Microbiology
FUNDAMENTALS

A Clinical Approach

SECOND EDITION

Marjorie Kelly Cowan

Miami University Middletown

WITH

Jennifer Bunn

RN, Clinical Advisor

Ronald M. Atlas

University of Louisville
Contributor

Heidi Smith

Front Range Community College
Digital Author

Mc
Graw
Hill
Education

MICROBIOLOGY FUNDAMENTALS: A CLINICAL APPROACH, SECOND EDITION
Published by McGraw-Hill Education, 2 Penn Plaza, New York, NY 10121. Copyright © 2016
by McGraw-Hill Education. All rights reserved. Printed in the United States of America.
Previous edition © 2013. No part of this publication may be reproduced or distributed in any
form or by any means, or stored in a database or retrieval system, without the prior written
consent of McGraw-Hill Education, including, but not limited to, in any network or other
electronic storage or transmission, or broadcast for distance learning.

Some ancillaries, including electronic and print components, may not be available to
customers outside the United States.

This book is printed on acid-free paper.

3 4 5 6 7 RMN 20 19 18 17 16

ISBN 978-0-07-802104-6
MHID 0-07-802104-9

Senior Vice President, Products & Markets: *Kurt L. Strand*
Vice President, General Manager, Products & Markets: *Marty Lange*
Vice President, Content Design & Delivery: *Kimberly Meriwether David*
Managing Director: *Michael Hackett*
Brand Manager: *Amy Reed/Marija Magner*
Director, Product Development: *Rose Koos*
Product Developer: *Darlene M. Schueller*
Marketing Manager: *Kristine Rellihan*
Digital Product Analyst: *Jake Theobald*
Director, Content Design & Delivery: *Linda Avenarius*
Program Manager: *Angela R. FitzPatrick*
Content Project Manager: *Sherry Kane*
Buyer: *Laura M. Fuller*
Design: *Trevor Goodman*
Content Licensing Specialists: *John Leland/Leonard Behnke*
Cover Image: *© Colin Anderson/Blend Images LLC © Janis Christie/Digital Vision/Gettyimages*
© Universal Images Group/Gettyimages © Eye of Science/Science Source
Compositor: *MPS Limited*
Printer: *R. R. Donnelley*

All credits appearing on page or at the end of the book are considered to be an extension of
the copyright page.

Library of Congress Cataloging-in-Publication Data

Cowan, M. Kelly, author.
 Microbiology fundamentals : a clinical approach / Marjorie Kelly Cowan,
Miami University with Jennifer Bunn, RN, clinical contributor, and with
contributions from Ronald M. Atlas -- Second edition.
    pages cm
 Includes index.
 ISBN 978-0-07-802104-6 (alk. paper)
 1. Microbiology. I. Bunn, Jennifer, RN, author. II. Atlas, Ronald M.,
1946- author. III. Title.
 QR41.2.C692 2016
 579--dc23
                                        2014031852

www.mhhe.com

Brief Contents

Contributed by Ronald M. Atlas

About the Authors

Kelly Cowan, PhD, has been a microbiologist at Miami University since 1993, where she teaches microbiology for pre-nursing/allied health students at the university's Middletown campus, a regional commuter campus that accepts first-time college students with a high school diploma or GED, at any age. She started life as a dental hygienist. She then went on to attain her PhD at the University of Louisville, and later worked at the University of Maryland's Center of Marine Biotechnology and the University of Groningen in The Netherlands. Kelly has published (with her students) 24 research articles stemming from her work on bacterial adhesion mechanisms and plant-derived antimicrobial compounds. But her first love is teaching—both doing it and studying how to do it better. She is past chair of the Undergraduate Education Committee of the American Society for Microbiology (ASM). When she is not teaching or writing, Kelly hikes, reads, and still tries to (s)mother her three grown kids.

Jennifer Bunn, RN, is a registered nurse, having spent most of her career in rural medicine, where she has had the opportunity to interact with patients of all ages. Her experience includes emergency medicine and critical care, pediatrics, acute care, long-term care, and labor and delivery. Currently, Jennifer works on an acute care unit. Over the span of her career, she has enjoyed mentoring and precepting LPN and RN students. Jennifer writes medical content for websites, apps, and blogs.

Ronald M. Atlas is Professor of Biology at the University of Louisville. He was a postdoctoral fellow at the Jet Propulsion Laboratory where he worked on Mars Life Detection. He has served as President of the American Society for Microbiology, as cochair of the American Society for Microbiology Biodefense Committee, as a member of the DHS Homeland Security Science and Technology Advisory Committee, and as chair of the Board of Directors of the One Health Commission. He is author of nearly 300 manuscripts and 20 books. His research on hydrocarbon biodegradation has helped pioneer the field of petroleum bioremediation. He has performed extensive studies on oil biodegradation and has worked for both Exxon and the U.S. EPA as a consultant on the *Exxon Valdez* spill and for BP on the *Deepwater Horizon* spill in the Gulf of Mexico.

Heidi Smith leads the microbiology department at Front Range Community College, Fort Collins, Colorado. Student success is a strategic priority at FRCC and a personal passion of Heidi's. Collaboration with other faculty across the nation, the development and implementation of new digital learning tools, and her focus on student learning outcomes have revolutionized her face-to-face and online teaching approaches and student performance in her classes. Outside of the classroom, Heidi served as the director of the FRCC Honors Program for six years, working with other faculty to build the program from the ground up. She is also an active member of the American Society for Microbiology and participated as a task force member for the development of their Curriculum Guidelines for Undergraduate Microbiology Education. Off campus, Heidi spends as much time as she can enjoying the beautiful Colorado outdoors with her husband and three young children.

Preface

Students:

Welcome! I am so glad you are here. I am very excited for you to try this book. I wrote it after years of frustration, teaching from books that didn't focus on the right things that my students needed. My students (and, I think, you) need a solid but not overwhelming introduction to microbiology and infectious diseases. I asked myself: What are the major concepts I want my students to remember 5 years from now? And then I worked backward from there, making sure everything pointed to the big picture.

While this book has enough detail to give you context, there is not so much detail that you will lose sight of the major principles. Biological processes are described right next to the illustrations that illustrate them. The format is easier to read than most books, because there is only one column of text on a page and wider margins. The margins gave me space to add interesting illustrations and clinical content. My coauthor, Jennifer Bunn, is a nurse who brings her years of experience to life on the page and shows you how this information will matter to you when you are working as a health care provider. We have interesting and up-to-the-moment Case Files, Medical Moments, Inside the Clinic selections, and NCLEX® questions in every chapter. Also be sure to use the Connect content—this is where you can really take control of your success in the class by making use of as many of the tools as you need.

I really wanted this to be a different kind of book. I started using it in my own classes and my students love it! Well, maybe they have to say that, but I hope that you truly do enjoy it and find that it is a refreshing kind of science book.

—Kelly Cowan

I dedicate this book to Ted.

LEARNSMART®
ADVANTAGE

McGraw-Hill LearnSmart® is one of the most effective and successful adaptive learning resources available on the market today. More than 2 million students have answered more than 1.3 billion questions in LearnSmart since 2009, making it the most widely used and intelligent adaptive study tool that's proven to strengthen memory recall, keep students in class, and boost grades. Students using LearnSmart are 13% more likely to pass their classes, and 35% less likely to drop out.

LearnSmart continuously adapts to each student's needs by building an individual learning path so students study smarter and retain more knowledge. Turnkey reports provide valuable insight to instructors, so precious class time can be spent on higher-level concepts and discussion.

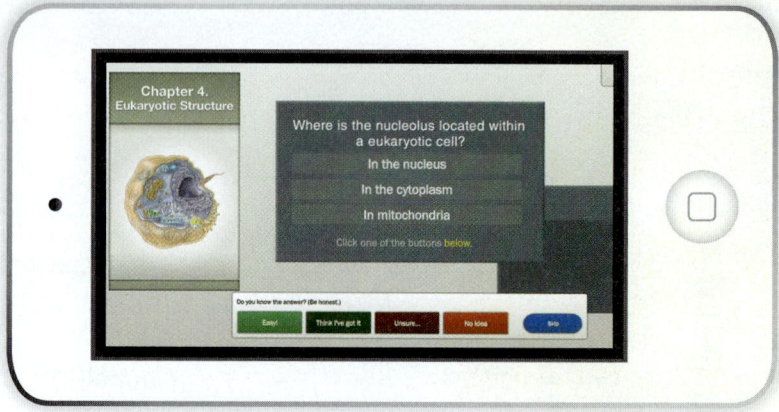

Fueled by LearnSmart—the most widely used and intelligent adaptive learning resource—**SmartBook®** is the first and only adaptive reading experience available today.

Distinguishing what students know from what they don't, and honing in on concepts they are most likely to forget, SmartBook personalizes content for each student in a continuously adapting reading experience. Reading is no longer a passive and linear experience, but an engaging and dynamic one where students are more likely to master and retain important concepts, coming to class better prepared.

As a result of the adaptive reading experience found in SmartBook, students are more likely to retain knowledge, stay in class, and get better grades.

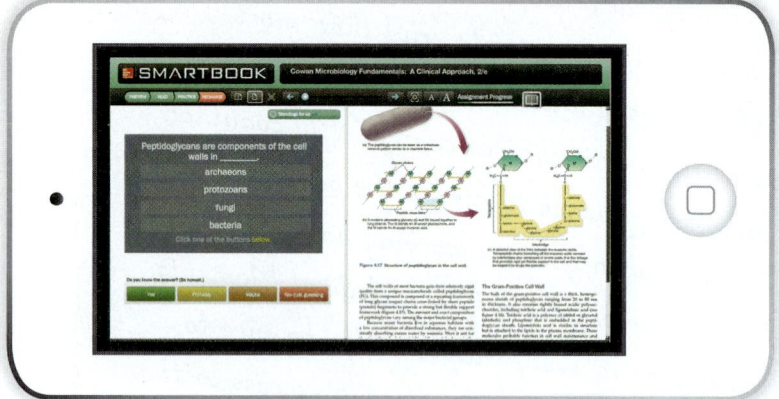

LearnSmart Labs® is an adaptive simulated lab experience that brings meaningful scientific exploration to students. Through a series of adaptive questions, LearnSmart Labs identifies a student's knowledge gaps and provides resources to quickly and efficiently close those gaps. Once students have mastered the necessary basic skills and concepts, they engage in a highly realistic simulated lab experience that allows for mistakes and the execution of the scientific method.

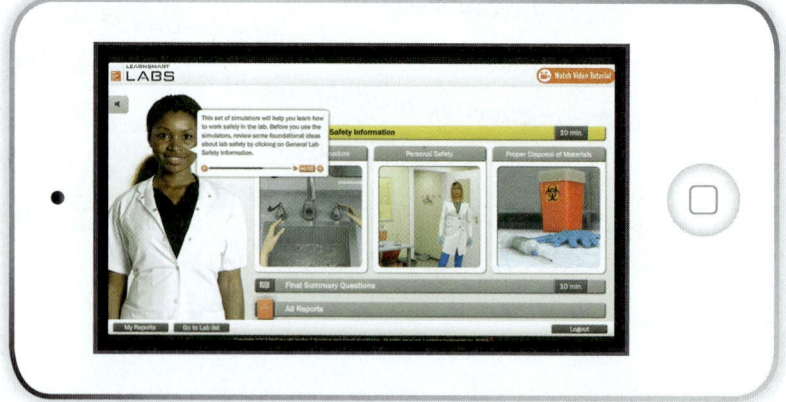

LearnSmart Prep® is designed to get students ready for a forthcoming course by quickly and effectively addressing prerequisite knowledge gaps that may cause problems down the road. LearnSmart Prep maintains a continuously adapting learning path individualized for each student, and tailors content to focus on what the student needs to master in order to have a successful start in the new class.

Digital efficacy study shows results!

Digital efficacy study final analysis shows students experience higher success rates when required to use LearnSmart.

- Passing rates increased by an average of **11.5%** across the schools and by a weighted average of **7%** across all students.
- Retention rates increased an average of **10%** across the schools and by a weighted average of **8%** across all students.

Study details:

- Included two state universities and four community colleges.
- Control sections assigned chapter assignments consisting of testbank questions and the experimental sections assigned LearnSmart, both through McGraw-Hill Connect®.
- Both types of assignments were counted as a portion of the grade, and all other course materials and assessments were consistent.
- 358 students opted into the LearnSmart sections and 332 into the sections where testbank questions were assigned.

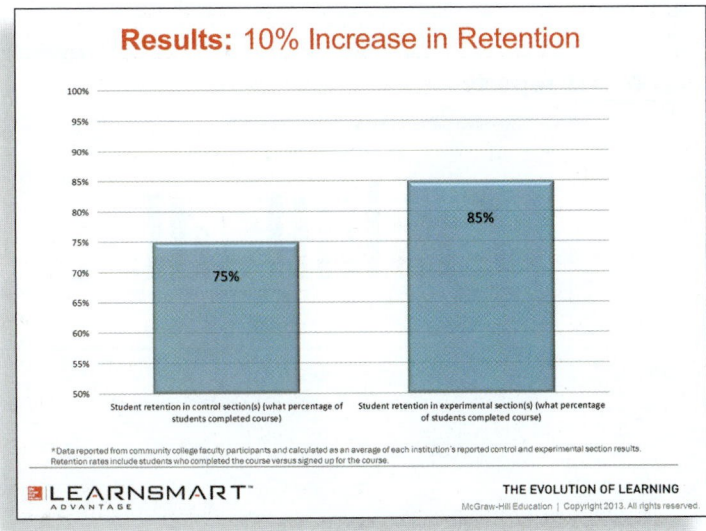

Results: 10% Increase in Retention

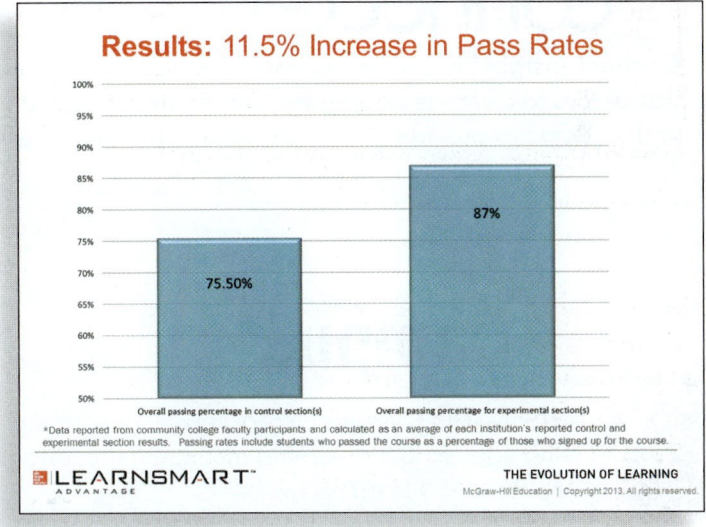

Results: 11.5% Increase in Pass Rates

"Use of technology, especially LearnSmart, assisted greatly in keeping on track and keeping up with the material."

— student, Triton College

"LearnSmart has helped me to understand exactly what concepts I do not yet understand. I feel like after I complete a module I have a deeper understanding of the material and a stronger base to then build on to apply the material to more challenging concepts."

— Student

"After collecting data for five semesters, including two 8-week intensive courses, the trend was very clear: students who used LearnSmart scored higher on exams and tended to achieve a letter grade higher than those who did not."

— Gabriel Guzman, Triton College

"This textbook was selected due to the LearnSmart online content as well as the fact that it is geared for an allied health student. This textbook has certainly enhanced the classroom experience and I see that my students are better prepared for class after they have worked within LearnSmart."

— Jennifer Bess, Hillsborough Community College

Connecting Instructors to Students

McGraw-Hill Connect® Microbiology

McGraw-Hill Connect Microbiology is a digital teaching and learning environment that saves students and instructors time while improving performance over a variety of critical outcomes.

- Instructors have access to a variety of resources including assignable and gradable interactive questions based on textbook images, case study activities, tutorial videos, and more.
- Digital images, PowerPoint® lecture outlines, and instructor resources are also available through Connect.
- All Connect questions are tagged to a learning outcome, specific section and topic, ASM topics and curriculum guidelines, and Bloom's level for easy tracking of assessment data.

Visit www.mcgrawhillconnect.com.

Connect Insight® is a powerful data analytics tool that allows instructors to leverage aggregated information about their courses and students to provide a more personalized teaching and learning experience.

Campus

McGraw-Hill Campus® integrates all of your digital products from McGraw-Hill Education with your school's learning management system for quick and easy access to best-in-class content and learning tools.

Through Innovative Digital Solutions

Unique Interactive Question Types in Connect, Tagged to ASM's Curriculum Guidelines for Undergraduate Microbiology

1 **Case Study:** Case studies come to life in a learning activity that is interactive, self-grading, and assessable. The integration of the cases with videos and animations adds depth to the content, and the use of integrated questions forces students to stop, think, and evaluate their understanding. Pre- and post-testing allow instructors and students to assess their overall comprehension of the activity.

2 **Concept Maps:** Concept maps allow students to manipulate terms in a hands-on manner in order to assess their understanding of chapter-wide topics. Students become actively engaged and are given immediate feedback, enhancing their understanding of important concepts within each chapter.

3 **What's the Diagnosis:** Specifically designed for the disease chapters of the text, this is an integrated learning experience designed to assess the student's ability to utilize information learned in the preceding chapters to successfully culture, identify, and treat a disease-causing microbe in a simulated patient scenario. This question type is true experiential learning and allows the students to think critically through a real-life clinical situation.

4 **Animations:** Animation quizzes pair our high-quality animations with questions designed to probe student understanding of the illustrated concepts.

5 **Tutorial Animation Learning Modules:** Animations, videos, audio, and text all combine to help students understand complex processes. These tutorials take a stand-alone, static animation and turn it into an interactive learning experience for your students with real-time remediation. Key topics have an Animated Learning Module assignable through Connect. An icon in the text indicates when these learning modules are available.

6 **Labeling:** Using the high-quality art from the textbook, check your students' visual understanding as they practice interpreting figures and learning structures and relationships.

7 **Classification:** Ask students to organize concepts or structures into categories by placing them in the correct "bucket."

8 **Sequencing:** Challenge students to place the steps of a complex process in the correct order.

9 **Composition:** Fill in the blanks to practice vocabulary, and then reorder the sentences to form a logical paragraph (these exercises may qualify as "writing across the curriculum" activities!).

All McGraw-Hill Connect content is tagged to Learning Outcomes for each chapter as well as topic, section, Bloom's Level, ASM topic, and ASM Curriculum Guidelines to assist you in customizing assignments and in reporting on your students' performance against these points. This will enhance your ability to assess student learning in your courses by allowing you to align your learning activities to peer-reviewed standards from an international organization.

NCLEX®

NCLEX® Prep Questions: Sample questions are available in Connect to assign to students, and there are questions throughout the book as well.

Presentation Tools allow you to customize your lectures.

Enhanced Lecture Presentations contain lecture outlines, art, photos, and tables, embedded where appropriate. Fully customizable, complete, and ready to use, these presentations will enable you to spend less time preparing for lecture!

Animations Over 100 animations bring key concepts to life, available for instructors and students.

Animation PPTs Animations are embedded in PowerPoint for ultimate ease of use! Just copy and paste into your custom slide show and you're done!

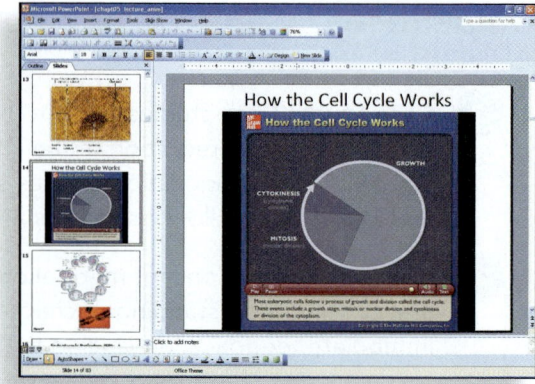

Take your course online—*easily*—with one-click Digital Lecture Capture.

McGraw-Hill Tegrity® is a fully automated lecture capture solution used in traditional, hybrid, "flipped classes," and online courses to record lesson, lectures, and skills.

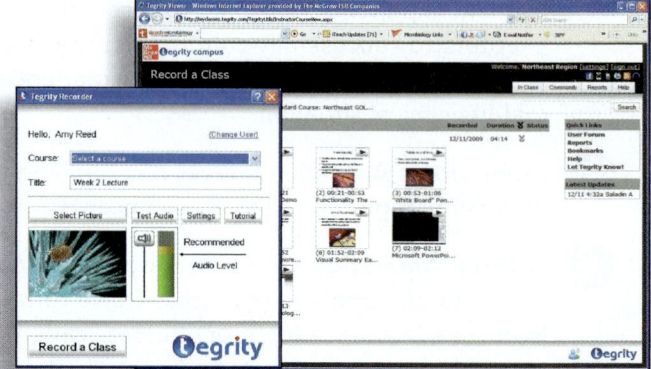

Create what you've only imagined.

Introducing **McGraw-Hill Create™**—a self-service website that allows you to create custom course materials—print and eBooks—by drawing upon McGraw-Hill Education's comprehensive, cross-disciplinary content. Add your own content quickly and easily. Tap into other rights-secured third-party sources as well. Then, arrange the content in a way that makes the most sense for your course. Even personalize your book with your course name and information! Choose the best format for your course: color print, black and white print, or eBook. The eBook is now even viewable on an iPad!® And, when you are done you will receive a free PDF review copy in just minutes!

Visit McGraw-Hill Create—www.mcgrawhillcreate.com—today and begin building your perfect book.

Microbiology Fundamentals Laboratory Manual, Second Edition

Steven Obenauf, Broward College
Susan Finazzo, Georgia Perimeter College

Written specifically for pre-nursing and allied health microbiology students, this manual features brief, visual exercises with a clinical emphasis.

Clinical applications help students see the relevance of microbiology.

Case File   Each chapter begins with a case written from the perspective of a former microbiology student.

These high-interest introductions provide a specific example of how the chapter content is relevant to real life and future health care careers.

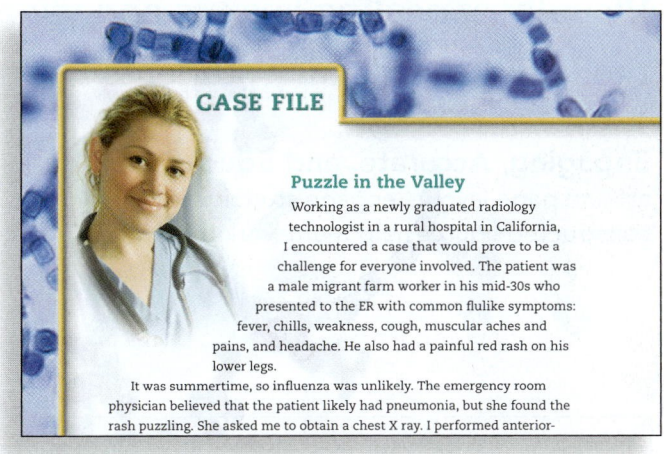

CASE FILE

Puzzle in the Valley

Working as a newly graduated radiology technologist in a rural hospital in California, I encountered a case that would prove to be a challenge for everyone involved. The patient was a male migrant farm worker in his mid-30s who presented to the ER with common flulike symptoms: fever, chills, weakness, cough, muscular aches and pains, and headache. He also had a painful red rash on his lower legs.

It was summertime, so influenza was unlikely. The emergency room physician believed that the patient likely had pneumonia, but she found the rash puzzling. She asked me to obtain a chest X ray. I performed anterior-

Clinical Contributor

This textbook features a clinical advisor, Jennifer Bunn, RN, who authored the following features, described on these pages:

▶ Added clinical relevance throughout the chapter
▶ Relevant case files
▶ Medical Moment boxes
▶ NCLEX® prep questions

"Jen added things that were fascinating to ME! And will enrich my own teaching. Pre-allied health students are so eager to start 'being' nurses, etc., they love these clinical details."

—Kelly Cowan

Medical Moment

These boxes give students a more detailed clinical application of a nearby concept in the chapter.

Medical Moment

Outsmarting Encapsulated Bacteria

Catheter-associated infections in critically ill patients requiring central venous access are unfortunately all too common. It has been estimated that blood-stream infection, a condition called **sepsis**, affects 3% to 8% of patients requiring an indwelling cathe-ter for a prolonged period of time. Sepsis increases morbidity and mortality and can increase the cost of a patient's care by approximately $30,000.

In order to colonize a catheter, microorgan-isms must first adhere to the surface of the tip on this medical device. Fimbriae and glycocalyces are bacterial structures most often used for this purpose.

NCLEX® PREP

1. The physician has ordered that a urine culture be taken on a client. What priority information should the nurse know in order to complete the collection of this specimen?
   a. Date and time of collection
   b. Method of collection
   c. Whether the client is NPO (to have nothing by mouth)
   d. Age of client

NCLEX® Prep Questions   Found throughout the chapter, these multiple-choice questions are application-oriented and designed to help students learn the microbiology information they will eventually need to pass the NCLEX examination. Students will begin learning to think critically, apply information, and over time, prep themselves for the examination.

Additional questions are available in Connect for homework and assessment.

Inside the Clinic   Each chapter ends with a reading that emphasizes the nursing aspect of microbiology.

Clinical Examples Throughout   Clinical insights and examples are woven throughout the chapter—not just in boxed elements.

Fever: To Treat or Not to Treat?

Inside the Clinic

Our immune system helps to protect us from invading microorganisms. One manner in which our body protects itself is by mounting a fever in response to microbes present in the body (body temperature can also rise in response to inflammation or injury).

The hypothalamus, located in the brain, serves as the temperature-control center of the body. Fever occurs when the hypothalamus actually resets itself at a higher temperature. The hypothalamus raises body temperature by shunting blood away from the skin and into the body's core. It also raises temperature by inducing shivering, which is a result of muscle contraction and serves to increase temperature. This is why people experience chills and shivering when they have a fever. Once the new, higher temperature is reached (warmer blood reaches the hypothalamus, the hypothalamus works to maintain this temperature. When the "thermostat" is reset once again to a lower level, the body reverses the process, shunting blood to the skin. This is why people become diaphoretic (sweaty) when a fever breaks.

When microorganisms gain entrance to the body and begin to proliferate, the body responds with an onslaught of macrophages and monocytes, whose pur-

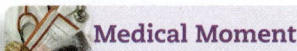

VISUAL

Visually appealing layouts and vivid art closely linked to narrative complement the way 21st-century students learn.

Engaging, Accurate, and Educational Art   Visually appealing art and page layouts engage students in the content, while carefully constructed figures help them work through difficult concepts.

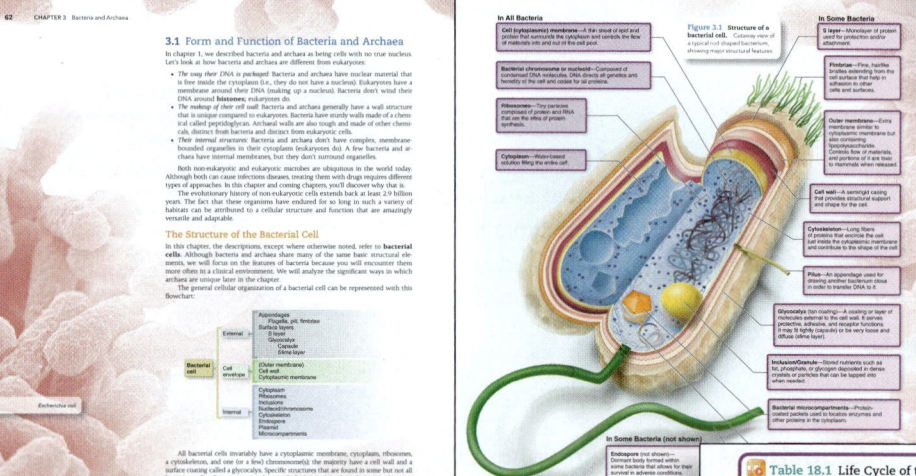

Visual Tables   The most important points explaining a concept are distilled into table format and paired with explanatory art.

Process Figures   Complex processes are broken into easy-to-follow steps. Numbered steps in the art coordinate with numbered text boxes to walk students through the figure.

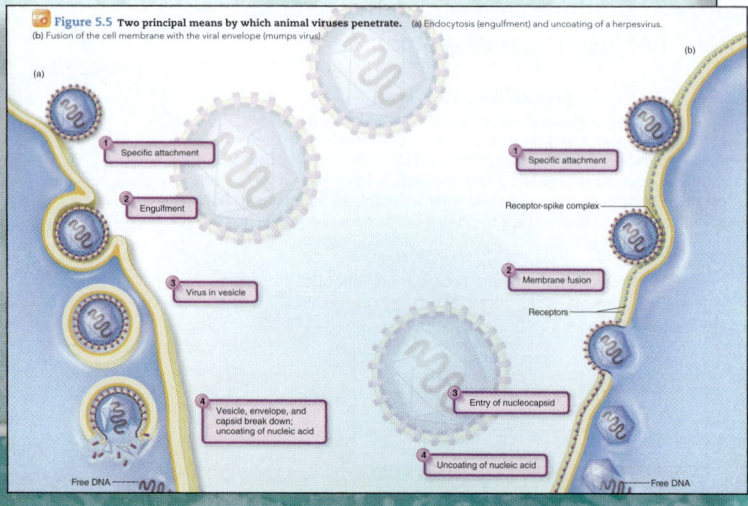

Streamlined coverage of core concepts help students retain the information they will need for advanced courses.

Chemistry topics required for understanding microbiology are combined with the foundation content found in chapter 1.

Genetics content is synthesized into one chapter covering the concepts that are key to microbiology students.

A chapter in microbiology textbooks that is often not used in health-related classes becomes relevant because it presents the 21st-century idea of "One Health"—that the environment and animals influence human health and infections.

Brief Contents

Contributed by Ronald M. Atlas

"The textbook is unique in that it was written with the health science student in mind. Unlike most texts, which just claim to be appropriate for nursing students, this textbook actually incorporates real world health care using the features such as Inside the Clinic and Case Files. The textbook also incorporates critical thinking and visual connections to illustrate how a student would 'function' in the field."

—Jill Roberts, University of South Florida

Duplication Eliminated Detail is incorporated into figures so students can learn in context with the art. This allows a more concise narrative flow while still retaining core information.

Tables Tables are used to further streamline content and help students understand relationships between concepts.

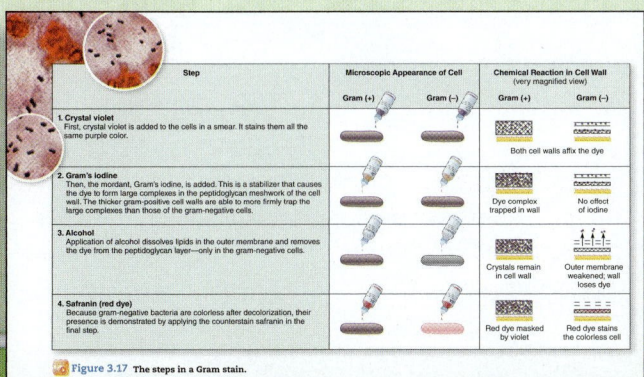

Figure 3.17 The steps in a Gram stain.

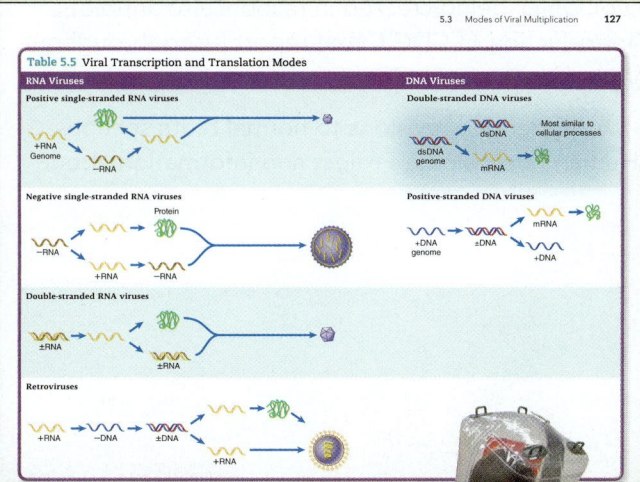

Changes to the Second Edition

Significant Changes

Epidemiological data (who, where, how common) are added for every organism in every disease table!

Twenty new chapter-opener case files include: a measles case, *C. diff*, Valley fever, Norwalk virus, gas gangrene, rheumatoid arthritis, UTI, and a bloodstream infection.

Throughout the Book

This edition has improved Learning Outcomes, new Critical Thinking questions, many new Medical Moments scattered throughout, and new Inside the Clinic scenarios at the ends of the chapters. Also, antibiotic-resistant bacteria are uniformly identified throughout the book according to CDC threat status, and neglected parasitic infections (NPIs) are highlighted.

Chapter Highlights

The Human Microbiome Project results have altered nearly every chapter. Other noteworthy changes are described here.

Chapters 1 and 4 Updates about origin of cells.

Chapter 2 New emphasis on nonculture methods.

Chapter 3 Much more information on biofilms; new material on S layers and microcompartments.

Chapter 6 Improved diffusion and osmosis discussion and exponential growth figures.

Chapter 9 Added concept of critical, semicritical disinfection.

Chapter 10 Significant changes and enhancements to the antibiotic-resistance section, incorporating information about resistance not ONLY being created in response to antibiotic presence; introduction of CDC threat report (used throughout disease chapters).

Chapter 11 Extensive revisions to normal biota sections based on Human Microbiome Project and information about normal biota in lungs, and so on; new information about polymicrobial infections, quorum sensing; added the built environment as a reservoir and the impact on epidemiology of Internet and social media.

Chapter 12 Updated to include gamma-delta T cells/NKT/NK cells as functioning in both specific and nonspecific immunity; added inflammasomes; updated discussion of interferon; complement section much clearer.

Chapter 13 Added detail on gamma-delta T cells and their important role; Medical Moment addresses Facebook group about pox parties.

Chapter 14 Updates on allergies and the microbiome.

Chapter 15 Many redrawn figures; new section titled "Breakthrough Methodologies" to discuss use of deep sequencing, mass spectrometry, and imaging as diagnostic techniques.

Chapter 16 Added *MRSA skin and soft-tissue infection* as first Highlight Disease; great emphasis on measles and recent outbreaks.

Chapter 18 Up-to-the-moment Inside the Clinic about the 2014 Ebola epidemic, including its presence in the United States.

Chapter 19 Extensive updates on influenza, TB, MDR-TB, and XDR-TB.

Chapter 20 Emphasis on neglected parasitic infections; addition of cysticercosis as a separate condition; addition of norovirus as a significant cause of diarrhea.

Chapter 21 UTI section completely rewritten to emphasize hospital and long-term-care infections.

Chapter 22 Completely new, revolutionary chapter by Ronald M. Atlas (One Health) which ties together the environment, animals and human health.

Acknowledgments

I am always most grateful to my students in my classes. They teach me every darned day how to do a better job helping them understand these concepts that are familiar to me but new to them. All the instructors who reviewed the manuscript for me were also great allies. I thank them for lending me some of their microbiological excellence. I had several contributors to the book and digital offerings—Hank Stevens, Andrea Rediske, Kimberly Harding, Kathleen Sandman and Heidi Smith chief among them. Jennifer Bunn, my coauthor, teaches me many things about many things. I would especially like to thank Ronald Atlas for the new chapter he wrote. I also am the beneficiary of the best copyediting on the planet delivered from the mind and keyboard of C. Jeanne Patterson. Amy Reed, Marija Magner, Sherry Kane, and Kristine Rellihan from McGraw-Hill Education make the wheels go round. Darlene Schueller, my day-to-day editor, is a wonderful human being and taskmaster, in that order. In short, I'm just a lucky girl surrounded by talented people.

—Kelly Cowan

Reviewers

Larry Barton, *University of New Mexico*
Jennifer Bess, *Hillsborough Community College*
Linda Bruslind, *Oregon State University*
Miranda Campbell, *Greenville Technical College*
Rudolph DiGirolamo, *St. Petersburg College*
Jason L. Furrer, *University of Missouri*
Kathryn Germain, *Southwest Tennessee Community College*
Ellen Gower, *Greenville Technical College*
Raymond L. Harris, *Prince George's Community College*
Ingrid Herrmann, *Santa Fe College*
John Jones, *Calhoun Community College*
Lara Kingeter, *Tarrant County College*
Suzanne Long, *Monroe Community College*
Margaret Major, *Georgia Perimeter College*
Matthew Morgan, *Greenville Technical College*
Laura O'Riorden, *Tallahassee Community College*
Karen L. Richardson, *Calhoun Community College*
Geraldine Rimstidt, *Daytona State College*
Seth Ririe, *BYU-Idaho*
Jill Roberts, *University of South Florida*
Meredith Rodgers, *Wright State University*
Rachael Romain, *Columbus State Community College*
Lindsey Shaw, *University of South Florida*
Tracey Steeno, *Northeast Wisconsin Technical College*
Cristina Takacs-Vesbach, *University of New Mexico*
John E. Whitlock, *Hillsborough Community College*
Michael Womack, *Gordon State College*
John M. Zamora, *Middle Tennessee State University*

Lance D. Bowen, *Truckee Meadows Community College*
David Brady, *Southwestern Community College*
Toni Brem, *Wayne County Community College District—Northwest Campus*
Lisa Burgess, *Broward College*
Elizabeth A. Carrington, *Tarrant County College District*
Joseph P. Caruso, *Florida Atlantic University*
Shima Chaudhary, *South Texas College*
Melissa A. Deadmond, *Truckee Meadows Community College*
Elizabeth Emmert, *Salisbury University*
Jason L. Furrer, *University of Missouri*
Chris Gan, *Highline Community College*
Zaida M. Gomez-Kramer, *University of Central Arkansas*
Brinda Govindan, *San Francisco State University*
Julianne Grose, *Brigham Young University*
Zafer Hatahet, *Northwestern State University*
James B. Herrick, *James Madison University*
James E. Johnson, *Central Washington University*
Laura Leverton, *Wake Tech Community College*
Philip Lister, *Central New Mexico Community College*
Suzanne Long, *Monroe Community College*
Tammy Lorince, *University of Arkansas*
Kimberly Roe Maznicki, *Seminole State College of Florida*
Amee Mehta, *Seminole State College of Florida*
Sharon Miles, *Itawamba Community College*
Rita B. Moyes, *Texas A&M University*
Ruth A. Negley, *Harrisburg Area Community College—Gettysburg Campus*
Julie A. Oliver, *Cosumnes River College*
Jean Revie, *South Mountain Community College*
Jackie Reynolds, *Richland College*
Donald L. Rubbelke, *Lakeland Community College*
George Shahla, *Antelope Valley College*
Sasha A. Showsh, *University of Wisconsin—Eau Claire*
Steven J. Thurlow, *Jackson College*
George Wawrzyniak, *Milwaukee Area Technical College*
Janice Webster, *Ivy Tech Community College*
John Whitlock, *Hillsborough Community College*

Focus Group Attendees

Corrie Andries, *Central New Mexico Community College*
John Bacheller, *Hillsborough Community College*
Michelle L. Badon, *University of Texas at Arlington*
David Battigelli, *University of North Carolina—Greensboro*
Cliff Boucher, *Tyler Junior College*

Contents

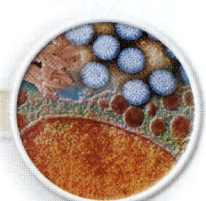

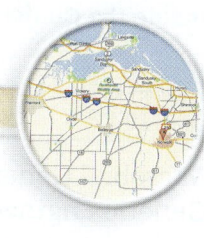

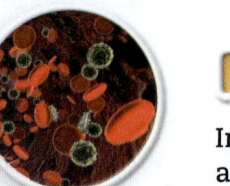
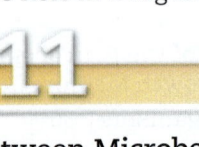
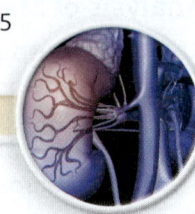
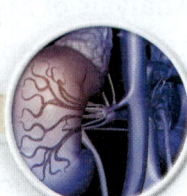

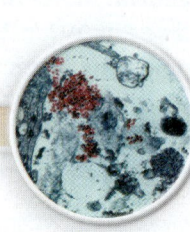
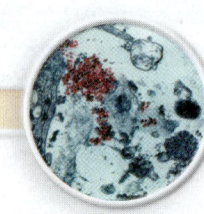

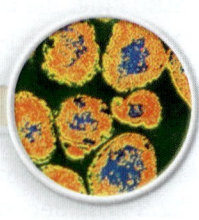

Microbiology
FUNDAMENTALS
A Clinical Approach

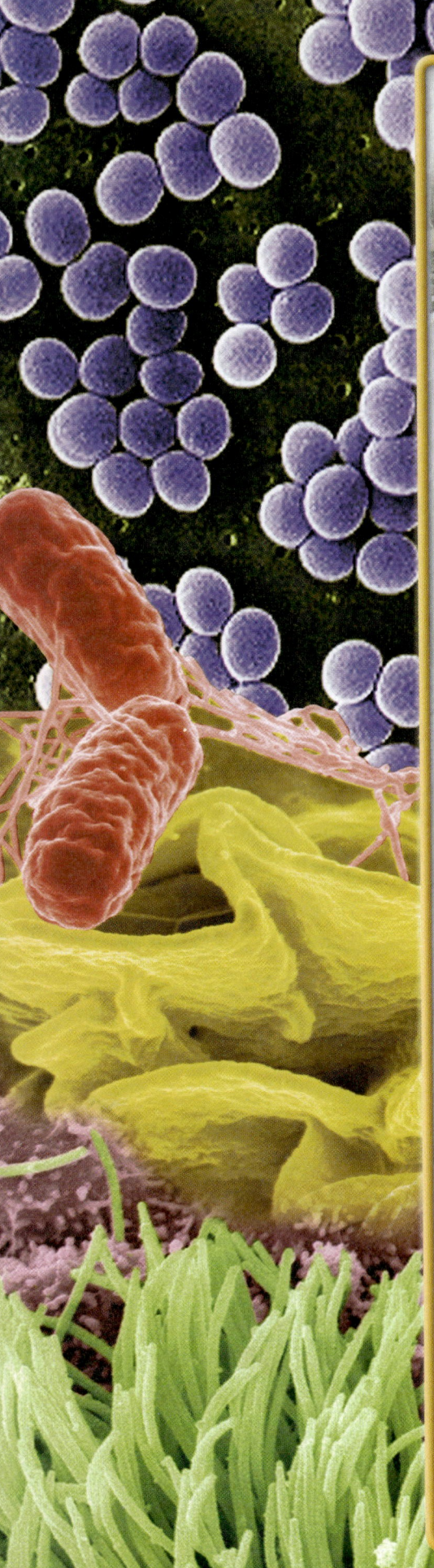

The Subject Is You!

At the beginning of every chapter in this book a different health care worker will tell you a story about something "microbiological" that happened to him or her in the line of duty. For this first chapter, though, I am claiming "dibs" as author and am going to introduce myself to you by telling you about the first day of class in my course.

Long ago I noticed that students have a lot of anxiety about their microbiology course. I know that starts you out with one strike against you, as attitudes are such powerful determinants of our success. So on the first day of class I often spend some time talking with students about how much they already know about microbiology.

Sometimes I start with "How many of you have taken your kids for vaccinations?" since in the classes I teach very many students are parents. Right away students will tell me why they or friends they know have not vaccinated their children and I can tell them there's a sophisticated microbiological concept they are referencing, even if they aren't naming it: *herd immunity,* discussed in chapter 11 of this book.

My favorite question (now that we're all warmed up) is "Who knows someone—whom you don't have to identify—who has had a really unusual or scary infection?" A surprising number of people have known someone who has had malaria, or leptospirosis, or endocarditis, or encephalitis. Then the conversation gets interesting as students begin to understand how much they already know about microbiology, and the class is not going to be as intimidating as they thought.

- Think about how many times you have taken antibiotics in the past few years. What is special about antibiotics that they are only given to treat infections?

- What is the most unusual infection you have ever encountered among family or friends or patients you have cared for?

Case File Wrap-Up appears on page 30.

Introduction to Microbes and Their Building Blocks

IN THIS CHAPTER...

1.1 Microbes: Tiny but Mighty

1. List the various types of microorganisms that can colonize humans.
2. Describe the role and impact of microbes on the earth.
3. Explain the theory of evolution and why it is called a theory.
4. Explain the ways that humans manipulate organisms for their own uses.
5. Summarize the relative burden of human disease caused by microbes.
6. Differentiate among bacteria, archaea, and eukaryotic microorganisms.
7. Identify a fourth type of microorganism.
8. Compare and contrast the relative sizes of the different microbes.

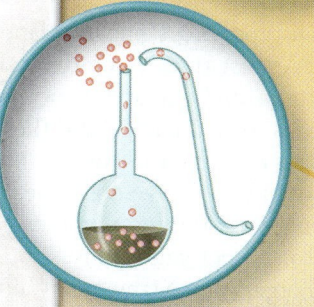

1.2 Microbes in History

9. Make a time line of the development of microbiology from the 1600s to today.
10. List some recent microbiology discoveries of great impact.

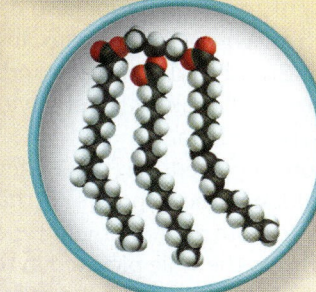

1.3 Macromolecules: Superstructures of Life

11. Name the four main families of biochemicals.
12. Provide examples of cell components made from each of the families of biochemicals.
13. Differentiate among primary, secondary, tertiary, and quaternary levels of protein structure.
14. List the three components of a nucleotide.
15. Name the nitrogen bases of DNA and RNA.
16. List the three components of ATP.
17. Recall three characteristics common to all cells.

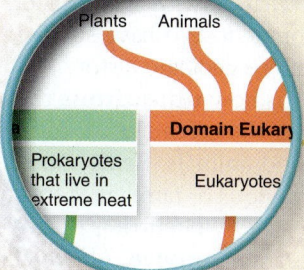

1.4 Naming, Classifying, and Identifying Microorganisms

18. Differentiate among the terms *nomenclature*, *taxonomy*, and *classification*.
19. Create a mnemonic device for remembering the taxonomic categories.
20. Correctly write the binomial name for a microorganism.
21. Draw a diagram of the three major domains.
22. Explain the difference between traditional and molecular approaches to taxonomy.

A rod-shaped bacterium with numerous flagella.

1.1 Microbes: Tiny but Mighty

Microbiology is a specialized area of biology that deals with living things ordinarily too small to be seen without magnification. Such **microscopic** organisms are collectively referred to as **microorganisms** (my″-kroh-or′-gun-izms), **microbes,** or several other terms depending on the kind of microbe or the purpose. There are several major groups of microorganisms that we'll be studying. They are **bacteria, archaea, protozoa, fungi, helminths,** and **viruses.** There is another very important group of organisms called **algae.** They are critical to the health of the biosphere but do not directly infect humans, so we will not consider them in this book. Each of the other six groups contains members that colonize humans, so we will focus on them.

The nature of microorganisms makes them both very easy and very difficult to study—easy because they reproduce so rapidly and we can quickly grow large populations in the laboratory, and difficult because we usually can't see them directly. We rely on a variety of indirect means of analyzing them in addition to using microscopes.

Microbes and the Planet

For billions of years, microbes have extensively shaped the development of the earth's habitats and the evolution of other life forms. It is understandable that scientists searching for life on other planets first look for signs of microorganisms.

Single-celled organisms that preceded our current cell types arose on this planet about 3.5 billion years ago, according to the fossil record. At that time, three types of cells arose from the original (now extinct) cell type: bacteria, archaea, and a specific cell type called a **eukaryote** (yoo-kar′-ee-ote). *Eu-kary* means "true nucleus," and these were the only cells containing a nucleus. Bacteria and archaea have no true nucleus. For that reason, they have traditionally been called **prokaryotes** (pro-kar′-ee-otes), meaning "prenucleus." But researchers are suggesting we no longer use the term *prokaryote* to lump them together because archaea and bacteria are so distinct genetically.

Bacteria and archaea are predominantly single-celled organisms. Many, many, eukaryotic organisms are also single-celled; but the eukaryotic cell type also developed into highly complex multicellular organisms, such as worms and humans. In terms of numbers, eukaryotic cells are a small minority compared to the bacteria and archaea; but their larger size (and our own status as eukaryotes!) makes us perceive them as dominant to—and more important than—bacteria and archaea.

For a long time, it was believed that eukaryotes evolved long after bacteria and archaea and actually derived *from* them. Most evidence today points to the near simultaneous rise of bacteria, archaea, and eukaryotes from an earlier cell type.

Figure 1.1 illustrates the history of life on earth. On the scale pictured in the figure, humans seem to have just appeared. Bacteria and archaea preceded even the earliest animals by more than 2 billion years. This is a good indication that humans are not likely to—nor should we try to—eliminate bacteria from our environment. We have survived and adapted to many catastrophic changes over the course of our geologic history.

Another indication of the huge influence bacteria exert is how **ubiquitous** they are. Microbes can be found nearly everywhere, from deep in the earth's crust, to the polar ice caps and oceans, to inside the bodies of plants and animals. Being mostly invisible, the actions of microorganisms are usually not as obvious or familiar as those of larger plants and animals. They make up for their small size by occurring in large numbers and living in places that many other organisms cannot survive. Above all, they play central roles in the earth's landscape that are essential to life.

When we point out that single-celled organisms have adapted to a wide range of conditions over the 3.5 billion years of their presence on this planet, we are talking about evolution. The presence of life in its present form would not be possible if the earliest life forms had not changed constantly, adapting to their environment and circumstances. Getting from the far left in figure 1.1 to the far right, where humans appeared, involved billions and billions of tiny changes, starting with the first cell that appeared about a billion years after the planet itself was formed.

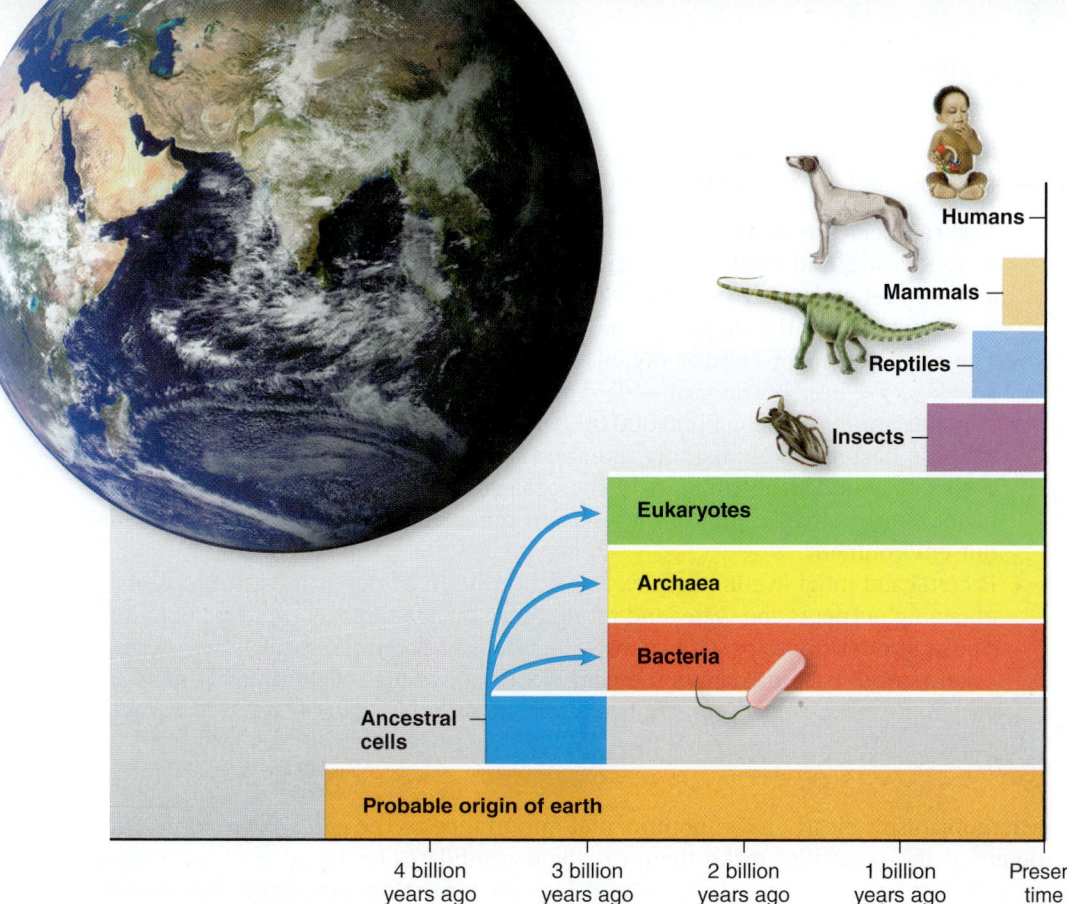

Figure 1.1 **Evolutionary time line.**

You have no doubt heard this concept described as the **theory of evolution.** Let's clarify some terms. **Evolution** is the accumulation of changes that occur in organisms as they adapt to their environments. It is documented every day in all corners of the planet, an observable phenomenon testable by science. Scientists use the term *theory* in a different way than the general public does, which often leads to great confusion. In science, a theory begins as a hypothesis, or an educated guess to explain an observation. By the time a hypothesis has been labeled a *theory* in science, it has undergone years and years of testing and not been disproved. It is taken as fact. This is much different from the common usage, as in "My theory is that he overslept and that's why he was late." The theory of evolution, like the germ theory and many other scientific theories, refers to a well-studied and well-established natural phenomenon, not just a random guess.

How Microbes Shape Our Planet

Microbes are deeply involved in the flow of energy and food through the earth's ecosystems. Most people are aware that plants carry out **photosynthesis,** which is the light-fueled conversion of carbon dioxide to organic material, accompanied by the formation of oxygen (called oxygenic photosynthesis). However, bacteria invented photosynthesis long before the first plants appeared, first as a process that did not produce oxygen (*anoxygenic photosynthesis*). This anoxygenic photosynthesis later evolved into oxygenic photosynthesis, which not only produced oxygen but also was much more efficient in extracting energy from sunlight. Hence, these ancient, single-celled microbes were responsible for changing the atmosphere of the earth from one without oxygen to one with oxygen. The production of oxygen also led to the use of oxygen for aerobic respiration and the formation of ozone, both of which set off an explosion in species diversification. Today, photosynthetic microorganisms (mainly bacteria and algae) account for more than 70% of the earth's photosynthesis, contributing the majority of the oxygen to the atmosphere **(figure 1.2).**

Figure 1.2 **A rich photosynthetic community.**
Summer pond with a thick mat of algae.

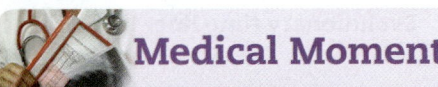

Medical Moment

Medications from Microbes

Penicillin is a worthy example of how microorganisms can be used to improve human life. Alexander Fleming, a Scottish bacteriologist, discovered penicillin quite by accident in 1928. While growing several bacterial cultures in Petri dishes, he accidentally forgot to cover them. They remained uncovered for several days; when Fleming checked the Petri dishes, he found them covered with mold. Just before Fleming went to discard the Petri dishes, he happened to notice that there were no bacteria to be seen around the mold—in other words, the mold was killing all of the bacteria in its vicinity.

Recognizing the importance of this discovery, Fleming experimented with the mold (of the genus *Penicillium*) and discovered that it effectively stopped or slowed the growth of several bacteria. The chemical that was eventually isolated from the mold—penicillin—became widely used during the Second World War and saved many soldiers' lives, in addition to cementing Fleming's reputation.

In the long-term scheme of things, microorganisms are the main forces that drive the structure and content of the soil, water, and atmosphere. For example:

- The temperature of the earth is regulated by gases, such as carbon dioxide, nitrous oxide, and methane, which create an insulation layer in the atmosphere and help retain heat. Many of these gases are produced by microbes living in the environment and the digestive tracts of animals.
- The most abundant cellular organisms in the oceans are not fish but bacteria. Think of a 2-liter bottle that soda comes in. Two liters of surface ocean water contains approximately 1,000,000,000,000,000,000 (1 quintillion) bacteria. Each of these bacteria likely harbors thousands of viruses inside of it, making viruses the most abundant inhabitants of the oceans. The bacteria and their viruses are major contributors to photosynthesis and other important processes that create our environment.
- Bacteria and fungi live in complex associations with plants that assist the plants in obtaining nutrients and water and may protect them against disease. Microbes form similar interrelationships with animals, notably, in the stomach of cattle, where a rich assortment of bacteria digests the complex carbohydrates of the animals' diets and causes the release of large amounts of methane into the atmosphere.

Microbes and Humans

Microorganisms clearly have monumental importance to the earth's operation. Their diversity and versatility make them excellent candidates for being used by humans for our own needs, and for them to "use" humans for their needs, sometimes causing disease along the way. We'll look at both of these kinds of microbial interactions with humans in this section.

By accident or choice, humans have been using microorganisms for thousands of years to improve life and even to shape civilizations. Baker's and brewer's yeasts, types of single-celled fungi, cause bread to rise and ferment sugar into alcohol to make wine and beers. Other fungi are used to make special cheeses such as Roquefort or Camembert. Historical records show that households in ancient Egypt kept moldy loaves of bread to apply directly to wounds and lesions. When humans manipulate microorganisms to make products in an industrial setting, it is called **biotechnology.** For example, some specialized bacteria have unique capacities to mine precious metals or to clean up human-created contamination.

Genetic engineering is an area of biotechnology that manipulates the genetics of microbes, plants, and animals for the purpose of creating new products and genetically modified organisms (GMOs). One powerful technique for designing GMOs is termed **recombinant DNA technology.** This technology makes it possible to transfer genetic material from one organism to another and to deliberately alter DNA. Bacteria and fungi were some of the first organisms to be genetically engineered. This was possible because they are single-celled organisms and they are so adaptable to changes in their genetic makeup. Recombinant DNA technology has unlimited potential in terms of medical, industrial, and agricultural uses. Microbes can be engineered to synthesize desirable products such as drugs, hormones, and enzymes.

Another way of tapping into the unlimited potential of microorganisms is the science of **bioremediation** (by′-oh-ree-mee-dee-ay″-shun). This term refers to the ability of microorganisms–ones already present or those introduced intentionally–to restore stability or to clean up toxic pollutants. Microbes have a surprising capacity to break down chemicals that would be harmful to other organisms **(figure 1.3).** This includes even human-made chemicals that scientists have developed and for which there are no natural counterparts.

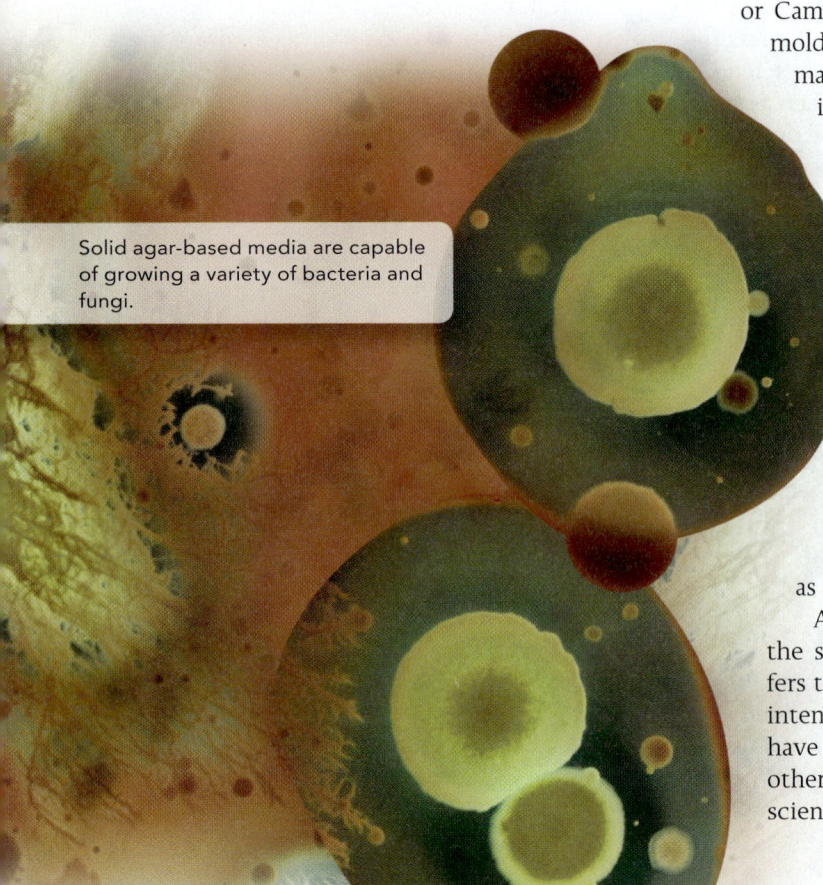

Solid agar-based media are capable of growing a variety of bacteria and fungi.

Microbes Harming Humans

One of the most fascinating aspects of the microorganisms with which we share the earth is that, despite all of the benefits they provide, they also contribute significantly to human misery as **pathogens** (path'-oh-jenz). The vast majority of microorganisms that associate with humans cause no harm. In fact, they provide many benefits to their human hosts. Note that a diverse microbial biota living in and on humans is an important part of human well-being. However, humankind is also plagued by nearly 2,000 different microbes that can cause various types of disease. Infectious diseases still devastate human populations worldwide, despite significant strides in understanding and treating them. The World Health Organization (WHO) estimates there are a total of 10 billion new infections across the world every year. Infectious diseases are also among the most common causes of death in much of humankind, and they still kill a significant percentage of the U.S. population. **Table 1.1** depicts the 10 top causes of death per year (by all causes, infectious and noninfectious) in the United States and worldwide.

Adding to the overload of infectious diseases, we are also witnessing an increase in the number of new (emerging) and older (reemerging) diseases. AIDS, hepatitis C, West Nile virus, and tuberculosis are examples. It is becoming clear that human actions in the form of reforestation, industrial farming techniques, and chemical and antibiotic usage can foster the emergence or reemergence of particular infectious diseases. These patterns will be discussed in chapter 22.

One of the most eye-opening discoveries in recent years is that many diseases that used to be considered noninfectious probably do involve microbial infection. One well-known example is that of gastric ulcers, now known to be caused by a bacterium called *Helicobacter*. But there are more. Associations have been established between certain cancers and bacteria and viruses, between diabetes and the Coxsackie virus, and between schizophrenia and a virus called the Borna agent. Diseases as different as multiple sclerosis, obsessive compulsive disorder, coronary artery disease, and even obesity have been linked to chronic infections with microbes. It seems that the golden age of microbiological discovery, during which all of the "obvious" diseases were characterized and cures or preventions were devised for them, should more accurately be referred to as the *first* golden age. We're now discovering the subtler side of microorganisms.

Another important development in infectious disease trends is the increasing number of patients with weakened defenses, who, because of welcome medical advances, are living active lives instead of enduring long-term disability or death from their conditions. They are subject to infections by common microbes that are not

Figure 1.3 **The 2011 Gulf oil spill.** There is evidence that ocean bacteria metabolized ("chewed up") some of the spilled oil.

NCLEX® PREP

1. For which of the following disease processes has microbial infection been implicated? Select all that apply.
   a. *gastric ulcers*
   b. *diabetes type 1*
   c. *renal artery stenosis*
   d. *schizophrenia*
   e. *obesity*
   f. *deep vein thrombosis*

Table 1.1 Top Causes of Death—All Diseases

United States	No. of Deaths	Worldwide	No. of Deaths
1. Heart disease	617,000	1. Heart disease	7 million
2. Cancer	565,000	2. Stroke	6.2 million
3. Chronic lower-respiratory disease	141,000	3. Lower-respiratory infections (influenza and pneumonia)	3.2 million
4. Cerebrovascular disease	134,000	4. Chronic obstructive pulmomary disease	3 million
5. Accidents (unintentional injuries)	122,000	5. Diarrheal diseases	1.9 million
6. Alzheimer's disease	82,000	6. HIV/AIDS	1.5 million
7. Diabetes	71,000	7. Trachea, bronchus, lung cancers	1.5 million
8. Influenza and pneumonia	56,000	8. Diabetes mellitus	1.4 million
9. Kidney disease	48,000	9. Road injury	1.3 million
10. Suicide	36,000	10. Prematurity	1.2 million

*Diseases in red are those most clearly caused by microorganisms.

Source: Data from the World Health Organization and the Centers for Disease Control and Prevention. Data published in 2014 representing final figures for the year 2011.

pathogenic to healthy people. There is also an increase in microbes that are resistant to drugs. It appears that even with the most modern technology available to us, microbes still have the "last word," as the great French scientist Louis Pasteur observed.

What Are They Exactly?

Cellular Organization

As discussed earlier, two basic cell types appeared during evolutionary history. The bacteria and archaea, along with **eukaryotic cells,** differ not only in the complexity of their cell structure but also in contents and function.

In general, bacterial and archaeal cells are about 10 times smaller than eukaryotic cells, and they lack many of the eukaryotic cell structures such as **organelles.** Organelles are small, double-membrane-bound structures in the eukaryotic cell that perform specific functions and include the nucleus, mitochondria, and chloroplasts. Examples of bacteria, archaea, and eukaryotic microorganisms are covered in more detail in chapters 3 and 4.

All bacteria and archaea are microorganisms, but only some eukaryotes are microorganisms **(figure 1.4).** Also, of course, humans are eukaryotes. Certain small eukaryotes–such as helminths (worms), many of which can be seen with the naked eye–are also included in the study of infectious diseases because of the way they are transmitted and the way the body responds to them, though they are not microorganisms.

Viruses are subject to intense study by microbiologists. They are not independently living cellular organisms. Instead, they are small particles that exist at the level of complexity somewhere between large molecules and cells. Viruses are much simpler than cells; outside their host, they are composed essentially of a small amount of hereditary material (either DNA or RNA but never both) wrapped up in a protein covering that is sometimes enveloped by a protein-containing lipid membrane. When inside their host

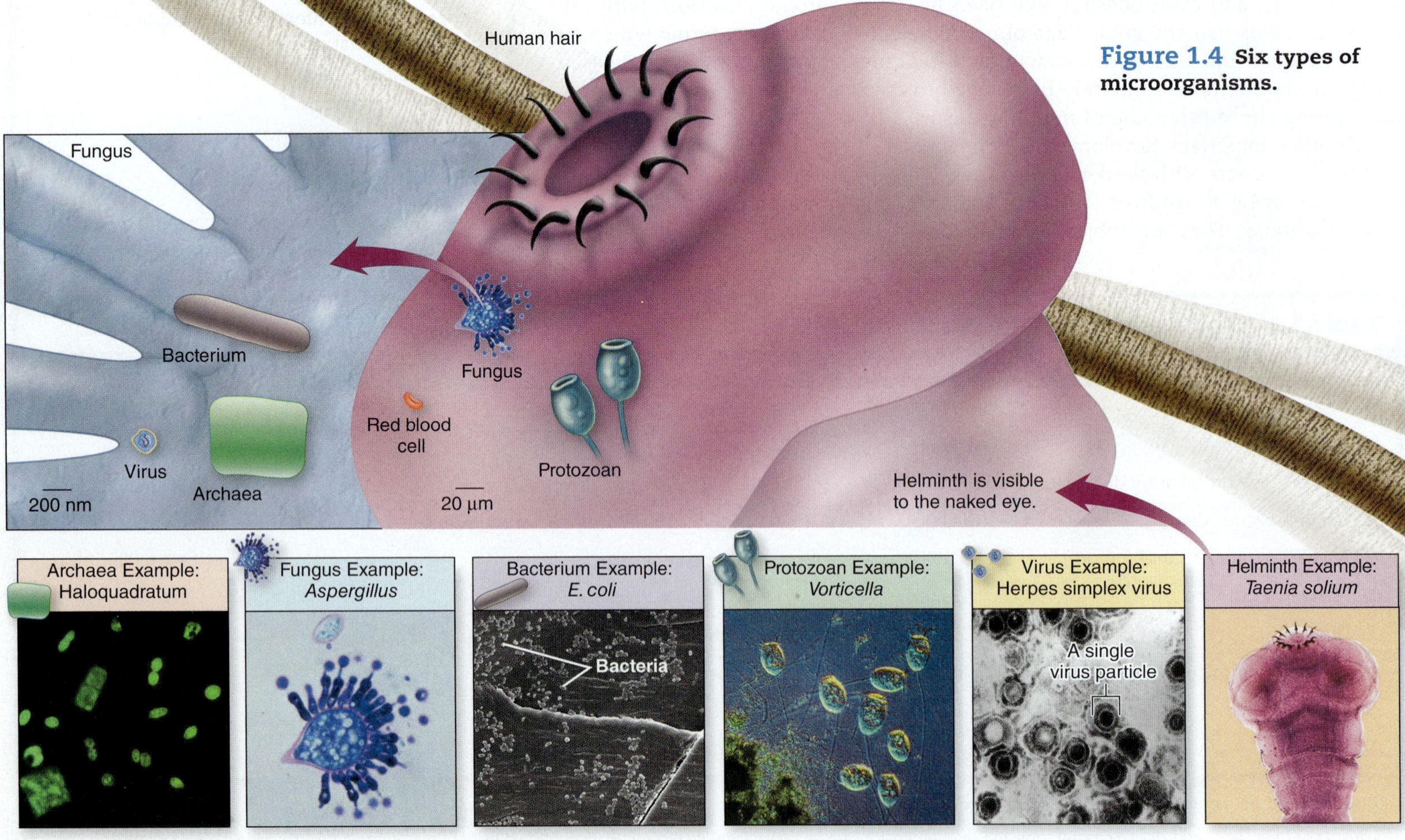

Figure 1.4 Six types of microorganisms.

Human hair

Fungus

Bacterium

Virus   Archaea

200 nm

Red blood cell

Fungus

Protozoan

20 μm

Helminth is visible to the naked eye.

| Archaea Example: Haloquadratum | Fungus Example: *Aspergillus* | Bacterium Example: *E. coli* | Protozoan Example: *Vorticella* | Virus Example: Herpes simplex virus | Helminth Example: *Taenia solium* |

Bacteria

A single virus particle

organism, in the intracellular state, viruses usually exist only in the form of genetic material that confers a partial genetic program on the host organisms.

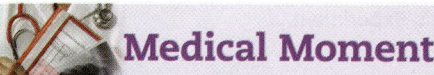

> ### 1.1 LEARNING OUTCOMES—Assess Your Progress
>
> 1. List the various types of microorganisms that can colonize humans.
> 2. Describe the role and impact of microbes on the earth.
> 3. Explain the theory of evolution and why it is called a theory.
> 4. Explain the ways that humans manipulate organisms for their own uses.
> 5. Summarize the relative burden of human disease caused by microbes.
> 6. Differentiate among bacteria, archaea, and eukaryotic microorganisms.
> 7. Identify a fourth type of microorganism.
> 8. Compare and contrast the relative sizes of the different microbes.

1.2 Microbes in History

If not for the extensive interest, curiosity, and devotion of thousands of microbiologists over the last 300 years, we would know little about the microscopic realm that surrounds us. Many of the discoveries in this science have resulted from the prior work of men and women who toiled long hours in dimly lit laboratories with the crudest of tools. Each additional insight, whether large or small, has added to our current knowledge of living things and processes. This section summarizes the prominent discoveries made in the past 300 years.

Spontaneous Generation

From very earliest history, humans noticed that when certain foods spoiled, they became inedible or caused illness, and yet other "spoiled" foods did no harm and even had enhanced flavor. Indeed, several centuries ago, there was already a sense that diseases such as the Black Plague and smallpox were caused by some sort of transmissible matter. But the causes of such phenomena were vague and obscure because, frankly, we couldn't *see* anything amiss. Consequently, they remained cloaked in mystery and regarded with superstition—a trend that led even well-educated scientists to believe in a concept called **spontaneous generation.** This was the belief that invisible vital forces present in matter led to the creation of life. The belief was continually reinforced as people observed that meat left out in the open soon "produced" maggots, that mushrooms appeared on rotting wood, that rats and mice emerged from piles of litter, and other similar phenomena. Though some of these early ideas seem quaint and ridiculous in light of modern knowledge, we must remember that, at the time, mysteries in life were accepted and the scientific method was not widely practiced.

Even after single-celled organisms were discovered during the mid-1600s, the idea of spontaneous generation continued to exist. Some scientists assumed that microscopic beings were an early stage in the development of more complex ones.

Over the subsequent 200 years, scientists waged an experimental battle over the two hypotheses that could explain the origin of simple life forms. Some tenaciously clung to the idea of **abiogenesis** (*a* = "without"; *bio* = "life"; *genesis* = "beginning"–"beginning in absence of life"), which embraced spontaneous generation. On the other side were

Medical Moment

Diabetes and the Viral Connection

Scientists have long believed that type 1 diabetes is triggered by an infection. Enteroviruses, such as Coxsackie virus B, have been the focus of intensive research. Several studies support this hypothesis. For example, a study published in 2010 showed that enteroviruses can play a role in the early development of type 1 diabetes through the infection of beta cells in the pancreas, which results in inflammation as a result of innate immunity. This study and others like it seem to suggest that many type 1 diabetic patients have persistent enterovirus infection, which is associated with inflammation in the gut mucosa.

Studies like these that are attempting to determine how diabetes develops are the first step in discovering a cure. If researchers definitively determine that type 1 diabetes is caused by a virus, perhaps one day there will be a vaccine to prevent the disease. Before this can be accomplished, however, researchers will need to determine why it is that not all individuals who become infected with the virus develop diabetes.

Source: 2010. *Nature Reviews Endocrinology* 6(5): 279–89.

Wine, cheese, and bread are each made using bacteria and fungi.

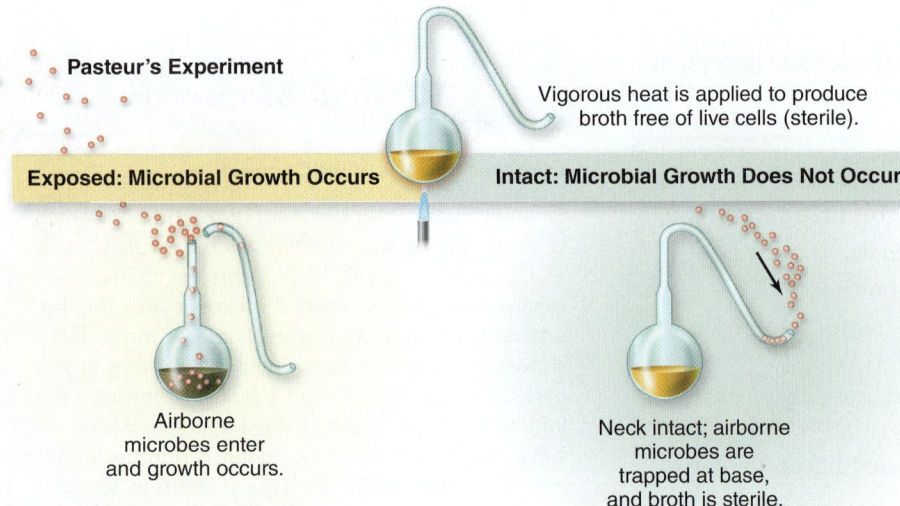

Pasteur's Experiment

Vigorous heat is applied to produce broth free of live cells (sterile).

Exposed: Microbial Growth Occurs   Intact: Microbial Growth Does Not Occur

Airborne microbes enter and growth occurs.

Neck intact; airborne microbes are trapped at base, and broth is sterile.

Figure 1.5 Pasteur's swan-neck flask experiment disproving spontaneous generation. He left the flask open to air but bent the neck so that gravity would trap any airborne microbes.

advocates of **biogenesis** ("beginning with life") saying that living things arise only from others of their same kind. There were serious advocates on both sides, and each side put forth what appeared on the surface to be plausible explanations of why its evidence was more correct. Finally in the mid-1800s, the acclaimed chemist and microbiologist Louis Pasteur entered the arena. He had recently been studying the roles of microorganisms in the fermentation of beer and wine, and it was clear to him that these processes were brought about by the activities of microbes introduced into the beverage from air, fruits, and grains. The methods he used to discount abiogenesis were simple yet brilliant.

To demonstrate that air and dust were the source of microbes, Pasteur filled flasks with broth and fashioned their openings into long, swan-neck-shaped tubes **(figure 1.5).** The flasks' openings were freely open to the air but were curved so that gravity would cause any airborne dust particles to deposit in the lower part of the necks. He heated the flasks to sterilize the broth and then incubated them. As long as the flask remained intact, the broth remained sterile; but if the neck was broken off so that dust fell directly down into the container, microbial growth immediately commenced.

Pasteur summed up his findings, "For I have kept from them, and am still keeping from them, that one thing which is above the power of man to make; I have kept from them the germs that float in the air, I have kept from them life."

The Role of the Microscope

True awareness of the widespread distribution of microorganisms and some of their characteristics was finally made possible by the development of the first microscopes. These devices revealed microbes as discrete entities sharing many of the characteristics of larger, visible plants and animals. The likely earliest record of microbes is in the works of Englishman Robert Hooke. In the 1660s, Hooke studied a great diversity of material from household objects, plants, and trees; described for the first time cellular structures in tree bark; and drew sketches of "little structures" that seemed to be alive. Hooke paved the way for even more exacting observations of microbes by Antonie van Leeuwenhoek (lay'-oo-wun-hook), a Dutch linen merchant and self-made microbiologist.

Figure 1.6 Leeuwenhoek's microscope.
A brass replica of a Leeuwenhoek microscope.

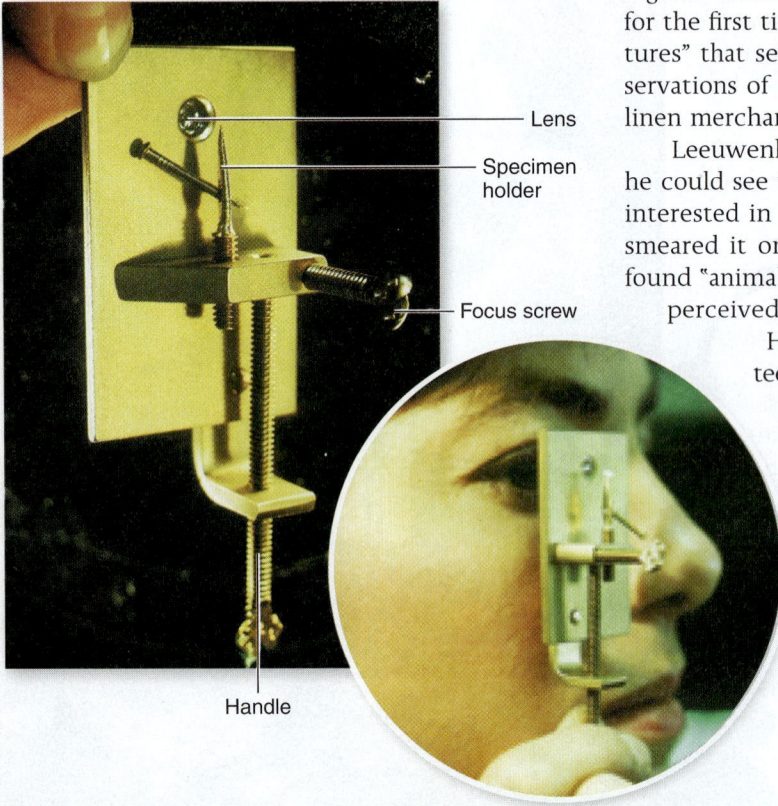

Lens

Specimen holder

Focus screw

Handle

Leeuwenhoek taught himself to grind glass lenses to ever-finer specifications so he could see with better clarity the threads in his fabrics. Eventually, he became interested in things other than thread counts. He took rainwater from a clay pot, smeared it on his specimen holder, and peered at it through his finest lens. He found "animals appearing to me ten thousand times less than those which may be perceived in the water with the naked eye."

He didn't stop there. He scraped the plaque from his teeth and from the teeth of some volunteers who had never cleaned their teeth in their lives and took a good close look at that. He recorded: "In the said matter there were many very little living animalcules, very prettily a-moving. . . . Moreover, the other animalcules were in such enormous numbers, that all the water . . . seemed to be alive." Leeuwenhoek started sending his observations to the Royal Society of London, and eventually he was recognized as a scientist of great merit.

Leeuwenhoek constructed more than 250 small, powerful microscopes that could magnify up to 300 times **(figure 1.6).** Considering that he had no formal training in science, his descriptions of bacteria and protozoa (which he called "animalcules") were astute and precise.

These events marked the beginning of our understanding of microbes and the diseases they can cause. Discoveries continue at a breakneck pace, however. In fact, the 2000s are being widely called the Century of Biology, fueled by our new abilities to study genomes and harness biological processes. Microbes have led the way in these discoveries and continue to play a large role in the new research.

To give you a feel for what has happened most recently, **table 1.2** provides a glimpse of some recent discoveries that have had huge impacts on our understanding of microbiology.

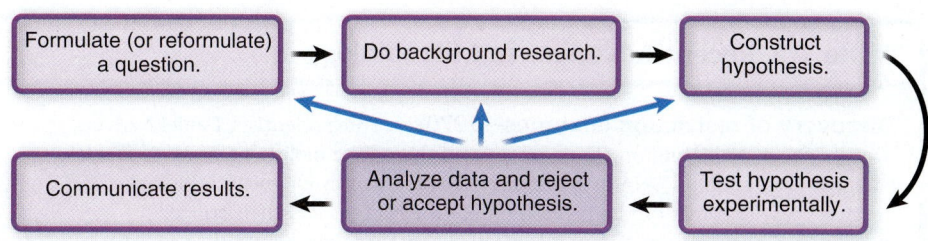

An overview of the scientific method.

The changes to our view of the role of RNAs that you see in table 1.2 highlight a feature of biology–and all of science–that is perhaps underappreciated. Because we have thick textbooks containing all kinds of assertions and "facts," many people think science is an ironclad collection of facts. Wrong! Science is an ever-evolving collection of new information, gleaned from observable phenomena and combined with old information to come up with the current understandings of nature. Some of the hypotheses explaining these observations have been confirmed so many times over such a long period of time that they are, if not "fact," very close to fact. Many other hypotheses will be altered over and over again as new findings emerge. And that is the beauty of science.

The Beginnings of Medical Microbiology

Early experiments on the sources of microorganisms led to the profound realization that microbes are everywhere: Not only are air and dust full of them, but the entire surface of the earth, its waters, and all objects are inhabited by them. This discovery led to immediate applications in medicine. Thus the seeds of medical microbiology were sown in the mid to latter half of the 19th century with the introduction of the germ theory of disease and the resulting use of sterile, aseptic, and pure culture techniques.

The Discovery of Spores and Sterilization

The discovery and detailed description of heat-resistant bacterial endospores by Ferdinand Cohn, a German botanist, clarified the reason that heat would sometimes fail to completely eliminate all microorganisms. The modern sense of the word **sterile,** meaning completely free of all life forms (including spores) and virus particles, was established from that point on (see chapter 9). The capacity to sterilize objects and materials is an absolutely essential part of microbiology, medicine, dentistry, and many industries.

The Development of Aseptic Techniques

At the same time that spontaneous generation was being hotly debated, a few physicians began to suspect that microorganisms could cause not only spoilage and decay but also infectious diseases. It occurred to these rugged individualists that even the human body itself was a source of infection. In 1843, Dr. Oliver Wendell Holmes, an American physician, published an article in which he observed that mothers who gave birth at home experienced fewer infections than did mothers who gave birth in the hospital; a few years later, the Hungarian Dr. Ignaz Semmelweis showed quite clearly that women became infected in the maternity ward after examinations by physicians coming directly from the autopsy room.

In the 1860s, the English surgeon Joseph Lister took notice of these observations and was the first to introduce **aseptic** (ay-sep'-tik) **techniques** aimed at reducing microbes in a medical setting and preventing wound infections. Lister's concept of asepsis was much more limited than our modern precautions. It mainly involved disinfecting the hands and the air with strong antiseptic chemicals, such as phenol, prior to surgery. It is hard for us to believe, but as recently as the late 1800s surgeons wore street clothes in the operating room and had little idea that hand washing was important **(figure 1.7).** Lister's techniques and the

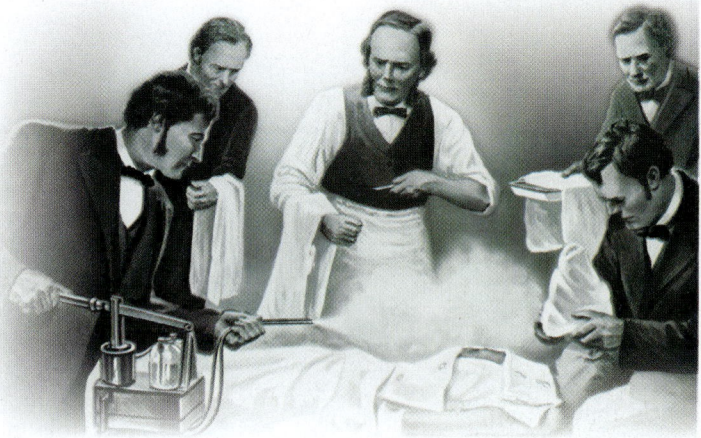

Figure 1.7 Joseph Lister's operating theater in the mid-1800s.

Table 1.2 Recent Advances in Microbiology

Discovery of restriction enzymes—1970s. Three scientists, Daniel Nathans, Werner Arber, and Hamilton Smith, discovered these little molecular "scissors" inside bacteria. They chop up DNA in specific ways. This was a major event in biology because these enzymes can be harvested from the bacteria and then utilized in research labs to cut up DNA in a controlled way that then allows us to splice the DNA pieces into vehicles that can carry them into other cells. This opened the floodgates to genetic engineering—and all that has meant for the treatment of diseases, the investigation into biological processes, and the biological "revolution" of the 21st century.

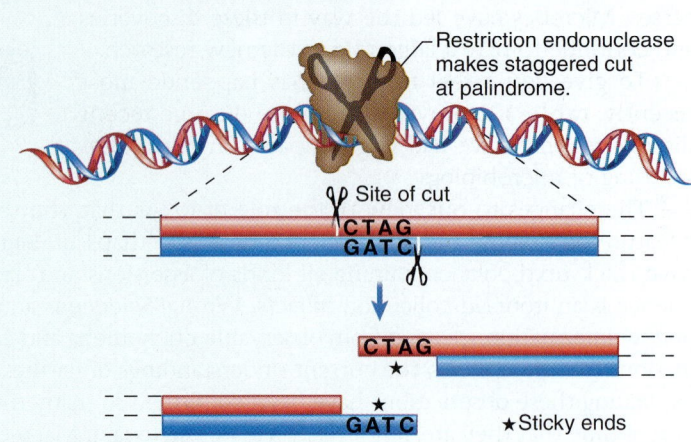

Restriction endonuclease makes staggered cut at palindrome.

Site of cut

CTAG
GATC

CTAG
★

★
GATC

★Sticky ends

The importance of biofilms in infectious diseases—1980s and beyond. Biofilms are accumulations of bacteria and other microbes on surfaces. Often there are multiple species in a single biofilm, and often they are several layers thick. They have been recognized in environmental microbiology for a long time. Biofilms on rocks, biofilms on ship hulls, even biofilms on ancient paintings have been well studied. We now understand that biofilms are relatively common in the human body (dental plaque is an example) and may be responsible for infections that are tough to conquer, such as some ear infections and recalcitrant infections of the prostate. Biofilms are also a danger to the success of any foreign body implanted in the body. Artificial hips, hearts, and even IUDs (intrauterine devices) have all been seen to fail due to biofilm colonization.

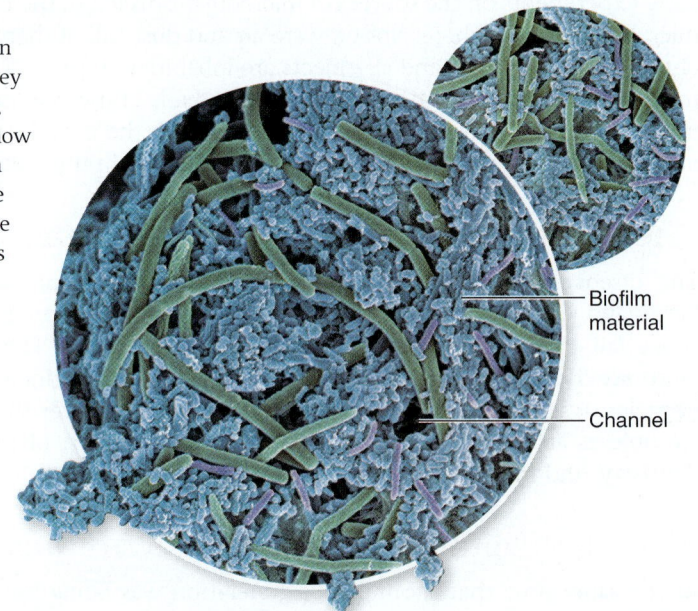

Biofilm material

Channel

application of heat for sterilization became the foundations for microbial control by physical and chemical methods, which are still in use today.

The Germ Theory of Disease

Louis Pasteur made enormous contributions to our understanding of the microbial role in wine and beer formation. He invented pasteurization and conducted some of the first studies showing that human diseases could arise from infection. These studies, supported by the work of other scientists, became known as the **germ theory of disease.** Pasteur's contemporary, Robert Koch, established *Koch's postulates*, a series of proofs, or logical steps, that verified the germ theory and could establish whether an organism was pathogenic and which disease it caused (see chapter 11). About 1875,

The importance of small RNAs—2000s. Once we were able to sequence entire genomes (another big move forward), scientists discovered something that turned a concept we literally used to call "dogma" on its head. The previously held "Central Dogma of Biology" was that DNA makes RNA which dictated the creation of proteins. Genome sequencing has revealed that perhaps only 2% of DNA leads to a resulting protein. Much RNA doesn't end up with a protein counterpart. These pieces of RNA are usually small. It now appears that they have critical roles in regulating what happens in the cell. It has led to new approaches to how diseases are treated. For example, if the small RNAs are important in bacteria that infect humans, they can be new targets for antimicrobial therapy.

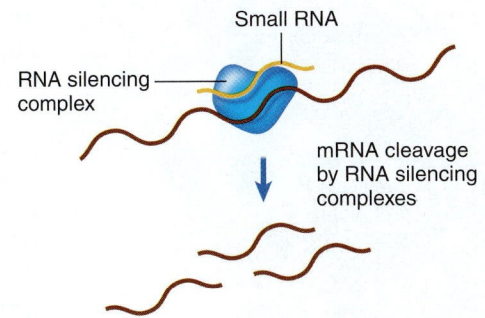

Small RNA

RNA silencing complex

mRNA cleavage by RNA silencing complexes

Genetic identification of the human microbiome—2010s and beyond. The first detailed information produced by the Human Microbiome Project (HMP) was astounding: For one thing, 90% of the cells in and on our body are not human at all but are microbial. For another, even though the exact types of microbes found in and on different people are highly diverse, the overall set of metabolic capabilities the bacterial communities possess is remarkably similar among people. This and other groundbreaking discoveries have set the stage for new knowledge of our microbial guests and their role in our overall health and disease.

Koch used this experimental system to show that anthrax was caused by a bacterium called *Bacillus anthracis*. So useful were his postulates that the causative agents of 20 other diseases were discovered between 1875 and 1900, and even today, they are the standard for identifying pathogens of plants and animals.

1.2 LEARNING OUTCOMES—Assess Your Progress

9. Make a time line of the development of microbiology from the 1600s to today.

10. List some recent microbiology discoveries of great impact.

The green specks are microorganisms in the stomach of a tube worm.

1.3 Macromolecules: Superstructures of Life

In this book, we won't be presenting the basics of chemistry, though of course it is important to understand chemical concepts to understand all of biology. But that is what chemistry textbooks are for! However, there will be so much emphasis on some important *bio*chemicals in this book and in your course that we want to present a concise description of cellular macromolecules.

All microorganisms–indeed, all organisms–are constructed from just a few major types of biological molecules, called **macromolecules,** because they are often very large. They include four main families: carbohydrates, lipids, proteins, and nucleic acids **(table 1.3).** All macromolecules except lipids are formed by polymerization, a process in which repeating subunits termed **monomers** are bound into chains of various lengths termed **polymers.** For example, proteins (polymers) are composed of a chain of amino acids (monomers). In the following section and in later chapters, we consider numerous concepts relating to the roles of macromolecules in cells. **Table 1.4** presents the important structural features of the four main macromolecules.

Carbohydrates: Sugars and Polysaccharides

The term **carbohydrate** originates from the composition of members of this class: They are combinations of carbon (*carbo-*) and water (*-hydrate*). Although carbohydrates can be generally represented by the formula $(CH_2O)_n$, in which *n* indicates the

Table 1.3 Macromolecules and Their Functions

Macromolecule	Basic Structure	Examples	Notes About the Examples
Carbohydrates			
Monosaccharides	3- to 7-carbon sugars	Glucose, fructose	Sugars involved in metabolic reactions; building block of disaccharides and polysaccharides
Disaccharide	Two monosaccharides	Maltose (malt sugar)	Composed of two glucoses; an important breakdown product of starch
		Lactose (milk sugar)	Composed of glucose and galactose
		Sucrose (table sugar)	Composed of glucose and fructose
Polysaccharides	Chains of monosaccharides	Starch, cellulose, glycogen	Cell wall, food storage
Lipids			
Triglycerides	Fatty acids + glycerol	Fats, oils	Major component of cell membranes; storage
Phospholipids	Fatty acids + glycerol + phosphate	Membrane components	
Waxes	Fatty acids, alcohols	Mycolic acid	Cell wall of mycobacteria
Steroids	Ringed structure	Cholesterol, ergosterol	In membranes of eukaryotes and some bacteria
Proteins			
	Chains of amino acids	Enzymes; part of cell membrane, cell wall, ribosomes, antibodies	Serve as structural components and perform metabolic reactions
Nucleic acids	Nucleotides (pentose sugar + phosphate + nitrogen base) Nitrogen bases Purines: adenine (A), guanine (G) Pyrimidines: cytosine (C), thymine (T), uracil (U)		
Deoxyribonucleic acid (DNA)	Contains deoxyribose sugar and thymine, not uracil	Chromosomes; genetic material of viruses	Mediate inheritance
Ribonucleic acid (RNA)	Contains ribose sugar and uracil, not thymine	Ribosomes; mRNA, tRNA, small RNAs, genetic material of viruses	Facilitate expression of genetic traits

Table 1.4 Macromolecules in the Cell

Carbohydrates. Another word for *sugar* is **saccharide.** A **monosaccharide** is a simple sugar containing from 3 to 7 carbons; a **disaccharide** is a combination of two monosaccharides; and a **polysaccharide** is a polymer of five or more monosaccharides bound in linear or branched chain patterns.

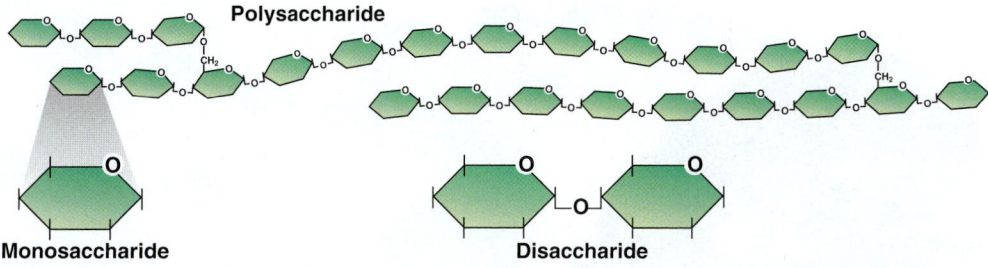

Lipids. The term **lipid,** derived from the Greek word *lipos,* meaning fat, is not a chemical designation but an operational term for a variety of substances that are not soluble in polar solvents such as water but will dissolve in nonpolar solvents such as benzene and chloroform. Here we see a model of a single molecule of a phospholipid. The phosphate-alcohol head leads a charge to one end of the molecule; its long trailing hydrocarbon chain is uncharged.

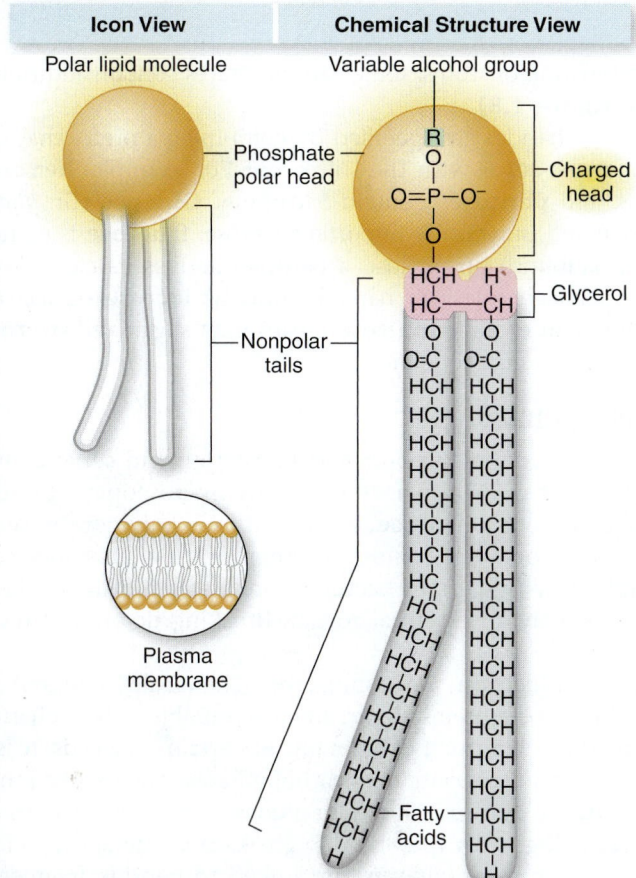

Proteins. Proteins are chains of amino acids. Amino acids have a basic skeleton consisting of a carbon (called the α carbon) linked to an amino group (NH_2), a carboxyl group (COOH), a hydrogen atom (H), and a variable R group. The variations among the amino acids occur at the R group, which is different in each amino acid and imparts the unique characteristics to the molecule and to the proteins that contain it. A covalent bond called a **peptide bond** forms between the amino group on one amino acid and the carboxyl group on another amino acid.

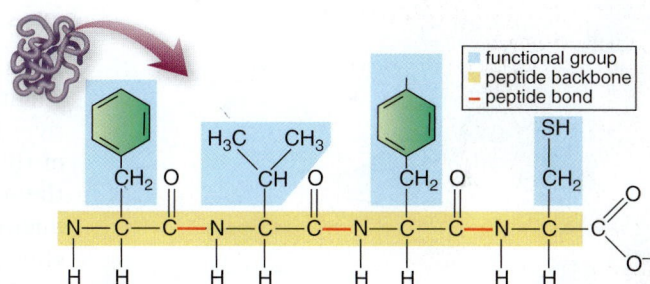

Nucleic acids. Both DNA and RNA are polymers of repeating units called **nucleotides,** each of which is composed of three smaller units: a **nitrogen base,** a **pentose** (5-carbon) sugar, and a **phosphate.** The nitrogen base is a cyclic compound that comes in two forms: *purines* (two rings) and *pyrimidines* (one ring). There are two types of purines—**adenine (A)** and **guanine (G)**—and three types of pyrimidines—**thymine (T), cytosine (C),** and **uracil (U).** The nitrogen base is covalently bonded to the sugar *ribose* in RNA and *deoxyribose* (because it has one less oxygen than ribose) in DNA. The backbone of a nucleic acid strand is a chain of alternating phosphate-sugar-phosphate-sugar molecules, and the nitrogen bases branch off the side of this backbone.

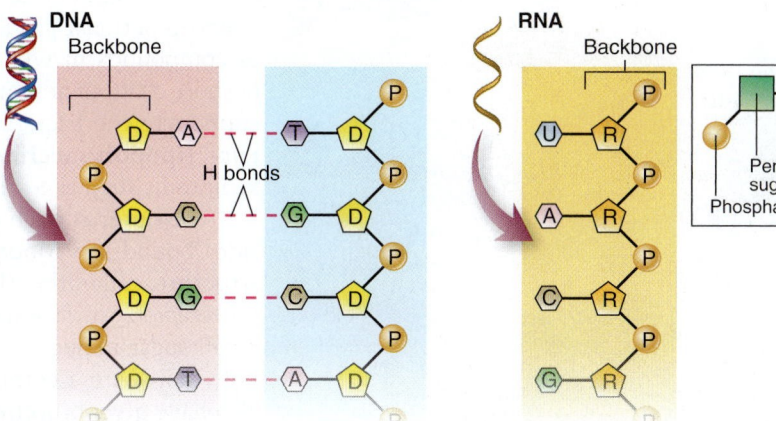

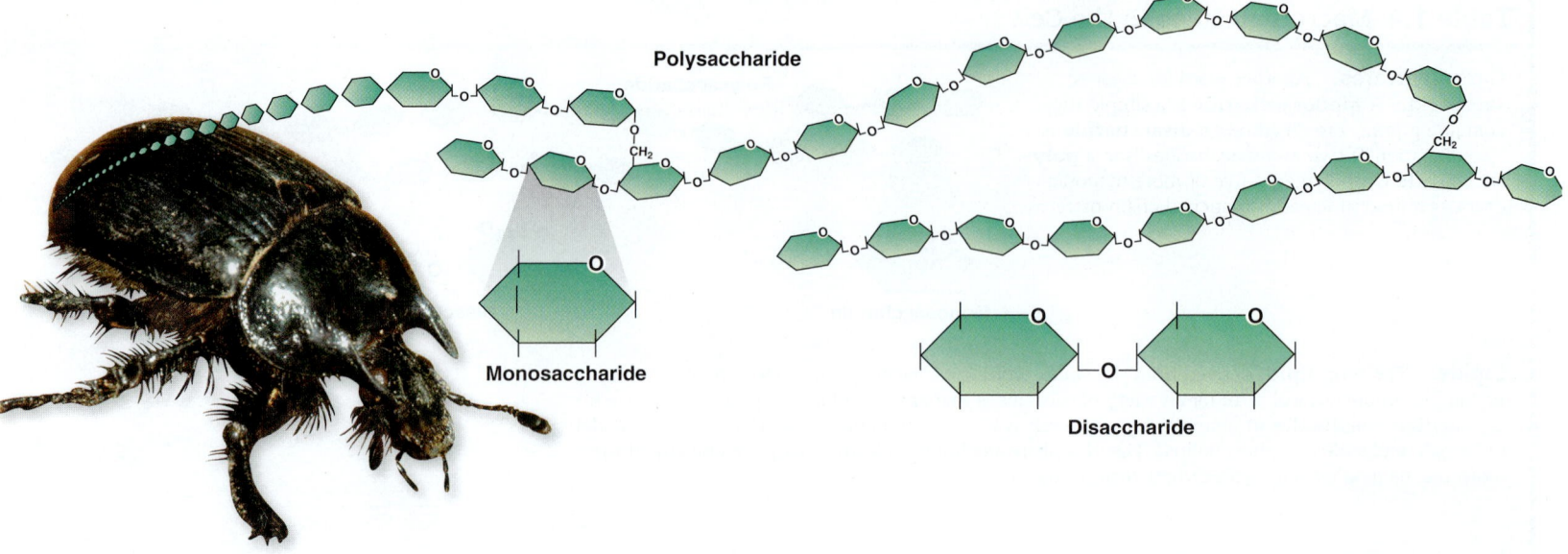

Figure 1.8 Carbohydrates. Polysaccharides are built of monomer sugars. They are present in many substances in nature, including chitin, which makes up the exoskeleton of some insects.

number of units of this combination of atoms, some carbohydrates contain additional atoms of sulfur or nitrogen **(figure 1.8).**

Monosaccharides and disaccharides are specified by combining a prefix that describes some characteristic of the sugar with the suffix *-ose.* For example, **hexoses** are composed of 6 carbons, and **pentoses** contain 5 carbons. **Glucose** (Gr. *glyko*, "sweet") is the most common and universally important hexose; **fructose** is named for fruit (one place where it is found); and xylose, a pentose, derives its name from the Greek word for wood. Disaccharides are named similarly: **lactose** (L. *lacteus*, "milk") is an important component of milk; **maltose** means malt sugar; and **sucrose** (Fr. *sucre*, "sugar") is common table sugar or cane sugar.

The Functions of Polysaccharides

Polysaccharides contribute to structural support and protection and serve as nutrient and energy stores. The cell walls in plants and many microscopic algae derive their strength and rigidity from **cellulose,** a long, fibrous polymer. Because of this role, cellulose is probably one of the most common organic substances on the earth, yet it is digestible only by certain bacteria, fungi, and protozoa. These microbes, called decomposers, play an essential role in breaking down and recycling plant materials.

Other structural polysaccharides can be conjugated (chemically bonded) to amino acids, nitrogen bases, lipids, or proteins. **Agar,** an indispensable polysaccharide in preparing solid culture media, is a natural component of certain seaweeds. It is a complex polymer of galactose and sulfur-containing carbohydrates. The exoskeletons of certain fungi contain **chitin** (ky'-tun), a polymer of glucosamine (a sugar with an amino functional group). **Peptidoglycan** (pep-tih-doh-gly'-kan) is one special class of compounds in which polysaccharides (glycans) are linked to peptide fragments (a short chain of amino acids). This molecule provides the main source of structural support to the bacterial cell wall. The cell wall of gram-negative bacteria also contains **lipopolysaccharide,** a complex of lipid and polysaccharide responsible for symptoms such as fever and shock (see chapters 3 and 11).

The outer surface of many cells has a "sugar coating" composed of polysaccharides bound in various ways to proteins (the combination is a glycoprotein). This structure, called the **glycocalyx,** functions in attachment to other cells or as a site for *receptors*–surface molecules that receive external stimuli or act as binding sites. Small sugar molecules account for the differences in human blood types, and carbohydrates are a component of large protein molecules called antibodies. Viruses also have glycoproteins on their surface with which they bind to and invade their host cells.

Lipids: Fats, Phospholipids, and Waxes

There are four main types of compounds classified as lipids: triglycerides, phospholipids, steroids, and waxes.

The **triglycerides** are an important storage lipid. This category includes fats and oils. Triglycerides are composed of a single molecule of glycerol bound to three fatty acids **(figure 1.9)**. **Glycerol** is a 3-carbon alcohol with three OH groups that serve as binding sites, and fatty acids are long-chain hydrocarbon molecules with a carboxyl group (COOH) at one end that is free to bind to the glycerol. The hydrocarbon portion of a fatty acid can vary in length from 4 to 24 carbons—and, depending on the fat, it may be saturated or unsaturated. If all carbons in the chain

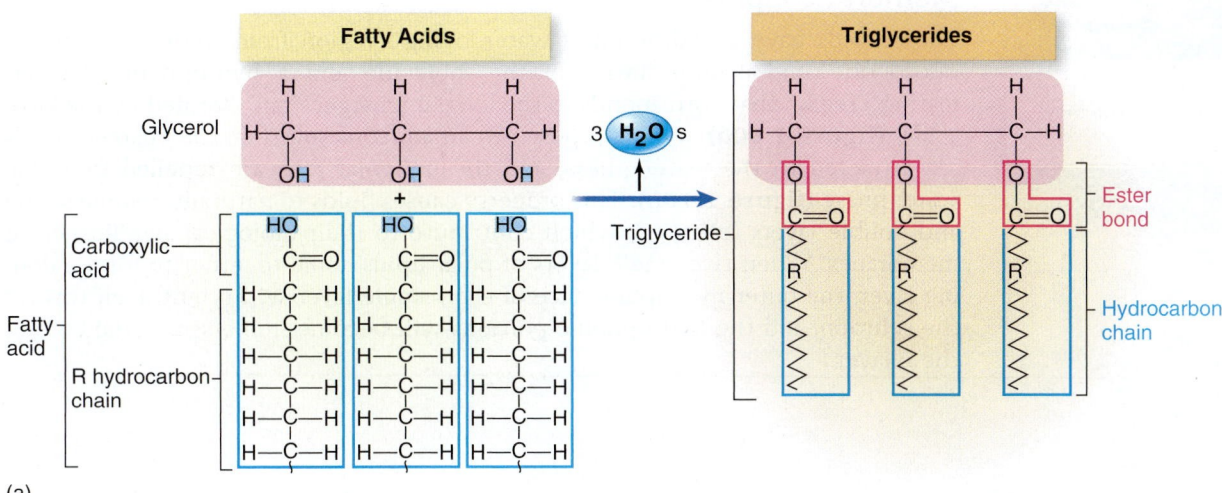

(a)

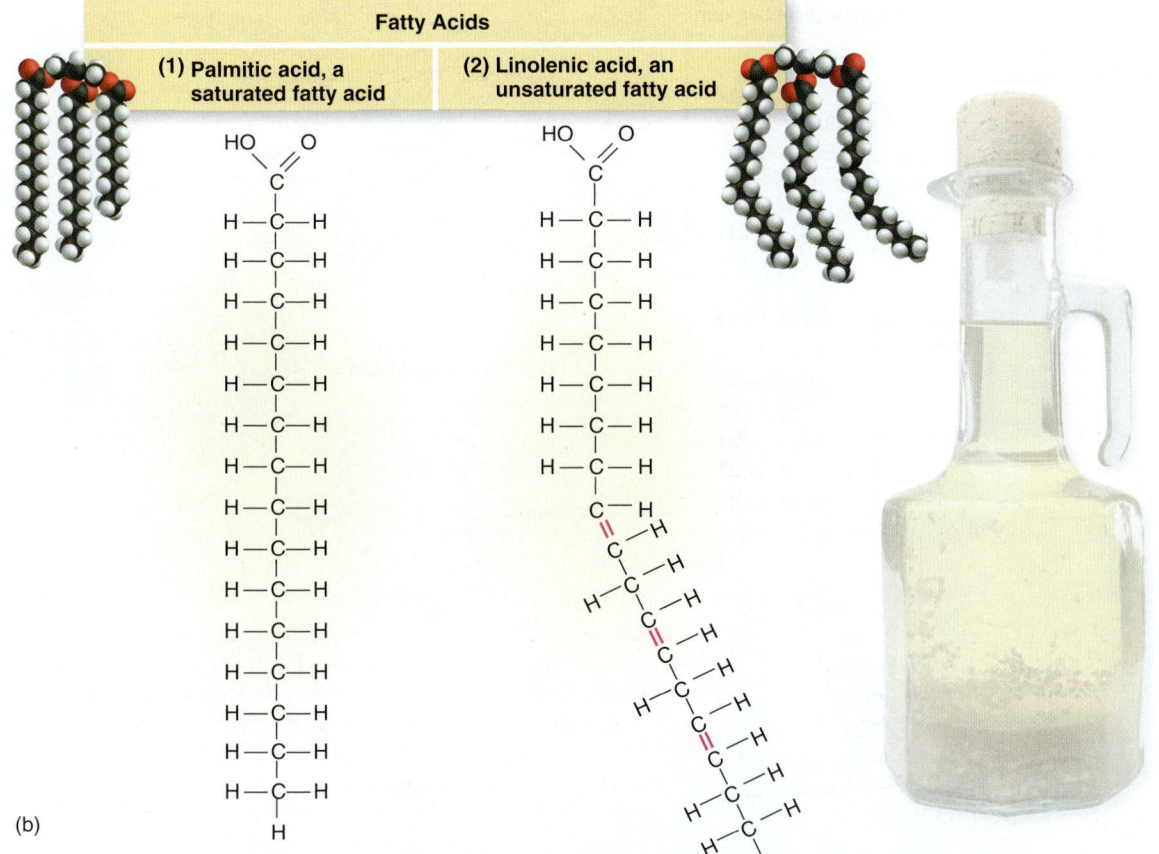

(b)

Figure 1.9 Synthesis and structure of a triglyceride.
(a) Because a water molecule is released at each ester bond, this is an example of dehydration synthesis. The jagged lines and R symbol represent the hydrocarbon chains of the fatty acids, which are commonly very long. (b) Structural and three-dimensional models of fatty acids and triglycerides. (1) A saturated fatty acid has long, straight chains that readily pack together and form solid fats. (2) An unsaturated fatty acid—here a polyunsaturated one with 3 double bonds—has a bend in the chain that prevents packing and produces oils (right).

Oils on duck feathers keep these two canvasback ducks insulated and dry, no matter how much time they spend in the water.

are single-bonded to 2 other carbons and 2 hydrogens, the fat is saturated; if there is at least one C=C double bond in the chain, it is unsaturated. The structure of fatty acids is what gives fats and oils (liquid fats) their greasy, insoluble nature. In general, solid fats (such as butter) are more saturated, and liquid fats (such as oils) are more unsaturated.

In most cells, triglycerides are stored in long-term concentrated form as droplets or globules. When they are acted on by digestive enzymes called lipases, the fatty acids and glycerol are freed to be used in metabolism. Fatty acids are a superior source of energy, yielding twice as much per gram as other storage molecules (starch). Soaps are K^+ or Na^+ salts of fatty acids whose qualities make them excellent grease removers and cleaners (see chapter 9).

Membrane Lipids

These lipids have a hydrophilic ("water-loving") region from the charge on the phosphoric acid-alcohol "head" of the molecule and a hydrophobic ("water-fearing") region that corresponds to the long, uncharged "tail" (formed by the fatty acids) **(figure 1.10a)**. When exposed to an aqueous solution, the charged heads are attracted to the water phase, and the nonpolar tails are repelled from the water phase **(figure 1.10b)**. This property causes lipids to naturally assume single and double layers (bilayers), which contribute to their biological significance in membranes. When two single layers of polar lipids come together to form a double layer, the outer hydrophilic face of each single layer will orient itself toward the solution, and the hydrophobic portions will become immersed in the core of the bilayer.

Figure 1.10 Phospholipids— membrane molecules.
(a) A model of a single molecule of a phospholipid. The phosphate-alcohol head lends a charge to one end of the molecule; its long, trailing hydrocarbon chain is uncharged. (b) The behavior of phospholipids in water-based solutions causes them to become arranged (1) in single layers called micelles, with the charged head oriented toward the water phase and the hydrophobic nonpolar tail buried away from the water phase, or (2) in double-layered phospholipid systems with the hydrophobic tails sandwiched between two hydrophilic layers.

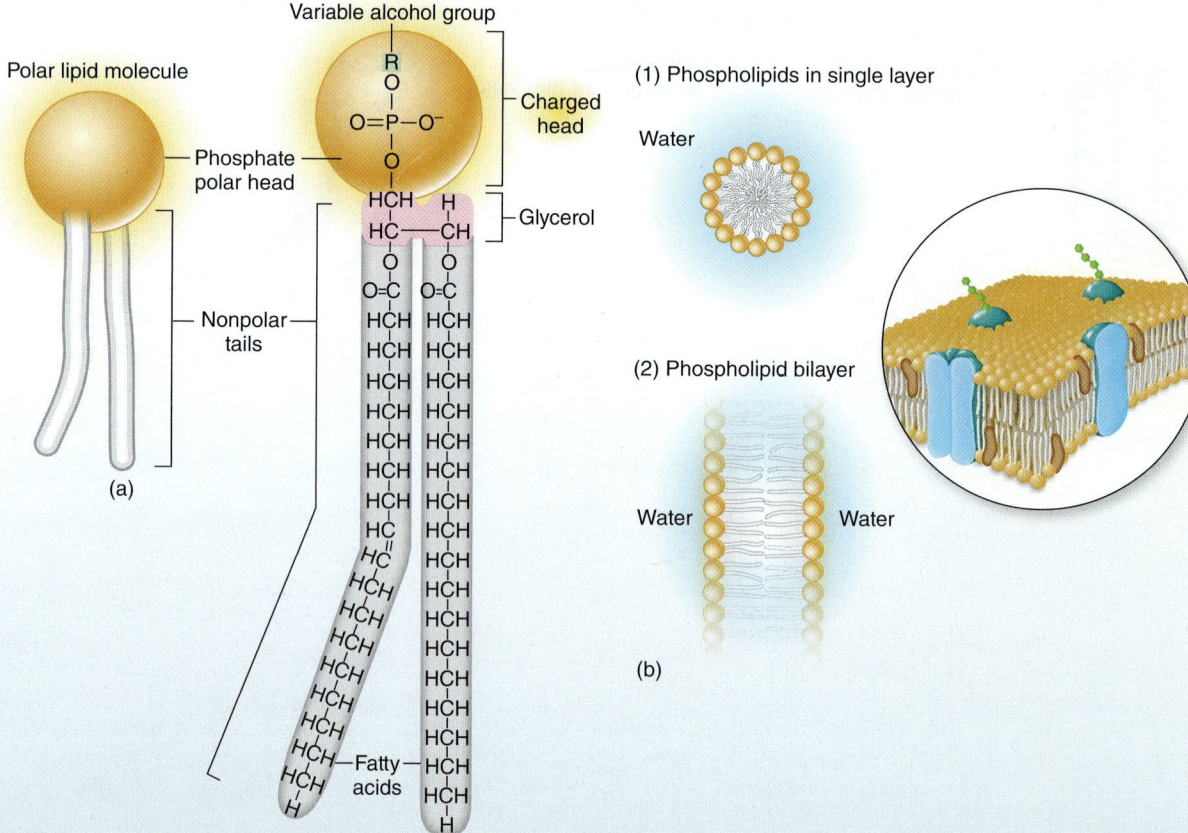

Steroids and Waxes

Steroids are complex ringed compounds commonly found in cell membranes and animal hormones. The best known of these is the sterol (meaning a steroid with an OH group) called **cholesterol (figure 1.11).** Cholesterol reinforces the structure of the cell membrane in animal cells and in an unusual group of cell-wall-deficient bacteria called the mycoplasmas (see chapter 3). The cell membranes of fungi also contain a sterol, called ergosterol.

Chemically, a *wax* is an ester formed between a long-chain alcohol and a saturated fatty acid. The resulting material is typically pliable and soft when warmed but hard and water resistant when cold (paraffin, for example). Among living things, fur, feathers, fruits, leaves, human skin, and insect exoskeletons are naturally waterproofed with a coating of wax. Bacteria that cause tuberculosis and leprosy produce a wax that repels ordinary laboratory stains and contributes to their pathogenicity.

Proteins: Shapers of Life

The predominant organic molecules in cells are **proteins.** To a large extent, the structure, behavior, and unique qualities of each living thing are a consequence of the proteins they contain. The building blocks of proteins are **amino acids,** which exist in 20 different naturally occurring forms **(table 1.5).** Various combinations of these amino acids account for the nearly infinite variety of proteins.

Various terms are used to denote the nature of proteins. **Peptide** usually refers to a molecule composed of short chains of amino acids, such as a dipeptide (two amino acids), a tripeptide (three), and a tetrapeptide (four). A **polypeptide** contains an unspecified number of amino acids but usually has more than 20 and is often a smaller subunit of a protein. A protein is the largest of this class of compounds and usually contains a minimum of 50 amino acids. It is common for the term *protein* to be used to describe all of these molecules. In chapter 8, we see that protein synthesis is not just a random connection of amino acids; it is directed by information provided in DNA.

Table 1.5 Twenty Amino Acids and Their Abbreviations

Acid	Abbreviation	Characteristic of R Groups
Alanine	Ala	nonpolar
Arginine	Arg	+
Asparagine	Asn	polar
Aspartic acid	Asp	−
Cysteine	Cys	polar
Glutamic acid	Glu	−
Glutamine	Gln	polar
Glycine	Gly	polar
Histidine	His	+
Isoleucine	Ile	nonpolar
Leucine	Leu	nonpolar
Lysine	Lys	+
Methionine	Met	nonpolar
Phenylalanine	Phe	nonpolar
Proline	Pro	nonpolar
Serine	Ser	polar
Threonine	Thr	polar
Tryptophan	Trp	nonpolar
Tyrosine	Tyr	polar
Valine	Val	nonpolar

+ = positively charged; − = negatively charged.

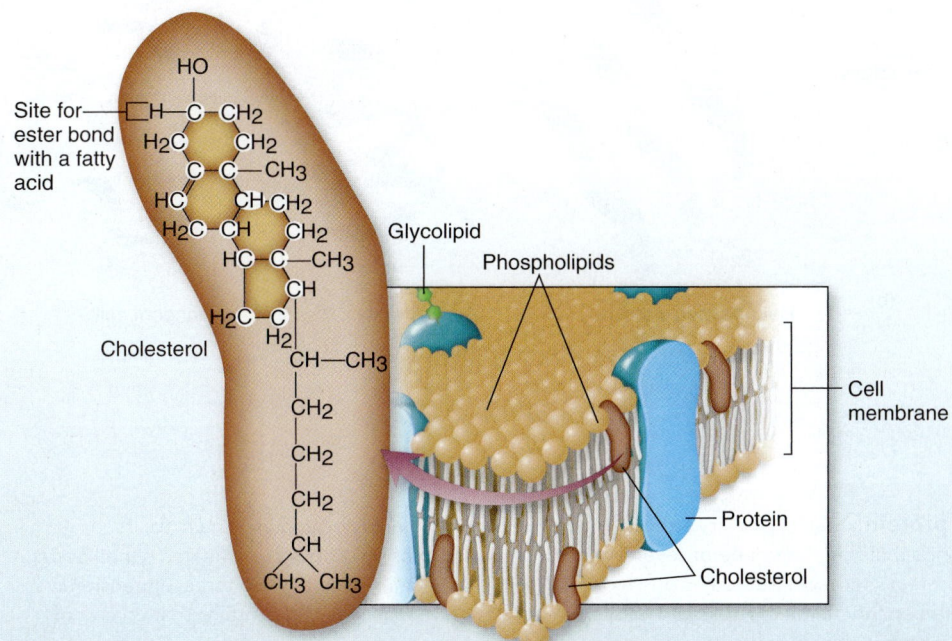

Figure 1.11 Cutaway view of a membrane with its bilayer of lipids. The primary lipid is phospholipid—however, cholesterol is inserted in some membranes. Other structures are protein and glycolipid molecules.

Protein Structure and Diversity

The reason that proteins are so varied and specific is that they do not function in the form of a simple straight chain of amino acids (called the primary structure). A protein has a natural tendency to assume more complex levels of organization, called the secondary, tertiary, and quaternary structures **(figure 1.12)**. The **primary (1°) structure** is the type, number, and order of amino acids in the chain, which varies extensively from protein to protein. The **secondary (2°) structure** arises when various functional groups exposed on the outer surface of the molecule interact by forming hydrogen bonds. This interaction causes the amino acid chain to twist into a coiled configuration called the *alpha helix* (α helix) or to fold into an accordion pattern called a *beta-pleated sheet* (β-pleated sheet). Many proteins contain both types of secondary configurations. Proteins at the secondary level undergo a third degree of torsion called the **tertiary (3°) structure** created by additional bonds between functional groups (figure 1.12c). In proteins with the sulfur-containing amino acid **cysteine,** considerable tertiary stability is achieved through covalent disulfide bonds between sulfur atoms on two different parts of the molecule. Some complex proteins assume a **quaternary (4°) structure,** in which more than one polypeptide forms a large, multiunit protein. This is typical of antibodies and some enzymes that act in cell synthesis.

Protein is a major component of meats, eggs, and nuts.

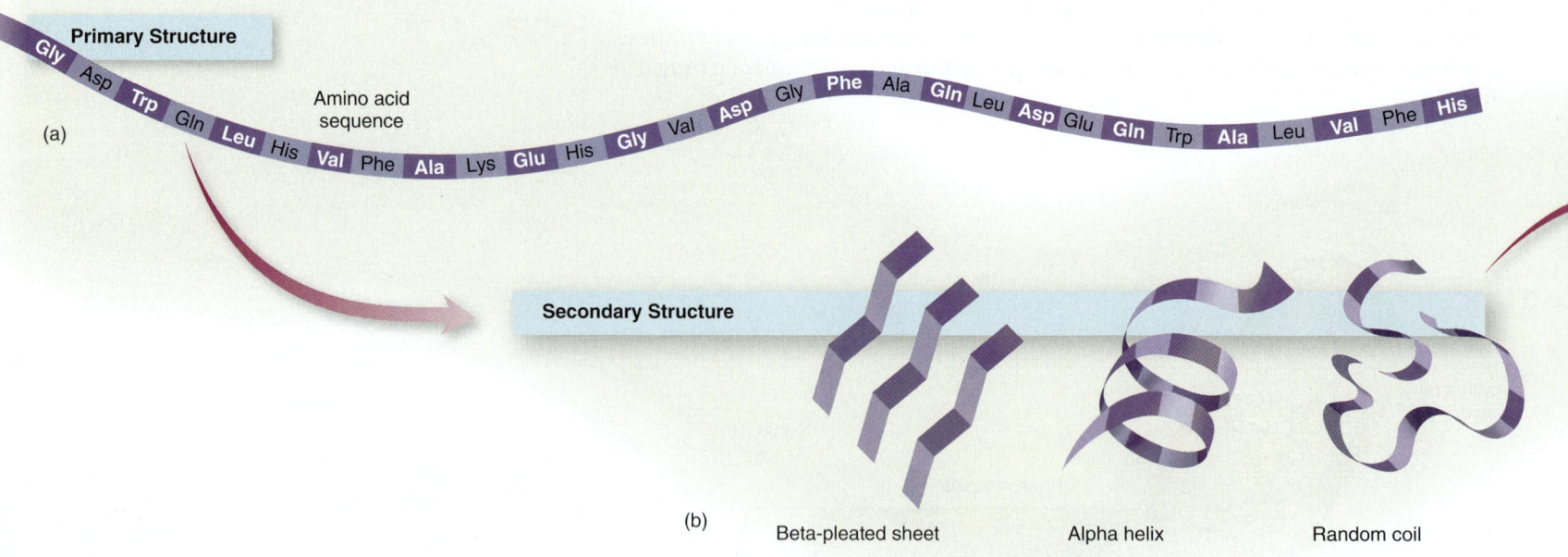

(a) **Primary Structure**

Gly Asp Trp Gln Leu His Val Phe Ala Lys Glu His Gly Val Asp Gly Phe Ala Gln Leu Asp Glu Gln Trp Ala Leu Val Phe His

Amino acid sequence

(b) **Secondary Structure**

Beta-pleated sheet   Alpha helix   Random coil

Figure 1.12 Stages in the formation of a functioning protein. (a) Its primary structure is a series of amino acids bound in a chain. (b) Its secondary structure develops when the chain forms hydrogen bonds that fold it into one of several configurations such as an α helix or β-pleated sheet. Some proteins have several configurations in the same molecule. (c) A protein's tertiary structure is due to further folding of the molecule into a three-dimensional mass that is stabilized by hydrogen, ionic, and disulfide bonds between functional groups. (d) The quaternary structure exists only in proteins that consist of more than one polypeptide chain. The chains in this protein each have a different color.

The most important outcome of the various forms of bonding and folding is that each different type of protein develops a unique shape, and its surface displays a distinctive pattern of pockets and bulges. As a result, a protein can react only with molecules that complement or fit its particular surface features like a lock and key. Such a degree of specificity can provide the functional diversity required for many thousands of different cellular activities. **Enzymes** serve as the catalysts for all chemical reactions in cells, and nearly every reaction requires a different enzyme (see chapter 7). This specificity comes from the architecture of the binding site, which determines which molecules fit it. The same is true of antibodies: **Antibodies** are complex glycoproteins with specific regions of attachment for bacteria, viruses, and other microorganisms; certain bacterial toxins (poisonous products) react with only one specific organ or tissue; and proteins embedded in the cell membrane have reactive sites restricted to a certain nutrient. The functional three-dimensional form of a protein is termed the *native state*, and if it is disrupted by some means, the protein is said to be *denatured*. Such agents as heat, acid, alcohol, and some disinfectants disrupt (and thus denature) the stabilizing bonds within the chains and cause the molecule to become nonfunctional, as described in chapter 9.

The Nucleic Acids: A Cell Computer and Its Programs

DNA, the master computer of cells, contains a special coded genetic program with detailed and specific instructions for each organism's heredity. It transfers the details of its program to RNA, "helper" molecules responsible for carrying out DNA's instructions and translating the DNA program into proteins that can perform life functions. For now, let us briefly consider the structure and some functions of DNA, RNA, and a close relative, adenosine triphosphate (ATP).

Curly hair is the result of particular protein folding patterns as described in figure 1.12.

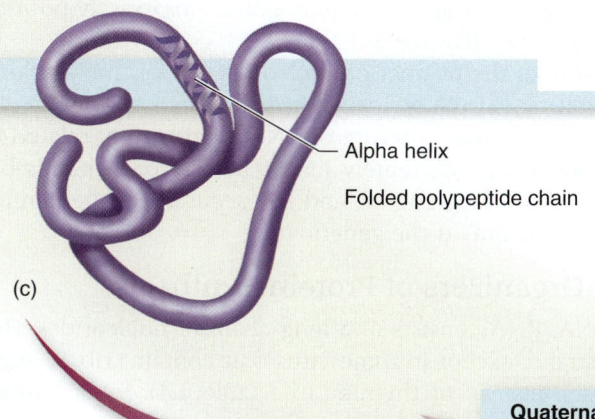

Tertiary Structure

Alpha helix

Folded polypeptide chain

(c)

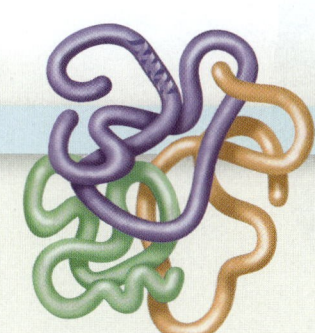

Quaternary Structure

Two or more polypeptide chains

(d)

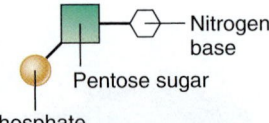

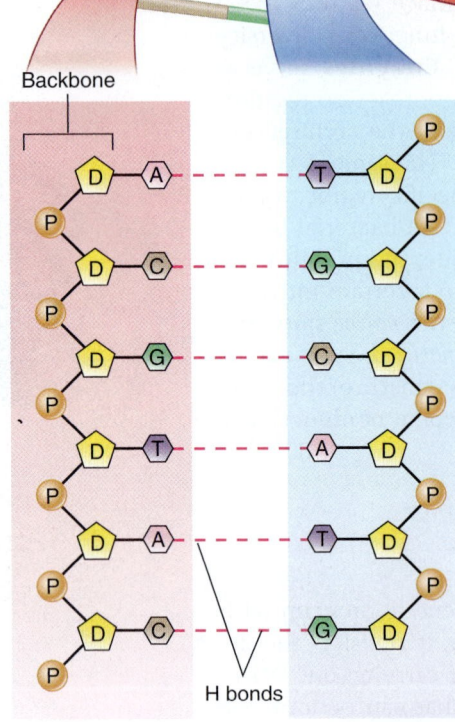

(a) A nucleotide, composed of a phosphate, a pentose sugar, and a nitrogen base (either A, T, C, G, or U) is the monomer of both DNA and RNA.

(b) DNA molecules are composed of alternating deoxyribose (D) and phosphate (P) with nitrogen bases (A, T, C, G) attached to the deoxyribose. DNA almost always exists in pairs of strands, oriented so that the bases are paired across the central axis of the molecule.

(c) RNA molecules are composed of alternating ribose (R) and phosphate (P) attached to nitrogen bases (A, U, C, G), but it is usually a single strand.

Figure 1.13 **The general structure of nucleic acids.**

The Double Helix of DNA

DNA is a huge molecule formed by two very long nucleotide strands linked along their length by hydrogen bonds between nitrogen bases. The pairing of the nitrogen bases occurs according to a predictable pattern: Adenine always pairs with thymine, and cytosine with guanine. The bases are attracted in this way because each pair shares oxygen, nitrogen, and hydrogen atoms exactly positioned to align perfectly for hydrogen bonds **(figure 1.13).**

Owing to the manner of nucleotide pairing and stacking of the bases, the actual configuration of DNA is a *double helix* that looks somewhat like a spiral staircase. As is true of protein, the structure of DNA is intimately related to its function. DNA molecules are usually extremely long. The hydrogen bonds between pairs break apart when DNA is being copied, and the accuracy of the complementary base-pairing is essential to maintain the genetic code.

RNA: Organizers of Protein Synthesis

Like DNA, RNA consists of a long chain of nucleotides. However, RNA is usually a single strand, except in some viruses. It contains ribose sugar instead of deoxyribose and uracil instead of thymine (see table 1.4). Several functional types of RNA are formed using the DNA template through a replication-like process. Three major types of RNA are important for protein synthesis. Messenger RNA (mRNA) is a copy of a gene (a single functional part of the DNA) that provides the order and type of amino acids in a protein; transfer RNA (tRNA) is a carrier that delivers the correct amino acids for protein assembly; and ribosomal RNA (rRNA) is a major component of ribosomes (described in chapter 3). A fourth type of RNA is the RNA that acts to regulate the genes and gene expression. More information on these important processes is presented in chapter 8.

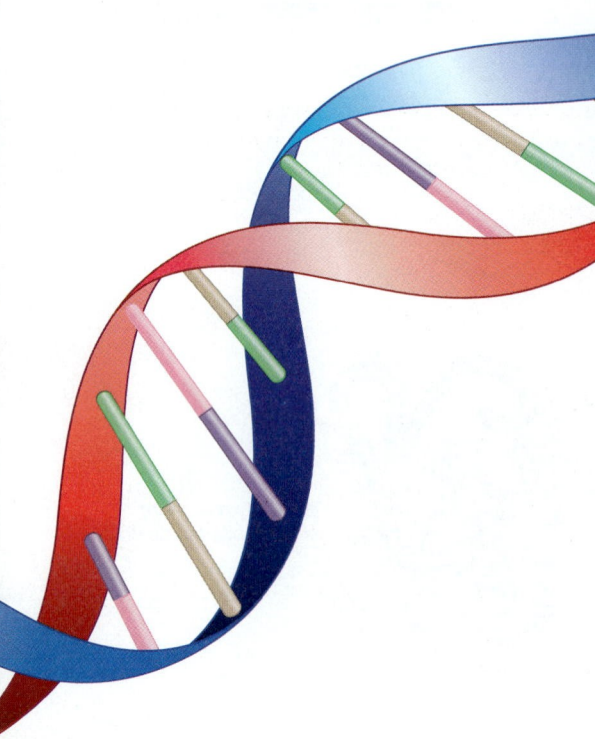

ATP: The Energy Molecule of Cells

A relative of RNA involved in an entirely different cell activity is **adenosine triphosphate (ATP)**. ATP is a nucleotide containing adenine, ribose, and three phosphates rather than just one **(figure 1.14)**. It belongs to a category of high-energy compounds (also including guanosine triphosphate [GTP]) that give off energy when the bond is broken between the second and third (outermost) phosphate. The presence of these high-energy bonds makes it possible for ATP to release and store energy for cellular chemical reactions. Breakage of the bond of the terminal phosphate releases energy to do cellular work and also generates adenosine diphosphate (ADP). ADP can be converted back to ATP when the third phosphate is restored, thereby serving as an energy depot. Carriers for oxidation-reduction activities (nicotinamide adenine dinucleotide [NAD], for instance) are also derivatives of nucleotides (see chapter 8).

Cells: Where Chemicals Come to Life

As we proceed in this chemical survey from the level of simple molecules to increasingly complex levels of macromolecules, at some point we cross a line from the realm of lifeless molecules and arrive at the fundamental unit of life called a **cell.** A cell is indeed a huge aggregate of carbon, hydrogen, oxygen, nitrogen, and many other atoms, and it follows the basic laws of chemistry and physics, but it is much more. The combination of these atoms produces characteristics, reactions, and products that can only be described as *living.*

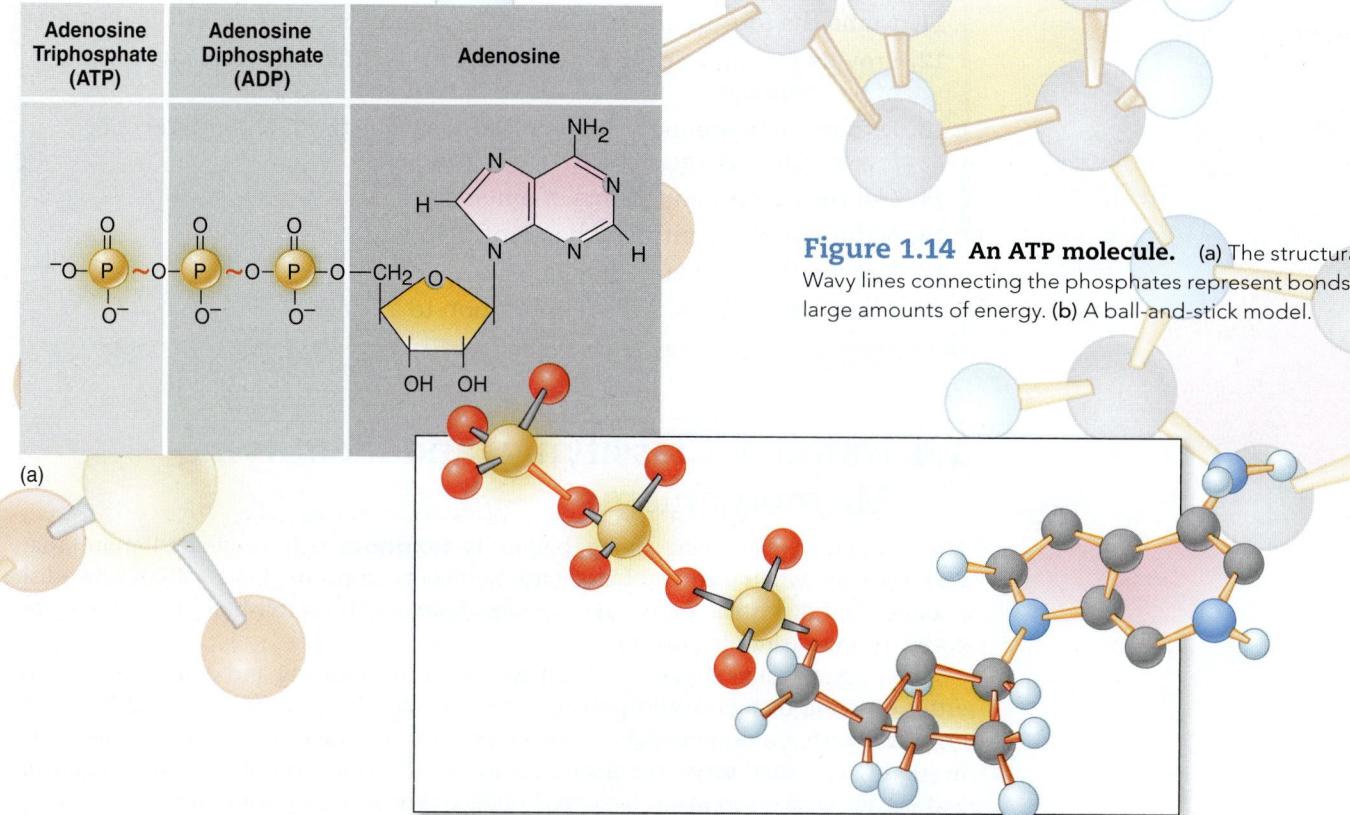

Figure 1.14 An ATP molecule. (a) The structural formula. Wavy lines connecting the phosphates represent bonds that release large amounts of energy. **(b)** A ball-and-stick model.

Adenosine Triphosphate (ATP)	Adenosine Diphosphate (ADP)	Adenosine

(a)

(b)

A poultry egg is a single large cell.

Fundamental Characteristics of Cells

The bodies of some living things, such as bacteria and protozoa, consist of only a single cell, whereas those of animals and plants contain trillions of cells. Regardless of the organism, all cells have a few common characteristics. They tend to be spherical, polygonal, cubical, or cylindrical; and their protoplasm (internal cell contents) is encased in a cell or cytoplasmic membrane. They have chromosomes containing DNA and ribosomes for protein synthesis, and they are exceedingly complex in function. Aside from these few similarities, the contents and structure of the three different cell types—bacterial, archaeal, and eukaryotic—differ significantly.

Animals, plants, fungi, and protozoa are all comprised of eukaryotic cells. Such cells contain a number of complex internal parts called organelles that perform useful functions for the cell involving growth, nutrition, or metabolism. Organelles are distinct cell components that perform specific functions and are enclosed by membranes. Organelles also partition the eukaryotic cell into smaller compartments. The most visible organelle is the nucleus, a roughly ball-shaped mass surrounded by a double membrane that surrounds the DNA of the cell. Other organelles include the Golgi apparatus, endoplasmic reticulum, vacuoles, and mitochondria.

Bacterial and archaeal cells may seem to be the cellular "have nots" because, for the sake of comparison, they are described by what they lack. They have no nucleus and generally no other organelles. This apparent simplicity is misleading, however, because the fine structure of these cells is complex. Overall, bacterial and archaeal cells can engage in nearly every activity that eukaryotic cells can, and many can function in ways that eukaryotes cannot. Chapters 3 and 4 delve deeply into the properties of bacterial, archaeal, and eukaryotic cells.

1.3 LEARNING OUTCOMES—Assess Your Progress

11. Name the four main families of biochemicals.

12. Provide examples of cell components made from each of the families of biochemicals.

13. Differentiate among primary, secondary, tertiary, and quaternary levels of protein structure.

14. List the three components of a nucleotide.

15. Name the three nitrogen bases of DNA and RNA.

16. List the three components of ATP.

17. Recall three characteristics common to all cells.

1.4 Naming, Classifying, and Identifying Microorganisms

The science of classifying living beings is **taxonomy.** It originated more than 250 years ago when Carl von Linné (also known as Linnaeus; 1701–1778), a Swedish botanist, laid down the basic rules for *classification* and established taxonomic categories, or **taxa** (singular, *taxon*).

Von Linné realized early on that a system for recognizing and defining the properties of living beings would prevent chaos in scientific studies by providing each organism with a unique name and an exact "slot" in which to catalog it. This classification would then serve as a means for future identification of that same organism and permit workers in many biological fields to know if they were indeed discussing the same organism.

The primary concerns of modern taxonomy are still naming, classifying, and identifying. These three areas are interrelated and play a vital role in keeping a dynamic inventory of the extensive array of living and extinct beings. In general,

Nomenclature (naming) is the assignment of scientific names to the various taxonomic categories and to individual organisms.

Classification is the orderly arrangement of organisms into a hierarchy.

Identification is the process of discovering and recording the traits of organisms so that they may be recognized or named and placed in an overall taxonomic scheme.

Nomenclature

Many macroorganisms are known by a common name suggested by certain dominant features. For example, a bird species may be called a "red-headed blackbird" or a flowering plant species a "black-eyed Susan." Some species of microorganisms are also called by informal names, including human pathogens such as "gonococcus" (*Neisseria gonorrhoeae*) or fermenters such as "brewer's yeast" (*Saccharomyces cerevisiae*), or the recent "Iraqabacter" (*Acinetobacter baumannii*), but this is not the usual practice. If we were to adopt common names such as the "little yellow coccus," the terminology would become even more cumbersome and challenging than scientific names.

The method of assigning a scientific or specific name is called the **binomial** (two-name) **system** of nomenclature. The scientific name is always a combination of the generic (genus) name followed by the species name. The generic part of the scientific name is capitalized, and the species part begins with a lowercase letter. Both should be italicized (or underlined if using handwriting), as follows:

The two-part name of an organism is sometimes abbreviated to save space, as in *E. coli*, but only if the genus name has already been stated. The inspiration for names is extremely varied and often rather imaginative. Some species have been named in honor of a microbiologist who originally discovered the microbe or who has made outstanding contributions to the field. Other names may designate a characteristic of the microbe (shape, color), a location where it was found, or a disease it causes. Some examples of specific names, their pronunciations, and their origins are

- *Staphylococcus aureus* (staf'-i-lo-kok'-us ah'-ree-us) Gr. *staphule*, "bunch of grapes," *kokkus*, "berry," and Gr. *aureus*, "golden." A common bacterial pathogen of humans.
- *Lactobacillus sanfrancisco* (lak'-toh-bass-ill'-us san-fran-siss'-koh) L. *lacto*, "milk," and *bacillus*, "little rod." A bacterial species used to make sourdough bread, for which San Francisco is known.
- *Giardia lamblia* (jee-ar'-dee-uh lam'-blee-uh) for Alfred Giard, a French microbiologist, and Vilem Lambl, a Bohemian physician, both of whom worked on the organism, a protozoan that causes a severe intestinal infection.

Here's a helpful hint: These names may seem difficult to pronounce and the temptation is to simply "slur over them." But when you encounter the names of microorganisms in the chapters ahead, it will be extremely useful to take the time to sound them out and repeat them until they seem familiar. Even experienced scientists stumble the first few times through new names. Stumbling out loud is a great way to figure them out and you are much more likely to remember them that way–they are less likely to end up in a tangled heap with all of the new language you will be learning.

Classification

Classification schemes are organized into several descending ranks, beginning with the most general all-inclusive taxonomic category and ending with the smallest and most specific category. This means that all members of the highest category share only one

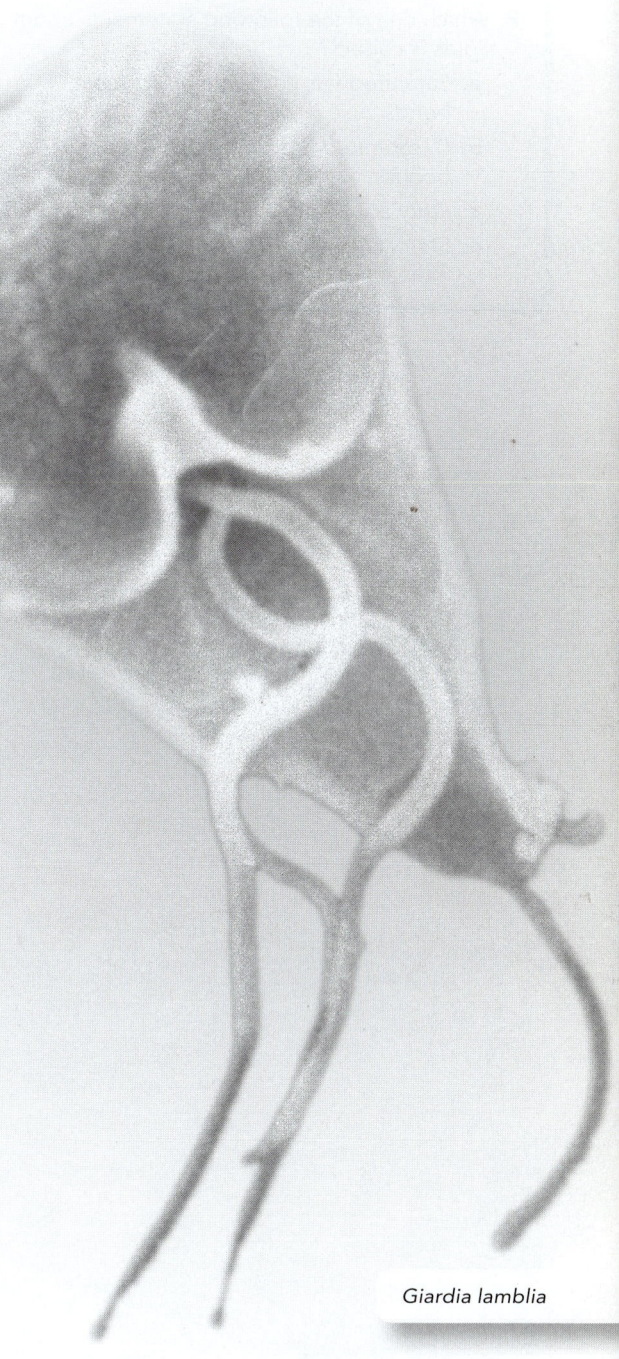

Giardia lamblia

or a few general characteristics, whereas members of the lowest category are essentially the same kind of organism—that is, they share the majority of their characteristics. The taxonomic categories from top to bottom are **domain, kingdom, phylum** or **division, class, order, family, genus,** and **species.** Thus, each kingdom can be subdivided into a series of phyla or divisions, each phylum is made up of several classes, each class contains several orders, and so on. Because taxonomic schemes are to some extent artificial, certain groups of organisms may not exactly fit into the main categories. In such a case, additional taxonomic levels can be imposed above (super) or below (sub) a taxon, giving us such categories as "superphylum" and "subclass."

Let's compare the taxonomic breakdowns of a human and a protozoan (pro-tuh-zoh'-un) to illustrate the fine points of this system **(figure 1.15).** Humans and protozoa are both organisms with nucleated cells (eukaryotes); therefore, they are in the same domain (Eukarya) but they are in different kingdoms. Humans are multicellular animals (kingdom Animalia), whereas protozoa are single-cellular organisms that, together with algae, belong to the kingdom Protista. To emphasize just how broad the category "kingdom" is, ponder the fact that we humans belong to the same kingdom as jellyfish. Of the several phyla within this kingdom, humans belong to the phylum Chordata, but even a phylum is rather all-inclusive, considering that humans share it with other vertebrates as well as with creatures called sea squirts. The next level, class Mammalia, narrows the field considerably by grouping only those vertebrates that have hair and suckle their young. Humans belong to the order Primates, a group that also includes apes, monkeys, and lemurs. Next comes the family Hominoidea, containing only humans and apes. The final levels are our genus, *Homo* (all modern and ancient humans), and our species, *sapiens* (meaning "wise"). Notice that for the human as well as the protozoan, the taxonomic categories in descending order become less inclusive and the individual members more closely related. In this text, we are usually concerned with only the most general (domain, kingdom, phylum) and specific (genus, species) taxonomic levels.

Identification

Discovering the identity of microbes we find in the environment or in diseases is an art and a science. The methods used in this process are extensively described in chapter 2 and in chapter 15.

The Origin and Evolution of Microorganisms

As we indicated earlier, *taxonomy,* the science of classification of biological species, is used to organize all of the forms of modern and extinct life. In biology today, there are different methods for deciding on taxonomic categories, but they all rely on the degree of relatedness among organisms. The scheme that represents the natural relatedness (relation by descent) between groups of living beings is called their *phylogeny* (Gr. *phylon,* "race or class"; L. *genesis,* "origin or beginning"). Biologists use phylogenetic relationships to refine the system of taxonomy.

To understand the natural history of and the relatedness among organisms, we must understand some fundamentals of the process of evolution. **Evolution** is an important theme that underlies all of biology, including the biology of microorganisms. As we said earlier, evolution states that the hereditary information in living beings changes gradually through time and that these changes result in various structural and functional changes through many generations. The process of evolution is selective in that those changes that most favor the survival and reproduction of a particular organism or group of organisms tend to be retained, whereas those that are less beneficial to survival tend to be lost. This is not always the case, but it often is. Charles Darwin called this process *natural selection.*

Usually, evolution progresses toward greater complexity but there are many examples of evolution toward lesser complexity (reductive evolution). This is because individual organisms never evolve in isolation but as populations of organisms in their specific environments, which exert the functional pressures of selection. Because

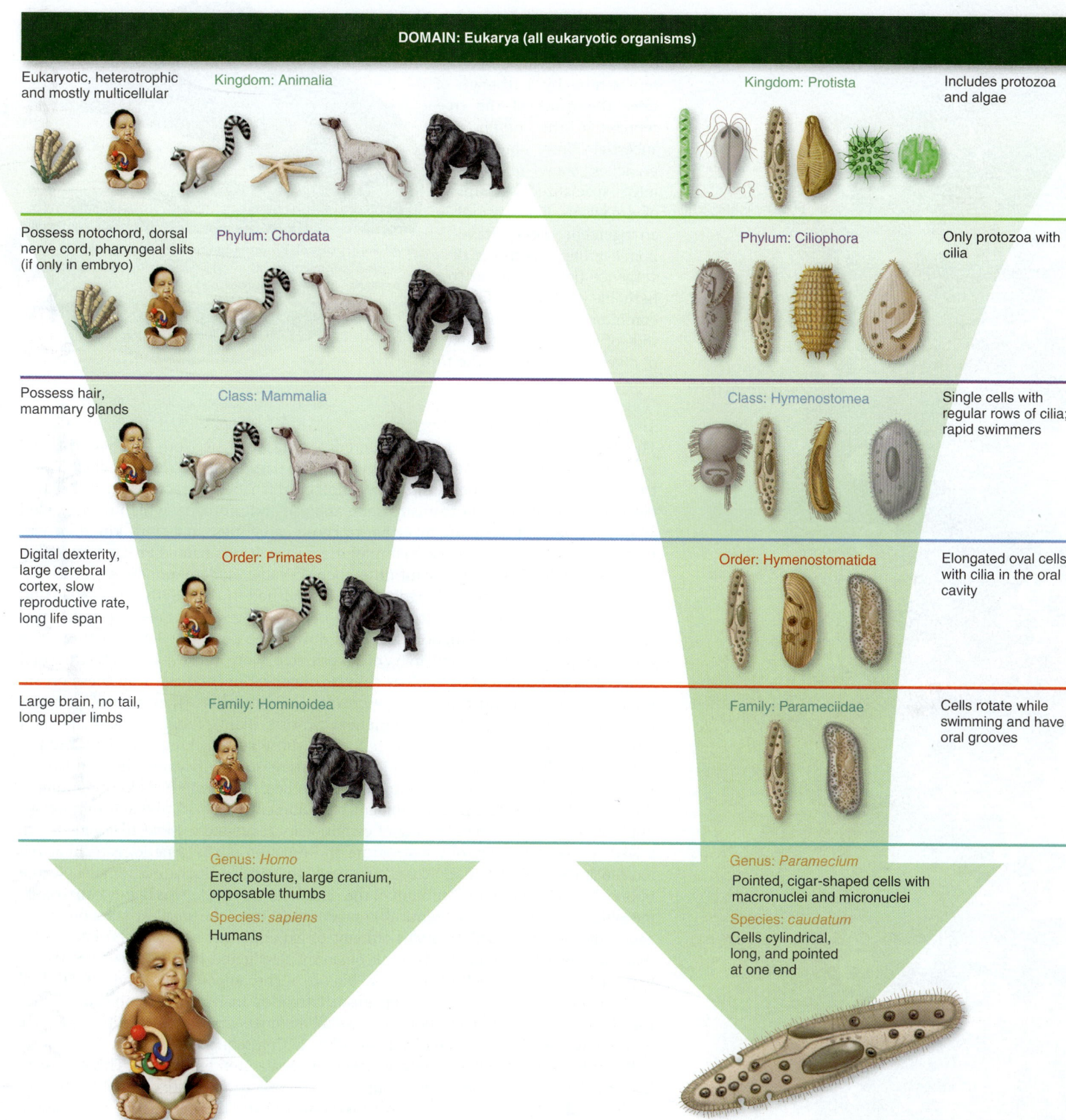

Figure 1.15 Sample taxonomy. Two organisms belonging to the Eukarya domain, traced through their taxonomic series. On the left, modern humans, *Homo sapiens.* On the right, a common protozoan, *Paramecium caudatum.*

of the divergent nature of the evolutionary process, the phylogeny, or relatedness by descent, of organisms is often represented by a diagram of a tree. The trunk of the tree represents the origin of ancestral lines, and the branches show offshoots into specialized groups of organisms. This sort of arrangement places taxonomic groups with less divergence (less change in the heritable information) from the common ancestor closer to the root of the tree and taxa with lots of divergence closer to the top.

A handful of soil is home to thousands of different kinds of organisms, including a wide diversity of fungi, bacteria, viruses, and protozoa.

A Universal Web of Life

The first trees of life were constructed a long time ago on the basis of just two kingdoms–plants and animals–by Charles Darwin and Ernst Haeckel. These trees were chiefly based on visible morphological (shape) characteristics. It became clear that certain (micro)organisms such as algae and protozoa, which only existed as single cells, did not truly fit either of those categories, so a third kingdom was recognized by Haeckel for these simpler organisms. It was named Protista (or Protozoa). Eventually, when significant differences became evident among even the unicellular organisms, a fourth kingdom was established in the 1870s by Haeckel and named Monera. Almost a century passed before Robert Whittaker extended this work and added a fifth kingdom for fungi during the period of 1959 to 1969. The relationships that were used in Whittaker's tree were those based on structural similarities and differences, such as cellular organization, and the way these organisms obtained their nutrition. These criteria indicated that there were five major taxonomic units, or kingdoms: the monera, protists, plants, fungi, and animals, all of which consisted of one of the two cell types, those cells lacking a nucleus and the eukaryotic cells. Whittaker's five-kingdom system quickly became the standard.

With the rise of genetics as a molecular science, newer methods for determining phylogeny have led to the development of a differently shaped tree–with important implications for our understanding of evolutionary relatedness. Molecular genetics allowed an in-depth study of the structure and function of the genetic material at the molecular level. In 1975, Carl Woese discovered that one particular macromolecule, the ribonucleic acid in the small subunit of the ribosome (ssu rRNA), was highly conserved– meaning that it was nearly identical in organisms within the smallest taxonomic category, the species. Based on a vast amount of experimental data and the knowledge that protein synthesis proceeds in all organisms facilitated by the ribosome, Woese hypothesized that ssu rRNA provides a "biological chronometer" or a "living record" of the evolutionary history of a given organism. Extended analysis of this molecule in prokaryotic and eukaryotic cells indicated that all members in a certain group of bacteria, then known as archaeabacteria, had ssu rRNA with a sequence that was significantly different from the ssu rRNA found in other bacteria and in eukaryotes. This discovery led Carl Woese and collaborator George Fox to propose a separate taxonomic unit for the archaeabacteria, which they named **Archaea.** Under the microscope, they resembled the structure of bacteria, but molecular biology has revealed that the archaea, though seemingly bacterial in nature, were actually more closely related to eukaryotic cells. To reflect these relationships, Carl Woese and George Fox proposed an entirely new system that assigned all known organisms to one of the three major taxonomic units, the **domains,** each being a different type of cell (**figure 1.16**).

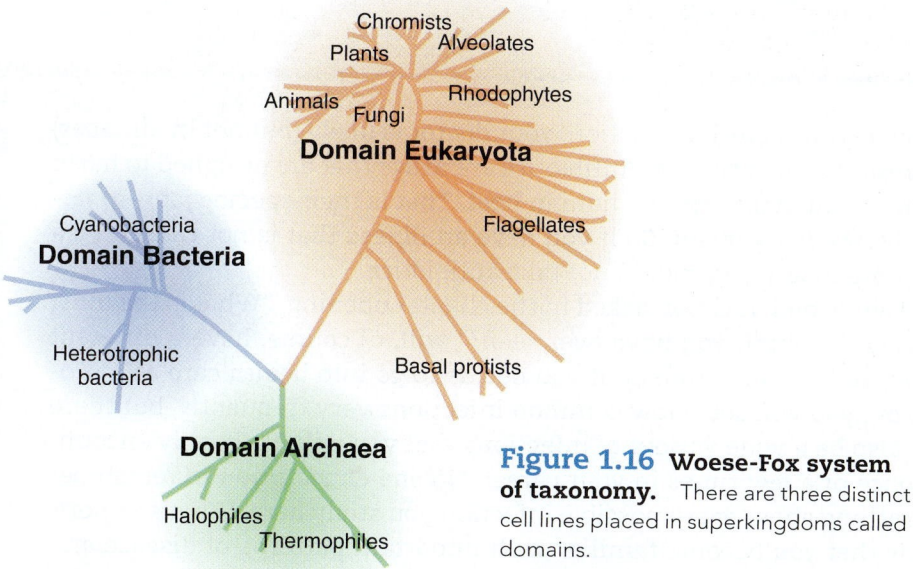

Figure 1.16 Woese-Fox system of taxonomy. There are three distinct cell lines placed in superkingdoms called domains.

The domains are the highest level in hierarchy and can contain many kingdoms and superkingdoms. Cell types lacking a nucleus are represented by the domains Archaea and **Bacteria,** whereas eukaryotes are all placed in the domain **Eukarya.** Analysis of the ssu rRNAs from all organisms in these three domains suggests that all modern and extinct organisms on earth arose from a common ancestor. Therefore, eukaryotes did not emerge from bacteria and archaea. Instead, it appears that bacteria, archaea, and eukaryotes all emerged separately from a different, now extinct, cell type.

To add another level of complexity, the most current data suggest that "trees" of life do not truly represent the relatedness—and evolution—of organisms in their totality. It has become obvious that genes travel horizontally—meaning from one species to another in nonreproductive ways—and that the neat generation-to-generation changes are combined with neighbor-to-neighbor exchanges of DNA. For example, it is estimated that 40% to 50% of human DNA has been carried to humans from other species (by viruses). For these reasons, most scientists like to think of a *web* as the proper representation of life these days. Nevertheless, this new scheme does not greatly affect our presentation of most microbes, because we will discuss them at the genus or species level. But be aware that biological taxonomy and, more important, our view of how organisms evolved on earth are in a period of transition. Keep in mind that our methods of classification or evolutionary schemes reflect our current understanding and will change as new information is uncovered.

Please note that viruses are not included in any of the classification or evolutionary schemes, because they are not cells or organisms, and their position in a "web of life" cannot be determined. The special taxonomy of viruses is discussed in chapter 5.

1.4 LEARNING OUTCOMES—Assess Your Progress

18. Differentiate among the terms *nomenclature, taxonomy,* and *classification.*
19. Create a mnemonic device for remembering the taxonomic categories.
20. Correctly write the binomial name for a microorganism.
21. Draw a diagram of the three major domains.
22. Explain the difference between traditional and molecular approaches to taxonomy.

CASE FILE WRAP-UP

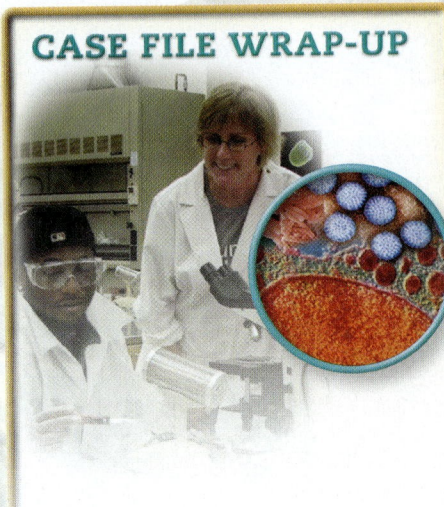

If you have a bacterial infection, your doctor is likely (but not in all cases) to prescribe an antibiotic. Antibiotics are drugs that are designed to harm microbes but not harm the human host. That is their specific job—to target the microorganism. So if you have an illness that is not caused by a microorganism, you should not take antibiotics.

The second question asked in the chapter opening, "What is the most unusual infection you have ever seen?" will, of course, have a different response for every student. If you decide to go into health care as a profession, you will see a few common infections very frequently, but there will also be a wide variety of infections that you will likely only encounter once or a few times in your career. No one expects you to remember everything about every possible infection you study here. What's important is that you become familiar with important patterns of disease and the ways that our body—and the treatments we apply—affect them.

The Vaccine Debate

Although we have the knowledge and the means to eradicate many diseases that threaten human life, in the recent past there has been a small but highly significant public movement in some developed countries (including the United States) against vaccinating children. Childhood immunization programs protect against infections that were once widespread and deadly, with high morbidity and mortality rates, such as measles, diphtheria, and whooping cough. Individuals who choose to not immunize their children generally do so for three main reasons: (1) They fear that immunizations are unsafe or will cause adverse side effects (i.e., the autism debate); (2) they do not believe immunizations are effective or necessary; or (3) they wrongly assume that, if everyone else vaccinates their children, their own children are safe from these illnesses. Other factors in choosing to not vaccinate include antigovernment sentiment, religious considerations, and cost.

Herd immunity is the term used to describe the concept of vaccines preventing illness in people who have not been vaccinated themselves or who have not been exposed to the natural disease. The crux of this theory is as follows: If most people around you are immune to a certain illness because they have been vaccinated, then they cannot become ill and infect you or others who have not been immunized. However, there is a catch: Herd immunity declines as immunization rates decline. For example, it is estimated that immunization rates for whooping cough must be 92% or higher to prevent outbreaks of the disease.

The result of this failure to vaccinate is the reappearance or resurgence of diseases that were once relatively rare. Measles is a prime example. In 2000, endemic transmission of the disease was eradicated in the United States and the Americas. It was eliminated in the United Kingdom in 1995. However, after the publication in the United Kingdom in 1998 of a misleading paper linking the vaccination to autism (that was later completely discredited), many parents stopped vaccinating against this deadly disease. Rates of the disease skyrocketed in the United Kingdom, and it is now considered endemic there once again. In the United States, measles rates in 2012 were the highest they had been since 1996.

Several studies have shown that the number of parents refusing to vaccinate their children is continuing to grow, a problem that is resulting in decreased herd immunity and a resurgence of diseases like measles and whooping cough. We'll investigate vaccine safety later in this book.

Source: Pertussis Outbreak Trends, Centers for Disease Control and Prevention. Updated March 2013.

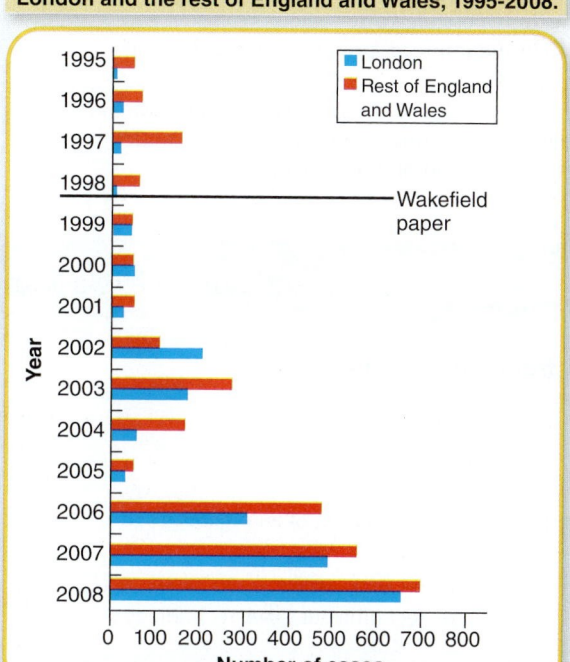

Percentage of 2-year-olds receiving MMR vaccine, England, Wales, and Scotland, 1994-2008.

Legend: England, Wales, Scotland

Wakefield paper

Y-axis: Year of second birthday (1994-95 through 2007-08)
X-axis: Percentage (75 to 100)

(a)

Annual laboratory-confirmed measles cases, London and the rest of England and Wales, 1995-2008.

Legend: London, Rest of England and Wales

Wakefield paper

Y-axis: Year (1995 through 2008)
X-axis: Number of cases (0 to 800)

(b)

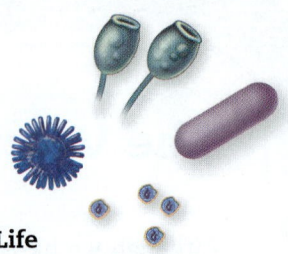

Chapter Summary

1.1 Microbes: Tiny but Mighty

- Microorganisms are defined as "living organisms too small to be seen with the naked eye." Members of this huge group of organisms are bacteria, archaea, protozoa, fungi, parasitic worms (helminths), and viruses.
- Microorganisms live nearly everywhere and influence many biological and physical activities on earth.
- There are many kinds of relationships between microorganisms and humans; most are beneficial, but some are harmful.
- Groups of organisms constantly evolve to produce new forms of life.
- Microbes are crucial to the cycling of nutrients and energy necessary for all life on earth.
- Humans have learned how to manipulate microbes to do important work for them in industry and medicine and in caring for the environment.
- In the last 160 years, microbiologists have identified the causative agents for many infectious diseases. They have discovered distinct connections between microorganisms and diseases whose causes were previously unknown.
- Excluding the viruses, there are three types of microorganisms: bacteria and archaea, which are small and lack a nucleus and (usually) organelles, and eukaryotes, which are larger and have both a nucleus and organelles.
- Viruses are not cellular and are therefore sometimes called particles rather than organisms. They are included in microbiology because of their small size and close relationship with cells.

1.2 Microbes in History

- The theory of spontaneous generation of living organisms from "vital forces" in the air was disproved finally by Louis Pasteur.
- Our current understanding of microbiology is the cumulative work of thousands of microbiologists, many of whom literally gave their lives to advance knowledge in this field.
- The microscope made it possible to see microorganisms and thus to identify their widespread presence, particularly as agents of disease.
- Medical microbiologists developed the germ theory of disease and introduced the critically important concept of aseptic technique to control the spread of disease agents.

1.3 Macromolecules: Superstructures of Life

- Macromolecules are very large organic molecules (polymers) usually built up by polymerization of smaller molecular subunits (monomers).
- Carbohydrates are biological molecules whose polymers are monomers linked together by glycosidic bonds. Their main functions are protection and support (in organisms with cell walls) and also nutrient and energy stores.
- Lipids are biological molecules such as fats that are insoluble in water. Their main functions are as cell components and nutrient and energy stores.
- Proteins are biological molecules whose polymers are chains of amino acid monomers linked together by peptide bonds.
- Proteins are called the "shapers of life" because of the many biological roles they play in cell structure and cell metabolism.
- Protein structure determines protein function. Structure and shape are dictated by amino acid composition and by the pH and temperature of the protein's immediate environment.
- Nucleic acids are biological molecules whose polymers are chains of nucleotide monomers linked together by phosphate-pentose sugar covalent bonds. Double-stranded nucleic acids are linked together by hydrogen bonds. Nucleic acids are information molecules that direct cell metabolism and reproduction. Nucleotides such as ATP also serve as energy-transfer molecules in cells.
- As the atom is the fundamental unit of matter, so is the cell the fundamental unit of life.

1.4 Naming, Classifying, and Identifying Microorganisms

- The taxonomic system has three primary functions: naming, classifying, and identifying species.
- The major groups in the most advanced taxonomic system are (in descending order): domain, kingdom, phylum or division, class, order, family, genus, and species.
- Evolutionary patterns show a treelike or weblike branching, thereby describing the diverging evolution of all life forms from the gene pool of a common ancestor.
- The Woese-Fox classification system places all organisms into three domains: Eukarya, Bacteria, and Archaea.

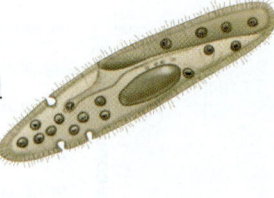

Multiple-Choice Questions   Bloom's Levels 1 and 2: Remember and Understand

Select the correct answer from the answers provided.

1. Which of the following is *not* considered a microorganism?
   - a. alga
   - b. bacterium
   - c. protozoan
   - d. flea

2. Which process involves the deliberate alteration of an organism's genetic material?
   - a. bioremediation
   - b. biotechnology
   - c. decomposition
   - d. recombinant DNA technology

3. Abiogenesis
   - a. refers to the spontaneous generation of organisms from nonliving matter.
   - b. explains the development of life forms from preexisting life forms.
   - c. only takes place in the absence of aseptic technique.
   - d. was supported by Pasteur's swan-necked flask experiments.

4. When a hypothesis has been thoroughly supported by long-term study and data, it is considered

   a. a law.
   b. a speculation.
   c. a theory.
   d. proved.

5. Which is the correct way to denote the scientific name of a microorganism?

   a. e. coli
   b. E. coli
   c. *E. coli*
   d. *e. Coli*

6. Which of the following is an acellular microorganism lacking a nucleus?

   a. bacterium
   b. helminth
   c. protozoan
   d. virus

7. Which is a correct statement about proteins?

   a. They are made up of nucleic acids.
   b. They contain fatty acids.
   c. They primarily serve as an energy source within the cell.
   d. Their shape determines their function.

8. DNA is a hereditary molecule that is composed of

   a. deoxyribose, phosphate, and nitrogen bases.
   b. deoxyribose, a pentose, and nucleic acids.
   c. sugar, proteins, and thymine.
   d. adenine, phosphate, and ribose.

Critical Thinking   Bloom's Levels 3, 4, and 5: Apply, Analyze, and Evaluate

Critical thinking is the ability to reason and solve problems using facts and concepts. These questions can be approached from a number of angles and, in most cases, they do not have a single correct answer.

1. Review figure 1.16 from this chapter and discuss the following.
   a. To which domain of life do humans belong?
   b. Most scientists believe that eukaryotic organisms are more closely related to archaea than to bacteria. Is this surprising? Why or why not?

2. Conduct additional research and discuss one current example in which microorganisms are used in the bioremediation of contaminated environments.

3. Discuss why it has been suggested that in the future obesity may be treated with antimicrobial drugs.

4. Compare and contrast how the maintenance of surgical suites and the use of basic surgical protocols have changed since the early 1800s.

5. Often when there is a local water main break, the town will post an advisory for everyone to boil their water before using for drinking or cooking. Discuss how this action would target the biological molecules discussed in this chapter, minimizing the microbial contaminants.

Visual Connections   Bloom's Level 5: Evaluate

This question connects previous images to a new concept.

1. **Figure 1.1.** Look at the red bar (the time that bacteria have been on earth) and at the time that humans appeared. Speculate on the probability that we will be able to completely eliminate all bacteria from our planet, and discuss whether or not this would even be a beneficial action.

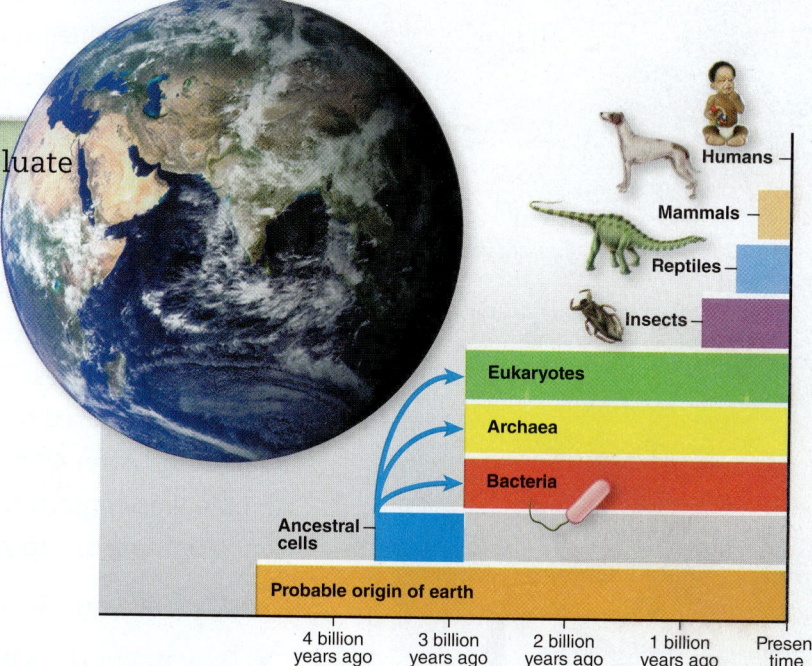

www.mcgrawhillconnect.com

Enhance your study of this chapter with study tools and practice tests. Also ask your instructor about the resources available through Connect, including the media-rich eBook, interactive learning tools, and animations.

Getting the Goods

As a nurse working in a busy obstetrics and gynecology practice, my job often included instructing pregnant women in collecting urine samples. Every expectant mother who attended the clinic provided a urine sample at every visit. A pregnant woman is at higher risk of developing urinary tract infections (UTIs) due to the increasing weight of her growing uterus, which compresses the bladder and prevents the bladder from draining completely. Urine left behind in the bladder becomes the perfect medium for bacterial growth.

I instructed a young mother how to properly collect a midstream urine sample. I told the patient to first wash her hands. I emphasized that she should ensure that her hands did not come in contact with the rim of the collection container. I further instructed her on how to cleanse the external genitalia with a disposable wipe saturated with povidone-iodine, a potent antimicrobial solution. I reminded her to wipe from front to back to prevent fecal contamination. I told her she was to void a small amount of urine into the toilet, then introduce the collection container into the urine stream, collecting the midstream portion of the urine. She was instructed to put the lid on the collection container, being careful not to touch the rim or the inside of the lid, and then wash her hands. I then donned gloves, wiped the outside of the container and delivered the specimen to the lab, after labeling it with the patient's name, the date and time of collection, and additional identification information.

The laboratory staff examined a small amount of urine under the microscope for the presence of bacteria, red blood cells, white blood cells, and other abnormalities. The lab staff identified the presence of bacteria, and the urine was cultured to identify the microorganism and to test its antibiotic sensitivity. After 48 hours, the culture result came back stating that the sample was contaminated. I informed the patient's physician, who asked that the patient return to provide another urine sample.

- What is a mixed culture? A contaminated culture?
- How might the sample have become contaminated during the collection process?

Case File Wrap-Up appears on page 56.

Tools of the Laboratory

Methods for the Culturing and Microscopic Analysis of Microorganisms

IN THIS CHAPTER...

2.1 How to Culture Microorganisms

1. Explain what the Five I's are and what each step entails.
2. Discuss three physical states of media and when each is used.
3. Compare and contrast selective and differential media, and give an example of each.
4. Provide brief definitions for *defined media* and *complex media*.

2.2 The Microscope

5. Convert among the different units of the metric system.
6. List and describe the three elements of good microscopy.
7. Differentiate between the principles of light microscopy and the principles of electron microscopy.
8. Give examples of simple, differential, and special stains.

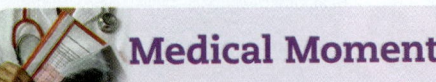

Medical Moment

The Making of the Flu Vaccine: An Example of a Live Growth Medium

Have you ever wondered why health care workers ask about allergic reactions to eggs prior to immunizing patients? Live attenuated vaccines are sometimes created by culturing a virus, such as the influenza virus, in live animals, often chick embryos. The virus is inoculated into fertilized eggs, which are then incubated to encourage the replication of large numbers of virus particles. The contents of the eggs are then collected and purified to create the vaccine.

Today, influenza vaccine preparations contain such low levels of egg protein that they can be safely administered even in most individuals with allergies though it is recommended that they be medically monitored after receiving the dose. There are also (egg-free) alternative forms of the vaccine available.

2.1 How to Culture Microorganisms

The Five I's

When you're trying to study microorganisms, you are confronted by some unique problems. First, most habitats (such as the soil and the human mouth) contain microbes in complex associations, so it is often necessary to separate the species from one another. Second, to maintain and keep track of such small research subjects, microbiologists usually have to grow them under artificial (and thus distorting) conditions. A third difficulty in working with microbes is that they are invisible. Fourth, microbes are everywhere, and undesirable ones can be introduced into your experiment, causing misleading results.

Microbiologists use five basic techniques to manipulate, grow, examine, and characterize microorganisms in the laboratory **(figure 2.1):**

1. inoculation,
2. incubation,
3. isolation,
4. inspection, and
5. identification.

Major Techniques Performed by Microbiologists to Locate, Grow, Observe, and Characterize Microorganisms

Specimen Collection:
Nearly any object or material can serve as a source of microbes. Common ones are body fluids and tissues, foods, water, or soil. Specimens are removed by some form of sampling device: a swab, syringe, or a special transport system that holds, maintains, and preserves the microbes in the sample.

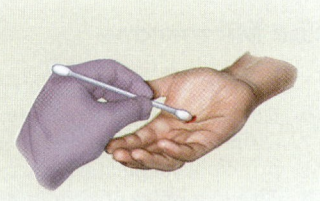

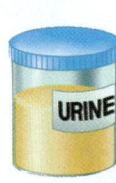

A GUIDE TO THE FIVE I's: How the Sample Is Processed and Profiled

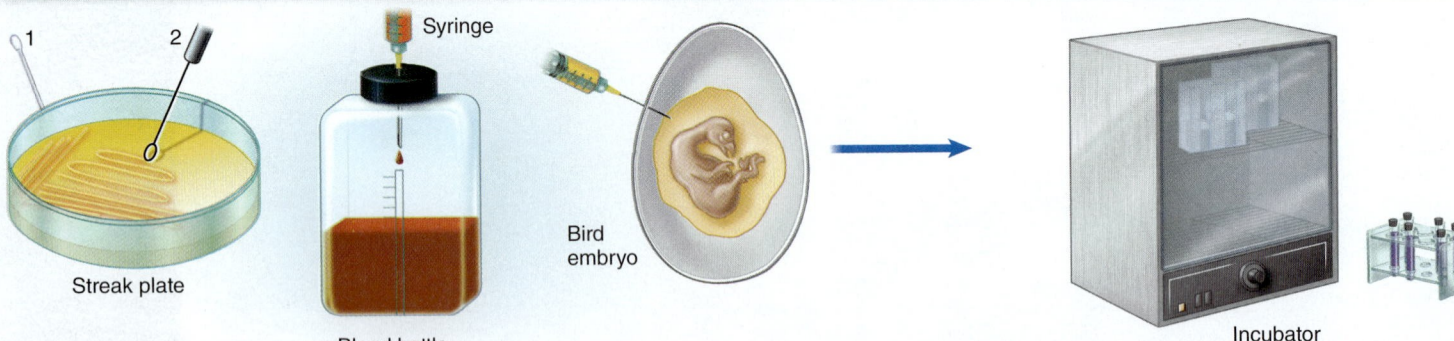

Syringe

Streak plate

Blood bottle

Bird embryo

Incubator

1 Inoculation:
The sample is placed into a container of sterile **medium** containing appropriate nutrients to sustain growth. Inoculation involves spreading the sample on the surface of a solid medium or introducing the sample into a flask or tube. Selection of media with specialized functions can improve later steps of isolation and identification. Some microbes may require a live organism (animal, egg) as the growth medium.

2 Incubation:
An incubator creates the proper growth temperature and other conditions. This promotes multiplication of the microbes over a period of hours, days, and even weeks. Incubation produces a culture—the visible growth of the microbe in or on the medium.

Figure 2.1 A summary of the general laboratory techniques carried out by microbiologists. It is not necessary to perform all the steps shown or to perform them exactly in this order, but all microbiologists participate in at least some of these activities. In some cases, one may proceed right from the sample to inspection, and in others, only inoculation and incubation on special media are required.

These procedures make it possible to handle and maintain microorganisms as discrete entities whose detailed biology can be studied and recorded. Having said that, keep in mind as we move through this chapter: It is not necessary to cultivate a microorganism to identify it anymore, though it still remains a very common method. You will read about noncultivation methods of identifying microbes in chapter 15. Sometimes growing microbes in isolated cultures can tell you very little about how they act in a mixed species environment, but being able to isolate them and study them is also valuable, as long as you keep in mind that it is an unnatural state for them.

Inoculation

To grow, or **culture,** microorganisms, one introduces a tiny sample (the inoculum) into a container of nutrient **medium** (plural, *media*), which provides an environment in which they multiply. This process is called **inoculation.** Any instrument used for sampling and inoculation must initially be *sterile.* The observable growth that appears in or on the medium after **incubation** is known as a culture. Clinical specimens for determining the cause of an infectious disease are obtained from body fluids (blood, cerebrospinal fluid), discharges (sputum, urine, feces), anatomical sites (throat, nose, ear, eye, genital tract), or diseased tissue (such as an abscess or wound). Other samples subject to microbiological analysis are soil, water, sewage, foods, air, and inanimate objects. Procedures for proper specimen collection are discussed in chapter 15.

Incubation

Once a container of medium has been inoculated, it is **incubated,** which means it is placed in a temperature-controlled chamber (incubator) to encourage multiplication. Although microbes have adapted to growth at temperatures ranging from freezing to boiling, the usual temperatures used in laboratory propagation fall between 20°C and 45°C. Incubators can also control the content of atmospheric

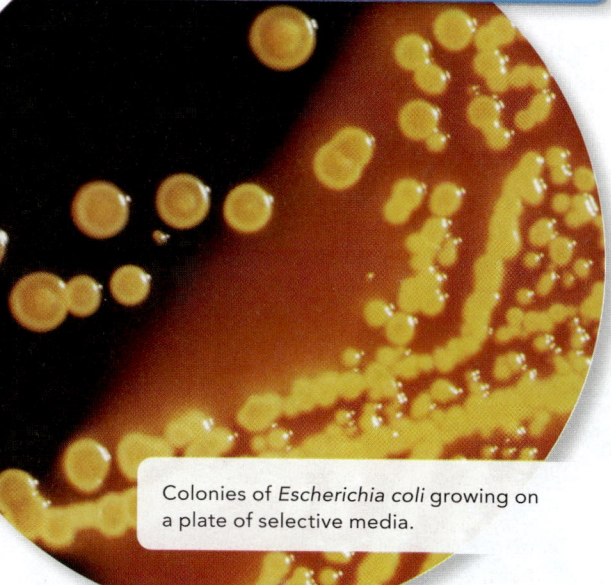

NCLEX® PREP

1. The physician has ordered that a urine culture be taken on a client. What priority information should the nurse know in order to complete the collection of this specimen?
   a. *Date and time of collection*
   b. *Method of collection*
   c. *Whether the client is NPO (to have nothing by mouth)*
   d. *Age of client*

Colonies of *Escherichia coli* growing on a plate of selective media.

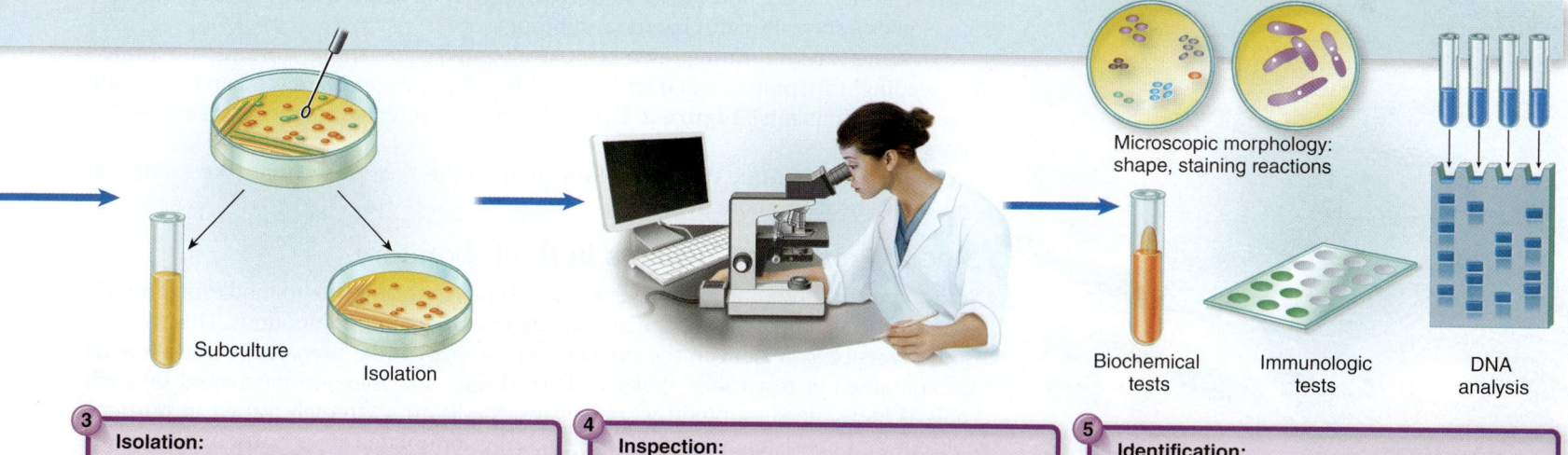

Subculture

Isolation

Microscopic morphology: shape, staining reactions

Biochemical tests

Immunologic tests

DNA analysis

3 **Isolation:**
One result of inoculation and incubation is **isolation** of the microbe. Isolated microbes may take the form of separate colonies (discrete mounds of cells) on solid media, or turbidity (free-floating cells) in broths. Further isolation by subculturing involves taking a bit of growth from an isolated colony and inoculating a separate medium. This is one way to make a pure culture that contains only a single species of microbe.

4 **Inspection:**
The colonies or broth cultures are observed macroscopically for growth characteristics (color, texture, size) that could be useful in analyzing the specimen contents. Slides are made to assess microscopic details such as cell shape, size, and motility. Staining techniques may be used to gather specific information on microscopic morphology.

5 **Identification:**
A major purpose of the Five I's is to determine the type of microbe, usually to the level of species. Information used in identification can include relevant data already taken during initial inspection and additional tests that further describe and differentiate the microbes. Specialized tests include biochemical tests to determine metabolic activities specific to the microbe, immunologic tests, and genetic analysis.

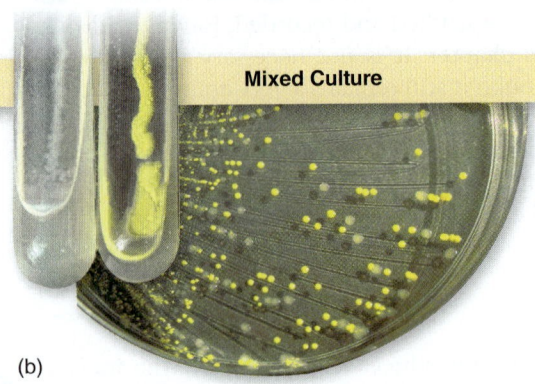

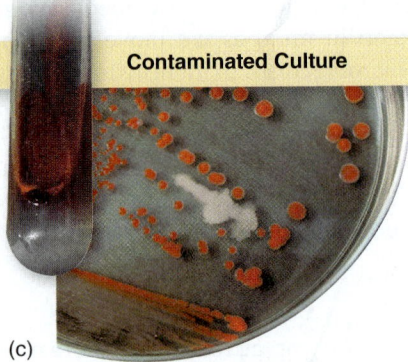

(a)

Pure Culture

Mixed Culture

Contaminated Culture

(b)

(c)

Figure 2.2 Various conditions of cultures.   **(a)** Three tubes containing pure cultures of *Escherichia coli* (white), *Micrococcus luteus* (yellow), and *Serratia marcescens* (red). A **pure culture** is a container of medium that grows only a single known species or type of microorganism. This type of culture is most frequently used for laboratory study, because it allows the systematic examination and control of one microorganism by itself.

(b) A **mixed culture** is a container that holds two or more *identified*, easily differentiated species of microorganisms, not unlike a garden plot containing both carrots and onions. Pictured here is a mixed culture of *M. luteus* (bright yellow colonies) and *E. coli* (faint white colonies).

(c) A **contaminated culture** was once pure or mixed (and thus a known entity) but has since had **contaminants** (unwanted microbes of uncertain identity) introduced into it, like weeds into a garden. Contaminants get into cultures when the lids of tubes or Petri dishes are left off for too long, allowing airborne microbes to settle into the medium. They can also enter on an incompletely sterilized inoculating loop or on an instrument that you have inadvertently reused or touched to the table or your skin.

This plate of *S. marcescens* was overexposed to room air, and it has developed a large, white colony. Because this intruder is not desirable and not identified, the culture is now contaminated.

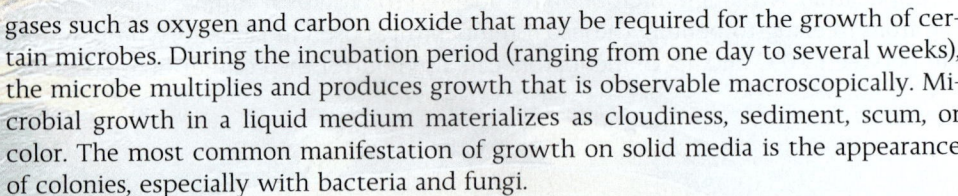

gases such as oxygen and carbon dioxide that may be required for the growth of certain microbes. During the incubation period (ranging from one day to several weeks), the microbe multiplies and produces growth that is observable macroscopically. Microbial growth in a liquid medium materializes as cloudiness, sediment, scum, or color. The most common manifestation of growth on solid media is the appearance of colonies, especially with bacteria and fungi.

In some ways, culturing microbes is analogous to gardening. Cultures are formed by "seeding" tiny plots (media) with microbial cells. Extreme care is taken to exclude weeds (contaminants). **Figure 2.2** provides an important summary of three different types of cultures.

Before we continue to cover information on the Five I's, we will take a side trip to look at media in more detail.

Media: Providing Nutrients in the Laboratory

Some microbes require only a very few simple inorganic compounds for growth; others need a complex list of specific inorganic and organic compounds. This tremendous diversity is evident in the types of media that can be prepared. Culture media are contained in test tubes, flasks, or Petri dishes, and they are inoculated by such tools as loops, needles, pipettes, and swabs. Media are extremely varied in nutrient content and consistency, and can be specially formulated for a particular purpose. Culturing microbes that cannot grow on artificial media (all viruses and certain bacteria) requires cell cultures or host animals. In this chapter, we will focus on artificial media, because these are the most frequently used type in clinical situations.

For an experiment to be properly controlled, sterile technique is necessary. This means that the inoculation must start with a sterile medium and inoculating tools with sterile tips must be used. Measures must be taken to prevent introduction of nonsterile materials, such as room air and fingers, into the media.

Agar, the main component of media, is commonly harvested from seaweed.

Types of Media

Media can be classified according to three properties **(table 2.1):**

1. physical state,
2. chemical composition, and
3. **functional type** (purpose).

Most media discussed here are designed for bacteria and fungi, though algae and some protozoa can be propagated in media.

Physical States of Media

Figure 2.3 provides a good summary of three physical types of media: liquid, semisolid, and solid.

Agar, a complex polysaccharide isolated from the alga *Gelidium*, is a critical tool in the microbiology lab. The benefits of agar are numerous. It is solid at room temperature, and it melts (liquefies) at the boiling temperature of water (100°C). Once liquefied, agar does not resolidify until it cools to 42°C, so it can be inoculated and poured in liquid form at temperatures (45°C to 50°C) that will not harm the microbes or the handler. Agar is flexible and moldable, and it provides a basic framework to hold moisture and nutrients. Importantly, it is not itself a digestible nutrient for most microorganisms.

Chemical Content of Media

Media whose compositions are precisely chemically defined are termed *defined* (also known as *synthetic*). Such media contain pure organic and inorganic compounds that vary little from one source to another and have a molecular content specified by means of an exact formula. Defined media may contain nothing more than a few essential compounds such as salts and amino acids dissolved in water or may

Table 2.1 Three Categories of Media Classification

Physical State	Chemical Composition	Functional Type
1. Liquid 2. Semisolid 3. Solid (can be converted to liquid) 4. Solid (cannot be liquefied)	1. Chemically defined 2. Complex; not chemically defined	1. General purpose 2. Enriched 3. Selective 4. Differential 5. Anaerobic growth 6. Specimen transport 7. Assay 8. Enumeration

Liquid   Semisolid   Solid/Reversible to Liquid

(a)   (b)   ① ② ③ ④   (c)

Figure 2.3 Media in different physical forms.   (a) **Liquid media** are water-based solutions that do not solidify at temperatures above freezing and that tend to flow freely when the container is tilted. Growth occurs throughout the container and can then present a dispersed, cloudy, or particulate appearance. Urea broth is used to show a biochemical reaction in which the enzyme urease digests urea and releases ammonium. This raises the pH of the solution and causes the dye to become increasingly pink. Left: uninoculated broth, pH 7; middle: weak positive, pH 7.5; right: strong positive, pH 8.0.

(b) **Semisolid media** have more body than liquid media but less body than solid media. They do not flow freely and have a soft, clotlike consistency at room temperature. Semisolid media are used to determine the motility of bacteria and to localize a reaction at a specific site. Here, sulfur indole motility (SIM) medium is pictured. The (1) medium is stabbed with an inoculum and incubated. Location of growth indicates nonmotility (2) or motility (3). If H_2S gas is released, a black precipitate forms (4).

(c) **Media containing 1%–5% agar** are solid enough to remain in place when containers are tilted or inverted. They are reversibly solid and can be liquefied with heat, poured into a different container, and resolidified. Solid media provide a firm surface on which cells can form discrete colonies. Nutrient gelatin contains enough gelatin (12%) to take on a solid consistency. The top tube shows it as a solid. The bottom tube indicates what happens when it is warmed or when microbial enzymes digest the gelatin and liquefy it.

be composed of a variety of defined organic and inorganic chemicals **(tables 2.2A and 2.2B).** Such standardized and reproducible media are most useful in research when the exact nutritional needs of the test organisms are known.

If even one component of a given medium is not chemically definable, the medium belongs in the *complex* category. Complex media contain extracts of animals, plants, or yeasts, including such materials as ground-up cells, tissues, and secretions. Examples are blood, serum, and meat extracts or infusions. Other possible ingredients are milk, yeast extract, soybean digests, and peptone. Nutrient broth, blood agar, and MacConkey agar, though different in function and appearance, are all complex media that present a rich mixture of nutrients for microbes that have complex nutritional needs.

Tables 2.2A and 2.2B provide a practical application of the two categories—*defined* and *complex* media—by comparing two different media for the growth of *Staphylococcus aureus*.

Media for Different Purposes

Microbiologists have many types of media at their disposal. Depending on what is added, a microbiologist can fine-tune a medium for nearly any purpose. Until recently, microbiologists knew of only a few species of bacteria or fungi that could not be cultivated artificially. However, newer DNA detection technologies have shown us that there are many more microbes that we don't know how to cultivate in the lab than those that we do. Although we can now study some vital traits of bacteria without actually growing the bacteria, developing new media is still important for growing the bacteria that we are discovering using genomic methods.

General-purpose media are designed to grow as broad a spectrum of microbes as possible. As a rule, they are of the complex variety and contain a mixture of nutrients that could support the growth of a variety of microbial life. Examples include nutrient agar and broth, brain-heart infusion, and trypticase soy agar (TSA). An **enriched medium** contains complex organic substances such as blood, serum, hemoglobin, or special **growth factors** (specific vitamins, amino acids) that certain species must have in order to grow. Bacteria that require growth factors and complex nutrients are termed **fastidious.** Blood agar, which is made by adding sterile sheep, horse, or rabbit blood to a sterile agar base **(figure 2.4a)** is widely used to grow fastidious streptococci and other pathogens. Pathogenic *Neisseria* (one species causes gonorrhea) are grown on either Thayer-Martin medium or "chocolate" agar, which is a blood agar with added hemin and nicotinamide adenine dinucleotide **(figure 2.4b).** Enriched media are also useful in the clinical laboratory to encourage the growth of pathogens that may be present in very low numbers, such as in urine or blood specimens.

Table 2.2A Defined Medium for Growth and Maintenance of Pathogenic *Staphylococcus aureus*

0.25 Grams Each of These Amino Acids	0.5 Grams Each of These Amino Acids	0.12 Grams Each of These Amino Acids
Cystine	Arginine	Aspartic acid
Histidine	Glycine	Glutamic acid
Leucine	Isoleucine	
Phenylalanine	Lysine	
Proline	Methionine	
Tryptophan	Serine	
Tyrosine	Threonine	
	Valine	

Additional ingredients

0.005 mole nicotinamide ⎤
0.005 mole thiamine ⎥ Vitamins
0.005 mole pyridoxine ⎥
0.5 micrograms biotin ⎦

1.25 grams magnesium sulfate ⎤
1.25 grams dipotassium hydrogen phosphate ⎥ Salts
1.25 grams sodium chloride ⎥
0.125 grams iron chloride ⎦

Ingredients dissolved in 1,000 milliliters of distilled water and buffered to a final pH of 7.0.

Table 2.2B Brain-Heart Infusion Broth: A Complex Medium for Growth and Maintenance of Pathogenic *Staphylococcus aureus*

27.5 grams brain, heart extract, peptone extract
2   grams glucose
5   grams sodium chloride
2.5 grams disodium hydrogen phosphate

Ingredients dissolved in 1,000 milliliters of distilled water and buffered to a final pH of 7.0.

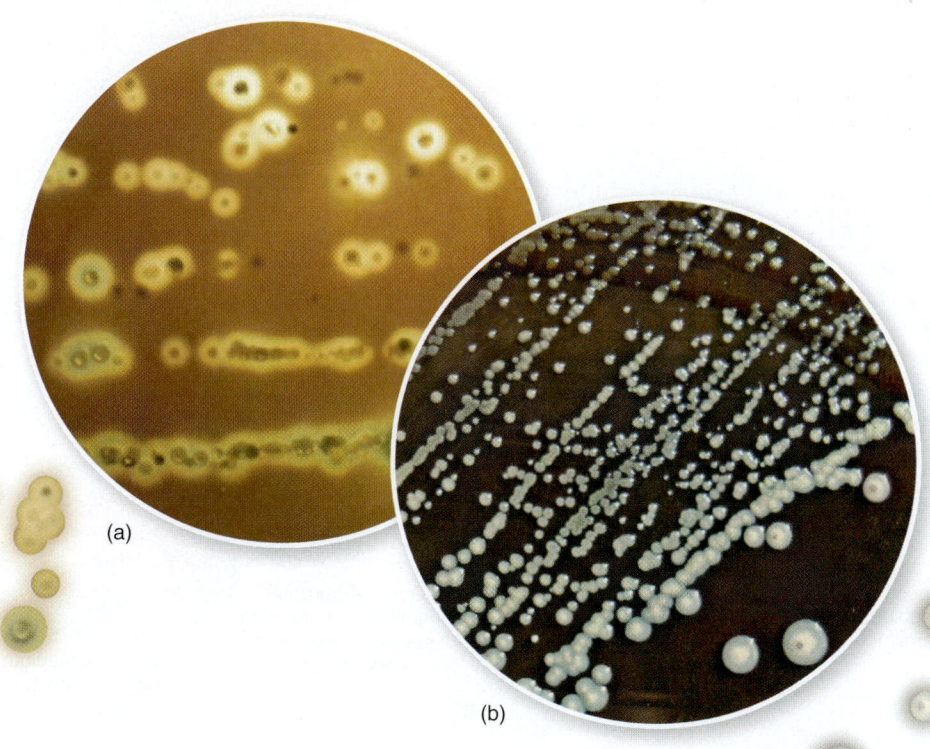

Figure 2.4 Examples of enriched media.
(a) Blood agar plate growing bacteria from the human throat. Note that this medium also differentiates among colonies by the zones of hemolysis (clear areas) they may show. (b) Culture of *Neisseria* sp. on chocolate agar. Chocolate agar gets its brownish color from cooked blood (not chocolate) and does not produce hemolysis.

Selective and Differential Media   These media are designed for special microbial groups, and they are extremely useful in isolation and identification. They can permit, in a single step, the preliminary identification of a genus or even a species.

A **selective medium** contains one or more agents that inhibit the growth of a certain microbe or microbes (call them A, B, and C) but not others (D) and thereby encourages, or *selects*, microbe D and allows it to grow. Selective media are very important in primary isolation of a specific type of microorganism from samples containing dozens of different species–for example, feces, saliva, skin, water, and soil. They speed up isolation by suppressing the unwanted background organisms and favoring growth of the desired ones.

Media for isolating intestinal pathogens (MacConkey agar, Hektoen enteric [HE] agar) contain bile salts as a selective agent. Other agents that have selective properties are dyes, such as methylene blue and crystal violet, and antimicrobial drugs. **Table 2.3** gives multiple examples of selective media and what they do.

Differential media allow multiple types of microorganisms to grow but are designed to display visible differences in how they grow. Differentiation shows up as variations in colony size or color **(figure 2.5),** in media color changes, or in the formation of gas bubbles and precipitates. These variations often come from the type of chemicals these media contain and the ways that microbes react

Table 2.3 Selective Media, Agents, and Functions		
Medium	**Selective Agent**	**Used For**
Enterococcus faecalis broth	Sodium azide, tetrazolium	Isolation of fecal enterococci
Tomato juice agar	Tomato juice, acid	Isolation of lacto-bacilli from saliva
MacConkey agar	Bile, crystal violet	Isolation of gram-negative enterics
Salmonella/ Shigella (SS) agar	Bile, citrate, brilliant green	Isolation of *Salmonella* and *Shigella*
Lowenstein-Jensen	Malachite green dye	Isolation and maintenance of *Mycobacterium*
Sabouraud's agar	pH of 5.6 (acid)	Isolation of fungi–inhibits bacteria

Figure 2.5 A medium that is both selective and differential.   MacConkey agar selects against gram-positive bacteria. Therefore, you will not see them here! It also differentiates between lactose-fermenting bacteria (indicated by a pink-red reaction in the center of the colony) and lactose-negative bacteria (indicated by an off-white colony with no dye reaction).

Figure 2.6 **Comparison of selective and differential media with general-purpose media.** **(a)** A mixed sample containing three different species is streaked onto plates of general-purpose nonselective medium and selective medium. **(b)** Another mixed sample containing three different species is streaked onto plates of general-purpose nondifferential medium and differential medium.

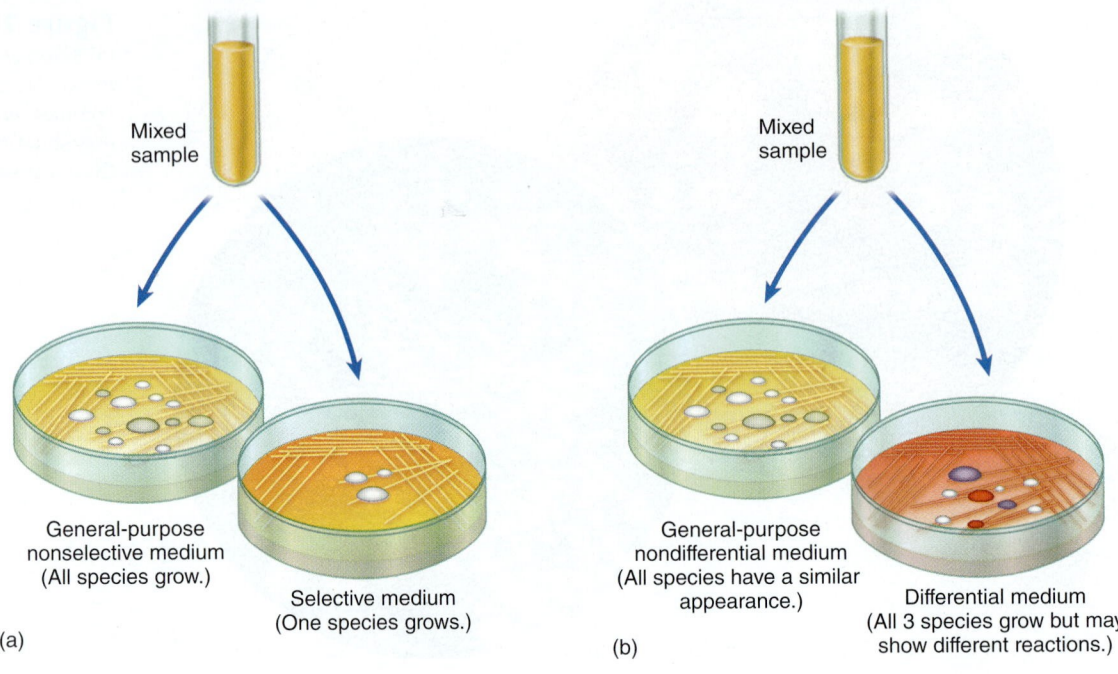

Mixed sample

Mixed sample

General-purpose nonselective medium (All species grow.)

Selective medium (One species grows.)

(a)

General-purpose nondifferential medium (All species have a similar appearance.)

Differential medium (All 3 species grow but may show different reactions.)

(b)

to them. For example, when microbe X metabolizes a certain substance not used by organism Y, then X will cause a visible change in the medium and Y will not **(figure 2.6)**. The simplest differential media show just two reaction types, such as the use or nonuse of a particular nutrient or a color change in some colonies but not in others. Some media are sufficiently complex to allow for three or four different reactions. A single medium can be both selective and differential, owing to its different ingredients. MacConkey agar, for example, appears in table 2.3 (selective media) and **table 2.4** (differential media) due to its ability to suppress the growth of some organisms while producing a visual distinction among the ones that do grow. The agar in figure 2.5 illustrates this activity; you just can't see the colonies that were suppressed. Media that are both selective and differential allow for microbial isolation and identification to occur at the same time, which can be very useful in the screening of patient specimens as well as food and water samples.

Dyes are frequently used as differential agents because many of them are pH indicators that change color in response to the production of an acid or a base. For example, MacConkey agar contains neutral red, a dye that is yellow when neutral

Table 2.4 Differential Media

Medium	Substances That Facilitate Differentiation	Differentiates Between or Among
Blood agar	Intact red blood cells	Types of hemolysis displayed by different species of *Streptococcus*
Mannitol salt agar	Mannitol, phenol red	Species of *Staphylococcus*
MacConkey agar	Lactose, neutral red	Bacteria that ferment lactose (lowering the pH) from those that do not
Urea broth	Urea, phenol red	Bacteria that hydrolyze urea to ammonia from those that do not
Sulfur indole motility (SIM)	Thiosulfate, iron	H_2S gas producers from nonproducers
Triple-sugar iron agar (TSIA)	Triple sugars, iron, and phenol red dye	Fermentation of sugars, H_2S production
Birdseed agar	Seeds from thistle plant	*Cryptococcus neoformans* and other fungi

Figure 2.7 Carbohydrate fermentation in broth. This medium is designed to show fermentation (acid production) and gas formation by means of a small, inverted Durham tube for collecting gas bubbles. The medium also changes color in the presence of acid.

and pink or red when acidic. A common intestinal bacterium such as *Escherichia coli* that gives off acid when it metabolizes the lactose in the medium develops red to pink colonies, and one like *Salmonella* that does not give off acid remains its natural color (off-white).

Miscellaneous Media   A **reducing medium** contains a substance (sodium thioglycollate or cystine) that absorbs oxygen or slows the penetration of oxygen in a medium, thus reducing its availability. Reducing media are important for growing anaerobic bacteria or for determining oxygen requirements of isolates (described in chapter 6). **Carbohydrate fermentation media** contain sugars that can be fermented (converted to acids) and a pH indicator to show this reaction **(figure 2.7)**.

Transport media are used to maintain and preserve specimens that have to be held for a period of time before clinical analysis or to sustain delicate species that die rapidly if not held under stable conditions. **Assay media** are used by technologists to test the effectiveness of antimicrobial drugs (see chapter 12) and by drug manufacturers to assess the effect of disinfectants, antiseptics, cosmetics, and preservatives on the growth of microorganisms. **Enumeration media** are used by industrial and environmental microbiologists to count the numbers of organisms in milk, water, food, soil, and other samples.

Isolation: Separating One Species from Another

Certain **isolation** techniques are based on the concept that if an individual bacterial cell is separated from other cells and provided adequate space on a nutrient surface, it will grow into a discrete mound of cells called a **colony (figure 2.8)**. If it was formed from a single cell, a colony consists of just that one species and no other. Proper isolation requires that a small number of cells be inoculated into a relatively large volume or over a large area of medium. It generally requires the following materials: a medium that has a relatively firm surface (see agar in "Physical States of Media," page 39), a Petri dish (a clear, flat dish with a cover), and inoculating tools. In the streak plate method, a small droplet of culture or sample is spread over the surface of the medium with an *inoculating loop* in a pattern that gradually thins out the sample and separates the cells spatially over several sections of the plate **(figure 2.9a)**. The goal here is to allow a single cell to grow into an isolated colony.

In the loop dilution, or pour plate, technique, the sample is inoculated serially into a series of cooled but still liquid agar tubes so as to dilute the number of cells

Figure 2.8 Isolation technique. Stages in the formation of an isolated colony, showing the microscopic events and the macroscopic result. Separation techniques such as streaking can be used to isolate single cells. After numerous cell divisions, a macroscopic mound of cells, or a colony, will be formed. This is a relatively simple yet successful way to separate different types of bacteria in a mixed sample.

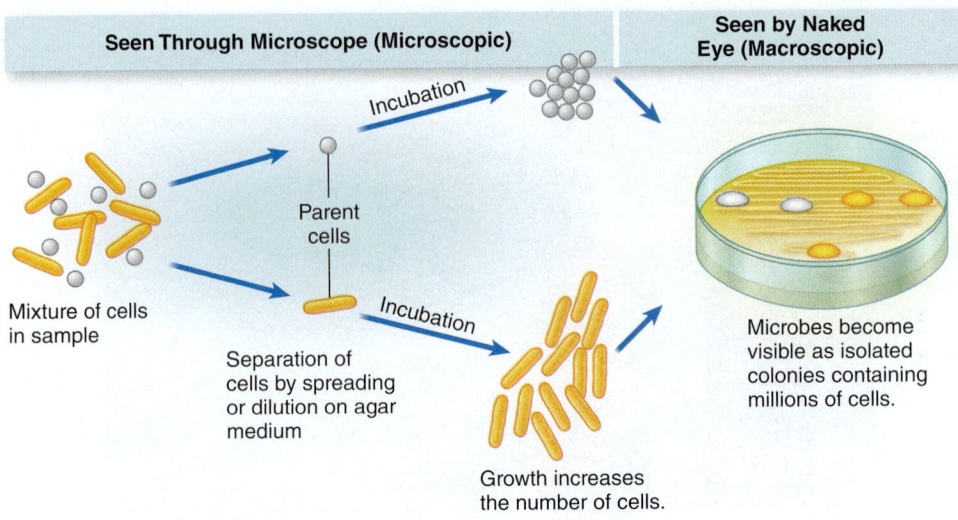

Seen Through Microscope (Microscopic)

Seen by Naked Eye (Macroscopic)

Mixture of cells in sample

Parent cells

Incubation

Separation of cells by spreading or dilution on agar medium

Incubation

Growth increases the number of cells.

Microbes become visible as isolated colonies containing millions of cells.

in each successive tube in the series **(figure 2.9b).** Inoculated tubes are then plated out (poured) into sterile Petri dishes and are allowed to solidify (harden). The end result (usually in the second or third plate) is that the number of cells per volume is so decreased that cells have ample space to grow into separate colonies. One difference between this and the streak plate method is that in this technique some of the colonies will develop deep in the medium itself and not just on the surface.

With the spread plate technique, a small volume of liquid, diluted sample is pipetted onto the surface of the medium and spread around evenly by a sterile spreading tool (sometimes called a "hockey stick" because of its shape). Like the streak plate, cells are pushed onto separate areas on the surface so that they can form individual colonies **(figure 2.9c).**

Rounding Out the Five I's: Inspection and Identification

How does one determine (i.e., identify) what sorts of microorganisms have been isolated in cultures? Certainly, microscopic appearance can be valuable in differentiating the smaller, simpler bacterial cells from the larger, more complex eukaryotic cells. Appearance can be especially useful in identifying eukaryotic microorganisms to the level of genus or species because of their distinctive morphological features; however, bacteria are generally not identifiable by these methods because very different species may appear quite similar. For them, we have to include other techniques, some of which characterize their cellular metabolism. These methods, called biochemical tests, can determine fundamental chemical characteristics such as nutrient requirements, products given off during growth, presence of enzymes, and mechanisms for deriving energy. Their genetic and immunologic characteristics are also used for identification. In chapter 15, we present more detailed examples of the most current genotypic and immunologic identification methods.

> ### 2.1 LEARNING OUTCOMES—Assess Your Progress
> 1. Explain what the Five I's are and what each step entails.
> 2. Discuss three physical states of media and when each is used.
> 3. Compare and contrast selective and differential media, and give an example of each.
> 4. Provide brief definitions for *defined media* and *complex media*.

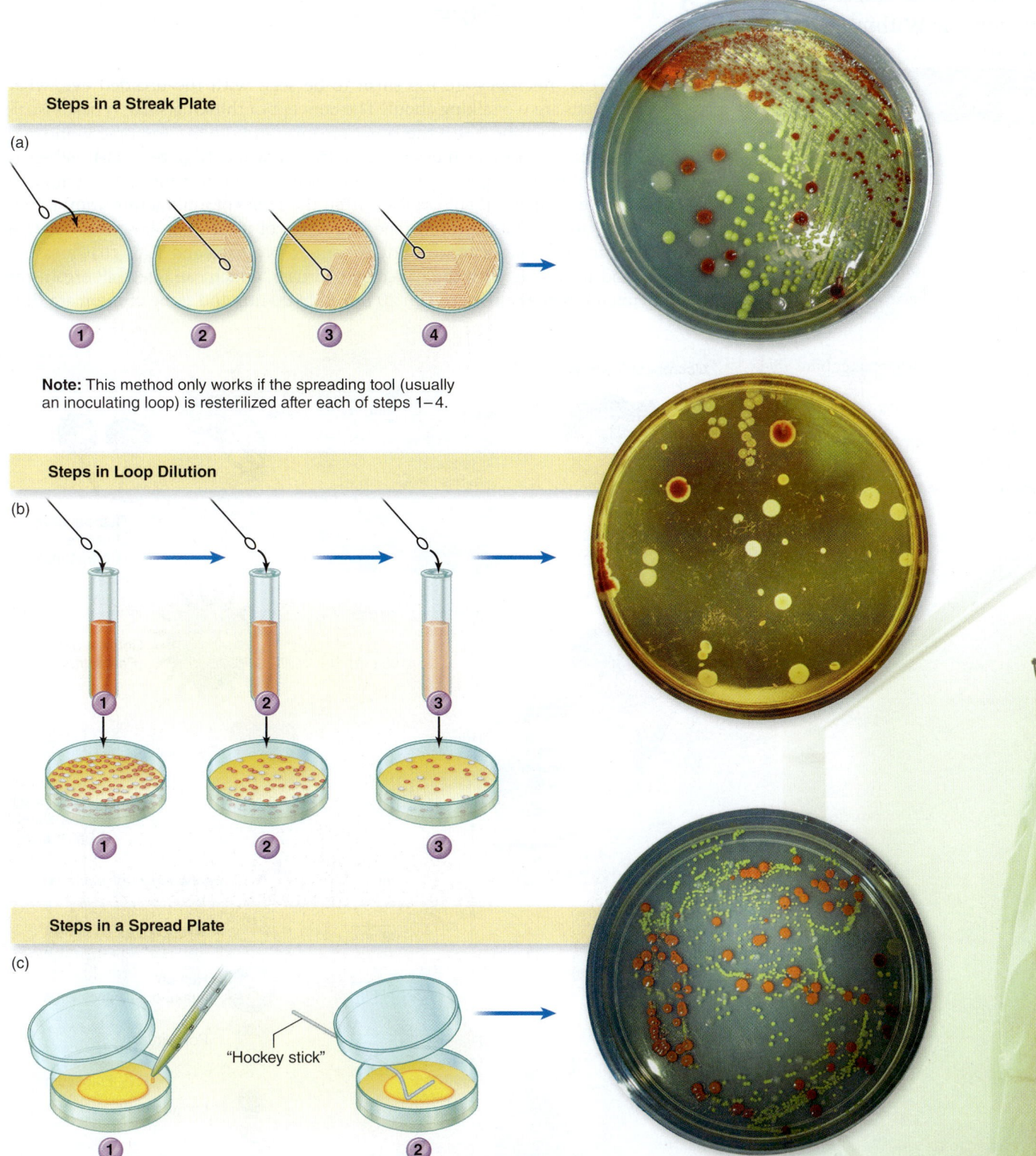

Steps in a Streak Plate

(a)

①   ②   ③   ④

Note: This method only works if the spreading tool (usually an inoculating loop) is resterilized after each of steps 1–4.

Steps in Loop Dilution

(b)

①   ②   ③

1   2   3

Steps in a Spread Plate

(c)

"Hockey stick"

1   2

Figure 2.9 **Methods for isolating bacteria.** (a) Steps in a quadrant streak plate and resulting isolated colonies of bacteria. (b) Steps in the loop dilution method and the appearance of plate 3. (c) Spread plate and its result.

Table 2.5 Conversions Within the Metric System

Log^{10} of Meters	Meters	Name
3	1,000	Kilometer (km)
0	1	Meter (m)
−1	0.1	Decimeter (dm)
−2	0.01	Centimeter (cm)
−3	0.001	Millimeter (mm)
−4	0.0001	–
−5	0.00001	–
−6	0.000001	Micrometer (μm)
−7	0.0000001	–
−8	0.00000001	–
−9	0.000000001	Nanometer (nm)

2.2 The Microscope

Microbial Size

When we say that microbes are too small to be seen with the unaided eye, what sorts of dimensions are we talking about? The concept of thinking small is best visualized by comparing microbes with the larger organisms of the macroscopic world and also with the atoms and molecules of the molecular world **(figure 2.10)**. Whereas the dimensions of macroscopic organisms are usually given in centimeters (cm) and meters (m), those of microorganisms fall within the range of millimeters (mm) to micrometers (μm) to nanometers (nm). The size range of most microbes extends from the smallest bacteria, measuring around 200 nm, to protozoa and algae that measure 3 to 4 mm and are visible with the naked eye. Viruses, which can infect all organisms including microbes, measure between 20 nm and 800 nm, and some of them are thus

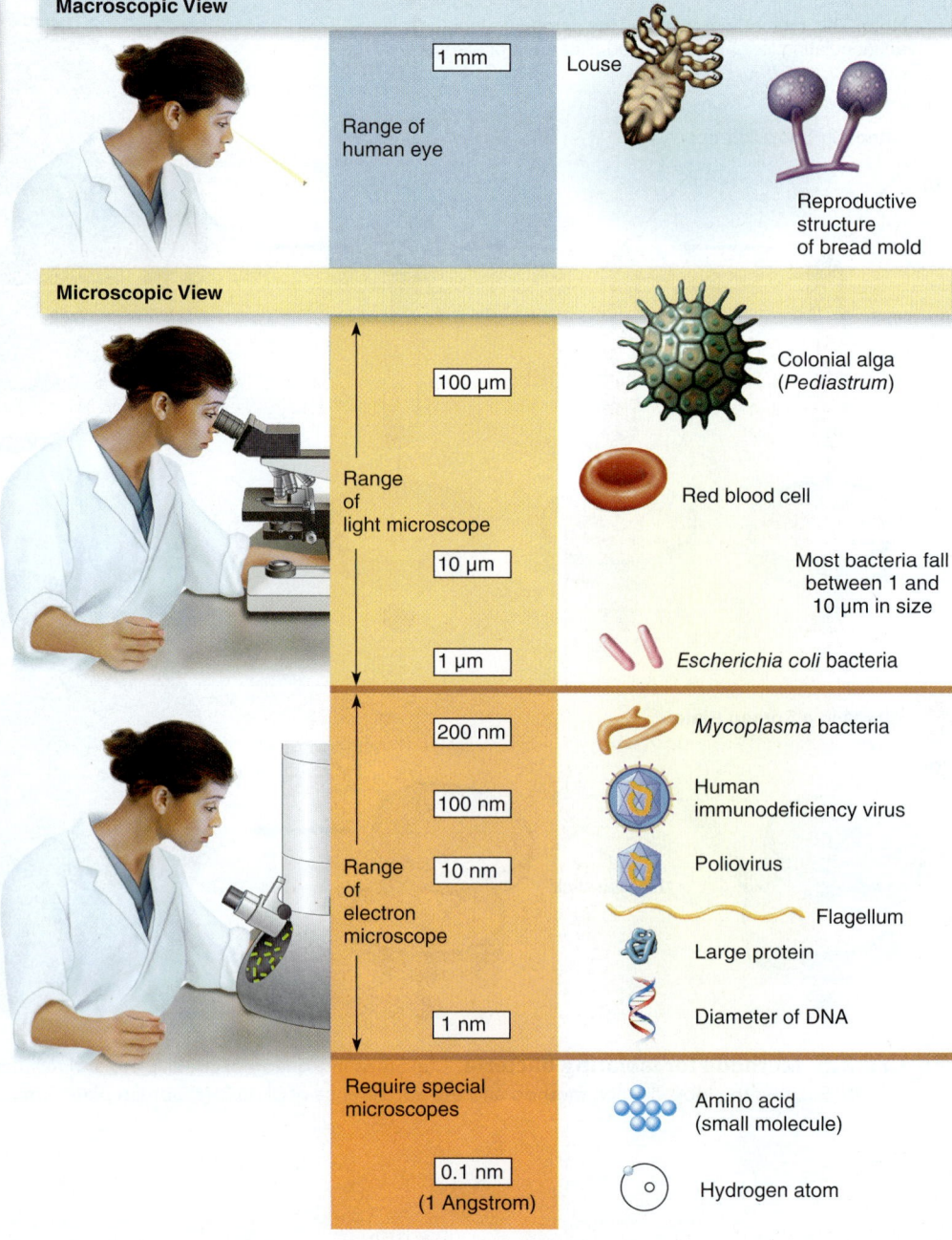

Figure 2.10. The size of things. Common measurements encountered in microbiology and a scale of comparison from the macroscopic to the microscopic, molecular, and atomic. Most microbes encountered in our studies will fall between 100 μm and 10 nm in overall dimensions. The microbes shown are more or less to scale within size zone but not between size zones.

not much bigger than large molecules, whereas others are just a tad larger than the smallest bacteria. Consult **table 2.5** for a reminder of relative size.

Magnification and Microscope Design

The microbial world is of obvious importance, but it would remain largely uncharted without an essential tool: the microscope. The fundamental parts of a modern compound light microscope are illustrated in **figure 2.11.**

Principles of Light Microscopy

Microscopes provide three important qualities:

- magnification,
- resolution,
- and contrast.

Magnification

Magnification occurs in two phases. The first lens in this system (the one closest to the specimen) is the objective lens, and the second (the one closest to the eye) is the ocular lens, or eyepiece **(figure 2.12).** The objective forms the initial image of the specimen, called the **real image.** When this image is projected up through the microscope body to the plane of the eyepiece, the ocular lens forms a second image, the **virtual image.** The virtual image is the one that will be received by the eye and converted to a retinal and visual image. The magnifying power of the objective lens usually ranges from 4× to 100×, and the power of the ocular lens is usually 10×.

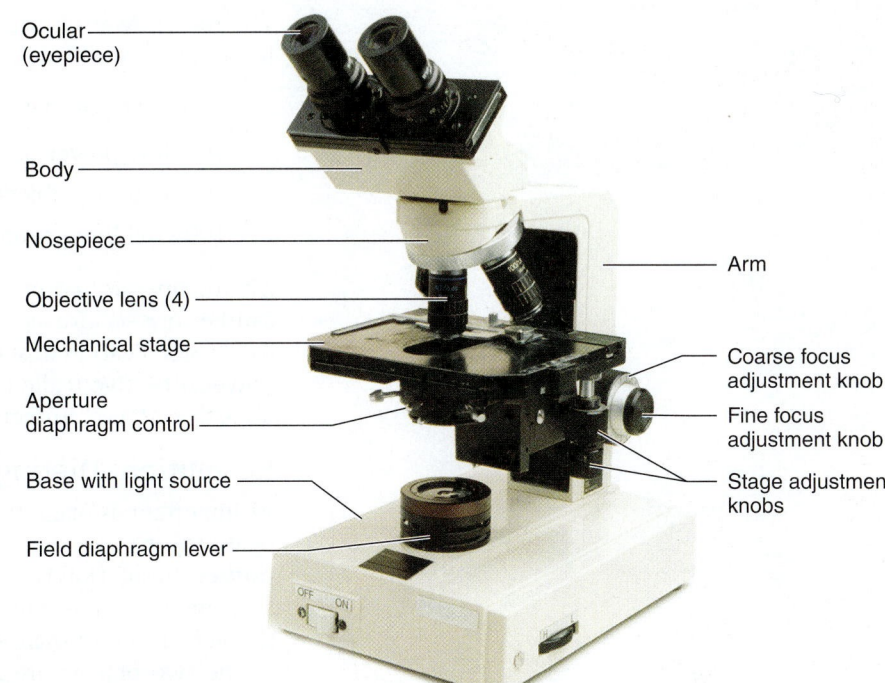

Figure 2.11 **The parts of a student laboratory microscope.** This microscope is a compound light microscope with two oculars (called binocular). It has four objective lenses.

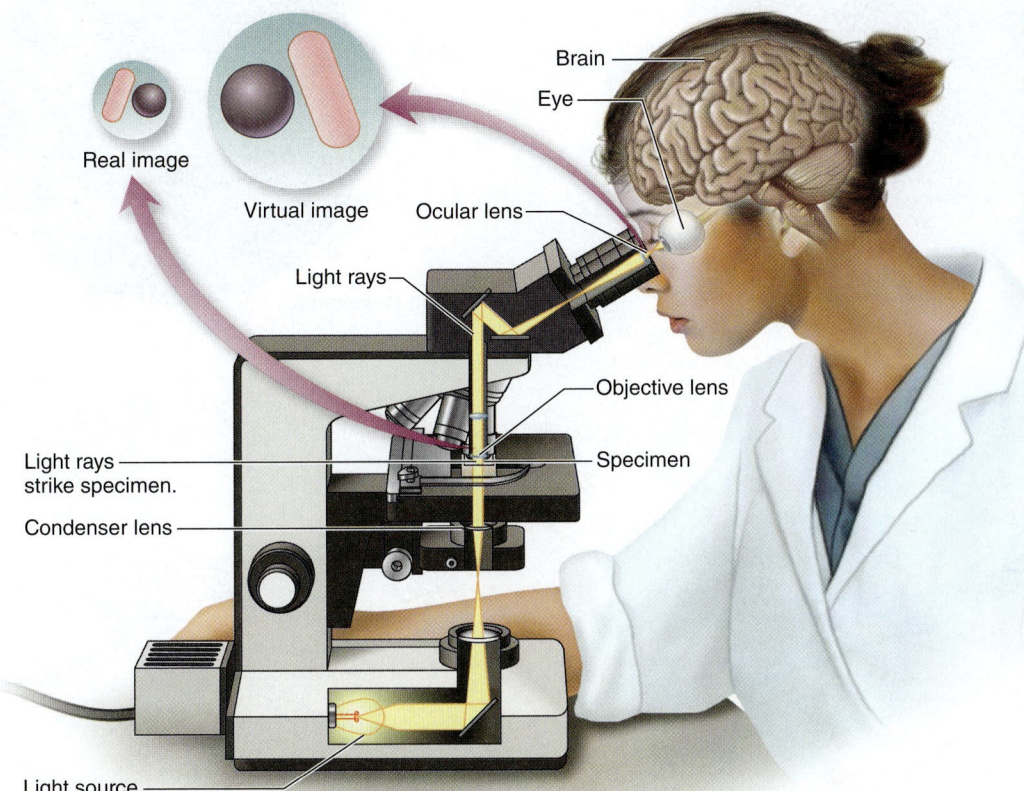

Figure 2.12 **The pathway of light and the two stages in magnification of a compound microscope.** As light passes through the condenser, it forms a solid beam that is focused on the specimen. Light leaving the specimen that enters the objective lens is refracted so that an enlarged primary image, the real image, is formed. One does not see this image, but its degree of magnification is represented by the smaller circle. The real image is projected through the ocular, and a second image, the virtual image, is formed by a similar process. The virtual image is the final magnified image that is received by the retina and perceived by the brain. Notice that the lens systems cause the image to be reversed.

The total power of magnification of the final image formed by the combined lenses is a product of the separate powers of the two lenses:

Power of objective	Power of ocular	Total magnification
10× low power objective	× 10×	= 100×
40× high dry objective	× 10×	= 400×
100× oil immersion objective	× 10×	= 1,000×

Microscopes are equipped with a nosepiece holding three or more objectives that can be rotated into position as needed. Depending on the power of the ocular, the total magnification of standard light microscopes can vary from 40× with the lowest power objective (called the scanning objective) to 2,000× with the highest power objective (the oil immersion objective).

Resolution: Distinguishing Magnified Objects Clearly

As important as magnification is for visualizing tiny objects or cells, an additional optical property is essential for seeing clearly. That property is resolution, or **resolving power.** Resolution is the capacity of an optical system to distinguish or separate two adjacent objects or points from one another. For example, at a certain fixed distance, the lens in the human eye can resolve two small objects as separate points as long as the two objects are no closer than 0.2 millimeters apart. The eye examination given by optometrists is in fact a test of the resolving power of the human eye for various-size letters read at a particular distance. **Figure 2.13** should help you understand the concept of resolution.

The oil immersion lens (100× magnification) uses oil to capture some of the light that would otherwise be lost to scatter **(figure 2.14).** Reducing this scatter increases

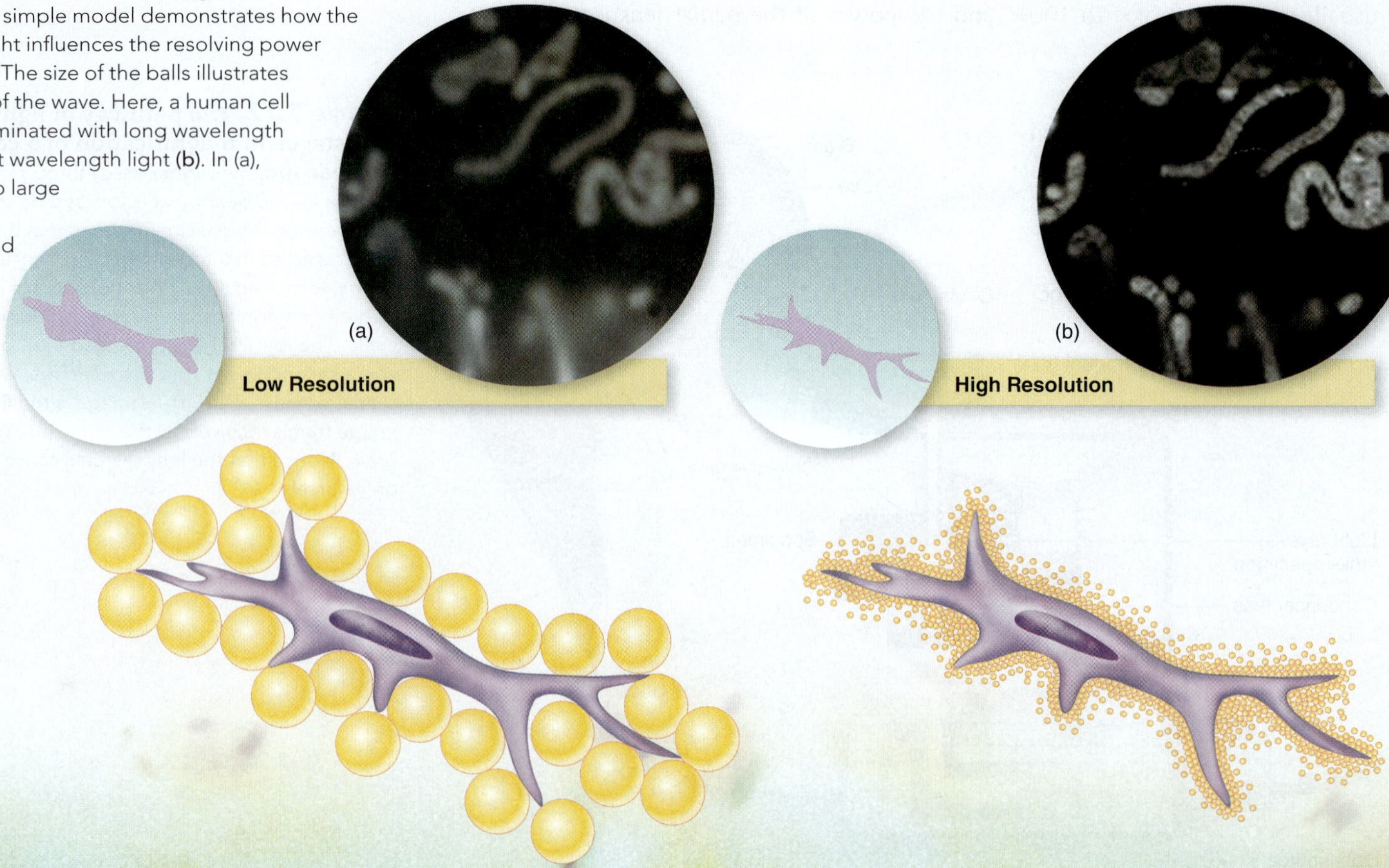

Figure 2.13 Effect of wavelength on resolution. A simple model demonstrates how the wavelength of light influences the resolving power of a microscope. The size of the balls illustrates the relative size of the wave. Here, a human cell (fibroblast) is illuminated with long wavelength light (a) and short wavelength light (b). In (a), the waves are too large to penetrate the tighter spaces and produce a fuzzy, undetailed image.

(a)

Low Resolution

(b)

High Resolution

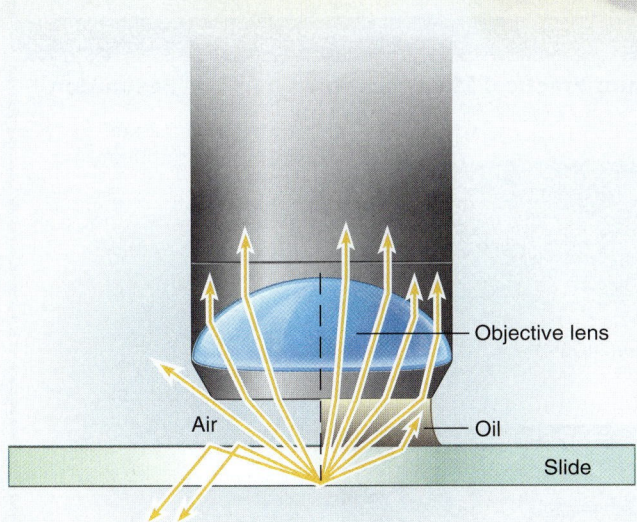

Figure 2.14 **Workings of an oil immersion lens.** Without oil, some of the peripheral light that passes through the specimen is scattered into the air or onto the glass slide; this scattering decreases resolution.

Appearance in Microscope	Appearance in Reality

Small bacterial cells

Eukaryotic cells

0.2 µm   2 µm   0.2 µm   2 µm

Figure 2.15 **The importance of resolution.** If a microscope has a resolving power of 0.2 µm, then the bacterial cells would not be resolvable as two separate cells. Likewise, the small specks inside the eukaryotic cell will not be visible.

resolution. In practical terms, the oil immersion lens can resolve any cell or cell part as long as it is at least 0.2 µm in diameter, and it can resolve two adjacent objects as long as they are at least 0.2 µm apart **(figure 2.15).** In general, organisms that are 0.5 µm or more in diameter are readily seen. This includes fungi and protozoa, some of their internal structures, and most bacteria. However, a few bacteria and most viruses are far too small to be resolved by the optical microscope and require electron micros-copy (discussed later in this chapter). In summary, then, the factor that most limits the clarity of a microscope's image is its resolving power. Even if a light microscope were designed to magnify several thousand times, its resolving power could not be increased, and the image it produced would simply be enlarged and fuzzy.

Contrast

The third quality of a well-magnified image is its degree of contrast from its sur-roundings. The contrast is measured by a quality called the **refractive index.** Re-fractive index refers to the degree of bending that light undergoes as it passes from one medium, such as water or glass, to another medium, such as bacterial cells. The higher the difference in refractive indexes (the more bending of light), the sharper the contrast that is registered by the microscope and the eye. Because too much light can reduce contrast and burn out the image, an adjustable iris diaphragm on most mi-croscopes controls the amount of light entering the condenser. The lack of contrast in cell components is compensated for by using special lenses (the phase-contrast microscope) and by adding dyes.

Different Types of Light Microscopes

Optical microscopes that use visible light can be described by the nature of their *field,* meaning the circular area viewed through the ocular lens. There are four types of visible-light microscopes: bright-field, dark-field, phase-contrast, and interference. A fifth type of optical microscope, the fluorescence microscope, uses ultraviolet radi-ation as the illuminating source; another, the confocal microscope, uses a laser beam. Each of these microscopes is adapted for viewing specimens in a particular way, as described in **table 2.6.**

NCLEX® PREP

3. The capacity of an optical system to distinguish or separate two adjacent objects or points from one another is known as
   a. *the real image.*
   b. *the virtual image.*
   c. *resolving power.*
   d. *numerical aperture.*
   e. *power.*

Table 2.6 Comparison of Types of Microscopy

Visible light as source of illumination

Microscope	Maximum Practical Magnification	Resolution
Bright Field The bright-field microscope is the most widely used type of light microscope. Although we ordinarily view objects like the words on this page with light reflected off the surface, a bright-field microscope forms its image when light is transmitted through the specimen. The specimen, being denser and more opaque than its surroundings, absorbs some of this light, and the rest of the light is transmitted directly up through the ocular. As a result, the specimen will produce an image that is darker than the surrounding brightly illuminated field. The bright-field microscope is a multipurpose instrument that can be used for both live, unstained material and preserved, stained material. 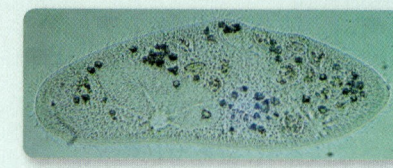 *Paramecium* (400×)	2,000×	0.2 µm (200 nm)
Dark Field A bright-field microscope can be adapted as a dark-field microscope by adding a special disc called a *stop* to the condenser. The stop blocks all light from entering the objective lens–except peripheral light that is reflected off the sides of the specimen itself. The resulting image is a particularly striking one: brightly illuminated specimens surrounded by a dark (black) field. The most effective use of dark-field microscopy is to visualize living cells that would be distorted by drying or heat or that cannot be stained with the usual methods. Dark-field microscopy can outline the organism's shape and permit rapid recognition of swimming cells that might appear in dental and other infections, but it does not reveal fine internal details.  *Paramecium* (400×)	2,000×	0.2 µm
Phase-Contrast If similar objects made of clear glass, ice, cellophane, or plastic are immersed in the same container of water, an observer would have difficulty telling them apart because they have similar optical properties. Internal components of a live, unstained cell also lack contrast and can be difficult to distinguish. But cell structures do differ slightly in density, enough that they can alter the light that passes through them in subtle ways. The phase-contrast microscope has been constructed to take advantage of this characteristic. This microscope contains devices that transform the subtle changes in light waves passing through the specimen into differences in light intensity. For example, denser cell parts such as organelles alter the pathway of light more than less dense regions (the cytoplasm). Light patterns coming from these regions will vary in contrast. The amount of internal detail visible by this method is greater than by either bright-field or dark-field methods. The phase-contrast microscope is most useful for observing intracellular structures such as bacterial endospores, granules, and organelles, as well as the locomotor structures of eukaryotic cells such as cilia. 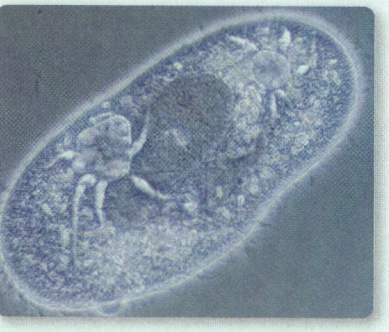 *Paramecium* (400×)	2,000×	0.2 µm
Differential Interference Like the phase-contrast microscope, the differential interference contrast (DIC) microscope provides a detailed view of unstained, live specimens by manipulating the light. But this microscope has additional refinements, including two prisms that add contrasting colors to the image and two beams of light rather than a single one. DIC microscopes produce extremely well-defined images that are vividly colored and appear three-dimensional.  *Amoeba proteus* (160×)	2,000×	0.2 µm

Table 2.6 (continued)

Ultraviolet rays as source of illumination

Microscope	Maximum Practical Magnification	Resolution
Fluorescence The fluorescence microscope is a specially modified compound microscope furnished with an ultraviolet (UV) radiation source and a filter that protects the viewer's eye from injury by these dangerous rays. The name of this type of microscopy originates from the use of certain dyes (acridine, fluorescein) and minerals that show **fluorescence.** The dyes emit visible light when bombarded by short ultraviolet rays. For an image to be formed, the specimen must first be coated or placed in contact with a source of fluorescence. Subsequent illumination by ultraviolet radiation causes the specimen to give off light that will form its own image, usually an intense yellow, orange, or red against a black field. Fluorescence microscopy has its most useful applications in diagnosing infections caused by specific bacteria, protozoans, and viruses.	2,000× 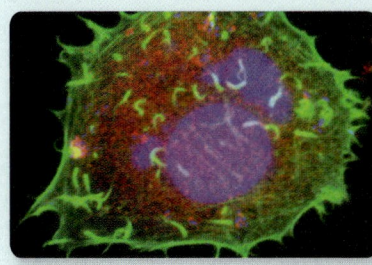 Fluorescence image of a eukaryotic cell.	0.2 μm
Confocal The scanning confocal microscope overcomes the problem of cells or structures being too thick, a problem resulting in other microscopes being unable to focus on all their levels. This microscope uses a laser beam of light to scan various depths in the specimen and deliver a sharp image focusing on just a single plane. It is thus able to capture a highly focused view at any level, ranging from the surface to the middle of the cell. It is most often used on fluorescently stained specimens but it can also be used to visualize live unstained cells and tissues.	2,000× 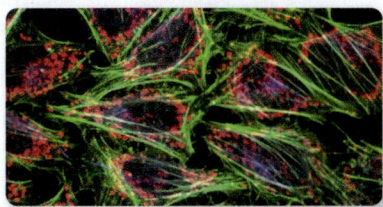 Myofibroblasts, cells involved in tissue repair (400×)	0.2 μm

Electron beam forms image of specimen

Microscope	Maximum Practical Magnification	Resolution
Transmission Electron Microscope (TEM) Transmission electron microscopes are the method of choice for viewing the detailed structure of cells and viruses. This microscope produces its image by transmitting electrons through the specimen. Because electrons cannot readily penetrate thick preparations, the specimen must be sectioned into extremely thin slices (20-100 nm thick) and stained or coated with metals that will increase image contrast. The darkest areas of TEM micrographs represent the thicker (denser) parts, and the lighter areas indicate the more transparent and less dense parts.	100,000,000× 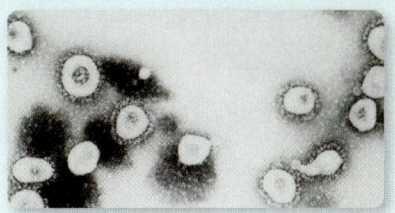 Coronavirus, causative agent of many respiratory infections (100,000×)	0.5 nm
Scanning Electron Microscope (SEM) The scanning electon microscope provides some of the most dramatic and realistic images in existence. This instrument is designed to create an extremely detailed three-dimensional view of all kinds of objects–from plaque on teeth to tapeworm heads. To produce its images, the SEM bombards the surface of a whole metal-coated specimen with electrons while scanning back and forth over it. A shower of electrons deflected from the surface is picked up with great fidelity by a sophisticated detector, and the electron pattern is displayed as an image on a television screen. You will often see these images in vivid colors. The color is always added afterward; the actual microscopic image is black and white.	100,000,000×  Algae showing cell walls made of calcium discs (10,000×)	10 nm

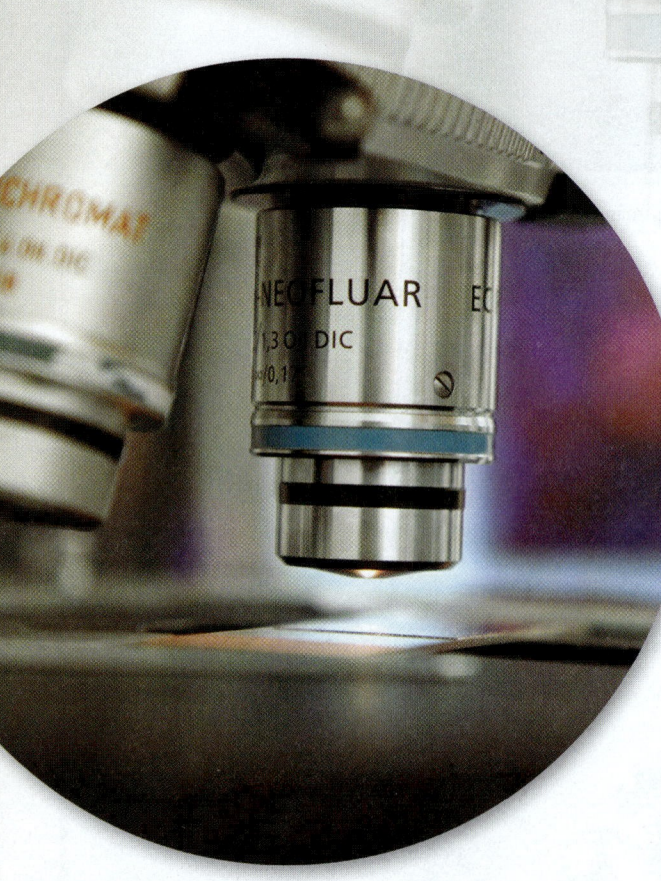

Preparing Specimens for the Microscope

A specimen for optical microscopy is generally prepared by mounting a sample on a suitable glass slide that sits on the stage between the condenser and the objective lens. The manner in which a slide specimen, or mount, is prepared depends upon (1) the condition of the specimen, either in a living or preserved state; (2) the aims of the examiner, whether to observe overall structure, identify the microorganisms, or see movement; and (3) the type of microscopy available, whether it is bright-field, dark-field, phase-contrast, or fluorescence.

Fresh, Living Preparations

Live samples of microorganisms are placed in wet mounts or in hanging drop mounts so that they can be observed as near to their natural state as possible. The cells are suspended in a suitable fluid (water, broth, saline) that temporarily maintains viability and provides space and a medium for locomotion. A wet mount consists of a drop or two of the culture placed on a slide and overlaid with a coverslip. The hanging drop preparation is made with a special concave (depression) slide, a Vaseline adhesive or sealant, and a coverslip from which a tiny drop of sample is suspended **(figure 2.16)**. These short-term mounts provide a true assessment of the size, shape, arrangement, color, and motility of cells. However, if you need to visualize greater cellular detail, you will have to use phase-contrast or interference microscopy.

Fixed, Stained Smears

A more permanent mount for long-term study can be obtained by preparing fixed, stained specimens. The smear technique, developed by Robert Koch more than 100 years ago, consists of spreading a thin film made from a liquid suspension of cells on a slide and air-drying it. Next, the air-dried smear is usually heated gently by a process called heat fixation that simultaneously kills the specimen and secures it to the slide.

Stains

Like images on undeveloped photographic film, the unstained cells of a fixed smear are quite indistinct, no matter how great the magnification or how fine the resolving power of the microscope. The process of "developing" a smear to create contrast and make inconspicuous features stand out requires staining techniques. Staining is any procedure that applies colored chemicals called *dyes* to specimens. Dyes impart a color to cells or cell parts by becoming affixed to them through a chemical reaction. Dyes can be classified as basic (cationic) dyes, which have a positive charge, or acidic (anionic) dyes, which have a negative charge. Because chemicals of opposite charge are attracted to each other, cell parts that are negatively charged will attract basic dyes, and those that are positively charged will attract acidic dyes. Many cells, especially those of bacteria, have numerous negatively charged acidic substances on

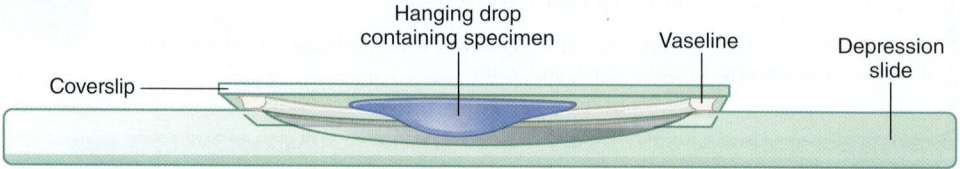

Figure 2.16 Hanging drop technique. Cross-section view of slide and coverslip. (Vaseline actually surrounds entire well of slide.)

Table 2.7 Comparison of Positive and Negative Stains

	Positive Staining	Negative Staining
Appearance of cell	Colored by dye	Clear and colorless
Background	Not stained (generally white)	Stained (dark gray or black)

Dyes employed	Basic dyes:
	    Crystal violet
	    Methylene blue
	    Safranin Malachite green
		Acidic dyes:
		    Nigrosin
		    India ink

Subtypes of stains	Several types:
	    Simple stain
	
	Differential stains
	    Gram stain
	    Acid-fast stain
	    Spore stain
	
	Special stains
	    Capsule
	    Flagella
	    Spore
	    Granules
	    Nucleic acid
		Few types:
		    Capsule
		    Spore

their surfaces and thus stain more readily with basic dyes. Acidic dyes, on the other hand, tend to be repelled by cells, so they are good for negative staining (discussed in the next section).

Negative Versus Positive Staining   Two basic types of staining technique are used, depending upon how a dye reacts with the specimen (summarized in **table 2.7**). Most procedures involve a **positive stain,** in which the dye actually sticks to the specimen and gives it color. A **negative stain,** on the other hand, is just the reverse (like a photographic negative). The dye does not stick to the specimen but settles some distance from its outer boundary, forming a silhouette. Nigrosin (blue-black) and India ink (a black suspension of carbon particles) are the dyes most commonly used for negative staining. The cells themselves do not stain because these dyes are negatively charged and are repelled by the negatively charged surface of the cells. The value of negative staining is its relative simplicity and the reduced shrinkage or distortion of cells, as the smear is not heat fixed. Negative staining is also used to accentuate the capsule that surrounds certain bacteria and yeasts.

Simple Versus Differential Staining   Positive staining methods are classified as simple, differential, or special. Whereas **simple stains** require only a single dye and an uncomplicated procedure, **differential stains** use two differently colored dyes, called the *primary dye* and the *counterstain,* to distinguish between cell types or parts. These staining techniques tend to be more complex and sometimes require additional chemical reagents to produce the desired reaction.

Simple stains cause all cells in a smear to appear more or less the same color, regardless of type, but they can still reveal bacterial characteristics such as shape, size, and arrangement **(figure 2.17).**

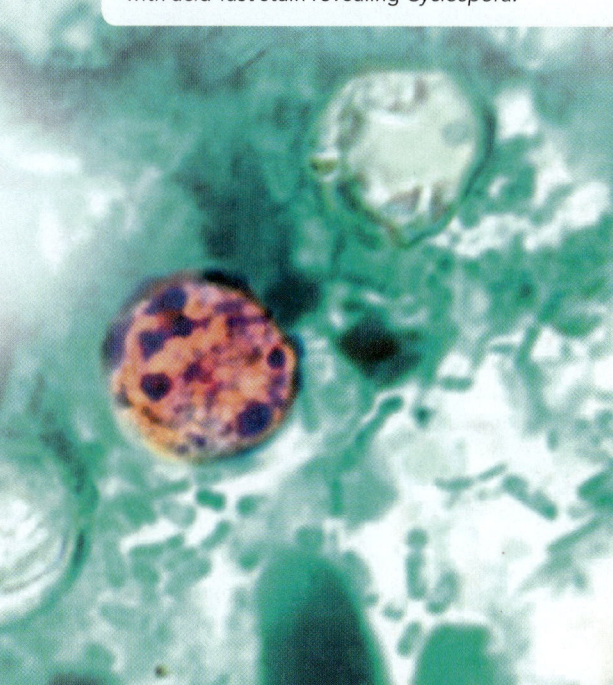

Photomicrograph of stool sample stained with acid-fast stain revealing *Cyclospora.*

Medical Moment

Gram-Positive Versus Gram-Negative Bacteria

The Gram stain is one type of differential stain that can help to identify bacterial species and guide treatment decisions. Differentiating between gram-positive and gram-negative organisms is important. One of the main differences between gram-positive and gram-negative bacteria is that gram-negative bacteria have an outer membrane containing LPS (lipopolysaccharide). This lipid portion acts as an endotoxin, which can cause a severe reaction if it enters the circulatory system, causing symptoms of shock (high fever, dangerously low blood pressure, and elevated respiratory rate). This is known as endotoxic shock.

Simple Stains

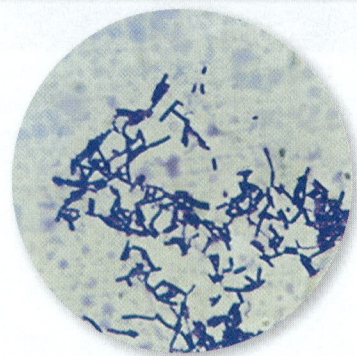

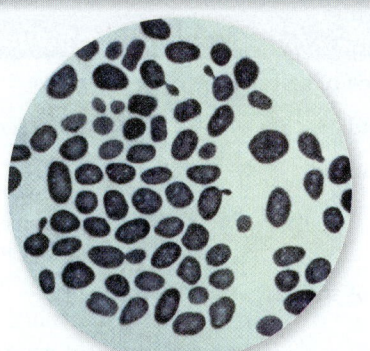

(a) Crystal violet stain of *Escherichia coli*　　　(b) Methylene blue stain of *Corynebacterium*

Figure 2.17 Simple stains.

Types of Differential Stains A satisfactory differential stain uses differently colored dyes to clearly contrast two cell types or cell parts. Common combinations are red and purple, red and green, or pink and blue **(figure 2.18).** Typical examples include Gram, acid-fast, and endospore stains. Some staining techniques (endospore, capsule), which are differential, are also in the "special" category.

The Gram Stain In 1884, Hans Christian Gram discovered a staining technique that could be used to make bacteria in infectious specimens more visible. His technique consisted of timed, sequential applications of crystal violet (the primary dye), Gram's iodine (the mordant), an alcohol rinse (decolorizer), and a contrasting counterstain. Bacteria that stain purple are called gram-positive, and those that stain red are called gram-negative. Gram-variable organisms produce both pink- and purple-staining cells.

The different results in the Gram stain are due to differences in the structure of the cell wall and how it reacts to the series of reagents applied to the cells. We will study it in more detail in chapter 3.

This century-old staining method remains the universal basis for bacterial classification and identification. The Gram stain can also be a practical aid in diagnosing infection and in guiding drug treatment. For example, Gram staining a fresh sputum or spinal fluid specimen can help pinpoint the possible cause of infection, and in some cases it is possible to begin drug therapy on the basis of this stain. Even in this day of elaborate and expensive medical technology, the Gram stain remains an important first tool in diagnosis.

 **Figure 2.18 Differential stains.**

Differential Stains

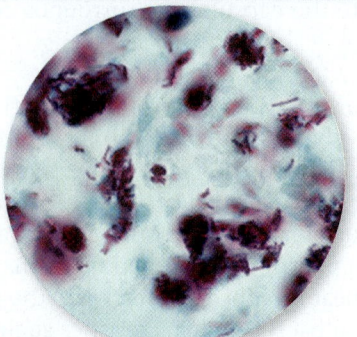

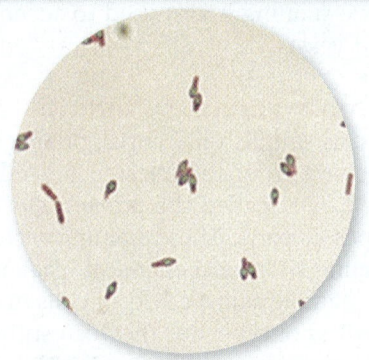

(a) Gram stain. Here both gram-negative (pink) rods and gram-positive (purple) cocci are visible.

(b) Acid-fast stain. Reddish-purple cells are acid-fast. Blue cells are non-acid-fast.

(c) Endospore stain, showing endospores (red) and vegetative cells (blue)

Special Stains

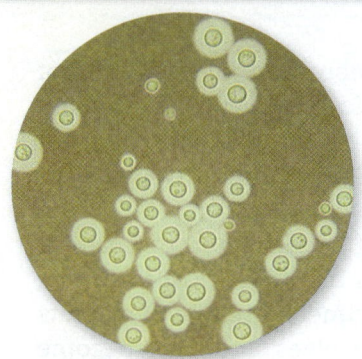

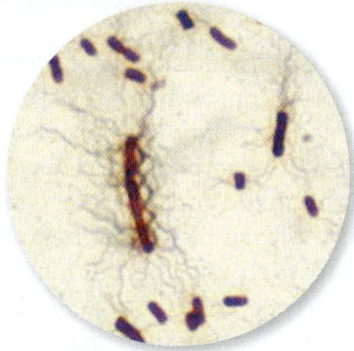

(a) India ink capsule stain of *Cryptococcus neoformans*

(b) Flagellar stain of *Proteus vulgaris*

Figure 2.19 Special stains.

The **acid-fast stain,** like the Gram stain, is an important diagnostic stain that differentiates acid-fast bacteria (pink) from non-acid-fast bacteria (blue). This stain originated as a specific method to detect *Mycobacterium tuberculosis* in specimens. It was determined that these bacterial cells have a particularly impervious outer wall that holds fast (tightly or tenaciously) to the dye (carbol fuchsin) even when washed with a solution containing acid or acid alcohol. This stain is used for other medically important bacteria, fungi, and protozoa; it is performed when a gram-variable result is seen in a specimen.

The **endospore** stain (spore stain) is similar to the acid-fast method in that a dye is forced by heat into resistant bodies called endospores (their formation and significance are discussed in chapter 3). This stain is designed to distinguish between endospores and the cells that they come from (so-called **vegetative** cells). Of significance in medical microbiology are the gram-positive, endospore-forming members of the genus *Bacillus* (the cause of anthrax) and *Clostridium* (the cause of botulism and tetanus)–dramatic diseases that we consider in later chapters.

Special stains are used to emphasize certain cell parts that are not revealed by conventional staining methods **(figure 2.19). Capsular staining** is a method of observing the microbial capsule, an unstructured protective layer surrounding the cells of some bacteria and fungi. Because the capsule does not react with most stains, it is often negatively stained with India ink, or it may be demonstrated by special positive stains. The fact that not all microbes exhibit capsules is a useful feature for identifying pathogens. One example is *Cryptococcus,* which causes a serious form of fungal meningitis in AIDS patients (see chapter 17).

Flagellar staining is a method of revealing **flagella** (singular, *flagellum*), the tiny, slender filaments used by bacteria for locomotion. Because the width of bacterial flagella lies beyond the resolving power of the light microscope, in order to be seen, they must be enlarged by depositing a coating on the outside of the filament and then staining it. Their presence, number, and arrangement on a cell are useful for identification of the bacteria.

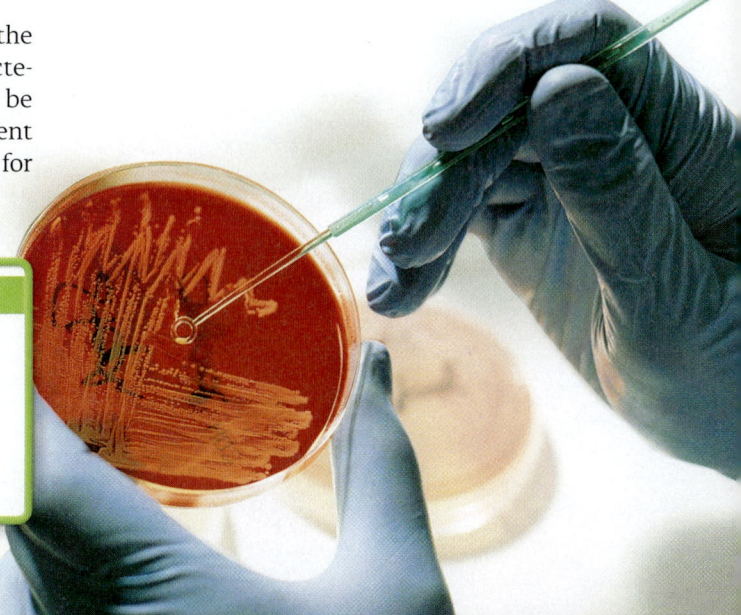

2.2 LEARNING OUTCOMES—Assess Your Progress

5. Convert among the different units of the metric system.

6. List and describe the three elements of good microscopy.

7. Differentiate between the principles of light microscopy and the principles of electron microscopy.

8. Give examples of simple, differential, and special stains.

CASE FILE WRAP-UP

This special medium is designed for urine cultures.

Mixed cultures are defined as containing two or more identifiable species of microorganisms. "Contaminated" is a designation given to cultures when unwanted (and usually unidentified) microbes are present. These intruders may have been introduced to the specimen through poor collection, handling, or storage technique.

In the case file at the beginning of this chapter, the patient was provided with verbal instructions regarding how to collect a midstream urine sample. Urine specimens are one of the few specimens collected by patients themselves and may become contaminated easily due to poor collection technique. Failure to wash hands, accidentally touching the rim or lid of the collection container, and failure to properly cleanse the external genitalia (in female patients) prior to specimen collection are some of the ways in which specimens may become contaminated.

In this case, the patient returned to provide another sample. Instructions were provided again, and the patient was asked whether she understood what was required of her. This time the sample yielded only one species, *Escherichia coli*, a bacterium that is a common causative agent of urinary tract infections. The patient was treated with antibiotics for 10 days, and a repeat culture was negative for any microorganisms.

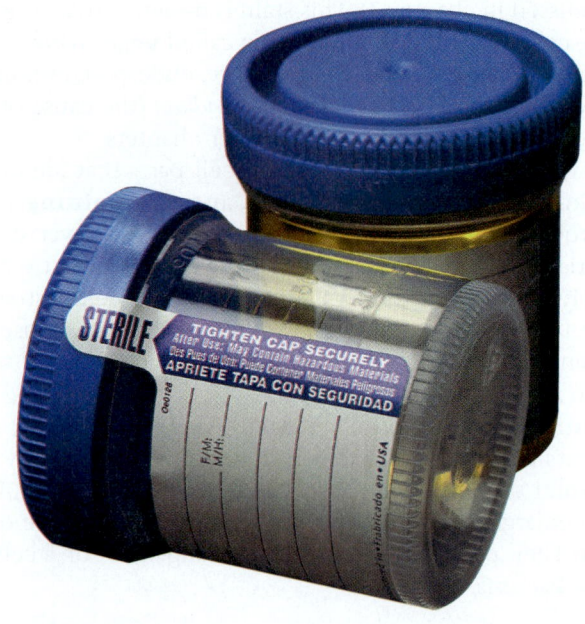

The Papanicolaou Stain

The *Papanicolaou test* (commonly referred to as a *Pap smear*) is a test used to screen for precancerous and cancerous conditions occurring in the female endocervical canal. It may also detect some vaginal and uterine infections caused by bacteria, fungi, or viruses. This staining technique was developed by Dr. George Papanicolaou in 1942 and is still widely used today, although it has been modified slightly over the years.

During a Pap smear, cells are collected from the cervical os (entrance to the uterus) using a swab, brush, or spatula. The procedure involved in collecting a Pap smear is not difficult but can cause some anxiety for patients. Patients are placed in the lithotomy position (lying on their back) on an examining table, and the patient's feet are placed in stirrups. This position allows the physician or nurse practitioner to visualize the external genitalia for signs of infection or other abnormalities and allows access to the vaginal canal. A speculum is used to gently open the walls of the vagina so that the cervical os can be visualized. A sample is taken from the cervical os using a small spatula or brush.

The sample is transferred immediately to a glass slide and fixed using an alcohol-based substance (usually ethanol). New liquid-based methods are currently available in which the sample is placed into a special liquid preservative and is later processed onto a glass slide. The sample is then stained and examined under a microscope in the usual fashion.

Pap smear staining uses a combination of four or five dyes. The slides are immersed in the dyes for established and specific periods of time. When properly performed, the stained specimen will display a variety of colors specific to different components of the cell. For example, the nuclei of the cell will appear blue to black, while cancerous cells will often appear pink and green within the same field of view.

The observation of abnormal cells and cell structures in a Pap smear has long been an indicator of infection with human papillomavirus (HPV), a known oncogenic (cancer-causing) virus. There is currently a great deal of debate surrounding the use of the Pap smear to screen for HPV infection versus more sensitive DNA-based tests for viral identification. These advances will be further discussed in chapters 15 and 21.

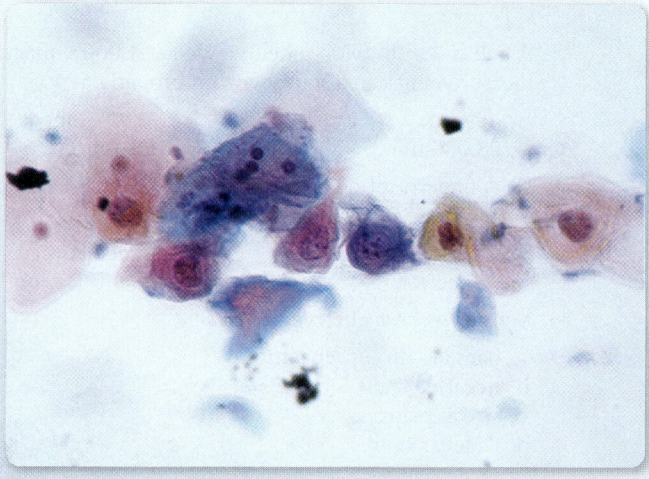

Pap smear of precancerous cervical cells. The cells with abnormally large nuclei indicate mild to moderate dysplasia.

Chapter Summary

2.1 How to Culture Microorganisms

- The Five I's–inoculation, incubation, isolation, inspection, and identification–summarize the kinds of laboratory procedures used in microbiology.
- Following *inoculation*, cultures are *incubated* at a specified temperature to encourage growth.
- Many microorganisms can be cultured on artificial media, but some can be cultured only in living tissue or in cells.
- Artificial media are classified by their *physical state* as either liquid, semisolid, liquefiable solid, or nonliquefiable solid.
- Artificial media are classified by their *chemical composition* as either *defined* or *complex*, depending on whether the exact chemical composition is known.
- Enriched, selective, differential, transport, assay, and enumerating media are all examples of media designed for specific purposes.
- *Isolated* colonies that originate from single cells are composed of large numbers of cells piled up together.
- A culture may be *pure*, containing only one species or type of microorganism; *mixed*, containing two or more known species; or *contaminated*, containing both known and unknown (unwanted) microorganisms.
- During *inspection*, the cultures are examined and evaluated macroscopically and microscopically.

- Microorganisms are *identified* in terms of their macroscopic or immunologic morphology, their microscopic morphology, their biochemical reactions, and their genetic characteristics.

2.2 The Microscope

- Magnification, resolving power, and contrast all influence the clarity of specimens viewed through the optical microscope.
- The maximum resolving power of the optical microscope is 200 nm, or 0.2 μm. This is sufficient to see the internal structures of eukaryotes and the morphology of most bacteria.
- There are six types of optical microscopes. Four types use visible light for illumination: bright-field, dark-field, phase-contrast, and interference microscopes. The fluorescence microscope uses UV light for illumination. The confocal microscope can use UV light or visible light reflected from specimens.
- Electron microscopes (EM) use electrons, not light waves, as an illumination source to provide high magnification (5,000× to 1,000,000×) and high resolution (0.5 nm).
- Specimens viewed through optical microscopes can be either alive or dead, depending on the type of specimen preparation, but all EM specimens are dead because they must be viewed in a vacuum.
- The Gram stain is an immensely useful differential stain that divides bacteria into two main groups, gram-positive and gram-negative. Some bacteria do not fall in either of these categories.
- Stains increase the contrast of specimens and they can be designed to differentiate cell shape, structure, and biochemical composition of the specimens being viewed.

Multiple-Choice Questions   Bloom's Levels 1 and 2: Remember and Understand

Select the correct answer from the answers provided.

1. A mixed culture is
   a. the same as a contaminated culture.
   b. one that has been adequately stirred.
   c. one that contains two or more known species.
   d. a pond sample containing algae and protozoa.

2. Resolution is ____ with a longer wavelength of light.
   a. improved
   b. worsened
   c. not changed
   d. not possible

3. A microscope that has a total magnification of 1,500× when using the oil immersion objective has an ocular of what power?
   a. 150×
   b. 1.5×
   c. 15×
   d. 30×

4. A cell is 25 μm wide when viewed at 1,000× magnification. This measurement can also be written properly as
   a. 25 mm.
   b. 25,000 mm.
   c. 0.025 mm.
   d. 2.5 mm.

5. DNA fingerprinting and antibody-based ELISA tests would be used during which step of microbial analysis?
   a. isolation
   b. inspection
   c. inoculation
   d. identification

6. Motility is best observed with a
   a. hanging drop preparation.
   b. negative stain.
   c. streak plate.
   d. flagellar stain.

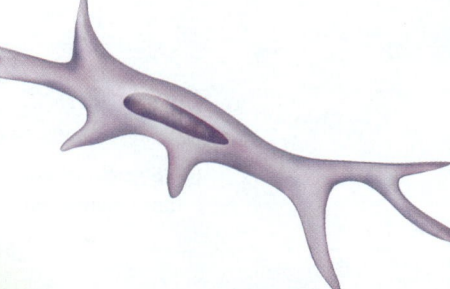

7. Bacteria tend to stain more readily with cationic (positively charged) dyes because bacteria

 a. contain large amounts of alkaline substances on their surfaces.
 b. contain large amounts of acidic substances on their surfaces.
 c. carry a neutral charge on their surfaces.
 d. have thick cell walls.

8. A fastidious organism must be grown on what type of medium?

 a. general-purpose medium
 b. differential medium
 c. defined medium
 d. enriched medium

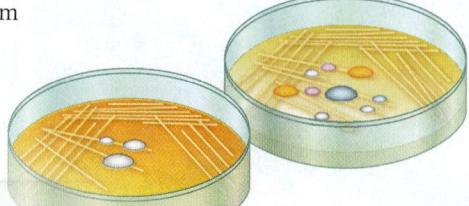

Critical Thinking   Bloom's Levels 3, 4, and 5: Apply, Analyze, and Evaluate

Critical thinking is the ability to reason and solve problems using facts and concepts. These questions can be approached from a number of angles and, in most cases, they do not have a single correct answer.

1. Your patient presents with a skin lesion that you believe to be impetigo, a bacterial infection. Please list the steps you will take to identify the pathogen(s) causing this infection, summarizing the tools and methods used in this process.

2. Which type(s) of medium would be used in each scenario?

 a. isolating the growth of *Streptococcus pyogenes* from a patient's throat swab
 b. isolating a pathogen from a patient's clean-catch urine sample
 c. isolating enteric bacteria such as *Escherichia coli* from a sample of organically grown spinach
 d. maintaining a patient's nasal swab specimen for further analysis and identification of possible respiratory syncytial virus (RSV) infection

3. a. Explain whether or not any of the methods in figure 2.9 could be used to determine the total number of cells present in a patient's specimen.
 b. After performing the streak plate method on a bacterial specimen, the culture was incubated for 48 hours at 37°C. Upon viewing the plate, there was heavy growth (with no isolated colonies) in the first quadrant, but no growth was apparent in the remaining quadrants. Please discuss errors in the procedure that could have produced this result.

4. a. Lactophenol cotton blue is utilized to stain the colorless cytoplasm of *Amoeba proteus*, a common pond protozoan. Please discuss which property of microscopy is enhanced by using this dye.
 b. Which type of microscopy would provide the best image in each scenario?
 • visualizing a viral pathogen in a patient's lung biopsy
 • visualizing the presence of multiple organisms within a specimen
 • visualizing the organelles within a eukaryotic cell

5. You have been told to obtain a sputum sample and to perform microbiological staining in order to determine the identity of the pathogen causing a patient's illness. You first perform a Gram stain, but upon microscopic analysis you visualize a mixture of pink and purple bacilli. Explain the results you have just observed, and discuss what you may now do in order to identify the pathogen.

Visual Connections   Bloom's Level 5: Evaluate

This question connects previous images to a new concept.

1. **Figure 2.9a.** If you were using the quadrant streak plate method to plate a very dilute broth culture (with many fewer bacteria than the broth used for the plate pictured to the right), would you expect to see single, isolated colonies in quadrant 4 or quadrant 3? Explain your answer.

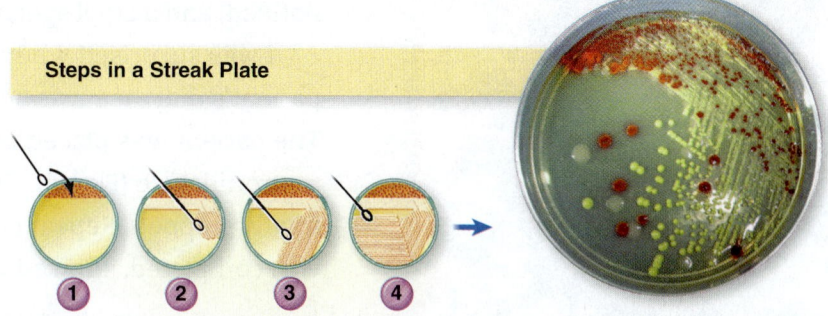

Steps in a Streak Plate

Note: This method only works if the spreading tool (usually an inoculating loop) is resterilized after each of steps 1–4.

www.mcgrawhillconnect.com

Enhance your study of this chapter with study tools and practice tests. Also ask your instructor about the resources available through Connect, including the media-rich eBook, interactive learning tools, and animations.

Extreme Endospores

While working as a newly graduated nurse, I was caring for an elderly female patient from a local nursing home who had been admitted for a hip replacement. The patient seemed to be recovering well until she developed redness, increased swelling, and purulent discharge at the surgical site. The wound was cultured and the patient was started on a cephalosporin antibiotic. The results from microbiological testing revealed that the infection was caused by *Staphylococcus aureus,* a pathogen known to be sensitive to the cephalosporin drug she was already taking.

The patient successfully completed the course of antibiotic therapy, and within a few days all signs of infection had subsided. The patient was progressing well with physiotherapy, and we were beginning to plan for discharge back to the nursing home when the patient suddenly began to experience diarrhea. At first I assumed that the diarrhea was because of an expected side effect from the antibiotic, but it soon became clear that this was something more than a general side effect. On the first day, the patient had two loose bowel movements. By the second day, the episodes of diarrhea were occurring every 2 to 3 hours. The stools were watery and foul-smelling and contained large amounts of mucus. The patient complained of mild abdominal pain and cramping, and she subsequently developed a fever. The physician was notified, and a stool specimen was collected for laboratory testing.

I was surprised when the stool culture came back showing that the patient's diarrhea was actually caused by the bacterium *Clostridium difficile.* The patient was placed on contact isolation and was started on intravenous metronidazole (Flagyl). With this treatment, the diarrhea gradually slowed and finally stopped. Repeat cultures, performed after the metronidazole therapy was completed, showed that the infection had been successfully cleared.

- How is C. *difficile* spread?

- What risk factors made this patient particularly vulnerable to infection with C. *difficile*?

Case File Wrap-Up appears on page 82.

Bacteria and Archaea

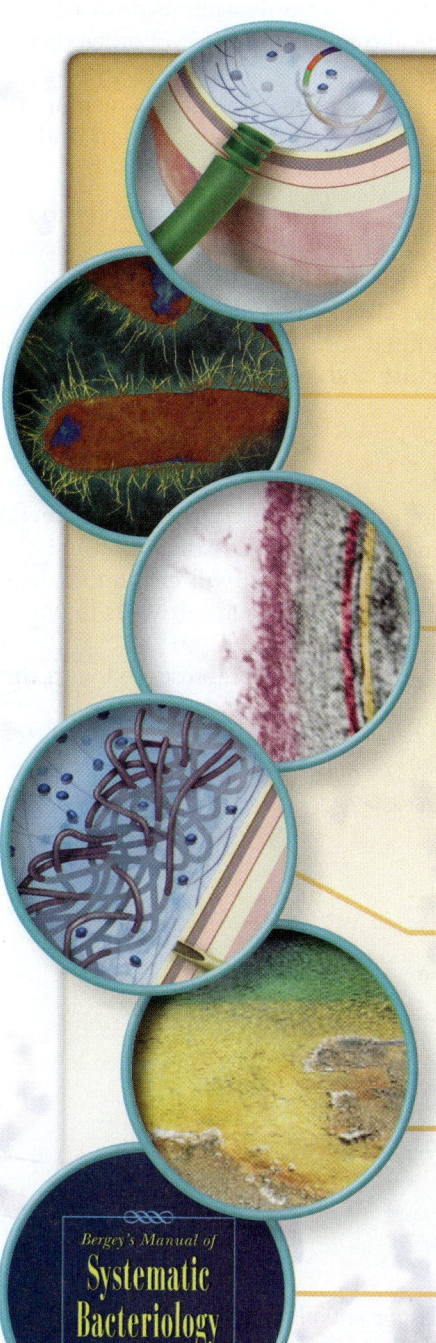

IN THIS CHAPTER...

3.1 Form and Function of Bacteria and Archaea

1. List the structures all bacteria possess.
2. Identify three structures some but not all bacteria possess.
3. Describe three major shapes of bacteria.
4. Provide at least four terms to describe bacterial arrangements.

3.2 External Structures

5. Describe the structure and function of four different types of bacterial appendages.
6. Explain how a flagellum works in the presence of an attractant.

3.3 The Cell Envelope: The Wall and Membrane(s)

7. Differentiate between the two main types of bacterial envelope structure.
8. Discuss why gram-positive cell walls are stronger than gram-negative cell walls.
9. Name a substance in the envelope structure of some bacteria that can cause severe symptoms in humans.

3.4 Bacterial Internal Structure

10. Identify five structures that may be contained in bacterial cytoplasm.
11. Detail the causes and mechanisms of sporogenesis and germination.

3.5 The Archaea: The Other "Prokaryotes"

12. Compare and contrast the major features of archaea, bacteria, and eukaryotes.

3.6 Classification Systems for Bacteria and Archaea

13. Differentiate between *Bergey's Manual of Systematic Bacteriology* and *Bergey's Manual of Determinative Bacteriology*.
14. Name four divisions ending in –*cutes* and describe their characteristics.
15. Define a *species* in terms of bacteria.

3.1 Form and Function of Bacteria and Archaea

In chapter 1, we described bacteria and archaea as being cells with no true nucleus. Let's look at how bacteria and archaea are different from eukaryotes:

- *The way their DNA is packaged:* Bacteria and archaea have nuclear material that is free inside the cytoplasm (i.e., they do not have a nucleus). Eukaryotes have a membrane around their DNA (making up a nucleus). Bacteria don't wind their DNA around **histones;** eukaryotes do.
- *The makeup of their cell wall:* Bacteria and archaea generally have a wall structure that is unique compared to eukaryotes. Bacteria have sturdy walls made of a chemical called peptidoglycan. Archaeal walls are also tough and made of other chemicals, distinct from bacteria and distinct from eukaryotic cells.
- *Their internal structures:* Bacteria and archaea don't have complex, membrane-bounded organelles in their cytoplasm (eukaryotes do). A few bacteria and archaea have internal membranes, but they don't surround organelles.

Both non-eukaryotic and eukaryotic microbes are ubiquitous in the world today. Although both can cause infections diseases, treating them with drugs requires different types of approaches. In this chapter and coming chapters, you'll discover why that is.

The evolutionary history of non-eukaryotic cells extends back at least 2.9 billion years. The fact that these organisms have endured for so long in such a variety of habitats can be attributed to a cellular structure and function that are amazingly versatile and adaptable.

The Structure of the Bacterial Cell

In this chapter, the descriptions, except where otherwise noted, refer to **bacterial cells.** Although bacteria and archaea share many of the same basic structural elements, we will focus on the features of bacteria because you will encounter them more often in a clinical environment. We will analyze the significant ways in which archaea are unique later in the chapter.

The general cellular organization of a bacterial cell can be represented with this flowchart:

Escherichia coli

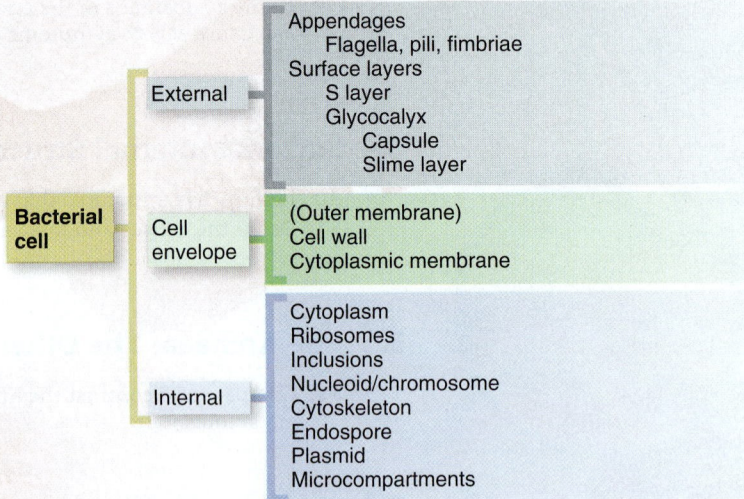

All bacterial cells invariably have a cytoplasmic membrane, cytoplasm, ribosomes, a cytoskeleton, and one (or a few) chromosome(s); the majority have a cell wall and a surface coating called a glycocalyx. Specific structures that are found in some but not all bacteria are flagella, an outer membrane, pili, fimbriae, plasmids, inclusions, endospores, and microcompartments. Most of these structures are observed in archaea as well.

Figure 3.1 presents a three-dimensional anatomical view of a generalized, rod-shaped bacterial cell. As we survey the principal anatomical features of this cell, we

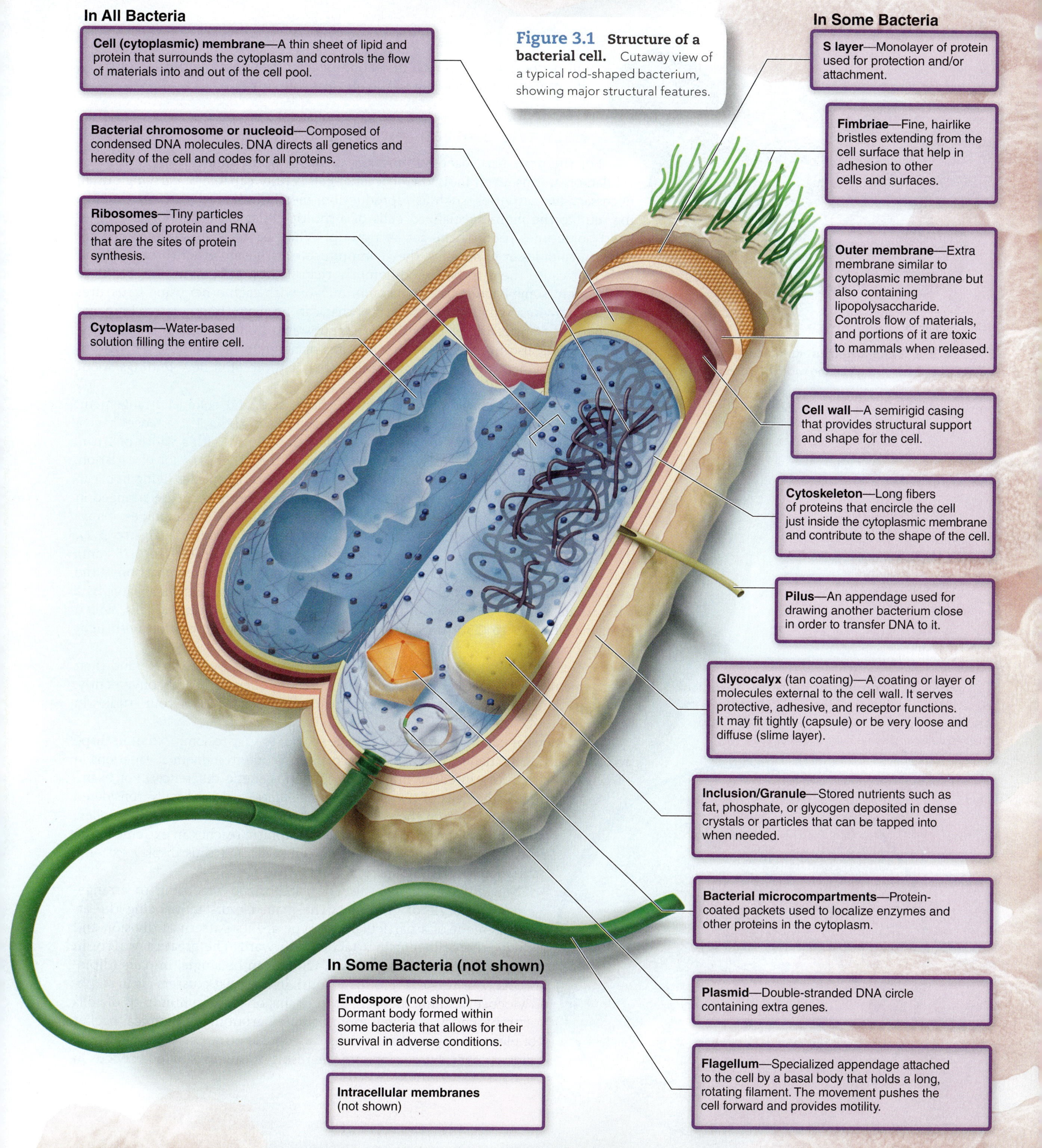

Figure 3.1 **Structure of a bacterial cell.** Cutaway view of a typical rod-shaped bacterium, showing major structural features.

In All Bacteria

Cell (cytoplasmic) membrane—A thin sheet of lipid and protein that surrounds the cytoplasm and controls the flow of materials into and out of the cell pool.

Bacterial chromosome or nucleoid—Composed of condensed DNA molecules. DNA directs all genetics and heredity of the cell and codes for all proteins.

Ribosomes—Tiny particles composed of protein and RNA that are the sites of protein synthesis.

Cytoplasm—Water-based solution filling the entire cell.

In Some Bacteria

S layer—Monolayer of protein used for protection and/or attachment.

Fimbriae—Fine, hairlike bristles extending from the cell surface that help in adhesion to other cells and surfaces.

Outer membrane—Extra membrane similar to cytoplasmic membrane but also containing lipopolysaccharide. Controls flow of materials, and portions of it are toxic to mammals when released.

Cell wall—A semirigid casing that provides structural support and shape for the cell.

Cytoskeleton—Long fibers of proteins that encircle the cell just inside the cytoplasmic membrane and contribute to the shape of the cell.

Pilus—An appendage used for drawing another bacterium close in order to transfer DNA to it.

Glycocalyx (tan coating)—A coating or layer of molecules external to the cell wall. It serves protective, adhesive, and receptor functions. It may fit tightly (capsule) or be very loose and diffuse (slime layer).

Inclusion/Granule—Stored nutrients such as fat, phosphate, or glycogen deposited in dense crystals or particles that can be tapped into when needed.

Bacterial microcompartments—Protein-coated packets used to localize enzymes and other proteins in the cytoplasm.

Plasmid—Double-stranded DNA circle containing extra genes.

Flagellum—Specialized appendage attached to the cell by a basal body that holds a long, rotating filament. The movement pushes the cell forward and provides motility.

In Some Bacteria (not shown)

Endospore (not shown)—Dormant body formed within some bacteria that allows for their survival in adverse conditions.

Intracellular membranes (not shown)

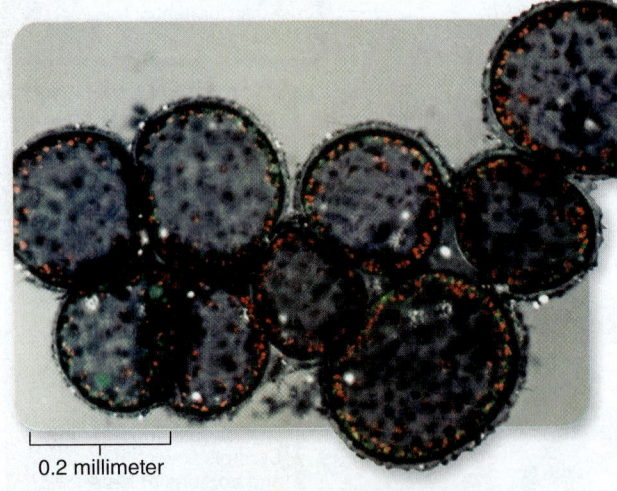

0.2 millimeter

Figure 3.2 *Thiomargarita namibiensis.* The bacteria are each about half the width of a common paper-clip.

Figure 3.3 **Pleomorphic bacteria.** If you look closely at this micrograph of stained *Rickettsia rickettsii* bacteria, you will see some coccoid cells, some rod-shaped cells, and some hybrid forms.

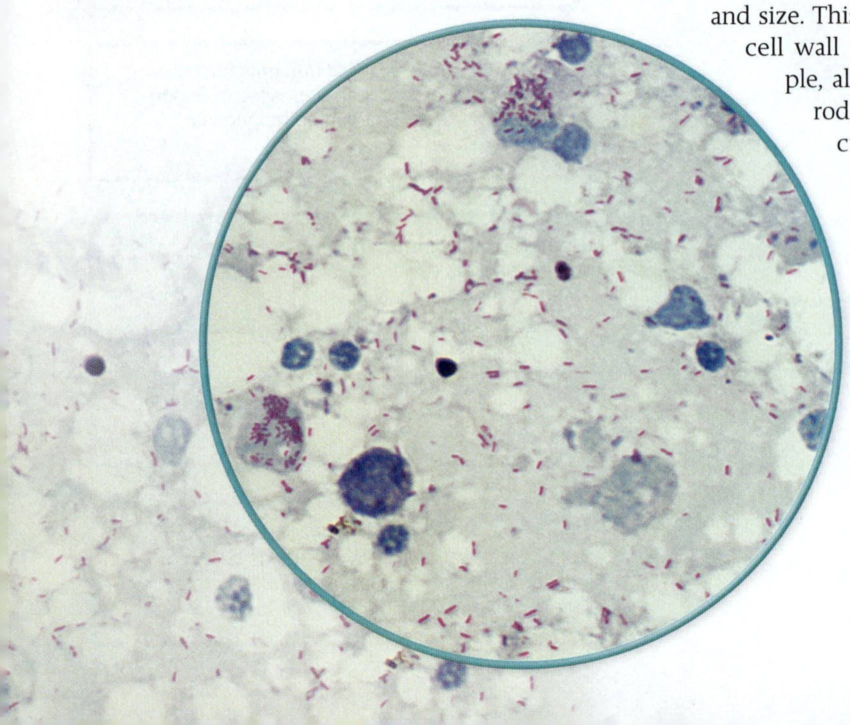

will perform a microscopic dissection of sorts, beginning with the outer cell structures and proceeding to the internal contents.

Bacterial Shapes and Arrangements

For the most part, bacteria function as independent single-celled, or unicellular, organisms. Each individual bacterial cell is fully capable of carrying out all necessary life activities, such as reproduction, metabolism, and nutrient processing, unlike the more specialized cells of a multicellular organism. On the other hand, sometimes bacteria *can* act as a group. When bacteria are close to one another in colonies or in biofilms, they communicate with each other through chemicals that cause them to behave differently than if they were living singly. More surprisingly, some bacteria seem to communicate with each other using structures called nanowires, which are appendages that can be many micrometers long and are used for transferring electrons or other substances outside the cell onto metals. The wires also intertwine with the wires of neighboring bacteria. This is not the same as being a multicellular organism, but it represents new findings about microbial cooperation.

Bacteria exhibit considerable variety in shape, size, and colonial arrangement. Let's start with size. Bacterial cells have an average size of about 1 μm. Cocci have a circumference of 1 μm, and rods may have a length of 2 μm with a width of 1 μm. But that's just the average. As with everything in nature, there is a lot of variation. One of the largest non-eukaryote yet discovered is a bacterial species living in ocean sediments near the African country of Namibia. These gigantic cocci are arranged in strands that look like pearls and contain hundreds of golden sulfur granules, inspiring their name, *Thiomargarita namibiensis* ("sulfur pearl of Namibia") **(figure 3.2).** The size of the individual cells ranges from 100 up to 750 μm (0.1 to 0.75 mm), and many are large enough to see with the naked eye. By way of comparison, if the average bacterium were the size of a mouse, *Thiomargarita* would be as large as a blue whale! On the other end of the spectrum, we have *Mycoplasma* cells, which are generally 0.15 to 0.30 μm, which is right at the limit of resolution with light microscopes.

One of the most important ways to describe bacteria is by the shape and their arrangement. **Table 3.1** presents these patterns comprehensively and conveniently. Gaining a familiarity with these will be a great help for the rest of your studies in this course.

It is somewhat common for cells of a single species to vary to some extent in shape and size. This phenomenon, called **pleomorphism,** is due to individual variations in cell wall structure caused by nutritional or slight genetic differences. For example, although the cells of *Corynebacterium diphtheriae* are generally considered rod-shaped, in culture they display variations such as club-shaped, swollen, curved, filamentous, and coccoid. Pleomorphism reaches an extreme in the mycoplasmas, which entirely lack cell walls and thus display extreme variations in shape **(figure 3.3).**

Bacterial cells can also be categorized according to arrangement, or style of grouping. The main factors influencing the arrangement of a particular cell type are its pattern of division and how the cells remain attached afterward. The greatest variety in arrangement occurs in cocci, which can be single, in pairs (diplococci), in **tetrads** (groups of four), in irregular clusters (as in staphylococci and micrococci), or in chains of a few to hundreds of cells (streptococci). An even more complex grouping is a cubical packet of eight, sixteen, or more cells called a **sarcina** (sar'-sih-nah). These different coccal groupings are the result of the division of a coccus in

Table 3.1 Bacterial Shapes

(a) Coccus

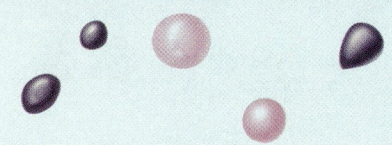

If the cell is spherical or ball-shaped, the bacterium is described as a **coccus** (kok′-us). Cocci (kok′-sie) can be perfect spheres, but they also can exist as oval, bean-shaped, or even pointed variants. This is a *Deinococcus* (2,000×).

(b) Rod/Bacillus

A cell that is cylindrical is termed a rod, or **bacillus** (bah-sil′-lus). There is also a genus named *Bacillus*. Rods are also quite varied in their actual form. Depending on the species, they can be blocky, spindle-shaped, round-ended, long and threadlike (filamentous), or even club-shaped or drumstick-shaped. Note: When a rod is short and plump, it is called a **coccobacillus**. This is a *Lactobacillus* (5,000×).

(c) Vibrio

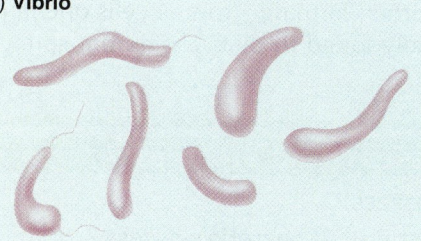

Singly occurring rods that are gently curved are called **vibrio** (vib′-ree-oh). This is a *Vibrio cholerae* (13,000×).

(d) Spirillum

A bacterium having a slightly curled or spiral-shaped cylinder is called a **spirillum** (spy-ril′-em), a rigid helix, twisted twice or more along its axis (like a corkscrew). This is an *Aquaspirillum* (7,500×).

(e) Spirochete

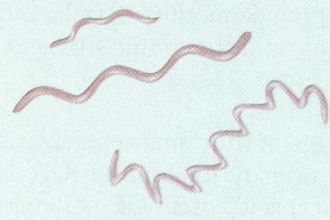

Another spiral cell containing periplasmic flagella is the **spirochete,** a more flexible form that resembles a spring. These are spirochetes (14,000×).

(f) Branching filaments

A few bacteria produce multiple branches off of a basic rod structure, a form called branching **filaments.** This is a *Streptomyces* (1,500×).

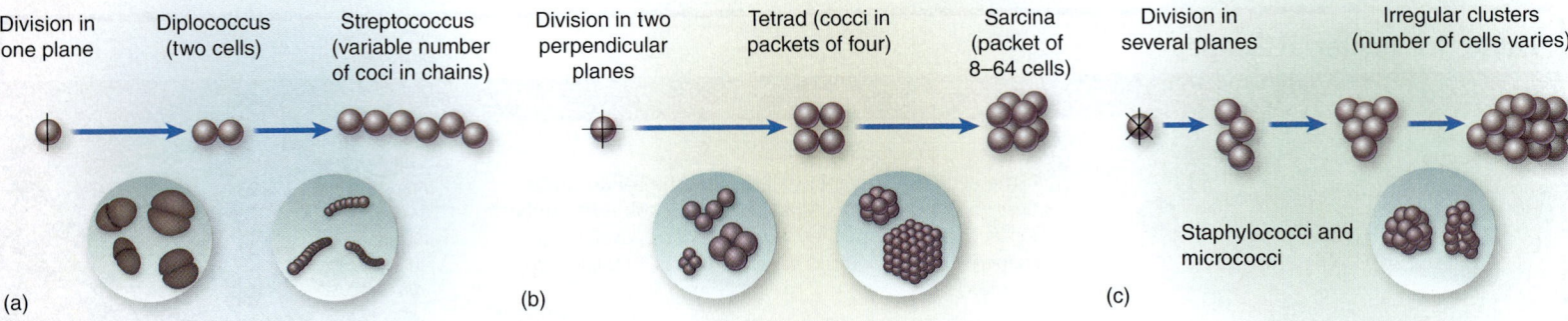

Division in one plane | Diplococcus (two cells) | Streptococcus (variable number of coci in chains)

(a)

Division in two perpendicular planes | Tetrad (cocci in packets of four) | Sarcina (packet of 8–64 cells)

(b)

Division in several planes | Irregular clusters (number of cells varies)

Staphylococci and micrococci

(c)

Figure 3.4 Arrangement of cocci resulting from different planes of cell division. (a) Division in one plane produces diplococci and streptococci. (b) Division in two or three planes at right angles produces tetrads and packets. (c) Division in several planes produces irregular clusters.

a single plane, in two perpendicular planes, or in several intersecting planes; after division, the resultant daughter cells remain attached **(figure 3.4)**.

Bacilli are less varied in arrangement because they divide only in the transverse plane (perpendicular to the axis). They occur either as single cells, as a pair of cells with their ends attached (diplobacilli), or as a chain of several cells (streptobacilli). A **palisades** (pal'-ih-saydz) arrangement, typical of the corynebacteria, is formed when the cells of a chain remain partially attached by a small hinge region at the ends. The cells tend to fold (snap) back upon each other, forming a row of cells oriented side by side **(figure 3.5)**. Spirilla are occasionally found in short chains, but spirochetes rarely remain attached after division.

> ### 3.1 LEARNING OUTCOMES—Assess Your Progress
>
> 1. List the structures all bacteria possess.
> 2. Identify three structures some but not all bacteria possess.
> 3. Describe three major shapes of bacteria.
> 4. Provide at least four terms to describe bacterial arrangements.

3.2 External Structures

Appendages: Cell Extensions

Several different types of accessory structures sprout from the surface of bacteria. These long **appendages** are common but are not present on all species. Appendages can be divided into two major groups: those that provide motility (flagella and axial filaments) and those that provide attachment points or channels (fimbriae and pili).

Flagella—Bacterial Propellers

The bacterial **flagellum** (flah-jel'-em), an appendage of truly amazing construction, is certainly unique in the biological world. The primary function of flagella is to confer **motility,** or self-propulsion–that is, the capacity of a cell to swim freely through an aqueous habitat. The flagellum has three distinct parts: the filament, the hook (sheath), and the basal body **(figure 3.6).** The **filament,** a helical structure composed of proteins, is approximately 20 nm in diameter and varies from 1 to 70 μm in length. It is inserted into a curved, tubular hook. The hook is anchored to the cell by the basal body, a stack of rings firmly anchored through the cell wall, to the cytoplasmic membrane and the outer membrane. This arrangement permits the hook with its filament to rotate 360°, rather than undulating back and forth like a whip as was once thought. Although many archaea possess flagella, recent studies have shown that the structure is quite different than the bacterial flagellum. It is called *archaellum* by some scientists.

Figure 3.5 *Corynebacterium* cells illustrating the palisades (stacking) arrangement.

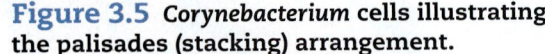

All spirilla, about half of the bacilli, and a small number of cocci are flagellated. Flagella vary both in number and arrangement according to two general patterns:

1. In a *polar* arrangement, the flagella are attached at one or both ends of the cell. Three subtypes of this pattern are
   - **monotrichous** (mah'-noh-trik'-us), with a single flagellum;
   - **lophotrichous** (lo'-foh-), with small bunches or tufts of flagella emerging from the same site; and
   - **amphitrichous** (am'-fee-), with flagella at both poles of the cell.
2. In a **peritrichous** (per'-ee-) arrangement, flagella are dispersed randomly over the surface of the cell **(figure 3.7).**

Motility is one piece of information used in the laboratory identification or diagnosis of pathogens. Flagella are hard to visualize in the laboratory, but often it is sufficient to know simply whether a bacterial species is motile. One way to detect motility is to stab a tiny mass of cells into a soft (semisolid) medium in a test tube. Growth spreading rapidly through the entire medium is indicative of motility. Alternatively, cells can be observed microscopically with a hanging drop slide. A truly motile cell will flit, dart, or wobble around the field, making some progress, whereas one that is nonmotile jiggles about in one place but makes no progress.

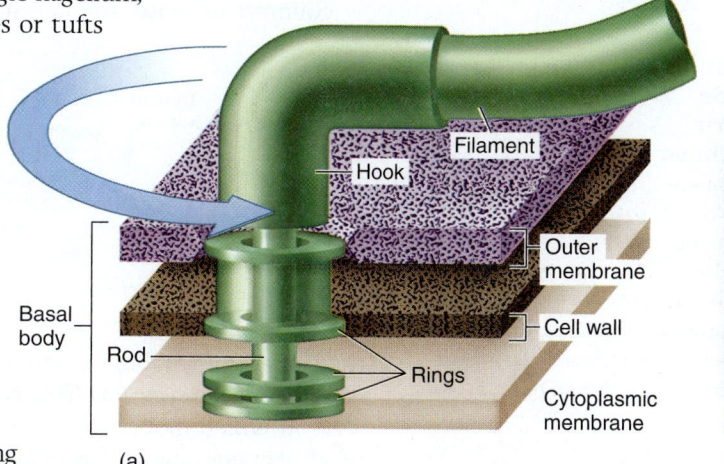

(a)   (b)

Figure 3.6   Details of the basal body of a flagellum in a gram-negative cell.   (a) The hook, rings, and rod function together as a tiny device that rotates the filament 360°. (b) An electron micrograph of the basal body of a bacterial flagellum.

Fine Points of Flagellar Function
Flagellated bacteria can perform some rather sophisticated feats. They can detect and move in response to chemical signals–a type of behavior called **chemotaxis** (ke'-moh-tak'-sis). Positive chemotaxis is movement of a cell in the direction of a favorable chemical stimulus (usually a nutrient); negative chemotaxis is movement away from a repellent (potentially harmful) compound.

The flagellum is effective in guiding bacteria through the environment primarily because the system for detecting chemicals is linked to the mechanisms that drive the flagellum. Located in the cytoplasmic membrane are clusters of receptors that bind specific molecules coming from the immediate environment. The attachment of sufficient numbers of these molecules transmits signals to the flagellum and sets

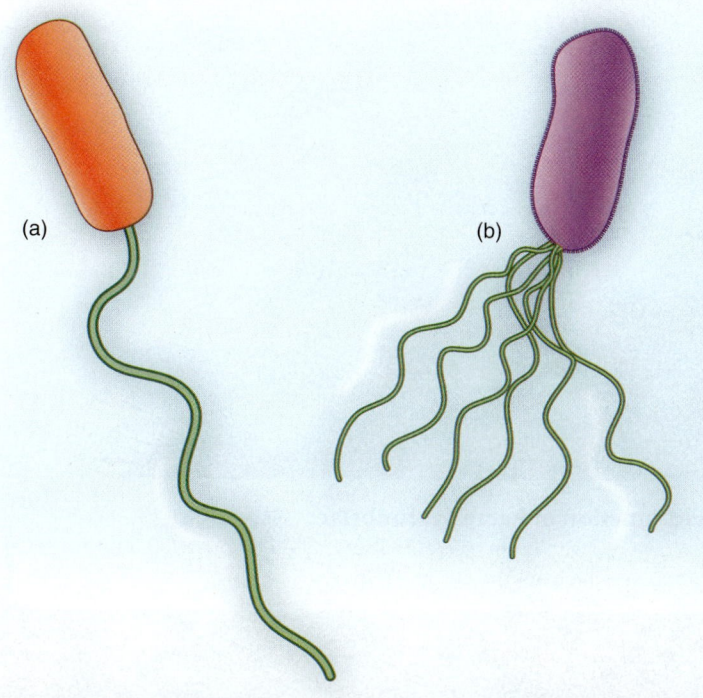

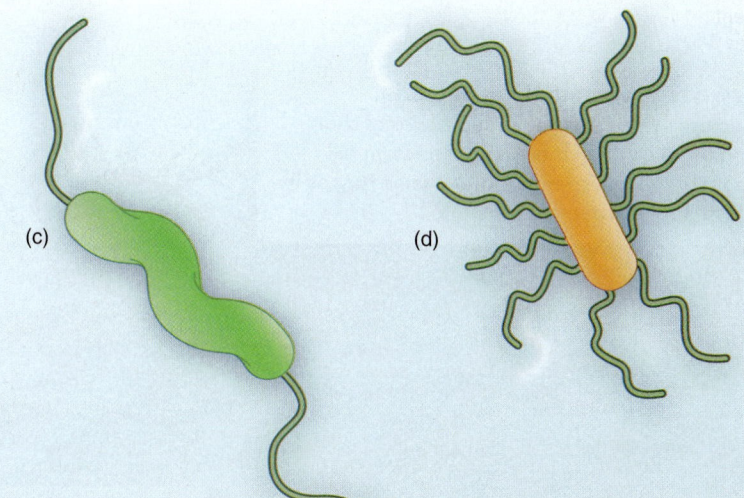

(a)   (b)   (c)   (d)

Figure 3.7   Electron micrographs depicting types of flagellar arrangements.   (a) Monotrichous polar flagellum on the bacterium *Bdellovibrio*. (b) Lophotrichous polar flagella on *Pseudomonas*. (c) Amphitrichous polar flagella on *Campylobacter*. (d) Peritrichous flagella on *Escherichia coli*.

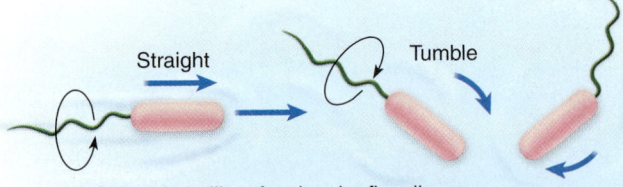

Straight   Tumble

(a) General motility of a singular flagellum

Figure 3.8 The operation of flagella and the mode of locomotion in bacteria with polar and peritrichous flagella. **(a)** In general, when a polar flagellum rotates in a counterclockwise direction, the cell swims forward. When the flagellum reverses direction and rotates clockwise, the cell stops and tumbles. **(b)** In peritrichous forms, all flagella sweep toward one end of the cell and rotate as a single group. During tumbles, the flagella lose coordination.

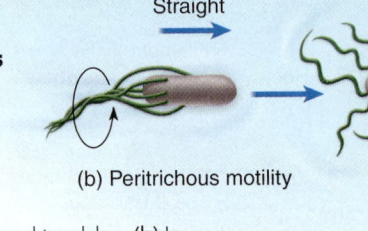

Straight   Tumble

(b) Peritrichous motility

Key

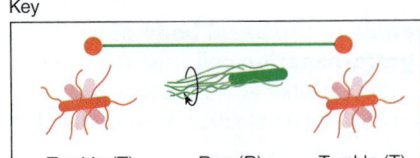

Tumble (T)   Run (R)   Tumble (T)

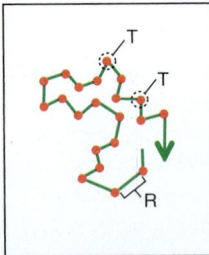 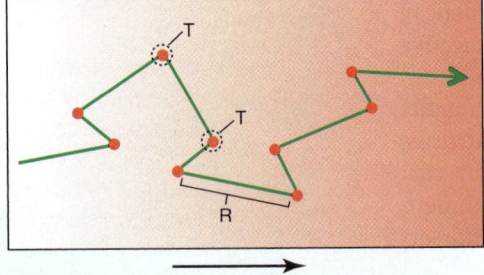

(a) No attractant or repellent   (b) Gradient of attractant concentration

Figure 3.9 Chemotaxis in bacteria. **(a)** A bacterium moves via a random series of short runs and tumbles when there is no attractant or repellent. **(b)** The cell spends more time on runs as it gets closer to the attractant.

it into rotary motion. The actual "fuel" for the flagellum to turn is a gradient of protons (hydrogen ions) that are generated by the metabolism of the bacterium and that bind to and detach from parts of the flagellar motor within the cytoplasmic membrane, causing the filament to rotate. If several flagella are present, they become aligned and rotate as a group **(figure 3.8).** As a flagellum rotates counterclockwise, the cell itself swims in a smooth linear direction toward the stimulus; this action is called a *run*. Runs are interrupted at various intervals by *tumbles*, during which the flagellum reverses direction and causes the cell to stop and change its course. Alternation between runs and tumbles generates what is termed a *random walk* form of motility in these bacteria. However, in response to a concentration gradient of an attractant molecule, the bacterium will begin to inhibit tumbles, permitting longer runs and overall progress toward the stimulus **(figure 3.9).** The movement now becomes a *biased* random walk in which movement is favored (biased) in the direction of the attractant. But what happens when a flagellated bacterium wants to run away from a toxic environment? In this case, the random walk then favors movement away from the concentration of repellent molecules. By delaying tumbles, the bacterium increases the length of its runs, allowing it to redirect itself away from the negative stimulus.

Periplasmic Flagella Corkscrew-shaped bacteria called **spirochetes** (spy'-roh-keets) show an unusual, wriggly mode of locomotion caused by two or more long, coiled threads, the periplasmic flagella or *axial filaments*. A periplasmic flagellum is a type of internal flagellum that is enclosed in the space between the cell wall and the cytoplasmic membrane.

Appendages for Attachment and Mating

Although their main function is motility, bacterial flagella can be used for attachment to surfaces in some species. The structures termed **pilus** (pil-us) and **fimbria** (fim'-bree-ah) are both bacterial surface appendages that provide some type of adhesion but not locomotion.

Fimbriae are small, bristlelike fibers sprouting off the surface of many bacterial cells **(figure 3.10).** Their exact composition varies, but most

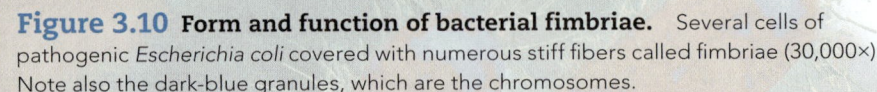

Figure 3.10 Form and function of bacterial fimbriae. Several cells of pathogenic *Escherichia coli* covered with numerous stiff fibers called fimbriae (30,000×). Note also the dark-blue granules, which are the chromosomes.

of them contain protein. Fimbriae have an inherent tendency to stick to each other and to surfaces. They may be responsible for the mutual clinging of cells that leads to biofilms and other thick aggregates of cells on the surface of liquids and for the microbial colonization of inanimate solids such as rocks and glass. Some pathogens can colonize and infect host tissues because of a tight adhesion between their fimbriae and epithelial cells. For example, the gonococcus (agent of gonorrhea) colonizes the genitourinary tract, and *Escherichia coli* colonizes the intestine by this means. Mutant forms of these pathogens that lack fimbriae are unable to cause infections.

A pilus is a long, rigid tubular structure made of a special protein, *pilin.* Pili are well-characterized in gram-negative bacteria but have more recently been identified in several gram-positive pathogens. Conjugation pili are utilized in a "mating" process between cells called **conjugation,** which involves partial transfer of DNA from one cell to another **(figure 3.11).** A conjugation pilus from the donor cell unites with a recipient cell, thereby providing a cytoplasmic connection for making the transfer. Production of these pili is controlled genetically, and conjugation takes place only between compatible gram-negative cells. The roles of pili and conjugation are further explored in chapter 8. There is a special type of structure in some bacteria called a type IV pilus. Like the pili described here, it can transfer genetic material. In addition, it can act like fimbriae and assist in attachment, and act like flagella and make a bacterium motile. Although conjugation does occur in gram-positive bacteria, it does not involve a conjugation pilus.

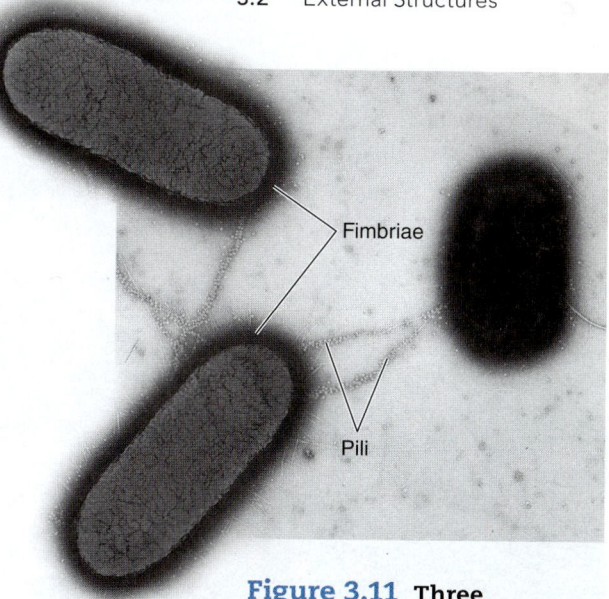

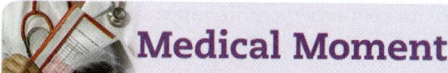

Figure 3.11   Three bacteria in the process of conjugating.   Clearly evident are the pili forming mutual conjugation bridges between a donor (middle cell) and two recipients (cells on the left side). Fimbriae can also be seen on the two left-hand cells.

Surface Coatings: The S Layer and the Glycocalyx

The bacterial cell surface is frequently exposed to severe environmental conditions. Bacterial cells protect themselves with either an **S layer** or a **glycocalyx** or both. **S layers** are single layers of thousands of copies of a single protein linked together like tiny chain mail. They are often called "the armor" of a bacterial cell **(figure 3.12).** It took scientists a long time to discover them because bacteria only produce them when they are in a hostile environment. The nonthreatening conditions of growing in a lab in a nutritious broth with no competitors around ensured that bacteria did not produce the layer. We now know that many different species have the ability to produce an S layer, including pathogens such as *Clostridium difficile* and *Bacillus anthracis.* Some bacteria use S layers to aid in attachment, as well. The **glycocalyx** develops as a coating of repeating polysaccharide or glycoprotein units. This protects the cell and, in some cases, helps it adhere to its environment. Glycocalyces differ among bacteria in thickness, organization, and chemical composition. Some bacteria are covered with a loose shield called a **slime layer** that evidently protects them from loss of water

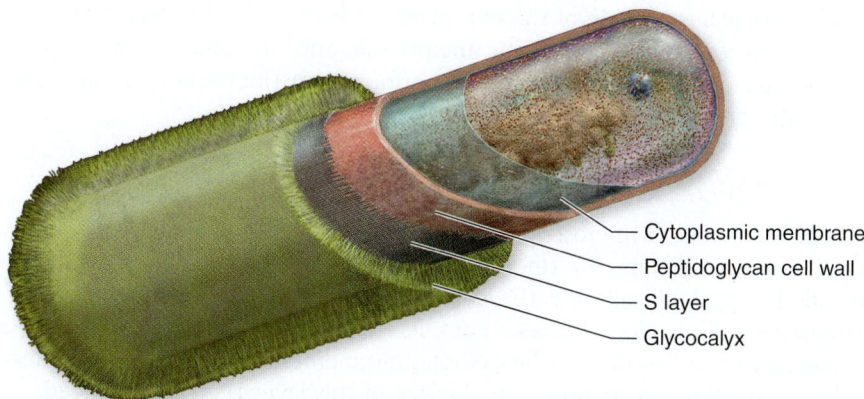

Cytoplasmic membrane

Peptidoglycan cell wall

S layer

Glycocalyx

Figure 3.12   Bacterial S layer, shown in purple.

Medical Moment

Outsmarting Encapsulated Bacteria

Catheter-associated infections in critically ill patients requiring central venous access are unfortunately all too common. It has been estimated that bloodstream infection, a condition called **sepsis,** affects 3% to 8% of patients requiring an indwelling catheter for a prolonged period of time. Sepsis increases morbidity and mortality and can increase the cost of a patient's care by approximately $30,000.

In order to colonize a catheter, microorganisms must first adhere to the surface of the tip on this medical device. Fimbriae and glycocalyces are bacterial structures most often used for this purpose. Researchers have now found a way to outsmart bacterial pathogens by creating catheters that are coated with antibacterial compounds. These agents prevent the bacteria from attaching to the device, eliminating their ability to colonize into thick biofilms capable of spreading infectious agents.

Catheters coated with a combination of rifampin and minocycline or chlorhexidine and silver sulfadiazine have been documented to reduce rates of infection. However, these agents can damage the catheter itself and may trigger drug resistance or tissue toxicity. New studies show that coating the tips in an antibiotic- and antiseptic-free polymer efficiently blocks bacterial colonization of the catheters and poses no threat to patient cells or tissues. To learn more about how biofilms can affect medical devices, see "Inside the Clinic" at the end of this chapter.

Source: 2013. *Biomaterials. 33*(28): 6593.

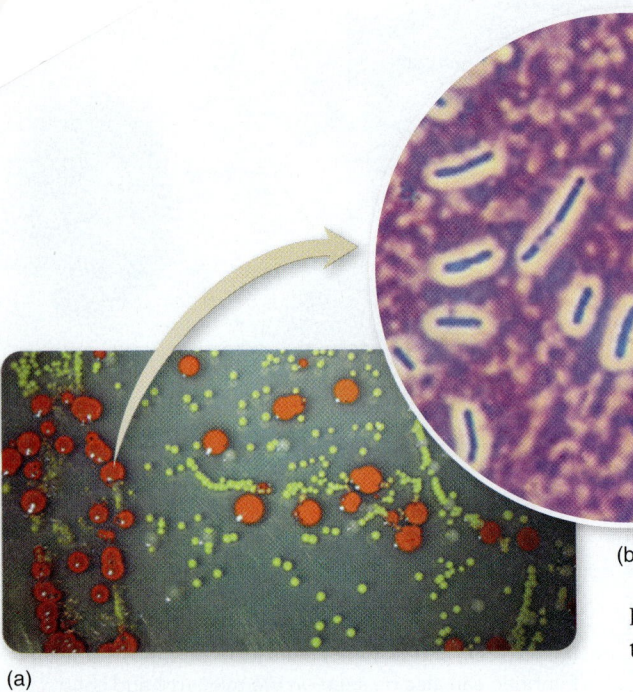

and nutrients. A glycocalyx is called a **capsule** when it is bound more tightly to the cell than a slime layer is and it is denser and thicker. Capsules are often visible in negatively stained preparations **(figure 3.13a)** and produce a prominently sticky (mucoid) character to colonies on agar **(figure 3.13b)**.

Specialized Functions of the Glycocalyx

Capsules are formed by many pathogenic bacteria, such as *Streptococcus pneumoniae* (a cause of pneumonia, an infection of the lung), *Haemophilus influenzae* (one cause of meningitis), and *Bacillus anthracis* (the cause of anthrax). Encapsulated bacterial cells generally have greater pathogenicity because capsules protect the bacteria against white blood cells called phagocytes. Phagocytes are a natural body defense that can engulf and destroy foreign cells through phagocytosis, thus preventing infection. A capsular coating blocks the mechanisms that phagocytes use to attach to and engulf bacteria. By escaping phagocytosis, the bacteria are free to multiply and infect body tissues. Encapsulated bacteria that mutate to nonencapsulated forms usually lose their ability to cause disease.

Glycocalyces can be important in formation of biofilms **(figure 3.14a)**. The thick, white plaque that forms on teeth comes in part from the surface slimes produced by certain streptococci in the oral cavity. This slime protects them from being dislodged from the teeth and provides a niche for other oral bacteria that, in time, can lead to dental disease. The glycocalyx of some bacteria is so highly adherent that it is responsible for persistent colonization of nonliving materials such as plastic catheters, intrauterine devices, and metal pacemakers that are in common medical use **(figure 3.14b)**.

(b)

Figure 3.13 Encapsulated bacteria.
(a) This agar plate contains two different species, indicated by the red and yellow pigments. But the characteristic of interest here is the relative *glossiness* of the colonies. Bacteria that produce capsules (the red ones) appear glossier on agar. They are called "smooth" by microbiologists, while those without capsules (the yellow here) are called "rough". (b) special stain of encapsulated bacteria.

3.2 LEARNING OUTCOMES—Assess Your Progress

5. Describe the structure and function of four different types of bacterial appendages.
6. Explain how a flagellum works in the presence of an attractant.

3.3 The Cell Envelope: The Wall and Membrane(s)

The majority of bacteria have a chemically complex external covering, termed the *cell envelope*, that lies outside of the cytoplasm. It is composed of two or three basic layers: the cell wall, the cytoplasmic membrane, and, in some bacteria, the outer membrane. Although each envelope layer performs a distinct function, together they act as a single protective unit.

Differences in Cell Envelope Structure

In gram-positive cells, a microscopic section **(figure 3.15)** resembles an open-faced sandwich with two layers: the thick cell wall, composed primarily of a unique molecule called peptidoglycan, and the cytoplasmic membrane. A similar section of a gram-negative cell envelope shows a complete sandwich with three layers: an outer membrane, a thin cell wall, and the cytoplasmic membrane. Although gram-negative cells contain peptidoglycan, note that the size of this layer is greatly reduced.

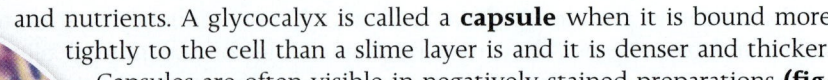

NCLEX® PREP

1. A client presents to the emergency room with a puncture wound. Which of the following procedures would be the *priority* intervention to help prevent wound contamination by bacterial spores in the clinical setting?
   a. *Give an injection of tetanus toxoid if indicated.*
   b. *Use sterile gloves while cleaning the wound.*
   c. *Use clean gloves while cleaning the wound.*
   d. *Medicate client with Tylenol (acetaminophen) if found to be febrile.*

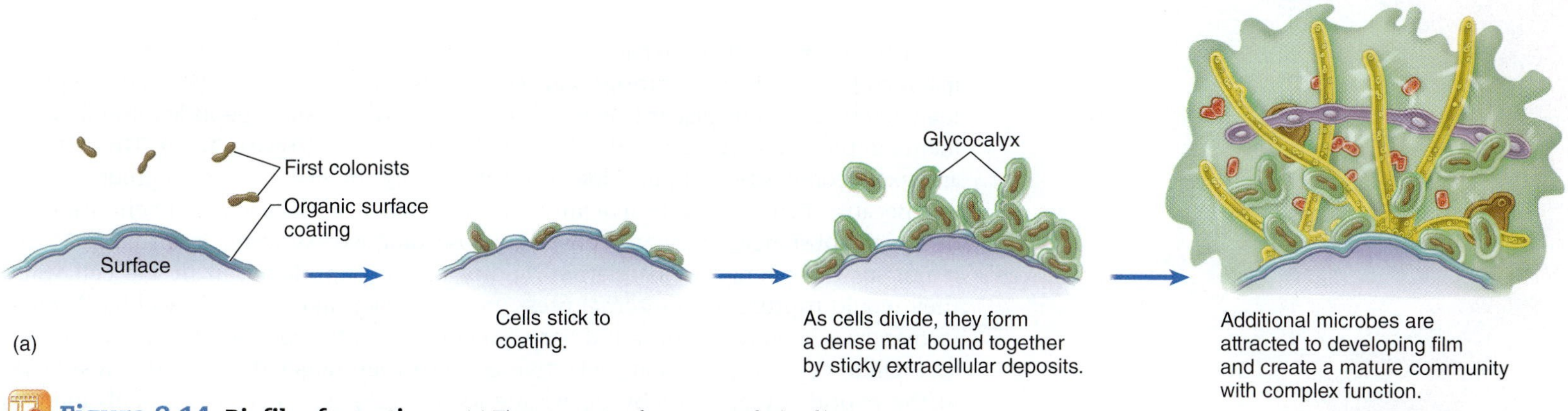

First colonists

Organic surface coating

Surface

Cells stick to coating.

Glycocalyx

As cells divide, they form a dense mat bound together by sticky extracellular deposits.

Additional microbes are attracted to developing film and create a mature community with complex function.

(a)

Figure 3.14 Biofilm formation.   (a) The step-wise formation of a biofilm on a surface. (b) Scanning electron micrograph of *Staphylococcus aureus* cells attached to a catheter by a slime secretion.

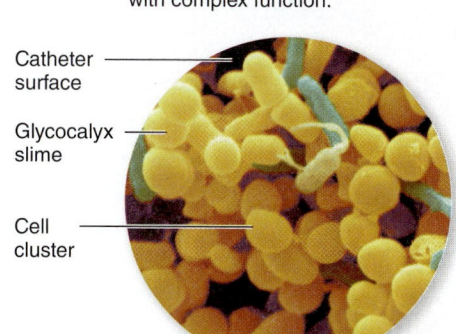

Catheter surface

Glycocalyx slime

Cell cluster

(b)

Moving from outside to in (see figure 3.1), the outer membrane (if present) lies just under the glycocalyx. Next comes the cell wall. Finally, the innermost layer is always the cytoplasmic membrane. Because only some bacteria have an outer membrane, we discuss the cell wall first.

The Cell Wall

The **cell wall** accounts for a number of important bacterial characteristics. In general, it helps determine the shape of a bacterium, and it also provides the kind of strong structural support necessary to keep a bacterium from bursting or collapsing because of changes in osmotic pressure.

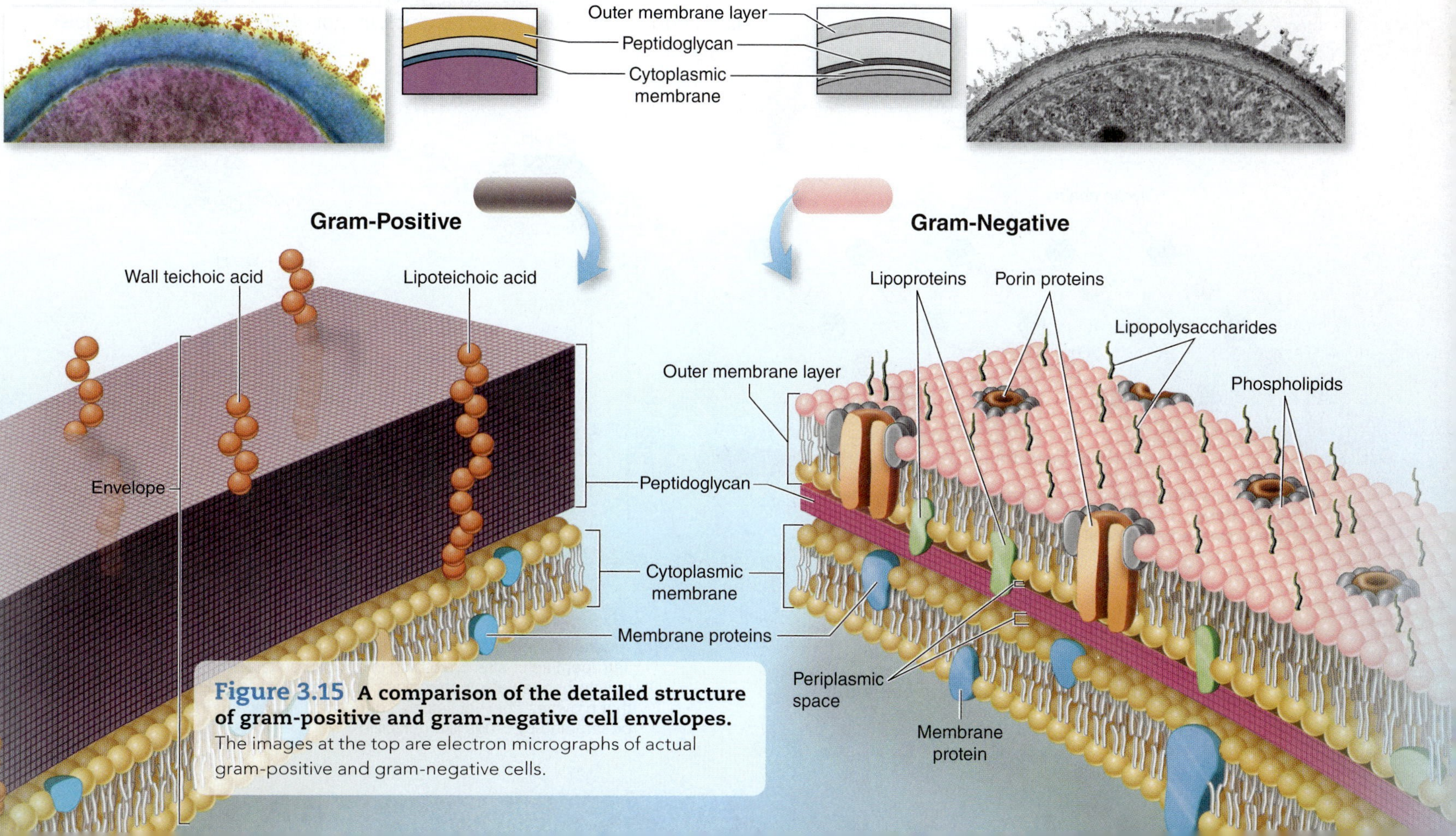

Outer membrane layer

Peptidoglycan

Cytoplasmic membrane

Gram-Positive

Gram-Negative

Wall teichoic acid

Lipoteichoic acid

Lipoproteins   Porin proteins

Lipopolysaccharides

Phospholipids

Outer membrane layer

Envelope

Peptidoglycan

Cytoplasmic membrane

Membrane proteins

Periplasmic space

Membrane protein

Figure 3.15 A comparison of the detailed structure of gram-positive and gram-negative cell envelopes. The images at the top are electron micrographs of actual gram-positive and gram-negative cells.

The cell walls of most bacteria gain their relatively rigid quality from a unique macromolecule called **peptidoglycan (PG)**. This compound is composed of a repeating framework of long *glycan (sugar)* chains cross-linked by short peptide (protein) fragments to provide a strong but flexible support framework **(figure 3.16)**. The amount and exact composition of peptidoglycan vary among the major bacterial groups.

Because many bacteria live in aqueous habitats with a low concentration of dissolved substances, they are constantly absorbing excess water by osmosis. Were it not for the strength and relative rigidity of the peptidoglycan in the cell wall, they would rupture from internal pressure. This function of the cell wall has been a tremendous boon to the drug industry. Several types of drugs used to treat infection (penicillin, cephalosporins) are effective because they target the peptide cross-links in the peptidoglycan, thereby disrupting its integrity. With their cell walls incomplete or missing, such cells have very little protection from **lysis** (ly'-sis), which is the disintegration or rupture of the cell. Lysozyme, an enzyme contained in tears and saliva, provides a natural defense against certain bacteria by hydrolyzing the bonds in the glycan chains and causing the wall to break down. (Chapter 9 discusses the actions of antimicrobial chemical agents.)

More than a hundred years ago, long before the detailed anatomy of bacteria was even remotely known, a Danish physician named Hans Christian Gram developed a staining technique, the **Gram stain,** that delineates two generally different groups of bacteria. The two major groups shown by this technique are the **gram-positive** bacteria and the **gram-negative** bacteria. The structural difference denoted by the designations *gram-positive* and *gram-negative* lies in large part within the peptidoglycan layer of the cell envelope, as you will see next.

The Gram-Positive Cell Wall

The bulk of the gram-positive cell wall is a thick, homogeneous sheath of peptidoglycan ranging from 20 to 80 nm in thickness. It also contains tightly bound acidic polysaccharides, including **teichoic acid** and **lipoteichoic acid** (see figure 3.15). Teichoic acid is a polymer of ribitol or glycerol (alcohols)

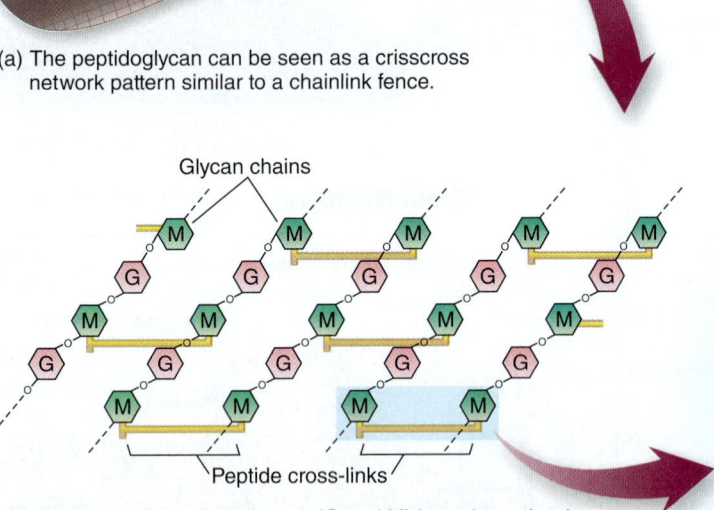

(a) The peptidoglycan can be seen as a crisscross network pattern similar to a chainlink fence.

(b) It contains alternating glycans (G and M) bound together in long strands. The G stands for *N*-acetyl glucosamine, and the M stands for *N*-acetyl muramic acid.

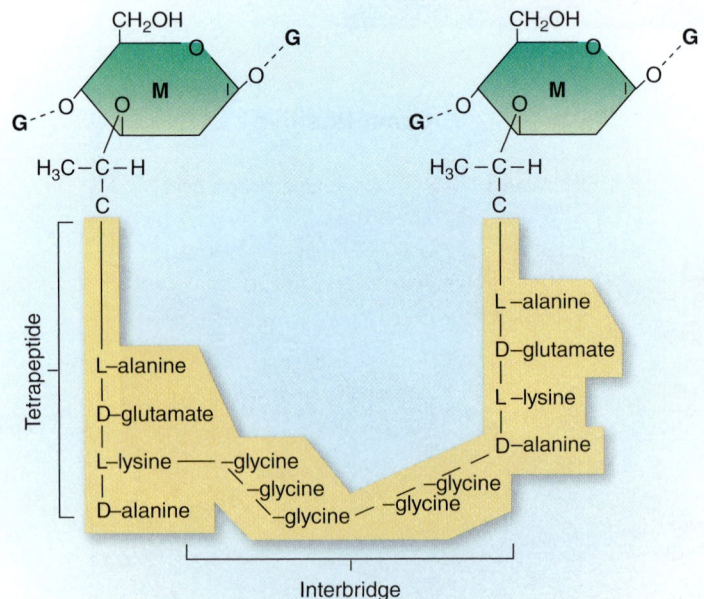

(c) A detailed view of the links between the muramic acids. Tetrapeptide chains branching off the muramic acids connect by interbridges also composed of amino acids. It is this linkage that provides rigid yet flexible support to the cell and that may be targeted by drugs like penicillin.

Figure 3.16 **Structure of peptidoglycan in the cell wall.**

and phosphate that is embedded in the peptidoglycan sheath. Lipoteichoic acid is similar in structure but is attached to the lipids in the plasma membrane. These molecules appear to function in cell wall maintenance and enlargement during cell division, and they also contribute to the acidic charge on the cell surface.

The Gram-Negative Cell Wall

The gram-negative cell wall is a single, thin (1–3 nm) sheet of peptidoglycan. Although it acts as a somewhat rigid protective structure as previously described, its thinness gives gram-negative bacteria a relatively greater flexibility–and sensitivity to lysis.

The Gram Stain

The technique of Hans Christian Gram consisted of timed, sequential applications of crystal violet (the primary dye), Gram's iodine (the mordant), an alcohol rinse (decolorizer), and a contrasting counterstain. Bacteria that stain purple are called gram-positive, and those that stain red are called gram-negative.

The different results in the Gram stain are due to differences in the structure of the cell wall and how it reacts to the series of reagents applied to the cells **(figure 3.17).**

This century-old staining method remains the universal basis for bacterial classification and identification. The Gram stain can also be a practical aid in diagnosing infection and in guiding drug treatment. For example, Gram staining a fresh urine

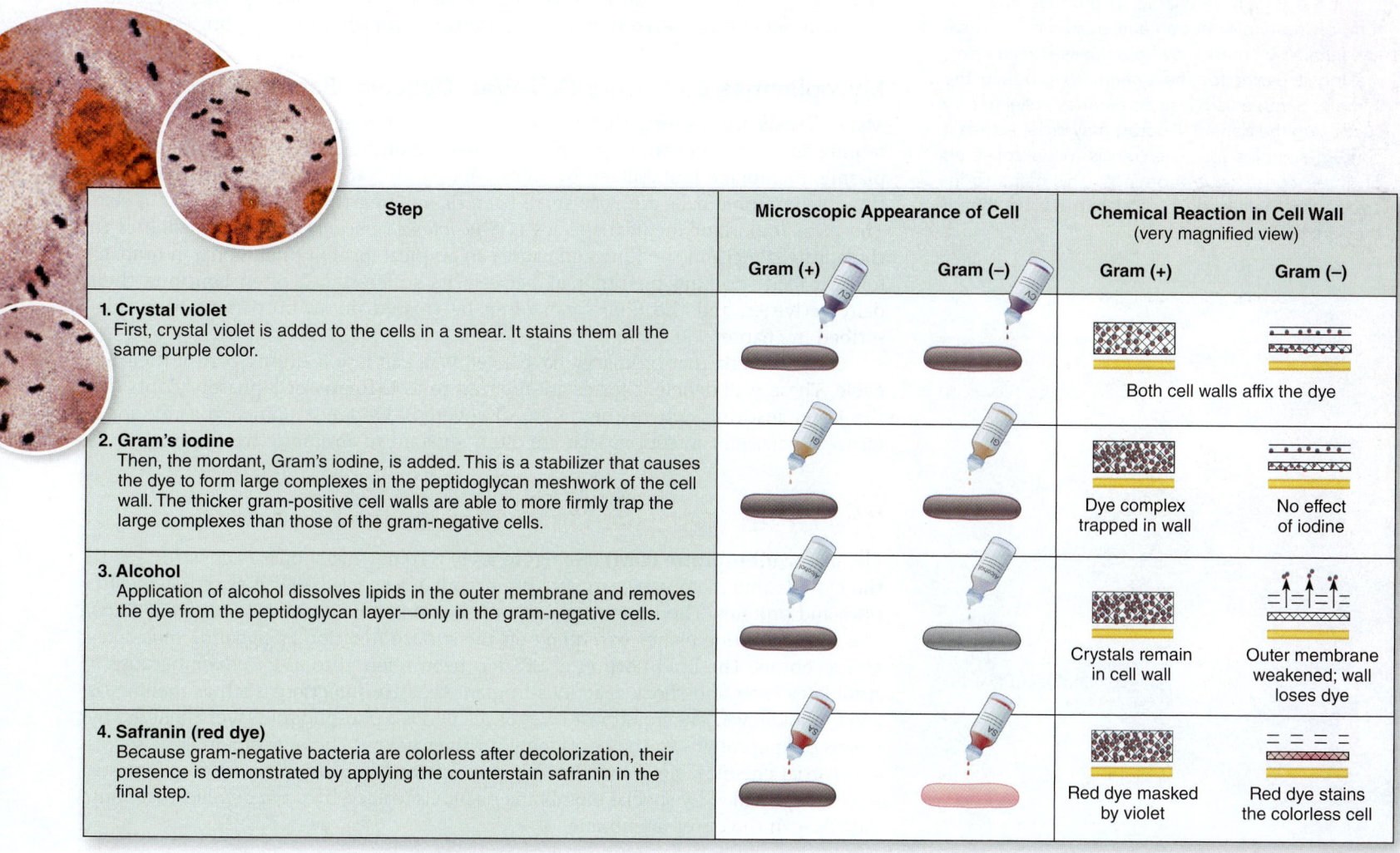

Step	Microscopic Appearance of Cell		Chemical Reaction in Cell Wall (very magnified view)	
	Gram (+)	Gram (−)	Gram (+)	Gram (−)
1. Crystal violet First, crystal violet is added to the cells in a smear. It stains them all the same purple color.				Both cell walls affix the dye
2. Gram's iodine Then, the mordant, Gram's iodine, is added. This is a stabilizer that causes the dye to form large complexes in the peptidoglycan meshwork of the cell wall. The thicker gram-positive cell walls are able to more firmly trap the large complexes than those of the gram-negative cells.			Dye complex trapped in wall	No effect of iodine
3. Alcohol Application of alcohol dissolves lipids in the outer membrane and removes the dye from the peptidoglycan layer—only in the gram-negative cells.			Crystals remain in cell wall	Outer membrane weakened; wall loses dye
4. Safranin (red dye) Because gram-negative bacteria are colorless after decolorization, their presence is demonstrated by applying the counterstain safranin in the final step.			Red dye masked by violet	Red dye stains the colorless cell

Figure 3.17 **The steps in a Gram stain.**

or throat specimen can help pinpoint the possible cause of infection, and in some cases it is possible to begin drug therapy on the basis of this stain. Even in this day of elaborate and expensive medical technology, the Gram stain remains an important and unbeatable first tool in diagnosis.

Nontypical Cell Walls

Several bacterial groups lack the cell wall structure of gram-positive or gram-negative bacteria, and some bacteria have no cell wall at all. Although these exceptional forms can stain positive or negative in the Gram stain, examination of their fine structure and chemistry shows that they do not really fit the descriptions for typical gram-negative or -positive cells. For example, the cells of *Mycobacterium* and *Nocardia* contain peptidoglycan and stain gram-positive, but the bulk of their cell wall is composed of unique types of lipids. One of these is a very-long-chain fatty acid called *mycolic acid*, or cord factor, that contributes to the pathogenicity of this group (see chapter 19). The thick, waxy nature imparted to the cell wall by these lipids is also responsible for a high degree of resistance to certain chemicals and dyes. Such resistance is the basis for the **acid-fast stain** used to diagnose tuberculosis and leprosy.

The archaea exhibit unusual and chemically distinct cell walls. In some, the walls are composed almost entirely of polysaccharides, and in others, the walls are pure protein; but as a group, they all lack the true peptidoglycan structure described previously. Because a few archaea lack a cell wall entirely, their cytoplasmic membrane must serve the dual functions of support and transport.

Mycoplasmas and Other Cell-Wall-Deficient Bacteria

Mycoplasmas are bacteria that naturally lack a cell wall. Although other bacteria require an intact cell wall to prevent the bursting of the cell, the mycoplasma cytoplasmic membrane is stabilized by sterols and is resistant to lysis. These extremely tiny, pleomorphic cells are very small bacteria, ranging from 0.1 to 0.5 μm in size. The most important medical species is *Mycoplasma pneumoniae*, which adheres to the epithelial cells in the lung and causes an atypical form of pneumonia in humans (often called "walking pneumonia" because its sufferers can often continue their daily activities, and the illness can often be treated on an outpatient basis) (described in chapter 19).

Some bacteria that ordinarily have a cell wall can lose it during part of their life cycle. These wall-deficient forms are referred to as **L forms** or L-phase variants (for the Lister Institute, where they were discovered). Evidence points to a role for L forms in persistent infections that are often resistant to antibiotic treatment.

The Gram-Negative Outer Membrane

The **outer membrane (OM)** (see figure 3.15) is somewhat similar in construction to the cytoplasmic membrane, except that it contains specialized types of polysaccharides and proteins. The uppermost layer of the OM contains *lipopolysaccharide* (LPS). The polysaccharide chains extending off the surface function as signaling molecules and receptors. The lipid portion of LPS has been referred to as *endotoxin* because it stimulates fever and shock reactions in gram-negative infections such as meningitis and typhoid fever. The innermost layer of the OM is a phospholipid layer anchored by means of lipoproteins to the peptidoglycan layer below. The outer membrane serves as a partial chemical sieve by allowing only relatively small molecules to penetrate. Access is provided by special membrane channels formed by *porin proteins* that completely span the outer membrane.

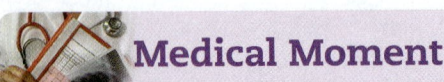

Medical Moment

Collecting Sputum

The nurse is often responsible for collecting sputum samples for acid-fast staining when a patient with a cough is suspected of having tuberculosis. A sterile container must be provided, and the patient should be instructed that early morning specimens are best, usually collected upon first awakening. This is due to the fact that sputum often "pools" in the bronchi when the patient is sleeping at night; therefore, it is easier to collect a larger sample in the morning after the patient has been lying down all night.

If the patient is unable to produce any sputum, giving him or her an aerosolized dose of saline inhaled by mask may help moisten secretions, making it easier for the patient to produce the sample. Samples are also sometimes collected by suctioning the patient. Doctors may order acid-fast sputum samples for tuberculosis to be collected on three consecutive mornings. This helps to increase the likelihood of identifying the bacteria if they are present.

Bacillus subtilis

Cytoplasmic Membrane Structure

Appearing just beneath the cell wall is the cell, or **cytoplasmic membrane,** a very thin (5–10 nm), flexible sheet molded completely around the cytoplasm. Its general composition is a lipid bilayer with proteins embedded to varying degrees. Bacterial cytoplasmic membranes have this typical structure, containing primarily phospholipids (making up about 30%–40% of the membrane mass) and proteins (contributing 60%–70%). Major exceptions to this description are the membranes of mycoplasmas, which contain high amounts of sterols–rigid lipids that stabilize and reinforce the membrane–and the membranes of archaea, which contain unique branched hydrocarbons rather than fatty acids.

Some environmental bacteria, including photosynthesizers and ammonia oxidizers, contain dense stacks of internal membranes. In many cases, they derive from the cytoplasmic membrane, and they are studded with enzymes or photosynthetic pigments. The inner membranes allow a higher concentration of these enzymes and pigments and also accomplish a compartmentalization that allows for higher energy production.

Functions of the Cytoplasmic Membrane

Because bacteria have none of the eukaryotic organelles, the cytoplasmic membrane provides a site for functions such as energy reactions, nutrient processing, and synthesis. A major action of the cytoplasmic membrane is to regulate *transport*, that is, the passage of nutrients into the cell and the discharge of wastes. Although water and small uncharged molecules can diffuse across the membrane unaided, the membrane is a *selectively permeable* structure with special carrier mechanisms for passage of most molecules (see chapter 6). The glycocalyx and cell wall can bar the passage of large molecules, but they are not the primary transport apparatuses.

The membranes of bacteria are an important site for a number of metabolic activities. Most enzymes of respiration and ATP synthesis reside in the cytoplasmic membrane since bacteria lack mitochondria (see chapter 7).

Practical Considerations of Differences in Cell Envelope Structure

Variations in cell envelope anatomy contribute to several other differences between the two cell types. The outer membrane contributes an extra barrier in gram-negative bacteria that makes them impervious to some antimicrobial chemicals such as dyes and disinfectants, so they are generally more difficult to inhibit or kill than are gram-positive bacteria. One exception is for alcohol-based compounds, which can dissolve the lipids in the outer membrane and therefore damage the cell. This is why alcohol swabs are often used to cleanse the skin prior to certain medical procedures, such as venipuncture. Treating infections caused by gram-negative bacteria often requires different drugs from gram-positive infections, especially drugs that can cross the outer membrane.

NCLEX® PREP

2. Walking pneumonia is most often caused by what type of bacterium?
   a. *Klebsiella*
   b. *Mycoplasma*
   c. *Corynebacterium*
   d. *Haemophilus*
   e. *Streptococcus pneumoniae*

3.3 LEARNING OUTCOMES—Assess Your Progress

7. Differentiate between the two main types of bacterial envelope structure.

8. Discuss why gram-positive cell walls are stronger than gram-negative cell walls.

9. Name a substance in the envelope structure of some bacteria that can cause severe symptoms in humans.

3.4 Bacterial Internal Structure

Contents of the Cell Cytoplasm

The **cytoplasm** is a gelatinous solution encased by the cytoplasmic membrane. Its major component is water (70%–80%), which serves as a solvent for the cell pool, a complex mixture of nutrients including sugars, amino acids, and salts. The components of this pool serve as building blocks for cell synthesis or as sources of energy.

Bacterial Chromosomes and Plasmids

The hereditary material of most bacteria exists in the form of a single circular strand of DNA designated as the **bacterial chromosome.** Some bacteria have multiple chromosomes. By definition, bacteria do not have a nucleus; that is, their DNA is not enclosed by a nuclear membrane but instead is aggregated in a dense area of the cell called the **nucleoid.** The chromosome is actually an extremely long molecule of double-stranded DNA that is tightly coiled around special basic protein molecules so as to fit inside the cell compartment. Arranged along its length are genetic units (genes) that carry information required for bacterial maintenance and growth.

Although the chromosome is the minimal genetic requirement for bacterial survival, many bacteria contain other, nonessential pieces of DNA called **plasmids** (refer to figure 3.1). These tiny strands exist as separate double-stranded circles of DNA, although at times they can become integrated into the chromosome. During conjugation, they may be duplicated and passed on to related nearby bacteria. During bacterial reproduction, they are duplicated and passed on to offspring. They are not essential to bacterial growth and metabolism, but they often confer protective traits such as resisting drugs and producing toxins and enzymes (see chapter 8). Because they can be readily manipulated in the laboratory and transferred from one bacterial cell to another, plasmids are an important agent in genetic engineering techniques.

Ribosomes: Sites of Protein Synthesis

A bacterial cell contains thousands of tiny **ribosomes,** the site of protein synthesis. When viewed even by very high magnification, ribosomes show up as fine, spherical specks dispersed throughout the cytoplasm that often occur in chains called polysomes. Many are also attached to the cytoplasmic membrane. Chemically, a ribosome is a combination of a special type of RNA called ribosomal RNA, or rRNA (about 60%), and protein (40%). Ribosomes are characterized by their density, designated by something called "S units." Ribosomes consist of a small subunit and a large subunit **(figure 3.18),** both of these made of a mixture of rRNA and protein. The small subunit has an S value of 30, and the large subunit has an S value of 50. Overall, the bacterial ribosome has a density of 70S. (It is not simply an additive property; that is why the total S value is not a product of the small and large subunits.) The two subunits fit together to form a miniature platform upon which protein synthesis is performed. Note that eukaryotic ribosomes are similar but different. Because of this, we can design drugs to target bacterial ribosomes that do not harm our own. Eukaryotic ribosomes are designated 80S. Although archaea possess 70S ribosomes, they are more similar in structure to that of 80S eukaryotic ribosomes!

Inclusion Bodies and Microcompartments

Bacteria manufacture inclusion bodies to respond to their environmental conditions. They can store nutrients in this way to respond to periods of low food availability. They can pack gas into vesicles to provide buoyancy in an aquatic environment. They can even store crystals of iron oxide with magnetic properties in inclusion bodies. These

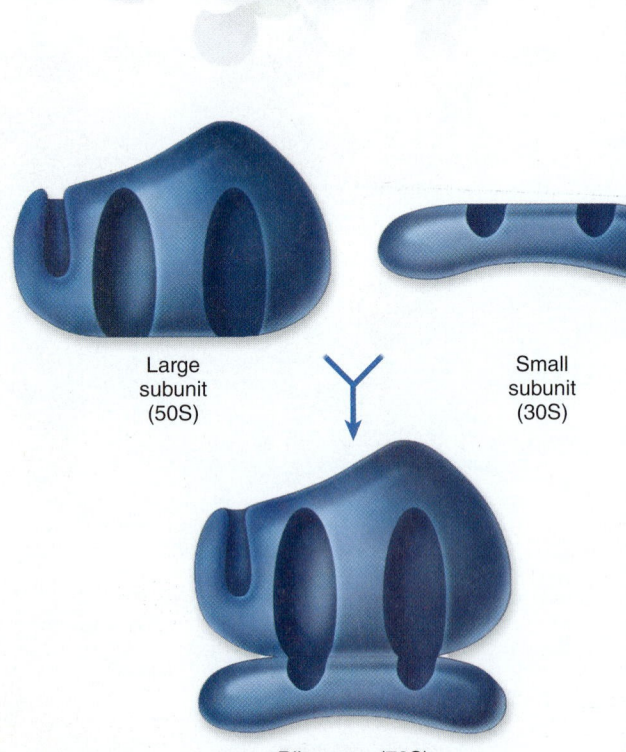

Large subunit (50S)

Small subunit (30S)

Ribosome (70S)

Figure 3.18 A model of a bacterial ribosome, showing the small (30S) and large (50S) subunits, both separate and joined.

magnetotactic bacteria use the granules to orient themselves in polar and gravitational fields to bring them to environments with the proper oxygen content. **Figure 3.19** illustrates a bacterium with an inclusion body packed with the energy-rich organic substance, poly-hydroxybutyrate (PHB).

In the early 2000s, new compartments inside bacterial cells were discovered. These were named *bacterial microcompartments* (BMCs). Their outer shells are made of protein, arranged geometrically, and are packed full of enzymes that are designed to work together in biochemical pathways, thereby ensuring that they are in close proximity to one another.

The Cytoskeleton

Until very recently, scientists thought that the shape of all bacteria was completely determined by the peptidoglycan layer (cell wall). Although this is true of many bacteria, particularly the cocci, other bacteria produce long polymers of proteins that are very similar to eukaryotic **actin.** These proteins are arranged in helical ribbons around the cell just under the cytoplasmic membrane. These fibers contribute to cell shape, perhaps by influencing the way peptidoglycan is manufactured, and also in cell division. Cytoskeletal proteins have also been identified in archaea. Because these proteins are unique to non-eukaryotic cells, they are a potentially powerful target for future antibiotic development.

Bacterial Endospores

The anatomy of bacteria helps them adjust rather well to adverse habitats. But of all microbial structures, nothing can compare to the bacterial **endospore** for withstanding hostile conditions and facilitating survival.

Endospores are dormant bodies produced by bacteria such as *Bacillus, Clostridium,* and *Sporosarcina.* These bacteria have a two-phase life cycle—a vegetative cell and an endospore **(figure 3.20).** The vegetative cell is a metabolically active and growing entity that can be induced by environmental conditions to undergo endospore formation, or **sporulation.** The endospore exists initially inside the cell, but eventually the cell disintegrates and the endospore is on its own. Both gram-positive and gram-negative bacteria can form endospores, but the medically relevant ones are all gram-positive. Most bacteria form only one endospore; therefore, this is not a reproductive function for them.

Bacterial endospores are the hardiest of all life forms, capable of withstanding extremes in heat, drying, freezing, radiation, and chemicals that would readily kill vegetative cells. Their survival under such harsh conditions is due to several factors. The heat resistance of endospores is due to their high content of calcium and dipicolinic acid. We know, for instance, that heat destroys cells by inactivating proteins and DNA and that this process requires a certain amount of water in the protoplasm. Because the deposition of calcium dipicolinate in the endospore removes water and leaves the endospore very dehydrated, it is less vulnerable to the effects of heat. The thick, impervious cortex and endospore coats also protect against radiation and chemicals. The longevity of bacterial endospores verges on immortality. Recently, microbiologists unearthed a viable endospore from a 250-million-year-old salt crystal. Initial analysis of this ancient microbe indicates it is a species of *Bacillus* that is genetically different from previously known species.

Endospore Formation: Sporulation

The depletion of nutrients, especially an adequate carbon or nitrogen source, is the stimulus for a vegetative cell to begin endospore formation. Once this stimulus has been received by the vegetative cell, it undergoes a conversion to become

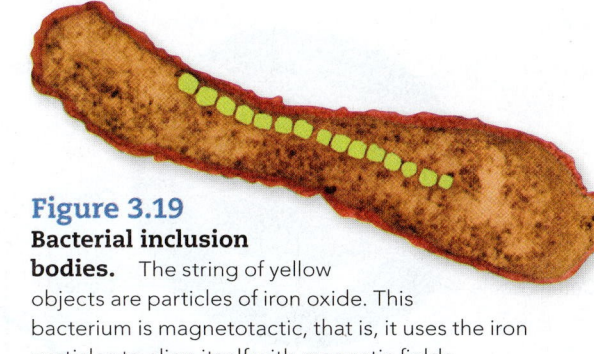

Figure 3.19
Bacterial inclusion bodies.   The string of yellow objects are particles of iron oxide. This bacterium is magnetotactic, that is, it uses the iron particles to align itself with magnetic fields.

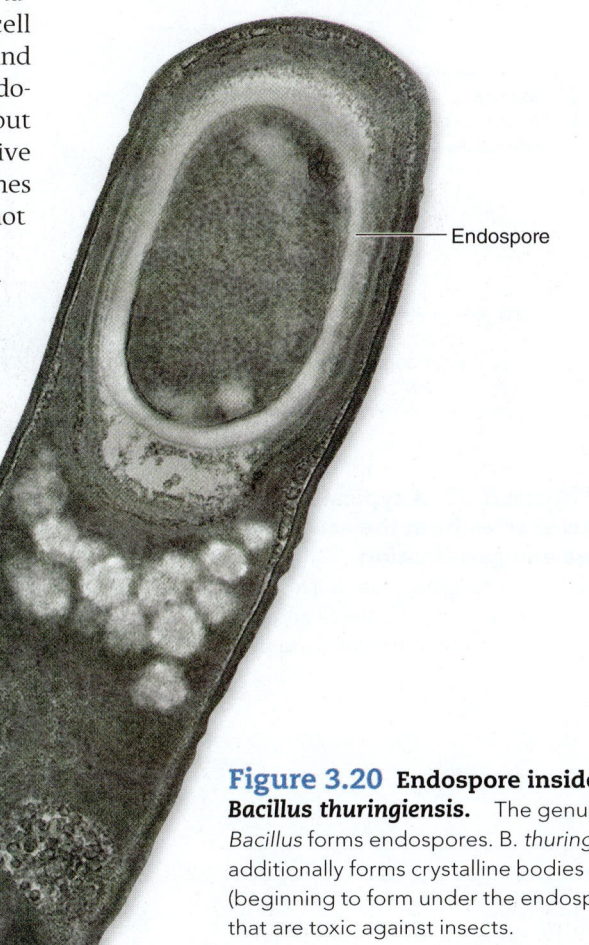

Endospore

Figure 3.20 Endospore inside *Bacillus thuringiensis.*   The genus *Bacillus* forms endospores. B. *thuringiensis* additionally forms crystalline bodies (beginning to form under the endospore) that are toxic against insects.

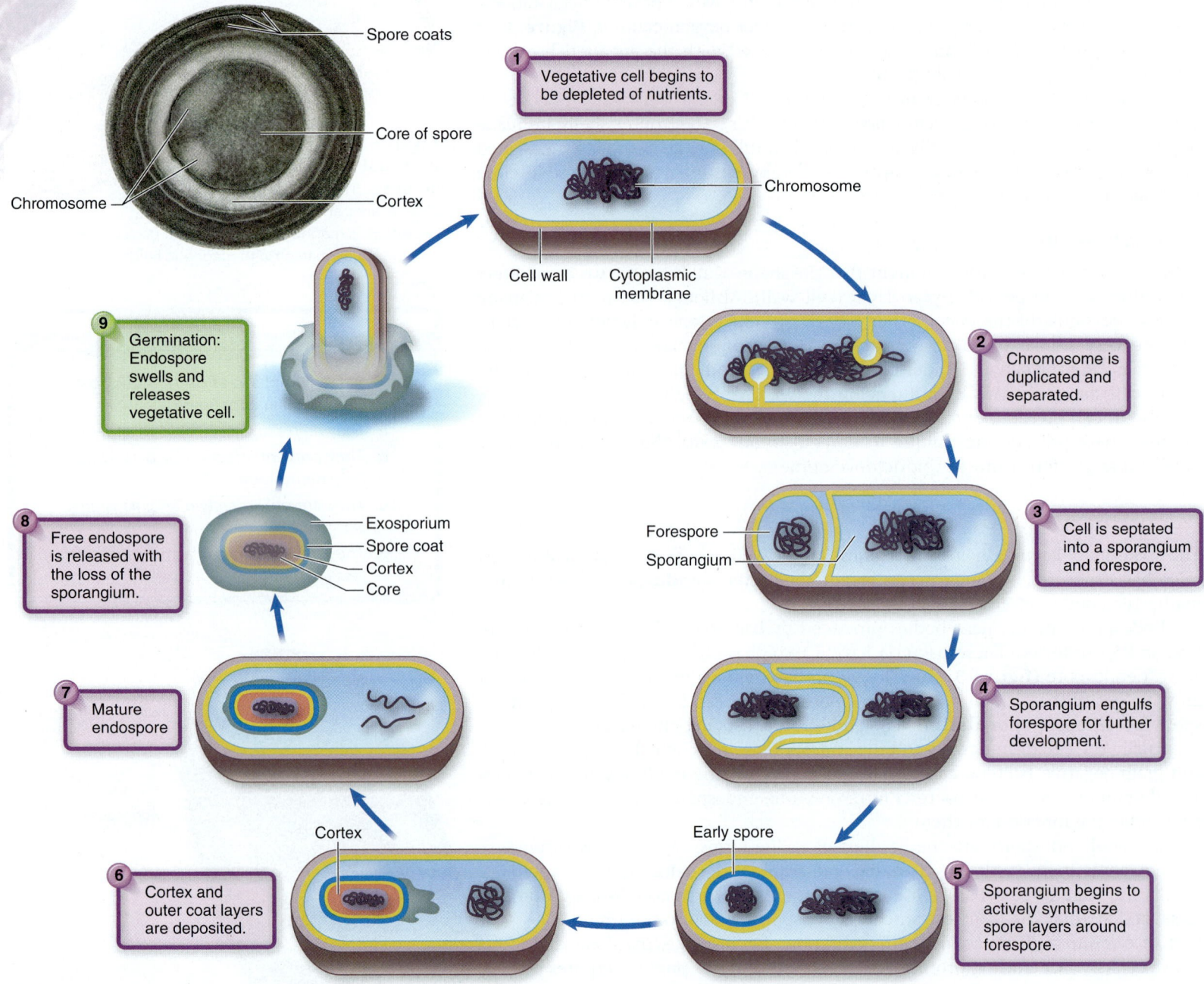

Spore coats

Core of spore

Chromosome

Cortex

1 Vegetative cell begins to be depleted of nutrients.

Chromosome

Cell wall   Cytoplasmic membrane

2 Chromosome is duplicated and separated.

Forespore

Sporangium

3 Cell is septated into a sporangium and forespore.

4 Sporangium engulfs forespore for further development.

Early spore

5 Sporangium begins to actively synthesize spore layers around forespore.

Cortex

6 Cortex and outer coat layers are deposited.

7 Mature endospore

8 Free endospore is released with the loss of the sporangium.

Exosporium
Spore coat
Cortex
Core

9 Germination: Endospore swells and releases vegetative cell.

Figure 3.21 A typical sporulation cycle in *Bacillus* species from the active vegetative cell to release and germination.   The process takes, on average, about 10 hours. Inset is a high magnification (10,000×) cross section of a single endospore showing the dense protective layers that surround the core with its chromosome.

a sporulating cell called a **sporangium.** Complete transformation of a vegetative cell into a sporangium and then into an endospore requires 6 to 8 hours in most endospore-forming species. **Figure 3.21** illustrates the major physical and chemical events in this process.

Return to the Vegetative State: Germination

After lying in a state of inactivity for an indefinite time, endospores can be revitalized when favorable conditions arise. Germination—the breaking of dormancy—happens in the presence of water and a specific chemical or environmental stimulus (germination agent). Once initiated, it proceeds to completion quite rapidly (1½ hours). Although the specific germination agent varies among species, it is generally a small organic molecule such as an amino acid or an inorganic salt. This agent stimulates the formation of hydrolytic (digestive) enzymes by the endospore membranes. These

enzymes digest the cortex and expose the core to water. As the core rehydrates and takes up nutrients, it begins to grow out of the endospore coats. In time, it reverts to a fully active vegetative cell, resuming the vegetative cycle.

Medical Significance of Bacterial Endospores

Although the majority of endospore-forming bacteria are relatively harmless, several bacterial pathogens are endospore formers. In fact, some aspects of the diseases they cause are related to the persistence and resistance of their spores. *Bacillus anthracis* is the agent of anthrax; its persistence in endospore form makes it an ideal candidate for bioterrorism. The genus *Clostridium* includes even more pathogens, such as *C. tetani,* the cause of tetanus (lockjaw), and *C. perfringens,* the cause of gas gangrene. When the endospores of these species are embedded in a wound that contains dead tissue, they can germinate, grow, and release potent toxins. Another toxin-forming species, *C. botulinum,* is the agent of botulism, a deadly form of food poisoning. (Each of these disease conditions is discussed in the infectious disease chapters, according to the organ systems it affects.)

Because they inhabit the soil and dust, endospores are constant intruders where sterility and cleanliness are important. They resist ordinary cleaning methods that use boiling water, soaps, and disinfectants; and they frequently contaminate cultures and media. Hospitals and clinics must take precautions to guard against the potential harmful effects of endospores, especially those of *Clostridium difficile,* the causative agent of a gastrointestinal disease commonly known as *C. diff.* Endospore destruction is a particular concern of the food-canning industry. Several endospore-forming species cause food spoilage or poisoning. Ordinary boiling (100°C) will usually not destroy such endospores, so canning is carried out in pressurized steam at 120°C for 20 to 30 minutes. Such rigorous conditions ensure that the food is sterile and free from viable bacteria.

> ### 3.4 LEARNING OUTCOMES—Assess Your Progress
>
> **10.** Identify five structures that may be contained in bacterial cytoplasm.
> **11.** Detail the causes and mechanisms of sporogenesis and germination.

3.5 The Archaea: The Other "Prokaryotes"

The discovery and characterization of novel cells resembling bacteria that have unusual anatomy, physiology, and genetics changed our views of microbial taxonomy and classification (see chapter 1). These single-celled, simple organisms, called **archaea,** are now considered a third cell type in a separate superkingdom (the domain Archaea). We include them in this chapter because they share many bacterial characteristics. But it has become clear that they are actually more closely related to domain Eukarya than to bacteria. For example, archaea and eukaryotes share a number of ribosomal RNA sequences that are not found in bacteria, and their protein synthesis and ribosomal subunit structures are similar. **Table 3.2** outlines selected points of comparison of the three domains.

Among the ways that the archaea differ significantly from other cell types are that they have entirely unique sequences in their rRNA. They exhibit a novel method of DNA compaction, and they contain unique membrane lipids, cell wall components, and pilin proteins. It is clear that the archaea are the most primitive of all life forms and are most closely related to the first cells that originated on the earth 4 billion years ago. The early earth is thought to have contained a hot, anaerobic "soup" with sulfuric gases and salts in abundance. The modern archaea still live in the remaining habitats on the earth that have these same ancient conditions–the most extreme habitats in nature. It is for this reason that they are often called *extremophiles,* meaning that they "love" extreme conditions in the environment.

Table 3.2 Comparison of Three Cellular Domains

Characteristic	Bacteria	Archaea	Eukarya
Chromosomes	Single or few, circular	Single, circular	Multiple, linear
Types of ribosomes	70S	70S but structure is similar to 80S	80S
Contains unique ribosomal RNA signature sequences	+	+	+
Protein synthesis similar to Eukarya	−	+	+
Presence of peptidoglycan in cell wall	+	−	−
Cytoplasmic membrane lipids	Fatty acids with ester linkages	Long-chain, branched hydrocarbons with ether linkages	Fatty acids with ester linkages
Sterols in membrane	− (some exceptions)	−	+
Nucleus and membrane-bound organelles	No	No	Yes
Flagellum	Bacterial flagellum	Archaellum	Eukaryotic flagellum

Some archaea thrive in extremely high temperatures. Others need extremely high concentrations of salt or acid to survive. Some archaea live on sulfur, reducing it to hydrogen sulfide to get their energy. Members of the group called *methanogens* can convert CO_2 and H_2 into methane gas (CH_4) through unusual and complex pathways. Archaea adapted to growth at very low temperatures are called *psychrophilic* (loving cold temperatures); those growing at very high temperatures are *hyperthermophilic* (loving high temperatures). Hyperthermophiles flourish at temperatures between 80°C and 113°C and cannot grow at 50°C. They live in volcanic waters and soils and submarine vents and are often salt- and acid-tolerant as well.

Archaea are not just environmental microbes. They have been isolated from human tissues such as the colon, the mouth, and the vagina. Recently, an association was found between the degree of severity of periodontal disease and the presence of archaeal RNA sequences in the gingiva, suggesting–but not proving–that archaea may be capable of causing human disease.

3.5 LEARNING OUTCOMES—Assess Your Progress

12. Compare and contrast the major features of archaea, bacteria, and eukaryotes.

3.6 Classification Systems for Bacteria and Archaea

Classification systems serve both practical and academic purposes. They aid in differentiating and identifying unknown species in medical and applied microbiology. They are also useful in organizing microorganisms and as a means of studying their relationships and origins. Since classification began around 200 years ago, several thousand species of bacteria and archaea have been identified, named, and cataloged.

There are two comprehensive databases compiled into books that help scientists classify bacteria and archaea. One, called *Bergey's Manual of Systematic Bacteriology*, presents a comprehensive view of bacterial and archaeal relatedness, combining phenotypic information with rRNA sequencing information to classify bacteria and archaea; it is a huge, five-volume set. (We need to remember that all bacteria and archaea classification systems are in a state of constant flux; no system is ever finished.)

Thermophilic archaea and cyanobacteria colonizing a thermal pool in Yellowstone National Park.

A separate book, called *Bergey's Manual of Determinative Bacteriology*, is based entirely on phenotypic characteristics. It is utilitarian in focus, categorizing bacteria by traits commonly assayed in clinical, teaching, and research labs. It is widely used by microbiologists who need to identify bacteria but need not know their evolutionary backgrounds. This phenotypic classification is more useful for students of medical microbiology, as well.

Taxonomic Scheme

Bergey's Manual of Determinative Bacteriology organizes the bacteria and archaea into four major divisions. These somewhat natural divisions are based on the nature of the cell wall. The **Gracilicutes** (gras'-ih-lik'-yoo-teez) have gram-negative cell walls and thus are thin-skinned; the **Firmicutes** have gram-positive cell walls that are thick and strong; the **Tenericutes** (ten'-er-ik'-yoo-teez) lack a cell wall and thus are soft; and the **Mendosicutes** (men-doh-sik'-yoo-teez) are the archaea (also called archaebacteria), primitive cells with unusual cell walls and nutritional habits. The first two divisions contain the greatest number of species. The 200 or so species that are so-far known to cause human and animal diseases can be found in four classes: the Scotobacteria, Firmibacteria, Thallobacteria, and Mollicutes. The system used in *Bergey's Manual* further organizes bacteria and archaea into subcategories such as classes, orders, and families, but these are not available for all groups.

Species and Subspecies in Bacteria and Archaea

Among most organisms, the species level is a distinct, readily defined, and natural taxonomic category. In animals, for instance, a species is a distinct type of organism that can produce viable offspring only when it mates with others of its own kind. This definition does not work for bacteria and archaea primarily because they do not exhibit a typical mode of sexual reproduction. Also, they can accept genetic information from unrelated forms, and they can alter their genetic makeup by a variety of mechanisms. Thus, it is necessary to hedge a bit when we define a *bacterial species*. Theoretically, it is a collection of bacterial cells, all of which share an overall similar pattern of traits, in contrast to other groups whose patterns differ significantly.

Members of a bacterial species should also share at least 70%–80% of their genes. Although the boundaries that separate two closely related species in a genus are in some cases arbitrary, this definition still serves as a method to separate the bacteria and archaea into various kinds that can be cultured and studied.

Individual members of given species can show variations, as well. Therefore, more categories within species exist, but they are not well defined. Microbiologists use terms like *subspecies, strain,* or *type* to designate bacteria of the same species that have differing characteristics. *Serotype* refers to representatives of a species that stimulate a distinct pattern of antibody (serum) responses in their hosts, because of distinct surface molecules.

3.6 LEARNING OUTCOMES—Assess Your Progress

13. Differentiate between *Bergey's Manual of Systematic Bacteriology* and *Bergey's Manual of Determinative Bacteriology.*

14. Name four divisions ending in –*cutes* and describe their characteristics.

15. Define a *species* in terms of bacteria.

CASE FILE WRAP UP

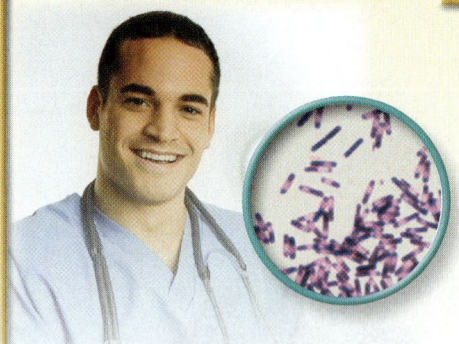

The circle contains an electron micrograph of Clostridium difficile, the endospore-forming bacterium that causes a common healthcare-associated intestinal infection.

Clostridium difficile is an endospore-forming bacterium that has gained attention over the last few years as the causative agent of a common (and potentially deadly) healthcare-associated infection. Often called "C. diff," this disease is spread by direct contact with an infected individual or the pathogen itself. In the hospital, the bacterium and more often its endospores can be present on bedrails, bedside tables, sinks, and even on surfaces such as stethoscopes and blood pressure cuffs. Endospores are most often the source of infection because they are extremely resistant to many cleaning agents.

Individuals at higher risk of contracting the disease include the elderly, individuals with weakened immune systems, people with intestinal disorders, and people who have recently taken antibiotics. The patient in the opening case file was elderly, had recently had major surgery, and was already battling another infection that was being treated with antibiotics, all risk factors for the development of C. diff. The disease can range from a mild infection to a life-threatening illness causing severe diarrhea up to 15 times a day. Note that some people are asymptomatic carriers of this pathogen, which makes controlling the disease that much more difficult, especially in health care settings.

Treatment of C. diff involves antibiotic therapy. For mild to moderate disease, metronidazole is used; vancomycin is used to treat severe infections. Probiotics can help to restore normal biota within the intestinal tract, because the overgrowth of C. *difficile* often occurs due to antibiotic-induced loss of these beneficial microbes. Unfortunately, approximately one-fourth of the individuals who recover from C. diff will experience a recurrence of the disease at some point—either due to regrowth of the initial pathogen or a new infection. Recurring bouts of C. diff often require treatment with different antibiotics. New studies indicate fecal transplants may be a beneficial option in some of these cases, as you will see in later chapters.

A Sticky Situation

A study published in the *Proceedings of the National Academy of Sciences* in 2013 revealed just how quickly biofilms can clog commonly used medical devices, such as cardiovascular stents. Researchers from Princeton utilized narrow tubes closely resembling those found in certain medical devices. Specific materials were chosen to replicate the surface of the equipment, and the tubes were then exposed to fluid under pressure in order to closely mimic conditions within the human body.

The researchers used microbes that are known to contaminate medical devices and engineered them to produce a green pigment that could be observed microscopically. After forcing a stream of these microbes through the experimental tubes for approximately 40 hours, microscopic analysis revealed the formation of a biofilm on the inside walls of the device.

Over the next few hours, the researchers then forced a stream of different microbes into the experimental tubes. These cells had been engineered so that they glowed red when viewed microscopically. Within a short period of time, red cells were noted adhering to the biofilm-coated inner walls of the tubes. Further analysis revealed that the flow within the narrow tubes nudged the trapped cells into threadlike "streamers" that rippled along with the moving fluid.

Initially, the formation of these microbial threads only slightly decreased the rate of fluid flow within the experimental tubes. However, after 55 hours, the streamers began to weave together, creating a net similar to a spider's web. This newly formed structure spanned the diameter of the narrow tube and trapped even more flowing cells, triggering a total blockage of the experimental tube within an hour.

This experiment revealed an important phenomenon that may explain why devices such as stents often fail. In addition, the researchers were able to identify which bacterial genes are likely involved in biofilm formation within a fluid environment. These data could lead to new strategies that maintain flow through medical devices, which could prevent unnecessary replacement of these devices or, in some cases, even death.

Source: 2013, February 11. *Proceedings of the National Academy of Sciences*. DOI: 1300321110.

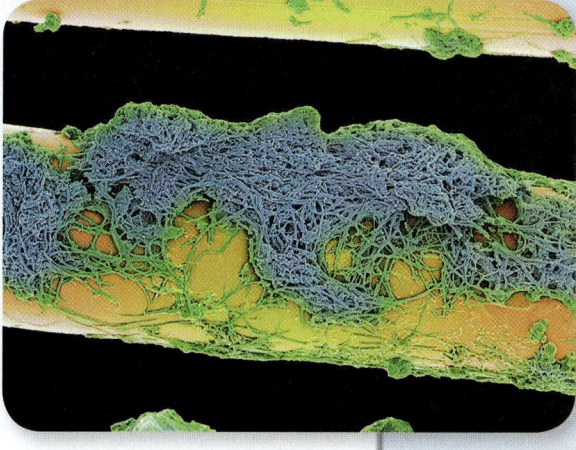

An accumulation of bacteria on a single fiber of a gauze bandage.

Chapter Summary

3.1 Form and Function of Bacteria and Archaea

- Bacteria and archaea are distinguished from eukaryotes by (a) the way their DNA is packaged, (b) their cell walls, and (c) their lack of membrane-bound internal structures.
  - Bacteria invariably have a cytoplasmic membrane, cytoplasm, ribosomes, a nucleoid, and a cytoskeleton.
  - Bacteria may also have a cell wall, a glycocalyx, flagella, an outer membrane, a pilus, plasmids, inclusions and/or microcompartments.
- Most bacteria have one of three general shapes: coccus (round), bacillus (rod), or spiral (spirochete, spirillum). Additional shapes are vibrio and branching filaments.
- Shape and arrangement of cells are key means of describing bacteria. Arrangements of cells are based on the number of planes in which a given species divides.

3.2 External Structures

- The external structures of bacteria include appendages (flagella, fimbriae, and pili) and the glycocalyx.
- Flagella vary in number and arrangement as well as in the type and rate of motion they produce.

3.3 The Cell Envelope: The Wall and Membrane(s)

- The cell envelope is the boundary between inside and outside in a bacterial cell. Gram-negative bacteria have an outer membrane, the cell wall, and the cytoplasmic membrane. Gram-positive bacteria have only the cell wall and cytoplasmic membrane.
- In a Gram stain, gram-positive bacteria retain the crystal violet and stain purple. Gram-negative bacteria lose the crystal violet and stain red from the safranin counterstain.
- The outer membrane of gram-negative cells contains lipopolysaccharide (LPS), which is toxic to mammalian hosts.
- The bacterial cytoplasmic membrane is typically composed of phospholipids and proteins, and it performs many metabolic functions as well as transport activities.

3.4 Bacterial Internal Structure

- The cytoplasm of bacterial cells serves as a solvent for materials used in all cell functions.
- The genetic material of bacteria is DNA, arranged on large, circular chromosomes. Additional genes can be carried on plasmids.
- Bacterial ribosomes are dispersed in the cytoplasm and are also embedded in the cytoplasmic membrane.
- Bacteria may store nutrients or other useful substances in their cytoplasm in either inclusions or microcompartments.
- Bacteria manufacture several types of proteins that help determine their cellular shape.
- A few families of bacteria produce dormant bodies called endospores, which are the hardiest of all life forms, surviving for hundreds or thousands of years.
- The genera *Bacillus* and *Clostridium* are endospore formers, and both contain deadly pathogens.

3.5 The Archaea: The Other "Prokaryotes"

- Archaea constitute the third domain of life. They superficially resemble bacteria but are most genetically related to eukaryotes.
- Although they exhibit similar external and internal structure, the unusual biochemistry and genetics of archaea set them apart from bacteria. Many members are adapted to extreme habitats with low or high temperature, salt, pressure, or acid.

3.6 Classification Systems for Bacteria and Archaea

- Bacteria and archaea are formally classified by phylogenetic relationships and phenotypic characteristics.
- Medical identification of pathogens uses an informal system of classification based on Gram stain, morphology, biochemical reactions, and metabolic requirements. It is summarized in *Bergey's Manual of Determinative Bacteriology*.
- A *bacterial species* is loosely defined as a collection of bacterial cells that shares an overall similar pattern of traits different from other groups of bacteria and that shares at least 70%-80% of its genes.
- Variant forms within a species (subspecies) include strains, types, and serotypes.

Multiple-Choice Questions   Bloom's Levels 1 and 2: Remember and Understand

Select the correct answer from the answers provided.

1. Which of the following is not found in all bacterial and archaeal cells?
   - a. cytoplasmic membrane
   - b. a nucleoid
   - c. ribosomes
   - d. flagellum

2. _____ refers to chains of spherical bacterial cells while clusters of spherical cells are called _____.
   - a. Diplococcus, streptococcus
   - b. Staphylococcus, streptococcus
   - c. Streptococcus, staphylococcus
   - d. Micrococcus, sarcina

3. Which structure plays a direct role in the exchange of genetic material between bacterial cells?
   - a. flagellum
   - b. pilus
   - c. capsule
   - d. fimbria

4. Which of the following is present in both gram-positive and gram-negative cell walls?
   - a. an outer membrane
   - b. peptidoglycan
   - c. teichoic acid
   - d. lipopolysaccharides

5. Bacterial endospores

   a. are visualized using the acid-fast stain.

   b. are a mechanism for survival.

   c. are used for nutrient storage.

   d. are easily inactivated by heat.

6. Which of the following would be used to identify an unknown bacterial culture in your nursing school laboratory exercise?

   a. *Gray's Anatomy*

   b. *Bergey's Manual of Systematic Bacteriology*

   c. *The Physicians' Desk Reference*

   d. *Bergey's Manual of Determinative Bacteriology*

7. Which stain is most frequently used to distinguish differences between the cell walls of medically important bacteria?

   a. simple stain   c. Gram stain

   b. acridine orange stain   d. negative stain

8. Archaea

   a. are most genetically related to bacteria.

   b. contain a nucleus.

   c. cannot cause disease in humans.

   d. lack peptidoglycan in their cell wall.

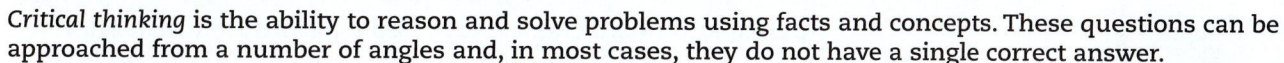

Critical Thinking   Bloom's Levels 3, 4, and 5: Apply, Analyze, and Evaluate

Critical thinking is the ability to reason and solve problems using facts and concepts. These questions can be approached from a number of angles and, in most cases, they do not have a single correct answer.

1. As a supervisor in the infection-control unit, you hire a local microbiologist to analyze samples from your hospital's hot-water tank for microbial contamination. Although she was unable to culture any microbes, she reports that basic microscopic analysis revealed the presence of cells 0.8 μm in diameter that lacked a nucleus. Transmission electron microscopy showed that the cells lacked membrane-bound organelles but did contain ribosomes. Which domain of life do you hypothesize these cells represent? Discuss any additional analysis that could be performed to determine this classification most accurately.

2. During your clinical diagnostic lab rotation, you are asked to perform a test to determine whether or not three patients are all infected with the same bacterial pathogen. Your results demonstrate that each patient's immune system produces a unique set of antibodies against his or her infectious agent.

   a. Based upon this information, discuss whether or not you can conclude that all three patients are infected with the same species of bacterium.

   b. Explain why each patient made different antibodies to the pathogen causing his or her disease.

3. Conduct additional research and discuss how bacterial endospores played a pivotal role in the 2001 anthrax attacks in the United States.

4. Create a chart to compare and contrast the known structure and functions of fimbriae, pili, flagella, and glycocalyces.

5. The results of your patient's wound culture just arrived, and Gram staining revealed the presence of pink, rod-shaped bacterial cells organized in pairs.

   a. Using terms from this chapter, describe the bacterium's arrangement.

   b. Based upon this information, summarize the Gram reaction displayed by this bacterium.

   c. You quickly realize that this patient could be at risk for developing fever and shock. Explain how the culture results indicated this potential risk.

Visual Connections   Bloom's Level 5: Evaluate

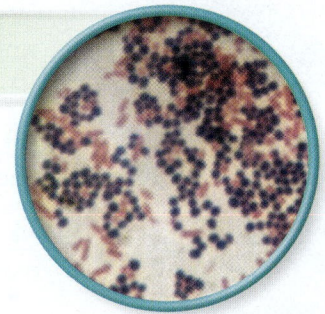

This question connects previous images to a new concept.

1. From chapter 2, **figure 2.18.** Explain why some cells are pink and others are purple in this image of a Gram-stained bacterial smear.

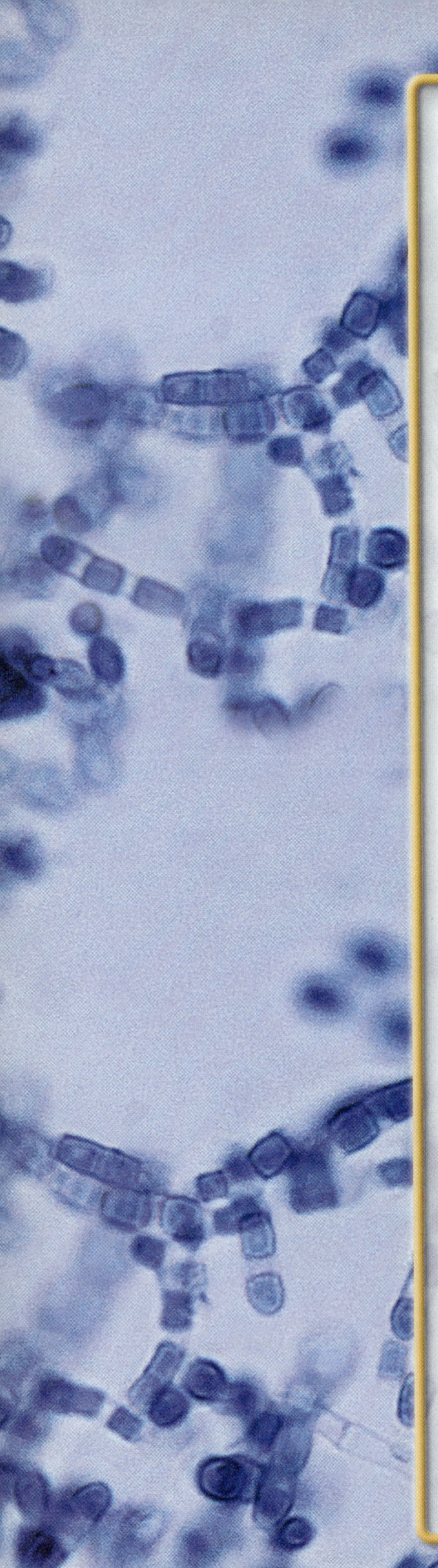

Puzzle in the Valley

Working as a newly graduated radiology technologist in a rural hospital in California, I encountered a case that would prove to be a challenge for everyone involved. The patient was a male migrant farm worker in his mid-30s who presented to the ER with common flulike symptoms: fever, chills, weakness, cough, muscular aches and pains, and headache. He also had a painful red rash on his lower legs.

It was summertime, so influenza was unlikely. The emergency room physician believed that the patient likely had pneumonia, but she found the rash puzzling. She asked me to obtain a chest X ray. I performed anterior-posterior and lateral views of the chest, which revealed two nodules approximately 2 cm in size in the patient's left upper lobe. The physician stated that the nodules were consistent with pneumonia, but the possibility of cancer could not be ruled out. The patient's age and the fact that he was a nonsmoker, however, made a diagnosis of lung cancer much less likely than pneumonia.

The patient was admitted to the hospital for IV antibiotic treatment. Before the antibiotic therapy was started, a sputum sample was collected and sent to a larger center for culture and sensitivity (C&S) testing. Despite IV fluids, rest, and broad-spectrum antibiotics targeting both gram-positive and gram-negative bacteria, the patient showed no improvement. After receiving the C&S report, I understood why the intravenous antibiotics were not working: The patient had a fungal infection, not a bacterial infection as first suspected. I notified the physician, who immediately started the patient on amphotericin B, a potent antifungal medication that would properly treat the patient's case of coccidioidomycosis.

- How might the patient have contracted this infection?

- Why did the initial antibiotic therapy fail to improve the patient's symptoms?

Case File Wrap-Up appears on page 110.

Eukaryotic Cells and Microorganisms

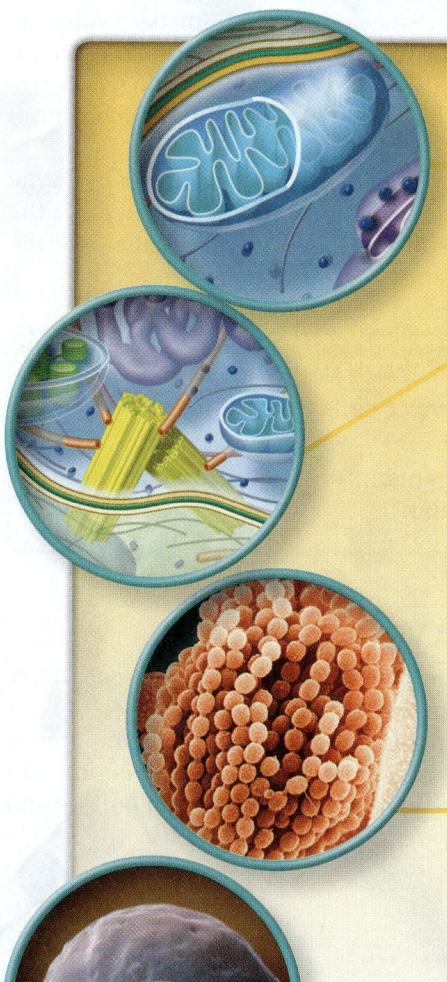

Table 4.1 Eukaryotic Organisms Studied in Microbiology

Always Unicellular	May Be Unicellular or Multicellular	Always Multicellular
Protozoa	Fungi Algae	Helminths (have unicellular egg or larval forms)

4.1 The History of Eukaryotes

Evidence from paleontology indicates that the first eukaryotic cells appeared on the earth approximately 2 to 3 billion years ago. While it used to be thought that eukaryotic cells evolved directly from ancient prokaryotic cells, we now believe that bacteria, archaea, and eukaryotes evolved from a different kind of cell, a precursor to both prokaryotes and eukaryotes that biologists call the *last common ancestor*. This ancestor was neither prokaryotic nor eukaryotic but gave rise to all three current cell types.

The first primitive eukaryotes were probably single-celled and independent, but, over time, some cells began to aggregate, forming colonies. With further evolution, some of the cells within colonies became *specialized*, or adapted to perform a particular function advantageous to the whole colony, such as movement, feeding, or reproduction. Complex multicellular organisms evolved as individual cells in the organism lost the ability to survive apart from the intact colony.

Only certain eukaryotes are traditionally studied by microbiologists–primarily the protozoa, the microscopic algae and fungi, and helminths **(table 4.1).** Because the vast majority of algae do not cause infections of humans, we will discuss only the other three eukaryotic microbes in this chapter.

4.1 LEARNING OUTCOMES—Assess Your Progress

1. Relate bacterial, archaeal, and eukaryotic cells to the *last common ancestor*.
2. List the types of eukaryotic microorganisms, and denote which are unicellular and which are multicellular.

4.2 Structures of the Eukaryotic Cell

In general, eukaryotic microbial cells have a cell membrane, nucleus, mitochondria, endoplasmic reticulum, Golgi apparatus, vacuoles, cytoskeleton, and glycocalyx. A cell wall, locomotor appendages, and chloroplasts are found only in some groups **(figure 4.1).** In the following sections, we cover the microscopic structure and functions of the eukaryotic cell. As with the bacteria, we begin on the outside and proceed inward through the cell.

Structure Flowchart

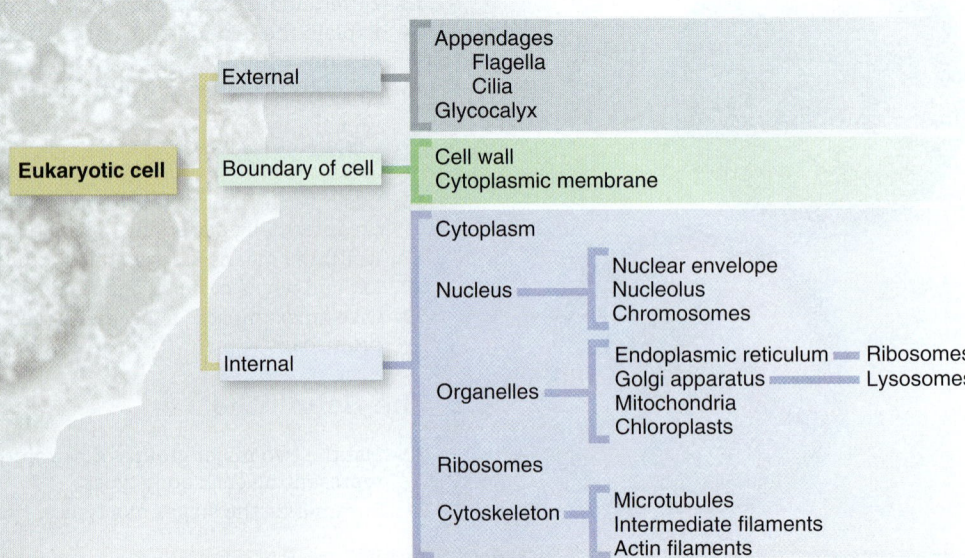

Eukaryotic cell
- External
  - Appendages
    - Flagella
    - Cilia
  - Glycocalyx
- Boundary of cell
  - Cell wall
  - Cytoplasmic membrane
- Internal
  - Cytoplasm
  - Nucleus
    - Nuclear envelope
    - Nucleolus
    - Chromosomes
  - Organelles
    - Endoplasmic reticulum — Ribosomes
    - Golgi apparatus — Lysosomes
    - Mitochondria
    - Chloroplasts
  - Ribosomes
  - Cytoskeleton
    - Microtubules
    - Intermediate filaments
    - Actin filaments

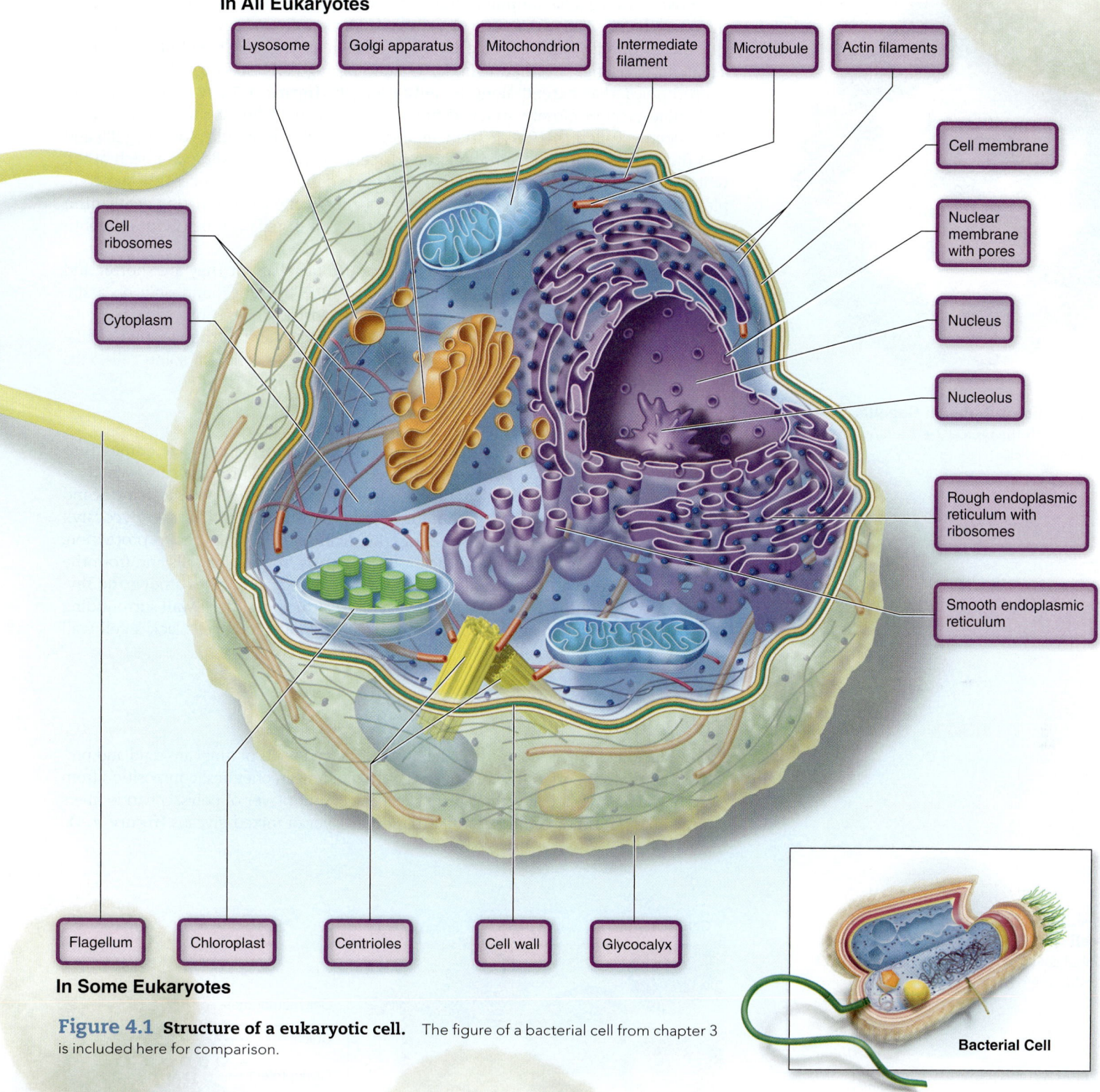

In All Eukaryotes

Lysosome | Golgi apparatus | Mitochondrion | Intermediate filament | Microtubule | Actin filaments

Cell membrane

Nuclear membrane with pores

Nucleus

Nucleolus

Rough endoplasmic reticulum with ribosomes

Smooth endoplasmic reticulum

Cell ribosomes

Cytoplasm

Flagellum | Chloroplast | Centrioles | Cell wall | Glycocalyx

In Some Eukaryotes

Bacterial Cell

Figure 4.1 **Structure of a eukaryotic cell.** The figure of a bacterial cell from chapter 3 is included here for comparison.

External Structures

Appendages for Moving: Cilia and Flagella

Motility allows a microorganism to locate nutrients and to migrate toward positive stimuli such as sunlight; it also enables them to avoid harmful substances and stimuli. Locomotion by means of flagella or cilia is common in protozoa, many algae, and a few fungal and animal cells.

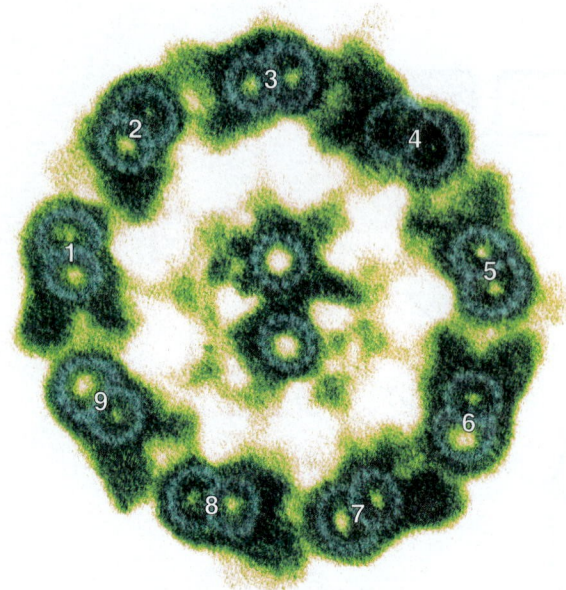

Figure 4.2 Microtubules in flagella. A cross section that reveals the typical 9 + 2 arrangement found in both flagella and cilia.

Eukaryotic flagella are much different from those of bacteria, though they share the same name. The eukaryotic flagellum is thicker (by a factor of 10), structurally more complex, and covered by an extension of the cell membrane. A single flagellum is a long, sheathed cylinder containing regularly spaced hollow tubules–microtubules–that extend along its entire length **(figure 4.2).** A cross section reveals nine pairs of closely attached microtubules surrounding a single central pair. This scheme, called the 9 + 2 arrangement, is the pattern of eukaryotic flagella and cilia (figure 4.2). During locomotion, the adjacent microtubules slide past each other, whipping the flagellum back and forth. Although details of this process are too complex to discuss here, it involves expenditure of energy and a coordinating mechanism in the cell membrane. The placement and number of flagella can be useful in identifying flagellated protozoa and certain algae.

Cilia are very similar in overall architecture to flagella, but they are shorter and more numerous (some cells have several thousand). They are found only on a single group of protozoa and in certain animal cells. In the ciliated protozoa, the cilia occur in rows over the cell surface, where they beat back and forth in regular oarlike strokes. Such protozoa are among the fastest of all motile cells. On some cells, cilia also function as feeding and filtering structures.

The Glycocalyx

Most eukaryotic cells have a **glycocalyx,** an outermost layer that comes into direct contact with the environment (see figure 4.1). This structure, which is sometimes called an extracellular matrix, is usually composed of polysaccharides and appears as a network of fibers, a slime layer, or a capsule much like the glycocalyx of prokaryotes. Because of its positioning, the glycocalyx contributes to protection, adherence of cells to surfaces, and reception of signals from other cells and from the environment. The nature of the layer beneath the glycocalyx varies among the several eukaryotic groups. Fungi and most algae have a thick, rigid cell wall surrounding a cell membrane, whereas protozoa, a few algae, and all animal cells lack a cell wall and have only a cell membrane.

Boundary Structures

The Cell Wall

Protozoa and helminths do not have cell walls. The cell walls of fungi are rigid and provide structural support and shape, but they are different in chemical composition from bacterial and archaeal cell walls. They have a thick, inner layer of polysaccharide fibers composed of chitin or cellulose, and a thin outer layer of mixed glycans **(figure 4.3).**

Figure 4.3 Cross-sectional views of fungal cell walls. (a) An electron micrograph of two fungal cells. (b) A drawing of the section of the wall inside the square in (a).

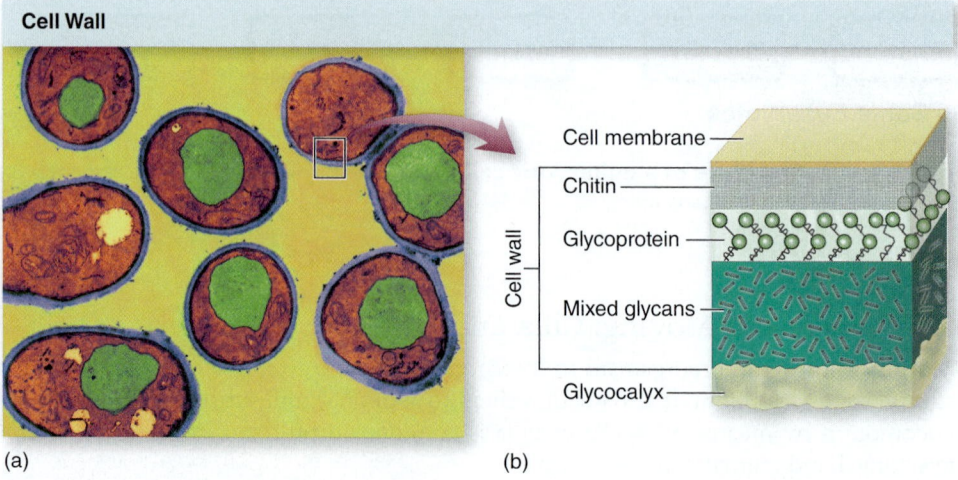

Cell Wall

Cell membrane

Chitin

Glycoprotein

Mixed glycans

Glycocalyx

Cell wall

(a)   (b)

The Cell Membrane

The cell (or cytoplasmic) membrane of eukaryotic cells is a typical bilayer of phospholipids in which protein molecules are embedded. In addition to phospholipids, eukaryotic membranes also contain *sterols* of various kinds. Sterols are different from phospholipids in both structure and behavior. Their relative rigidity makes eukaryotic membranes more stable than those of non-eukaryotic cells. This strengthening feature is extremely important in those cells that don't have a cell wall. Cytoplasmic membranes of eukaryotes are functionally similar to those of bacteria and archaea, serving as selectively permeable barriers.

Internal Structures

Unlike bacteria and archaea, eukaryotic cells contain a number of individual membrane-bound organelles that are extensive enough to account for 60% to 80% of their volume.

The Nucleus

The nucleus is a compact sphere that is the most prominent organelle of eukaryotic cells. It is separated from the cell cytoplasm by an external boundary called a nuclear envelope. The envelope has a unique architecture. It is composed of two parallel membranes separated by a narrow space, and it is perforated with small, regularly spaced openings, or pores, formed at sites where the two membranes unite **(figure 4.4).** The nuclear pores are passageways through which macromolecules migrate from the nucleus to the cytoplasm and vice versa. The nucleus contains a matrix called the nucleoplasm and a granular mass, the **nucleolus,** that stains more intensely than the immediate surroundings because of its RNA content. The nucleolus is the site for ribosomal RNA synthesis and a collection area for ribosomal subunits. The subunits are transported through the nuclear pores into the cytoplasm for final assembly into ribosomes.

A prominent feature of the nucleoplasm in stained preparations is a network of dark fibers known as **chromatin.** Chromatin is made of linear DNA, which, of course, is the genetic material of the cell. When it is wound around histone proteins, chromatin forms structures called chromosomes. Elaborate processes have evolved for transcription and duplication of this genetic material.

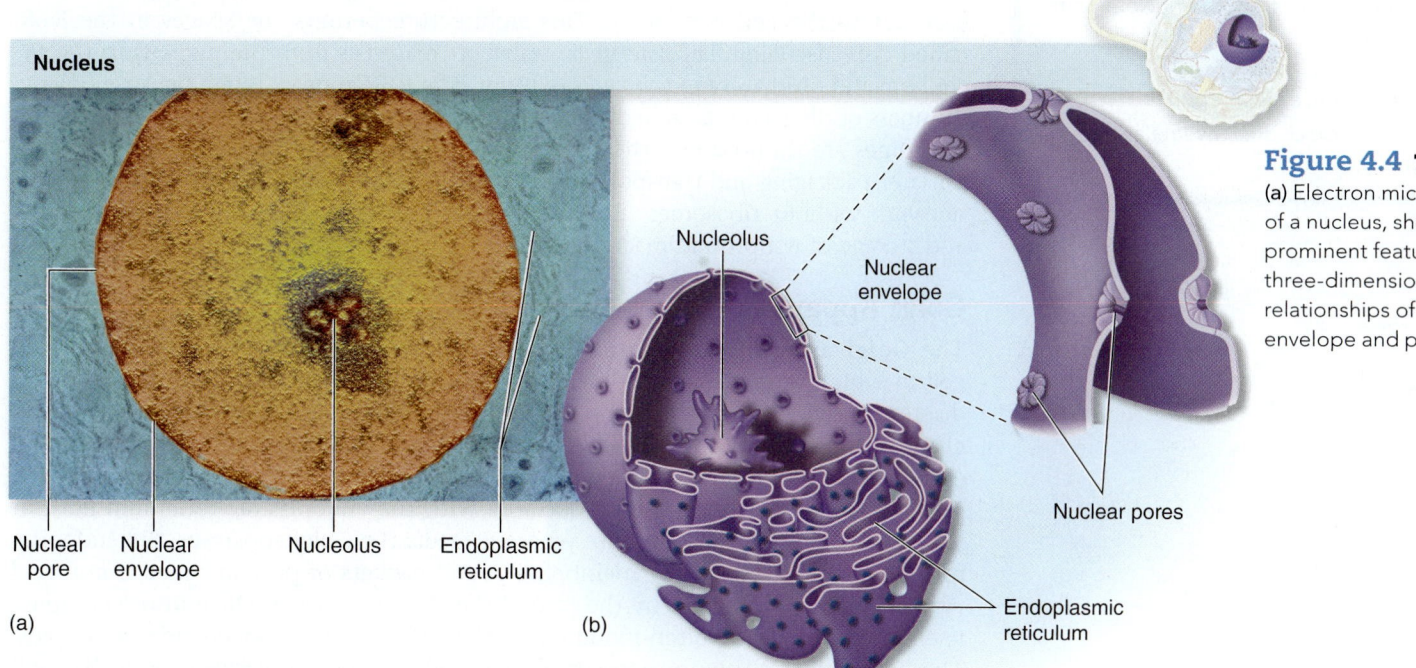

Nucleus

Nucleolus

Nuclear envelope

Nuclear pores

Endoplasmic reticulum

Nuclear pore   Nuclear envelope   Nucleolus   Endoplasmic reticulum

(a)

(b)

Figure 4.4 The nucleus.
(a) Electron micrograph section of a nucleus, showing its most prominent features. (b) Cutaway three-dimensional view of the relationships of the nuclear envelope and pores.

Endoplasmic Reticulum

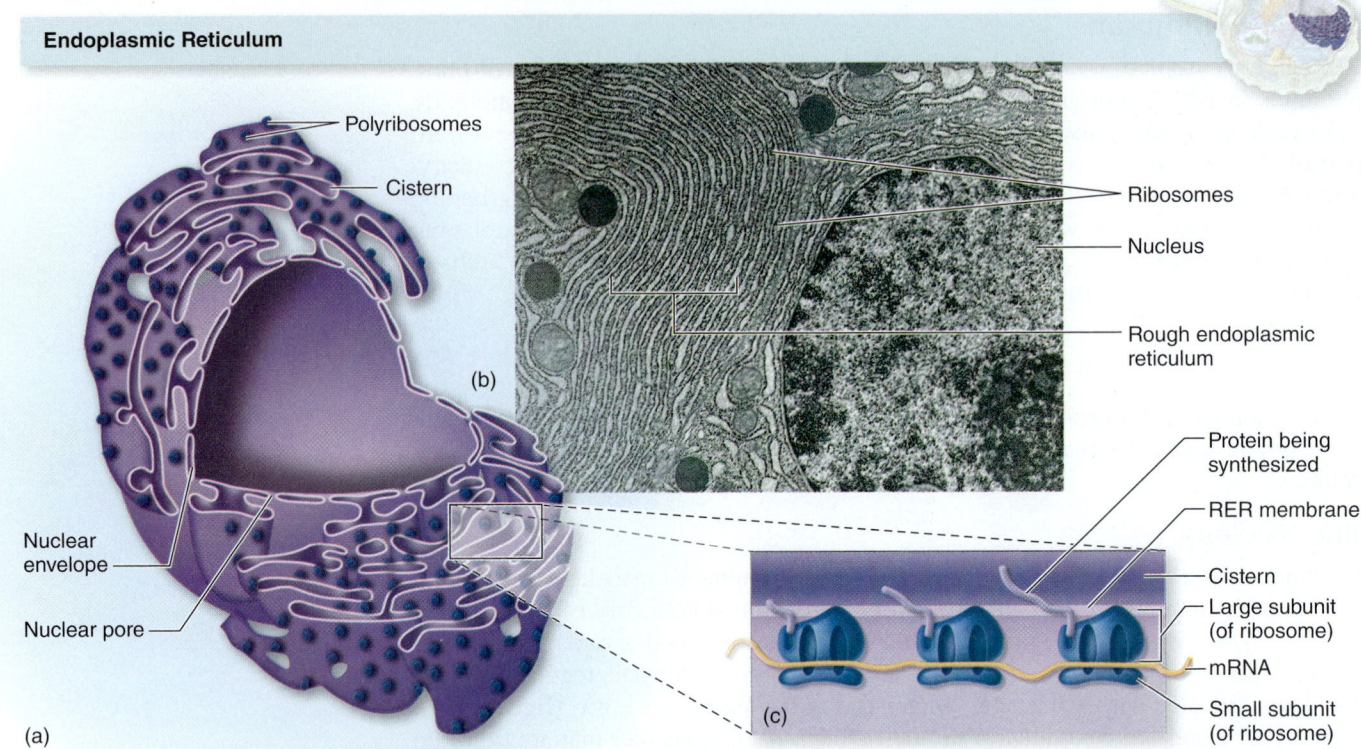

Figure 4.5 The origin and detailed structure of the rough endoplasmic reticulum (RER). **(a)** Schematic view of the origin of the RER from the outer membrane of the nuclear envelope. **(b)** Electron micrograph of the RER. **(c)** Detail of the orientation of a ribosome on the RER membrane.

Endoplasmic Reticulum

The **endoplasmic reticulum (ER)** is a series of microscopic tunnels used in transport and storage. There are two kinds of endoplasmic reticulum: the **rough endoplasmic reticulum (RER) (figure 4.5)** and the **smooth endoplasmic reticulum (SER).** The RER originates from the outer membrane of the nuclear envelope and extends in a continuous network through the cytoplasm, even all the way out to the cell membrane. This architecture permits the spaces in the RER, called cisternae (singular, *cistern*), to transport materials from the nucleus to the cytoplasm and ultimately to the cell's exterior. The RER appears rough because of large numbers of ribosomes attached to its membrane surface. Proteins synthesized on the ribosomes are shunted into the inside space (the lumen) of the RER and held there for later packaging and transport. In contrast to the RER, the SER is a closed tubular network without ribosomes that functions in nutrient processing and in synthesis and storage of nonprotein macromolecules such as lipids.

Golgi Apparatus

The **Golgi apparatus,** also called the Golgi complex or body, is the site in the cell in which proteins are modified and then sent to their final destinations. It is a discrete organelle consisting of a stack of several flattened, disc-shaped sacs called cisternae. These sacs have outer limiting membranes and cavities like those of the endoplasmic reticulum, but they do not form a continuous network **(figure 4.6).** This organelle is always closely associated with the endoplasmic reticulum both in its location and function. At a site where it meets the Golgi apparatus, the endoplasmic reticulum buds off tiny membrane-bound packets of protein called *transitional vesicles* that are picked up by the face of the Golgi apparatus. Once in the complex itself, the proteins are often modified by the addition of polysaccharides and lipids. The final action of this apparatus is to pinch off finished *condensing vesicles* that will

Golgi Apparatus

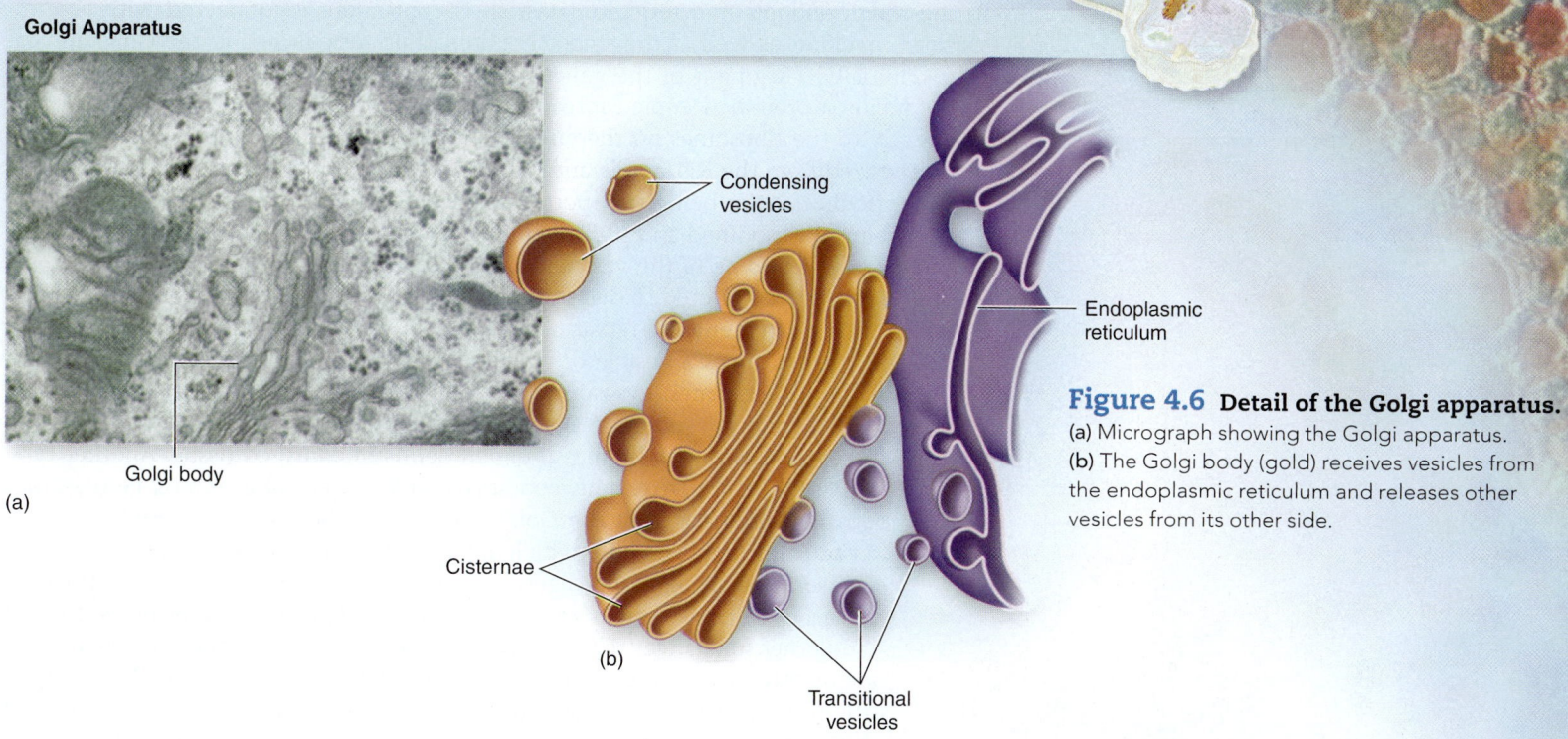

(a)

Golgi body

Condensing vesicles

Endoplasmic reticulum

Cisternae

(b)

Transitional vesicles

Figure 4.6 Detail of the Golgi apparatus.
(a) Micrograph showing the Golgi apparatus.
(b) The Golgi body (gold) receives vesicles from the endoplasmic reticulum and releases other vesicles from its other side.

be conveyed to organelles such as lysosomes or transported outside the cell as secretory vesicles **(figure 4.7)**.

Nucleus, Endoplasmic Reticulum, and Golgi Apparatus: Nature's Assembly Line

As the keeper of the eukaryotic genetic code, the nucleus ultimately governs and regulates all cell activities. But, because the nucleus remains fixed in a specific cellular site, it must direct these activities through a structural and chemical network (figure 4.7). This network includes ribosomes, which originate in the nucleus, and

Figure 4.7 The transport process.   The cooperation of organelles in protein synthesis and transport: Nucleus → RER → Golgi apparatus → vesicles → secretion.

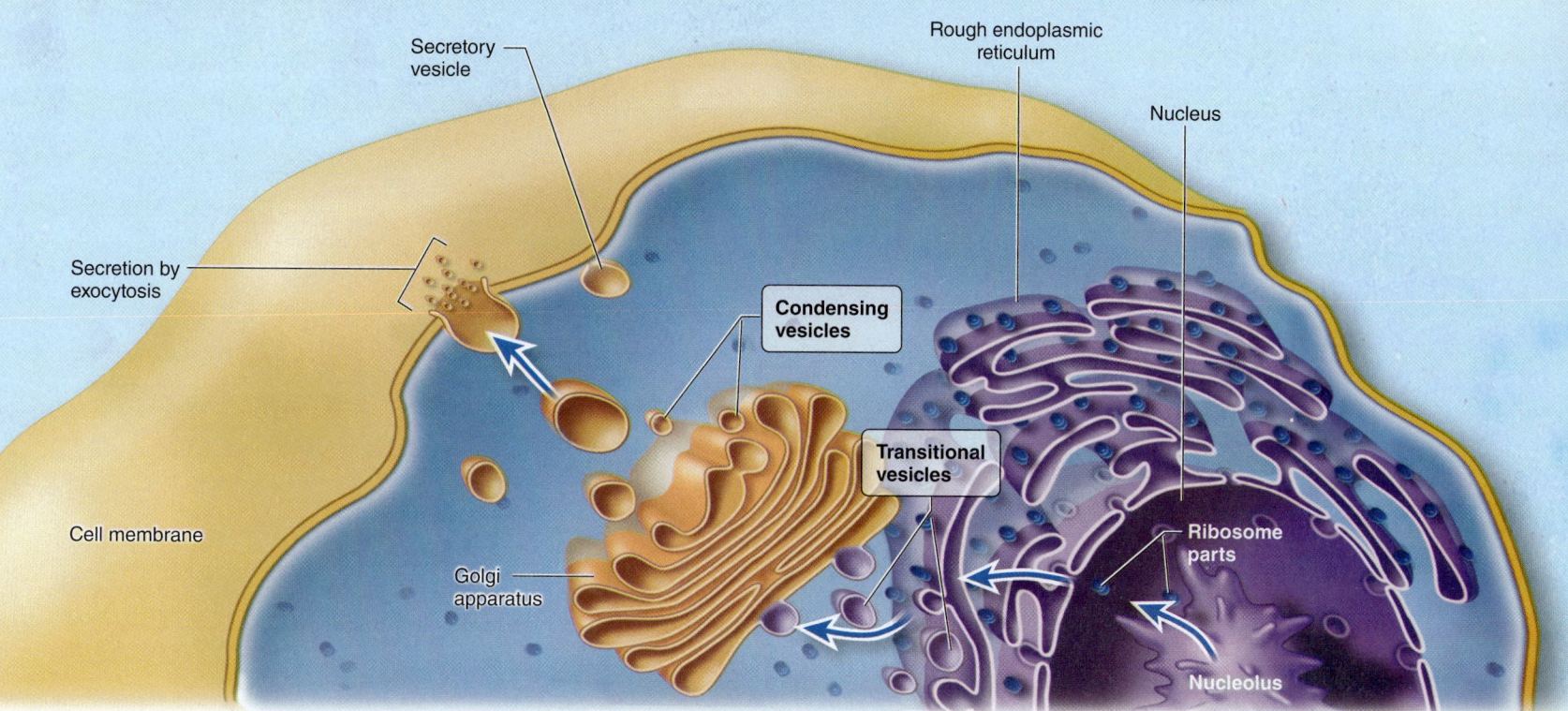

Secretory vesicle

Rough endoplasmic reticulum

Nucleus

Secretion by exocytosis

Condensing vesicles

Cell membrane

Transitional vesicles

Golgi apparatus

Ribosome parts

Nucleolus

This colorized transmission electron micrograph of a mast cell, a cell type of of the immune system, shows the nucleus as a large orange oval.

the rough endoplasmic reticulum, which is continuously connected with the nuclear envelope, as well as the smooth endoplasmic reticulum and the Golgi apparatus. Initially, a segment of the genetic code of DNA containing the instructions for producing a protein is copied into RNA and passed out through the nuclear pores directly to the ribosomes on the endoplasmic reticulum. Here, specific proteins are synthesized from the RNA code and deposited in the lumen (space) of the endoplasmic reticulum. After being transported to the Golgi apparatus, the protein products are chemically modified and packaged into vesicles that can be used by the cell in a variety of ways. Some of the vesicles contain enzymes to digest food inside the cell; other vesicles are secreted to digest materials outside the cell, and others are important in the enlargement and repair of the cell wall and membrane.

A **lysosome** is a vesicle originating from the Golgi apparatus that contains a variety of enzymes. Lysosomes are involved in intracellular digestion of food particles and in protection against invading microorganisms. They also participate in digestion and removal of cell debris in damaged tissue. Another type of vesicle, the peroxisome, contains a wide variety of enzymes. Peroxisomes do not originate from the Golgi apparatus. Other types of vesicles include **vacuoles** (vak'-yoo-ohl), which are membrane-bound sacs containing fluids or solid particles to be digested, excreted, or stored. They are formed in phagocytic cells (certain white blood cells and protozoa) in response to food and other substances that have been engulfed. The contents of a food vacuole are digested through the merger of the vacuole with a lysosome. This merged structure is called a phagosome **(figure 4.8).** Other types of vacuoles are used in storing reserve food such as fats and glycogen. Protozoa living in freshwater

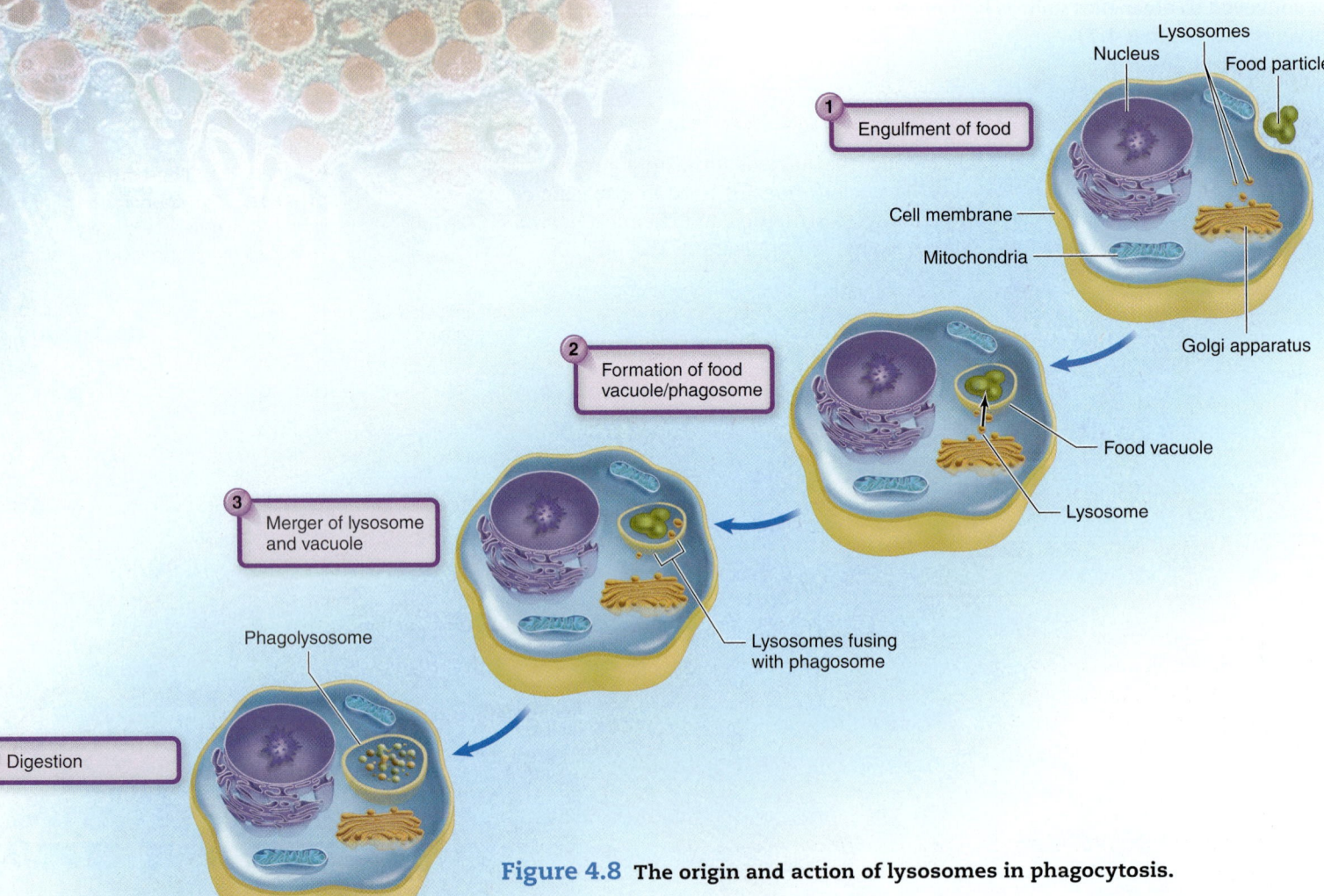

Figure 4.8 The origin and action of lysosomes in phagocytosis.

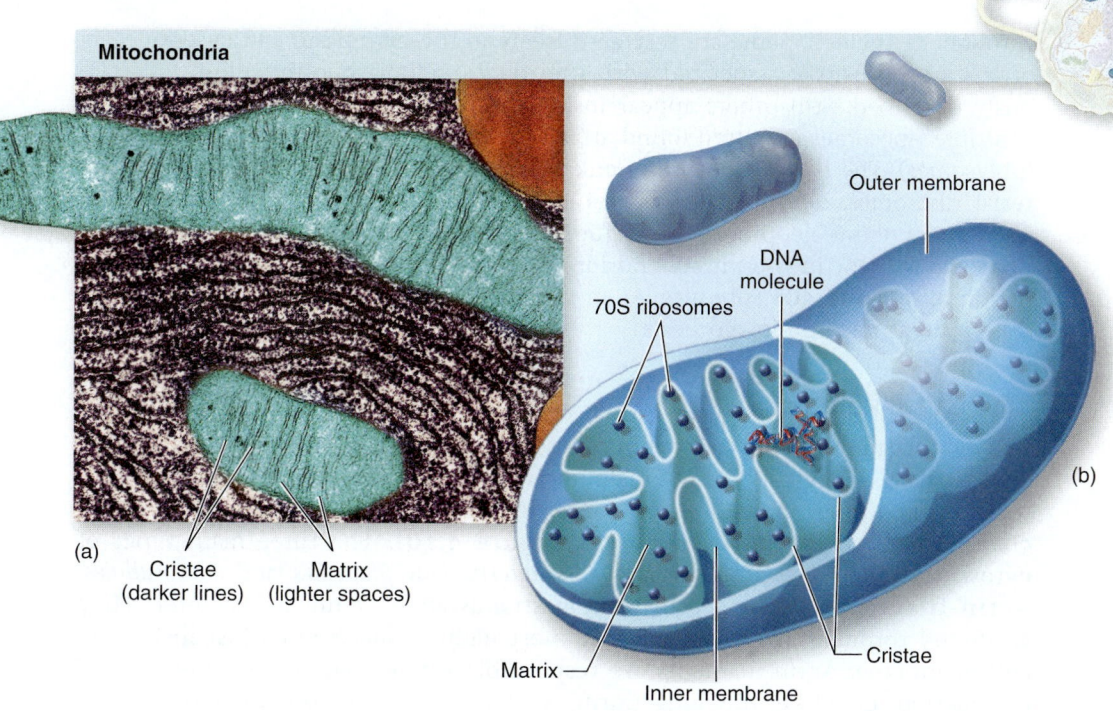

Mitochondria

(a) Cristae (darker lines)   Matrix (lighter spaces)

70S ribosomes

DNA molecule

Outer membrane

Matrix

Inner membrane

Cristae

(b)

Figure 4.9 **General structure of a mitochondrion.**

habitats regulate osmotic pressure by means of contractile vacuoles, which regularly expel excess water that has diffused into the cell (described later).

Mitochondria

Although the nucleus is the cell's control center, none of the cellular activities it commands could proceed without a constant supply of energy, the bulk of which is generated in most eukaryotes by **mitochondria** (my″-toh-kon′-dree-uh). When viewed with light microscopy, mitochondria appear as round or elongated particles scattered throughout the cytoplasm. The internal ultrastructure reveals that a single mitochondrion consists of a smooth, continuous outer membrane that forms the external contour, and an inner, folded membrane nestled neatly within the outer membrane **(figure 4.9a).** The folds on the inner membrane, called **cristae** (kris′-te), may be tubular, like fingers, or folded into shelflike bands.

The cristae membranes hold the enzymes and electron carriers of aerobic respiration. This is an oxygen-using process that extracts chemical energy contained in nutrient molecules and stores it in the form of high-energy molecules, or ATP. Mitochondria (along with chloroplasts) are unique among organelles in that they divide independently of the cell, contain circular molecules of DNA, and have bacteria-sized 70S ribosomes. These characteristics have caused scientists to suggest that mitochondria were once bacterial cells that developed into eukaryotic organelles over time.

Chloroplasts

Chloroplasts are remarkable organelles found in algae and plant cells that are capable of converting the energy of sunlight into chemical energy through photosynthesis. Another important photosynthetic product of chloroplasts is oxygen gas. Although chloroplasts resemble mitochondria, chloroplasts are larger, contain special pigments, and are much more varied in shape.

Ribosomes

In an electron micrograph of a eukaryotic cell, ribosomes are numerous, tiny particles that give a "dotted" appearance to the cytoplasm. Ribosomes are distributed

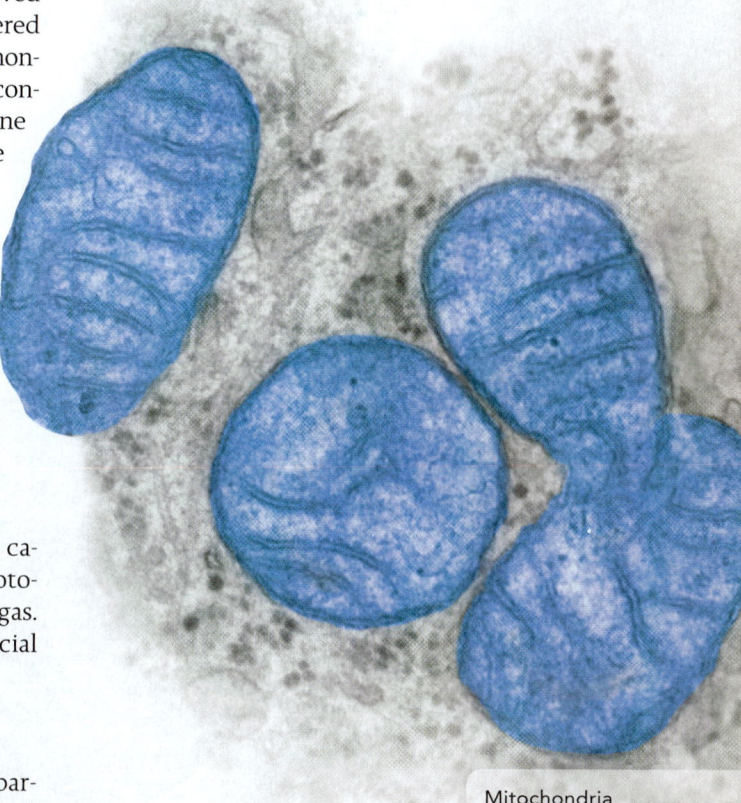

Mitochondria

throughout the cell: Some are scattered freely in the cytoplasm and cytoskeleton; others are intimately associated with the rough endoplasmic reticulum as previously described. Still others appear inside the mitochondria and in chloroplasts. Multiple ribosomes are often found arranged in short chains called polyribosomes (polysomes). The basic structure of eukaryotic ribosomes is similar to that of bacterial ribosomes, described in chapter 3. Both are composed of large and small subunits of ribonucleoprotein (see figure 4.5). By contrast, however, the eukaryotic ribosome (except in the mitochondrion) is the larger 80S variety that is a combination of 60S and 40S subunits. As in the bacteria, eukaryotic ribosomes are the staging areas for protein synthesis.

The Cytoskeleton

The cytoplasm of a eukaryotic cell is criss-crossed by a flexible framework of molecules called the cytoskeleton. This framework appears to have several functions, such as anchoring organelles, moving RNA and vesicles, and permitting shape changes and movement in some cells **(figure 4.10)**. The three main types of cytoskeletal elements are *actin filaments, intermediate filaments,* and *microtubules.* **Actin filaments** are long thin protein strands about 7 nm in diameter. They are found throughout the cell but are most highly concentrated just inside the cell membrane. Actin filaments are responsible for cellular movements such as contraction, crawling, pinching during cell division, and formation of cellular extensions. **Microtubules** are long, hollow tubes that maintain the shape of

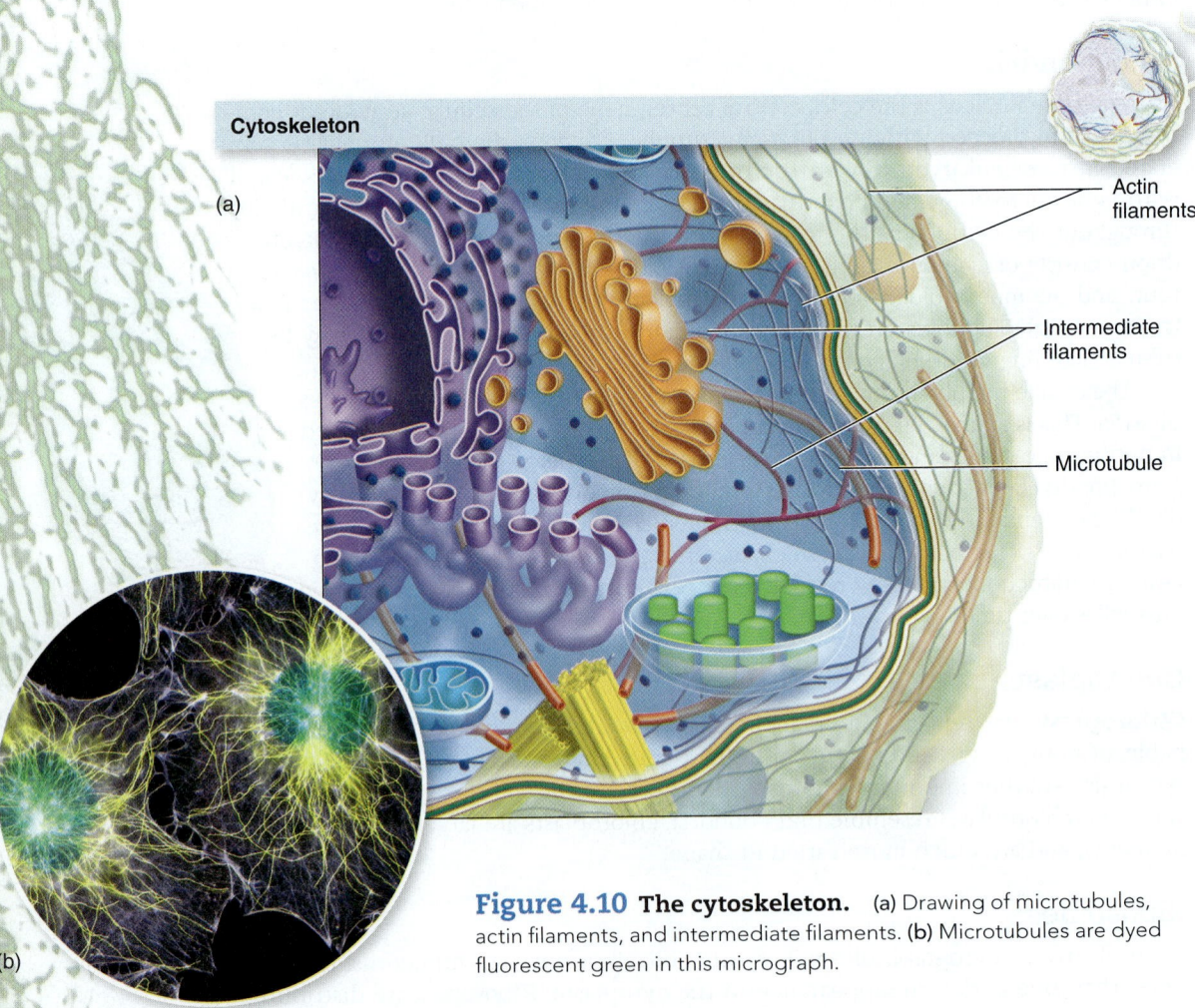

Cytoskeleton

(a)

Actin filaments

Intermediate filaments

Microtubule

(b)

Figure 4.10 The cytoskeleton. (a) Drawing of microtubules, actin filaments, and intermediate filaments. (b) Microtubules are dyed fluorescent green in this micrograph.

eukaryotic cells without walls and transport substances from one part of a cell to another. The spindle fibers that play an essential role in mitosis are actually microtubules that attach to chromosomes and separate them into daughter cells. As indicated earlier, microtubules are also responsible for the movement of cilia and flagella. **Intermediate filaments** are ropelike structures that are about 10 nm in diameter. (Their name comes from their intermediate size, between actin filaments and microtubules.) Their main role is in structural reinforcement to the cell and to organelles. For example, they support the structure of the nuclear envelope.

Table 4.2 summarizes the differences between eukaryotic and bacterial and archaeal cells. Viruses (discussed in chapter 5) are included as well.

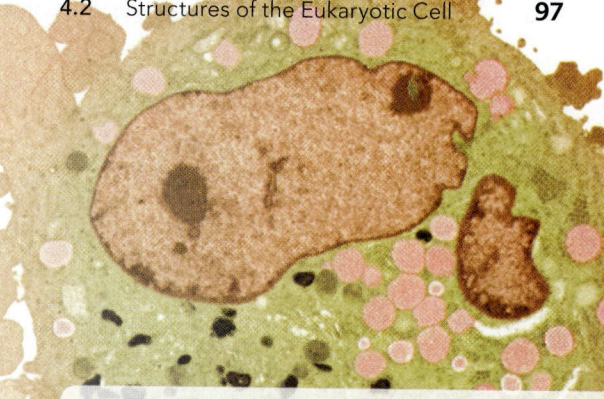

This human epithelial cell has turned cancerous. It has an irregular surface and an enlarged nucleus.

Table 4.2 A General Comparison of Cells and Viruses*

Function or Structure	Characteristic	Bacterial/Archaeal Cells	Eukaryotic Cells	Viruses**
Genetics	Nucleic acids	+	+	+
	Chromosomes	+	+	−
	True nucleus	−	+	−
	Nuclear envelope	−	+	−
Reproduction	Mitosis	−	+	−
	Production of sex cells	+/−	+	−
	Binary fission	+	+	−
Biosynthesis	Independent	+	+	−
	Golgi apparatus	−	+	−
	Endoplasmic reticulum	−	+	−
	Ribosomes	+***	+	−
Respiration	Mitochondria	−	+	−
Photosynthesis	Pigments	+/−	+/−	−
	Chloroplasts	−	+/−	−
Motility/locomotor structures	Flagella	+/−***	+/−	−
	Cilia	−	+/−	−
Shape/protection	Membrane	+	+	+/− (called "envelope" when present)
	Cell wall	+***	+/−	− (have capsids instead)
	Glycocalyx	+/−	+/−	−
Complexity of function		+	+	+/−
Size (in general)		0.5-3 μm****	2-100 μm	<0.2 μm

°+ Means most members of the group exhibit this characteristic; − means most lack it; +/− means some members have it and some do not.

**Viruses cannot participate in metabolic or genetic activity outside their host cells.

***The bacterial/archaeal type is functionally similar to the eukaryotic but structurally unique.

****Much smaller and much larger bacteria exist.

4.3 The Fungi

The kingdom Fungi is large and filled with a great variety and complexity of forms. For practical purposes, the approximately 50,000 species of fungi identified to date can be divided into two groups: the *macroscopic fungi* (mushrooms, puffballs, gill fungi) and the *microscopic fungi* (molds, yeasts). Although the majority of fungi are either unicellular or colonial, a few complex forms such as mushrooms and puffballs are considered multicellular. Cells of the microscopic fungi exist in two basic forms: yeasts and hyphae. A yeast cell is distinguished by its round to oval shape and by its mode of asexual reproduction. It grows swellings on its surface called buds, which then become separate cells. **Hyphae** (hy′-fee) are long, threadlike cells found in the bodies of

Aspergillus hyphae and spores

filamentous fungi, or molds **(figure 4.11).** Some species form a **pseudohypha,** a chain of yeast cells formed when buds remain attached in a row **(figure 4.12).** Because of its manner of formation, it is not a true hypha like that of molds. While some fungal cells exist only in a yeast form and others occur primarily as hyphae, a few are classified as **dimorphic.** This means they can take either form, depending on growth conditions, such as changing temperature. This variability in growth form is particularly characteristic of some fungi that cause human disease.

Humans are generally quite resistant to fungal infection, yet nearly 300 species of fungi can still cause human disease. This number is only expected to rise with the continuing climate changes seen worldwide today. The Centers for Disease Control and Prevention currently monitor three types of fungal disease in humans: community-acquired infections caused by environmental pathogens in the general population; hospital-associated infections caused by fungal pathogens in clinical settings; and opportunistic infections caused by pathogens infecting already weakened individuals **(table 4.3).**

Mycoses (fungal infections) vary in the way the pathogen enters the body and the degree of tissue involvement they display. Even so-called harmless species found in the air and dust around us may be able to cause infections, especially in individuals who already have AIDS, cancer, or diabetes. Fungi can cause other dangerous medical conditions without establishing an actual infection. Fungal cell walls give off chemical substances that can trigger allergies. The toxins produced by poisonous mushrooms can induce neurological disturbances and even death. The mold *Aspergillus flavus* synthesizes a potentially lethal poison called aflatoxin. The consumption of grain contaminated with this mold has led to increased cases of liver cancer in developing nations.

Fungi pose an ever-present economic hindrance to the agricultural industry. A number of species are pathogenic to field plants such as corn and grain, which reduces crop production but can also cause disease in domestic animals consuming the contaminated feed crops. Fungi also rot fresh produce during shipping and storage. It has been estimated that as much as 40% of the yearly fruit crop is consumed not by humans but by fungi. On the beneficial side, however, fungi play an essential role in decomposing organic matter and returning essential minerals to the soil. They form stable associations with plant roots (mycorrhizae) that increase the ability of the roots to absorb water and nutrients. Industry has tapped the biochemical potential of fungi to produce large quantities of antibiotics, alcohol, organic acids, and vitamins. Some fungi are eaten or used to impart flavorings to food. The yeast *Saccharomyces* produces the alcohol in beer and wine and the gas that causes bread to rise. Blue cheese, soy sauce, and cured meats derive their unique flavors from the actions of fungi.

Fungal Nutrition

All fungi are **heterotrophic.** They acquire nutrients from a wide variety of organic materials called **substrates.** Most fungi are **saprobes,** meaning that they obtain these substrates from the remnants of dead plants and animals in soil or aquatic habitats. Fungi can also be **parasites** on the bodies of living animals or plants, although very few fungi absolutely require a living host. In general, the fungus penetrates the substrate and secretes enzymes that reduce it to small molecules that can be absorbed by the cells. Fungi have enzymes for digesting an incredible array of substances, including feathers, hair, cellulose, petroleum products, wood, and rubber. Fungi are often found in nutritionally poor or adverse environments. Various fungi thrive in substrates with high salt or sugar content, at relatively high temperatures, and even in snow and glaciers.

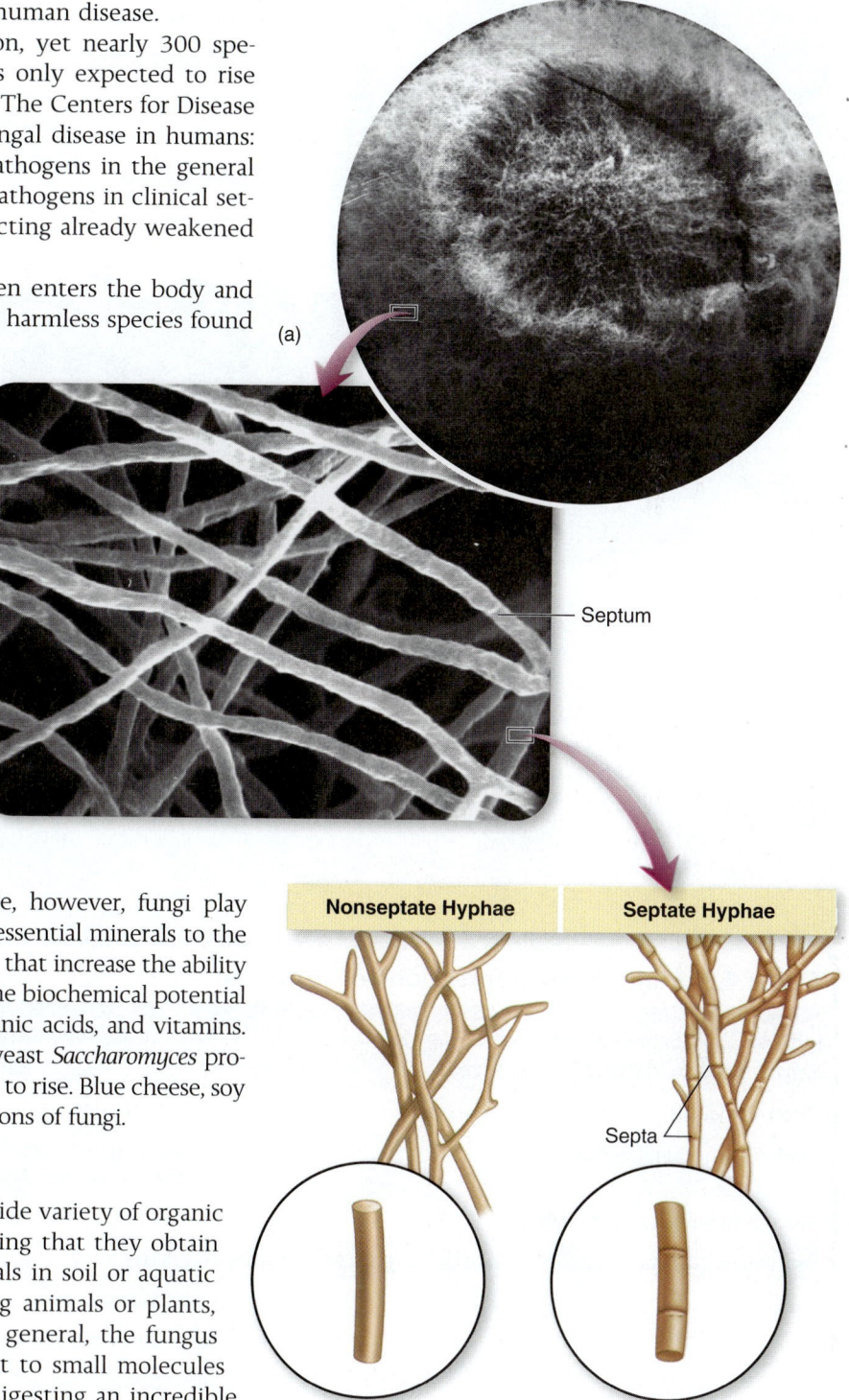

Figure 4.11 *Diplodia maydis,* **a pathogenic fungus of corn plants.** (a) Scanning electron micrograph of a single colony showing its filamentous texture (24×). (b) Close-up of hyphal structure (1,200×). (c) Basic structural types of hyphae.

(a)

Septum

(b)

Nonseptate Hyphae	Septate Hyphae

Septa

as in *Rhizopus*   as in *Penicillium*

(c)

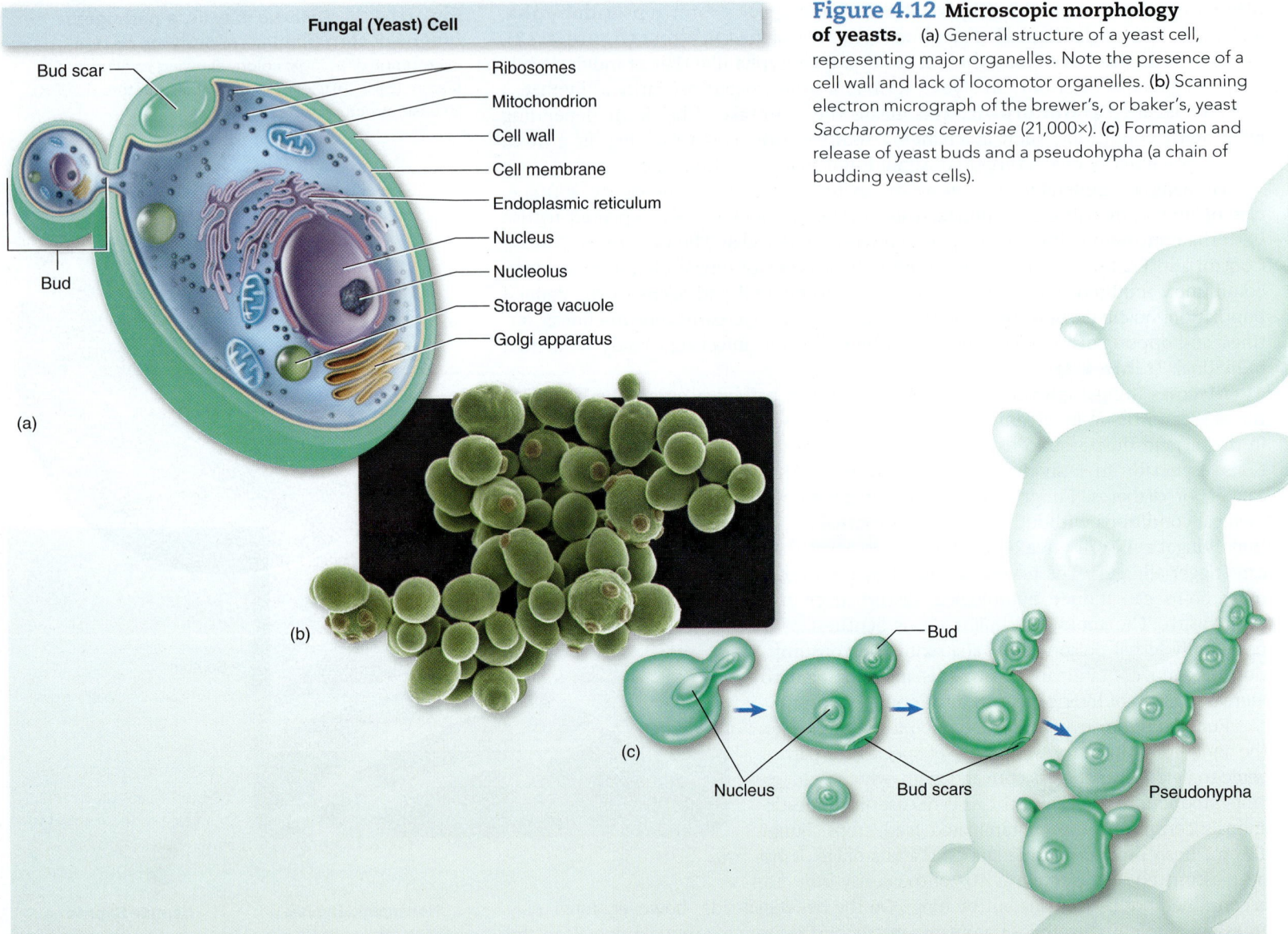

Fungal (Yeast) Cell

Bud scar
Ribosomes
Mitochondrion
Cell wall
Cell membrane
Endoplasmic reticulum
Nucleus
Nucleolus
Storage vacuole
Golgi apparatus
Bud
(a)

(b)

(c)
Bud
Nucleus
Bud scars
Pseudohypha

Figure 4.12 Microscopic morphology of yeasts. (a) General structure of a yeast cell, representing major organelles. Note the presence of a cell wall and lack of locomotor organelles. (b) Scanning electron micrograph of the brewer's, or baker's, yeast *Saccharomyces cerevisiae* (21,000×). (c) Formation and release of yeast buds and a pseudohypha (a chain of budding yeast cells).

Table 4.3 Major Fungal Infections of Humans

Degree of Tissue Involvement and Area Affected	Name of Infection	Name of Causative Fungus
Superficial (not deeply invasive)		
Outer epidermis	Tinea versicolor	*Malassezia furfur*
Epidermis, hair, and dermis	Dermatophytosis, also called tinea or ringworm of the scalp, body, feet (athlete's foot), toenails	*Microsporum, Trichophyton,* and *Epidermophyton*
Mucous membranes, skin, nails	Candidiasis, or yeast infection	*Candida albicans*
Systemic (deep; organism enters lungs; can invade other organs)		
Lung	Coccidioidomycosis (San Joaquin Valley fever)	*Coccidioides immitis*
	North American blastomycosis (Chicago disease)	*Blastomyces dermatitidis*
	Histoplasmosis (Ohio Valley fever)	*Histoplasma capsulatum*
	Cryptococcosis	*Cryptococcus neoformans*
Lung, skin	Paracoccidioidomycosis (South American blastomycosis)	*Paracoccidioides brasiliensis*

Morphology of Fungi

The cells of most microscopic fungi grow in loose associations or colonies. The colonies of yeasts are much like those of bacteria in that they have a soft, uniform texture and appearance. The colonies of filamentous fungi are noted for the striking cottony, hairy, or velvety textures that arise from their microscopic organization and morphology. The woven, intertwining mass of hyphae that makes up the body or colony of a mold is called a **mycelium.**

Although hyphae contain the usual eukaryotic organelles, they also have some unique organizational features. In most fungi, the hyphae are septate, or divided into segments by cross walls called **septa** (singular, *septum;* see figure 4.11*c*). The nature of the septa varies from solid partitions with no communication between the compartments to partial walls with small pores that allow the flow of organelles and nutrients between adjacent compartments. Nonseptate hyphae consist of one long, continuous cell *not* divided into individual compartments by cross walls. With this construction, the cytoplasm and organelles move freely from one region to another, and each hyphal element can have several nuclei.

Hyphae can also be classified according to their particular function. Vegetative hyphae (mycelia) are responsible for the visible mass of growth that appears on the surface of a substrate and penetrates it to digest and absorb nutrients. During the development of a fungal colony, the vegetative hyphae give rise to structures called reproductive, or fertile, hyphae, which branch off a vegetative mycelium. These hyphae are responsible for the production of fungal reproductive bodies called **spores.**

Reproductive Strategies and Spore Formation

Fungi have many complex and successful reproductive strategies. Most can propagate by the simple outward growth of existing hyphae or by fragmentation, in which a separated piece of mycelium can generate a whole new colony. But the primary reproductive mode of fungi involves the production of various types of spores. (Do not confuse fungal spores with the more resistant, nonreproductive bacterial spores.) Spores help the fungus disperse throughout the environment. Because of their compactness and relatively light weight, spores are dispersed widely through the environment by air, water, and living things. Upon encountering a favorable substrate, a spore will germinate and produce a new fungus colony in a very short time.

Fungal spores are explicitly responsible for multiplication. The fungi have such a wide variety of different spores that they are largely classified and identified by their spores and spore-forming structures, but we won't cover this information. Instead, we will focus on the most general subdivision, which is based on the way the spores arise. Asexual spores are the products of mitotic division of a single parent cell, and sexual spores are formed through a process involving the fusing of two parental nuclei followed by meiosis. An important consequence of meiosis and sexual reproduction is a resulting increase in genetic variation among spores.

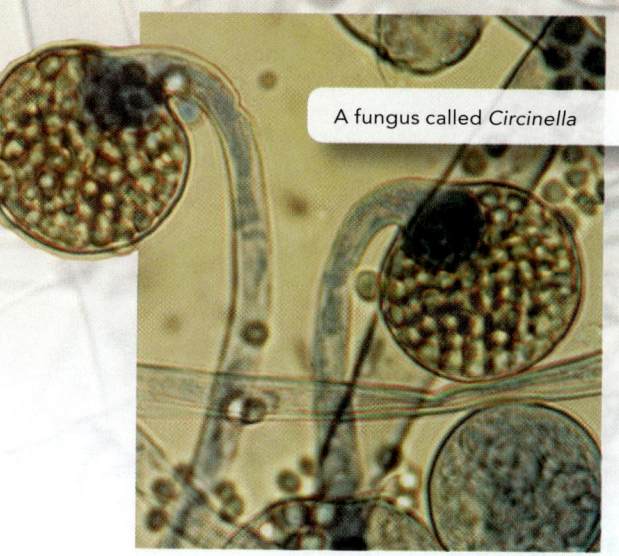

A fungus called *Circinella*

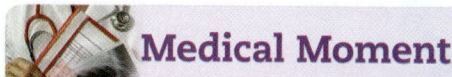

Medical Moment

Vaginal Candidiasis

Almost every woman will experience a vaginal yeast infection, caused by an overgrowth of *Candida albicans*, at some time in her life. Although uncomfortable due to irritation, itching, and vaginal discharge, the infection is easily treatable with antifungal medication in the form of creams, oral medications, or vaginal suppositories.

The female reproductive system is quite amazing, and a very delicate balance is maintained within this environment. A small amount of *C. albicans* is nearly always present within the vagina, but its growth is limited by the acidic pH of the vaginal canal. Interestingly, the acid-producing bacteria also living within the vagina help to tightly control this pH level. When something disrupts the vaginal pH, *C. albicans* takes the opportunity to proliferate, leading to an overgrowth of this microbe and in many cases to a subsequent yeast infection.

But what causes disruption of the normal vaginal pH? Although pregnancy, diabetes, obesity, and monthly hormonal changes can cause pH levels to fluctuate, by far the most common cause that leads to the development of a yeast infection is antibiotic therapy. This is due to the fact that along with the pathogen it is really trying to target, the drug kills the protective bacteria (normal biota) of the vagina that help to keep yeast in check. So, Ladies, remember to eat your yogurt during and after antibiotic therapy, as the live active bacterial cultures within this fermented product may actually help to boost the levels of these beneficial bacteria within the vaginal canal.

Source: www.cdc.gov/fungal/Candidiasis/genital. Accessed August 5, 2013.

Figure 4.13 Types of asexual mold spores.
(a) Sporangiospores: (1) *Absidia,* (2) *Syncephalastrum.*
(b) Conidial variations: (1) arthrospores (e.g.,
Coccidioides), (2) chlamydospores and blastospores (e.g.,
Candida albicans), (3) phialospores (e.g., *Aspergillus*),
(4) macroconidia and microconidia (e.g., *Microsporum*),
and (5) porospores (e.g., *Alternaria*).

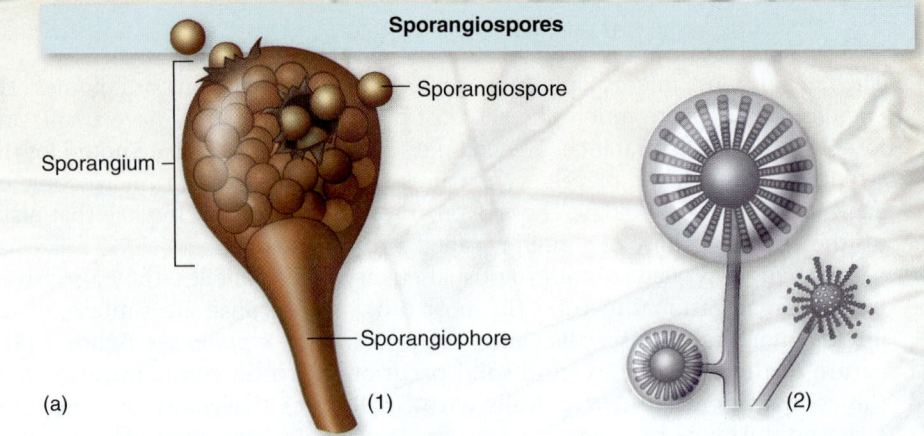

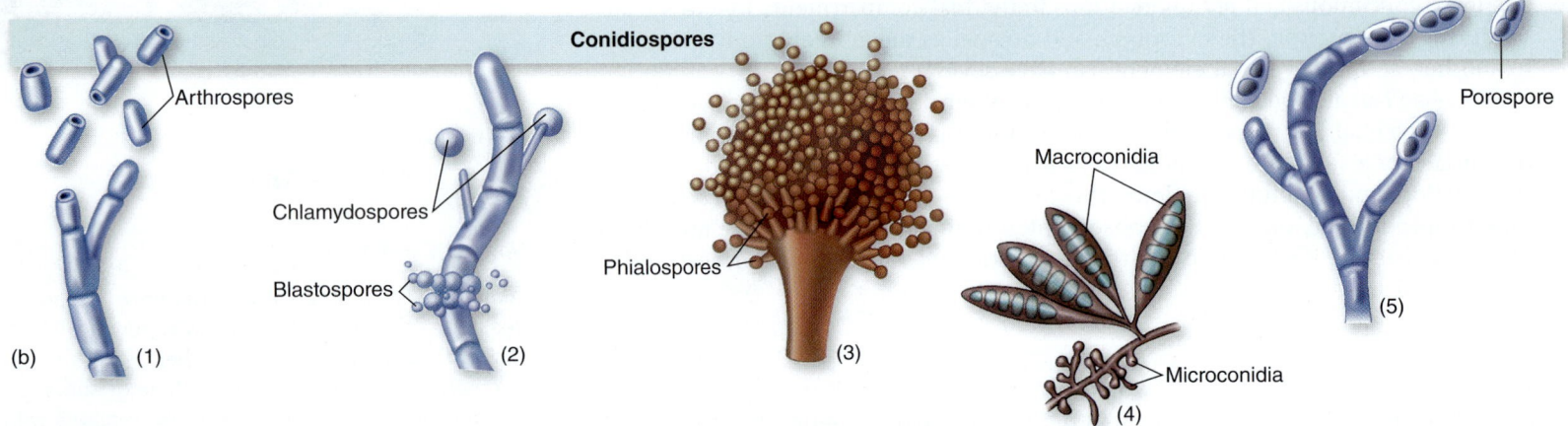

Medical Moment

Toxoplasmosis and Pregnancy

It's not a myth, nor just an excuse to avoid an unpleasant chore: Women who are pregnant (or who are trying to conceive) should avoid changing their cat's litter box. Women who come in contact with *Toxoplasma gondii* for the first time while pregnant risk passing the pathogen on to their unborn baby, which can result in serious problems including blindness and mental disability.

Cats, especially those who are allowed outdoors, can come in contact with the parasite when they kill and eat small mammals (i.e., mice) and birds that are infected with the protozoan. Although many cats may never show any signs of illness, they can still shed *T. gondii* in their stool for several weeks after becoming infected.

However, there is no need to give up the family feline. Instead, a woman who is pregnant should ask someone else in her household to take on litter box duties for the duration of the pregnancy. If there is no one else to perform this chore, wearing gloves and washing hands thoroughly after changing the litter should sufficiently protect the expecting mother and her baby-to-be. Daily removal of feces from the litter and frequent changing of the litter help as well, because *T. gondii* does not become infectious until 1 to 5 days after defecation.

Source: www.cdc.gov/parasites/toxoplasmosis. Accessed August 5, 2013.

Asexual Spore Formation

There are two types of asexual spores, **sporangiospores** and **conidiospores,** also called conidia **(figure 4.13):**

1. Sporangiospores **(figure 4.13a)** are formed by successive cleavages within a saclike head called a **sporangium,** which is attached to a stalk, the sporangiophore.
2. Conidiospores, or **conidia,** are free spores not enclosed by a spore-bearing sac. They develop either by the pinching off of the tip of a special fertile hypha or by the segmentation of a pre-existing vegetative hypha. There are many different forms of conidia, illustrated in **figure 4.13b.**

Sexual Spore Formation

Fungi can propagate themselves successfully with their millions of asexual spores. That being the case, why is the production of sexual spores necessary? The answer lies in important variations that occur when fungi of different genetic makeup combine their genetic material. Just as in plants and animals, this mixing of DNA from two parents creates offspring with combinations of genes different from that of either parent. The offspring from such a union can have slight variations in form and function that are potentially advantageous in the adaptation and survival of their species.

The majority of fungi produce sexual spores at some point. The nature of this process varies from the simple fusion of fertile hyphae of two different strains to a complex union of differentiated male and female structures and the development of special fruiting structures. It may be a surprise to discover that the fleshy part of a mushroom is actually a fruiting body designed to protect and help disseminate its sexual spores.

4.4 The Protozoa

Although their name comes from the Greek for "first animals," protozoa are far from being simple, primitive organisms. The protozoa constitute a very large group (about 12,000 species) of creatures that, although single-celled, have startling properties when it comes to movement, feeding, and behavior. Although most members of this group are harmless, free-living inhabitants of water and soil, a few species are parasites collectively responsible for hundreds of millions of infections of humans each year. Remember that the term *protozoan* is more of a convenience than an accurate taxonomic designation. As we next describe them, you will see why protozoa are categorized together. As it turns out, it is because of their similar physical characteristics rather than their genetic relatedness.

Protozoan Form and Function

Most protozoan cells are single cells containing all the major eukaryotic organelles. Their organelles can be highly specialized for feeding, reproduction, and locomotion. The cytoplasm is usually divided into a clear outer layer called the **ectoplasm** and a granular inner region called the endoplasm. Ectoplasm is involved in locomotion, feeding, and protection. Endoplasm houses the nucleus, mitochondria, and food and contractile vacuoles. Some protozoa even have organelles that work somewhat like a primitive nervous system to coordinate movement. Protozoa can move through fluids by means of *pseudopods* ("false feet"), flagella, or cilia. Because protozoa lack a cell wall, they have a certain amount of flexibility. Their outer boundary is a cell membrane that regulates the movement of food, wastes, and secretions. Cell shape can remain constant (as in most ciliates) or can change constantly (as in amoebas). Certain amoebas encase themselves in hard shells made of calcium carbonate. The size of most protozoan cells falls within the range of 3 to 300 µm. Some notable exceptions are giant amoebas and ciliates that are large enough (3 to 4 mm in length) to be seen swimming in pond water.

Nutritional and Habitat Range

The protozoa we will be interested in are typically heterotrophic and usually require their food in a complex organic form. Free-living species scavenge dead plant or animal debris and even graze on live bacteria and algae. Some species have special feeding structures, such as oral grooves, which carry food particles into a passageway or gullet that packages the captured food into vacuoles for digestion. Some protozoa absorb food directly through the cell membrane. Parasitic species live on the fluids of their host, such as plasma and digestive juices, or they can actively feed on tissues.

Euglena is a type of flagellated protozoan that can be found in a drop of pond water or even aquarium water.

An amoeba exhibiting pseudopod formation

Although protozoa have adapted to a wide range of habitats, their main limiting factor is the availability of moisture. Their predominant habitats are fresh and marine water, soil, plants, and animals. Even extremes in temperature and pH are not a barrier to their existence; hardy species are found in hot springs, ice, and habitats with low or high pH. Many protozoa can convert to a resistant, dormant stage called a cyst.

Life Cycles and Reproduction

Most protozoa can be recognized in their motile feeding stage called the **trophozoite.** This is a stage that requires ample food and moisture to remain active. A large number of species are also capable of entering into a dormant, resting stage called a **cyst** when conditions in the environment become unfavorable for growth and feeding. During *encystment*, the trophozoite cell rounds up into a sphere, and its ectoplasm secretes a tough, thick cuticle around the cell membrane **(figure 4.14).** Because cysts are more resistant than ordinary cells to heat, drying, and chemicals, they can survive adverse periods. They can be dispersed by air currents and may even be an important factor in the spread of diseases such as amoebic dysentery. If provided with moisture and nutrients, a cyst breaks open and releases the active trophozoite.

The life cycles of protozoans vary from simple to complex. Several protozoan groups exist only in the trophozoite state. Many alternate between a trophozoite and a cyst stage, depending on the conditions of the habitat. The life cycle of a parasitic protozoan dictates its mode of transmission to other hosts. For example, the flagellate *Trichomonas vaginalis* causes a common sexually transmitted infection. Because it does not form cysts, it is more delicate and must be transmitted by intimate contact between sexual partners. In contrast, intestinal pathogens such as *Entamoeba histolytica* and *Giardia lamblia* form cysts and are readily transmitted in contaminated water and foods.

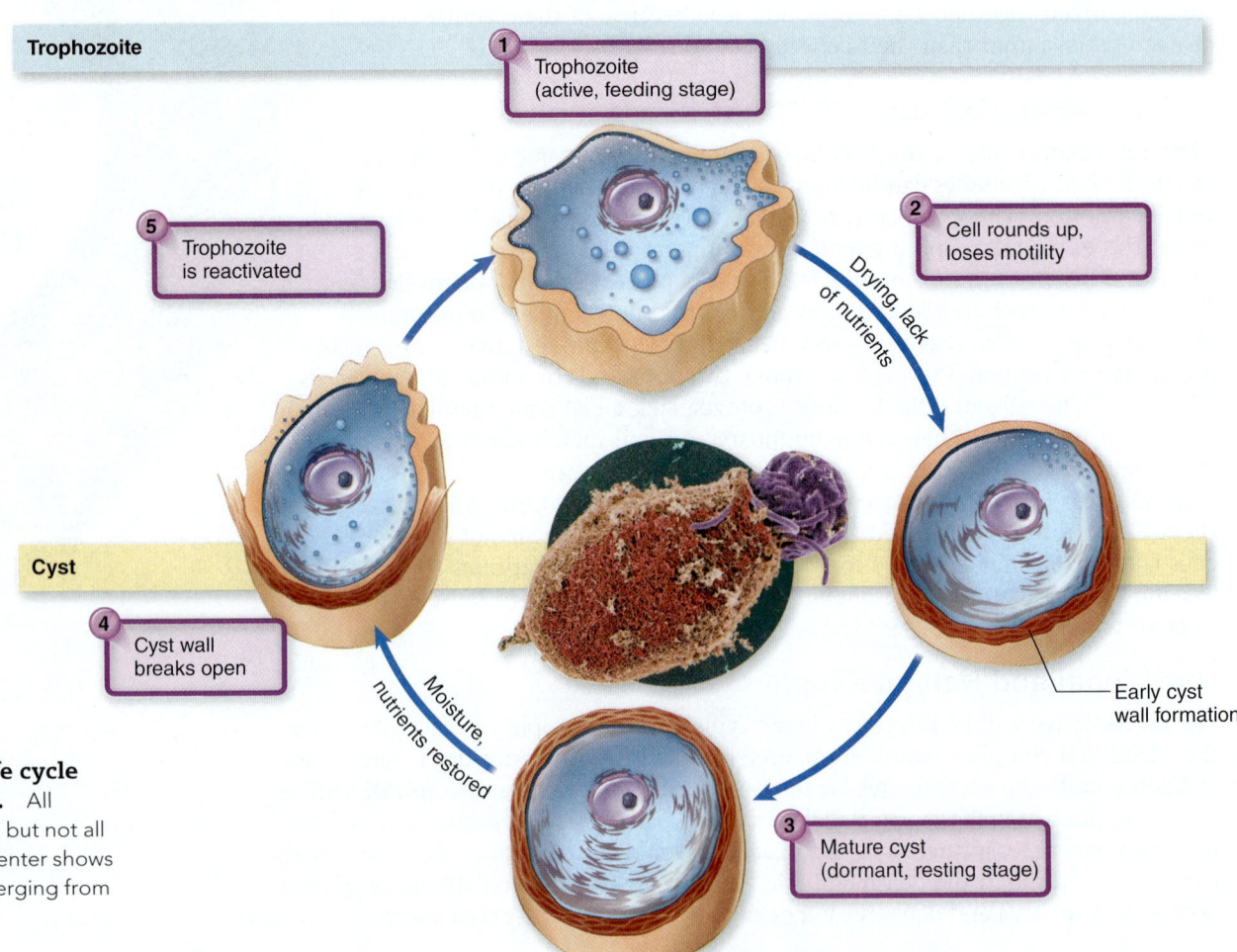

Trophozoite

1 Trophozoite (active, feeding stage)

5 Trophozoite is reactivated

2 Cell rounds up, loses motility

Drying, lack of nutrients

Cyst

4 Cyst wall breaks open

Moisture, nutrients restored

Early cyst wall formation

3 Mature cyst (dormant, resting stage)

Figure 4.14 The general life cycle exhibited by many protozoa. All protozoa have a trophozoite form, but not all produce cysts. The photo in the center shows a *Giardia* trophozoite (purple) emerging from its cyst form (orange).

All protozoa reproduce by relatively simple, asexual methods, usually mitotic cell division. Several parasitic species, including the causative agents of malaria and toxoplasmosis, reproduce asexually by multiple rounds of division inside a host cell. Sexual reproduction also occurs during the life cycle of most protozoa. Ciliates participate in **conjugation,** a form of genetic exchange in which two cells fuse temporarily and exchange micronuclei. This process of sexual recombination yields new and different genetic combinations that can be advantageous in evolution.

Classification of Selected Medically Important Protozoa

As has been stated, taxonomists have problems classifying protozoa. They are very diverse and frequently frustrate attempts to generalize or place them in neat groupings. We will use a common and simple system of four groups, based on their method of motility: Sarcodina (pseudopods), Ciliophora (cilia), Mastigophora (flagella), and Sporozoa (gliding motility) **(table 4.4).**

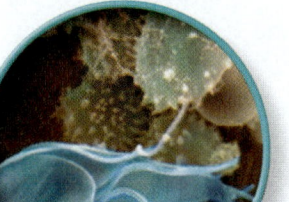
Giardia lamblia

Table 4.4 Major Pathogenic Protozoa

Protozoan	Disease	Reservoir/Source	
Amoeboid Protozoa (Sarcodina)			
Entamoeba histolytica	Amoebiasis (intestinal and other symptoms)	Humans, water, and food	
Naegleria, Acanthamoeba	Brain infection	Water	
Ciliated Protozoa (Ciliophora)			
Balantidium coli (photo to right is of *Stentor*)	Balantidiosis (intestinal and other symptoms)	Pigs, cattle	
Flagellated Protozoa (Mastigophora)			
Giardia lamblia	Giardiasis (intestinal distress)	Animals, water, and food	
Trichomonas vaginalis	Trichomoniasis (vaginal symptoms)	Human	
Trypanosoma brucei, T. cruzi	Trypanosomiasis (intestinal distress and widespread organ damage)	Animals, vector-borne	
Leishmania donovani, L. tropica, L. brasiliensis	Leishmaniasis (either skin lesions or widespread involvement of internal organs)	Animals, vector-borne	
Apicomplexan Protozoa (Sporozoa)			
Plasmodium vivax, P. falciparum, P. malariae	Malaria (cardiovascular and other symptoms)	Human, vector-borne	
Toxoplasma gondii	Toxoplasmosis (flulike illness or silent infection)	Animals, vector-borne	
Cryptosporidium	Cryptosporidiosis (intestinal and other symptoms)	Water, food	
Cyclospora cayetanensis	Cyclosporiasis (intestinal and other symptoms)	Water, fresh produce	

4.5 The Helminths

Tapeworms, flukes, and roundworms are collectively called *helminths*, from the Greek word meaning "worm." Adult specimens are usually large enough to be seen with the naked eye, and they range from the longest tapeworms, measuring up to about 25 m in length, to roundworms less than 1 mm in length. Helminths are included in the study of microbes mainly due to their infective abilities and the production of microscopic eggs and larvae.

On the basis of body type, the two major groups of parasitic helminths are the flatworms (phylum Platyhelminthes) and the roundworms (phylum Aschelminthes, also called **nematodes**). Flatworms have a very thin, often segmented body plan **(figure 4.15),** and roundworms have an elongated, cylindrical, unsegmented

Figure 4.15 Parasitic flatworms.   (a) A cestode (tapeworm), showing the scolex; long, tapelike body; and magnified views of immature and mature proglottids (body segments). The photo shows an actual tapeworm. (b) The structure of a trematode (liver fluke). Note the suckers that attach to host tissue and the dominance of reproductive and digestive organs.

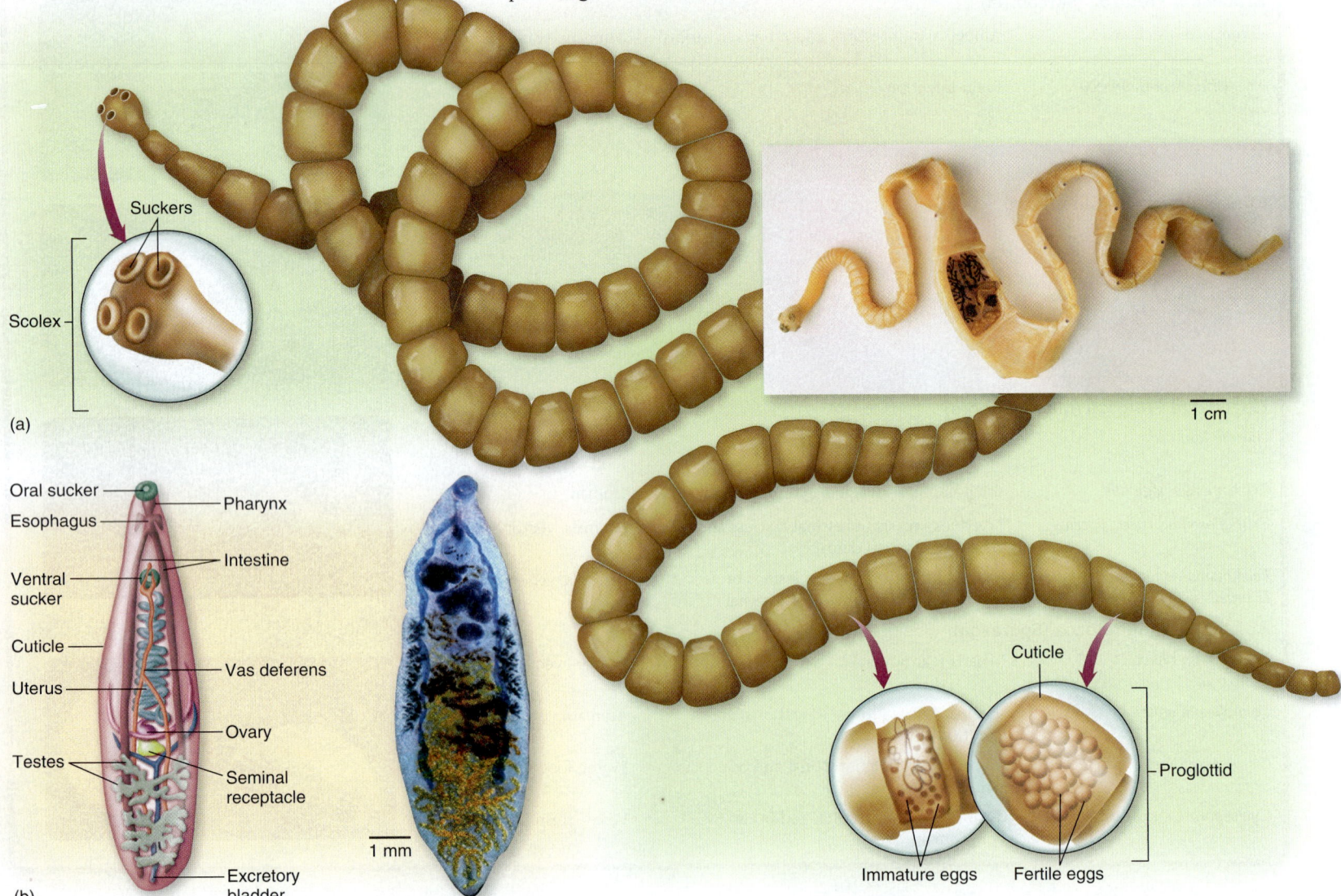

body **(figure 4.16).** The flatworm group is subdivided into the **cestodes,** or tapeworms, named for their long, ribbonlike arrangement, and the **trematodes,** or flukes, characterized by flat, ovoid bodies. Not all flatworms and roundworms are parasites by nature; many live free in soil and water. Because most disease-causing helminths spend part of their lives in the gastrointestinal tract, they are discussed in chapter 20.

NCLEX® PREP

3. Which group of helminths is subdivided into cestodes and trematodes?
   - a. *roundworms*
   - b. *tapeworms*
   - c. *flatworms*
   - d. *pinworms*

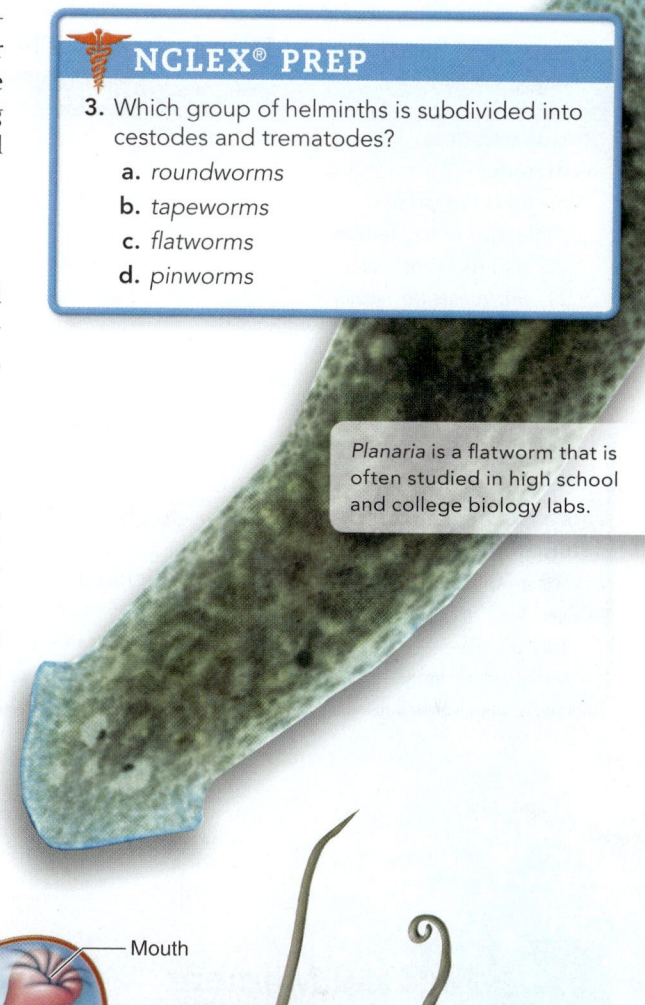

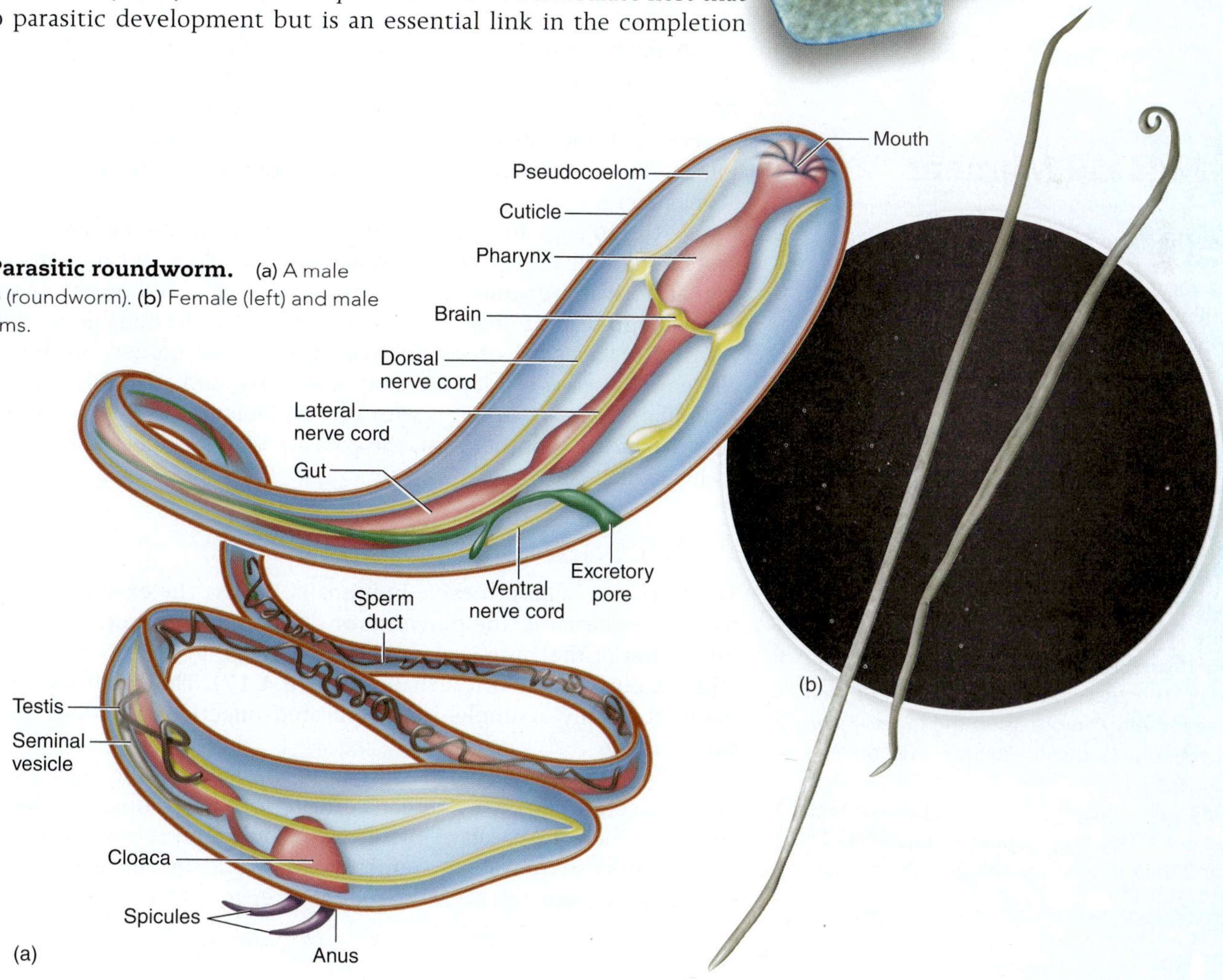

Planaria is a flatworm that is often studied in high school and college biology labs.

General Worm Morphology

All helminths are multicellular animals equipped to some degree with organs and organ systems. In parasitic helminths, the most developed organs are those of the reproductive tract, with some degree of reduction in the digestive, excretory, nervous, and muscular systems.

Life Cycles and Reproduction

The complete life cycle of helminths includes the fertilized egg (embryo), larval, and adult stages. In the majority of helminths, adults derive nutrients and reproduce sexually in a host's body. In nematodes, the sexes are separate and usually different in appearance; in trematodes, the sexes can be either separate or **hermaphroditic,** meaning that male and female sex organs are in the same worm; cestodes are generally hermaphroditic. Helminths must complete the life cycle by transmitting an infective form, usually an egg or larva, to the body of another host, either of the same or a different species. The host in which larval development occurs is the intermediate (secondary) host, and adulthood and mating occur in the **definitive (final) host.** A *transport host* is an intermediate host that experiences no parasitic development but is an essential link in the completion of the cycle.

Mouth
Pseudocoelom
Cuticle
Pharynx
Brain
Dorsal nerve cord
Lateral nerve cord
Gut
Ventral nerve cord
Excretory pore
Sperm duct
Testis
Seminal vesicle
Cloaca
Spicules
Anus
(a)
(b)

Figure 4.16 Parasitic roundworm. (a) A male *Ascaris* nematode (roundworm). (b) Female (left) and male (right) *Ascaris* worms.

Table 4.5 Examples of Helminths and Their Modes of Transmission

Common Name	Disease or Worm	Life Cycle Requirement	Spread to Humans By
Roundworms			
Nematodes			
Intestinal Nematodes			Ingestion
Infective in egg (embryo) stage			
Ascaris lumbricoides	Ascariasis	Humans	Fecal pollution of soil with eggs
Enterobius vermicularis	Pinworm	Humans	Close contact
Infective in larval stage			
Trichinella spiralis	Trichina worm	Pigs, wild mammals	Consumption of meat containing larvae
Tissue Nematodes			Burrowing of larva into tissue
Onchocerca volvulus	River blindness	Humans, black flies	Fly bite
Dracunculus medinensis	Guinea worm	Humans and *Cyclops* (an aquatic invertebrate)	Ingestion of water containing *Cyclops*
Flatworms			
Trematodes			
Schistosoma japonicum	Blood fluke	Humans and snails	Skin penetration of larval stage
Cestodes			
Taenia solium	Pork tapeworm	Humans, swine	Consumption of undercooked or raw pork
Diphyllobothrium latum	Fish tapeworm	Humans, fish	Consumption of undercooked or raw fish

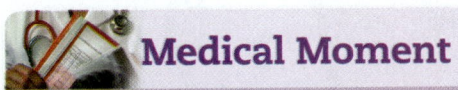

Medical Moment

Pinworms: The Tape Test

The roundworm *Enterobius vermicularis* can cause a significant amount of discomfort for its tiny size! The anal itching associated with these tiny helminths can result in nights of lost sleep. The severe itching is due to the fact that the female of the species emerges at night to lay her eggs around the anus.

One of the easiest tests to perform for detection of *Enterobius vermicularis* is the tape test. To perform the tape test, a piece of cellophane tape is applied to the skin surrounding the anus. The tape is then removed and placed, sticky side down, on a glass slide, which can then be examined under the microscope for the presence of the roundworm. The test should be performed first thing in the morning before using the toilet, showering, or bathing, as these activities may remove the "evidence."

These tiny helminths can actually be seen with the naked eye. They may appear as tiny white filaments, and their movement will give them away!

In general, sources for human infection are contaminated food, soil, and water or infected animals; routes of infection are by oral intake or penetration of unbroken skin. Humans are the definitive hosts for many of the parasites listed in **table 4.5.** In about half the diseases, they are also the sole biological reservoir. In other cases, animals or insect vectors serve as reservoirs or are required to complete worm development.

Fertilized eggs are usually released to the environment and are provided with a protective shell and extra food to aid their development into larvae. Even so, most eggs and larvae are vulnerable to heat, cold, drying, and predators and are destroyed or unable to reach a new host. To counteract this formidable mortality rate, certain worms have adapted a reproductive capacity that borders on the incredible: A single female *Ascaris* can lay 200,000 eggs a day, and a large female can contain over 25 million eggs at varying stages of development! If only a tiny number of these eggs makes it to another host, the parasite will have been successful in completing its life cycle.

A Helminth Cycle: The Pinworm

To illustrate a helminth cycle in humans, we use the example of a roundworm, *Enterobius vermicularis*, the pinworm or seatworm. This worm causes a very common infestation of the large intestine. Worms range from 2 to 12 mm long and have a tapered, curved cylindrical shape **(figure 4.17).** The condition they cause, enterobiasis, is usually a simple, uncomplicated infection that does not spread beyond the intestine.

A cycle starts when a person swallows microscopic eggs picked up from another infected person by direct contact or by touching articles that person has touched. The eggs hatch in the intestine and then release larvae that mature into adult worms within about 1 month. Male and female worms mate, and the female migrates out to the anus to deposit eggs, which cause intense itchiness that

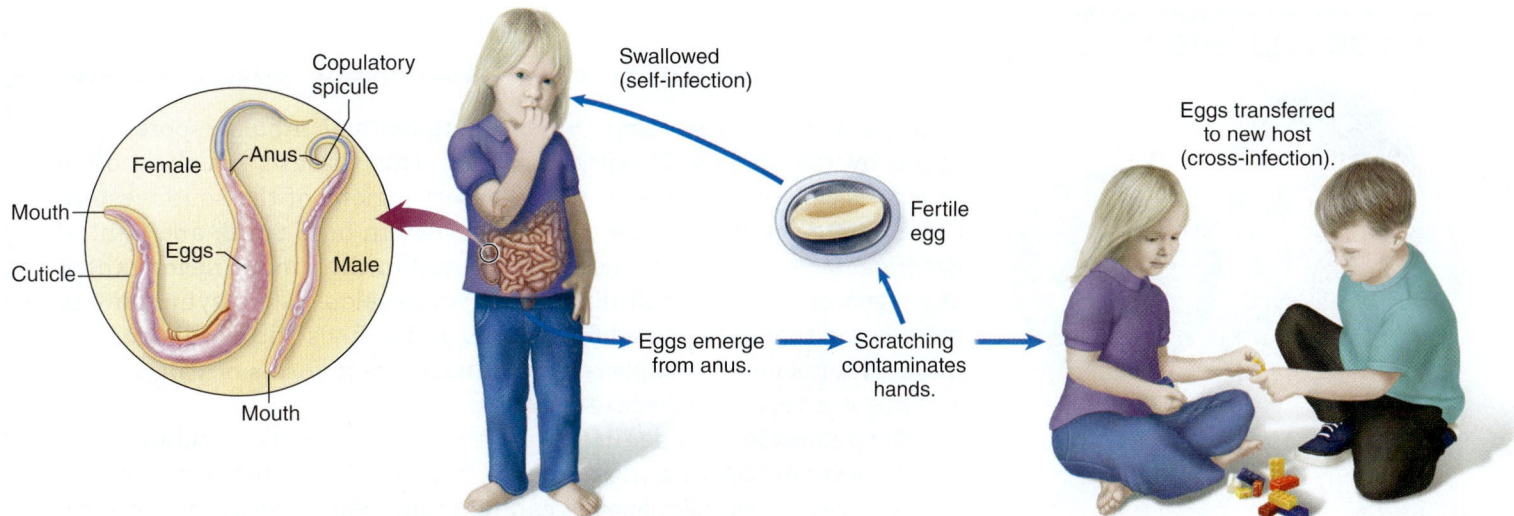

Figure 4.17 The life cycle of the pinworm, a roundworm. Eggs are the infective stage and are transmitted by contaminated hands. Children frequently reinfect themselves and also pass the parasite on to others.

is relieved by scratching. Herein lies a significant means of dispersal: Scratching contaminates the fingers, which, in turn, transfer eggs to bedclothes and other inanimate objects. This person becomes a host and a source of eggs, and can spread them to others in addition to reinfecting him- or herself. Enterobiasis occurs most often among families and in other close living situations. Its distribution is worldwide among all socioeconomic groups, but it seems to attack younger people more frequently than older ones.

Distribution and Importance of Parasitic Worms

About 50 species of helminths parasitize humans. They are distributed in all areas of the world that support human life. Some worms are restricted to a given geographic region, and many have a higher incidence in tropical areas. This knowledge must be tempered with the realization that jet-age travel, along with human migration, is gradually changing the patterns of helminth infections, especially of those species that do not require alternate hosts or special climatic conditions for development. The yearly estimate of worldwide cases numbers in the billions, and these are not confined to developing countries. A conservative estimate places 50 million helminth infections in North America alone.

4.5 LEARNING OUTCOMES—Assess Your Progress

20. List the two major groups of helminths, and provide examples representing each body type.

21. Summarize the stages of a typical helminth life cycle.

CASE FILE WRAP-UP

Coccidioidomycosis develops when an individual inhales spores produced by the fungus *Coccidioides immitis*. This disease, which is often called *Valley fever*, is endemic to the desert regions of the southwestern United States, and cases are commonly seen in both South and Central America. Most people who become exposed to the fungus never exhibit any signs or symptoms of illness. Others develop flulike symptoms or pneumonia that may persist for months. Individuals with weakened immune systems tend to experience the most severe forms of Valley fever, and in some cases the disease is fatal.

The patient in the opening case file was a farm worker and likely inhaled spores while working outdoors, because California is one of the states where *Coccidioides immitis* can be found. His chest X ray revealed the lung nodules typical of this disease (which may be mistaken for cancer), and the patient's blood tested positive for the fungus. Microscopic evidence of the fungus was identified in his sputum sample. Once the patient was started on an antifungal medication, he began to gradually improve and fully recovered from the disease. Due to misinterpretation of the initial symptoms, the medication the patient was initially started on was an antibiotic drug. Although effective against bacteria, antibiotics will not lead to targeted destruction of this fungus.

Deadly Bite: Malaria

Malaria is currently one of the world's most dreaded diseases, affecting poor tropical countries most heavily. Around 220 million cases of malaria are diagnosed each year worldwide, with one child dying every minute of every day from this preventable disease. Almost all of the nearly 1,500 cases of malaria that occur in the United States yearly are acquired from foreign travel.

Malaria is caused by parasitic protozoa of the genus *Plasmodium*. Malaria is contracted via the bite of the female *Anopheles* mosquito. The female mosquito passes the infectious parasites through its saliva to the host on which it is feeding. They enter the bloodstream of the host and eventually enter red blood cells, feeding on proteins and hemoglobin contained within the cells. The protozoa multiply rapidly within the red blood cells; these cells eventually burst and release toxins into the bloodstream, causing the symptoms associated with malaria.

Symptoms occur in a cyclical nature every 48 or 72 hours (depending on species) and include joint and muscle aches, headache, malaise, and fever and chills. Anemia and jaundice may also occur due to destruction and hemolysis of red blood cells. Kidney failure, liver disease, coma, and death may occur with untreated disease, particularly with *P. falciparum*, the most deadly of the *Plasmodium* species.

Nearly 90% of all malaria deaths today occur in rural sub-Saharan Africa. This is due to the biology of African mosquito species, the climate in this area, and low immunity to the pathogen in certain populations. Interestingly, the sickle-cell trait seen in many individuals within this region confers some immunity against malaria, particularly the *P. falciparum* form.

Quinine has been the standard of treatment for many years, but resistance to the drug is a growing problem. Artesunate (artemesinin) is a newer drug in the fight against malaria, but its use in the United States is very limited and elsewhere in the world resistance to the drug has already appeared. Whether treatment is given on an outpatient or inpatient basis is dependent on the condition of the patient and the severity of the infection and availability of suitable medical facilities. Most people who receive adequate treatment will recover completely. Being simultaneously infected with HIV may complicate recovery from malaria, creating terrible problems in central and southern African countries where HIV/AIDS prevalence is especially high.

Travelers should keep their skin covered as much as possible to prevent contracting malaria when traveling to countries where malaria is known to be endemic. *Anopheles* mosquitos are most active between the hours of dusk and dawn. Mosquito nets can diminish the risk of being bitten. Quinine and other drugs are sometimes recommended prophylactically for people traveling to countries where malaria is prevalent.

Anopheles albimanus is one species of mosquito that carries malaria. This one is found in Central America.

Chapter Summary

4.1 The History of Eukaryotes

- Eukaryotes are cells with a nucleus and organelles compartmentalized by membranes. They, like bacteria, originated from a primitive cell referred to as the *last common ancestor*. Eukaryotic cell structure enabled eukaryotes to diversify from single cells into a huge variety of complex multicellular forms.

4.2 Structures of the Eukaryotic Cell

- The cell structures common to most eukaryotes are the cell membrane, nucleus, vacuoles, mitochondria, endoplasmic reticulum, Golgi apparatus, and a cytoskeleton. Cell walls, chloroplasts, and locomotor organs are present in some eukaryote groups.
- Microscopic eukaryotes use locomotor organs such as flagella or cilia for moving themselves or their food.
- The glycocalyx is the outermost boundary of most eukaryotic cells. Its functions are protection, adherence, and reception of chemical signals from the environment or from other organisms. The glycocalyx is supported by either a cell wall or a cell membrane.
- The cell membrane of eukaryotes is similar in function to that of bacteria, but it differs in composition, possessing sterols as additional stabilizing agents.
- Fungi have a cell wall that is composed of glycans and chitin or cellulose.
- The genome of eukaryotes is located in the nucleus, a spherical structure surrounded by a double membrane. The nucleus contains the nucleolus, the site of ribosome synthesis. DNA is organized into chromosomes in the nucleus.
- The endoplasmic reticulum (ER) is an internal network of membranous passageways extending throughout the cell.
- The Golgi apparatus is a packaging center that receives materials from the ER and then forms vesicles around them for storage or for transport to the cell membrane for secretion.
- The mitochondria generate energy in the form of ATP to be used in numerous cellular activities.

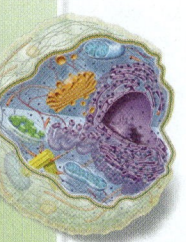

- Chloroplasts, membranous packets found in plants and algae, are used in photosynthesis.
- Ribosomes are the sites for protein synthesis present in both eukaryotes and bacteria.
- The cytoskeleton maintains the shape of cells and produces movement of cytoplasm within the cell, movement of chromosomes at cell division, and, in some groups, movement of the cell as a unit.

4.3 The Fungi

- The fungi are a nonphotosynthetic species with cell walls. They are either saprobes or parasites and may be unicellular, colonial, or multicellular.
- There are two categories of fungi that cause human disease: the primary pathogens, which infect healthy persons, and the opportunistic pathogens, which cause disease only in compromised hosts.
- All fungi are heterotrophic.
- Fungi use both asexual and sexual reproductive strategies.
- Fungi can produce asexual spores called sporangiospores and conidiospores.
- Fungal sexual spores enable the organisms to incorporate variations in form and function.

4.4 The Protozoa

- Protozoa are mostly unicellular eukaryotes that lack specialized tissues.
- Disease-causing protozoa are typically heterotrophic and usually display some form of locomotion. All have a trophozoite form, and many produce a resistant stage, or cyst.

4.5 The Helminths

- The kingdom Animalia has only one group that contains members that are studied in microbiology. These are the helminths, or worms. Parasitic members include flatworms and roundworms that are able to invade and reproduce in human tissues.

Multiple-Choice Questions   Bloom's Levels 1 and 2: Remember and Understand

Select the correct answer from the answers provided.

1. Pseudopodia are used for motility by
   - a. helminths.
   - b. protozoa.
   - c. fungi.
   - d. algae.

2. The Golgi apparatus
   - a. receives vesicles from the mitochondrion.
   - b. packages products into transitional vesicles.
   - c. modifies proteins.
   - d. synthesizes proteins and sterols.

3. Yeasts are __ fungi, and molds are __ fungi.
   - a. macroscopic; microscopic
   - b. unicellular; filamentous
   - c. motile; nonmotile
   - d. water; terrestrial

4. Fungi produce which structures for reproduction and multiplication?
   - a. endospores
   - b. cysts
   - c. spores
   - d. eggs

5. The protozoa that cause malaria belong to the following group:
   - a. Sarcodina
   - b. Ciliophora
   - c. Mastigophora
   - d. Sporozoa

6. Parasitic helminths reach adulthood and mate within a
   - a. intermediate host.
   - b. temporary host.
   - c. definitive host.
   - d. multiplicative host.

7. Mitochondria likely originated from
   a. archaea.
   b. invaginations of the cell membrane.
   c. bacteria.
   d. chloroplasts.

8. Helminths
   a. are unicellular animals.
   b. are all parasitic.
   c. can be hermaphroditic.
   d. reproduce by means of spores.

Critical Thinking   Bloom's Levels 3, 4, and 5: Apply, Analyze, and Evaluate

Critical thinking is the ability to reason and solve problems using facts and concepts. These questions can be approached from a number of angles and, in most cases, they do not have a single correct answer.

1. Thinking of a eukaryotic cell as a little city, write a short essay describing the role that each organelle plays and its importance to the overall function of the city itself. Be sure to analyze what would happen to the city if the organelle stopped working.

2. a. Analysis of your patient's specimen just revealed the presence of cells containing a cell wall. Discuss the type of microbe(s) that could be causing the infection at hand.

   b. Further analysis showed the presence of a nucleus and 80S ribosomes. Discuss how this changes your hypothesis regarding the type of microbe(s) causing the patient's infection.

   c. Final biochemical analysis of the microbe showed that its cell wall was composed of chitin. Provide a final assessment of the pathogen's microbial type based upon all of this evidence.

3. a. *Candida albicans* normally causes superficial infections of the mucous membranes that are typically resolved with topical drugs. However, infection within the lungs of immunocompromised patients can develop if *C. albicans* is aspirated and treatment is usually just as toxic to the patient than to the pathogen itself. Based upon your understanding of fungal biology, explain how this pathogen is able to cause a more severe form of disease within this area of the body and why such an infection is difficult to treat.

   b. Conduct additional research, and summarize how fungi played a role in Alexander Fleming's discovery of the first antibiotic compound. Draw a sketch of his famous Petri dish, and describe what aspect of the plate revealed to him that the fungal contaminant was actually providing to him a very valuable biochemical clue.

4. a. Compare and contrast the methods used by bacteria and protozoa to stay alive even when conditions within the environment are unfavorable to the growth and survival of the organism.

   b. Based upon the presenting symptoms, you hypothesize that your patient is suffering from amoebiasis. However, when the laboratory analysis of the pathogen returns, you find out that the microbe is ciliated. Explain how this information may change your thoughts on the patient's disease diagnosis.

5. a. Many tapeworms exhibit the ability to regenerate, or grow new proglottids, when their body segments are broken. Based upon your understanding of flatworm structure, describe what portion of the tapeworm's body would be best targeted by an antihelminthic drug in order to effectively treat the infection.

   b. Conduct additional research, and summarize the role of helminths in the burden of "neglected tropical diseases" worldwide today.

Visual Connections   Bloom's Level 5: Evaluate

This question connects previous images to a new concept.

1. **From chapter 2, figure 2.1.** Discuss how the techniques of the Five I's of microbiology would be completed if your patient's infection was due to a protozoan, a eukaryotic microbe.

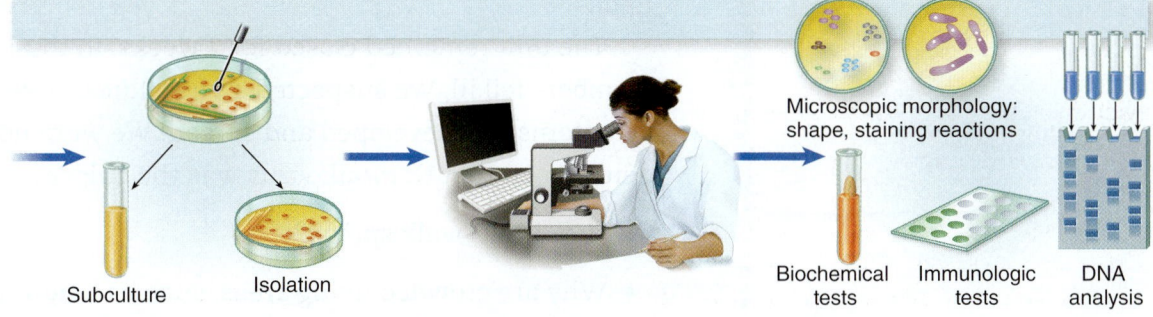

Subculture   Isolation   Microscopic morphology: shape, staining reactions   Biochemical tests   Immunologic tests   DNA analysis

|MICROBIOLOGY

www.mcgrawhillconnect.com

Enhance your study of this chapter with study tools and practice tests. Also ask your instructor about the resources available through Connect, including the media-rich eBook, interactive learning tools, and animations.

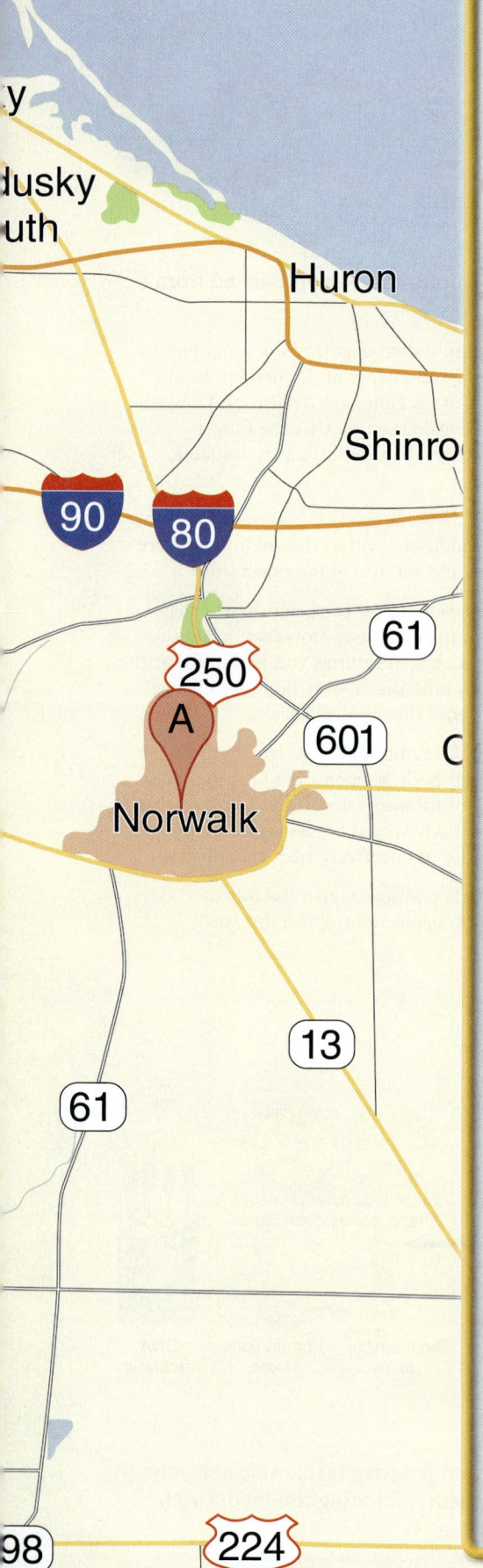

CASE FILE

The Domino Effect

I was an LPN working in a long-term care unit. My duties included supervising the certified nursing aides (CNAs), administering medications, changing dressings, and providing any other required treatments that fell within my scope of practice.

One day, one of the CNAs reported that two of the residents were experiencing vomiting and diarrhea. I checked on them and determined that their symptoms had started suddenly, without warning. One of the residents had a low-grade fever, and both patients were weak and experiencing abdominal discomfort.

I reported their condition to my supervising nurse, who saw the patients herself. She was concerned that their symptoms could be contagious. We placed the affected patients on contact isolation, and the infection-control nurse was consulted. She recommended that we obtain stool samples for culture and maintain isolation precautions until the stool sample reports came back. We notified the physician that one of the patients was becoming dehydrated due to vomiting and diarrhea. The patient was started on intravenous fluids.

When I returned to work the next day, three more residents had fallen ill with the same symptoms. At that time, we closed the unit to visitors and unnecessary personnel, suspecting an infectious cause, and worried that more residents could become ill. By the next day, two staff members and other residents had also become ill. It seemed we had an epidemic on our hands!

The unit remained closed for 2 weeks. In total, 11 residents and 4 staff members fell ill. We suspected a viral illness, based on how quickly symptoms had developed and spread. We were not surprised when it was determined that Norwalk virus was the culprit.

- How is Norwalk spread?

- Why are crowded living areas, such as long-term care units, prone to the spread of viral illnesses?

Case File Wrap-Up appears on page 136.

Viral Structure and Life Cycles

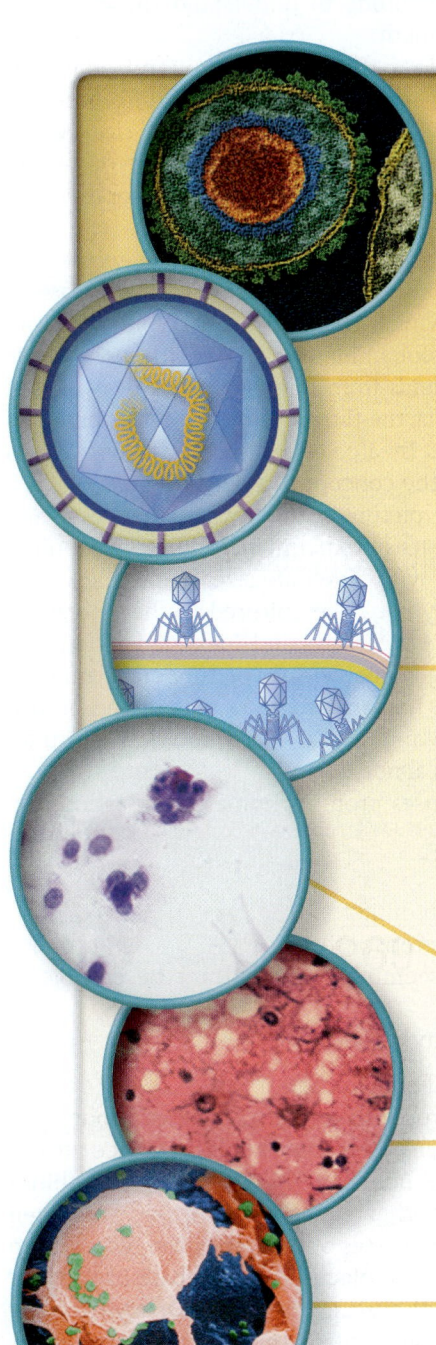

IN THIS CHAPTER...

5.1 The Position of Viruses in the Biological Spectrum

1. Explain what it means when viruses are described as *filterable*.
2. Identify better terms for viruses than *alive* or *dead*.

5.2 The General Structure of Viruses

3. Discuss the size of viruses relative to other microorganisms.
4. Describe the function and structure(s) of viral capsids.
5. Distinguish between enveloped and naked viruses.
6. Explain the importance of viral surface proteins, or spikes.
7. Diagram the possible configurations that nucleic acid viruses may possess.

5.3 Modes of Viral Multiplication

8. Diagram the five-step life cycle of animal viruses.
9. Define the term *cytopathic effect* and provide one example.
10. Discuss both persistent and transforming infections.
11. Provide thorough descriptions of both lysogenic and lytic bacteriophage infections.

5.4 Techniques in Cultivating and Identifying Animal Viruses

12. List the three principal purposes of cultivating viruses.
13. Describe three ways in which viruses are cultivated.

5.5 Other Noncellular Infectious Agents

14. Name two noncellular infectious agents besides viruses.

5.6 Viruses and Human Health

15. Analyze the relative importance of viruses in human infection and disease.
16. Discuss the primary reason that antiviral drugs are more difficult to design than antibacterial drugs.

This variegated tulip gets its beautiful colors from a viral infection.

5.1 The Position of Viruses in the Biological Spectrum

Viruses are a unique group of biological entities known to infect every type of cell, including bacteria, algae, fungi, protozoa, plants, and animals. Viruses are extremely abundant on our planet. For example, it is documented that seawater can contain 10 million viruses per milliliter, and human feces probably contain 100 times that many. It is estimated that the sum of viruses in the ocean represents 270 million metric tons of organic matter. We are just beginning to understand the impact of these huge numbers of viruses on our environment.

For many years, the cause of viral infections such as smallpox and polio was unknown, even though it was clear that the diseases were transmitted from person to person. The French scientist Louis Pasteur was certainly on the right track when he postulated that rabies was caused by a "living thing" smaller than bacteria, and in 1884 he was able to develop the first vaccine for rabies. Pasteur also proposed the term **virus** (L. "poison") to denote this special group of infectious agents.

The first substantial revelations about the unique characteristics of viruses occurred in the 1890s. First, D. Ivanovski and M. Beijerinck showed that a disease in tobacco was caused by a virus (tobacco mosaic virus). Then, Friedrich Loeffler and Paul Frosch discovered an animal virus that causes foot-and-mouth disease in cattle. These early researchers found that when infectious fluids from host organisms were passed through porcelain filters designed to trap bacteria, the filtrate remained infectious. This result proved that an infection could be caused by a cell-free fluid containing agents smaller than bacteria and thus first introduced the concept of a *filterable virus*.

Over the succeeding decades, a remarkable picture of the physical, chemical, and biological nature of viruses has taken form. Years of experimentation were required to show that viruses were noncellular particles with a definite size, shape, and chemical composition. Using special techniques, they could be cultured in the laboratory. Thanks to new genomic techniques, including DNA arrays and "next-generation" nucleic acid sequencing techniques, we are getting a much clearer picture of the number and variety of viruses on earth. Studies of the human virome (a part of the human microbiome) and of the world's oceans are showing us that there are vast multitudes of viruses that have roles we cannot even guess about.

The exceptional and curious nature of viruses prompts numerous questions, including the following:

1. Are they organisms; that is, are they alive?
2. What role did viruses play in the evolution of life?
3. What are their distinctive biological characteristics?
4. How can particles so small, simple, and seemingly insignificant be capable of causing disease and death?
5. What is the connection between viruses and cancer?

In this chapter, we address these questions and many others.

The unusual structure and behavior of viruses have led to debates about their connection to the rest of the microbial world. One viewpoint holds that since viruses are unable to multiply independently from the host cell, they are not living things but should be called infectious molecules. Another viewpoint proposes that even though viruses do not exhibit most of the life processes of cells, they can direct them and thus are certainly more than inert and lifeless molecules. This debate has greater philosophical than practical importance when discussing disease because viruses are agents of disease and must be dealt with through control, therapy, and prevention, whether we regard them as living or not. In keeping with their special position in the biological spectrum, it is best to describe viruses as either *active* or *inactive* (rather than alive or dead).

Recent discoveries suggest that viruses have been absolutely vital in forming cells and other life forms as they are today. By infecting other cells, and sometimes

> ### Table 5.1 Properties of Viruses
>
> - Are obligate intracellular parasites of bacteria, protozoa, fungi, algae, plants, and animals
> - Estimated 10^{31} virus particles on earth, approximately 10 times the number of prokaryotes
> - Are ubiquitous in nature and have had major impact on development of biological life
> - Are utramicroscopic in size, ranging from 20 nm up to 450 nm (diameter)
> - Are not cells; structure is very compact and economical
> - Do not independently fulfill the characteristics of life
> - Basic structure consists of protein shell (capsid) surrounding nucleic acid core
> - Nucleic acid can be either DNA or RNA, but not both
> - Nucleic acid can be double-stranded DNA, single-stranded DNA, single-stranded RNA, or double-stranded RNA
> - Molecules on virus surface impart high specificity for attachment to host cell
> - Multiply by taking control of host cell's genetic material and regulating the synthesis and assembly of new viruses
> - Lack enzymes for most metabolic processes
> - Lack machinery for synthesizing proteins

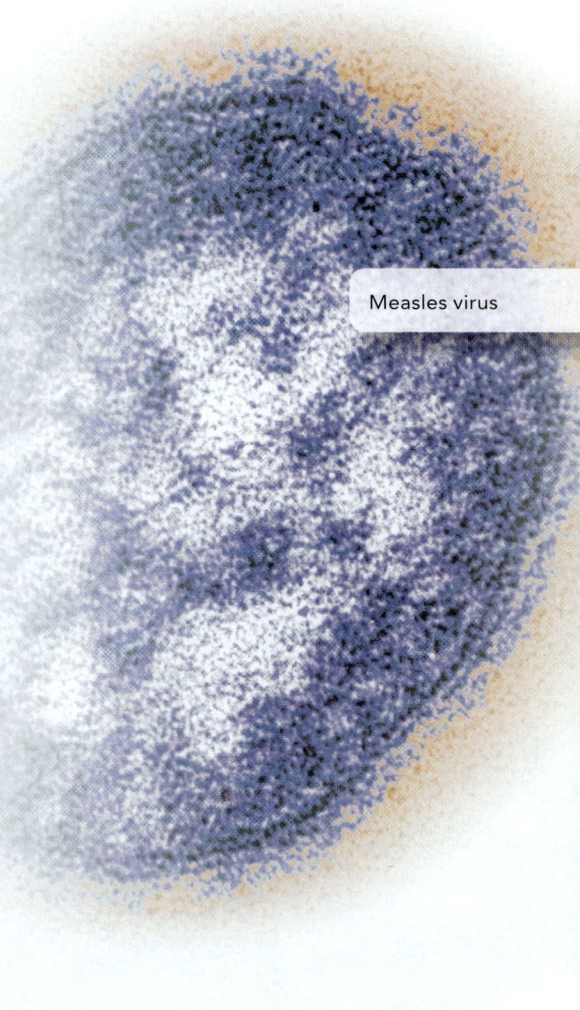

Measles virus

influencing their genetic makeup, they have shaped the way cells, tissues, bacteria, plants, and animals have evolved to their present forms. For example, scientists think that approximately 10% of the human genome consists of sequences that come from viruses that have incorporated their genetic material permanently into human DNA. Bacterial DNA also contains 10% to 20% viral sequences. As you learn more about how viruses work, you will see how this could happen.

Viruses are different from their host cells in size, structure, behavior, and physiology. They are a type of *obligate intracellular parasites* that cannot multiply unless they invade a specific host cell and instruct its genetic and metabolic machinery to make and release quantities of new viruses. Other unique properties of viruses are summarized in **table 5.1.**

How Viruses Are Classified and Named

In an informal and general way, we have already begun classifying viruses—as animal, plant, or bacterial viruses; enveloped or naked viruses; DNA or RNA viruses; and helical or icosahedral viruses. For many years, the animal viruses were classified mainly on the basis of their hosts and the kind of diseases they caused. Newer systems for naming viruses also take into account the actual nature of the virus particles themselves, with only partial emphasis on host and disease. The main criteria presently used to group viruses are structure, chemical composition, and similarities in genetic makeup.

In 2012, the International Committee on the Taxonomy of Viruses issued a report on the classification of viruses. The committee listed 7 orders with 23 families and 143 genera, plus another 71 families with 278 genera not yet assigned to any order. These numbers will only continue to grow as we discover more and more about life on earth. Previous to 2000, there had been only a single recognized order of viruses. Usually clinicians use the common vernacular names (for example, poliovirus and rabies virus), and this book will do the same.

> ### 5.1 LEARNING OUTCOMES—Assess Your Progress
> **1.** Explain what it means when viruses are described as *filterable.*
> **2.** Identify better terms for viruses than *alive* or *dead.*

5.2 The General Structure of Viruses

Size Range

As a group, viruses represent the smallest infectious agents (with some unusual exceptions to be discussed later in this chapter). They are dwarfed by their host cells: More than 2,000 bacterial viruses could fit into an average bacterial cell, and more than 50 million polioviruses could be accommodated by an average human cell. Animal viruses range in size from the small parvoviruses (around 20 nm [0.02 μm] in diameter) to the newly discovered pandoraviruses, that are about the same size as a common bacterial cell (1 μm) **(figure 5.1)**. Some cylindrical viruses are relatively long (800 nm [0.8 μm] in length) but so narrow in diameter (15 nm [0.015 μm]) that their visibility is still limited without the high magnification and resolution of an electron microscope. Figure 5.1 compares the sizes of several viruses with bacterial and eukaryotic cells and molecules.

Viral architecture is most readily observed through special stains in combination with electron microscopy **(figure 5.2)**.

Viral Components: Capsids, Nucleic Acids, and Envelopes

It is important to realize that viruses bear no real resemblance to cells, and that they lack any of the protein-synthesizing machinery found in even the simplest cells. Their molecular structure is composed of regular, repeating subunits that give

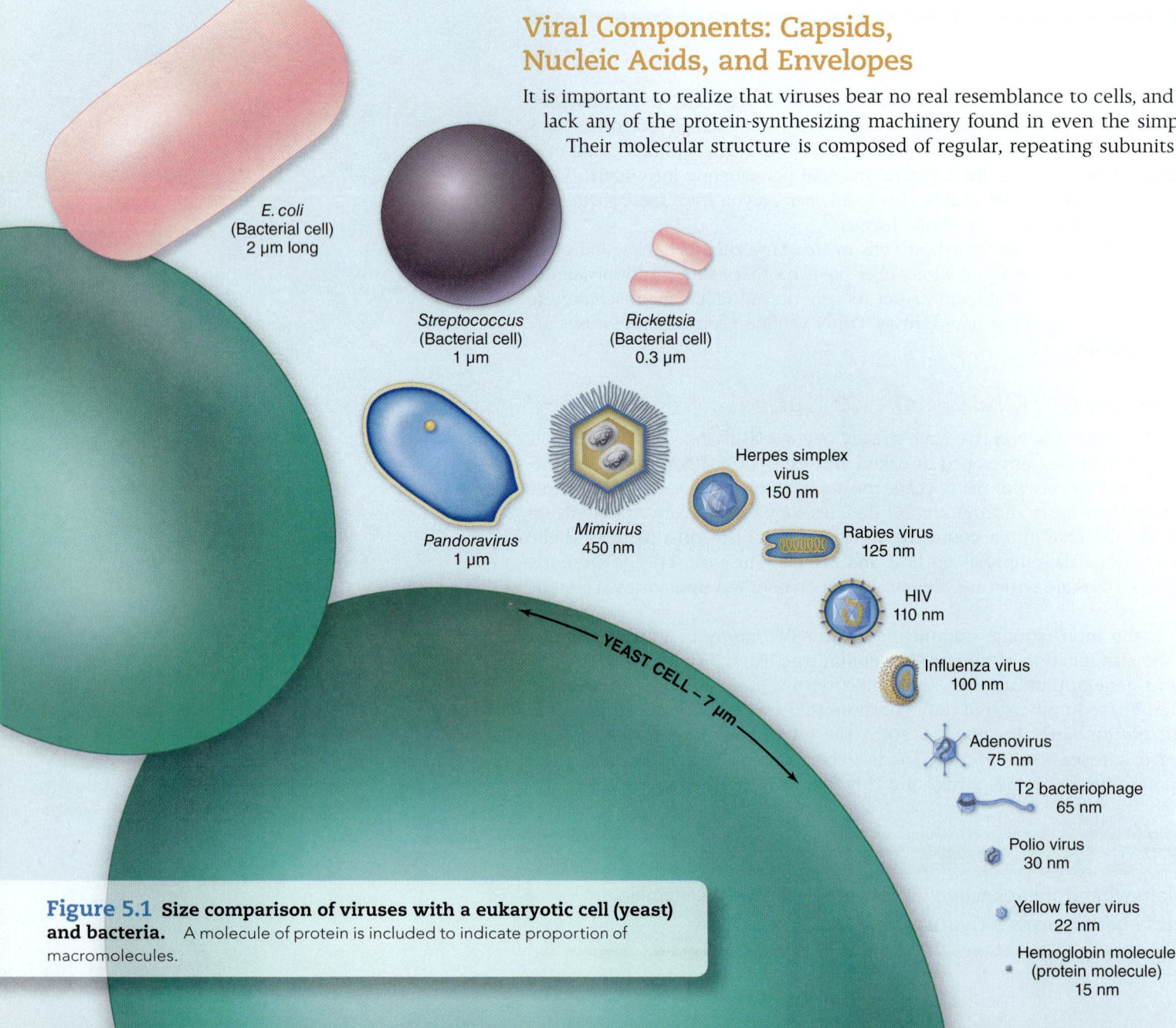

E. coli (Bacterial cell) 2 μm long

Streptococcus (Bacterial cell) 1 μm

Rickettsia (Bacterial cell) 0.3 μm

Pandoravirus 1 μm

Mimivirus 450 nm

Herpes simplex virus 150 nm

Rabies virus 125 nm

HIV 110 nm

Influenza virus 100 nm

Adenovirus 75 nm

T2 bacteriophage 65 nm

Polio virus 30 nm

Yellow fever virus 22 nm

Hemoglobin molecule (protein molecule) 15 nm

YEAST CELL – 7 μm

Figure 5.1 Size comparison of viruses with a eukaryotic cell (yeast) and bacteria. A molecule of protein is included to indicate proportion of macromolecules.

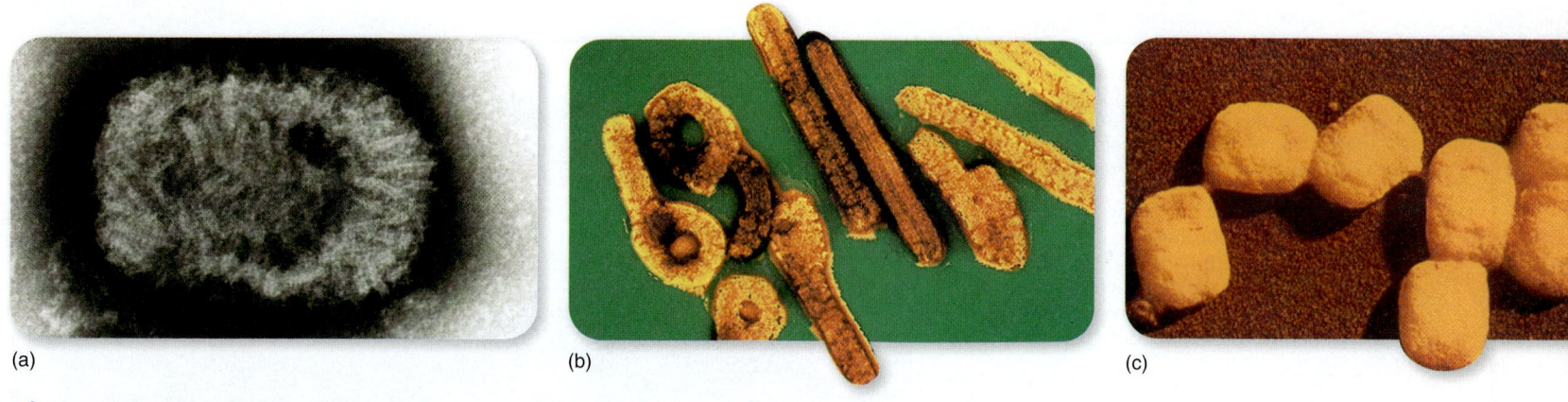

(a)   (b)   (c)

Figure 5.2 Methods of viewing viruses. (a) Negative staining of an orfvirus (a type of poxvirus), revealing details of its outer coat. (b) Positive stain of the Ebola virus. Note the textured capsid. (c) Shadowcasting image of a vaccinia virus.

rise to their crystalline appearance. The general plan of virus organization is the utmost in simplicity and compactness. Viruses contain only those parts needed to invade and control a host cell: an external coating and a core containing one or more nucleic acid strands of either DNA or RNA, and sometimes one or two enzymes. This pattern of organization can be represented with a flowchart:

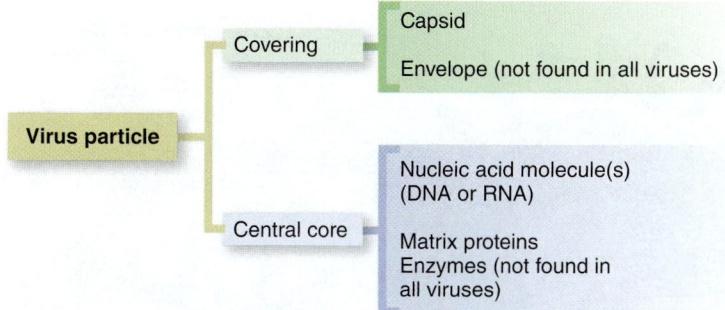

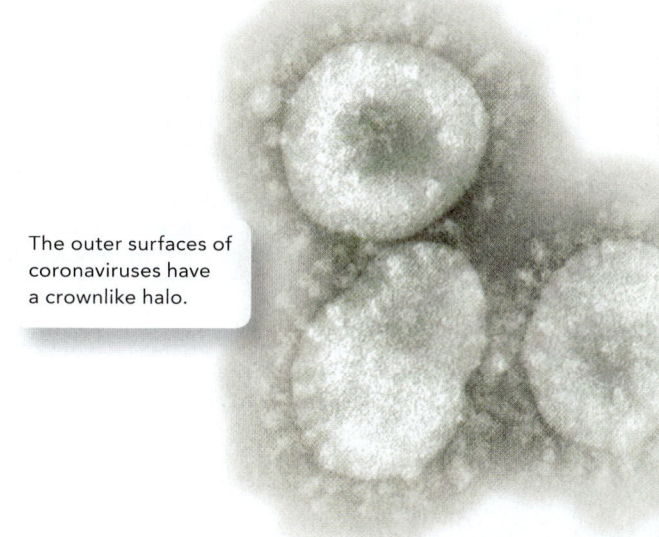

The outer surfaces of coronaviruses have a crownlike halo.

All viruses have a protein **capsid,** or shell, that surrounds the nucleic acid in the central core. Together the capsid and the nucleic acid are referred to as the **nucleocapsid (figure 5.3).** Members of 13 of the 20 families of animal viruses possess an additional covering external to the capsid called an envelope, which is usually a modified piece of the host's cell membrane **(figure 5.3b).** Viruses that consist of only a nucleocapsid are considered *naked viruses* **(figure 5.3a).** Both naked and enveloped viruses possess proteins on their outer surfaces that project from either the nucleocapsid or the envelope. They are the molecules that allow viruses to dock with their host cells and are called *spikes.* As we shall see later, the enveloped viruses differ from the naked viruses in the way that they enter and leave a host cell. A fully formed virus that is able to establish an infection in a host cell is often called a **virion.**

The Viral Capsid and Envelope

When a virus particle is magnified several hundred thousand times, the capsid appears as the most prominent geometric feature. In general, each capsid is constructed from identical subunits called **capsomers** that are constructed from protein molecules. The capsomers spontaneously self-assemble into the finished capsid. Depending on how the capsomers are shaped and arranged, this assembly results in two different types: helical and icosahedral. **Table 5.2** depicts the variations on these two themes.

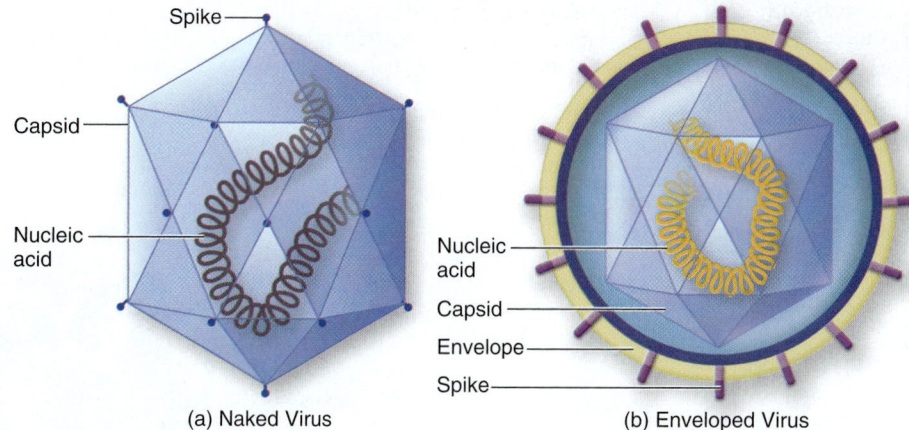

(a) Naked Virus   (b) Enveloped Virus

Figure 5.3 Generalized structure of viruses. (a) The simplest virus is a naked virus (nucleocapsid), consisting of a geometric capsid assembled around a nucleic acid strand or strands. (b) An enveloped virus is composed of a nucleocapsid surrounded by a flexible membrane called an envelope. The envelope usually has special receptor spikes inserted into it.

Table 5.2 Capsid Structure

Helical Capsids	The simpler **helical capsids** have rod-shaped capsomers that bond together to form a series of hollow discs resembling a bracelet. During the formation of the nucleocapsid, these discs link with other discs to form a continuous helix into which the nucleic acid strand is coiled.
Naked	The nucleocapsids of naked helical viruses are very rigid and tightly wound into a cylinder-shaped package. An example is the *tobacco mosaic virus*, which attacks tobacco leaves (right). 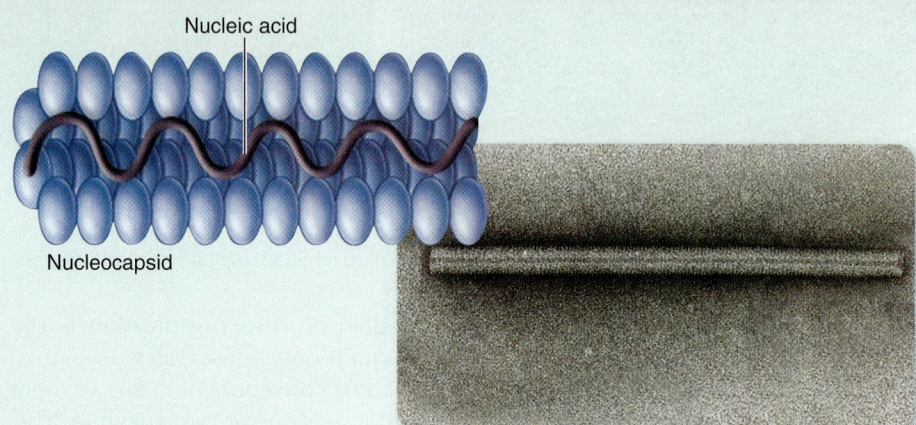
Enveloped	Enveloped helical nucleocapsids are more flexible and tend to be arranged as a looser helix within the envelope. This type of morphology is found in several enveloped human viruses, including influenza, measles, and rabies. 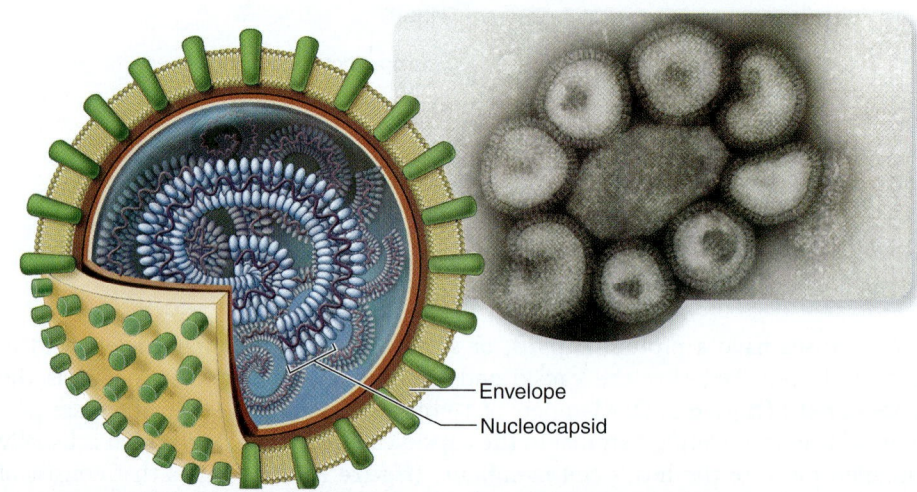
Icosahedral Capsids	These capsids form an **icosahedron** (eye″-koh-suh-hee′-drun)—a three-dimensional, 20-sided figure with 12 evenly spaced corners. The arrangements of the capsomers vary from one virus to another. Some viruses construct the capsid from a single type of capsomer, while others may contain several types of capsomers. There are major variations in the number of capsomers; for example, a poliovirus has 32, and an adenovirus has 252 capsomers.
Naked	Adenovirus is an example of a naked icosahedral virus. In the photo you can clearly see the spikes, some of which have broken off. 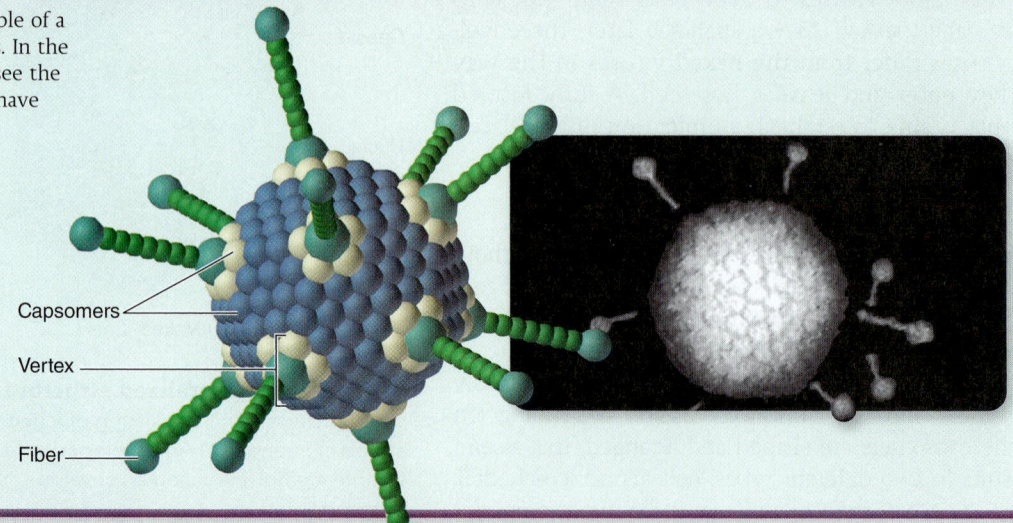

Table 5.2 (continued)

Icosahedral Capsids (*continued*)

Enveloped

Two very common viruses, hepatitis B virus (left) and the herpes simplex virus (right), possess enveloped icosahedrons.

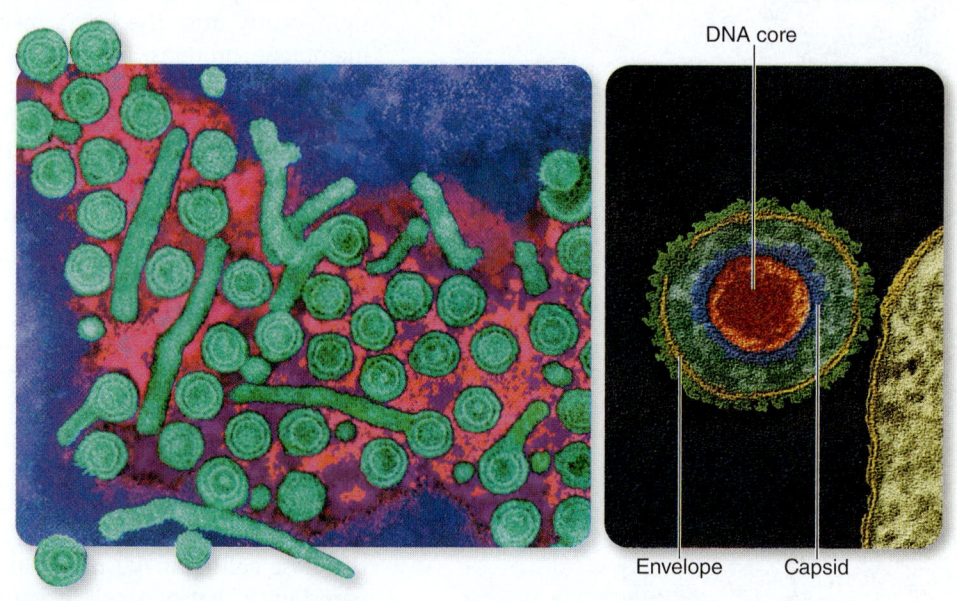

DNA core

Envelope   Capsid

Complex Capsids

Complex capsids, found in the viruses that infect bacteria, may have multiple types of proteins and take shapes that are not symmetrical. They are never enveloped. The one pictured on the right is a T4 bacteriophage.

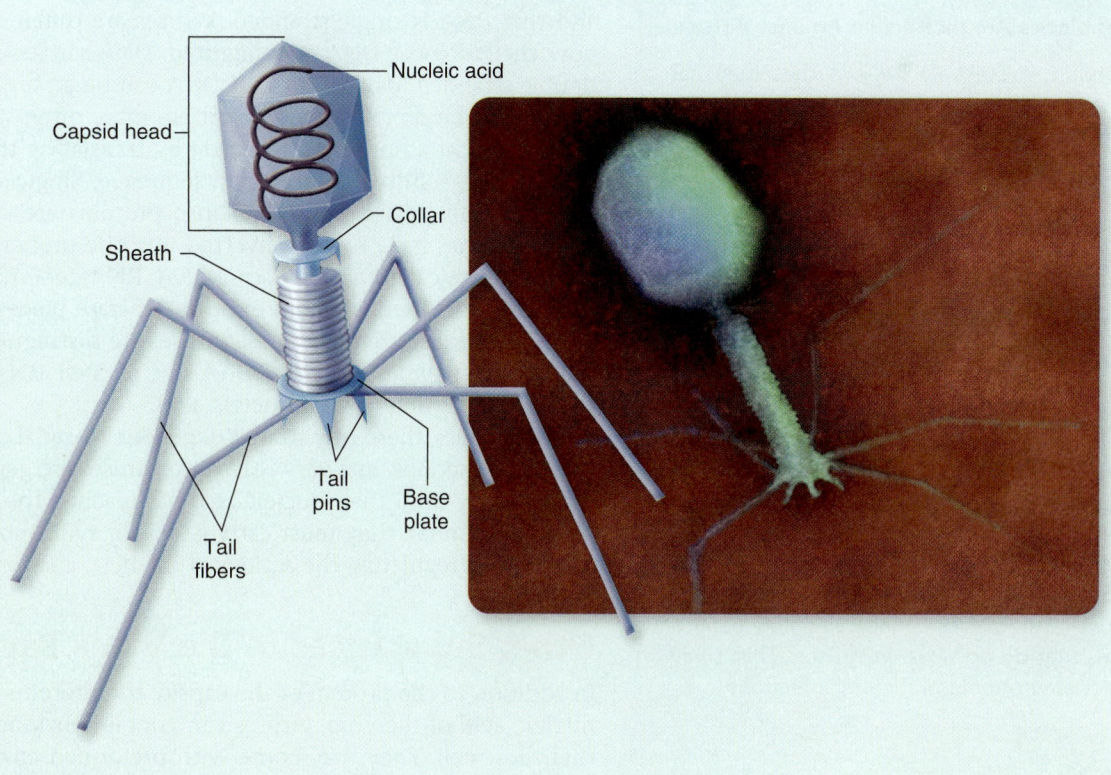

Nucleic acid

Capsid head

Collar

Sheath

Tail pins

Base plate

Tail fibers

A human herpesvirus (blue and gold) released from its host cell.

When **enveloped viruses** (mostly animal) are released from the host cell, they take with them a bit of its membrane system in the form of an envelope, as described later. Some viruses bud off the cell membrane; others leave via the nuclear envelope or the endoplasmic reticulum. Whichever avenue of escape, the viral envelope differs significantly from the host's membranes. In the envelope, some or all of the regular membrane proteins are replaced with special viral proteins. Some of the envelope proteins attach to the capsid of the virus, and glycoproteins (proteins bound to a carbohydrate) remain exposed on the outside of the envelope. These protruding molecules, called spikes when they are on enveloped viruses, are essential for the attachment of viruses to the next host cell. Because the envelope is more supple than the capsid, enveloped viruses are pleomorphic (of variable shape) and range from spherical to filamentous in shape.

Nucleic Acids: At the Core of a Virus

The sum total of the genetic information carried by an organism is called its **genome.** We know that the genetic information of living cells is carried by nucleic acids (DNA, RNA). Viruses, although neither alive nor cells, are no exception to this rule, but there is a significant difference. Unlike cells, which contain both DNA and RNA, viruses contain either DNA or RNA *but not both.*

Because viruses pack into a tiny space all of the genes necessary to instruct the host cell to make new viruses, the number of viral genes is quite small compared with that of a cell. It varies from four genes in hepatitis B virus to hundreds of genes in some herpesviruses. Viruses possess only the genes needed to invade host cells and redirect their activity. By comparison, the bacterium *Escherichia coli* has approximately 4,000 genes, and a human cell has approximately 20,000 to 30,000 genes. These additional genes allow cells to carry out the complex metabolic activity necessary for independent life.

In chapter 1, you learned that DNA usually exists as a double-stranded molecule and that RNA is single-stranded. Viruses are different; they exhibit wide variety in how their RNA or DNA is configured. DNA viruses can have single-stranded (ss) or double-stranded (ds) DNA; the dsDNA can be arranged linearly or in ds circles. RNA viruses can be double-stranded but are more often single-stranded. You will learn in chapter 8 that all proteins are made by translating the nucleic acid code on a single strand of RNA into an amino acid sequence. Single-stranded RNA genomes that are ready for immediate translation into proteins are called *positive-sense* RNA. Other RNA genomes have to be converted into the proper form to be made into proteins, and these are called *negative-sense* RNA. RNA genomes may also be *segmented,* meaning that the individual genes exist on separate pieces of RNA. A special type of RNA virus is called a *retrovirus.* These viruses are distinguished by the fact that they carry their own enzymes to create DNA out of their RNA. **Table 5.3** gives examples of each configuration of viral nucleic acid.

In all cases, these tiny strands of genetic material carry the blueprint for viral structure and functions. In a very real sense, viruses are genetic parasites because they cannot multiply until their nucleic acid has reached the internal habitat of the host cell. At the minimum, they must carry genes for synthesizing the viral capsid and genetic material, for regulating the actions of the host, and for packaging the mature virus.

Other Substances in the Virus Particle

In addition to the protein of the capsid, the proteins and lipids of envelopes, and the nucleic acid of the core, viruses can contain enzymes for specific operations within their host cell. They may come with preformed enzymes that are required for viral replication. Examples include *polymerases* (pol-im′-ur-ace-uz) that synthesize DNA and RNA, and replicases that copy RNA. The AIDS virus comes equipped with *reverse transcriptase* for synthesizing DNA from RNA. However, viruses completely lack the

Medical Moment

Why Antibiotics Are Ineffective Against Viruses

Many people mistakenly believe that they can take an antibiotic to cure a viral infection, such as the common cold. Why are antibiotics ineffective against viruses? To understand why antibiotics do not work on viral infections, we need to think about what antibiotics do, as well as the properties of viruses that make them unique (see table 5.1).

Most antibiotics target specific functions or processes within bacteria. Antibiotics may inhibit cell wall synthesis, protein synthesis, nucleic acid synthesis, or the synthesis of specific proteins required for the bacteria to survive and reproduce. They may also cause injury to the cytoplasmic membrane.

Viruses lack cytoplasmic membranes, are unable to synthesize proteins and contain either DNA or RNA (but not both). Viruses can only reproduce by hijacking their host's genetic material to create new viruses. Antibiotics cannot alter functions or processes that do not exist in viruses. This is why antibiotics are not helpful for viral infections.

Table 5.3 Viral Nucleic Acid

	Diagram	Virus Name	Disease It Causes
DNA Viruses			
Double-stranded DNA		Variola virus	Smallpox
		Herpes simplex II	Genital herpes
Single-stranded DNA		Parvovirus	Erythema infectiosum (skin condition)
RNA Viruses			
Single-stranded (+) polarity		Poliovirus	Poliomyelitis
Single-stranded (−) polarity		Influenza virus	Influenza
Double-stranded RNA		Rotavirus	Gastroenteritis
Single-stranded RNA reverse transcriptase		HIV	AIDS

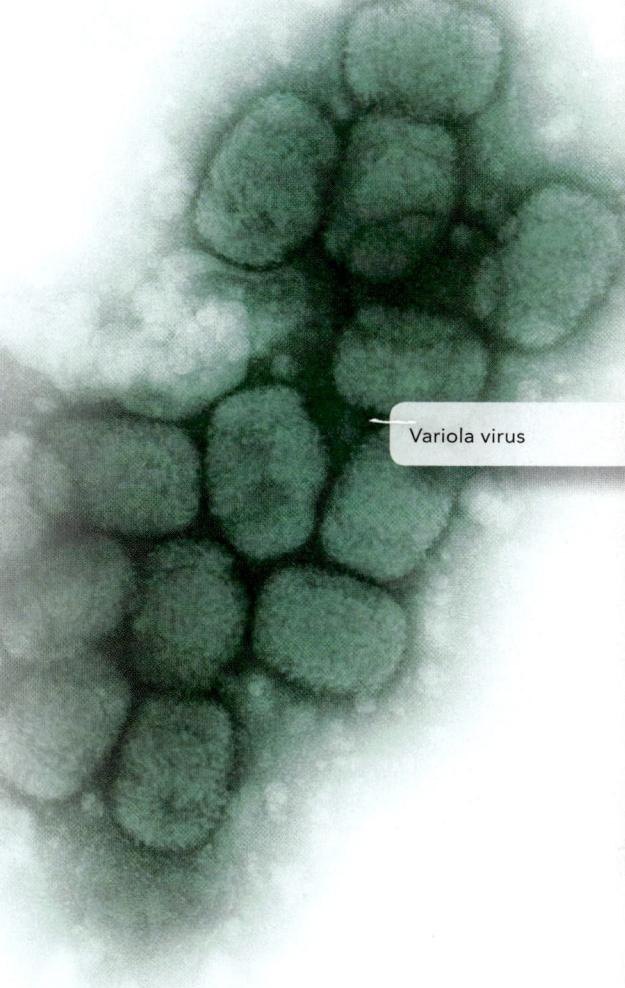

Variola virus

genes for synthesis of metabolic enzymes. As we shall see, this deficiency has little consequence, because viruses have adapted to completely take over their hosts' metabolic resources. Some viruses can actually carry away substances from their host cell. For instance, arenaviruses pack along host ribosomes, and retroviruses "borrow" the host's tRNA molecules.

5.2 LEARNING OUTCOMES—Assess Your Progress

3. Discuss the size of viruses relative to other microorganisms.
4. Describe the function and structure(s) of viral capsids.
5. Distinguish between enveloped and naked viruses.
6. Explain the importance of viral surface proteins, or spikes.
7. Diagram the possible configurations that nucleic acid viruses may possess.

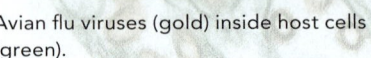

Avian flu viruses (gold) inside host cells (green).

5.3 Modes of Viral Multiplication

The process of viral multiplication is an extraordinary biological phenomenon. Viruses are minute parasites that seize control of the synthetic and genetic machinery of cells. The nature of this cycle dictates the way the virus is transmitted and what it does to its host, the responses of the immune defenses, and human measures to control viral infections.

Multiplication Cycles in Animal Viruses

The general phases in the life cycle of animal viruses are **adsorption, penetration and uncoating, synthesis, assembly,** and **release** from the host cell. The length of the entire multiplication cycle varies from 8 hours in polioviruses to 36 hours in some herpesviruses. **Table 5.4** walks through the major phases of the viral life cycle, using a + strand RNA virus (of which rubella virus is an example) as a model.

Notes on the Multiplication Cycle

Adsorption   Because a virus can invade its host cell only through making an exact fit with a specific host molecule, the range of hosts it can infect is limited **(figure 5.4).** This limitation, known as the **host range,** may be highly restricted as in the case of hepatitis B, which infects only liver cells of humans; moderately restrictive like the poliovirus, which infects intestinal and nerve cells of primates (humans, apes, and monkeys); or broad like the rabies virus, which can infect various cells of all mammals. Cells that lack compatible virus receptors are resistant to adsorption and invasion by that virus. This explains why, for example, human liver cells are not infected by the canine hepatitis virus and dog liver cells cannot host the human hepatitis A virus. It also explains why viruses usually have tissue specificities called *tropisms* (troh'-pizmz) for certain cells in the body. The hepatitis B virus targets the liver, and the mumps virus targets salivary glands.

Figure 5.4 The viral attachment process.   An enveloped coronavirus with prominent spikes. The configuration of the spike has a complementary fit for cell receptors. The process in which the virus lands on the cell and plugs into receptors is termed *docking*.

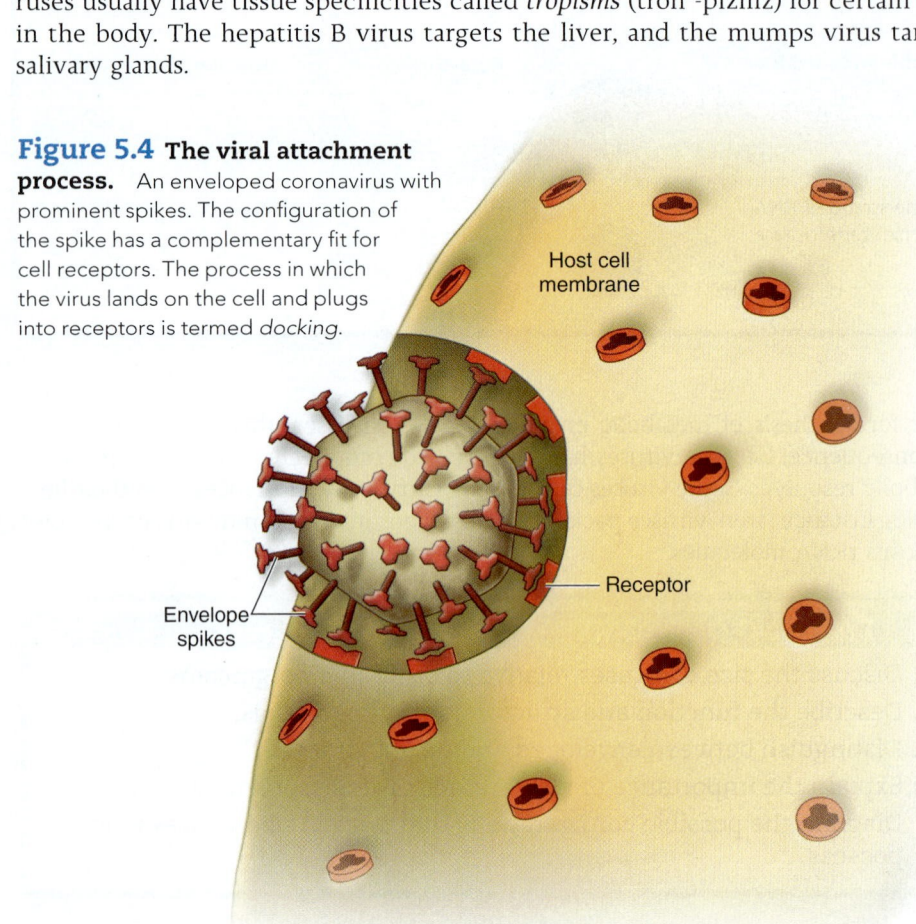

Host cell membrane

Receptor

Envelope spikes

Table 5.4 Life Cycle of Animal Viruses

1 **Adsorption.** The virus encounters a susceptible host cell and adsorbs specifically to receptor sites on the cell membrane. The membrane receptors that viruses attach to are usually glycoproteins that the cell requires for its normal function. Glycoprotein spikes on the envelope (or on the capsid of naked viruses) bind to the cell membrane receptors.

2 **Penetration and Uncoating.** In this example, the entire virus is engulfed (endocytosed) by the cell and enclosed in a vacuole or vesicle. When enzymes in the vacuole dissolve the envelope and capsid, the virus is said to be uncoated, a process that releases the viral nucleic acid into the cytoplasm.

3 **Synthesis: Replication and Protein Production.** Almost immediately, the viral nucleic acid begins to synthesize the building blocks for new viruses. First, the + ssRNA, which can serve immediately upon entry as mRNA, starts being translated into viral proteins, especially those useful for further viral replication. The + strand is then replicated into – ssRNA. This RNA becomes the template for the creation of many new + ssRNAs, used as the viral genomes for new viruses. Additional + ssRNAs are synthesized and used for late-stage mRNAs. Some viruses come equipped with the necessary enzymes for synthesis of viral components; others utilize those of the host. Proteins for the capsid, spikes, and viral enzymes are synthesized on the host's ribosomes using its amino acids.

4 **Assembly.** Toward the end of the cycle, mature virus particles are constructed from the growing pool of parts. In most instances, the capsid is first laid down as an empty shell that will serve as a receptacle for the nucleic acid strand. One important event leading to the release of enveloped viruses is the insertion of viral spikes into the host's cell membrane so they can be picked up as the virus buds off with its envelope.

5 **Release.** Assembled viruses leave their host in one of two ways. Nonenveloped and complex viruses that reach maturation in the cell nucleus or cytoplasm are released when the cell lyses or ruptures. Enveloped viruses are liberated by budding from the membranes of the cytoplasm, nucleus, endoplasmic reticulum, or vesicles. During this process, the nucleocapsid binds to the membrane, which curves completely around it and forms a small pouch. Pinching off the pouch releases the virus with its envelope.

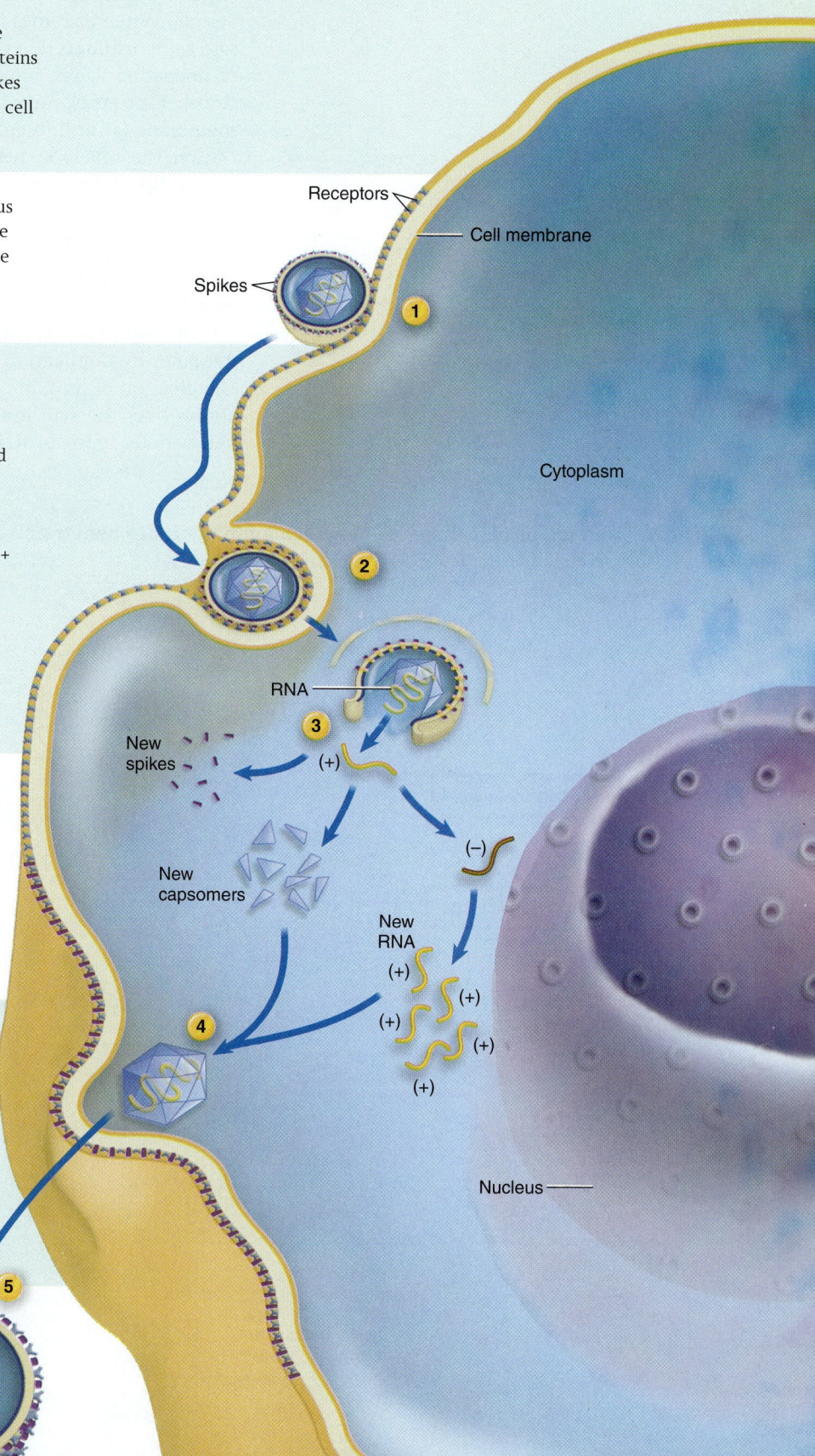

Penetration and Uncoating

Animal viruses exhibit some impressive mechanisms for entering a host cell. The flexible cell membrane of the host is penetrated by the whole virus or its nucleic acid **(figure 5.5)**. In penetration by **endocytosis (figure 5.5a)**, the entire virus is engulfed by the cell and enclosed in a vacuole or vesicle. When enzymes in the vacuole dissolve the envelope and capsid, the virus is said to be **uncoated,** a process that releases the viral nucleic acid. The exact manner of uncoating varies, but in most cases, the virus fuses with the wall of the vesicle. Another means of entry involves direct fusion of the viral envelope with the host cell membrane (as in influenza and mumps viruses) **(figure 5.5b)**. In this form of penetration, the envelope merges directly with the cell membrane, thereby liberating the nucleocapsid into the cell's interior.

Synthesis: Replication and Protein Production

In general, the DNA viruses (except poxviruses) enter the host cell's nucleus and are replicated and assembled there. With few exceptions (such as retroviruses), RNA viruses are replicated and assembled in the cytoplasm.

In chapter 8 you will learn that cellular organisms make new copies of their new genomes by duplicating their DNA. They also use DNA to make mRNA that directs the creation of proteins. These processes can be very different in viruses. **Table 5.5** shows how the synthesis of new genomes and mRNAs for translation differ among the various types of RNA and DNA viruses. Note that the retroviruses turn their

Figure 5.5 Two principal means by which animal viruses penetrate. (a) Endocytosis (engulfment) and uncoating of a herpesvirus. (b) Fusion of the cell membrane with the viral envelope (mumps virus).

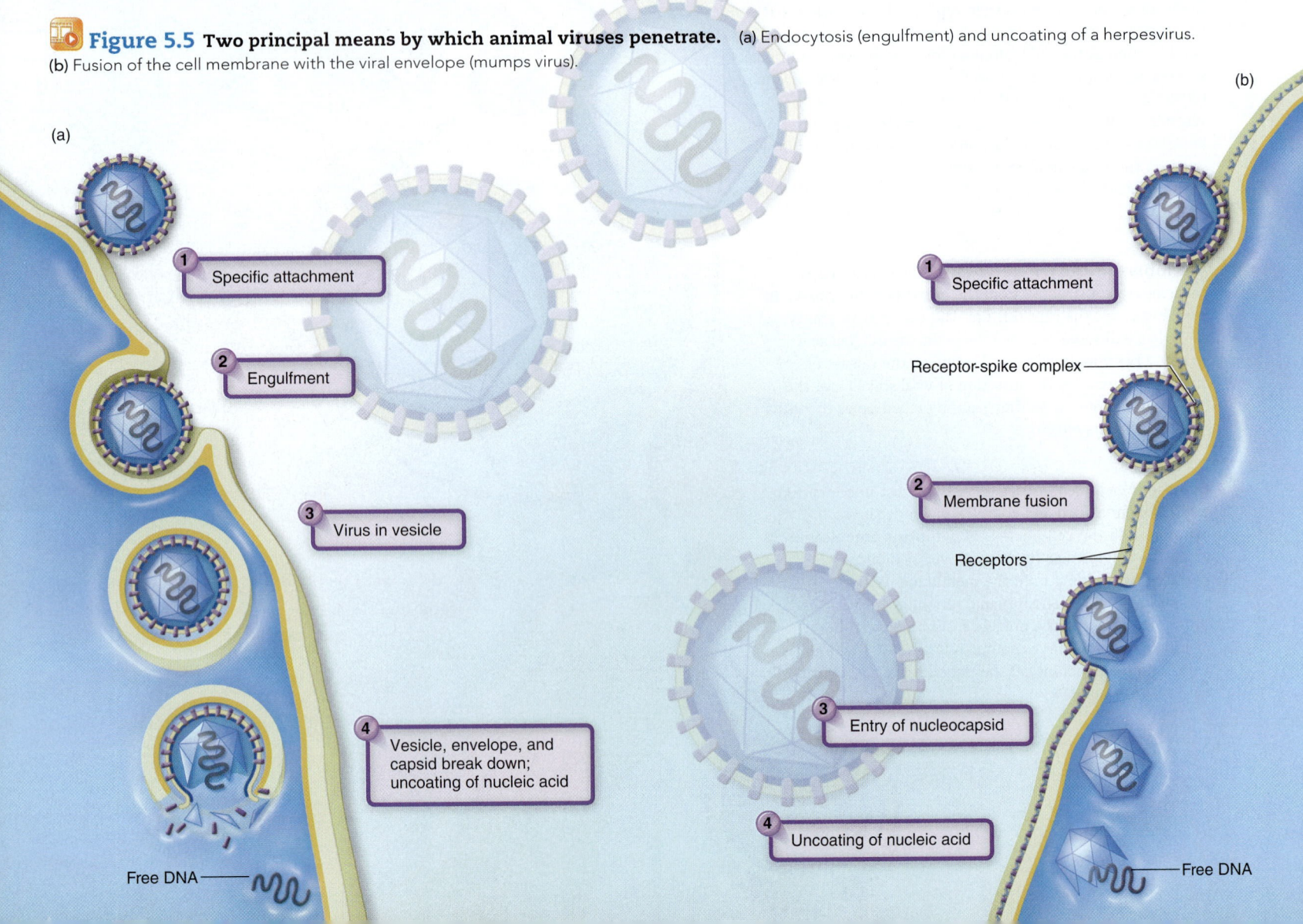

(a)

1. Specific attachment
2. Engulfment
3. Virus in vesicle
4. Vesicle, envelope, and capsid break down; uncoating of nucleic acid

Free DNA

(b)

1. Specific attachment

Receptor-spike complex

2. Membrane fusion

Receptors

3. Entry of nucleocapsid

4. Uncoating of nucleic acid

Free DNA

Table 5.5   Viral Transcription and Translation Modes

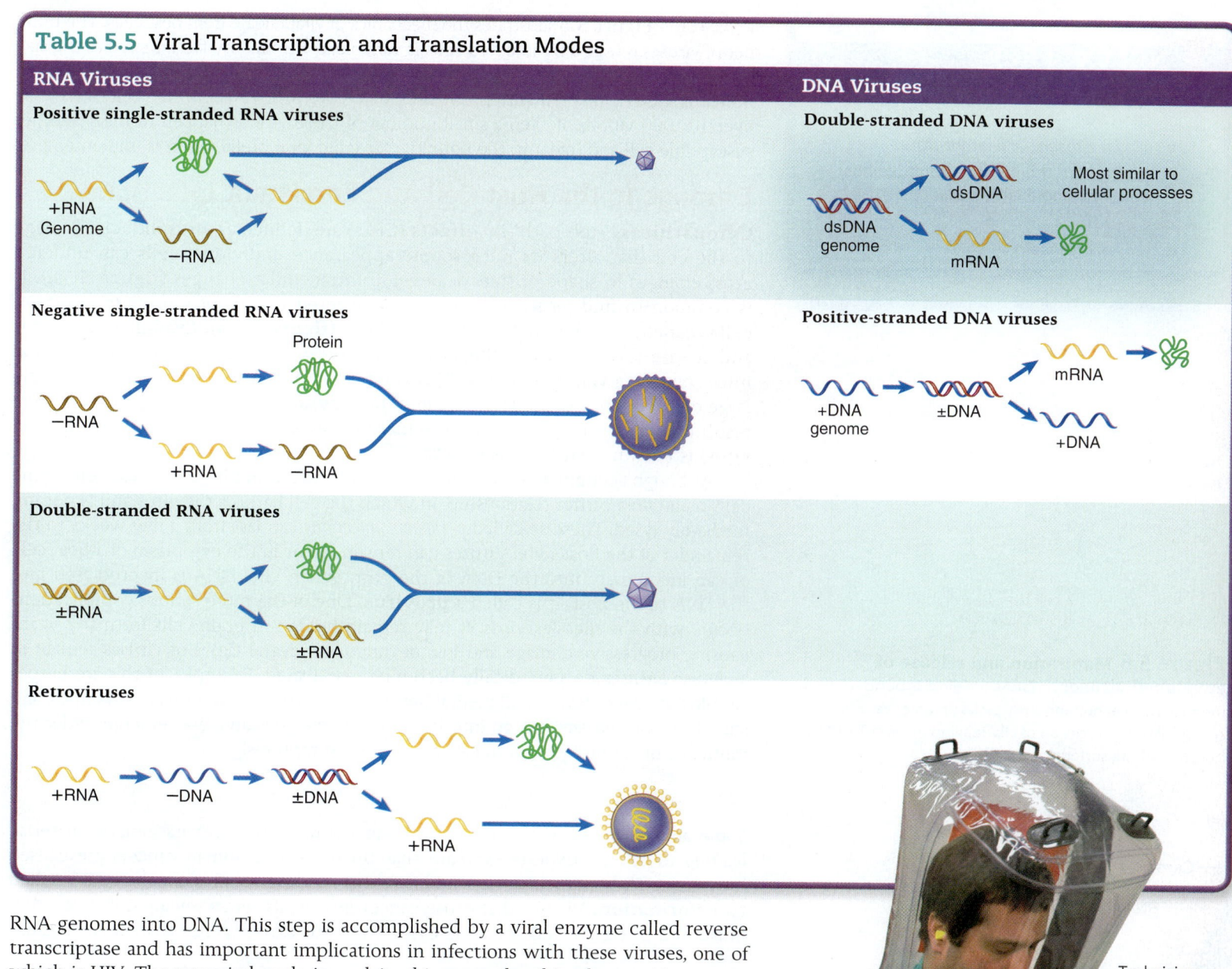

RNA genomes into DNA. This step is accomplished by a viral enzyme called reverse transcriptase and has important implications in infections with these viruses, one of which is HIV. The retroviral cycle is explained in more detail in chapter 18.

In the life cycle of dsDNA viruses, the synthesis phase is divided into two parts. During the early phase, viral DNA enters the nucleus, where several genes are transcribed into a messenger RNA. The newly synthesized RNA transcript then moves into the cytoplasm to be translated into viral proteins (enzymes) needed to replicate the viral DNA; this replication occurs in the nucleus. The host cell's own DNA polymerase is often involved, though some viruses (herpes, for example) have their own. During the late phase, other parts of the viral genome are transcribed and translated into proteins required to form the capsid and other structures. The new viral genomes and capsids are assembled, and the mature viruses are released by budding or cell disintegration. In some viruses, the viral DNA becomes silently *integrated* into the host's genome by insertion at a particular site on the host genome. This integration may later lead to the transformation of the host cell into a cancer cell and the production of a tumor.

Assembly   As illustrated in table 5.5, this step actually puts together the new viruses using the "parts" manufactured in the synthesis process: new capsids and new nucleic acids.

Technicians who work on certain viruses take special precautions.

Release **Figure 5.6** illustrates the mechanics of viral release from host cells. The number of viruses released by infected cells is variable, controlled by factors such as the size of the virus and the health of the host cell. About 3,000 to 4,000 virions are released from a single cell infected with poxviruses, whereas a poliovirus-infected cell can release over 100,000 virions. If even a small number of these virions happen to meet another susceptible cell and infect it, the potential for rapid viral proliferation is immense.

Damage to the Host Cell and Persistent Infections

Cytopathic (sy″-toh-path′-ik) **effects** (CPEs) are defined as virus-induced damage to the cell that alters its microscopic appearance. Individual cells can undergo gross changes in shape or size, or develop intracellular changes **(figure 5.7a)**. It is common to find *inclusion bodies,* or compacted masses of viruses or damaged cell organelles, in the nucleus and cytoplasm **(figure 5.7b).** Examination of cells and tissues for cytopathic effects is an important part of the diagnosis of viral infections. One very common CPE is the fusion of multiple host cells into single large cells containing multiple nuclei. These **syncytia** (singular, *syncytium*) are a result of some viruses' ability to fuse membranes. One virus (respiratory syncytial virus) is even named for this effect.

Although accumulated damage from a virus infection kills most host cells, some cells maintain a carrier relationship, in which the cell harbors the virus and is not immediately lysed. These so-called *persistent infections* can last from a few weeks to the remainder of the host's life. Viruses can remain latent in the cytoplasm of a host cell, or can incorporate into the DNA of the host. When viral DNA is incorporated into the DNA of the host, it is called a **provirus.** One of the more serious complications occurs with the measles virus. It may remain hidden in brain cells for many years, causing progressive damage and loss of function. Several types of viruses remain in a *chronic latent state,* periodically becoming reactivated. Examples of this are herpes simplex virus (cold sores and genital herpes) and herpes zoster virus (chickenpox and shingles). Both viruses can go into latency in nerve cells and later emerge under the influence of various stimuli to cause recurrent symptoms.

Viruses and Cancer

Some animal viruses enter a host cell and permanently alter its genetic material, leading to cancer. Experts estimate that up to 20% of human cancers are caused by viruses. These viruses are termed *oncogenic,* and their effect on the cell is called **transformation.** Viruses that cause cancer in animals act in several different ways, illustrated in **figure 5.8.** In some cases, the virus carries genes that directly cause

Figure 5.6 Maturation and release of enveloped viruses. **(a)** As the virus is budded off the membrane, it simultaneously picks up an envelope and spikes. **(b)** A micrograph of HIV leaving its host T cell by budding off its surface.

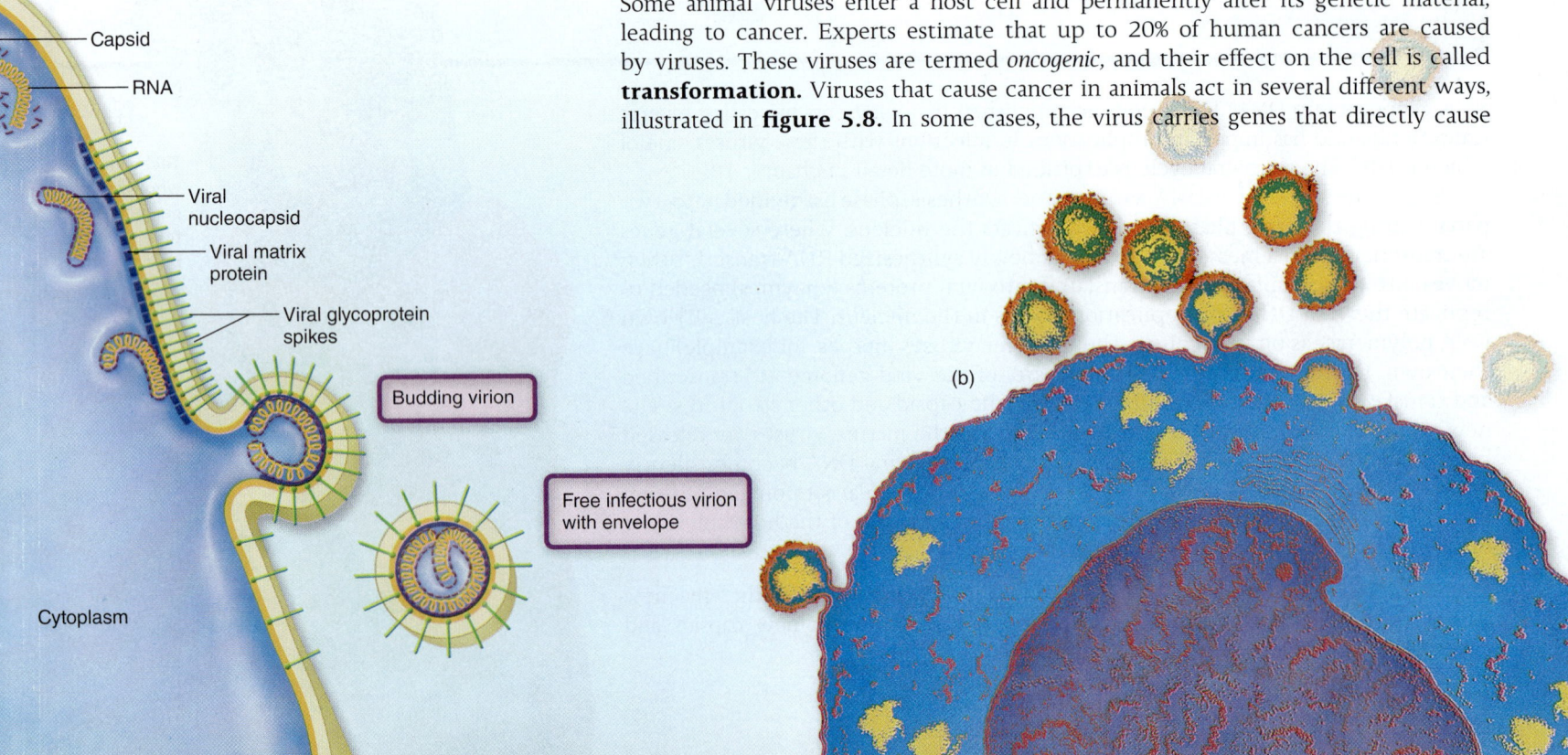

(a)

Capsid

RNA

Viral nucleocapsid

Viral matrix protein

Viral glycoprotein spikes

Budding virion

Free infectious virion with envelope

Cytoplasm

(b)

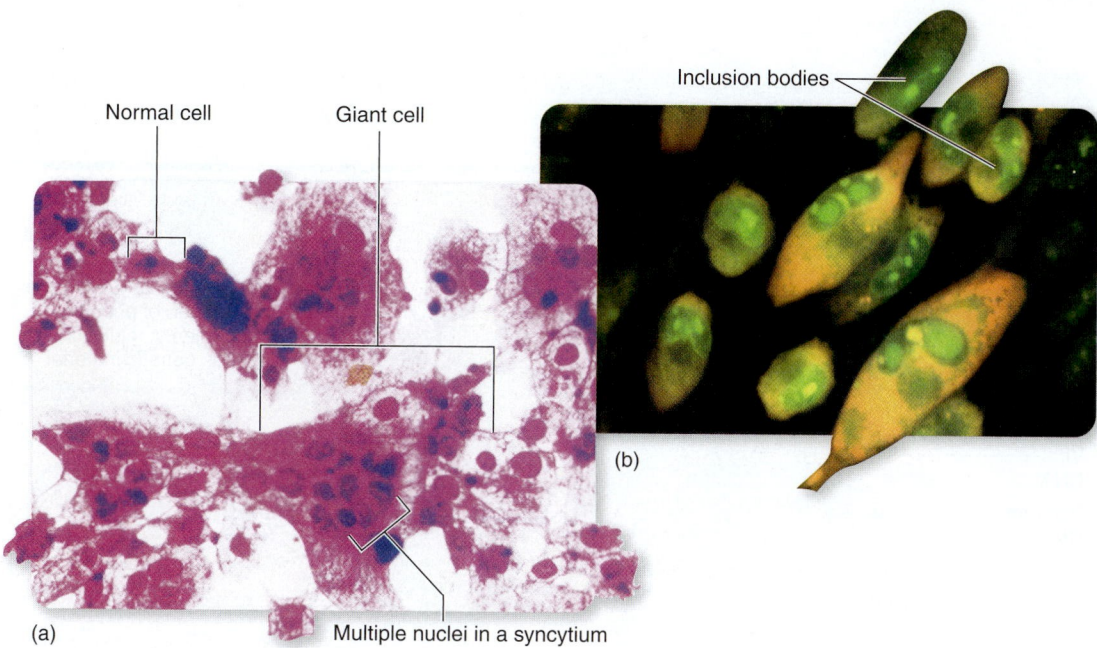

Normal cell · Giant cell · Inclusion bodies · Multiple nuclei in a syncytium · (a) · (b)

Figure 5.7 Cytopathic changes in cells and cell cultures infected by viruses. (a) Human epithelial cells infected by herpes simplex virus demonstrate giant cells with multiple nuclei. (b) Fluorescent-stained human cells infected with cytomegalovirus. Note the inclusion bodies (arrows). Note also that both viruses disrupt the cohesive junctions between cells, which would ordinarily be arranged side by side in neat patterns.

the cancer. In other cases, the virus produces proteins that induce a loss of growth regulation in the cell, leading to cancer. Transformed cells have an increased rate of growth; alterations in chromosomes; changes in the cell's surface molecules; and the capacity to divide for an indefinite period, unlike normal animal cells. Mammalian viruses capable of initiating tumors are called **oncoviruses.** Some of these are DNA viruses such as papillomavirus (genital warts are associated with cervical cancer), herpesviruses (one herpesvirus, Epstein-Barr virus, causes Burkitt's lymphoma), and hepatitis B virus (liver cancer). A virus related to HIV–HTLV I–is also involved in human cancers. These findings have spurred a great deal of speculation on the possible involvement of viruses in cancers and other diseases such as multiple sclerosis.

Viruses That Infect Bacteria

We now turn to the life cycle of another type of virus called **bacteriophage.** When Frederick Twort and Felix d'Herelle discovered bacterial viruses in 1915, it first appeared that the bacterial host cells were being eaten by some unseen parasite; hence, the name *bacteriophage* was used (*phage* coming from the Greek word for "eating"). Most bacteriophages–a term often shortened to *phage*–contain double-stranded DNA, although single-stranded DNA and RNA types exist as well.

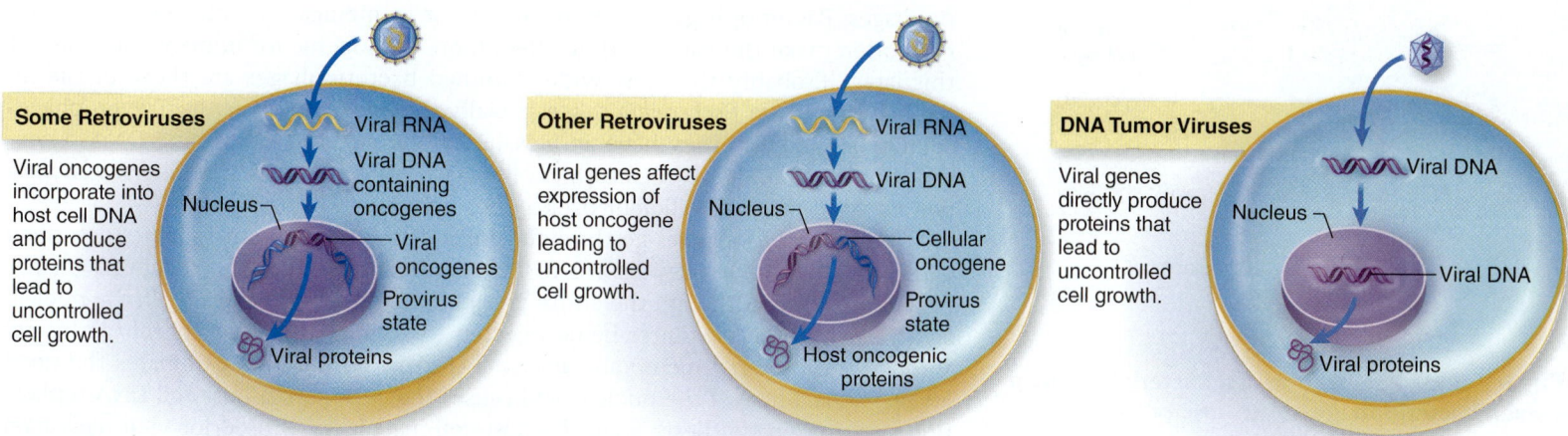

Some Retroviruses
Viral oncogenes incorporate into host cell DNA and produce proteins that lead to uncontrolled cell growth.
Nucleus · Viral RNA · Viral DNA containing oncogenes · Viral oncogenes · Provirus state · Viral proteins

Other Retroviruses
Viral genes affect expression of host oncogene leading to uncontrolled cell growth.
Nucleus · Viral RNA · Viral DNA · Cellular oncogene · Provirus state · Host oncogenic proteins

DNA Tumor Viruses
Viral genes directly produce proteins that lead to uncontrolled cell growth.
Nucleus · Viral DNA · Viral DNA · Viral proteins

Figure 5.8 Three mechanisms for viral induction of cancer.

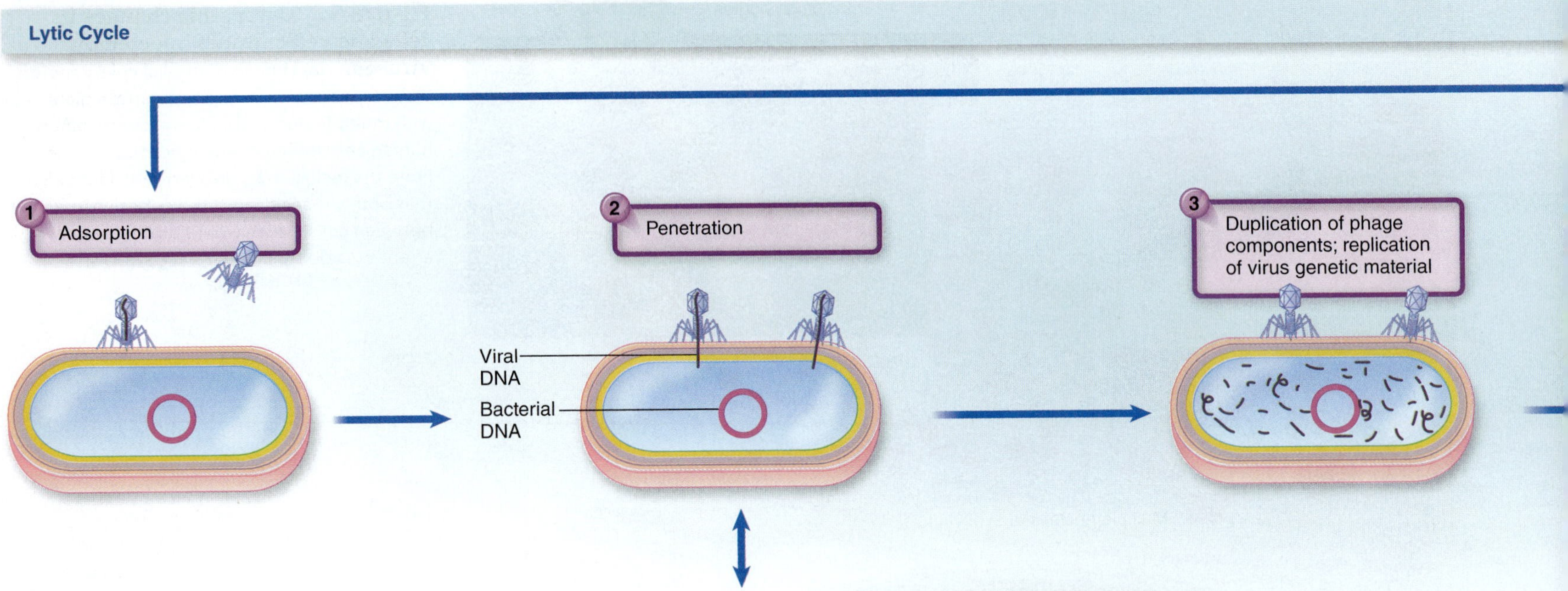

Lytic Cycle

① Adsorption

② Penetration

Viral DNA

Bacterial DNA

③ Duplication of phage components; replication of virus genetic material

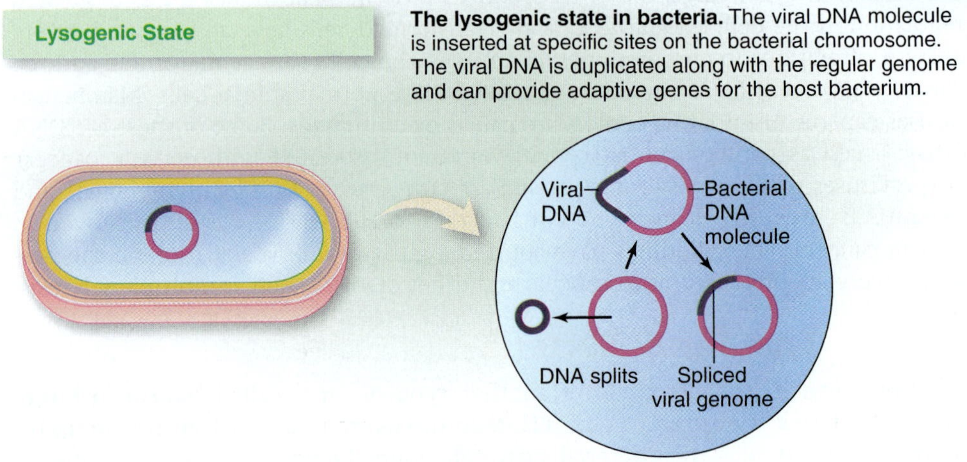

▶ **Figure 5.9 Events in the lytic cycle of T-even bacteriophages.**   The lytic cycle (1–6) involves full completion of viral infection through lysis and release of virions. Occasionally the virus enters a reversible state of lysogeny (pictured to the right) and is incorporated into the host's genetic material.

Lysogenic State

The lysogenic state in bacteria. The viral DNA molecule is inserted at specific sites on the bacterial chromosome. The viral DNA is duplicated along with the regular genome and can provide adaptive genes for the host bacterium.

Viral DNA

Bacterial DNA molecule

DNA splits   Spliced viral genome

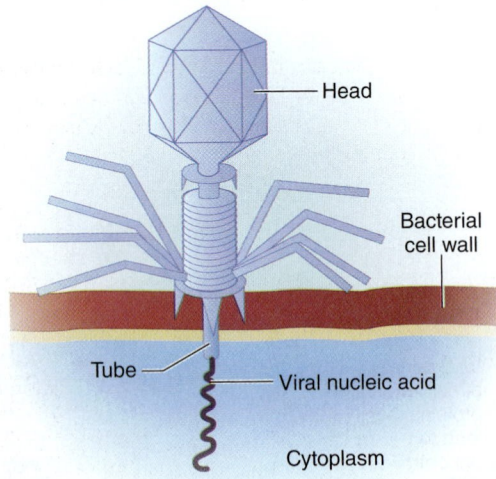

Head

Bacterial cell wall

Tube

Viral nucleic acid

Cytoplasm

Figure 5.10 Penetration of a bacterial cell by a T-even bacteriophage.   After adsorption, the phage plate becomes embedded in the cell wall and the sheath contracts, pushing the tube through the cell wall and releasing the nucleic acid into the interior of the cell.

So far as is known, every bacterial species is parasitized by various specific bacteriophages. Bacteriophages are of great interest to medical microbiologists because they often make the bacteria they infect more pathogenic for humans (more about this later). Probably the most widely studied bacteriophages are those of the intestinal bacterium *Escherichia coli*–especially the ones known as the T-even phages such as T2 and T4. They have an icosahedral capsid head containing DNA, a central tube (surrounded by a sheath), collar, base plate, tail pins, and fibers, which in combination make an efficient package for infecting a bacterial cell .

T-even bacteriophages go through similar stages as the animal viruses described earlier **(figure 5.9).** They *adsorb* to host bacteria using specific receptors on the bacterial surface. Although the entire phage does not enter the host cell, the nucleic acid *penetrates* the host after being injected through a rigid tube the phage inserts through the bacterial membrane and wall **(figure 5.10).** This eliminates the need for *uncoating*. Entry of the nucleic acid causes the cessation of host cell DNA replication and protein synthesis. Soon the host cell machinery is used for viral *replication* and synthesis of viral proteins. As the host cell produces new phage parts, the parts spontaneously *assemble* into bacteriophages.

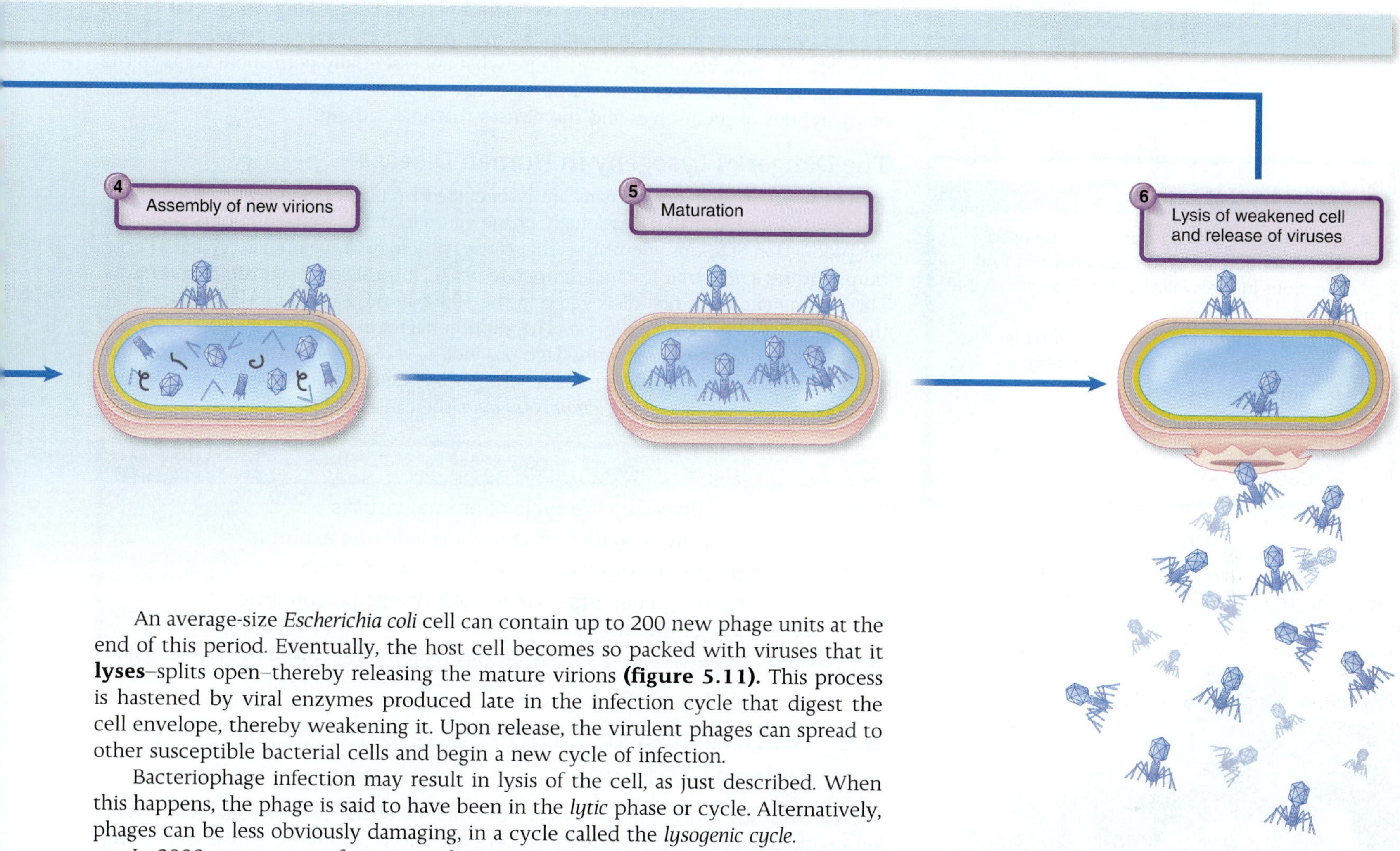

4	5	6
Assembly of new virions	Maturation	Lysis of weakened cell and release of viruses

An average-size *Escherichia coli* cell can contain up to 200 new phage units at the end of this period. Eventually, the host cell becomes so packed with viruses that it **lyses**–splits open–thereby releasing the mature virions **(figure 5.11).** This process is hastened by viral enzymes produced late in the infection cycle that digest the cell envelope, thereby weakening it. Upon release, the virulent phages can spread to other susceptible bacterial cells and begin a new cycle of infection.

Bacteriophage infection may result in lysis of the cell, as just described. When this happens, the phage is said to have been in the *lytic* phase or cycle. Alternatively, phages can be less obviously damaging, in a cycle called the *lysogenic cycle.*

In 2008, a new type of virus was discovered. These have been named *virophages.* They parasitize other viruses that are infecting the same host cell they infect, using genes from other (usually larger) viruses for their own replication and production. Even though these are parasites of viruses, note that they must be in a host cell, along with their "host" virus.

Lysogeny: The Silent Virus Infection

While special DNA phages, called **temperate phages,** can participate in a lytic phase, they also have the ability to undergo adsorption and penetration into the bacterial host and not undergo replication or release immediately. Instead, the viral DNA enters an inactive **prophage** state, reminiscent of the provirus state in animal viruses, during which it is inserted into the bacterial chromosome. This viral DNA will be retained by the bacterial cell and copied during its normal cell division so that the cell's progeny will also have the temperate phage DNA (see figure 5.9). This condition, in which the host chromosome carries bacteriophage DNA, is termed **lysogeny** (ly-soj′-uhn-ee). Because viral particles are not produced, the bacterial cells carrying temperate phages do not lyse, and they appear entirely normal. On occasion, in a process called **induction,** the prophage in a lysogenic cell will be activated and progress directly into viral replication and the lytic cycle. Lysogeny is a less deadly form of parasitism than the full lytic cycle and is thought to be an advancement that allows the virus to spread without killing the host.

Bacteriophages are just now receiving their due as important shapers of biological life. Scientists believe that there are more bacteriophages than all other forms of

Figure 5.11   A weakened bacterial cell, crowded with viruses.   The cell has ruptured and released numerous virions that can then attack nearby susceptible host cells. Note the empty heads of "spent" phages lined up around the ruptured wall.

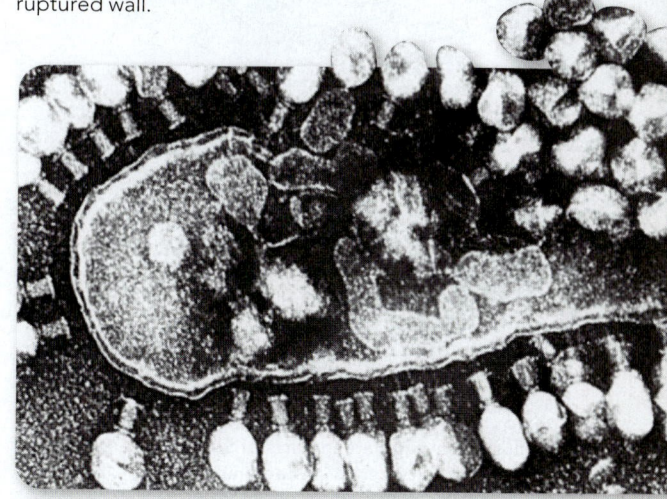

life in the biosphere combined. As we mentioned in the opening paragraphs of this chapter, viral genes linger in human, animal, plant, and bacterial genomes in huge numbers. As such, viruses can contribute what are essentially permanent traits to the bacteria, so much so that it could be said that all bacteria—indeed all organisms—are really hybrids of themselves and the viruses that infect them.

The Danger of Lysogeny in Human Disease

Many bacteria that infect humans are lysogenized by phages. Sometimes that is very bad news for the human: Occasionally phage genes in the bacterial chromosome cause the production of toxins or enzymes that cause pathology in the human. When a bacterium acquires a new trait from its temperate phage, it is called **lysogenic conversion.** The phenomenon was first discovered in the 1950s in the bacterium that causes diphtheria, *Corynebacterium diphtheriae.* The diphtheria toxin responsible for the deadly nature of the disease is a bacteriophage product. *C. diphtheriae* without the phage are harmless. Other bacteria that are made virulent by their prophages are *Vibrio cholerae,* the agent of cholera, and *Clostridium botulinum,* the cause of botulism.

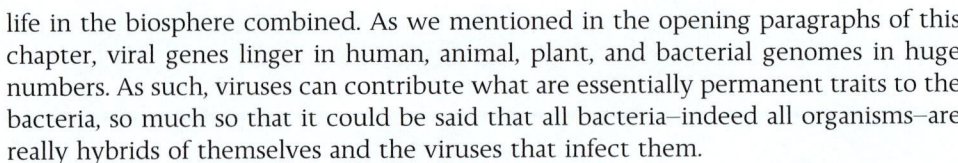

NCLEX® PREP

3. What diagnostic methods would be used to differentiate between bacterial and viral infections in the clinical setting? Select all that apply.
   a. *determination of whether patient is experiencing nausea or vomiting*
   b. *blood cultures*
   c. *rapid strep testing*
   d. *initiation of antibiotic therapy*
   e. *CBC with differential*

5.3 LEARNING OUTCOMES—Assess Your Progress

8. Diagram the five-step life cycle of animal viruses.
9. Define the term *cytopathic effect* and provide one example.
10. Discuss both persistent and transforming infections.
11. Provide thorough descriptions of both lysogenic and lytic bacteriophage infections.

5.4 Techniques in Cultivating and Identifying Animal Viruses

In order to study viruses, it is necessary to cultivate them. This presents many problems with organisms that require living cells as their "medium." Scientists have developed methods, which include inoculation of laboratory-bred animals and embryonic bird tissues (such methods are termed *in vivo*) and cell (or tissue) culture methods (called *in vitro*).

The primary purposes of viral cultivation are to

1. isolate and identify viruses in clinical specimens;
2. prepare viruses for vaccines; and
3. do detailed research on viral structure, multiplication cycles, genetics, and effects on host cells.

Using Live Animal Inoculation

Specially bred strains of white mice, rats, hamsters, guinea pigs, and rabbits are the usual choices for animal cultivation of viruses. Invertebrates (insects) or nonhuman primates are occasionally used as well. Because viruses can exhibit host specificity, certain animals can propagate a given virus more readily than others.

Avian flus often originate in parts of Southeast Asia where contact between avians and humans is commonplace.

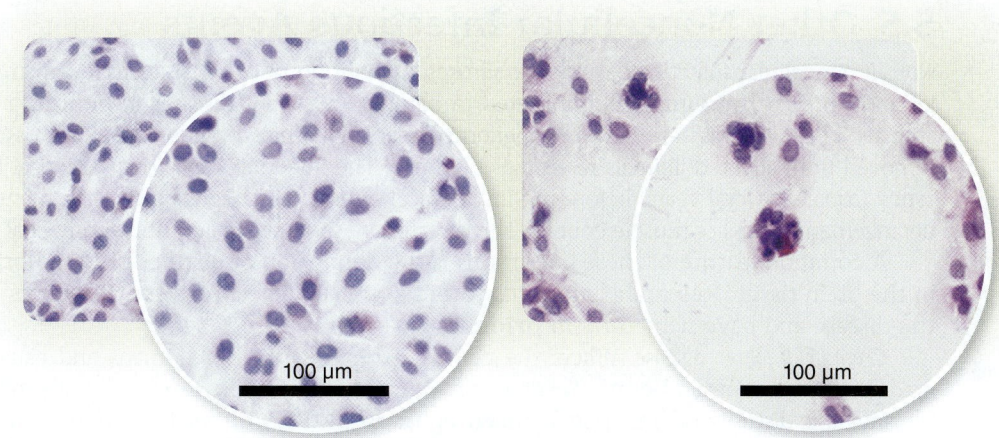

Figure 5.12 Appearance of normal and infected cell culture. Microscopic view of a layer of animal cells before infection with the appropriate virus (left), and after (right).

Using Bird Embryos

A bird egg containing an embryo provides an intact and self-supporting unit, complete with its own sterile environment and nourishment. Furthermore, it furnishes several embryonic tissues that readily support viral multiplication. Chicken, duck, and turkey eggs are the most common choices for inoculation. The virus must be injected through the egg shell, usually by drilling a hole or making a small window.

Using Cell (Tissue) Culture Techniques

The most important early discovery that led to easier cultivation of viruses in the laboratory was the development of a simple and effective way to grow populations of isolated animal cells in culture. These types of *in vitro* cultivation systems are termed *cell culture*, or *tissue culture*. Animal cell cultures are grown in sterile chambers with special media that contain the correct nutrients required by animal cells to survive. The cultured cells grow in the form of a *monolayer*, a single, confluent sheet of cells that supports viral multiplication and permits close inspection of the culture for signs of infection **(figure 5.12)**.

The recent avian flu worries have prompted scientists to look for faster and more efficient ways to grow the vaccine strains of influenza virus, which has been grown in chicken eggs since the 1950s. Scientists have succeeded in propagating the viruses in a continuous cell line derived from dog kidney cells. There were plans to produce flu vaccine in cell culture beginning in 2009, but they were mostly thwarted. In 2012, for the first time, the FDA approved for general use a cell-culture-based vaccine for the seasonal influenza virus.

One way to detect the growth of a virus in culture is to observe degeneration and lysis of infected cells in the monolayer of cells. The areas where virus-infected cells have been destroyed show up as clear, well-defined patches in the cell sheet called **plaques** (figure 5.12). Plaques are essentially the macroscopic manifestation of cytopathic effects (CPEs), discussed earlier. This same technique is used to detect and count bacteriophages, because they also produce plaques when grown in soft agar cultures of their host cells (bacteria). A plaque develops when the viruses released by an infected host cell radiate out to adjacent host cells. As new cells become infected, they die and release more viruses, and so on. As this process continues, the infection spreads gradually and symmetrically from the original point of infection, causing the macroscopic appearance of round, clear spaces that correspond to areas of dead cells.

5.4 LEARNING OUTCOMES—Assess Your Progress

12. List the three principal purposes of cultivating viruses.
13. Describe three ways in which viruses are cultivated.

 NCLEX® PREP

4. Bird embryos are used to cultivate viruses because they
    a. *provide a sterile environment.*
    b. *provide nourishment needed for replication.*
    c. *are self-supporting units.*
    d. *all of the above*

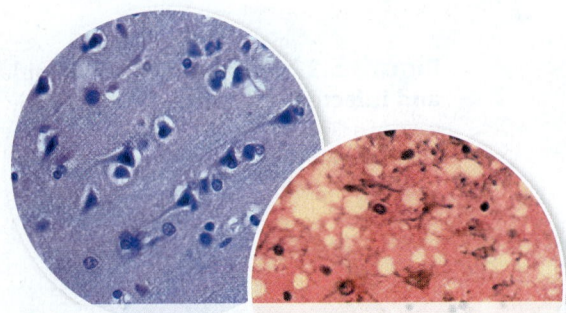

The damage inflicted on brain tissue by Creutzfeldt-Jakob disease. Diseased tissue (*right*) shows sponge-like holes not seen in healthy brains (*left*).

Medical Moment

Differentiating Between Bacterial and Viral Infections

Viral and bacterial diseases can share many of the same symptoms. How can physicians tell the difference? This is an important question, as treatment for bacterial infections often requires antibiotic therapy, whereas treatment for viral illnesses is often supportive—antibiotic therapy is ineffective against viruses.

Many viral illnesses will cause milder symptoms than their bacterial counterpart. For example, viral meningitis is typically a less serious disease than bacterial meningitis, and pharyngitis caused by a bacterium will cause more severe symptoms than viral pharyngitis. Therefore, doctors will take into account how sick the patient is when trying to determine whether a patient is suffering from a viral or bacterial infection. They will also look at duration of symptoms, time of year, and known illnesses circulating in the community.

However, none of the above are hard and fast rules. Doctors become very proficient at recognizing various illnesses, but sometimes even they cannot tell the difference. When in doubt, blood work, cultures, and other diagnostic tests can help them make the diagnosis. Advances in laboratory testing, such as rapid strep testing, have made life easier for physicians attempting to differentiate between viral and bacterial illnesses.

NCLEX® PREP

5. Which of the following is a known association between viruses and cancers?
   a. *Papillomavirus causes brain cancer.*
   b. *Infection with herpesvirus leads to AIDS.*
   c. *Hepatitis B is associated with liver cancer.*
   d. *Papillomavirus is associated with gastric cancer.*

5.5 Other Noncellular Infectious Agents

Not all noncellular infectious agents are viruses. One group of unusual forms, even smaller and simpler than viruses, is implicated in chronic, persistent diseases in humans and animals. These diseases are called spongiform encephalopathies because the brain tissue removed from affected animals resembles a sponge. The infection has a long period of latency (usually several years) before the first clinical signs appear. Signs range from mental derangement to loss of muscle control. The diseases are progressive and universally fatal.

A common feature of these conditions is the deposition of distinct protein fibrils in the brain tissue. Researchers have hypothesized that these fibrils are the agents of the disease and have named them **prions** (pree′-onz).

Creutzfeldt-Jakob disease afflicts the central nervous system of humans and causes gradual degeneration and death. Several animals (sheep, mink, elk) are victims of similar transmissible diseases. Bovine spongiform encephalopathy (BSE), or "mad cow disease," was recently the subject of fears and a crisis in Europe when researchers found evidence that the disease could be acquired by humans who consumed contaminated beef. This was the first incidence of prion disease transmission from animals to humans. Several hundred Europeans developed symptoms of a variant form of Creutzfeldt-Jakob disease, leading to strict governmental controls on exporting cattle and beef products. In 2003, isolated cows with BSE were found in Canada and in the United States. Precautionary measures have been taken to protect North American consumers. As of 2011, only three BSE-positive cows have been found in the United States, compared to over 184,000 in the United Kingdom. (This disease is described in more detail in chapter 17.)

The exact mode of prion infection is currently being investigated. The fact that prions are composed primarily of protein (no nucleic acid) has certainly revolutionized our ideas of what can constitute an infectious agent. One of the most compelling questions is just how a prion could be replicated, because all other infectious agents require some nucleic acid.

Other fascinating viruslike agents in human disease are defective forms called satellite viruses that are actually dependent on other viruses for replication. Two remarkable examples are the adeno-associated virus (AAV), so named because it was originally thought that it could replicate only in cells infected with adenovirus. But it can also infect cells that are infected with other viruses or that have had their DNA disrupted through other means. Another satellite virus, called the delta agent, is a naked circle of RNA that is expressed only in the presence of the hepatitis B virus and can worsen the severity of liver damage.

Plants are also parasitized by viruslike agents called **viroids** that differ from ordinary viruses by being very small (about one-tenth the size of an average virus) and being composed of only naked strands of RNA, lacking a capsid or any other type of coating. Viroids are significant pathogens in several economically important plants, including tomatoes, potatoes, cucumbers, citrus trees, and chrysanthemums.

> **5.5 LEARNING OUTCOMES—Assess Your Progress**
> 14. Name two noncellular infectious agents besides viruses.

5.6 Viruses and Human Health

The number of viral infections that occur on a worldwide basis is nearly impossible to measure accurately. Certainly, viruses are extremely common causes of acute infections such as colds, hepatitis, chickenpox, influenza, herpes, and warts. If one also takes into account prominent viral infections found only in certain regions of the world, such as Dengue fever, Rift Valley fever, and yellow fever, the total could easily exceed several billion cases each year. Although most viral infections do not result in death, some, such as rabies, AIDS, and Ebola, have very high mortality rates, and others can lead to long-term debility (polio, neonatal rubella). Current research is focused on the possible connection of viruses to chronic afflictions of unknown cause, such as type 1 diabetes, multiple sclerosis, various cancers, Alzheimer's, and even obesity. Additionally, as mentioned earlier, several cancers have their origins in viral infection.

Table 5.6 provides a list of the most common viruses causing diseases in humans.

Table 5.6 Important Human Virus Families, Genera, Common Names, and Types of Diseases

Family	Genus of Virus	Common Name of Genus Members	Name of Disease
DNA Viruses			
Poxviridae	*Orthopoxvirus*	Variola and vaccinia	Smallpox, cowpox
Herpesviridae	*Simplexvirus*	Herpes simplex 1 virus (HSV)	Fever blister, cold sores
		Herpes simplex 2 virus (HSV)	Genital herpes
	Varicellovirus	Varicella zoster virus (VZV)	Chickenpox, shingles
	Cytomegalovirus	Human cytomegalovirus (CMV)	CMV infections
Adenoviridae	*Mastadenovirus*	Human adenoviruses	Adenovirus infection
Papovaviridae	*Papillomavirus*	Human papillomavirus (HPV)	Several types of warts
	Polyomavirus	JC virus (JCV)	Progressive multifocal leukoencephalopathy (PML)
Hepadnaviridae	*Orthohepadnavirus*	Hepatitis B virus (HBV or Dane particle)	Serum hepatitis
Parvoviridae	*Erythrovirus*	Parvovirus B19	Erythema infectiosum
RNA Viruses			
Picornaviridae	*Enterovirus*	Poliovirus	Poliomyelitis
		Coxsackie virus	Hand-foot-mouth disease
	Hepatovirus	Hepatitis A virus (HAV)	Short-term hepatitis
	Rhinovirus	Human rhinovirus	Common cold, bronchitis
Caliciviridae	*Norovirus*	Norwalk virus	Viral diarrhea, Norwalk virus syndrome
Togaviridae	*Alphavirus*	Eastern equine encephalitis virus	Eastern equine encephalitis (EEE)
		Western equine encephalitis virus	Western equine encephalitis (WEE)
		St. Louis encephalitis virus	St. Louis encephalitis
	Rubivirus	Rubella virus	Rubella (German measles)
Flaviviridae	*Flavivirus*	Dengue fever virus	Dengue fever
		West Nile fever virus	West Nile fever
		Yellow fever virus	Yellow fever
Coronaviridae	*Coronavirus*	Infectious bronchitis virus (IBV)	Bronchitis
		Enteric corona virus	Coronavirus enteritis
	Betacoronavirus	SARS virus	Severe acute respiratory syndrome
Filoviridae	*Ebolavirus* *Marburgvirus*	Ebola, Marburg virus	Ebola fever
Orthomyxoviridae	*Influenza A virus*	Influenza virus, type A (Asian, Hong Kong, and swine influenza viruses)	Influenza or "flu"
Paramyxoviridae	*Respirovirus* *Rubulavirus*	Parainfluenza virus, types 1–5	Parainfluenza
		Mumps virus	Mumps
	Morbillivirus	Measles virus	Measles
	Pneumovirus	Respiratory syncytial virus (RSV)	Common cold syndrome
Rhabdoviridae	*Lyssavirus*	Rabies virus	Rabies
Bunyaviridae	*Orthobunyavirus*	Bunyamwera viruses	California encephalitis
	Hantavirus	Sin Nombre virus	Respiratory distress syndrome
	Phlebovirus	Rift Valley fever virus	Rift Valley fever
	Nairovirus	Crimean–Congo hemorrhagic fever virus (CCHF)	Crimean–Congo hemorrhagic fever
Reoviridae	*Coltivirus*	Colorado tick fever virus	Colorado tick fever
	Rotavirus	Human rotavirus	Rotavirus gastroenteritis
Retroviridae	*Deltaretrovirus*	Human T-lymphotropic virus 1 (HTLV-1)	T-cell leukemia
	Lentivirus	HIV (human immunodeficiency viruses 1 and 2)	Acquired immunodeficiency syndrome (AIDS)
Arenaviridae	*Arenavirus*	Lassa virus	Lassa fever

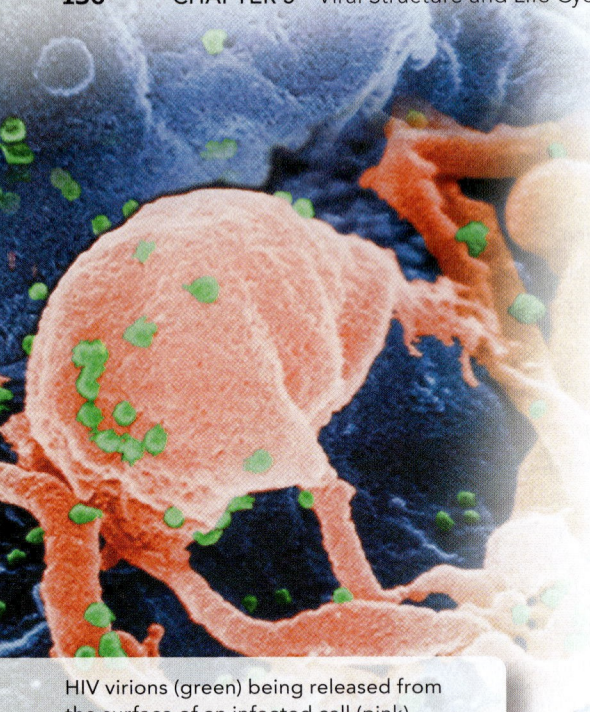

HIV virions (green) being released from the surface of an infected cell (pink).

Treatment of Animal Viral Infections

The nature of viruses makes it difficult to design effective therapies against them. Because viruses are not bacteria, antibiotics aimed at disrupting bacterial cells do not work on them. Until now, most antiviral drugs were designed to block virus replication by targeting the function of host cells and therefore could cause severe side effects. Almost all antiviral drugs so far licensed have been designed to target one of the steps in the viral life cycle you learned about earlier in this chapter. The integrase inhibitor class of HIV drugs interrupts the ability of HIV genetic information to incorporate into the host cell DNA.

A breakthrough was made in 2011 in the development of antiviral drugs. A molecule called a *double-stranded RNA activated caspase oligomizer* (affectionately called DRACO) was developed. It causes virus-infected cells to destroy themselves, no matter what the virus is. Even though this strategy has great promise, it will be several years before it can be used clinically.

Vaccines that stimulate immunity are an extremely valuable tool but are available for only a limited number of viral diseases.

5.6 LEARNING OUTCOMES—Assess Your Progress

15. Analyze the relative importance of viruses in human infection and disease.
16. Discuss the primary reason that antiviral drugs are more difficult to design than antibacterial drugs.

CASE FILE WRAP-UP

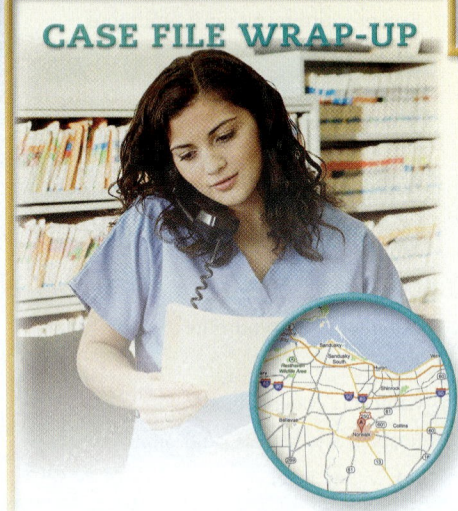

Norwalk virus (*Norovirus*, from the *Caliciviridae* family) is responsible for approximately 90% of nonbacterial epidemic gastroenteritis outbreaks worldwide. The virus is extremely contagious—only 20 virus particles are needed to cause illness (the infectious dose), which explains how the virus spreads so easily. The virus may be spread by direct contact or through ingestion of contaminated water or food (salads and shellfish are often implicated). The virus can also be aerosolized (i.e., when an individual in close proximity to an infected person who is vomiting breathes in virus particles).

Symptoms include nausea, vomiting, abdominal pain or cramping, watery diarrhea, weakness, headache, muscle aches, and low-grade fever. Symptoms occur 24 to 48 hours after exposure to the virus and subside within 24 to 60 hours. The elderly, the very young, and individuals with weakened immune systems can quickly become very dehydrated. Although death is rare, Norwalk virus is responsible for approximately 300 deaths per year in the United States. Outbreaks occur in closed communities where people interact in close proximity, such as schools, long-term care facilities, camps, prisons, and cruise ships.

Shingles

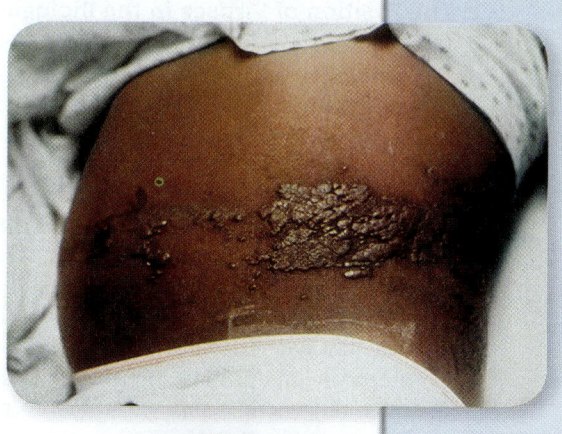

Shingles is an example of a disease caused by a virus that remains in the body in a chronic latent state, only to reappear years later. The disease is caused by the varicella zoster virus, the same virus that causes chickenpox. Once an individual has recovered from chickenpox, the virus can "hide" in the nerves for years. When the virus is triggered again, perhaps by changes in immunity, it becomes active and causes the disease known as shingles. In some people, emotional or physical stress seems to trigger reactivation. Shingles occurs most commonly in older people who had chickenpox at some point in their lives, usually during childhood. Most people experience shingles only once, but a few unlucky people may suffer from shingles more than once. A person who has shingles can pass the virus on to someone who has never had it, but that person will get chickenpox, not shingles.

Shingles initially causes a tingling, burning, or painful sensation. Discomfort precedes the rash, which starts as reddened areas on the skin that eventually form small blisters. The blisters eventually break and crusted areas of skin remain, which are shed in 2 to 3 weeks. The rash usually follows dermatomes, areas of the skin supplied by sensory fibers of the spinal cord. Typically, the rash starts on the back and extends around to the skin on the chest or abdomen on one side of the body. The pain of shingles can sometimes be severe. Some people develop *postherpetic neuralgia*, resulting from damage to the nerves, which causes chronic pain. Other symptoms associated with shingles include fever, headache, malaise, abdominal pain, and joint pain. Shingles affecting the eye or ear may result in vision or hearing loss.

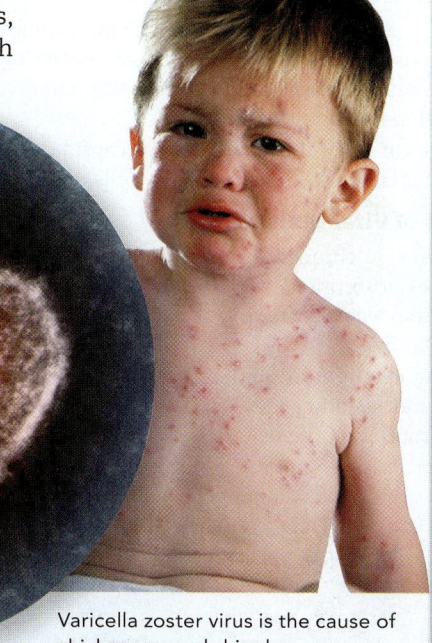

The diagnosis can usually be made based on the appearance of the rash. Leukocytosis (an elevated white blood cell count) and antibodies to the chickenpox virus can help to confirm the virus when there is any doubt as to the cause of the rash. Most of the time, however, the appearance of the rash is enough to make the diagnosis. Antiviral drugs may be used; although they can't cure shingles, they may shorten the course of the disease. However, antivirals need to be started within 72 hours of the start of symptoms and preferably before the blisters that accompany the rash appear. Acyclovir, famciclovir, and valacyclovir are the antivirals typically used. They are given in higher doses than would typically be used for herpes simplex or genital herpes. Antiviral medications may be given intravenously to people who are immunocompromised or at risk for disseminated disease. It is important for individuals to visit their physician immediately upon noticing symptoms of shingles so that antiviral therapy can be commenced.

A live vaccine, called Zostavax, is now recommended in the United States for adults aged 50 and older who have had chickenpox. The vaccine has been shown to prevent up to 50% of cases of shingles and has also been shown to reduce the occurrence of postherpetic neuralgia, which can be very debilitating and can lead to lifelong pain that is very difficult to treat.

Varicella zoster virus is the cause of chickenpox and shingles.

137

Chapter Summary

5.1 The Position of Viruses in the Biological Spectrum

- Viruses are noncellular entities whose properties have been identified through technological advances in microscopy and tissue culture.
- Viruses are infectious particles that invade every known type of cell. They are not alive, yet they are able to redirect the metabolism of living cells to reproduce virus particles.
- Viruses have a profound influence on the genetic makeup of the biosphere.
- The International Committee on the Taxonomy of Viruses oversees naming and classification of viruses. Viruses are classified into orders, families, and genera.
- Viruses are grouped in various ways. This textbook uses their structure, genetic composition, and host range to categorize them.

5.2 The General Structure of Viruses

- Virus size range is from 20 to 450 nm (diameter). Viruses are composed of an outer protein capsid enclosing either DNA or RNA plus a variety of enzymes. Some viruses also exhibit an envelope around the capsid.
- Spikes on the surface of the virus capsid or envelope are critical for their attachment to host cells.

5.3 Modes of Viral Multiplication

- Viruses go through a multiplication cycle that generally involves adsorption, penetration (sometimes followed by uncoating), viral synthesis and assembly, and viral release by lysis or budding.
- These events turn the host cell into a factory solely for making and shedding new viruses. This results in the ultimate destruction of the cell.
- Animal viruses can cause acute infections or can persist in host tissues as chronic latent infections that can reactivate periodically throughout the host's life. Some persistent animal viruses can cause cancer.

- Bacteriophages vary significantly from animal viruses in their methods of adsorption, penetration, site of replication, and method of exit from host cells.
- Lysogeny is a condition in which viral DNA is inserted into the bacterial chromosome and remains inactive for an extended period. The viral DNA is replicated with the chromosome every time the bacterium divides.
- Some bacteria express virulence traits that are coded for by the bacteriophage DNA in their chromosomes. This phenomenon is called lysogenic conversion.

5.4 Techniques in Cultivating and Identifying Animal Viruses

- Animal viruses must be studied in some sort of living cell or tissue.
- Viruses are grouped in various ways. This textbook uses their structure, genetic composition, and host range to categorize them.
- Cell and tissue cultures are cultures of host cells grown in special sterile chambers using aseptic techniques to exclude unwanted microorganisms.
- Virus growth in cell culture and bacteriophage growth on bacterial lawns are detected by the appearance of plaques.

5.5 Other Noncellular Infectious Agents

- Other noncellular agents of disease are the prions, which are not viruses at all but protein fibers; viroids, extremely small lengths of naked nucleic acid; and satellite viruses, which require the presence of larger viruses to cause disease.

5.6 Viruses and Human Health

- Viruses are easily responsible for several billion infections each year. It is conceivable that many chronic diseases of unknown cause will eventually be connected to viral agents.
- Viral infections are difficult to treat because the drugs that attack viral replication also cause side effects in the host.

Multiple-Choice Questions   Bloom's Levels 1 and 2: Remember and Understand

Select the correct answer from the answers provided.

1. When phage nucleic acid is incorporated into the nucleic acid of its host cell and is replicated when the host DNA is replicated, this is considered part of which cycle?

   a. lytic cycle
   b. virulence cycle
   c. lysogenic cycle
   d. cell cycle
   e. multiplication cycle

2. A virus that undergoes lysogeny is a/an

   a. temperate phage.
   b. intemperate phage.
   c. T-even phage.
   d. animal virus.
   e. DNA virus.

3. The nucleic acid of a virus is

   a. DNA only.
   b. RNA only.
   c. both DNA and RNA.
   d. either DNA or RNA.

4. The general steps in a viral multiplication cycle are

   a. adsorption, penetration, synthesis, assembly, and release.
   b. endocytosis, uncoating, replication, assembly, and budding.
   c. adsorption, uncoating, duplication, assembly, and lysis.
   d. endocytosis, penetration, replication, maturation, and exocytosis.

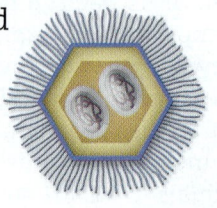

5. A prophage is an early stage in the development of a/an
   a. bacterial virus.
   b. poxvirus.
   c. lytic virus.
   d. enveloped virus.

6. In general, RNA viruses multiply in the cell _____, and DNA viruses multiply in the cell _____.
   a. nucleus; cytoplasm
   b. cytoplasm; nucleus
   c. vesicles; ribosomes
   d. endoplasmic reticulum; nucleolus

7. Viruses cannot be cultivated in
   a. tissue culture.
   b. bird embryos.
   c. live mammals.
   d. blood agar.

8. Clear patches in cell cultures that indicate sites of virus infection are called
   a. plaques.
   b. pocks.
   c. colonies.
   d. prions.

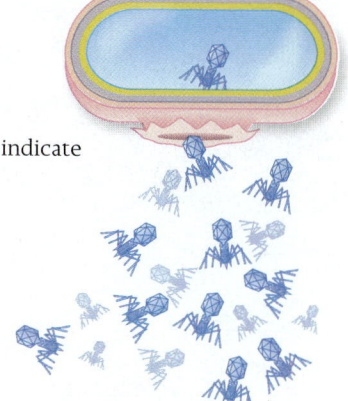

Critical Thinking   Bloom's Levels 3, 4, and 5: Apply, Analyze, and Evaluate

Critical thinking is the ability to reason and solve problems using facts and concepts. These questions can be approached from a number of angles and, in most cases, they do not have a single correct answer.

1. a. What characteristics of viruses could be used to characterize them as life forms?
   b. What makes them more similar to lifeless molecules?

2. a. If viruses that normally form envelopes were prevented from budding, would they still be infectious? Why or why not?
   b. If the RNA of an influenza virus were injected into a cell by itself, could it cause an active infection?

3. a. If you were involved in developing an antiviral drug, what would be some important considerations? (Can a drug "kill" a virus?)
   b. How could multiplication be blocked?

4. Is there such a thing as a "good virus"? Explain why or why not. Consider both bacteriophages and viruses of eukaryotic organisms.

5. Discuss some advantages and disadvantages of using bacteriophage therapy in treating bacterial infections of humans.

Visual Connections   Bloom's Level 5: Evaluate

This question connects previous images to a new concept.

1. **From chapter 1, table 1.1.** This chart from chapter 1 identified diseases most clearly caused by microorganisms. Considering what you have learned in this chapter, are there more deaths caused by microorganisms than might be accounted for by the red-labeled diseases? Can you make a rough guess of how many total deaths might be caused by viruses?

Table 1.1 Top Causes of Death—All Diseases

United States	No. of Deaths	Worldwide	No. of Deaths
1. Heart disease	617,000	1. Heart disease	7 million
2. Cancer	565,000	2. Stroke	6.2 million
3. Chronic lower-respiratory disease	141,000	3. Lower-respiratory infections (influenza and pneumonia)	3.2 million
4. Cerebrovascular disease	134,000	4. Chronic obstructive pulmomary disease	3 million
5. Accidents (unintentional injuries)	122,000	5. Diarrheal diseases	1.9 million
6. Alzheimer's disease	82,000	6. HIV/AIDS	1.5 million
7. Diabetes	71,000	7. Trachea, bronchus, lung cancers	1.5 million
8. Influenza and pneumonia	56,000	8. Diabetes mellitus	1.4 million
9. Kidney disease	48,000	9. Road injury	1.3 million
10. Suicide	36,000	10. Prematurity	1.2 million

*Diseases in red are those most clearly caused by microorganisms.

Source: Data from the World Health Organization and the Centers for Disease Control and Prevention. Data published in 2014 representing final figures for the year 2011.

|MICROBIOLOGY

www.mcgrawhillconnect.com

Enhance your study of this chapter with study tools and practice tests. Also ask your instructor about the resources available through Connect, including the media-rich eBook, interactive learning tools, and animations.

CASE FILE

Wound Care

I was an RN working in a large city hospital on a medical floor. A lot of our patients had diabetes and were suffering various complications of the disease, particularly diabetic wounds caused by poor circulation. Wound care was a large part of my job. After 2 years on the unit, I decided to pursue wound care certification. Once I became a wound care specialist, I continued to work in the same hospital and saw patients with complicated and/or chronic wounds.

Mr. Jones was one of the first patients I consulted about after I became certified. He was an elderly gentleman who had lost his sight due to diabetes. When I met Mr. Jones, he had a chronic wound on his lower leg that had been present for months. The wound was circumferential, taking up half of his lower leg. It was also grossly infected. Mr. Jones had been admitted to the hospital for antibiotics to treat his infection. It was clear that if the antibiotics failed to improve his wound, Mr. Jones was in danger of losing his leg.

Within a day of admission, we realized that antibiotic therapy alone was not going to be enough. Mr. Jones developed signs of gas gangrene. Wound cultures were positive for *Clostridium perfringens,* which produces toxins that destroy muscle tissue and results in sepsis and death if untreated.

Mr. Jones was taken immediately to surgery where his wound was debrided, meaning that dead or devitalized tissue was removed. Following, he was given large doses of penicillin in an effort to stop the spread of the infection.

The next day, Mr. Jones was started on daily hyperbaric oxygen therapy, with sessions lasting for 45 minutes. Slowly, his wound began to improve. The wound was debrided twice more under anesthesia, and the patient remained on antibiotics until wound cultures came back free of *C. perfringens*. Although the wound took several months to heal, Mr. Jones kept his leg.

- What is hyperbaric oxygen therapy, and why is it used to treat wounds infected with *C. perfringens*?

- Is *C. perfringens* considered an aerobe or an anaerobe?

Case File Wrap-Up appears on page 162.

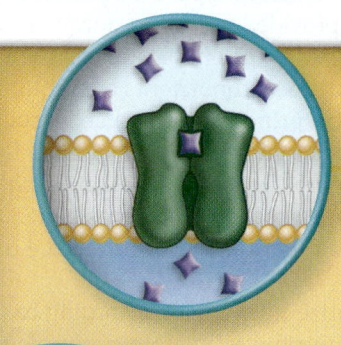

Microbial Nutrition and Growth

6

IN THIS CHAPTER...

6.1 Microbial Nutrition

1. List the essential nutrients of a bacterial cell.
2. Differentiate between macronutrients and micronutrients.
3. List and define four different terms that describe an organism's sources of carbon and energy.
4. Define *saprobe* and *parasite*, and provide microbial examples of each.
5. Compare and contrast the processes of diffusion and osmosis.
6. Identify the effects of isotonic, hypotonic, and hypertonic conditions on a cell.
7. Name two types of passive transport and one type of active transport.

6.2 Environmental Factors That Influence Microbes

8. List and define five terms used to express a microbe's optimal growth temperature.
9. Summarize three ways in which microorganisms function in the presence of differing oxygen conditions.
10. Identify three important environmental factors (other than temperature and oxygen) with which organisms must cope.
11. List and describe the five types of associations microbes can have with their hosts.
12. Discuss characteristics of biofilms that differentiate them from planktonic bacteria.

6.3 The Study of Bacterial Growth

13. Summarize the steps of cell division used by most bacteria.
14. Define *doubling time*, and describe how it leads to exponential growth.
15. Compare and contrast the four phases of growth in a bacterial growth curve.
16. Identify one quantitative and one qualitative method used for analyzing bacterial growth.

6.1 Microbial Nutrition

With respect to nutrition, microbes are not really so different from humans. Bacteria living in mud on a diet of inorganic sulfur, or protozoa digesting wood in a termite's intestine, seem to live radical lifestyles, but even these organisms require a constant influx of certain substances from their habitat. In general, all living things require a source of elements such as carbon, hydrogen, oxygen, phosphorus, potassium, nitrogen, sulfur, calcium, iron, sodium, chlorine, magnesium, and certain other elements. But the ultimate source of a particular element, its chemical form, and how much of it the microbe needs are all points of variation between different types of organisms.

The pristine waters of this beautiful coral reef depend on keeping microbial nutrients very low so that harmful bacteria are not able to outcompete phytoplankton or cause coral diseases.

Any substance that must be provided to an organism is called an **essential nutrient.** Two categories of essential nutrients are **macronutrients** and **micronutrients.** Macronutrients are required in relatively large quantities and play principal roles in cell structure and metabolism. Examples of macronutrients are carbon, hydrogen, and oxygen. Micronutrients, or **trace elements,** such as manganese, zinc, and nickel, are present in much smaller amounts and are involved in enzyme function and maintenance of protein structure.

Another way to categorize nutrients is according to their carbon content. An inorganic nutrient is an atom or simple molecule that contains a combination of atoms other than carbon and hydrogen. The natural reservoirs of inorganic compounds are mineral deposits in the crust of the earth, bodies of water, and the atmosphere. Examples include metals and their salts (magnesium sulfate, ferric nitrate, sodium phosphate), gases (oxygen, carbon dioxide), and water. In contrast, the molecules of organic nutrients contain carbon and hydrogen atoms and are usually the products of living things. They range from the simplest organic molecule, methane (CH_4), to large polymers (carbohydrates, lipids, proteins, and nucleic acids). The source of nutrients is extremely varied: Some microbes obtain their nutrients entirely from inorganic sources, and others require a combination of organic and inorganic sources.

Chemical Analysis of Microbial Cytoplasm

Table 6.1 lists the major contents of the bacterium *Escherichia coli.* Some of these components are absorbed in a ready-to-use form, and others must be synthesized by the cell from simple nutrients. The important features of cell composition can be summarized as follows:

- Water is the most abundant of all the components (70%).
- Proteins are the next most prevalent chemical.
- About 97% of the dry cell weight is composed of organic compounds.
- About 96% of the dry cell weight is composed of six elements (represented by CHONPS and shown later in table 6.3).
- Chemical elements are needed in the overall scheme of cell growth, but most of them are available to the cell as compounds and not as pure elements.
- A cell as "simple" as *E. coli* contains on the order of 5,000 different compounds, yet it needs to absorb only a few types of nutrients to synthesize this great diversity. These include ammonium sulfate (($NH_4)_2SO_4$), iron chloride ($FeCl_2$), sodium chloride (NaCl), trace elements, glucose, potassium phosphate (KH_2PO_4), magnesium sulfate ($MgSO_4$), calcium phosphate ($CaHPO_4$), and water.

What Microbes Eat

The earth's limitless habitats and microbial adaptations are matched by an elaborate menu of microbial nutritional schemes. Fortunately, most organisms show consistent trends and can be described by a few general categories **(table 6.2)** and a few

Table 6.1 The Chemical Composition of an *Escherichia coli* Cell

Organic Compounds	% Dry Weight	Elements	% Dry Weight
Proteins	50	Carbon (C)	50
Nucleic Acids		Oxygen (O)	20
RNA	20	Nitrogen (N)	14
DNA	3	Hydrogen (H)	8
Carbohydrates	10	Phosphorus (P)	3
Lipids	10	Sulfur (S)	1
Miscellaneous	4	Potassium (K)	1
Inorganic Compounds		Sodium (Na)	1
Water	(—)	Calcium (Ca)	0.5
All others	3	Magnesium (Mg)	0.5
		Chlorine (Cl)	0.5
		Iron (Fe)	0.2
		Trace metals	0.3

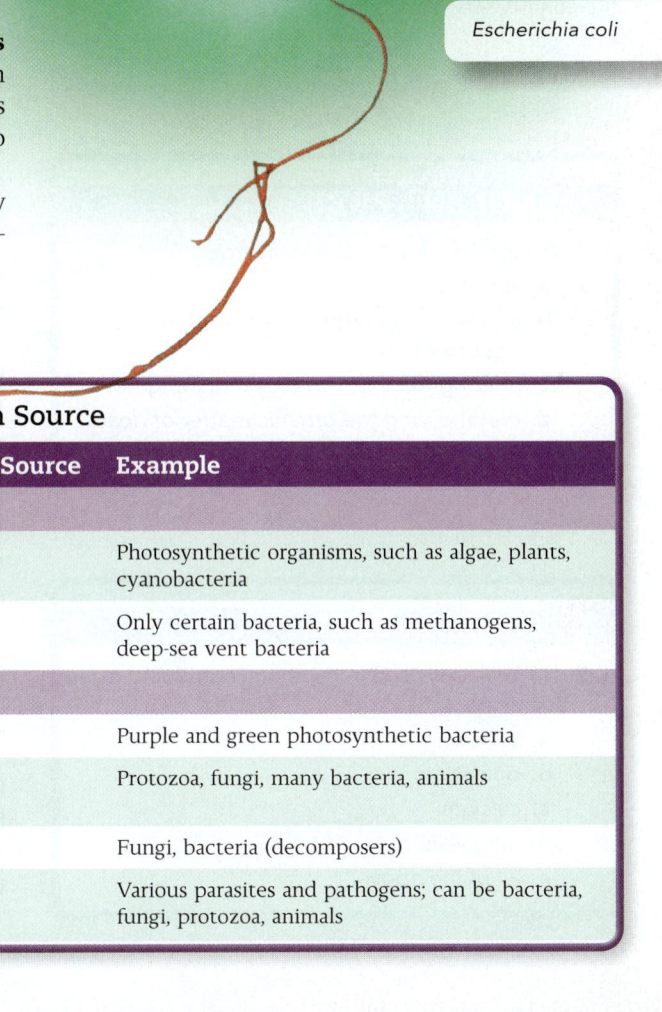

Escherichia coli

selected terms. The main determinants of a microbe's nutritional type are its sources of carbon and energy.

We'll start with an organism's carbon source: Microbes are either **heterotrophs** or **autotrophs.** A heterotroph is an organism that must obtain its carbon in an organic form. An autotroph ("self-feeder") is an organism that uses inorganic CO_2 as its carbon source. Because autotrophs have the special capacity to convert CO_2 into organic compounds, they are not nutritionally dependent on other living things.

The next way that microbes are categorized is via their energy source. They are either **phototrophs** or **chemotrophs.** Microbes that photosynthesize are phototrophs, and those that gain energy from chemical compounds are chemotrophs.

Table 6.2 Nutritional Categories of Microbes by Energy and Carbon Source

Category	Energy Source	Carbon Source	Example
Autotroph	**Nonliving Environment**		
Photoautotroph	Sunlight	CO_2	Photosynthetic organisms, such as algae, plants, cyanobacteria
Chemoautotroph	Simple inorganic	CO_2	Only certain bacteria, such as methanogens, deep-sea vent bacteria
Heterotroph	**Other Organisms or Sunlight**		
Photoheterotroph	Sunlight	Organic	Purple and green photosynthetic bacteria
Chemoheterotroph	Metabolic conversion of the nutrients from other organisms	Organic	Protozoa, fungi, many bacteria, animals
Saprobe	Metabolizing the organic matter of dead organisms	Organic	Fungi, bacteria (decomposers)
Parasite	Utilizing the tissues, fluids of a live host	Organic	Various parasites and pathogens; can be bacteria, fungi, protozoa, animals

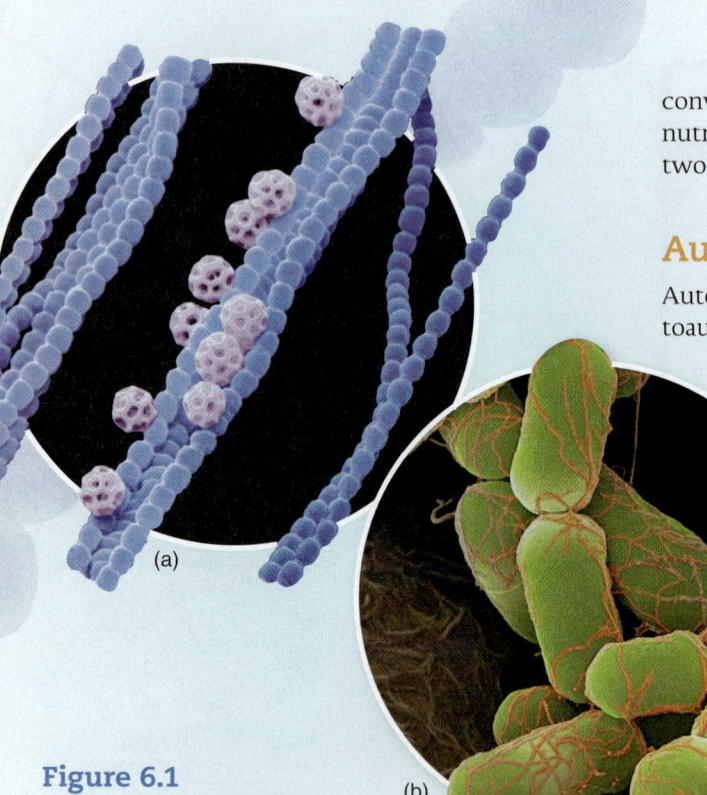

Figure 6.1

A photoautotroph and a chemoheterotroph.

(a) *Cyanobacterium,* in blue, a photosynthetic autotroph. (b) *Escherichia coli,* a chemoheterotroph.

The terms for carbon and energy source are often merged into a single word for convenience. The categories described here are meant to describe only the major nutritional groups and do not include unusual exceptions. **Figure 6.1** illustrates two examples.

Autotrophs and Their Energy Sources

Autotrophs derive energy from one of two possible nonliving sources: sunlight (photoautotrophs) and chemical reactions involving simple chemicals (chemoautotrophs). **Photoautotrophs** are photosynthetic–that is, they capture the energy of light rays and transform it into chemical energy that can be used in cell metabolism. Because photosynthetic organisms (algae, plants, some bacteria) produce organic molecules that can be used by themselves and by heterotrophs, they form the basis for most food webs. Their role as primary producers of organic matter is discussed in chapter 22.

Chemoautotrophs are of two types: One of these is the group called chemoorganic autotrophs. These use organic compounds for energy and inorganic compounds as a carbon source. The second type of chemoautotroph is a group called **lithoautotrophs,** which require neither sunlight nor organic nutrients, relying totally on inorganic minerals. These bacteria derive energy in diverse and rather amazing ways. In very simple terms, they remove electrons from inorganic substrates–such as hydrogen gas, hydrogen sulfide, sulfur, or iron–and combine them with carbon dioxide and hydrogen.

Heterotrophs and Their Energy Sources

The majority of heterotrophic microorganisms are **chemoheterotrophs** that derive both carbon and energy from organic compounds. Processing these organic molecules by respiration or fermentation releases energy in the form of ATP. Chemoheterotrophic microorganisms belong to one of two main categories that differ in how they obtain their organic nutrients: **Saprobes** are free-living microorganisms that feed primarily on organic detritus from dead organisms, and **parasites** ordinarily derive nutrients from the cells or tissues of a living host.

Saprobes occupy a niche as decomposers of plant litter, animal matter, and dead microbes. If not for the work of decomposers, the earth would gradually fill up with organic material, and the nutrients it contains would not be recycled.

Parasites live in or on the body of a host, which they harm to some degree. Because parasites cause damage to tissues (disease) or even death, they are also called **pathogens.** Parasites range from viruses to helminths (worms), and they can live on the body (ectoparasites), in the organs and tissues (endoparasites), or even within cells (intracellular parasites, the most extreme type). *Obligate parasites* (for example, the leprosy bacillus and the syphilis spirochete) are unable to grow outside of a living host. Parasites that are less strict can be cultured artificially if provided with the correct nutrients and environmental conditions. Bacteria such as *Streptococcus pyogenes* (the cause of strep throat) and *Staphylococcus aureus* can grow on artificial media.

The vast majority of microbes causing human disease are chemoheterotrophs.

Essential Nutrients

Chemicals that are necessary for particular organisms, which they cannot manufacture by themselves, are called **essential.** For microbes, the essential nutrients are carbon, hydrogen, oxygen, nitrogen, phosphate, and sulfur. Biologists remember these elements with the acronym CHONPS **(table 6.3).**

NCLEX® PREP

1. A saprobe derives its energy from
   a. *sunlight.*
   b. *conversion of nutrients from other organisms.*
   c. *utilizing the tissues/fluids of a living host.*
   d. *metabolizing the organic matter of dead organisms.*

NCLEX® PREP

2. Mineral ions used in microbial nutrition include
   a. *sodium.*
   b. *potassium.*
   c. *calcium.*
   d. *magnesium.*
   e. *all of the above.*

Table 6.3 Essential Nutrients

Carbon	Among the common organic molecules that can satisfy this requirement are proteins, carbohydrates, lipids, and nucleic acids. In most cases, these molecules provide several other nutrients as well.
Hydrogen	Hydrogen is a major element in all organic and several inorganic compounds, including water (H_2O), salts ($Ca[OH]_2$), and certain naturally occurring gases (H_2S, CH_4, and H_2). These gases are both used and produced by microbes. Hydrogen helps cells maintain their pH, is useful for forming hydrogen bonds between molecules, and also serves as a source of free energy in respiration.
Oxygen	Because oxygen is a major component of organic compounds such as carbohydrates, lipids, nucleic acids, and proteins, it plays an important role in the structural and enzymatic functions of the cell. Oxygen is likewise a common component of inorganic salts such as sulfates, phosphates, nitrates, and water. Free gaseous oxygen (O_2) makes up 20% of the atmosphere.
Nitrogen	The main reservoir of nitrogen is nitrogen gas (N_2), which makes up 79% of the earth's atmosphere. This element is indispensable to the structure of proteins, DNA, RNA, and ATP. Such compounds are the primary nitrogen source for heterotrophs, but to be useful, they must first be degraded into their basic building blocks (proteins into amino acids; nucleic acids into nucleotides). Some bacteria and algae utilize inorganic nitrogenous nutrients (NO_3^-, NO_2^-, or NH_3). A small number of bacteria and archaea can transform N_2 into compounds usable by other organisms through the process of nitrogen fixation. Regardless of the initial form in which the inorganic nitrogen enters the cell, it must first be converted to NH_3, the only form that can be directly combined with carbon to synthesize amino acids and other compounds.
Phosphate	The main inorganic source of phosphorus is phosphate (PO_4^{3-}), derived from phosphoric acid (H_3PO_4) and found in rocks and oceanic mineral deposits. Phosphate is a key component of nucleic acids and is therefore essential to the genetics of cells and viruses. Because it is also found in ATP, it serves in cellular energy transfers. Other phosphate-containing compounds are phospholipids in cytoplasmic membranes and coenzymes such as NAD^+.
Sulfur	Sulfur is widely distributed throughout the environment in mineral form. Rocks and sediments (such as gypsum) can contain sulfate (SO_4^{2-}), sulfides (FeS), hydrogen sulfide gas (H_2S), and elemental sulfur (S). Sulfur is an essential component of some vitamins (vitamin B_1) and the amino acids methionine and cysteine; the latter help determine shape and structural stability of proteins by forming unique linkages called disulfide bonds.

Other Important Nutrients

Mineral ions are also important components in microbial metabolism. Potassium is essential to protein synthesis and membrane function. Sodium is important for certain types of cell transport. Calcium is a stabilizer of the cell wall and endospores of bacteria. Magnesium is a component of chlorophyll and a stabilizer of membranes and ribosomes. Iron is an important component of the cytochrome proteins of cell respiration. Zinc is an essential regulatory element for eukaryotic genetics. It is a major component of "zinc fingers"–binding factors that help enzymes adhere to specific sites on DNA. Copper, cobalt, nickel, molybdenum, manganese, silicon, iodine, and boron are needed in small amounts by some microbes but not others. On the other hand, in chapter 9 you will see that metals can also be very toxic to microbes. The concentration of metal ions can even influence the diseases microbes cause. For example, the bacteria that cause gonorrhea and meningitis grow more rapidly in the presence of iron ions.

How Microbes Eat: Transport Mechanisms

A microorganism's habitat provides necessary nutrients–some abundant, others scarce– that must still be taken into the cell. Survival also requires that cells transport waste materials out of the cell and into the environment. Whatever the direction, transport occurs across the cytoplasmic membrane, the structure specialized for this role. This is true even in organisms with cell walls (bacteria, algae, and fungi), because the cell wall is usually too nonselective to screen the entrance or exit of molecules.

The driving force of transport is atomic and molecular movement–the natural tendency of atoms and molecules to be in constant random motion. This phenomenon of molecular movement, in which atoms or molecules move in a gradient from an area of higher density or concentration to an area of lower density or concentration, is **diffusion.**

Medical Moment

Osmosis and IV Fluids

The provision of intravenous (IV) solutions is a very common practice in medicine. Keeping in mind that the osmotic movement of water occurs as the body attempts to create a balance between the different solute concentrations that exist on either side of a semipermeable membrane, let's look at different types of IV solutions commonly used in medicine.

Isotonic solutions have a tonicity that is the same as the body's plasma. When isotonic solutions are administered, there will be very little movement, if any, between the cells and the blood vessels.

Hypertonic solutions have a tonicity that is higher than the body plasma. Administering hypertonic solutions will cause water to shift from the extravascular spaces into the bloodstream to increase the intravascular volume. This is how the body attempts to dilute the higher concentration of electrolytes in the IV fluid.

Hypotonic solutions have a tonicity that is lower than the body plasma, causing water to shift from the intravascular to the extravascular space, and eventually into the cells of the tissues. In this case, the body moves water from the intravascular space to the cells in order to dilute the electrolytes in the solution.

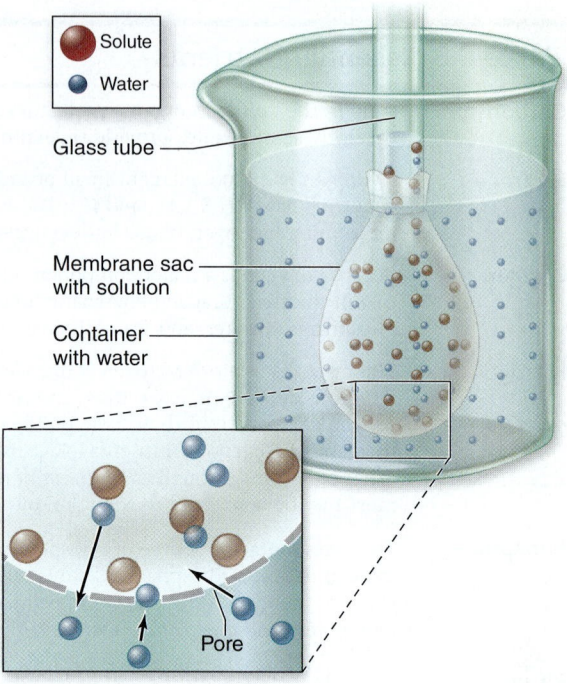

Figure 6.2 Model system to demonstrate osmosis. Here we have a solution enclosed in a sack-shaped membrane and attached to a hollow tube. The membrane is permeable to water (solvent) but not to solute. The sack is immersed in a container of pure water. In the inset, you see that the net direction of water diffusion is into the sac.

Solute
Water
Glass tube
Membrane sac with solution
Container with water
Pore

NCLEX® PREP

3. A physician has ordered hypotonic parenteral therapy for a post-operative client over a 24-hour period. Based on this order, what is the *priority* nursing action?
   a. *Verify the order with the physician and pharmacy and hang the fluid as ordered.*
   b. *Obtain a pump for the infusion.*
   c. *Place the client on intake and output measures.*
   d. *Begin intravenous hypotonic fluids.*

The Movement of Water: Osmosis

Diffusion of water through a selectively permeable membrane, a process called **osmosis,** is a physical phenomenon that is easily demonstrated in the laboratory with nonliving materials. It provides a model of how cells deal with various solute concentrations in aqueous solutions. In an osmotic system, the membrane is *selectively,* or *differentially, permeable,* having passageways that allow free diffusion of water but can block certain other dissolved molecules **(figure 6.2).** When this membrane is placed between solutions of differing concentrations and the solute cannot pass through the membrane, then under the laws of diffusion, water will diffuse at a faster rate from the side that has more water to the side that has less water. As long as the concentrations of the solutions differ, one side will experience a net loss of water and the other a net gain of water, until equilibrium is reached and the rate of diffusion is equalized.

Osmosis in living systems is similar to the model shown in **figure 6.3.** Living membranes generally block the entrance and exit of larger molecules and permit free diffusion of water. Because most cells are surrounded by some free water, the amount of water entering or leaving has a far-reaching impact on cellular activities and survival. This osmotic relationship between cells and their environment is determined by the relative concentrations of the solutions on either side of the cytoplasmic membrane (figure 6.3). Such systems can be compared using the terms *isotonic,* *hypotonic,* and *hypertonic.* (The root *-tonic* means "tension." *Iso-* means "the same," *hypo-* means "less," and *hyper-* means "more" or "greater.")

Under **isotonic** conditions, the environment is equal in solute concentration to the cell's internal environment, and because diffusion of water proceeds at the same rate in both directions, there is no net change in cell volume. Isotonic solutions are generally the most stable environments for cells, because they are already in an osmotic steady state with the cell. Parasites living in host tissues are most likely to be living in isotonic habitats.

Under **hypotonic** conditions, the solute concentration of the external environment is lower than that of the cell's internal environment. Pure water provides the most hypotonic environment for cells because it has no solute. The net direction of

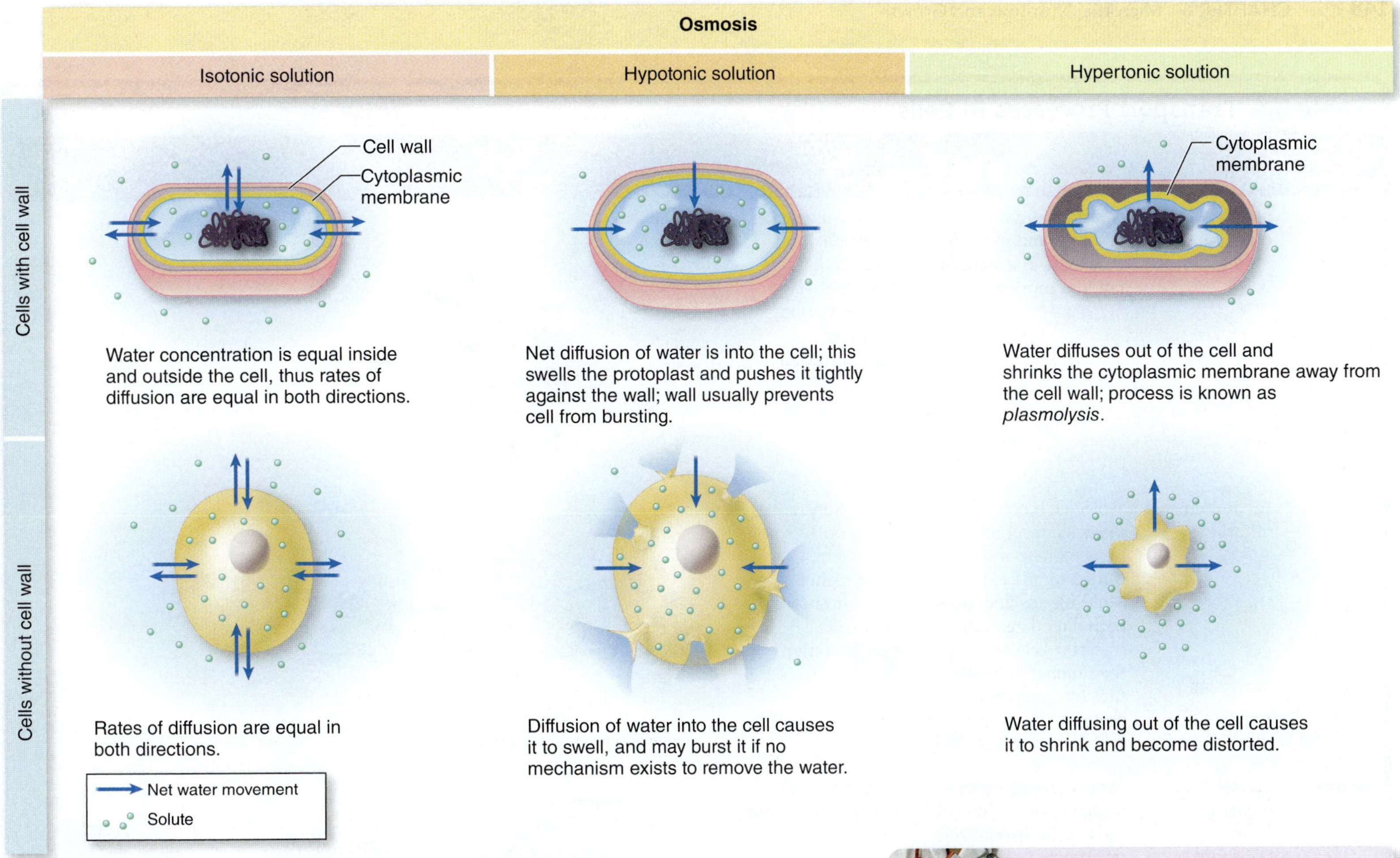

Isotonic solution	Hypotonic solution	Hypertonic solution

Cells with cell wall

Cell wall
Cytoplasmic membrane

Cytoplasmic membrane

Water concentration is equal inside and outside the cell, thus rates of diffusion are equal in both directions.

Net diffusion of water is into the cell; this swells the protoplast and pushes it tightly against the wall; wall usually prevents cell from bursting.

Water diffuses out of the cell and shrinks the cytoplasmic membrane away from the cell wall; process is known as *plasmolysis*.

Cells without cell wall

Rates of diffusion are equal in both directions.

Diffusion of water into the cell causes it to swell, and may burst it if no mechanism exists to remove the water.

Water diffusing out of the cell causes it to shrink and become distorted.

→ Net water movement

°₀ °₀ Solute

Figure 6.3 Cell responses to solutions of differing osmotic content.   Note that, unlike in figure 6.2, there is no tube into which the extra fluid can rise.

osmosis is from the hypotonic solution into the cell, and cells without walls swell and can burst.

A slightly hypotonic environment can be quite favorable for bacterial cells. The constant slight tendency for water to flow into the cell keeps the cytoplasmic membrane fully extended and the cytoplasm full. This is the optimum condition for the many processes occurring in and on the membrane. Slight hypotonicity is tolerated quite well by most bacteria because of their rigid cell walls.

Hypertonic conditions are also out of balance with the tonicity of the cell's cytoplasm, but in this case, the environment has a higher solute concentration than the cytoplasm. Because a hypertonic environment will force water to diffuse out of a cell, it is said to have high *osmotic pressure*, or potential. The growth-limiting effect of hypertonic solutions on microbes is the principle behind using concentrated salt and sugar solutions as preservatives for food, such as in salted hams.

Movement of Solutes

So far, the discussion of passive or simple diffusion has not included the added complexity of membranes or cell walls, which hinder simple diffusion by adding a physical barrier. Therefore, simple diffusion is limited to small nonpolar molecules like oxygen or lipid-soluble molecules that may pass through the membranes. But it is imperative that a cell be able to move polar molecules and ions across the plasma membrane as well, and this is impossible via simple diffusion. So microbes have developed multiple mechanisms to move substances across membranes. We look at these (and simple diffusion) in **table 6.4.**

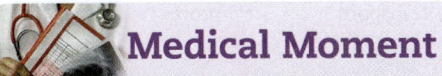

Medical Moment

Active Transport and Ion Channels

Active transport is defined as the transport of substances from low to high concentration against the diffusion gradient. Metal ions, such as calcium, require ion channels to cross membranes. Ion channels are proteins that form pores and are responsible for controlling small voltage gradients across the plasma membrane of cells. Ion channels are important in a wide variety of processes at the cellular level that involve rapid changes in cells; thus, ion channels are also the target of many drug therapies.

Calcium channel blockers are some of the most widely prescribed medications in the United States. They are used to treat hypertension and certain arrhythmias. How do they work? Calcium channel blockers inhibit calcium from passing through calcium ion channels, reducing the amount of calcium available for muscle contraction. Calcium channel blockers work on smooth muscle cells and myocytes (muscle cells) of the heart. Thus, calcium channel blockers can reduce blood pressure by relaxing smooth muscles in the arteries (they do not relax venous smooth muscle) and can also reduce conduction of impulses, resulting in a slowing of the heart rate.

Table 6.4 Transport Processes in Cells

	Examples	Description	Energy Requirements	
Passive	Simple diffusion	A fundamental property of atoms and molecules that exist in a state of random motion	None. Substances move on a gradient from higher concentration to lower concentration.	
	Facilitated diffusion	Molecule binds to a specific receptor in membrane and is carried to other side. Molecule-specific. Goes both directions. Rate of transport is limited by the number of binding sites on transport proteins.	None. Substances move on a gradient from higher concentration to lower concentration.	Membrane Extracellular Intracellular Extracellular Intracellular
Active	Carrier-mediated active transport	Atoms or molecules are pumped into or out of the cell by specialized receptors.	Driven by ATP or the proton motive force	Membrane ATP Extracellular Intracellular

As you see in table 6.4, very often energy is required to move molecules into or out of cells. In that case, the process is more accurately called transport and is seen as "active." Features inherent in **active transport** systems are

1. the transport of nutrients against the diffusion gradient or in the same direction as the natural gradient but at a rate faster than by diffusion alone;
2. the presence of specific membrane proteins (permeases and pumps); and
3. the expenditure of energy.

Examples of substances transported actively are monosaccharides, amino acids, organic acids, phosphates, and metal ions.

Endocytosis: Eating and Drinking by Cells

Some eukaryotic cells transport large molecules, particles, liquids, or even other cells across the cell membrane. Because the cell usually expends energy to carry out this transport, it is also a form of active transport. The substances transported do not pass physically through the membrane but are carried into the cell by **endocytosis.** First the cell encloses the substance in its membrane, simultaneously forming a vacuole and engulfing it. Amoebas and certain white blood cells ingest whole cells or

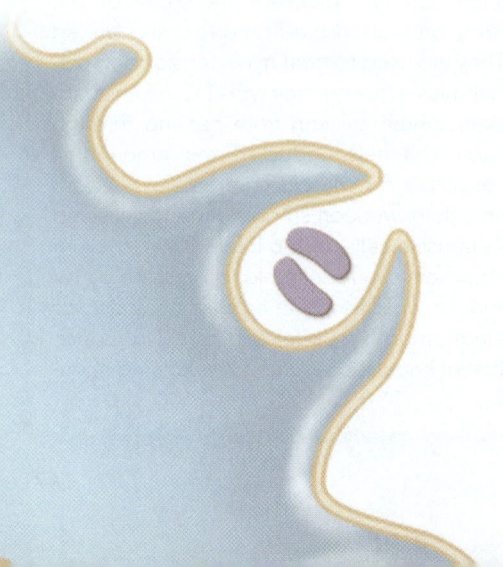

large solid matter by a type of endocytosis called **phagocytosis.** Liquids, such as oils or molecules in solution, enter the cell through a type of endocytosis called **pinocytosis.**

6.2 Environmental Factors That Influence Microbes

Microbes are exposed to a wide variety of environmental factors in addition to nutrients. These include such factors as heat, cold, gases, acid, radiation, osmotic and hydrostatic pressures, and even the effects of other microbes. For most microbes, environmental factors fundamentally affect the function of metabolic enzymes. Thus, survival in a changing environment is largely a matter of whether the enzyme systems of microorganisms can adapt to alterations in their habitat.

Temperature

Microbial cells are unable to control their temperature and therefore take on the ambient temperature of their natural habitats. Their survival is dependent on adapting to whatever temperature variations are encountered in that habitat. The range of temperatures for the growth of a given microbial species can be expressed as three *cardinal temperatures*. The **minimum temperature** is the lowest temperature that permits a microbe's continued growth and metabolism; below this temperature, its activities are inhibited. The **maximum temperature** is the highest temperature at which growth and metabolism can proceed. If the temperature rises slightly above maximum, growth will stop, but if it continues to rise beyond that point, the enzymes and nucleic acids will eventually become permanently inactivated (otherwise known as denaturation), and the cell will die. This is why heat works so well as an agent in microbial control. The **optimum temperature** covers a small range, intermediate between the minimum and maximum, which promotes the fastest rate of growth and metabolism (rarely is the optimum a single point).

Depending on their natural habitats, some microbes have a narrow cardinal range, others a broad one. Some strict parasites will not grow if the temperature varies more than a few degrees below or above the host's body temperature. For instance, the typhus bacterium multiplies only in the range of 32°C to 38°C, and rhinoviruses (one cause of the common cold) multiply most successfully in tissues that are slightly below normal body temperature (33°C to 35°C). Other organisms are not so limited. Strains of *Staphylococcus aureus* grow within the range of 6°C to 46°C, and the intestinal bacterium *Enterococcus faecalis* grows within the range of 0°C to 44°C.

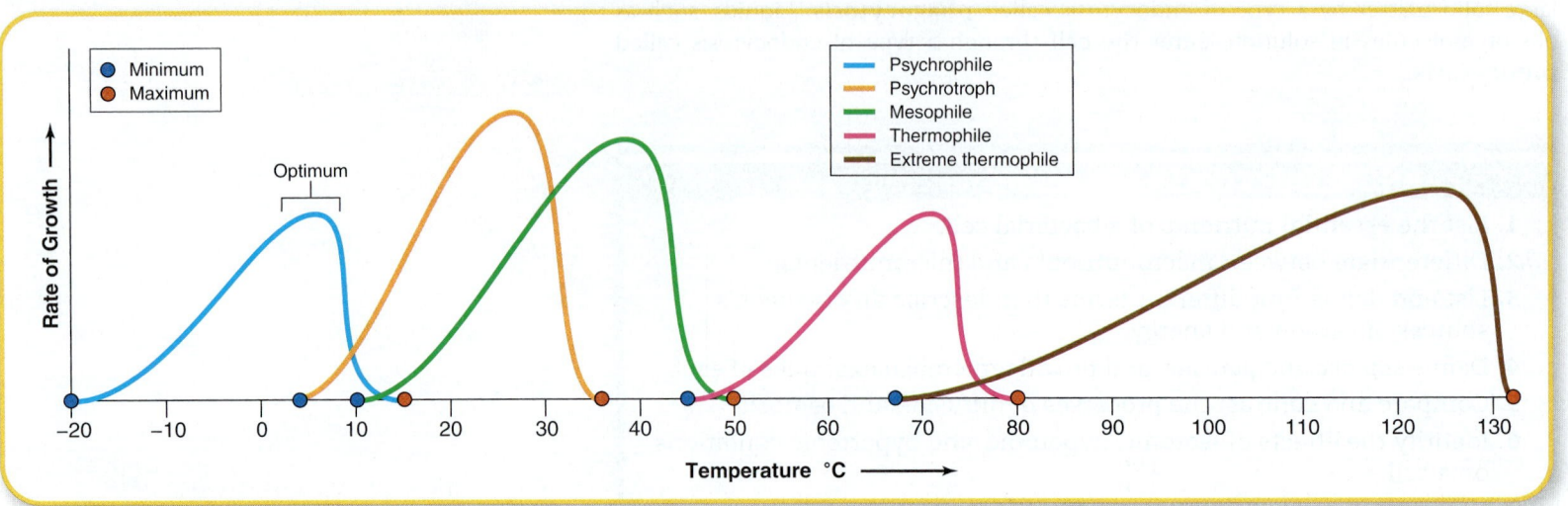

Figure 6.4 **Ecological groups by temperature range.** Psychrophiles can grow at or below 0°C and have an optimum below 15°C. Psychrotrophs have an optimum of from 15°C to 30°C. As a group, mesophiles can grow between 10°C and 50°C, but their optima usually fall between 20°C and 40°C. Generally speaking, thermophiles require temperatures above 45°C and grow optimally between this temperature and 80°C. Extreme thermophiles have optima above 80°C. Note that the ranges can overlap to an extent.

(a)

(b)

Figure 6.5 **Red snow.** (a) An early summer snowbank provides a perfect habitat for psychrophilic photosynthetic organisms like *Chlamydomonas nivalis*. (b) Microscopic view of this snow alga (actually classified as a green alga, although a red pigment dominates at this stage of its life cycle).

Another way to express temperature adaptation is to describe whether an organism grows optimally in a cold, moderate, or hot temperature range. The terms used for these ecological groups are *psychrophile, mesophile,* and *thermophile* **(figure 6.4),** respectively.

A **psychrophile** (sy'-kroh-fyl)–the blue line in figure 6.4–is a microorganism that has an optimum temperature below 15°C and is capable of growth at 0°C. It is obligate with respect to cold and generally cannot grow above 20°C. Unlike most laboratory cultures, storage in the refrigerator incubates, rather than inhibits, them. As one might predict, the habitats of psychrophilic bacteria, fungi, and algae are lakes and rivers, snowfields **(figure 6.5),** polar ice, and the deep ocean. Rarely, if ever, are they pathogenic. True psychrophiles must be distinguished from the less extreme *psychrotrophs* (the gold line in figure 6.4) that grow slowly in cold but have an optimum temperature between 15°C and 30°C. Bacteria such as *Staphylococcus aureus* and *Listeria monocytogenes* are a concern because they can grow in refrigerated food and cause food-borne illness.

The majority of medically significant microorganisms are **mesophiles** (mez'-oh-fylz; the green line in figure 6.4), organisms that grow at intermediate temperatures. The optimum growth temperatures (optima) of most mesophiles fall into the range of 20°C to 40°C. Organisms in this group inhabit animals and plants as well as soil and water in temperate, subtropical, and tropical regions. Most human pathogens have optima somewhere between 30°C and 40°C (human body temperature is 37°C). *Thermoduric* microbes, which can survive short exposure to high temperatures but are normally mesophiles, are common contaminants of heated or pasteurized foods. Examples include heat-resistant endospore formers such as *Bacillus* and *Clostridium*.

A **thermophile** (thur'-moh-fyl; the pink line in figure 6.4) is a microbe that grows optimally at temperatures greater than 45°C. Such heat-loving microbes live in soil and water associated with volcanic activity, in compost piles, and in habitats directly exposed to the sun. Thermophiles vary in heat requirements, with a general range of growth of 45°C to 80°C. Most eukaryotic forms cannot survive above 60°C, but a few thermophilic bacteria, called extreme thermophiles (the brown line in figure 6.4), grow between 80°C and 121°C. Strict thermophiles are so heat tolerant that researchers may use an autoclave to isolate them in culture.

Gases

The atmospheric gases that most influence microbial growth are O_2 and CO_2. Of these, oxygen gas has the greatest impact on microbial growth. Not only is it an important respiratory gas, but it is also a powerful oxidizing agent that exists in many toxic forms. In general, microbes fall into one of three categories:

- those that use oxygen and can detoxify it,
- those that can neither use oxygen nor detoxify it, and
- those that do not use oxygen but can detoxify it.

How Microbes Process Oxygen

As oxygen enters into cellular reactions, it is transformed into several toxic products. *Singlet oxygen* (O) is an extremely reactive molecule. Notably, it is one of the substances produced by phagocytes to kill invading bacteria. The buildup of singlet oxygen and the oxidation of membrane lipids and other molecules can damage and destroy a cell. The highly reactive *superoxide ion* (O_2^-), *hydrogen peroxide* (H_2O_2), and *hydroxyl radicals* (OH^-) are other destructive metabolic by-products of oxygen. To protect themselves against damage, most cells have developed enzymes that go about the business of scavenging and neutralizing these chemicals. The complete conversion of superoxide ion into harmless oxygen requires a two-step process and at least two enzymes:

$$\text{Step 1. } 2O_2^- + 2H^+ \xrightarrow{\text{Superoxide dismutase}} H_2O_2 \text{ (hydrogen peroxide)} + O_2$$

$$\text{Step 2. } 2H_2O_2 \xrightarrow{\text{Catalase}} 2H_2O + O_2$$

In this series of reactions (essential for aerobic organisms), the superoxide ion is first converted to hydrogen peroxide and normal oxygen by the action of an enzyme called superoxide dismutase. Because hydrogen peroxide is also toxic to cells (after all, it is used as a disinfectant and antiseptic), it must be degraded by the enzyme catalase into water and oxygen. If a microbe is not capable of dealing with toxic oxygen by these or similar mechanisms, it is forced to live in habitats free of oxygen.

Because oxygen requirements differ so dramatically and are so important clinically, microbes are grouped into several general categories **(table 6.5).** A photo of tubes depicting three of the gas-utilizing categories can be found in **figure 6.6.**

Carbon Dioxide

Although all microbes require some carbon dioxide in their metabolism, *capnophiles* grow best at a higher CO_2 tension than is normally present in the atmosphere. This becomes important in the initial isolation of some pathogens from clinical specimens, notably *Neisseria* (gonorrhea, meningitis), *Brucella* (undulant fever), and *Streptococcus pneumoniae*.

pH

The term **pH** is defined as the degree of acidity or alkalinity (basicity) of a solution. It is expressed by the pH scale, a series of numbers ranging from 0 to 14. The pH of pure water (7.0) is neutral, neither acidic nor basic. As the pH value decreases toward 0, the acidity increases, and as the pH increases toward 14, the alkalinity increases. The majority of organisms live or grow in habitats between pH 6 and 8 because strong acids and bases can be highly damaging to enzymes and other cellular substances.

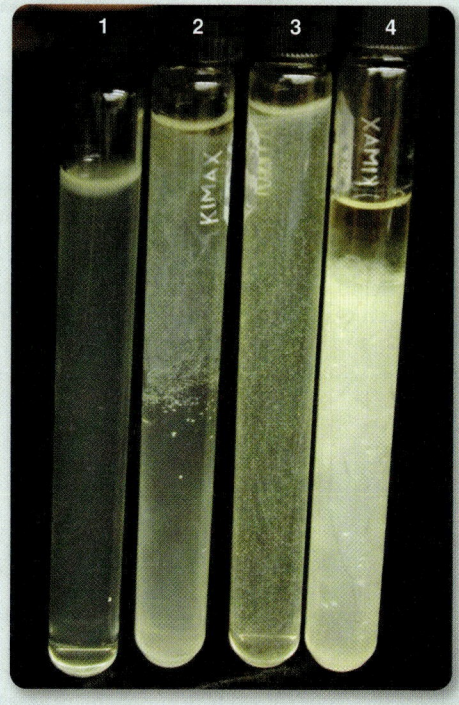

Figure 6.6 Four tubes showing three different patterns of oxygen utilization. These tubes use thioglycollate, which reduces oxygen to water, to restrict oxygen diffusion through the agar. So, whereas there is an oxygen-rich layer at the top of the agar, the oxygen concentration rapidly decreases deeper in the agar. In tube 1, the obligately aerobic *Pseudomonas aeruginosa* grows only at the very top of the agar. Tubes 2 and 3 contain two different examples of facultatively anaerobic bacteria. Many facultatives, though able to grow both aerobically and anaerobically, grow more efficiently in the aerobic mode. This is more obvious in tube 2 (*Staphylococcus aureus*) and less obvious in tube 3 (*Escherichia coli*). Tube 4 contains *Clostridium butyricum*, an obligate anaerobe.

Communities of extreme thermophilic microorganisms live around hot deep-sea ocean vents.

Table 6.5 Oxygen Usage and Tolerance Patterns in Microbes

Culture Appearance

In all cases, bacteria are grown in a medium called thioglycollate, which allows anaerobic bacteria to grow in tubes exposed to air. Oxygen concentration is highest at the top of the tube.

Aerobes

Can use gaseous oxygen in their metabolism and possess the enzymes needed to process toxic oxygen products. An organism that cannot grow without oxygen is an *obligate aerobe*.

Examples: Most fungi, protozoa, and many bacteria, such as *Bacillus* species and *Mycobacterium tuberculosis*

(Obligate aerobe)

Growth

Microaerophiles

Do not grow at normal atmospheric concentrations of oxygen but require a small amount of it in metabolism.

Examples: Organisms that live in soil or water or in mammalian hosts, not directly exposed to atmosphere; *Helicobacteri pylori*, *Borrelia burgdorferi*

Growth

Facultative anaerobes

Do not require oxygen for metabolism but use it when it is present. Can also perform anaerobic metabolism. In the tube to the right, the bacteria are growing throughout, but there is heavier growth in the aerobic portion of the tube (upper) because aerobic growth can proceed more quickly in some facultative anaerobes.

Examples: Many gram-negative intestinal bacteria, staphylococci

Growth

Anaerobes

Lack the metabolic enzyme systems for using oxygen in respiration. Obligate anaerobes also lack the enzymes for processing toxic oxygen and die in its presence.

Examples: Many oral bacteria, intestinal bacteria

(Obligate anaerobe)

Growth

Aerotolerant anaerobes

Do not utilize oxygen but can survive and grow to a limited extent in its presence. They are not harmed by oxygen, mainly because they possess alternate mechanisms for breaking down peroxides and superoxide.

Examples: Certain lactobacilli and streptococci, clostridial species

Growth

A few microorganisms live at pH extremes. Obligate *acidophiles* include *Euglena mutabilis*, an alga that grows in acid pools between 0 and 1.0 pH, and *Thermoplasma*, an archaea that lacks a cell wall, lives in hot coal piles at a pH of 1 to 2, and will die if exposed to pH 7. *Picrophilus* thrives at a pH of 0.7, and can grow at a pH of 0. Because many molds and yeasts tolerate moderate acid, they are the most common spoilage agents of pickled foods. *Alkalinophiles*, such as *Natromonas* species, live in hot pools and soils that contain high levels of basic minerals (up to pH 12.0). Bacteria that decompose urine create alkaline conditions, because ammonium (NH_4^+) can be produced when urea (a component of urine) is digested. Metabolism of urea is one way that *Proteus* spp. can neutralize the acidity of the urine to colonize and infect the urinary system.

Osmotic Pressure

Although most microbes exist under hypotonic or isotonic conditions, a few, called **osmophiles**, live in habitats with a high solute concentration. One common type of osmophile prefers high concentrations of salt; these organisms are called **halophiles** (hay'-loh-fylz). Obligate halophiles such as *Halobacterium* and *Halococcus* inhabit salt lakes, ponds, and other hypersaline habitats. They grow optimally in solutions of 25% NaCl but require at least 9% NaCl (combined with other salts) for growth. These archaea have significant modifications in their cell walls and membranes and will lyse in hypotonic habitats. Facultative halophiles are remarkably resistant to salt, even though they do not normally reside in high-salt environments. For example, *Staphylococcus aureus* can grow on NaCl media ranging from 0.1% up to 20%.

Radiation and Hydrostatic/Atmospheric Pressure

Radiation

Some microbes (phototrophs) can use visible light rays as an energy source, but nonphotosynthetic microbes tend to be damaged by the toxic oxygen products produced by contact with light. Some microbial species produce yellow carotenoid pigments to protect against the damaging effects of light by absorbing and dismantling toxic oxygen. Other types of radiation that can damage microbes are ultraviolet and ionizing rays (X rays and cosmic rays). In chapter 9, you will see just how these types of energy are applied in microbial control.

Pressure

The ocean depths subject organisms to increasing hydrostatic pressure. Deep-sea microbes called **barophiles** exist under pressures that range from a few times to over 1,000 times the pressure of the atmosphere. These bacteria are so strictly adapted to high pressures that they will rupture when exposed to normal atmospheric pressure.

Other Organisms

Up to now, we have considered the importance of nonliving environmental influences on the growth of microorganisms. Another profound influence comes from other organisms that share (or sometimes *are*) their habitats. In all but the rarest instances, microbes live in shared habitats, which give rise to complex and fascinating associations. Some associations are between similar or dissimilar types of microbes; others involve multicellular organisms such as animals or plants. Interactions can have beneficial, harmful, or no particular effects on the organisms involved; they can be obligatory or nonobligatory to the members; and they often involve nutritional interactions. This outline provides an overview of the major types of microbial associations:

Cucumbers placed in a hypertonic solution turn into smaller, denser pickles.

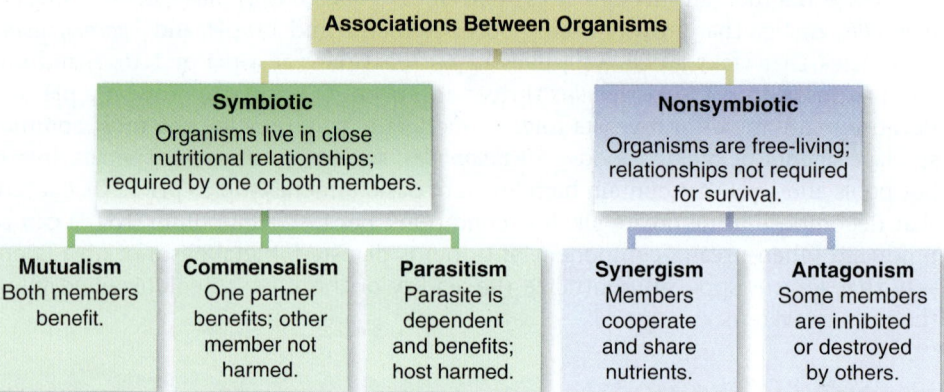

Associations Between Organisms

Symbiotic
Organisms live in close nutritional relationships; required by one or both members.

Nonsymbiotic
Organisms are free-living; relationships not required for survival.

Mutualism
Both members benefit.

Commensalism
One partner benefits; other member not harmed.

Parasitism
Parasite is dependent and benefits; host harmed.

Synergism
Members cooperate and share nutrients.

Antagonism
Some members are inhibited or destroyed by others.

Oxpecker birds and herd ungulates may have a mutualistic relationship, in which the birds benefit from a steady source of ticks and the ungulates have blood-sucking ticks removed.

Strong Partnerships: Symbioses

A general term used to denote a situation in which two organisms live together in a close partnership is **symbiosis,** and the members are termed *symbionts*. Three main types of symbiosis occur:

- **Mutualism** exists when organisms live in an obligatory and mutually beneficial relationship. Many microbes that live in or on humans fall in this category. The microbes receive necessary nutrients, and the host gets a variety of benefits, ranging from protection from transient, pathogenic microbes to the healthy development of the immune system that is only possible when a robust resident microbiome is present.
- In a relationship known as **commensalism,** the member called the commensal receives benefits, while its partner is neither harmed nor benefited. One example is the relationship between cattle egrets (birds) and cattle and other livestock. As the large animals graze in fields, they stir up insects, which can then be more easily eaten by the birds. The birds benefit, and the livestock are not harmed or helped. Most "normal biota" microbes used to be considered commensals, with the assumption that they provided neither harm nor benefit to their hosts. We now know that the normal biota contribute greatly to the health of their human host, so they are more accurately classified as mutualists.
- **Parasitism** is a relationship in which the host organism provides the parasitic microbe with nutrients and a habitat. Microbes that make humans sick fall in this category.

Associations but Not Partnerships: Antagonism and Synergism

Even when organisms are not engaged in symbiotic relationships, they are interacting. Relationships between free-living species can either have negative or positive results. **Antagonism** is an association between free-living species that arises when members of a community compete. In this interaction, one microbe secretes chemical substances into the surrounding environment that inhibit or destroy another microbe in the same habitat. The first microbe may gain a competitive advantage by increasing the space and nutrients available to it. Interactions of this type are common in the soil, where mixed communities often compete for space and food. *Antibiosis*—the production of inhibitory compounds such as antibiotics—is actually a form of antagonism. Hundreds of naturally occurring antibiotics have been isolated from bacteria and fungi and used as drugs to control diseases.

Synergism is an interrelationship between two or more free-living organisms that benefits them but is not necessary for their survival. Together, the participants cooperate to produce a result that none of them could do alone. Gum disease, dental caries, and some bloodstream infections involve mixed infections by bacteria interacting synergistically.

Biofilms: The Epitome of Synergy Biofilms are mixed communities of different kinds of bacteria and other microbes that are attached to a surface and to each other, forming a multilayer conglomerate of cells and intracellular material. Usually there is a "pioneer" colonizer, a bacterium that initially attaches to a surface, such as a tooth or the lung tissue **(figure 6.7).** Other microbes then attach either to those bacteria or to the polymeric sugar and protein substance that inevitably is secreted by microbial colonizers of surfaces. In many cases, once the cells are attached, they are stimulated to release chemicals that accumulate as the cell population grows. By this means, they can monitor the size of their own population. This is a process called **quorum sensing.** Bacteria can use quorum sensing to interact with other members of the same species, as well as members of other species that are close by. Eventually large complex communities are formed, which have different physical and biological characteristics in different locations of the community. The bottom of a biofilm may have very different pH and oxygen conditions than the surface of a biofilm, for example. It is now clearly established that microbes in a biofilm, as opposed to those in a planktonic (free-floating) state, behave and respond very differently to their environments. Different genes are even activated in the two situations. At any rate, a single biofilm is actually a partnership among multiple microbial inhabitants and thus cannot be eradicated by traditional methods targeting individual infections. This kind of synergism has led to the necessity of rethinking treatment of a great many different conditions.

Figure 6.7 Steps in the formation of a biofilm. The photograph is an SEM of a biofilm formed on a gauze bandage. The blue bacteria are methicillin-resistant *Staphylococcus aureus* (MRSA); the orange substance is the extracellular matrix; and the green tubes are fibers of the gauze. This is probably a single-species biofilm.

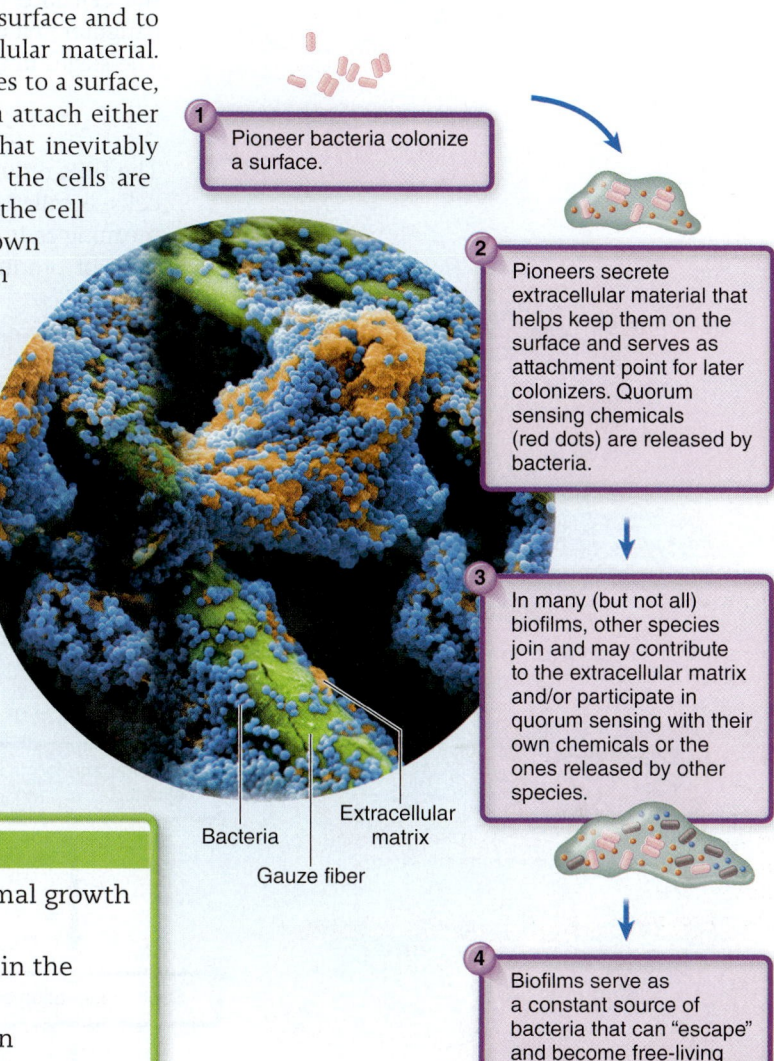

1. Pioneer bacteria colonize a surface.

2. Pioneers secrete extracellular material that helps keep them on the surface and serves as attachment point for later colonizers. Quorum sensing chemicals (red dots) are released by bacteria.

3. In many (but not all) biofilms, other species join and may contribute to the extracellular matrix and/or participate in quorum sensing with their own chemicals or the ones released by other species.

4. Biofilms serve as a constant source of bacteria that can "escape" and become free-living again.

Bacteria

Extracellular matrix

Gauze fiber

6.2 LEARNING OUTCOMES—Assess Your Progress

8. List and define five terms used to express a microbe's optimal growth temperature.

9. Summarize three ways in which microorganisms function in the presence of differing oxygen conditions.

10. Identify three important environmental factors (other than temperature and oxygen) with which organisms must cope.

11. List and describe the five types of associations microbes can have with their hosts.

12. Discuss characteristics of biofilms that differentiate them from planktonic bacteria.

6.3 The Study of Bacterial Growth

The growth of eukaryotic microorganisms can occur in various ways, but bacteria most often grow using a unique process called **binary fission**. When we discuss bacterial growth, we are referring to the growth in their population size. Individual cell size does increase before the actual fission event, but the most relevant aspect of growth is in their *numbers*.

Binary Fission

Binary fission refers to the fact that one cell becomes two. During binary fission, the parent cell enlarges, duplicates its chromosome, and then starts to pull its cell envelope together in the center of the cell using a band of protein that is made of substances that resemble actin and tubulin–the protein component of microtubules in eukaryotic cells. The cell wall eventually forms a complete central septum. This process divides the cell into two daughter cells. This process is repeated at intervals by each new daughter cell in turn; with each successive round of division, the population increases. The stages in this continuous process are shown in greater detail in **figure 6.8.**

The Rate of Population Growth

The time required for a complete fission cycle–from parent cell to two new daughter cells–is called the **generation,** or **doubling, time.** The term *generation* has a similar meaning as it does in humans. It is the period between an individual's birth and the time of producing offspring. In bacteria, each new fission cycle, or generation, increases the population by a factor of 2, or doubles it. Thus, the initial parent stage consists of 1 cell, the first generation consists of 2 cells, the second 4, the third 8, then 16, 32, 64, and so on. As long as the environment remains favorable, this doubling effect can continue at a constant rate. With the passing of each generation, the population will double, over and over again.

The length of the generation time is a measure of the growth rate of an organism. Compared with the growth rates of most other living things, bacteria are notoriously rapid. The average generation time is 30 to 60 minutes under optimum conditions. The shortest generation times can be 10 to 12 minutes, although some bacteria have generation times of days. For example, *Mycobacterium leprae,* the cause of Hansen's disease, has a generation time of 10 to 30 days–as long as that of some animals. Environmental bacteria commonly have generation times measured in months. Most pathogens have relatively short doubling times. *Salmonella enteritidis* and *Staphylococcus aureus,* bacteria that cause food-borne illness, double in 20 to 30 minutes, which is why leaving food at room temperature even for a short period has caused many cases of food-borne disease. In a few hours, a population of these bacteria can easily grow from a small number of cells to several million.

Figure 6.9 shows several quantitative characteristics of growth: The cell population size can be represented by the number 2 with an exponent (2^1, 2^2, 2^3, 2^4); the exponent increases by one in each generation; and the number of the exponent is also the number of the generation. This growth pattern is termed **exponential.** Because these populations often contain very large numbers of cells, it is useful to express them by means of exponents or logarithms. The data from a growing bacterial population are graphed by plotting the number of cells as a function of time. Plotting the logarithm number over time provides a straight line indicative of exponential growth. Plotting the data arithmetically gives a constantly curved slope. In general, logarithmic graphs are preferred because an accurate cell number is easier to read, especially during early growth phases.

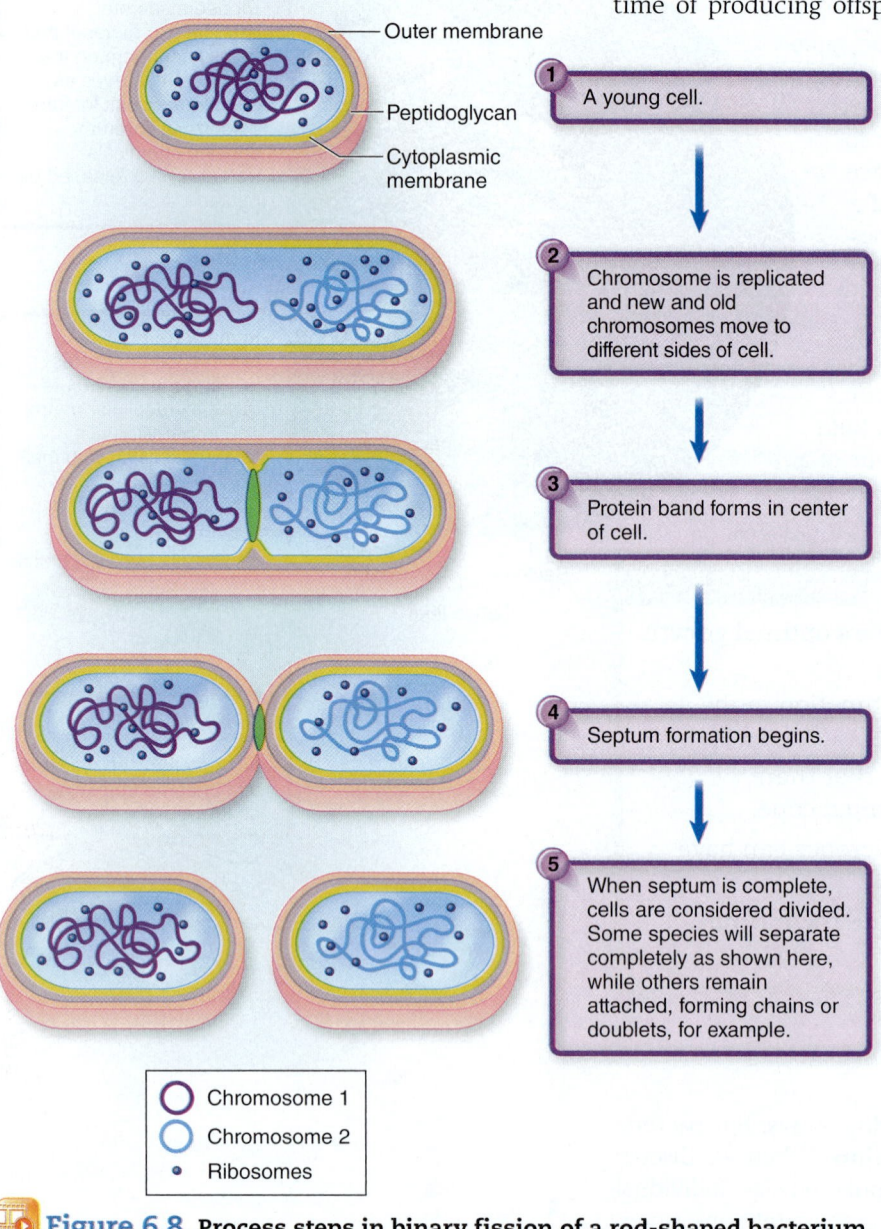

Outer membrane
Peptidoglycan
Cytoplasmic membrane

1 A young cell.

2 Chromosome is replicated and new and old chromosomes move to different sides of cell.

3 Protein band forms in center of cell.

4 Septum formation begins.

5 When septum is complete, cells are considered divided. Some species will separate completely as shown here, while others remain attached, forming chains or doublets, for example.

○ Chromosome 1
○ Chromosome 2
• Ribosomes

Figure 6.8 Process steps in binary fission of a rod-shaped bacterium. Note that even though the two chromosomes are colored differently, the new one is an exact copy of the old one (with some mistakes that you will learn about later).

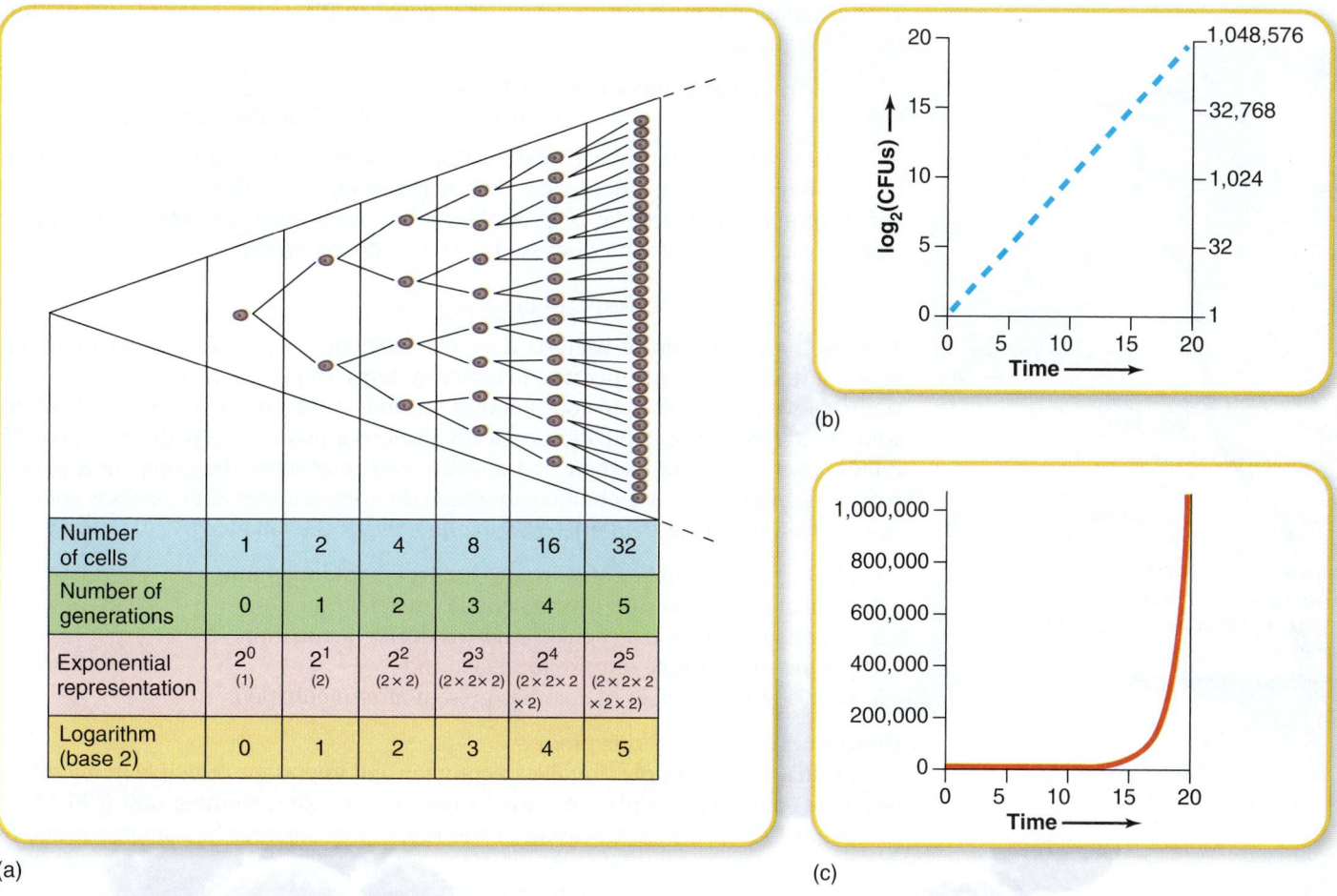

(a)

(b)

(c)

Figure 6.9 **The mathematics of population growth.**   (a) Starting with a single cell, if each product of reproduction goes on to divide by binary fission, the population doubles with each new cell division or generation. This process can be represented by logarithms using exponents or by simple numbers. **(b)** Plotting the logarithm of the cells produces a straight line indicative of exponential growth, whereas **(c)** plotting the cell numbers arithmetically gives a curved slope.

Clusters of *Staphylococcus aureus* bacteria that have divided by binary fission

Predicting the number of cells that will arise during a long growth period (yielding millions of cells) is based on a relatively simple concept. One could use the method of addition (2 + 2 = 4; 4 + 4 = 8; 8 + 8 = 16; 16 + 16 = 32, and so on) or a method of multiplication (for example, $2^5 = 2 \times 2 \times 2 \times 2 \times 2$), but it is easy to see that for 20 or 30 generations, this calculation could be very tedious. An easier way to calculate the size of a population over successive generations is to use this equation:

$$N_t = (N_i)2^n$$

Here, N_t is the total number of cells in the population (the "t" denotes "at some point in time t"). N_i is the starting number, the exponent n denotes the generation number, and 2^n represents the number of cells in that generation. If we know any two of the values, the other values can be calculated. Let us use the example of *Staphylococcus aureus* to calculate how many cells (N_t) will be present in an egg salad sandwich after it sits in a warm car for 4 hours. We will assume that N_i is 10 (number of cells deposited in the sandwich while it was being prepared). To derive n, we need to divide 4 hours (240 minutes) by the generation time (we will use 20 minutes).

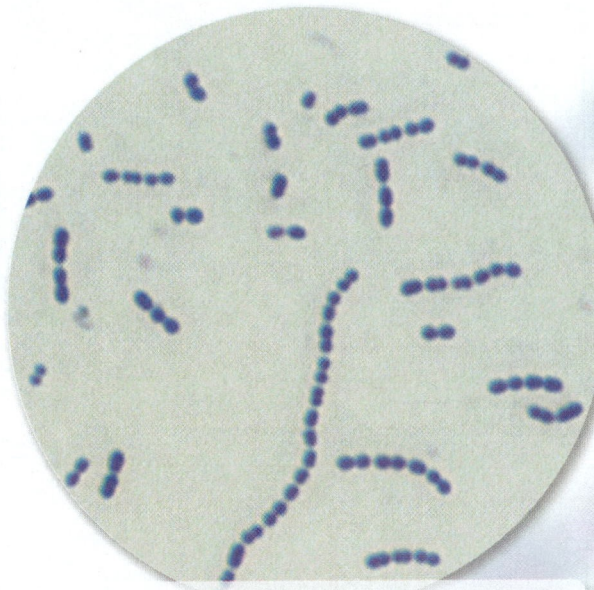

"Strips" of *Streptococcus* bacteria that have divided by binary fission. When mature cells stay attached like this, it means they stop their binary fission before completely separating.

This calculation comes out to 12, so 2^n is equal to 2^{12}. Using a calculator, we find that 2^{12} is 4,096.

$$\text{Final number } (N_t) = 10 \times 4,096$$
$$= 40,960 \text{ bacterial cells in the sandwich}$$

This same equation, with modifications, is used to determine the generation time, a more complex calculation that requires knowing the number of cells at the beginning and end of a growth period. Such data are obtained through actual testing by a method discussed in the following section.

The Population Growth Curve

In reality, a population of bacteria does not maintain its potential growth rate and does not double endlessly, because in closed systems (called batch cultures) numerous factors prevent the cells from continuously dividing at their maximum rate. Laboratory studies indicate that a population typically displays a predictable pattern, or **growth curve,** over time. The method traditionally used to observe the population growth pattern is a viable count technique, in which the total number of live cells is counted over a given time period. In brief, this method entails the following:

1. placing a tiny number of cells into a sterile liquid medium,
2. incubating this culture over a period of several hours,
3. sampling the broth at regular intervals during incubation,
4. plating each sample onto solid media, and
5. counting the number of colonies present after incubation.

Figure 6.10 illustrates this process.

Evaluating the samples involves a common and important principle in microbiology: One colony on the plate represents one cell or colony-forming unit (CFU) from the original sample. Multiplication of the number of colonies in a single sample by

Equally spaced time intervals	60 min	120 min	180 min	240 min	300 min	360 min	420 min	480 min	540 min	600 min
0.1 mL sample added to tube										
Sample is diluted in liquid agar medium and poured or spread over surface of solidified medium										
Plates are incubated, colonies are counted	None									
Number of colonies (CFU) per 0.1 mL	<1*	2	4	7	13	23	45	80	135	230
Total estimated cell population in flask	<5,000	10,000	20,000	35,000	65,000	115,000	225,000	400,000	675,000	1,150,000

500 mL inoculated flask

*Only means that too few cells are present to be assayed.

Figure 6.10 **Steps in a viable plate count: batch culture method.**

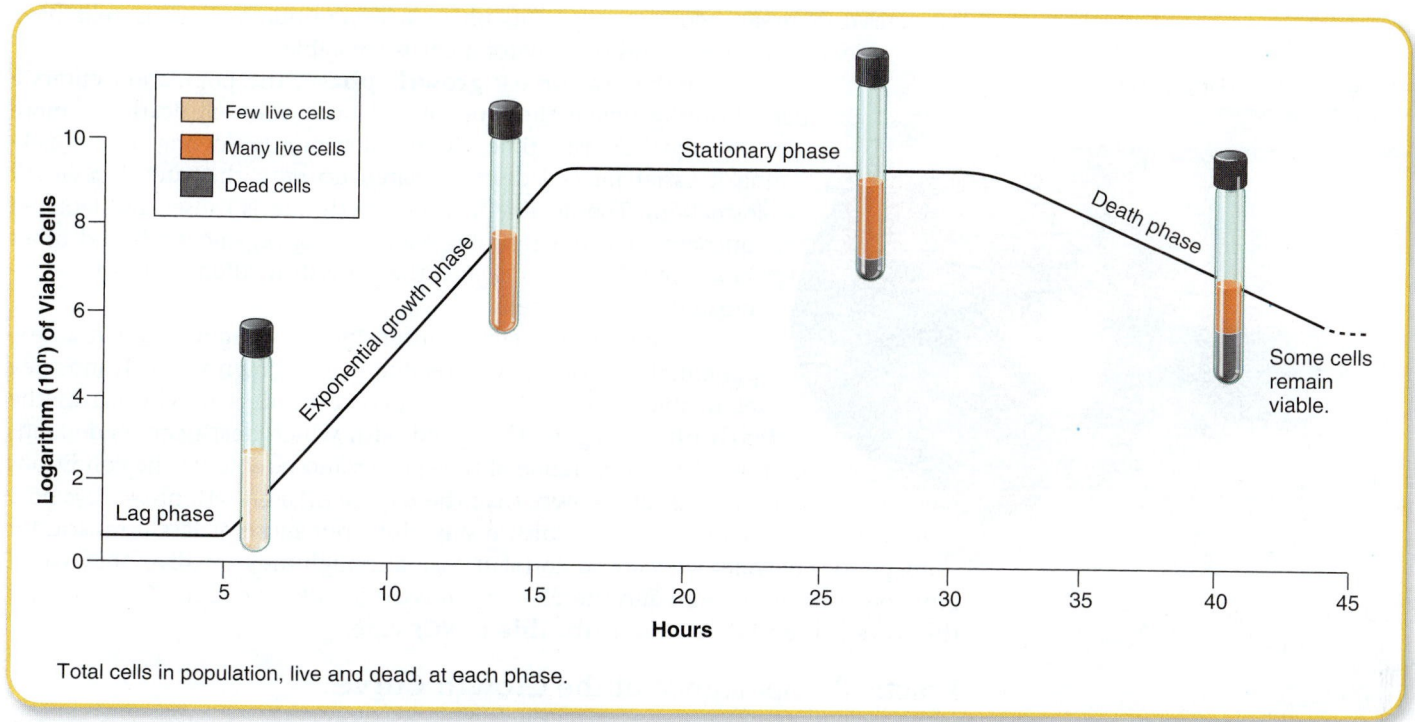

Total cells in population, live and dead, at each phase.

Figure 6.11   The growth curve in a bacterial culture.   On this graph, the number of viable cells expressed as a logarithm (log) is plotted against time. See text for discussion of the various phases. Note that with a generation time of 35 minutes, the population has risen from 10 (10^1) cells to over 1,000,000,000 (10^9) cells in only 16 hours.

the container's volume gives a fair estimate of the total population size (number of cells) at any given point. The growth curve is determined by graphing the number for each sample in sequence for the whole incubation period.

Because of the scarcity of cells in the early stages of growth, some samples can give a zero reading even if there are viable cells in the culture. Also, the sampling itself can remove enough viable cells to alter the tabulations, but since the purpose is to compare relative trends in growth, these factors do not significantly change the overall pattern.

Stages in the Normal Growth Curve

The system of batch culturing just described is *closed*, meaning that nutrients and space are finite and there is no mechanism for the removal of waste products. Data from an entire growth period typically produce a curve with a series of phases termed the *lag phase*, the *exponential growth (log) phase*, the *stationary phase*, and the *death phase* (**figure 6.11**).

The **lag phase** is a relatively "flat" period on the graph when the population appears not to be growing or is growing at less than the exponential rate. Growth lags primarily because

1. the newly inoculated cells require a period of adjustment, enlargement, and synthesis;
2. the cells are not yet multiplying at their maximum rate; and
3. the population of cells is so sparse or dilute that the sampling misses them.

The length of the lag period varies somewhat from one population to another. It is important to note that even though the population of cells is not increasing (growing), individual cells are metabolically active as they increase their contents and prepare to divide.

The cells reach the maximum rate of cell division during the **exponential growth (logarithmic or log) phase,** a period during which

The bacteria in cold yogurt are probably in stationary phase. Once inside a warm body, the bacteria can re-enter lag and then exponential phase and increase their numbers in the gut.

the curve increases geometrically. This phase will continue as long as cells have adequate nutrients and the environment is favorable.

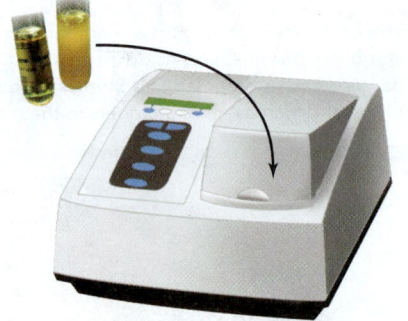

The cloudiness in fish bowls is due in large part to the growth of bacteria in the water.

At the **stationary growth phase,** the population enters a period during which the rates of cell birth and cell death are more or less equal. At this time, the division rate is slowing down (making it easier for cell death to catch up with the rate of new cell formation). The decline in the growth rate is caused by depleted nutrients and oxygen plus excretion of organic acids and other biochemical pollutants into the growth medium, due to the increased density of cells.

As the limiting factors intensify, cells begin to die at an exponential rate (literally perishing in their own wastes), and they are unable to multiply. The curve now dips downward as the **death phase** begins. The speed with which death occurs depends on the relative resistance of the species and how toxic the conditions are, but it is usually slower than the exponential growth phase. It is now clear that many cells in a culture stay alive, but more or less dormant, for long periods of time. They are so dormant that, although they are alive, they won't grow on culture medium and therefore are missed in colony counts. The name for this state is the **viable nonculturable (VNC)** state.

Practical Importance of the Growth Curve

The tendency for populations to exhibit phases of rapid growth, slow growth, and death has important implications in microbial control, infection, food microbiology, and culture technology. Antimicrobial agents such as heat and disinfectants rapidly accelerate the death phase in all populations, but microbes in the exponential growth phase are more vulnerable to these agents than are those that have entered the stationary phase. In general, actively growing cells are more vulnerable to conditions that disrupt cell metabolism and binary fission.

Growth patterns in microorganisms can account for the stages of infection. A person shedding bacteria in the early and middle stages of an infection is more likely to spread it to others than is a person in the late stages. The course of an infection is also influenced by the relatively faster rate of multiplication of the microbe, which can overwhelm the slower growth rate of the host's own cellular defenses.

For certain research or industrial applications, closed batch culturing with its four phases is inefficient. The alternative is an automatic growth chamber called the **chemostat,** or continuous culture system. This device can admit a steady stream of

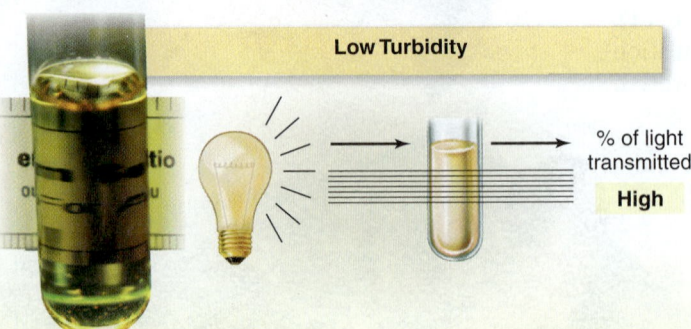

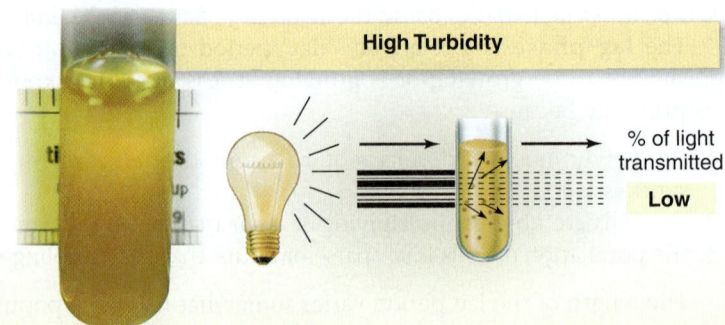

Figure 6.12 Turbidity measurements as indicators of growth. Holding a broth to the light is one method of checking for gross differences in cloudiness (turbidity). The broth on the left is transparent, indicating little or no growth; the broth on the right is cloudy and opaque, indicating heavy growth. The eye is not sensitive enough to pick up fine degrees in turbidity; more sensitive measurements can be made with a spectrophotometer. On the left you will see that a tube with no growth will allow light to easily pass. Therefore more light will reach the photodetector and give a higher transmittance value. In a tube with growth (as on the right), the cells scatter the light, resulting in less light reaching the photodetector and, therefore, giving a lower transmittance value.

new nutrients and siphon off used media and old bacterial cells, thereby stabilizing the growth rate and cell number.

Analyzing Population Size Without Culturing

Turbidity

Microbiologists have developed several alternative ways of analyzing bacterial growth qualitatively and quantitatively. One of the simplest methods for estimating the size of a population is through turbidometry. This technique relies on the simple observation that a tube of clear nutrient solution becomes cloudy, or **turbid,** as microbes grow in it. In general, the greater the turbidity, the larger the population size, which can be measured by means of sensitive instruments **(figure 6.12).**

Counting

If you know how many cells are present in a specified amount of fluid, you can multiply that to determine total numbers. It is possible to conduct a **direct cell count** microscopically **(figure 6.13).** This technique, very similar to that used in blood cell counts, employs a special microscope slide (cytometer) calibrated to accept a tiny sample that is spread over a premeasured grid. One inherent inaccuracy in this method, as well as in spectrophotometry, is that no distinction can be made between dead and live cells, both of which are included in the count.

Counting can be automated by sensitive devices such as the *Coulter counter,* which electronically scans a fluid as it passes through a tiny pipette. As each cell flows by, it is detected and registered on an electronic sensor **(figure 6.14).** A *flow cytometer* works on a similar principle, but in addition to counting, it can measure cell size and even differentiate between live and dead cells. When used in conjunction with fluorescent dyes and antibodies to tag cells, it has been used to differentiate between gram-positive and gram-negative bacteria. It has been adapted for use as a rapid method to identify pathogens in patient specimens and to differentiate blood cells. More sophisticated forms of the flow cytometer can actually sort cells of different types into separate compartments of a collecting device.

Although flow cytometry can be used to count bacteria in natural samples without the need for culturing them, it requires fluorescent labeling of the cells you are interested in detecting, which is not always possible.

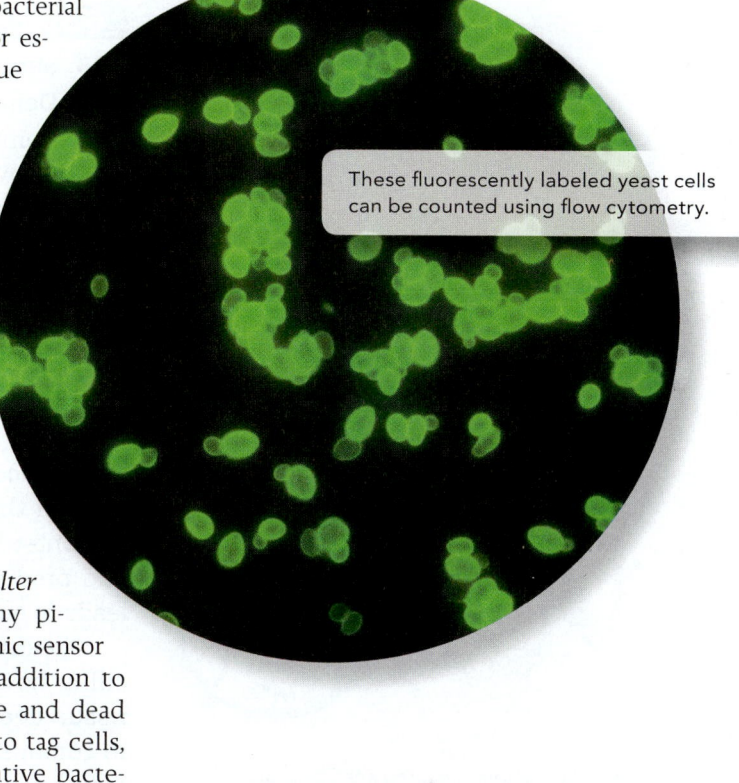

These fluorescently labeled yeast cells can be counted using flow cytometry.

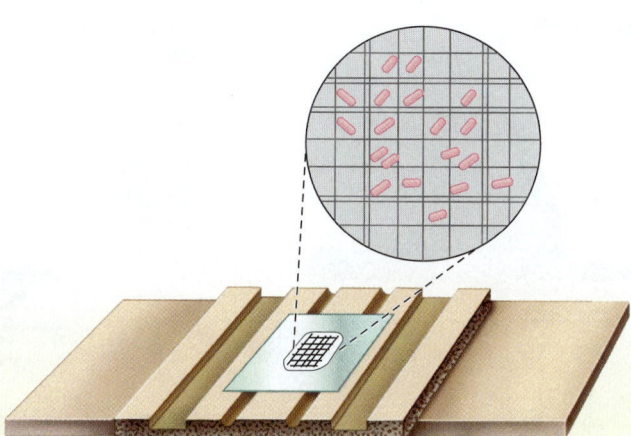

Figure 6.13 Direct microscopic count of bacteria. A small sample is placed on the grid under a cover glass. Individual cells, both living and dead, are counted. This number can be used to calculate the total cell count of a sample.

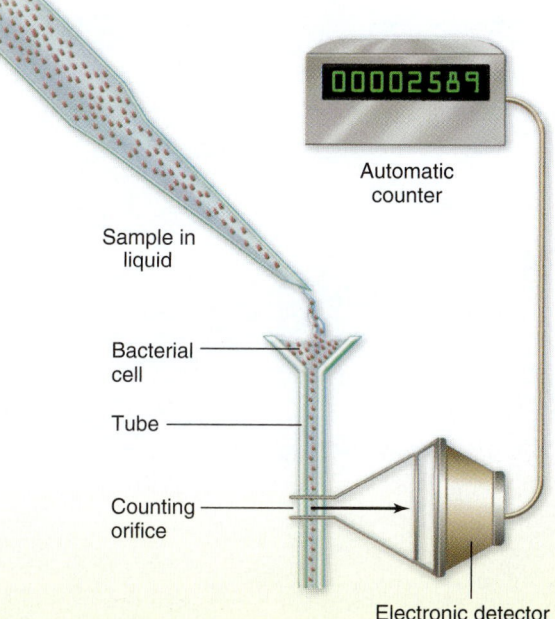

Automatic counter

00002589

Sample in liquid

Bacterial cell

Tube

Counting orifice

Electronic detector

Figure 6.14 Coulter counter. As cells pass through this device, they trigger an electronic sensor that tallies their numbers.

Genetic Probing

A variation of the **polymerase chain reaction (PCR),** called real-time PCR, allows scientists to quantify bacteria and other microorganisms that are present in environmental or tissue samples without isolating them and without culturing them.

> ### 6.3 LEARNING OUTCOMES—Assess Your Progress
>
> **13.** Summarize the steps of cell division used by most bacteria.
> **14.** Define *doubling time*, and describe how it leads to exponential growth.
> **15.** Compare and contrast the four phases of growth in a bacterial growth curve.
> **16.** Identify one quantitative and one qualitative method used for analyzing bacterial growth.

CASE FILE WRAP-UP

The bacteria pictured are Clostridium perfringens.

C. perfringens is commonly found in soil but is also found in the normal biota of the intestines. It produces a toxin known as alpha toxin, which can enter into muscle tissue through a wound. The toxin destroys tissue and produces gas, hence the term *gas gangrene*. Massive infection often leads to sepsis and death.

Treatment of gas gangrene includes wound debridement and penicillin; however, penicillin alone is unable to penetrate infected muscle tissue deeply enough to kill all of the bacteria; thus, penicillin is used in combination with surgical treatment. Amputation is often required to definitively treat *C. perfringens*.

Hyperbaric oxygen therapy (HBOT) has been used to successfully treat wound infections, including infections caused by *C. perfringens*. Anaerobic bacteria such as *C. perfringens* are susceptible to increased concentrations of oxygen. With HBOT, large quantities of oxygen-free radicals are generated, which renders bacteria vulnerable to oxidative death. HBOT has also been found to inhibit the release of alpha toxins. Lastly, HBOT enhances the effects of antibiotics. HBOT in cases of *C. perfringens* has resulted in avoidance of amputation, increased patient survival, and shorter hospital stays.

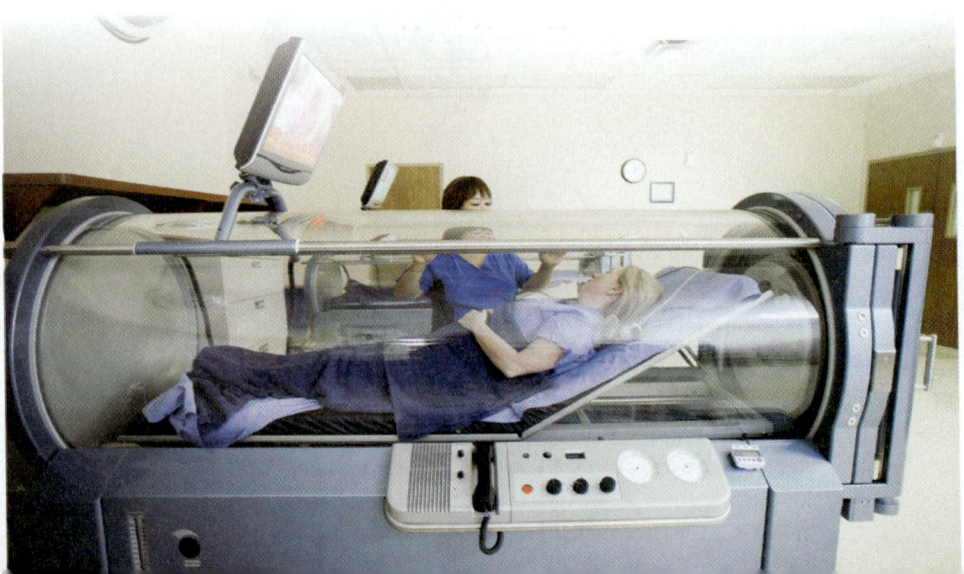

Hyperbaric chamber

Fever: To Treat or Not to Treat?

Our immune system helps to protect us from invading microorganisms. One manner in which our body protects itself is by mounting a fever in response to microbes present in the body (body temperature can also rise in response to inflammation or injury).

The hypothalamus, located in the brain, serves as the temperature-control center of the body. Fever occurs when the hypothalamus actually resets itself at a higher temperature. The hypothalamus raises body temperature by shunting blood away from the skin and into the body's core. It also raises temperature by inducing shivering, which is a result of muscle contraction and serves to increase temperature. This is why people experience chills and shivering when they have a fever. Once the new, higher temperature is reached (warmer blood reaches the hypothalamus), the hypothalamus works to maintain this temperature. When the "thermostat" is reset once again to a lower level, the body reverses the process, shunting blood to the skin. This is why people become diaphoretic (sweaty) when a fever breaks.

When microorganisms gain entrance to the body and begin to proliferate, the body responds with an onslaught of macrophages and monocytes, whose purpose is to destroy microorganisms. This immune response induces fever.

Fever is often one of the first symptoms a patient with an illness will experience, prompting the individual with fever to take stock of his or her symptoms. Many people, including physicians, routinely treat fever with fever-reducing agents such as acetaminophen or other NSAIDs (nonsteroidal anti-inflammatory drugs). Is it a good idea to reduce fever if fever is a normal response to an abnormal process occurring in the body, such as an infection? Not all experts agree.

We know that microorganisms thrive at different temperatures (see "Environmental Factors That Influence Microbes"). For example, rhinoviruses, responsible for causing the common cold, thrive at temperatures slightly below normal human body temperature. If this is the case, fever can be seen as the body's attempt to make the internal environment less hospitable to the virus, and lowering body temperature may allow the virus to proliferate. Therefore, fever can be seen as a natural and useful method of curbing the growth of microorganisms.

For most people, fever is not harmful. It may cause unpleasant symptoms such as chills, headache, and muscle and joint pain, which is why people tend to want to treat it. A small segment of the population may experience adverse effects of a high fever, for example, children who experience febrile seizures; however, most people tolerate fever well without any ill effects. Because a high fever may sometimes be caused by serious illness, the following guidelines regarding fever should be kept in mind:

- Children under the age of 6 months should be examined by a physician if they develop a high fever.
- Fever should be treated if it rises to 40°C/104°F, regardless of age.
- A patient of any age who has neck stiffness, difficulty breathing or labored/rapid breathing, altered level of consciousness (i.e., confusion), persistent/severe abdominal pain, or severe headache with photophobia (aversion to light) or who experiences a febrile seizure should be seen by a physician, as these symptoms may be indicative of a serious illness.

Chapter Summary

6.1 Microbial Nutrition

- Nutrition is a process by which all living organisms obtain substances from their environment to convert to metabolic uses.
- Nutrients are categorized by the amount required (macronutrients or micronutrients), by chemical structure (organic or inorganic), and by their importance to the organism's survival (essential or nonessential).
- Microorganisms are classified both by the chemical form of their nutrients and the energy sources they utilize.
- Although the chemical form of nutrients varies widely, all organisms require six elements—carbon, hydrogen, oxygen, nitrogen, phosphorus, and sulfur—to survive, grow, and reproduce.
- Nutrients are transported into microorganisms by two kinds of processes: active transport that expends energy and passive transport that does not need energy input.

6.2 Environmental Factors That Influence Microbes

- The environmental factors that control microbial growth are temperature; gases; pH; osmotic, hydrostatic and atmospheric pressure; radiation; and other organisms in their habitats.
- Environmental factors control microbial growth mainly by their influence on microbial enzymes.
- Three cardinal temperatures for a microorganism describe its temperature range and the temperature at which it grows best. These are the minimum temperature, the maximum temperature, and the optimum temperature.

- Microorganisms are classified by their temperature requirements as psychrophiles, mesophiles, or thermophiles.
- Most eukaryotic microorganisms are aerobic, whereas bacteria vary widely in their oxygen requirements from obligately aerobic to anaerobic.
- Microorganisms live in association with other species that range from mutually beneficial symbiosis to parasitism and antagonism.
- Biofilms are examples of complex synergistic communities of microbes that behave differently than free-living microorganisms.

6.3 The Study of Bacterial Growth

- The splitting of a parent bacterial cell to form a pair of similar-size daughter cells is known as binary fission.
- Microbial growth refers both to increase in cell size and increase in number of cells in a population.
- The generation time is a measure of the growth rate of a microbial population. It varies in length according to environmental conditions.
- Microbial cultures in a nutrient-limited batch environment exhibit four distinct stages of growth: the lag phase, the exponential growth (log) phase, the stationary phase, and the death phase.
- Microbial cell populations show distinct phases of growth in response to changing nutrient and waste conditions.
- Population growth can be quantified by measuring colony numbers, the turbidity of a solution, and direct cell counts.

Multiple-Choice Questions   Bloom's Levels 1 and 2: Remember and Understand

Select the correct answer from the answers provided.

1. An organism that can synthesize all its required organic components from CO_2 using energy from the sun is a
   a. photoautotroph.
   b. photoheterotroph.
   c. chemoautotroph.
   d. chemoheterotroph.

2. Which of the following is not one of the six major elements microbes need to survive, grow, and reproduce?
   a. oxygen
   b. calcium
   c. sulfur
   d. nitrogen
   e. carbon

3. A microbe that does not require oxygen for metabolism but will use it if available is a/an
   a. microaerophile.
   b. facultative anaerobe.
   c. obligate anaerobe.
   d. aerotolerant anaerobe.
   e. obligate aerobe.

4. A pathogen would most accurately be described as a
   a. parasite.
   b. commensal.
   c. saprobe.
   d. symbiont.

5. Which of the following is true of passive transport?
   a. It requires a gradient.
   b. It uses the cell wall.
   c. It includes endocytosis.
   d. It only moves water.

6. A cell exposed to a hypertonic environment will _____ by osmosis.
   a. gain water
   b. lose water
   c. neither gain nor lose water
   d. burst

7. Psychrophiles would be expected to grow
  a. in hot springs.
  b. on the human body.
  c. at refrigeration temperatures.
  d. at low pH.

8. In a viable plate count, each _____ represents a _____ from the sample population.
  a. cell; colony
  b. colony; cell
  c. hour; generation
  d. cell; generation

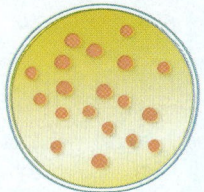

Critical Thinking   Bloom's Levels 3, 4, and 5: Apply, Analyze, and Evaluate

Critical thinking is the ability to reason and solve problems using facts and concepts. These questions can be approached from a number of angles and, in most cases, they do not have a single correct answer.

1. a. Compare and contrast passive and active forms of transport, using examples of what is being transported and the requirements for each.

   b. How are phagocytosis and pinocytosis similar? How are they different?

2. Compare the effects of isotonic, hypotonic, and hypertonic solutions on an amoeba and on a bacterial cell. If a cell lives in a hypotonic environment, what will occur if it is placed in a hypertonic one?

3. How can you explain the observation that unopened milk will spoil even while refrigerated?

4. a. If an egg salad sandwich sitting in a warm car for 4 hours develops 40,960 bacterial cells, how many more cells would result in just one more generation?

   b. What would the cell count be after 4 hours if the initial bacterial dose was 100?

   c. What do your answers tell you about using clean techniques in food preparation and storage (other than esthetic considerations)?

5. Consider figure 6.6, and note that the medium at the bottom of tube 2 is clear containing no growth, whereas tube 3 contains growth throughout. Explain what is probably causing this difference.

Visual Connections   Bloom's Level 5: Evaluate

This question connects previous images to a new concept.

1. **From chapter 5, figure 5.10.** What type of symbiotic relationship is pictured here?

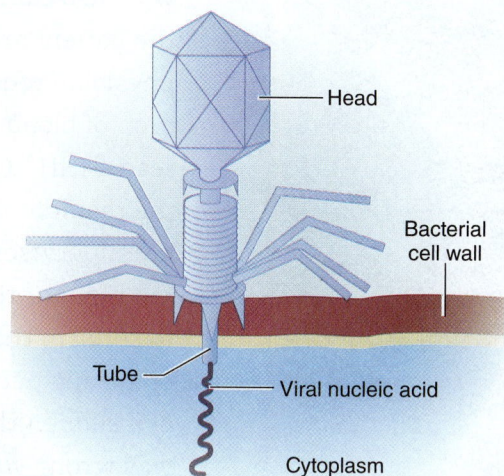

www.mcgrawhillconnect.com
Enhance your study of this chapter with study tools and practice tests. Also ask your instructor about the resources available through Connect, including the media-rich eBook, interactive learning tools, and animations.

Not So Sweet

I was working in triage in the emergency room of a pediatrics hospital during my final nursing practicum when a mother and her 8-year-old daughter came in. The mother stated that her daughter had been ill for 3 days with nausea and vomiting. She had been unable to keep any food or fluids down during this time, although she complained of constant thirst and begged for water and juice. The mother was concerned that her daughter was becoming dehydrated.

I quickly assessed the little girl and found that she did not have a fever (although her skin was flushed and hot), her heart rate was fast (125 beats per minute), and her blood pressure was low at 88/56 mmHg. Her respirations were fast, deep, and labored. She was very pale and her lips were cracked. She complained that her "tummy" hurt and she was quite drowsy, seeming to fall asleep whenever I was not touching or talking to her. Her breath had an odd odor—it was almost sweet, which seemed at odds with a 3-day history of vomiting.

Concerned because the patient appeared to be very ill, I notified my preceptor, who quickly came to assess the little girl. We notified the doctor, who also came quickly. We started an intravenous line so we could rehydrate the patient and drew blood for laboratory tests at the same time. The physician requested a spot glucose, so I poked the patient's finger to obtain a drop of blood and applied it to a glucose monitor. The result on the monitor read "HHH." Concerned that I had not performed the test properly, I showed the result to the physician, who said that that was the result he expected. My supervisor explained that "HHH" meant that the patient's blood sugar was too high for the monitor to read and that was all the information we needed to confirm the patient's diagnosis. She asked me if I had noticed the patient's breathing pattern and her odd sweet-smelling breath. I said that I had, and all of a sudden the patient's symptoms came together in my mind. I knew what was wrong with the patient—it was diabetic ketoacidosis.

- What is catabolism, and how does it relate to the patient's symptoms?

- When the body lacks insulin and cannot burn glucose for energy, what does the body use as fuel instead?

Case File Wrap-Up appears on page 188.

Microbial Metabolism

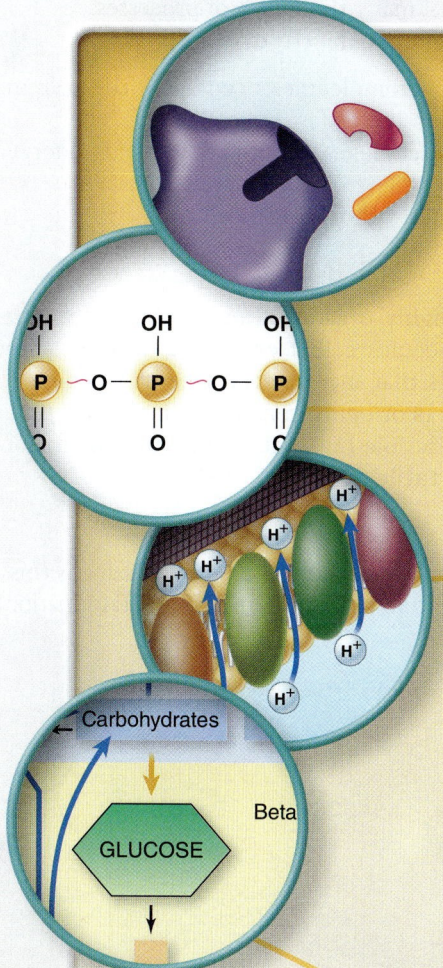

7.1 Metabolism and the Role of Enzymes

Metabolism, from the Greek term *metaballein,* meaning "change," pertains to all chemical reactions and physical workings of the cell. Although metabolism entails thousands of different reactions, most of them fall into one of two general categories. **Anabolism,** sometimes also called *biosynthesis,* is any process that results in synthesis of cell molecules and structures. It is a building and bond-making process that forms larger macromolecules from smaller ones, and it usually requires the input of energy. **Catabolism** is the opposite of anabolism. Catabolic reactions break the bonds of larger molecules into smaller molecules and often release energy. In a cell, linking anabolism to catabolism ensures the efficient completion of many thousands of processes.

In summary, metabolism accomplishes the following **(figure 7.1):**

1. assembles smaller molecules into larger macromolecules needed for the cell; in this process, ATP (energy) is utilized to form bonds (anabolism);
2. degrades macromolecules into smaller molecules, a process that yields energy (catabolism); and
3. stores energy in the form of ATP (adenosine triphosphate).

Enzymes: Catalyzing the Chemical Reactions of Life

The chemical reactions of life, even when highly organized and complex, cannot proceed without a special class of macromolecules called **enzymes.** Enzymes are a remarkable example of **catalysts,** chemicals that increase the rate of a chemical reaction without becoming part of the products or being consumed in the reaction. It is easy to think that an enzyme creates a reaction, but that is not true. The major characteristics of enzymes are summarized in **table 7.1.**

How Do Enzymes Work?

An enzyme speeds up the rate of a metabolic reaction, but just how does it do this? During a chemical reaction, reactants are converted to products either by bond formation

Figure 7.1 Simplified model of metabolism. Cellular reactions fall into two major categories. Catabolism (yellow) involves the breakdown of complex organic molecules to extract energy and form simpler end products. Anabolism (blue) uses the energy to synthesize necessary macromolecules and cell structures from precursors.

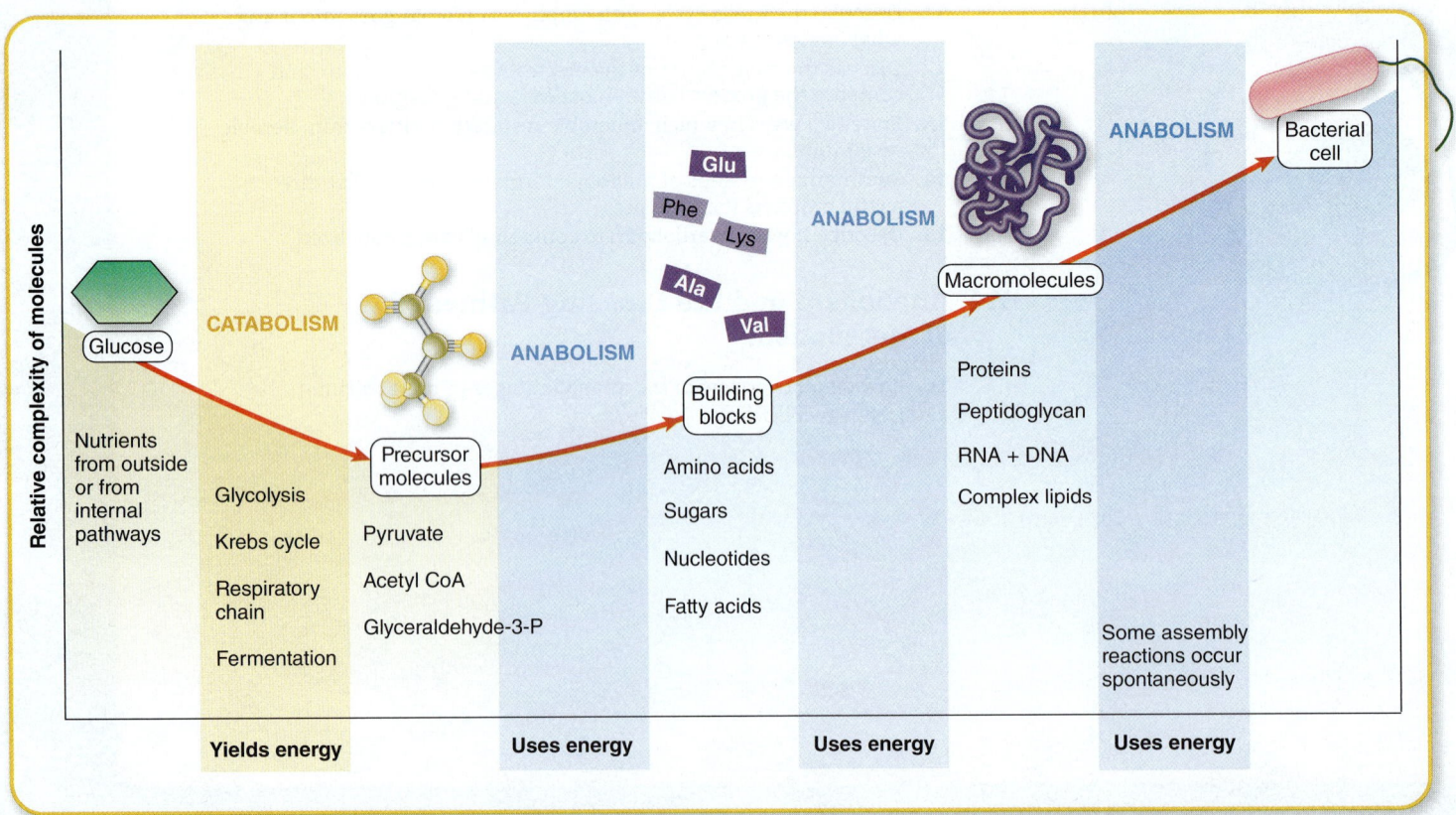

Table 7.1 Checklist of Enzyme Characteristics

- Most composed of protein; may require cofactors
- Act as organic catalysts to speed up the rate of cellular reactions
- Have unique characteristics such as shape, specificity, and function
- Enable metabolic reactions to proceed at a speed compatible with life
- Have an active site for target molecules called substrates
- Are much larger in size than their substrates
- Associate closely with substrates but do not become integrated into the reaction products
- Are not used up or permanently changed by the reaction
- Can be recycled, thus function in extremely low concentrations
- Are greatly affected by temperature and pH
- Can be regulated by feedback and genetic mechanisms

or by bond breakage. A certain amount of energy is required to initiate every such reaction, which limits its rate. While the rate could be sped up by increased heat or other means, biological cells use enzymes to vastly increase the speed of important reactions.

At the molecular level, an enzyme promotes a reaction by serving as a physical site upon which the reactant molecules, called **substrates,** can be positioned for various interactions. The enzyme is much larger in size than its substrate, and it presents a unique active (or catalytic) site that fits only that particular substrate. Although an enzyme binds to the substrate and participates directly in changes to the substrate, it does not become a part of the products, is not used up by the reaction, and can function over and over again. Enzyme speed, defined as the number of substrate molecules converted per enzyme per second, is well documented. Speeds range from several million for catalase to a thousand for lactate dehydrogenase.

Enzyme Structure

Most enzymes are proteins—although there is a special class that are made of RNA—and they can be classified as simple or conjugated. Simple enzymes consist of protein alone, whereas conjugated enzymes **(figure 7.2)** contain protein and nonprotein molecules. A conjugated enzyme, sometimes referred to as a **holoenzyme,** is a combination of a protein and one or more **cofactors.** In this situation, the actual protein portion is called the **apoenzyme.** Cofactors are either organic molecules, called **coenzymes,** or inorganic elements (metal ions). For example, catalase, an enzyme that we studied in chapter 6, breaks down hydrogen peroxide and requires iron as a metallic cofactor.

Enzyme-Substrate Interactions

For a reaction to take place, a temporary enzyme-substrate union must occur at the active site **(figure 7.3).** The fit is so specific that it is often described as a "lock-and-key" fit in which the substrate is inserted into the active site's pocket.

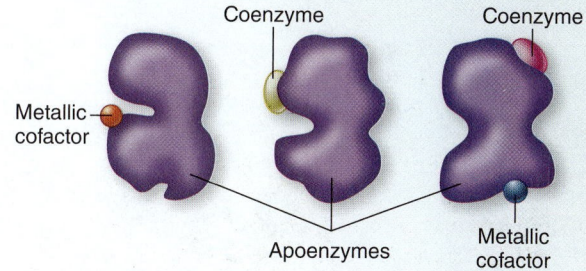

Figure 7.2 Conjugated enzyme structure.
Conjugated enzymes have an apoenzyme (polypeptide or protein) component and one or more cofactors.

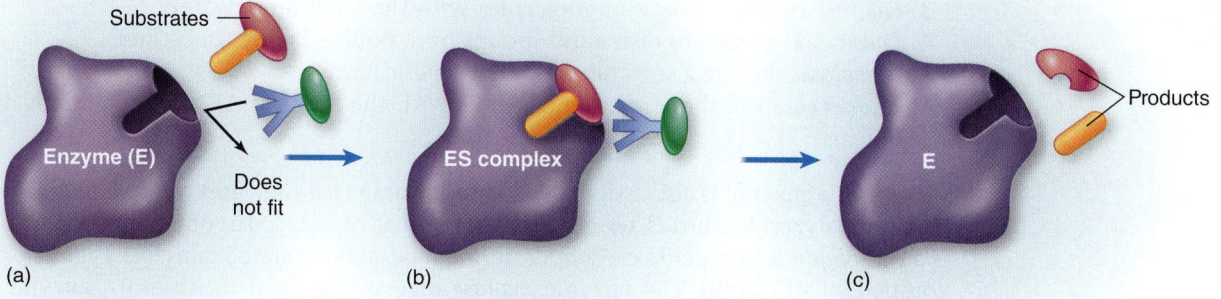

Figure 7.3 Enzyme-substrate reactions.   (a) When the enzyme and substrate come together, the substrate (S) must show the correct fit and position with respect to the enzyme (E). (b) When the ES complex is formed, it enters a transition state. During this temporary but tight interlocking union, the enzyme participates directly in breaking or making bonds. (c) Once the reaction is complete, the enzyme releases the products.

The bonds formed between the substrate and enzyme are weak and, of necessity, easily reversible. Once the enzyme-substrate complex has formed, appropriate reactions occur on the substrate, often with the aid of a cofactor, and a product is formed and released. The enzyme can then attach to another substrate molecule and repeat this action. Although enzymes can potentially catalyze reactions in both directions, most examples in this chapter depict them working in one direction only.

Cofactors: Supporting the Work of Enzymes

In chapter 6, you learned that microorganisms require specific metal ions called trace elements and certain organic growth factors. In many cases, the need for these substances arises from their roles as cofactors for enzymes. The metallic cofactors, including iron, copper, magnesium, manganese, zinc, cobalt, selenium, and many others, participate in precise functions between the enzyme and its substrate. In general, metals activate enzymes, help bring the active site and substrate close together, and participate directly in chemical reactions with the enzyme-substrate complex.

Coenzymes are a type of cofactor. They are organic compounds that work in conjunction with an apoenzyme to perform a necessary alteration of a substrate. The general function of a coenzyme is to remove a chemical group from one substrate molecule and add it to another substrate, thereby serving as a transient carrier of this group. In a later section of this chapter, we shall see that coenzymes carry and transfer hydrogen atoms, electrons, carbon dioxide, and amino groups. Many coenzymes are derived from **vitamins,** which explains why vitamins are important to nutrition and may be required as growth factors for living things. Vitamin deficiencies prevent the complete holoenzyme from forming. Consequently, both the chemical reaction and the structure or function dependent upon that reaction are compromised.

Classification of Enzyme Functions

Enzymes are classified and named according to characteristics such as site of action, type of action, and substrate.

In general, an enzyme name is composed of two parts: a prefix or stem word derived from a certain characteristic—usually the substrate acted upon, the type of reaction catalyzed, or both—followed by the ending -ase.

The system classifies the enzyme in one of the following six classes, on the basis of its general biochemical action:

1. *Oxidoreductases* transfer electrons from one substrate to another, and *dehydrogenases* transfer a hydrogen from one compound to another.
2. *Transferases* transfer functional groups from one substrate to another.
3. *Hydrolases* cleave bonds on molecules with the addition of water.
4. *Lyases* add groups to or remove groups from double-bonded substrates.
5. *Isomerases* change a substrate into its isomeric form.
6. *Ligases* catalyze the formation of bonds with the input of ATP and the removal of water.

Each enzyme is also assigned a common name that indicates the specific reaction it catalyzes. With this system, an enzyme that digests a carbohydrate substrate is a *carbohydrase;* a specific carbohydrase, *amylase,* acts on starch (amylose is a major component of starch). The enzyme *maltase* digests the sugar maltose. An enzyme that hydrolyzes peptide bonds of a protein is a *proteinase, protease,* or *peptidase.* Some fats and other lipids are digested by *lipases.* DNA is hydrolyzed by *deoxyribonuclease,* generally shortened to *DNase.* A *synthetase* or *polymerase* bonds together many small molecules into large molecules.

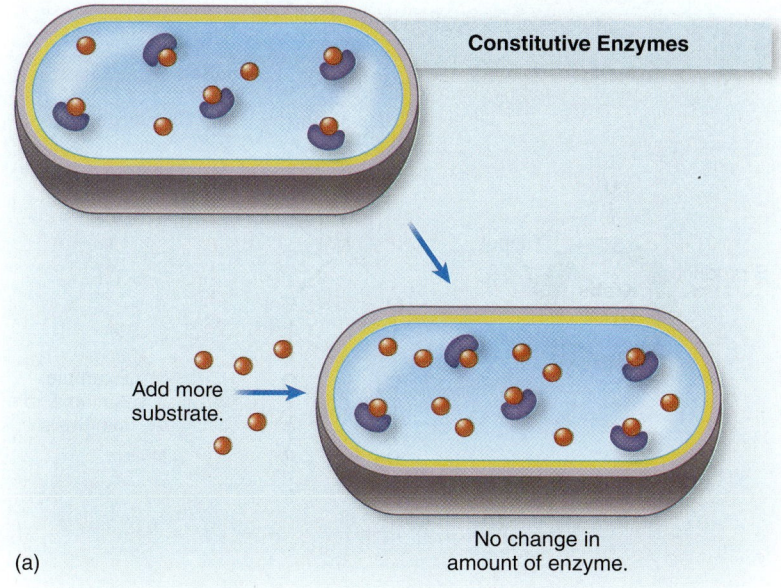

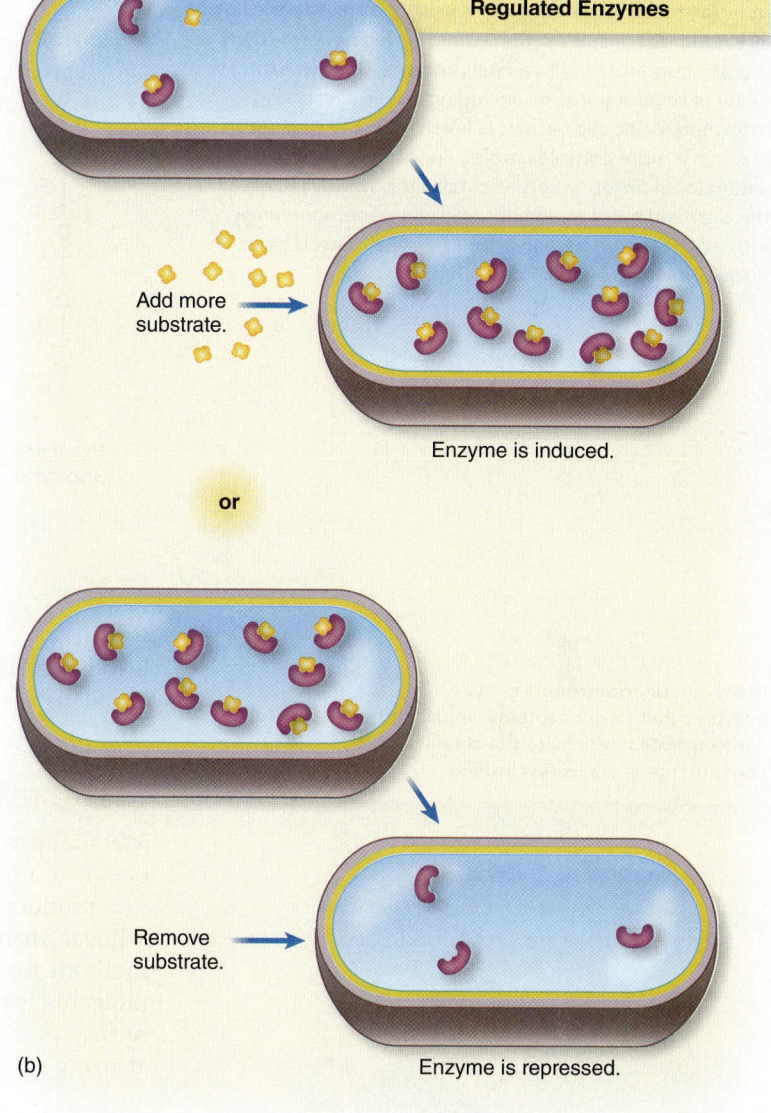

Figure 7.4 Constitutive and regulated enzymes.
(a) Constitutive enzymes are present in constant amounts in a cell. The addition of more substrate does not increase the numbers of these enzymes. (b) The concentration of regulated enzymes in a cell increases or decreases in response to substrate levels.

Regulation of Enzyme Action

Enzymes are not all produced in equal amounts or at equal rates. Some, called **constitutive enzymes (figure 7.4a),** are always present and in relatively constant amounts, regardless of the amount of substrate. The enzymes involved in utilizing glucose, for example, are very important in metabolism and thus are constitutive. Other enzymes are **regulated enzymes (figure 7.4b),** the production of which is either turned on (induced) or turned off (repressed) in response to changes in concentration of the substrate.

The activity of an enzyme is highly influenced by the cell's environment. In general, enzymes operate only under the natural temperature, pH, and osmotic pressure of an organism's habitat. When enzymes are subjected to changes in these normal conditions, they tend to be chemically unstable, or **labile.** Low temperatures inhibit catalysis, and high temperatures denature the apoenzyme. **Denaturation** is a process by which the weak bonds that collectively maintain the native shape of the apoenzyme are broken. This disruption causes extreme distortion of the enzyme's shape and prevents the substrate from attaching to the active site. Such nonfunctional enzymes block metabolic reactions and thereby can

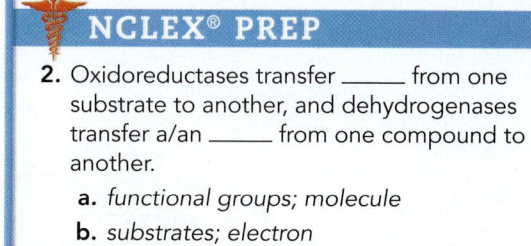

NCLEX® PREP

2. Oxidoreductases transfer _____ from one substrate to another, and dehydrogenases transfer a/an _____ from one compound to another.
   a. *functional groups; molecule*
   b. *substrates; electron*
   c. *electrons; hydrogen*
   d. *double-bonded substrates; bond*

Figure 7.5 Patterns of metabolism. In general, metabolic pathways consist of a linked series of individual chemical reactions that produce intermediary metabolites and lead to a final product. These pathways occur in several patterns, including linear, cyclic, and branched. Anabolic pathways involved in biosynthesis result in a more complex molecule, each step adding on a functional group, whereas catabolic pathways involve the dismantling of molecules and can generate energy. Virtually every reaction in a series—represented by an arrow—involves a specific enzyme.

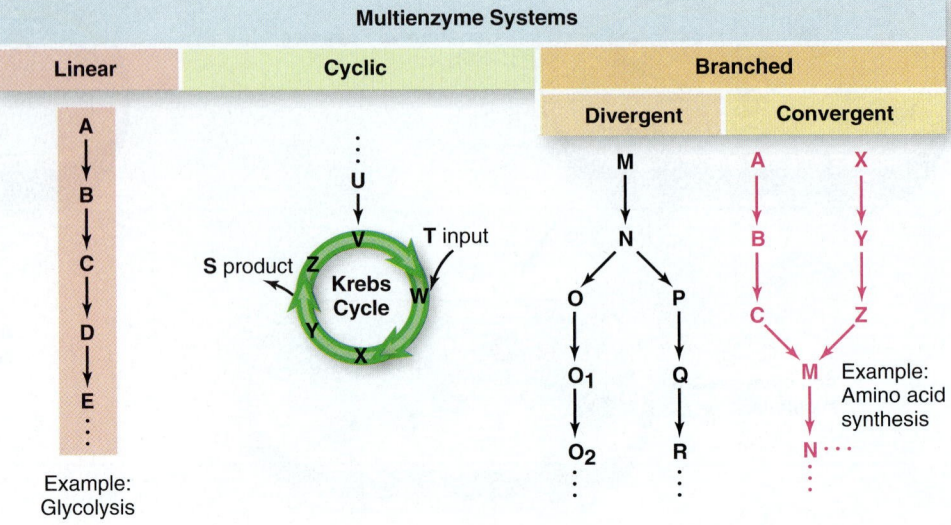

Many laundry detergents contain enzymes that target proteins, lipids, and carbohydrates, which are the chemical constituents of stains on clothing.

lead to cell death. Low or high pH or certain chemicals (heavy metals, alcohol) are also denaturing agents.

Metabolic Pathways

Metabolic reactions rarely consist of a single action or step. More often, they occur in a multistep series or pathway, with each step catalyzed by an enzyme. The product of one reaction is often the reactant (substrate) for the next, forming a linear chain of reactions. Many pathways have branches that provide alternate methods for nutrient processing. Others take a cyclic form, in which the starting molecule is regenerated to initiate another turn of the cycle **(figure 7.5)**. On top of that, pathways generally do not stand alone; they are interconnected and merge at many sites.

Direct Controls on the Action of Enzymes

The bacterial cell has many ways of directly influencing the activity of its enzymes. It can inhibit enzyme activity by supplying a molecule that resembles the enzyme's normal substrate. The "mimic" can then occupy the enzyme's active site, preventing the actual substrate from binding there. Because the mimic cannot actually be acted on by the enzyme or function in the way the product would have, the enzyme is effectively shut down. This form of inhibition is called **competitive inhibition,** because the mimic is competing with the substrate for the binding site **(figure 7.6).** (In chapter 10, you will see that some antibiotics use the same strategy of competing with enzymatic active sites to shut down metabolic processes.)

Another form of inhibition can occur with special types of enzymes that have two binding sites—the active site and another area called the regulatory site (figure 7.6). These enzymes are regulated by the binding of molecules other than the substrate in their regulatory sites. Often the regulatory molecule is the product of the enzymatic reaction itself. This provides a negative feedback mechanism that can slow down enzymatic activity once a certain concentration of product is produced. This is **noncompetitive inhibition,** because the regulator molecule does not bind in the same site as the substrate.

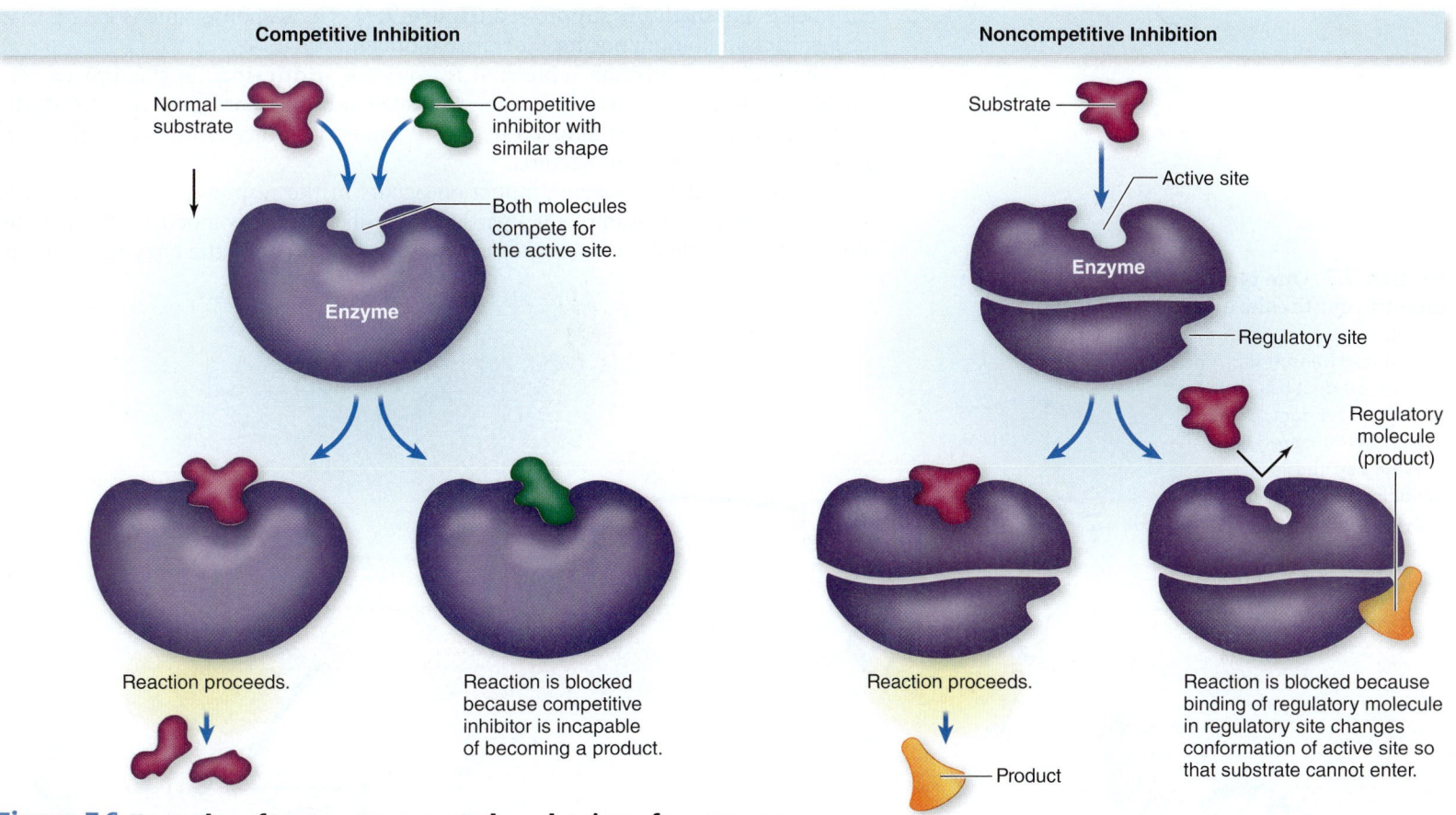

| Competitive Inhibition | Noncompetitive Inhibition |

Normal substrate

Competitive inhibitor with similar shape

Both molecules compete for the active site.

Enzyme

Reaction proceeds.

Reaction is blocked because competitive inhibitor is incapable of becoming a product.

Substrate

Active site

Enzyme

Regulatory site

Regulatory molecule (product)

Reaction proceeds.

Product

Reaction is blocked because binding of regulatory molecule in regulatory site changes conformation of active site so that substrate cannot enter.

Figure 7.6 **Examples of two common control mechanisms for enzymes.**

Controls on Enzyme Synthesis

Controlling enzymes by controlling their synthesis is another effective mechanism, because enzymes do not last indefinitely. Some wear out, some are deliberately degraded, and others are diluted with each cell division. For catalysis to continue, enzymes eventually must be replaced. This cycle works into the scheme of the cell, where replacement of enzymes can be regulated according to cell demand. The mechanisms of this system are genetic in nature; that is, they require regulation of DNA and the protein synthesis machinery–topics we shall encounter once again in chapter 8.

Enzyme repression is a means to stop further synthesis of an enzyme some-where along its pathway. As the level of the end product from a given enzymatic reaction has built to excess, the genetic apparatus responsible for replacing these

Enzymes isolated from microbes are used by denim manufacturers to make the fabric softer and impart different colors to it.

enzymes is automatically suppressed **(figure 7.7).** The response time is longer than for feedback inhibition, but its effects are more enduring.

The inverse of enzyme repression is **enzyme induction.** In this process, enzymes appear (are induced) only when suitable substrates are present—that is, the synthesis of an enzyme is induced by its substrate. Both mechanisms are important genetic control systems in bacteria.

A classic model of enzyme induction occurs in the response of *Escherichia coli* to certain sugars. For example, if a particular strain of *E. coli* is inoculated into a medium whose principal carbon source is lactose, it will produce the enzyme lactase to

Figure 7.7 **One type of genetic control of enzyme synthesis: enzyme repression.**
(1)–(5), The enzyme is synthesized continuously via uninhibited transcription and translation until enough product has been made. (6), (7), Excess product reacts with a site on DNA that regulates the enzyme's synthesis, thereby inhibiting further enzyme production.

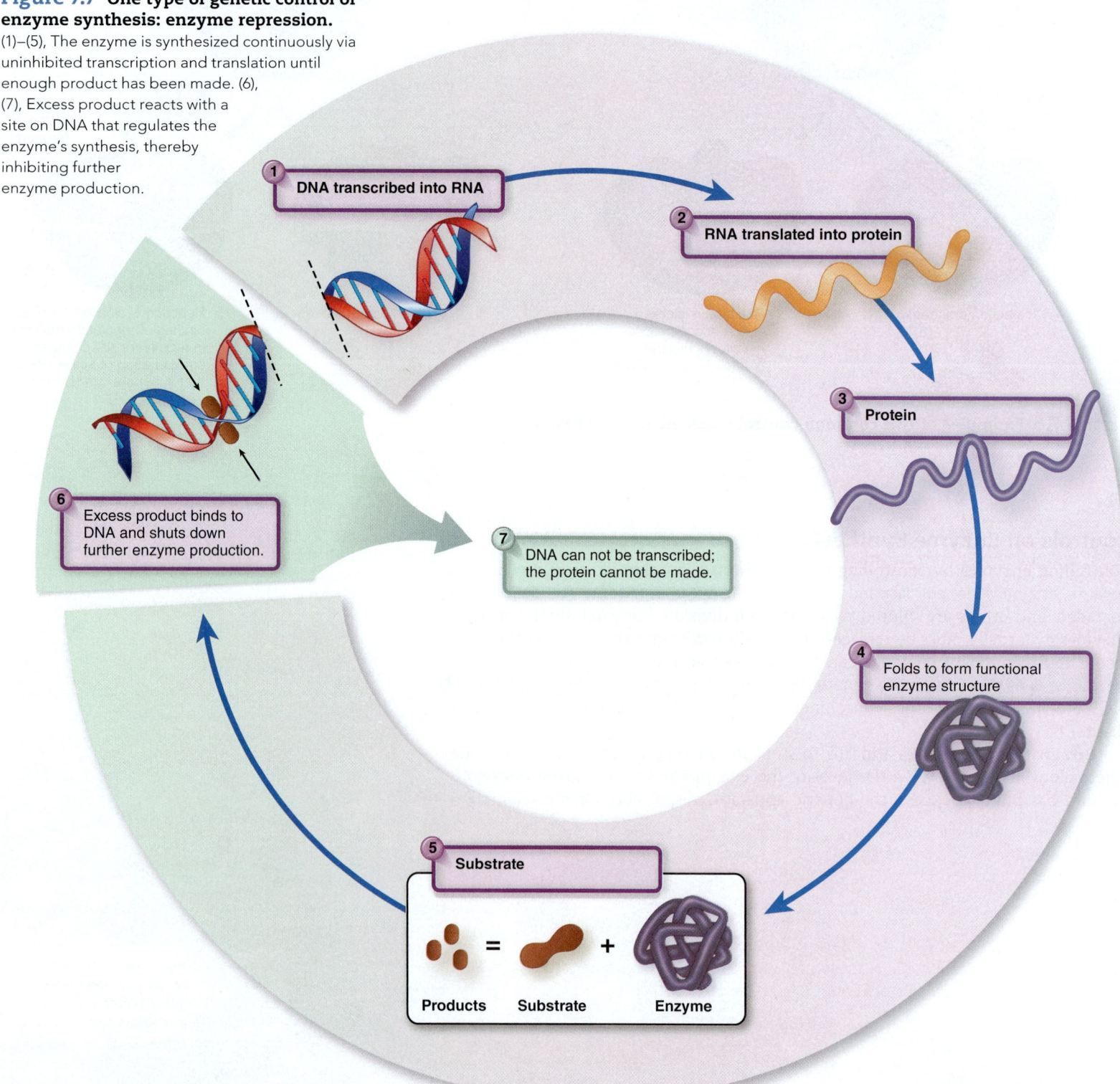

1. DNA transcribed into RNA
2. RNA translated into protein
3. Protein
4. Folds to form functional enzyme structure
5. Substrate

   Products = Substrate + Enzyme
6. Excess product binds to DNA and shuts down further enzyme production.
7. DNA can not be transcribed; the protein cannot be made.

hydrolyze it into glucose and galactose. If the bacterium is subsequently inoculated into a medium containing only sucrose as a carbon source, it will cease synthesizing lactase and begin synthesizing sucrase. This response enables the organism to utilize a variety of nutrients, and it also prevents a microbe from wasting energy making enzymes for which no substrates are present.

> ### 7.1 LEARNING OUTCOMES—Assess Your Progress
>
> 1. Describe the relationship among metabolism, catabolism, and anabolism.
> 2. Fully describe the structure and function of enzymes.
> 3. Differentiate between constitutive and regulated enzymes.
> 4. Diagram four major patterns of metabolism.
> 5. Describe how enzymes are controlled.

7.2 The Pursuit and Utilization of Energy

In order to carry out the work of an array of metabolic processes, cells require constant input and expenditure of some form of usable energy. The energy comes directly from light or is contained in chemical bonds and released when substances are catabolized, or broken down. The energy is mostly stored in ATP.

Energy in Cells

Cells manage energy in the form of chemical reactions that change molecules. This often involves activities such as the making or breaking of bonds and the transfer of electrons. Not all cellular reactions are equal with respect to energy. Some release energy, and others require it to proceed. For example, a reaction that proceeds as follows:

$$X + Y \xrightarrow{\text{Enzyme}} Z + \text{Energy}$$

releases energy as it goes forward. This type of reaction is an **exergonic** (ex-er-gon'-ik) **reaction.** Energy of this type is available for doing cellular work. Energy transactions such as the following:

$$\text{Energy} + A + B \xrightarrow{\text{Enzyme}} C$$

are called **endergonic** (en-der-gon'-ik) **reactions,** because they require the addition of energy. In cells, exergonic and endergonic reactions are often coupled, so that released energy is immediately put to use.

Summaries of metabolism may make it seem that cells "create" energy from nutrients, but they do not. What they actually do is extract chemical energy already present in nutrient fuels and apply that energy toward useful work in the cell, much like a gasoline engine releases energy as it burns fuel. The engine does not actually produce energy, but it converts some of the potential energy to do work.

At the simplest level, cells possess specialized enzyme systems that trap the energy present in the bonds of nutrients as they are progressively broken. During exergonic reactions, energy released by bonds is stored in certain high-energy phosphate bonds, such as in ATP. The ability of ATP to temporarily store and release the energy of chemical bonds fuels endergonic cell reactions. Before discussing ATP, we examine the process behind electron transfer: redox reactions.

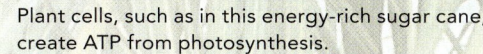
Plant cells, such as in this energy-rich sugar cane, create ATP from photosynthesis.

NCLEX® PREP

3. A reaction that requires energy as it goes forward is termed
   a. *exergonic.*
   b. *endergonic.*
   c. *oxidative.*
   d. *reductive.*

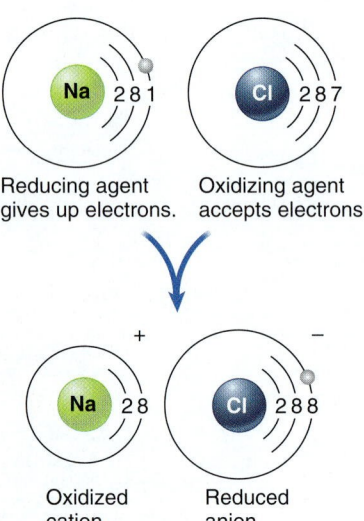

Reducing agent gives up electrons. Oxidizing agent accepts electrons.

Oxidized cation   Reduced anion

Figure 7.8 Redox pairs.

Oxidation and Reduction

Some atoms and compounds readily give or receive electrons and participate in oxidation (the loss of electrons) or reduction (the gain of electrons). The compound that loses the electrons is **oxidized,** and the compound that receives the electrons is **reduced (figure 7.8).** Such **oxidation-reduction (redox)** reactions are common in the cell and indispensable to the required energy transformations. Important components of cellular redox reactions are oxidoreductases, which remove electrons from one substrate and add them to another. Their coenzyme carriers are nicotinamide adenine dinucleotide (NAD) (**figure 7.9**) and flavin adenine dinucleotide (FAD).

Redox reactions always occur in pairs, with an electron donor and an electron acceptor, which constitute a *redox pair*. Oxidation-reduction reactions salvage electrons along with the energy they contain.

This changes the energy balance, leaving the previously reduced compound with less energy than the now oxidized one. The energy now present in the electron acceptor can be captured to **phosphorylate** (add an inorganic phosphate) to ADP or to some other compound. This process stores the energy in a high-energy molecule (ATP, for example). In many cases, the cell does not handle electrons as separate entities but rather as parts of an atom such as hydrogen, which contains a proton and an electron. For simplicity's sake, we will continue to use the term *electron transfer*, but keep in mind that hydrogens are often involved in the transfer process. The removal of hydrogens from a compound during a redox reaction is called dehydrogenation. The job of handling these protons and electrons falls to one or more carriers, which function as short-term repositories for the electrons until they can be transferred.

Electron Carriers: Molecular Shuttles

Electron carriers resemble shuttles that are alternately loaded and unloaded, repeatedly accepting and releasing electrons and hydrogens to facilitate the transfer of redox energy. In catabolic pathways, electrons are extracted and carried through a series of redox reactions until the final electron acceptor at the end of a particular

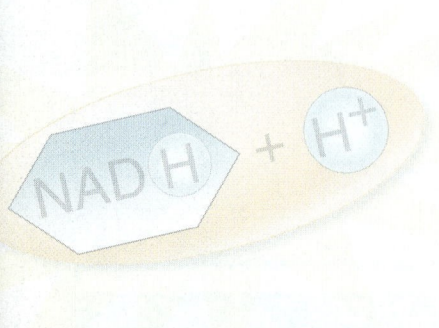

Figure 7.9 Details of NAD reduction. The coenzyme NAD contains the vitamin nicotinamide (niacin) and the purine adenine attached to double ribose phosphate molecules (a dinucleotide). The principal site of action is on the nicotinamide (boxed areas). Hydrogens and electrons donated by a substrate interact with a carbon on the top of the ring. One hydrogen bonds there, carrying two electrons, and the other hydrogen is carried in solution as H$^+$ (a proton).

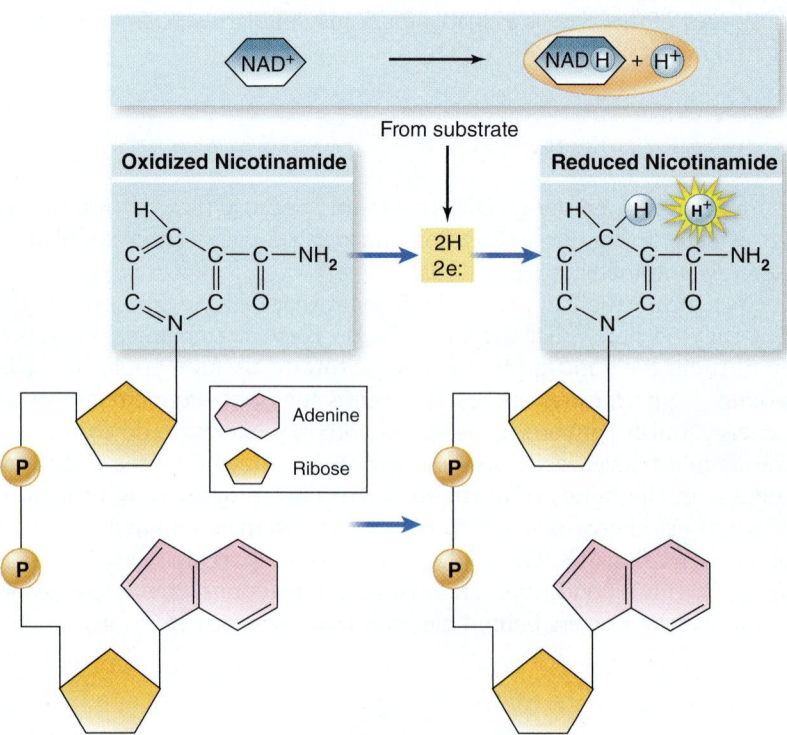

pathway is reached. In aerobic metabolism, this acceptor is molecular oxygen; in anaerobic metabolism, it is some other inorganic or organic compound.

Adenosine Triphosphate: Metabolic Money

Let's look more closely at the powerhouse molecule, adenosine triphosphate. ATP has also been described as metabolic money because it can be earned, banked, saved, spent, and exchanged. As a temporary energy repository, ATP provides a connection between energy-yielding catabolism and the other cellular activities that require energy. Some clues to its energy-storing properties lie in its unique molecular structure.

The Molecular Structure of ATP

ATP is a three-part molecule consisting of a nitrogen base (adenine) linked to a 5-carbon sugar (ribose), with a chain of three phosphate groups bonded to the ribose **(figure 7.10)**. The high energy of ATP comes from the orientation of the phosphate groups, which are relatively bulky and carry negative charges. The proximity of these repelling electrostatic charges imposes a strain that is most acute on the bonds between the last two phosphate groups. The strain on the phosphate bonds accounts for the energetic quality of ATP because removal of the terminal phosphates releases free energy.

Breaking the bonds between the two outermost phosphates of ATP yields adenosine diphosphate (ADP), which is then converted to adenosine monophosphate (AMP). AMP derivatives help form the backbone of RNA and are also a major component of certain coenzymes (NAD, FAD, and coenzyme A).

The Metabolic Role of ATP

ATP is the primary energy currency of the cell, and when it is used in a chemical reaction, it must then be replaced. Therefore, ATP utilization and replenishment make up an ongoing cycle. Often, the energy released during ATP hydrolysis drives biosynthesis by activating individual substrates before they are enzymatically linked together. ATP is also used to prepare molecules for catabolism, such as when a 6-carbon sugar is phosphorylated during the early stages of glycolysis:

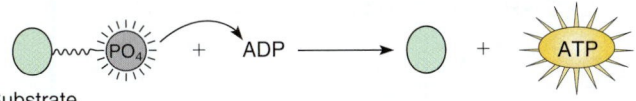

$$\text{Glucose} \xrightarrow[\qquad\qquad\qquad\qquad]{\text{ATP} \quad \text{ADP}} \text{Glucose-6-phosphate}$$

When ATP is utilized, by the removal of the terminal phosphate to release energy plus ADP, ATP then needs to be re-created. The reversal of this process–that is, adding the terminal phosphate to ADP–will replenish ATP, but it requires an input of energy:

$$\text{Substrate} \sim \text{PO}_4 + \text{ADP} \longrightarrow \bigcirc + \text{ATP}$$

In heterotrophs, the energy infusion that regenerates a high-energy phosphate comes from certain steps of catabolic pathways, in which nutrients such as carbohydrates are degraded and yield energy. ATP is formed when substrates or electron carriers provide a high-energy phosphate that becomes bonded to ADP.

7.2 LEARNING OUTCOMES—Assess Your Progress

6. Name the chemical in which energy is stored in cells.

7. Create a general diagram of a redox reaction.

8. Identify electron carriers used by cells.

Figure 7.10 The structure of adenosine triphosphate (ATP). Removing the left-most phosphate group yields ADP; removing the next one yields AMP.

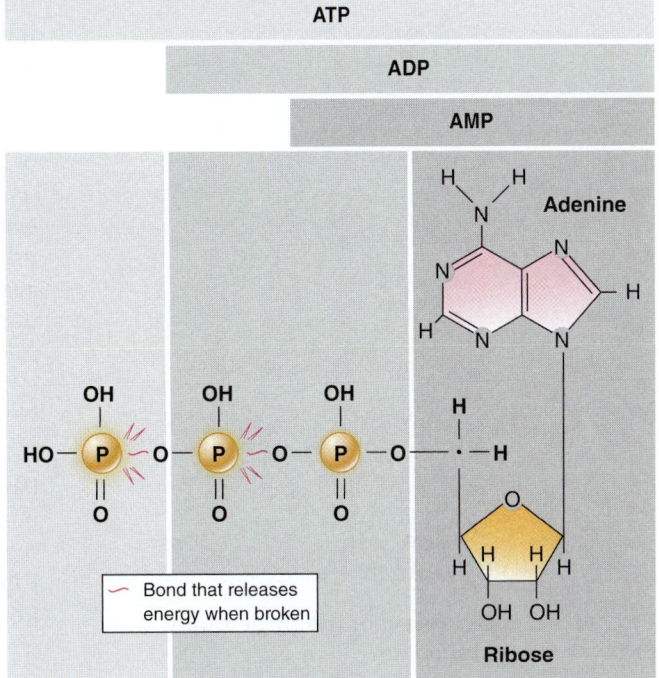

ATP

ADP

AMP

Adenine

OH   OH   OH

HO—P—O—P—O—P—O—•—H

O   O   O

H

Ribose

OH OH

⌇ Bond that releases energy when broken

ATP is used for energy in all cells, including human cells.

7.3 Catabolism

Now you have an understanding of all the tools a cell needs to *metabolize*. Metabolism uses *enzymes* to catalyze reactions that break down (*catabolize*) organic molecules to materials (*precursor molecules*) that cells can then use to build (*anabolize*) larger, more complex molecules that are particularly suited to them. This process is presented symbolically in figure 7.1. Another very important point about metabolism is that *reducing power* (the electrons available in NADH and FADH$_2$) and *energy* (stored in the bonds of ATP) are needed in large quantities for the anabolic parts of metabolism (the blue bars in our figure). They are produced during the catabolic part of metabolism (the yellow bar).

Metabolism starts with "nutrients" from the environment, usually discarded molecules from other organisms. Cells have to get the nutrients inside; to do this, they use the mechanisms discussed in chapter 6. Some of these require energy, which is available from catabolism already occurring in the cell. In the next step, intracellular nutrients have to be broken down to the appropriate precursor molecules. These catabolic pathways are discussed next.

Getting Materials and Energy

Nutrient processing is extremely varied, especially in bacteria, yet in most cases it is based on three basic catabolic pathways. Frequently, the nutrient is glucose. In previous discussions, microorganisms were categorized according to their requirement for oxygen gas; this designation is related directly to these nutrient processing pathways. **Figure 7.11** provides an overview of the three major pathways for producing the needed precursors and energy (i.e., catabolism).

Figure 7.11 Overview of the three main pathways of catabolism.

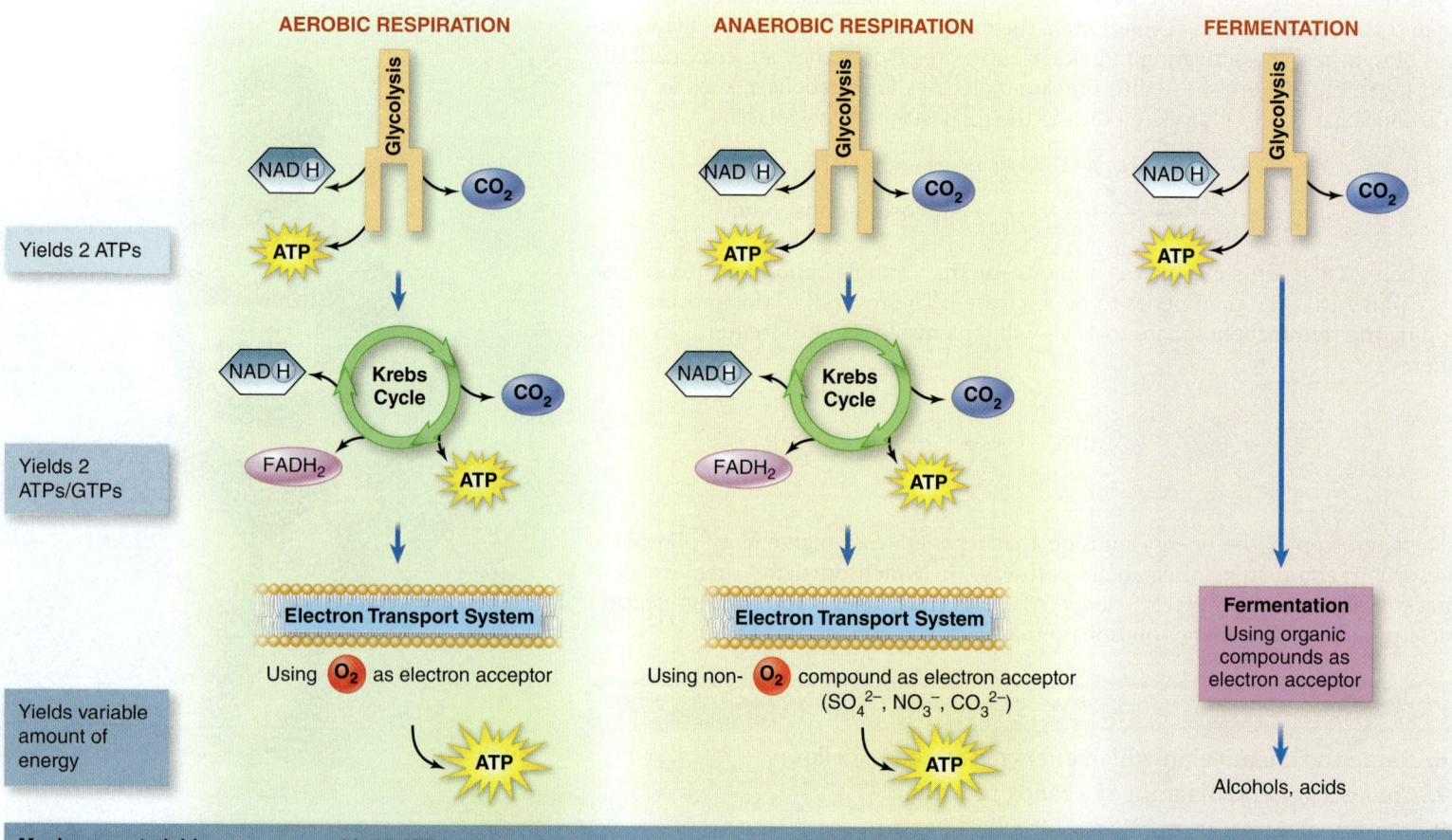

| Maximum net yield | 36–38 ATPs | 2–36 ATPs | 2 ATPs |

Aerobic respiration is a series of reactions (glycolysis, the Krebs cycle, and the respiratory chain) that converts glucose to CO_2 and allows the cell to recover significant amounts of energy. Aerobic respiration relies on free oxygen as the final acceptor for electrons and hydrogens and produces a relatively large amount of ATP. Aerobic respiration is characteristic of many bacteria, fungi, protozoa, and animals.

Anaerobic respiration is the metabolic strategy used by many microorganisms, some strictly anaerobic and others who are able to metabolize with or without oxygen. This system involves the same three pathways as aerobic respiration, but it does not use oxygen as the final electron acceptor; instead, NO_3^-, SO_4^{2-}, CO_3^{3-}, and other oxidized compounds are utilized.

Fermentation is an adaptation used by facultative and aerotolerant anaerobes to incompletely oxidize (ferment) glucose. In this case, oxygen is not required, organic compounds are the final electron acceptors, and a relatively small amount of ATP is produced.

Microbes often obtain nutrients from dead plants and organisms.

Glycolysis

All three of the metabolic pathways begin with **glycolysis,** which turns glucose into two copies of a chemical uniquely capable of yielding energy in the pathways that follow. **Table 7.2** illustrates glycolysis.

Pyruvic Acid—Central to All Three Metabolic Strategies

In strictly aerobic organisms and some anaerobes, pyruvic acid enters the Krebs cycle for further processing and energy release. Facultative anaerobes can use fermentation, in which pyruvic acid is re-reduced into acids or other products.

After Pyruvic Acid I: Aerobic and Anaerobic Respiration

Bacterial respiration, whether done aerobically or anaerobically, utilizes the Krebs cycle and the electron transport system to harvest the energy and products needed to build cell parts.

The Krebs cycle operates similarly in aerobic and anaerobic respiration and is covered next.

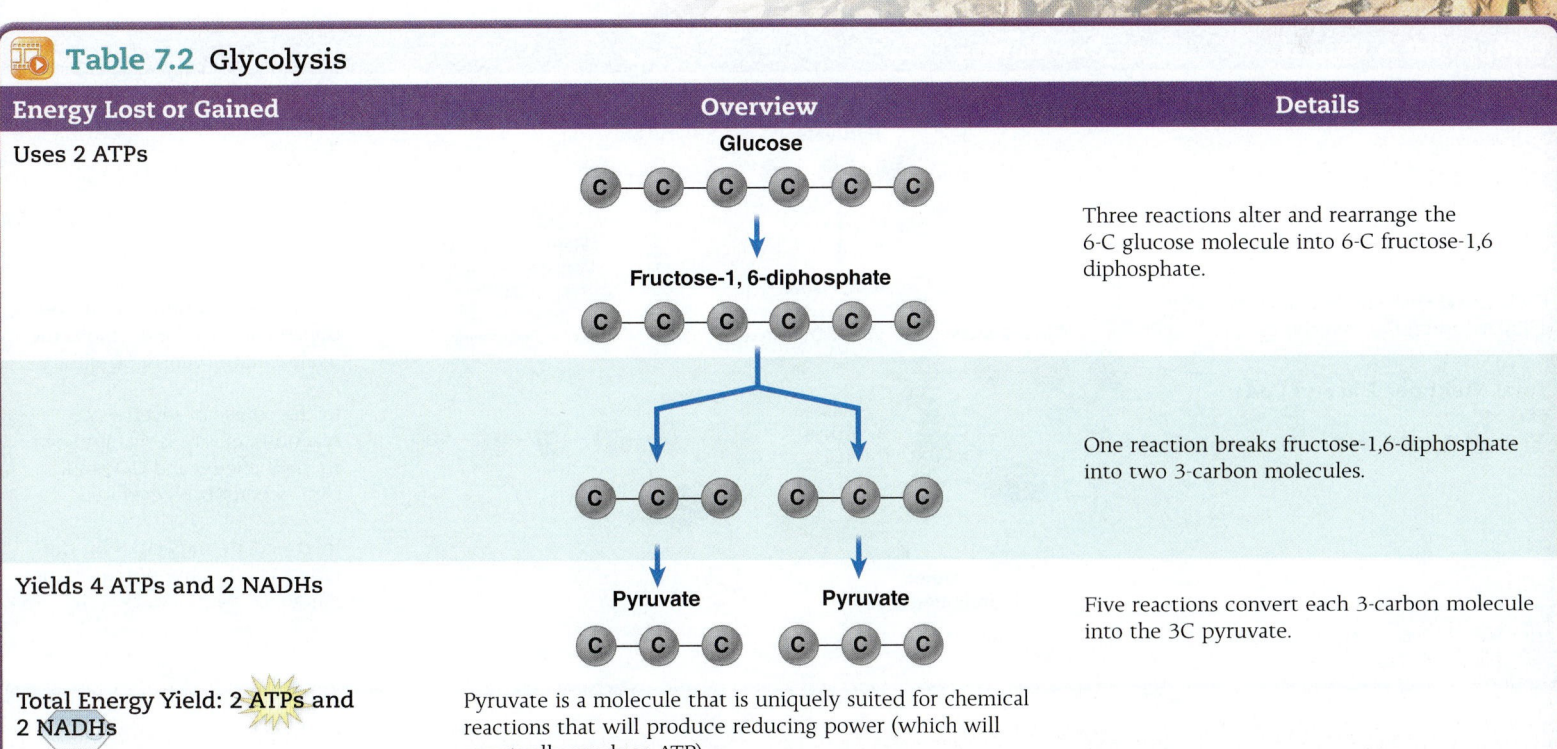

Table 7.2 Glycolysis

Energy Lost or Gained	Overview	Details
Uses 2 ATPs	Glucose	Three reactions alter and rearrange the 6-C glucose molecule into 6-C fructose-1,6 diphosphate.
	Fructose-1, 6-diphosphate	One reaction breaks fructose-1,6-diphosphate into two 3-carbon molecules.
Yields 4 ATPs and 2 NADHs	Pyruvate Pyruvate	Five reactions convert each 3-carbon molecule into the 3C pyruvate.
Total Energy Yield: 2 ATPs and 2 NADHs	Pyruvate is a molecule that is uniquely suited for chemical reactions that will produce reducing power (which will eventually produce ATP).	

The Krebs Cycle—A Carbon and Energy Wheel

As you have seen, the oxidation of glucose yields a comparatively small amount of energy and gives off pyruvic acid. Pyruvic acid is still energy-rich, containing a number of extractable hydrogens and electrons to power ATP synthesis, but this can be achieved only through the work of the second and third phases of respiration, in which pyruvic acid's hydrogens are transferred to oxygen. In the following section, we examine the next phase of this process, which takes place in the cytoplasm of bacteria and in the mitochondrial matrix in eukaryotes. **Table 7.3** illustrates that step and the Krebs cycle.

The **Krebs cycle** (also known as the TCA cycle) serves to transfer the energy stored in acetyl CoA to NAD⁺ and FAD by reducing them (transferring hydrogen ions to them). Thus, the main products of the Krebs cycle are these reduced molecules (as well as 2 ATPs for each glucose molecule). The reduced coenzymes NADH and $FADH_2$ are vital to the energy production that will occur in electron transport. Along the way, the 2-carbon acetyl CoA joins with a 4-carbon compound, oxaloacetic acid, and then participates in seven additional chemical transformations while "spinning off" the NADH and $FADH_2$. That's why we sometimes call the Krebs cycle the "carbon and energy wheel."

The Respiratory Chain: Electron Transport

We now come to the energy chain, which is the final "processing mill" for electrons and hydrogen ions and the major generator of ATP. It is the final step in both aerobic and anaerobic respiration. Overall, the electron transport system (ETS) consists of a chain of special redox carriers that receives electrons from reduced carriers (NADH, $FADH_2$) generated by glycolysis and the Krebs cycle and passes

Table 7.3 The Krebs Cycle

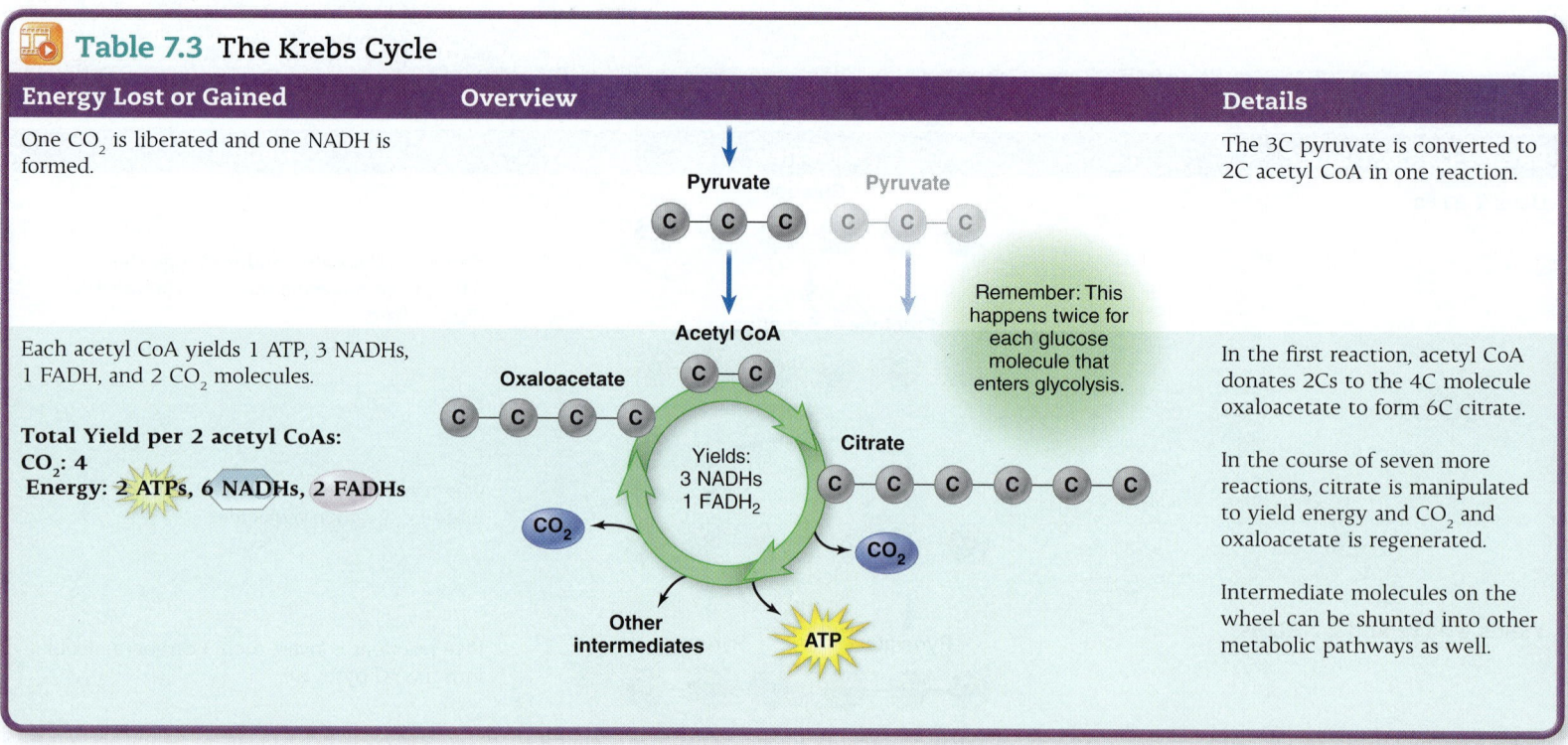

Energy Lost or Gained	Overview	Details
One CO_2 is liberated and one NADH is formed.		The 3C pyruvate is converted to 2C acetyl CoA in one reaction.
Each acetyl CoA yields 1 ATP, 3 NADHs, 1 FADH, and 2 CO_2 molecules. **Total Yield per 2 acetyl CoAs:** **CO_2:** 4 **Energy: 2 ATPs, 6 NADHs, 2 FADHs**		In the first reaction, acetyl CoA donates 2Cs to the 4C molecule oxaloacetate to form 6C citrate. In the course of seven more reactions, citrate is manipulated to yield energy and CO_2 and oxaloacetate is regenerated. Intermediate molecules on the wheel can be shunted into other metabolic pathways as well.

them in a sequential and orderly fashion from one redox molecule to the next. The flow of electrons down this chain is highly energetic and allows the active transport of hydrogen ions to the outside of the membrane where the respiratory chain is located. The step that finalizes the transport process is the acceptance of electrons and hydrogen by oxygen, producing water. This process consumes oxygen. Some variability exists from one organism to another, but the principal compounds that carry out these complex reactions are NADH dehydrogenase, flavoproteins, coenzyme Q (ubiquinone), and **cytochromes** (sy'-toh-krohm). The cytochromes contain a tightly bound metal atom at their center that is actively involved in accepting electrons and donating them to the next carrier in the series. The highly compartmentalized structure of the respiratory chain is an important factor in its function. Note in **table 7.4** that the electron transport carriers and enzymes are embedded in the cytoplasmic membrane in bacteria. The equivalent structure for housing them in eukaryotes is the inner mitochondrial membranes pictured in **figure 7.12.**

Table 7.4 The Respiratory (Electron Transport) Chain

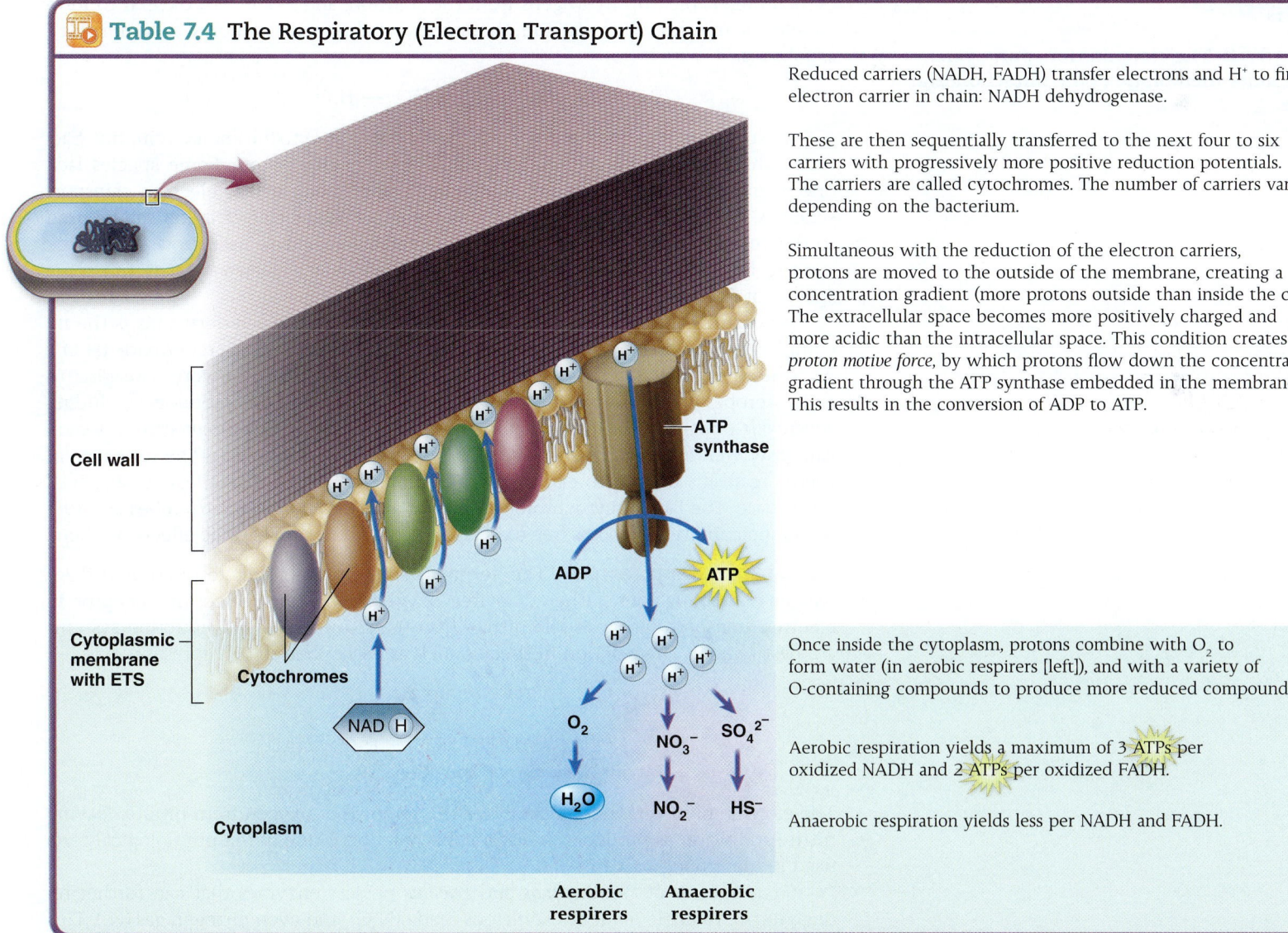

Reduced carriers (NADH, FADH) transfer electrons and H⁺ to first electron carrier in chain: NADH dehydrogenase.

These are then sequentially transferred to the next four to six carriers with progressively more positive reduction potentials. The carriers are called cytochromes. The number of carriers varies, depending on the bacterium.

Simultaneous with the reduction of the electron carriers, protons are moved to the outside of the membrane, creating a concentration gradient (more protons outside than inside the cell). The extracellular space becomes more positively charged and more acidic than the intracellular space. This condition creates the *proton motive force*, by which protons flow down the concentration gradient through the ATP synthase embedded in the membrane. This results in the conversion of ADP to ATP.

Once inside the cytoplasm, protons combine with O_2 to form water (in aerobic respirers [left]), and with a variety of O-containing compounds to produce more reduced compounds.

Aerobic respiration yields a maximum of 3 ATPs per oxidized NADH and 2 ATPs per oxidized FADH.

Anaerobic respiration yields less per NADH and FADH.

Conveyance of the NADHs from glycolysis and the Krebs cycle to the first carrier sets in motion the remaining steps. With each redox exchange, the energy level of the reactants is lessened. The released energy is channeled through the **ATP synthase** complex, stationed along the membrane in close association with the ETS carriers. Each NADH that enters electron transport can give rise to a maximum of 3 ATPs (though actual numbers are probably lower due to inefficiencies in the pathways). This coupling of ATP synthesis to electron transport is termed **oxidative phosphorylation.** Because the electrons from $FADH_2$ from the Krebs cycle enter the cycle at a later point than the NAD reactions, there is less energy to release, and only 2 ATPs are the result.

The Terminal Step

Aerobic Respiration The terminal step in aerobic respiration, during which oxygen accepts the electrons, is catalyzed by cytochrome aa_3, also called cytochrome oxidase. This large enzyme complex is specifically adapted to receive electrons from cytochrome c, pick up hydrogens from the solution, and react with oxygen to form a molecule of water. This reaction, though in actuality more complex, is summarized as follows:

$$2H^+ + 2e^- + \tfrac{1}{2}O_2 \rightarrow H_2O$$

Most eukaryotic aerobes have a fully functioning cytochrome system, but bacteria exhibit wide-ranging variations in this part of the system. Some species lack one or more of the redox steps; others have several alternative electron transport schemes. Because many bacteria lack cytochrome oxidase, this variation can be used to differentiate among certain genera of bacteria. An oxidase detection test can be used to help identify members of the genera *Neisseria* and *Pseudomonas* and some species of *Bacillus*.

A potential side reaction of the respiratory chain in aerobic organisms is the incomplete reduction of oxygen to superoxide ion (O_2^-) and hydrogen peroxide (H_2O_2). As mentioned in chapter 6, these toxic oxygen products can be very damaging to cells. Aerobes have neutralizing enzymes to deal with these products, including *superoxide dismutase* and *catalase*. One exception is the genus *Streptococcus*, which can grow well in oxygen yet lacks both cytochromes and catalase. The tolerance of these organisms to oxygen can be explained by the neutralizing effects of a special peroxidase. The lack of cytochromes, catalase, and peroxidases in anaerobes as a rule limits their ability to process free oxygen and contributes to its toxic effects on them.

Anaerobic Respiration The terminal step in anaerobic respiration utilizes oxygen-containing ions, rather than free oxygen, as the final electron acceptor in electron transport. Of these, the nitrate (NO_3^-) and nitrite (NO_2^-) reduction systems are best known. The reaction in species such as *Escherichia coli* is represented as:

$$\overset{\text{Nitrate reductase}}{\underset{}{\downarrow}}$$
$$\underset{\text{nitrate}}{NO_3^-} + NADH \rightarrow \underset{\text{nitrite}}{NO_2^-} + H_2O + NAD^+$$

The enzyme nitrate reductase catalyzes the removal of oxygen from nitrate, leaving nitrite and water as products. A test for this reaction is one of the physiological tests used in identifying bacteria.

Some species of *Pseudomonas* and *Bacillus* possess enzymes that can further reduce nitrite to nitric oxide (NO), nitrous oxide (N_2O), and even nitrogen gas (N_2). This process, called **denitrification,** is a very important step in recycling nitrogen in the biosphere. Other oxygen-containing nutrients reduced anaerobically by various

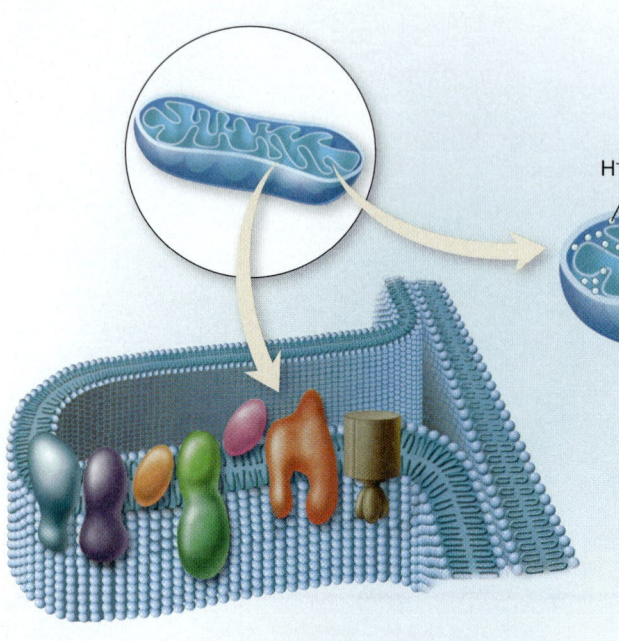

Figure 7.12 **The electron transport system on the inner membrane of the mitochondrial cristae.**

H+ ions

Intermembrane space

Cristae

bacteria are carbonates and sulfates. None of the anaerobic pathways produce as much ATP as aerobic respiration.

After Pyruvic Acid II: Fermentation

The definition of *fermentation* is the incomplete oxidation of glucose or other carbohydrates in the absence of oxygen. This process uses organic compounds as the terminal electron acceptors and yields a small amount of ATP (see figure 7.11). This pathway is used by organisms that do not have an electron transport chain and therefore cannot respire. Other organisms repress the production of electron transport chain proteins when oxygen is lacking in their environment. They can then revert to fermentation.

Without an electron transport chain to churn out large quantities of ATP from reduced carriers, it may seem that fermentation would yield only meager amounts of energy (2 ATPs maximum per glucose), and that would slow down growth. What actually happens, however, is that many bacteria can grow as fast as they would in the presence of oxygen. This rapid growth is made possible by an increase in the rate of glycolysis. From another standpoint, fermentation permits independence from molecular oxygen and allows colonization of anaerobic environments. It also enables microorganisms with a versatile metabolism to adapt to variations in the availability of oxygen. For them, fermentation provides a means to grow even when oxygen levels are too low for aerobic respiration.

Bacteria that digest cellulose in the rumens of cattle are largely fermentative. After initially hydrolyzing cellulose to glucose, they ferment the glucose to organic acids, which are then absorbed as the bovine's principal energy source. Even human muscle cells can undergo a form of fermentation that permits short periods of activity after the oxygen supply in the muscle has been exhausted. Muscle cells convert pyruvic acid into lactic acid, which allows anaerobic production of ATP to proceed for a time. But this cannot go on indefinitely, and after a few minutes, the accumulated lactic acid causes muscle fatigue. **Table 7.5** gives an overview of fermentation.

Yeasts turn the sugar in grapes into alcohol through fermentation.

Table 7.5 Fermentation

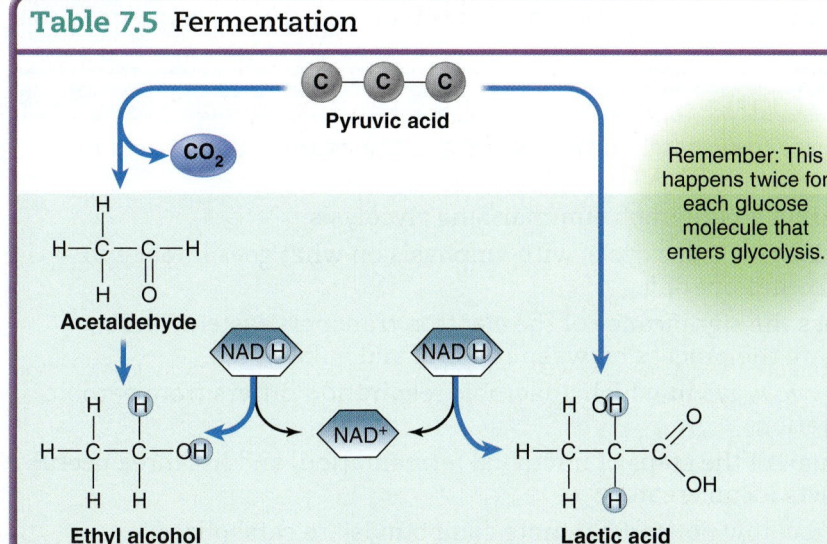

Pyruvic acid from glycolysis can itself become the electron acceptor.

Pyruvic acid can also be enzymatically altered and then serve as the electron acceptor.

The NADs are recycled to reenter glycolysis.

The organic molecules that became reduced in their role as electron acceptors are extremely varied, and often yield useful products such as ethyl alcohol, lactic acid, propionic acid, butanol, and others.

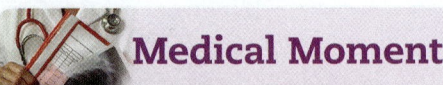

Medical Moment

Muscle Metabolism

The burning sensation felt in the muscles during intense exercise has always been blamed on the buildup of lactic acid in the muscles. It has been thought for many years that when muscles deplete their supply of oxygen during exercise, they convert pyruvic acid to lactic acid so that production of ATP can continue to proceed for a short period of time. As the level of lactic acid builds, muscle pain and fatigue are experienced.

New research has suggested that lactic acid may have gotten a bad rap all these years. Instead of being viewed as the cause of muscle pain, it is now thought that lactic acid may be simply another fuel source. Lactic acid is created from glucose and is used by the muscles as another energy source, particularly during high-intensity exercise. It is now thought that the body creates additional proteins whose role is to convert lactic acid into energy. If this is indeed the case, what is to blame for the muscle pain we have all experienced during bouts of intense exercise? Some experts now state that muscle pain is simply the result of microscopic tears and muscle trauma.

Products of Fermentation in Microorganisms

Alcoholic beverages (wine, beer, whiskey) are perhaps the most prominent among fermentation products. Note that the products of alcoholic fermentation are not only ethanol but also CO_2, a gas that accounts for the bubbles in champagne and beer and the rising of bread dough.

Other fermentation products are solvents (acetone, butanol), organic acids (lactic, acetic), dairy products, and many other foods. Derivatives of proteins, nucleic acids, and other organic compounds are fermented to produce vitamins, antibiotics, and even hormones such as hydrocortisone.

We have provided only a brief survey of fermentation products, but it is worth noting that microbes can be harnessed to synthesize a variety of other substances by varying the raw materials provided them. In fact, so broad is the colloquial meaning of the word *fermentation* that the large-scale industrial syntheses by microorganisms often utilize entirely different mechanisms from those described here, and they even occur aerobically, particularly in antibiotic, hormone, vitamin, and amino acid production.

Catabolism of Noncarbohydrate Compounds

We have given you one version of events for catabolism, using glucose, a carbohydrate, as our example. Other compounds serve as fuel, as well. The more complex polysaccharides are easily broken down into their component sugars, which can enter glycolysis at various points. Microbes also break down other molecules for their own use, of course. Two other major sources of energy and building blocks for microbes are lipids (fats) and proteins. Both of these must be broken down to their component parts to produce precursor metabolites and energy.

Recall from chapter 1 that fats are fatty acids joined to glycerol. Enzymes called **lipases** break these apart. The glycerol is then converted to dihydroxyacetone phosphate (DHAP), which can enter a step midway through glycolysis. The fatty acid component goes through a process called **beta oxidation.** Fatty acids have a variable number of carbons; in beta oxidation, 2-carbon units are successively transferred to coenzyme A, creating acetyl CoA, which enters the Krebs cycle. This process can yield a large amount of energy. Oxidation of a 6-carbon fatty acid yields 50 ATPs, compared with 38 for a 6-carbon sugar.

Proteins are chains of amino acids. Enzymes called **proteases** break proteins down to their amino acid components, after which the amino groups are removed by a reaction called **deamination.** This leaves a carbon compound, which is easily converted to one of several Krebs cycle intermediates.

7.3 LEARNING OUTCOMES—Assess Your Progress

9. List three basic catabolic pathways and the estimated ATP yield for each.
10. Construct a paragraph summarizing glycolysis.
11. Describe the Krebs cycle, with emphasis on what goes into it and what comes out of it.
12. Discuss the significance of the electron transport system, and compare the process between bacteria and eukaryotes.
13. State two ways in which anaerobic respiration differs from aerobic respiration.
14. Summarize the steps of microbial fermentation, and list three useful products it can create.
15. Describe how noncarbohydrate compounds are catabolized.

7.4 Anabolism and the Crossing Pathways of Metabolism

Our discussion now turns from catabolism and energy extraction to anabolic functions and biosynthesis. In this section, we present aspects of intermediary metabolism, including amphibolic pathways, the synthesis of simple molecules, and the synthesis of macromolecules.

The Frugality of the Cell

It must be obvious by now that cells have mechanisms for careful management of carbon compounds. Rather than being dead ends, most catabolic pathways contain strategic molecular intermediates (metabolites) that can be diverted into anabolic pathways. In this way, a given molecule can serve multiple purposes, and the maximum benefit can be derived from all nutrients and metabolites of the cell pool. The ability of a system to integrate catabolic and anabolic pathways to improve cell efficiency is termed **amphibolism** (am-fee-bol′-izm).

At this point in the chapter, you can appreciate a more complex view of metabolism than that presented at the beginning, in figure 7.1. **Table 7.6** demonstrates the amphibolic nature of intermediary metabolism.

Table 7.6 Amphibolic Pathways of Glucose Metabolism

Anabolic Pathways

Intermediates from glycolysis are fed into the amino acid synthesis pathway. From there, the compounds are formed into proteins. Amino acids can then contribute nitrogenous groups to nucleotides to form nucleic acids.

Glucose and related simple sugars are made into additional sugars and polymerized to form complex carbohydrates.

The glycolysis product acetyl CoA can be oxidized to form fatty acids, critical components of lipids.

Catabolic Pathways

In addition to the respiration and fermentation pathways already described, bacteria can deaminate amino acids, which leads to the formation of a variety of metabolic intermediates, including pyruvate and acetyl CoA.

Also, fatty acids can be oxidized to form acetyl CoA.

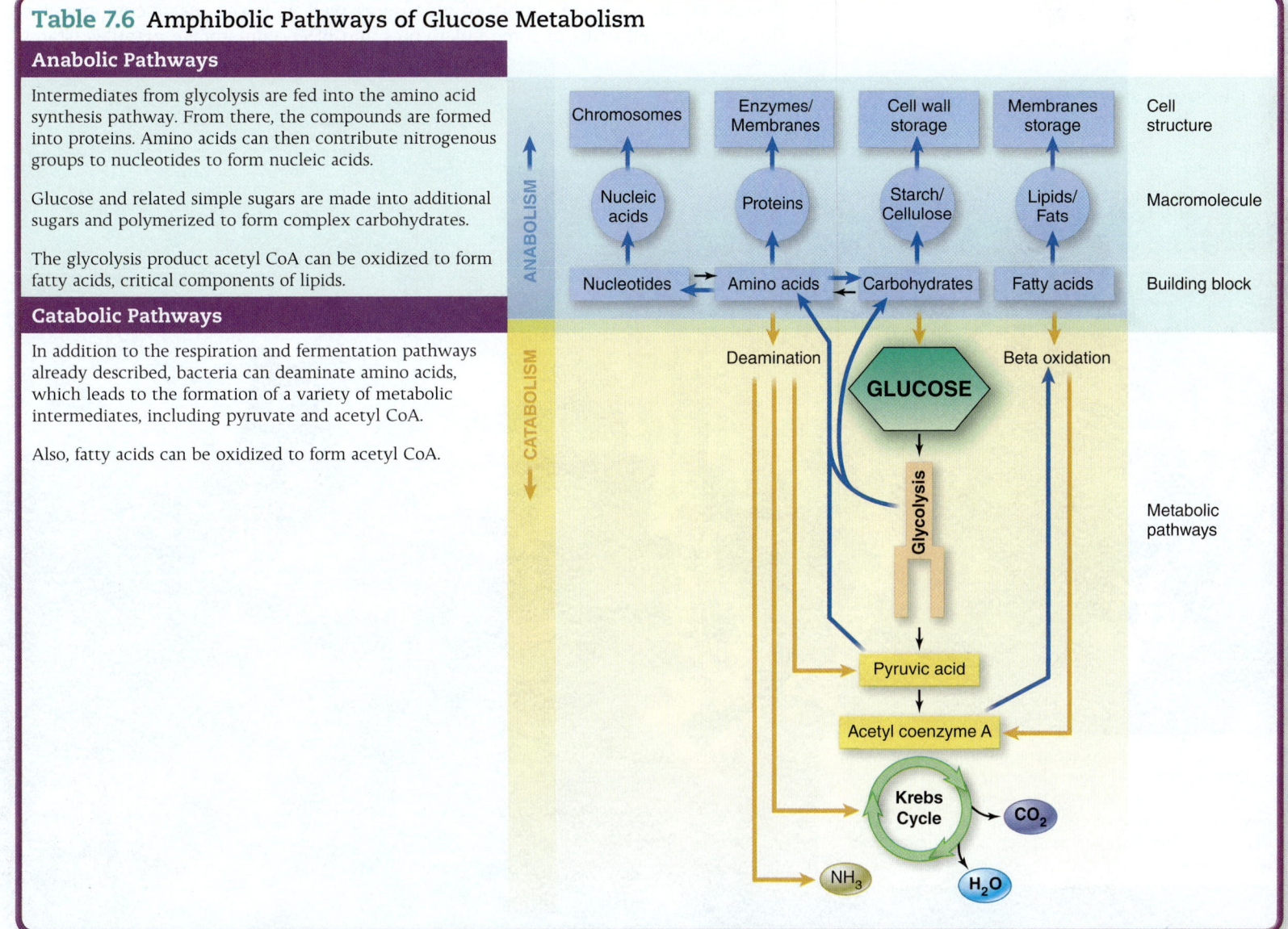

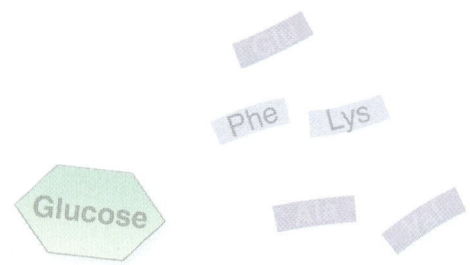

Anabolism: Formation of Macromolecules

Monosaccharides, amino acids, fatty acids, nitrogen bases, and vitamins–the building blocks that make up the various macromolecules and organelles of the cell–come from two possible sources. They can enter the cell from the outside "ready to use," or they can be synthesized through various cellular pathways. The degree to which an organism can synthesize its own building blocks (simple molecules) is determined by its genetic makeup, a factor that varies tremendously from group to group. In chapter 6, you learned that autotrophs require only CO_2 as a carbon source, a few minerals to synthesize all cell substances, and no organic nutrients. Some heterotrophic organisms (*E. coli*, yeasts) are also very efficient in that they can synthesize all cellular substances from minerals and one organic carbon source such as glucose. Compare this with a strict parasite that has few synthetic abilities of its own and derives most precursor molecules from the host.

Whatever their source, once these building blocks are added to the metabolic pool, they are available for synthesis of polymers by the cell.

Carbohydrate Biosynthesis

The role of glucose in metabolism and energy utilization is so crucial that its biosynthesis is ensured by several alternative pathways. Certain structures in the cell depend on an adequate supply of glucose as well. It is the major component of the cellulose cell walls of some eukaryotes and of certain storage granules (starch,

In 2008 researchers found a bacterium living alone—with no other organisms—in water found in a South African gold mine 2 miles underneath the earth's surface. It is the first instance of an organism living alone in an ecosystem, obtaining all of its nutrients from minerals in the surrounding rock.

glycogen). One of the intermediaries in glycolysis, glucose-6-P, is used to form glycogen. Monosaccharides other than glucose are important in the synthesis of bacterial cell walls. Peptidoglycan contains a linked polymer of muramic acid and glucosamine. Fructose-6-P from glycolysis is used to form these two sugars. Carbohydrates (deoxyribose, ribose) are also essential building blocks in nucleic acids. Polysaccharides are the predominant components of cell surface structures such as capsules and the glycocalyx, and they are commonly found in slime layers.

Amino Acids, Protein Synthesis, and Nucleic Acid Synthesis

Proteins account for a large proportion of a cell's constituents. They are essential components of enzymes, the cytoplasmic membrane, the cell wall, and cell appendages. As a general rule, 20 amino acids are needed to make these proteins. Although some organisms (*E. coli,* for example) have pathways that will synthesize all 20 amino acids, others, including animals, lack some or all of the pathways for amino acid synthesis and must acquire the essential ones from their diets. Protein synthesis itself is a complex process that requires a genetic blueprint and the operation of intricate cellular machinery, as you will see in chapter 8.

DNA and RNA are responsible for the hereditary continuity of cells and the overall direction of protein synthesis. Because nucleic acid synthesis is a major topic of genetics and is closely allied to protein synthesis, it will likewise be covered in chapter 8.

Assembly of the Cell

The component parts of a bacteria cell are synthesized on a continuous basis, and catabolism is also taking place, as long as nutrients are present and the cell is in a nondormant state. When anabolism produces enough macromolecules to serve two cells, and when DNA replication produces duplicate copies of the cell's genetic material, the cell undergoes binary fission, which results in two cells from one parent cell. The two cells will need twice as many ribosomes, twice as many enzymes, and so on. The cell has created these during the anabolic phases we have described. Before cell division, the membrane(s) and the cell wall will have increased in size to create a cell that is almost twice as big as a "newborn" cell. Once synthesized, the phospholipid bilayer components of the membranes assemble themselves spontaneously with no energy input. Other assembly reactions require the input of energy. Proteins and other components must be added to the membranes. Growth of the cell wall, accomplished by the addition and coupling of sugars and peptides, requires energy input. The energy acquired during catabolic processes provides all the energy for these complex building reactions.

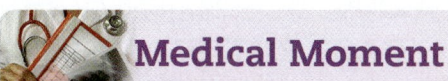

Medical Moment

Amino Acids: Essential, Nonessential, and Conditionally Essential Amino Acids

Essential amino acids are those amino acids that must be obtained from our diet, as our bodies lack the ability to synthesize them, while nonessential amino acids are those that can be synthesized by our bodies.

However, as is often the case, there are exceptions to every rule just to keep things interesting! *Conditionally essential* amino acids are not normally required in the diet, except in specific populations that are unable to synthesize them in adequate amounts. For example, individuals with PKU (phenylketonuria), a genetic condition, must keep only very low levels of phenylalanine in their bodies to prevent complications such as mental retardation; however, because they cannot synthesize enough tyrosine from the phenylalanine, tyrosine becomes an essential amino acid for these individuals.

7.4 LEARNING OUTCOMES—Assess Your Progress

16. Provide an overview of the anabolic stages of metabolism.

17. Define *amphibolism.*

CASE FILE WRAP-UP

Diabetic ketoacidosis (DKA) is a life-threatening condition that occurs only in people with diabetes—usually type 1 diabetes, but occasionally in people with type 2 diabetes, as well. It is caused by a lack of insulin in the body and is sometimes the first indication that someone is diabetic. In DKA, the body is forced into a catabolic state. Insulin is required to break down glucose for energy. When insulin is lacking, the cells of the body cannot access glucose; thus the body begins to burn muscle and fat for energy. When this occurs, fatty acids are produced that are released into the blood, resulting in metabolic acidosis.

The young patient in the case file opener presented with all of the typical signs of DKA: altered breathing (fast and deep breathing is the body's attempt to "blow off" carbon dioxide), altered level of consciousness (which may range from drowsiness to coma), fruity odor to the breath (ketone breath), vomiting and abdominal pain, weakness and excessive thirst accompanied by frequent urination.

The patient was admitted to the pediatric ICU and was treated with IV fluids and insulin. She required frequent blood work to monitor her electrolytes because ketones released in the urine have a tendency to pull potassium and other electrolytes with them, causing severe electrolyte imbalance. Once she was stabilized, she and her family were provided with intensive guidance for handling her condition before being discharged home.

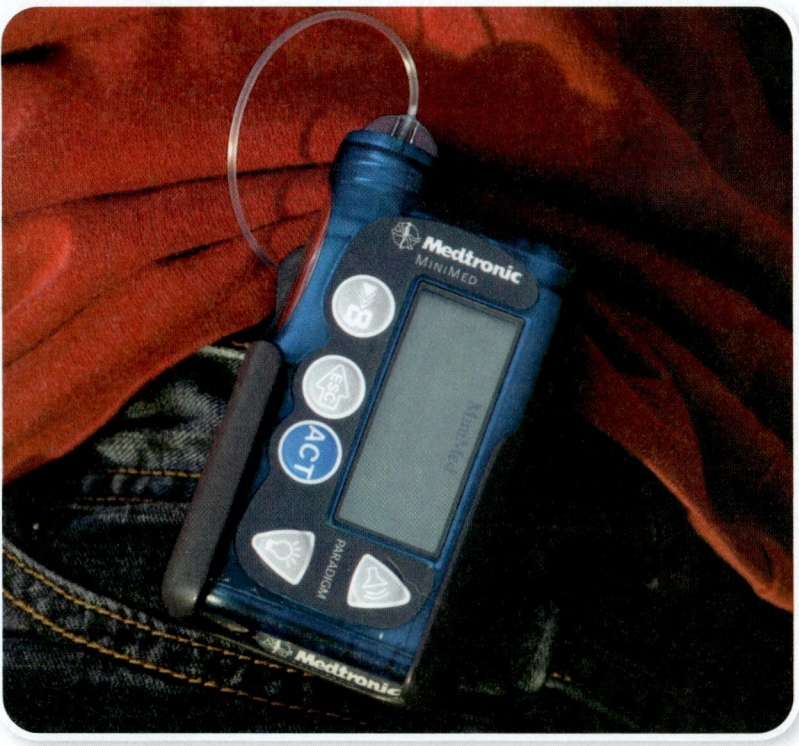

Insulin pump

Vitamin D Deficiency

Vitamins are required for normal metabolism. Many vitamins cannot be synthesized in the human body or are synthesized in inadequate amounts; therefore, they must be obtained from the diet. Vitamins are necessary because many coenzymes are derived from vitamins. Inadequate intake of a vitamin can lead to a deficiency state, in which the complete holoenzyme (the combination of a protein apoenzyme plus one or more cofactors) is prevented from forming.

Vitamin D has been studied extensively in recent years. Deficiency of vitamin D has been linked to multiple sclerosis (MS), cancer, hypertension, autoimmune disease, Alzheimer's disease, dementia, premature labor, and certain infectious diseases. Deficiency has long been known to predispose to rickets in children and to osteomalacia, osteoporosis, and bone fractures in adults.

Few foods contain vitamin D in more than negligible amounts, and although milk and a few other foods are fortified with vitamin D, our major source of vitamin D is exposure to sunlight (which is why vitamin D is often called the "sunshine vitamin").

Unfortunately, excessive exposure to sunlight, especially UV-B radiation, causes cell and molecular damage, so we are left wondering how to balance the good and the bad aspects of sunlight. We have received the message that we should protect ourselves from skin cancer by using sunscreen whenever we are out in the sun. We slather ourselves and our children with sunscreen as recommended by dermatologists; as a consequence, skin cancer rates are declining even in countries such as Australia, which once boasted the highest rate of skin cancer. Unfortunately, although skin cancer rates are declining, rates of vitamin D deficiency are climbing.

Individuals with increased skin pigmentation (i.e., people of African-American descent) have a reduced ability to synthesize vitamin D from sunlight. A sunscreen with an SPF of 15 will decrease synthesis of vitamin D in the skin by 97% to 99%. African Americans with very dark skin have the equivalent of an SPF of 15, so their ability to synthesize vitamin D is decreased by as much as 99%. Those who cover their skin completely for religious or cultural reasons are also at high risk of becoming deficient in vitamin D. Obesity and aging also decrease our ability to synthesize vitamin D.

So what should we do? It's a predictable—and boring—answer: We can enjoy the sun in moderation.

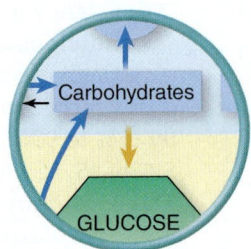

Chapter Summary

7.1 Metabolism and the Role of Enzymes

- Metabolism is the sum of cellular chemical and physical activities. It consists of anabolism, synthetic reactions that convert small molecules into large molecules, and catabolism, in which large molecules are degraded and energy is produced.
- Metabolism is made possible by organic catalysts, or enzymes, that speed up reactions to rates compatible with biological processes.
- Enzymes are not consumed and can be reused. Each enzyme acts specifically upon its matching molecule or substrate.
- Substrate attachment occurs in the special pocket called the active, or catalytic, site.
- Enzymes are labile (unstable) and function only within narrow operating ranges of temperature, osmotic pressure, and pH, and they are especially vulnerable to denaturation.
- Enzymes are frequently the targets for physical and chemical agents used in control of microbes.
- Regulatory controls can act on enzymes directly or on the process that gives rise to the enzymes.

7.2 The Pursuit and Utilization of Energy

- Energy is the capacity of a system to perform work. It is consumed in endergonic reactions and is released in exergonic reactions.
- Extracting energy requires a series of electron carriers arrayed in a redox chain between electron donors and electron acceptors.

7.3 Catabolism

- Carbohydrates, such as glucose, are energy-rich because when catabolized they can yield a large number of electrons per molecule.
- Glycolysis is a pathway that degrades glucose to pyruvic acid without requiring oxygen.

- Pyruvic acid is processed in aerobic respiration via the Krebs cycle and its associated electron transport chain.
- Pyruvic acid undergoes further oxidation and decarboxylation in the Krebs cycle, which generates ATP, CO_2, and large amounts of reduced carriers (NADH and $FADH_2$).
- The respiratory chain completes energy extraction.
- The final electron acceptor in aerobic respiration is oxygen. In anaerobic respiration, compounds such as sulfate, nitrate, or nitrite serve this function.
- Fermentation is anaerobic respiration in which both the electron donor and final electron acceptors are organic compounds.
- Production of alcohol, vinegar, and certain industrial solvents relies upon fermentation.
- Glycolysis and the Krebs cycle are central pathways that link catabolic and anabolic pathways, allowing cells to break down different classes of molecules in order to synthesize compounds required by the cell.
- Intermediates, such as pyruvic acid, are convertible into amino acids, and amino acids can in turn be used as precursors to glucose and other carbohydrates.
- Two-carbon acetyl molecules from pyruvate can be used in fatty acid synthesis.

7.4 Anabolism and the Crossing Pathways of Metabolism

- The ability of a cell or system to integrate catabolic and anabolic pathways to improve efficiency is called amphibolism.
- Macromolecules, such as proteins, carbohydrates, and nucleic acids, are made of building blocks from two possible sources: from outside the cell (preformed) or via synthesis in one of the anabolic pathways.

Multiple-Choice Questions   Bloom's Levels 1 and 2: Remember and Understand

Select the correct answer from the answers provided.

1. Anabolism is the form of metabolism in which energy is _____ important molecules.

   a. used to break down
   b. used to build
   c. released through the breakdown of
   d. released through the assembly of

2. An enzyme

   a. becomes part of the final products.
   b. is nonspecific for substrate.
   c. is consumed by the reaction.
   d. is heat and pH labile.

3. Which of these molecules cannot be broken down to release energy?

   a. sugar
   b. starch
   c. fat
   d. protein
   e. none of the above (all of these *can* release energy)

4. Many coenzymes are formed from

   a. metals.
   b. vitamins.
   c. proteins.
   d. substrates.

5. Energy is carried from catabolic to anabolic reactions in the form of
   a. ADP.
   b. high-energy ATP bonds.
   c. coenzymes.
   d. inorganic phosphate.

6. A product or products of glycolysis is/are
   a. ATP.
   b. pyruvate.
   c. CO_2
   d. both a and b.

7. Complete oxidation of glucose in aerobic respiration can yield a maximum net output of _____ ATPs.
   a. 40
   b. 6
   c. 38
   d. 2

8. ATP synthase complexes can generate _____ ATPs for each NADH that enters electron transport.
   a. 1
   b. 2
   c. 3
   d. 4

Critical Thinking   Bloom's Levels 3, 4, and 5: Apply, Analyze, and Evaluate

Critical thinking is the ability to reason and solve problems using facts and concepts. These questions can be approached from a number of angles and, in most cases, they do not have a single correct answer.

1. Show diagrammatically the interaction of a holoenzyme and its substrate and general products that can be formed from a reaction.

2. Explain how oxidation of a substrate proceeds without oxygen.

3. a. Describe the roles played by ATP and NAD in metabolism.

   b. What particular features of the structure of ATP lend themselves to these functions?

4. a. What is meant by the concept of the "final electron acceptor"?

   b. What are the final electron acceptors in aerobic, anaerobic, and fermentative metabolism?

5. What is the fate of NADH in a fermentative organism?

Visual Connections   Bloom's Level 5: Evaluate

This question connects previous images to a new concept.

1. **From chapter 3, figure 3.15.** On these depictions of the gram-positive and gram-negative envelopes, draw protons in the proper compartment in such a way that creates a proton motive force.

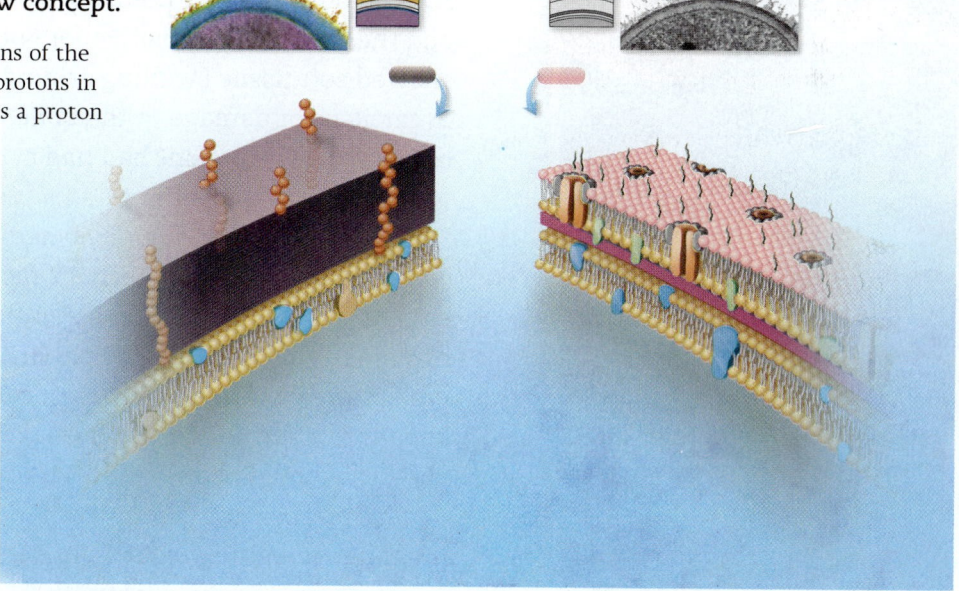

www.mcgrawhillconnect.com

Enhance your study of this chapter with study tools and practice tests. Also ask your instructor about the resources available through Connect, including the media-rich eBook, interactive learning tools, and animations.

CASE FILE

A Body Attacking Itself

A 57-year-old woman was admitted to the acute care unit where I was working as an RN. Her admitting diagnosis was rheumatoid arthritis (RA). Her past medical history was significant for gastric ulcers and insulin-dependent diabetes. She had been seen by her primary care physician in the clinic, who decided to admit her after examining her and listening to her history of symptoms.

This patient was unique in that her symptoms had begun quite suddenly 2 days prior, rather than insidiously as is usually the case in RA. She had awoken with moderately severe pain in her knees, ankles, hands, and feet. She also complained of extreme fatigue and stated that she felt as though she had the flu. She had a low-grade fever on admission. Her other vital signs were normal. Her hands, knees, and ankles were mildly swollen and were warm to the touch. She had difficulty moving due to joint pain and was having difficulty managing at home.

Blood work was ordered, including a complete blood count (CBC), chemistry panel, anti-CPP, ESR (erythrocyte sedimentation rate), rheumatoid factor (RF), and C-reactive protein. X rays were also ordered of the affected joints. Test results indicated that the patient was mildly anemic. The ESR was elevated and the patient was positive for both rheumatoid factor and anti-CPP. X rays showed soft tissue swelling around the affected joints, but little destruction of cartilage or damage to the bone was seen, which was not surprising considering the patient had had symptoms for such a short time.

Given that the patient had a history of GI (gastrointestinal) bleeding and had brittle diabetes, the physician decided to start the patient on Enbrel (entanercept), a tumor necrosis factor (TNF) inhibitor. I questioned why the patient was not started on a steroid, as is usually the case. The physician explained that starting the patient on anti-inflammatory drugs might have precipitated another stomach ulcer, and steroids such as prednisone are known to affect blood glucose levels; therefore, the physician felt that the patient would do better on a TNF inhibitor such as Enbrel. Before starting the patient on the drug, the physician screened her for hepatitis and tuberculosis. All of her screening tests were negative, and within 2 weeks the patient was in remission and feeling great.

- How is tumor necrosis factor significant in rheumatoid arthritis?

- How do TNF inhibitors reduce inflammation?

Case File Wrap-Up appears on page 228.

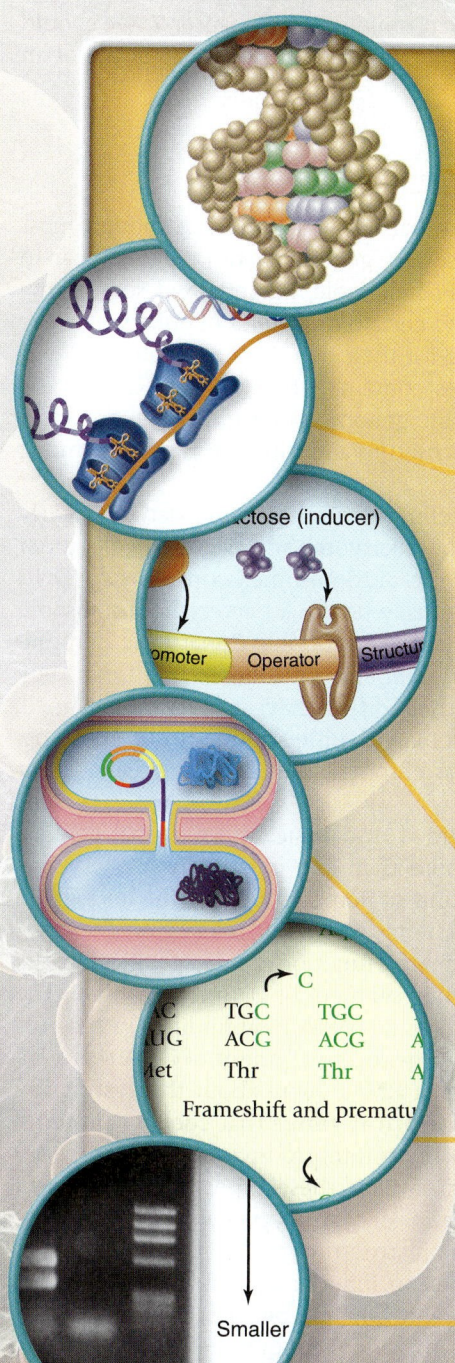

CHAPTER

8

Microbial Genetics and Genetic Engineering

IN THIS CHAPTER...

8.1 Introduction to Genetics and Genes

1. Define the terms *genome* and *gene*.
2. Differentiate between genotype and phenotype.
3. Draw a segment of DNA, labeling all important chemical groups within the molecule.
4. Summarize the steps of bacterial DNA replication, and identify the enzymes used in this process.
5. Compare and contrast the synthesis of leading and lagging strands during DNA replication.

8.2 Transcription and Translation

6. Provide an overview of the relationship among DNA, RNA, and proteins.
7. Identify important structural and functional differences between RNA and DNA.
8. Draw a picture of the process of transcription.
9. List the three types of RNA directly involved in translation.
10. Define the terms *codon* and *anticodon,* and list three start and stop codons.
11. Identify the locations of the promoter, the start codon, and the A and P sites during translation.
12. Indicate how eukaryotic transcription and translation differ from these processes in bacteria.

8.3 Genetic Regulation of Protein Synthesis

13. Define the term *operon*, and explain one advantage it provides to a bacterial cell.
14. Highlight the main points of *lac* operon operation.

8.4 DNA Recombination Events

15. Explain the defining characteristics of a recombinant organism.
16. Describe three forms of horizontal gene transfer used in bacteria.

8.5 Mutations: Changes in the Genetic Code

17. Define the term *mutation,* and discuss one positive and one negative example of it in microorganisms.
18. Differentiate among frameshift, nonsense, silent, and missense mutations.

8.6 Genetic Engineering

19. Explain the importance of restriction endonucleases to genetic engineering.
20. List the steps in the polymerase chain reaction.
21. Describe how you can clone a gene into a bacterium.

8.1 Introduction to Genetics and Genes

Genetics is the study of the inheritance, or **heredity,** of living things. It is a wide-ranging science that explores

- the transmission of biological properties (traits) from parent to offspring,
- the expression and variation of those traits,
- the structure and function of the genetic material, and
- how this material changes.

This chapter will explore DNA, which is the genetic material, and the proteins and other products that it gives rise to in a cell. Coming out of chapter 7, we should point out that the production of new DNA, RNA, and proteins is an example of an anabolic process.

The Nature of the Genetic Material

The **genome** is the sum total of genetic material of an organism. Although most of the genome exists in the form of chromosomes, genetic material can appear in nonchromosomal sites as well **(figure 8.1).** For example, bacteria and some fungi contain tiny extra pieces of DNA (plasmids), and certain organelles of eukaryotes (the mitochondria and chloroplasts) are equipped with their own DNA. Genomes of cells are composed exclusively of DNA, but viruses contain either DNA or RNA as the principal genetic material. Although the specific genome of an individual organism is unique, the general pattern of nucleic acid structure and function is similar among all organisms.

In general, a **chromosome** is a discrete cellular structure composed of a neatly packaged DNA molecule. The chromosomes of eukaryotes and bacterial cells differ in several respects. The structure of eukaryotic chromosomes consists of a DNA molecule tightly wound around histone proteins, whereas a bacterial chromosome is condensed into a packet by means of histonelike proteins. Eukaryotic chromosomes are located in the nucleus, they vary in number from a few to hundreds, they can occur in pairs (diploid) or singles (haploid), and they have a linear appearance. In contrast, most bacteria have a single, circular (double-stranded) chromosome, although many bacteria have multiple circular chromosomes and some have linear chromosomes.

The chromosomes of all cells are subdivided into basic informational packets called genes. A **gene** can be defined from more than one perspective. In classical genetics, the term refers to the fundamental unit of heredity responsible for a given trait in an organism. In the molecular and biochemical sense, it is a site on the chromosome that provides information for a certain cell function. More specifically still, it has traditionally been characterized as a certain segment of DNA that contains the necessary code to make a **protein** or RNA molecule. With new findings in the area of gene expression, we now prefer to speak of a gene as a segment of DNA that contains code to make a group of related proteins or RNAs. More about this distinction later. Genes fall into three basic categories: **structural genes** that code for proteins, genes that code for the RNA machinery used in protein production, and regulatory genes that control gene expression. The sum of all of these types of genes constitutes an organism's distinctive genetic makeup, or **genotype** (jee'-noh-typ). The expression of the genotype creates traits (certain structures or functions) referred to as the **phenotype** (fee'-noh-typ). Just as a person inherits a combination of genes (genotype) that gives a certain eye color or height (phenotype), a bacterium inherits genes that direct the formation of a flagellum, and a virus contains genes for its capsid structure. All organisms contain more genes in their genotypes than are manifested as a phenotype at any given time. In other words, the phenotype can change depending on which genes are "turned on" (expressed).

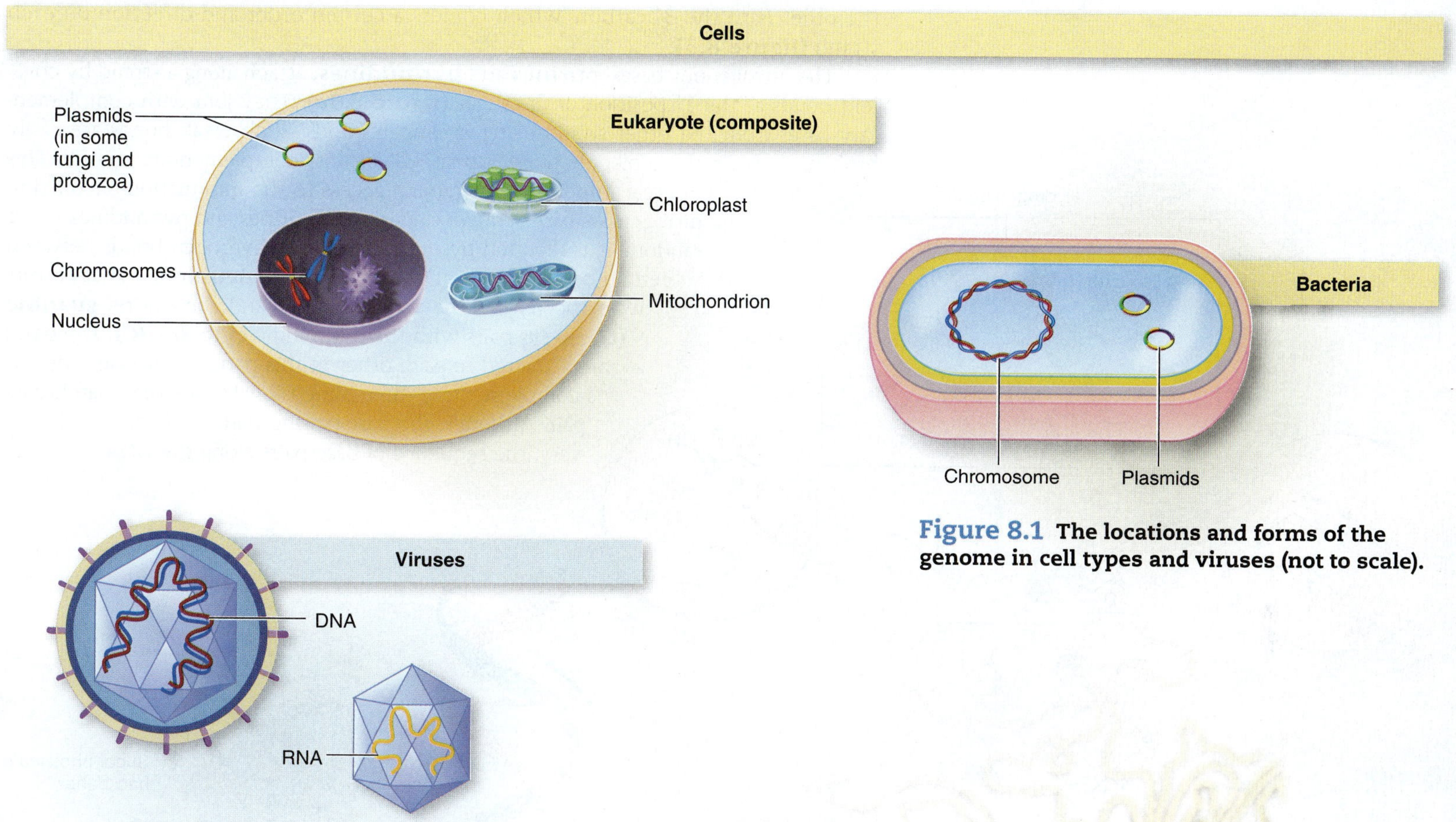

Cells

Eukaryote (composite)

Plasmids
(in some
fungi and
protozoa)

Chloroplast

Chromosomes

Nucleus

Mitochondrion

Bacteria

Chromosome   Plasmids

Figure 8.1 The locations and forms of the genome in cell types and viruses (not to scale).

Viruses

DNA

RNA

The Size and Packaging of Genomes

Genomes vary greatly in size. The smallest viruses have four or five genes; the bacterium *Escherichia coli* has a single chromosome containing 4,288 genes, and a human cell has about 25,000 genes on 46 chromosomes. The chromosome of *E. coli* would measure about 1 mm if unwound and stretched out linearly, and yet this fits within a cell that measures just over 1 μm across, making the stretched-out DNA 1,000 times longer than the cell **(figure 8.2).** Still, the bacterial chromosome takes up only about one-third to one-half of the cell's volume. Likewise, if the sum of all DNA contained in the 46 human chromosomes were unraveled and laid end to end, it would measure about 6 feet.

The DNA Code

The general structure of DNA is universal, except in some viruses that contain single-stranded DNA. The basic unit of DNA structure is a **nucleotide,** and a chromosome in a typical bacterium consists of several million nucleotides linked end to end. Each nucleotide is composed of a **phosphate,** a **deoxyribose,** and a **nitrogenous base.** The nucleotides covalently bond to each other in a sugar-phosphate linkage that becomes the backbone of each strand. Each sugar attaches in a repetitive pattern to two phosphates. One of the bonds is to the number 5′ (read "five prime") carbon on deoxyribose, and

Figure 8.2 An *Escherichia coli* cell disrupted to release its DNA molecule. The cell has spewed out its single, uncoiled DNA strand into the surrounding medium.

the other is to the 3′ carbon, which confers a certain order and direction on each strand **(figure 8.3).**

The nitrogenous bases, **purines** and **pyrimidines,** attach along a strand by covalent bonds at the 1′ position of the sugar **(figure 8.3a).** They join with complementary bases from the other strand using hydrogen bonds. Such weak bonds are easily broken, allowing the molecule to be "unzipped" into its complementary strands. This feature is of great importance in gaining access to the information encoded in the nitrogenous base sequence. Pairing of purines and pyrimidines is not random; it is dictated by the formation of hydrogen bonds between certain bases. Thus, in DNA, the purine **adenine (A)** always pairs with the pyrimidine **thymine (T),** and the purine **guanine (G)** always pairs with the pyrimidine **cytosine (C).** The bases are attracted to each other in this pattern because each has a complementary three-dimensional shape that matches its pair. Although the base-pairing partners generally do not vary, the sequence of base pairs along the DNA molecule

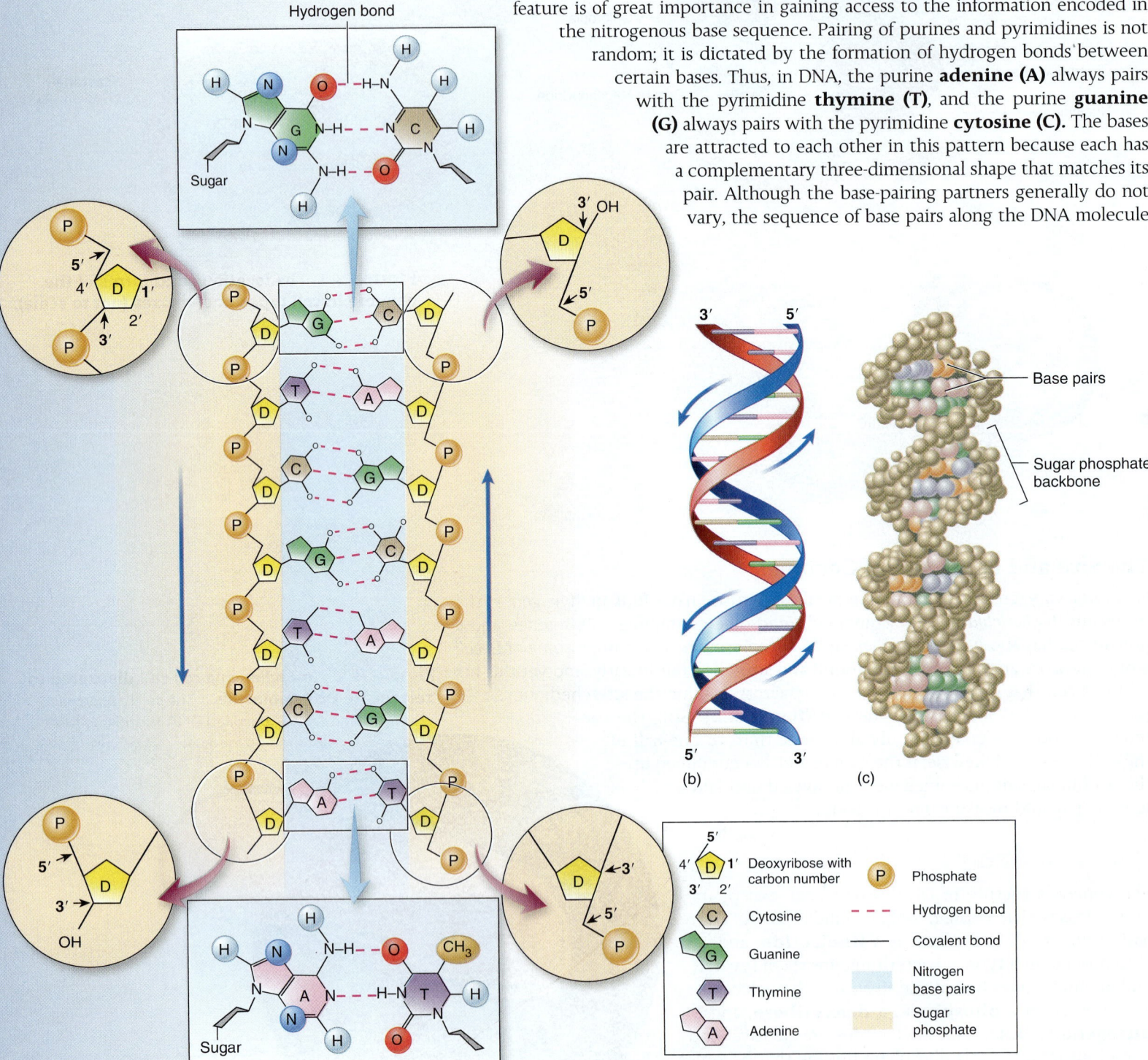

Figure 8.3 Three views of DNA structure.   (a) A schematic nonhelical model, to show the arrangement of the molecules it is made of. Note that the order of phosphate and sugar bonds differs between the two strands, going from the #5 carbon to the #3 carbon on one strand, and from the #3 carbon to the #5 carbon on the other strand. Insets show details of the nitrogen *bases.* (b) Simplified model that highlights the antiparallel arrangement. (c) Space-filling model that more accurately depicts the three-dimensional structure of DNA.

can assume any order, resulting in a nearly infinite number of possible nucleotide sequences.

Other important considerations of DNA structure concern the nature of the double helix itself. The halves are not oriented in the same direction. One side of the helix runs in the opposite direction of the other, in what is called an **antiparallel** arrangement **(figure 8.3b).** The order of the bond between the carbon on deoxyribose and the phosphates is used to keep track of the direction of the two sides of the helix. Thus, one helix runs from the 5′ to 3′ direction, and the other runs from the 3′ to 5′ direction. This characteristic is a significant factor in DNA synthesis and protein production.

The Significance of DNA Structure

The English language, based on 26 letters, can create an infinite variety of words, but how can an apparently complex genetic language such as DNA be based on just four nitrogen base "letters"? A mathematical example can explain the possibilities. For a segment of DNA that is 1,000 nucleotides long, there are $4^{1,000}$ different sequences possible. Carried out, this number would approximate 1.5×10^{602}, a number so huge that it provides nearly endless degrees of variation.

In this chapter, we will address two separate reactions DNA enters into: its own replication (facilitating cell division) and its role in producing proteins.

DNA Replication

The process of duplicating DNA is called *DNA replication*. In the following example, we will show replication in bacteria; but with some exceptions, it also applies to the process as it works in eukaryotes and some viruses.

The Overall Replication Process

DNA replication requires a careful orchestration of the actions of 30 different enzymes (partial list in **table 8.1),** which separate the strands of the existing DNA molecule, copy one strand, and produce two complete daughter molecules. A critical feature of DNA replication is that each daughter molecule will be identical to the parent in composition but neither one is completely new; the strand that serves as a template is an original parental DNA strand. The preservation of the parent molecule in this way, termed **semiconservative replication**–*semi-* meaning "half" as in *semicircle*– helps explain the reliability and fidelity of replication.

Refinements and Details of Replication

The process of synthesizing a new daughter strand of DNA using the parental strand as a template is carried out by the enzyme DNA polymerase III. The entire process of replication does, however, depend on several enzymes and can be most easily understood by keeping in mind a few points concerning both the structure of the DNA molecule and the limitations of DNA polymerase III:

1. DNA polymerase III is unable to *begin* synthesizing a chain of nucleotides but can only continue to add nucleotides to an already existing chain.
2. DNA polymerase III can add nucleotides in only one direction, so a new strand is always synthesized 5′ to 3′.

With these constraints in mind, we outline the process in **table 8.2.**

Table 8.1 Some Enzymes Involved in DNA Replication and Their Functions

Enzyme	Function
Helicase	Unzipping the DNA helix
Primase	Synthesizing an RNA primer
DNA polymerase III	Adding bases to the new DNA chain; proofreading the chain for mistakes
DNA polymerase I	Removing primer, closing gaps, repairing mismatches
Ligase	Final binding of nicks in DNA during synthesis and repair
Topoisomerases I and II	Supercoiling and untangling

Table 8.2 DNA Replication

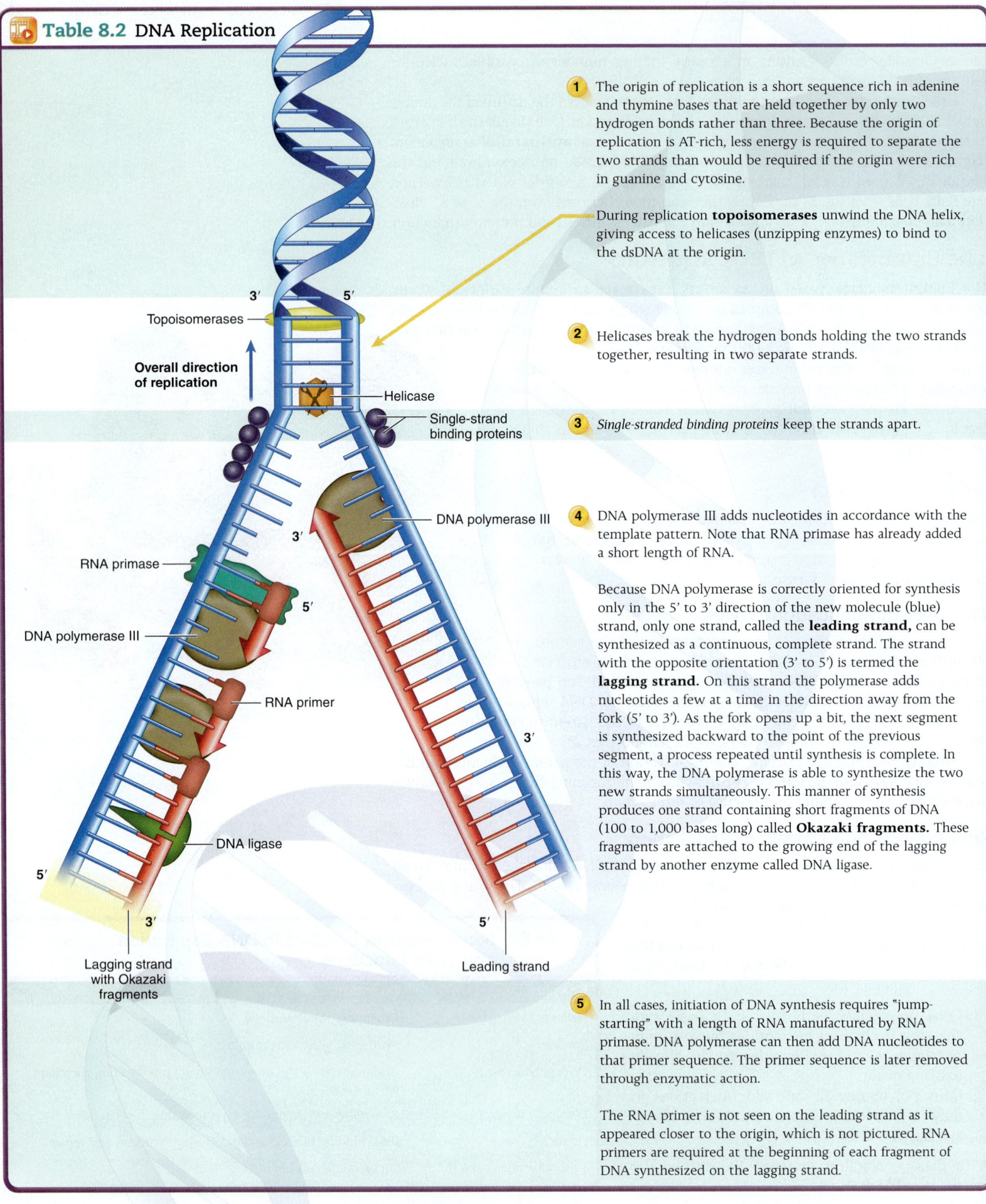

Topoisomerases

3′   5′

Overall direction of replication

Helicase

Single-strand binding proteins

RNA primase

DNA polymerase III

DNA polymerase III

3′

5′

RNA primer

3′

DNA ligase

5′

3′

5′

Lagging strand with Okazaki fragments

3′

5′

Leading strand

1 The origin of replication is a short sequence rich in adenine and thymine bases that are held together by only two hydrogen bonds rather than three. Because the origin of replication is AT-rich, less energy is required to separate the two strands than would be required if the origin were rich in guanine and cytosine.

During replication **topoisomerases** unwind the DNA helix, giving access to helicases (unzipping enzymes) to bind to the dsDNA at the origin.

2 Helicases break the hydrogen bonds holding the two strands together, resulting in two separate strands.

3 *Single-stranded binding proteins* keep the strands apart.

4 DNA polymerase III adds nucleotides in accordance with the template pattern. Note that RNA primase has already added a short length of RNA.

Because DNA polymerase is correctly oriented for synthesis only in the 5′ to 3′ direction of the new molecule (blue) strand, only one strand, called the **leading strand,** can be synthesized as a continuous, complete strand. The strand with the opposite orientation (3′ to 5′) is termed the **lagging strand.** On this strand the polymerase adds nucleotides a few at a time in the direction away from the fork (5′ to 3′). As the fork opens up a bit, the next segment is synthesized backward to the point of the previous segment, a process repeated until synthesis is complete. In this way, the DNA polymerase is able to synthesize the two new strands simultaneously. This manner of synthesis produces one strand containing short fragments of DNA (100 to 1,000 bases long) called **Okazaki fragments.** These fragments are attached to the growing end of the lagging strand by another enzyme called DNA ligase.

5 In all cases, initiation of DNA synthesis requires "jump-starting" with a length of RNA manufactured by RNA primase. DNA polymerase can then add DNA nucleotides to that primer sequence. The primer sequence is later removed through enzymatic action.

The RNA primer is not seen on the leading strand as it appeared closer to the origin, which is not pictured. RNA primers are required at the beginning of each fragment of DNA synthesized on the lagging strand.

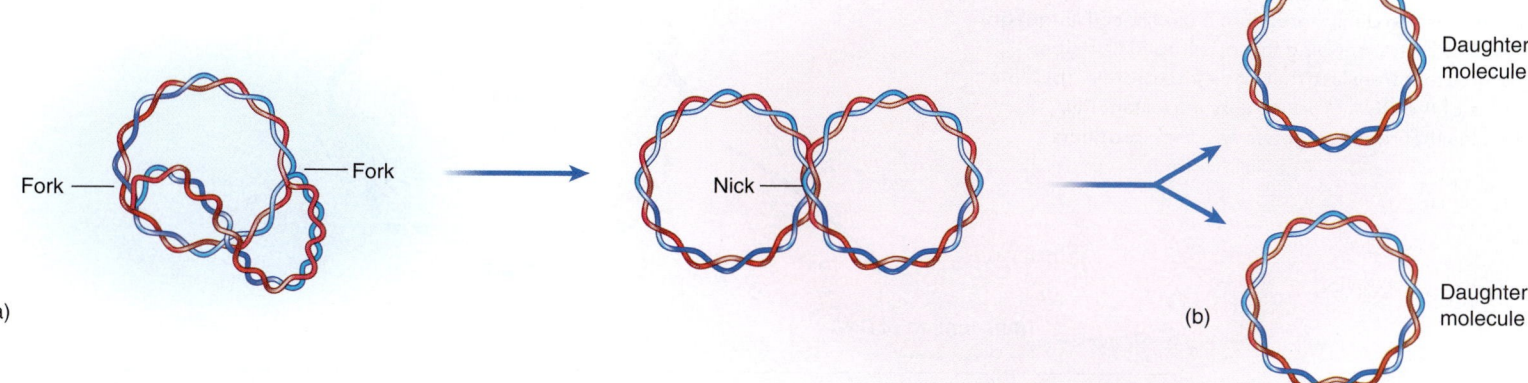

(a)

(b) Daughter molecule

Daughter molecule

Nick

Fork

Fork

Figure 8.4 Completion of chromosome replication in bacteria. (a) As replication proceeds, one double strand loops away from the other. (b) Final separation is achieved through repair and the release of two completed molecules. The daughter cells receive these during binary fission.

Elongation and Termination of the Daughter Molecules The addition of nucleotides proceeds at an astonishing pace, estimated in some bacteria to be 750 bases per second at each fork! As replication proceeds, the newly produced double strand loops down **(figure 8.4a).** DNA polymerase I removes the RNA primers used to initiate DNA synthesis and replaces them with DNA. When the forks come full circle and meet, ligases move along the lagging strand to begin the initial linking of the fragments and to complete synthesis and separation of the two circular daughter molecules **(figure 8.4b).**

Like any language, DNA is occasionally "misspelled" when an incorrect base is added to the growing chain. Studies have shown that such mistakes are made once in approximately 10^8 to 10^9 bases, but most of these are corrected. If not corrected, they are referred to as *mutations* (covered later in this chapter). Because continued cellular integrity is very dependent on accurate replication, cells have evolved their own proofreading function for DNA. DNA polymerase III, the enzyme that elongates the molecule, can detect incorrect, unmatching bases; excise them; and replace them with the correct base. DNA polymerase I can also proofread the molecule and repair damaged DNA.

8.1 LEARNING OUTCOMES—Assess Your Progress

1. Define the terms *genome* and *gene*.
2. Differentiate between genotype and phenotype.
3. Draw a segment of DNA, labeling all important chemical groups within the molecule.
4. Summarize the steps of bacterial DNA replication, and identify the enzymes used in this process.
5. Compare and contrast the synthesis of leading and lagging strands during DNA replication.

8.2 Transcription and Translation

Although the genome is full of critical information, the molecule itself does not perform cell processes directly. Its stored information is conveyed to RNA molecules, which carry out the instructions. The concept that genetic information flows from DNA to RNA to protein is a central theme of biology **(figure 8.5a).** More precisely, it states that the master code of DNA is first used to synthesize an RNA molecule via a process called **transcription,** and the information contained in the RNA is then used to produce proteins in a process known as **translation.** The principal exceptions to this pattern are found in RNA viruses, which convert RNA to other RNA, and in retroviruses, which convert RNA to DNA.

NCLEX® PREP

1. A central theme of biology is that the master code of DNA is used to synthesize an RNA molecule via a process called _____, and the information contained in the RNA is used to produce proteins in a process known as _____.

   a. *translation; transcription*
   b. *transcription; translation*
   c. *posttranslational modification; redundancy*
   d. *transcription; termination*

Figure 8.5 Summary of the flow of genetic information in cells. DNA is the ultimate storehouse and distributor of genetic information. (a) DNA must be deciphered into a usable cell language. It does this by transcribing its code into RNA helper molecules that translate that code into protein. (b) Other sections of the DNA produce very important RNA molecules that regulate genes and their products.

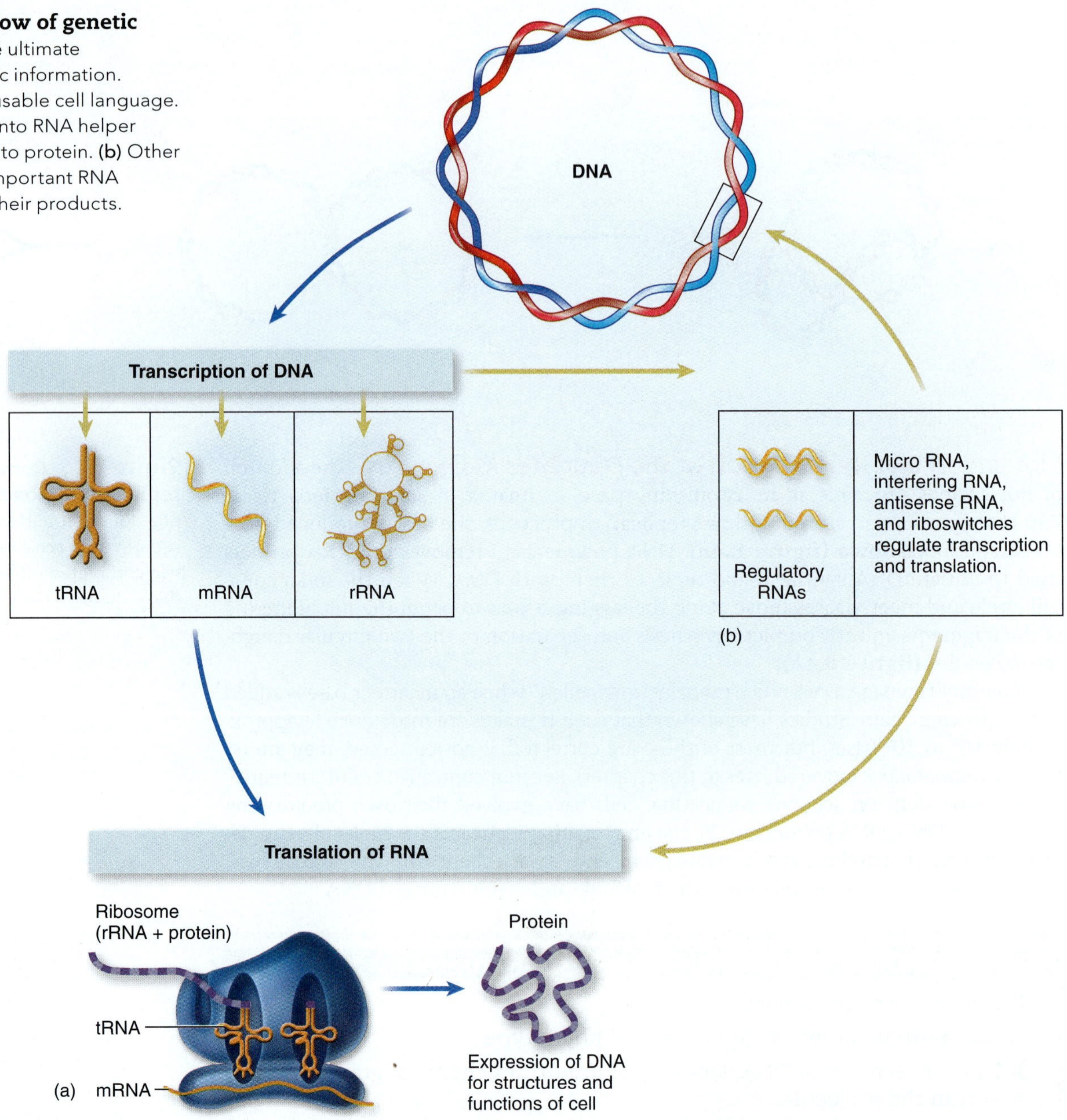

DNA

Transcription of DNA

tRNA

mRNA

rRNA

Regulatory RNAs

Micro RNA, interfering RNA, antisense RNA, and riboswitches regulate transcription and translation.

(b)

Translation of RNA

Ribosome (rRNA + protein)

Protein

tRNA

(a) mRNA

Expression of DNA for structures and functions of cell

In addition to the RNA that is used to produce proteins, a wide variety of RNAs are used to regulate gene function. This means that vast amounts of DNA sequences code for RNAs that never get made into proteins (**figure 8.5b**). The DNA that codes for these very crucial RNA molecules was called "junk" DNA until very recently, because we didn't understand their function.

Transcription: The First Stage of Gene Expression

During transcription, the DNA code is converted to RNA through several stages, directed by a large and very complex enzyme system, **RNA polymerase.** Only one strand of the DNA–the **template strand**–contains meaningful instructions for synthesis of a functioning polypeptide.

Table 8.3 describes transcription.

▶ **Table 8.3** **Transcription**

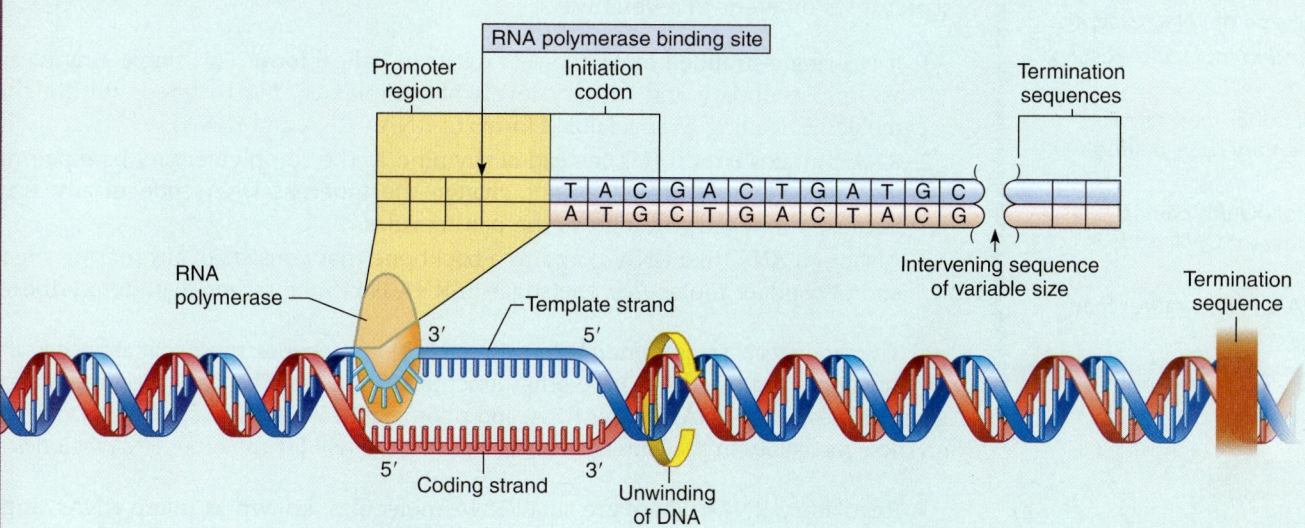

RNA polymerase binding site

Promoter region | Initiation codon | Termination sequences

| T | A | C | G | A | C | T | G | A | T | G | C |
| A | T | G | C | T | G | A | C | T | A | C | G |

RNA polymerase

Template strand

3′   5′

Intervening sequence of variable size

Termination sequence

5′   3′

Coding strand

Unwinding of DNA

1   *Initiation.* Transcription is initiated when RNA polymerase recognizes a segment of the DNA called the **promoter** region. This region consists of two sequences of DNA just prior to the beginning of the gene to be transcribed. These promoter sequences provide the signal for RNA polymerase to bind to the DNA. Then there is a special codon called the initiation codon, which is where the RNA polymerase begins its transcription. As the DNA helix unwinds, the polymerase first pulls the early parts of the DNA into itself, a process called "DNA scrunching," and then, having acquired energy from the scrunching process, begins to advance down the DNA strand to continue synthesizing an RNA molecule complementary to the template strand of DNA. The nucleotide sequence of promoters differs only slightly from gene to gene, with all promoters being rich in adenine and thymine.

Only one strand of DNA, called the **template strand,** is copied by RNA polymerase.

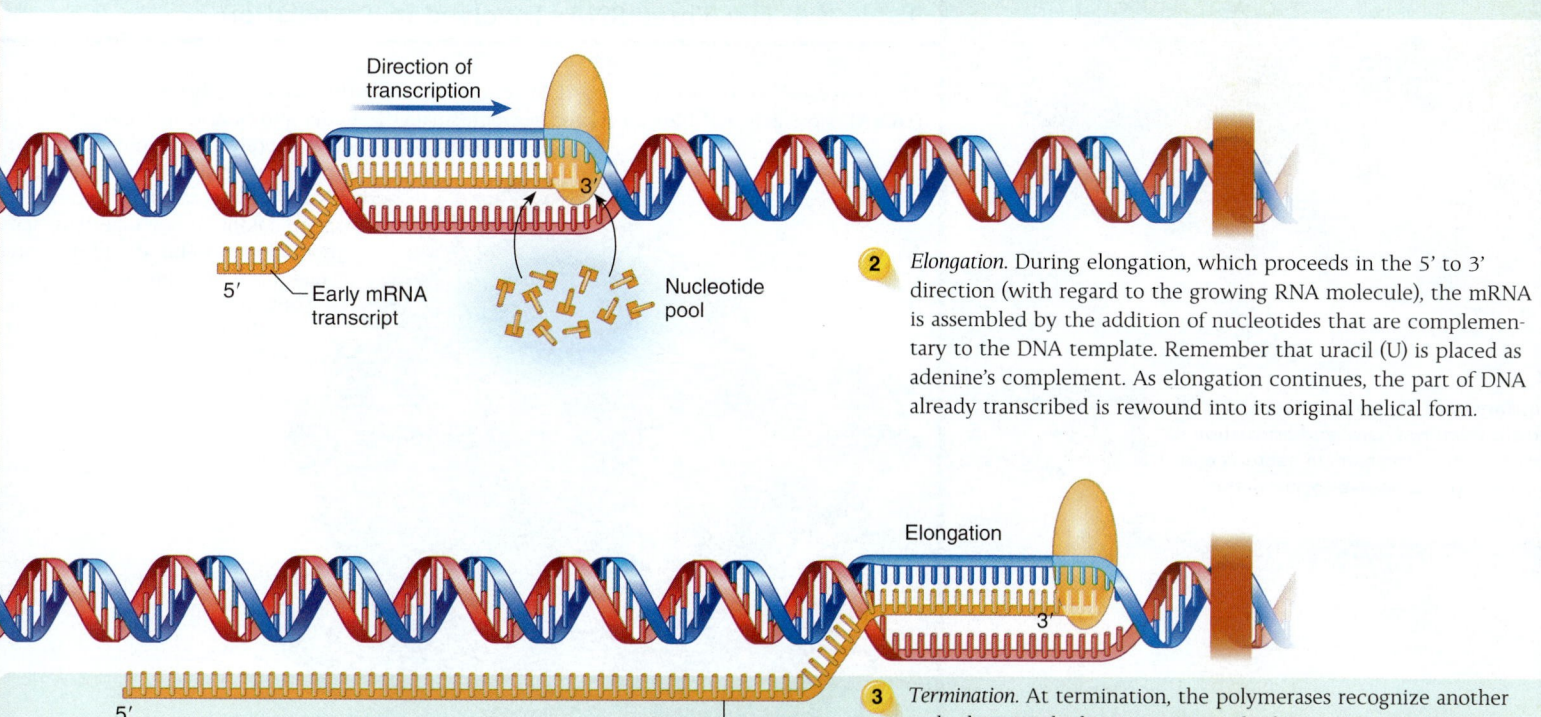

Direction of transcription

5′ Early mRNA transcript

Nucleotide pool

2   *Elongation.* During elongation, which proceeds in the 5' to 3' direction (with regard to the growing RNA molecule), the mRNA is assembled by the addition of nucleotides that are complementary to the DNA template. Remember that uracil (U) is placed as adenine's complement. As elongation continues, the part of DNA already transcribed is rewound into its original helical form.

Elongation

3′

5′ Late mRNA transcript

3   *Termination.* At termination, the polymerases recognize another code that signals the separation and release of the mRNA strand, or transcript. The smallest mRNA might consist of 100 bases; an average-size mRNA might consist of 1,200 bases; and a large one might consist of several thousand.

The RNAs

In terms of its general properties, ribonucleic acid is similar to DNA, but its general structure is different in several ways:

1. It is a single-stranded molecule that exists in helical form. This single strand can assume secondary and tertiary levels of complexity due to bonds within the molecule, leading to specialized forms of RNA (tRNA and rRNA).
2. RNA contains **uracil (U),** instead of thymine, as the complementary base-pairing mate for adenine. This does not change the inherent DNA code in any way because the uracil still follows the pairing rules.
3. Although RNA, like DNA, contains a backbone that consists of alternating sugar and phosphate molecules, the sugar in RNA is **ribose** rather than deoxyribose.

The products of transcription belong in two major categories: those that are necessary for translation, and those that have other functions in the cell. The translation machinery includes messenger RNA, transfer RNA, and ribosomal RNA. We will spend a lot of time on these molecules in the coming section, but first we will list the other RNA varieties:

- Regulatory RNAs: These are small RNA molecules, known as micro RNAs, antisense RNAs, riboswitches, and small interfering RNAs. They are important regulators of gene expression in bacteria and eukaryotes and also act in the coiling of chromatin in eukaryotic cells.
- Primer RNA: This RNA is laid down in DNA replication, as a template of sorts for the DNA sequence. Primer RNAs are operative in both bacterial and eukaryotic cells.

In eukaryotes, transcription occurs in the nucleus (pictured here) and translation occurs in the cytoplasm. In bacteria and archaea, both processes occur in the cytoplasm.

Table 8.4 The Three RNAs Involved in Translation

The bacterial (70S) ribosome is a particle composed of tightly packaged **ribosomal RNA (rRNA)** (large gray and blue areas) and protein (smaller lavender and purple areas). A metabolically active bacterial cell can contain up to 20,000 of these minuscule factories—all actively engaged in reading the genetic program, taking in raw materials, and producing proteins at an impressive rate. Each ribosome has a large subunit and small subunit, each of which is composed of rRNA and protein.

- Ribozymes: These enzymes are made of RNA and, in eukaryotes, remove un-needed sequences from other RNAs.

All of these–and tRNA, mRNA, and rRNA–are products of the transcription of distinct genes in the chromosome.

Now, on to translation.

After Transcription: Translation

Three different RNA products of transcription are needed for the final step of protein expression. They are described in **table 8.4.**

The ribosomes of bacteria and eukaryotes are different sizes. Ribosomes in bacteria, as well as the ribosomes in chloroplasts and mitochondria of eukaryotes, are of a 70S size, made up of a 50S (large) subunit and a 30S (small) subunit. The "S" is a measurement of sedimentation rates, which is how ribosomes are characterized. It is a nonlinear measure; therefore, 30S and 50S add up to 70S. Eukaryotic ribosomes are 80S (a large subunit of 60S and a 40S small subunit). The small subunit binds to the 5′ end of the mRNA, and the large subunit supplies enzymes for making peptide bonds on the protein.

Translation: The Second Stage of Gene Expression

In translation, all of the elements needed to synthesize a protein, from the mRNA to the amino acids, are brought together on the ribosomes **(figure 8.6).**

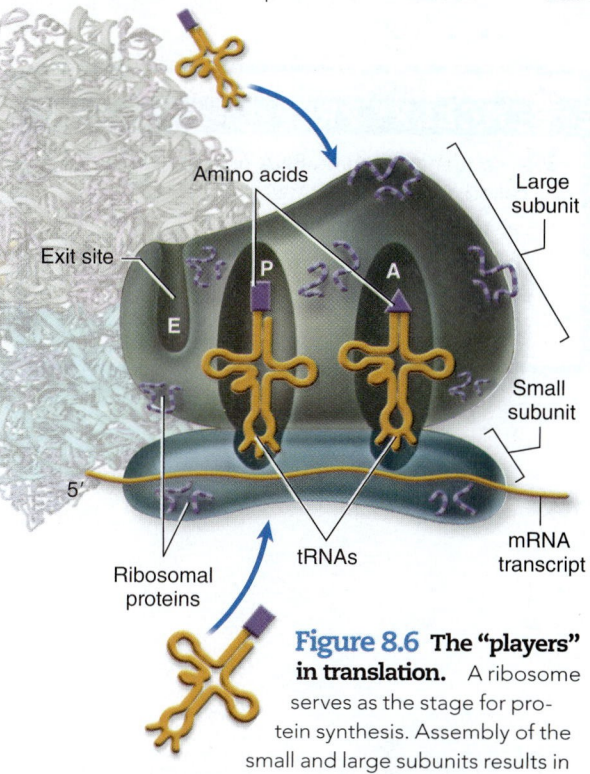

Figure 8.6 The "players" in translation. A ribosome serves as the stage for protein synthesis. Assembly of the small and large subunits results in specific sites for holding the mRNA and two tRNAs with their amino acids. This depiction of the ribosome matches the depiction on the left-hand page so you can see the connection between the molecular image and the image we will use in the book.

Transfer RNA (tRNA) is also a copy of a specific region of DNA; however, it differs from mRNA. Each one is 75 to 95 nucleotides long, and contains sequences of bases that form hydrogen bonds with complementary sections within the same tRNA strand. At these points, the molecule bends back upon itself into several hairpin loops, giving the molecule a secondary cloverleaf structure that folds even further into a complex, three-dimensional helix. This compact molecule is an adaptor that converts RNA language into protein language. The bottom loop of the cloverleaf exposes a triplet, the **anticodon,** that both designates the specificity of the tRNA and complements mRNA's codons. At the opposite end of the molecule is a binding site for the amino acid that is specific for that tRNA's anticodon. For each of the 20 amino acids, there is at least one specialized type of tRNA to carry it.

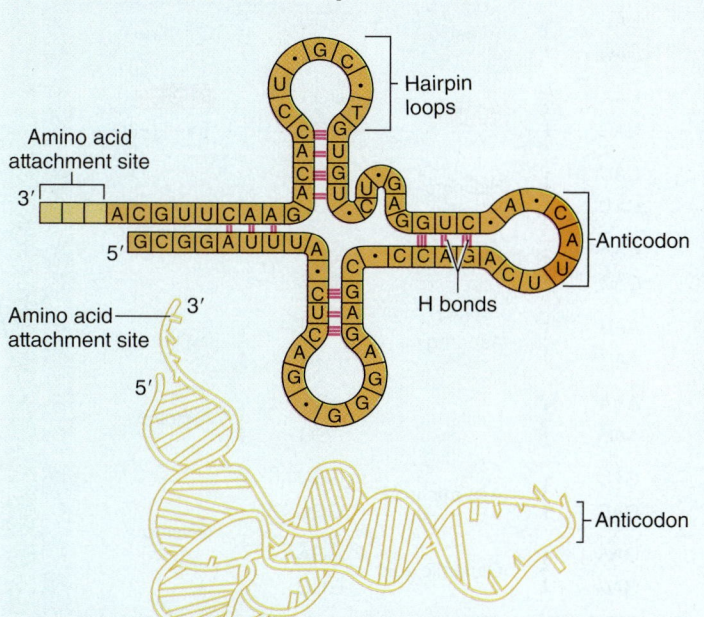

Messenger RNA (mRNA) is a transcript (copy) of a structural gene or genes in the DNA. The complementary base-pairing rules ensure that the code will be faithfully copied in the mRNA transcript. The message of this transcribed strand is later read as a series of triplets called **codons.** The length of the mRNA molecule varies from about 100 nucleotides to several thousand. It carries the sequence that will dictate the eventual amino acid sequence of the protein.

> **NCLEX® PREP**
>
> 3. A sequence of three RNA nucleotides is called a
>   a. *wobble.*
>   b. *polyribosomal complex.*
>   c. *codon.*
>   d. *peptide chain.*

The Master Genetic Code: The Message in Messenger RNA

Translation relies on a central principle: The mRNA nucleotides are read in groups of three. Three nucleotides are called a **codon,** and it is the codon that dictates which amino acid is added to the growing peptide chain. In **figure 8.7,** the mRNA codons and their corresponding amino acid specificities are given. Because there are 64 different triplet codes and only 20 different amino acids, it is not surprising that some amino acids are represented by several codons. For example, leucine and serine can each be represented by any of six different triplets, and only tryptophan and methionine are represented by a single codon. This property is called **redundancy** and allows for the insertion of correct amino acids (sometimes) even when mistakes occur in the DNA sequence, as they do with regularity. Also, in codons such as leucine, only the first two nucleotides are required to encode the correct amino acid, and the third nucleotide does not change its sense. This property, called **wobble,** is thought to permit some variation or mutation without altering the message. **Figure 8.8** shows the relationship between DNA sequence, RNA codons, tRNA, and amino acids.

Before newly made proteins can carry out their structural or enzymatic roles, they often require finishing touches. Even before the peptide chain is released from the ribosome, it begins folding upon itself to achieve its biologically active tertiary conformation. Other alterations, called **posttranslational** modifications, may be necessary. Some proteins must have the starting amino acid (formyl methionine) clipped off; proteins destined to become complex enzymes have cofactors added; and some join with other completed proteins to form quaternary levels of structure.

In bacteria, the translation of mRNA starts while transcription is still occurring **(figure 8.9).** A single mRNA is long enough to be fed through more than one ribosome simultaneously. This permits the synthesis of hundreds of protein molecules from the same mRNA transcript arrayed along a chain of ribosomes. This **polyribosomal complex** is like an assembly line for mass production of proteins. It occurs in bacteria, but not in eukaryotic cells, because there is no nucleus; and transcription and translation both occur in the cytoplasm. (In eukaryotes, transcription occurs in the nucleus.) Remember that all of the processes involved in gene expression are anabolic processes; protein synthesis consumes an enormous

Figure 8.7 The genetic code: codons of mRNA that specify a given amino acid. The master code for translation is found in the mRNA codons.

First Base Position	Second Base Position: U		C		A		G		Third Base Position
U	UUU	Phenylalanine	UCU	Serine	UAU	Tyrosine	UGU	Cysteine	U
	UUC	Phenylalanine	UCC	Serine	UAC	Tyrosine	UGC	Cysteine	C
	UUA	Leucine	UCA	Serine	UAA	STOP**	UGA	STOP**	A
	UUG	Leucine	UCG	Serine	UAG	STOP**	UGG	Tryptophan	G
C	CUU	Leucine	CCU	Proline	CAU	Histidine	CGU	Arginine	U
	CUC	Leucine	CCC	Proline	CAC	Histidine	CGC	Arginine	C
	CUA	Leucine	CCA	Proline	CAA	Glutamine	CGA	Arginine	A
	CUG	Leucine	CCG	Proline	CAG	Glutamine	CGG	Arginine	G
A	AUU	Isoleucine	ACU	Threonine	AAU	Asparagine	AGU	Serine	U
	AUC	Isoleucine	ACC	Threonine	AAC	Asparagine	AGC	Serine	C
	AUA	Isoleucine	ACA	Threonine	AAA	Lysine	AGA	Arginine	A
	AUG	START f-Methionine*	ACG	Threonine	AAG	Lysine	AGG	Arginine	G
G	GUU	Valine	GCU	Alanine	GAU	Aspartic acid	GGU	Glycine	U
	GUC	Valine	GCC	Alanine	GAC	Aspartic acid	GGC	Glycine	C
	GUA	Valine	GCA	Alanine	GAA	Glutamic acid	GGA	Glycine	A
	GUG	Valine	GCG	Alanine	GAG	Glutamic acid	GGG	Glycine	G

* This codon initiates translation.
**For these codons, which give the orders to stop translation, there are no corresponding tRNAs and no amino acids.

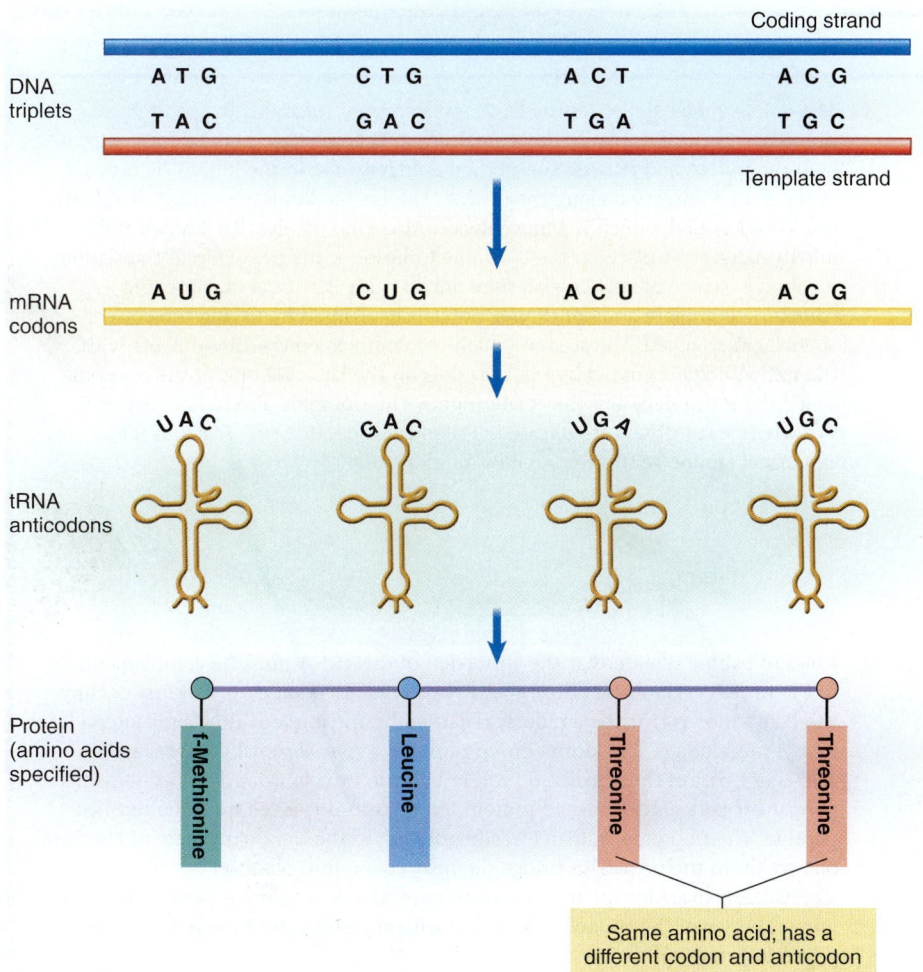

Figure 8.8 Interpreting the DNA code. If the DNA sequence is known, the mRNA codon can be surmised. If a codon is known, the anticodon and, finally, the amino acid sequence can be determined. The reverse is not as straightforward (determining the exact codon or anticodon from amino acid sequence) due to the redundancy of the code.

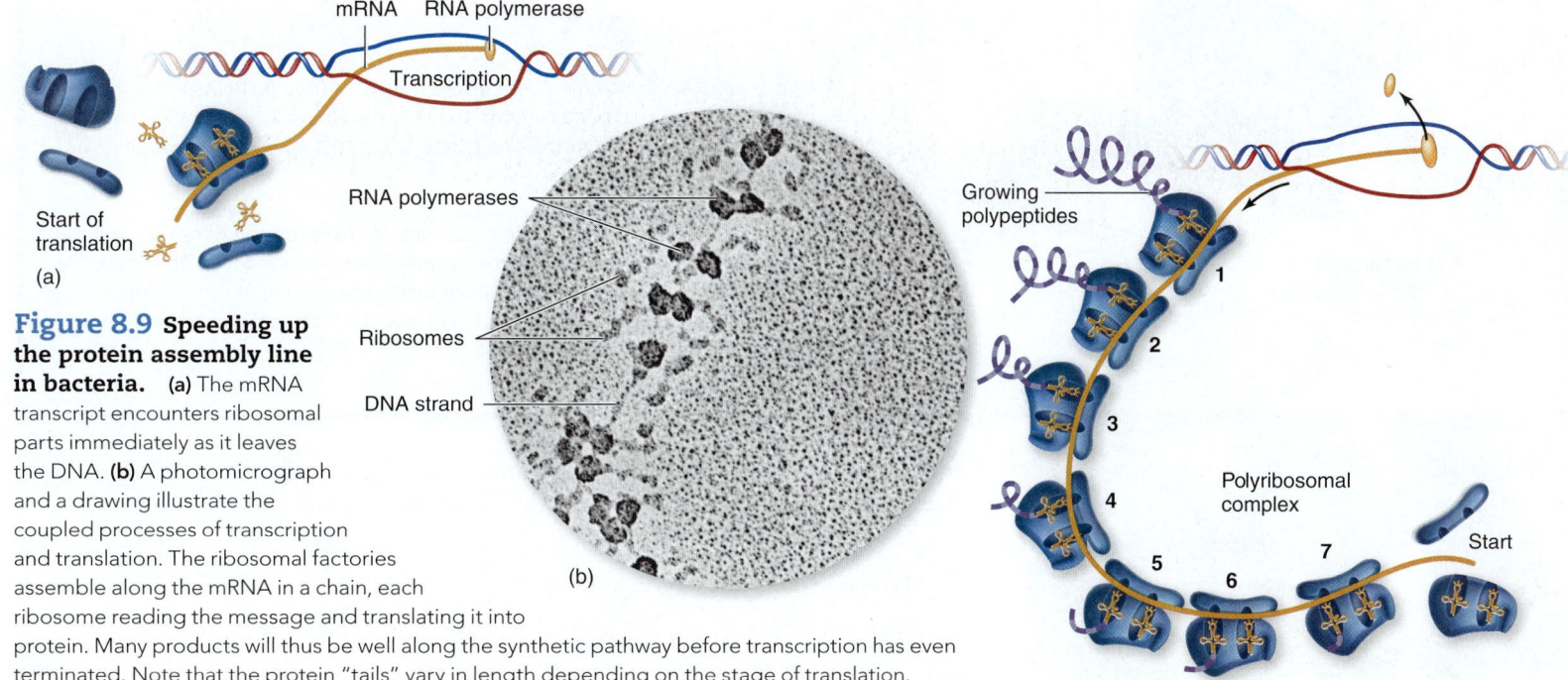

Figure 8.9 Speeding up the protein assembly line in bacteria. **(a)** The mRNA transcript encounters ribosomal parts immediately as it leaves the DNA. **(b)** A photomicrograph and a drawing illustrate the coupled processes of transcription and translation. The ribosomal factories assemble along the mRNA in a chain, each ribosome reading the message and translating it into protein. Many products will thus be well along the synthetic pathway before transcription has even terminated. Note that the protein "tails" vary in length depending on the stage of translation.

▶ Table 8.5 Translation

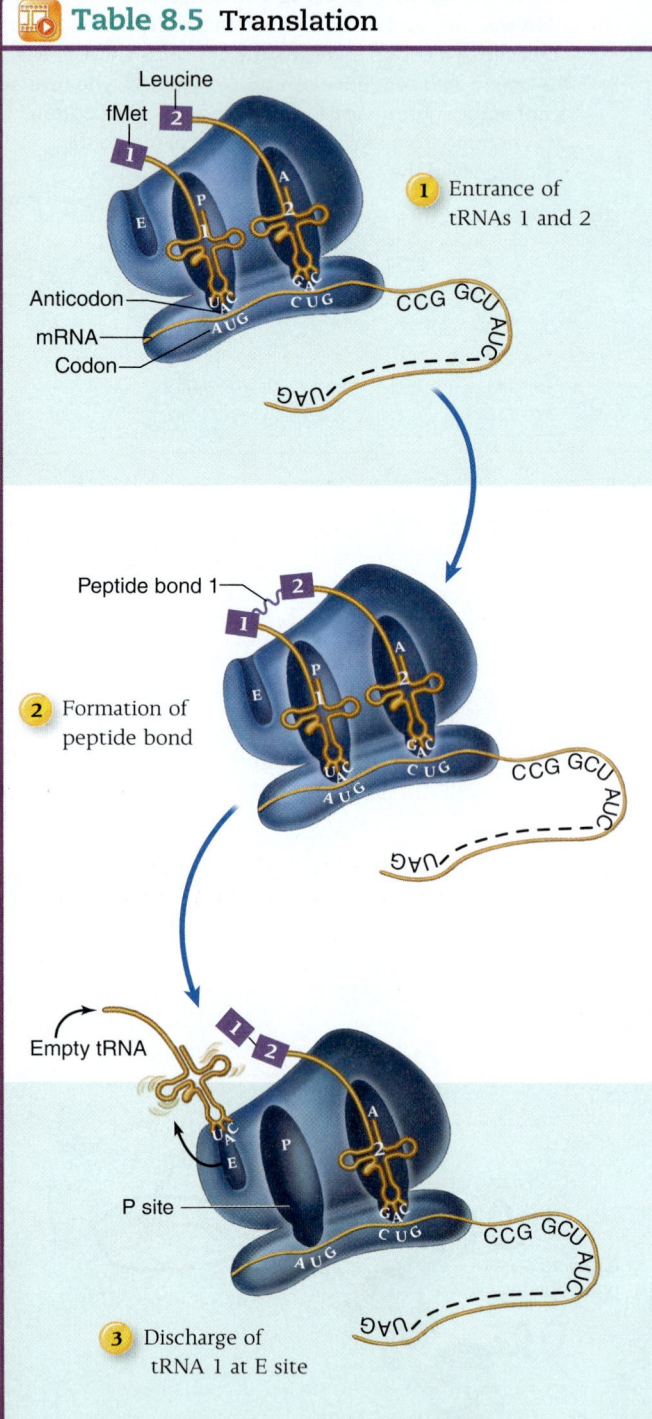

1. Entrance of tRNAs 1 and 2

Leucine
fMet
Anticodon
mRNA
Codon

2. Formation of peptide bond

Peptide bond 1
Empty tRNA

P site

3. Discharge of tRNA 1 at E site

1 The mRNA molecule leaves the DNA transcription site and is transported to ribosomes in the cytoplasm. Ribosomal subunits come together and form sites to hold the mRNA and tRNAs. The ribosome begins to scan the mRNA by moving in the 5' to 3' direction along the mRNA. The first codon it encounters is called the START codon, which is almost always AUG (and, rarely, GUG). With the mRNA message in place on the assembled ribosome, the next step in translation involves entrance of tRNAs with their amino acids. The pool of cytoplasm contains a complete array of tRNAs, previously charged by having the correct amino acid attached. The step in which the complementary tRNA meets with the mRNA code is guided by the two sites on the large subunit of the ribosome called the P site (left) and the A site (right). The ribosome also has an exit or E site where used tRNAs are released. (P stands for peptide site; A stands for aminoacyl (amino acid) site; E stands for exit site.)

2 Rules of pairing dictate that the anticodon of this tRNA must be complementary to the mRNA codon AUG; thus, the tRNA with anticodon UAC will first occupy site P. It happens that the amino acid carried by the initiator tRNA in bacteria is formyl methionine. The formyl group provides a special signal that this amino acid is not part of the translated protein because usually fMet does not remain a permanent part of the finished protein but instead is cleaved from the finished peptide. The ribosome shifts its "reading frame" to the right along the mRNA from one codon to the next. This brings the next codon into place on the ribosome and makes a space for the next tRNA to enter the A position. A peptide bond is formed between the amino acids on the adjacent tRNAs, and the polypeptide grows in length.

Elongation begins with the filling of the A site by a second tRNA. The identity of this tRNA and its amino acid is dictated by the second mRNA codon.

3 The entry of tRNA 2 into the A site brings the two adjacent tRNAs in favorable proximity for a peptide bond to form between the amino acids (aa) they carry. The fMet is transferred from the first tRNA to aa 2, resulting in two coupled amino acids called a dipeptide.

For the next step to proceed, some room must be made on the ribosome, and the next codon in sequence must be brought into position for reading. This process is accomplished by translocation, the enzyme-directed shifting of the ribosome to the right along the mRNA strand, which causes the blank tRNA 1 to be discharged from the ribosome at the E site.

amount of energy. Nearly 1,200 ATPs are required just for synthesis of an average-size protein. **Table 8.5** contains the details of translation.

Differences Between Eukaryotic and Bacterial Transcription and Translation

Eukaryotes and bacteria share many similarities in protein synthesis. The start codon in eukaryotes is also AUG, but it codes for a different form of methionine.

Table 8.5 Translation (continued)

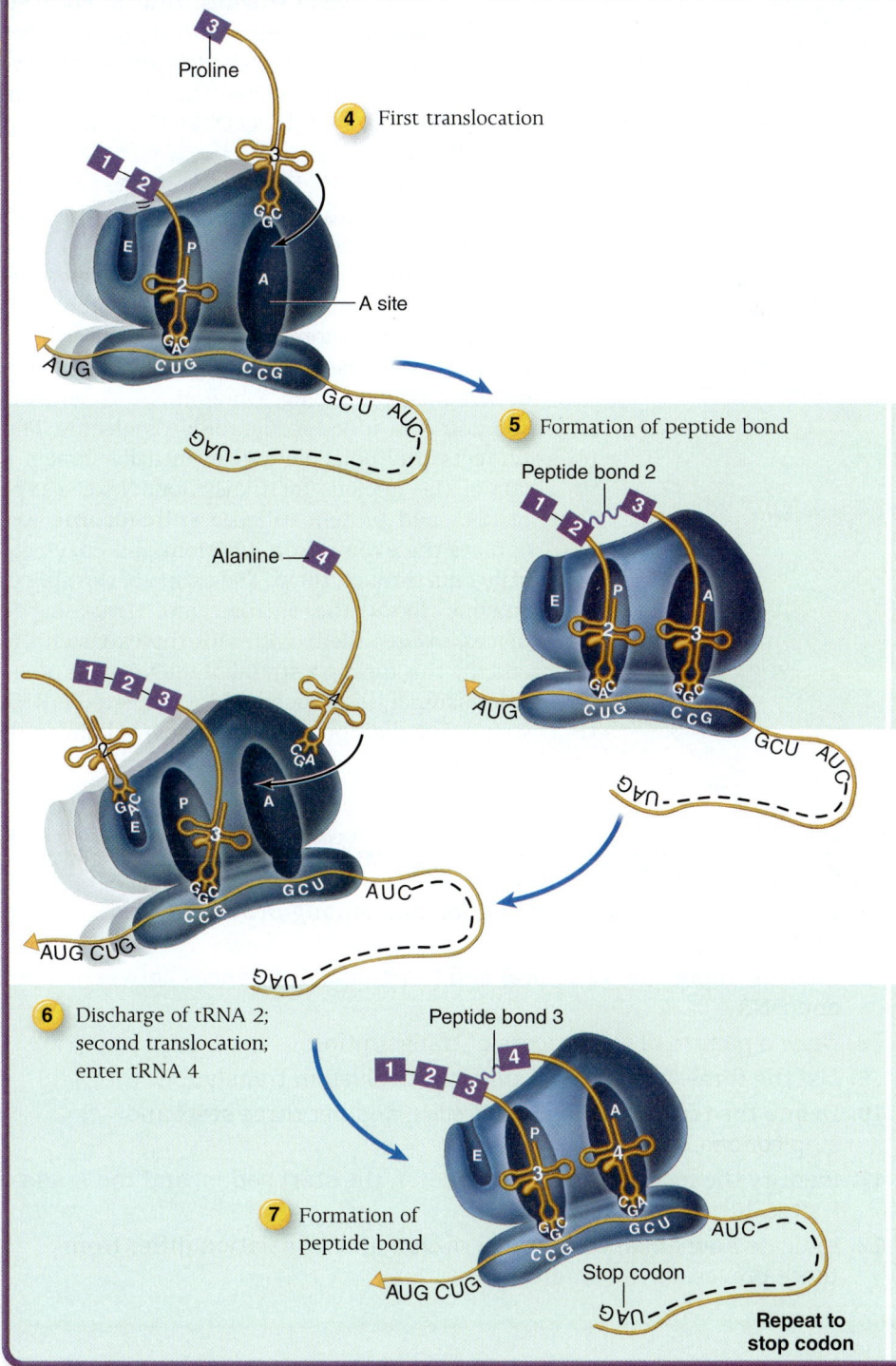

4 First translocation

Proline

A site

5 Formation of peptide bond

Peptide bond 2

Alanine — 4

6 Discharge of tRNA 2; second translocation; enter tRNA 4

Peptide bond 3

7 Formation of peptide bond

Stop codon

Repeat to stop codon

4 This also shifts the tRNA holding the dipeptide into P position. Site A is temporarily left empty. The tRNA that has been released is now free to drift off into the cytoplasm and become recharged with an amino acid for later additions to this or another protein.

The stage is now set for the insertion of tRNA 3 at site A as directed by the third mRNA codon. This insertion is followed once again by peptide bond formation between the dipeptide and amino acid 3 (making a tripeptide), splitting of the peptide from tRNA 2, and translocation.

5 This releases tRNA 2, shifts mRNA to the next position, moves tRNA 3 to position P, and opens position A for the next tRNA (which will be called tRNA 4).

6 From this point on, peptide elongation proceeds repetitively by this same series of actions out to the end of the mRNA.

7 The termination of protein synthesis is not simply a matter of reaching the last codon on mRNA. It is brought about by the presence of at least one special codon occurring just after the codon for the last amino acid. Termination codons–UAA, UAG, and UGA–are codons for which there is no corresponding tRNA. Although they are often called nonsense codons, they carry a necessary and useful message: Stop here. When this codon is reached, a special enzyme breaks the bond between the final tRNA and the finished polypeptide chain, releasing it from the ribosome.

Another difference is that eukaryotic mRNAs code for just one protein, unlike bacterial mRNAs, which often contain information from several genes in series.

As just mentioned, the presence of the DNA in a separate compartment (the nucleus) means that eukaryotic transcription and translation cannot be simultaneous. The mRNA transcript must pass through pores in the nuclear membrane and be carried to the ribosomes in the cytoplasm for translation.

We have given the simplified definition of a gene that works well for bacteria, but most eukaryotic genes (and, surprisingly, archaeal genes) do *not* exist as an uninterrupted

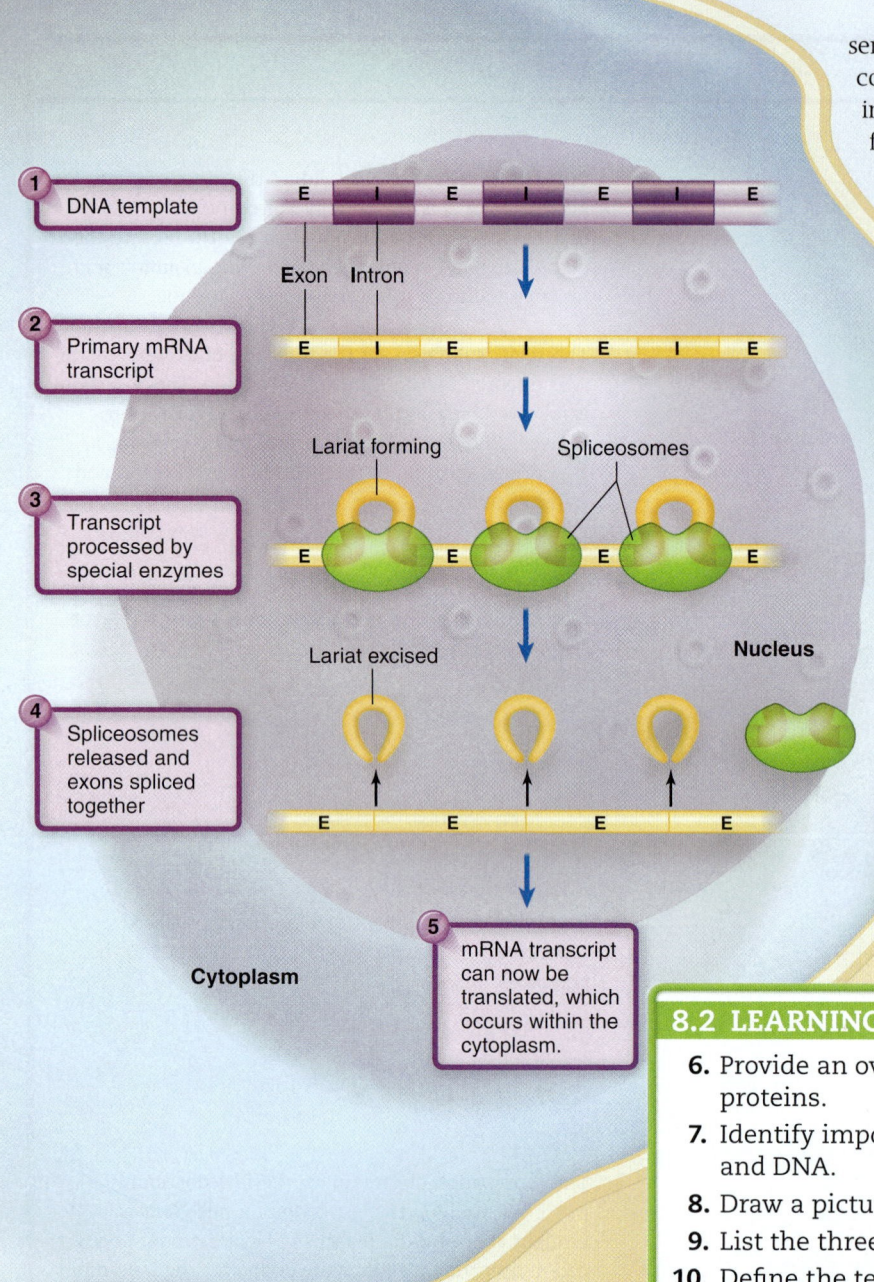

① **DNA template**

Exon   Intron

② **Primary mRNA transcript**

Lariat forming   Spliceosomes

③ **Transcript processed by special enzymes**

Lariat excised   **Nucleus**

④ **Spliceosomes released and exons spliced together**

Cytoplasm

⑤ **mRNA transcript can now be translated, which occurs within the cytoplasm.**

Figure 8.10 The split gene of eukaryotes. Eukaryotic genes have an additional complicating factor in their translation. Their coding sequences, or exons (E), are interrupted at intervals by segments called introns (I) that are not part of that protein's code. Introns are transcribed but not translated, which necessitates their removal by RNA splicing enzymes before translation.

series of triplets coding for a protein. A eukaryotic gene contains the code for a protein, but located along the gene are one to several intervening sequences of bases, called **introns,** that do not code for protein. Introns are interspersed between coding regions, called **exons,** that will be translated into protein **(figure 8.10).** We can use words as examples. A short section of colinear bacterial gene might read TOM SAW OUR DOG DIG OUT; a eukaryotic gene that codes for the same portion would read TOM SAW XZKP FPL OUR DOG QZWVP DIG OUT. The recognizable words are the exons, and the nonsense letters represent the introns.

This unusual genetic architecture, sometimes called a split gene, requires further processing for eukaryotes before translation. Transcription of the entire gene with both exons and introns occurs first, producing a pre-mRNA. A series of adenosines is added to the mRNA molecule. This protects the molecule and eventually directs it out of the nucleus for translation. Next, a type of RNA and protein called a **spliceosome** recognizes the exon-intron junctions and enzymatically cuts through them. The action of this splicer enzyme loops the introns into lariat-shaped pieces, excises them, and joins the exons end to end. By this means, a strand of mRNA with no intron material is produced. This completed mRNA strand can then proceed to the cytoplasm to be translated.

8.2 LEARNING OUTCOMES—Assess Your Progress

6. Provide an overview of the relationship among DNA, RNA, and proteins.
7. Identify important structural and functional differences between RNA and DNA.
8. Draw a picture of the process of transcription.
9. List the three types of RNA directly involved in translation.
10. Define the terms *codon* and *anticodon,* and list three start and stop codons.
11. Identify the locations of the promoter, the start codon, and the A and P sites during translation.
12. Indicate how eukaryotic transcription and translation differ from these processes in bacteria.

8.3 Genetic Regulation of Protein Synthesis

In chapter 7, we surveyed the metabolic reactions in cells and the enzymes involved in those reactions. At that time, we mentioned that some enzymes are regulated and that one form of regulation occurs at the genetic level. Control mechanisms ensure that genes are active only when their products are required. In this way, enzymes will be produced as they are needed and prevent the waste of energy and materials in dead-end synthesis. Antisense RNAs, micro RNAs, and riboswitches provide regulation in bacteria, archaea, and eukaryotes. Bacteria and archaea have an additional strategy: They organize collections of genes into **operons.** Operons consist of a coordinated set of genes, all of which are regulated as a single unit. Operons are described as either inducible or repressible. Many catabolic operons, or operons encoding enzymes that act in catabolism, are inducible, meaning that the operon is turned on (induced) by the substrate of the enzyme(s) for which the structural genes code. In this way, the enzymes needed to metabolize a nutrient (lactose, for example) are produced only when that nutrient is present in the environment. Repressible operons often contain genes coding for anabolic enzymes, such as those used to synthesize amino acids. In the case of these operons, several genes in series are turned off (repressed) by the product synthesized by the enzyme.

The Lactose Operon: A Model for Inducible Gene Regulation in Bacteria

The best understood cell system for explaining control through genetic induction is the **lactose (*lac*) operon.** This system, first described in 1961 by François Jacob and Jacques Monod, accounts for the regulation of lactose metabolism in *Escherichia coli*. Many other operons with similar modes of action have since been identified, and together they show us that the environment of a cell can have great impact on gene expression.

The lactose operon has three important features **(table 8.6):**

1. the **regulator,** composed of the gene that codes for a protein capable of repressing the operon (a **repressor**);
2. the control locus, composed of two areas, the **promoter** (recognized by RNA polymerase) and the **operator,** a sequence that acts as an on/off switch for transcription; and
3. the structural locus, made up of three genes, each coding for a different enzyme needed to catabolize lactose.

One of the enzymes, β-galactosidase, hydrolyzes the lactose into its monosaccharides; another, permease, brings lactose across the cytoplasmic membrane.

The operon provides an efficient strategy that permits genes for a particular metabolic pathway to be induced or repressed in unison by a single regulatory element. The promoter, operator, and structural components usually lie adjacent to one another, but the regulator can be at a distant site.

Table 8.6 supplies the details of how the *lac* operon works.

A fine but important point about the *lac* operon is that it functions only in the absence of glucose or if the cell's energy needs are not being met by the available glucose. Glucose is the preferred carbon source because it can be used immediately in growth and does not require induction of an operon. When glucose is present, a second regulatory system ensures that the *lac* operon is inactive, regardless of lactose levels in the environment.

Phase Variation

When bacteria turn on or off a complement of genes that leads to obvious phenotypic changes, it is sometimes called **phase variation.** Phase variation is a type of phenotypic variation, but it has its own name because it has some special characteristics, the most important of which is that the phenotype is heritable, meaning it is passed down to subsequent generations. The process of turning on genes is often mediated by regulatory proteins, as described with operons. The term *phase variation* is most often applied to traits affecting the bacterial cell surface and was originally coined to describe the ability of bacteria to change components of their surface that marked them for targeting by the host's immune system. Since these surface molecules also influenced the bacterium's ability to attach to surfaces, the ability to undergo phase variation allowed the microbes to adapt to–and stick in–different environments. Examples of phase variation include the ability of *Neisseria gonorrhoeae* strains to produce attachment fimbriae, and the ability of *Streptococcus pneumoniae* to produce a capsule.

NCLEX® PREP

4. Which of the following is true of operons?
   a. *They consist of a coordinated set of genes, all of which are regulated as a single unit.*
   b. *They may be inducible or repressible.*
   c. *Only bacteria and archaea utilize operons.*
   d. *All of the above are true.*

Lactose used by bacteria is the same disaccharide found in the milk we drink.

Table 8.6 The *lac* Operon

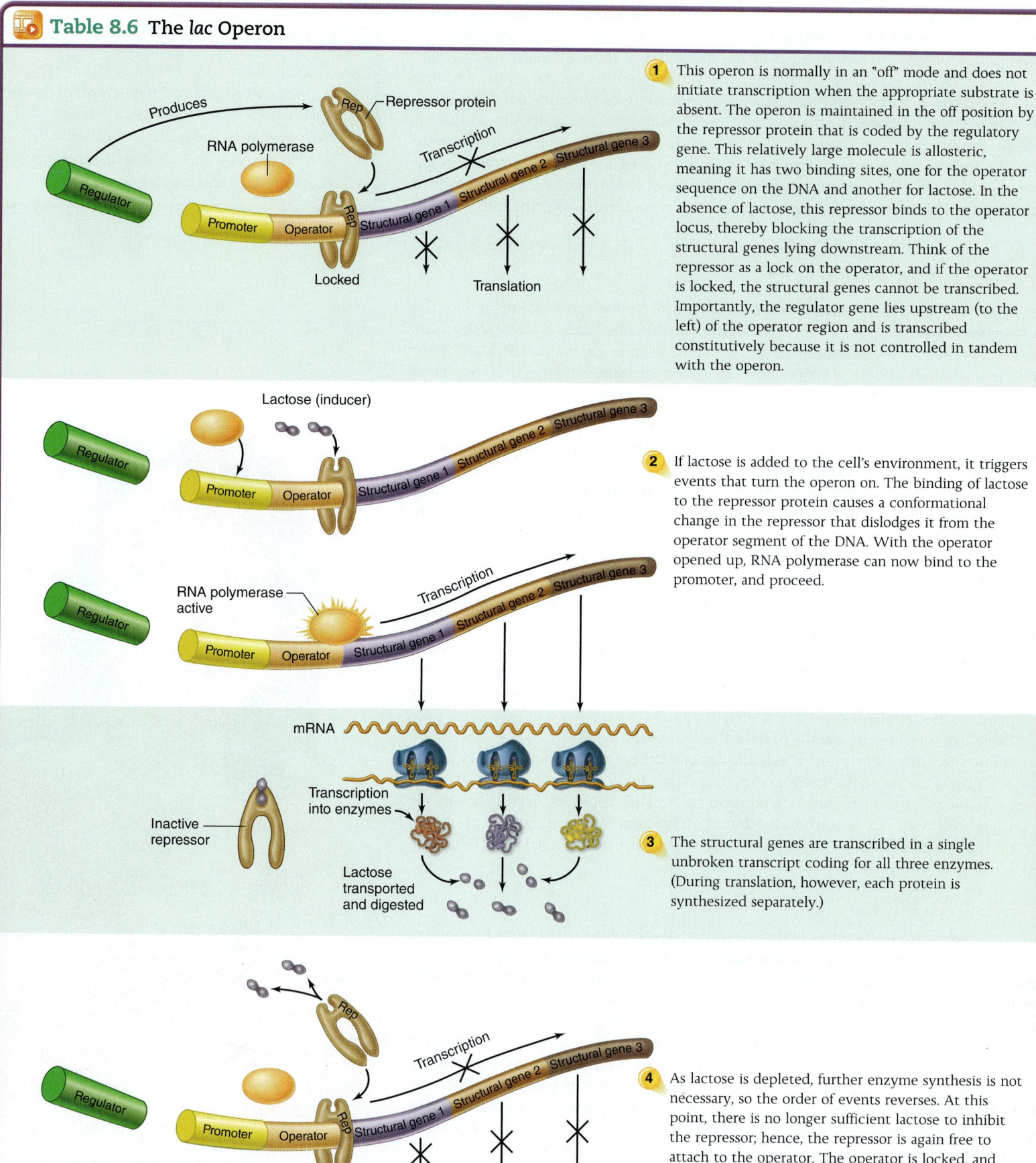

1 This operon is normally in an "off" mode and does not initiate transcription when the appropriate substrate is absent. The operon is maintained in the off position by the repressor protein that is coded by the regulatory gene. This relatively large molecule is allosteric, meaning it has two binding sites, one for the operator sequence on the DNA and another for lactose. In the absence of lactose, this repressor binds to the operator locus, thereby blocking the transcription of the structural genes lying downstream. Think of the repressor as a lock on the operator, and if the operator is locked, the structural genes cannot be transcribed. Importantly, the regulator gene lies upstream (to the left) of the operator region and is transcribed constitutively because it is not controlled in tandem with the operon.

2 If lactose is added to the cell's environment, it triggers events that turn the operon on. The binding of lactose to the repressor protein causes a conformational change in the repressor that dislodges it from the operator segment of the DNA. With the operator opened up, RNA polymerase can now bind to the promoter, and proceed.

3 The structural genes are transcribed in a single unbroken transcript coding for all three enzymes. (During translation, however, each protein is synthesized separately.)

4 As lactose is depleted, further enzyme synthesis is not necessary, so the order of events reverses. At this point, there is no longer sufficient lactose to inhibit the repressor; hence, the repressor is again free to attach to the operator. The operator is locked, and transcription of the structural genes and enzyme synthesis related to lactose both stop.

8.4 DNA Recombination Events

Genetic recombination through sexual reproduction is an important means of genetic variation in eukaryotes. Although bacteria have no exact equivalent to sexual reproduction, they exhibit a primitive means for sharing or recombining parts of their genome. An event in which one bacterium donates DNA to another bacterium is a type of genetic transfer termed **recombination,** the end result of which is a new strain different from both the donor and the original recipient strain. Recombination in bacteria depends in part on the fact that bacteria contain extrachromosomal DNA–that is, plasmids–and are adept at interchanging genes. Genetic exchanges have tremendous effects on the genetic diversity of bacteria. They provide additional genes for resistance to drugs and metabolic poisons, new nutritional and metabolic capabilities, and increased virulence and adaptation to the environment.

In general, any organism that contains genes that originated in another organism is called a **recombinant.**

Horizontal Gene Transfer in Bacteria

Any transfer of DNA that results in organisms acquiring new genes that did not come directly from parent organisms is called **horizontal gene transfer.** (Acquiring genes from parent organisms during reproduction would be vertical gene transfer.) Bacteria have been known to engage in horizontal gene transfer for decades. It is now becoming clear that eukaryotic organisms–including humans–also engage in horizontal gene transfer, often aided and abetted by microbes such as viruses. This revelation has upended traditional views about taxonomy and even "human-ness."

DNA transfer between bacterial cells typically involves small pieces of DNA in the form of plasmids or chromosomal fragments. Plasmids are small, circular pieces of DNA that contain their own origin of replication and therefore can replicate independently of the bacterial chromosome. Plasmids are found in many bacteria (as well as some fungi) and typically contain, at most, only a few dozen genes. Although plasmids are not necessary for bacterial survival, they often carry useful traits, such as antibiotic resistance. Chromosomal fragments that have escaped from a lysed bacterial cell are also commonly involved in the transfer of genetic information between cells. An important difference between plasmids and fragments is that while a plasmid has its own origin of replication and is stably replicated and inherited, chromosomal fragments must integrate themselves into the bacterial chromosome in order to be replicated and eventually passed to progeny cells. While the process of genetic recombination is relatively rare in nature, its frequency can be increased in the laboratory, where the ability to shuffle genes between organisms is highly prized.

Depending on the mode of transmission, the means of genetic recombination in bacteria is called conjugation, transformation, or transduction. **Conjugation** requires the attachment of two related species and the formation of a bridge that can transport DNA. **Transformation** entails the transfer of naked DNA and requires no special vehicle. **Transduction** is the transfer of DNA from one bacterium to another via a bacterial virus **(table 8.7).**

Researchers have found snake DNA in the cow genome, undoubtedly transferred there by a virus that infected both—or that infected bacteria colonizing both.

Table 8.7 Types of Horizontal Gene Transfer in Bacteria

Examples of Mode	Factors Involved	Direct or Indirect*	Genes Commonly Transferred in Nature**
Conjugation	Donor cell with pilus Fertility plasmid in donor Both donor and recipient alive Bridge forms between cells to transfer DNA.	Direct	Drug resistance; resistance to metals; toxin production; enzymes; adherence molecules
Transformation	Free donor DNA (fragment) Live, competent recipient cell	Indirect	Polysaccharide capsule
Transduction	Donor is lysed bacterial cell. Defective bacteriophage is carrier of donor DNA. Live recipient cell of same species as donor	Indirect	Toxins; enzymes for sugar fermentation; drug resistance

*Direct means the donor and recipient are in contact during exchange; indirect means they are not.
**In the lab almost any gene can be transferred.

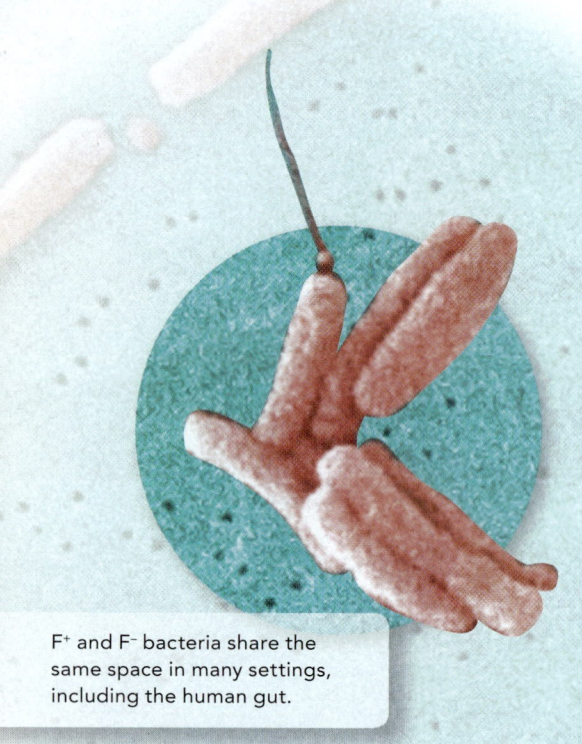

F+ and F− bacteria share the same space in many settings, including the human gut.

Conjugation: Exchanging Genes

Conjugation is a mode of genetic exchange in which a plasmid or other genetic material is transferred by a donor to a recipient cell via a direct connection **(figure 8.11)**. Both gram-negative and gram-positive cells can conjugate. In gram-negative cells, the donor's plasmid (called a **fertility, or F factor**) allows the synthesis of a conjugative **pilus.** The recipient cell has a recognition site on its surface. A cell's role in conjugation is denoted by F+ for the cell that has the F plasmid and by F− for the cell that lacks it. Contact is made when a pilus grows out from the F+ cell, attaches to the surface of the F− cell, contracts, and draws the two cells together. In gram-positive cells, an opening is created between two adjacent cells, and the replicated DNA passes across from one cell to the other. Conjugation is a conservative process, in that the donor bacterium generally retains ("conserves") a copy of the genetic material being transferred.

There are hundreds of conjugative plasmids with some variations in their properties. One of the best understood plasmids is the F factor in *E. coli*, which can do either of two things:

1. The donor (F+) cell makes a copy of its F factor and transmits this to a recipient (F−) cell. The F− cell is thereby changed into an F+ cell capable of producing a pilus and conjugating with other cells. No additional donor genes are transferred at this time.

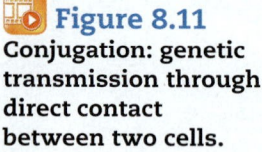 **Figure 8.11**
Conjugation: genetic transmission through direct contact between two cells.

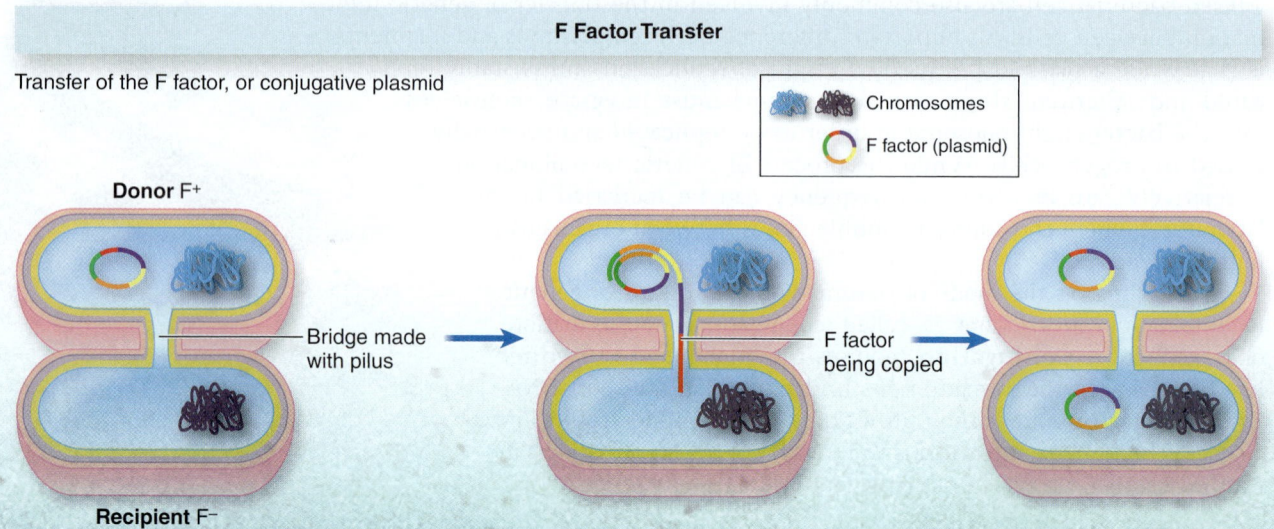

2. In high-frequency recombination (Hfr) donors, the plasmid becomes integrated into the F⁺ donor chromosome, which, when replicated, begins to transfer to the recipient cell. This means that some chromosomal genes get transferred to the recipient. Plasmid genes may or may not be transferred.

Conjugation has great biomedical importance. Special **resistance (R) plasmids,** or **factors,** that bear genes for resisting antibiotics and other drugs are commonly shared among bacteria through conjugation. Transfer of R factors can confer multiple resistance to antibiotics such as tetracycline, chloramphenicol, streptomycin, sulfonamides, and penicillin. Other types of R factors carry genetic codes for resistance to heavy metals (nickel and mercury) or for synthesizing virulence factors (toxins, enzymes, and adhesion molecules) that increase the pathogenicity of the bacterial strain.

Transformation: Capturing DNA from Solution

We now know that a chromosome released by a lysed cell breaks into fragments small enough to be accepted by a recipient cell and that DNA, even from a dead cell, retains its genetic sequence. This nonspecific acceptance by a bacterial cell of small fragments of soluble DNA from the surrounding environment is termed **transformation.** Transformation is apparently facilitated by special DNA-binding proteins on the cell wall that capture DNA from the surrounding medium. Cells that are capable of accepting genetic material through this means are termed **competent.** The new DNA is transported into the cytoplasm, where some of it is inserted into the bacterial chromosome. Transformation is a natural event found in several groups of gram-positive and gram-negative bacterial species.

Because transformation requires no special appendages, and the donor and recipient cells do not have to be in direct contact, the process is useful for certain types of recombinant DNA technology. With this technique, foreign genes from a completely unrelated organism are inserted into a plasmid, which is then introduced into a competent bacterial cell through transformation in the same way that small pieces are taken up naturally. These recombinations can be carried out in a test tube, and human genes can be experimented upon and even expressed outside the human body by placing them in a microbial cell. This same phenomenon in eukaryotic cells, termed **transfection,** is an essential aspect of genetically engineered yeasts, plants, and mice, and it has been proposed as a future technique for curing genetic diseases in humans.

Transduction: The Case of the Piggyback DNA

Bacteriophages (bacterial viruses) have been previously described as bacterial parasites. Viruses can in fact serve as genetic vectors (an entity that can bring foreign DNA into a cell). The process by which a bacteriophage serves as the carrier of DNA

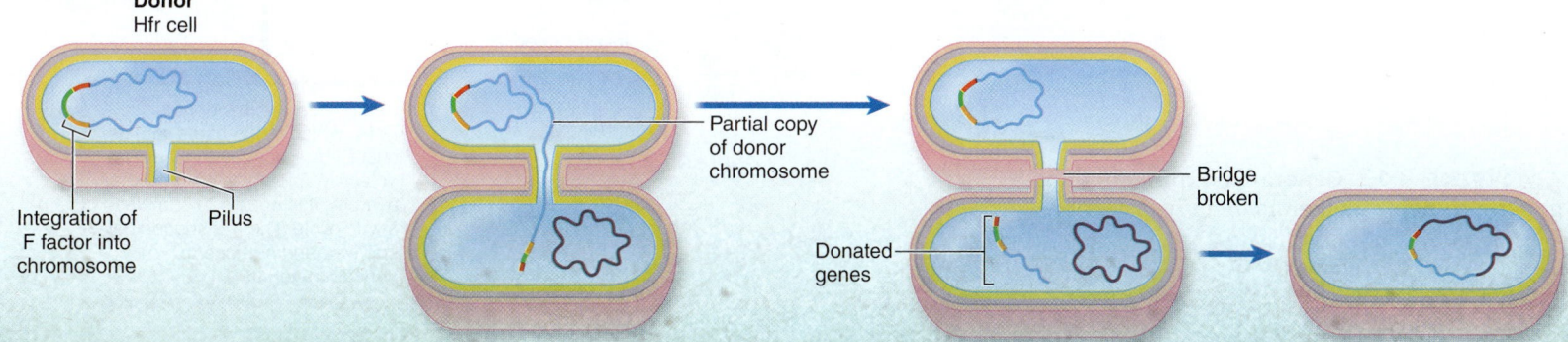

Hfr Transfer

High-frequency (Hfr) transfer involves transmission of chromosomal genes from a donor cell to a recipient cell. The plasmid jumps into the chromosome, and when the chromosome is duplicated the plasmid and part of the chromosome are transmitted to a new cell through conjugation. This plasmid/chromosome hybrid then incorporates into the recipient chromosome.

Donor
Hfr cell

Partial copy of donor chromosome

Bridge broken

Integration of F factor into chromosome

Pilus

Donated genes

from a donor cell to a recipient cell is *transduction*. It occurs naturally in a broad spectrum of bacteria. The participating bacteria in a single transduction event must be the same species because of the specificity of viruses for host cells.

There are two versions of transduction. In *generalized transduction* (**figure 8.12**), random fragments of disintegrating host DNA are taken up by the phage during assembly.

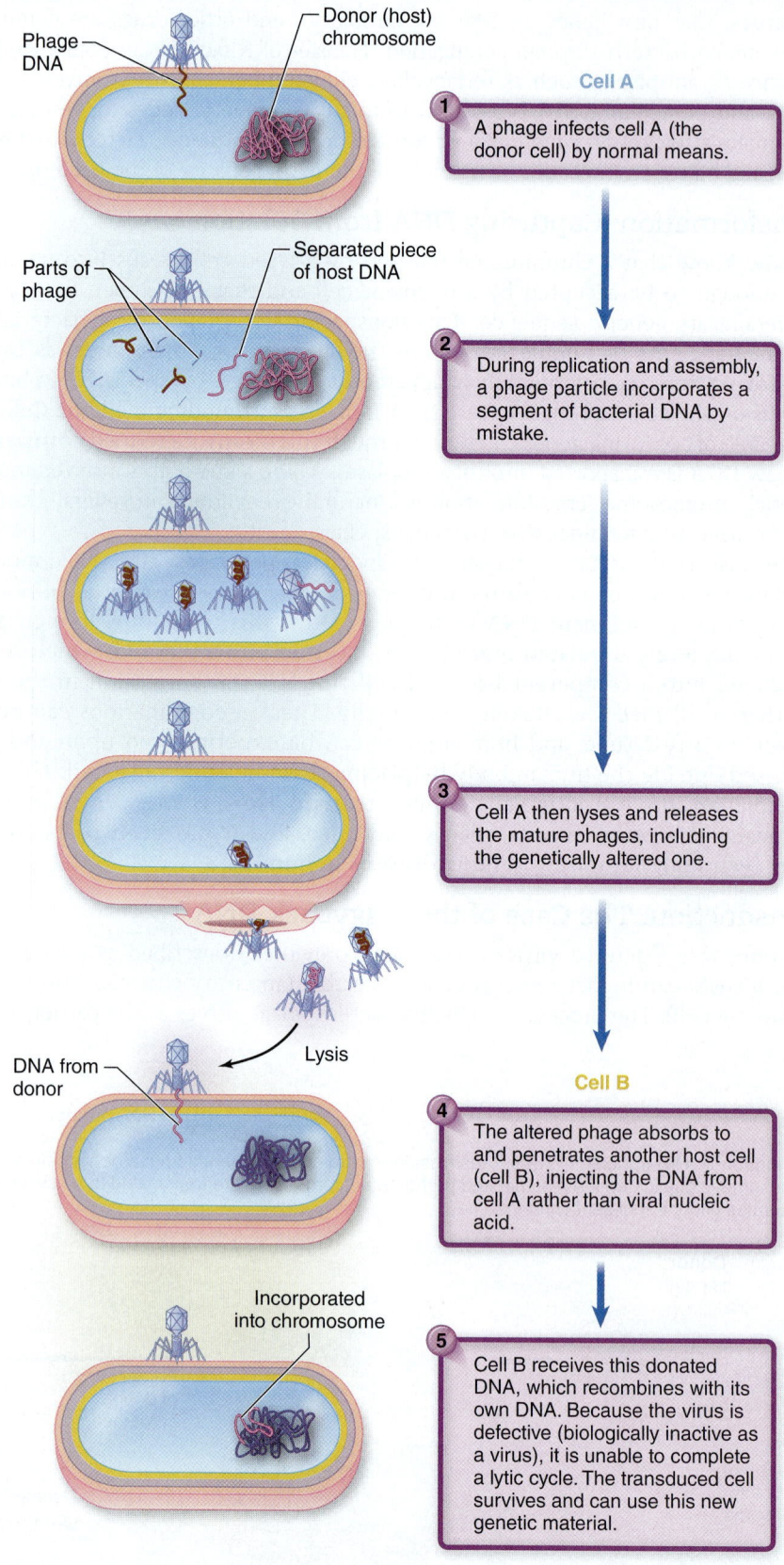

Cell A

1. A phage infects cell A (the donor cell) by normal means.

2. During replication and assembly, a phage particle incorporates a segment of bacterial DNA by mistake.

3. Cell A then lyses and releases the mature phages, including the genetically altered one.

Cell B

4. The altered phage absorbs to and penetrates another host cell (cell B), injecting the DNA from cell A rather than viral nucleic acid.

5. Cell B receives this donated DNA, which recombines with its own DNA. Because the virus is defective (biologically inactive as a virus), it is unable to complete a lytic cycle. The transduced cell survives and can use this new genetic material.

Figure 8.12 Generalized transduction: genetic transfer by means of a virus carrier.

Virtually any gene from the bacterium can be transmitted through this means. In *specialized transduction* **(figure 8.13),** a highly specific part of the host genome is regularly incorporated into the virus. This specificity is explained by the prior existence of a temperate prophage inserted in a fixed site on the bacterial chromosome. When activated, the prophage DNA separates from the bacterial chromosome, carrying a small segment of host genes with it. During a lytic cycle, these specific viral-host gene combinations are incorporated into the viral particles and carried to another bacterial cell.

Several cases of specialized transduction have biomedical importance. The virulent strains of bacteria such as *Corynebacterium diphtheriae*, *Clostridium* spp.,

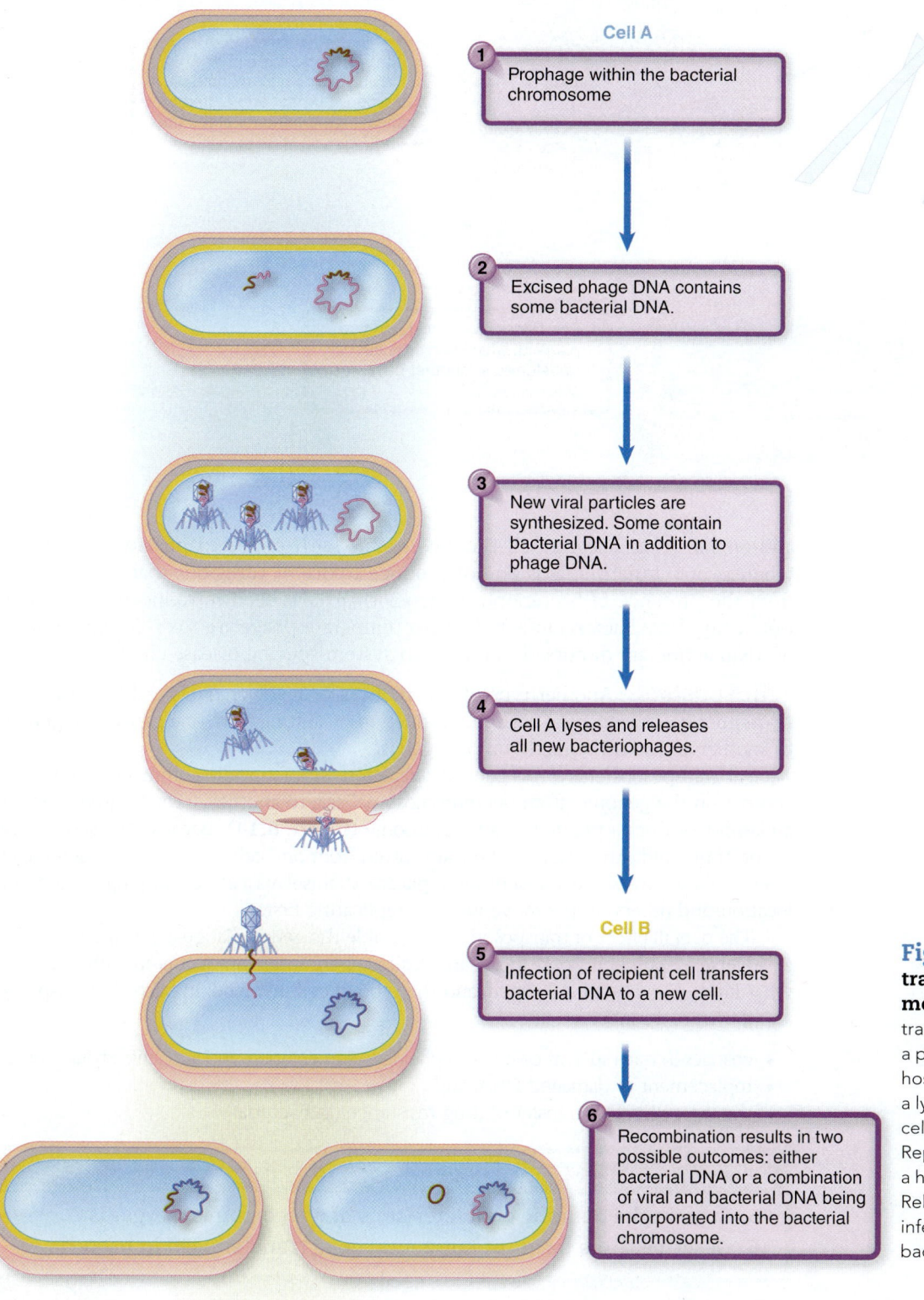

Cell A

1 Prophage within the bacterial chromosome

2 Excised phage DNA contains some bacterial DNA.

3 New viral particles are synthesized. Some contain bacterial DNA in addition to phage DNA.

4 Cell A lyses and releases all new bacteriophages.

Cell B

5 Infection of recipient cell transfers bacterial DNA to a new cell.

6 Recombination results in two possible outcomes: either bacterial DNA or a combination of viral and bacterial DNA being incorporated into the bacterial chromosome.

Figure 8.13 Specialized transduction: transfer of specific genetic material by means of a virus carrier. Specialized transduction begins with a cell that contains a prophage (a viral genome integrated into the host cell chromosome). Rarely, the virus enters a lytic cycle and, as it excises itself from its host cell, inadvertently includes some bacterial DNA. Replication and assembly result in production of a hybrid virus, containing some bacterial DNA. Release of the recombinant virus and subsequent infection of a new host result in transfer of bacterial DNA between cells.

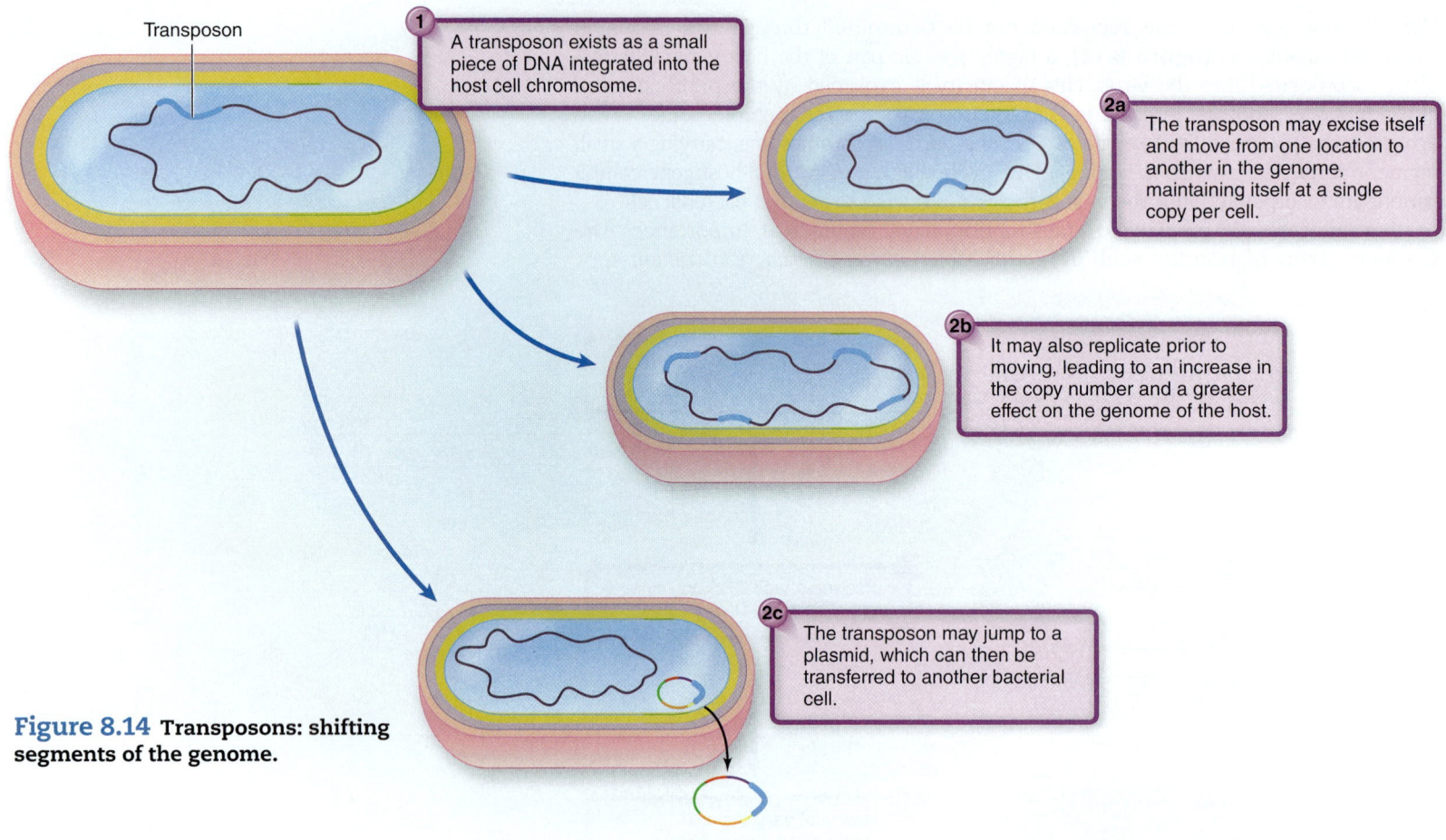

Figure 8.14 Transposons: shifting segments of the genome.

and *Streptococcus pyogenes* all produce toxins with profound physiological effects, whereas nonvirulent strains do not produce toxins. It turns out that toxicity arises from the presence of bacteriophage genes that have been introduced by transduction. Only those bacteria infected with a temperate phage are toxin formers. (Details of toxin action are discussed in the organ system–specific disease chapters.)

Transposons Another type of genetic transfer involves transposable elements, or **transposons.** Transposons have the distinction of shifting from one part of the genome to another and so are termed "jumping genes."

All transposons share the general characteristic of traveling from one location to another on the genome—from one chromosomal site to another, from a chromosome to a plasmid, or from a plasmid to a chromosome **(figure 8.14).** Because transposons can occur in plasmids, they can also be transmitted from one cell to another in bacteria and a few eukaryotes. Some transposons replicate themselves before jumping to the next location, and others simply move without replicating first.

The overall effect of transposons—to scramble the genetic language—can be beneficial or adverse, depending upon such variables as where insertion occurs in a chromosome, what kinds of genes are relocated, and the type of cell involved. In bacteria, transposons are known to be involved in

- changes in traits such as colony morphology, pigmentation, and antigenic characteristics;
- replacement of damaged DNA; and
- the intermicrobial transfer of drug resistance (in bacteria).

8.4 LEARNING OUTCOMES—Assess Your Progress

15. Explain the defining characteristics of a recombinant organism.

16. Describe three forms of horizontal gene transfer used in bacteria.

8.5 Mutations: Changes in the Genetic Code

As precise and predictable as the rules of genetic expression seem, permanent changes do occur in the genetic code. Indeed, genetic change is the driving force of evolution. Any change to the nucleotide sequence in the genome is called a mutation. Mutations are most noticeable when the genotypic change leads to a change in phenotype. Mutations can involve the loss of base pairs, the addition of base pairs, or a rearrangement in the order of base pairs. Do not confuse this with genetic recombination, in which microbes transfer whole segments of genetic information among themselves.

A microorganism that exhibits a natural, nonmutated characteristic is known as a **wild type,** or wild strain with respect to that trait. If a microorganism bears a mutation, it is called a **mutant strain.** Mutant strains can show variance in morphology, nutritional characteristics, genetic control mechanisms, resistance to chemicals, temperature preference, and nearly any type of enzymatic function.

Causes of Mutations

Mutations can be spontaneous or induced, depending upon their origin. A **spontaneous mutation** is a random change in the DNA arising from errors in replication that occur randomly. The frequency of spontaneous mutations has been measured for a number of organisms. Mutation rates vary tremendously, from one mutation in 10^5 replications (a high rate) to one mutation in 10^{10} replications (a low rate). The rapid rate of bacterial reproduction allows these mutations to be observed more readily in bacteria than in most eukaryotes.

Induced mutations result from exposure to known **mutagens,** which are primarily physical or chemical agents that interact with DNA in a disruptive manner. Examples of mutagens are some types radiation (UV light, X rays) and certain chemicals such as nitrous acid.

Categories of Mutations

Mutations range from large mutations, in which large genetic sequences are gained or lost, to small ones that affect only a single base on a gene. These latter mutations, which involve addition, deletion, or substitution of single bases, are called **point mutations.**

To understand how a change in DNA influences the cell, remember that the DNA code appears in a particular order of triplets (three bases) that is transcribed into mRNA codons, each of which specifies an amino acid. A permanent alteration in the DNA that is copied faithfully into mRNA and translated can change the structure of the protein. A change in a protein can likewise change the morphology and physiology of a cell. Some mutations have a harmful effect on the cell, leading to cell dysfunction or death; these are called lethal mutations. Neutral mutations produce neither adverse nor helpful changes. A small number of mutations are beneficial in that they provide the cell with a useful change in structure or physiology.

Any change in the code that leads to placement of a different amino acid is called a **missense mutation.** A missense mutation can do one of the following:

- create a faulty, nonfunctional (or less functional) protein;
- produce a protein that functions in a different manner; or
- cause no significant alteration in protein function (see **table 8.8** to see how missense mutations look).

A **nonsense mutation,** on the other hand, changes a normal codon into a stop codon that does not code for an amino acid and stops the production of the protein wherever it occurs. A nonsense mutation almost always results in a nonfunctional protein. (**Table 8.8, row d,** shows a nonsense mutation resulting

Barbara McClintock won the Nobel Prize for Physiology or Medicine in 1983 for the discovery of mobile genetic elements, or transposons, in the corn plant.

Table 8.8 Categories of Point Mutations and Their Effects

(a)	DNA RNA Protein	TAC AUG Met	TGG ACC Thr	CTG GAC Asp	CTC GAG Glu	TAC AUG Met	TTT... AAA... Lys...	Normal gene
(b)	DNA RNA Protein	TAC AUG Met	TGG ACC Thr	CTT GAA Glu	CTC GAG Glu	TAC AUG Met	TTT... AAA... Lys...	Missense mutation: leading to amino acid switch (may or may not function well)
(c)	DNA RNA Protein	TAC AUG Met	TGG ACC Thr	CTA GAU Asp	CTC GAG Glu	TAC AUG Met	TTT... AAA... Lys...	Base substitution: silent (no change in function)
(d)	DNA RNA Protein	TAC AUG Met	TGC ↗G ACG Thr	TGC ACG Thr	TCT AGA Arg	ACT UGA STOP	TT AAA...	**Frameshift mutation**
			Frameshift and premature stop					Deletion mutation (d) ↓ Both lead to frameshifts and can lead to premature stop codons and/or poorly functioning protein ↑ Insertion mutation (e)
(e)	DNA RNA Protein	TAC AUG Met	TGG ACC Thr	GCT CGA Arg	GCT CGA Arg	CTA GAU Asp	CTT... GAA... Glu...	
			Frameshift					

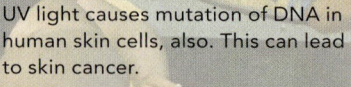

UV light causes mutation of DNA in human skin cells, also. This can lead to skin cancer.

from a frameshift [described below].) A **silent mutation (table 8.8, row c)** alters a base but does not change the amino acid and thus has no effect. For example, because of the redundancy of the code, ACU, ACC, ACG, and ACA all code for threonine, so a mutation that changes only the last base will not alter the sense of the message in any way. A **back-mutation** occurs when a gene that has undergone mutation reverses (mutates back) to its original base composition.

Mutations also occur when one or more bases are inserted into or deleted from a newly synthesized DNA strand. This type of mutation, known as a **frameshift (table 8.8, rows d and e),** is so named because the reading frame of the mRNA has been changed. Frameshift mutations nearly always result in a nonfunctional protein because every amino acid after the mutation is different from what was coded for in the original DNA. Also note that insertion or deletion of bases in multiples of three (3, 6, 9, etc.) results in the addition or deletion of amino acids but does not disturb the reading frame.

Repair of Mutations

Earlier we indicated that DNA has a proofreading mechanism to repair mistakes in replication that might otherwise become permanent. Because mutations are potentially life-threatening, the cell has additional systems for finding and repairing DNA that has been damaged by various mutagenic agents and processes. Most ordinary DNA damage is resolved by enzymatic systems specialized for finding and fixing such defects.

DNA that has been damaged by ultraviolet radiation can be restored by photoactivation or

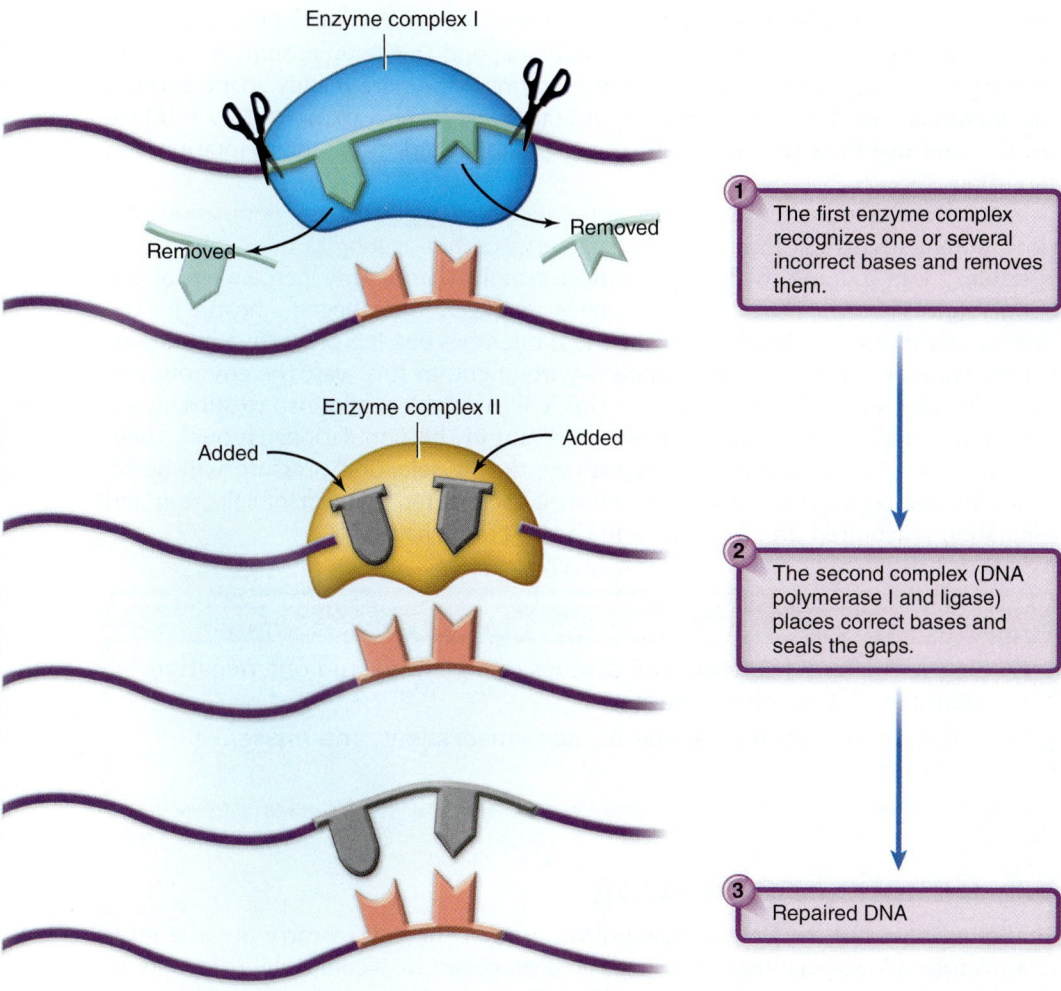

Figure 8.15 Excision repair of mutation by enzymes.

1. The first enzyme complex recognizes one or several incorrect bases and removes them.

2. The second complex (DNA polymerase I and ligase) places correct bases and seals the gaps.

3. Repaired DNA

light repair. This repair mechanism requires visible light and a light-sensitive enzyme, DNA photolyase, which can detect and attach to the damaged areas (sites of abnormal pyrimidine binding). Ultraviolet repair mechanisms are successful only for a relatively small number of UV mutations. Cells cannot repair severe, widespread damage and will die.

Mutations can be excised by a series of enzymes that remove the incorrect bases and add the correct ones. This process is known as excision repair. First, enzymes break the bonds between the bases and the sugar-phosphate strand at the site of the error. A different enzyme subsequently removes the defective bases one at a time, leaving a gap that will be filled in by DNA polymerase I and ligase **(figure 8.15).** A repair system can also locate mismatched bases that were missed during proofreading—for example, C mistakenly paired with A, or G with T. The base must be replaced soon after the mismatch is made, or it will not be recognized by the repair enzymes.

Positive and Negative Effects of Mutations

Many mutations are not repaired. How the cell copes with them depends on the nature of the mutation and the strategies available to that organism. Mutations are permanent and heritable and will be passed on to the offspring of organisms and new viruses and become a long-term part of the gene pool. Many mutations are harmful to organisms; others provide adaptive advantages.

Although most spontaneous mutations are not beneficial, a small number contribute to the success of the individual and the population by creating variant strains

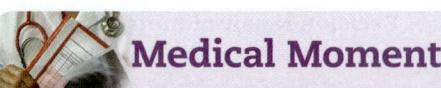

Medical Moment

Is There Hope for Combating Antibiotic-Resistant Organisms?

There has been increasing concern in the scientific community for the past several years as the occurrence of antibiotic-resistant organisms has increased. "The bugs are winning" seems to be the general consensus, and it is indeed cause for concern.

Not all antibiotic-resistant microbes do well once they have acquired the resistance. For example, researchers have found that rifampicin-resistant meningococci did not multiply as rapidly and were less effective at invading cells than their rifampicin-susceptible counterparts. In other words, the resistant organisms were weaker. This is known as a "fitness cost"—meaning that resistant bacteria may be less "fit" than their nonresistant counterparts.

Other research has found just the opposite—that antibiotic-resistant strains of some bacteria display no fitness costs whatsoever. Further research will help us understand the consequences when bacteria become antibiotic-resistant.

with alternate ways of expressing a trait. Microbes are not "aware" of this advantage and do not direct these changes; they simply respond to the environment they encounter. Those organisms with beneficial mutations can more readily adapt, survive, and reproduce. In the long-range view, mutations and the variations they produce are the raw materials for change in the population and, thus, for adaptation and evolution.

Mutations that create variants occur frequently enough that any population contains mutant strains for a number of characteristics, but as long as the environment is stable, these mutants will never comprise more than a tiny percentage of the population. When the environment changes, however, it can become hostile for the survival of certain individuals, and only those microbes bearing protective mutations will be equipped to survive in the new environment. In this way, the environment naturally selects certain mutant strains that will reproduce, give rise to subsequent generations, and, in time, be the dominant strain in the population. Through these means, any change that confers an advantage during selection pressure will be retained by the population. One of the clearest models for this sort of selection and adaptation is acquired drug resistance in bacteria (see chapter 10).

8.5 LEARNING OUTCOMES—Assess Your Progress

17. Define the *term mutation,* and discuss one positive and one negative example of it in microorganisms.

18. Differentiate among frameshift, nonsense, silent, and missense mutations.

8.6 Genetic Engineering

The knowledge of how DNA is manipulated within the cell to carry out the goals of a microbe allows scientists to utilize these processes to accomplish goals more to the liking of human beings. Since the 1970s, discoveries and advances have led to an explosion of new capabilities and, as a result, an explosion of new knowledge about microbes and about biology in general. In this section, we will highlight a few techniques that have relevance for microbiology and in particular for infectious diseases.

Enzymes for Dicing and Splicing Nucleic Acids

The groundbreaking discovery in 1971 of **restriction endonucleases** made almost everything we discuss in this section possible. These enzymes come from bacterial cells. They recognize foreign DNA and are capable of breaking the phosphodiester bonds between adjacent nucleotides on both strands of DNA, leading to a break in the DNA strand. In the bacterial cell, this ability protects against the incompatible DNA of bacteriophages or plasmids. In the biotechnologist's lab, the enzymes can be used to cleave DNA at desired sites and are necessary for the techniques of recombinant DNA technology.

Hundreds of restriction endonucleases have been discovered in bacteria. Each type has a known sequence of 4 to 10 base pairs as its target, so sites of cutting can be finely controlled. These enzymes have the unique property of recognizing and clipping at base sequences called **palindromes (figure 8.16).** Palindromes are sequences of DNA that are identical when read from the 5′ to 3′ direction on one strand and the 5′ to 3′ direction on the other strand.

Endonucleases are usually named by combining the first letter of the bacterial genus, the first two letters of the species, and the endonuclease number. Thus, *Eco*RI is the first endonuclease found

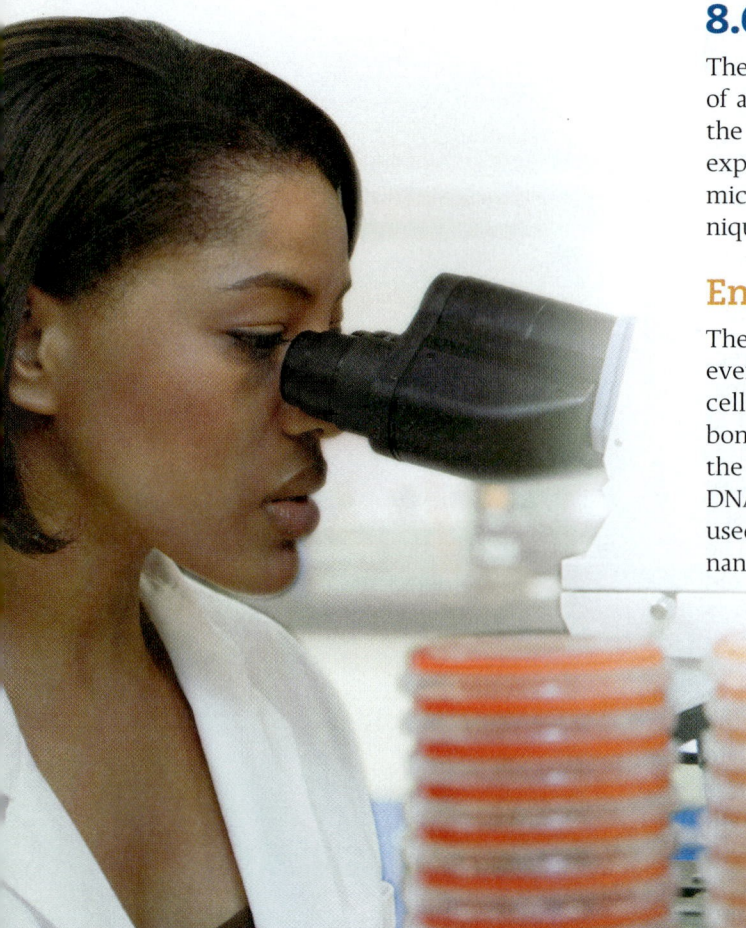

Figure 8.16 **Some useful properties of DNA.**

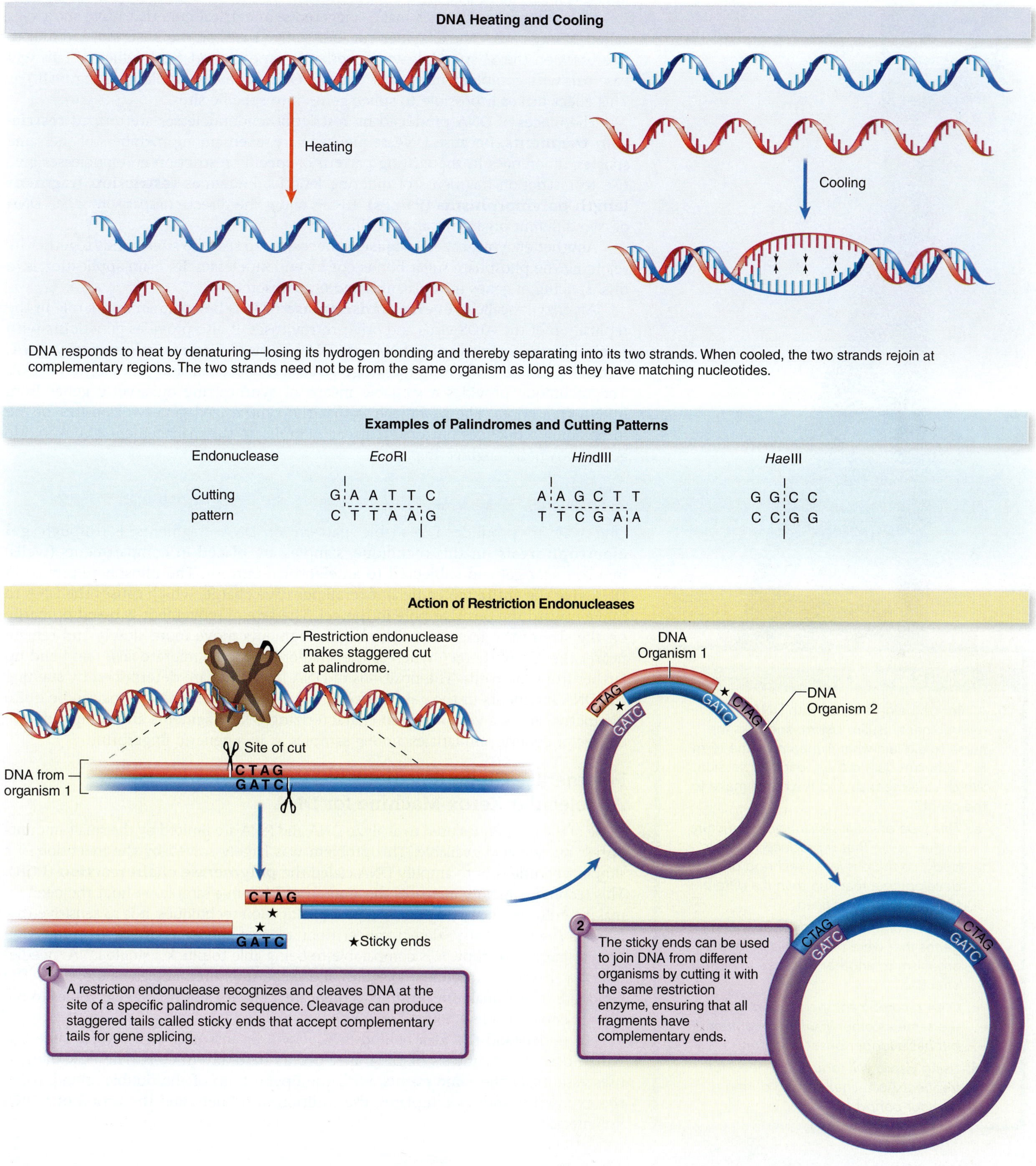

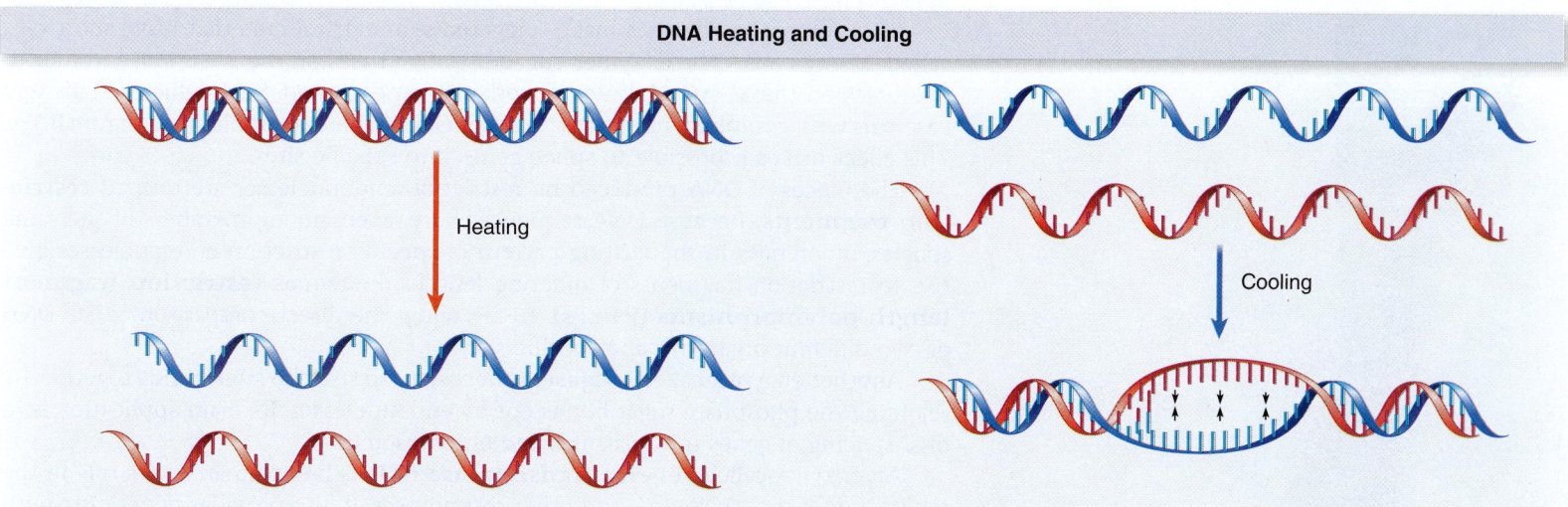

DNA Heating and Cooling

Heating

Cooling

DNA responds to heat by denaturing—losing its hydrogen bonding and thereby separating into its two strands. When cooled, the two strands rejoin at complementary regions. The two strands need not be from the same organism as long as they have matching nucleotides.

Examples of Palindromes and Cutting Patterns

Endonuclease	*Eco*RI	*Hind*III	*Hae*III
Cutting pattern	G A A T T C C T T A A G	A A G C T T T T C G A A	G G C C C C G G

Action of Restriction Endonucleases

Restriction endonuclease makes staggered cut at palindrome.

DNA Organism 1

DNA Organism 2

DNA from organism 1

Site of cut

C T A G
G A T C

C T A G

★

G A T C   ★Sticky ends

1 A restriction endonuclease recognizes and cleaves DNA at the site of a specific palindromic sequence. Cleavage can produce staggered tails called sticky ends that accept complementary tails for gene splicing.

2 The sticky ends can be used to join DNA from different organisms by cutting it with the same restriction enzyme, ensuring that all fragments have complementary ends.

CTAG
GATC

CTAG
GATC

in *Escherichia coli* (in the R strain), and *Hind*III is the third endonuclease discovered in *Haemophilus influenzae* type d (figure 8.16).

Most often, the enzymes make staggered symmetrical cuts that leave short tails called "sticky ends." The enzymes cut four to five bases on the 3′ strand, and four to five bases on the 5′ strand, leaving overhangs on each end. Such adhesive tails will base-pair with complementary tails on other DNA fragments or plasmids (figure 8.16). This effect makes it possible to splice genes into specific sites.

The pieces of DNA produced by restriction endonucleases are termed **restriction fragments.** Because DNA sequences vary, even among members of the same species, differences in the cutting pattern of specific restriction endonucleases give rise to restriction fragments of differing lengths, known as **restriction fragment length polymorphisms (RFLPs).** RFLPs allow the direct comparison of the DNA of two different organisms at a specific site.

Another enzyme, called a **ligase,** is necessary to seal the sticky ends together by rejoining the phosphate-sugar bonds cut by endonucleases. Its main application is in final splicing of genes into plasmids and chromosomes.

An enzyme called **reverse transcriptase (RT)** is best known for its role in the replication of the AIDS virus and other retroviruses. It also provides geneticists with a valuable tool for converting RNA into DNA. Copies called **complementary DNA,** or **cDNA,** can be made from messenger, transfer, ribosomal, and other forms of RNA. The technique provides a valuable means of synthesizing eukaryotic genes from mRNA transcripts. The advantage is that the synthesized gene will be free of the intervening sequences (introns) that can complicate the management of eukaryotic genes in genetic engineering.

Analysis of DNA

One way to produce a readable pattern of DNA fragments is through **gel electrophoresis.** In this technique, samples are placed in compartments (wells) in a soft agar gel and subjected to an electrical current. The phosphate groups in DNA give the entire molecule an overall negative charge, which causes the DNA to move toward the positive pole in the gel. The rate of movement is based primarily on the size of the fragments. The larger fragments move more slowly and remain nearer the top of the gel, whereas the smaller fragments migrate faster and end up farther from the wells. The positions of DNA fragments are determined by staining the DNA fragments in the gel **(figure 8.17).** Electrophoresis patterns can be quite distinctive and are very useful in characterizing DNA fragments and comparing the degree of genetic similarities among samples as in a genetic fingerprint.

Polymerase Chain Reaction: A Molecular Xerox Machine for DNA

Some of the techniques used to analyze DNA and RNA are limited by the small amounts of test nucleic acid available. This problem was largely solved by the invention of a simple, versatile way to amplify DNA called the **polymerase chain reaction (PCR).** This technique rapidly increases the amount of DNA in a sample without the need for making cultures or carrying out complex purification techniques. It is so sensitive that it holds the potential to detect cancer from a single cell or to diagnose an infection from a single gene copy. It is comparable to being able to pluck a single DNA "needle" out of a "haystack" of other molecules and make unlimited copies of the DNA. The rapid rate of PCR makes it possible to replicate a target DNA from a few copies to billions of copies in a few hours.

To understand the idea behind PCR, it will be instructive to review table 8.2, which describes synthesis of DNA as it occurs naturally in cells. The PCR method uses essentially the same events, with the opening up of the double strand, using the exposed strands as templates, the addition of primers, and the action of a DNA polymerase.

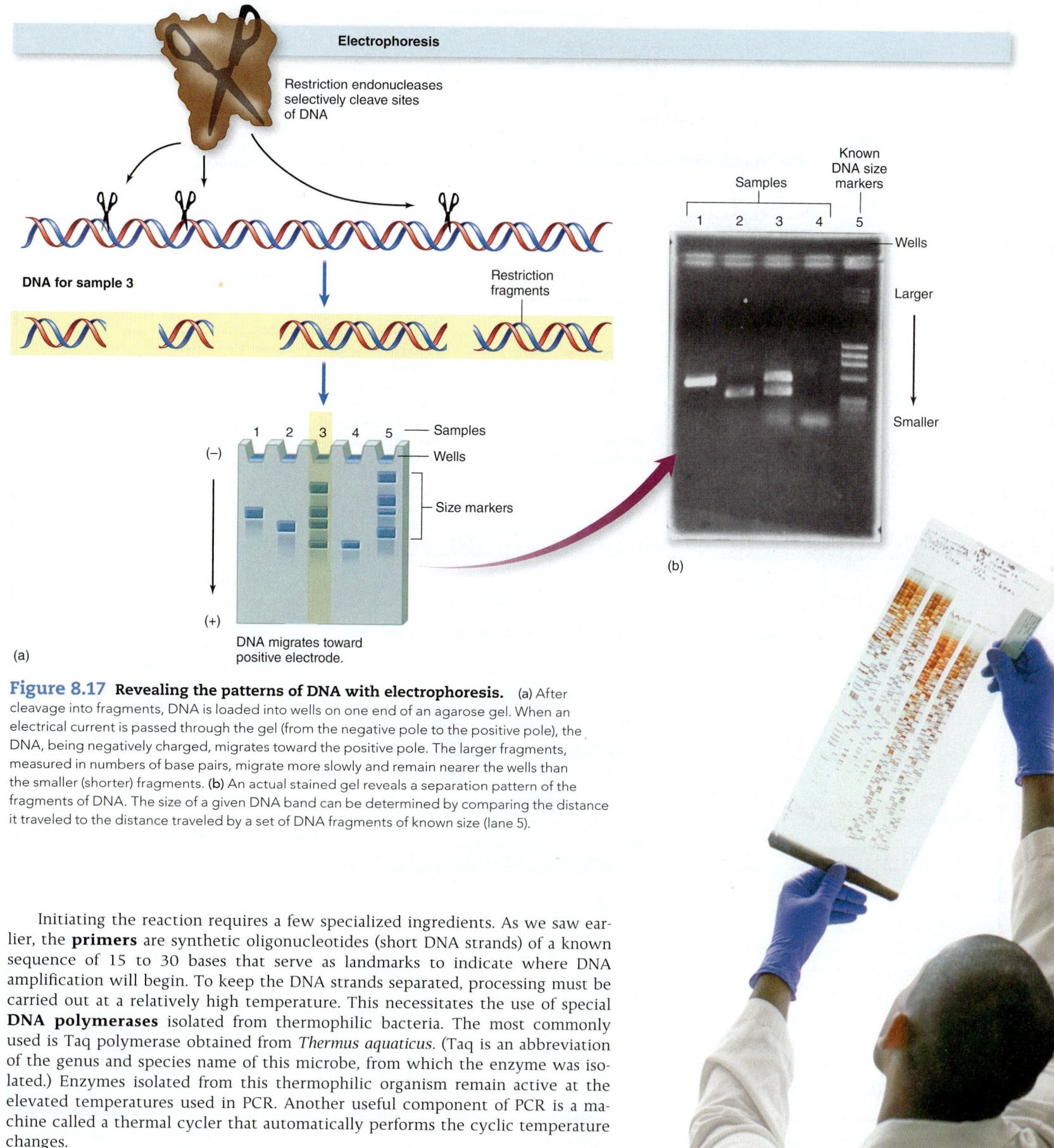

Figure 8.17 Revealing the patterns of DNA with electrophoresis.   (a) After cleavage into fragments, DNA is loaded into wells on one end of an agarose gel. When an electrical current is passed through the gel (from the negative pole to the positive pole), the DNA, being negatively charged, migrates toward the positive pole. The larger fragments, measured in numbers of base pairs, migrate more slowly and remain nearer the wells than the smaller (shorter) fragments. **(b)** An actual stained gel reveals a separation pattern of the fragments of DNA. The size of a given DNA band can be determined by comparing the distance it traveled to the distance traveled by a set of DNA fragments of known size (lane 5).

Initiating the reaction requires a few specialized ingredients. As we saw earlier, the **primers** are synthetic oligonucleotides (short DNA strands) of a known sequence of 15 to 30 bases that serve as landmarks to indicate where DNA amplification will begin. To keep the DNA strands separated, processing must be carried out at a relatively high temperature. This necessitates the use of special **DNA polymerases** isolated from thermophilic bacteria. The most commonly used is Taq polymerase obtained from *Thermus aquaticus*. (Taq is an abbreviation of the genus and species name of this microbe, from which the enzyme was isolated.) Enzymes isolated from this thermophilic organism remain active at the elevated temperatures used in PCR. Another useful component of PCR is a machine called a thermal cycler that automatically performs the cyclic temperature changes.

Table 8.9 Polymerase Chain Reaction

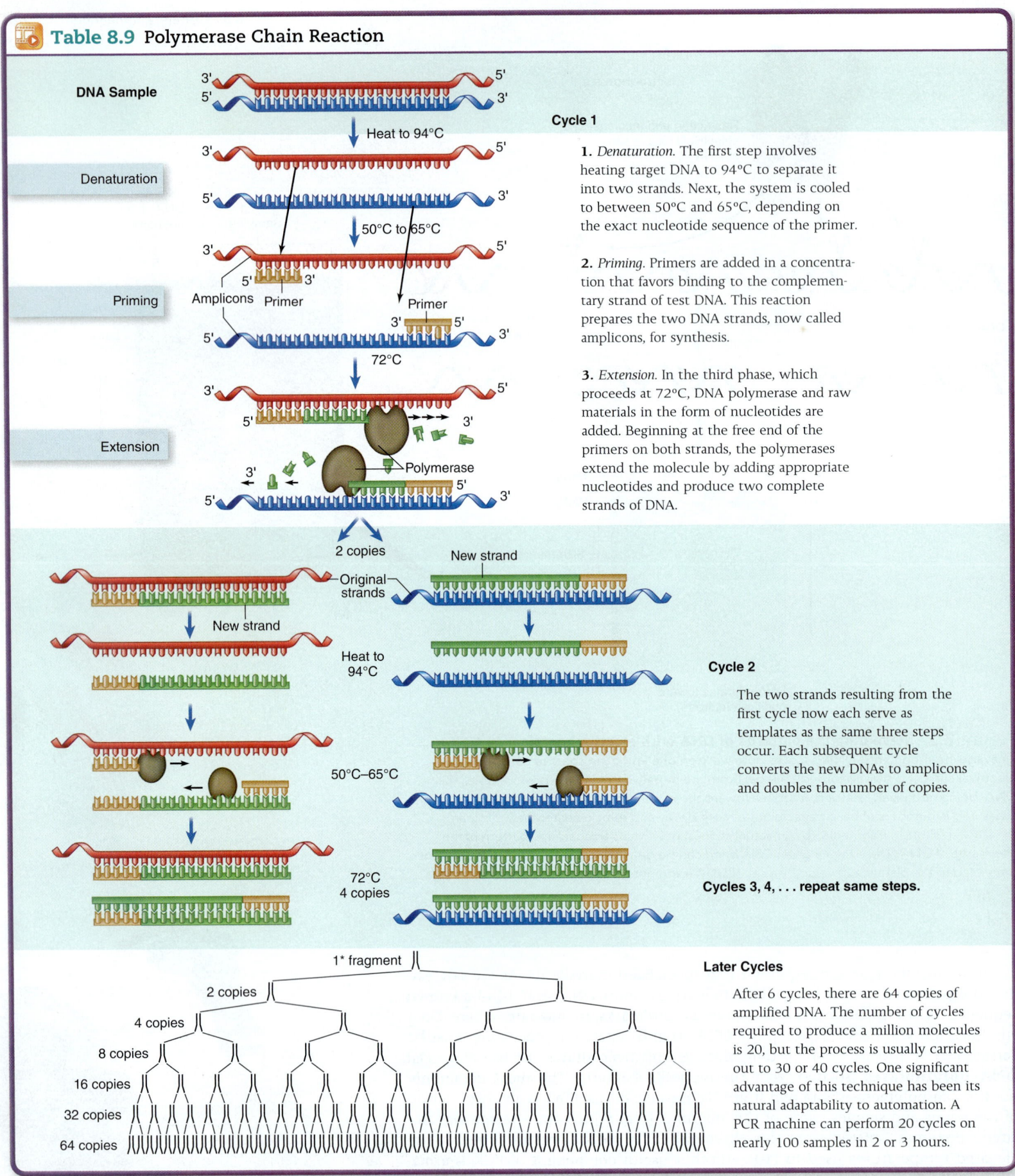

DNA Sample

Denaturation

Heat to 94°C

Priming

Amplicons   Primer   Primer

50°C to 65°C

Extension

Polymerase

72°C

2 copies

New strand

Original strands

New strand

Heat to 94°C

50°C–65°C

72°C
4 copies

1* fragment

2 copies

4 copies

8 copies

16 copies

32 copies

64 copies

Cycle 1

1. *Denaturation.* The first step involves heating target DNA to 94°C to separate it into two strands. Next, the system is cooled to between 50°C and 65°C, depending on the exact nucleotide sequence of the primer.

2. *Priming.* Primers are added in a concentration that favors binding to the complementary strand of test DNA. This reaction prepares the two DNA strands, now called amplicons, for synthesis.

3. *Extension.* In the third phase, which proceeds at 72°C, DNA polymerase and raw materials in the form of nucleotides are added. Beginning at the free end of the primers on both strands, the polymerases extend the molecule by adding appropriate nucleotides and produce two complete strands of DNA.

Cycle 2

The two strands resulting from the first cycle now each serve as templates as the same three steps occur. Each subsequent cycle converts the new DNAs to amplicons and doubles the number of copies.

Cycles 3, 4, . . . repeat same steps.

Later Cycles

After 6 cycles, there are 64 copies of amplified DNA. The number of cycles required to produce a million molecules is 20, but the process is usually carried out to 30 or 40 cycles. One significant advantage of this technique has been its natural adaptability to automation. A PCR machine can perform 20 cycles on nearly 100 samples in 2 or 3 hours.

*For simplicity's sake, we have omitted the elongation of the complete original parent strand during the first cycles. Ultimately, templates that correspond only to the smaller fragments dominate and become the primary population of replicated DNA.

The PCR technique operates by repetitive cycling of three basic steps: denaturation, priming, and extension.

Table 8.9 outlines the steps of PCR.

The polymerase chain reaction quickly became prominent as a powerful workhorse of molecular biology, medicine, and biotechnology. It is taking on an important role in diagnosis of infectious diseases, as well.

Recombinant DNA Technology

The primary intent of **recombinant DNA technology** is to deliberately remove genetic material from one organism and combine it with that of a different organism. Its origins can be traced to 1970, when microbiologists first began to duplicate the clever tricks bacteria do naturally with bits of extra DNA such as plasmids, transposons, and proviruses. As mentioned earlier, humans have been trying to artificially influence genetic transmission of traits for centuries. The discovery that bacteria can readily accept, replicate, and express foreign DNA made them powerful agents for studying the genes of other organisms in isolation. The practical applications of this work were soon realized by biotechnologists. Bacteria could be genetically engineered to mass produce substances such as hormones, enzymes, and vaccines that were difficult to synthesize by the usual industrial methods.

An important objective of this technique is to form genetic **clones.** Cloning involves the removal of a selected gene from an animal, plant, or microorganism (the genetic donor) followed by its propagation in a different host organism. Cloning requires that the desired donor gene first be selected, excised by restriction endonucleases, and isolated. The gene is next inserted into a **vector** (usually a plasmid or a virus) that will insert the DNA into a **cloning host.** The cloning host is usually a bacterium or a yeast that can replicate the gene and translate it into the protein product for which it codes. In the next section, we examine the elements of gene isolation, vectors, and cloning hosts and show how they participate in a complete recombinant DNA procedure.

Technical Aspects of Recombinant DNA and Gene Cloning

The first hurdles in cloning a target gene are to locate its exact site on the genetic donor's chromosome and to isolate it. Among the most common strategies for obtaining genes in an isolated state are the following:

1. The DNA is removed from cells and separated into fragments by endonucleases. Each fragment is then inserted into a vector and cloned. The cloned fragments are probed to identify desired sequences. This is a long and tedious process, because each fragment of DNA must be examined for the cloned gene.
2. A gene can be synthesized from isolated mRNA transcripts using reverse transcriptase. In this process, the reverse transcriptase creates DNA out of the mRNA.
3. A gene can be amplified using PCR in many cases.

Although gene cloning and isolation can be very laborious, a fortunate outcome is that, once isolated, genes can be maintained in a cloning host and vector just like a microbial pure culture. **Genomic libraries** are collections of DNA clones that represent the entire genome of numerous organisms.

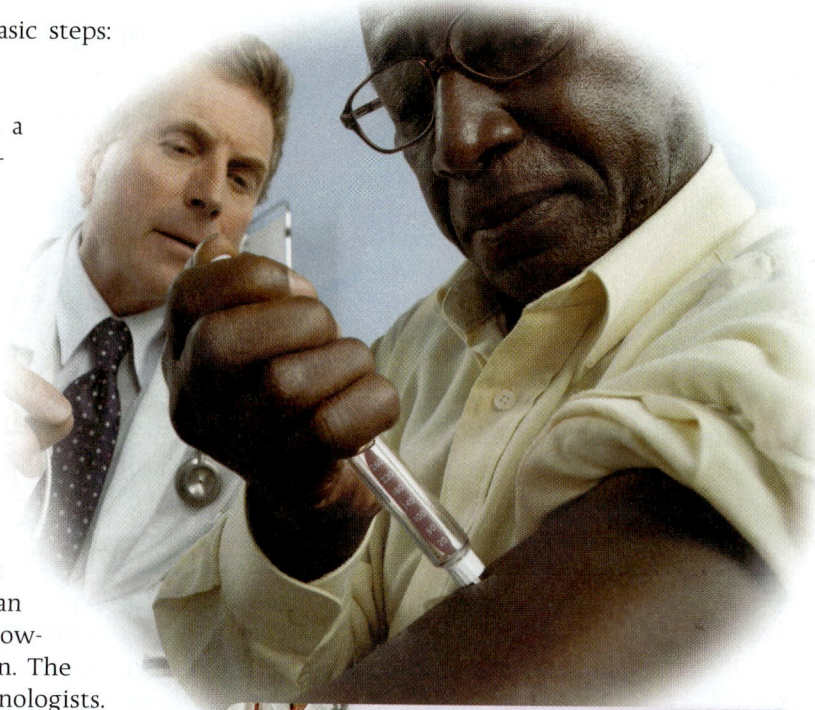

Medical Moment

Bactofection—Direct Gene Transfer

Bactofection is a technique that directly transfers genes into a tissue, organ, or organism using bacteria. In other words, transformed bacteria deliver genes that are located on plasmids into the target cells for the purpose of gene therapy. Bacteria that are nonpathogenic to humans are used for these purposes in order to avoid causing unwanted disease. Bacteria used for the purposes of bactofection are generally modified—for example, the bacteria may be altered so as not to induce an immune response. However, despite alteration of the bacteria, unwanted side effects do sometimes occur, including infection and autoimmune reactions.

The steps in bactofection are as follows:

1. Bacteria that have been transformed and contain plasmids carrying the transgene are introduced into the target host.
2. The genetically modified bacteria penetrate the target cells.
3. The bacteria undergo lysis or destruction in the host cell cytoplasm.
4. The released plasmids penetrate the target cell nucleus.
5. The therapeutic transgene is expressed via the eukaryotic processes of transcription and translation.

Bactofection has been studied in various disease models, including immune disease, cancer, and cystic fibrosis.

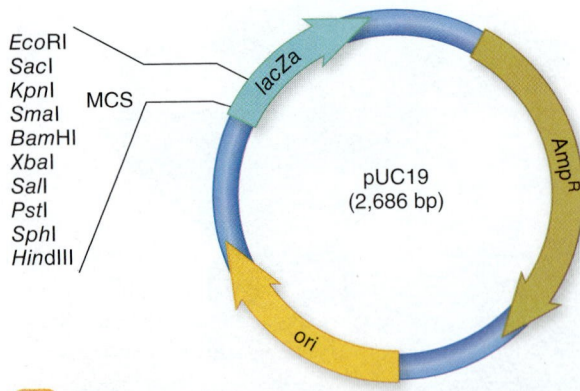

EcoRI
SacI
KpnI
SmaI   MCS
BamHI
XbaI
SalI
PstI
SphI
HindIII

lacZα

pUC19
(2,686 bp)

Amp^R

ori

Figure 8.18 **The cloning vector pUC19.**
The ampicillin-resistance gene is in tan.

Cloning Vectors

Genes in isolation are not easily manipulated in the lab. They are typically spliced into a cloning vector, using restriction enzymes. Plasmids are excellent vectors because they are small, well characterized, easy to manipulate, and can be transferred into appropriate host cells through transformation. Bacteriophages also serve well because they have the natural ability to inject DNA into bacterial hosts through transduction.

Vectors typically contain a gene that confers drug resistance to their cloning host. In this way, cells can be grown on drug-containing media, and only those cells that harbor a plasmid will be selected for growth **(figure 8.18).**

Construction of a Recombinant, Insertion into a Cloning Host, and Genetic Expression

Table 8.10 is a step-by-step guide to cloning a gene.

8.6 LEARNING OUTCOMES—Assess Your Progress

19. Explain the importance of restriction endonucleases to genetic engineering.
20. List the steps in the polymerase chain reaction.
21. Describe how you can clone a gene into a bacterium.

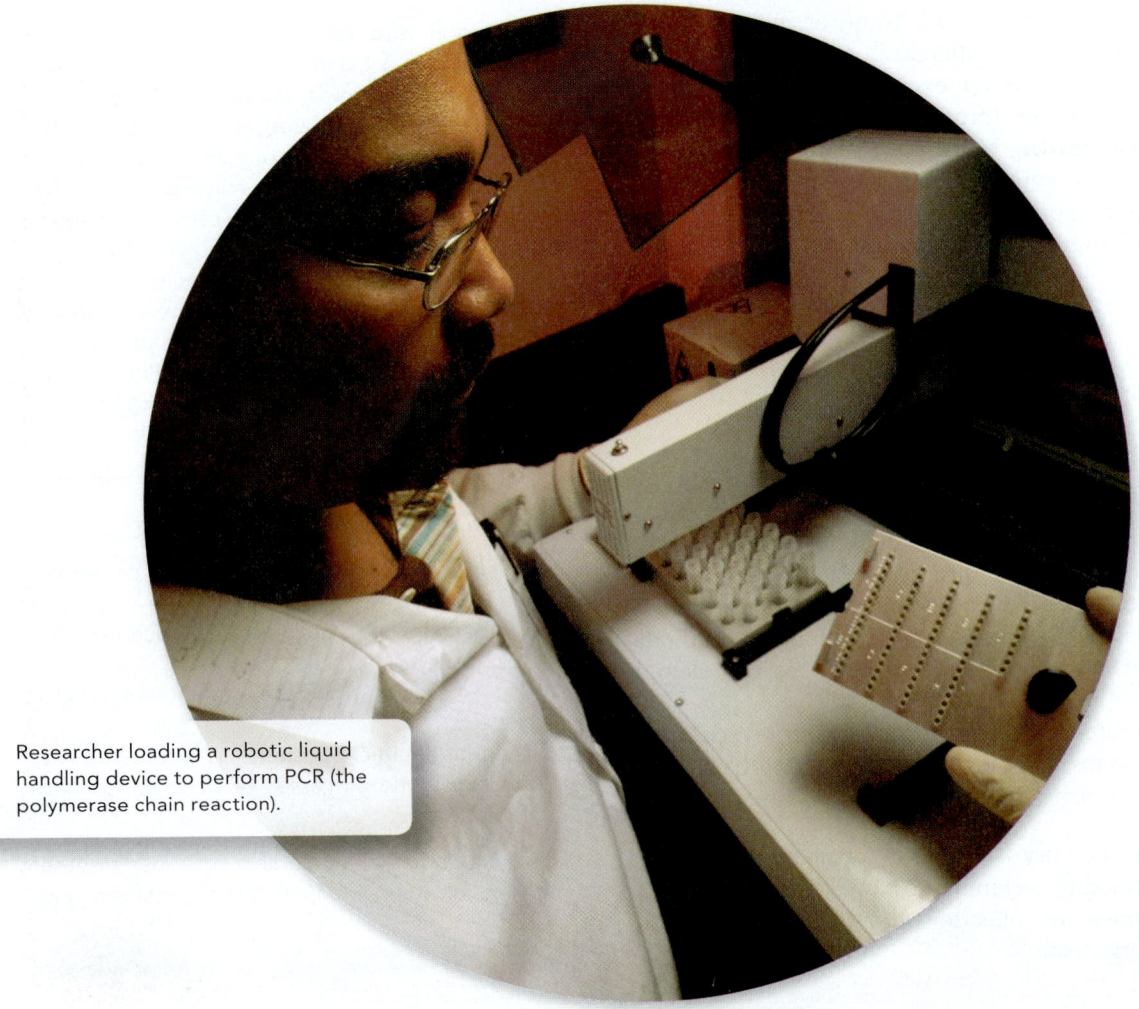

Researcher loading a robotic liquid handling device to perform PCR (the polymerase chain reaction).

Table 8.10 Gene Cloning

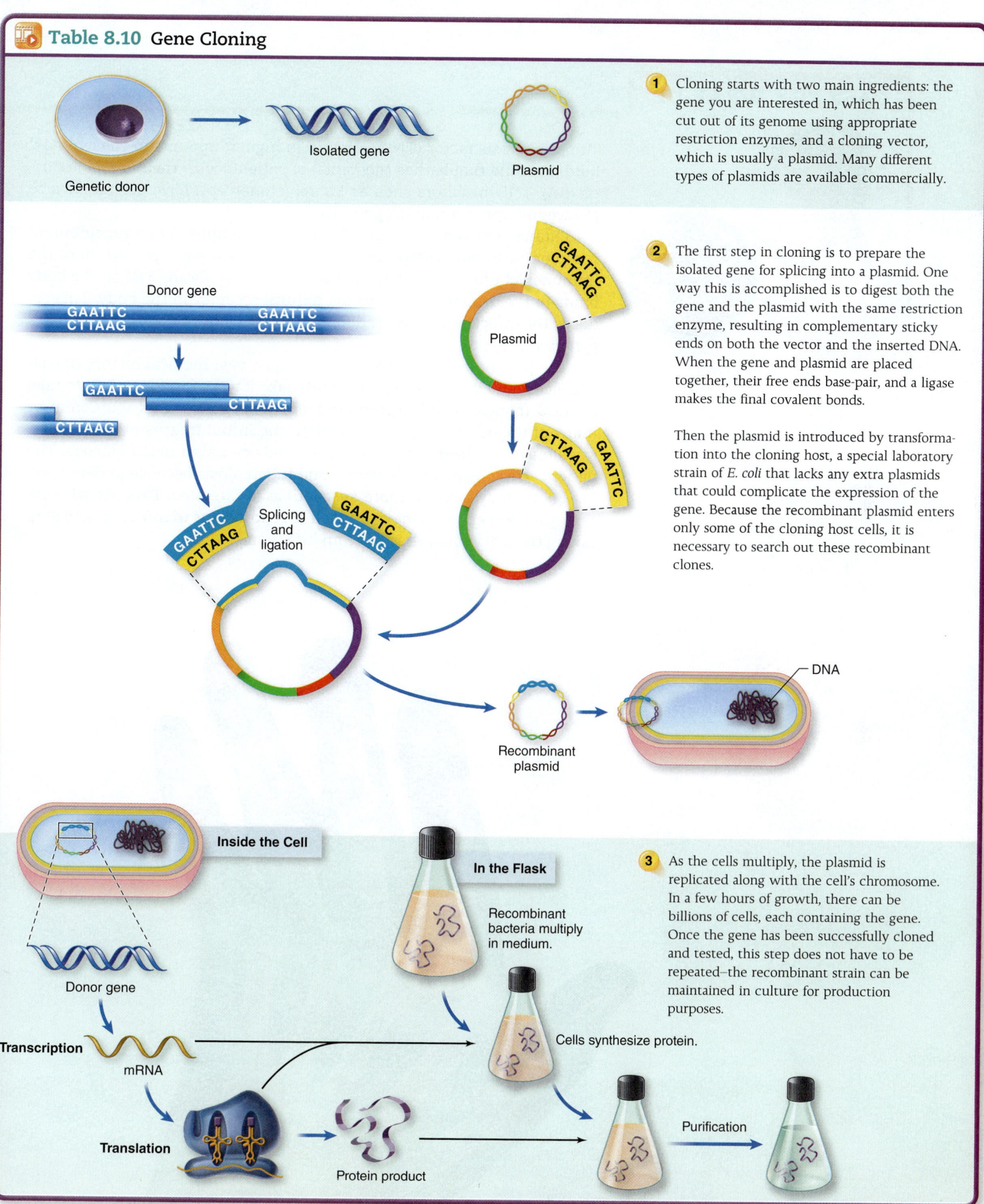

Genetic donor → Isolated gene   Plasmid

1 Cloning starts with two main ingredients: the gene you are interested in, which has been cut out of its genome using appropriate restriction enzymes, and a cloning vector, which is usually a plasmid. Many different types of plasmids are available commercially.

Donor gene

GAATTC   GAATTC
CTTAAG   CTTAAG

GAATTC
CTTAAG

GAATTC
CTTAAG

CTTAAG

GAATTC   GAATTC
CTTAAG   CTTAAG
        Splicing
        and
        ligation

GAATTC
CTTAAG

Plasmid

CTTAAG   GAATTC

2 The first step in cloning is to prepare the isolated gene for splicing into a plasmid. One way this is accomplished is to digest both the gene and the plasmid with the same restriction enzyme, resulting in complementary sticky ends on both the vector and the inserted DNA. When the gene and plasmid are placed together, their free ends base-pair, and a ligase makes the final covalent bonds.

Then the plasmid is introduced by transformation into the cloning host, a special laboratory strain of *E. coli* that lacks any extra plasmids that could complicate the expression of the gene. Because the recombinant plasmid enters only some of the cloning host cells, it is necessary to search out these recombinant clones.

Recombinant
plasmid

DNA

Inside the Cell

Donor gene

Transcription

mRNA

Translation

Protein product

In the Flask

Recombinant
bacteria multiply
in medium.

Cells synthesize protein.

Purification

3 As the cells multiply, the plasmid is replicated along with the cell's chromosome. In a few hours of growth, there can be billions of cells, each containing the gene. Once the gene has been successfully cloned and tested, this step does not have to be repeated—the recombinant strain can be maintained in culture for production purposes.

CASE FILE WRAP-UP

Tumor necrosis factor (TNF) is found in higher amounts in the synovial fluid (the fluid that bathes the joints) of patients with rheumatoid arthritis (RA). TNF inhibitors such as Enbrel (etanercept) have been shown to reduce inflammation in RA patients.

Entanercept is a pharmaceutical that is composed of a recombinant receptor for human tumor necrosis factor, fused to the Fc portion of the immunoglobulin IgG. How does it work? The circulating TNF in the body binds to the drug instead of to the natural receptor in the body. This reduces the effective concentration of TNF acting in the body and reduces symptoms due to TNF.

The patient in the case file opener had a past medical history of gastric ulcers and diabetes, which would make it risky for the patient to take NSAIDS (non-steroidal anti-inflammatory drugs, such as ibuprofen) or many of the other drugs that are often the initial treatment for RA. Steroids may have been effective, but steroids can alter blood glucose. The patient was screened for hepatitis and tuberculosis because patients taking Enbrel may become more susceptible to infection. Patients who are taking etanercept must be cautioned to report signs of infection and stop taking the drug when they have an infection.

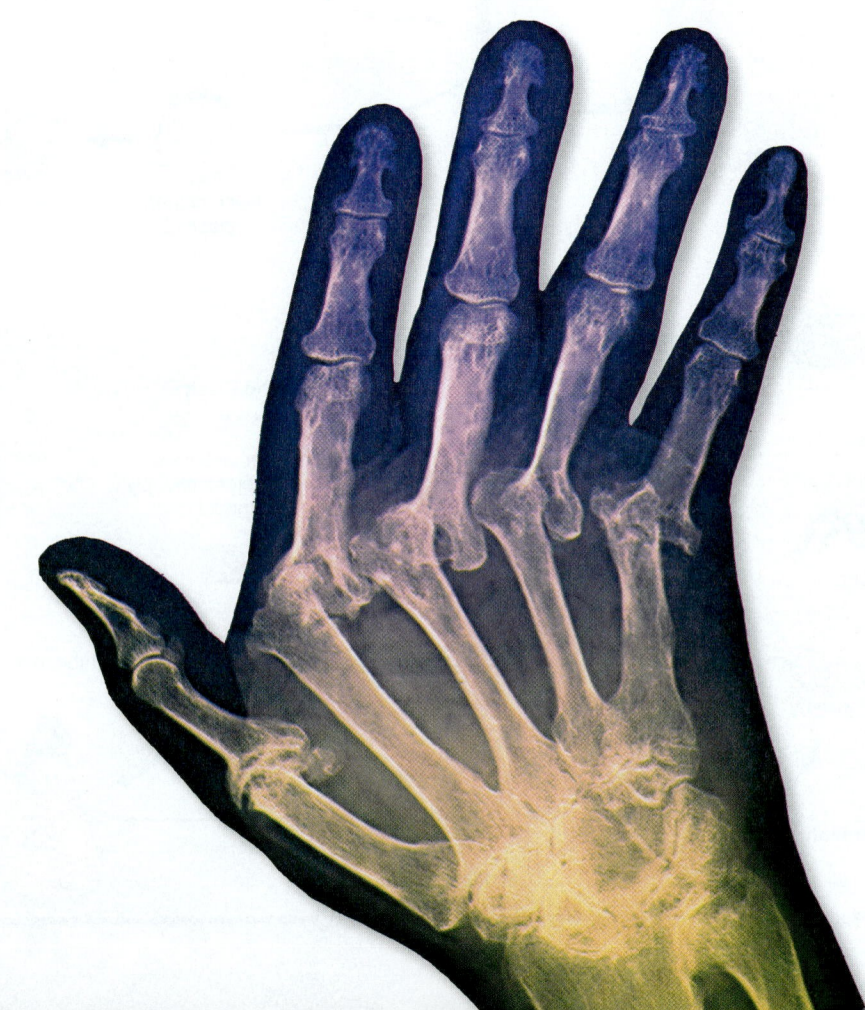

Using Recombinant DNA to Produce Insulin

Prior to January of 1922, when Canadian physican Frederick Banting and colleagues administered the first dose of insulin to a 14-year-old Canadian boy, people with diabetes were essentially doomed to die. The only treatment available was to restrict their diet so stringently that many diabetics died of malnutrition. Even on a strict diet entirely devoid of sugar and extremely low in carbohydrates, people with diabetes could expect to live a few years at best after diagnosis.

The first insulin was made by extracting the pancreas from cattle and pigs, grinding it up, and then purifying it. Unfortunately, insulin derived from animal sources caused reactions in some patients; there was also concern about the long-term effects of injecting a "foreign" substance.

In 1955, Frederick Sanger (who would later win the Nobel Prize for his work) studied insulin, eventually discovering that is was made of a specific sequence of amino acids. (It was the first protein to be sequenced.) Once insulin's exact protein sequence was known, it became possible to recreate it synthetically. It was a very important discovery and paved the way for many new technologies. There was a problem, however. Researchers could not produce enough of it at one time to be of much use.

In the 1970s, researchers found a way to synthesize human insulin in the laboratory in the large amounts needed. How did they accomplish this? They used recombinant DNA technology to "highjack" bacterial cells and force them to produce human insulin. Here's how it works:

1. The human chromosomal gene responsible for insulin production is isolated and then copied over and over so that there are plenty of insulin genes with which to work.
2. Restriction endonucleases are used to cut a plasmid open. (The plasmids come from microbes but are made available to purchase for purposes such as these.)
3. Once the plasmid ring has been opened, the human gene for insulin is inserted into the plasmid and it is closed up again. Thus, human insulin genes are "recombined" with the bacterial plasmid DNA. Many copies of the plasmid are constructed this way.
4. The plasmids, with the human insulin gene inside, are inserted into a suitable host bacterial (or fungal) species.
5. The bacteria are provided with everything they need to grow and multiply. Bacterial cell processes "turn on" the gene for human insulin and begin to produce human insulin within the bacterial cell. When the bacterial cells divide, the human insulin gene is also produced in the new cells.
6. The human insulin proteins produced by the bacteria are collected and purified.

The human insulin produced is not a protein from a different species, reducing any chance of autoimmune reaction. The human body does not distinguish between this synthetic insulin and natural insulin produced by the pancreas.

Both bacteria and yeast have been utilized to produce human insulin. Millions of people with diabetes now lead full lives thanks to the miracle of recombinant DNA technology.

A ribbon diagram of the insulin protein.

Chapter Summary

8.1 Introduction to Genetics and Genes

- Nucleic acids contain the blueprints of life in the form of genes. DNA is the blueprint molecule for all cellular organisms. The blueprints of viruses can be either DNA or RNA.
- The total amount of DNA in an organism is termed its genome (also genotype). Not all genes are expressed all the time; the ones that are expressed determine an organism's phenotype.
- Bacterial DNA consists of a few thousand genes in one circular chromosome. Eukaryotic genomes range from thousands to tens of thousands of genes.
- DNA copies itself just before cellular division by the process of semiconservative replication. Semiconservative replication means that each "old" DNA strand is the template upon which each "new" strand is synthesized.
- The circular bacterial chromosome is replicated at two forks as directed by DNA polymerase III. At each fork, two new strands are synthesized—one continuously and one in short fragments called Okazaki fragments.

8.2 Transcription and Translation

- Information in DNA is converted to proteins by the processes of transcription and translation.
- DNA also contains a great number of non-protein-coding sequences. These sequences are often transcribed into RNA that serves to regulate cell function.
- Eukaryotes transcribe DNA in the nucleus, remove its introns, and then translate it in the cytoplasm. Bacteria transcribe and translate simultaneously because the DNA is not sequestered in a nucleus and the bacterial DNA is free of introns.

8.3 Genetic Regulation of Protein Synthesis

- Operons are collections of genes in bacteria that code for products with a coordinated function.
- Nutrients can combine with regulator gene products to turn a set of structural genes on (inducible genes) or off (repressible genes). The *lac* (lactose) operon is an example of an inducible operon.

8.4 DNA Recombination Events

- Genetic recombination occurs in eukaryotes through sexual reproduction and through horizontal gene transfer.
- In bacteria, recombination occurs only through horizontal gene transfer.
- The three main types of horizontal gene transfer in bacteria are transformation, conjugation, and transduction.
- Transposons are genes that can relocate from one part of the genome to another, causing rearrangement of genetic material.

8.5 Mutations: Changes in the Genetic Code

- Changes in the genetic code can occur by two means: mutation and recombination.
- Mutations are changes in the nucleotide sequence of the organism's genome.
- Mutations can be either spontaneous or induced by exposure to some external mutagenic agent.
- All cells have enzymes that repair damaged DNA. When the degree of damage exceeds the ability of the enzymes to make repairs, mutations occur.

8.6 Genetic Engineering

- Genetic engineering utilizes a wide range of methods that physically manipulate DNA for purposes of visualization, sequencing, hybridizing, and identifying specific sequences.
- The tools of genetic engineering include restriction endonucleases, gel electrophoresis, and DNA sequencing.
- The polymerase chain reaction (PCR) technique amplifies small amounts of DNA into much larger quantities for further analysis.
- Cloning is the process by which genes are removed from the original host and duplicated for transfer into a cloning host by means of cloning vectors.

Multiple-Choice Questions   Bloom's Levels 1 and 2: Remember and Understand

Select the correct answer from the answers provided.

1. Which of the following is a characteristic of RNA?
   a. RNA is double stranded.
   b. RNA contains thymine, which pairs with adenine.
   c. RNA contains deoxyribose.
   d. RNA molecules are necessary for translation and gene regulation.
   e. All of the above are true.

2. Which of the following groups of start and stop codons is the complete and correct set?
   a. Start: AUG (f-Methionine). Stop: UGA.
   b. Start: AUG (f-Methionine). Stop: UAA, UAG, or UGA.
   c. Start: AUG (f-Methionine) or AUU (f-Isoleucine). Stop: UGA.
   d. Start: AUG (f-Methionine) or AUU (f-Isoleucine). Stop: UAA, UAG, or UGA.

3. DNA replication is semiconservative because the _____ strand will become half of the _____ molecule.
   a. RNA; DNA
   b. template; finished
   c. sense; mRNA
   d. codon; anticodon

4. In DNA, adenine is the complementary base for _____, and cytosine is the complement for _____.

   a. guanine; thymine   c. thymine; guanine

   b. uracil; guanine   d. thymine; uracil

5. As a general rule, during transcription, the template strand on DNA will always begin with

   a. TAC.   c. ATG.

   b. AUG.   d. UAC.

6. The *lac* operon is usually in the _____ position and is activated by a/an _____ molecule.

   a. on; repressor   c. on; inducer

   b. off; inducer   d. off; repressor

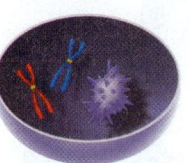

7. Which of the following is *not* a mechanism of horizontal gene transfer?

   a. spontaneous mutations

   b. transformation

   c. transduction

   d. conjugation

8. When genes are turned on differently under different environmental conditions, this represents a change in

   a. species.

   b. genotype.

   c. phenotype.

   d. growth rate.

Critical Thinking   Bloom's Levels 3, 4, and 5: Apply, Analyze, and Evaluate

Critical thinking is the ability to reason and solve problems using facts and concepts. These questions can be approached from a number of angles and, in most cases, they do not have a single correct answer.

1. Describe what is meant by the antiparallel arrangement of DNA.

2. On paper, replicate the following segment of DNA:

   5′ A T C G G C T A C G T T C A C 3′

   3′ T A G C C G A T G C A A G T G 5′

   a. Show the direction of replication of the new strands and explain what the lagging and leading strands are.

   b. Explain how this is semiconservative replication. Are the new strands identical to the original segment of DNA?

3. The following sequence represents triplets on DNA:

   TAC CAG ATA CAC TCC CCT GCG ACT

   a. Give the mRNA codons and tRNA anticodons that correspond with this sequence, and then give the sequence of amino acids in the polypeptide.

   b. Provide another mRNA strand that can be used to synthesize this same protein.

4. Using the piece of DNA in question 3, show a deletion, an insertion, a substitution, and nonsense mutations. Which ones are frameshift mutations? Are any of your mutations nonsense? Missense? (Use the universal code to determine this.)

5. a. If gene probes, profiling, and mapping could make it possible for you to know of future genetic diseases in you or one of your children, would you wish to use this technology to find out?

   b. Explain the risks and benefits if these technologies were used to gather information on employees. How would it make you feel if you were such an employee?

   c. Explain the risks and benefits if these technologies were used by health insurance companies to gather information on beneficiaries. How would it make you feel if you were one of those benficiaries?

Visual Connections   Bloom's Level 5: Evaluate

This question connects previous images to a new concept.

1. **From chapter 3, figure 3.10a.** Speculate on why these cells contain two chromosomes (shown in blue).

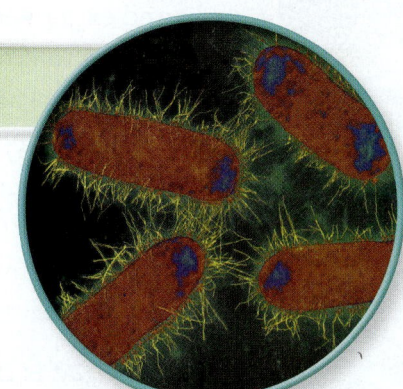

www.mcgrawhillconnect.com

Enhance your study of this chapter with study tools and practice tests. Also ask your instructor about the resources available through Connect, including the media-rich eBook, interactive learning tools, and animations.

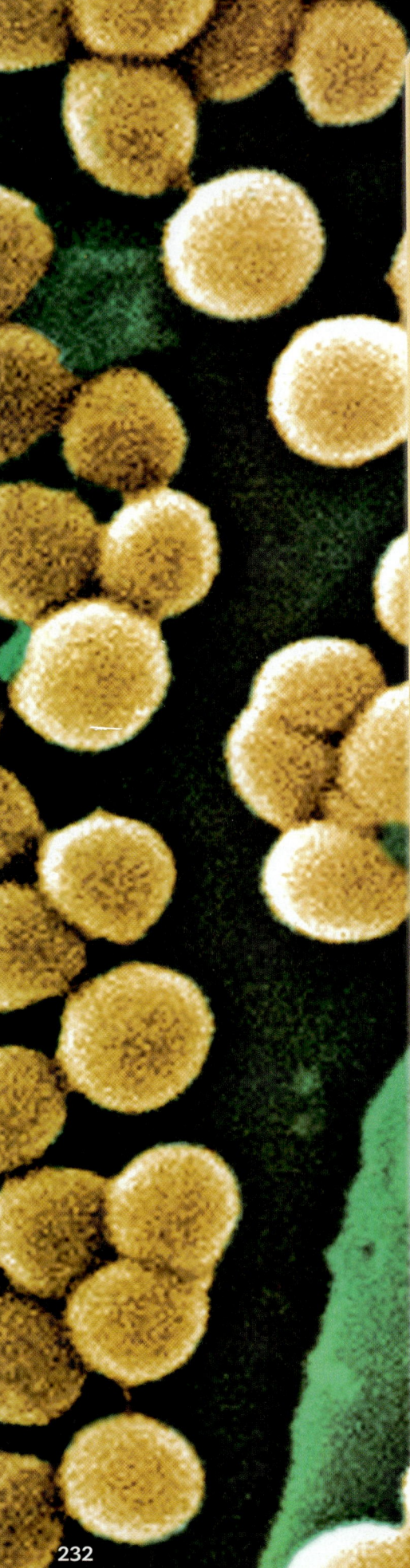

Preparing the Skin

I was just starting to work in the emergency room of a hospital located in a rural community where farming was one of the mainstays of the local economy. I learned that during the spring we would see many patients who had injured themselves in their fields during seeding, suffering lacerations of their fingers, hands, arms, and legs. Many of these injuries were sustained while working with various types of machinery, so in addition to a tetanus shot, most of these lacerations required suturing.

My job as an LPN was to assist the physician during suturing. I would first set up the sterile suture tray while the physician washed his hands and donned sterile gloves. The physician would tell me what type of suture material to add to the tray, what he or she wanted to use as an anesthetic, and what he or she wanted to use to clean the wound prior to suturing.

The first time I assisted with suturing, the physician requested that I add povidone iodine (Betadine) to the small sterile cup on the suture tray, to be used to thoroughly cleanse the wound before suturing took place. The next time we had a patient requiring suturing, I was working with a different physician and automatically added Betadine to the cup, assuming that all of the physicians would use the same agent as a skin disinfectant. Needless to say, the physician was unhappy and stated that he only ever used sterile saline solution to cleanse the wound. I quickly switched out the Betadine for sterile saline.

When we were finished with the suturing, I sought out my preceptor and explained what had happened. She felt badly that she had neglected to tell me that many of the physicians had different preferences for cleansing solutions. I asked her why the second physician only used sterile saline solution to cleanse his patient's wound, as it seemed to me that Betadine would do a better job of killing any bacteria in the wound. She told me that habit often trumped best practice. She asked me to go online on the computer at the nursing station and see what I could find in support of using normal saline versus Betadine in wound cleansing.

- What is the mechanism of action of Betadine on bacteria?

- Why might the second physician have chosen sterile normal saline over Betadine to cleanse the patient's wound?

Case File Wrap-Up appears on page 254.

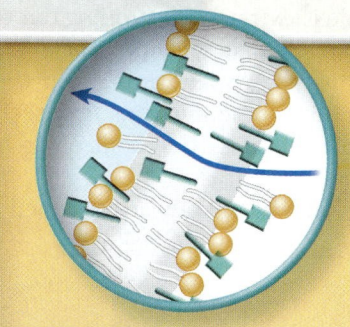

Physical and Chemical Control of Microbes

IN THIS CHAPTER...

9.1 Controlling Microorganisms

1. Clearly define the terms *sterilization*, *disinfection*, *decontamination*, *sanitization*, *antisepsis*, and *degermation*.
2. Identify the microorganisms that are most resistant and least resistant to control measures.
3. Compare the action of microbicidal and microbistatic agents, providing an example of each.
4. Name four categories of cellular targets for physical and chemical agents.

9.2 Methods of Physical Control

5. Name six methods of physical control of microorganisms.
6. Discuss both moist and dry heat methods, and identify multiple examples of each.
7. Define *thermal death time* and *thermal death point*.
8. Explain methods of moist heat control.
9. Explain two methods of dry heat control.
10. Identify advantages and disadvantages of cold and desiccation.
11. Differentiate between the two types of radiation control methods.
12. Explain how filtration and osmotic pressure function as control methods.

9.3 Chemical Agents in Microbial Control

13. Name the desirable characteristics of chemical control agents.
14. Discuss chlorine and iodine and their uses.
15. List advantages and disadvantages to phenolic compounds.
16. Explain the mode of action of chlorhexidine.
17. Explain the applications of hydrogen peroxide agents.
18. Identify some heavy metal control agents.
19. Discuss the disadvantages of aldehyde agents.
20. Identify applications for ethylene oxide sterilization.

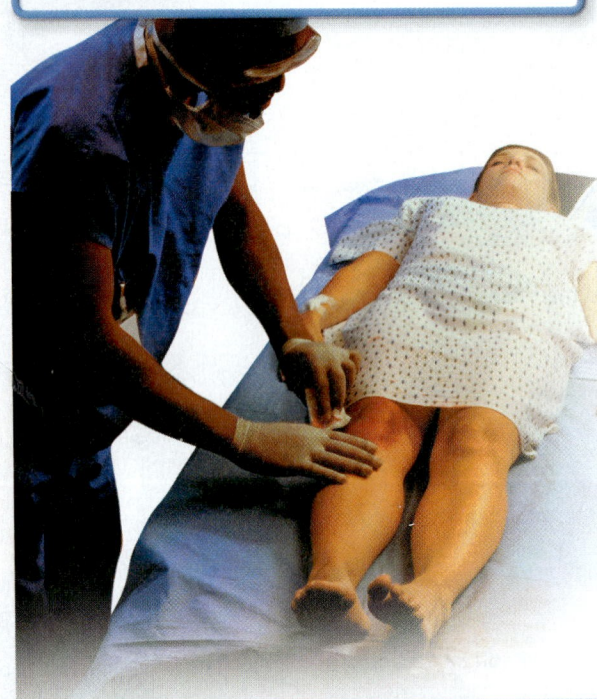

9.1 Controlling Microorganisms

Much of the time in the developed world, we take for granted tap water that is drinkable, food that is not spoiled, shelves full of products to eradicate "germs," and drugs to treat infections. Controlling our degree of exposure to potentially harmful microbes is a monumental concern in our lives. A prime example of this occurred when letters containing endospores of *Bacillus anthracis* (anthrax) were delivered and opened in the Hart Office Building of the U.S. Senate in 2001. The EPA (Environmental Protection Agency) had to determine a method of removing all traces of a lethal, highly infectious, endospore-forming bacterium from an enormous space, which presented a huge challenge. Although this example is extreme, controlling our exposure to harmful microbes is always an ongoing concern. The ancient Greeks learned to burn corpses and clothing during epidemics; the Egyptians embalmed the bodies of their dead, using strong salts and pungent oils. These methods may seem rather archaic by modern measures, but these examples illustrate that controlling microbes has been a concern for several centuries.

The methods of microbial control used outside of the body are designed to result in four possible outcomes:

1. sterilization,
2. disinfection,
3. decontamination (also called sanitization), or
4. antisepsis (also called degermation).

These terms are differentiated in **table 9.1.** While it may seem clumsy to have more than one word for some of these processes, it is important that you recognize them when you hear them used so we include them here. To complicate matters, the everyday use of some of these terms can at times be vague and inexact. For example, occasionally one may be directed to "sterilize" or "disinfect" a patient's skin, even though this usage does not fit the technical definition of either term. We also provide a flowchart **(figure 9.1)** to summarize the major applications and aims in microbial control. NOTE: In this table and the others in this chapter,

Table 9.1 Concepts in Antimicrobial Control

Techniques and chemicals that are capable of sterilizing are highlighted with a pink background.

Term	Definition	Key Points	Examples of Agents
Sterilization	Process that destroys or removes all viable microorganisms (including viruses)	The term *sterile* should be used only in the strictest sense to refer to materials that have been subjected to the process of sterilization (there is no such thing as slightly sterile). Generally reserved for inanimate objects as it would be impractical or dangerous to sterilize parts of the human body Common uses: surgical instruments, syringes, commercially packaged food	Heat (autoclave) Sterilants (chemical agents capable of destroying endospores)
Disinfection	Physical process or a chemical agent to destroy vegetative pathogens but not bacterial endospores Removes harmful products of microorganisms (toxins) from material	Normally used on inanimate objects because the concentration of disinfectants required to be effective is harmful to human tissue Common uses: boiling food utensils, applying 5% bleach solution to an examining table, immersing thermometers in an iodine solution between uses	Bleach Iodine Heat (boiling)
Decontamination/ Sanitization	Cleansing technique that mechanically removes microorganisms as well as other debris to reduce contamination to safe levels	Important to restaurants, dairies, breweries, and other commercial entities that handle large numbers of soiled utensils/containers Common uses: Cooking utensils, dishes, bottles, and cans must be sanitized for reuse.	Soaps Detergents Commercial dishwashers
Antisepsis/ Degermation	Reduces the number of microbes on the human skin A form of decontamination but on living tissues	Involves scrubbing the skin (mechanical friction) or immersing it in chemicals (or both)	Alcohol Surgical hand scrubs

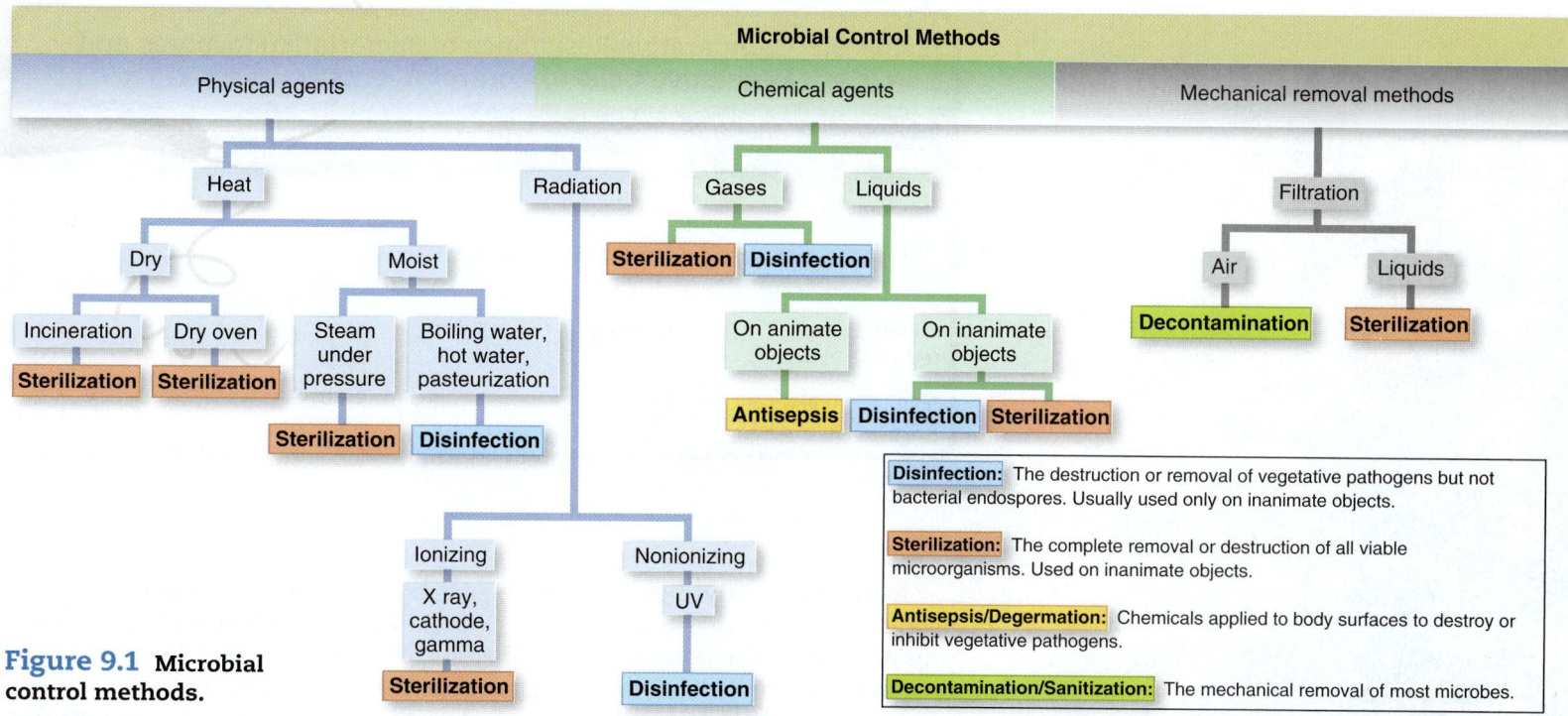

Figure 9.1 Microbial control methods.

Disinfection: The destruction or removal of vegetative pathogens but not bacterial endospores. Usually used only on inanimate objects.

Sterilization: The complete removal or destruction of all viable microorganisms. Used on inanimate objects.

Antisepsis/Degermation: Chemicals applied to body surfaces to destroy or inhibit vegetative pathogens.

Decontamination/Sanitization: The mechanical removal of most microbes.

techniques and chemicals that are capable of sterilizing are highlighted with a pink background.

Relative Resistance of Microbial Forms

The primary targets of microbial control are microorganisms capable of causing infection or spoilage that are constantly present in the external environment and on the human body. This targeted population is rarely simple or uniform; in fact, it often contains mixtures of microbes with extreme differences in resistance and harmfulness. **Figure 9.2** compares the general resistance these forms have to physical and chemical methods of control.

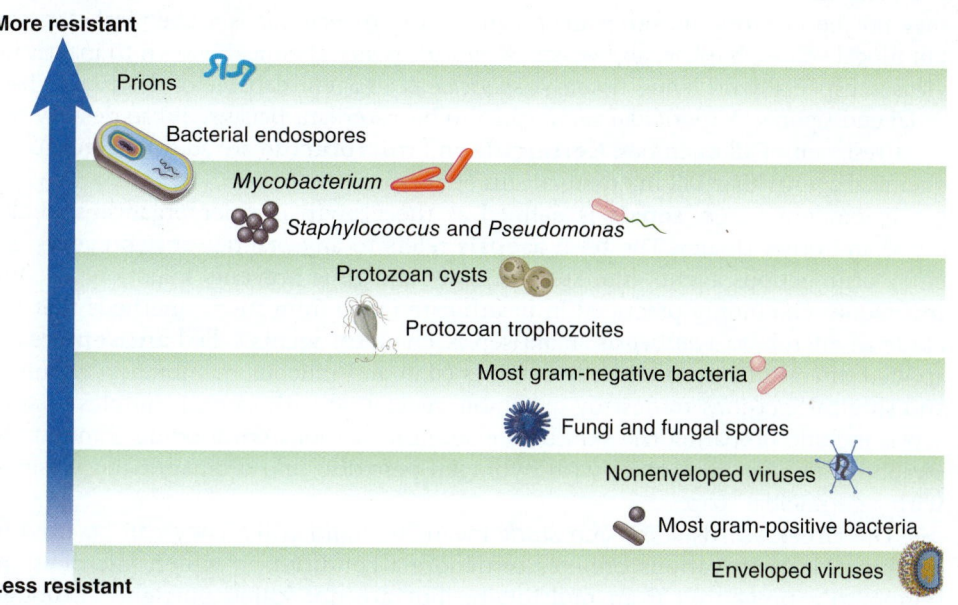

Figure 9.2 Relative resistance of different microbial types to microbial control agents. This is a very general hierarchy; different control agents are more or less effective against the various microbes.

A Note About Prions

Prions are in a class of their own when it comes to "sterilization" procedures. This chapter defines "sterile" as the absence of all viable microbial life—but none of the procedures described in this chapter are necessarily sufficient to destroy prions. Prions are extraordinarily resistant to heat and chemicals. If instruments or other objects become contaminated with these unique agents, they must either be discarded as biohazards or, if this is not possible, a combination of chemicals and heat must be applied in accordance with the Centers for Disease Control and Prevention (CDC) guidelines. The guidelines themselves are constantly evolving as new information becomes available. In the meantime, this chapter discusses sterilization using bacterial endospores as the toughest form of microbial life. When tissues, fluids, or instruments are suspected of containing prions, consultation with infection control experts and/or the CDC is recommended when determining effective sterilization conditions. Chapter 17 describes prions in detail.

Table 9.2 Comparative Resistance of Bacterial Endospores and Vegetative Cells to Control Agents

Method	Endospores	Vegetative Forms	Endospores Are ___× More Resistant
Heat (moist)	120°C	80°C	1.53
Radiation (X-ray) dosage	4,000 Grays	1,000 Grays	43
Sterilizing gas (ethylene oxide)	1,200 mg/L	700 mg/L	1.73
Sporicidal liquid (2% glutaraldehyde)	3 h	10 min	183

Actual comparative figures on the requirements for destroying various groups of microorganisms are shown in **table 9.2.** Bacterial endospores have traditionally been considered the most resistant microbial entities, being as much as 18 times harder to destroy than their counterpart vegetative cells. Because of their resistance to microbial control, their destruction is the goal of *sterilization* because any process that kills endospores will invariably kill all less resistant microbial forms. Other methods of control (disinfection, antisepsis) act primarily upon microbes that are less hardy than endospores.

Agents Versus Processes

The terms *sterilization, disinfection,* and so on refer to processes. You will encounter other terms that describe the agents used in the process. Two examples of these are the terms *bactericidal* and *bacteristatic.* The root *-cide,* meaning "having the capacity to kill," can be combined with other terms to define an antimicrobial agent aimed at destroying a certain group of microorganisms. For example, a **bactericide** is a chemical that destroys bacteria except for those in the endospore stage. It may or may not be effective on other microbial groups. A *fungicide* is a chemical that can kill fungal spores, hyphae, and yeasts. A *virucide* is any chemical known to inactivate viruses, especially on living tissue. A *sporicide* is an agent capable of destroying bacterial endospores. A sporicidal agent can also be a sterilant because it can destroy the most resistant of all microbes. **Germicide** and **microbicide** are additional terms for chemical agents that kill microorganisms.

In modern usage, **sepsis** is defined as the growth of microorganisms in the blood and other tissues. The term **asepsis** refers to any practice that prevents the entry of infectious agents into sterile tissues and thus prevents infection. Aseptic techniques commonly practiced in health care range from sterile methods that exclude all microbes to *antisepsis.* In antisepsis, chemical agents called **antiseptics** are applied directly to exposed body surfaces (skin and mucous membranes), wounds, and surgical incisions to destroy or inhibit vegetative pathogens. Examples of antisepsis include preparing the skin before surgical incisions with iodine compounds, swabbing an open root canal with hydrogen peroxide, and ordinary hand washing with a germicidal soap.

The Greek words *stasis* and *static* mean "to stand still." They can be used in combination with various prefixes to denote a condition in which microbes are temporarily prevented from multiplying but are not killed outright. Although killing or permanently inactivating microorganisms is the usual goal of microbial control, microbistasis does have meaningful applications. **Bacteristatic** agents prevent the growth of bacteria on tissues or on objects in the environment, and *fungistatic* chemicals inhibit fungal growth. Materials used to control microorganisms in the body (antiseptics and drugs) often have **microbistatic** effects because many microbicidal compounds can be highly toxic to human cells. Note that even a *-cidal* agent doesn't necessarily result in sterilization, depending on how it is used.

Practical Matters in Microbial Control

Numerous considerations govern the selection of a workable method of microbial control. One useful framework for determining how devices that come in contact with patients should be handled is whether they are considered *critical*, *semicritical*, or *noncritical*. *Critical* medical devices are those that are expected to come into contact with sterile tissues. A good example of this would be a syringe needle or an artificial hip. These must be sterilized before use. *Semicritical* devices are those that come into contact with mucosal membranes. An endoscopy tube is an example. These must receive at least high-level disinfection and preferably should be sterilized. *Noncritical* items are those that do not touch the patient or are only expected to touch intact skin, such as blood pressure cuffs or crutches. They require only low-level disinfection unless they become contaminated with blood or body fluids.

A remarkable variety of substances can require sterilization. They range from durable solids such as rubber to sensitive liquids such as serum, and even to entire office buildings, as seen in 2001 when the Hart Senate Office Building was contaminated with *Bacillus anthracis* endospores. Hundreds of situations requiring sterilization confront the network of persons involved in health care, whether technician, nurse, doctor, or manufacturer, and no universal method works well in every case.

Considerations such as cost, effectiveness, and method of disposal are all important. For example, disposable plastic items such as catheters and syringes that are used in invasive medical procedures have the potential for infecting the tissues. These must be sterilized during manufacture by a nonheating method (gas or radiation), because heat can damage plastics. After these items have been used, it is often necessary to destroy or decontaminate them before they are discarded because of the potential risk to the handler. Steam sterilization, which is quick and sure, is a sensible choice at this point, because it does not matter if the plastic is destroyed.

What Is Microbial Death?

Death is a phenomenon that involves the permanent termination of an organism's vital processes. Signs of life in complex organisms such as animals are self-evident, and death is made clear by loss of nervous function, respiration, or heartbeat. In contrast, death in microscopic organisms that are composed of just one or a few cells is often hard to detect, because they reveal no conspicuous vital signs to begin with.

At present, the most practical way to detect this damage is to determine if a microbial cell can still reproduce when exposed to a suitable environment. If the microbe has sustained metabolic or structural damage to such an extent that it can no longer reproduce, even under ideal environmental conditions, then it is no longer viable. The permanent loss of reproductive capability, even under optimum growth conditions, has become the accepted microbiological definition of death.

Factors Affecting Death Rate

The cells of a culture show marked variation in susceptibility to a given microbicidal agent. Death of the whole population is not instantaneous but begins when a certain threshold of microbicidal agent (some combination of time and concentration) is met. Death continues in a logarithmic manner as the time or concentration of the agent is increased.

Because many microbicidal agents target the cell's metabolic processes, active cells (younger, rapidly dividing) tend to die more quickly than those that are less metabolically active (older, inactive). Eventually, a point is reached at which survival of any cells is highly unlikely; this point is equivalent to sterilization.

The effectiveness of a particular agent is governed by several factors besides time. These additional factors influence the action of antimicrobial agents:

1. The number of microorganisms. A higher load of contaminants requires more time to destroy.
2. The nature of the microorganisms in the population. In most actual circumstances of disinfection and sterilization, the target population is not a single species of microbe but a mixture of bacteria, fungi, endospores, and viruses, presenting a broad spectrum of microbial resistance.
3. The temperature and pH of the environment.
4. The concentration (dosage, intensity) of the agent. For example, ultraviolet (UV) radiation is most effective at 260 nanometers (nm), and most disinfectants are more active at higher concentrations.
5. The mode of action of the agent. How does it kill or inhibit the microorganism?
6. The presence of solvents, interfering organic matter, and inhibitors. Saliva, blood, and feces can inhibit the actions of disinfectants and even of heat.

The influence of these factors is discussed in greater detail in subsequent sections.

Modes of Action of Antimicrobial Agents

An antimicrobial agent's adverse effect on cells is known as its *mode* (or *mechanism*) *of action.* Agents affect one or more cellular targets, inflicting damage progressively until the cell is no longer able to survive. Antimicrobials have a range of cellular targets, with the agents that are least selective in their targeting tending to be effective against the widest range of microbes (examples include heat and radiation). More selective agents (drugs, for example) tend to target only a single cellular component and are much more restricted as to the microbes they are effective against.

The cellular targets of physical and chemical agents fall into four general categories:

1. the cell wall,
2. the cytoplasmic membrane,
3. cellular synthetic processes (DNA, RNA), and
4. proteins.

Table 9.3 depicts the effects of various agents on cellular structures and processes. **Figure 9.3** illustrates what happens to a membrane when it is exposed to surfactants.

Table 9.3	Actions of Various Physical and Chemical Agents upon the Cell	
Cellular Target	**Effects of Agents**	**Examples of Agents Used**
Cell wall	Chemical agents can damage the cell wall by <ul><li>blocking its synthesis, or</li><li>digesting the cell wall.</li></ul>	Chemicals Detergents Alcohol
Cytoplasmic membrane	Agents physically bind to lipid layer of the cytoplasmic membrane, opening up the cytoplasmic membrane and allowing injurious chemicals to enter the cell and important ions to exit the cell.	Detergents
Cellular synthesis	Agents can interrupt the synthesis of proteins via the ribosomes, inhibiting proteins needed for growth and metabolism and preventing multiplication. Agents can change genetic codes (mutation).	Formaldehyde Radiation Ethylene oxide
Proteins	Some agents are capable of denaturing proteins (breaking of protein bonds, which results in breakdown of the protein structure). Agents may attach to the active site of a protein, preventing it from interacting with its chemical substrate.	Moist heat Alcohol Phenolics

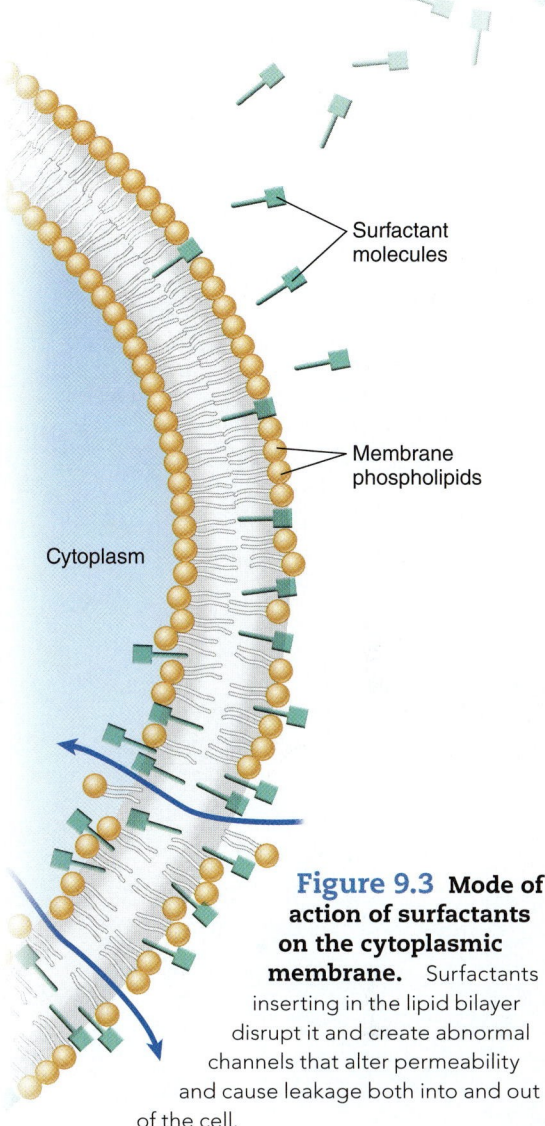

Figure 9.3 Mode of action of surfactants on the cytoplasmic membrane. Surfactants inserting in the lipid bilayer disrupt it and create abnormal channels that alter permeability and cause leakage both into and out of the cell.

Labels: Surfactant molecules; Membrane phospholipids; Cytoplasm

9.1 LEARNING OUTCOMES—Assess Your Progress

1. Clearly define the terms *sterilization, disinfection, decontamination, sanitization, antisepsis,* and *degermation.*
2. Identify the microorganisms that are most resistant and least resistant to control measures.
3. Compare the action of microbicidal and microbistatic agents, providing an example of each.
4. Name four categories of cellular targets for physical and chemical agents.

9.2 Methods of Physical Control

We can divide our methods of controlling microorganisms into two broad categories: physical and chemical. We'll start with physical methods. Microorganisms have adapted to the tremendous diversity of habitats the earth provides, even severe conditions of temperature, moisture, pressure, and light. For microbes that normally withstand such extreme physical conditions, our attempts at control would probably have little effect. Fortunately for us, we are most interested in controlling microbes that flourish in the same environment in which humans live. The vast majority of these microbes are readily controlled by abrupt changes in their environment. Most prominent among antimicrobial physical agents is heat. Other less widely used agents include radiation, filtration, ultrasonic waves, and even cold. The following sections examine some of these methods and explore their practical applications in medicine, commerce, and the home.

Heat

As a rule, elevated temperatures (exceeding the maximum growth temperature) are microbicidal, whereas lower temperatures (below the minimum growth temperature) are microbistatic. We'll start with heat. Heat can be applied in either moist or dry forms. *Moist heat* occurs in the form of hot water, boiling water, or steam (vaporized water). In practice, the temperature of moist heat usually ranges from 60°C to 135°C. As we shall see, the temperature of steam can be regulated by adjusting its pressure in a closed container. *Dry heat* refers to hot air (such as in an oven) or an open flame. In practice, the temperature of dry heat ranges from 160°C to several thousand degrees Celsius.

Mode of Action and Relative Effectiveness of Heat

Moist heat and dry heat differ in their modes of action as well as in their efficiency. Moist heat operates at lower temperatures and shorter exposure times to achieve the same effectiveness as dry heat **(table 9.4)**. Although many cellular structures are damaged by moist heat, its most microbicidal effect is the coagulation and denaturation of proteins, which quickly and permanently halts cellular metabolism.

Dry heat dehydrates the cell, removing the water necessary for metabolic reactions, and it also denatures proteins. However, the lack of water actually increases the stability of some protein conformations, necessitating the use of higher temperatures when dry heat is employed as a method of microbial control. At very high temperatures, dry heat oxidizes cells, burning them to ashes. This method is the one used in the laboratory when a loop is flamed or in industry when medical waste is incinerated.

Heat Resistance and Thermal Death: Endospores and Vegetative Cells

Bacterial endospores exhibit the greatest resistance and vegetative states of bacteria and fungi are the least resistant to both moist and dry heat. Destruction of endospores usually requires temperatures above boiling, although resistance varies widely.

Vegetative cells also vary in their sensitivity to heat. Among bacteria, the death times with moist heat range from 50°C for 3 minutes (*Neisseria gonorrhoeae*) to 60°C

Table 9.4 Comparison of Times and Temperatures to Achieve Sterilization with Moist and Dry Heat

	Temperature (°C)	Time to Sterilize (Min)
Moist heat	121	15
	125	10
	134	3
Dry heat	121	600
	140	180
	160	120
	170	60

Pasteurization increases the shelf life of dairy products.

for 60 minutes (*Staphylococcus aureus*). It is worth noting that vegetative cells of endospore formers are just as susceptible as vegetative cells of non-endospore-formers, and that pathogens are neither more nor less susceptible than nonpathogens. Other microbes, including fungi, protozoa, and worms, are rather similar in their sensitivity to heat. Viruses are surprisingly resistant to heat, with a tolerance range extending from 55°C for 2 to 5 minutes (adenoviruses) to 60°C for 600 minutes (hepatitis A virus).

Susceptibility of Microbes to Heat: Thermal Death Measurements

As we have seen, higher temperatures allow shorter exposure times, and lower temperatures require longer exposure times. A combination of these two variables constitutes the **thermal death time,** or **TDT,** defined as the shortest length of time required to kill all test microbes at a specified temperature. The TDT has been experimentally determined for the microbial species that are common or important contaminants in various heat-treated materials. Another way to compare the susceptibility of microbes to heat is the **thermal death point (TDP),** defined as the lowest temperature required to kill all microbes in a sample in 10 minutes.

Many perishable substances are processed with moist heat. Some of these products are intended to remain on the shelf at room temperature for several months or even years. The chosen heat treatment must render the product free of agents of spoilage or disease. At the same time, the quality of the product and the speed and cost of processing must be considered. For example, in the commercial preparation of canned green beans, one of the manufacturer's greatest concerns is to prevent growth of the agent of botulism. From several possible TDTs (i.e., combinations of time and temperature) for *Clostridium botulinum* endospores, the factory must choose one that kills all endospores but does not turn the beans to mush. Out of these many considerations emerges an optimal TDT for a given processing method. Commercial canneries heat low-acid foods at 121°C for 30 minutes, a treatment that sterilizes these foods. Because of such strict controls in canneries, cases of botulism due to commercially canned foods are rare.

Frequently Used Approaches to Moist Heat Control

The three ways that moist heat is employed to control microbes are described in **table 9.5.**

Canned sardines stay preserved for 2–3 years, preserved by the osmotic pressure of high salt.

Table 9.5 Moist Heat Methods

Techniques and chemicals that are capable of sterilizing are highlighted with a pink background.

Method	Applications

Boiling Water: Disinfection   A simple boiling water bath or chamber can quickly decontaminate items in the clinic and home. Because a single processing at 100°C will not kill all resistant cells, this method can be relied on only for disinfection and not for sterilization. Exposing materials to boiling water for 30 minutes will kill most non-endospore-forming pathogens, including resistant species such as the tubercle bacillus and staphylococci. Probably the greatest disadvantage with this method is that the items can be easily recontaminated when removed from the water.

Useful in the home for disinfection of water, materials for babies, food and utensils, bedding, and clothing from the sickroom

Pasteurization: Disinfection of Beverages   Fresh beverages such as milk, fruit juices, beer, and wine are easily contaminated during collection and processing. Because microbes have the potential for spoiling these foods or causing illness, heat is frequently used to reduce the microbial load and destroy pathogens. **Pasteurization** is a technique in which heat is applied to liquids to kill potential agents of infection and spoilage, while at the same time retaining the liquid's flavor and food value.

Ordinary pasteurization techniques require special heat exchangers that expose the liquid to 71.6°C for 15 seconds (flash method) or to 63°C to 66°C for 30 minutes (batch method). The first method is preferable because it is less likely to change flavor and nutrient content, and it is more effective against certain resistant pathogens such as *Coxiella* and *Mycobacterium*. Although these treatments inactivate most viruses and destroy the vegetative stages of 97% to 99% of bacteria and fungi, they do not kill endospores or particularly heat-resistant microbes (mostly nonpathogenic lactobacilli, micrococci, and yeasts). Milk is not sterile after regular pasteurization. In fact, it can contain 20,000 microbes per milliliter or more, which explains why even an unopened carton of milk will eventually spoil. (Newer techniques can also produce *sterile milk* that has a storage life of 3 months. This milk is processed with ultrahigh temperature [UHT]–134°C–for 1 to 2 seconds.) This is not generally considered pasteurization, so we don't consider pasteurization a sterilization method.

Milk, wine, beer, other beverages

Dry Heat: Hot Air and Incineration

Dry heat is not as versatile or as widely used as moist heat, but it has several important sterilization applications. The temperatures and times employed in dry heat vary according to the particular method, but in general, they are greater than with moist heat. **Table 9.6** describes the two methods.

Table 9.5 (continued)

Method	Applications
Steam Under Pressure: Autoclaving At sea level, normal atmospheric pressure is 15 pounds per square inch (psi), or 1 atmosphere. At this pressure, water will boil (change from a liquid to a gas) at 100°C, and the resultant steam will remain at exactly that temperature, which is unfortunately too low to reliably kill all microbes. In order to raise the temperature of steam, the pressure at which it is generated must be increased. As the pressure is increased, the temperature at which water boils and the temperature of the steam produced both rise. For example, at a pressure of 20 psi (5 psi above normal), the temperature of steam is 109°C. As the pressure is increased to 10 psi above normal, the steam's temperature rises to 115°C, and at 15 psi above normal (a total of 2 atmospheres), it will be 121°C. It is not the pressure by itself that is killing microbes but the increased temperature it produces.	Heat-resistant materials such as glassware, cloth (surgical dressings), metallic instruments, liquids, paper, some media, and some heat-resistant plastics. If items are heat-sensitive (plastic Petri dishes) but will be discarded, the autoclave is still a good choice. However, it is ineffective for sterilizing substances that repel moisture (oils, waxes), or for those that are harmed by it (powders).

Such pressure-temperature combinations can be achieved only with a special device that can subject pure steam to pressures greater than 1 atmosphere. Health and commercial industries use an **autoclave** for this purpose, and a comparable home appliance is the pressure cooker. The most efficient pressure-temperature combination for achieving sterilization is 15 psi, which yields 121°C. It is important to avoid overpacking or haphazardly loading the chamber, which prevents steam from circulating freely around the contents and impedes the full contact that is necessary. The duration of the process is adjusted according to the bulkiness of the items in the load (thick bundles of material or large flasks of liquid) and how full the chamber is. The range of holding times varies from 10 minutes for light loads to 40 minutes for heavy or bulky ones; the average time is 20 minutes.

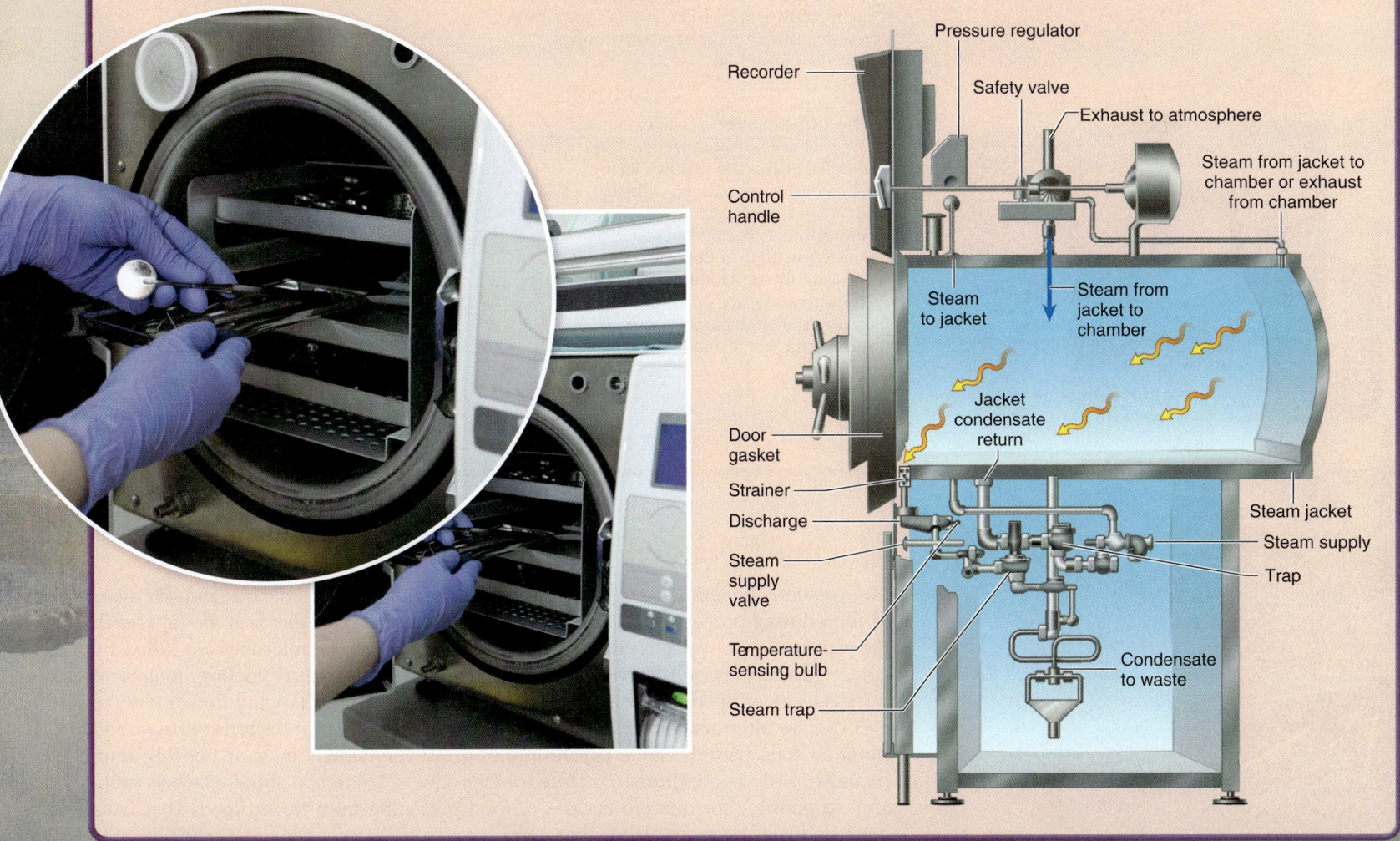

Table 9.6 Dry Heat Methods

Techniques and chemicals that are capable of sterilizing are highlighted with a pink background.

Method		Applications
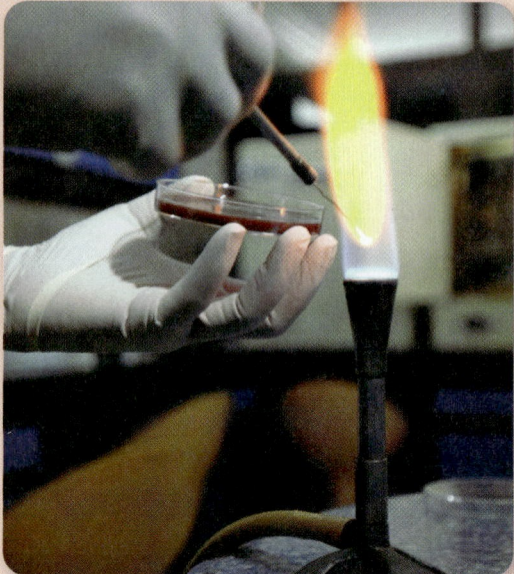	**Incineration** in a flame is perhaps the most rigorous of all heat treatments. The flame of a Bunsen burner reaches 1,870°C at its hottest point, and furnaces/incinerators operate at temperatures of 800°C to 6,500°C. Direct exposure to such intense heat ignites and reduces microbes and other substances to ashes and gas. Incineration of microbial samples on inoculating loops and needles using a Bunsen burner is a very common practice in the microbiology laboratory. This method is fast and effective, but it is also limited to metals and heat-resistant glass materials. This method also presents hazards to the operator (an open flame) and to the environment (contaminants on needle or loop often spatter when placed in flame). Tabletop infrared incinerators have replaced Bunsen burners in many labs for these reasons. Large incinerators are regularly employed in hospitals and research labs for complete destruction of infectious materials.	Bunsen burners/small incinerators: laboratory instruments such as inoculating loops. Large incinerators: syringes, needles, culture materials, dressings, bandages, bedding, animal carcasses, and pathology samples.
	**The hot-air oven** provides another means of dry-heat sterilization. The so-called *dry oven* is usually electric (occasionally gas) and has coils that radiate heat within an enclosed compartment. Heated, circulated air transfers its heat to the materials in the oven. Sterilization requires exposure to 150°C to 180°C for 2 to 4 hours, which ensures thorough heating of the objects and destruction of endospores.	Glassware, metallic instruments, powders, and oils that steam does not penetrate well. Not suitable for plastics, cotton, and paper, which may burn at the high temperatures, or for liquids, which will evaporate.

The Effects of Cold and Desiccation

The principal benefit of cold treatment is to slow growth of cultures and microbes in food during processing and storage. *It must be emphasized that cold merely retards the activities of most microbes.* Although it is true that some microbes are killed by cold temperatures, most are not adversely affected by gradual cooling, long-term refrigeration, or deep-freezing. In fact, freezing temperatures, ranging from −70°C to −135°C, are often used in research labs to preserve cultures of bacteria, viruses, and fungi for long periods. Some psychrophiles grow very slowly even at freezing temperatures and can continue to secrete toxic products. Ignorance of these facts is probably responsible for numerous cases of food poisoning from frozen foods that have been defrosted at room temperature and then inadequately cooked. Pathogens able

A wide variety of foods are now irradiated to control microbial growth.

to survive several months in the refrigerator are *Staphylococcus aureus, Clostridium* species (endospore formers), *Streptococcus* species, and several types of yeasts, molds, and viruses. Outbreaks of *Salmonella* food infection traced backed to refrigerated foods such as ice cream, eggs, and tiramisu are testimony to the inability of freezing temperatures to reliably kill pathogens.

Vegetative cells directly exposed to normal room air gradually become dehydrated, or **desiccated.** Delicate pathogens such as *Streptococcus pneumoniae,* the spirochete of syphilis, and *Neisseria gonorrhoeae* can die after a few hours of air drying, but many others are not killed and some are even preserved. Endospores of *Bacillus* and *Clostridium* are viable for thousands of years under extremely arid conditions. Staphylococci and streptococci in dried secretions and the tubercle bacillus surrounded by sputum can remain viable in air and dust for lengthy periods. Many viruses (especially nonenveloped) and fungal spores can also withstand long periods of desiccation. Desiccation can be a valuable way to preserve foods because it greatly reduces the amount of water available to support microbial growth.

It is interesting to note that a combination of freezing and drying–**lyophilization** (ly-off″-il-ih-za′-shun)–is a common method of *preserving* microorganisms and other cells in a viable state for many years. Pure cultures are frozen instantaneously and exposed to a vacuum that rapidly removes the water (it goes right from the frozen state into the vapor state). This method avoids the formation of ice crystals that would damage the cells. Although not all cells survive this process, enough of them do to permit future reconstitution of that culture.

As a general rule, chilling, freezing, and desiccation should not be construed as methods of disinfection or sterilization because their antimicrobial effects are erratic and uncertain, and one cannot be sure that pathogens subjected to them have been killed.

Radiation

Another way in which energy can serve as an antimicrobial agent is through the use of radiation. **Radiation** is defined as energy emitted from atomic activities and dispersed at high velocity through matter or space. **Figure 9.4** illustrates the different wavelengths of radiation. In our discussion, we consider only those types suitable for microbial control: gamma rays, X rays, and ultraviolet radiation. There are several

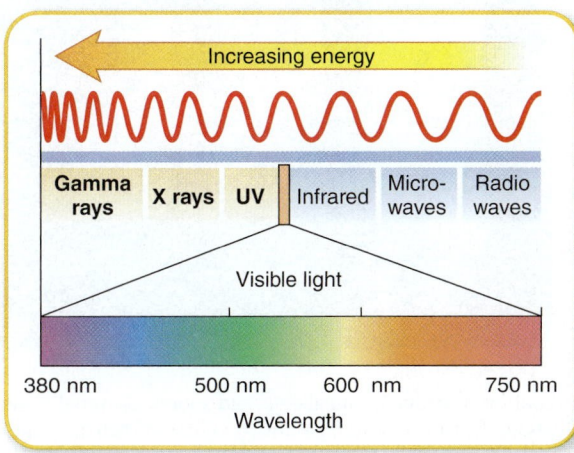

Figure 9.4 The electromagnetic spectrum, showing different types of radiation.

Table 9.7 Radiation Methods

Techniques and chemicals that are capable of sterilizing are highlighted with a pink background.

Method	Applications

Ionizing Radiation: Gamma Rays and X Rays   Ionizing radiation is a highly effective alternative for sterilizing materials that are sensitive to heat or chemicals. Devices that emit ionizing rays include gamma-ray machines containing radioactive cobalt, X-ray machines similar to those used in medical diagnosis, and cathode-ray machines. Items are placed in these machines and irradiated for a short time with a carefully chosen dosage. The dosage of radiation is measured in *Grays* (which has replaced the older term, *rads*). Depending on the application, exposure ranges from 5 to 50 kiloGrays (kGray; a kiloGray is equal to 1,000 Grays). Although all ionizing radiation can penetrate liquids and most solid materials, gamma rays are most penetrating, X rays are intermediate, and cathode rays are least penetrating.

Foods have been subject to **irradiation** in limited circumstances for more than 50 years. From flour to pork and ground beef, to fruits and vegetables, radiation is used to kill not only bacterial pathogens but also insects and worms and even to inhibit the sprouting of white potatoes. Irradiated food has been extensively studied, and found to be safe and nonradioactive.

Irradiation may lead to a small decrease in the amount of thiamine (vitamin B_1) in food, but this change is small enough to be inconsequential. The irradiation process does produce short-lived free radical oxidants, which disappear almost immediately (this same type of chemical intermediate is produced through cooking as well). Certain foods do not irradiate well and are not good candidates for this type of antimicrobial control. The white of eggs becomes milky and liquid, grapefruit gets mushy, and alfalfa seeds do not germinate properly. Lastly, it is important to remember that food is not made radioactive by the irradiation process, and many studies, in both animals and humans, have concluded that there are no ill effects from eating irradiated food. In fact, NASA relies on irradiated meat for its astronauts.

Drugs, vaccines, medical instruments (especially plastics), syringes, surgical gloves, tissues such as bone and skin, and heart valves for grafting.

After the anthrax attacks of 2001, mail delivered to certain Washington, D.C., ZIP Codes was irradiated with ionizing radiation. Its main advantages include speed, high penetrating power (it can sterilize materials through outer packages and wrappings), and the absence of heat. Its main disadvantages are potential dangers to radiation machine operators from exposure to radiation and possible damage to some materials.

Nonionizing Radiation: Ultraviolet Rays   Ultraviolet (UV) **radiation** ranges in wavelength from approximately 100 to 400 nm. It is most lethal from 240 to 280 nm (with a peak at 260 nm). Owing to its lower energy state, UV radiation is not as penetrating as ionizing radiation. Because UV radiation passes readily through air, slightly through liquids, and only poorly through solids, the object to be disinfected must be directly exposed to it for full effect.

Ultraviolet rays are a powerful tool for destroying fungal cells and spores, bacterial vegetative cells, protozoa, and viruses. Bacterial spores are about 10 times more resistant to radiation than are vegetative cells, but they can be killed by increasing the time of exposure. Even though it is possible to sterilize with UV, it is so technically challenging that we don't regularly call it a sterilizing technology.

Usually directed at disinfection rather than sterilization. Germicidal lamps can cut down on the concentration of airborne microbes as much as 99%. They are used in hospital rooms, operating rooms, schools, food preparation areas, and dental offices. Ultraviolet disinfection of air has proved effective in reducing postoperative infections, preventing the transmission of infections by respiratory droplets, and curtailing the growth of microbes in food-processing plants and slaughterhouses.

Ultraviolet irradiation of liquids requires special equipment to spread the liquid into a thin, flowing film that is exposed directly to a lamp. This method can be used to treat drinking water and to purify other liquids (milk and fruit juices) as an alternative to heat. The photo shows a UV treatment system for the disinfection of water.

especially useful practices that take advantage of these forms of radiation **(table 9.7)**. Given the ubiquitous nature of UV radiation in the lab and in your daily life, it is important to understand its mode of action **(figure 9.5)**.

Other Physical Methods: Filtration

Filtration is an effective method to remove microbes from air and liquids. In practice, a fluid is strained through a filter with openings large enough for the fluid to pass through but too small for microorganisms to pass through **(figure 9.6a)**.

Most modern microbiological filters are thin membranes of cellulose acetate, polycarbonate, and a variety of plastic materials (Teflon, nylon) whose pore size can be carefully controlled and standardized. Ordinary substances such as charcoal, diatomaceous earth, or unglazed porcelain are also used in some applications. Viewed microscopically, most filters are perforated by very precise, uniform pores **(figure 9.6b)**. The pore diameters vary from coarse (8 μm) to ultrafine (0.02 μm), permitting selection of the minimum particle size to be trapped. Those with even smaller pore diameters permit true sterilization by removing viruses, and some will even remove large proteins. A sterile liquid filtrate is typically produced by suctioning the liquid through a sterile filter into a presterilized container. These filters are also used to separate mixtures of microorganisms and to enumerate bacteria in water analysis.

Filtration is used to prepare liquids that cannot withstand heat, including serum and other blood products, vaccines, drugs, IV fluids, enzymes, and media. Filtration has been employed as an alternative method for decontaminating milk and beer without altering their flavor. It is also an important step in water purification. Its use extends to filtering out particulate impurities (crystals, fibers, and so on) that can cause severe reactions in the body. It has the disadvantage of not removing soluble molecules (toxins) that can cause disease.

Filtration is also an efficient means of removing airborne contaminants that are a common source of infection and spoilage. High-efficiency particulate air (HEPA) filters are widely used to provide a flow of decontaminated air to hospital rooms and sterile rooms. A vacuum

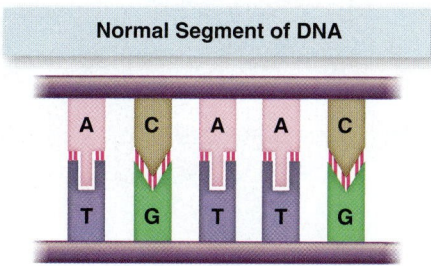

Normal Segment of DNA

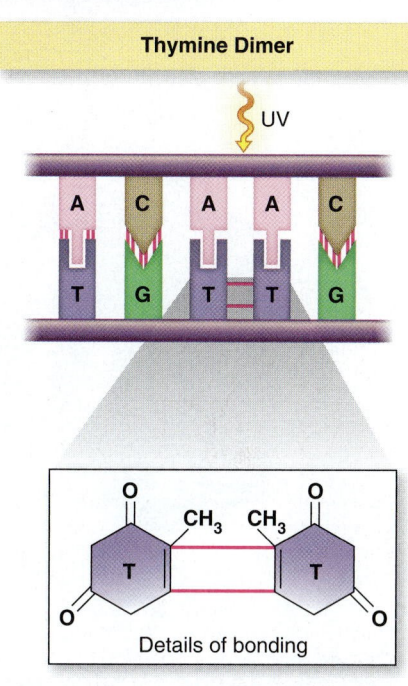

Thymine Dimer

Details of bonding

Figure 9.5 Formation of pyrimidine dimers by the action of ultraviolet (UV) radiation. This shows what occurs when two adjacent thymine bases on one strand of DNA are induced by UV rays to bond laterally with each other. The result is a thymine dimer (shown in greater detail). Dimers can also occur between adjacent cytosines, and thymine and cytosine bases. If they are not repaired, dimers can prevent that segment of DNA from being correctly replicated or transcribed. Massive dimerization is lethal to cells.

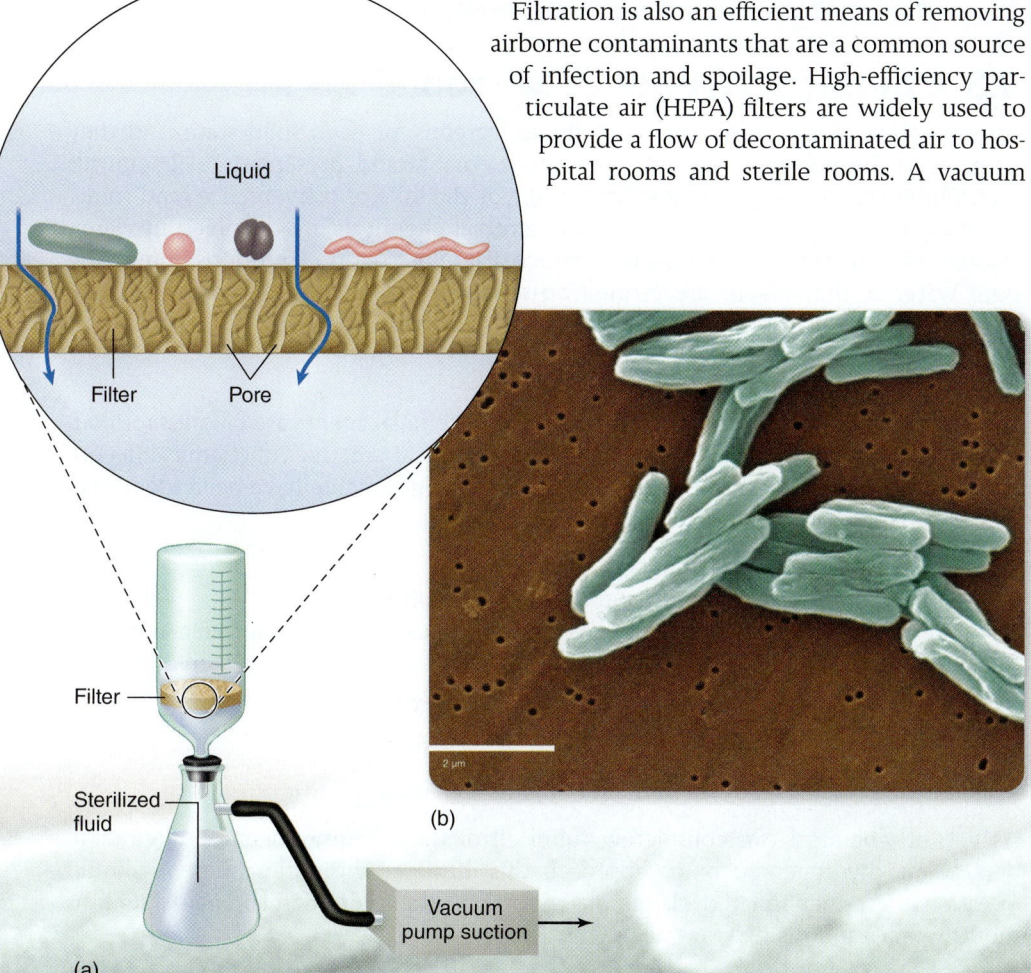

Liquid

Filter   Pore

Filter

Sterilized fluid

(a)

(b)

2 μm

Vacuum pump suction →

Figure 9.6 Membrane filtration. (a) Vacuum assembly for achieving filtration of liquids through suction. The surface of the filter is shown magnified in the blown-up section, with tiny passageways (pores) too small for the microbial cells to enter but large enough for liquid to pass through. (b) Scanning electron micrograph (5,900×) of filter, showing relative size of pores and bacteria trapped on its surface.

with a HEPA filter was even used to remove anthrax endospores from the Senate offices most heavily contaminated after the terrorist attack in late 2001.

Osmotic Pressure

In chapter 6, you learned about the effects of osmotic pressure on cells. This fact has long been exploited as a means of preserving food. Adding large amounts of salt or sugar to foods creates a hypertonic environment for bacteria in the foods, causing plasmolysis and making it impossible for the bacteria to multiply. People knew that these techniques worked long before the discovery of bacteria. Even in ancient times, people used pickling, smoking, and drying of foods to control growth of microorganisms. This is why meats are "cured," or treated with high salt concentrations so they can be kept for long periods without refrigeration. High sugar concentrations in foods like jellies have the same effect. Osmotic pressure is never a sterilizing technique.

9.2 LEARNING OUTCOMES—Assess Your Progress

5. Name six methods of physical control of microorganisms.
6. Discuss both moist and dry heat methods, and identify multiple examples of each.
7. Define *thermal death time* and *thermal death point*.
8. Explain methods of moist heat control.
9. Explain two methods of dry heat control.
10. Identify advantages and disadvantages of cold and desiccation.
11. Differentiate between the two types of radiation control methods.
12. Explain how filtration and osmotic pressure function as control methods.

9.3 Chemical Agents in Microbial Control

Antimicrobial chemicals occur in the liquid, gaseous, or even solid state, and they range from disinfectants and antiseptics to sterilants and preservatives (chemicals that inhibit the deterioration of substances). For the sake of convenience (and sometimes safety), many solid or gaseous antimicrobial chemicals are dissolved in water, alcohol, or a mixture of the two to produce a liquid solution. Solutions containing pure water as the solvent are termed **aqueous,** whereas those dissolved in pure alcohol or water-alcohol mixtures are termed **tinctures.**

Selecting a Microbicidal Chemical

The choice and appropriate use of antimicrobial chemical agents are of constant concern in medicine and dentistry. Although actual clinical practices of chemical decontamination vary widely, some desirable qualities in a germicide have been identified, including the following:

- rapid action even in low concentrations
- solubility in water or alcohol and long-term stability
- broad-spectrum microbicidal action without being toxic to human and animal tissues
- penetration of inanimate surfaces to sustain a cumulative or persistent action
- resistance to becoming inactivated by organic matter
- noncorrosive or nonstaining properties
- sanitizing and deodorizing properties
- affordability and ready availability

As yet, no chemical can completely fulfill all of those requirements, but glutaraldehyde and hydrogen peroxide approach this ideal. At the same time, we should question the rather inflated claims made about certain commercial agents such as mouthwashes and disinfectant air sprays.

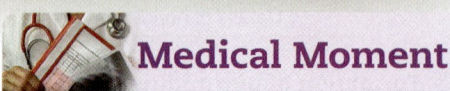

Medical Moment

The Use of Alcohol-Based Hand Cleansers

Hand washing is recognized as being one of the most important methods of preventing infection in hospitals, yet studies have shown that many health care workers (HCWs) do not wash their hands as often as they should, or fail to wash their hands in accordance with guidelines. For this reason, the use of alcohol-based hand antiseptics between hand washes has become common practice. Most alcohol-based hand antiseptics contain isopropanol, n-propanol, or ethanol. Some contain combinations of two of these substances. A concentration of 60% to 80% alcohol is deemed most effective due to the fact that proteins require water in order for denaturation to occur, so higher concentrations of alcohol are less effective.

Studies have shown that many HCWs do not apply enough alcohol to their hands, resulting in the survival of some microorganisms on the skin. To use alcohol-based hand antiseptics properly, apply a liberal amount to the palm of one hand, then rub the hands together until the product has dissipated. Pay attention to the areas between the fingers and the base of the thumb—these areas are often missed.

Germicides are evaluated in terms of their effectiveness in destroying microbes in medical and dental settings. The three levels of chemical decontamination procedures are *high*, *intermediate*, and *low*. High-level germicides kill endospores and, if properly used, are sterilants. Materials that necessitate high-level control are medical devices–for example, catheters, heart-lung equipment, and implants–that are not heat-sterilizable and are intended to enter body tissues during medical procedures. Intermediate-level germicides kill fungal (but not bacterial) spores, resistant pathogens such as the tubercle bacillus, and viruses. They are used to disinfect items (respiratory equipment, thermometers) that come into intimate contact with the mucous membranes but are noninvasive. Low levels of disinfection eliminate only vegetative bacteria, vegetative fungal cells, and some viruses. They are used to clean materials such as electrodes, straps, and pieces of furniture that touch the skin surfaces but not the mucous membranes.

Factors Affecting the Germicidal Activity of Chemicals

Factors that control the effect of a germicide include the nature of the microorganisms being treated, the nature of the material being treated, the degree of contamination, the time of exposure, and the strength and chemical action of the germicide. The variations in concentration and time needed can be quite wide **(table 9.8).**

The Chemical Categories

The modes of action of most germicides are to attack the cellular targets discussed earlier: proteins, nucleic acids, the cell wall, and the cytoplasmic membrane. **Table 9.9** on the following pages provides details about the most commonly used chemicals and their modes of action.

Table 9.8 Required Concentrations and Times for Chemical Destruction of Selected Microbes

Organism	Concentration	Time
Agent: Chlorine		
Mycobacterium tuberculosis	50 ppm	50 sec
Entamoeba cysts (protozoa)	0.1 ppm	150 min
Hepatitis A virus	3 ppm	30 min
Agent: Ethyl Alcohol		
Staphylococcus aureus	70%	10 min
Escherichia coli	70%	2 min
Poliovirus	70%	10 min
Agent: Hydrogen Peroxide		
Staphylococcus aureus	3%	12.5 sec
Neisseria gonorrhoeae	3%	0.3 sec
Herpes simplex virus	3%	12.8 sec
Agent: Quaternary Ammonium Compound		
Staphylococcus aureus	450 ppm	10 min
Salmonella typhi	300 ppm	10 min
Agent: Ethylene Oxide Gas		
Streptococcus faecalis	500 mg/L	2–4 min
Influenza virus	10,000 mg/L	25 h

The main antimicrobial ingredient in many mouthwashes is alcohol.

Table 9.9 Germicidal Categories According to Chemical Group
Techniques and chemicals that are capable of sterilizing are highlighted with a pink background.

Agent	Target Microbes	Form(s)	Mode of Action	Indications for Use	Limitations
Halogens: chlorine	Can kill endospores (slowly); all other microbes	Liquid/gaseous chlorine (Cl_2), hypochlorites (OCl), chloramines (NH_2Cl)	In solution, these compounds combine with water and release hypochlorous acid (HOCl); denature enzymes permanently and suspend metabolic reactions	Chlorine kills bacteria, endospores, fungi, and viruses; gaseous/liquid chlorine: used to disinfect drinking water, sewage and waste water; hypochlorites: used in health care to treat wounds, disinfect bedding and instruments, sanitize food equipment and in restaurants, pools, and spas; chloramines: alternative to pure chlorine in treating drinking water; also used to treat wounds and skin surfaces	Less effective if exposed to light, alkaline pH, and excess organic matter
Halogens: iodine	Can kill endospores (slowly); all other microbes	Free iodine in solution (I_2) Iodophors (complexes of iodine and alcohol)	Penetrates cells of microorganisms where it interferes with a variety of metabolic functions; interferes with the hydrogen and disulfide bonding of proteins	2% iodine, 2.4% sodium iodide (aqueous iodine) used as a topical antiseptic 5% iodine, 10% potassium iodide used as a disinfectant for plastic and rubber instruments, cutting blades, etc. Iodophor products contain 2% to 10% of available iodine, which is released slowly; used to prepare skin for surgery, in surgical scrubs, to treat burns, and as a disinfectant	Can be extremely irritating to the skin and is toxic when absorbed
Hydrogen peroxide (H_2O_2)	Kills endospores and all other microbes	Colorless, caustic liquid Decomposes in the presence of light metals or catalase into water, and oxygen gas	Oxygen forms free radicals (−OH), which are highly toxic and reactive to cells	As an antiseptic, 3% hydrogen peroxide used for skin and wound cleansing, mouth washing, bedsore care Used to treat infections caused by anaerobic bacteria 35% hydrogen peroxide used in low temperature sterilizing cabinets for delicate instruments	Sporicidal only in high concentrations
Aldehydes	Kill endospores and all other microbes	Organic substances bearing a −CHO functional group on the terminal carbon	Glutaraldehyde can irreversibly disrupt the activity of enzymes and other proteins within the cell Formaldehyde is a sharp irritating gas that readily dissolves in water to form an aqueous solution called formalin; attaches to nucleic acids and functional groups of amino acids	Glutaraldehyde kills rapidly and is broad-spectrum; used to sterilize respiratory equipment, scopes, kidney dialysis machines, dental instruments Formaldehyde kills more slowly than glutaraldehyde; used to disinfect surgical instruments	Glutaraldehyde is somewhat unstable, especially with increased pH and temperature Formaldehyde is extremely toxic and is irritating to skin and mucous membranes
Gaseous sterilants/disinfectants	Ethylene oxide kills endospores; other gases less effective	Ethylene oxide is a colorless substance that exists as a gas at room temperature	Ethylene oxide reacts vigorously with functional groups of DNA and proteins, blocking both DNA replication and enzymatic actions Chlorine dioxide is a strong alkylating agent	Ethylene oxide is used to disinfect plastic materials and delicate instruments; can also be used to sterilize syringes, surgical supplies, and medical devices that are prepackaged	Ethylene oxide is explosive–it must be combined with a high percentage of carbon dioxide or fluorocarbon It can damage lungs, eyes, and mucous membranes if contacted directly Ethylene oxide is rated as a carcinogen by the government

Table 9.9 (continued)

Agent	Target Microbes	Form(s)	Mode of Action	Indications for Use	Limitations
Phenol (carbolic acid)	Some bacteria, viruses, fungi	Derived from the distillation of coal tar Phenols consist of one or more aromatic carbon rings with added functional groups	In high concentrations, they are cellular poisons, disrupting cell walls and membranes, proteins In lower concentrations, they inactivate certain critical enzyme systems	Phenol remains one standard against which other (less toxic) phenolic disinfectants are rated; the phenol coefficient quantitatively compares a chemical's antimicrobial properties to those of phenol Phenol is now used only in certain limited cases, such as in drains, cesspools, and animal quarters	Toxicity of many phenolics makes them dangerous to use as antiseptics
Chlorhexidine	Most bacteria, viruses, fungi	Complex organic base containing chlorine and two phenolic rings	Targets both bacterial membranes, where selective permeability is lost, and proteins, resulting in denaturation	Mildness, low toxicity and rapid action make chlorhexidine a popular choice of agents Used in hand scrubs, prepping skin for surgery, as an obstetric antiseptic, as a mucous membrane irrigant, etc.	Effects on viruses and fungi are variable
Alcohol	Most bacteria, viruses, fungi	Colorless hydrocarbons with one or more −OH functional groups Ethyl and isopropyl alcohol are suitable for antimicrobial control	Concentrations of 50% and higher dissolve membrane lipids, disrupt cell surface tension, and compromise membrane integrity	Germicidal, nonirritating, and inexpensive Routinely used as skin degerming agents (70% to 95% solutions)	Rate of evaporation decreases effectiveness Inhalation of vapors can affect the nervous system
Detergents	Some bacteria, viruses, fungi	Polar molecules that act as surfactants Anionic detergents have limited microbial power Cationic detergents, such as quaternary ammonium compounds ("quats"), are much more effective antimicrobials	Positively charged end of the molecule binds well with the predominantly negatively charged bacterial surface proteins Long, uncharged hydrocarbon chain allows the detergent to disrupt the cytoplasmic membrane Cytoplasmic membrane loses selective permeability, causing cell death	Effective against viruses, algae, fungi, and gram-positive bacteria Rated only for low-level disinfection in the clinical setting Used to clean restaurant utensils, dairy equipment, equipment surfaces, restrooms	Ineffective against tuberculosis bacterium, hepatitis virus, *Pseudomonas*, and endospores Activity is greatly reduced in presence of organic matter Detergents function best in alkaline solutions
Heavy metal compounds	Some bacteria, viruses, fungi	Heavy metal germicides contain either an inorganic or an organic metallic salt; may come in tinctures, soaps, ointment, or aqueous solution	Mercury, silver, and other metals exert microbial effects by binding onto functional groups of proteins and inactivating them	Organic mercury tinctures are fairly effective antiseptics Organic mercurials serve as preservatives in cosmetics, ophthalmic solutions, and other substances Silver nitrate solutions are used for topical germicides and ointments	Microbes can develop resistance to metals Not effective against endospores Can be toxic if inhaled, ingested, or absorbed May cause allergic reactions in susceptible individuals

Medical Moment

Silver in Wound Care: Silver-Impregnated Dressings

Colonization and infection of wounds has always been a problem for clinicians. Wounds colonized with bacteria may heal more slowly, particularly in patients whose immune systems are compromised or in patients whose wounds are poorly perfused (i.e., leg ulcers in diabetics).

For this reason, dressings containing antimicrobials have been developed, with silver being the most common agent used. Silver has been used in the past to treat burns in the form of silver sulfadiazine; however, in this form, staining of the skin and toxicity may develop. In addition, dressings must be changed more frequently, causing increased pain for patients.

New dressings containing silver are more convenient in their application—they may be applied and left in place for several days. Silver has broad-spectrum antimicrobial activity, even against antibiotic-resistant organisms such as methicillin-resistant *Staphylococcus aureus* (MRSA) and vancomycin-resistant enterococci (VRE). However, dressings containing silver vary widely in the nature of their content and their release of silver, which leads to much confusion for clinicians choosing among these dressings. Although there have been concerns voiced that the use of silver may give rise to increased resistance of organisms against biocides, this has not been shown to be the case, and dressings impregnated with silver continue to be widely used.

A chemical's strength or concentration is expressed in various ways, depending on convention and the method of preparation. In dilutions, a small volume of the liquid chemical (solute) is diluted in a larger volume of solvent to achieve a certain ratio. For example, a common laboratory phenolic disinfectant such as Lysol is usually diluted 1:200; that is, one part of chemical has been added to 200 parts of water by volume. Solutions such as chlorine that are effective in very diluted concentrations are expressed in parts per million (ppm). In percentage solutions, the solute is added to water by weight or volume to achieve a certain percentage in the solution. Alcohol, for instance, is used in percentages ranging from 50% to 95%.

Another factor that contributes to germicidal effectiveness is the length of exposure. Most compounds require adequate contact time to allow the chemical to penetrate and to act on the microbes present. The composition of the material being treated must also be considered. Smooth, solid objects are more reliably disinfected than are those with pores or pockets that can trap soil. An item contaminated with common biological matter such as serum, blood, saliva, pus, fecal material, or urine presents a problem in disinfection. Large amounts of organic material can hinder the penetration of a disinfectant and, in some cases, can form bonds that reduce its activity. Adequate cleaning of instruments and other reusable materials must precede use of the germicide or sterilant. Otherwise, there is no way to predict whether your procedure will be effective.

Figure 9.7 provides an illustration of the effects of both heat and heavy metals on one important cellular target: protein.

The structure of detergents is depicted in **figure 9.8**.

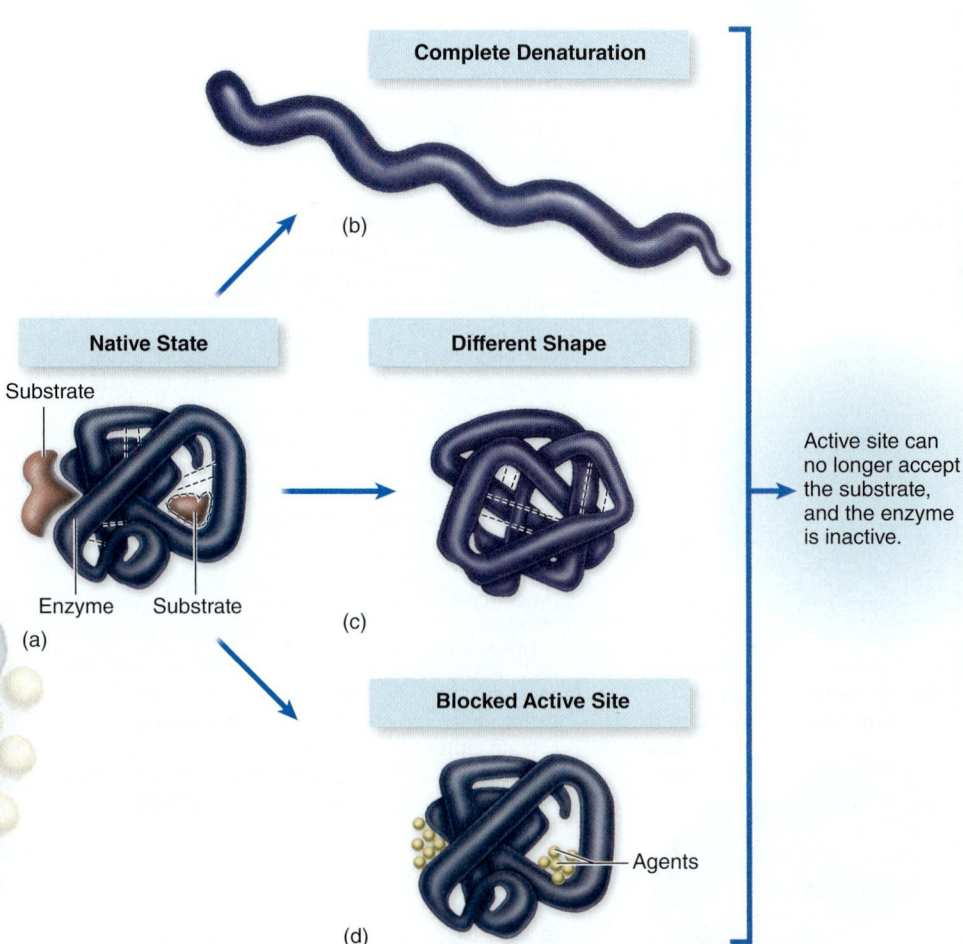

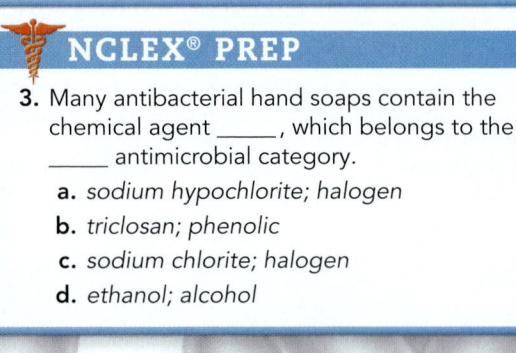

Figure 9.7 Modes of action affecting protein function. (a) The native (functional) state is maintained by bonds that create active sites to fit the substrate. Some agents denature the protein by breaking all or some secondary and tertiary bonds. Results are (b) complete unfolding or (c) random bonding and incorrect folding. (d) Some agents react with functional groups on the active site and interfere with bonding.

Other Antimicrobial Agents: Dyes, Acids, and Alkalis

Dyes, of course, have their primary usage in staining techniques and as selective and differential agents in media; they are also a primary source of certain drugs used in chemotherapy. But because aniline dyes such as crystal violet and malachite green are very active against gram-positive species of bacteria and various fungi, they are sometimes incorporated into solutions and ointments to treat skin infections (ringworm, for example). The yellow acridine dyes, acriflavine and proflavine, are sometimes utilized for antisepsis and wound treatment in medical and veterinary clinics. For the most part, dyes will continue to have limited applications because they stain and have a narrow spectrum of activity.

Acids and Alkalis

Conditions of very low or high pH can destroy or inhibit microbial cells, but they are limited in applications due to their corrosive, caustic, and hazardous nature. Aqueous solutions of ammonium hydroxide remain a common component of detergents, cleansers, and deodorizers. Organic acids are widely used in food preservation because they prevent endospore germination and bacterial and fungal growth and because they are generally regarded as safe to eat. Acetic acid (in the form of vinegar) is a pickling agent that inhibits bacterial growth; propionic acid is commonly incorporated into breads and cakes to retard molds; lactic acid is added to sauerkraut and olives to prevent growth of anaerobic bacteria (especially the clostridia); and benzoic and sorbic acids are added to beverages, syrups, and margarine to inhibit yeasts.

For a look at the antimicrobial chemicals found in some common household products, see **table 9.10.**

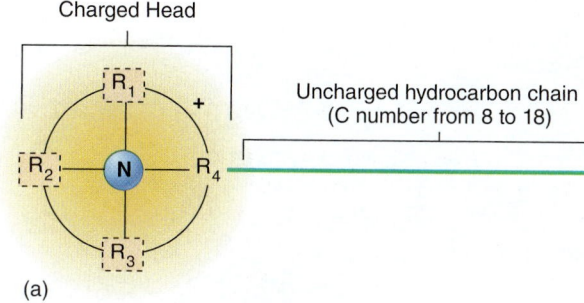

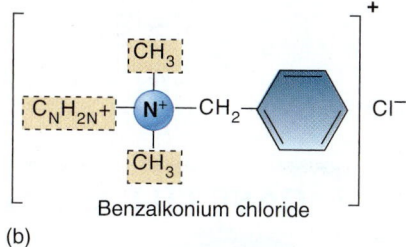

Figure 9.8 **The structure of detergents.**
(a) In general, detergents are polar molecules with a positively charged head and at least one long, uncharged hydrocarbon chain. The head contains a central nitrogen nucleus with various alkyl (R) groups attached. (b) A common quaternary ammonium detergent, benzalkonium chloride.

Table 9.10 **Active Ingredients of Various Commercial Antimicrobial Products**

Product	Specific Chemical Agent	Antimicrobial Category
Lysol® Sanitizing Wipes	Dimethyl benzyl ammonium chloride	Detergent (quat)
Clorox® Disinfecting Wipes	Dimethyl benzyl ammonium chloride	Detergent (quat)
Tilex® Mildew Remover	Sodium hypochlorites	Halogen
Lysol® Mildew Remover	Sodium hypochlorites	Halogen
Ajax® Antibacterial Hand Soap	Triclosan	Phenolic
Dawn® Antibacterial Hand Soap	Triclosan	Phenolic
Dial® Antibacterial Hand Soap	Triclosan	Phenolic
Lysol® Disinfecting Spray	Alkyl dimethyl benzyl ammonium saccharinate/ethanol	Detergent (quats)/alcohol
ReNu® Contact Lens Solution	Polyaminopropyl biguanide	Chlorhexidine
Wet Ones® Antibacterial Moist Towelettes	Benzethonium chloride	Detergents (quat)
Noxzema® Triple Clean	Triclosan	Phenolic
Scope® Mouthwash	Ethanol	Alcohol
Purell® Instant Hand Sanitizer	Ethanol	Alcohol
Pine-Sol®	Phenolics and surfactant	Mixed
Allergan® Eye Drops	Sodium chlorite	Halogen

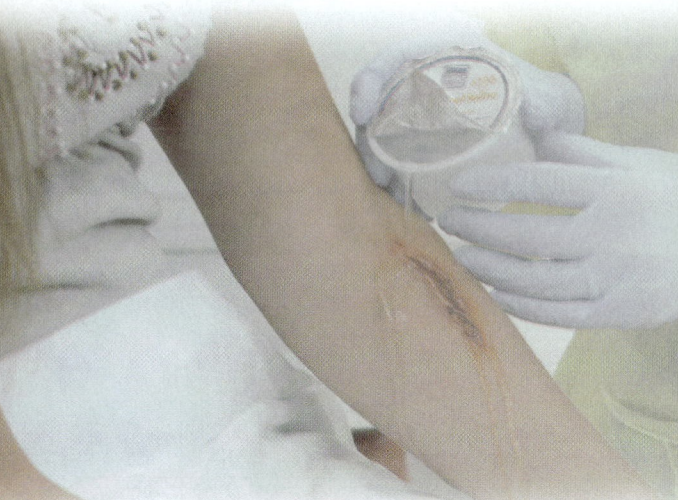

9.3 LEARNING OUTCOMES—Assess Your Progress

13. Name the desirable characteristics of chemical control agents.

14. Discuss chlorine and iodine and their uses.

15. List advantages and disadvantages to phenolic compounds.

16. Explain the mode of action of chlorhexidine.

17. Explain the applications of hydrogen peroxide agents.

18. Identify some heavy metal control agents.

19. Discuss the disadvantages of aldehyde agents.

20. Identify applications for ethylene oxide sterilization.

CASE FILE WRAP-UP

In wound care, the goal of antiseptic use is to destroy bacteria that may be present in the wound without affecting the wound healing process. Although antiseptics can reduce the bioburden (amount of harmful bacteria) present in wounds, antiseptics may also impede wound healing and can be toxic to healthy human cells. Numerous studies have shown an adverse effect on wound healing when antiseptics are used to cleanse wounds, with many authors now supporting sterile normal saline as a preferable alternative. In fact, the U.S. Agency for Health Care Research and Policy (2008) states: "Do not use povidone iodine, iodophor, sodium hypochlorite solution, hydrogen peroxide and acetic acid as they have been shown to be cytotoxic. Use normal saline at a pressure between 4 and 15 pounds per square inch (psi)."

Normal saline is isotonic and also causes less discomfort during wound cleansing. When combined with irrigation (a syringe with a needle or pulsed lavage), bacterial bioburden is diminished just as effectively as if an antiseptic were used, without cytotoxic effects on healthy cells.

Fresh Air and Sunshine: The Low-Tech Is Cutting Edge Again

At the turn of the 20th century, tuberculosis was a dreaded and common disease. Families with money sent their sick loved ones to "take the cure" in what were termed "solar clinics." These were often resortlike facilities in the countryside, where patients spent significant amounts of time outside in their mobile beds or wheelchairs, soaking up the sunshine and fresh air.

When antibiotics became available in the middle of the 20th century, that idea became nothing more than a quaint historical artifact. For several decades, the effectiveness of antibiotics left little apparent need for other mechanisms of infection fighting. It was a short-lived era, however. In current times, we know that several microbes are resistant to most or all antibiotics. Only now are some of these older methods being reexamined.

Researchers are discovering that exposure to circulating air and to UV light—such as that found in sunlight—can help people avoid new infections and can speed up recovery from tuberculosis. One component in circulating air found to be helpful are the hydroxyl radicals ($\bullet$OH). These radicals are continually produced in environmental air as a result of reactions between ozone and water. They then oxidize organic molecules that can cause harm to bacteria. It is well known that certain wavelengths of ultraviolet light have microbicidal action.

Before the days of air conditioning, of course, hospital windows were made to open. Nowadays, that is considered a safety hazard. Hospitals and buildings of all types have been designed to be airtight and well insulated as a direct result of our quest for energy conservation and efficiency. Although it is probably still not advisable to expose hospital patients to urban air pollution (to say nothing of noise pollution), scientists are considering whether restricting fresh air and sunlight in dwellings and hospitals has robbed us of valuable natural antimicrobial influences.

Chapter Summary

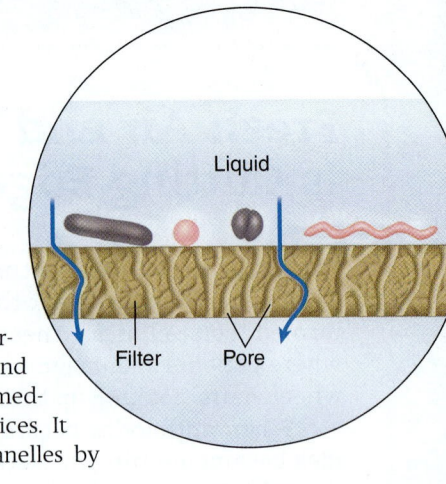

9.1 Controlling Microorganisms

- Microbial control methods involve the use of physical and chemical agents to eliminate or reduce the numbers of microorganisms from a specific environment to prevent the spread of infectious agents, retard spoilage, and keep commercial products safe.
- The population of microbes that cause spoilage or infection varies widely, so microbial control methods must be adjusted to fit individual situations.
- The type of microbial control is indicated by the terminology used. Sterilization agents destroy all viable organisms, including viruses. Antisepsis, disinfection, decontamination/sanitization, and antisepsis/degermation reduce the numbers of viable microbes to a specified level.
- Antimicrobial agents are described according to their ability to destroy or inhibit microbial growth. Microbicidal agents cause microbial death. They are described by what they are -cidal for: sporicides, bactericides, fungicides, and viricides.
- An antiseptic agent is applied to living tissue to destroy or inhibit microbial growth.
- A disinfectant agent is used on inanimate objects to destroy vegetative pathogens but not bacterial endospores.
- Decontamination/sanitization reduces microbial numbers on inanimate objects to safe levels by physical or chemical means.
- Microbial death is defined as the permanent loss of reproductive capability in microorganisms.
- Antimicrobial agents attack specific cell sites to cause microbial death or damage. The four major cell targets are the cell wall, the cytoplasmic membrane, biosynthesis pathways for DNA or RNA, or protein (enzyme) function.

9.2 Methods of Physical Control

- Physical methods of microbial control include heat, cold, radiation, drying, filtration, and osmotic pressure.
- Heat is the most widely used method of microbial control. It is used in combination with water (moist heat) or as dry heat (oven, flames).
- The thermal death time (TDT) is the shortest length of time required to kill all microbes at a specific temperature.
- The thermal death point (TDP) is the lowest temperature at which all microbes are killed in a specified length of time (10 minutes).
- Autoclaving, or steam sterilization, is the process by which steam is heated under pressure to sterilize a wide range of materials in a comparatively short time (minutes to hours).
- Boiling water and pasteurization of beverages disinfect but do not sterilize materials.
- Dry heat is microbicidal under specified times and temperatures. Flame heat, or incineration, is microbicidal.
- Chilling, freezing, and desiccation are microbistatic but not microbicidal. They are not considered true

methods of disinfection because they are not consistent in their effectiveness.

- Ionizing radiation (cold sterilization) by gamma rays and X rays is used to sterilize medical products, meats, and spices. It damages DNA and cell organelles by producing disruptive ions.
- Ultraviolet light, or nonionizing radiation, has limited penetrating ability. It is therefore usually restricted to disinfecting air and certain liquids.
- Decontamination by filtration removes microbes from heat-sensitive liquids and circulating air. The pore size of the filter determines what kinds of microbes are removed.
- The addition of high amounts of salt or sugar to food results in preservation through osmotic pressure.

9.3 Chemical Agents in Microbial Control

- Chemical agents of microbial control are classified by their physical state and chemical nature.
- Chemical agents can be either microbicidal or microbistatic. They are also classified as high-, medium-, or low-level germicides.
- Factors that determine the effectiveness of a chemical agent include the type and numbers of microbes involved, the material involved, the strength of the agent, and the exposure time.
- Halogens are effective chemical agents at both microbicidal and microbistatic levels. Chlorine, iodine, and iodophors are examples.
- Phenols are strong microbicidal agents used in general disinfection.
- Alcohols dissolve membrane lipids and destroy cell proteins. Their action depends upon their concentration, but they are generally only microbistatic.
- Hydrogen peroxide is a versatile microbicide that can be used as an antiseptic for wounds and a disinfectant for utensils. A high concentration is an effective sporicide.
- Detergents reduce cytoplasmic membrane surface tension, causing membrane rupture. Cationic detergents, or quats, are low-level germicides limited by the amount of organic matter present and the microbial load.
- Aldehydes are potent sterilizing agents and high-level disinfectants that irreversibly disrupt microbial enzymes.
- Ethylene oxide and chlorine dioxide are gaseous sterilants that work by alkylating protein and DNA.

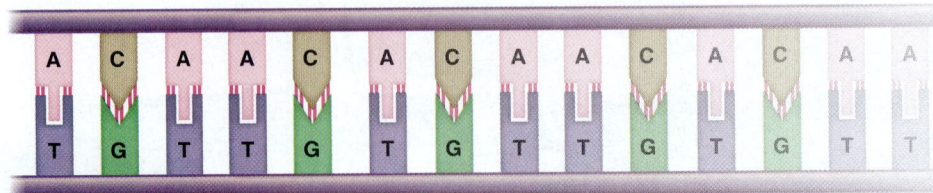

Multiple-Choice Questions   Bloom's Levels 1 and 2: Remember and Understand

Select the correct answer from the answers provided.

1. Microbial control methods that kill ____ are able to sterilize.
   a. viruses
   b. the tubercle bacillus
   c. endospores
   d. cysts

2. Sanitization is a process by which
   a. the microbial load on objects is reduced.
   b. objects are made sterile with chemicals.
   c. utensils are scrubbed.
   d. skin is debrided.

3. An example of an agent that lowers the surface tension of cells is
   a. phenol.
   b. chlorine.
   c. alcohol.
   d. formalin.

4. High temperatures ___ and low temperatures ___.
   a. sterilize; disinfect
   b. kill cells; inhibit cell growth
   c. denature proteins; burst cells
   d. speed up metabolism; slow down metabolism

5. Microbe(s) that is/are the target(s) of pasteurization include
   a. *Clostridium botulinum.*
   b. *Mycobacterium* species.
   c. *Salmonella* species.
   d. both b and c.

6. The primary mode of action of nonionizing radiation is to
   a. produce superoxide ions.
   b. make pyrimidine dimers.
   c. denature proteins.
   d. break disulfide bonds.

7. The most versatile method of sterilizing heat-sensitive liquids is
   a. UV radiation.
   b. exposure to ozone.
   c. beta propiolactone.
   d. filtration.

8. Select the correct sequence, from least to most resistant.
   a. most gram-negative bacteria, enveloped viruses, prions
   b. most gram-negative bacteria, staphylococcus, enveloped viruses
   c. most gram-negative bacteria, most gram-positive bacteria, prions
   d. enveloped viruses, protozoan cysts, bacterial endospores
   e. enveloped viruses, most gram-negative bacteria, most gram-positive bacteria

Critical Thinking   Bloom's Levels 3, 4, and 5: Apply, Analyze, and Evaluate

Critical thinking **is the ability to reason and solve problems using facts and concepts. These questions can be approached from a number of angles and, in most cases, they do not have a single correct answer.**

1. Briefly explain how the type of microorganisms present will influence the effectiveness of exposure to antimicrobial agents.

2. In the emergency room, an RN has cleaned an open wound with normal saline and hydrogen peroxide. You make a microscope slide of material from the cleaned wound, and at high magnification you observe bacterial cells. Explain whether these cells are likely to be dead or living.

3. Can you think of situations in which the same microbe would be considered a serious contaminant in one case and completely harmless in another?

4. Devise an experiment that will differentiate between bactericidal and bacteristatic effects.

5. Most antimicrobials that arrest protein function are nonselective as to the microbes they affect. Why would this be? What would the effect of these agents be on human skin or tissue, if applied there?

Gram-Negative

Visual Connections   Bloom's Level 5: Evaluate

This question connects previous images to a new concept.

1. **From chapter 3, figure 3.15.** Study this illustration of a gram-negative cell envelope. In what ways could alcohol damage these two membranes? How would that harm the cell?

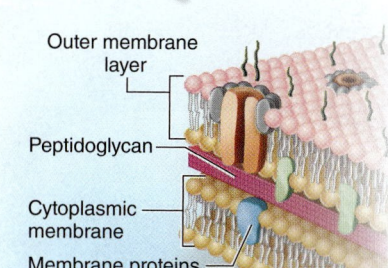

Outer membrane layer

Peptidoglycan

Cytoplasmic membrane

Membrane proteins

www.mcgrawhillconnect.com

Enhance your study of this chapter with study tools and practice tests. Also ask your instructor about the resources available through Connect, including the media-rich eBook, interactive learning tools, and animations.

CASE FILE

Not What We Were Expecting

I was working in a pediatric hospital in the emergency room. A 2-year-old girl with an obvious rash was brought in by her mother, who wondered whether the rash might be chickenpox. I took the child and her mother back to a cubicle and began to take the child's history. The child was healthy and had no major health problems. Her vaccinations were all current, including the chickenpox vaccine, which made a diagnosis of chickenpox unlikely. She had recently had otitis media, an ear infection, and was on her second-to-last day of antibiotic therapy with Ceclor, a cephalosporin antibiotic. The child had taken Ceclor on one other occasion for an ear infection.

The vital signs were normal. She did not have a fever. The rash was maculopapular, a flat red rash with tiny pimplelike eruptions in the center. The rash covered her face, chest, back, arms, and legs. In fact, it covered almost her entire body except the palms of her hands and soles of her feet. The child was clearly uncomfortable and was scratching exposed areas of skin.

The mother reported that she had taken the child for a haircut in the afternoon and noticed a few spots on the child's neck. She did not think much of the spots at the time. She dropped the little girl off at her mother-in-law's for child care while she went to work as a waitress. When she returned to her mother-in-law's house to pick the child up after her shift, the child was awake, irritable, and covered in the rash. Alarmed, the mother brought the child immediately to the emergency room.

After recording the history and the child's vital signs, I went to find the physician and reported my findings. He told me that he felt he knew what the problem was but would quickly examine the child first before telling me his diagnosis. The doctor looked at the patient's rash and told the mother, "Just as I thought. The rash is a reaction to the Ceclor your daughter has been taking. Stop the Ceclor and the rash will go away." The mother was surprised because she had always thought that an allergic reaction to a drug would start with the first dose. The physician told her that an allergic reaction could begin at any time, even after taking the same drug numerous times.

The child's ears were checked and there was no sign of infection. The girl and her mother were discharged after receiving a prescription for an antihistamine to help control the itching.

- What category of antibiotic does Ceclor fall under?
- What is the mechanism underlying the allergic response to an antibiotic?

Case File Wrap-Up appears on page 284.

CHAPTER 10

Antimicrobial Treatment

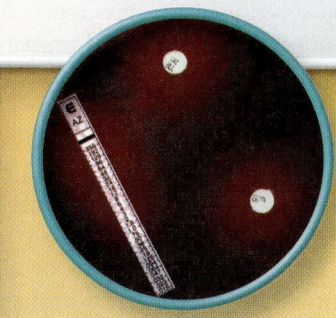

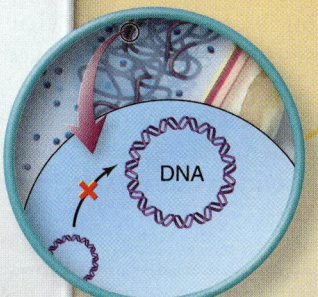

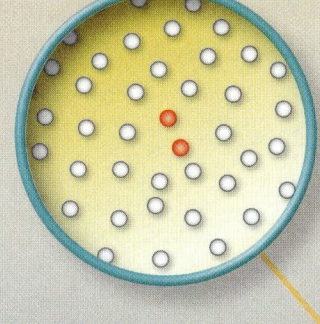

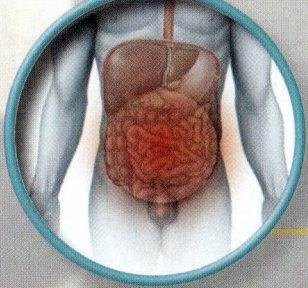

IN THIS CHAPTER...

10.1 Principles of Antimicrobial Therapy

1. State the main goal of antimicrobial treatment.
2. Identify the sources for the most commonly used antimicrobials.
3. Describe two methods for testing antimicrobial susceptibility.
4. Define *therapeutic index*, and identify whether a high or a low index is preferable.

10.2 Interactions Between Drug and Microbe

5. Explain the concept of selective toxicity.
6. List the five major targets of antimicrobial agents.
7. Identify which categories of drugs are most selectively toxic and why.
8. Distinguish between broad-spectrum and narrow-spectrum antimicrobials, and explain the significance of the distinction.
9. Identify the microbes against which the various penicillins are effective.
10. Explain the mode of action of penicillinases and their importance in treatment.
11. Identify two antimicrobials that act by inhibiting protein synthesis.
12. Explain how drugs targeting folic acid synthesis work.
13. Identify one example of a fluoroquinolone.
14. Describe the mode of action of drugs that target the cytoplasmic or cell membrane.
15. Discuss how treatments of biofilm and nonbiofilm infections differ.
16. Name the four main categories of antifungal agents.
17. Explain why antiprotozoal and antihelminthic drugs are likely to be more toxic than antibacterial drugs.
18. List the three major targets of action of antiviral drugs.

10.3 Antimicrobial Resistance

19. Discuss two possible ways that microbes acquire antimicrobial resistance.
20. List five cellular or structural mechanisms that microbes use to resist antimicrobials.
21. Discuss at least two novel antimicrobial strategies that are under investigation.

10.4 Interactions Between Drug and Host

22. Distinguish between drug toxicity and allergic reactions to drugs.
23. Explain what a superinfection is and how it occurs.

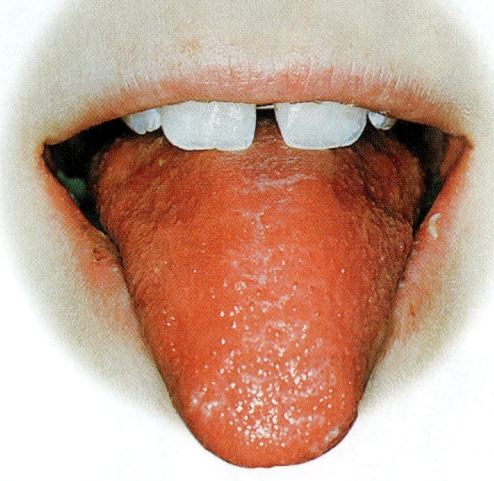

Scarlet fever, once common, is now treatable with antibiotics. Strawberry tongue is one of its symptoms.

10.1 Principles of Antimicrobial Therapy

A hundred years ago in the United States, one out of three children was expected to die of an infectious disease before the age of 5. Early death or severe lifelong debilitation from scarlet fever, diphtheria, tuberculosis, meningitis, and many other bacterial diseases was a fearsome yet undeniable fact of life to most of the world's population. The introduction of modern drugs to control infections in the 1930s was a medical revolution that has added significantly to the life span and health of humans. It is no wonder that, for many years, antibiotics in particular were regarded as miracle drugs. Although antimicrobial drugs have greatly reduced the incidence of certain infections, they have definitely not eradicated infectious disease and probably never will. In fact, many doctors are now warning that we are dangerously close to a post-antibiotic era, where the drugs we have are no longer effective.

The goal of antimicrobial chemotherapy is deceptively simple: Administer a drug to an infected person, which destroys the infective agent without harming the host's cells. In actuality, this goal is rather difficult to achieve, because many (often contradictory) factors must be taken into account. The ideal drug should be easy to administer, yet be able to reach the infectious agent anywhere in the body; be absolutely toxic to the infectious agent, while being nontoxic to the host; and remain active in the body as long as needed, yet be safely and easily broken down and excreted. Additionally, microbes in biofilms often require different drugs than when they are not in biofilms. In short, the perfect drug does not exist–but by balancing drug characteristics against one another, a satisfactory compromise can usually be achieved **(table 10.1)**.

Chemotherapeutic agents are described with regard to their origin, range of effectiveness, and whether they are naturally produced or chemically synthesized. A few of the more important terms you will encounter are found in **table 10.2**.

Table 10.1 Characteristics of the Ideal Antimicrobial Drug
• Selectively toxic to the microbe but nontoxic to host cells
• Microbicidal rather than microbistatic
• Relatively soluble; functions even when highly diluted in body fluids
• Remains potent long enough to act and is not broken down or excreted prematurely
• Does not lead to the development of antimicrobial resistance
• Complements or assists the activities of the host's defenses
• Remains active in tissues and body fluids
• Readily delivered to the site of infection
• Reasonably priced
• Does not disrupt the host's health by causing allergies or predisposing the host to other infections

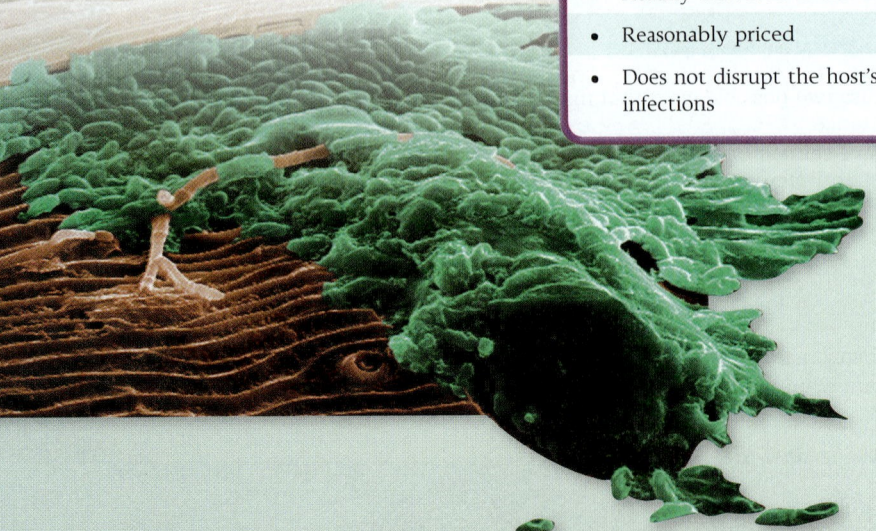

Bacterial biofilm formed on the surface of a spider.

Table 10.2 Terminology of Chemotherapy	
Chemotherapeutic Drug	Any chemical used in the treatment, relief, or prophylaxis of a disease
Prophylaxis	Use of a drug to prevent imminent infection of a person at risk
Antimicrobial Chemotherapy	The use of chemotherapeutic drugs to control infection
Antimicrobials	All-inclusive term for any antimicrobial drug, regardless of its origin
Antibiotics	Substances produced by the natural metabolic processes of some microorganisms that can inhibit or destroy other microorganisms
Semisynthetic Drugs	Drugs that are chemically modified in the laboratory after being isolated from natural sources
Synthetic Drugs	Drugs produced entirely by chemical reactions
Narrow-Spectrum (Limited Spectrum)	Antimicrobials effective against a limited array of microbial types—for example, a drug effective mainly against gram-positive bacteria
Broad-Spectrum (Extended Spectrum)	Antimicrobials effective against a wide variety of microbial types—for example, a drug effective against both gram-positive and gram-negative bacteria

A Note About Chemotherapy

The word **chemotherapy** is commonly associated with the treatment of cancer. As you see in table 10.2, its official meaning is broader than that and can also be applied to antimicrobial treatment.

In this chapter, we describe different types of antibiotic drugs, their mechanism of action, and the types of microbes on which they are effective. The organ system chapters 16 through 21 list specific disease agents and the drugs used to treat them.

The Origins of Antimicrobial Drugs

Nature is a prolific producer of antimicrobial drugs. Antibiotics, after all, are natural metabolic products of aerobic bacteria and fungi. By inhibiting the growth of other microorganisms in the same habitat (antagonism), antibiotic producers presumably enjoy less competition for nutrients and space. The greatest numbers of current antibiotics are derived from bacteria in the genera *Streptomyces* and *Bacillus* and from molds in the genera *Penicillium* and *Cephalosporium.* Not only have chemists created new drugs by altering the structure of naturally occurring antibiotics, they are actively searching for metabolic compounds with antimicrobial effects in species other than bacteria and fungi.

Identifying the Microbe and Starting Treatment

Before actual antimicrobial therapy can begin, it is important that at least three factors be known:

1. the nature of the microorganism causing the infection,
2. the degree of the microorganism's susceptibility (also called sensitivity) to various drugs, and
3. the overall medical condition of the patient.

Identification of infectious agents from body specimens should be attempted as soon as possible. It is especially important that such specimens be taken before any antimicrobial drug is given, before the drug reduces the numbers of the infectious agent. Direct examination of body fluids, sputum, or stool is a rapid initial method for detecting and perhaps even identifying bacteria or fungi. A doctor often begins the therapy on the basis of such immediate findings, or even on the basis of an informed best guess. For instance, if a sore throat appears to be caused by *Streptococcus pyogenes,* the physician might prescribe penicillin, because this species

First mass-produced in 1944, penicillin saved many lives in WWII.

seems to be almost universally sensitive to it so far. If the infectious agent is not or cannot be isolated, epidemiological statistics may be required to predict the most likely agent in a given infection. For example, *Streptococcus pneumoniae* accounts for the majority of cases of bacterial meningitis in children, followed by *Neisseria meningitidis* (discussed in detail in chapter 17).

Testing for the Drug Susceptibility of Microorganisms

Some infectious agents require antimicrobial sensitivity testing and some do not. Testing is essential in those groups of bacteria commonly showing resistance, such as *Staphylococcus* species, *Neisseria gonorrhoeae, Streptococcus pneumoniae, Enterococcus faecalis,* and the aerobic gram-negative intestinal bacilli. On the other hand, drug testing in fungal or protozoal infections is difficult and is often unnecessary because the antimicrobial agents generally target all representatives of these groups. Lastly, when certain groups, such as group A streptococci and all anaerobes (except *Bacteroides*), are known to be uniformly susceptible to Penicillin G, testing may not be necessary unless the patient is allergic to penicillin.

Selection of a proper antimicrobial agent begins by demonstrating the *in vitro* activity of several drugs against the infectious agent by means of standardized methods. In general, these tests involve exposing a pure culture of the bacterium to several different drugs and observing the effects of the drugs on growth.

The *Kirby-Bauer* technique is an agar diffusion test that provides useful data on antimicrobial susceptibility. In this test, the surface of a plate of special medium is spread with the test bacterium, and small discs containing a premeasured amount of antimicrobial are dispensed onto the bacterial lawn. After incubation, the zone of inhibition surrounding the discs is measured and compared with a standard for each drug **(table 10.3** and **figure 10.1).** This profile of antimicrobial sensitivity is called an antibiogram. The Kirby-Bauer procedure is less effective for bacteria that are

Penicillins damage bacterial cell walls.

Table 10.3 Results of a Sample Kirby-Bauer Test

Drug	Zone Sizes (mm) Required for Susceptibility (S)	Resistance (R)	Example Results (mm) for *Staphylococcus aureus*	Evaluation
Bacitracin	>13	<8	15	S
Chloramphenicol	>18	<12	20	S
Erythromycin	>18	<13	15	I
Gentamicin	>13	<12	16	S
Kanamycin	>18	<13	20	S
Neomycin	>17	<12	12	R
Penicillin G	>29	<20	10	R
Polymyxin B	>12	<8	10	R
Streptomycin	>15	<11	11	R
Vancomycin	>12	<9	15	S
Tetracycline	>19	<14	25	S

R = resistant, I = intermediate, S = sensitive

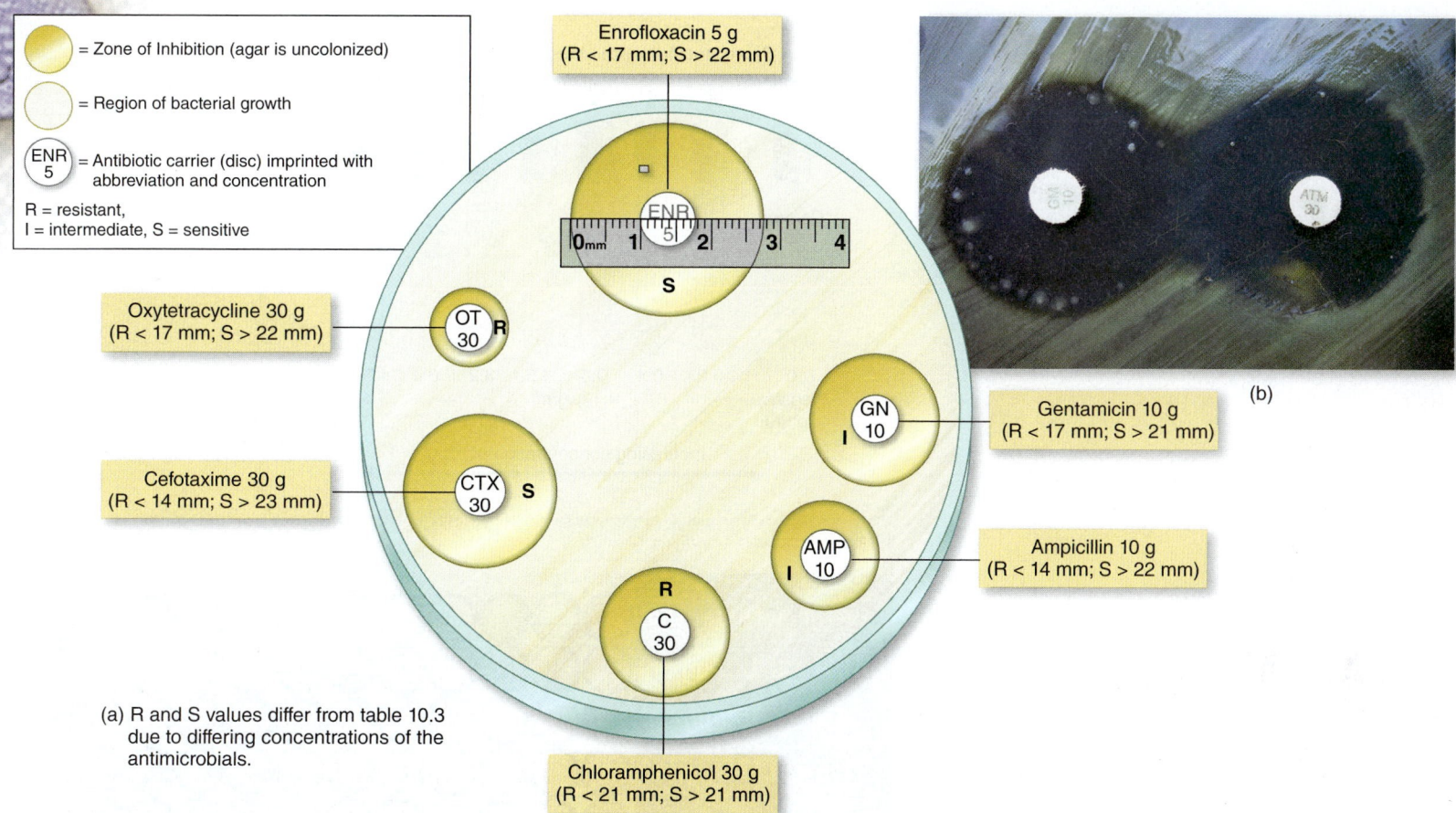

Legend:
- = Zone of Inhibition (agar is uncolonized)
- = Region of bacterial growth
- ENR 5 = Antibiotic carrier (disc) imprinted with abbreviation and concentration

R = resistant,
I = intermediate, S = sensitive

Enrofloxacin 5 g
(R < 17 mm; S > 22 mm)

Oxytetracycline 30 g
(R < 17 mm; S > 22 mm)

Cefotaxime 30 g
(R < 14 mm; S > 23 mm)

Gentamicin 10 g
(R < 17 mm; S > 21 mm)

Ampicillin 10 g
(R < 14 mm; S > 22 mm)

Chloramphenicol 30 g
(R < 21 mm; S > 21 mm)

(a) R and S values differ from table 10.3 due to differing concentrations of the antimicrobials.

(b)

Figure 10.1 Technique for preparation and interpretation of disc diffusion tests. (a) Standardized methods are used to spread a lawn of bacteria over the medium. A dispenser delivers several drugs onto a plate, followed by incubation. Interpretation of results: During incubation, antimicrobials become increasingly diluted as they diffuse out of the disc into the medium. If the test bacterium is sensitive to a drug, a zone of inhibition develops around its disc. Roughly speaking, the larger the size of this zone, the greater is the bacterium's sensitivity to the drug. The diameter of each zone is measured in millimeters and evaluated for susceptibility or resistance by means of a comparative standard (see table 10.3). (b) Results of test with *Escherichia hermannii* indicate a synergistic effect between two different antibiotics (note the expanded zone between these two drugs).

anaerobic, highly fastidious, or slow-growing (*Mycobacterium*). An alternative diffusion system that provides additional information on drug effectiveness is the E-test **(figure 10.2).**

More sensitive and quantitative results can be obtained with tube dilution tests. First the antimicrobial is diluted serially in tubes of broth, and then each tube is inoculated with a small uniform sample of pure culture, incubated, and examined for growth (turbidity). The smallest concentration (highest dilution) of drug that visibly inhibits growth is called the **minimum inhibitory concentration,** or **MIC.** The MIC is useful in determining the smallest effective dosage of a drug and in providing a comparative index against other antimicrobials

Figure 10.2 Alternative to the Kirby-Bauer procedure.
Another diffusion test is the E-test, which uses a strip to produce the zone of inhibition. The advantage of the E-test is that the strip contains a gradient of drug calibrated in micrograms. This way, the MIC can be measured by observing the mark on the strip that corresponds to the edge of the zone of inhibition.

Figure 10.3 Tube dilution test for determining the minimum inhibitory concentration (MIC). (a) The antibiotic is diluted serially through tubes of liquid nutrient from right to left. All tubes are inoculated with an identical amount of a test bacterium and then incubated. The first tube on the left is a control that lacks the drug and shows maximum growth. The dilution of the first tube in the series that shows no growth (no turbidity) is the MIC. (b) Microbroth dilution in a multiwell plate adapted for eukaryotic pathogens. Here, amphotericin B, flucytosine, and several azole drugs are tested on a pathogenic yeast. Pink indicates growth and blue, no growth. Numbers indicate the dilution of the MIC, and Xs show the first well without growth.

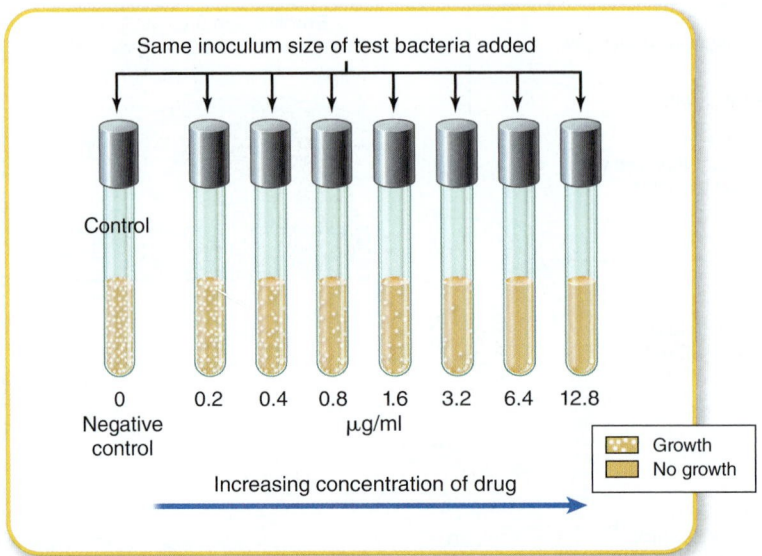

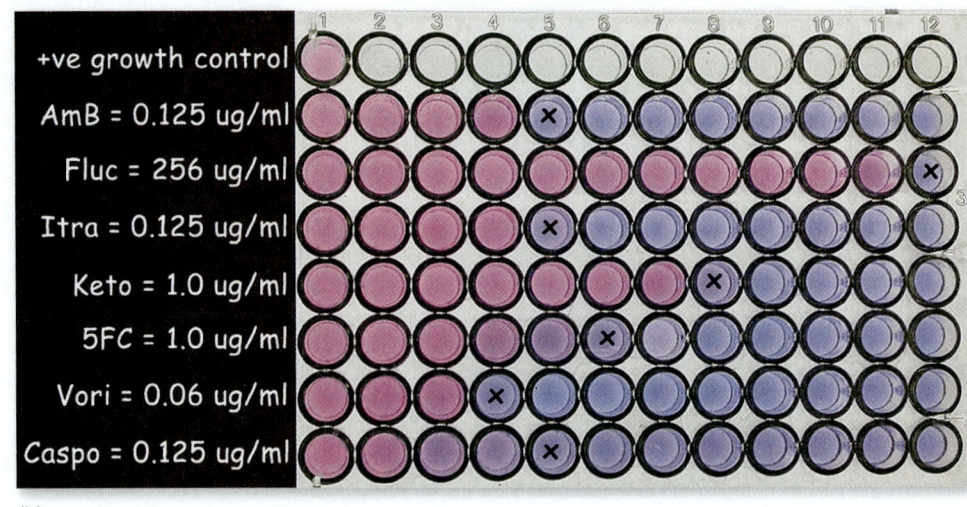

(figure 10.3). In many clinical laboratories, these antimicrobial testing procedures are performed in automated machines that can test dozens of drugs simultaneously.

The MIC and Therapeutic Index

The results of antimicrobial sensitivity tests guide the physician's choice of a suitable drug. If therapy has already commenced, it is imperative to determine if the tests bear out the use of that particular drug. Once therapy has begun, it is important to observe the patient's clinical response, because the *in vitro* activity of the drug is not always correlated with its *in vivo* effect. When antimicrobial treatment fails, the failure is due to one or more the following:

- the inability of the drug to diffuse into that body compartment (the brain, joints, skin);
- resistant microbes in the infection that didn't make it into the sample collected for testing; or
- an infection caused by more than one pathogen (mixed), some of which are resistant to the drug.

If therapy does fail, a different drug, combined therapy, or a different method of administration must be considered.

Because drug toxicity to the host is of concern, it is best to choose the one with high selective toxicity for the infectious agent and low human toxicity. The **therapeutic index (TI)** is defined as the ratio of the dose of the drug that is toxic to humans as compared to its minimum effective (therapeutic) dose. The closer these two figures are (the smaller the ratio), the greater is the potential for toxic drug reactions. For example, a drug that has a therapeutic index of

$$\frac{10\ \mu g/mL\ \text{(toxic dose)}}{9\ \mu g/mL\ \text{(MIC)}}\ TI = 1.1$$

is a riskier choice than one with a therapeutic index of

$$\frac{10\ \mu g/mL}{1\ \mu g/mL}\ TI = 10$$

When a series of drugs being considered for therapy have similar MICs, the drug with the highest therapeutic index usually has the widest margin of safety.

The physician must also take a careful history of the patient to discover any preexisting medical conditions that will influence the activity of the drug or the response of the patient. A history of allergy to a certain class of drugs precludes the use of that drug and any drugs related to it. Underlying liver or kidney disease will ordinarily require changing the drug therapy, because these organs play such an important part in metabolizing or excreting the drug. Infants, the elderly, and pregnant women require special precautions. For example, age can diminish gastrointestinal absorption and organ function, and most antimicrobial drugs cross the placenta and could affect fetal development.

Patients must be asked about other drugs they are taking, because incompatibilities can result in increased toxicity or failure of one or more of the drugs. For example, the combination of aminoglycosides and cephalosporins can be toxic to kidneys; antacids reduce the absorption of isoniazid; and the interaction of tetracycline or rifampin with oral contraceptives can abolish the contraceptive's effect. Some drug combinations (penicillin with certain aminoglycosides, or amphotericin B with flucytosine) act synergistically, so that reduced doses of each can be used in combined therapy. Other concerns in choosing drugs include any genetic or metabolic abnormalities in the patient, the site of infection, the route of administration, and the cost of the drug.

The Art and Science of Choosing an Antimicrobial Drug

Even when all the information is in, the final choice of a drug is not always easy or straightforward. Consider the hypothetical case of an elderly alcoholic patient with pneumonia caused by *Klebsiella* and complicated by diminished liver and kidney function. All drugs must be given by injection because of prior damage to the gastrointestinal lining and poor absorption. Drug tests show that the infectious agent is sensitive to third-generation cephalosporins, gentamicin, imipenem, and azlocillin. The patient's history shows previous allergy to the penicillins, so these would be ruled out. Drug interactions occur between alcohol and the cephalosporins, which are also associated with serious bleeding in elderly patients, so this may not be a good choice. Aminoglycosides such as gentamicin are toxic to the kidneys and poorly cleared by damaged kidneys. Imipenem causes intestinal discomfort, but it has less toxicity and would be a viable choice.

In the case of a cancer patient with severe systemic *Candida* infection, there will be fewer criteria to weigh. Intravenous amphotericin B or fluconazole are the only possible choices, despite drug toxicity and other possible adverse side effects. In a life-threatening situation in which a dangerous chemotherapy is perhaps the only chance for survival, the choices are reduced and the priorities are different.

While choosing the right drug is an art and a science, requiring the consideration of many different things, the process has been made simpler–or at least more portable–with the advent of smartphones and applications ("apps"). Most doctors now have the information literally at their fingertips, when they pull their smartphones out of their pockets.

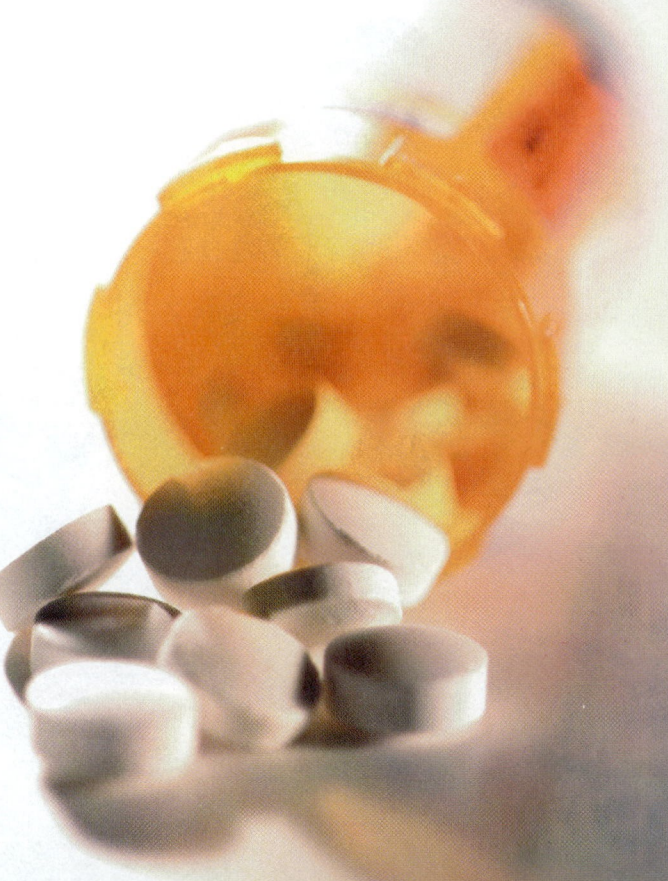

10.2 Interactions Between Drug and Microbe

The goal of antimicrobial drugs is either to disrupt the cell processes or structures of bacteria, fungi, and protozoa or to inhibit virus replication. Most of the drugs used in chemotherapy interfere with the function of enzymes required to synthesize or assemble macromolecules, or they destroy structures already formed in the cell. Above all, drugs should be **selectively toxic,** which means they should kill or inhibit microbial cells without simultaneously damaging host tissues. This concept of selective toxicity is central to antibiotic treatment, and the best drugs in current use are those that block the actions or synthesis of molecules in microorganisms but not in vertebrate cells. Examples of drugs with excellent selective toxicity are those that block the synthesis of the cell wall in bacteria (penicillins). They have low toxicity and few direct effects on human cells because human cells lack the chemical peptidoglycan and are thus unaffected by this action of the antibiotic. Among the most toxic to human cells are drugs that act upon a structure common to both the infective agent and the host cell, such as the cytoplasmic membrane (e.g., amphotericin B, used to treat fungal infections). As the characteristics of the infectious agent become more and more similar to those of the host cell, selective toxicity becomes more difficult to achieve, and undesirable side effects are more likely to occur.

Mechanisms of Drug Action

If the goal of chemotherapy is to disrupt the structure or function of an organism to the point where it can no longer survive, then the first step toward this goal is to identify the structural and metabolic needs of a living cell. Once

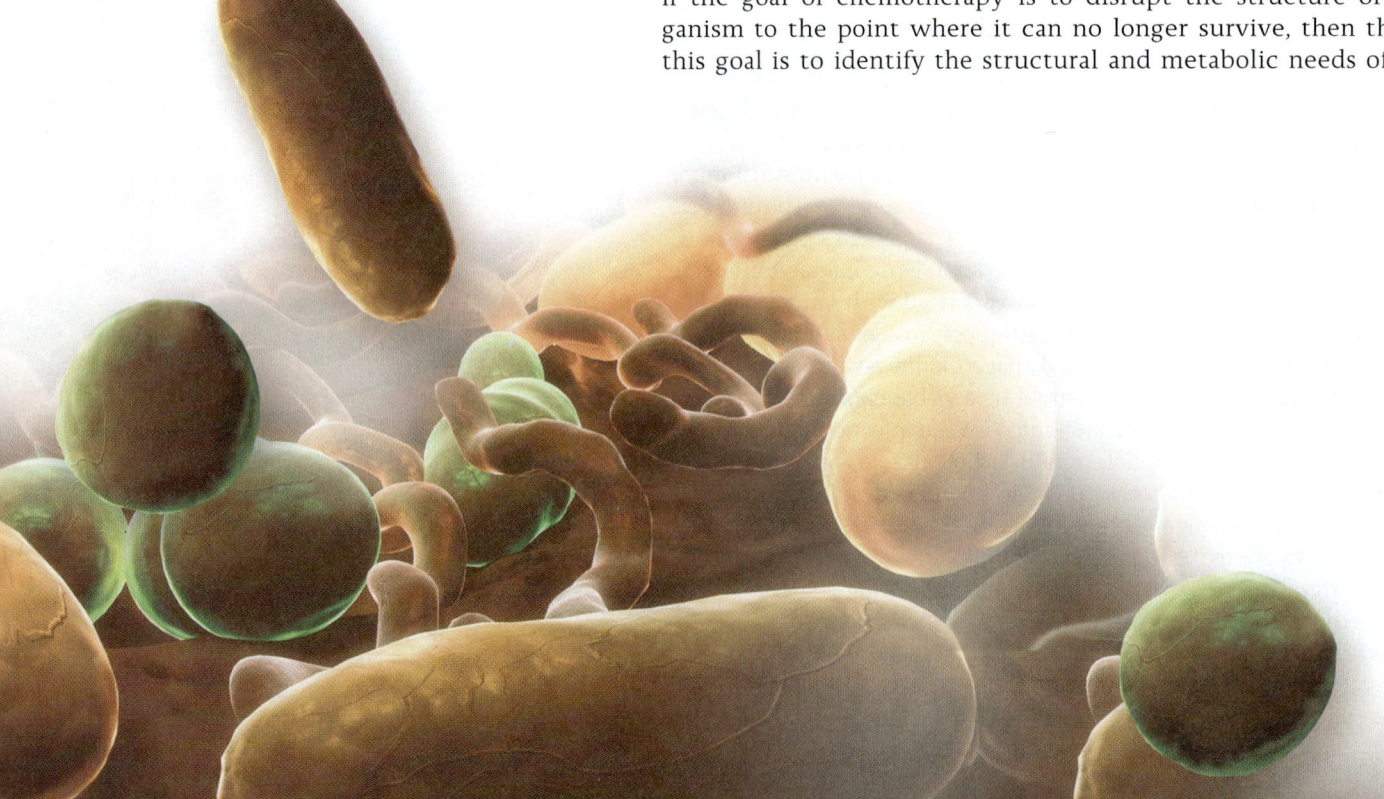

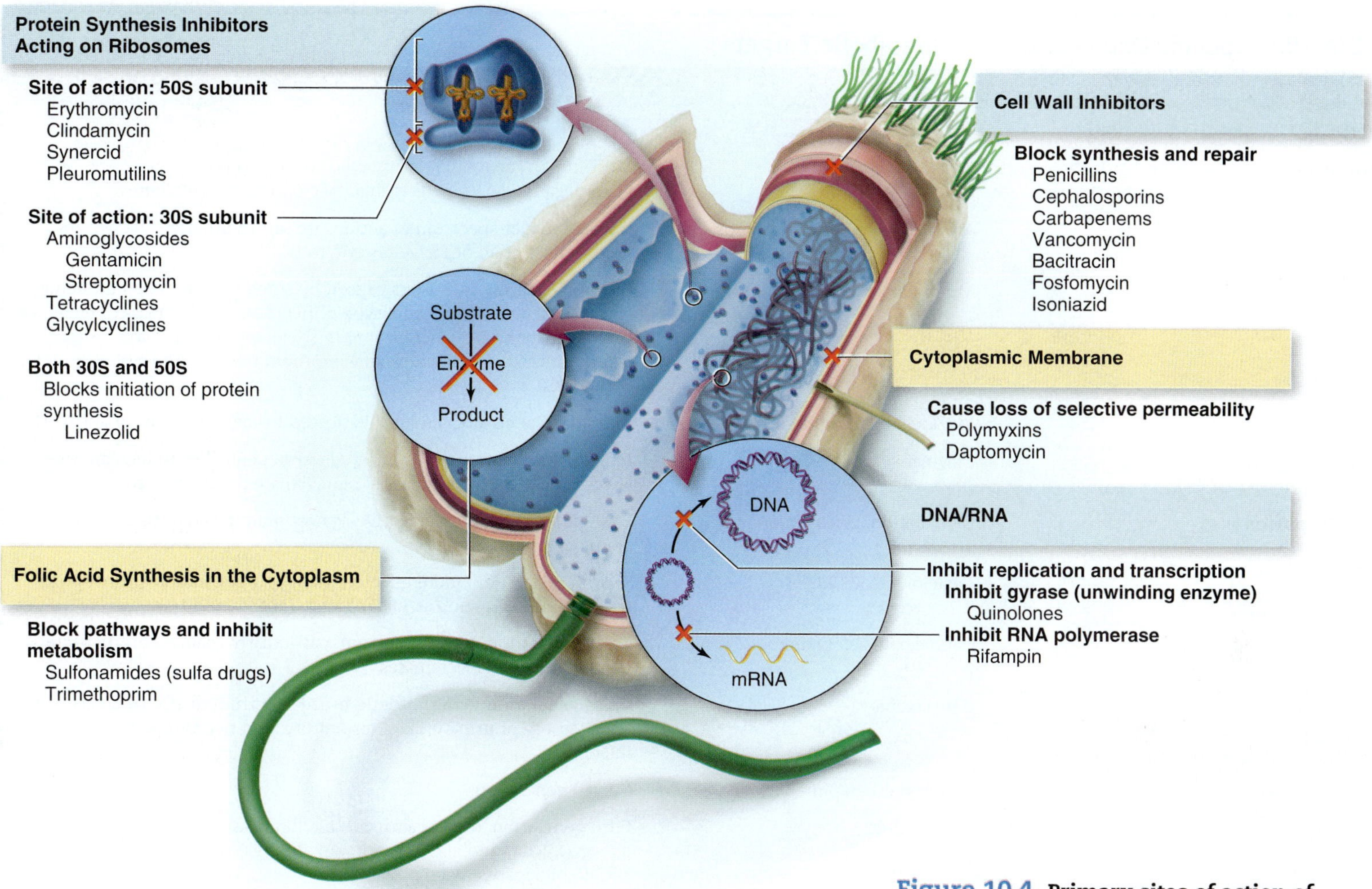

**Protein Synthesis Inhibitors
Acting on Ribosomes**

Site of action: 50S subunit
Erythromycin
Clindamycin
Synercid
Pleuromutilins

Site of action: 30S subunit
Aminoglycosides
Gentamicin
Streptomycin
Tetracyclines
Glycylcyclines

Both 30S and 50S
Blocks initiation of protein
synthesis
Linezolid

Folic Acid Synthesis in the Cytoplasm

**Block pathways and inhibit
metabolism**
Sulfonamides (sulfa drugs)
Trimethoprim

Substrate

Enzyme

Product

Cell Wall Inhibitors

Block synthesis and repair
Penicillins
Cephalosporins
Carbapenems
Vancomycin
Bacitracin
Fosfomycin
Isoniazid

Cytoplasmic Membrane

Cause loss of selective permeability
Polymyxins
Daptomycin

DNA

mRNA

DNA/RNA

**Inhibit replication and transcription
Inhibit gyrase (unwinding enzyme)**
Quinolones
Inhibit RNA polymerase
Rifampin

Figure 10.4 **Primary sites of action of
antimicrobial drugs on bacterial cells.**

the requirements of a living cell have been determined, methods of removing, disrupting, or interfering with these requirements can be used as potential chemotherapeutic strategies. The metabolism of an actively dividing cell is marked by the production of new cell wall components (in most cells), DNA, RNA, proteins, and cytoplasmic membrane. Consequently, antimicrobial drugs are divided into categories based on which of these metabolic targets they affect. These categories are outlined in **figure 10.4** and include the following:

1. inhibition of cell wall synthesis,
2. inhibition of nucleic acid (RNA and DNA) structure and function,
3. inhibition of protein synthesis,
4. interference with cytoplasmic membrane structure or function, and
5. inhibition of folic acid synthesis.

As you will see, these categories are not completely discrete, and some effects can overlap. **Table 10.4** describes these categories, as well as common drugs comprising each of these categories.

NCLEX® PREP

2. In evaluating a treatment plan, the therapeutic index (TI) is calculated as 1.0. Based on this result, how would the nurse interpret this information?

   a. *The medication can be utilized as there is less potential for a toxic reaction.*

   b. *The medication can be used as long as the dosage is within therapeutic range.*

   c. *There is no chance of a drug reaction occurring based on this result.*

   d. *A different medication should be considered for use in the treatment plan.*

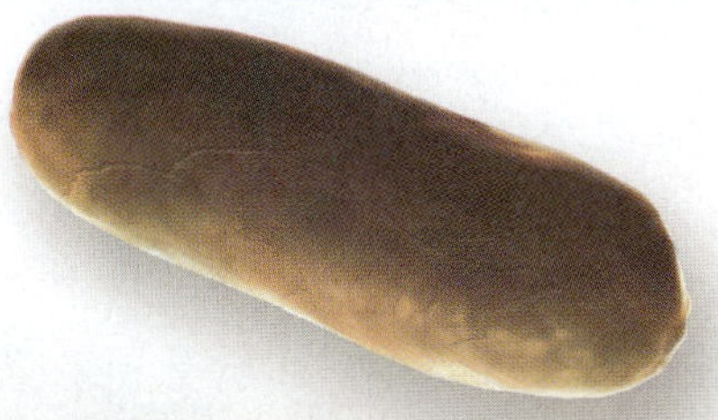

Table 10.4 Specific Drugs and Their Metabolic Targets

Drug Class/Mechanism of Action	Subgroups	Uses/Characteristics
Drugs That Target the Cell Wall		
Penicillins	Penicillins G and V	Most important natural forms used to treat gram-positive cocci, some gram-negative bacteria (meningococci, syphilis, spirochetes)
	Ampicillin, carbenicillin, amoxicillin	Have a broader spectrum of action, are semisynthetic; used against gram-negative enteric rods
	Methicillin, nafcillin, cloxacillin	Useful in treating infections caused by some penicillinase-producing bacteria (penicillinase is one type of beta-lactamase, a class of enzymes that destroy the beta-lactam ring in some antibiotics; some bacteria can produce these enzymes, making them resistant to these types of antibiotics)
	Mezlocillin, azlocillin	Extended spectrum; can be substituted for combinations of antibiotics
	Clavulanic acid	Inhibits beta-lactamase enzymes; added to penicillins to increase their effectiveness in the presence of penicillinase-producing bacteria
Cephalosporins	Cephalothin, cefazolin	First generation°; most effective against gram-positive cocci, few gram-negative bacteria
	Cefaclor, cefonicid	Second generation; more effective than first generation against gram-negative bacteria such as *Enterobacter, Proteus,* and *Haemophilus*
	Cephalexin, cefotaxime	Third generation; broad-spectrum, particularly against enteric bacteria that produce beta-lactamases
	Ceftriaxone	Third generation; semisynthetic broad-spectrum drug that treats wide variety of urinary, skin, respiratory, and nervous system infections
	Cefpirome, cefepime	Fourth generation
	Ceftobiprole	Fifth generation; used against methicillin-resistant *Staphylococcus aureus* (MRSA) and also against penicillin-resistant gram-positive and gram-negative bacteria
Carbapenems	Doripenem, imipenem	Powerful but potentially toxic; reserved for use when other drugs are not effective
	Aztreonam	Narrow-spectrum; used to treat gram-negative aerobic bacilli causing pneumonia, septicemia, and urinary tract infections; effective for those who are allergic to penicillin
Miscellaneous Drugs That Target the Cell Wall	Bacitracin	Narrow-spectrum; used to combat superficial skin infections caused by streptococci and staphylococci; main ingredient in Neosporin
	Isoniazid	Used to treat *Mycobacterium tuberculosis*, but only against growing cells; used in combination with other drugs in active tuberculosis
	Vancomycin	Narrow-spectrum of action; used to treat staphylococcal infections in cases of penicillin and methicillin resistance or in patients with an allergy to penicillin
	Fosfomycin tromethamine	Phosphoric acid agent; effective as an alternative treatment for urinary tract infection caused by enteric bacteria
Drugs That Target Protein Synthesis		
Aminoglycosides Insert on sites on the 30S subunit and cause the misreading of the mRNA, leading to abnormal proteins	Streptomycin	Broad-spectrum; used to treat infections caused by gram-negative rods, certain gram-positive bacteria; used to treat bubonic plague, tularemia, and tuberculosis; vancomycin also targets protein synthesis as well as cell walls

°New improved versions of drugs are referred to as new "generations."

Table 10.4 (continued)

Drug Class/Mechanism of Action	Subgroups	Uses/Characteristics
Drugs That Target Protein Synthesis (continued)		
Tetracyclines Block the attachment of tRNA on the A acceptor site and stop further protein synthesis	Tetracycline, terramycin	Effective against gram-positive and gram-negative rods and cocci, aerobic and anaerobic bacteria, mycoplasmas, rickettsias, and spirochetes
Glycylcyclines	Tigecycline	Newer derivative of tetracycline; effective against bacteria that have become resistant to tetracyclines
Macrolides Inhibit translocation of the subunit during translation (erythromycin)	Erythromycin, clarithromycin, azithromycin	Relatively broad-spectrum, semisynthetic; used in treating ear, respiratory, and skin infections, as well as *Mycobacterium* infections in AIDS patients
Miscellaneous Drugs That Target Protein Synthesis	Clindamycin	Broad-spectrum antibiotic used to treat penicillin-resistant staphylococci, serious anaerobic infections of the stomach and intestines unresponsive to other antibiotics
	Quinupristin and dalfopristin (Synercid)	A combined antibiotic from the streptogramin group of drugs; effective against *Staphylococcus* and *Enterococcus* species causing endocarditis and surgical infections, including resistant strains
	Linezolid	Synthetic drug from the oxazolidinones; a novel drug that inhibits the initiation of protein synthesis; used to treat antibiotic-resistant organisms such as MRSA and VRE
Drugs That Target Folic Acid Synthesis		
Sulfonamides Interfere with folate metabolism by blocking enzymes required for the synthesis of tetrahydrofolate, which is needed by the cells for folic acid synthesis and eventual production of DNA, RNA, and amino acids	Sulfasoxazole	Used to treat shigellosis, acute urinary tract infections, certain protozoal infections
	Silver sulfadiazine	Used to treat burns, eye infections (in ointment and solution forms)
	Trimethoprim	Inhibits the enzymatic step in an important metabolic pathway that comes just before the step inhibited by sulfonamides; trimethoprim often given in conjunction with sulfamethoxazole because of this synergistic effect; used to treat *Pneumocystis jiroveci* in AIDS patients
Drugs That Target DNA or RNA		
Fluoroquinolones Inhibit DNA unwinding enzymes or helicases, thereby stopping DNA transcription	Nalidixic acid	First generation; rarely used anymore
	Ciprofloxacin, ofloxacin	Second generation
	Levofloxacin	Third generation; used against gram-positive organisms, including some that are resistant to other drugs
	Trovafloxacin	Fourth generation; effective against anaerobic organisms
Miscellaneous Drugs That Target DNA or RNA	Rifamycin (altered chemically into rifampin)	Limited in spectrum because it cannot pass through the cell envelope of many gram-negative bacilli; mainly used to treat infections caused by gram-positive rods and cocci and a few gram-negative bacteria; used to treat leprosy and tuberculosis
Drugs That Target Cytoplasmic or Cell Membranes		
Polymyxins Interact with membrane phospholipids; distort the cell surface and cause leakage of protein and nitrogen bases, particularly in gram-negative bacteria	Polymyxin B and E	Used to treat drug-resistant *Pseudomonas aeruginosa* and severe urinary tract infections caused by gram-negative rods
	Daptomycin	Most active against gram-positive bacteria

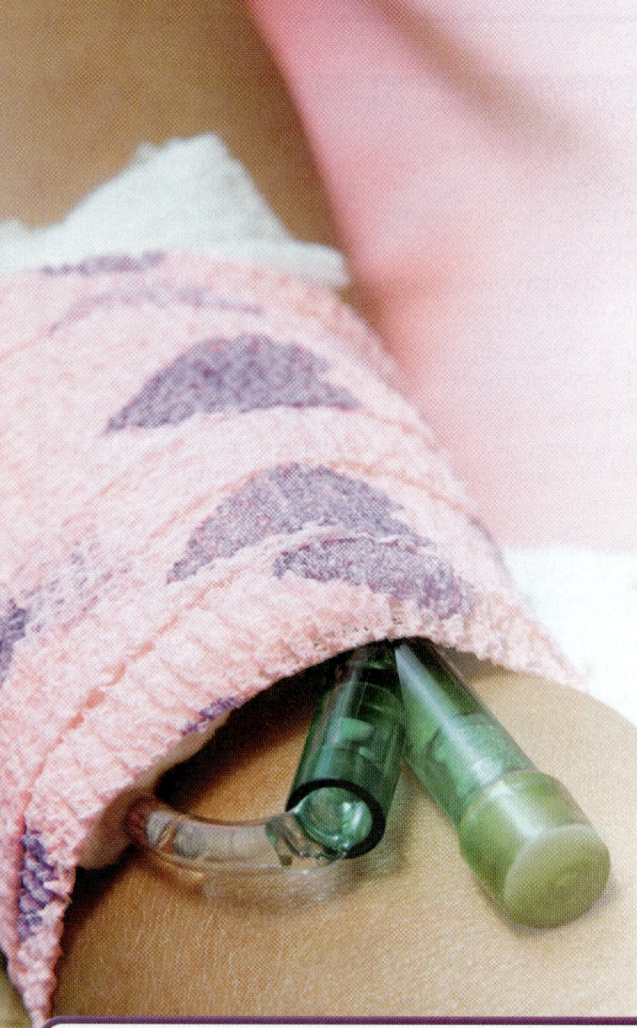

Spectrum of Activity

Scores of antimicrobial drugs are marketed in the United States. Although the medical and pharmaceutical literature contains a wide array of names for antimicrobials, most of them are variants of a small number of drug families. One of the most useful ways of categorizing antimicrobials, which you have already encountered in the previous section, is to designate them as either **broad-spectrum** or **narrow-spectrum.** Broad-spectrum drugs are effective against more than one group of bacteria, whereas narrow-spectrum drugs generally target a specific group. **Table 10.5** demonstrates that tetracyclines are broad-spectrum, whereas polymyxin and even penicillins are narrow-spectrum agents.

Since penicillin is such a familiar antibiotic, and since the alterations in the molecule over the years illustrate how antibiotics are developed and improved upon, we provide an overview in **table 10.6**. Here you will see that original penicillin was narrow-spectrum and susceptible to microbial counterattacks. Later penicillins were developed to overcome those two limitations.

Referring back to table 10.4, you can view details about various antimicrobial drugs based on which of the five major mechanisms they target.

Antibiotics and Biofilms

As you read in chapter 6, biofilm inhabitants behave differently than their free-living counterparts. One of the major ways they differ—at least from a medical perspective—is that they are often unaffected by the same antimicrobials that work against them when they are free-living. When this was first recognized, it was assumed that it was a problem of penetration, that the (often ionically charged) antimicrobial drugs could not penetrate the sticky extracellular material surrounding biofilm organisms. While that is a factor, there is something more important contributing to biofilm resistance: the different phenotype expressed by biofilm bacteria. When secured to surfaces, they express different genes and therefore have different antibiotic susceptibility profiles.

Table 10.5 Spectrum of Activity for Antibiotics

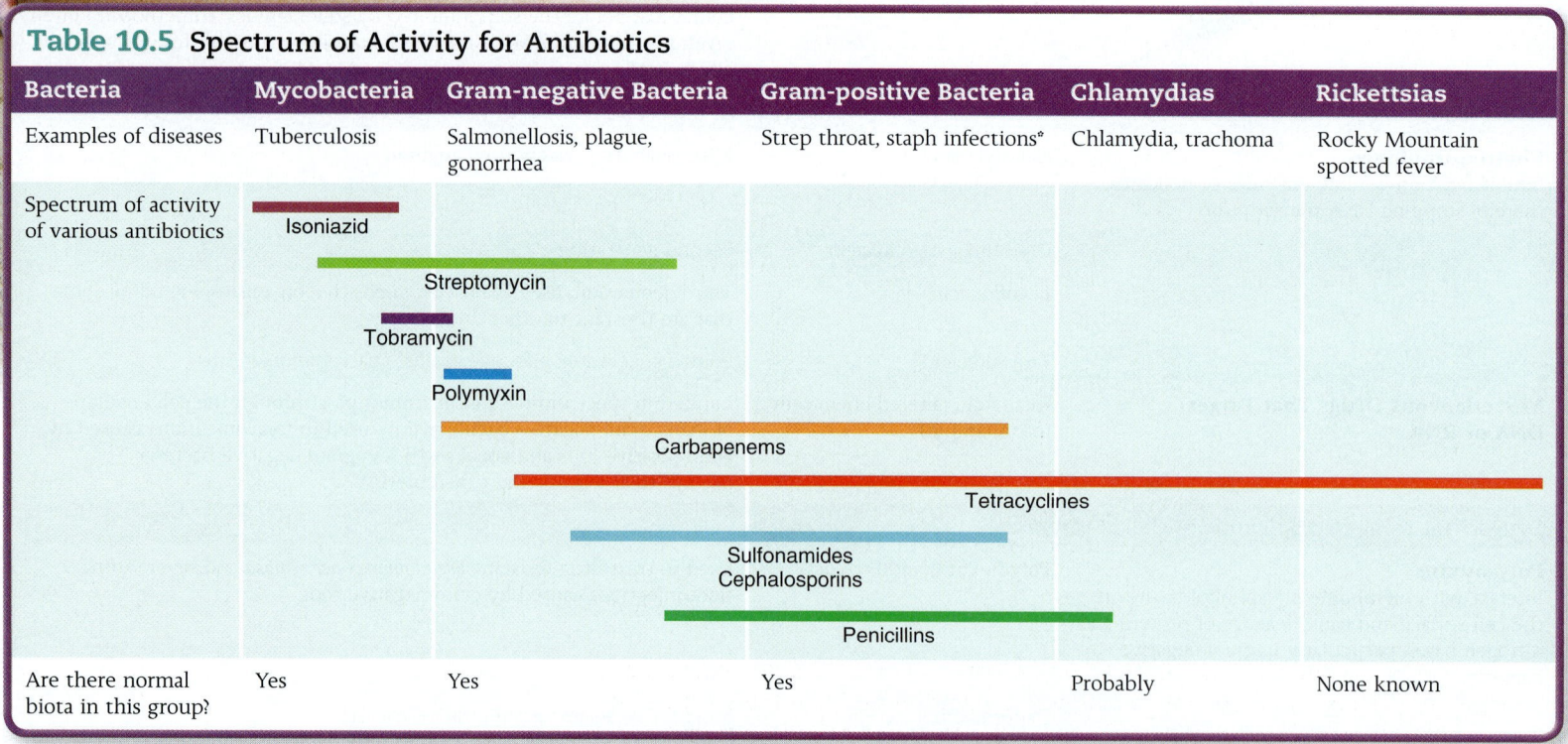

Bacteria	Mycobacteria	Gram-negative Bacteria	Gram-positive Bacteria	Chlamydias	Rickettsias
Examples of diseases	Tuberculosis	Salmonellosis, plague, gonorrhea	Strep throat, staph infections°	Chlamydia, trachoma	Rocky Mountain spotted fever
Spectrum of activity of various antibiotics	Isoniazid	Streptomycin — Tobramycin — Polymyxin — Carbapenems	Tetracyclines — Sulfonamides — Cephalosporins — Penicillins		
Are there normal biota in this group?	Yes	Yes	Yes	Probably	None known

°Note that some members of a bacterial group may not be affected by the antibiotics indicated, due to acquired or natural resistance. In other words, exceptions do exist.

Table 10.6 Characteristics of Selected Penicillin Drugs

Name		Spectrum of Action	Uses, Advantages	Disadvantages
Penicillin G	Beta-lactam ring — S, H_2, CH$_2$—CO—N, CH$_3$, CH$_3$, O, N, COOH	Narrow	Best drug of choice when bacteria are sensitive; low cost; low toxicity	Can be hydrolyzed by penicillinase; allergies occur; requires injection
Penicillin V		Narrow	Good absorption from intestine; otherwise, similar to Penicillin G	Hydrolysis by penicillinase; allergies
Methicillin, nafcillin	CO—N, S, CH$_3$, CH$_3$, O, N, COOH	Narrow	Not usually susceptible to penicillinase	Poor absorption; allergies; growing resistance
Ampicillin		Broad	Works on gram-negative bacilli	Can be hydrolyzed by penicillinase; allergies; only fair absorption
Amoxicillin		Broad	Gram-negative infections; good absorption	Hydrolysis by penicillinase; allergies
Azlocillin, mezlocillin, ticarcillin	CH—CO—N, S, CH$_3$, CH$_3$, COONa, O, N, COOH, S	Very broad	Effective against *Pseudomonas* species; low toxicity compared with aminoglycosides	Allergies; susceptible to many beta-lactamases

Years of research have so far not yielded an obvious solution to this problem, though there are several partially successful strategies. One of these involves interrupting the quorum-sensing pathways that mediate communication between cells and may change phenotypic expression. Daptomycin, a lipopeptide that is effective in deep tissue infections with resistant bacteria, has also shown some success in biofilm infection treatment. Also, some researchers have found that adding DNase to their antibiotics can help with penetration of the antibiotic through the extracellular debris—apparently some of which is DNA from lysed cells.

Many biofilm infections can be found on biomaterials inserted in the body, such as cardiac or urinary catheters. These can be impregnated with antibiotics prior to insertion to prevent colonization. This, of course, cannot be done with biofilm infections of natural tissues, such as the prostate or middle ear.

Interestingly, it appears that chemotherapy with some antibiotics–notably aminogly-cosides–can cause bacteria to form biofilms at a higher rate than they otherwise would. Obviously there is much more to come in understanding biofilms and their control.

Agents to Treat Fungal Infections

Because the cells of fungi are eukaryotic, they present special problems in che-motherapy. For one, the great majority of chemotherapeutic drugs are designed to act on bacteria and are generally ineffective in combating fungal infections. For another, the similarities between fungal and human cells often mean that drugs toxic to fungal cells are also capable of harming human tissues. A few agents with special antifungal properties have been developed for treating systemic and super-ficial fungal infections. Four main drug groups currently in use are the macrolide polyene antibiotics, the azoles, the echinocandins, and flucytosine. **Table 10.7** describes in further detail the antifungal drug groups and their actions.

Agents to Treat Protozoal Infections

The enormous diversity among protozoal and helminthic parasites and their cor-responding therapies reach far beyond the scope of this textbook; however, a few of the more common drugs are surveyed here and described again for particular diseases in the organ systems chapters.

Antimalarial Drugs: Quinine and Its Relatives

Quinine, extracted from the bark of the cinchona tree, was the principal treatment for malaria for hundreds of years, but it has been replaced by the synthesized quinolones, mainly chloroquine and primaquine, which have less toxicity to humans. Because there are several species of *Plasmodium* (the malaria parasite) and many stages in its life cycle, no single drug is universally effective for every species and stage, and each drug is restricted in application. For instance, primaquine eliminates the liver phase of infec-tion, and chloroquine suppresses acute attacks associated with infection of red blood cells. Artemisinin combination therapy (ACT) is now recom-mended for the treatment of certain types of malaria today; it employs the use of artemisinin with quinine derivatives or other drugs.

Chemotherapy for Other Protozoal Infections

A widely used amoebicide, metronidazole (Flagyl), is effective in treating mild and severe intestinal infections and hepatic disease caused by *Entamoeba his-tolytica.* Given orally, it also has applications for infections by *Giardia lamblia* and *Trichomonas vaginalis* (described in chapters 20 and 21, respectively). Other drugs with antiprotozoal activities are quinacrine (a quinine-based drug), sulfonamides, and tetracyclines.

Agents to Treat Helminthic Infections

Treating helminthic infections has been one of the most difficult and challenging of all chemotherapeutic tasks. Flukes, tapeworms, and roundworms are much larger parasites than other microorganisms and, being animals, have greater similarities to human physiology. Also, the usual strategy of using drugs to block their reproduction is usually not successful in eradicating the adult worms. The most effective drugs immobilize, disintegrate, or inhibit the metabolism in all stages of the life cycle.

Mebendazole and albendazole are broad-spectrum antiparasitic drugs used in several roundworm intestinal infestations. These drugs work locally in the intestine to inhibit the function of the microtubules of worms, eggs, and larvae. This means the parasites can no longer utilize glucose, which leads to their demise. The com-pound pyrantel paralyzes the muscles of intestinal roundworms. Consequently, the worms are unable to maintain their grip on the intestinal wall and are expelled

Antimalarial quinine is extracted from the bark of the cinchona tree.

Table 10.7 Agents Used to Treat Fungal Infections

Drug Group	Drug Examples	Action
Macrolide polyenes	Amphotericin B (shown above in gray)	• Bind to fungal membranes, causing loss of selective permeability; extremely versatile • Can be used to treat skin, mucous membrane lesions caused by *Candida albicans* • Injectable form of the drug can be used to treat histoplasmosis and *Cryptococcus* meningitis
Azoles	Ketoconazole, fluconazole, miconazole, and clotrimazole	• Interfere with sterol synthesis in fungi • Ketoconazole–cutaneous mycoses, vaginal and oral candidiasis, systemic mycoses • Fluconazole–AIDS-related mycoses (aspergillosis, *Cryptococcus* meningitis) • Clotrimazole and miconazole–used to treat infections in the skin, mouth, and vagina
Echinocandins	Micafungin, caspofungin	• Inhibit fungal cell wall synthesis • Used against *Candida* strains and aspergillosis
Nucleotide cytosine analog	Flucytosine	• Rapidly absorbed orally, readily dissolves in the blood and CSF (cerebrospinal fluid) • Used to treat cutaneous mycoses • Usually combined with amphotericin B to treat systemic mycoses because many fungi are resistant to this drug

Tapeworm

along with the feces by the normal peristaltic action of the bowel. Two newer antihelminthis drugs are praziquantel, a treatment for various tapeworm and fluke infections, and ivermectin, a veterinary drug now used for strongyloidiasis and onocercosis in humans. Helminthic diseases are described in chapter 20 because these organisms spend a large part of their life cycles in the digestive tract.

Agents to Treat Viral Infections

The chemotherapeutic treatment of viral infections presents unique problems. With a virus, we are dealing with an infectious agent that relies upon the host cell for the vast majority of its metabolic functions. With currently used drugs, disrupting viral metabolism requires that we disrupt the metabolism of the host cell to a much greater extent than is desirable. Put another way, selective toxicity with regard to viral infection is difficult to achieve because a single metabolic system is responsible for the well-being of both virus and host. Although viral diseases such as measles, mumps, and hepatitis are routinely prevented by the use of effective vaccinations, epidemics of AIDS, influenza, and even the "commonness" of the common cold attest to the need for more effective medications for the treatment of viral pathogens.

The currently used antiviral drugs were developed to target specific points in the infectious cycle of viruses. Three major modes of action are as follows:

1. barring penetration of the virus into the host cell,
2. blocking the transcription and translation of viral molecules, and
3. preventing the maturation of viral particles.

Table 10.8 presents an overview of antivirals from each of these categories. Meanwhile, researchers continue to work on additional drugs. A breakthrough treatment for viral infection is currently being tested in the laboratory. It is called DRACO (standing for "double-stranded RNA-activated caspase oligomerizer"). Viruses of nearly every type create long, double-stranded RNAs at some point in their life cycle, and cells do not, so DRACO goes after cells containing dsRNA and causes their destruction. Researchers believe this may result in a broad-spectrum antiviral, once it has been thoroughly tested.

Table 10.8 Actions of Antiviral Drugs

Mode of Action	Examples	Effects of Drug
Inhibition of Virus Entry Receptor/fusion/uncoating inhibitors	Enfuvirtide (Fuzeon®)	Blocks **HIV** infection by preventing the binding of viral proteins to cell receptor, thereby preventing fusion of virus with cell
	Amantadine and its relatives, zanamivir (Relenza®), oseltamivir (Tamiflu©)	Block entry of **influenza virus** by interfering with fusion of virus with cell membrane (also release); stop the action of influenza neuraminidase, required for entry of virus into cell (also assembly)
Inhibition of Nucleic Acid Synthesis	Acyclovir (Zovirax®), other "cyclovirs," vidarabine	Purine analogs that terminate DNA replication in **herpesviruses**
	Ribavirin	Purine analog, used for **respiratory syncytial virus (RSV)** and some **hemorrhagic fever viruses**
	Zidovudine (AZT), lamivudine (3TC), didanosine (ddI), zalcitabine (ddC), and stavudine (d4T)	Nucleotide analog reverse transcriptase (RT) inhibitors; stop the action of reverse transcriptase in **HIV**, blocking viral DNA production
	Nevirapine, efavirenz, delavirdine	Nonnucleotide analog reverse transcriptase inhibitors; attach to **HIV** RT binding site, stopping its action
Inhibition of Viral Assembly/Release	Indinavir, saquinavir	Protease inhibitors; insert into **HIV** protease, stopping its action and resulting in inactive noninfectious viruses

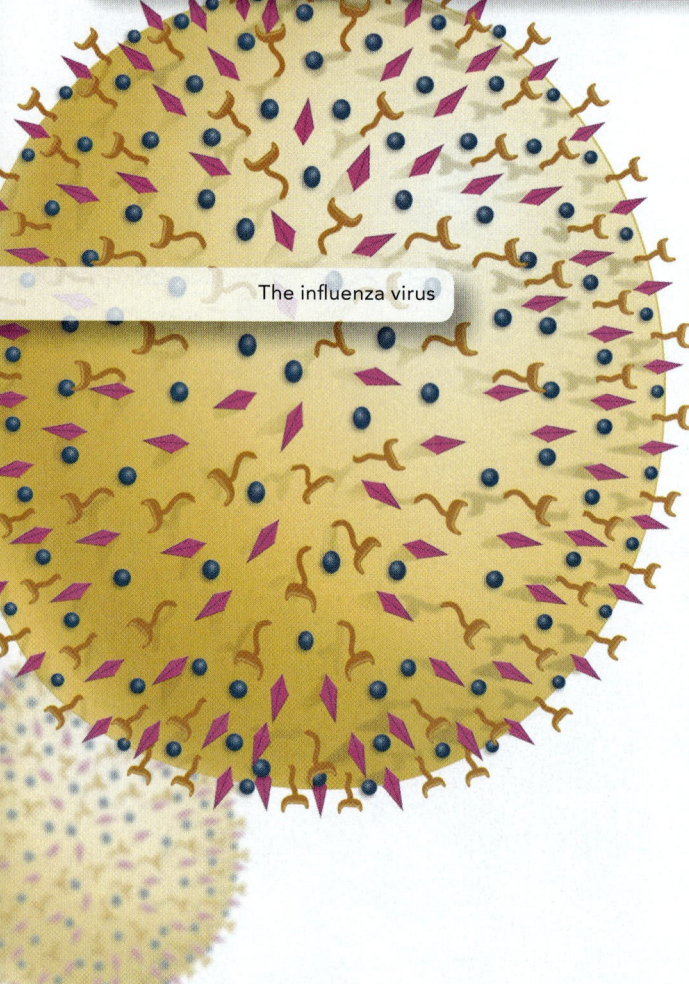

The influenza virus

10.2 LEARNING OUTCOMES—Assess Your Progress

5. Explain the concept of selective toxicity.
6. List the five major targets of antimicrobial agents.
7. Identify which categories of drugs are most selectively toxic and why.
8. Distinguish between broad-spectrum and narrow-spectrum antimicrobials, and explain the significance of the distinction.
9. Identify the microbes against which the various penicillins are effective.
10. Explain the mode of action of penicillinases and their importance in treatment.
11. Identify two antimicrobials that act by inhibiting protein synthesis.
12. Explain how drugs targeting folic acid synthesis work.
13. Identify one example of a fluoroquinolone.
14. Describe the mode of action of drugs that target the cytoplasmic or cell membrane.
15. Discuss how treatments of biofilm and nonbiofilm infections differ.
16. Name the four main categories of antifungal agents.
17. Explain why antiprotozoal and antihelminthic drugs are likely to be more toxic than antibacterial drugs.
18. List the three major targets of action of antiviral drugs.

10.3 Antimicrobial Resistance

One unfortunate outcome of the use of antimicrobials is the development of microbial **drug resistance,** an adaptive response in which microorganisms begin to tolerate an amount of drug that would ordinarily be inhibitory. The ability to circumvent or inactivate antimicrobial drugs is due largely to the genetic versatility and adaptability of microbial populations. The property of drug resistance can be intrinsic as well as acquired. Intrinsic drug resistance can best be exemplified by the fact that bacteria must, of course, be resistant to any antibiotic that they themselves produce. Of much greater importance is the acquisition of resistance to a drug by a microbe that was previously sensitive to the drug. In our context, the term **drug resistance** will refer to this last type of acquired resistance.

How Does Drug Resistance Develop?

Contrary to popular belief, antibiotic resistance is an ancient phenomenon. In 2012, 93 bacterial species were discovered in a cave in New Mexico that had been cut off from the surface for millions of years. Most of these species were found to have resistance to multiple antibiotics–antibiotics naturally produced by other microbes. Because most of our oldest therapeutically used antibiotics are natural products from fungi and bacteria, resistance to them has been a survival strategy for *other* microbes for as long as the microbes have been around. The scope of the problem in terms of using the antibiotics as treatments for humans became apparent in the 1980s and 1990s, when scientists and physicians observed treatment failures on a large scale. What the New Mexico data and other recent findings tell us is that the acquisition of drug resistance is not always a result of exposure to the drug. This adds another dimension to the efforts to prolong antibiotic effectiveness, which so far have focused on limiting the amount of antibiotic in the environment. We see now that this is important but not enough to prevent microorganisms from developing resistance altogether.

Whether antibiotics are present or not, microbes become newly resistant to a drug after one of the following two events occurs:

1. spontaneous mutations in critical chromosomal genes, or
2. acquisition of entire new genes or sets of genes via horizontal transfer from another species.

Drug resistance that is found on chromosomes usually results from spontaneous random mutations in bacterial populations. The chance that such a mutation will be advantageous is minimal, and the chance that it will confer resistance to a specific drug is lower still. Nevertheless, given the huge numbers of microorganisms in any population and the constant rate of mutation, such mutations do occur. The end result varies from slight changes in microbial sensitivity, which can be over-come by larger doses of the drug, to complete loss of sensitivity. There may be a third mechanism of acquiring resistance to a drug, which is a pheno-typic, not a genotypic, adaptation. Recent studies suggest that bacteria can "go to sleep" when exposed to antibiotics, meaning they will slow or stop their metabolism so that they cannot be harmed by the antibiotic. They can then rev back up after the antibiotic concentration decreases. Sometimes these bacteria are called "persisters." (This is one reason bio-film bacteria are less susceptible to antibiotics than free-living bacteria are.) In the next sections, we will focus on the two genetic changes that can result in acquired resistance.

Resistance occurring through horizontal transfer originates from plas-mids called **resistance (R) factors** that are transferred through conjugation, transformation, or transduction. Such traits are "lying in wait" for an opportu-nity to be expressed and to confer adaptability on the species. Many bacteria also maintain transposable drug resistance sequences (transposons) that are duplicated and inserted from one plasmid to another or from a plasmid to the chromosome. Chromosomal genes and plasmids containing codes for drug resistance are faithfully replicated and inherited by all subsequent progeny. This sharing of resistance genes accounts for the rapid proliferation of drug-resistant species. As you have read in earlier chapters, gene transfers are extremely frequent in nature, with genes coming from totally unrelated bacteria, viruses, and other organisms living in the body's normal biota and the environment.

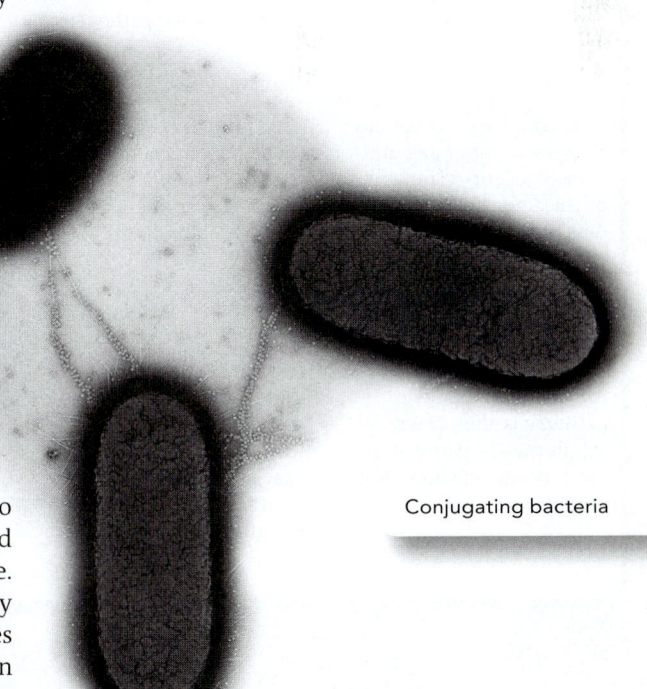

Conjugating bacteria

Specific Mechanisms of Drug Resistance

Mutations and horizontal transfer, just described, result in mutants acquiring one of several mechanisms of drug resistance. **Table 10.9** lists the most common mechanisms of drug resistance and provides specific examples of each.

Table 10.9 Mechanisms of Drug Resistance

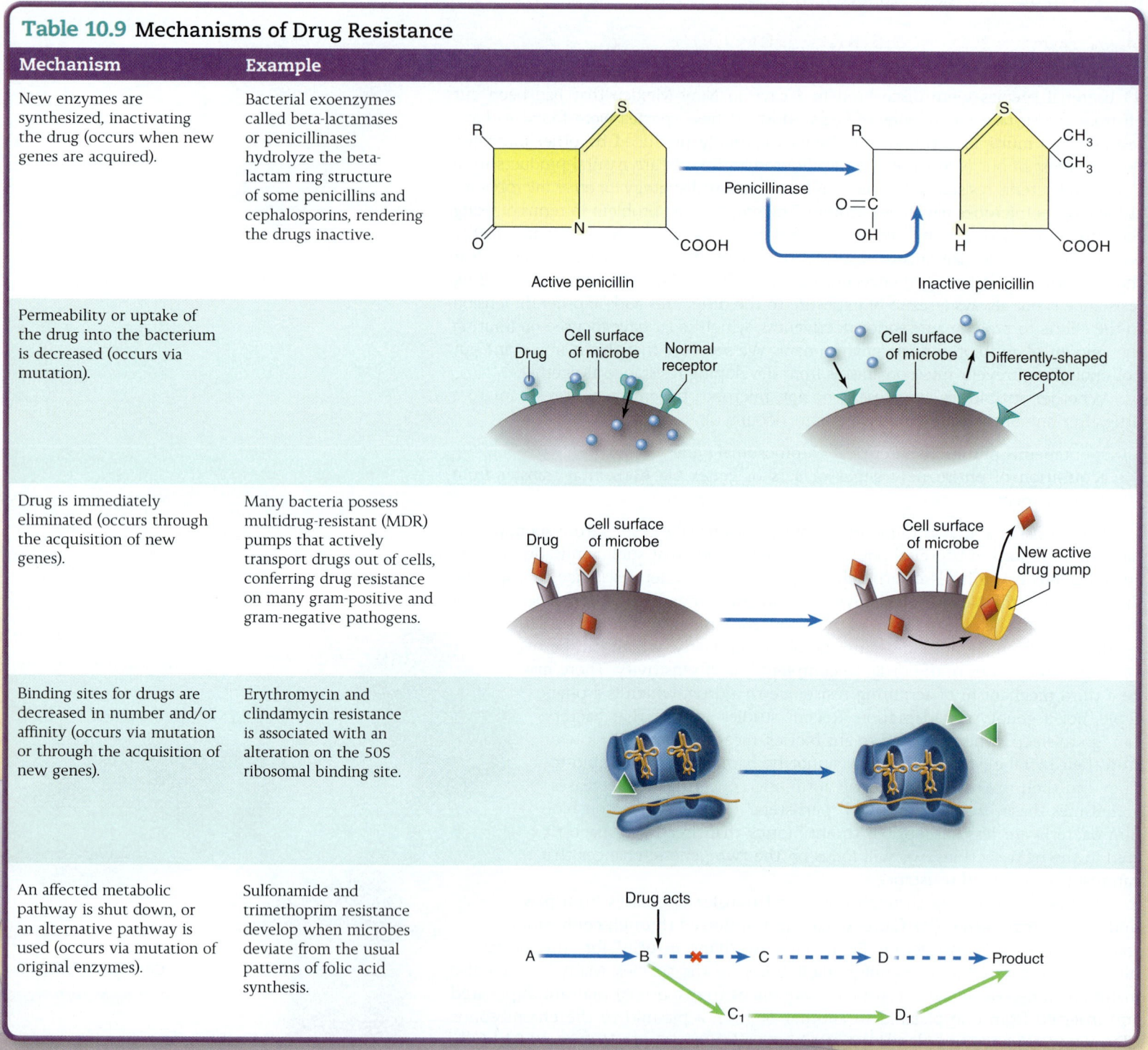

Mechanism	Example
New enzymes are synthesized, inactivating the drug (occurs when new genes are acquired).	Bacterial exoenzymes called beta-lactamases or penicillinases hydrolyze the beta-lactam ring structure of some penicillins and cephalosporins, rendering the drugs inactive.
Permeability or uptake of the drug into the bacterium is decreased (occurs via mutation).	
Drug is immediately eliminated (occurs through the acquisition of new genes).	Many bacteria possess multidrug-resistant (MDR) pumps that actively transport drugs out of cells, conferring drug resistance on many gram-positive and gram-negative pathogens.
Binding sites for drugs are decreased in number and/or affinity (occurs via mutation or through the acquisition of new genes).	Erythromycin and clindamycin resistance is associated with an alteration on the 50S ribosomal binding site.
An affected metabolic pathway is shut down, or an alternative pathway is used (occurs via mutation of original enzymes).	Sulfonamide and trimethoprim resistance develop when microbes deviate from the usual patterns of folic acid synthesis.

Natural Selection and Drug Resistance

So far, we have been considering drug resistance at the cellular and molecular levels, but its full impact is felt only if this resistance occurs throughout the cell population. Let us examine how this might happen and its long-term therapeutic consequences.

Any large population of microbes is likely to contain a few individual cells that are already drug resistant because of prior mutations or transfer of plasmids **(figure 10.5a).** While we now know that many things can cause these "odd balls" to start overtaking the population, one of the most reliable ways to make this happen is for the correct antibiotic to be present **(figure 10.5b).** Sensitive individuals are inhibited or destroyed, and resistant forms survive and proliferate. During subsequent population growth, offspring of these resistant microbes will inherit this drug resistance. In time, the replacement population will have a preponderance of the drug-resistant forms and can eventually become completely resistant **(figure 10.5c).** In ecological terms, the environmental factor (in this case, the drug) has put selection pressure on the population, allowing the more "fit" microbe (the drug-resistant one) to survive, and the population has evolved to a condition of drug resistance.

An Urgent Problem

Textbooks generally avoid using superlatives and exclamation points. But the danger of antibiotic resistance can hardly be overstated. The Centers for Disease Control and Prevention (CDC) issued a "Threat Report" about this issue for the first time in 2013, and they continue to monitor the situation, which they label "potentially catastrophic."

Even though the antibiotic era began less than 70 years ago, we became so confident it would be permanent that we may have forgotten what it was like before antibiotics were available. Certain types of pneumonia had a 50% fatality rate. Strep throat could turn deadly overnight. Infected wounds often required amputations or led to death. Yet the effectiveness of our currently available antibiotics is declining, in some cases very rapidly. There is a real possibility that we will enter a post-antibiotic era, in which some infections will be untreatable.

New and effective antibiotics have been slow to come to market. There are a variety of reasons for this, including the economic reality that antibiotics (taken in short courses) are not as lucrative for drug manufacturers as drugs for chronic diseases, which must often be taken for life, even though they are just as time-consuming and expensive to develop. Policy-makers are starting to create incentives for the discovery and manufacture of new antibiotics, although we should keep in mind that even new drugs will eventually become less effective over time as bacteria adapt to them.

The CDC has categorized resistant bacteria into three groups, termed "hazard levels". The three hazard levels are *concerning, serious,* and *urgent.* We will look at them individually in the disease chapters later in the book.

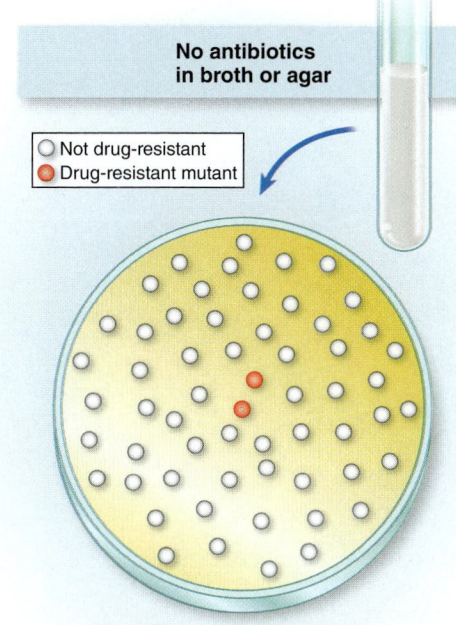

No antibiotics in broth or agar

○ Not drug-resistant
● Drug-resistant mutant

(a) Population of microbial cells

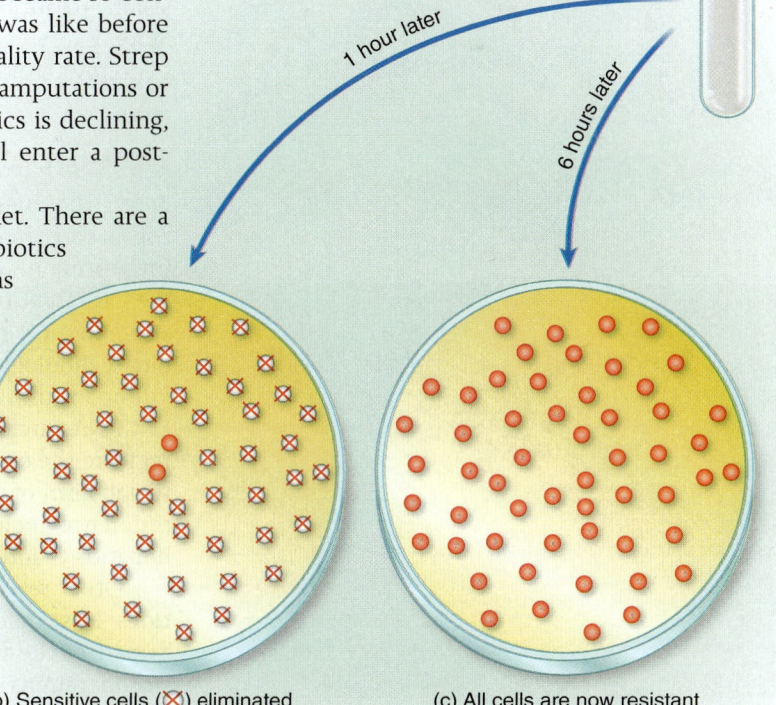

Antibiotics added to broth and agar; same bacterial population as above

1 hour later

6 hours later

(b) Sensitive cells (⊗) eliminated by drug; resistant mutants survive

(c) All cells are now resistant

Figure 10.5 The events in natural selection for drug resistance. (a) Populations of microbes can harbor some members with a prior mutation that confers drug resistance. (b) Environmental pressure (here, the presence of the drug) selects for survival of these mutants. (c) They eventually become the dominant members of the population.

Urgent Threats

- Carbapenem-resistant Enterobacteriaceae (CRE)
- Drug-resistant *Neisseria gonorrhoeae*

Serious Threats

- Multidrug-resistant *Acinetobacter*
- Drug-resistant *Campylobacter*
- Fluconazole-resistant *Candida* (a fungus)
- Extended spectrum β-lactamase producing Enterobacteriaceae (ESBLs)
- Vancomycin-resistant *Enterococcus* (VRE)
- Multidrug-resistant *Pseudomonas aeruginosa*
- Drug-resistant non-typhoidal *Salmonella*
- Drug-resistant *Salmonella* typhi
- Drug-resistant *Shigella*
- Methicillin-resistant *Staphylococcus aureus* (MRSA)
- Drug-resistant *Streptococcus pneumoniae*
- Drug-resistant tuberculosis

Concerning Threats

- Vancomycin-resistant *Staphylococcus aureus* (VRSA)
- Erythromycin-resistant Group A *Streptococcus*
- Clindamycin-resistant Group B *Streptococcus*

In the United States alone, 2 million people a year become infected with resistant bacteria, and at least 23,000 deaths are attributed to them. (The CDC also considers *Clostridium difficile* in the "urgent" category, even though it is not particularly resistant to antibiotic treatment itself. Instead, it causes 14,000 deaths in the United States every year because extensive antibiotic treatments for other infections lead to overgrowth of this bacterium, which then causes severe disease.)

New Approaches to Antimicrobial Therapy

Often, the quest for new antimicrobial strategies focuses on finding new targets in the bacterial cell and custom-designing drugs that aim for them. There are many interesting new strategies that have not yet resulted in a marketable drug—for example, (1) targeting iron-scavenging capabilities of bacteria; (2) using RNA interference strategies; (3) mimicking molecules called defense peptides; and (4) exploiting an old technology, using bacteriophages, the natural enemies of bacteria, to do the killing for us.

RNA interference, you recall from chapter 8, refers to small pieces of RNA that regulate the expression of genes. This is being exploited in attempts to shut down the metabolism of pathogenic microbes. There have been several human trials of RNA interference, including trials to evaluate the effectiveness of synthetic RNAs in treating hepatitis C and respiratory syncytial virus.

Other researchers are looking into proteins called host or bacterial defense peptides. Host defense peptides are peptides of 20 to 50 amino acids that are secreted as part of the mammalian innate immune system. They have names such as defensin, magainins, and protegrins. Some bacteria produce similar peptides. These are called bacteriocins and lantibiotics. Both host and bacterial defense peptides have multiple activities against bacteria—inserting in their membranes and also targeting other structures in the cells. For this reason, researchers believe they may be more effective than narrowly targeted drugs in current use and will be much less likely to foster resistance.

Sometimes the low-tech solution can be the best one. Eastern European countries have gained a reputation for using mixtures of bacteriophages as medicines for bacterial infections. There is little argument about the effectiveness of these treatments, though they have never been approved for use in the West. One recent human trial used a mixture of bacteriophages specific for *Pseudomonas aeruginosa* to treat ear infections caused by the bacterium. These infections are found in the form of biofilms and have been extremely difficult to treat. The phage preparation called Biophage-PA successfully treated patients who had experienced long-term antibiotic-resistant infections. Other researchers are incorporating phages into wound dressings. One clear advantage to bacteriophage treatments is the extreme specificity of the phages–only one species of bacterium is affected, leaving the normal inhabitants of the body, and the body itself, alone.

Helping Nature Along

Other novel approaches to controlling infections include the use of **probiotics** and **prebiotics.** Probiotics are preparations of live microorganisms that are fed to animals and humans to improve the intestinal biota. This can serve to replace microbes lost during antimicrobial therapy or simply to augment the biota that is already there. This is a slightly more sophisticated application of methods that have long been used in an empiric fashion, for instance, by people who consume yogurt because of the beneficial microbes it contains. Recent years have seen a huge increase in the numbers of probiotic products sold in ordinary grocery stores **(figure 10.6).** Experts generally find these products safe, and in some cases they can be effective. Probiotics are thought to be useful for the management of food allergies; their role in the stimulation of mucosal immunity is also being investigated.

Prebiotics are nutrients that encourage the growth of beneficial microbes in the intestine. For instance, certain sugars such as fructans are thought to encourage the growth of the beneficial *Bifidobacterium* in the large intestine and to discourage the growth of potential pathogens.

A technique that is gaining mainstream acceptance is the use of fecal transplants in the treatment of recurrent *Clostridium difficile* infection and ulcerative colitis. This procedure involves the transfer of feces from healthy patients via colonoscopy. This is, in fact, just an adaptation of probiotics. But instead of a few beneficial bacterial species being given orally with the hope that they will establish themselves in the intestines, a rich microbiota is administered directly to the site it must colonize–the large intestine. Work is also underway to develop a pill

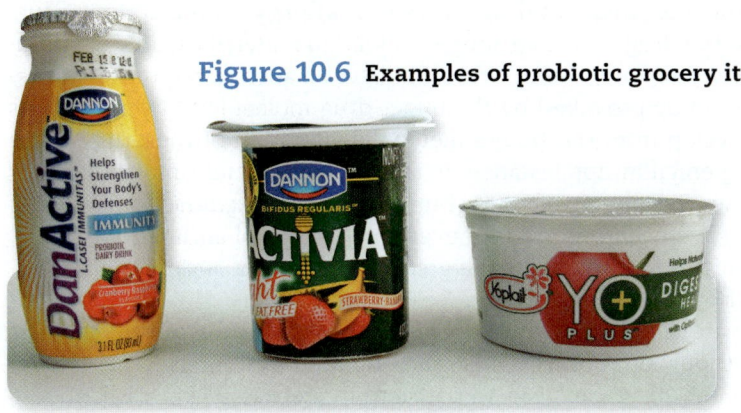

Figure 10.6 Examples of probiotic grocery items.

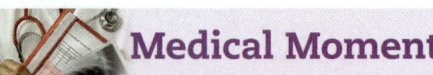

Medical Moment

Why Do Antibiotics Cause Diarrhea?

You are prescribed an antibiotic for strep throat. You take it as prescribed. Then, suddenly, you have diarrhea to go along with your fever and sore throat—just what you didn't need!

Why do we often get diarrhea when we take antibiotics? We have resident microbial biota in our intestines. These bacteria serve a useful purpose; they help us to keep numbers of harmful bacteria in check. We can refer to these helpful bacteria as "good" bacteria and the potentially illness-causing bacteria as "bad" bacteria. When we take antibiotics, we upset the delicate balance between numbers of good and bad bacteria so that the bad begin to outnumber the good. This may result in diarrhea, an unpleasant side effect of many antibiotics.

Having diarrhea while taking antibiotics is not considered an allergy (an allergic response results in activation of the immune system) but is considered an unpleasant side effect. If diarrhea is severe or prolonged, you should consult your physician, because superinfection with *C. difficile* sometimes occurs after antibiotic treatment (see chapter 20).

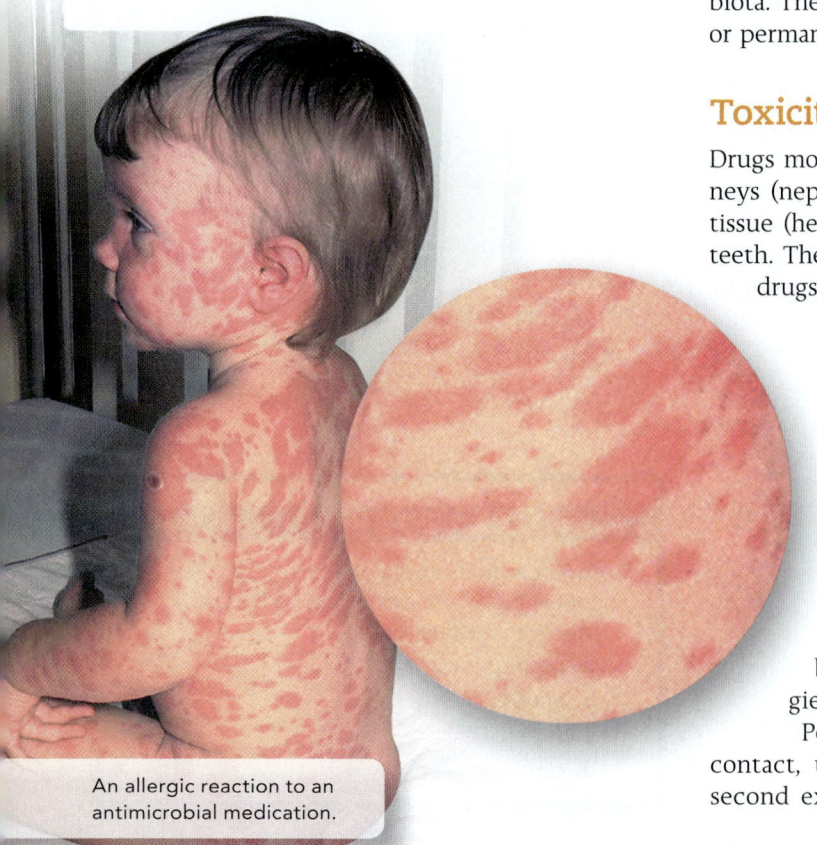

An allergic reaction to an antimicrobial medication.

containing the appropriate species, with a coating that will enable it to remain intact as it traverses the stomach and small intestine and releases the bacteria in the lower intestine. Clearly, the use of these agents is a different type of antimicrobial strategy than we are used to, but it may have its place in a future in which traditional antibiotics are more problematic.

10.3 LEARNING OUTCOMES—Assess Your Progress

19. Discuss two possible ways that microbes acquire antimicrobial resistance.
20. List five cellular or structural mechanisms that microbes use to resist antimicrobials.
21. Discuss at least two novel antimicrobial strategies that are under investigation.

10.4 Interactions Between Drug and Host

Until now, this chapter has focused on the interaction between antimicrobials and the microorganisms they target. During an infection, the microbe is living in or on a host; therefore, the drug is administered to the host though its target is the microbe. Therefore, the effect of the drug on the host must always be considered.

Although selective antimicrobial toxicity is the ideal constantly being sought, chemotherapy by its very nature involves contact with foreign chemicals that can harm human tissues. In fact, estimates indicate that at least 5% of all persons taking an antimicrobial drug experience some type of serious adverse reaction to it. The major side effects of drugs fall into one of three categories: direct damage to tissues through toxicity, allergic reactions, and disruption in the balance of normal microbial biota. The damage incurred by antimicrobial drugs can be short term and reversible or permanent, and it ranges in severity from cosmetic to lethal.

Toxicity to Organs

Drugs most often adversely affect the following organs: the liver (hepatotoxic), kidneys (nephrotoxic), gastrointestinal tract, cardiovascular system and blood-forming tissue (hemotoxic), nervous system (neurotoxic), respiratory tract, skin, bones, and teeth. The potential toxic effects of drugs on the body, along with the responsible drugs, are detailed in **table 10.10.**

Allergic Responses to Drugs

One of the most frequent drug reactions is **allergy.** This reaction occurs because the drug acts as an antigen (a foreign material capable of stimulating the immune system) and stimulates an allergic response. This response can be provoked by the intact drug molecule or by substances that develop from the body's metabolic alteration of the drug. In the case of penicillin, for instance, it is not the penicillin molecule itself that causes the allergic response but a product, *benzylpenicilloyl.* Allergic reactions have been reported for every major type of antimicrobial drug, but the penicillins account for the greatest number of antimicrobial allergies, followed by the sulfonamides.

People who are allergic to a drug become sensitized to it during the first contact, usually without symptoms. Once the immune system is sensitized, a second exposure to the drug can lead to a reaction such as a skin rash (hives),

Table 10.10 Major Adverse Toxic Reactions to Common Drug Groups

Antimicrobial Drug	Primary Damage or Abnormality Produced
Antibacterials	
Penicillin G	Rash, hives, watery eyes
Carbenicillin	Abnormal bleeding
Ampicillin	Diarrhea and enterocolitis
Cephalosporins	Inhibition of platelet function
	Decreased circulation of white blood cells; nephritis
Tetracyclines	Diarrhea and enterocolitis
	Discoloration of tooth enamel
	Reactions to sunlight (photosensitivity)
Chloramphenicol	Injury to red and white blood cell precursors
Aminoglycosides (streptomycin, gentamicin, amikacin)	Diarrhea and enterocolitis
	Malabsorption
	Loss of hearing, dizziness, kidney damage
Isoniazid	Hepatitis (liver inflammation)
	Seizures
	Dermatitis
Sulfonamides	Formation of crystals in kidney; blockage of urine flow
	Hemolysis
	Reduction in number of red blood cells
Polymyxin	Kidney damage
	Weakened muscular responses
Quinolones (ciprofloxacin, norfloxacin)	Headache, dizziness, tremors, GI distress
Rifampin	Damage to hepatic cells
	Dermatitis
Antifungals	
Amphotericin B	Disruption of kidney function
Flucytosine	Decreased number of white blood cells
Antiprotozoal Drugs	
Metronidazole	Nausea, vomiting
Chloroquine	Vomiting
	Headache
	Itching
Antihelminthics	
Niclosamide	Nausea, abdominal pain
Pyrantel	Intestinal irritation
	Headache, dizziness
Antivirals	
Acyclovir	Seizures, confusion
	Rash
Amantadine	Nervousness, light-headedness
	Nausea
AZT	Immunosuppression, anemia

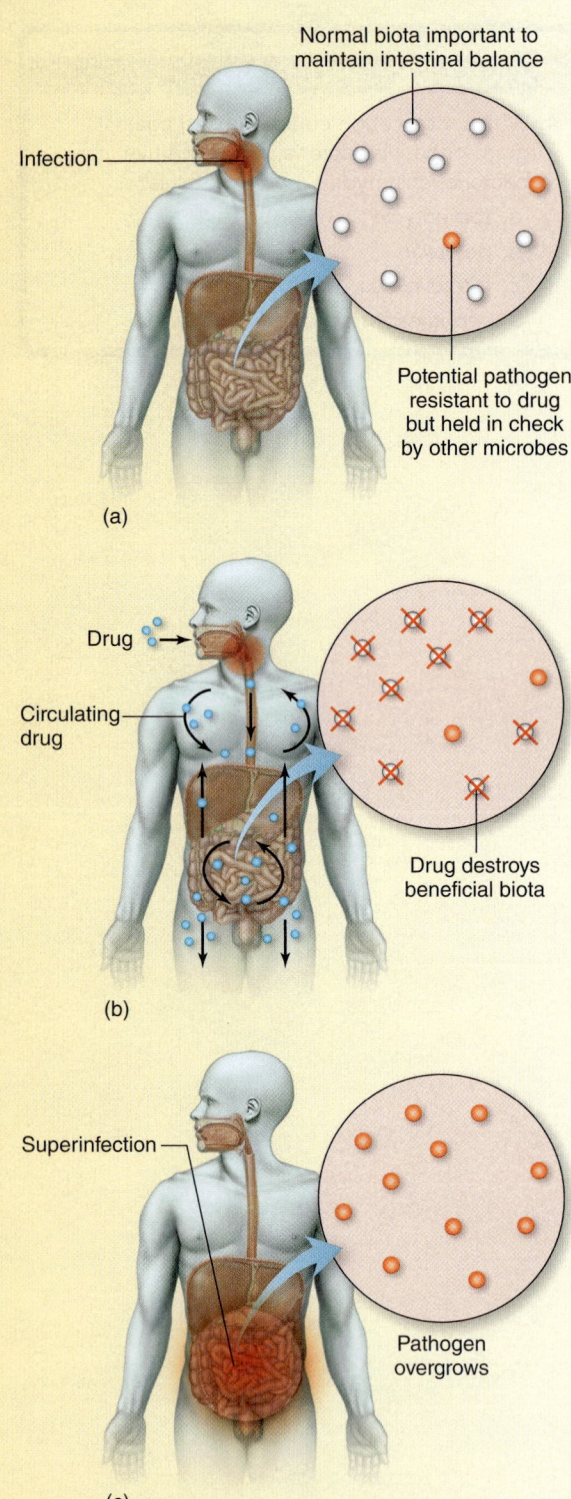

Normal biota important to maintain intestinal balance

Infection

Potential pathogen resistant to drug but held in check by other microbes

(a)

Drug

Circulating drug

Drug destroys beneficial biota

(b)

Superinfection

Pathogen overgrows

(c)

Figure 10.7 **The role of antimicrobials in disrupting microbial biota and causing superinfections.** (a) A primary infection in the throat is treated with an oral antibiotic. (b) The drug is carried to the intestine and is absorbed into the circulation. (c) The primary infection is cured, but drug-resistant pathogens have survived and create an intestinal superinfection.

respiratory inflammation, and, rarely, anaphylaxis, an acute, overwhelming allergic response that develops rapidly and can be fatal. (This topic is discussed in greater detail in chapter 14.)

Suppression and Alteration of the Microbiota by Antimicrobials

Most normal, healthy body surfaces, such as the skin, large intestine, outer openings of the urogenital tract, and oral cavity, provide numerous habitats for a virtual "garden" of microorganisms. These normal colonists, or residents, called the **biota,** or microbiota, consist mostly of harmless or beneficial bacteria, but a small number can potentially be pathogens. Although we defer a more detailed discussion of this topic to chapter 11 and later chapters, here we focus on the general effects of drugs on this population.

If a broad-spectrum antimicrobial is introduced into a host to treat infection, it will destroy microbes regardless of their roles as normal biota, affecting not only the targeted infectious agent but also many others in sites far removed from the original infection **(figure 10.7).** When this therapy destroys beneficial resident species, other microbes that were once in small numbers begin to overgrow and cause disease. This complication is called a **superinfection.**

Some common examples demonstrate how a disturbance in microbial biota leads to replacement biota and superinfection. A broad-spectrum cephalosporin used to treat a urinary tract infection by *Escherichia coli* will cure the infection, but it will also destroy the lactobacilli in the vagina that normally maintain a protective acidic environment there. The drug has no effect, however, on *Candida albicans,* a yeast that also resides in normal vaginas. Released from the inhibitory environment provided by lactobacilli, the yeasts proliferate and cause symptoms. *Candida* can cause similar superinfections of the oropharynx (thrush) and the large intestine.

Oral therapy with tetracyclines, clindamycin, and broad-spectrum penicillins and cephalosporins is associated with a serious and potentially fatal condition known as *antibiotic-associated colitis* (pseudomembranous colitis). This condition is due to the overgrowth in the bowel of *Clostridium difficile,* an endospore-forming bacterium that is resistant to the antibiotic. It invades the intestinal lining and releases toxins that induce diarrhea, fever, and abdominal pain. (You'll learn more about infectious diseases of the gastrointestinal tract, including *C. difficile,* in chapter 20.)

An Antimicrobial Drug Dilemma

The remarkable progress in treating many infectious diseases has spawned a view of antimicrobials as a "cure-all" for infections as diverse as the common cold and acne. And, although it is true that few things are as dramatic as curing an infectious disease with the correct antimicrobial drug, in many instances, drugs have no effect or can be harmful. For example, roughly 200 million prescriptions for antimicrobials are written in the United States every year. The CDC estimates that up to 50% of them are not needed or not "optimally prescribed."

In the past, many physicians tended to use a "shotgun" antimicrobial therapy for minor infections, which involves administering a broad-spectrum drug instead of a more specific narrow-spectrum one. This practice led to superinfections and other adverse reactions. Importantly, it also caused the development of resistance in "bystander" microbes (normal biota) that were exposed to the drug as well. This helped to spread antibiotic resistance to pathogens. With growing awareness of the problems of antibiotic resistance, this practice is much less frequent.

Tons of excess antimicrobial drugs produced in this country are exported to other countries, where controls are not as strict. Nearly 200 different antibiotics are sold over the counter in Latin America and Asian countries. It is common for people in these countries to self-medicate without understanding the correct medical indication. Drugs used in this way are largely ineffectual, but, worse yet, they are known to be responsible for emergence of drug-resistant bacteria that subsequently cause epidemics.

In the final analysis, every allied health professional should be critically aware not only of the admirable and utilitarian nature of antimicrobials but also of their limitations.

10.4 LEARNING OUTCOMES—Assess Your Progress

22. Distinguish between drug toxicity and allergic reactions to drugs.

23. Explain what a superinfection is and how it occurs.

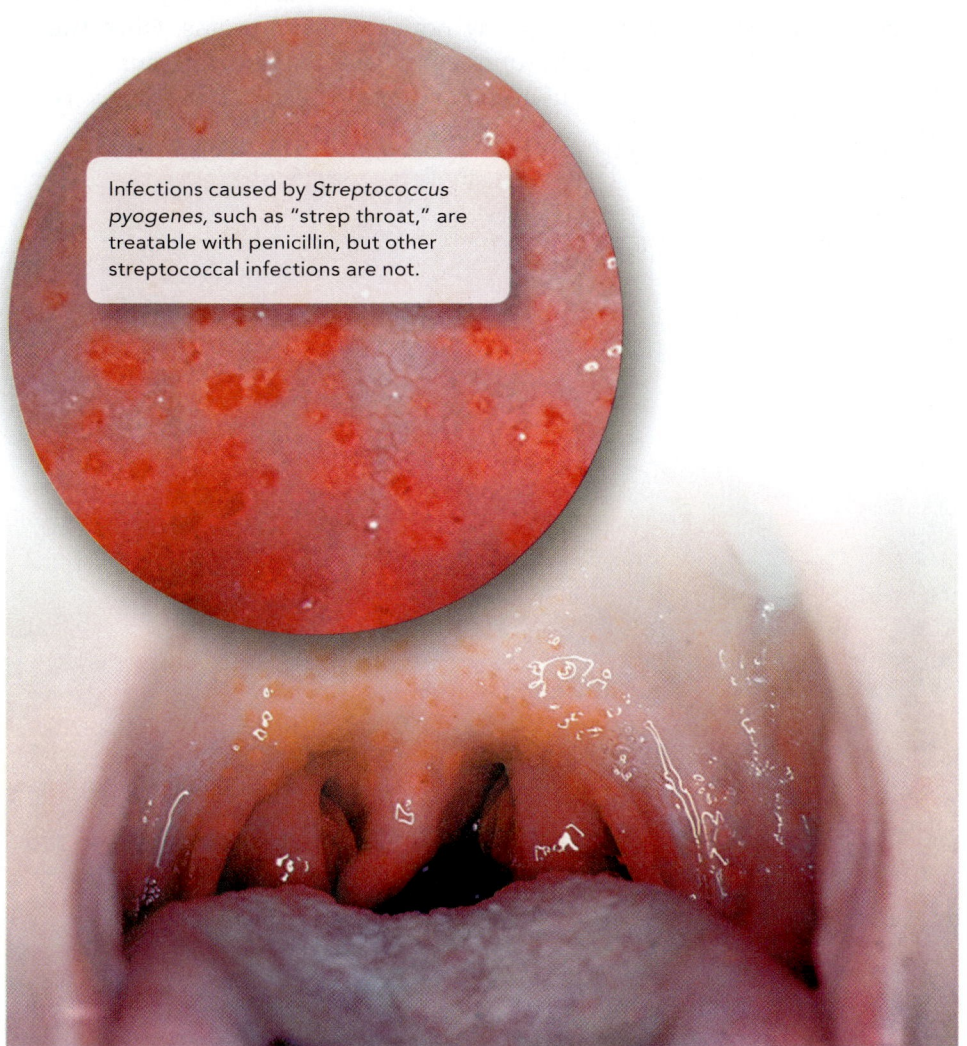

Infections caused by *Streptococcus pyogenes,* such as "strep throat," are treatable with penicillin, but other streptococcal infections are not.

Medical Moment

Side Effect or Allergy?

Medical professionals must often ask patients about their medication allergies. Patients will often report being allergic to a drug, when in actual fact they probably experienced an unpleasant side effect of the drug in question. What's the difference, and why does it matter?

A side effect is an unintended effect caused by taking a medication. For example, you may experience nausea and stomach upset when you take codeine. This is not a true drug allergy—a drug allergy involves activation of your immune system.

Why does it matter? Sometimes side effects can be avoided by giving another medication simultaneously with the first drug, by giving the drug with food, or by giving a lower dosage, for example. If the patient suffered a true allergic response, the drug cannot be given again and another drug must be chosen.

NCLEX® PREP

5. Mary has a urinary tract infection and is prescribed cephalexin for 10 days. Toward the end of her course of treatment, Mary develops a vaginal yeast infection. The yeast infection is an example of a/an

  a. *superinfection.*

  b. *expected complication.*

  c. *allergic reaction.*

  d. *toxic reaction.*

CASE FILE WRAP-UP

Cefaclor, which goes by a variety of brand names, including Ceclor, is a second-generation cephalosporin antibiotic used to treat gram-negative bacteria. When it first came out, it was popular among physicians for treating otitis media infections; however, cefaclor caused rash in a large number of patients. It has now fallen out of favor as newer cephalosporins have come along.

People with an allergy to penicillin may not be able to take cefaclor, as there is a possibility of a cross-reaction occurring. This is due to a similarity in the side chain structure of penicillins and some cephalosporins. The choice of whether to avoid the use of cephalosporins in individuals who are allergic to penicillin is often based on the allergic manifestations and the drug under consideration. Some people are able to take cephalosporins without suffering any adverse effects but should be aware of the possibility of reaction, however remote.

Allergic response to an antibiotic occurs because the drug acts as an antigen, a foreign agent that stimulates the immune response. People who are allergic to antibiotics usually become sensitized during the first contact, usually without suffering any noticeable symptoms. Once the body has become sensitized, subsequent exposure to the drug leads to an allergic response. Each subsequent exposure will result in more severe symptoms.

Demanding Antibiotics: The Consumer's Role in Drug Resistance

There have been many reasons cited for the rise of antibiotic resistance, including the use of antibiotics in livestock to improve health and size of livestock, the indiscriminate use of antibiotics in developing countries (particularly the sale of antibiotics without a prescription), and inappropriate prescribing of antibiotics by physicians (e.g., antibiotics prescribed to treat viral infections).

Most physicians have become more aware that prescribing practices for antibiotics must be tightened. However, many of their patients have yet to learn this important lesson. Many people continue to visit their physician with a viral infection, such as the common cold, and demand a prescription for an antibiotic. Society has become accustomed to being provided with an antibiotic prescription for whatever ails them, and health care consumers often demand antibiotics even when their condition does not warrant one. Putting pressure on their physicians sometimes yields the coveted prescription, a dangerous practice for the individual patient and society as a whole.

Health care education is the responsibility not only of physicians but also of nurses, pharmacists, and other professionals who deal directly with patients. Patients demanding antibiotics for viral infections often require an explanation as to why antibiotics are not appropriate for use against viruses and why this practice is irresponsible. Hearing this information from trusted health care professionals may have a bigger impact on the public than hearing the same information via government education ads.

The following are some suggestions on instructions that can be given to patients to decrease the spread of antibiotic-resistant organisms:

- Finish all antibiotics as prescribed—do not stop taking antibiotics partway through, even if you feel better. Antibiotics should be stopped only if your doctor instructs you to quit taking them (i.e., in the event of an allergic reaction).
- Don't ask your physician to prescribe antibiotics for viral infections. Your doctor will know whether you require an antibiotic, and it can be dangerous to take antibiotics when they are not necessary. Antibiotics are not effective against viruses.
- Never share antibiotics with others.
- Do not flush unused antibiotics down the toilet or dispose of them in your garbage disposal system. Do not throw out unused antibiotics in the garbage. Antibiotics can end up in the water supply, increasing the problem of antibiotic resistance. Instead, take them to your pharmacy and ask them to dispose of the medication for you.
- If you are a parent, ensure that your children are given or take antibiotics as prescribed by a physician, and be sure they finish the entire course.
- Avoid illness in the first place—be sure you are fully immunized against preventable diseases. Wash your hands frequently to prevent the spread of disease. Hand washing is the most effective means of preventing illness. Store, handle, and prepare food safely.

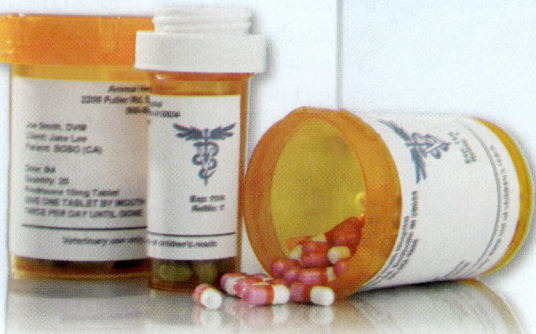

Chapter Summary

10.1 Principles of Antimicrobial Therapy

- Antimicrobial chemotherapy involves the use of drugs to control infection on or in the body.
- Antimicrobial drugs are produced either synthetically or from natural sources.
- Broad-spectrum antimicrobials are effective against many types of microbes. Narrow-spectrum antimicrobials are effective against a limited group of microbes.
- Bacteria and fungi are the primary sources of most currently used antibiotics. The molecular structures of these compounds can be chemically altered or mimicked in the laboratory.
- The three major considerations necessary to choose an effective antimicrobial are the identity of the infecting microbe, the microbe's sensitivity to available drugs, and the overall medical status of the infected host.
- The Kirby-Bauer test identifies antimicrobials that are effective against a specific infectious bacterial isolate.
- The MIC (minimum inhibitory concentration) identifies the smallest effective dose of an antimicrobial toxic to the infecting microbe.
- The therapeutic index is a ratio of the amount of drug toxic to the infected host and the MIC. The smaller the ratio, the greater the potential for toxic host-drug reactions.

10.2 Interactions Between Drug and Microbe

- Antimicrobials are classified into approximately 20 major drug families, based on chemical composition, source or origin, and their site of action.
- There are a great number of antibacterial drugs but a limited number that are effective against protozoa, helminths, fungi, and viruses.
- There are five main cellular targets for antibiotics in microbes: cell wall synthesis, nucleic acid structure and function, protein synthesis, cytoplasmic membranes, and folic acid synthesis.
- Penicillins, cephalosporins, carbapenems, and vancomycin block cell wall synthesis.
- Aminoglycosides, tetracyclines, oxazolidinone, and pleuromutilins block protein synthesis in bacteria.
- Sulfonamides, trimethoprim, and the fluoroquinolones are synthetic antimicrobials effective against a broad range of microorganisms. They block steps in the synthesis of nucleic acids.
- Polymyxins and daptomycin are the major drugs that disrupt cell membranes.
- Bacteria in biofilms respond differently to antibiotics than when they are free-floating. It is therefore difficult to eradicate biofilms in the human body.

- Fungal antimicrobials, such as macrolide polyenes, azoles, echinocandins, and allylamines, must be monitored carefully because of the potential toxicity to the infected host.
- There are fewer antiprotozoal drugs than antibacterial drugs because protozoa are eukaryotes like their human hosts, and they have several life stages, some of which can be resistant to the drug.
- Antihelminthic drugs immobilize or disintegrate infesting helminths or inhibit their metabolism in some manner.
- Antiviral drugs interfere with viral replication by blocking viral entry into cells, blocking the replication process, or preventing the assembly of viral subunits into complete virions.
- Many antiviral agents are analogs of nucleotides. They inactivate the replication process when incorporated into viral nucleic acids. HIV antivirals interfere with reverse transcriptase or proteases to prevent the maturation of viral particles.

10.3 Antimicrobial Resistance

- Microorganisms are termed *drug resistant* when they are no longer inhibited by an antimicrobial to which they were previously sensitive.
- Microbes acquire genes that code for methods of inactivating or escaping the antimicrobial, or acquire mutations that affect the drug's impact.
- Mechanisms of microbial drug resistance include drug inactivation, decreased drug uptake, decreased drug receptor sites, and modification of metabolic pathways formerly attacked by the drug.
- Widespread indiscriminate use of antimicrobials is one factor that has resulted in an explosion of microorganisms resistant to all common drugs.
- Probiotics and prebiotics are methods of crowding out pathogenic bacteria and providing a favorable environment for the growth of beneficial bacteria.

10.4 Interactions Between Drug and Host

- The three major side effects of antimicrobials are toxicity to organs, allergic reactions, and problems resulting from alteration of normal biota.
- Antimicrobials that destroy most but not all normal biota can allow the unaffected normal biota to overgrow, causing a superinfection.

Multiple-Choice Questions   Bloom's Levels 1 and 2: Remember and Understand

Select the correct answer from the answers provided.

1. A compound synthesized by bacteria or fungi that destroys or inhibits the growth of other microbes is a/an
   - a. synthetic drug.
   - b. antibiotic.
   - c. interferon.
   - d. competitive inhibitor.

2. The main consideration(s) in selecting an effective antimicrobial is/are
   - a. the identity of the infecting microbe.
   - b. the microbe's sensitivity to available drugs.
   - c. the overall medical status of the infected host.
   - d. a and b.
   - e. a, b, and c.

3. Drugs that prevent the formation of the bacterial cell wall are
   - a. quinolones.
   - b. penicillins.
   - c. tetracyclines.
   - d. aminoglycosides.

4. Microbial resistance to drugs is acquired through
   - a. conjugation.
   - b. transformation.
   - c. transduction.
   - d. all of these.

5. Antimalarial treatments are difficult because
   - a. the protozoal parasite (*Plasmodium*) is eukaryotic and therefore similar to human cells.
   - b. there are several species of *Plasmodium*.
   - c. no single drug can target all the life stages of *Plasmodium*.
   - d. all of the above are true.

6. Most antihelminthic drugs function by
   - a. weakening the worms so they can be flushed out by the intestine.
   - b. inhibiting worm metabolism.
   - c. blocking the absorption of nutrients.
   - d. inhibiting egg production.

7. The MIC is the _____ of a drug that is required to inhibit growth of a microbe.
   - a. largest concentration
   - b. standard dose
   - c. smallest concentration
   - d. lowest dilution

8. An antimicrobial drug with a _____ therapeutic index is a better choice than one with a _____ therapeutic index.
   - a. low; high
   - b. high; low

Critical Thinking   Bloom's Levels 3, 4, and 5: Apply, Analyze, and Evaluate

Critical thinking is the ability to reason and solve problems using facts and concepts. These questions can be approached from a number of angles and, in most cases, they do not have a single correct answer.

1. Can you think of a situation in which it would be better for a drug to be microbistatic rather than microbicidal? Discuss thoroughly.

2. Why does the penicillin group of drugs have milder toxicity than other antibiotics?

3. Explain the phenomenon of drug resistance from the standpoint of microbial genetics (include a description of R factors).

4. You have been directed to take a sample from a growth-free portion of the zone of inhibition in the Kirby-Bauer test and inoculate it onto a plate of nonselective medium.
   - a. What does it mean if growth occurs on the new plate?
   - b. What if there is no growth?

5. a. Explain the basis for combined therapy.
   - b. Give reasons why it could be helpful to use combined therapy in treating HIV infection.

Visual Connections   Bloom's Level 5: Evaluate

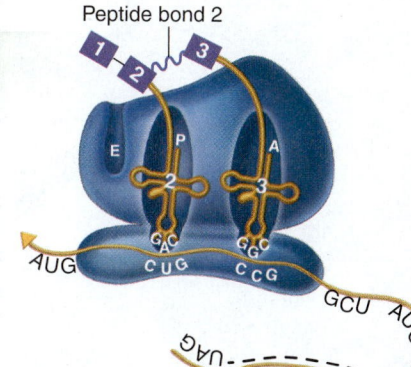

This question connects previous images to a new concept.

1. **From chapter 8, table 8.5.** Place Xs over this figure in places where bacterial protein synthesis might be inhibited by drugs.

MICROBIOLOGY

www.mcgrawhillconnect.com

Enhance your study of this chapter with study tools and practice tests. Also ask your instructor about the resources available through Connect, including the media-rich eBook, interactive learning tools, and animations.

CASE FILE

A Permanent Fix

When I was an ultrasound technician in an urban hospital, I met Jaelyn, a little girl with vesicoureteral reflux (VUR). Vesicoureteral reflux is a congenital condition of the urinary tract system in which the ureters are attached to the wall of the bladder at an angle that allows urine to "reflux" backward from the bladder to the kidneys. Children with VUR can experience frequent kidney infections, which can damage the kidneys, sometimes permanently.

Many children with moderate VUR require constant antibiotic suppression therapy to prevent episodes of pyelonephritis (kidney infections). These children remain on antibiotics for a few years and will sometimes outgrow the problem as their ureters grow. Jaelyn was no different than most children with moderate VUR—she was on antibiotics continuously from the time she was 7 months old when her condition was discovered. I would see Jaelyn and perform an ultrasound on her kidneys every 6 months to monitor her kidneys—and more often if she developed an infection requiring hospitalization and intravenous antibiotics.

In addition to kidney infections, Jaelyn was a sickly child who seemed to catch every bug that went around. She had constant colds, ear infections, and gastrointestinal viruses, possibly because her immune system was constantly working to fight off urinary tract infections. The fact that she was constantly taking antibiotics might have also contributed to her frequent infections.

When Jaelyn was 3 years old, she was hospitalized with her fifth kidney infection. I was called to perform another ultrasound. On ultrasound, her kidneys appeared dilated, and cultures of her urine came back showing *Pseudomonas aeruginosa*, a gram-negative bacterium that is an opportunistic pathogen.

The discovery of *P. aeruginosa* in Jaelyn's urine led to the decision to perform surgery to correct the angle of Jaelyn's ureters so that urine could no longer reflux into the kidneys. Following surgery, Jaelyn continued to take suppressive antibiotic therapy for 1 month, after which she was able to stop taking antibiotics. I saw Jaelyn once more after her surgery to recheck her kidneys. At her last ultrasound appointment, Jaelyn's kidneys were normal size and functioning well.

- What is an opportunistic pathogen?
- Why did the discovery of *P. aeruginosa* in Jaelyn's urine lead to the decision to perform surgery?

Case File Wrap-Up appears on page 318.

Interactions Between Microbes and Humans

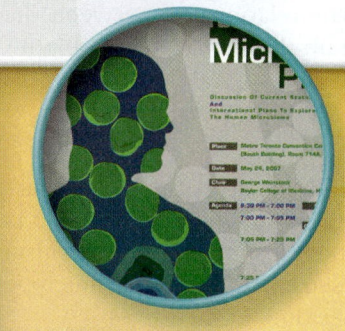

IN THIS CHAPTER...

11.1 The Human Host

1. Differentiate among the terms *colonization, infection,* and *disease.*
2. Enumerate the sites where normal biota is found in humans.
3. Discuss how the Human Microbiome Project is changing our understanding of normal biota.

11.2 The Progress of an Infection

4. Differentiate between a microbe's pathogenicity and its virulence.
5. Define *opportunism,* and list examples of common opportunistic pathogens.
6. List the steps a microbe has to take to get to the point where it can cause disease.
7. List several portals of entry and exit.
8. Define *infectious dose,* and explain its role in establishing infection.
9. Describe three ways microbes cause tissue damage.
10. Compare and contrast major characteristics of endotoxin and exotoxins.
11. Provide a definition of *virulence factors.*
12. Draw a diagram of the stages of disease in a human.
13. Differentiate among the various types of reservoirs, providing examples of each.
14. List several different modes of transmission of infectious agents.
15. Define *healthcare-associated infection,* and list the three most common types.
16. List Koch's postulates, and discuss when they might not be appropriate in establishing causation.

11.3 Epidemiology: The Study of Disease in Populations

17. Summarize the goals of epidemiology, and differentiate it from traditional medical practice.
18. Explain what is meant by a disease being "notifiable" or "reportable," and provide examples.
19. Define *incidence* and *prevalence,* and explain the difference between them.
20. Discuss the three major types of epidemics, and identify the epidemic curve associated with each.

11.1 The Human Host

It is easy to think of humans and other mammals as discrete, stand-alone organisms that are also colonized by some nice, nonpathogenic microorganisms. In fact, that's what scientists thought for the last 150 years or so. But the truer picture is that humans and other mammals have the form and the physiology that they have due to *having been formed in intimate contact with their microbes.* Do you see the difference? The **human microbiome,** the sum total of all microbes found on and in a normal human, is critically important to the health and functioning of its host organism. This chapter describes the relationship between the human and microorganisms, both the ones that make up the human's microbiome and the ones that are harmful.

The Human Microbiome

When you consider the evolutionary time line (refer to figure 1.1) of bacteria and humans, it is quite clear that humans evolved in an environment that had long been populated by bacteria and single-celled eukaryotes. It should not be surprising, therefore, that humans do not do well if they are separated from their microbes, either during growth and development or at any other time in their lives.

The extent to which this is true has been surprising even to the scientists studying it. Since 2007, a worldwide research effort has been underway that utilizes the powerful techniques of genome sequencing and "big data" tools. The American effort is called the **Human Microbiome Project (HMP),** and there are similar projects occurring around the world. The aim has been not only to characterize the microbes living on human bodies when they are healthy but also to determine how the microbiome differs in various diseases. Previous to this international project, scientists and clinicians mainly relied on culture techniques to determine what the "normal biota" consisted of. That meant we only knew about bacteria and fungi that we could grow in the laboratory, which vastly undercounts the actual number and variety, since many—even the majority of—microbes cannot be cultured in the laboratory, though they grow quite happily on human tissues.

Viruses are not traditionally discussed in the context of normal biota. However, they are most certainly present in healthy humans in vast quantities. Throughout evolutionary history, viral infections (of cells of all types) have influenced the way cells and organisms and communities and, yes, the entire ecosystem have developed. The critical contributions of viruses is just now being rigorously studied.

The information about the human microbiome presented in this chapter reflects the new findings, which should still be considered preliminary. We will try to show you the differences between the old picture of normal biota in various organ systems and the new, emerging picture. At this point in medical history, it will be important to appreciate the transitioning view.

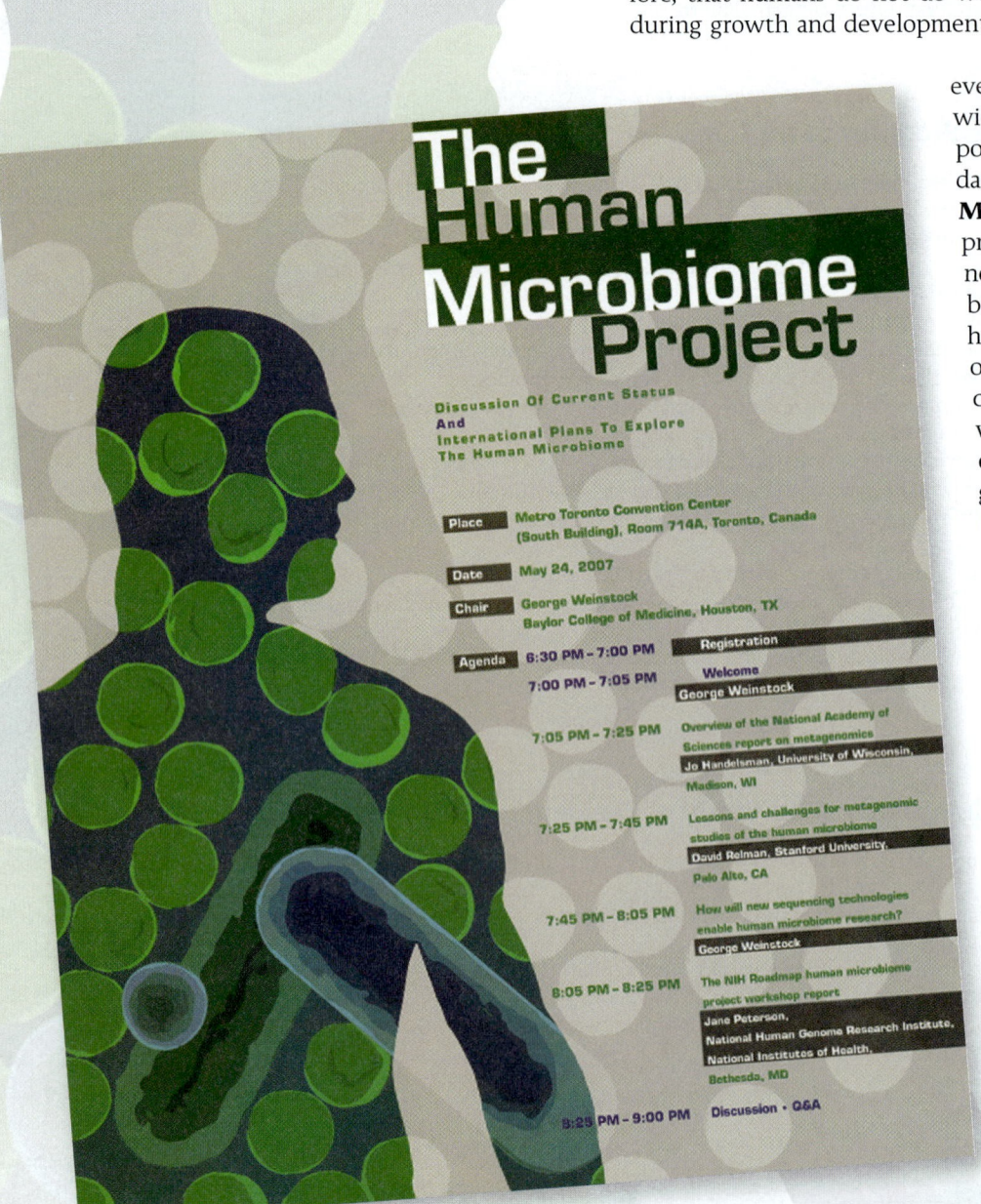

The Human Microbiome Project

Discussion Of Current Status And International Plans To Explore The Human Microbiome

Place	Metro Toronto Convention Center (South Building), Room 714A, Toronto, Canada
Date	May 24, 2007
Chair	George Weinstock Baylor College of Medicine, Houston, TX
Agenda	6:30 PM – 7:00 PM Registration
	7:00 PM – 7:05 PM Welcome George Weinstock
	7:05 PM – 7:25 PM Overview of the National Academy of Sciences report on metagenomics Jo Handelsman, University of Wisconsin, Madison, WI
	7:25 PM – 7:45 PM Lessons and challenges for metagenomic studies of the human microbiome David Relman, Stanford University, Palo Alto, CA
	7:45 PM – 8:05 PM How will new sequencing technologies enable human microbiome research? George Weinstock
	8:05 PM – 8:25 PM The NIH Roadmap human microbiome project workshop report Jane Peterson, National Human Genome Research Institute, National Institutes of Health, Bethesda, MD
	8:25 PM – 9:00 PM Discussion • Q&A

Acquiring Resident Biota

The human body offers a seemingly endless variety of environmental niches, with wide variations in temperature, pH, nutrients, and oxygen tension occurring from one area to another. Because the body provides such a range of habitats, it should not be surprising that the body supports a wide range of microbes. **Table 11.1** provides a breakdown of our current understanding of the microbiota living in and on a healthy host. The uppermost row contains the set of sites that microbiologists have long known to host a normal biota. The middle row presents some new sites recently found to harbor microbiota in a healthy human. The bottom row reports that two sites, the brain and the bloodstream, have both been found to contain DNA from multiple species of bacteria. Their exact role there is not entirely clear yet.

The vast majority of microbes that come in contact with the body are removed or destroyed by the host's defenses long before they are able to colonize a particular area. Of those microbes able to establish an ongoing presence, an even smaller number are able to remain without attracting the unwanted attention of the body's defenses. This last group of organisms has evolved, along with its human hosts, to produce a complex relationship in which its effects are generally not damaging to the host. Recall from chapter 6 that microbes exist in different kinds of relationships with their hosts. Normal biota are generally either in a commensal or a mutualistic association with their hosts.

The generally antagonistic effect "good" microbes have against intruder microorganisms is called **microbial antagonism.** Normal biota exist in a steady established relationship with the host and are unlikely to be displaced by incoming microbes. This antagonistic protection is partly the result of a limited number of attachment sites in the host site, all of which are stably occupied by normal biota. This antagonism is also enabled by the chemical or physiological environment created by the resident biota, which is hostile to most other microbes. There are often members of the "normal" biota that would be pathogenic if they were allowed to multiply to larger numbers. Microbial antagonism is also responsible for keeping them in check.

Characterizing the normal biota as beneficial or, at worst, commensal to the host presupposes that the host is in good health, with a fully functioning immune system, and that the biota is present only in its natural microhabitat within the body. Hosts with compromised immune systems could very easily experience disease caused by their (previously normal) biota. Factors that weaken host defenses and increase susceptibility to infection include the following:

- old age and extreme youth (infancy, prematurity);
- genetic defects in immunity, and acquired defects in immunity (AIDS);
- surgery and organ transplants;
- underlying disease: cancer, liver malfunction, diabetes;
- chemotherapy/immunosuppressive drugs;
- physical and mental stress;
- pregnancy; and
- other infections.

Initial Colonization of the Newborn

Until 2013, the uterus and its contents were thought to be sterile during embryonic and fetal development and remain essentially germ-free until just before birth. We do know that comprehensive exposure

Table 11.1 Sites Previously Known to Harbor Normal Microbiota	
Skin and adjacent mucous membranes	External genitalia
Upper respiratory tract	Vagina
Gastrointestinal tract, including mouth	External ear canal
Outer portion of urethra	External eye (lids, conjunctiva)
Additional Sites Now Thought to Harbor At Least Some Normal Microbiota (or Their DNA)	
Lungs (lower respiratory tract) Bladder (and urine) Breast milk Amniotic fluid and fetus	
Sites in Which DNA from Microbiota Has Been Detected	
Brain Bloodstream	

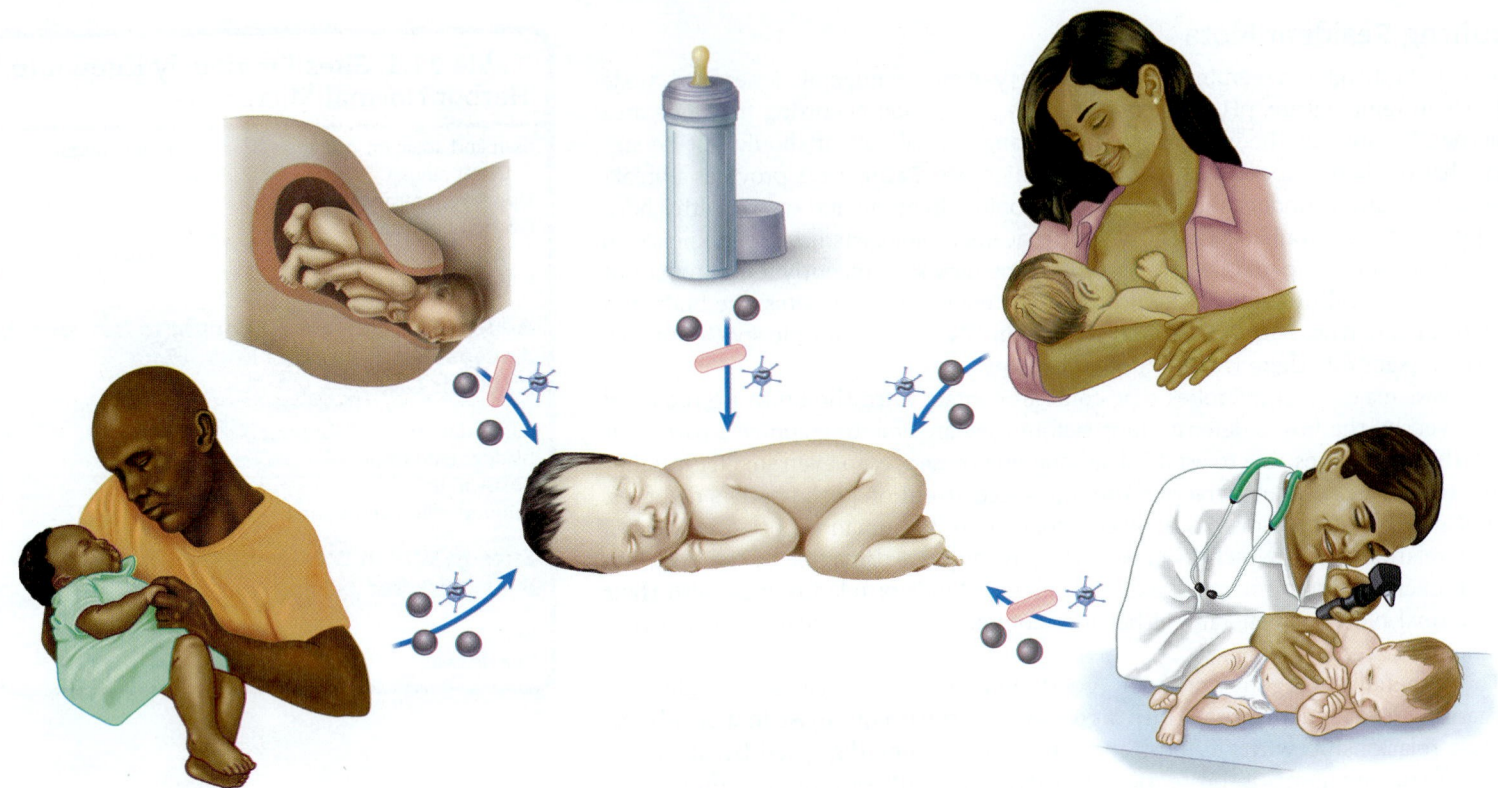

Figure 11.1 **The origins of microbiota in newborns.** From the moment of birth, the infant will begin to acquire microbes from its environment.

occurs during the birth process itself, when the baby becomes colonized with the mother's vaginal biota **(figure 11.1).** Many scientists now believe that the womb is not a sterile environment. Research in 2010 found that healthy newborns' stools, sampled before they have their first meal, contain a diversity of bacteria, indicating that their intestines are colonized *in utero*. These findings have been confirmed by other studies since then.

Within 8 to 12 hours after delivery, the vaginally delivered newborn typically has been colonized by bacteria such as *Lactobacillus*, *Prevotella*, and *Sneathia*, acquired primarily from the birth canal. Data from the Human Microbiome Project revealed that the microbial composition of the vagina changes significantly in pregnant women. Early on, a *Lactobacillus* species that digests milk begins to populate the vagina. Immediately prior to delivery, additional bacterial species colonize the birth canal. Scientists suggest that the lactobacilli provide the newborn baby with the enzymes necessary to digest milk, and that the later colonizers are better equipped to protect a newborn baby from skin disorders and other conditions. After the baby is born, the mother's vaginal microbiota returns to its former state. The baby continues to acquire resident microbiota from the environment, notably from its diet; throughout most of evolutionary history, of course, that means human breast milk. Scientists have found that human milk contains around 600 species of bacteria and a lot of sugars that babies cannot digest. The sugars are used by healthy gut bacteria, suggesting a role for breast milk in maintaining a healthy gut microbiome in the baby. The skin, gastrointestinal tract, and portions of the respiratory and genitourinary tracts all continue to be colonized as contact continues with family members, health care personnel, the environment, and food.

The Human Microbiome Project has shown that among healthy adults, the normal microbiota varies significantly. For instance, the microbiota on a person's right hand was found to be significantly different than that on the same person's left hand. What seemed to be more important than the exact microbial profile of any given body site was the profile of proteins, especially the enzymatic capabilities. That profile remained stable across subjects, though the microbes that were supplying those enzymes could differ broadly. Scientists are in the process of cataloging other microorganisms besides bacteria via metagenomics—and just beginning to appreciate their numbers in the human microbiome. For example, we now know that at least 100 types of fungi reside in the intestine and as many as a billion viruses are present per gram of feces.

11.1 LEARNING OUTCOMES—Assess Your Progress

1. Differentiate among the terms *colonization*, *infection*, and *disease*.
2. Enumerate the sites where normal biota is found in humans.
3. Discuss how the Human Microbiome Project is changing our understanding of normal biota.

11.2 The Progress of an Infection

A microbe whose relationship with its host is parasitic and results in infection and disease is termed a **pathogen**. A disease is defined as any deviation from health. There are hundreds of different diseases caused by such factors as infections, diet, genetics, and aging. In this chapter, however, we discuss only **infectious disease**—the disruption of a tissue or organ caused by microbes or their products.

The pattern of the host-parasite relationship can be viewed as a series of stages that begins with contact, progresses to infection, and ends in disease. Because of numerous factors relating to host resistance and degree of pathogenicity, not all contacts lead to colonization, not all colonizations lead to infection, and not all infections lead to disease. In fact, contamination without colonization and colonization without disease are the rules. The type and severity of infection depend both on the pathogenicity of the organism and the condition of the host. **Figure 11.2** puts this in graphic form. It explains all those questions you have always had about why *you* got the disease but your friend did not. Spend some time with this figure. It contains a wealth of information about why a certain microbe will cause diseases in only certain individuals.

Various aspects of the host influence whether a microbe will have severe, mild, or no effects. Variation in the genes coding for components of the immune system—or even the anatomy of infection sites—is one of these factors. Gender, hormone levels, and overall health also play a role. **Pathogenicity**, you will recall, is a broad concept that describes an organism's potential to cause disease and is used to divide pathogenic microbes into one of two groups. **True pathogens** (primary pathogens) are capable of causing disease in healthy persons with normal immune defenses. They are generally associated with a specific, recognizable disease, which may vary in severity from mild (colds) to severe (malarial) to fatal (rabies). Examples of true pathogens include the influenza virus, plague bacillus, and malarial protozoan.

Opportunistic pathogens cause disease when the host's defenses are compromised or when the pathogens become established in a part of the body that is not natural to them. Opportunists are not considered pathogenic to a normal healthy person and, unlike primary pathogens, do not generally possess well-developed virulence properties. Examples of opportunistic pathogens include *Pseudomonas* species and *Candida albicans*.

The relative severity of the disease caused by a particular microorganism depends on the **virulence** of the microbe. Although the terms *pathogenicity* and

NCLEX® PREP

2. What is the difference between a true pathogen and an opportunistic pathogen?
   a. *True pathogens cause a disease in the presence of immunosuppression whereas opportunistic pathogens do not.*
   b. *Opportunistic pathogens develop virulence properties whereas true pathogens do not.*
   c. *The diseases associated with true pathogens may vary in presentation ranging from mild to severe infections whereas opportunistic pathogens always present in severe form.*
   d. *True pathogens cause disease in healthy individuals whereas opportunistic pathogens typically cause disease in clients who are immunocompromised.*

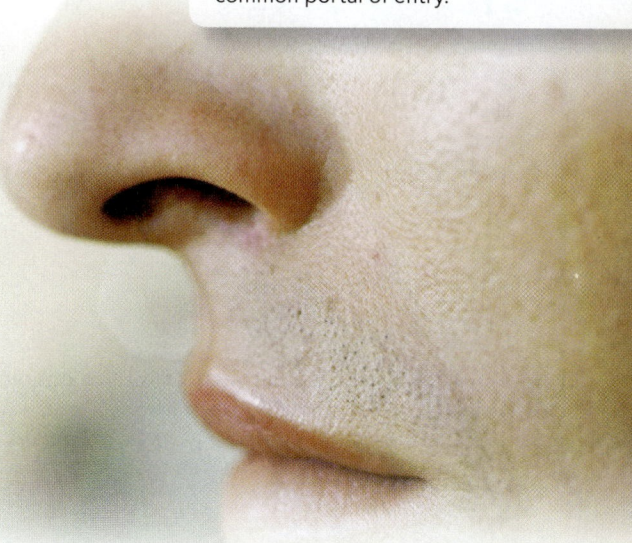

The respiratory tract is the most common portal of entry.

Microbe X			Host			Outcomes
Virulence	Percentage of optimal infectious dose	Correct portal of entry	Genetic profile that resists Microbe X (nonspecific defenses)	Previous exposure to Microbe X (specific immunity)	General level of health	
High — Low	100 — 0	Off		On		Microbe passes through unnoticed.
High — Low	100 — 0	On		On		Microbe passes through unnoticed. *or* Microbe becomes established without disease (colonization or infection).
High — Low	100 — 0	Off		Off		Microbe passes through unnoticed. *or* Microbe becomes established without disease (colonization or infection).
High — Low	100 — 0	On		Off		Microbe causes disease.

Figure 11.2 Will disease result from an encounter between a (human) host and a microorganism? In most cases, all of the slider bars must be in the correct ranges and the microbe's toggle switch must be in the "on" position, while the host's toggle switch must be in the "off" position in order for disease to occur. These are just a few examples and not the only options. For instance, you can see from the third row that even when the host has no specific immunity, for example, the microbe does not have enough advantages to cause disease.

virulence are often used interchangeably, *virulence* is the accurate term for describing the degree of pathogenicity. The virulence of a microbe is determined by its ability to

1. establish itself in the host, and
2. cause damage.

There is much involved in both of these steps. To establish themselves in a host, microbes must enter the host, attach firmly to host tissues, and survive the host defenses. To cause damage, microbes produce toxins or induce a host response that is actually injurious to the host. Any characteristic or structure of the microbe that contributes to the preceding activities is called a **virulence factor.** Virulence can be due to single or multiple factors. In some microbes, the causes of virulence are clearly established, but in others they are not. There is also an increasing appreciation of **polymicrobial** infections, in which the disease symptoms are influenced by more than one colonizer. In the following section, we examine the effects of virulence factors, while outlining the stages in the progress of an infection.

Step One: Becoming Established—Portals of Entry

To initiate an infection, a microbe enters the tissues of the body by a characteristic route, the **portal of entry,** usually the skin or a mucous membrane. The source of the infectious agent can be **exogenous,** originating from a source outside the body (the environment or another person or animal), or endogenous, already existing on or in the body (normal biota or a previously silent infection).

The majority of pathogens have adapted to a specific portal of entry, one that provides a habitat for further growth and spread. This adaptation can be so restrictive that if certain pathogens enter the "wrong" portal, they will not be infectious. For instance, inoculation of the nasal mucosa with the influenza virus is likely to give rise to the flu, but if this virus contacts only the skin, no infection will result. Occasionally, an infective agent can enter by more than one portal. For instance, *Mycobacterium tuberculosis* enters through both the respiratory and gastrointestinal tracts, and pathogens in the genera *Streptococcus* and *Staphylococcus* have adapted to invasion through several portals of entry such as the skin, urogenital tract, and respiratory tract. **Table 11.2** outlines common portals of entry, the organisms and diseases associated with these portals, and methods of entry.

The Size of the Inoculum

Another factor crucial to the course of an infection is the quantity of microbes in the inoculating dose. For most agents, infection will proceed only if a minimum number, called the *infectious dose* (*ID*), is present. This number has been determined experimentally for many microbes. In general, microorganisms with smaller infectious doses have greater virulence. On the low end of the scale, the ID for *Coxiella burnetii*, the causative agent of Q fever, is only a single cell, and the ID is only about 10 infectious cells in tuberculosis, giardiasis, and coccidioidomycosis. The ID is 1,000 bacteria for gonorrhea and 10,000 bacteria for typhoid fever, in contrast to 1,000,000,000 bacteria in cholera. Numbers below an infectious dose will generally not result in an infection. But if the quantity is far in excess of the ID, the onset of disease can be extremely rapid.

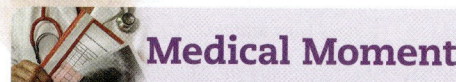

Medical Moment

When the Portal of Entry Is Compromised

Different portals of entry have protective mechanisms to prevent infectious agents from gaining entry. For example, the eye produces tears, which not only rinse pathogens out of the eye but also contain pathogen-fighting chemicals. The skin acts as a physical barrier, providing it is intact.

What happens when there is a failure to protect at a portal of entry? The respiratory tract is lined with cilia, fingerlike projections that protrude from cells that sweep back and forth to move particles toward the throat so that they can be swallowed rather than remain in the respiratory tract. In *primary ciliary dyskinesia,* affected individuals lack properly functioning cilia. These individuals have frequent respiratory tract infections beginning in early childhood. They may even experience breathing problems at birth. Chronic respiratory infections lead to bronchiectasis, which results from damage affecting the bronchial tubes leading to the lungs. This condition affects approximately one in 16,000 individuals and is passed down from two parents who have the defective gene but do not have the disease themselves (autosomal recessive pattern).

Table 11.2 Portals of Entry and Organisms Typically Involved

Portal of Entry	Organism/Disease	How Access Is Gained
Skin	*Staphylococcus aureus, Streptococcus pyogenes, Clostridium tetani*	Via nicks, abrasions, punctures, areas of broken skin
	Herpes simplex (type 1)	Via mucous membranes of the lips
	Helminth worms	Burrow through the skin
	Viruses, rickettsias, protozoa (i.e., malaria, West Nile virus)	Via insect bites
	Haemophilus aegyptius, Chlamydia trachomatis, Neisseria gonorrhoeae	Via the conjunctiva of the eye
Gastrointestinal tract	*Salmonella, Shigella, Vibrio, Escherichia coli*, poliovirus, hepatitis A, echovirus, rotavirus, enteric protozoans (*Giardia lamblia, Entamoeba histolytica*)	By eating/drinking contaminated foods and fluids. Via fomites (inanimate objects contaminated with the infectious organism)
Respiratory tract	Bacteria causing meningitis, influenza, measles, mumps, rubella, chickenpox, common cold, *Streptococcus pneumoniae, Klebsiella, Mycoplasma, Cryptococcus, Pneumocystis, Mycobacterium tuberculosis, Histoplasma*	Via inhalation of offending organism
Urogenital tract	HIV, *Trichomonas*, hepatitis B, syphilis, *Treponema pallidum, Neisseria gonorrhoeae, Chlamydia trachomatis*, herpes, genital warts	Enter through the skin/mucosa of penis, external genitalia, vagina/cervix, urethra; may enter through an unbroken surface or through a cut or abrasion

Step Two: Becoming Established—Attaching to the Host

Adhesion is a process by which microbes gain a more stable foothold on host tissues. Because adhesion is dependent on binding between specific molecules on both the host and pathogen, a particular pathogen is limited to only those cells (and organisms) to which it can bind. Once attached, the pathogen is poised advantageously to invade the body compartments. Bacterial, fungal, and protozoal pathogens attach most often by mechanisms such as fimbriae (pili), surface proteins, and adhesive slimes or capsules; viruses attach by means of specialized receptors. In addition, parasitic worms are mechanically fastened to the portal of entry by suckers, hooks, and barbs. There are many different methods in which microbes can attach themselves to host tissues. Firm attachment to host tissues is almost always a prerequisite for causing disease since the body has so many mechanisms for flushing microbes and foreign materials from its tissues.

Step Three: Becoming Established—Surviving Host Defenses

Microbes that are not established in a normal biota relationship in a particular body site in a host are likely to encounter resistance from host defenses when first entering, especially from certain white blood cells called **phagocytes.** These cells ordinarily engulf and destroy pathogens by means of enzymes and antimicrobial chemicals (see chapter 12).

Antiphagocytic factors are a type of virulence factor used by some pathogens to avoid phagocytes. The antiphagocytic factors of microorganisms help them to circumvent some part of the phagocytic process **(figure 11.3c).** The most aggressive strategy involves bacteria that kill phagocytes outright. Species of both *Streptococcus* and *Staphylococcus* produce **leukocidins,** substances that are toxic to white blood

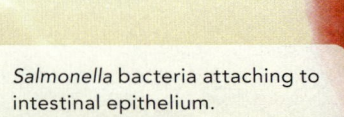
Salmonella bacteria attaching to intestinal epithelium.

cells. Some microorganisms secrete an extracellular surface layer (slime or capsule) that makes it physically difficult for the phagocyte to engulf them. *Streptococcus pneumoniae, Salmonella typhi, Neisseria meningitidis,* and *Cryptococcus neoformans* are notable examples. Some bacteria are well adapted to survival inside phagocytes after ingestion. For instance, pathogenic species of *Legionella, Mycobacterium,* and many rickettsias are readily engulfed but are capable of avoiding further destruction. The ability to survive intracellularly in phagocytes has special significance because it provides a place for the microbes to hide, grow, and be spread throughout the body.

Step Four: Causing Disease
How Virulence Factors Contribute to Tissue Damage

Virulence factors are structures or capabilities that allow a pathogen to cause infection in a host. From a microbe's perspective, they are simply adaptations it uses to invade and establish itself in the host. The effects of a pathogen's virulence factors on tissues vary greatly. Cold viruses, for example, invade and multiply but cause relatively little damage to their host. At the other end of the spectrum, pathogens such as *Clostridium tetani* or HIV severely damage or kill their host. There are three major ways that microorganisms damage their host:

1. directly through the action of enzymes **(figure 11.3*a*),**
2. directly through the action of toxins (both endotoxin and exotoxins), **(figure 11.3*b*),** and
3. indirectly by inducing the host's defenses to respond excessively or inappropriately (figure 11.3c).

It is obvious that enzymes, endotoxin and exotoxins are virulence factors, but other characteristics of microbes that lead to host overreaction are also considered virulence factors. The capsule of *Streptococcus pneumoniae* is a good example. Its presence prevents the bacterium from being cleared from the lungs by phagocytic cells, leading to a continuous influx of fluids into the lung spaces, and the condition we know as pneumonia (figure 11.3c).

Extracellular Enzymes   Many pathogenic bacteria, fungi, protozoa, and worms secrete **exoenzymes** that break down and inflict damage on tissues. Other enzymes dissolve the host's defense barriers and promote the spread of microbes to deeper tissues.
Examples of enzymes are

1. mucinase, which digests the protective coating on mucous membranes and is a factor in amoebic dysentery; and
2. hyaluronidase, which digests hyaluronic acid, the ground substance that cements animal cells together. This enzyme is an important virulence factor in staphylococci, clostridia, streptococci, and pneumococci.

Some enzymes react with components of the blood. Coagulase, an enzyme produced by pathogenic staphylococci, causes clotting of blood or plasma. By contrast, the bacterial kinases (streptokinase, staphylokinase) do just the opposite, dissolving fibrin clots and expediting the invasion of damaged tissues. In fact, one form of streptokinase is a therapy to dissolve blood clots in patients who have problems with thrombi and embolisms.

Bacterial Toxins: A Potent Source of Cellular Damage   A **toxin** is a specific chemical product of microbes that is poisonous to other organisms. A toxin is named according to its specific target of action: Neurotoxins act on the nervous system; enterotoxins act on the intestine; hemotoxins lyse red blood cells; and nephrotoxins damage the kidneys.
There are two broad categories of bacterial toxins. **Exotoxins** are proteins with a strong specificity for a target cell and extremely powerful, sometimes deadly, effects.

Many gastrointestinal (GI) diseases are caused by bacterial toxins that affect the GI tract. These are called enterotoxins.

Enzymes

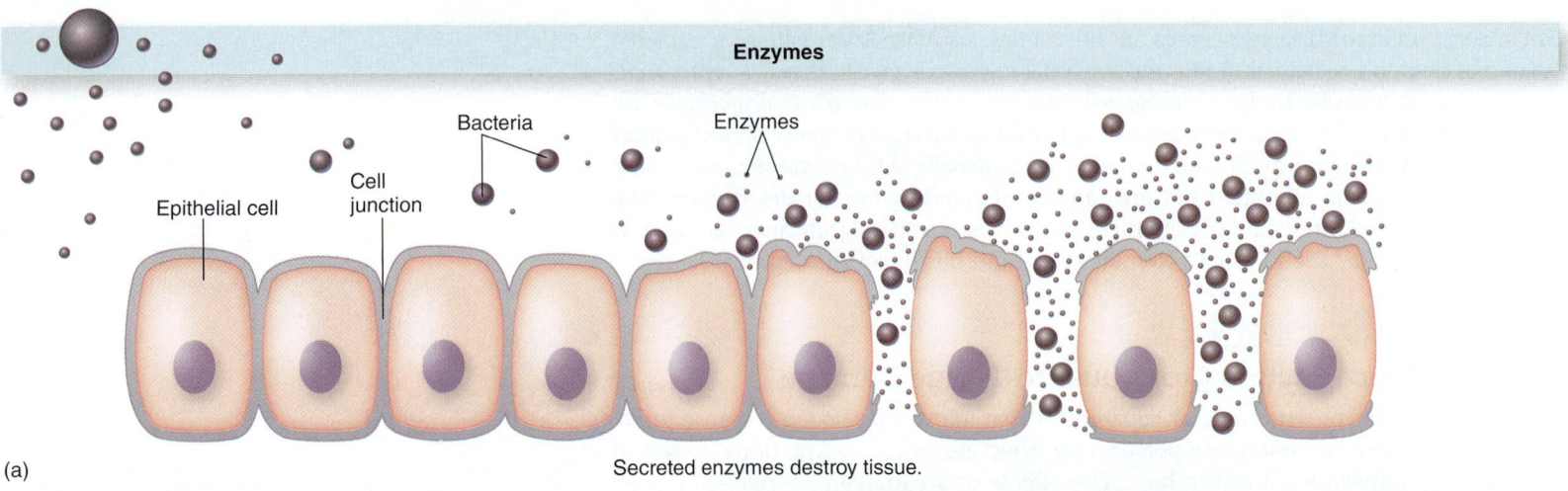

Epithelial cell

Cell junction

Bacteria

Enzymes

(a)   Secreted enzymes destroy tissue.

Toxins

| **Exotoxins** | **Endotoxin** |

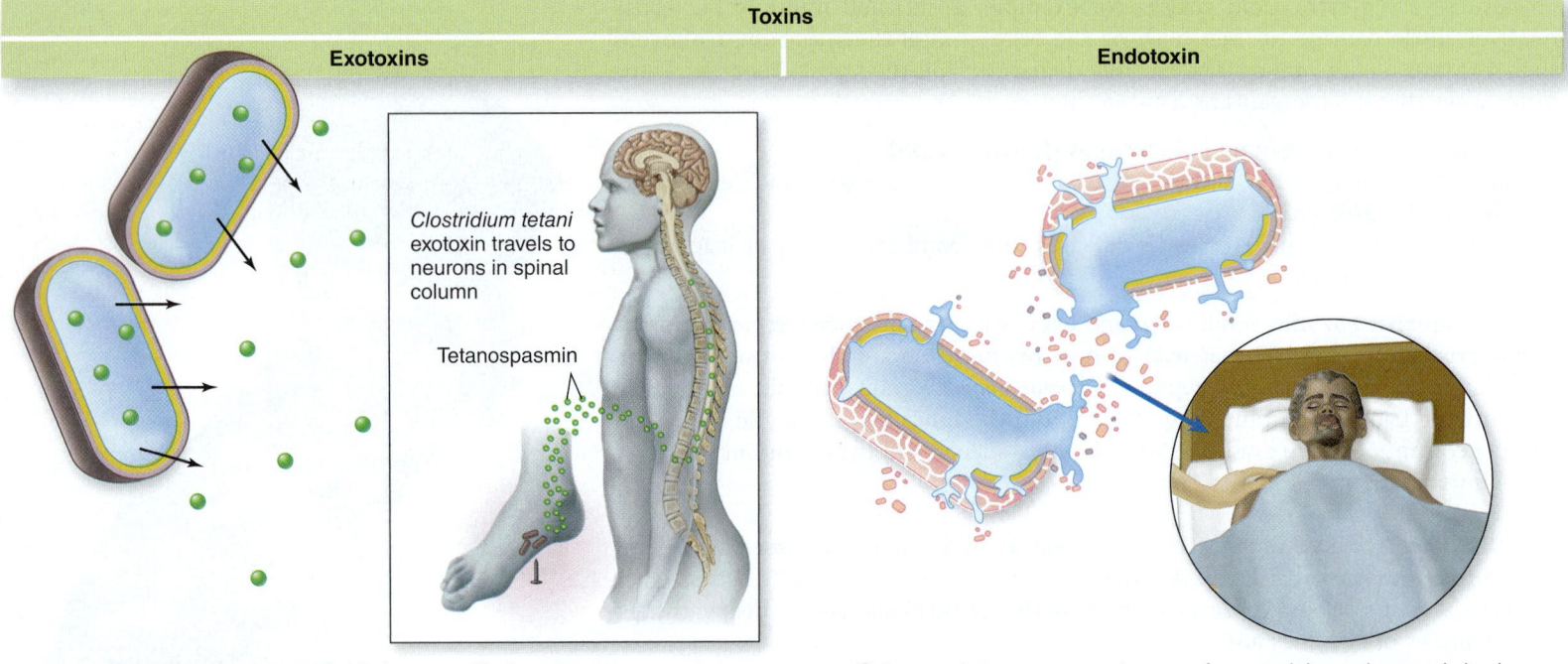

Clostridium tetani exotoxin travels to neurons in spinal column

Tetanospasmin

(b)   Specific secreted protein binds to specific tissue target.   Outer membrane component causes fever, malaise, aches, and shock.

Induction of Host Defenses

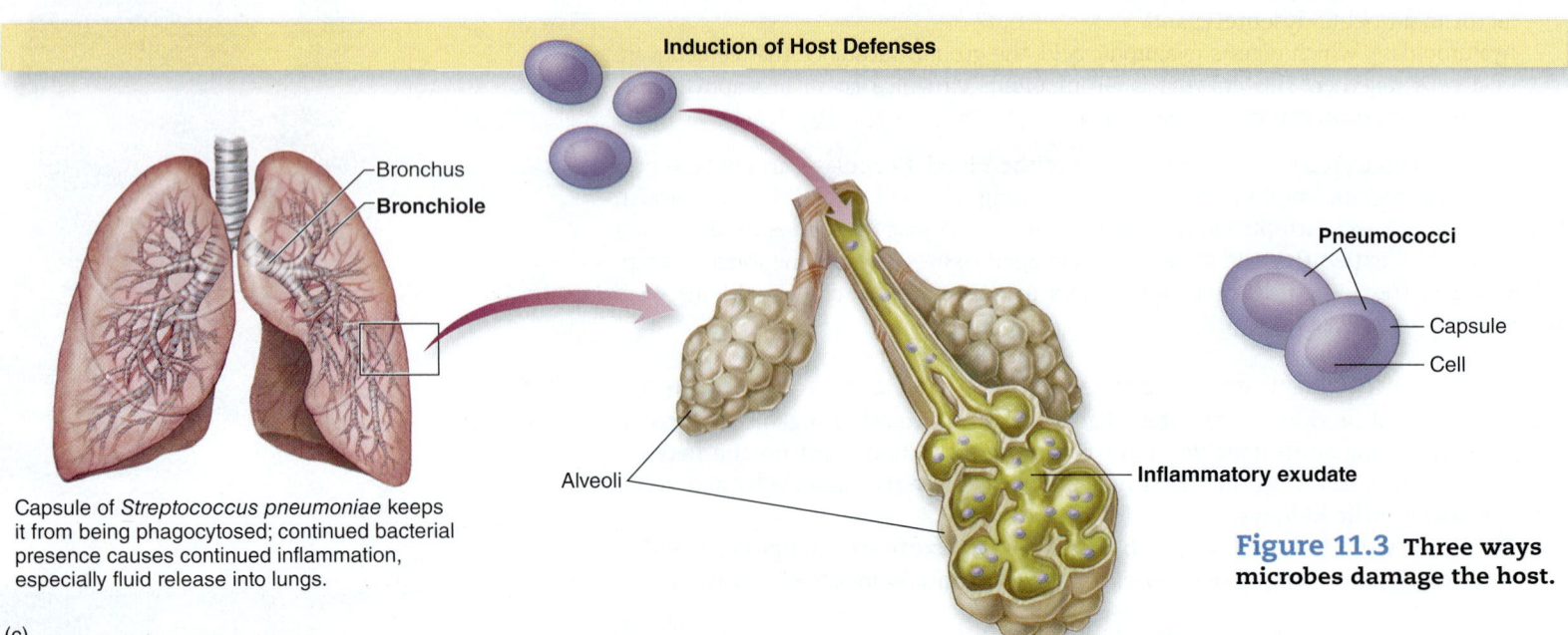

Bronchus

Bronchiole

Pneumococci

Capsule

Cell

Alveoli

Inflammatory exudate

Capsule of *Streptococcus pneumoniae* keeps it from being phagocytosed; continued bacterial presence causes continued inflammation, especially fluid release into lungs.

Figure 11.3 Three ways microbes damage the host.

(c)

They generally affect cells by damaging the cell membrane and initiating lysis or by disrupting intracellular function. **Hemolysins** (hee-mahl′-uh-sinz) are a class of bacterial exotoxin that disrupts the cell membrane of red blood cells (and some other cells, too). This damage causes the red blood cells to **hemolyze**–to burst and release hemoglobin pigment. Hemolysins that increase pathogenicity include the streptolysins of *Streptococcus pyogenes* and the alpha (α) and beta (β) toxins of *Staphylococcus aureus*. When colonies of bacteria growing on blood agar produce hemolysin, distinct zones appear around the colony. The pattern of hemolysis is often used to identify bacteria and determine their degree of virulence **(figure 11.4)**.

In contrast to the category *exotoxin*, which contains many different examples, the word *endotoxin* refers to a single substance. Endotoxin is actually a chemical called lipopolysaccharide (LPS), which is part of the outer membrane of gram-negative cell walls. Gram-negative bacteria shed these LPS molecules into tissues or into the circulation. Endotoxin differs from exotoxins in having a variety of systemic effects on tissues and organs. Depending upon the amounts present, endotoxin can cause fever, inflammation, hemorrhage, and diarrhea. Blood infection by gram-negative bacteria such as *Salmonella*, *Shigella*, *Neisseria meningitidis*, and *Escherichia coli* are particularly dangerous, in that it can lead to fatal endotoxic shock. **Table 11.3** contains important information about exotoxins and endotoxin.

Inducing an Injurious Host Response   Despite the extensive discussion on direct virulence factors, such as enzymes and toxins, it is probably the case that more microbial diseases are the result of indirect damage, or the host's excessive or inappropriate response to a microorganism. This is an extremely important point because it means that pathogenicity is not a trait inherent in microorganisms but is really a consequence of the interplay between microbe and host.

The Process of Infection and Disease

Establishment, Spread, and Pathologic Effects

Aided by virulence factors, microbes eventually settle in a particular target organ and cause damage at the site. The type and scope of injuries inflicted during this process account for the typical stages of an infection, the patterns of the infectious disease, and its manifestations in the body.

In addition to the adverse effects of enzymes, toxins, and other factors, multiplication by a pathogen frequently weakens host tissues. Pathogens can obstruct tubular structures such as blood vessels, lymphatic channels, fallopian tubes, and

Figure 11.4 Beta-hemolysis and alpha-hemolysis by different bacteria on blood agar. Beta-hemolysis, in the lower right, results in complete clearing of the red blood cells incorporated in the agar. Alpha-hemolysis, on the lower left, refers to incomplete lysis of the red blood cells, leaving a greenish tinge to the colonies and the area surrounding them.

Table 11.3 Differential Characteristics of Bacterial Exotoxins and Endotoxin

Characteristic	Exotoxins	Endotoxin
Toxicity	Toxic in minute amounts	Toxic in high doses
Effects on the body	Specific to a cell type (blood, liver, nerve)	Systemic: fever, inflammation
Chemical composition	Small proteins	Lipopolysaccharide of cell wall
Denatured by heat (60°C)	Yes	No
Toxoid formation	Can be converted to toxoid	Cannot be converted to toxoid
Immune response	Stimulate antitoxins	Does not stimulate antitoxins
Fever stimulation	Usually not	Yes
Manner of release	Secreted from live cell	Released by cell via shedding or during lysis
Typical sources	A few gram-positive and gram-negative	All gram-negative bacteria

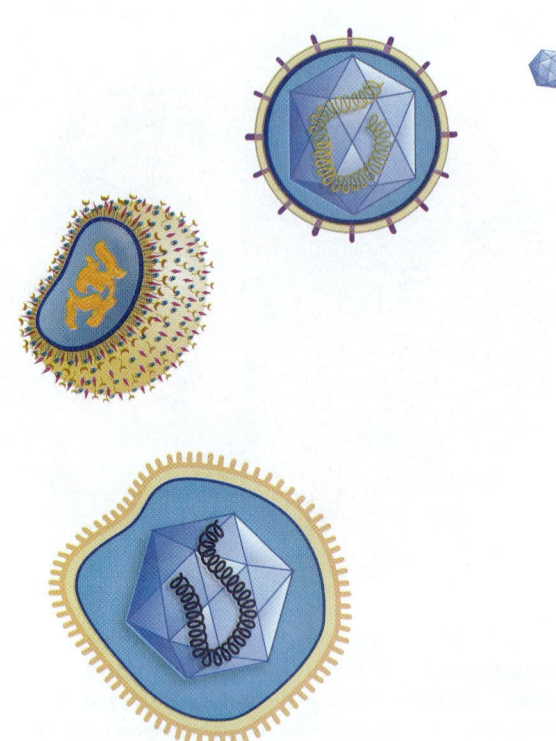

Table 11.4 Definitions of Infection Types

Type of Infection	Definition	Example
Localized infection	Microbes enter the body, remain confined to a specific tissue	Boils, warts, fungal skin infections
Systemic infection	Infection spreads to several sites and tissue fluids (usually via the bloodstream), but may travel by other means such as nerves (rabies) and cerebrospinal fluid (meningitis)	Mumps, rubella, chickenpox, AIDS, anthrax, typhoid, syphilis
Focal infection	Infectious agent spreads from a local site and is carried to other tissues	Tuberculosis, streptococcal pharyngitis
Mixed infection (polymicrobial infection)	Several agents establish themselves simultaneously at the infection site	Human bite infections, wound infections, gas gangrene
Primary infection	The initial infection	Can be any infection
Secondary infection	A second infection caused by a different microbe, which complicates a primary infection; often a result of lowered host immune defenses	Influenza complicated by pneumonia, common cold complicated by bacterial otitis media
Acute infection	Infection comes on rapidly, with severe but short-lived effects	Influenza
Chronic infection	Infection that progresses and persists over a long period of time	HIV

bile ducts. Accumulated damage can lead to cell and tissue death, a condition called **necrosis.** Although viruses do not produce toxins or destructive enzymes, they destroy cells by multiplying in and lysing them. Many of the cytopathic effects of viral infection arise from the impaired metabolism and death of cells (see chapter 5).

Patterns of Infection Patterns of infection are many and varied. **Table 11.4** describes various terms used to describe infection.

Figure 11.5 is a summary of the pathway a microbe follows when it causes disease.

Signs and Symptoms: Warning Signals of Disease

When an infection causes pathologic changes leading to disease, it is often accompanied by a variety of signs and symptoms. A **sign** is any objective evidence of disease as noted by an observer; a **symptom** is the subjective evidence of disease as sensed

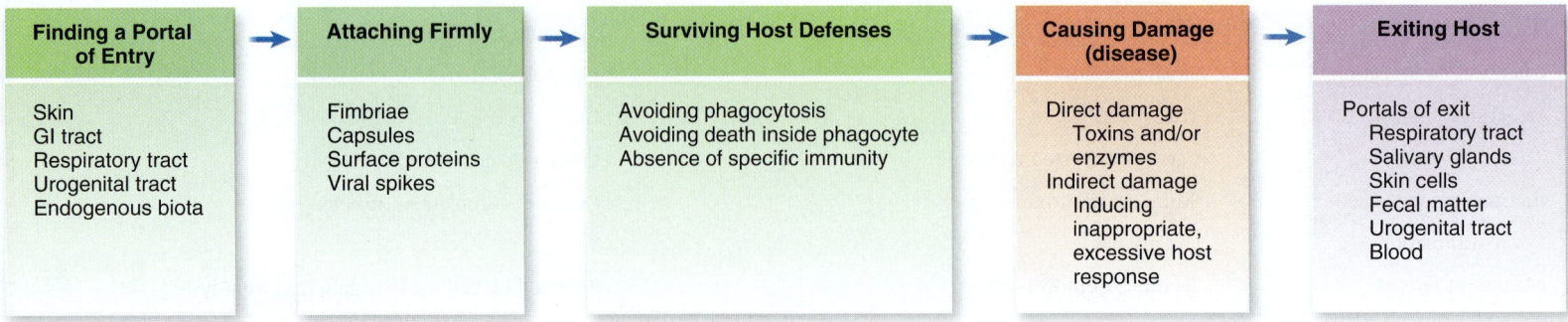

Finding a Portal of Entry	Attaching Firmly	Surviving Host Defenses	Causing Damage (disease)	Exiting Host
Skin GI tract Respiratory tract Urogenital tract Endogenous biota	Fimbriae Capsules Surface proteins Viral spikes	Avoiding phagocytosis Avoiding death inside phagocyte Absence of specific immunity	Direct damage Toxins and/or enzymes Indirect damage Inducing inappropriate, excessive host response	Portals of exit Respiratory tract Salivary glands Skin cells Fecal matter Urogenital tract Blood

Figure 11.5 **The steps involved when a microbe causes disease in a host.**

by the patient. In general, signs are more precise than symptoms, though both can have the same underlying cause. For example, an infection of the brain might present with the sign of bacteria in the spinal fluid and symptom of headache. When a disease can be identified or defined by a certain complex of signs and symptoms, it is termed a **syndrome.** Specific signs and symptoms for particular infectious diseases are covered in chapters 16 through 21.

Signs and Symptoms of Inflammation

The earliest symptoms of disease result from the activation of the body defense process called **inflammation.** The inflammatory response includes cells and chemicals that respond nonspecifically to disruptions in the tissue. This subject is discussed in greater detail in chapter 12, but as noted earlier, many signs and symptoms of infection are caused by the mobilization of this system. Some common symptoms of inflammation include fever, pain, soreness, and swelling. Signs of inflammation include **edema,** the accumulation of fluid in an afflicted tissue; **granulomas** and **abscesses,** walled-off collections of inflammatory cells and microbes in the tissues; and **lymphadenitis,** swollen lymph nodes.

Signs of Infection in the Blood

Changes in the number of circulating white blood cells, as determined by special counts, are considered to be signs of possible infection. **Leukocytosis** (loo″-koh′-sy-toh′-sis) is an increase in the level of white blood cells, whereas **leukopenia** (loo″-koh-pee′-nee-uh) is a decrease. Other signs of infection revolve around the occurrence of a microbe or its products in the blood. The clinical term for blood infection, **septicemia,** refers to a general state in which microorganisms are multiplying in the blood and are present in large numbers. When small numbers of bacteria or viruses are found in the blood, the correct terminology is **bacteremia,** or **viremia,** which means that these microbes are present in the blood but are not necessarily multiplying.

During infection, a normal host will invariably show signs of an immune response in the form of antibodies in the serum. This fact is the basis for several serological tests used in diagnosing infectious diseases such as AIDS or syphilis. Such specific immune reactions indicate the body's attempt to develop specific immunities against pathogens. We concentrate on this role of the host defenses in chapters 12 and 13.

Infections That Go Unnoticed

It is rather common for an infection to produce no noticeable symptoms, even though the microbe is active in the host tissue. In other words, although infected, the host does not manifest the disease. An infection of this nature is known as **asymptomatic** or **subclinical** (inapparent) because the patient experiences no symptoms or disease and does not seek medical attention.

Step Five: Vacating the Host—Portals of Exit

Earlier, we introduced the idea that a parasite is considered *unsuccessful* if it does not have a provision for leaving its host and moving to other susceptible hosts. With few exceptions, pathogens depart by a specific avenue called the **portal of exit (figure 11.6).** In most cases, the pathogen is shed or released from the body through secretion, excretion, discharge, or sloughed tissue. The usually very high number of infectious agents in these materials increases the likelihood that the pathogen will reach other hosts. In many cases, the portal of exit is the same as the portal of entry, but some pathogens use a different route. As we see in the next section, the portal of exit concerns epidemiologists because it greatly influences the dissemination of infection in a population.

Humans shed about 1 million skin cells—and the microbes on them—every day.

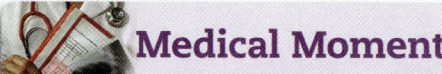

Medical Moment

Differentiating Between Signs and Symptoms

Many health care professionals find it difficult to differentiate between signs and symptoms when they are beginning practitioners. A simple way to think about the difference between signs and symptoms is to think of a symptom as something that the patient experiences, and signs as something the health care professional can see, hear, feel, or smell. For example, a patient who visits his or her doctor may complain of chills, fever, cough, and a sore throat. The chills, fever, cough, and sore throat are the patient's *symptoms*. The nurse evaluating the patient may observe that the patient's throat is red, there is nasal discharge present, and the lungs sound congested when auscultated by stethoscope. These are the *signs* of the patient's illness.

Of course, nothing in medicine is that simple! Some manifestations of disease can be both a sign and a symptom. Fever is one example. The patient may report symptoms of fever, such as feeling chilled or excessively warm; fever can also be observed objectively by taking the patient's temperature with a thermometer. Another example is epistaxis, or nosebleed—the patient may complain of a bleeding nose, and the condition can also be observed by others.

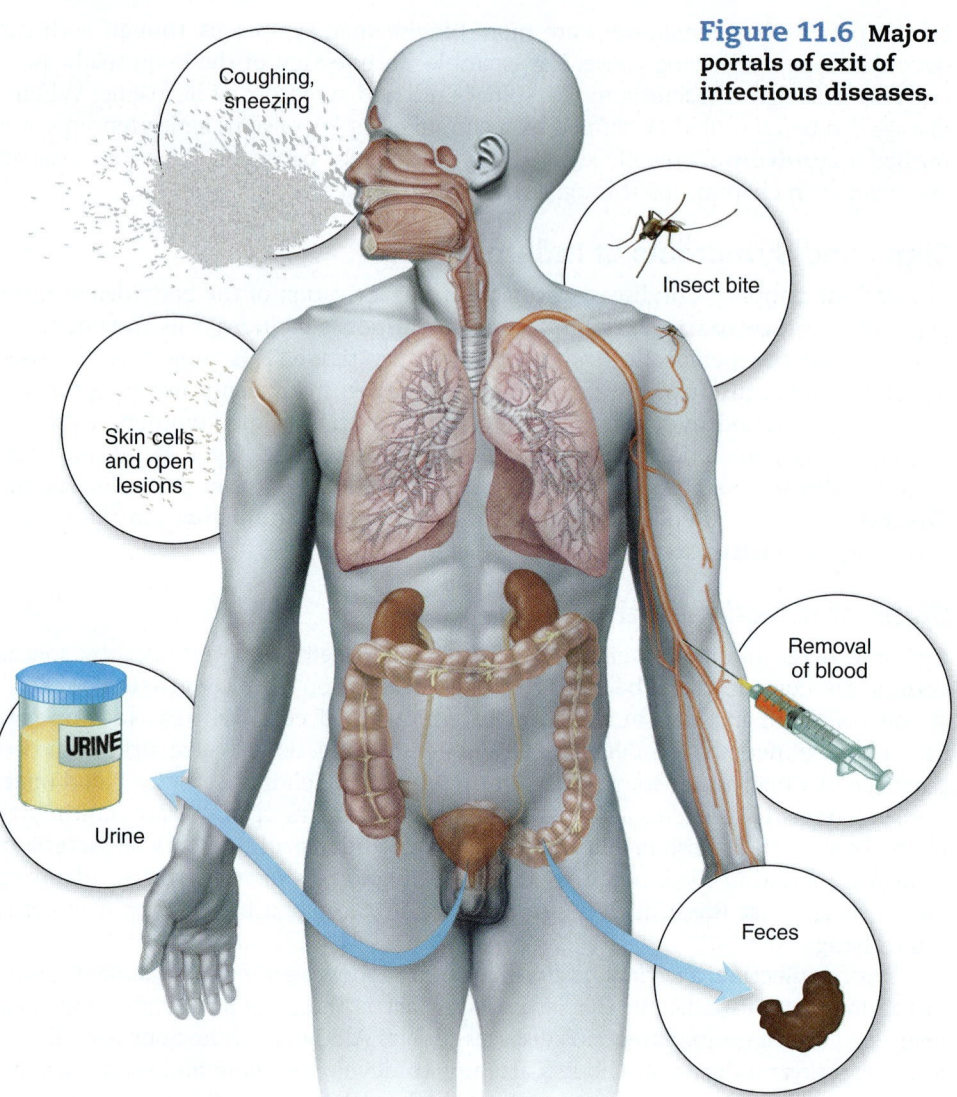

Figure 11.6 Major portals of exit of infectious diseases.

Coughing, sneezing

Insect bite

Skin cells and open lesions

Removal of blood

URINE

Urine

Feces

The Persistence of Microbes and Pathologic Conditions

The apparent recovery of the host does not always mean that the microbe has been completely removed or destroyed by the host defenses. After the initial symptoms in certain chronic infectious diseases, the infectious agent retreats into a dormant state called **latency.** Throughout this latent state, the microbe can periodically become active and produce a recurrent disease. The viral agents of herpes simplex, herpes zoster, hepatitis B, AIDS, and Epstein-Barr can persist in the host for long periods. The agents of syphilis, typhoid fever, tuberculosis, and malaria can also enter into latent stages. The person harboring a persistent infectious agent may or may not shed it during the latent stage. If it is shed, such persons are chronic carriers who serve as sources of infection for the rest of the population.

Some diseases leave **sequelae** in the form of long-term or permanent damage to tissues or organs. For example, meningitis can result in deafness, strep throat can lead to rheumatic heart disease, Lyme disease can cause arthritis, and polio can produce paralysis.

There are four distinct phases of infection and disease: the incubation period, the prodrome, the period of invasion, and the convalescent period.

What Happens in Your Body

The **incubation period** is the time from initial contact with the infectious agent (at the portal of entry) to the appearance of the first symptoms. During the incubation period, the agent is multiplying at the portal of entry but has not yet caused enough damage to elicit symptoms. Although this period is relatively well defined and predictable for each microorganism, it does vary according to host resistance, degree of virulence, and distance between the target organ and the portal of entry (the farther apart, the longer the incubation period). Overall, an incubation period can range from several hours in pneumonic plague to several years in leprosy. The majority of infections, however, have incubation periods ranging between 2 and 30 days.

The earliest notable symptoms of most infections appear as a vague feeling of discomfort, such as head and muscle aches, fatigue, upset stomach, and general malaise. This short period (1–2 days) is known as the **prodromal stage.** Some diseases have very specific prodromal symptoms. Next, the infectious agent enters a **period of invasion,** during which it multiplies at high levels, exhibits its greatest virulence, and becomes well established in its target tissue. This period is often marked by fever and other prominent and more specific signs and symptoms, which can include cough, rashes, diarrhea, loss of muscle control, swelling, jaundice, discharge of exudates, or severe pain, depending on the particular infection. The length of this period is extremely variable.

As the patient begins to respond to the infection, the symptoms decline–sometimes dramatically, other times slowly. During the recovery that follows, called the **convalescent period,** the patient's strength and health gradually return owing to the healing nature of the immune response. During this period, many patients stop taking their antibiotics, even though there are still pathogens in their system. Think about it–the ones still alive at this stage of treatment are the ones in the population with the most resistance to the antibiotic. In most cases, continuing the antibiotic dosing will take care of them. But stop taking the drug now and the bacteria that are left to repopulate are the ones with the higher resistance.

The transmissibility of the microbe during these four stages is different for each microorganism. A few agents are released mostly during incubation (measles, for example); many are released primarily during the invasive period (*Shigella*); and others can be transmitted during all of these periods (hepatitis B).

Reservoirs: Where Pathogens Persist

In order for an infectious agent to continue to exist and be spread, it must have a permanent place to reside. The **reservoir** is the primary habitat in the natural world from which a pathogen originates. Often it is a human or animal carrier, although soil, water, and plants are also reservoirs. The reservoir can be distinguished from the infection *transmitter*, which is the individual or object from which an infection is actually acquired. In diseases such as syphilis, the reservoir and the transmitter are the same (the human body). In the case of hepatitis A, the reservoir (a human carrier) is usually different from the mode of transmission (contaminated food). **Table 11.5** shows how reservoirs and transmission are interrelated.

Living Reservoirs

The list of living reservoirs is presented in table 11.5, but you may surmise (correctly) that a great number of infections that affect humans have their reservoirs in other humans. Persons or animals with obvious symptomatic infection are obvious sources

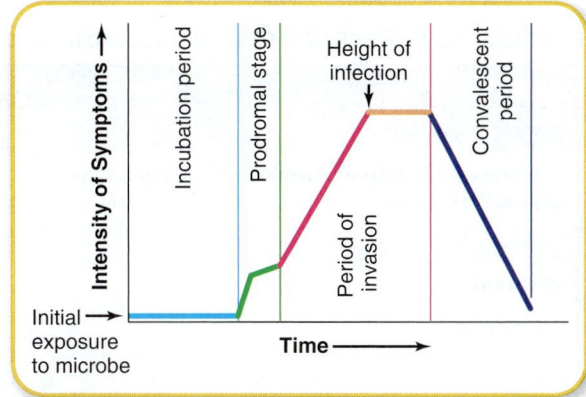

Figure 11.7 Stages in the course of infection and disease.   The stages have different durations in different infections.

NCLEX® PREP

4. Which of the following characteristics is/are associated with endotoxin? Select all that apply.

   a. *toxicity in minimal concentration*

   b. *gram-negative bacteria*

   c. *presence of fever*

   d. *denaturation*

   e. *released by cell as a result of shedding*

Table 11.5 Reservoirs and Transmitters

Reservoirs	Transmission Examples
Living Reservoirs	
Animals (Other than humans and arthropods) Mammals, birds, reptiles, etc.	Animals harboring pathogens can directly transmit them to humans (bats transmitting rabies to humans); vectors can transmit the pathogens from animals to humans (fleas passing the plague from rats to people); vehicles such as water can transmit pathogens which originated in animals, as in the case of leptospirosis.
Humans Actively ill	 A person suffering from a cold contaminates a pen, which is then picked up by a healthy person. That is indirect transmission. Alternatively, a sick person can transmit the pathogen directly by sneezing on a healthy person.
Carriers	A person who is fully recovered from his hepatitis but is still shedding hepatitis A virus in his feces may use suboptimal hand-washing technique. He contaminates food, which a healthy person ingests (indirect transmission). Carriers can also transmit through direct means, as when an incubating carrier of HIV, who does not know she is infected, transmits the virus through sexual contact.
Arthropods Biological vectors	 When an arthropod is the host (and reservoir) of the pathogen, it is also the mode of transmission.
Nonliving Reservoirs	
Soil **Water** **Air** **The built environment**	 Some pathogens, such as the TB bacterium, can survive for long periods in nonliving reservoirs. They are then directly transmitted to humans when they come in contact with the contaminated soil, water, or air.

of infection, but a **carrier** is, by definition, an individual who *inconspicuously* shelters a pathogen and spreads it to others without any notice. The duration of the carrier state can be short or long term, and it is important to remember that the carrier may or may not have experienced disease due to the microbe.

Several situations can produce the carrier state. **Table 11.6** describes the various carrier states and provides examples of each.

Table 11.6 Carrier States

Carrier State	Explanation	Example
Asymptomatic carriers	Infected but show no symptoms of disease	Gonorrhea, genital herpes with no lesions, human papillomavirus
Incubating carriers	Spread the infectious agent during the incubation period	Infectious mono-nucleosis
Convalescent carriers	Recuperating patients without symptoms; they continue to shed viable microbes and convey the infection to others	Hepatitis A
Chronic carriers	Individuals who shelter the infectious agent for a long period after recovery because of the latency of the infectious agent	Tuberculosis, typhoid fever
Passive carriers	Medical and dental personnel who must constantly handle patient materials that are heavily contaminated with patient secretions and blood risk picking up pathogens mechanically and accidentally transferring them to other patients	Various healthcare-associated infections

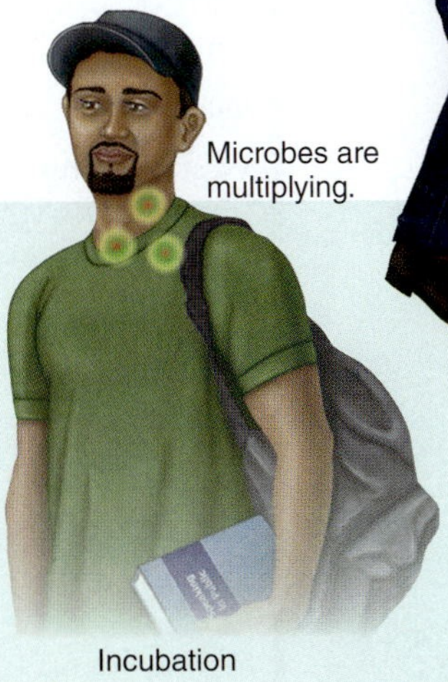

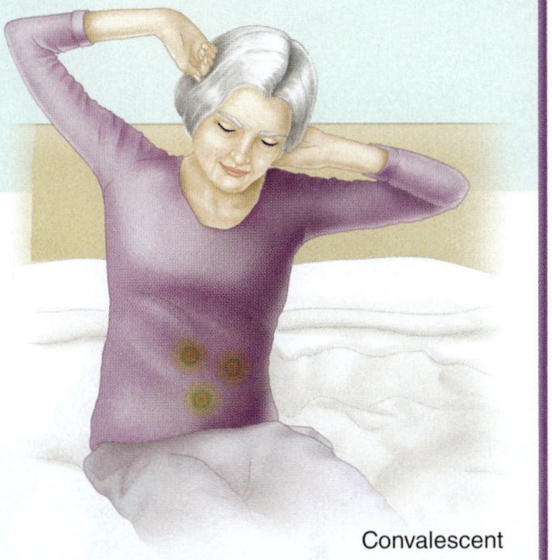

Microbes are multiplying.

Asymptomatic STD

Incubation

Convalescent

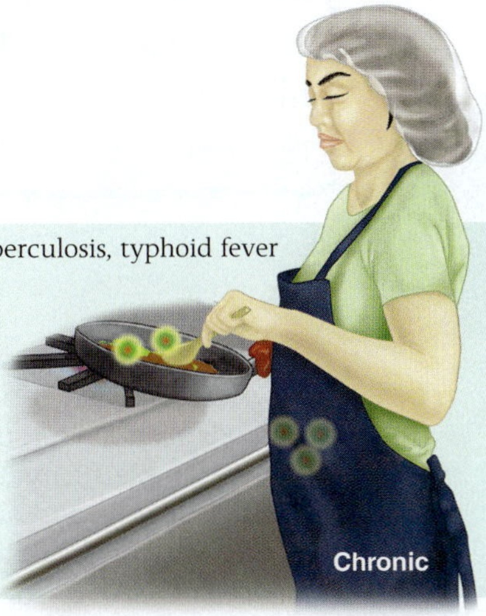

Chronic

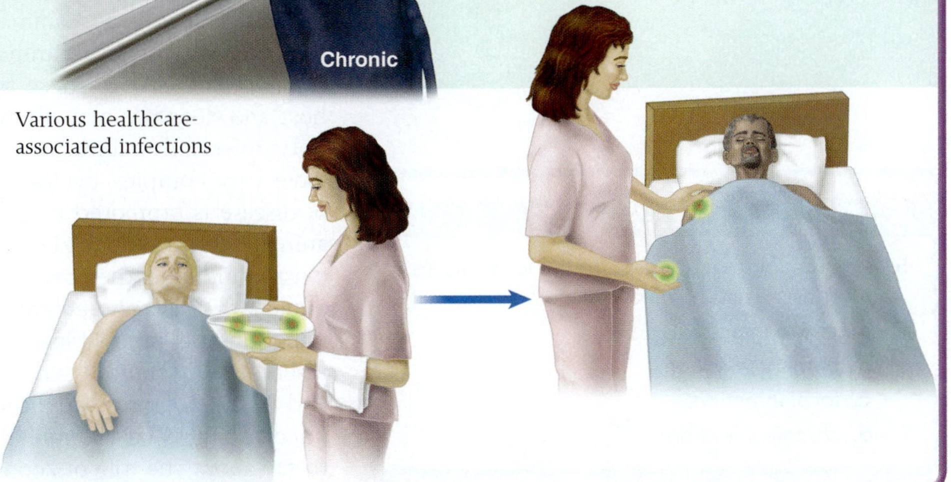

305

Table 11.7 Common Zoonotic Infections

Disease	Primary Animal Reservoirs
Viruses	
Rabies	Mammals
Yellow fever	Wild birds, mammals, mosquitoes
Viral fevers	Wild mammals
Hantavirus	Rodents
Influenza	Chickens, birds, swine
West Nile virus	Wild birds, mosquitoes
Bacteria	
Rocky Mountain spotted fever	Dogs, ticks
Psittacosis	Birds
Leptospirosis	Domestic animals
Anthrax	Domestic animals
Brucellosis	Cattle, sheep, pigs
Plague	Rodents, fleas
Salmonellosis	Mammals, birds, reptiles, and rodents
Tularemia	Rodents, birds, arthropods
Miscellaneous	
Ringworm	Domestic mammals
Toxoplasmosis	Cats, rodents, birds
Trypanosomiasis	Domestic and wild mammals
Trichinosis	Swine, bears
Tapeworm	Cattle, swine, fish

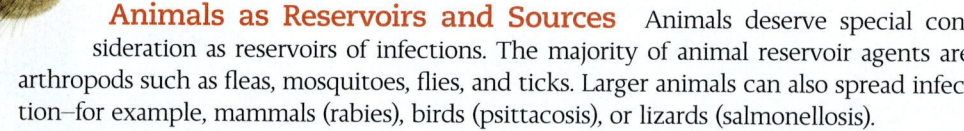

Animals as Reservoirs and Sources Animals deserve special consideration as reservoirs of infections. The majority of animal reservoir agents are arthropods such as fleas, mosquitoes, flies, and ticks. Larger animals can also spread infection—for example, mammals (rabies), birds (psittacosis), or lizards (salmonellosis).

Many vectors and animal reservoirs spread their own infections to humans. An infection indigenous to animals but naturally transmissible to humans is a **zoonosis** (zoh″-uh-noh′-sis). In these types of infections, the human is essentially a dead-end host and does not contribute to the natural persistence of the microbe. Some zoonotic infections (rabies, for instance) can have multihost involvement, and others can have very complex cycles in the wild (see plague in chapter 18). Zoonotic spread of disease is promoted by close associations of humans with animals, and people in animal-oriented or outdoor professions are at greatest risk. At least 150 zoonoses exist worldwide; the most common ones are listed in **table 11.7.** Zoonoses make up a full 70% of all new emerging diseases worldwide. It is worth noting that zoonotic infections are impossible to completely eradicate without also eradicating the animal reservoirs. Attempts have been made to eradicate mosquitoes and certain rodents, and in 2004 China slaughtered tens of thousands of civet cats who were thought (incorrectly) to be a source of the respiratory disease SARS. Chapter 22 has a lot more to say about the role of zoonoses in human infections.

Nonliving Reservoirs

Clearly, microorganisms have adapted to nearly every habitat in the biosphere. They thrive in soil and water and often find their way into the air. They also colonize what is known as "the built environment," surfaces in homes, office buildings, and structures of all kinds. Although most of these microbes are saprobic and cause little harm and considerable benefit to humans, some are opportunists and a few are regular pathogens. Because human hosts are in regular contact with these environmental sources, acquisition of pathogens from natural habitats is of diagnostic and epidemiological importance.

The Acquisition and Transmission of Infectious Agents

Infectious diseases can be categorized on the basis of how they are acquired. A disease is **communicable** when an infected host can transmit the infectious agent to another host and establish infection in that host. (Although this terminology is standard, one must realize that it is not the disease that is communicated but the microbe. Also be aware that the word *infectious* is sometimes used interchangeably with the word *communicable*, but this is not precise usage.) The transmission of the agent can be direct or indirect, and the ease with which the disease is transmitted varies considerably from one agent to another. If the agent is highly communicable, especially through direct contact, the disease is **contagious.** Influenza and measles move readily from host to host and thus are contagious, whereas Hansen's disease (leprosy) is only weakly communicable. Because they can be spread through the population, communicable diseases are our main focus in the following sections.

In contrast, a **noncommunicable** infectious disease does *not* arise through transmission of the infectious agent from host to host. The infection and disease are acquired through some other special circumstance. Noncommunicable infections occur primarily when a compromised person is invaded by his or her own microbiota (as with certain pneumonias, for example) or when an individual has accidental contact with a microbe that exists in a nonliving reservoir such as soil. Some examples are certain mycoses, acquired through inhalation of fungal spores, and tetanus, in which *Clostridium tetani* spores from a soiled object enter a cut or wound. Persons thus infected do not become a source of disease to others.

Patterns of Transmission in Communicable Diseases

The routes or patterns of disease transmission are many and varied. The spread of diseases is by direct or indirect contact with animate or inanimate objects and can be horizontal or vertical. The term *horizontal* means the disease is spread through a population from one infected individual to another; *vertical* signifies transmission from parent to offspring via the ovum, sperm, placenta, or milk. The extreme complexity of transmission patterns among microorganisms makes it very difficult to generalize. However, for easier organization, we will divide transmission into two major groups, as shown in **table 11.8:** transmission by some form of direct contact or transmission by indirect routes, in which some vehicle is involved.

Arthropods, like other reservoirs, can also serve as transmitters of infection. A biological vector communicates the infectious agent to the human host by biting, aerosol formation, or touch. In the case of biting vectors, the animal can

1. inject infected saliva into the blood (the mosquito),
2. defecate around the bite wound (the flea), or
3. regurgitate blood into the wound (the tsetse fly).

Airplanes can play a role in spreading diseases—but not in the way you might think. Studies have shown that airborne diseases are *not* more easily spread in airplane cabins. However, airplanes can move sick people from one continent to another quickly and thus widen an epidemic into a pandemic.

Table 11.8 Patterns of Transmission in Communicable Diseases

Mode of Transmission	Definition
Vertical	Transmission is from parent to offspring via the ovum, sperm, placenta, or milk
Horizontal	Disease is spread through a population from one infected individual to another

Direct (contact) transmission

Droplets (colds, chickenpox)

Involves physical contact between infected person and that of the new infectee

Types:

- Touching, kissing, sex
- Droplet contact, in which fine droplets are sprayed directly upon a person during sneezing or coughing
- **Parenteral** transmission via intentional or unintentional injection into deeper tissues (needles, knives, branches, broken glass, etc.)

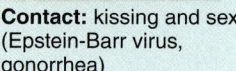

Contact: kissing and sex (Epstein-Barr virus, gonorrhea)

Indirect transmission

Infectious agent must pass from an infected host to an intermediate conveyor (a vehicle) and from there to another host

Infected individuals contaminate objects, food, or air through their activities

Types:

- **Fomite**—inanimate object that harbors and transmits pathogens (doorknobs, telephone receivers, faucet handles)

- **Vehicle**—a natural, nonliving material that can transmit infectious agents

  - **Air**—smaller particles evaporate and remain in the air and can be encountered by a new host; **aerosols** are suspensions of fine dust or moisture particles in the air that contain live pathogens

  - **Water**—some pathogens survive for long periods in water and can infect humans long after they were deposited in the water

  - **Soil**—microbes resistant to drying live in and can be transmitted from soil

  - **Food**—meats may contain pathogens with which the animal was infected; foods can also be contaminated by food handlers

Special Category: oral-fecal route—using either vehicles or fomites. A fecal carrier with inadequate personal hygiene contaminates food during handling, and an unsuspecting person ingests it; alternatively a person touches a surface that has been contaminated with fecal material and touches his or her mouth, leading to ingestion of fecal microbes

Vector transmission

Types:

- **Mechanical vector**—insect carries microbes to host on its body parts

- **Biological vector**—insect injects microbes into host; part of microbe life cycle completed in insect

Mechanical vectors are not necessary to the life cycle of an infectious agent and merely transport it without being infected. The external body parts of these animals become contaminated when they come into physical contact with a source of pathogens. The agent is subsequently transferred to humans indirectly by an intermediate such as food or, occasionally, by direct contact (as in certain eye infections). Houseflies are noxious mechanical vectors. They feed on decaying garbage and feces, and while they are feeding, their feet and mouthparts easily become contaminated.

Healthcare-Associated Infections

Infectious diseases that are acquired or develop during a hospital or health care facility stay are known as **healthcare-associated** or **nosocomial** (nohz″-oh-koh′-mee-al) **infections.** This concept seems incongruous at first thought, because a hospital is regarded as a place to get treatment for a disease, not a place to acquire a disease. Yet it is not uncommon for a surgical patient's incision to become infected or a burn patient to develop a case of pneumonia in the clinical setting. The rate of healthcare-associated infections can be as low as 0.1% or as high as 20% of all admitted patients, depending on the clinical setting, with an average of about 5%. In light of the number of admissions, this adds up to 2 to 4 million cases a year, which result in nearly 90,000 deaths. Healthcare-associated infections cost time and money as well as suffering. By one estimate, they amount to 8 million additional days of hospitalization a year and an increased cost of $5 to $10 billion.

So many factors unique to the hospital or extended-care facility environment are tied to healthcare-associated infections that a certain number of infections are virtually unavoidable. After all, the hospital both attracts and creates compromised patients, and it serves as a collection point for pathogens. Some patients become infected when surgical procedures or lowered defenses permit resident biota to invade their tissues. Other patients acquire infections directly or indirectly from fomites, medical equipment, other patients, medical personnel, visitors, air, and water.

The health care process itself increases the likelihood that infectious agents will be transferred from one patient to another. Treatments using reusable instruments such as respirators and thermometers constitute a possible source of infectious agents. Indwelling devices such as catheters, prosthetic heart valves, grafts, drainage tubes, and tracheostomy tubes form ready portals of entry and habitats for infectious agents. Because such a high proportion of the hospital population receives antimicrobial drugs during their stay, drug-resistant microbes are selected for at a much greater rate than is the case outside the hospital.

The most common healthcare-associated infections involve the urinary tract, the respiratory tract, and surgical incisions **(figure 11.8).** Gram-negative intestinal biota (*Escherichia coli, Klebsiella, Pseudomonas*) are cultured in more than half of patients with healthcare-associated infections. The gram-positive bacteria staphylococci and streptococci, and yeasts make up most of the remainder. True pathogens such as *Mycobacterium tuberculosis, Salmonella*, hepatitis B, and influenza virus can be transmitted in the clinical setting as well.

The federal government has taken steps to incentivize hospitals to control healthcare-associated transmission. In the fall of 2008, the Medicare and Medicaid programs announced they would not reimburse hospitals for healthcare-associated catheter-associated urinary tract infections, vascular catheter-associated bloodstream infections, and surgical site infections.

Hospitals generally employ an *infection control officer* who not only implements proper practices and procedures throughout the hospital but is also charged with tracking potential outbreaks, identifying breaches in asepsis, and training other health care workers in aseptic technique. Among those most in need of this training are nurses and other caregivers whose work, by its very nature, exposes

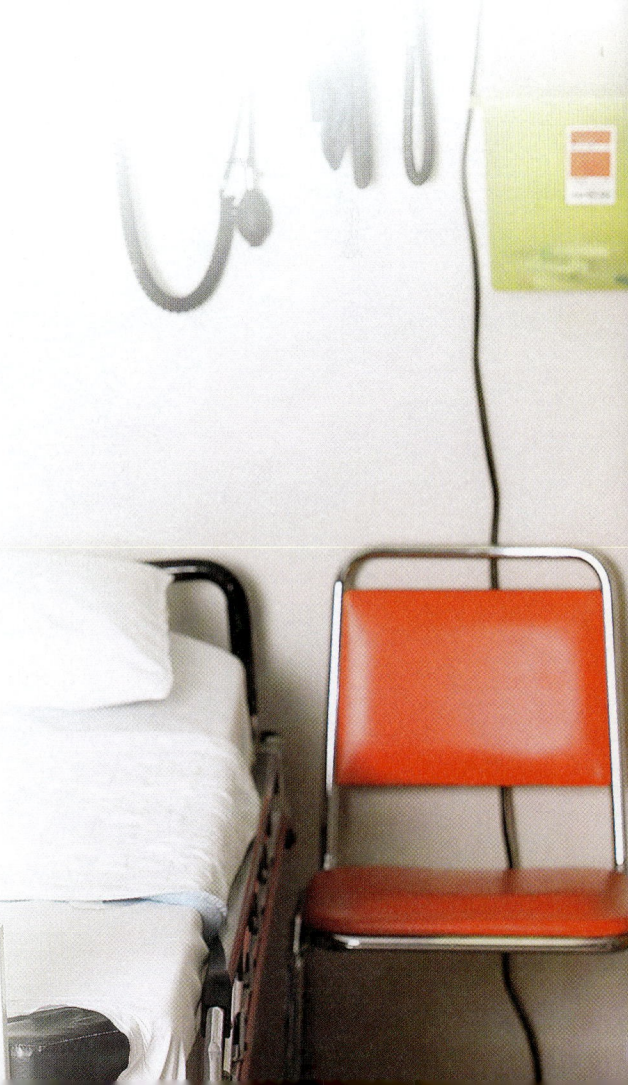

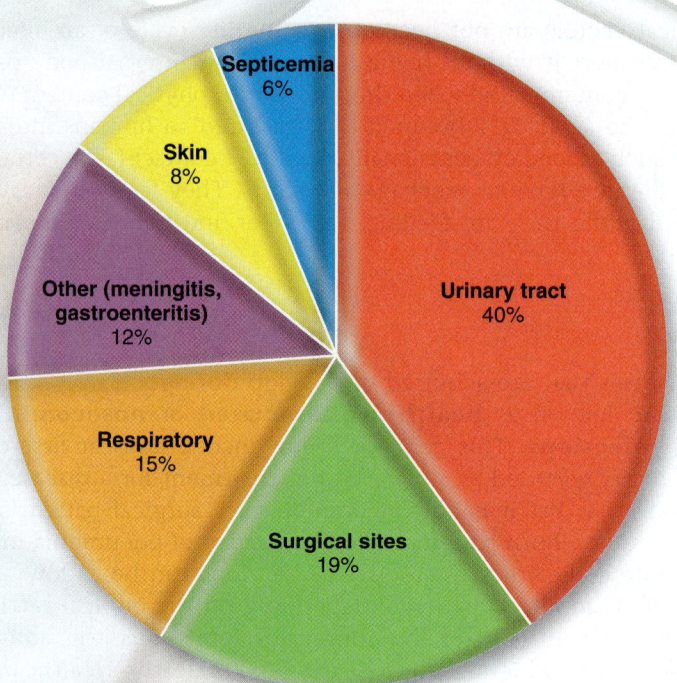

Figure 11.8 Most common healthcare-associated infections. Relative frequency by target area.

them to needlesticks, infectious secretions, blood, and physical contact with the patient. The same practices that interrupt the routes of infection in the patient can also protect the health care worker. It is for this reason that most hospitals have adopted universal precautions that recognize that all secretions from all persons in the clinical setting are potentially infectious and that transmission can occur in either direction.

Universal Blood and Body Fluid Precautions

Medical and dental settings require stringent measures to prevent the spread of healthcare-associated infections from patient to patient, from patient to worker, and from worker to patient. Even with precautions, the rate of such infections is rather high. Recent evidence indicates that more than one-third of healthcare-associated infections could be prevented by consistent and rigorous infection control methods.

Previously, control guidelines were disease-specific, and clearly identified infections were managed with particular restrictions and techniques. With this arrangement, personnel tended to handle materials labeled *infectious* with much greater care than those that were not so labeled. The AIDS epidemic spurred a reexamination of that policy. Because of the potential for increased numbers of undiagnosed HIV-infected patients, the Centers for Disease Control and Prevention laid down more stringent guidelines for handling patients and body substances. These guidelines have been termed **universal precautions (UPs),** because they are based on the assumption that all patients could harbor infectious agents and so must be treated with the same degree of care. They also include body substance isolation (BSI) techniques to be used in known cases of infection.

These precautions are designed to protect all individuals in the clinical setting—patients, workers, and the public alike. In general, they include techniques designed to prevent contact with pathogens and contamination and, if prevention is not possible, to take purposeful measures to decontaminate potentially infectious materials.

The universal precautions recommended for all health care settings follow.

1. Barrier precautions, including masks and gloves, should be taken to prevent contact of skin and mucous membranes with patients' blood or other body fluids. Because gloves can develop small invisible tears, double gloving decreases the risk further. For protection during surgery, venipuncture, or emergency procedures, gowns, aprons, and other body coverings should be worn. Dental workers should wear eyewear and face shields to protect against splattered blood and saliva.

2. More than 10% of health care personnel are pierced each year by sharp (and usually contaminated) instruments. These accidents carry risks not only for HIV (or HIV infection) but also for hepatitis B, hepatitis C, and other diseases. Preventing inoculation infection requires vigilant observance of proper techniques. All disposable needles, scalpels, or sharp devices from invasive procedures must immediately be placed in puncture-proof containers for sterilization and final discard. Under no circumstances should a worker attempt to recap a syringe, remove a needle from a syringe, or leave unprotected used syringes where they pose a risk to others. Reusable needles or other sharp devices must be heat-sterilized in a puncture-proof holder before they are handled.

3. Dental handpieces should be sterilized between patients, but if this is not possible, they should be thoroughly disinfected with a high-level disinfectant (peroxide, hypochlorite). Blood and saliva should be removed completely from all contaminated dental instruments and intraoral devices prior to sterilization.

4. Hands and other skin surfaces that have been accidently contaminated with blood or other fluids should be scrubbed immediately with a germicidal soap. Hands should likewise be washed after removing rubber gloves, masks, or other barrier devices.

5. Because saliva can be a source of some types of infections, barriers should be used in all mouth-to-mouth resuscitations.

6. Health care workers with active, draining skin or mucous membrane lesions must refrain from handling patients or equipment that will come into contact with other patients. Pregnant health care workers risk infecting their fetuses and must pay special attention to these guidelines. Personnel should be protected by vaccination whenever possible.

Isolation procedures for known or suspected infections should still be instituted on a case-by-case basis.

Which Agent Is the Cause? Using Koch's Postulates to Determine Etiology

An essential aim in the study of infection and disease is determining the precise **etiologic,** or causative, **agent** of a newly recognized condition. In our modern technological age, we take for granted that a certain infection is caused by a certain microbe, but such has not always been the case. More than a century ago, Robert Koch realized that in order to prove the germ theory of disease he would have to develop a standard for determining causation that would stand the test of scientific scrutiny. Out of his experimental observations on the transmission of anthrax in cows came a series of proofs, called **Koch's postulates,** that established the principal criteria for etiologic studies. **Table 11.9** demonstrates the principles of Koch's postulates.

Koch's postulates continue to play an essential role in modern epidemiology. Every decade, new diseases challenge the scientific community and require application of the postulates.

Koch's postulates are reliable for many infectious diseases, but they cannot be completely fulfilled in certain situations. For example, some infectious agents are not readily isolated or grown in the laboratory. If one cannot elicit an infection similar to that seen in humans by inoculating it into an animal, it is very difficult to prove the etiology. It is difficult to satisfy Koch's postulates for viral diseases because viruses usually have a very narrow host range. Human viruses may cause disease only in humans,

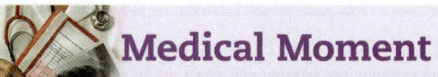

Medical Moment

Eye on Careers: Infection Control Practitioner

An infection control practitioner (ICP) holds an integral position in many hospitals and health care organizations. An ICP's biggest role is to reduce the spread of healthcare-associated infections, limiting the spread of infectious disease. Many different professionals may be designated as ICPs, including nurses, doctors, epidemiologists, or others who have taken specialized training to prepare them for this very important role.

An ICP may be responsible for the following:

- tracking positive cultures to ensure treatment has been implemented;
- following up actual and potential exposures to communicable diseases;
- training and education of staff regarding infection control practices and protocols;
- updating infection control manuals; and
- communicating with government entities such as the CDC (the Centers for Disease Control and Prevention), state public health departments, workers' compensation boards, and OSHA (the Occupational Safety and Health Administration).

Infection control practitioners may be trained by others in similar roles or may choose to become certified. Certification in the United States necessitates having at least 2 years of relevant clinical experience; the candidate must also be licensed in his or her chosen profession or hold at least a bachelor's degree in a health-related field.

Table 11.9 Koch's Postulates

Postulate #1

Find evidence of a particular microbe in every case of a disease.

Postulate #2

Isolate that microbe from an infected subject and cultivate it in pure culture in the laboratory; perform full microscopic and biological characterization.

Antibody
Antigens

ID   C T

Control Test   Sample well

NCLEX® PREP

6. A client reported onset of these symptoms a few days ago: low-grade fever, arthralgia, and fatigue. The client was seen in the outpatient clinic by a health care provider, diagnosed with a bacterial infection, and prescribed a course of antibiotic therapy. At present, the client has 3 days of antibiotic treatment remaining. The client would be in which stage of infection?

   a. *period of invasion*
   b. *convalescent period*
   c. *incubation period*
   d. *prodromal stage*

or perhaps in primates, though the disease symptoms in apes will often be different. To address this, T. M. Rivers proposed modified postulates for viral infections. These were used in 2003 to definitively determine the coronavirus cause of SARS.

It is also usually not possible to use Koch's postulates to determine causation in polymicrobial diseases. Diseases such as periodontitis and soft tissue abscesses are caused by complex mixtures of microbes. While it is theoretically possible to isolate each member and to re-create the exact proportions of individual cultures for the third step in Koch's postulates, it is not attempted in practice.

11.2 LEARNING OUTCOMES—Assess Your Progress

4. Differentiate between a microbe's pathogenicity and its virulence.
5. Define *opportunism*, and list examples of common opportunistic pathogens.
6. List the steps a microbe has to take to get to the point where it can cause disease.
7. List several portals of entry and exit.
8. Define *infectious dose*, and explain its role in establishing infection.
9. Describe three ways microbes cause tissue damage.
10. Compare and contrast major characteristics of endotoxin and exotoxins.
11. Provide a definition of *virulence factors*.
12. Draw a diagram of the stages of disease in a human.
13. Differentiate among the various types of reservoirs, providing examples of each.
14. List several different modes of transmission of infectious agents.
15. Define *healthcare-associated infection*, and list the three most common types.
16. List Koch's postulates, and discuss when they might not be appropriate in establishing causation.

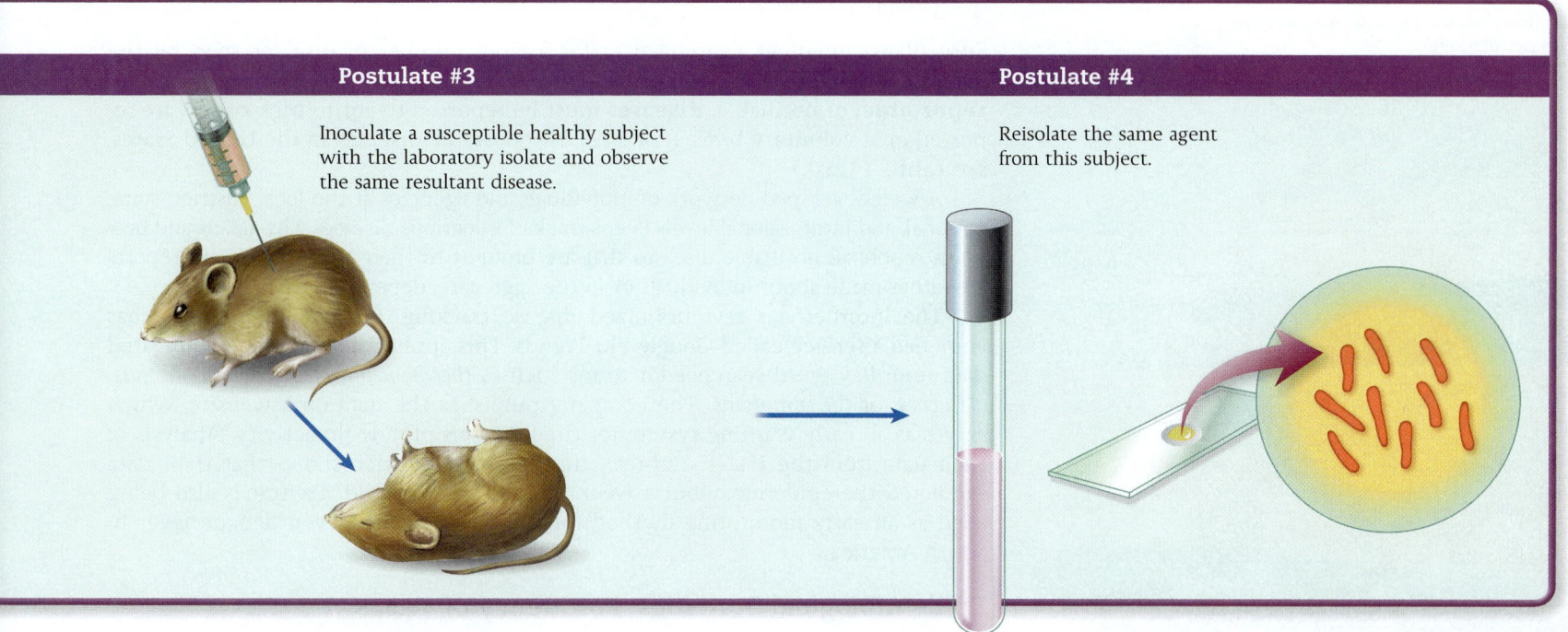

Postulate #3

Inoculate a susceptible healthy subject with the laboratory isolate and observe the same resultant disease.

Postulate #4

Reisolate the same agent from this subject.

11.3 Epidemiology: The Study of Disease in Populations

So far, our discussion has revolved primarily around the impact of an infectious disease in a single individual. Let us now turn our attention to the effects of diseases on the community–the realm of **epidemiology.** By definition, this term involves the study of the frequency and distribution of disease and other health-related factors in defined populations. It involves many disciplines–not only microbiology but also anatomy, physiology, immunology, medicine, psychology, sociology, ecology, and statistics–and it considers all forms of disease, including heart disease, cancer, drug addiction, and mental illness.

A groundbreaking British nurse named Florence Nightingale helped to lay the foundations of modern epidemiology. She arrived in the Crimean war zone in Turkey in the mid-1850s, where the British were fighting and dying at an astonishing rate. Estimates suggest that 20% of the soldiers there died (by contrast, 2.6% of U.S. soldiers in the Vietnam war died). Even though this was some years before the discovery of the germ theory, Nightingale understood that filth contributed to disease and instituted methods that had never been seen in military field hospitals. She insisted that separate linens and towels be used for each patient, and that the floors be cleaned and the pipes of sewage unclogged. She kept meticulous notes of what was killing the patients and was able to demonstrate that many more men died of disease than of their traumatic injuries. She used statistical analysis to convince government officials that these patterns were real. This was indeed one of the earliest forays into epidemiology–trying to understand how diseases were being transmitted and using statistics to do so.

The techniques of epidemiology are also used to track behaviors, such as exercise or smoking. The epidemiologist is a medical sleuth who collects clues on the causative agent, pathology, sources, and modes of transmission and tracks the numbers and distribution of cases of disease in the community. The outcomes of these studies help public health departments develop prevention and treatment programs and establish a basis for predictions.

Epidemiology is the study of disease in populations.

Tracking Disease in the Population

Surveillance involves keeping data for a large number of diseases seen by the medical community and reported to public health authorities. By law, certain **reportable,** or notifiable, **diseases** must be reported to authorities; others are reported on a voluntary basis. (For a list of reportable diseases in the United States, see **table 11.10.**)

A well-developed network of individuals and agencies at the local, district, state, national, and international levels keeps track of infectious diseases. Physicians and hospitals report all notifiable diseases that are brought to their attention. These reports are either made about individuals or in the aggregate, depending on the disease.

The Internet has revolutionized disease tracking. For example, Google has launched a service called Google Flu Trends. This application compiles aggregated data from key word searches for terms such as *thermometer, chest congestion, muscle aches,* or *flu symptoms.* The company publishes the data on a website, which serves as an early warning system for the locations of new flu activity. Analysis of their data from the H1N1 outbreak that began in Mexico shows that their data predicted the epidemic about a week before CDC data did. Twitter is also being used as an early monitoring method for flu epidemics and even dengue fever in South America.

Epidemiological Statistics: Frequency of Cases

The **prevalence** of a disease is the total number of existing cases with respect to the entire population. It is often thought of as a snapshot and is usually reported as the percentage of the population having a particular disease at any given time. Disease **incidence** measures the number of *new* cases over a certain time period. This statistic, also called the case, or morbidity, rate, indicates both the rate and the risk of infection. The equations used to figure these rates are

$$\text{Prevalence} = \frac{\text{Total number of cases in population}}{\text{Total number of persons in population}} \times 100 = \%$$

$$\text{Incidence} = \frac{\text{Number of new cases in a designated time period}}{\text{Total number of susceptible persons}} \quad \text{(Usually reported per 100,000 persons)}$$

Changes in incidence and prevalence are usually followed over a seasonal, yearly, and long-term basis and are helpful in predicting trends **(figure 11.9).** Statistics of concern to the epidemiologist are the rates of disease with regard to sex, race, or geographic region. Also of importance is the **mortality rate,** which measures the total number of deaths in a population due to a certain disease. Over the past century, the overall death rate from infectious diseases in the developed world has dropped, although the number of persons afflicted with infectious diseases (the **morbidity rate**) has remained relatively high.

When there is an increase in disease in a particular geographic area, it can be helpful to examine the epidemic curve (incidence over time) to determine if the infection is a **point-source, common-source,** or **propagated epidemic.** A point-source epidemic, illustrated in **figure 11.10a,** is one in which the infectious agent came from a single source, and all of its "victims" were exposed to it from that source. The classic example of this is food illnesses brought on by exposure to a contaminated food item at a potluck dinner or restaurant. Common-source epidemics or outbreaks result from common exposure to a single source of infection that can occur over a period of time **(figure 11.10b).** Think of a contaminated water plant that infects multiple people over the course of a week, or even of a single restaurant worker who is a carrier of hepatitis A and does not practice good hygiene. Lastly, a propagated epidemic **(figure 11.10c)** results from an infectious agent that is communicable from

Table 11.10 Reportable Diseases in the United States*

- *Anaplasma phagocytophilum*
- Anthrax
- Babesiosis
- Botulism
- Brucellosis
- California serogroup virus neuroinvasive disease
- Cancer†
- Chancroid
- *Chlamydia trachomatis* infections
- Cholera
- Coccidioidomycosis
- Cryptosporidiosis
- Cyclosporiasis
- Dengue fever
- Diphtheria
- Ehrlichiosis
- Encephalitis/meningitis, arboviral
  - Encephalitis/meningitis, California serogroup viral
  - Encephalitis/meningitis, eastern equine
  - Encephalitis/meningitis, Powassan
  - Encephalitis/meningitis, St. Louis
  - Encephalitis/meningitis, western equine
  - Encephalitis/meningitis, West Nile
- Food-borne disease outbreak
- Giardiasis
- Gonorrhea
- *Haemophilus influenzae* invasive disease
- Hansen's disease (leprosy)
- Hantavirus pulmonary syndrome
- Hemolytic uremic syndrome
- Hepatitis, viral, acute
- Hepatitis A, acute
- Hepatitis B, acute
- Hepatitis B virus, perinatal infection
- Hepatitis C, acute
- Hepatitis, viral, chronic
- Chronic hepatitis B
- Hepatitis C virus infection (past or present)
- HIV infection
- Influenza-associated pediatric mortality
- Lead poisoning†
- Legionellosis
- Leptospirosis
- Listeriosis
- Lyme disease
- Malaria
- Measles
- Meningococcal disease
- Mumps

- Novel influenza A infections
- Pertussis
- Pesticide poisoning†
- Plague
- Poliomyelitis, paralytic
- Poliovirus infection
- Powassan virus diseases
- Psittacosis
- Q fever
- Rabies
  - Rabies, animal
  - Rabies, human
- Rubella
- Rubella, congenital syndrome
- Salmonellosis
- Severe acute respiratory syndrome–associated coronavirus (SARS-CoV) disease
- Shiga toxin–producing *Escherichia coli* (STEC)
- Shigellosis
- Silicosis†
- Smallpox
- Spotted fever rickettsiosis
- Streptococcal toxic shock syndrome
- *Streptococcus pneumoniae*, invasive disease
- Syphilis
- Syphilis, congenital
- Tetanus
- Toxic shock syndrome
- Trichinellosis
- Tuberculosis
- Tularemia
- Typhoid fever
- Vancomycin-intermediate *Staphylococcus aureus* (VISA)
- Vancomycin-resistant *Staphylococcus aureus* (VRSA)
- Varicella
- Vibriosis
- Viral hemorrhagic fevers
- Yellow fever

*Reportable to the CDC; other diseases may be reportable to state departments of health.

†Diseases not of infectious origin (in case of cancer, may have been initiated by microbes–as in cervical cancer–but category includes all cancers)

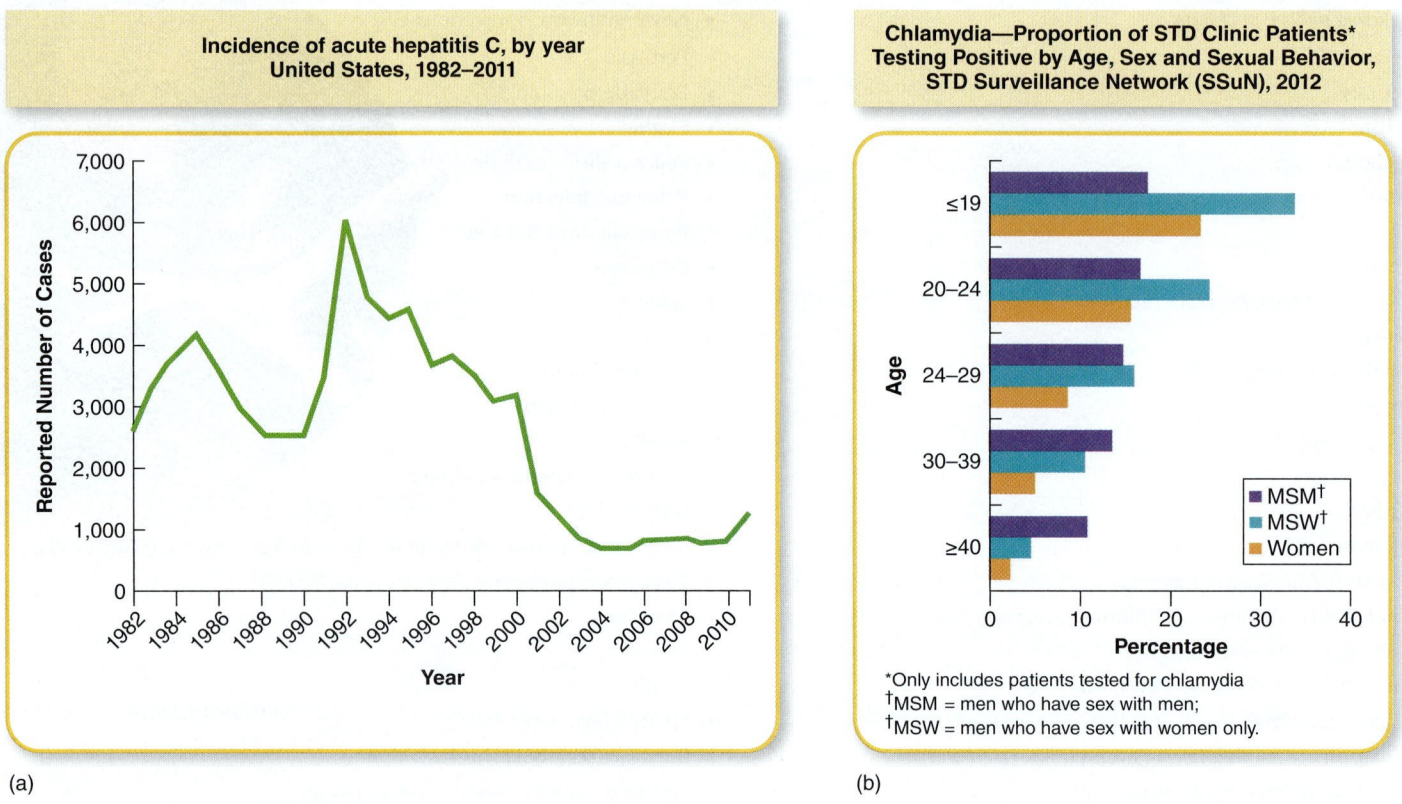

(a)

(b)

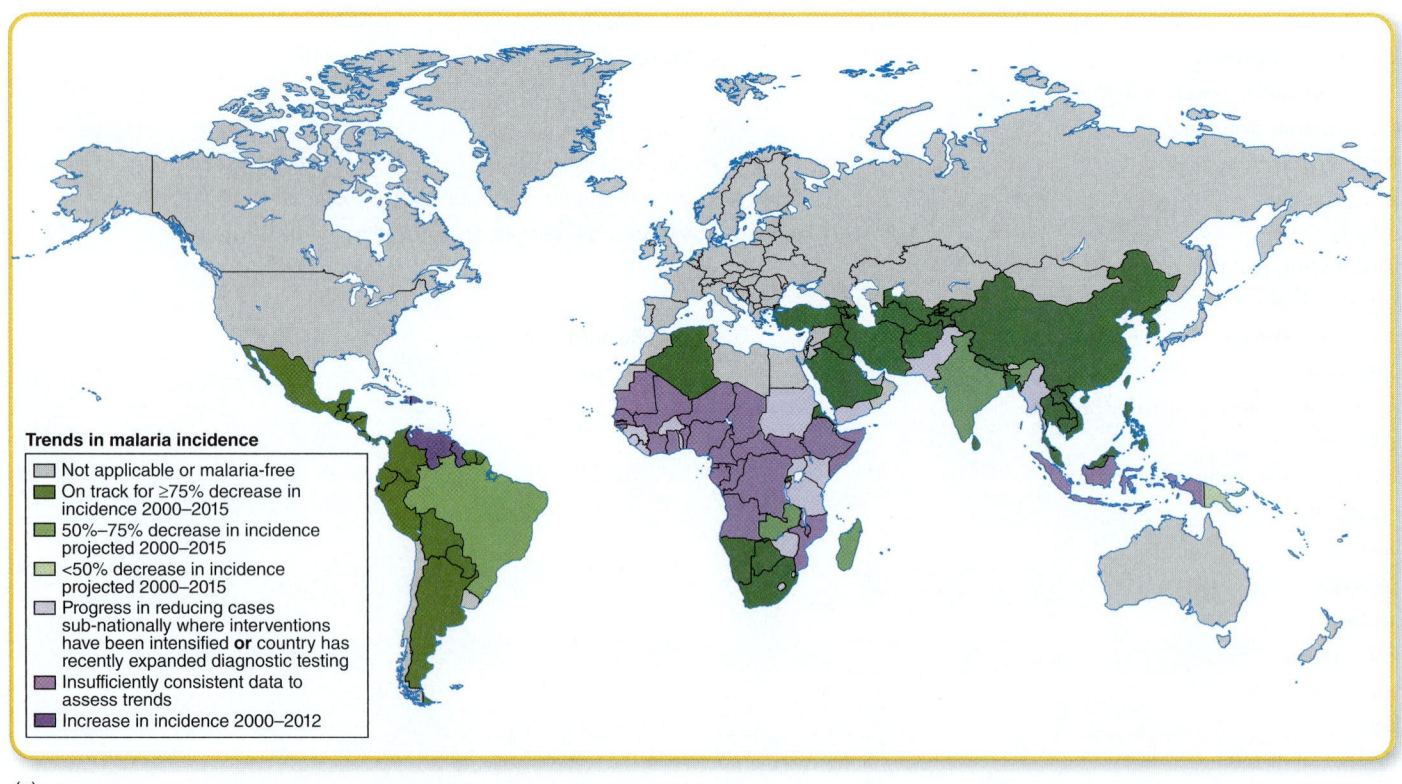

(c)

Figure 11.9 Graphical representation of epidemiological data. The Centers for Disease Control and Prevention collect epidemiological data that are analyzed with regard to (a) time frame and (b) age. In (c), both geographic and time frame data are presented.

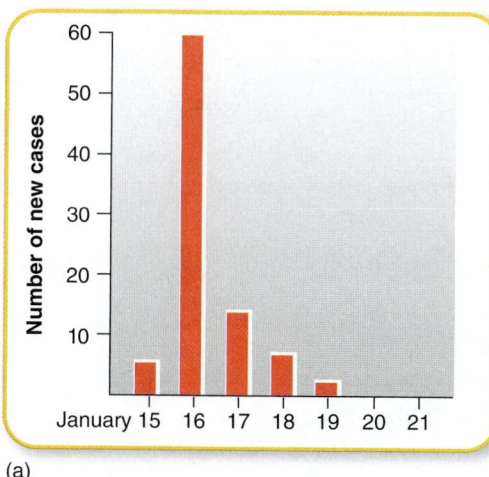

(a)

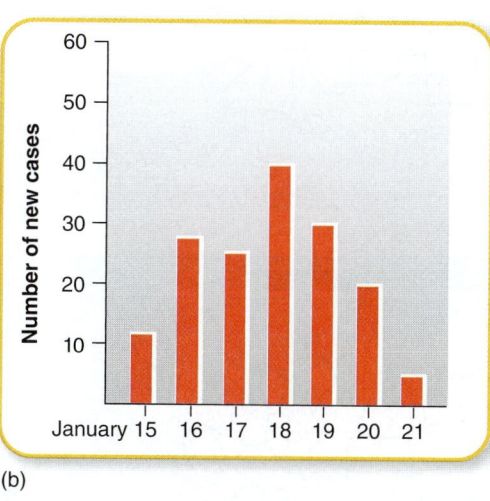

(b)

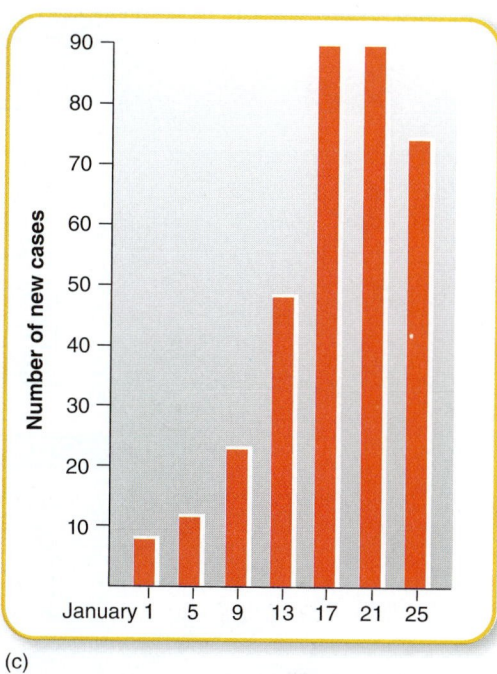

(c)

Figure 11.10 Different outbreak or epidemic curves with different shapes. (a) Point-source epidemic, (b) common-source epidemic, and (c) propagated epidemic.

person to person and therefore is sustained—propagated—over time in a population. Influenza is the classic example of this. The point is that each of these types of spread become apparent from the shape of the outbreak or epidemic curves (determined by plotting numbers of new cases over time on a graph).

An additional term, the **index case,** refers to the first patient found in an epidemiological investigation. How the cases unfurl from this case helps explain the type of epidemic it is. The index case may not turn out to be the first case—as the investigation continues, earlier cases may be found—but the index case is the case that brought the epidemic to the attention of officials. Monitoring statistics also makes it possible to define the frequency of a disease in the population. An infectious disease that exhibits a relatively steady frequency over a long time period in a particular geographic locale is **endemic (figure 11.11a).** For example, Lyme disease is endemic to certain areas of the United States where the tick vector is found. A certain number of new cases are expected in these areas every year. When a disease is **sporadic,** occasional cases are reported at irregular intervals in random locales **(figure 11.11b).** Tetanus and diphtheria are reported sporadically in the United States (fewer than 50 cases a year).

When statistics indicate that the prevalence of an endemic or sporadic disease is increasing beyond what is expected for that population, the pattern is described as an **epidemic (figure 11.11c).** Several epidemics occur every year in the United States, most recently among STDs such as chlamydia and gonorrhea. The spread of an epidemic across continents is a **pandemic,** as exemplified by AIDS and influenza **(figure 11.11d).**

11.3 LEARNING OUTCOMES—Assess Your Progress

17. Summarize the goals of epidemiology, and differentiate it from traditional medical practice.

18. Explain what is meant by a disease being "notifiable" or "reportable," and provide examples.

19. Define *incidence* and *prevalence*, and explain the difference between them.

20. Discuss the three major types of epidemics, and identify the epidemic curve associated with each.

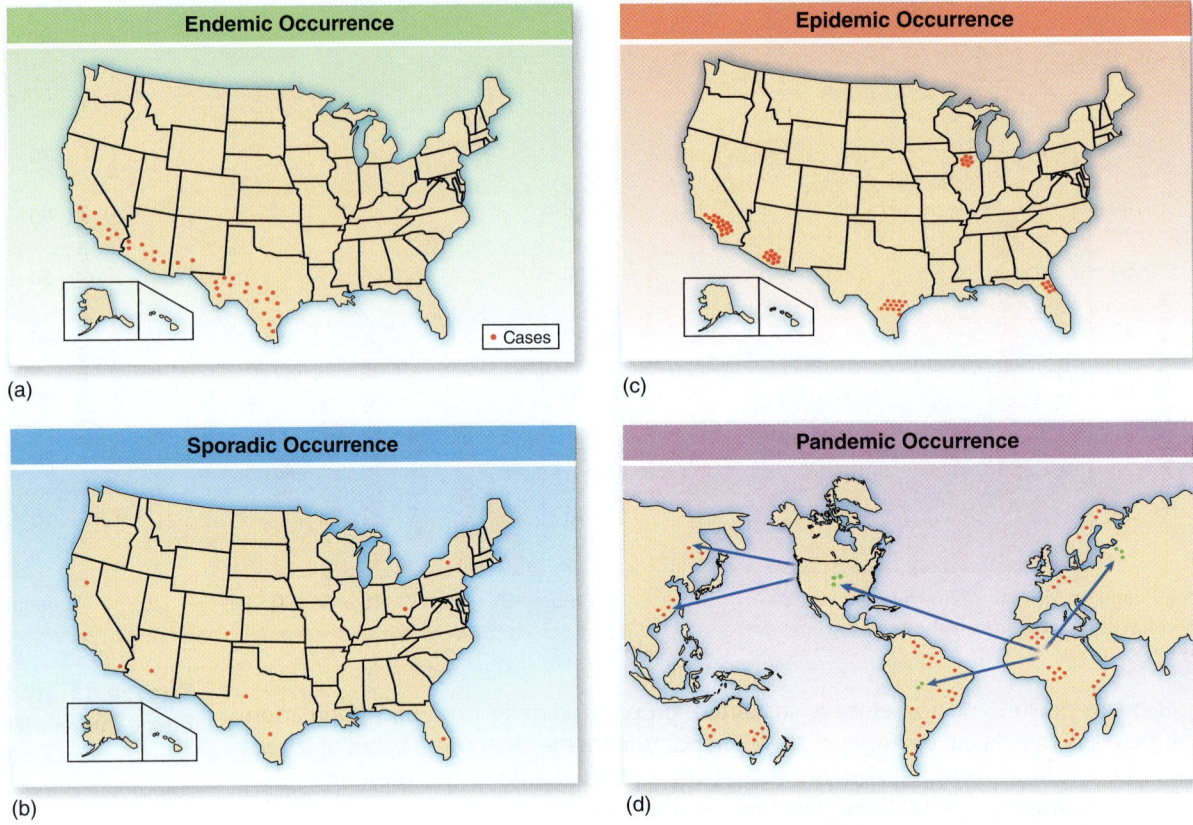

Figure 11.11 Patterns of infectious disease occurrence. **(a)** In endemic occurrence, cases are concentrated in one area at a relatively stable rate. **(b)** In sporadic occurrence, a few cases occur randomly over a wide area. **(c)** An epidemic is an increased number of cases that often appear in geographic clusters. The clusters may be local, as in the case of a restaurant-related food-borne epidemic, or nationwide, as is the case with *Chlamydia*. **(d)** Pandemic occurrence means that an epidemic ranges over more than one continent.

CASE FILE WRAP-UP

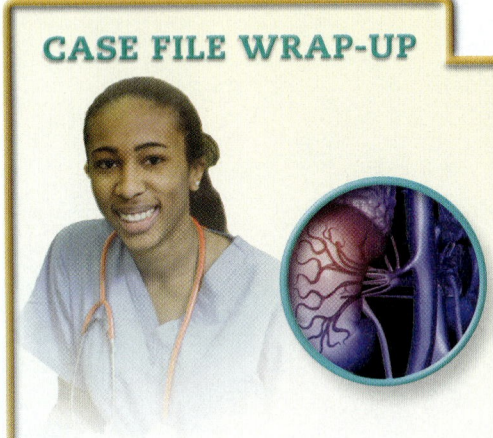

Pseudomonas aeruginosa is a common bacterium implicated in human opportunistic infections. It thrives in water, soil, on the skin, and in most man-made environments. It is a common cause of hospital-acquired infections in susceptible individuals. It is estimated that 10% of all hospital-acquired infections are caused by *P. aeruginosa*.

P. aeruginosa is of concern in vulnerable patients (immunocompromised patients) because some strains are highly resistant to many antibiotics. This resistance can make infection with *P. aeruginosa* very difficult to treat.

Jaelyn was already taking daily antibiotic therapy when she was found to be infected with *P. aeruginosa*. This pathogen posed an unacceptable risk for Jaelyn, whose immune system was already compromised by frequent infections and constant antibiotic use. Infection with *P. aeruginosa* is an undesirable outcome for children with VUR, as the bacterium can quickly become resistant to antibiotic therapy and can lead to a life-threatening kidney infection. This was the reason that surgery was chosen, because surgery for VUR is curative. When the ureters are attached at the correct angle in the bladder, there is no further reflux and no more risk of infection of the kidneys.

Fecal Transplants

Clostridium difficile infection is an extreme example of what happens when the normal microbiota of the bowel is destroyed. *C. difficile* is a gram-positive organism that forms endospores. It grows slowly and can be challenging to isolate.

C. difficile is one of the most common, and most serious, healthcare-associated infections. The microbe, in its worst manifestation, causes a condition known as pseudomembranous colitis, in which pseudomembranous plaques can often be seen on the colon mucosa. Antibiotic therapy is one of the most common causes of *C. difficile*. Antibiotics so drastically alter the normal microbiota of the colon that *C. difficile* is allowed to proliferate, releasing toxins that attack the mucosa of the bowel, resulting in cramping and diarrhea that can be so severe that susceptible individuals (particularly elderly and immunosuppressed people) sometimes die from the infection.

Treatment of *C. difficile* infection frequently entails potent drugs known to be effective against *C. difficile*, namely metronidazole (Flagyl) or fidaxomicin (Dificid). Dificid is used in more severe cases of the disease. However, relapse can still occur. Patients who relapse generally do so within 3 days to 3 weeks of finishing antibiotic therapy.

One therapy that is gaining attention is fecal transplant, or human probiotic infusion (HPI). The theory behind fecal transplant is to combat bacterial infection of the colon with harmless bacteria in an effort to displace the pathogenic microbes, restoring the normal balance in the colon. The treatment may sound distasteful but has proven to be effective.

Fecal transplants involve administering enemas containing donor feces to individuals suffering from bacterial infections of the bowel such as *C. difficile*. Donors are generally family members, and the donor feces are extensively tested to be sure that the donor is free of any bacterial, viral, or other harmful infections. Recipients may be administered the fecal enemas for 5 to 10 days, although sometimes one treatment is sufficient.

A study published in the *NEJM* (*New England Journal of Medicine*) in January of 2013 determined a 94% cure rate in patients treated with fecal microbiota transplants, compared to only a 31% cure rate with vancomycin. The results were so overwhelmingly positive that the study was stopped early, as it was considered unethical not to offer fecal microbiota transplants to all of the study participants. In June of 2013, the FDA approved the use of fecal transplants in patients with *C. difficile* who do not respond to standard medical therapy.

For those who have not yet developed *C. difficile* but are at risk, ARGF (autologous restoration of gastrointestinal flora) may be the answer. In this procedure, the patient provides a sample of his or her own feces, which is stored in a refrigerated environment. Should the patient become infected, the sample is extracted with saline and is run through a filter. The filtered material is then freeze-dried and encapsulated in enteric-coated capsules. The patient then swallows the capsules, which help to restore the patient's normal biota.

Chapter Summary

11.1 The Human Host

- Humans coexist with microorganisms from the moment of birth onward and possibly even *in utero*.
- Normal biota are easily cultured from the skin and in the respiratory tract, the gastrointestinal tract, the outer parts of the urethra, the vagina, the eye, and the external ear canal. Molecular techniques have also revealed normal biota to be present in the lungs, bladder, breast milk, and amniotic fluid.
- The Human Microbiome Project is finding a much wider array of normal biota than known previously.

11.2 The Progress of an Infection

- The pathogenicity of a microbe refers to its ability to cause disease. Its virulence is the degree of damage it can inflict.
- The virulence of a microbe is determined by its ability to establish itself in the host and then do damage. Any characteristic or structure that enhances its ability to do these two things is called a virulence factor.
- True pathogens cause infectious disease in healthy hosts; opportunistic pathogens cause damage only when the host immune system is compromised in some way.
- The site at which a microorganism first contacts host tissue is called the portal of entry.
- The reservoir of the microbe may or may not be the same as the transmitter of the microbe.
- The infectious dose, or ID, refers to the minimum number of microbial cells required to initiate infection in the host. The ID is often influenced by quorum-sensing chemicals.
- Fimbriae and adhesive capsules allow pathogens to physically attach to host tissues.
- Secreted enzymes and toxins and the ability to induce injurious host responses are the three main types of virulence factors pathogens utilize to damage host tissue.
- Exotoxins and endotoxin differ in their chemical composition and tissue specificity.
- Inappropriate or extreme host responses are a major factor in most infectious diseases.
- Patterns of infection vary with the pathogen or pathogens involved. Examples are local, focal, and systemic.

- The stages of infections in humans parallel what we know about how microorganisms grow.
- Mixed infections are more common than previously appreciated.
- The portal of exit by which a pathogen leaves its host is often but not always the same as the portal of entry.
- The portals of exit and entry determine how pathogens spread in a population.
- Some pathogens persist in the body in a latent state.
- There are four distinct phases of infection and disease: the incubation period, the prodrome, the period of invasion, and the convalescent period.
- A communicable disease can be transmitted from an infected host to others, but not all infectious diseases are communicable.
- The spread of infectious disease from person to person is called horizontal transmission. The spread from parent to offspring is called vertical transmission.
- Infectious diseases are spread by direct, indirect, or vector routes of transmission. Vehicles of indirect transmission include soil, water, food, air, and fomites (inanimate objects).
- Healthcare-associated infections are acquired in a hospital or health care facility from surgical procedures, equipment, personnel, and exposure to drug-resistant microorganisms.
- Causative agents of infectious disease may be identified according to Koch's postulates.

11.3 Epidemiology: The Study of Disease in Populations

- Epidemiology is the study of the determinants and distribution of infectious and noninfectious diseases in populations.
- Data on specific, reportable diseases are collected by local, national, and worldwide agencies.
- The prevalence of a disease is the percentage of existing cases in a given population. The disease incidence, or morbidity rate, is the number of newly infected members in a population during a specified time period.
- Outbreaks and epidemics are described as point-source, common-source, or propagated based on the source of the pathogen.
- Disease frequency is described as sporadic, epidemic, pandemic, or endemic.

Multiple-Choice   Bloom's Levels 1 and 2: Remember and Understand

Select the correct answer from the answers provided.

1. Normal resident microbes in our bodies interact with us as
   a. commensals.
   b. mutualists.
   c. pathogens.
   d. a and b.
   e. a, b, and c.

2. Resident biota is absent from the
   a. pharynx.
   b. heart.
   c. intestine.
   d. hair follicles.

3. Virulence factors include
   a. toxins.
   b. enzymes.
   c. capsules.
   d. a and b.
   e. a, b, and c.

4. The _____ is the time that lapses between an encounter with a pathogen and the first symptoms.
   a. prodrome
   b. period of invasion
   c. period of convalescence
   d. period of incubation

5. A short period early in a disease that may manifest with general malaise and achiness is the
   a. period of incubation.
   b. prodrome.
   c. sequela.
   d. period of invasion.

6. A/an _____ is a passive animal transporter of pathogens.
   a. zoonosis
   b. biological vector
   c. mechanical vector
   d. asymptomatic carrier

7. A positive antibody test for HIV would be a _____ of infection.
   a. sign
   b. symptom
   c. syndrome
   d. sequela

8. A seasonal outbreak of influenza would be an example of what type of outbreak?
   a. point-source
   b. common-source
   c. propagated
   d. a and b
   e. a, b, and c

Critical Thinking   Bloom's Levels 3, 4, and 5: Apply, Analyze, and Evaluate

Critical thinking is the ability to reason and solve problems using facts and concepts. These questions can be approached from a number of angles and, in most cases, they do not have a single correct answer.

1. Differentiate between contamination, infection, and disease. What are the possible outcomes in each?

2. How are infectious diseases different from other diseases?

3. Describe the course of infection from contact with the pathogen to its exit from the host.

4. The day after she skinned her knee climbing a tree, your little sister complains of feeling hot and dizzy. Upon examining her, you find she does have an elevated body temperature and the scrape on her knee looks bright red and is oozing fluid. Explain whether she has a local or systemic infection.

5. a. List the main features of Koch's postulates.
   b. Why is it so difficult to prove them for some diseases?

Visual Connections   Bloom's Level 5: Evaluate

This question connects previous images to a new concept.

1. **From chapter 2, figure 2.4a.** What chemical is the organism in this illustration producing? How does this add to an organism's pathogenicity?

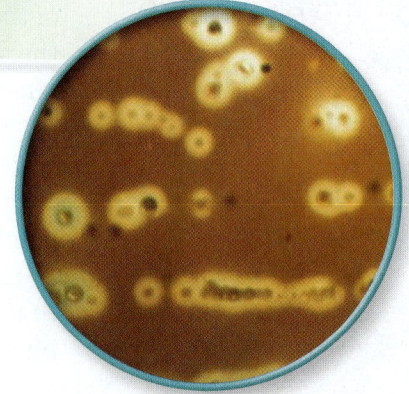

www.mcgrawhillconnect.com
Enhance your study of this chapter with study tools and practice tests. Also ask your instructor about the resources available through Connect, including the media-rich eBook, interactive learning tools, and animations.

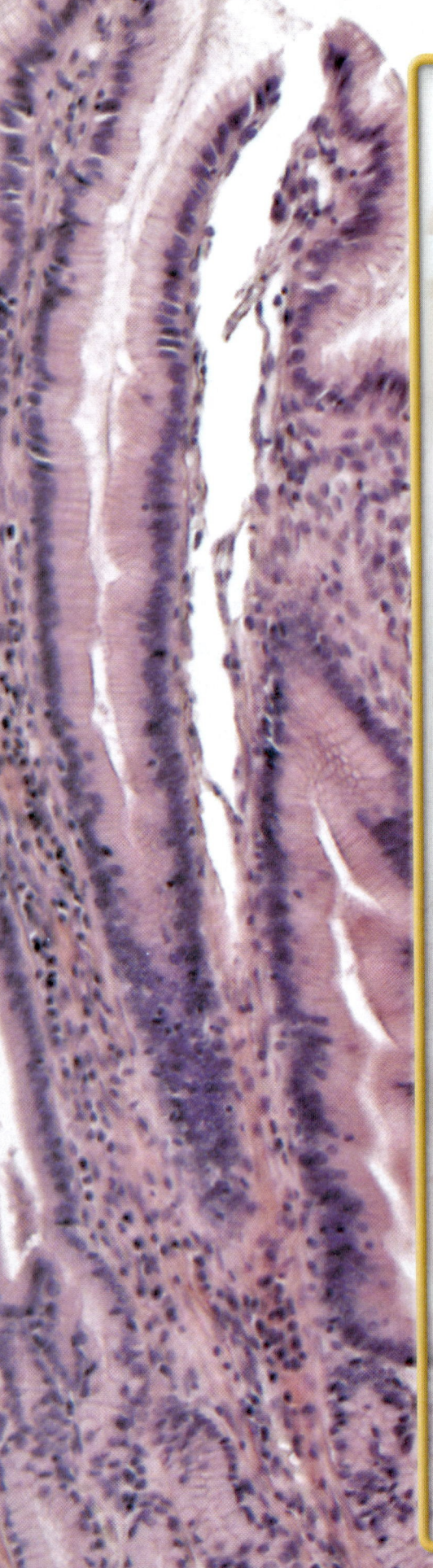

CASE FILE

Bacteria Cause That?

I was working in an endoscopy unit as a registered nurse when I met Natasha, a 20-year-old college student. Natasha was having a gastroscopy to determine the cause of her persistent stomach pain. Natasha had been having episodes of epigastric pain intermittently for 5 years. Approximately every 4 to 6 weeks, she would experience intense upper abdominal pain that left her unable to engage in her usual activities, including her classes and work. She would experience what she described as intense hunger, but pain would worsen immediately after eating so she learned to avoid eating during the worst of her "attacks," which could last as long as 5 days.

She had been admitted to the emergency room after a particularly intense bout of pain. She was dehydrated and weak and her pain was constant. Lab studies revealed an elevated hemoglobin level, indicative of dehydration, and a normal white blood cell count. She had no fever. Her epigastric area was moderately tender on palpation. Her heart rate was elevated, likely as a response to the pain she was experiencing. The ER physician who examined her had admitted her to the hospital and arranged an urgent gastroscopy.

Once in the endoscopy unit, Natasha was given medications to induce conscious sedation. The back of her throat was anesthetized and the gastroscopy scope was passed down her throat. The esophagus was visualized as normal. The stomach, however, was inflamed, with evidence of prior healed ulcerations. One small unhealed lesion remained.

Just as the ER physician had expected, Natasha had a gastric ulcer. Biopsies of the inflamed area were obtained, as well as photographs of the lesion, and the scope was removed. Natasha was then given medication to reverse the effects of the sedatives. Once fully recovered, Natasha was released to go home.

A week later, Natasha had an appointment with the gastroenterologist, who told Natasha that biopsies had revealed the presence of *Helicobacter pylori* in her stomach, which had caused the ulcers. She was prescribed a combination of antibiotics and an acid reducer and was given an appointment to follow up after she had completed the therapy to confirm the efficacy of the treatment. She was also instructed to contact them again if she experienced any more episodes of stomach pain.

- How does *H. pylori* survive in the acidic environment of the stomach?

- What classic response to injury did Natasha display?

Case File Wrap-Up appears on page 344.

CHAPTER 12

Host Defenses I

Overview and Nonspecific Defenses

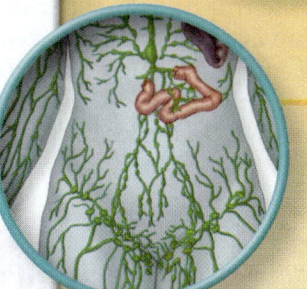

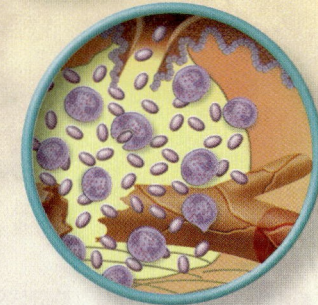

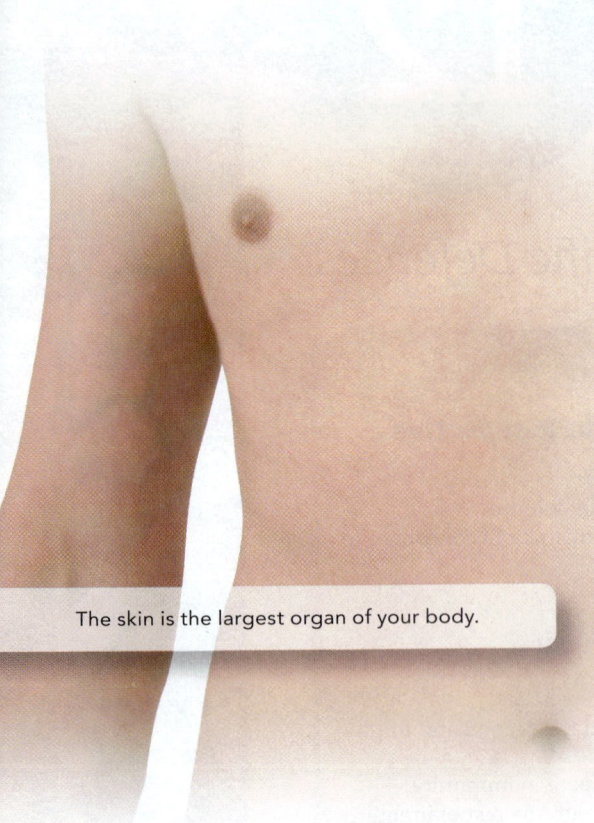

The skin is the largest organ of your body.

12.1 Defense Mechanisms of the Host in Perspective

The host defenses are a multilevel network of innate, nonspecific protections and specific **immunities** referred to as the first, second, and third lines of defense **(figure 12.1).** The interaction and cooperation of these three levels of defense normally provide complete protection against infection. The *first line of defense* includes any barrier that blocks invasion at the portal of entry. This nonspecific line of defense limits access to the internal tissues of the body. However, it is not considered a true immune response because it does not involve recognition of a specific foreign substance but is very general in action. The *second line of defense*, also nonspecific, is a more internal system of protective cells, fluids, and processes that includes inflammation and phagocytosis. It acts rapidly at both the local and systemic levels once the first line of defense has been circumvented. The highly specific *third line of defense* is acquired only as each foreign substance is encountered by white blood cells called lymphocytes. The reaction with each different foreign microbe produces unique protective substances and cells that can come into play if that microbe is encountered again. The third line of defense provides long-term immunity. It is discussed in detail in chapter 13. This chapter focuses on the first and second lines of defense.

Human systems are armed with various levels of defense that do not operate in a completely separate fashion; most defenses overlap and are even redundant in some of their effects. This bombards microbial invaders with an entire assault force, making their survival unlikely. Figure 12.1 provides an overview of the three systems. The light blue box at the bottom that connects both the second and third lines of defense is a good example of how the systems overlap.

Barriers: A First Line of Defense

The inborn, nonspecific defenses are the physical and chemical barriers that impede the entry of not only microbes but any foreign agent, whether living or not **(figure 12.2).**

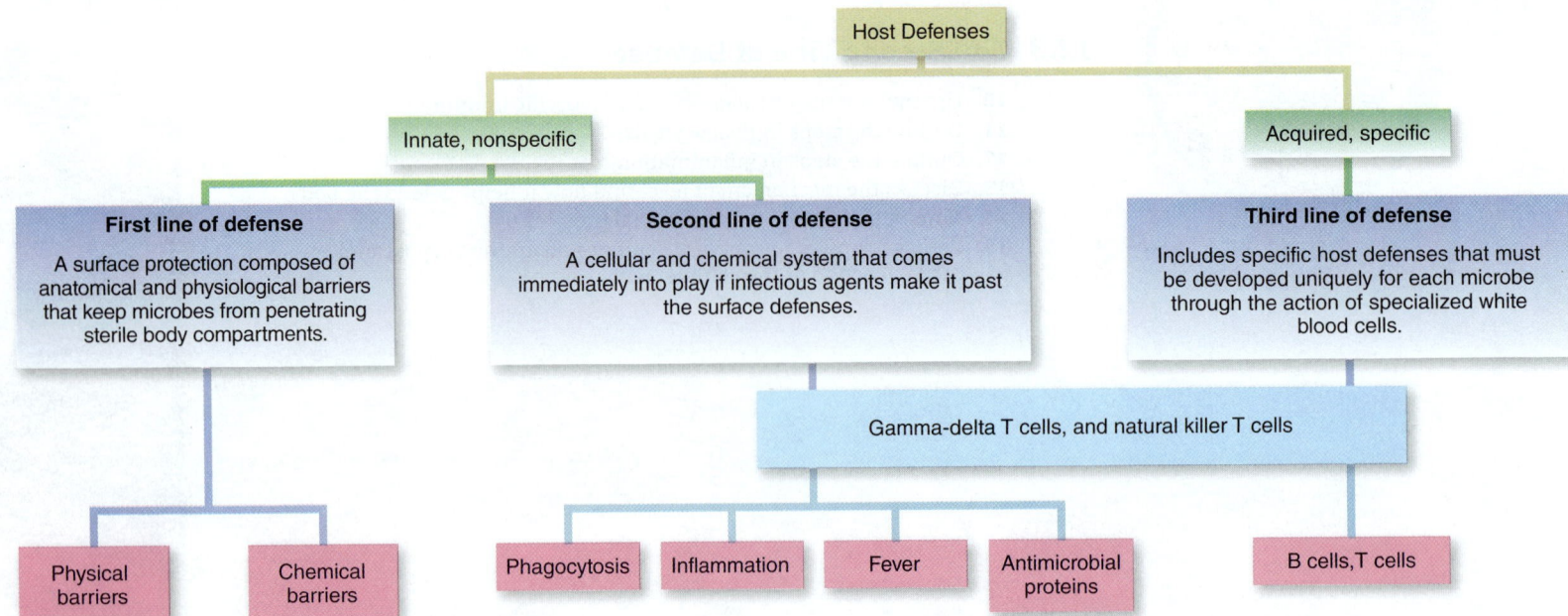

Figure 12.1 Flowchart summarizing the major components of the host defenses. Defenses are classified into one of two general categories: (1) innate and nonspecific or (2) acquired and specific. These can be further subdivided into the first, second, and third lines of defense, each being characterized by a different level and type of protection. There is also a set of cells that straddle the categories of innate and acquired defenses (light blue box).

Physical or Anatomical Barriers at the Body's Surface

The skin and mucous membranes of the respiratory and digestive tracts have several built-in defenses. The outermost layer (stratum corneum) of the skin is composed of epithelial cells that have become compacted, cemented together, and impregnated with an insoluble protein, keratin. The result is a thick, tough layer that is highly impervious and waterproof. Few pathogens can penetrate this unbroken barrier, especially in regions such as the soles of the feet or the palms of the hands, where the stratum corneum is much thicker than on other parts of the body. In addition, outer layers of skin are constantly sloughing off, taking associated microbes with them. Other cutaneous barriers include hair follicles and skin glands. The hair shaft is periodically shed, and the follicle cells are **desquamated** (des'-kwuh-mayt-ud). The flushing effect of sweat glands also helps remove microbes.

The mucous membranes of the digestive, urinary, and respiratory tracts and of the eye are moist and permeable. They do provide barrier protection but without a keratinized layer. The mucous coat on the free surface of some membranes impedes the entry and attachment of bacteria. Blinking and tear production (lacrimation) flush the eye's surface with tears and rid it of irritants. The constant flow of saliva helps carry microbes into the harsh conditions of the stomach. Vomiting and defecation also evacuate noxious substances or microorganisms from the body.

The respiratory tract is constantly guarded from infection by elaborate and highly effective adaptations. Nasal hair traps larger particles. The copious flow of mucus and fluids that occurs in allergy and colds exerts a flushing action. In the respiratory tree (primarily the trachea and bronchi), a ciliated epithelium (called the ciliary escalator) conveys foreign particles entrapped in mucus toward the pharynx to be removed **(figure 12.3).** Irritation of the nasal passage reflexively initiates a sneeze, which expels a large volume of air at high velocity. Similarly, the acute sensitivity of the bronchi, trachea, and larynx to foreign matter triggers coughing, which ejects irritants.

The genitourinary tract derives partial protection via the continuous trickle of urine through the ureters and from periodic bladder emptying that flushes the urethra. Vaginal secretions provide cleansing of the lower reproductive tract in females. Older women who are postmenopausal are sometimes prone to vaginal infections due to lack of vaginal secretions resulting from a decrease in estrogen production.

The composition of resident microbiota and its protective effect were discussed in chapter 11. Even though the resident biota does not constitute an anatomical barrier, its presence can block the access of pathogens to epithelial surfaces and can create an unfavorable environment for pathogens by competing for limited nutrients or by altering the local pH.

New research stemming from the Human Microbiome Project has continued to highlight the importance of the normal resident biota on the development of nonspecific defenses (described in this chapter) as well as specific immunity (described in the next chapter). The presence of a robust commensal biota "trains" host defenses in such a way that commensals are kept in check and pathogens are eliminated. Evidence suggests that inflammatory bowel diseases, including Crohn's disease and ulcerative colitis, may well be a result of our overzealous attempts to free our environment of microbes and to overtreat ourselves with antibiotics. The result is an "ill-trained" gut defense system that responds inappropriately to commensal biota.

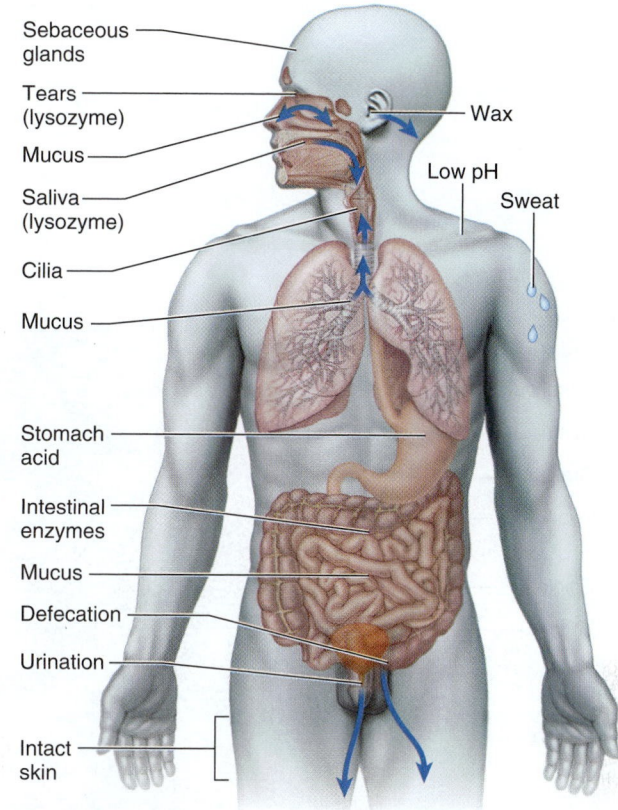

Figure 12.2 **The primary physical and chemical defense barriers.**

Tears cleanse the eyes of foreign particles.

NCLEX® PREP

1. The presence of intestinal microbiota is considered
   a. *a first line of defense.*
   b. *a second line of defense.*
   c. *a third line of defense.*
   d. *none of the above.*

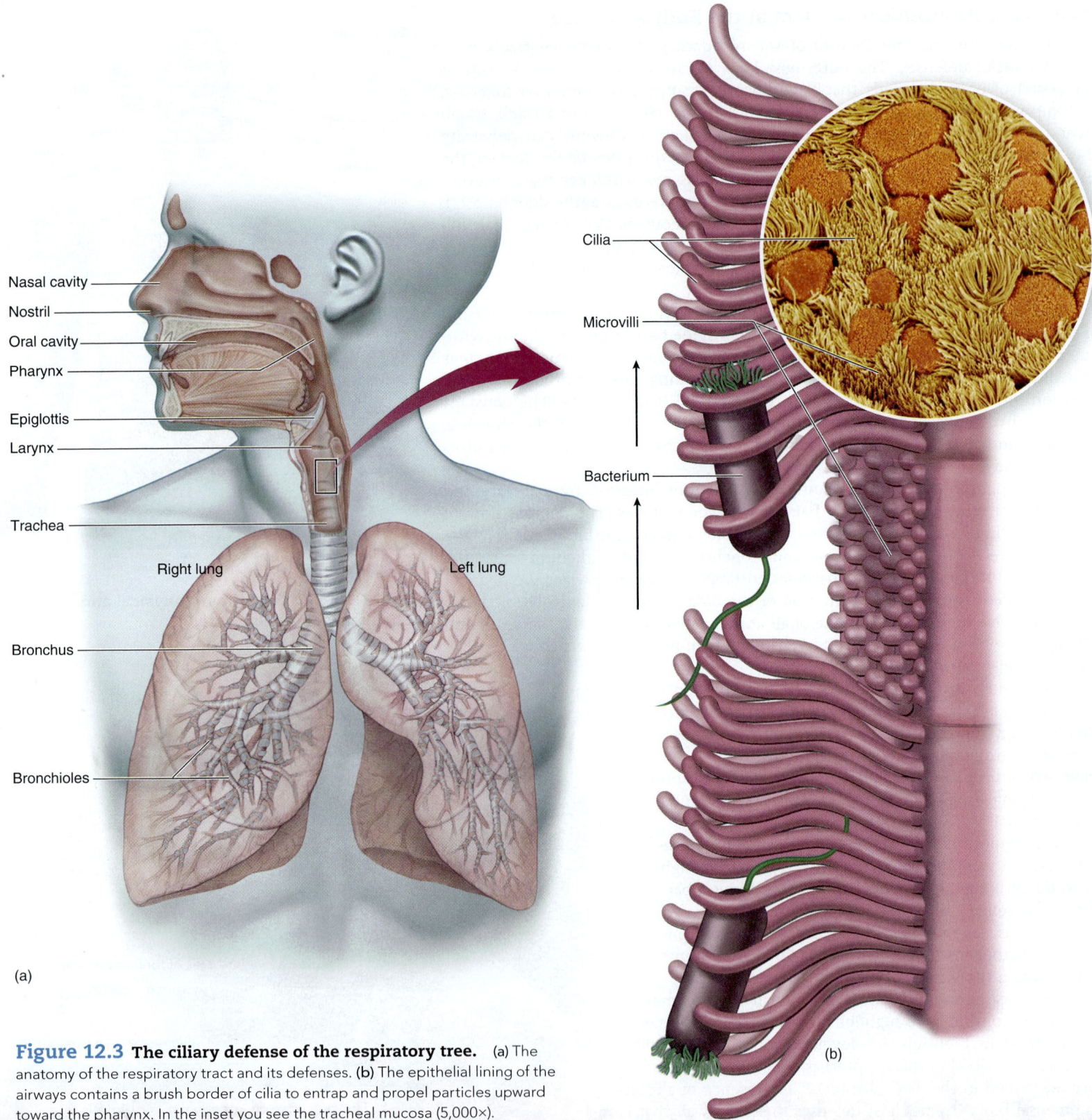

Nasal cavity

Nostril

Oral cavity

Pharynx

Epiglottis

Larynx

Trachea

Right lung

Left lung

Bronchus

Bronchioles

(a)

Cilia

Microvilli

Bacterium

(b)

Figure 12.3 The ciliary defense of the respiratory tree. (a) The anatomy of the respiratory tract and its defenses. (b) The epithelial lining of the airways contains a brush border of cilia to entrap and propel particles upward toward the pharynx. In the inset you see the tracheal mucosa (5,000×).

Nonspecific Chemical Defenses

The skin and mucous membranes offer a variety of chemical defenses. Sebaceous secretions exert an antimicrobial effect, and specialized glands of the eyelids lubricate the conjunctiva with an antimicrobial secretion. An additional defense in tears and saliva is **lysozyme,** an enzyme that hydrolyzes the peptidoglycan in the cell wall of bacteria. The high lactic acid and electrolyte concentrations of sweat and the skin's acidic pH and fatty acid content are also inhibitory to many microbes. Likewise, the hydrochloric acid in the stomach renders protection against many pathogens that are swallowed, and the intestine's digestive juices and bile are potentially destructive to microbes. Even semen contains an antimicrobial chemical that inhibits bacteria, and the vagina has a protective acidic pH maintained by normal biota.

The vital contribution of all types of barriers is clearly demonstrated in people who have lost them or never had them. Patients with severe skin damage due to burns are extremely susceptible to infections; those with blockages in the salivary glands, tear ducts, intestine, and urinary tract are also at greater risk for infection. But as important as it is, the first line of defense alone is not sufficient to protect against infection. Because many pathogens find a way to circumvent the barriers by using their virulence factors (discussed in chapter 11), a whole new set of defenses—inflammation, phagocytosis, specific immune responses—are brought into play.

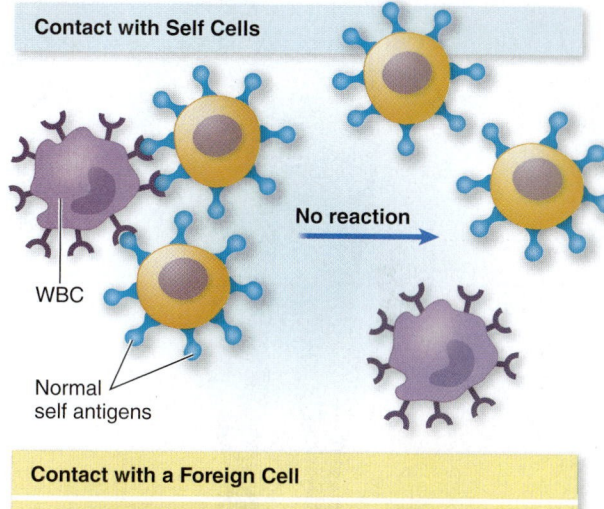

Contact with Self Cells

No reaction

WBC

Normal self antigens

12.1 LEARNING OUTCOMES—Assess Your Progress

1. Summarize the three lines of host defenses.
2. Identify three components of the first line of defense.
3. Describe two examples of how the normal microbiota contribute to the first line of defense.

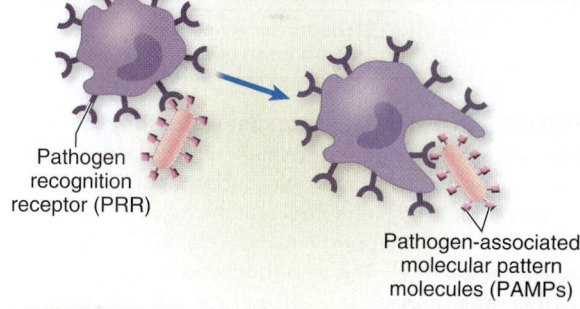

Contact with a Foreign Cell

1. Detection and recognition of foreign cell or virus

Pathogen recognition receptor (PRR)

Pathogen-associated molecular pattern molecules (PAMPs)

12.2 The Second and Third Lines of Defense: An Overview

Immunology encompasses the study of all features of the body's second and third lines of defense. Although this chapter is concerned, not surprisingly, with infectious microbial agents, be aware that immunology is central to the study of fields as diverse as cancer and allergy.

In the body, the mandate of the immune system can be easily stated. A healthy functioning immune system is responsible for the following:

1. surveillance of the body,
2. recognition of foreign material, and
3. destruction of entities deemed to be foreign **(figure 12.4).**

All cells (microbial and otherwise), as well as some particles such as pollen, display chemicals that the immune system "senses" to determine if they are foreign or not. These chemicals are called antigens. Because foreign cells or particles could potentially enter through any number of portals, the cells of the immune system constantly move about the body, searching for potential pathogens. This process is carried out primarily by white blood cells, which have been trained to recognize body cells (so-called **self**) and differentiate them from any foreign material in the body, such as an invading bacterial cell **(nonself).** The ability to evaluate cells and macromolecules as either self or nonself is central to the functioning of the immune system. While foreign substances must be recognized as a potential threat and dealt with appropriately, self cells and chemicals must not come under attack by the immune defenses. Many autoimmune

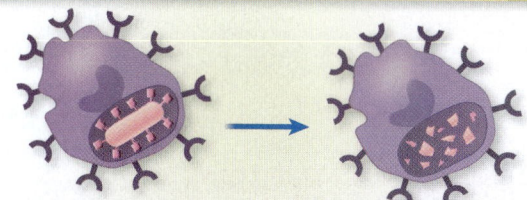

2. Destruction

Figure 12.4 *Search, recognize, and destroy* **is the mandate of the immune system.** White blood cells are equipped with a very sensitive sense of "touch." As they travel through the tissues, they feel surface markers that help them determine what is self and what is not. When self markers are recognized, no response occurs. However, when nonself is detected, a reaction to destroy it is mounted.

disorders are a result of the immune system mistakenly attacking the body's own tissues and organs. For example, in rheumatoid arthritis, the body attacks its own joints and tissues, causing pain and loss of function.

The surface chemicals that the immune system evaluates are called **markers.** These markers, which generally consist of proteins and/or sugars, can be thought of as the cellular equivalent of facial characteristics in humans and allow the cells of the immune system to identify whether or not a newly discovered cell poses a threat. While cells deemed to be self are left alone, cells and other objects designated as foreign are marked for destruction by a number of methods, the most common of which is phagocytosis. Markers that many different kinds of microbes have in common are called **pathogen-associated molecular patterns (PAMPs).** Host cells with important roles in the innate immunity of the second line of defense use **pattern recognition receptors (PRRs)** to recognize PAMPs. There is a middle ground as well. Nonself proteins that are not harmful—such as those found in food we ingest and on commensal microorganisms—are generally recognized as such and the immune system is signaled not to react.

Unlike many systems, the immune system does not exist in a single, well-defined site; rather, it is a large, complex, and diffuse network of cells and fluids that permeate every organ and tissue. It is this arrangement that promotes the surveillance and recognition processes that help screen the body for harmful substances.

The body is partitioned into several fluid-filled spaces called the intracellular, extracellular, lymphatic, cerebrospinal, and circulatory compartments. Although these compartments are physically separated, they have numerous connections.

The Communicating Body Compartments

For effective immune responsiveness, the activities in one fluid compartment must be communicated to other compartments. At the microscopic level, clusters of tissue cells are in direct contact with the **mononuclear phagocyte system (MPS),** which is described shortly, and the extracellular fluid (ECF). Blood and lymphatic capillaries penetrate into these tissues. This close association allows cells and chemicals that originate in the MPS and ECF to diffuse or migrate into the blood and lymphatics; any products of a lymphatic reaction can be transmitted directly into the blood through the connection between these two systems; and certain cells and chemicals originating in the blood can move through the vessel walls into the extracellular spaces and migrate into the lymphatic system.

The Mononuclear Phagocyte System

The tissues of the body are permeated by a support network of connective tissue fibers, or a reticulum, that interconnects nearby cells and meshes with the massive connective tissue network surrounding all organs. The phagocytic cells enmeshed in this network are collectively called the mononuclear phagocyte system (MPS; **figure 12.5).** (It is also sometimes called the reticuloendothelial system, or the RES.) This system is intrinsic to the immune function because it provides a passageway within and between tissues and organs. The MPS is found in the thymus, where important white blood cells mature, and the lymph nodes, tonsils, spleen, and lymphoid tissue in the mucosa of the gut and respiratory tract, where most of the MPS "action" takes place. The MPS is loaded with white blood cells called macrophages waiting to attack passing foreign intruders as they arrive in the skin, lungs, liver, lymph nodes, spleen, and bone marrow.

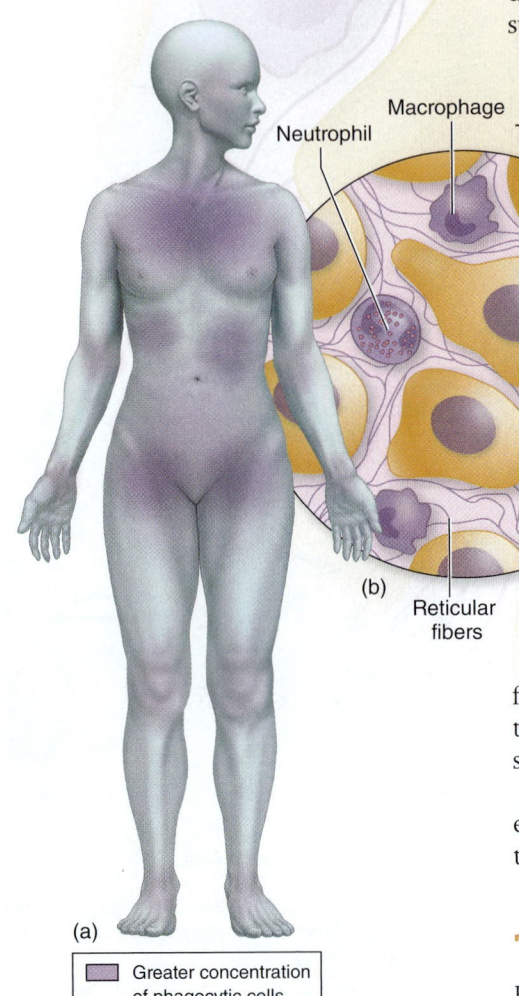

Neutrophil · Macrophage · Tissue cell · Reticular fibers · Dendritic cell · (b) · (a)

Greater concentration of phagocytic cells

Figure 12.5 The mononuclear phagocyte system. (a) The degrees of shading in the body indicate variations in phagocyte concentration (darker = greater). (b) This system begins at the microscopic level with a fibrous support network (reticular fibers) enmeshing each cell. This web connects one cell to another within a tissue or organ and provides a niche for phagocytic white blood cells, which can crawl within and between tissues.

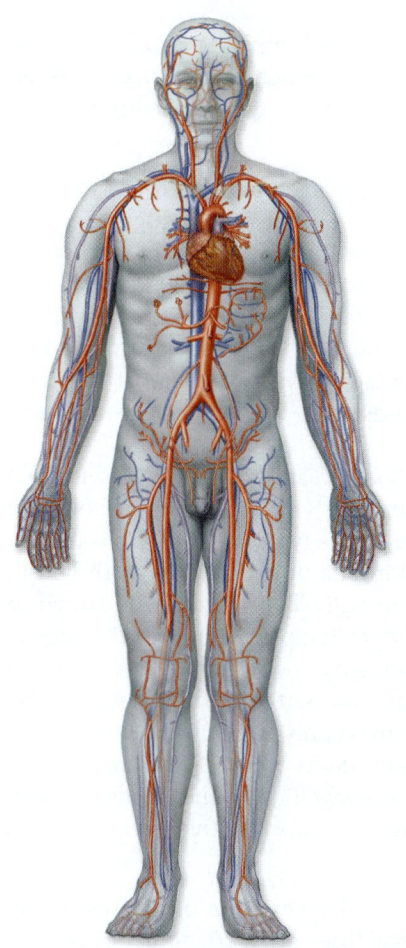

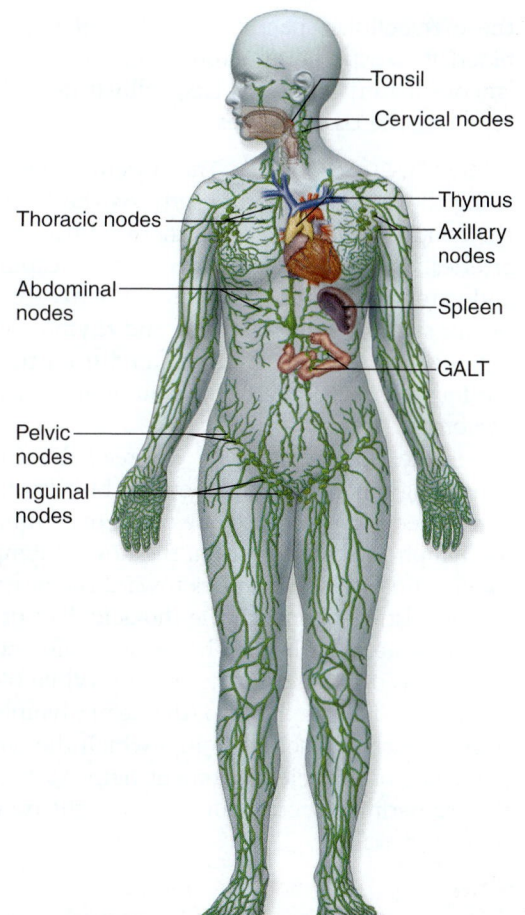

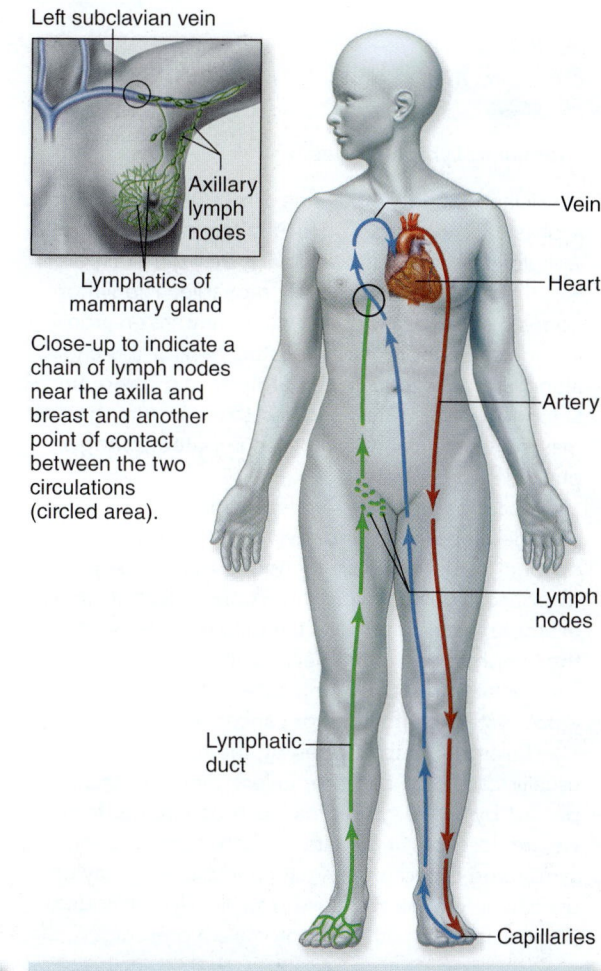

The Circulatory System	**The Lymphatic System**	**The Lymphatic and Circulatory Systems**
Body compartments are screened by circulating WBCs in the cardiovascular system.	The lymphatic system consists of a branching network of vessels that extend into most body areas. Note the higher density of lymphatic vessels in the "dead-end" areas of the hands, feet, and breast, which are frequent contact points for infections. Other lymphatic organs include the lymph nodes, spleen, gut-associated lymphoid tissue (GALT), the thymus, and the tonsils.	Comparison of the generalized circulation of the lymphatic system and the blood. Although the lymphatic vessels parallel the regular circulation, they transport in only one direction unlike the cyclic pattern of blood. Direct connection between the two circulations occurs at points near the heart where large lymph ducts empty their fluid into veins (circled area).

Figure 12.6 **The circulatory and lymphatic systems.**

The Lymphatic System

The **lymphatic system** is a compartmentalized network of vessels, cells, and specialized accessory organs **(figure 12.6)**. It begins in the farthest reaches of the tissues as tiny capillaries that transport a special fluid (lymph) through an increasingly larger tributary system of vessels and filters (lymph nodes), and it leads to major vessels that drain back into the regular circulatory system. Some major functions of the lymphatic system are as follows:

1. to provide a route for the return of extracellular fluid to the circulatory system proper;
2. to act as a "drain-off" system for the inflammatory response; and
3. to render surveillance, recognition, and protection against foreign materials through a system of lymphocytes, phagocytes, and antibodies.

Lymphatic Fluid   Lymph is a plasmalike liquid carried by the lymphatic circulation. It is formed when certain blood components move out of the blood vessels into

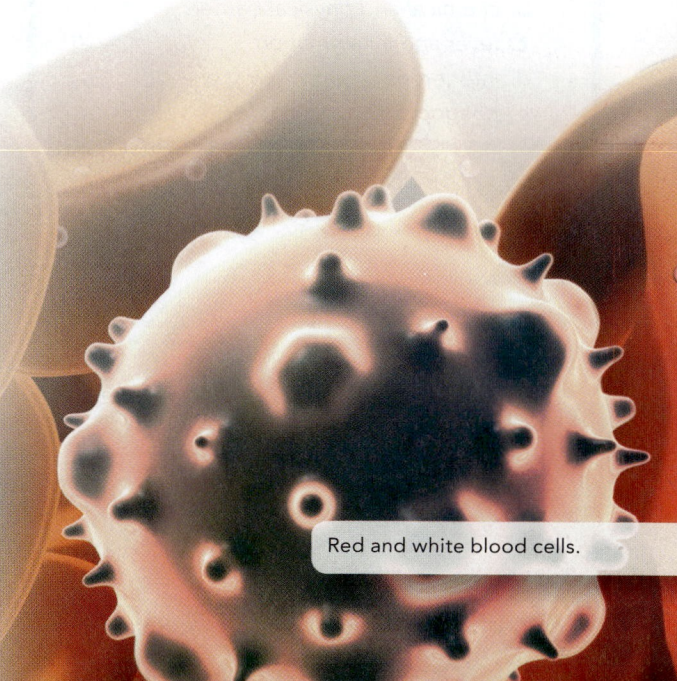

Red and white blood cells.

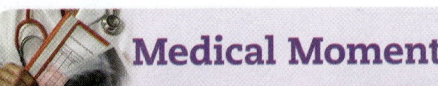

Medical Moment

Examining Lymph Nodes

Physicians routinely examine lymph nodes during a physical examination. Healthy lymph nodes feel well defined and slightly rubbery; they generally are less than 2 cm and are easily movable. The lymph nodes that can easily be felt when they are enlarged include the lymph nodes located behind the ears, along the neck, and in the armpits and the groin. Deeper lymph nodes, such as those in the chest, may be visible on an X ray or computed tomography (CT) scan.

Location of the swollen lymph node may provide clues as to underlying pathology. An individual node or chain of nodes may enlarge in response to a relatively innocuous infection. For example, a cold may result in mild enlargement of the lymph nodes of the neck. Diffuse and generalized enlargement may indicate autoimmune disease, systemic infection, or cancer.

Enlarged lymph nodes by themselves are not usually cause for concern, unless they are accompanied by other symptoms, such as fatigue, fever, weight loss, night sweats, or other constitutional symptoms. Blood tests, X rays, and biopsies may be used to determine the cause of swollen lymph nodes, particularly when other symptoms are present.

NCLEX® PREP

2. Which component(s) is/are associated with the second line of defense? Select all that apply.
   a. *maternal antibodies*
   b. *chemical and physical barriers*
   c. *fever and inflammation*
   d. *interferons*
   e. *phagocytosis*

Lymph nodes in the neck swell when the immune system is active.

the extracellular spaces and diffuse or migrate into the lymphatic capillaries. Like blood, it transports numerous white blood cells (especially lymphocytes) and miscellaneous materials such as fats, cellular debris, and infectious agents that have gained access to the tissue spaces.

Lymphatic Vessels   The system of vessels that transports lymph is constructed along the lines of blood vessels. As the lymph is never subjected to high pressure, the lymphatic vessels appear more similar to thin-walled veins than to thicker-walled arteries. The tiniest vessels, lymphatic capillaries, accompany the blood capillaries and extend into all parts of the body except the central nervous system and certain organs such as bone, placenta, and thymus. Their thin walls are easily permeated by extracellular fluid that has escaped from the circulatory system. Lymphatic vessels are found in particularly high numbers in the hands, feet, and around the areola of the breast.

In the next section, you will read about the bloodstream and blood vessels. Two overriding differences between the bloodstream and the lymphatic system should be mentioned. First, because one of the main functions of the lymphatic system is returning lymph to the circulation, the flow of lymph is in one direction only, with lymph moving from the extremities toward the heart. Eventually, lymph will be returned to the bloodstream through the thoracic duct or the right lymphatic duct to the subclavian vein near the heart. The second difference concerns how lymph travels through the vessels of the lymphatic system. While blood is transported through the body by means of a dedicated pump (the heart), lymph is moved only through the contraction of the skeletal muscles through which the lymphatic ducts wend their way. This dependence on muscle movement helps to explain the swelling of the hands and feet that sometimes occurs during the night (when muscles are inactive) yet dissipates soon after waking.

The Thymus: Site of T-Cell Maturation   The **thymus** originates in the embryo as two lobes in the lower neck region that fuse into a triangular structure. Under the influence of thymic hormones, thymus cells develop specificity and are released into the circulation as mature T cells. The T cells subsequently migrate to and settle in other lymphoid organs (e.g., the lymph nodes and spleen), where they occupy the specific sites described previously.

Lymph Nodes   Lymph nodes are small, encapsulated, bean-shaped organs stationed, usually in clusters, along lymphatic channels and large blood vessels of the thoracic and abdominal cavities (see figure 12.6). Major aggregations of nodes occur in the loose connective tissue of the armpit (axillary nodes), groin (inguinal nodes), and neck (cervical nodes). Their job is to filter out materials in the lymph and to provide appropriate cells for immune reactions. Enlargement of the lymph nodes can provide physicians with important clues as to a patient's condition. Generalized lymph node enlargement may indicate the presence of a systemic illness, while enlargement of an individual lymph node may be evidence of a localized infection.

The Spleen   The spleen is a lymphoid organ in the upper left portion of the abdominal cavity. It is somewhat similar to a lymph node except that it serves as a filter for blood instead of lymph. While the spleen's primary function is to remove worn-out red blood cells from circulation, its most important immunologic function is the filtering of pathogens from the blood and their subsequent phagocytosis by macrophages in the spleen. Although adults whose spleens have been surgically removed can live a relatively normal life, asplenic children are severely immunocompromised. The spleen also acts as a storehouse of blood that can be released in the event of hemorrhage. It can hold up to 1 cup of blood; for this reason, injury to the spleen can result in profuse bleeding.

Miscellaneous Lymphoid Tissue   At many sites on or just beneath the mucosa of the gastrointestinal and respiratory tracts lie discrete bundles of lymphocytes. The positioning of this diffuse system provides an effective first-strike potential against the constant influx of microbes and other foreign materials in food and air. In the pharynx, a ring of tissues called the tonsils provides an active source of lymphocytes. The breasts of pregnant and lactating women also become temporary sites of antibody-producing lymphoid tissues. The intestinal tract houses the best-developed collection of lymphoid tissue, called **gut-associated lymphoid tissue,** or **GALT.** Examples of GALT include the appendix, the lacteals (special lymphatic vessels stationed in each intestinal villus), and **Peyer's patches,** compact aggregations of lymphocytes in the ileum of the small intestine.

Blood appears red because there are approximately 1,000× more red cells than "white" cells in this fluid.

The Blood

The circulatory system consists of the heart, arteries, veins, and capillaries that circulate the blood, and the lymphatic system, which includes lymphatic vessels and lymphatic organs (lymph nodes) that circulate lymph. As you will see, these two circulations parallel, interconnect with, and complement one another.

The substance that courses through the arteries, veins, and capillaries is **whole blood,** a liquid consisting of **blood cells** (formed elements) suspended in **plasma.** One can visualize these two components with the naked eye when a tube of unclotted blood is allowed to sit or is spun in a centrifuge. The cells' density causes them to settle into an opaque layer at the bottom of the tube, leaving the plasma, a clear, yellowish fluid, on top. Serum is essentially the same as plasma, except it is the clear fluid from clotted blood. Serum is often used in immune testing and therapy.

A Survey of Blood Cells   The production of blood cells is called **hematopoiesis** (hee″-mat-o-poy-ee′-sis). The primary precursor of new blood cells is a pool of undifferentiated cells called pluripotential **stem cells** maintained in the marrow. During development, these stem cells proliferate and differentiate—meaning that immature or unspecialized cells develop the specialized form and function of mature cells. The primary lines of cells that arise from this process produce red blood cells (RBCs, or erythrocytes), white blood cells (WBCs, or leukocytes), and platelets (thrombocytes). **Figure 12.7** outlines the origin and development of the various cells found in blood—most of which are of vital importance to nonspecific and specific defenses.

The white blood cell lines are programmed to develop into several secondary lines of cells during the final process of differentiation. These committed lines of WBCs are largely responsible for immune function.

The **white blood cells** are also called **leukocytes.** One subtype of leukocyte is the **lymphocyte.**

NCLEX® PREP

3. Which of the following act/acts as a filter for blood, removing worn-out red cells from circulation?
   a. *Peyer's patches*
   b. *thymus*
   c. *spleen*
   d. *tonsils*

Medical Moment

The Tonsils

The tonsils, located on either side of the tongue at the back of the throat, and the adenoids, extending down from the back of the nasal cavity, are composed of lymphoid tissue. They are thought to protect from disease by trapping and killing foreign invaders that enter through the nose or mouth.

Once routinely removed, tonsils are nowadays left untouched unless they result in recurrent infection. Symptoms of tonsillitis include an extremely sore throat, difficulty swallowing, ear pain, headache, fever, and changes in the voice. Some children experience recurrent episodes of tonsillitis and experience significant symptoms as a result of chronic, subacute infection of the tonsils. In addition, infection with *Streptococcus* that is treated inadequately may lead to serious complications, such as peritonsillar abscess or kidney disease.

However, removal of the tonsils and/or adenoids is not an innocuous procedure. Hemorrhage can occur and may be severe. In addition, there are always risks associated with general anesthetic. Postsurgical infection may complicate healing. The old adage "if it ain't broke, don't fix it" may apply to the routine removal of tonsils.

12.2 LEARNING OUTCOMES—Assess Your Progress

4. Define *marker,* and discuss its importance in the second and third lines of defense.
5. Name four body compartments that participate in immunity.
6. Connect the mononuclear phagocyte system to the rest of innate immunity.
7. Describe the structure and function of the lymphatic system.
8. Name three kinds of blood cells that function in nonspecific immunity and the most important function of each.
9. Name two kinds of lymphocytes involved in specific immunity.

Figure 12.7 The development of blood cells and platelets.

Undifferentiated stem cells in the red marrow differentiate to give rise to several cell lines that become increasingly specialized until mature cells are released into circulation. Some cells migrate into the tissues to achieve fully functional status. The shaded areas indicate mature leukocytes (white blood cells).

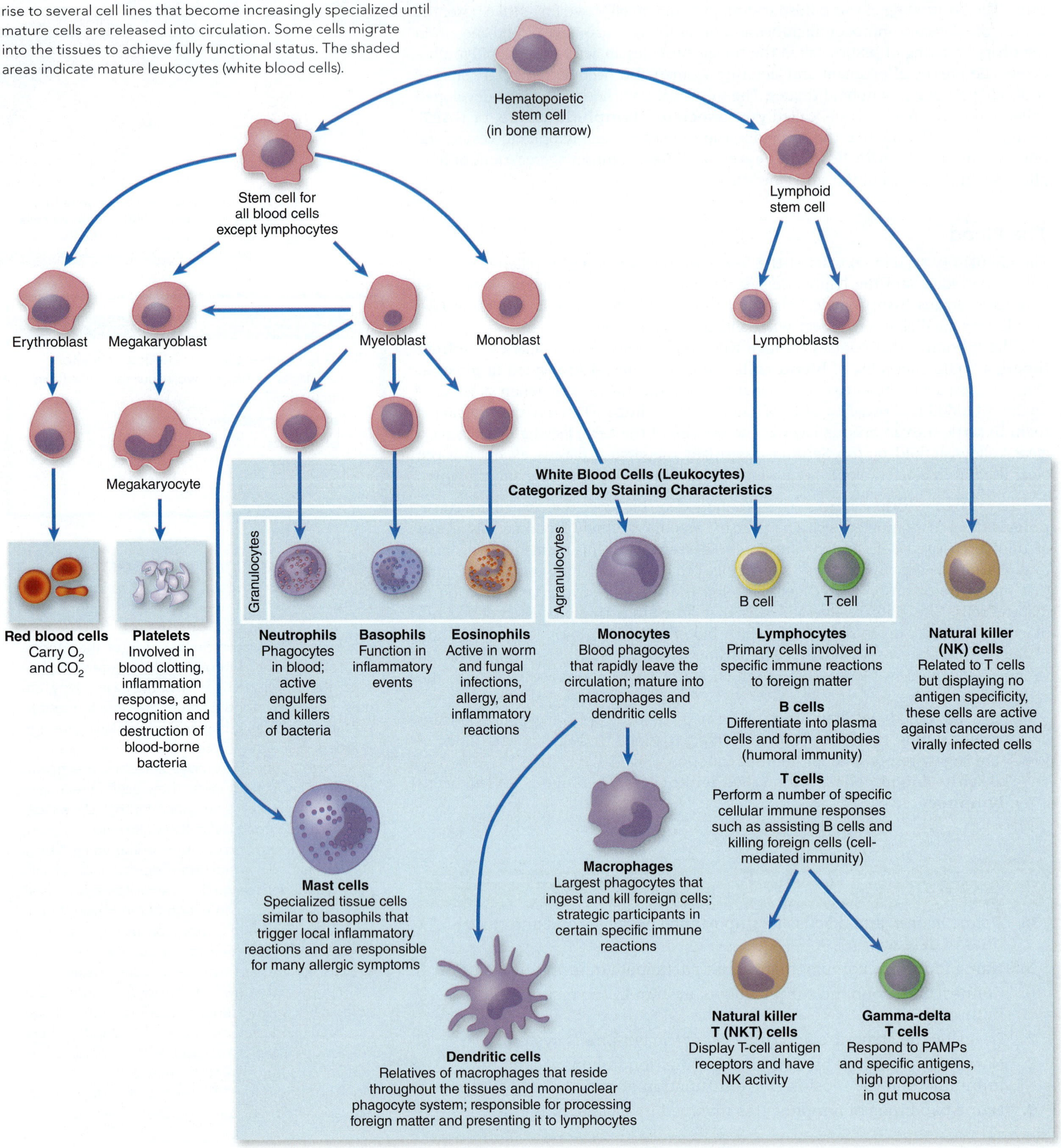

Hematopoietic stem cell (in bone marrow)

Stem cell for all blood cells except lymphocytes

Lymphoid stem cell

Erythroblast

Megakaryoblast

Myeloblast

Monoblast

Lymphoblasts

Megakaryocyte

White Blood Cells (Leukocytes) Categorized by Staining Characteristics

Granulocytes

Agranulocytes

B cell

T cell

Red blood cells
Carry O_2 and CO_2

Platelets
Involved in blood clotting, inflammation response, and recognition and destruction of blood-borne bacteria

Neutrophils
Phagocytes in blood; active engulfers and killers of bacteria

Basophils
Function in inflammatory events

Eosinophils
Active in worm and fungal infections, allergy, and inflammatory reactions

Monocytes
Blood phagocytes that rapidly leave the circulation; mature into macrophages and dendritic cells

Lymphocytes
Primary cells involved in specific immune reactions to foreign matter

B cells
Differentiate into plasma cells and form antibodies (humoral immunity)

T cells
Perform a number of specific cellular immune responses such as assisting B cells and killing foreign cells (cell-mediated immunity)

Natural killer (NK) cells
Related to T cells but displaying no antigen specificity, these cells are active against cancerous and virally infected cells

Mast cells
Specialized tissue cells similar to basophils that trigger local inflammatory reactions and are responsible for many allergic symptoms

Macrophages
Largest phagocytes that ingest and kill foreign cells; strategic participants in certain specific immune reactions

Natural killer T (NKT) cells
Display T-cell antigen receptors and have NK activity

Gamma-delta T cells
Respond to PAMPs and specific antigens, high proportions in gut mucosa

Dendritic cells
Relatives of macrophages that reside throughout the tissues and mononuclear phagocyte system; responsible for processing foreign matter and presenting it to lymphocytes

12.3 The Second Line of Defense

Now that we have introduced the principal anatomical and physiological framework of the immune system, we address some mechanisms that play important roles in host defenses: (1) phagocytosis, (2) inflammation, (3) fever, and (4) antimicrobial proteins. Because of the generalized nature of these defenses, they are primarily nonspecific in their effects, but they also support and interact with the specific immune responses described in chapter 13.

Phagocytosis: Cornerstone of Inflammation and Specific Immunity

By any standard, a phagocyte represents an impressive piece of living machinery, meandering through the tissues to seek, capture, and destroy a target. The general activities of phagocytes are as follows:

1. to survey the tissue compartments and discover microbes, particulate matter (dust, carbon particles, antigen-antibody complexes), and injured or dead cells;
2. to ingest and eliminate these materials; and
3. to extract immunogenic information (antigens) from foreign matter.

It is generally accepted that all cells have some capacity to engulf materials, but professional phagocytes do it for a living. The three main types of phagocytes are neutrophils, monocytes, and macrophages.

Neutrophils

As previously stated, **neutrophils** are general-purpose phagocytes that react early in the inflammatory response to bacteria and other foreign materials and to damaged tissue. A common sign of bacterial infection is a high neutrophil count in the blood (neutrophilia), and neutrophils are also a primary component of pus.

Monocytes and Macrophages: Kings of the Phagocytes

After emigrating out of the bloodstream into the tissues, **monocytes** are transformed by various inflammatory mediators into **macrophages.** This process is marked by an increase in size and by enhanced development of lysosomes and other organelles. Specialized macrophages called **histiocytes** live in a certain tissue and remain there during their life span. Examples are alveolar (lung) macrophages; the Kupffer cells in the liver; dendritic cells in the skin **(figure 12.8);** and macrophages in the spleen, lymph nodes, bone marrow, kidney, bone, and brain. Other macrophages do not reside permanently in a particular tissue and drift nomadically throughout the MPS. Not only are macrophages dynamic scavengers, but they also process foreign substances and prepare them for reactions with B and T lymphocytes.

Mechanisms of Phagocytic Recognition, Engulfment, and Killing

The term **phagocyte** literally means "eating cell." But phagocytosis (the name for what phagocytes do) is more than just the physical process of engulfment, because phagocytes also actively attack and dismantle foreign cells with a wide array of antimicrobial substances. Phagocytosis can occur as an isolated event performed by a lone phagocytic cell responding to a minor irritant in its area or as part of the orchestrated events of inflammation described in the next section. The events in phagocytosis include chemotaxis, ingestion, phagolysosome formation, destruction, and excretion. **Table 12.1** provides the details of phagocytosis.

Phagocytes and other defensive cells are able to recognize some microorganisms as foreign because of signal molecules that the microbes have on their surfaces. These pathogen-associated molecular patterns (PAMPs) are molecules shared by many microorganisms–but not present in mammals–and therefore serve as "red flags" for phagocytes and other cells of innate immunity.

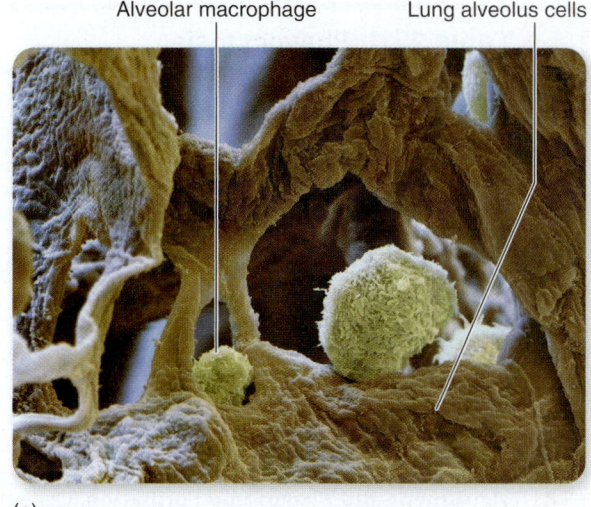
(a)

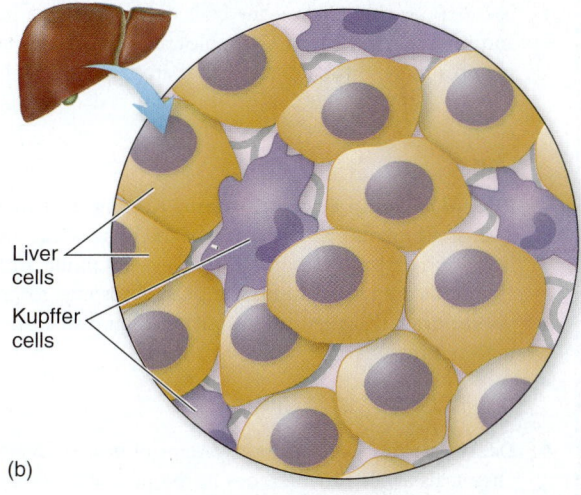
(b)

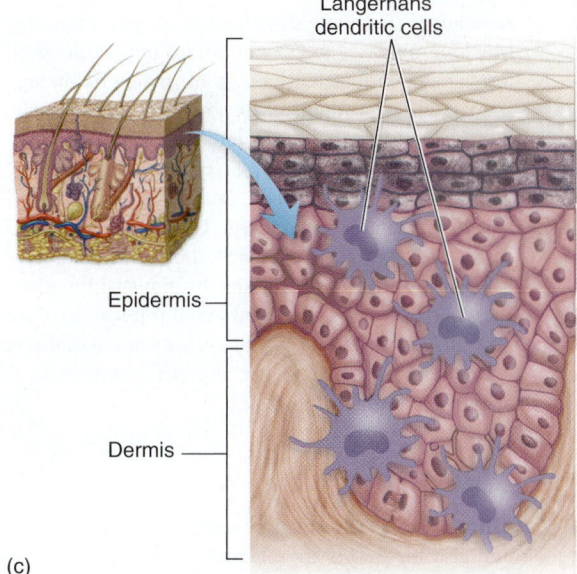
(c)

Figure 12.8 Macrophages. (a) Scanning electron micrograph of an alveolar macrophage inside a lung. (b) Liver tissue with Kupffer cells. (c) Langerhans cells deep in the epidermis.

Table 12.1 Phagocytosis

1 **Chemotaxis.** Phagocytes migrate into a region of inflammation with a deliberate sense of direction, attracted by a gradient of stimulant products from the parasite and host tissue at the site of injury.

2 **Adhesion.** Phagocytes use pattern recognition to identify and stick to foreign cells.

3 **Engulfment and Phagosome Formation.** Once the
4 phagocyte has made contact with its prey, it extends pseudopods that enclose the cells or particles in a pocket and internalize them in a vacuole called a phagosome. It also secretes more cytokines to further amplify the innate response.

5 **Phagolysosome Formation and Killing.** In a short time, lysosomes migrate to the scene of the phagosome and fuse with it to form a phagolysosome. Granules containing antimicrobial chemicals are released into the phagolysosome, forming a potent brew designed to poison and then dismantle the ingested material.

6 **Destruction.** Two separate systems of destructive chemicals await the microbes in the phagolysosome. The oxygen-dependent system (known as the respiratory burst, or oxidative burst) involves several substances that were described in chapters 6 and 9. Myeloperoxidase, an enzyme found in granulocytes, forms halogen ions (OCl^-) that are strong oxidizing agents. Other products of oxygen metabolism such as hydrogen peroxide, the superoxide anion (O_2^-), activated or so-called singlet oxygen ($^\bullet O_2$), and the hydroxyl free radical ($OH^\bullet$) separately and together have formidable killing power. Other mechanisms that come into play are the liberation of lactic acid, lysozyme, and nitric oxide (NO), a powerful mediator that kills bacteria and inhibits viral replication. Cationic proteins that injure bacterial cytoplasmic membranes and a number of proteolytic and other hydrolytic enzymes complete the job.

7 **Elimination.** The small bits of undigestible debris are released from the macrophage by exocytosis.

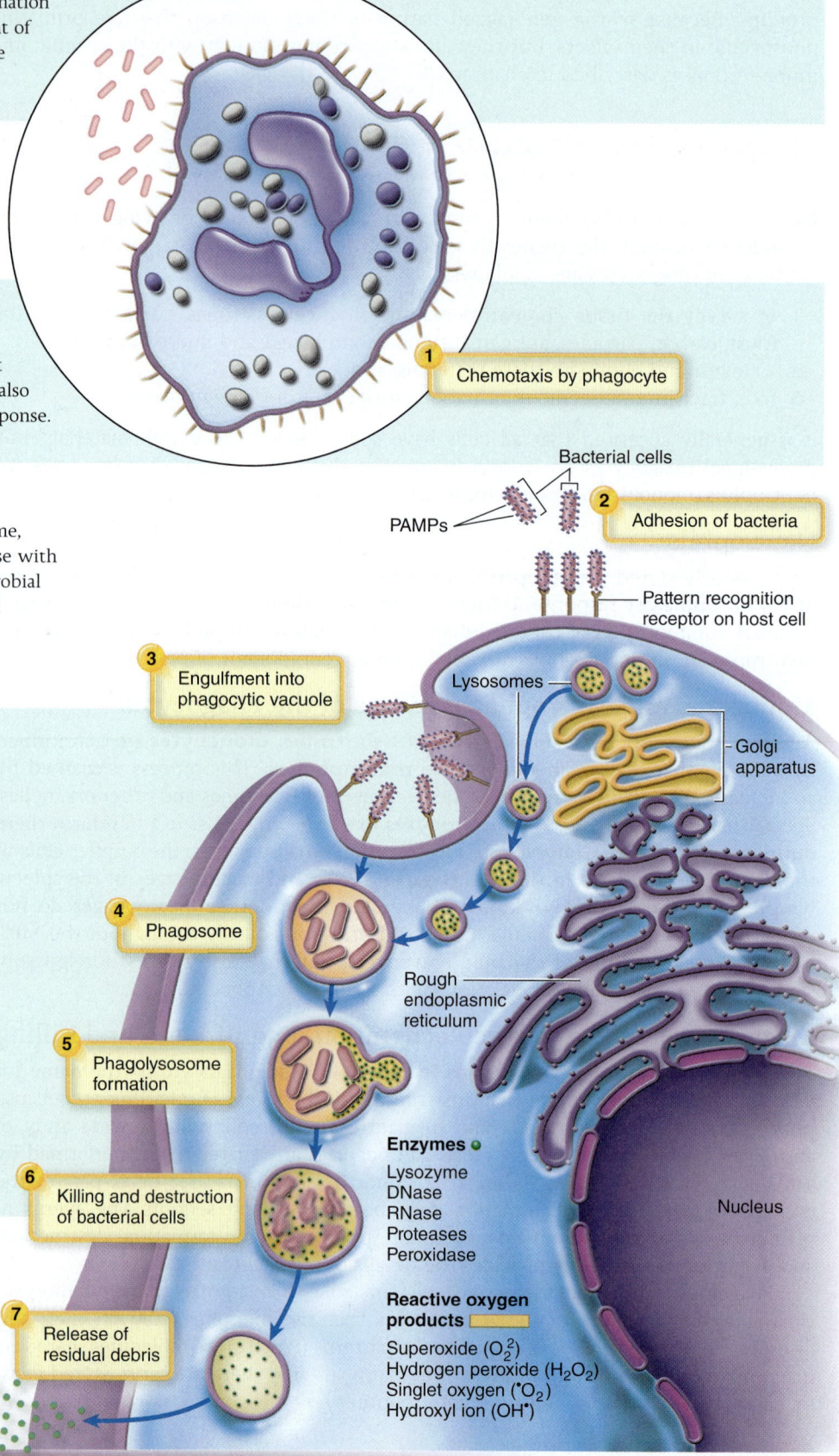

1 Chemotaxis by phagocyte

Bacterial cells

PAMPs

2 Adhesion of bacteria

Pattern recognition receptor on host cell

3 Engulfment into phagocytic vacuole

Lysosomes

Golgi apparatus

4 Phagosome

Rough endoplasmic reticulum

5 Phagolysosome formation

Enzymes
Lysozyme
DNase
RNase
Proteases
Peroxidase

6 Killing and destruction of bacterial cells

Nucleus

Reactive oxygen products
Superoxide (O_2^2)
Hydrogen peroxide (H_2O_2)
Singlet oxygen ($^\bullet O_2$)
Hydroxyl ion ($OH^\bullet$)

7 Release of residual debris

Bacterial PAMPs include peptidoglycan and lipopolysaccharide. Double-stranded RNA, which is found only in some viruses, is also a PAMP. On the host side, phagocytes, dendritic cells, endothelial cells, and even lymphocytes possess pattern recognition receptors (PRRs) on their surfaces that recognize and bind PAMPs. The cells possess these PRRs all the time, whether or not they have encountered PAMPs before. Many phagocytic cells of the innate immune system contain PRRs inside their cytoplasm. These have a special name—**inflammasomes**—and they recognize microbial PAMPs as well as markers from damaged host cells once they have been phago-cytosed. Recognition leads to the release of signals that initiate and regulate inflammation. (There are a lot of acronyms in immunology. Don't let them get away from you; keep up with them. If you know what all the acronyms stand for and what they do, you are halfway there in understanding host defenses!)

Figure 12.9 is an electron micrograph of a phagocyte engulfing bacteria.

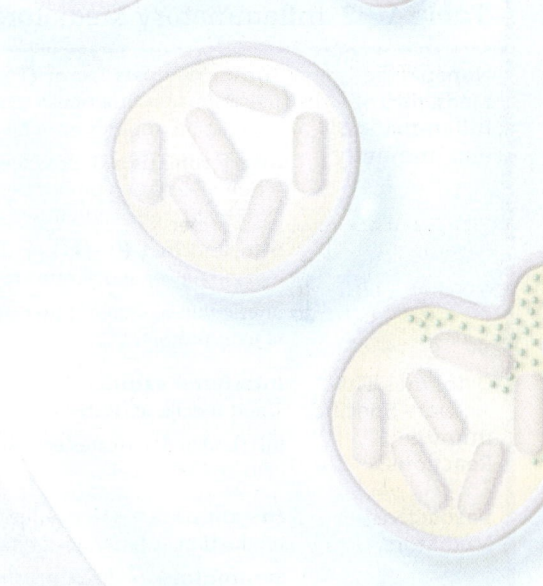

Figure 12.9 Scanning electron micrograph of a macrophage. This macrophage is devouring bacteria (10,000×).

The Inflammatory Response: A Complex Concert of Reactions to Injury

At its most general level, the inflammatory response is a reaction to any traumatic event in the tissues. It is so commonplace that all of us manifest inflammation in some way every day. It appears in the nasty flare of a cat scratch, the blistering of a burn, the painful lesion of an infection, and the symptoms of allergy. When close to our external surfaces, it is readily identifiable by a classic series of signs and symptoms characterized succinctly by four Latin terms: *rubor, calor, tumor,* and *dolor*. Rubor (redness) is caused by increased circulation and vasodilation in the injured tissues; calor (warmth) is the heat given off by the increased flow of blood; tumor (swelling) is caused by increased fluid escaping into the tissues; and dolor (pain) is caused by the stimulation of nerve endings. A fifth symptom, loss of function, has been added to give a complete picture of the effects of inflammation.

It is becoming increasingly clear that some chronic diseases, such as cardiovascular disease, can be caused by chronic inflammation. While we speak of inflammation at a local site (such as a finger), inflammation can affect an entire system—such as blood vessels, lungs, skin, the joints, and so on. Some researchers believe that the phenomenon of aging is a consequence of increasing inflammation in multiple body systems.

Factors that can elicit inflammation include trauma from infection (the primary emphasis here), tissue injury or necrosis due to physical or chemical agents, and specific immune reactions. Although the details of inflammation are very complex, its chief functions can be summarized as follows:

1. to mobilize and attract immune components to the site of the injury,
2. to set in motion mechanisms to repair tissue damage and localize and clear away harmful substances, and
3. to destroy microbes and block their further invasion.

The inflammatory response is a powerful defensive reaction, a means for the body to maintain stability and restore itself after an injury. But it has the potential to actually *cause* tissue injury, destruction, and disease.

The Role of Inflammatory Mediators

Before we dive into the details of the inflammatory process, we need to understand the role of some potent chemicals that influence inflammatory events, as well as much of the defensive response.

Hundreds of small, active molecules are constantly being secreted to regulate, stimulate, suppress, and otherwise control the many aspects of cell development, inflammation, and immunity. These substances are the products of several types

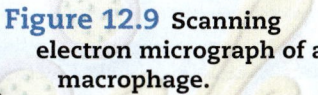

NCLEX® PREP

4. In order to assess immune function, the nurse anticipates that the physician will *first* order a
   a. *white blood cell (WBC) count with differential.*
   b. *red blood cell (RBC) count with differential.*
   c. *chemistry profile.*
   d. *coagulation profile.*

of cells, including monocytes, macrophages, lymphocytes, fibroblasts, mast cells, platelets, and endothelial cells of blood vessels. Their effects may be local or systemic, short term or long lasting, nonspecific or specific, protective or pathologic.

These molecules are generally termed **cytokines.** The major functional types can be categorized into the following:

1. cytokines that mediate nonspecific immune reactions such as inflammation and phagocytosis,
2. cytokines that activate immune reactions during inflammation,
3. vasoactive mediators,
4. cytokines that regulate the growth and activation of lymphocytes,
5. hematopoiesis factors for white blood cells, and
6. miscellaneous inflammatory mediators.

Table 12.2 provides important examples of many of these mediators.

The Stages of Inflammation

The process leading to inflammation is a dynamic, predictable sequence of events that can be acute, lasting from a few minutes or hours, or chronic, lasting for days, weeks, or years. Once the initial injury has occurred, a chain reaction takes place at the site of damaged tissue, summoning beneficial cells and fluids into the injured area. As an example, we will look at an injury at the microscopic level and observe the flow of major events in **table 12.3.**

NCLEX® PREP

5. During phagocytosis, which action is associated with destruction?
   a. *exocytosis*
   b. *adhesion*
   c. *oxidative burst*
   d. *chemotaxis*

Itchiness in animals and humans can be caused by the release of cytokines.

Table 12.2 Inflammatory Mediators and Other Cytokines

Nonspecific Mediators of Inflammation and Immunity	**Tumor necrosis factor (TNF),** a substance from macrophages, lymphocytes, and other cells that increases chemotaxis and phagocytosis and stimulates other cells to secrete inflammatory cytokines. It also serves as an endogenous pyrogen that induces fever, increases blood coagulation, suppresses bone marrow, and suppresses appetite.
	Interferons (IFNs), produced by leukocytes, fibroblasts, and other cells, inhibit virus replication and cell division and increase the action of certain lymphocytes that kill other cells. Interferons also reduce the amount of cholesterol in the body. Because cholesterol is used by bacteria and viruses as a nutrient, this provides more innate protection.
	Interleukin-1 (IL-1), a product of macrophages and dendritic cells that has many of the same biological activities as TNF, such as inducing fever and activation of certain white blood cells.
	Interleukin-6, secreted by macrophages and T cells. Its primary effects are to stimulate the growth of B cells and to increase the synthesis of liver proteins.
Cytokines That Activate Specific Immune Reactions	**Interferon gamma,** a T-cell-derived mediator whose primary function is to activate macrophages. It also promotes the differentiation of T and B cells, activates neutrophils, and stimulates diapedesis.
	Interleukin-5 activates eosinophils and B cells; interleukin-10 inhibits macrophages and stimulates B cells; and interleukin-12 activates T cells and killer cells.
Vasoactive Mediators	**Histamine,** a vasoactive mediator produced by mast cells and basophils, causes vasodilation, increased vascular permeability, and mucus production. It functions primarily in inflammation and allergy.
	Serotonin, a mediator produced by platelets and intestinal cells, causes smooth muscle contraction, inhibits gastric secretion, and acts as a neurotransmitter.
	Bradykinin, a vasoactive amine from the blood or tissues, stimulates smooth muscle contraction and increases vascular permeability, mucus production, and pain. It is particularly active in allergic reactions.
Cytokines That Regulate Lymphocyte Growth and Activation	Interleukin-2, the primary growth factor from T cells. Interestingly, it acts on the same cells that secrete it. It stimulates mitosis and secretion of other cytokines. In B cells, it is a growth factor and stimulus for antibody synthesis.
	Macrophage colony-stimulating factor (M-CSF), produced by a variety of cells. M-CSF promotes the growth and development of macrophages from undifferentiated precursor cells.
Miscellaneous Inflammatory Mediators	**Prostaglandins,** produced by most body cells; complex chemical mediators that can have opposing effects (e.g., dilation or constriction of blood vessels) and are powerful stimulants of inflammation and pain.
	Leukotrienes stimulate the contraction of smooth muscle and enhance vascular permeability. They are implicated in the more severe manifestations of immediate allergies (constriction of airways).
	Platelet-activating factor, a substance released from basophils, causes the aggregation of platelets and the release of other chemical mediators during immediate allergic reactions.

Table 12.3 Inflammation

① Injury/Immediate Reactions

Following an injury, early changes occur in the vasculature (arterioles, capillaries, venules) in the vicinity of the damaged tissue. These changes are controlled by nervous stimulation, **chemical mediators,** and **cytokines** released by blood cells, tissue cells, and platelets in the injured area. Some mediators are **vasoactive**—that is, they affect the endothelial cells and smooth muscle cells of blood vessels, and others are **chemotactic factors,** also called **chemokines,** that affect white blood cells.

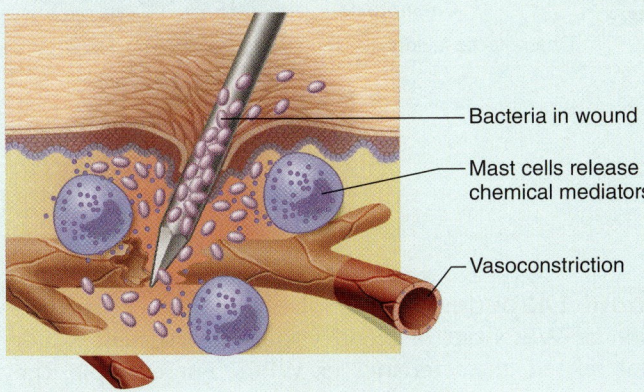

Bacteria in wound

Mast cells release chemical mediators

Vasoconstriction

② Vascular Reactions

Although the constriction of arterioles is stimulated first, it lasts for only a few seconds or minutes and is followed in quick succession by the opposite reaction, vasodilation. The overall effect of vasodilation is to increase the flow of blood into the area, which facilitates the influx of immune components and also causes redness and warmth.

Some vasoactive substances cause the endothelial cells in the walls of postcapillary venules to contract and form gaps through which blood-borne components exude into the extracellular spaces. The fluid part that escapes is called the **exudate.**

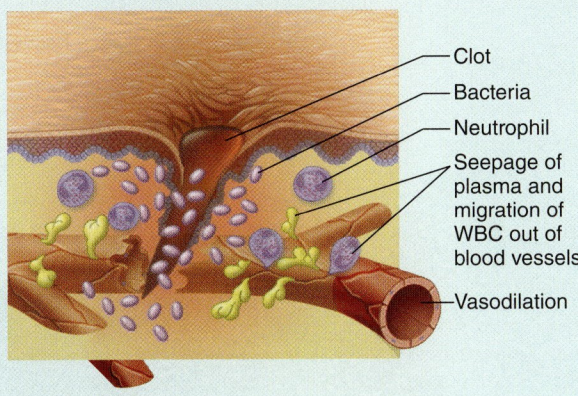

Clot

Bacteria

Neutrophil

Seepage of plasma and migration of WBC out of blood vessels

Vasodilation

③ Edema and Pus Formation

Accumulation of this fluid in the tissues gives rise to local swelling and hardness called **edema.** The fluid contains varying amounts of plasma proteins, such as globulins, albumin, the clotting protein fibrinogen, blood cells, and cellular debris. Depending on its content, the fluid may be clear (called **serous**), or it may contain red blood cells or pus. Pus is composed mainly of white blood cells and the debris generated by phagocytosis. In some types of edema, the fibrinogen is converted to fibrin threads that enmesh the injury site. Within an hour, multitudes of neutrophils responding chemotactically to special signaling molecules converge on the injured site.

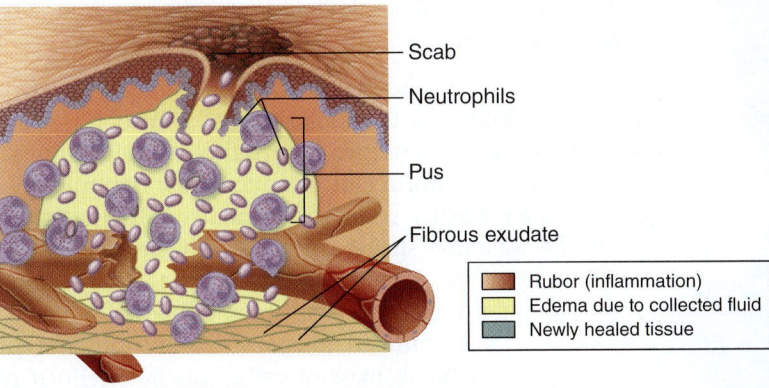

Scab

Neutrophils

Pus

Fibrous exudate

	Rubor (inflammation)
	Edema due to collected fluid
	Newly healed tissue

④ Resolution/Scar Formation

Repair is the last step and results either in complete resolution to healthy tissue, or in formation of scar tissue, depending on the tissue type and the extent of the damage. Note here that macrophages are pictured leaving the blood vessels in a process called **diapedesis** (dye″-ah-puh-dee′-sis).

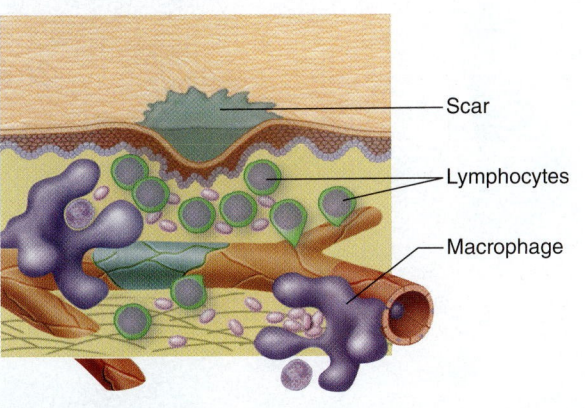

Scar

Lymphocytes

Macrophage

Figure 12.10 Diapedesis and chemotaxis of leukocytes. (a) View of a venule depicts white blood cells squeezing themselves between spaces in the blood vessel wall via diapedesis. This process, shown in cross section, indicates how the pool of leukocytes adheres to the endothelial wall. From this site, they are poised to migrate out of the vessel into the tissue space. (b) This micrograph captures neutrophils in the process of diapedesis.

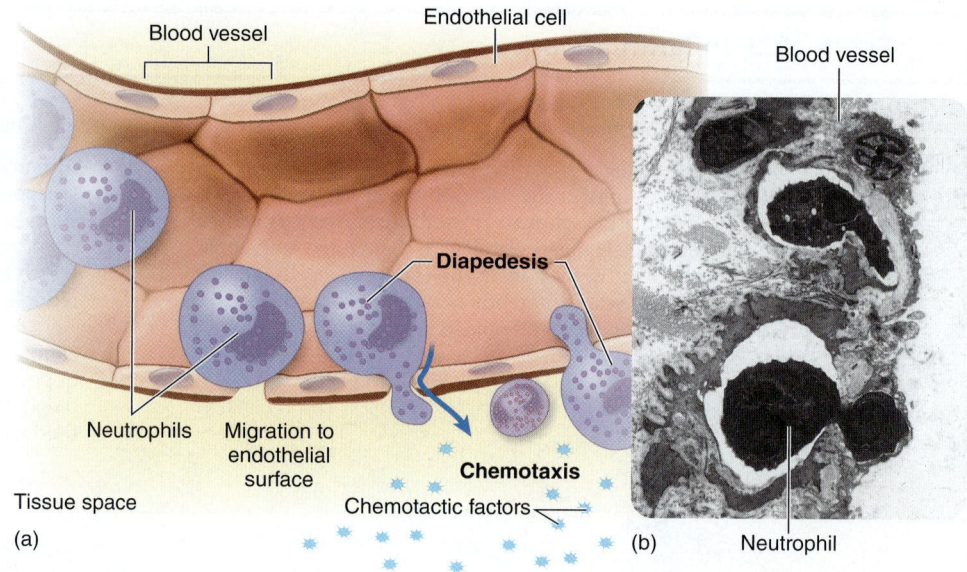

The pain of a sore throat is caused by (1) a combination of damage inflicted by microbes; (2) the inflammatory response that can lead to pressure on surrounding nerves; and (3) the release of pain-inducing cytokines.

More About Diapedesis and Chemotaxis Diapedesis, the migration of WBCs out of blood vessels into tissues, is aided by several related characteristics of WBCs. For example, they are actively motile and readily change shape. This phenomenon is also assisted by the nature of the endothelial cells lining the venules. They contain complex adhesive receptors that capture the WBCs and participate in their transport from the venules into the extracellular spaces **(figure 12.10).**

Another factor in the migratory habits of these WBCs is **chemotaxis,** or the tendency of cells to migrate in response to a specific chemical stimulus given off at a site of injury or infection. Through this means, cells swarm from many compartments to the site of infection and remain there to perform general and specific immune functions. These basic properties are absolutely essential for the sort of intercommunication and deployment of cells required for most immune reactions.

The Benefits of Edema and Leaky Vessels Both the secretion of fluids and the infiltration of neutrophils are physiologically beneficial activities. The influx of fluid dilutes toxic substances, and the fibrin clot can effectively trap microbes and prevent their further spread. The neutrophils that aggregate in the inflamed site are immediately involved in phagocytosing and destroying bacteria, dead tissues, and particulate matter. In some types of inflammation, accumulated phagocytes contribute to **pus,** a whitish mass of cells, liquefied cellular debris, and bacteria. Certain bacteria (streptococci, staphylococci, gonococci, and meningococci) are especially powerful attractants for neutrophils and are thus termed **pyogenic,** or pusforming, bacteria.

Fever: An Adjunct to Inflammation

An important systemic component of inflammation—and innate immunity in general—is fever, defined as an abnormally elevated body temperature. Although fever is very common in infection, it is also associated with certain allergies, cancers, and other organic illnesses. Fevers whose causes are unknown are called fevers of unknown origin, or FUO.

The body temperature is normally maintained by a control center in the hypothalamus region of the brain. This thermostat regulates the body's heat production and heat loss and sets the core temperature at around 37°C (98.6°F) with slight fluctuations during a daily cycle. Fever is initiated when circulating substances called **pyrogens** (py'-rohjenz) reset the hypothalamic thermostat to a higher setting. This change signals the musculature to increase heat production and peripheral arterioles to decrease heat loss through vasoconstriction. Fevers range in severity from low grade (37.7°C to 38.3°C, or 100°F to 101°F) to high (40.0°C to 41.1°C, or 104°F to 106°F). Pyrogens are described as **exogenous** (coming from outside the body) or **endogenous** (originating internally). Exogenous pyrogens are products of infectious agents such as viruses, bacteria, protozoa, and fungi. One well-characterized exogenous pyrogen is endotoxin, the lipopolysaccharide found in the cell walls of gram-negative bacteria. Blood, blood products, vaccines, or injectable solutions can also contain exogenous pyrogens. Endogenous pyrogens are liberated by monocytes, neutrophils, and macrophages during the process of phagocytosis and appear to be a natural part of the immune response. Two potent pyrogens released by macrophages are the cytokines interleukin-1 (IL-1) and tumor necrosis factor (TNF).

For thousands of years, people believed fever was part of an innate protective response. Hippocrates offered the idea that it was the body's attempt to burn off a noxious agent. Sir Thomas Sydenham wrote in the seventh century: "Why, fever itself is Nature's instrument!" So widely held was the view that fever could be therapeutic that pyretotherapy (treating disease by inducing an intermittent fever) was once used to treat syphilis, gonorrhea, leishmaniasis (a protozoal infection), and cancer. This attitude fell out of favor when drugs for relieving fever (aspirin) first came into use in the early 1900s, and an adverse view of fever began to dominate.

Benefits of Fever

Aside from its practical and medical importance as a sign of disease, increased body temperature has additional benefits:

- Fever inhibits multiplication of temperature-sensitive microorganisms such as the poliovirus, cold viruses, herpes zoster virus, and systemic and subcutaneous fungal pathogens.
- Fever impedes the nutrition of bacteria by reducing the availability of iron.
- Fever increases metabolism and stimulates immune reactions and naturally protective physiological processes. It speeds up hematopoiesis, phagocytosis, and specific immune reactions and helps specific lymphocytes home in on sites of infection.

Treatment of Fever

With this revised perspective on fever, whether to suppress it or not can be a difficult decision. Some advocates feel that a slight to moderate fever in an otherwise healthy person should be allowed to run its course, in light of its potential benefits and minimal side effects. Side effects of fever include tachycardia (rapid heart rate), tachypnea (elevated respiratory rate), and, in some individuals, a lowering of seizure threshold; therefore, all medical experts do agree that high and prolonged fevers or fevers in patients with cardiovascular disease, head trauma, seizures, or respiratory ailments are risky and must be treated immediately with fever-reducing drugs.

Antimicrobial Proteins (1): Interferons

Interferons (IFNs) are small proteins produced naturally by certain white blood and tissue cells. Although the interferon system was originally thought to be directed exclusively against viruses, it is now known to be involved also in defenses against other microbes and in immune regulation and intercommunication. Three major types are interferons alpha and beta, products of many cells, including lymphocytes, fibroblasts, and macrophages; and **interferon gamma,** a product of T cells.

Their biological activities are extensive. In all cases, interferons bind to cell surfaces and induce changes in genetic expression, but the exact results vary. In addition to antiviral effects discussed in the next section, all three IFNs can inhibit the expression of cancer genes and have tumor suppressor effects. IFN alpha and beta stimulate phagocytes, and IFN gamma is an immune regulator of macrophages and T and B cells.

Characteristics of Interferons

When viruses, and sometimes other microbes or their component parts, bind to the receptors on a host cell, a signal is sent to the nucleus that directs the cell to synthesize interferons. After transcribing and translating the interferon genes, newly synthesized interferon molecules are rapidly secreted by the cell into the extracellular space, where they bind to other host cells. The binding of interferons to a second cell induces the production of proteins in that cell that inhibit viral multiplication through a variety of mechanisms, such as by degrading the viral RNA or by preventing the translation of viral proteins **(figure 12.11).** Interferons can also induce the production of other proteins that combat infection, such as an enzyme that produces reactive oxygen chemicals that damage pathogens. Interferons are not

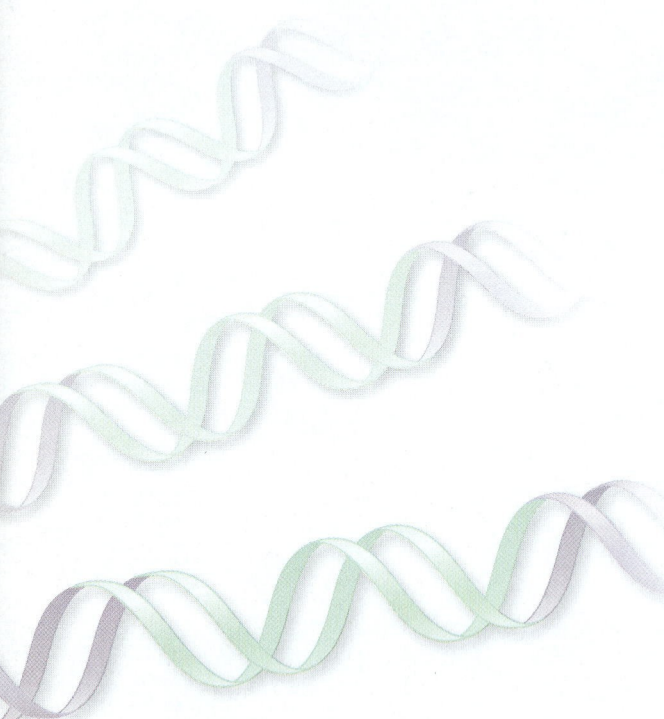

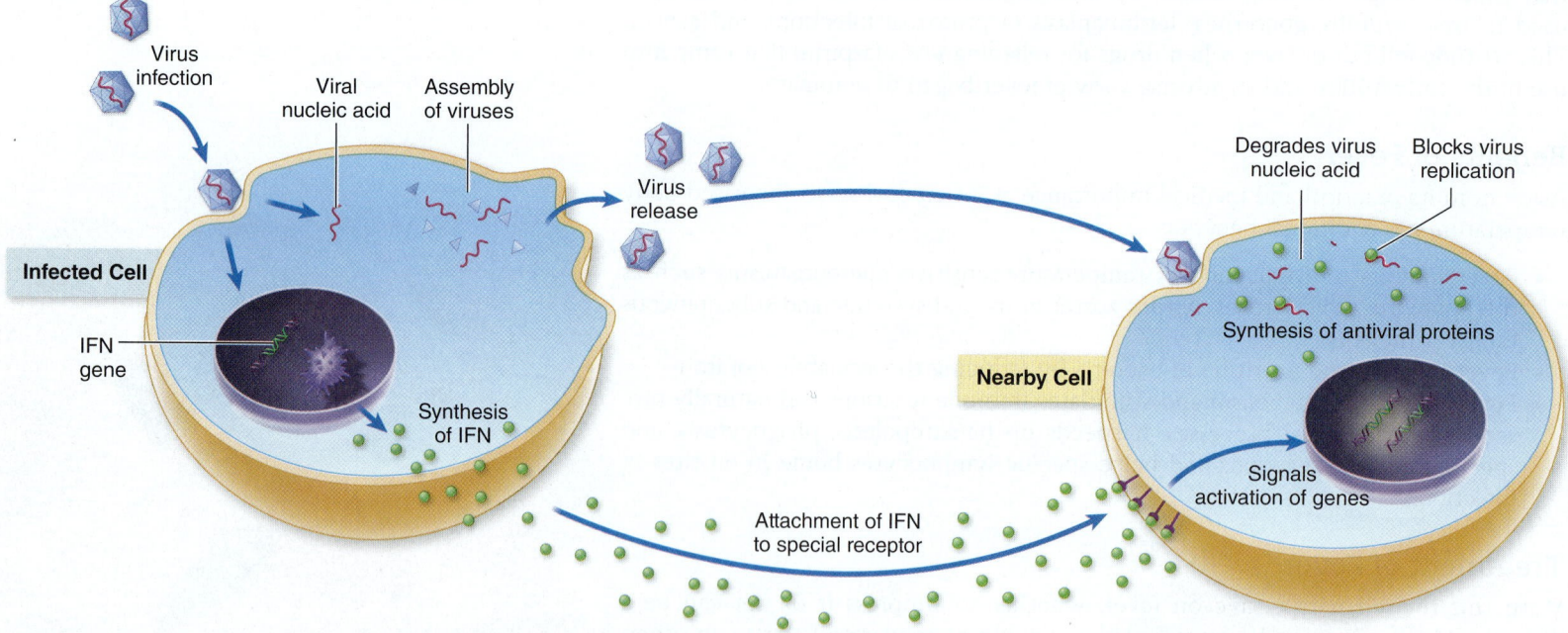

Figure 12.11 The antiviral activity of interferons. When a cell is infected by a virus (and sometimes other microbes), its nucleus is triggered to transcribe and translate the interferon (IFN) genes. Interferons diffuse out of the cell and bind to IFN receptors on nearby uninfected cells, where they induce production of proteins that eliminate genes from foreign organisms. Note that the original cell is not protected by IFNs and that IFNs do not prevent viruses (or other organisms) from invading the protected cells.

microbe-specific, so their synthesis in response to one type of microbe will also protect against other types. Because these proteins are inhibitors of viruses, they have been a valuable treatment for a number of virus infections.

Other Roles of Interferons

Interferons are also important immune regulatory cytokines that activate or instruct the development of white blood cells. See table 12.2 for more information about interferons.

Antimicrobial Proteins (2): Complement

The immune system has another complex and multiple-duty system called **complement** that, like inflammation and phagocytosis, is brought into play at several levels. The complement system, named for its property of "complementing" immune reactions, consists of over 30 different blood proteins that work in concert to destroy bacteria and certain viruses.

The concept of a cascade reaction is helpful in understanding how complement functions. A cascade reaction is a sequential physiological response like that of blood clotting, in which the first substance in a chemical series activates the next substance, which activates the next, and so on, until a desired end product is reached. There are three different complement pathways, distinguished by how they become activated. The final stages of the three pathways are the same and yield a similar end result. For our discussion, we will focus on what is known as the alternative pathway and point out how the others differ afterward.

Overall Stages in the Complement Cascade

In general, the complement cascade includes the four stages of initiation, amplification and cascade, polymerization, and membrane attack. At the outset, an initiator (in this case, the surface of a bacterium or a virus) reacts with the first complement protein, C3, which propels the reaction on its course. There is a recognition site on the surface of the target cell where the initial C components will bind. Through a stepwise series, each component reacts with another on or near the recognition site. The functioning end product of all complement pathways is a large ring-shaped protein termed the membrane attack complex. This complex can digest holes in the cytoplasmic membranes of bacteria, cells, and enveloped viruses, thereby destroying them **(figure 12.12).**

The alternative pathway pictured here is truly a nonspecific defense mechanism because it is activated by the simple presence of pathogen membranes. It can act quickly. The classical complement pathway receives help from the specific immune response covered in chapter 13. It is initiated by host antibody (produced by specific immunity) bound to a pathogen. In that case, other complement proteins begin the complement cascade when they bind to the antibodies. Several reactions take place, leading into the C3 part of the pathway seen in figure 12.12. Because it depends on antibody, the classical pathway takes longer to get started after a new infection. The lectin-mediated pathway is similar to the alternative pathway except that mannose-binding proteins (lectins) must bind to mannose residues on the surface of the pathogen in order for the pathway to proceed.

Overall inflammation will be amplified by the action of complement. In recent years, the excessive actions of complement have been implicated as aggravators of several autoimmune diseases, such as lupus, rheumatoid arthritis, and myasthenia gravis.

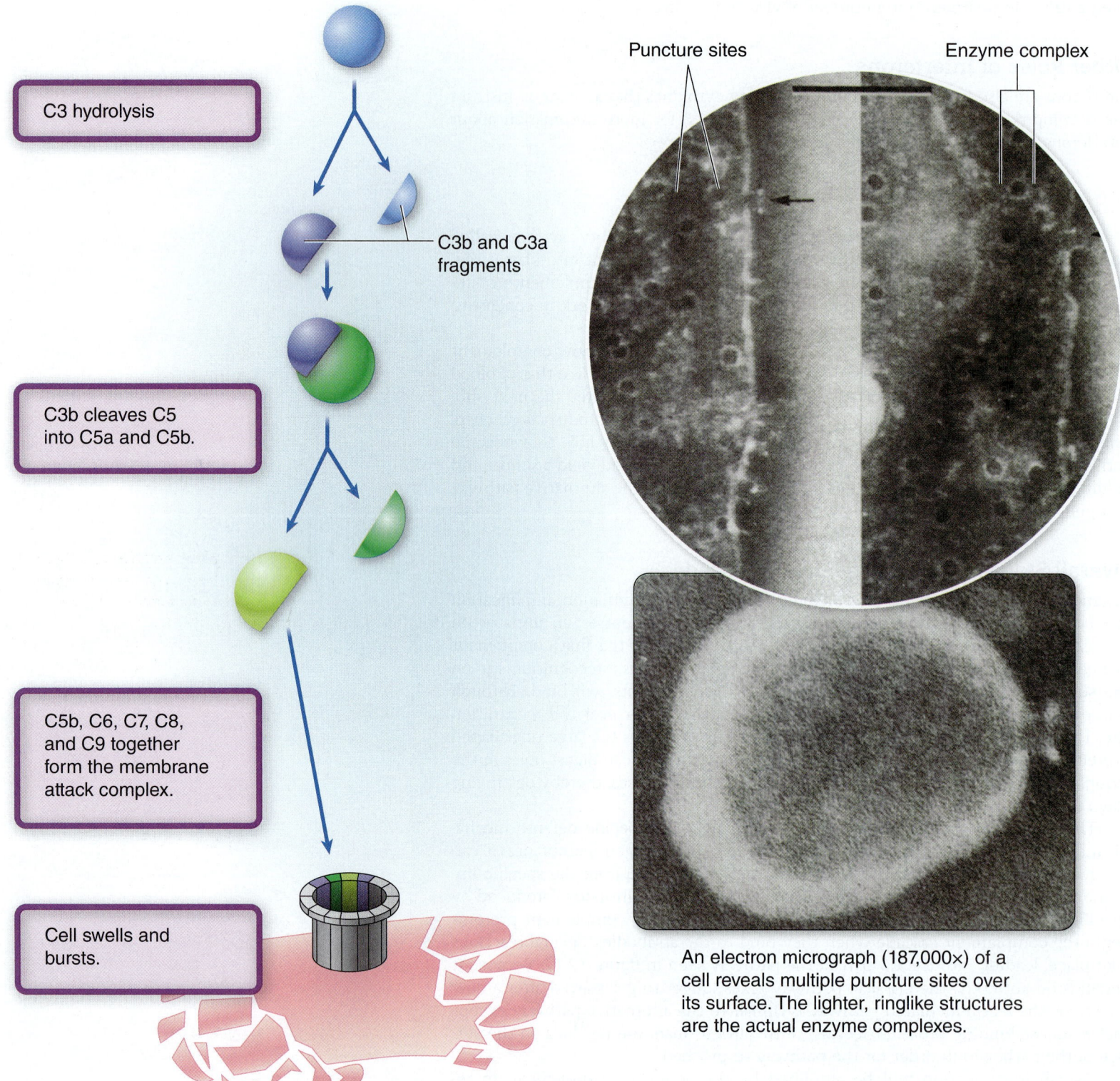

C3 hydrolysis

C3b and C3a fragments

C3b cleaves C5 into C5a and C5b.

C5b, C6, C7, C8, and C9 together form the membrane attack complex.

Cell swells and bursts.

Puncture sites

Enzyme complex

An electron micrograph (187,000×) of a cell reveals multiple puncture sites over its surface. The lighter, ringlike structures are the actual enzyme complexes.

Figure 12.12 Steps in the alternative complement pathway at a single site. All complement pathways function in a similar way, but the details differ. The alternative pathway is illustrated here.

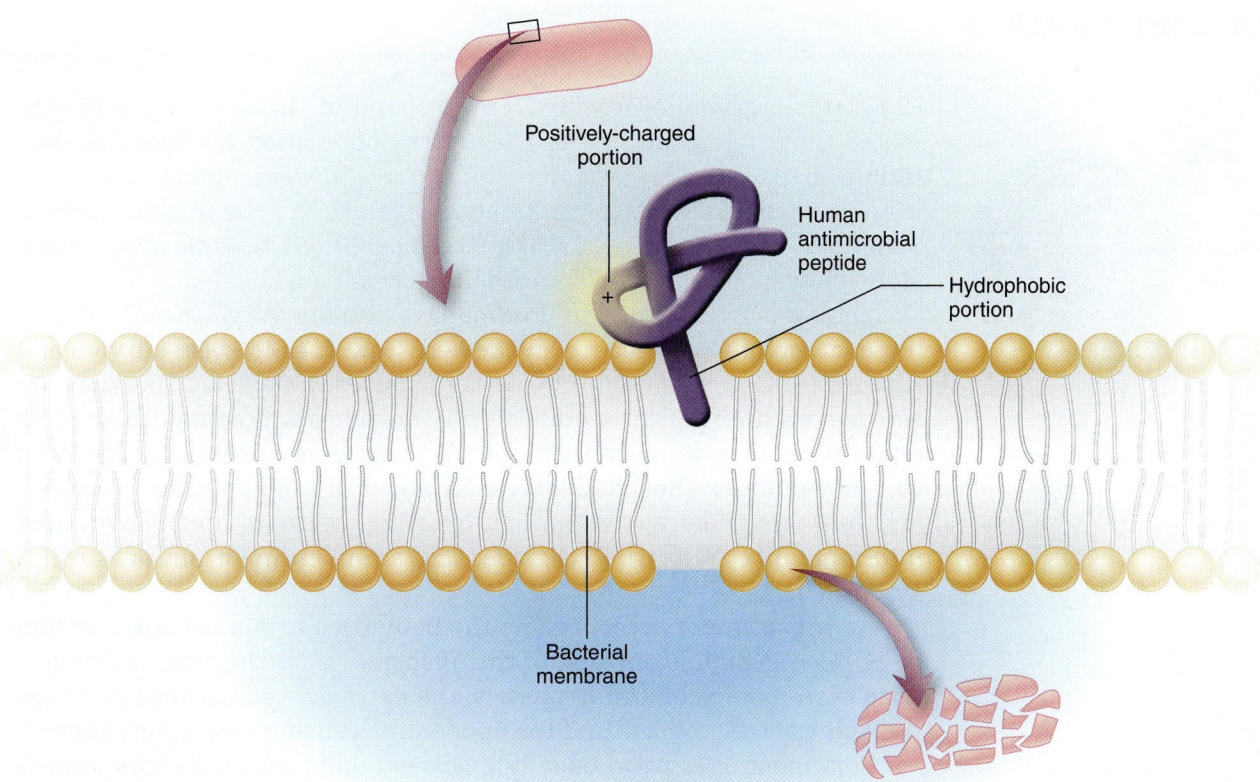

Figure 12.13 Antimicrobial peptides. These peptides have various mechanisms, but a very common one is to insert into pathogen membranes using a positive charge plus a hydrophobic tail.

Antimicrobial Proteins (3): Antimicrobial Peptides

Antimicrobial peptides are short proteins, of between 12 and 50 amino acids, that have the capability of inserting themselves into bacterial membranes **(figure 12.13)**. Through this mechanism and others, they kill the microbes. They have names like *defensin, magainins,* and *protegrins.* They are part of the innate immune system and also have an effect on other actions of nonspecific and specific immunity. Many researchers are trying to turn these antimicrobial peptides into practical use as therapeutic drugs. Their ability to modulate immune responses would distinguish them from other antibiotics on the market and may represent a new weapon in the war against microbial drug resistance. Although their clinical use has not fully developed, researchers are discovering ways to utilize computer programs to design the most effective antimicrobial peptides in the laboratory.

12.3 LEARNING OUTCOMES—Assess Your Progress

10. List the four major categories of nonspecific immunity.
11. Outline the steps in phagocytosis.
12. Outline the steps in inflammation.
13. Discuss the mechanism of fever and how it helps defend the body.
14. Name three types of antimicrobial proteins.
15. Compose one good overview sentence about the purpose and the mode of action of the complement system.

CASE FILE WRAP-UP

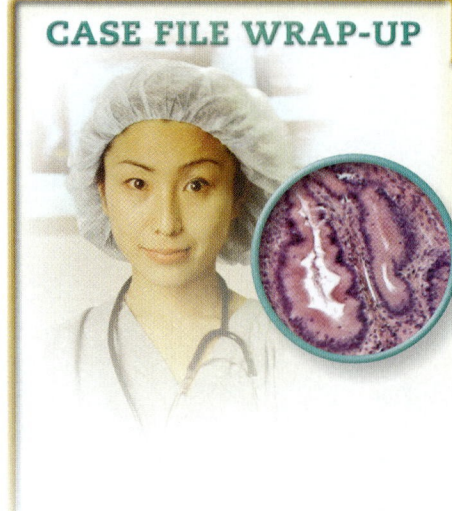

H. *pylori* is a gram-negative bacterium found in the stomach. It is estimated that more than half of the world's population harbors this bacterium in their upper gastrointestinal tract. However, most people are asymptomatic and may never know that they carry the microorganism. Symptoms occur in about 20% of infected people. Symptoms may include stomach pain, bloating, belching, and nausea.

The stomach's acidic environment is usually very efficient at killing microorganisms that are swallowed. How does H. *pylori* survive and thrive in this harsh environment? H. *pylori* uses its flagella to burrow into the mucous lining of the stomach until it reaches a more neutral area. In a process known as chemotaxis, H. *pylori* can sense areas of acidity and can move to more hospitable areas. It also has the ability to neutralize acid in its immediate environment by using urease to break down urea into ammonia and carbon dioxide. The ammonia neutralizes harmful stomach acid.

An inflammatory response by the body tries to combat the invasion by H. *pylori*. A significant part of this response is for the stomach to produce even more acid. Unfortunately, the extra acid sometimes damages the lining of the stomach and the duodenum, causing ulcerations (ulcers).

Treatment of H. *pylori* usually involves taking the antibiotics amoxicillin and clarithromycin, along with a proton pump inhibitor, which reduces the production of excess stomach acid. This is commonly called *triple therapy* and is usually effective at eradicating H. *pylori*, as it did in Natasha's case.

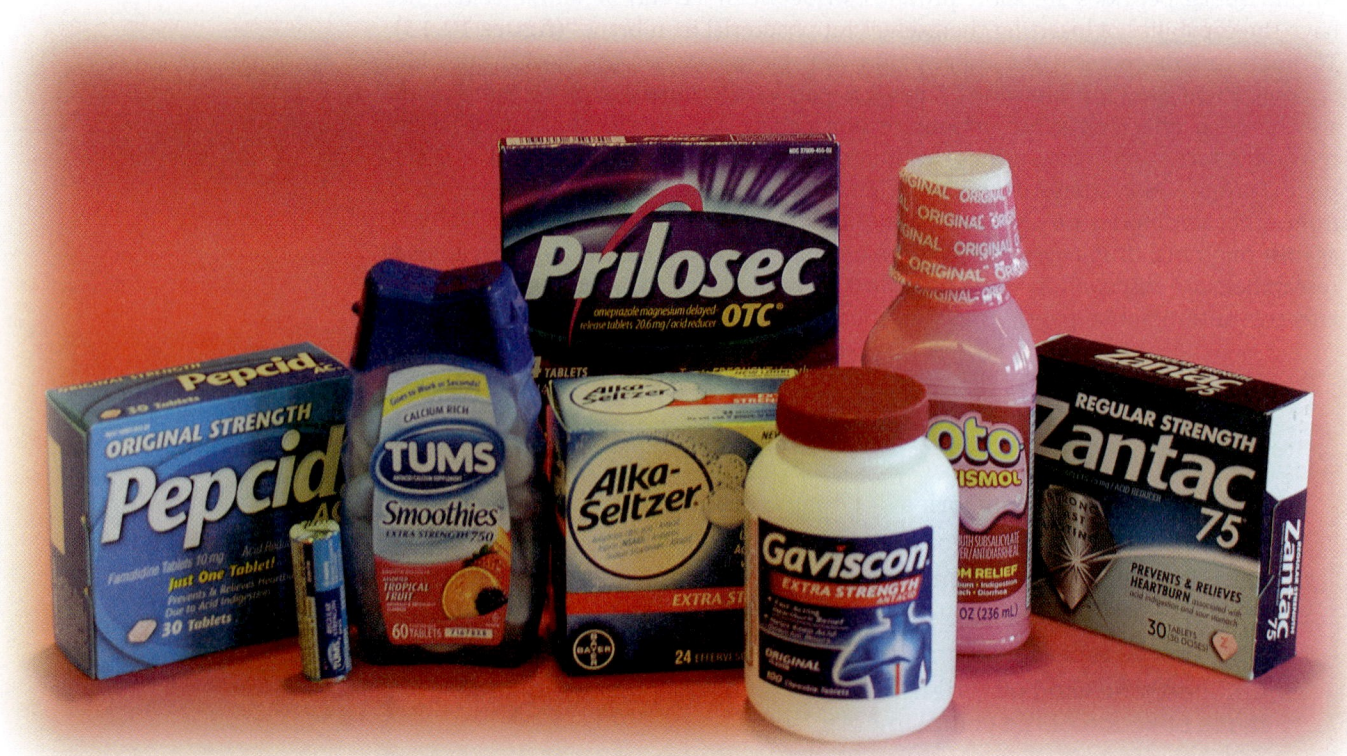

Fever of Unknown Origin: Medical Mystery

Fever that has no apparent cause can be worrisome for both patient and doctor. Fever of unknown origin (FUO) is a medical diagnosis, defined as the presence of fever (greater than 38.3°C [101°F]) off and on for at least 3 weeks and failure to determine a cause despite 1 week of inpatient testing.

Despite exhaustive investigations, approximately 5% to 15% of patients never have a cause of their fever diagnosed. In the rest of patients, FUOs are caused by the following:

- 30% to 40% by infections,
- 20% to 30% by cancers,
- 10% to 20% by collagen vascular diseases, and
- 15% to 20% by other miscellaneous diseases.

Discovering the cause of FUO is detective work. Obtaining a detailed history is often the first and most important step in determining cause. The history should include a detailed analysis of symptoms, including onset and associated symptoms. Special attention should be paid to the presence of constitutional symptoms, such as weight loss, night sweats, and fatigue. Occupational history, travel history, drug use (prescribed, over-the-counter, and illegal), exposure to animals (domestic and wild), family history, sexual history, and nutrition should all be carefully examined and may provide important clues. Demographics and geography can also provide important clues, as certain diseases may be more prevalent in some geographic areas (e.g., tuberculosis, which is most common in developing nations).

Physical examinations should include all organ symptoms, with particular emphasis on the lymph nodes, skin, heart, abdomen, spleen, and eyes. The genitalia should not be ignored even in the elderly or the young. Joint pain and neurological symptoms should also be questioned.

Tests to uncover the underlying cause of FUO generally start with the obvious and proceed to the less obvious, guided by results. To start, a CBC (complete blood count), chemistry studies, urinalysis, chest X ray, and blood cultures should be obtained. Other tests are then ordered according to the results of initial testing and possible diagnoses depending on factors uncovered in the history and a physical examination.

Specialized examinations may include the following:

- Computerized tomography (CT) scanning examinations;
- bone scans;
- magnetic resonance imaging (MRI) examinations;
- radionucleotide studies;
- ultrasonography (ultrasound) studies;
- positron emission tomography (PET) scanning;
- endoscopic examination;
- bone marrow biopsy; and
- serology, including immune function testing.

Generally, the sophisticated tests available today can accurately diagnose the cause of FUO. However, as previously mentioned, up to 15% of patients with FUO never receive a diagnosis. Watchful waiting may be necessary, allowing an illness time to present itself so that it can be treated appropriately. Sometimes patients experience fever for a long period of time and do not develop any other symptoms. In some cases, the fever disappears after a period of weeks or months, leaving behind questions with no answers. Research shows that there is little reason to treat a FUO empirically, that is, administer antibiotics without cause, especially with the current concern of antibiotic-resistant organisms.

Chapter Summary

12.1 Defense Mechanisms of the Host in Perspective

- The interconnecting network of host protection against microbial invasion is organized into three lines of defense.
- The first line of defense consists of physical and chemical barricades associated with the skin and mucous membranes.
- The second line encompasses all the nonspecific cells and chemicals found in the tissues and blood.

12.2 The Second and Third Lines of Defense: An Overview

- The immune system operates as a surveillance system that discriminates between the host's self identity markers and the nonself identity markers of foreign cells.
- The immune system is a complex collection of fluids and cells that penetrate every organ, tissue space, fluid compartment, and vascular network of the body. The four major subdivisions of this system are the MPS, the ECF, the lymphatic system, and the blood vascular system.
- The mononuclear phagocyte system consists of cells distributed throughout the body's network of connective tissue fibers. They stand ready to attack and ingest microbes that have managed to bypass the first line of defense.
- The ECF, or extracellular fluid, compartment surrounds all tissue cells and is penetrated by both blood and lymph vessels, which bring all components of the second and third lines of defense to attack infectious microbes.
- The lymphatic system has three functions: (1) It returns tissue fluid to general circulation; (2) it carries away excess fluid in inflamed tissues; and (3) it concentrates and processes foreign invaders and initiates the specific immune response. Important sites of lymphoid tissues are lymph nodes, spleen, thymus, tonsils, and GALT.
- The blood contains both specific and nonspecific defenses. Nonspecific cellular defenses include the granulocytes, macrophages, and dendritic cells. The two major components of the specific immune response are the T lymphocytes, which provide specific cell-mediated immunity, and the B lymphocytes, which produce specific antibody-mediated immunity.

12.3 The Second Line of Defense

- Nonspecific immune reactions are generalized responses to invasion, regardless of the type. These include phagocytosis, inflammation, fever, and an array of antimicrobial proteins.
- Phagocytosis is accomplished by macrophages along with neutrophils and a few other cell types.
- The four symptoms of inflammation are rubor (redness), calor (heat), tumor (swelling), and dolor (pain). Loss of function often accompanies these.
- Fever is another component of nonspecific immunity. It is caused by both endogenous and exogenous pyrogens. Fever increases the rapidity of the host immune responses and reduces the viability of many microbial invaders.
- There are three main types of antimicrobial proteins: the complement system, interferons, iron-binding proteins, and antimicrobial peptides.

Multiple-Choice Questions   Bloom's Levels 1 and 2: Remember and Understand

Select the correct answer from the answers provided.

1. An example of a nonspecific chemical barrier to infection is
   a. unbroken skin.
   b. lysozyme in saliva.
   c. cilia in respiratory tract.
   d. all of these

2. Host defenses include both innate, nonspecific components and acquired specific components. Which of the following is/are exclusive to the acquired, specific defenses?
   a. leukocytes
   b. antimicrobial proteins
   c. inflammation
   d. fever
   e. lymphocytes

3. What is included in GALT?
   a. thymus
   b. Peyer's patches
   c. tonsils
   d. breast lymph nodes

4. Which of the following is a specialized tissue cell similar to basophils that triggers local inflammatory reactions and is responsible for many allergic symptoms?
   a. mast cell
   b. macrophage
   c. dendritic cell
   d. natural killer cell
   e. neutrophil

5. An example of an exogenous pyrogen is
   a. interleukin-1.
   b. complement.
   c. interferon.
   d. endotoxin.

6. Which of the following components has a role in the third line of defense as well as in the second line of defense?

   a. fever
   b. complement
   c. lysosome
   d. all of the above

7. Which of the following substances is/are not produced by phagocytes to destroy engulfed microorganisms?

   a. hydroxyl radicals
   b. superoxide anion
   c. hydrogen peroxide
   d. bradykinin

8. Which of the following is the end product of the complement system?

   a. properdin
   b. cascade reaction
   c. membrane attack complex
   d. complement factor C9

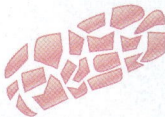

Critical Thinking   Bloom's Levels 3, 4, and 5: Apply, Analyze, and Evaluate

Critical thinking is the ability to reason and solve problems using facts and concepts. These questions can be approached from a number of angles and, in most cases, they do not have a single correct answer.

1. Suggest some reasons for so much redundancy of action and so many interacting aspects of immune responses.

2. a. Describe the main elements of the process through which the immune system distinguishes self from nonself.
   b. How is surveillance of the tissues carried out?
   c. What is responsible for this surveillance?
   d. What does the term *foreign* mean in reference to the immune system?

3. A cut in the skin breaches the first level of host defense. However, explain how a fully functioning first level may help reduce the negative effect of a cut.

4. In what ways is a phagocyte a tiny container of disinfectants?

5. One beautiful warm day in June, you are on a picnic and are stung on your finger by a bee. Although it hurt a little, you are soon enjoying the early summer day. The next morning you wake up to find that your finger is swollen, quite stiff, red, and warmer than the rest of your fingers. In addition, you seem to have a bit of a fever.

   a. Which parts of the immune system seem to be activated?
   b. Which classic responses to injury did you experience?
   c. Explain which cytokines and cell types were probably most responsible for many of your symptoms.

Visual Connections   Bloom's Level 5: Evaluate

This question connects previous images to a new concept.

1. **From chapter 11, figure 11.3.** Relate specific events in inflammation to the symptoms of pneumonia pictured in this drawing.

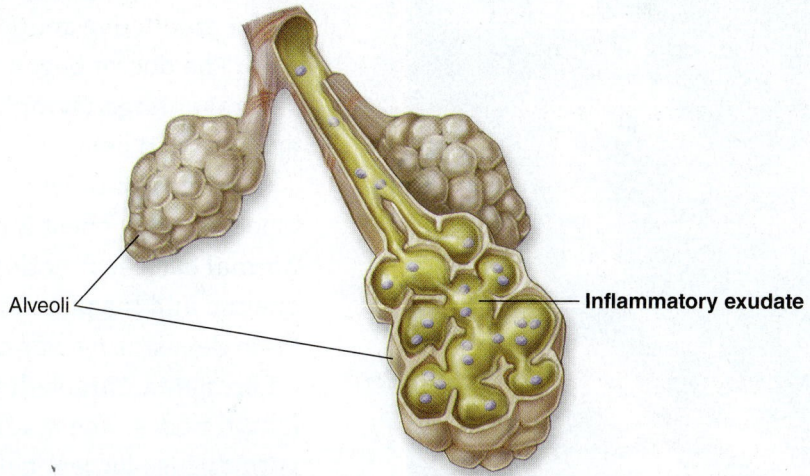

Alveoli

Inflammatory exudate

www.mcgrawhillconnect.com
Enhance your study of this chapter with study tools and practice tests. Also ask your instructor about the resources available through Connect, including the media-rich eBook, interactive learning tools, and animations.

Immune Trade-Off

I was working in a cardiologist's office when I met Mr. Campbell, a 65-year-old patient who had had a heart transplant three years previously following a viral illness that had damaged his heart irreparably. Mr. Campbell had done very well following his transplant and was a model patient. He never complained about the handfuls of drugs he had to take in order to suppress his immune system to prevent organ rejection. Everyone in the clinic looked forward to Mr. Campbell's office visits. His joy in his renewed health was infectious. He always had a huge smile and a joke to share with us.

Mr. Campbell came in one November morning 3 weeks earlier than his usual standing appointment. Although he smiled and joked, it was easy to tell that he was not feeling well. He told the doctor and me that he had been feeling weak and tired and had been experiencing a low-grade fever off and on for a couple of weeks. He was presently taking antibiotics for a chest infection. His family doctor had urged him to visit us sooner rather than later, concerned about his infection when Mr. Campbell was taking immunosuppressant drugs.

We took Mr. Campbell into an examining room. His vital signs were normal except for a moderate fever and a slightly elevated heart rate. He had a productive cough and his lungs sounded congested. He looked very pale. The doctor began to examine Mr. Campbell and discovered that he had several enlarged lymph nodes in his neck. Mr. Campbell admitted that his appetite had been "off" and he had lost some weight.

The doctor ordered a battery of tests: blood work, including a complete blood count, a chest X ray and an abdominal CT scan, in addition to his normal cardiac function tests. The results were worrisome. Mr. Campbell was anemic and his platelets were low. His white blood cells were abnormally low. More devastating was a tumor that was discovered during the abdominal CT scan. Mr. Campbell was scheduled immediately for biopsy of his swollen lymph nodes. The results were as we feared: Mr. Campbell was diagnosed with diffuse large B-cell lymphoma.

- What factor put Mr. Campbell at increased risk of developing lymphoma?

- Why is infection common in lymphoma?

Case File Wrap-Up appears on page 375.

Host Defenses II

Specific Immunity and Immunization

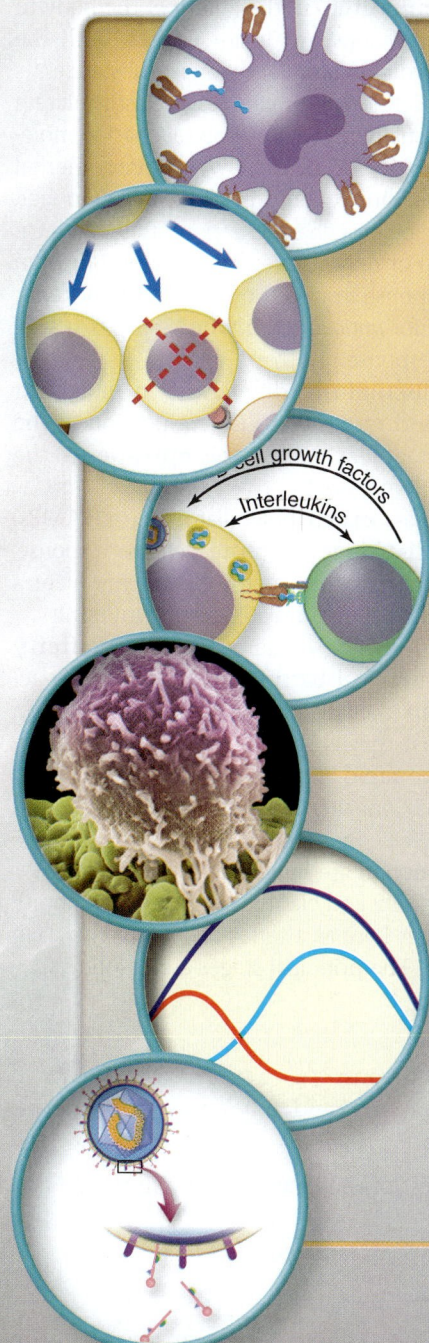

13.1 Specific Immunity: The Third and Final Line of Defense

When host barriers and nonspecific defenses fail to control an infectious agent, a person with a normally functioning immune system has a mechanism to resist the pathogen–the third, specific line of immunity. This sort of immunity is not innate but adaptive; it is acquired only after an immunizing event such as an infection.

Acquired specific immunity is the product of a dual system that we have previously mentioned–the B and T lymphocytes. During development, these lymphocytes undergo a selective process that prepares them for reacting only to one specific antigen or immunogen. During this time **immunocompetence,** the ability of the body to react with countless foreign substances, develops. An infant is born with the theoretical potential to react to an extraordinary array of different substances.

Antigens or **immunogens** are defined as molecules that stimulate a response by T and B cells. They are usually protein or polysaccharide molecules on or inside all cells and viruses, including our own. (Environmental chemicals can also be antigens. These are covered in chapter 14.) In fact, any exposed or released protein or polysaccharide is potentially an antigen, even those on our own cells. For reasons we discuss later, our own antigens do not usually evoke a response from our own immune systems. The term *immunogen* is a synonym for *antigen.* Both words refer to a substance that can elicit an immune response.

A lymphocyte's capacity to discriminate differences in molecular shape is so fine that it recognizes and responds to only a portion of the antigen molecule. This molecular fragment, called the **epitope,** is the primary signal that the molecule is foreign.

In chapter 12, we discussed pathogen-associated molecular patterns (PAMPs) that stimulate responses by phagocytic cells during an innate defense response. While PAMPs are molecules shared by many types of microbes that stimulate a nonspecific response, antigens are highly individual and stimulate specific immunity.

Two features that most characterize this third line of defense are **specificity** and **memory.** Unlike mechanisms such as anatomical barriers or phagocytosis, acquired immunity is specific. In general, the antibodies produced during an infection against the chickenpox virus will function against that virus and not against the measles virus. The property of memory refers to the rapid mobilization of lymphocytes that have been programmed to "recall" their first engagement with the invader and rush to the attack once again.

The elegance and complexity of immune function are largely due to lymphocytes working closely together with phagocytes. To simplify and clarify the network of immunologic development and interaction, we present it here as a series of stages, with each stage covered in a separate section **(figure 13.1).** The principal stages are as follows:

I. lymphocyte development and differentiation;
II. the presentation of antigens;
III. the challenge of B and T lymphocytes by antigens;
IV. T-lymphocyte response: cell-mediated immunity; and B-lymphocyte response: the production and activities of antibodies.

We will give an overview here and spend the rest of the chapter filling in the details.

A Brief Overview of the Immune Response

Lymphocyte Development

Although all lymphocytes arise from the same basic stem cell type, at some point in development they diverge into two distinct types. Final maturation of B cells occurs in specialized bone marrow sites, and that of T cells occurs in the thymus. Both cell types subsequently migrate to separate areas in the lymphoid organs (for instance, nodes and spleen). B and T cells constantly recirculate through the circulatory system and lymphatics, migrating into and out of the lymphoid organs.

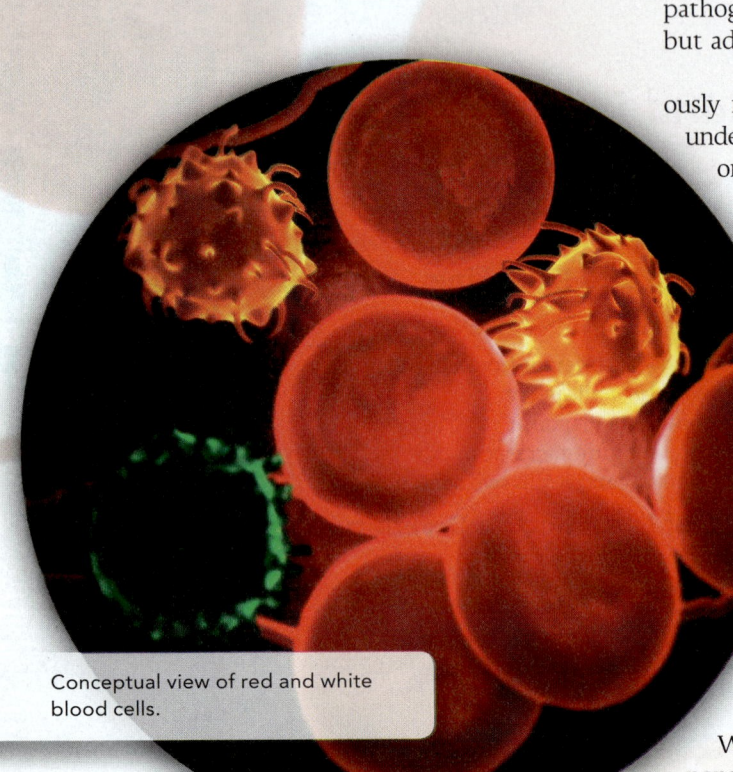

Conceptual view of red and white blood cells.

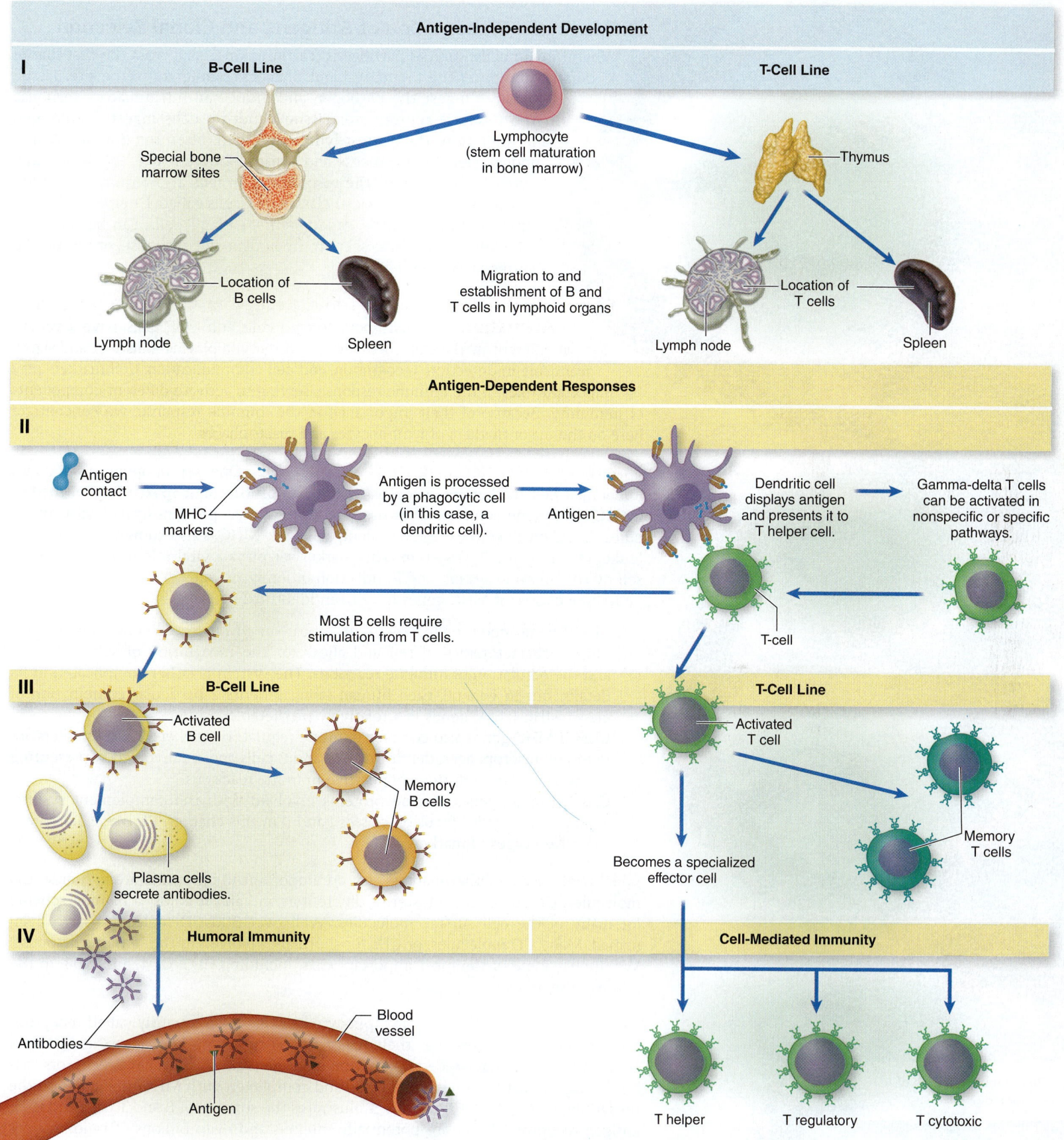

Figure 13.1 Overview of the stages of lymphocyte development and function. I. Development of B- and T-lymphocyte specificity and migration to lymphoid organs. II. Antigen processing by dendritic cell and presentation to lymphocytes; assistance to B cells by T cells. III. Lymphocyte activation, clonal expansion, and formation of memory B and T cells. IV. End result of lymphocyte activation. Left-hand side: antibody release; right-hand side: cell-mediated immunity. Details of these processes are covered in each corresponding section heading.

Entrance and Presentation of Antigens and Clonal Selection

When foreign cells, such as pathogens (carrying antigens), cross the first line of defense and enter the tissue, resident phagocytes migrate to the site. Tissue macrophages ingest the pathogen and induce an inflammatory response in the tissue if appropriate. Tissue dendritic cells ingest the antigen and migrate to the nearest lymphoid organ (often the draining lymph nodes). Here they process and present antigen to T and B lymphocytes. In most cases, the response of B cells also requires the additional assistance of special classes of T cells called T helper cells. One special class of T cells, called gamma-delta T cells, can be activated quickly by PAMPs, as seen in the nonspecific response, or by specific antigens as seen here.

The Role of Markers and Receptors in Presentation and Activation

All cells–both foreign cells and "self" cells–have a variety of different markers on their surfaces, each type playing a distinct and significant role in detection, recognition, and cell communication. Cell markers play important roles in the immune response, serving to activate different components of immunity. Because of their importance in the immune response, we concentrate here on the major markers of lymphocytes and macrophages.

The Major Histocompatibility Complex

One set of genes that codes for human cell markers or receptors is the **major histocompatibility complex (MHC).** This gene complex gives rise to a series of glycoproteins (called MHC molecules) found on all cells except red blood cells. The MHC is also known as the human leukocyte antigen (HLA) system. This marker set plays a vital role in recognition of self by the immune system and in rejection of foreign tissue.

Three classes of MHC genes have been identified:

1. Class I genes code for markers that appear on all nucleated cells. They display unique characteristics of self and allow for the recognition of self molecules and the regulation of immune reactions. The system is rather complicated in its details, but in general, each human being inherits a particular combination of class I MHC (HLA) genes in a relatively predictable fashion.
2. Class II MHC genes also code for immune regulatory markers. These markers are found on macrophages, dendritic cells, and B cells and are involved in presenting antigens to T cells during cooperative immune reactions.
3. Class III MHC genes encode proteins involved with the complement system, among others. We'll focus on classes I and II in this chapter. **Figure 13.2** shows these two types of markers.

CD Molecules

Other markers that are important in immunity are particular CD molecules. *CD* stands for "cluster of differentiation," and it is just a naming scheme for many of the cell surface molecules. Well over 300 CD molecules have been named. Many CD molecules, or CDs for short, are involved in the immune response. We discuss some of the most important CDs, including CD3, CD4, and CD8, in the following sections.

Lymphocyte Receptors

Lymphocyte markers are frequently called receptors, a name that emphasizes that their major role is to "accept" or "grasp" antigens in some form. B cells have receptors that bind antigens, and T cells have receptors that bind antigens that have been processed and complexed with MHC molecules on the presenting cell surface. **Figure 13.3** illustrates the surfaces of B and T cells and their antigen receptors. There are potentially millions and even billions of unique types of antigens. The many sources of antigens include microorganisms as well as an endless array of chemical compounds in the environment. We will soon see how T and B cells recognize so many different antigens.

Genes are the source of immunologic diversity.

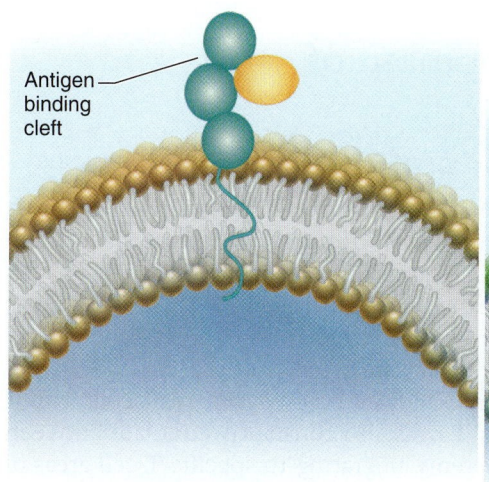
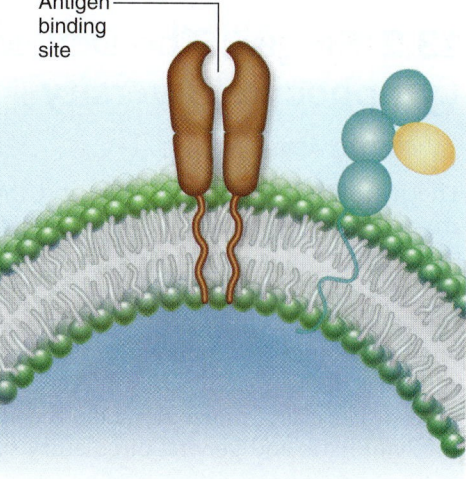

Antigen binding cleft

Antigen binding site

Class I MHC molecule found on all nucleated human cells.

Class II MHC found on some types of white blood cells (class I molecules here also, of course).

Figure 13.2 **Classes I and II of molecules of the human major histocompatibility complex.**

Challenging B and T Cells with Antigen

When challenged by antigen, both B cells and T cells proliferate and differentiate. The multiplication of a particular lymphocyte creates a clone, or group of genetically identical cells, some of which are memory cells that will ensure future reactiveness against that antigen. Because the B-cell and T-cell responses differ significantly from this point on in the sequence, they are summarized separately.

How T Cells Respond to Antigen: Cell-Mediated Immunity (CMI)

T-cell types and responses are extremely varied. When activated (sensitized) by antigen, a T cell gives rise to one of three different types of progeny, each involved in a cell-mediated immune function. The three main functional types of T cells are as follows:

1. helper T cells that activate macrophages, assist B-cell processes, and help activate cytotoxic T cells;
2. regulatory T cells that control the T-cell response; and
3. cytotoxic T cells that lead to the destruction of infected host cells and other "foreign" cells.

Although T cells secrete cytokines that help destroy pathogens and regulate immune responses, they do not produce antibodies.

How B Cells Respond to Antigen: Release of Antibodies

When a B cell is activated, or sensitized, by an antigen, it divides, giving rise to plasma cells, each with the same reactive profile. Plasma cells release antibodies into the tissue and blood. When these antibodies attach to the antigen for which they are specific, the antigen is marked for destruction or neutralization.

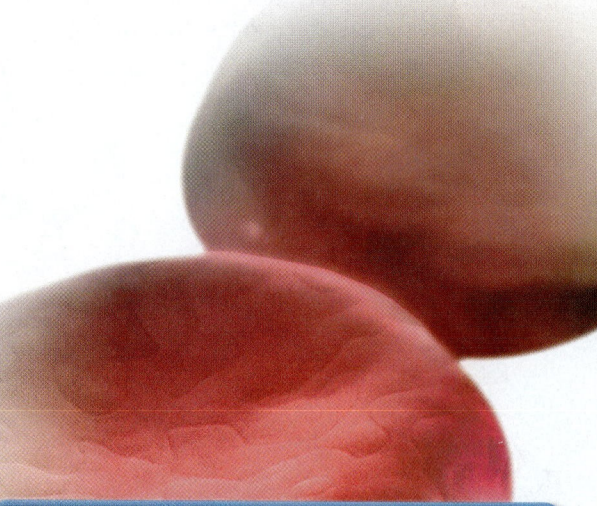

13.1 LEARNING OUTCOMES—Assess Your Progress

1. Describe how the third line of defense is different from the other two.
2. Compare the terms *antigen*, *immunogen*, and *epitope*.
3. List the four stages of a specific immune response.
4. Discuss the role of cell markers in the immune response.
5. Describe the major histocompatibility complex in two sentences.

NCLEX® PREP

1. The major histocompatibility complex (MHC) is a gene complex that gives rise to a series of glycoproteins (MHC molecules) found on all cells *except*
   a. *red blood cells.* c. *plasma cells.*
   b. *white blood cells.* d. *B cells.*

Medical Moment

The Thymus

The thymus is a small gland situated partly in the neck and partly in the thorax, behind the sternum. It weighs about 15 grams at birth, achieves its largest mass around puberty, and then atrophies slowly thereafter.

What is the main function of the thymus? Within the thymus, thymocytes (hematopoietic precursors from the bone marrow) mature into T cells. Once the T cells have become fully mature, they travel from the thymus and comprise the stock of circulating T cells responsible for many aspects of immunity. What happens when the thymus malfunctions, or when it is not working properly due to underlying disease? The following conditions may occur:

- **DiGeorge syndrome.** This rare genetic condition results in thymus deficiency. Patients with the syndrome may present with severe immunodeficiency.
- **Severe combined immunodeficiency (SCID) syndromes.** These are a group of genetic disorders resulting from defective hematopoietic precursor cells, causing a deficiency of both B lymphocytes and T lymphocytes.
- **Cancer.** Thymomas, tumors that originate from thymic epithelial cells, and thymic lymphomas, tumors originating from thymocytes, are the primary tumors affecting the thymus.

13.2 Stage I: The Development of Lymphocyte Diversity

Specific Events in T-Cell Maturation

The maturation of most T cells and the development of their specific receptors are directed by the thymus and its hormones **(table 13.1)**. Other T cells reach full maturity in the gastrointestinal tract. In addition to the antigen-specific T-cell receptor, all mature T lymphocytes express coreceptors called CD3. CD3 molecules surround the T-cell receptor and assist in binding. T cells also express either a CD4 or a CD8 coreceptor (figure 13.3). CD4 is an accessory receptor protein on T helper cells that binds to MHC class II molecules. CD8 is found on cytotoxic T cells, and it binds MHC class I molecules. Like B cells, T cells also constantly circulate between the lymphatic and general circulatory systems, migrating to specific T-cell areas of

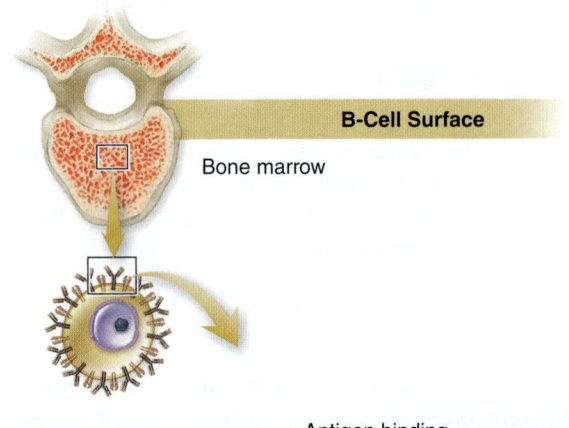

Figure 13.3 **The surfaces of T cells and B cells.**

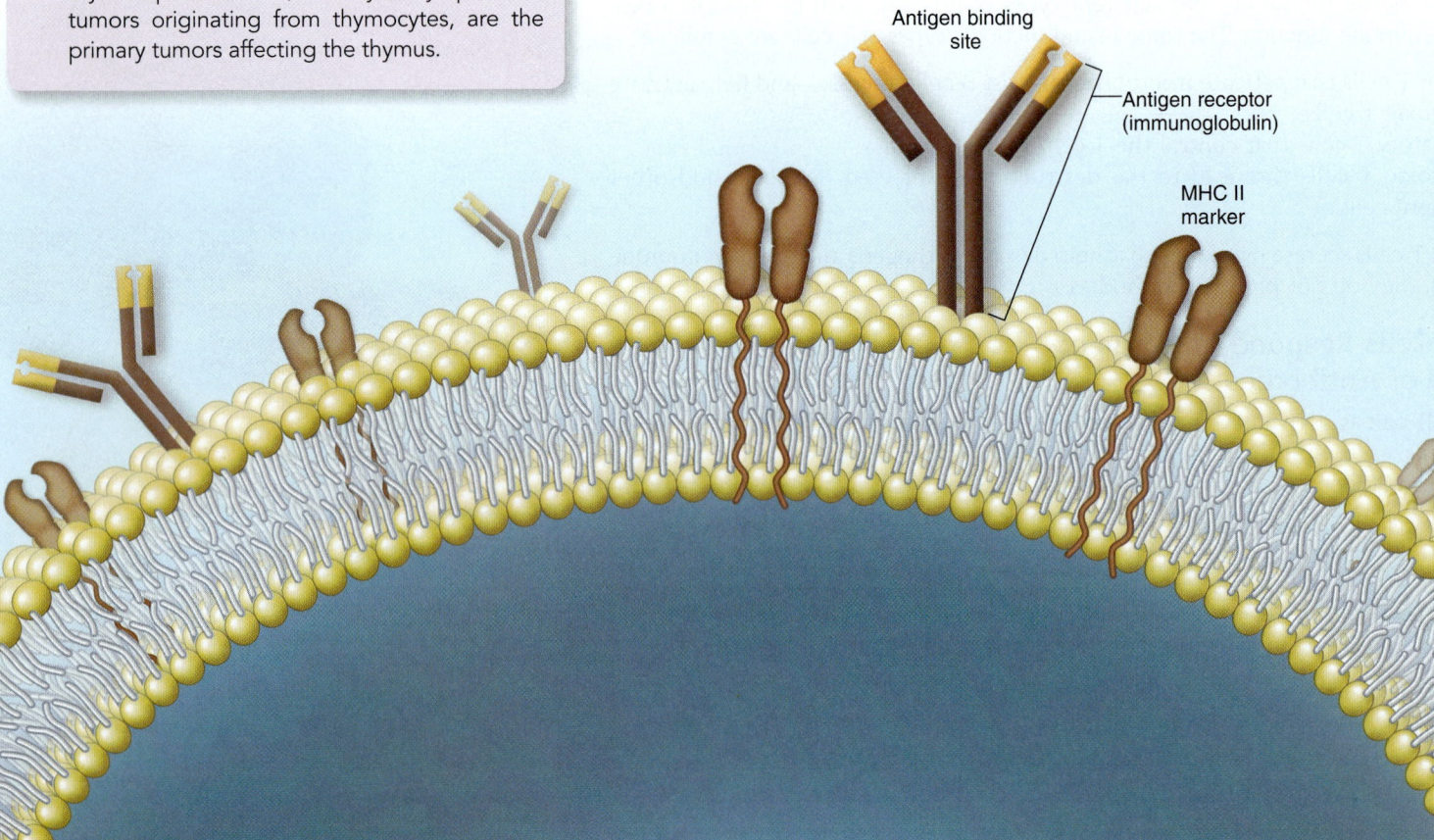

the lymph nodes and spleen. It has been estimated that more than 10^9 T cells pass between the lymphatic and general circulations per day.

Specific Events in B-Cell Maturation

B cells develop in the bone marrow. As a result of gene modification and selection, hundreds of millions of distinct B cells develop. These naive lymphocytes circulate through the blood, "homing" to specific sites in the lymph nodes, spleen, and other lymphoid tissue, where they adhere to specific binding molecules. Here they will come into contact with antigens throughout life.

Table 13.1 Contrasting Properties of B Cells and T Cells

	B Cells	T Cells
Site of Maturation	Bone marrow	Thymus
Specific Surface Markers	Immunoglobulin	T-cell receptor Several CD molecules
Circulation in Blood	Low numbers	High numbers
Receptors for Antigen	B-cell receptor (immunoglobulin)	T-cell receptor
Distribution in Lymphatic Organs	Cortex (in follicles)	Paracortical sites (interior to the follicles)
Require Antigen Presented with MHC	No	Yes°
Product of Antigenic Stimulation	Plasma cells and memory cells	Several types of activated T cells and memory cells
General Functions	Production of antibodies to inactivate, neutralize, target antigens	Cells activated to help other immune cells; suppress or kill abnormal cells; mediate hypersensitivity; synthesize cytokines

°Gamma-delta T cells can be activated differently.

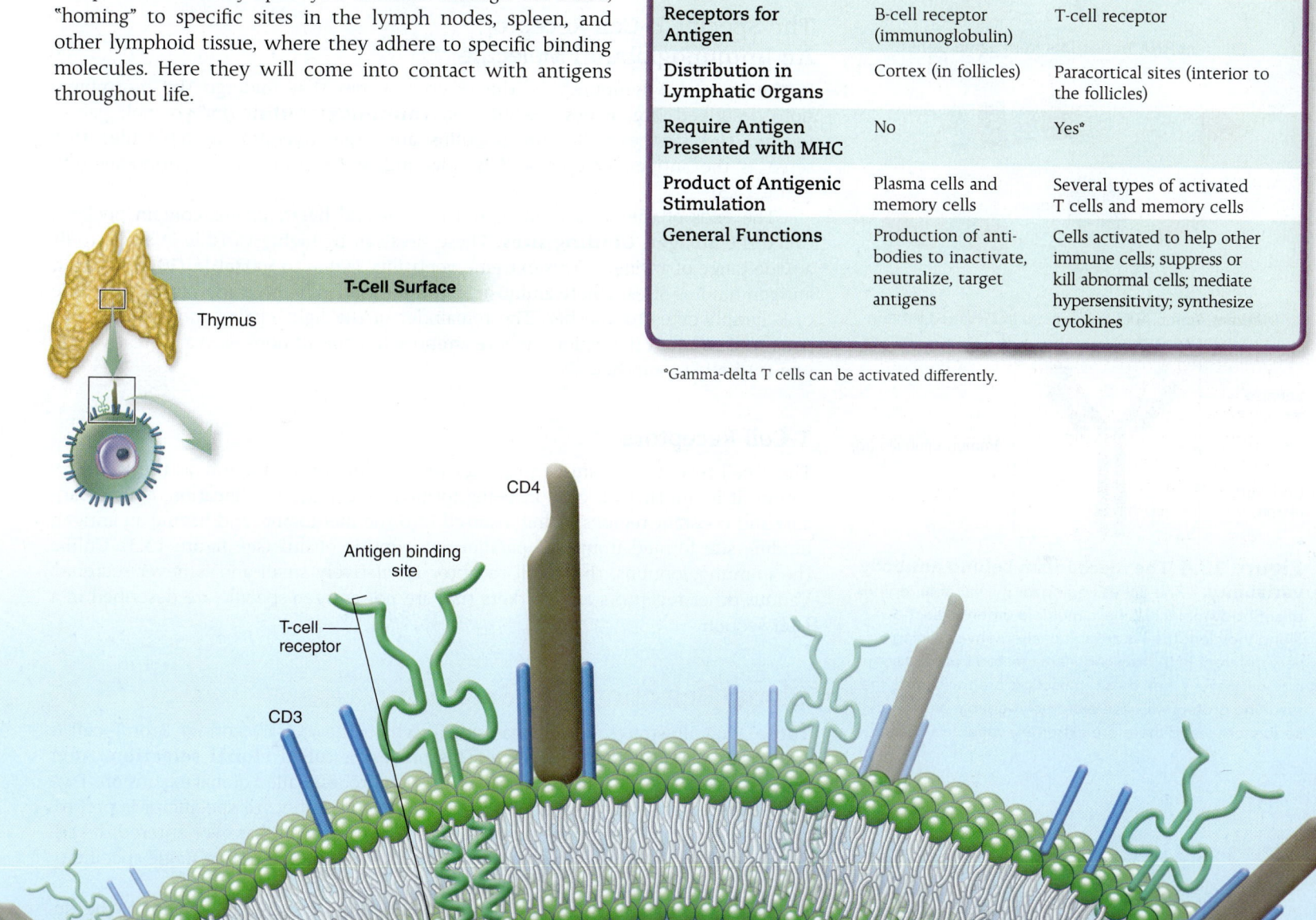

Thymus

T-Cell Surface

CD4

Antigen binding site

T-cell receptor

CD3

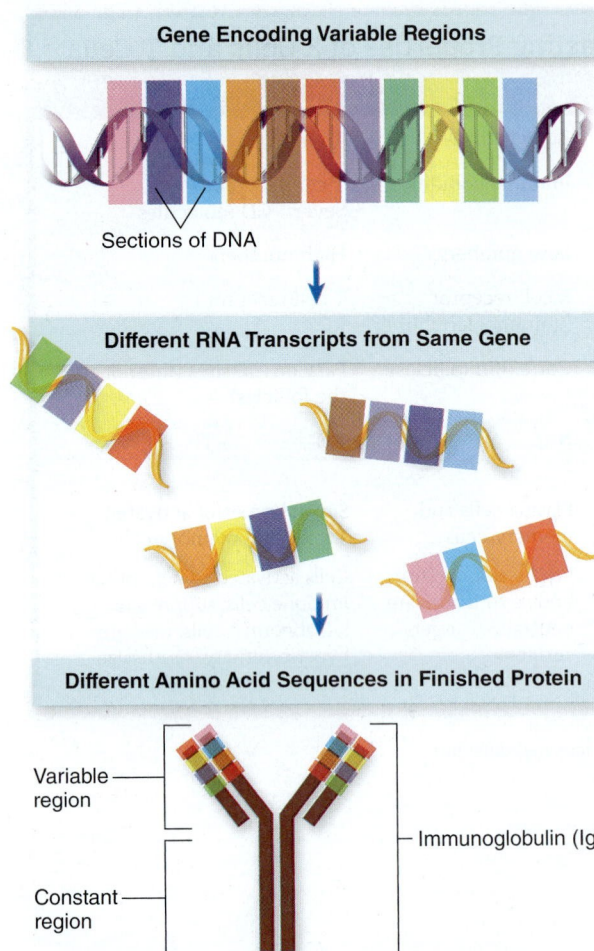

Gene Encoding Variable Regions

Sections of DNA

Different RNA Transcripts from Same Gene

Different Amino Acid Sequences in Finished Protein

Variable region

Immunoglobulin (Ig)

Constant region

Figure 13.4 The mechanism behind antibody variability.   The genes coding for the variable regions of antibody molecules have multiple different sections along their lengths. As a result of alternative splicing, very different RNA transcripts are created from the same original gene. When those transcripts are translated, the resulting protein will have extremely variable amino acid sequences—and therefore extremely variable shapes.

The Origin of Immunologic Diversity

By the time T and B cells reach the lymphoid tissues, each one is already equipped to respond to a unique antigen. This amazing specificity is generated by extensive rearrangements of more than 500 different gene segments that code for the antigen receptors on the T and B cells **(figure 13.4).** In time, every possible recombination occurs, leading to a huge assortment of lymphocytes. It is estimated that each human produces antibodies with 10 trillion different specificities.

The Specific B-Cell Receptor: An Immunoglobulin Molecule

In the case of B lymphocytes, the receptor genes that undergo the recombination described are genes coding for **immunoglobulin** (im″-yoo-noh-glahb′-yoo-lin) **(Ig)** synthesis. Immunoglobulins are large glycoprotein molecules that serve as the antigen receptors of B cells and, when secreted, as antibodies (see figure 13.3).

The ends of the forks formed by the light and heavy chains contain pockets, called the **antigen binding sites.** These sites can be highly variable in shape to fit a wide range of antigens. This extreme versatility is due to **variable (V) regions** in antigen binding sites, where amino acid composition is highly varied from one clone of B lymphocytes to another. The remainder of the light chains and heavy chains consist of constant (C) regions, whose amino acid content does not vary greatly from one antibody to another.

T-Cell Receptors

The T-cell receptor for antigen belongs to the same protein family as the B-cell receptor. It is similar to B cells in being formed by genetic modification, having variable and constant regions, being inserted into the membrane, and having an antigen binding site formed from two parallel polypeptide chains (see figure 13.3). Unlike the immunoglobulins, the T-cell receptor is relatively small and is never secreted. Various other receptors and markers that are not antigen-specific are described in a later section.

Clonal Selection and Expansion

Table 13.2 illustrates the mechanism by which the exactly correct B or T cell is activated by any incoming antigen. This process is called **clonal selection.** After activation, the B or T cell multiplies rapidly in a process called clonal expansion. Two important features of clonal selection are that (1) lymphocyte specificity is preprogrammed, existing in the genetic makeup before an antigen has ever entered the tissues; and (2) each genetically distinct lymphocyte expresses only a single specificity and can react to that chemical epitope.

One potentially problematic outcome of random genetic assortment is the development of clones of lymphocytes able to react to *self.* This outcome can lead to severe damage if the immune system actually perceives self molecules as foreign and mounts a harmful response against the host's tissues. Any such clones are destroyed during development through clonal *deletion.* The removal of such potentially harmful clones is the basis of immune tolerance or tolerance to self. Since humans are exposed to many new antigenic substances during their lifetimes, T cells and B cells in the periphery of the body have mechanisms for *not* reacting to innocuous antigens. Some diseases (autoimmunity) are thought to be caused by the loss of immune tolerance through the survival of certain "forbidden clones" or failure of these other systems (see chapter 14).

Table 13.2 Clonal Selection and Expansion of B and T Cells

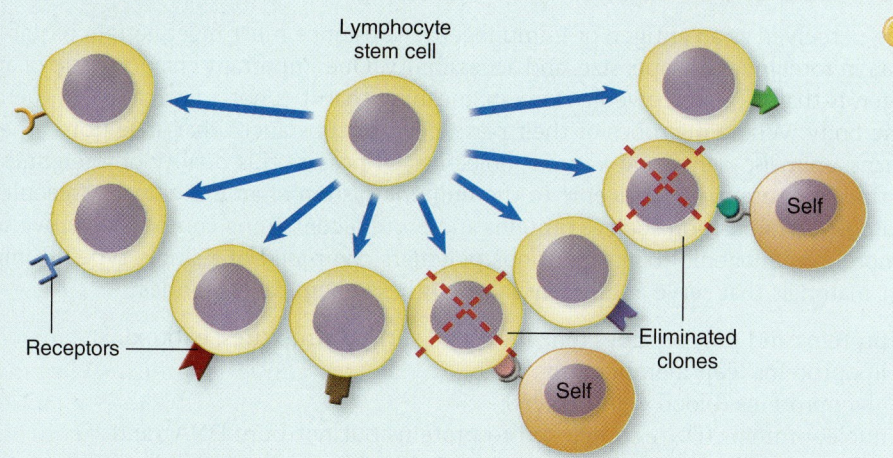

I

Repertoire of lymphocyte clones, each with unique receptor display

1. Each genetically unique line of lymphocytes arising from extensive recombinations of surface proteins is termed a **clone**. This proliferative stage of lymphocyte development does not require the actual presence of foreign antigens.

   At the same time, any lymphoctyes that develop a specificity for self molecules and could be harmful are eliminated from the pool of cells. This is called **immune tolerance.**

2. The specificity for a single epitope is programmed into the lymphocyte and is set for the life of a given cell. The end result is an enormous pool of mature but naive lymphocytes that are ready to further differentiate under the influence of certain organs and immune stimuli.

II–IV

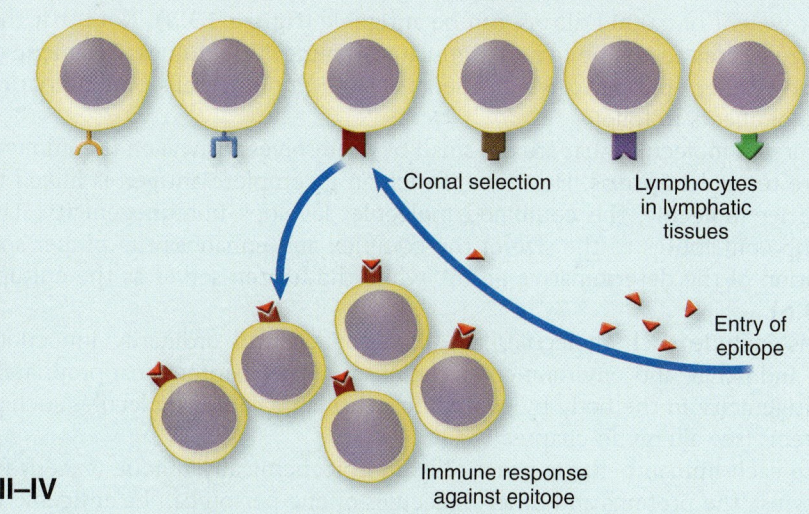

3. When any epitope enters the immune surveillance system, it encounters specific lymphocytes ready to recognize it. Such contact stimulates that clone to undergo mitotic divisions and expands it into a larger population of lymphocytes, all bearing the same specificity.

13.2 LEARNING OUTCOMES—Assess Your Progress

6. Summarize the maturation process of both B cells and T cells.

7. Draw a diagram showing how lymphocytes are capable of responding to nearly any epitope imaginable.

8. Describe the structures of the B-cell receptor and the T-cell receptor.

9. Outline the processes of clonal selection and expansion.

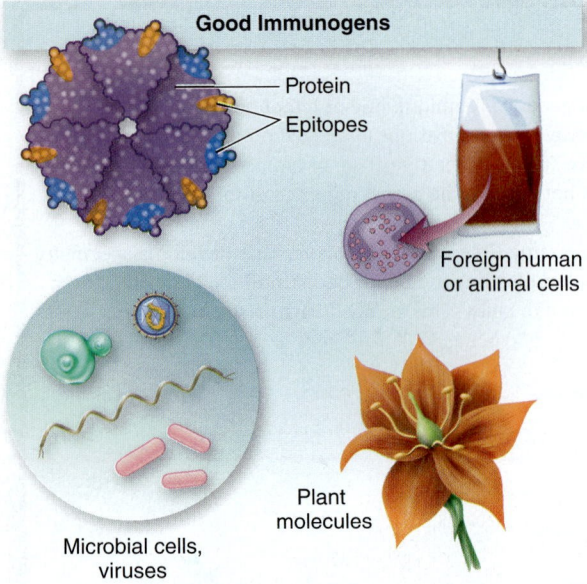

Good Immunogens

Protein

Epitopes

Foreign human
or animal cells

Plant
molecules

Microbial cells,
viruses

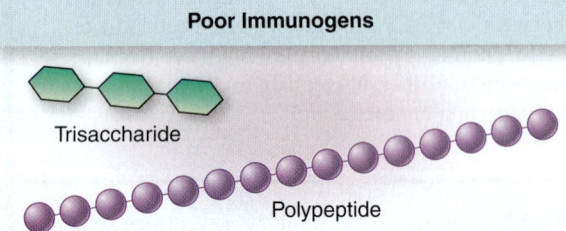

Poor Immunogens

Trisaccharide

Polypeptide

Figure 13.5 **A comparison of good immunogens and poor immunogens.** Top: Good immunogens are large and complex. Bottom: Small molecules and linear molecules are less likely to be good immunogens.

13.3 Stage II: Presentation of Antigens

Entrance of Antigens

To be perceived as an antigen or immunogen, a substance must meet certain requirements in foreignness, shape, size, and accessibility. One important characteristic of an antigen is that it be perceived as foreign, meaning that it is not a normal constituent of the body. Whole microbes or their parts, cells, or substances that arise from other humans, animals, plants, and various molecules all possess this quality of foreignness and thus are potentially antigenic to the immune system of an individual. Molecules of complex composition such as proteins and protein-containing compounds prove to be more immunogenic than repetitious polymers composed of a single type of unit. Most materials that serve as antigens fall into these chemical categories:

- proteins and polypeptides (enzymes, cell surface structures, exotoxins);
- lipoproteins (cell membranes);
- glycoproteins (blood cell markers);
- nucleoproteins (DNA complexed to proteins but not pure DNA); and
- polysaccharides (certain bacterial capsules) and lipopolysaccharides.

Effects of Molecular Shape and Size

To initiate an immune response, a substance must also be large enough to "catch the attention" of the surveillance cells. Large, complex macromolecules approaching a molecular weight (MW) of 100,000 Daltons (a unit of molecular weight) are the most immunogenic. Note that large size alone is not sufficient for antigenicity; glycogen, a polymer of glucose with a highly repetitious structure, has a molecular weight over 100,000 Daltons and is not normally antigenic, whereas insulin, a protein with a molecular weight of 6,000 Daltons, can be antigenic (**figure 13.5**). Note that this aspect of size relates to the ability to stimulate immune "wakefulness." Remember that it is the smaller epitope to which the awakened immune cells bind, initiating the cascade of effects to follow.

Small foreign molecules that are too small by themselves to awaken the immune response are termed **haptens.** However, if such an incomplete antigen is linked to a larger carrier molecule, the combined molecule develops immunogenicity. The carrier group contributes to the size of the complex and enhances the proper spatial orientation of the determinative group, while the hapten serves as the epitope **(figure 13.6).**

Haptens include such molecules as drugs, metals, and ordinarily innocuous household, industrial, and environmental chemicals. Many haptens inappropriately develop antigenicity in the body by combining with large carrier molecules such as serum proteins (see allergy in chapter 14).

Because each human being is genetically and biochemically unique (except for identical twins), the proteins and other molecules of one person can be antigenic to another. **Alloantigens** are cell surface markers and molecules that occur in some members of the same species but not in others. Alloantigens are the basis for an individual's blood group and major histocompatibility profile, and they are responsible for incompatibilities that can occur in blood transfusion or organ grafting.

Figure 13.6 **Haptens.** Some small molecules are poor immunogens but can be made more immunogenic by complexing them with a carrier.

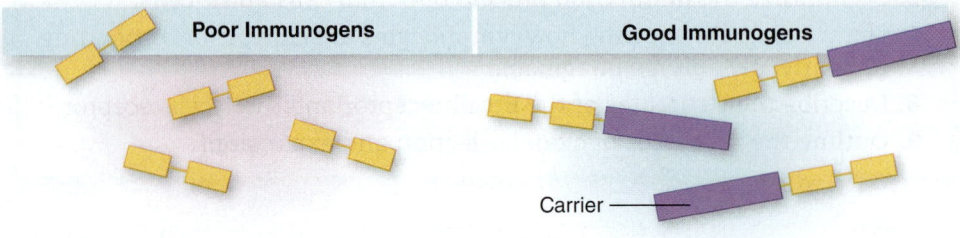

Poor Immunogens | **Good Immunogens**

Carrier

Some bacterial toxins, which belong to a group of immunogens called **superantigens,** are potent stimuli for T cells. Their presence in an infection activates T cells at a rate 100 times greater than ordinary antigens. The result can be an overwhelming release of cytokines and cell death. Such diseases as toxic shock syndrome and certain autoimmune diseases are associated with this class of antigens.

Antigens that evoke allergic reactions, called **allergens,** are characterized in detail in chapter 14.

Antigen Processing and Presentation

In most immune reactions, the antigen must be further acted upon and formally presented to lymphocytes by cells called **antigen-presenting cells (APCs).** Three different cells can serve as APCs: macrophages, B cells, and **dendritic** (den′-drih-tik) **cells.** After processing is complete, the antigen is bound to the MHC receptor and moved to the surface of the APC so that it will be readily accessible to T lymphocytes during presentation. **Table 13.3** illustrates how antigen is presented to T cells. The activated T cell that is the product of this reaction is central to all of T-cell immunity and most of B-cell immunity.

Most antigens must be presented first to T cells, even though they will eventually activate both the T-cell and B-cell systems. However, a few antigens can trigger a response directly from B lymphocytes without the cooperation of APCs or T helper cells. These are called T-cell-independent antigens. They are usually simple molecules such

Table 13.3 How Antigen Is Presented to T Cells

Microbes

Antigen-presenting cell (APC)

MHC-II receptor

Processed antigen

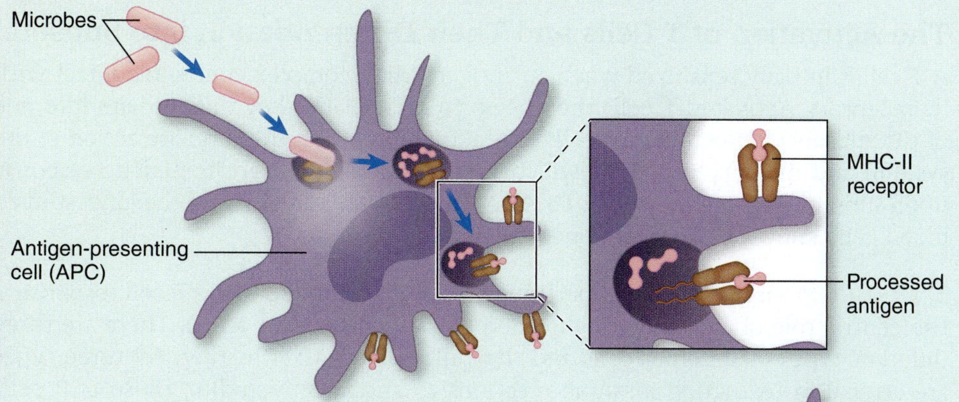

Antigen-preseting cells (APCs) engulf a microbe, take it into intracellular vesicles, and degrade it into smaller peptides or pieces. The antigen pieces complexed with MHC-II receptors are transported to the APC membrane (inset). From this surface location the antigens are presented to a T helper cell, which is specific for the antigen being presented.

First, the MHC-II antigen on the APC binds to the T-cell receptor.

Next, a coreceptor on the T cell (typically CD4) hooks itself to a position on the MHC-II receptor. This combination ensures the simultaneous recognition of the antigen (nonself) and the MHC receptor (self). Once identification has occurred, the APC activates this T helper (T$_H$) cell. The APC also secretes **interleukin-1 (IL-1).** The T$_H$ cell, in turn, produces **IL-2,** which is a growth factor for the T helper cells and cytotoxic T cells. These T helper cells can then help activate B cells.

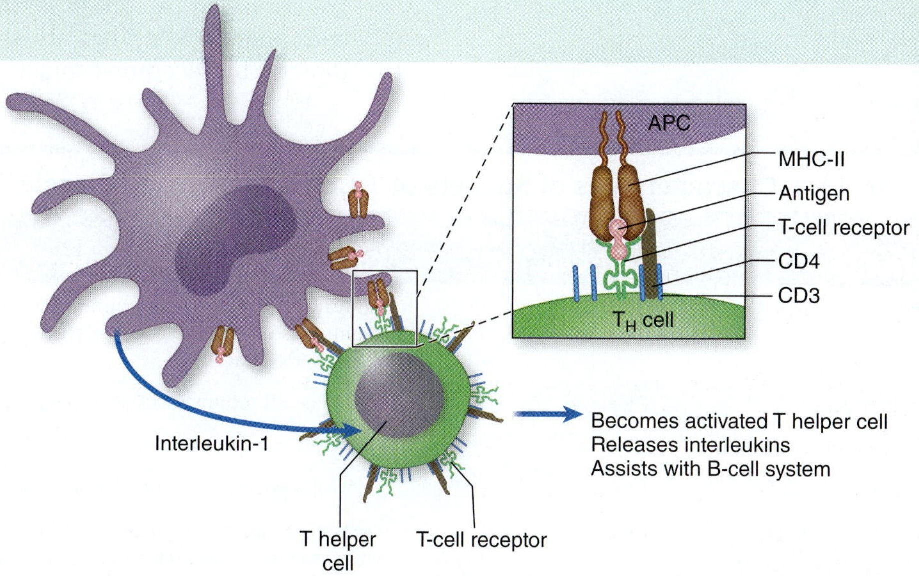

APC
MHC-II
Antigen
T-cell receptor
CD4
CD3
T$_H$ cell

Interleukin-1

Becomes activated T helper cell
Releases interleukins
Assists with B-cell system

T helper cell

T-cell receptor

A Note About Epitopes and Antigens

While up to now we have been calling the immunogenic substance "the antigen," it is more precisely termed the epitope. You could say, for instance, "the antigenic portion of the protein on a microbe is the epitope." You will also note that in practice, clinicians, and even other parts of this book, use the word "antigen" when the precise term is "epitope." You will know, however, that the part of the molecule that is actually recognized by the immune system is the epitope. This means, also, that every epitope can be recognized by B- and T-cell receptors that were formed during genetic reassortment. The particular tertiary structure and shape of this determinant must conform like a key to the receptor "lock" of the lymphocyte, which then responds to it. Certain amino acids accessible at the surface of proteins or protruding carbohydrate side chains are typical examples. Many foreign cells and molecules are very complex antigenically, with numerous component parts, each of which will elicit a separate and different lymphocyte response. Examples of these multiple, or *mosaic,* antigens include bacterial cells containing cell wall; membrane; and flagellar, capsular, and toxin antigens, as well as viruses. T-cell antigen receptors recognize these small pieces of antigens—epitopes—in combination with MHC molecules.

as carbohydrates with many repeating and invariable determinant groups. Examples include lipopolysaccharide from the cell wall of *Escherichia coli*, polysaccharide from the capsule of *Streptococcus pneumoniae*, and molecules from rabies and Epstein-Barr virus. Because so few antigens are of this type, most B-cell reactions require T helper cells.

13.3 LEARNING OUTCOMES—Assess Your Progress

10. List characteristics of antigens that optimize their immunogenicity.

11. Describe how the immune system responds to alloantigens, superantigens, and allergens.

12. List the types of cells that can act as antigen-presenting cells.

13.4 Stages III and IV: T-Cell Response

Cell-Mediated Immunity (CMI)

T-cell reactions are among the most complex and diverse in the immune system and involve several subsets of T cells, whose particular actions are dictated by the APCs that activate them. We refer to T cells as "restricted"–that is, they require some type of MHC (self) recognition before they can be activated, and all produce cytokines with a spectrum of biological effects.

The end result of T-cell stimulation is the mobilization of other T cells, B cells, and phagocytes.

The Activation of T Cells and Their Differentiation into Subsets

A T cell is initially sensitized when an antigen/MHC complex comes in contact with its receptors. Activated T cells then begin to divide, and they differentiate into one of the subsets of effector cells (cells that actually perform the ultimate action of the system) and memory cells that will be available to mount an immediate response upon subsequent contact **(table 13.4)**. Memory T cells are some of the longest-lived blood cells known (70 years in one well-documented case).

T Helper (T$_H$) Cells The first thing you will notice about the T-cell response is the central role of a type of T cell called helper cells **(table 13.5)**. There are three different types of T helper cells and they all bear the CD4 marker. All three types are critical in regulating immune reactions to antigens, including those of B cells and other T cells. They are also involved in activating macrophages. They do this directly by receptor contact and indirectly by releasing cytokines like interferon

Table 13.4 Characteristics of Subsets of T-Cell Types in the Classic T-Cell Response

Types	Primary Receptors on T Cell	Functions/Important Features
T helper cell 1 (T$_H$1)	CD4	Activates the cell-mediated immunity pathway; secretes tumor necrosis factor and interferon gamma; also responsible for delayed hypersensitivity (allergy occurring several hours or days after contact)
T helper cell 2 (T$_H$2)	CD4	Drives B-cell proliferation; secretes IL-4, IL-5, IL-13
T helper cell 17 (T$_H$17)	CD4	Promotes inflammation; secretes IL-17
T regulatory cell (T$_{reg}$)	CD4	Controls specific immune response; prevents autoimmunity
T cytotoxic cell (T$_C$)	CD8	Destroys a target foreign cell by lysis; important in destruction of complex microbes, cancer cells, virus-infected cells; graft rejection; requires MHC-I for function

gamma (IFNγ). T helper cells secrete interleukin-2, which stimulates the primary growth and activation of many types of T cells, including cytotoxic T cells. Some T helper cells secrete interleukins-4, -5, and -6, which stimulate various activities of B cells. T helper cells are the most prevalent type of T cell in the blood and lymphoid organs, making up about 65% of this population. The severe depression of this class of T cells (with CD4 receptors) by HIV is what largely accounts for the immunopathology of AIDS.

The details of the T-cell reaction to antigen can seem complex. Spend some time with table 13.5 and the mechanism will start to be clear.

Cytotoxic T (T_C) Cells: Cells That Kill Other Cells   Cytotoxic T cells have one job: to destroy other cells. Target cells that T_C cells can destroy include the following:

- Virally infected cells. Cytotoxic cells recognize these because of telltale virus peptides expressed on their surface. Cytotoxic defenses are an essential protection against viruses.
- Cancer cells. T cells constantly survey the tissues and immediately attack any abnormal cells they encounter **(figure 13.7)**. The importance of this function is clearly demonstrated in the susceptibility of T-cell-deficient people to cancer (see chapter 14).
- Cells from other animals and humans. Cytotoxic CMI is the most important factor in graft rejection. In this instance, the T_C cells attack the foreign tissues that have been implanted into a recipient's body.

Gamma-Delta T Cells   The subcategory of T cells called gamma-delta T cells is distinct from other T cells. They do have T-cell receptors that are rearranged to recognize a wide range of antigens, but they frequently respond to certain kinds of PAMPs on microorganisms the way WBCs in the nonspecific system do. This allows them to respond more quickly. However, they still produce memory cells when they are activated. For these reasons, they are considered to bridge the nonspecific and specific immune responses. They are particularly responsive to certain types of phospholipids and can recognize and react against tumor cells.

Additional Cells with Orders to Kill   Natural killer (NK) cells are a type of lymphocyte related to T cells that lack specificity for antigens. They circulate through the spleen, blood, and lungs, and are probably the first killer cells to attack cancer cells and virus-infected cells. They destroy such cells by similar mechanisms as T cells. They are not considered part of specific cell-mediated immunity, although their activities are acutely sensitive to cytokines such as interleukin-12 and interferon. Finally, there is a hybrid kind of cell that is part killer cell and part T cell, with T-cell receptors for antigen and the ability to release large amounts of cytokines very quickly, leading to cell death. These cells are called natural killer T cells, or NKT cells. Because they have T-cell receptors but respond very quickly, they are considered to be another important link between nonspecific and specific immunity.

As you can see, the T-cell system is very complex. In summary, T cells differentiate into five different types of cells (and also memory cells), each of which contributes to the orchestrated immune response, under the influence of a multitude of cytokines.

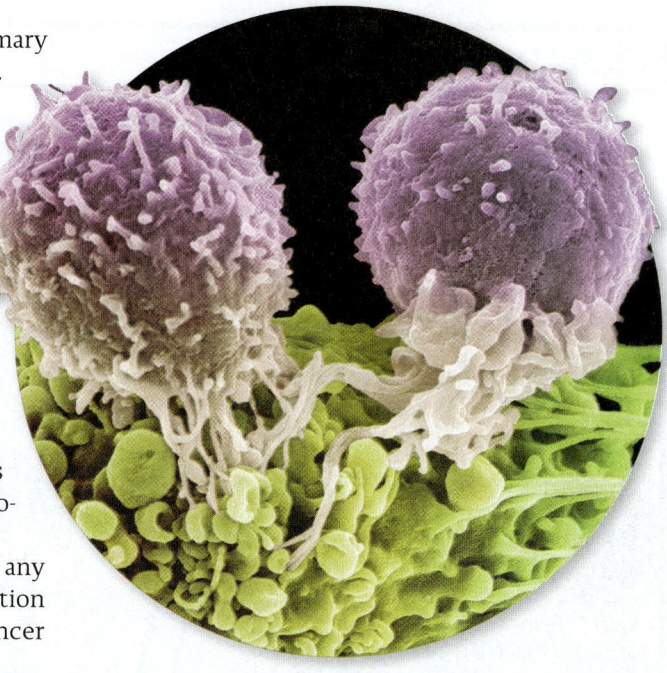

Figure 13.7   T cells attacking a cancer cell.
We've been showing drawings of the markers and receptors bringing immune cells and foreign cells together. That is how these T cells bound to this cancer cell.

Fish—and, in fact, all jawed vertebrates—share many features of the specific immune system.

> ## 13.4 LEARNING OUTCOMES—Assess Your Progress
>
> **13.** List the three major types of cells that T cells will differentiate into after stimulation.
> **14.** Describe the main functions of these three types of T cells.
> **15.** Note the similarities and differences between gamma-delta T cells and the other T cells.

Table 13.5 T-Cell Activation

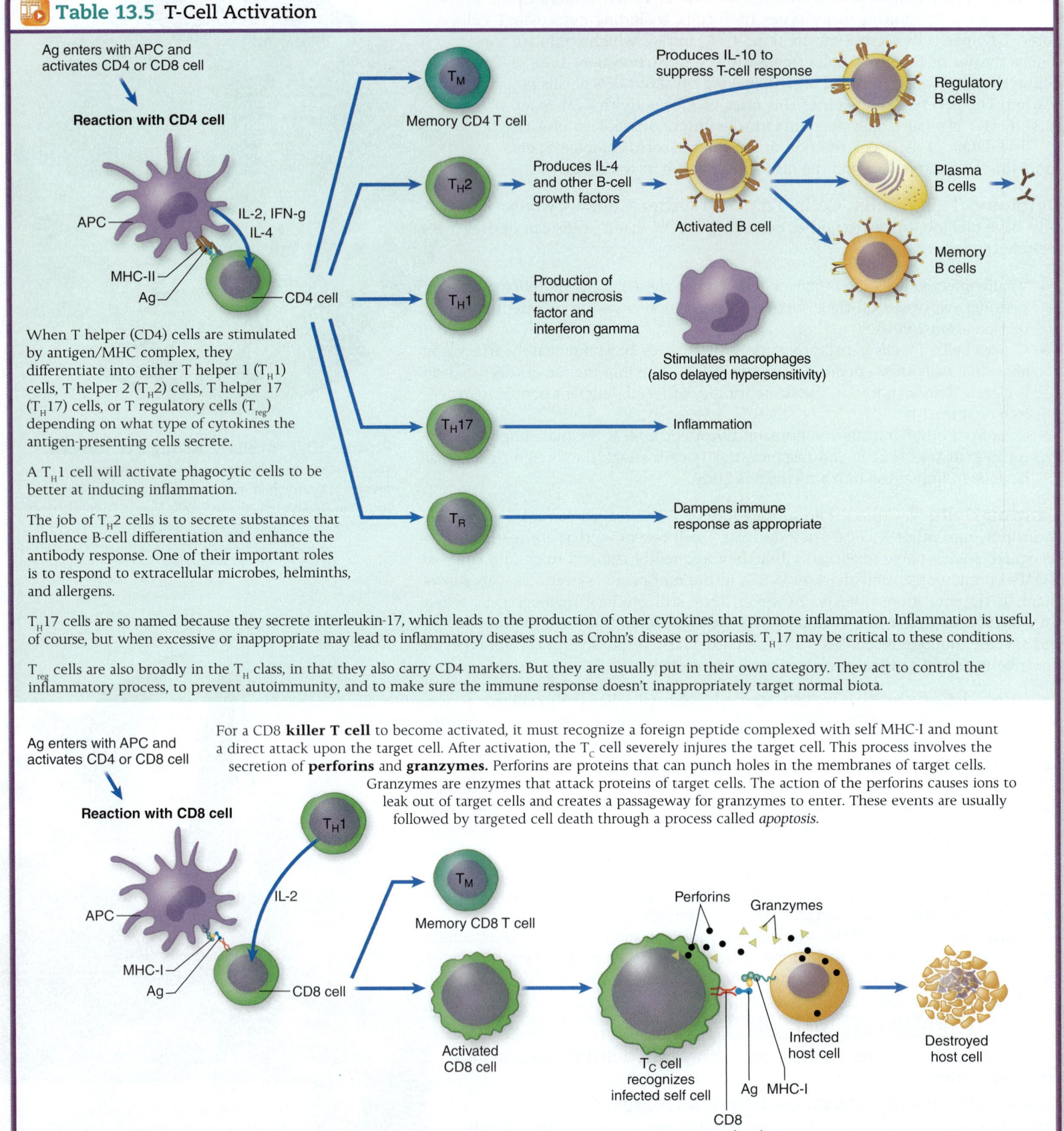

Ag enters with APC and activates CD4 or CD8 cell

Reaction with CD4 cell

APC

IL-2, IFN-g
IL-4

MHC-II

Ag

CD4 cell

T$_M$ — Memory CD4 T cell

T$_H$2 — Produces IL-4 and other B-cell growth factors → Activated B cell

Produces IL-10 to suppress T-cell response → Regulatory B cells

Plasma B cells

Memory B cells

T$_H$1 — Production of tumor necrosis factor and interferon gamma → Stimulates macrophages (also delayed hypersensitivity)

T$_H$17 — Inflammation

T$_R$ — Dampens immune response as appropriate

When T helper (CD4) cells are stimulated by antigen/MHC complex, they differentiate into either T helper 1 (T$_H$1) cells, T helper 2 (T$_H$2) cells, T helper 17 (T$_H$17) cells, or T regulatory cells (T$_{reg}$) depending on what type of cytokines the antigen-presenting cells secrete.

A T$_H$1 cell will activate phagocytic cells to be better at inducing inflammation.

The job of T$_H$2 cells is to secrete substances that influence B-cell differentiation and enhance the antibody response. One of their important roles is to respond to extracellular microbes, helminths, and allergens.

T$_H$17 cells are so named because they secrete interleukin-17, which leads to the production of other cytokines that promote inflammation. Inflammation is useful, of course, but when excessive or inappropriate may lead to inflammatory diseases such as Crohn's disease or psoriasis. T$_H$17 may be critical to these conditions.

T$_{reg}$ cells are also broadly in the T$_H$ class, in that they also carry CD4 markers. But they are usually put in their own category. They act to control the inflammatory process, to prevent autoimmunity, and to make sure the immune response doesn't inappropriately target normal biota.

For a CD8 **killer T cell** to become activated, it must recognize a foreign peptide complexed with self MHC-I and mount a direct attack upon the target cell. After activation, the T$_C$ cell severely injures the target cell. This process involves the secretion of **perforins** and **granzymes**. Perforins are proteins that can punch holes in the membranes of target cells. Granzymes are enzymes that attack proteins of target cells. The action of the perforins causes ions to leak out of target cells and creates a passageway for granzymes to enter. These events are usually followed by targeted cell death through a process called *apoptosis*.

Ag enters with APC and activates CD4 or CD8 cell

Reaction with CD8 cell

APC

IL-2

MHC-I

Ag

CD8 cell

T$_H$1

T$_M$ — Memory CD8 T cell

Activated CD8 cell

T$_C$ cell recognizes infected self cell

CD8 molecule

Ag   MHC-I

Perforins   Granzymes

Infected host cell

Destroyed host cell

13.5 Stages III and IV: B-Cell Response

Activation of B Lymphocytes: Clonal Expansion and Antibody Production

At the same time that the T-cell system is being activated by antigen, B cells are being stimulated as well. The immunologic activation of most B cells also requires a series of events. **Table 13.6** contains the details.

Products of B Lymphocytes: Antibody Structure and Functions

Earlier we saw how a basic immunoglobulin (Ig) molecule can have so many different variable regions. Let us view this structure once again, using an IgG molecule as a model. There are two functionally distinct segments called *fragments*. The two "arms" that bind antigen are termed *antigen binding fragments (Fabs)*, and the rest of the molecule is the crystallizable fragment (Fc), so called because it was the first to be crystallized in pure form. The basic immunoglobulin molecule is a composite of four polypeptide chains: a pair of identical heavy (H) chains and a pair of identical light (L) chains (**figure 13.8**). One light chain is bonded to one heavy chain, and the two heavy chains are bonded to one another with disulfide bonds, creating a symmetrical, Y-shaped arrangement.

The end of each Fab fragment (consisting of the variable regions of the heavy and light chains) folds into a groove that will accommodate one epitope. The presence of a special region at the site of attachment between the Fab and Fc fragments allows swiveling of the Fab fragments. In this way, they can change their angle to accommodate nearby antigen sites that vary slightly in distance and position. Figure 13.8 shows two views of antibody structure.

Don't Give Up!
Specific immunity is very complex. Don't be overwhelmed with all the illustrations and tables in this chapter. Seeing the processes with the use of drawings is better than just reading about them, but they are still very intimidating. Even scientists studying immunology get overwhelmed. Look at the end results of each process and think about what is important. Ask your instructor what the key concepts are for T-cell activation, B-cell activation, and so on. Just don't ask them "Will it be on the test?"!

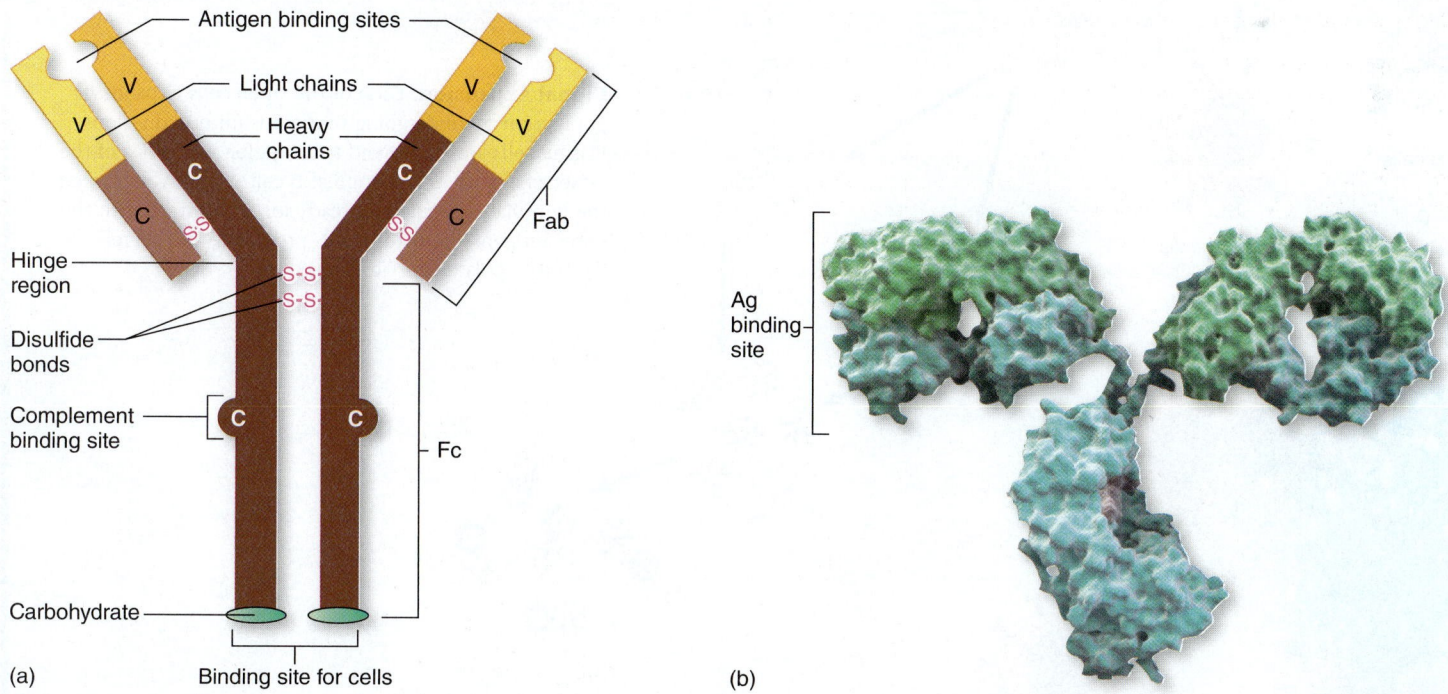

(a)

(b)

Figure 13.8 Working models of antibody structure. **(a)** Diagrammatic view of IgG depicts the principal functional areas (Fabs and Fc) of the molecule. Each Fab contains a hypervariable region (V) and a constant, nonvariable region (C). **(b)** Realistic model of immunoglobulin shows the tertiary and quaternary structure achieved by additional intrachain and interchain bonds.

Table 13.6 B-Cell Activation

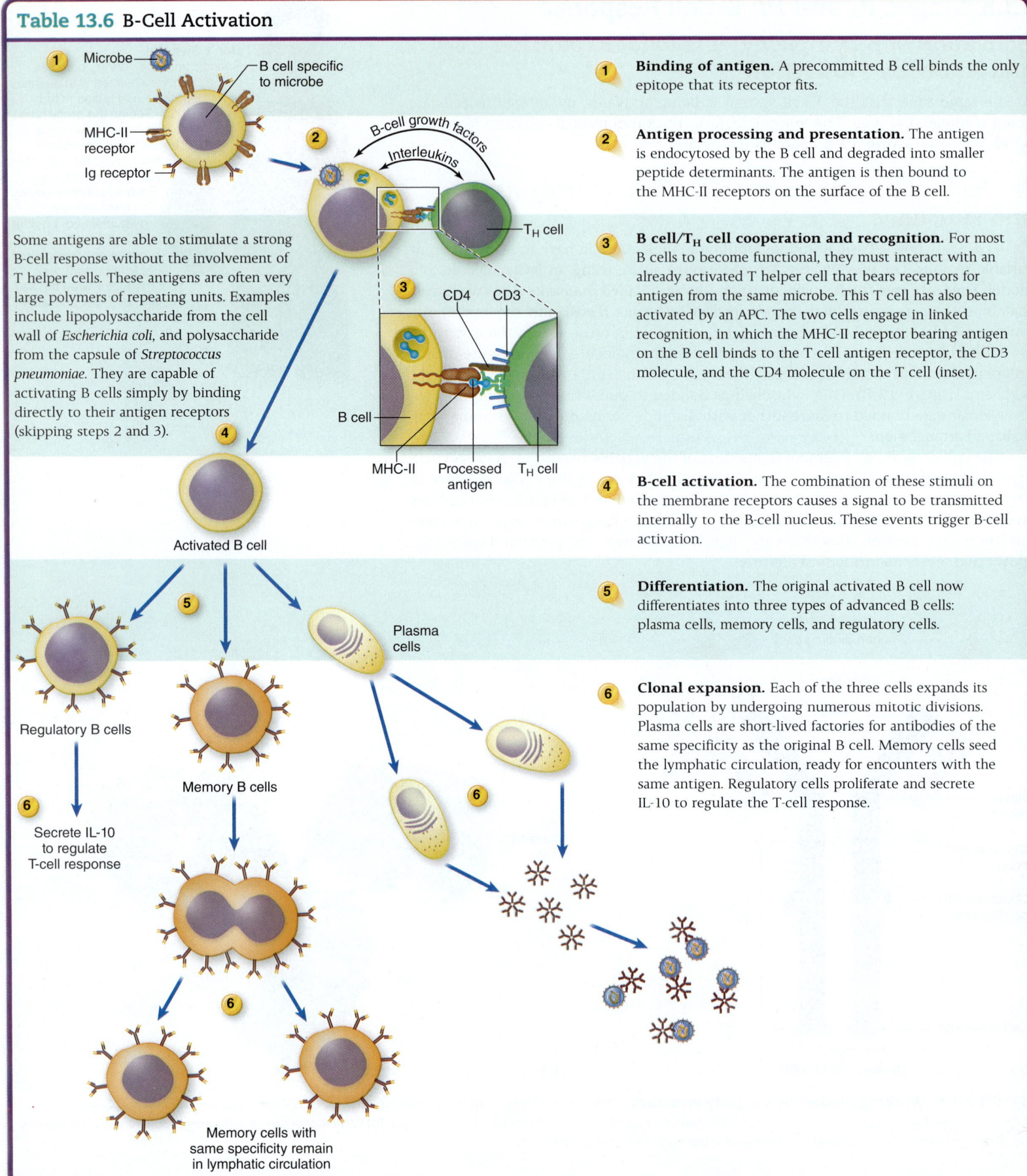

Some antigens are able to stimulate a strong B-cell response without the involvement of T helper cells. These antigens are often very large polymers of repeating units. Examples include lipopolysaccharide from the cell wall of *Escherichia coli*, and polysaccharide from the capsule of *Streptococcus pneumoniae*. They are capable of activating B cells simply by binding directly to their antigen receptors (skipping steps 2 and 3).

1 **Binding of antigen.** A precommitted B cell binds the only epitope that its receptor fits.

2 **Antigen processing and presentation.** The antigen is endocytosed by the B cell and degraded into smaller peptide determinants. The antigen is then bound to the MHC-II receptors on the surface of the B cell.

3 **B cell/T$_H$ cell cooperation and recognition.** For most B cells to become functional, they must interact with an already activated T helper cell that bears receptors for antigen from the same microbe. This T cell has also been activated by an APC. The two cells engage in linked recognition, in which the MHC-II receptor bearing antigen on the B cell binds to the T cell antigen receptor, the CD3 molecule, and the CD4 molecule on the T cell (inset).

4 **B-cell activation.** The combination of these stimuli on the membrane receptors causes a signal to be transmitted internally to the B-cell nucleus. These events trigger B-cell activation.

5 **Differentiation.** The original activated B cell now differentiates into three types of advanced B cells: plasma cells, memory cells, and regulatory cells.

6 **Clonal expansion.** Each of the three cells expands its population by undergoing numerous mitotic divisions. Plasma cells are short-lived factories for antibodies of the same specificity as the original B cell. Memory cells seed the lymphatic circulation, ready for encounters with the same antigen. Regulatory cells proliferate and secrete IL-10 to regulate the T-cell response.

Table 13.7 Summary of Antibody Functions

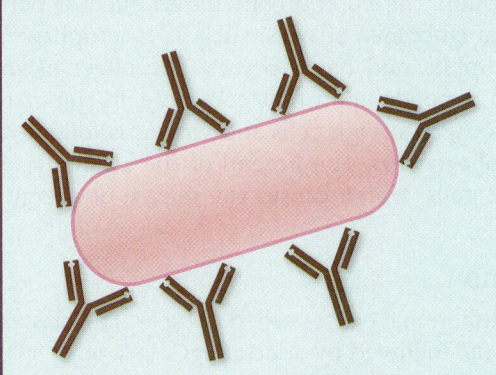

Antibodies coat the surface of a bacterium, preventing its normal function and reproduction in various ways.

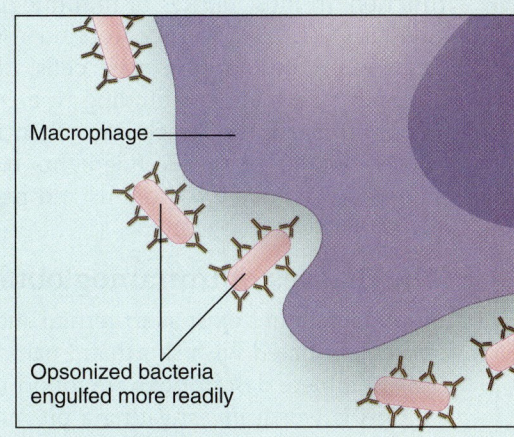

Macrophage

Opsonized bacteria engulfed more readily

Antibodies called opsonins stimulate **opsonization** (ahp″-son-uh-zaz′-shun), a process that makes microbes more readily recognized by phagocytes, which dispose of them. Opsonization has been likened to putting handles on a slippery object to provide phagocytes a better grip.

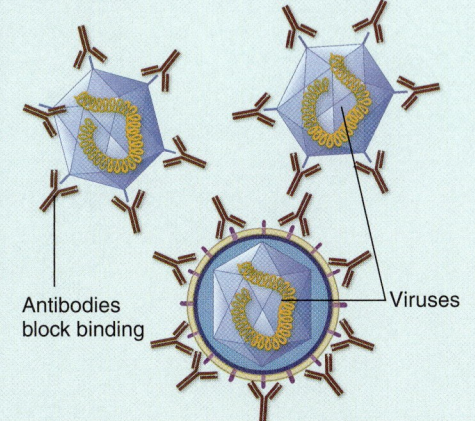

Antibodies block binding

Viruses

In **neutralization** reactions, antibodies fill the surface receptors on a virus or the active site on a microbial enzyme to prevent it from attaching normally.

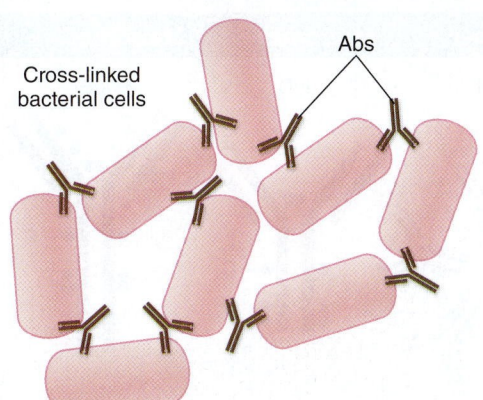

Cross-linked bacterial cells

Abs

The capacity for antibodies to aggregate, or **agglutinate,** antigens is the consequence of their cross-linking cells or particles into large clumps. Agglutination renders microbes immobile and enhances their phagocytosis. This is a principle behind certain immune tests discussed in chapter 15.

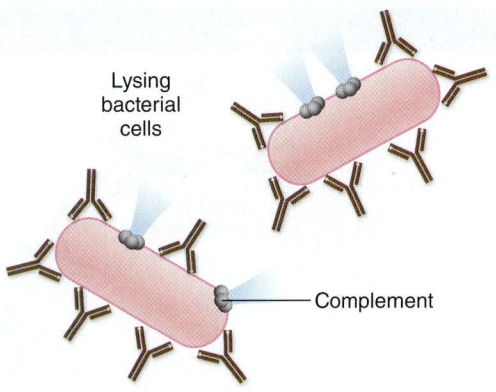

Lysing bacterial cells

Complement

The interaction of an antibody with complement can result in the specific rupturing of cells and some viruses.

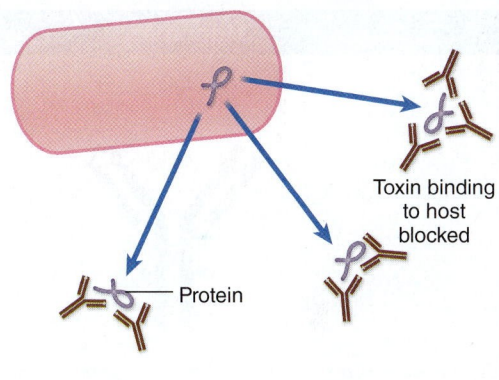

Toxin binding to host blocked

Protein

An **antitoxin** is a special type of antibody that neutralizes bacterial exotoxins.

Antibody-Antigen Interactions and the Function of the Fab

The site on the antibody where the epitope binds is composed of a *hypervariable region,* whose amino acid content can be extremely varied. The specificity of antigen binding sites for antigens is very similar to enzymes and substrates. Because the specificity of the two Fab sites is identical, an Ig molecule can bind epitope on the same cell or on two separate cells and thereby link them.

The principal activity of an antibody is to unite with, immobilize, call attention to, or neutralize the antigen for which it was formed. Follow along in **table 13.7** to see how each of these works.

Functions of the Fc Fragment

Although the Fab fragments bind antigen, the Fc fragment has a different binding function. In most classes of immunoglobulin, the Fc end contains an effector portion that can bind to receptors on the membranes of cells, such as macrophages, neutrophils, eosinophils, mast cells, basophils, and lymphocytes. The effect of an antibody's Fc fragment binding to a cell depends upon that cell's role. In the case of opsonization, the attachment of antibody to foreign cells and viruses is followed by the binding of the Fc fragments to phagocytes. The Fc end of the antibody of allergy (IgE) binds to basophils and mast cells, which causes the release of allergic mediators such as histamine.

The Classes of Immunoglobulins

Immunoglobulins exist as structural and functional classes called *isotypes*. The classes are differentiated with shorthand names (Ig, followed by a letter: IgG, IgA, IgM, IgD, IgE). Complete descriptions are found in **table 13.8.**

IgA is worth investigating a bit more. The two forms of IgA are (1) a monomer that circulates in small amounts in the blood and (2) a dimer that is a significant component of the mucous and serous secretions of the salivary glands, intestine, nasal membrane, breast, lung, and genitourinary tract. The dimer, called secretory IgA, is formed by two monomers held together by a J chain. To facilitate the transport of IgA

Table 13.8 Characteristics of the Immunoglobulin (Ig) Classes

	Monomer	Dimer, Monomer	Pentamer	Monomer	Monomer
	IgG	IgA	IgM	IgD	IgE
Number of Antigen Binding Sites	2	4, 2	10	2	2
Molecular Weight	150,000	170,000–385,000	900,000	180,000	200,000
Percentage of Total Antibody in Serum	80%	13%	6%	1%	0.002%
Average Half-Life in Serum (Days)	23	6	5	3	2.5
Crosses Placenta?	Yes	No	No	No	No
Fixes Complement?	Yes	No	Yes	No	No
Fc Binds to	Phagocytes				Mast cells and basophils
Biological Function	Monomer produced by plasma cells in a primary response and by memory cells responding the second time to a given antigenic stimulus; most prevalent antibody circulating throughout the tissue fluids and blood; neutralizes toxins, opsonizes, fixes complement	Dimer is secretory antibody on mucous membranes; monomer in small quantities in blood	Produced at first response to antigen; can serve as B-cell receptor	Receptor on B cells; triggering molecule for B-cell activation	Antibody of allergy; worm infections; mediates anaphylaxis, asthma, etc.

across membranes, a secretory piece is later added. IgA coats the surface of mucous membranes and also is suspended in saliva, tears, colostrum, and mucus. It provides the most important specific local immunity to enteric, respiratory, and genitourinary pathogens. During lactation, the breast becomes a site for the proliferation of lymphocytes that produce IgA. The very earliest secretion of the breast, a thin, yellow milk called **colostrum,** is very high in IgA. These antibodies form a protective coating in the gastrointestinal tract of a nursing infant that guards against infection by a number of enteric pathogens (*Escherichia coli*, *Salmonella*, poliovirus, rotavirus). This is one of the reasons that new mothers are encouraged to nurse their newborns, at least for a few weeks, to take advantage of the immune properties of this first milk. Protection at this level is especially critical because an infant's own IgA and natural intestinal barriers are not yet developed. As with immunity *in utero*, the necessary antibodies will be donated only if the mother herself has active immunity to the microbe through a prior infection or vaccination.

Monitoring Antibody Production over Time: Primary and Secondary Responses to Antigens

We can learn a great deal about how the immune system reacts to an antigen by studying the levels of antibodies in serum over time. This level is expressed quantitatively as the **titer** (ty′-tur), or concentration of antibodies. **Table 13.9** illustrates one of the most important features of specific immunity: memory.

Table 13.9 Primary and Secondary Response to Antigens

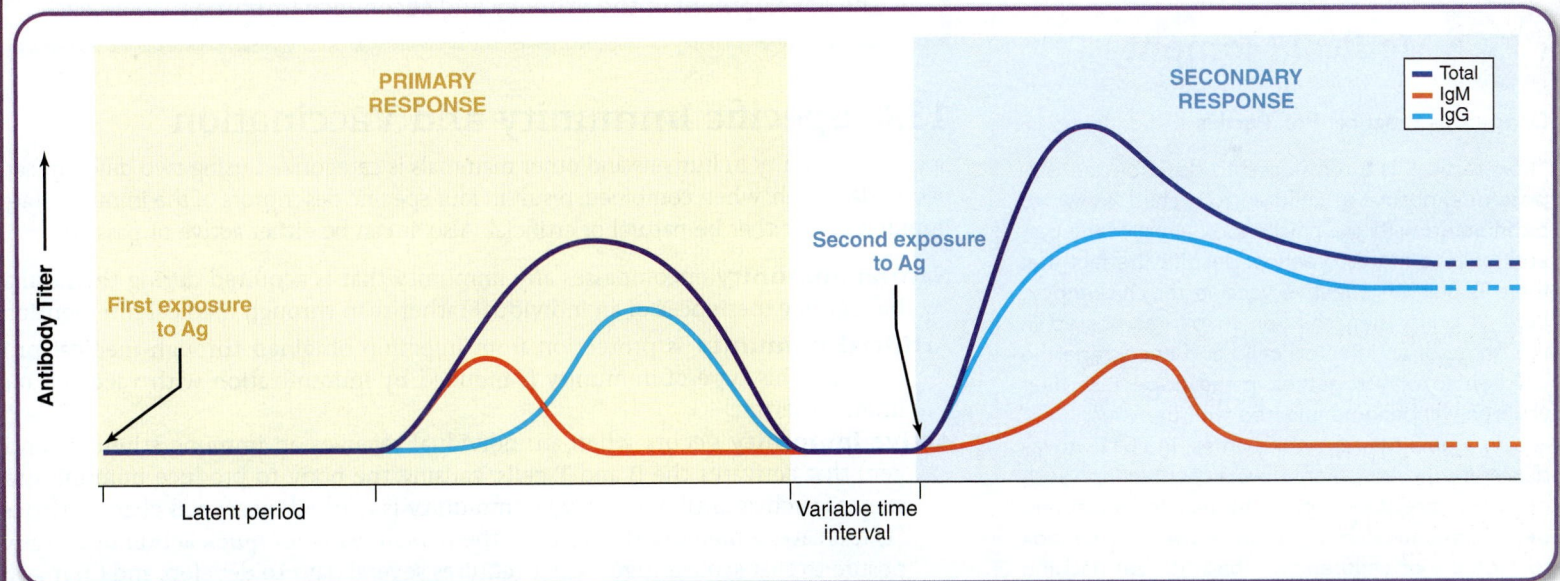

Upon the first exposure to an antigen, the system undergoes a **primary response.** The earliest part of this response, the *latent period*, is marked by a lack of antibodies for that antigen, but much activity is occurring. During this time, the antigen is being concentrated in lymphoid tissue and is being processed by the correct clones of B lymphocytes. As plasma cells synthesize antibodies, the serum titer increases to a certain plateau and then tapers off to a low level over a few weeks or months. Early in the primary response, most of the antibodies are the IgM type, which is the first class to be secreted by plasma cells. Later, the class of the antibodies (but not their specificity) is switched to IgG or some other class (IgA or IgE).

After the initial response, there is no activity, but memory cells of the same specificity are seeded throughout the lymphatic system.

When the immune system is exposed again to the same immunogen within weeks, months, or even years, a **secondary response** occurs. The rate of antibody synthesis, the peak titer, and the length of antibody persistence are greatly increased over the primary response. The speed and intensity seen in this response are attributable to the memory B cells that were formed during the primary response. The secondary response is also called the **anamnestic response** (from the Greek word for "memory"). The advantage of this response is evident: It provides a quick and potent strike against subsequent exposures to infectious agents.

In certain settings, the oral polio vaccine is dispensed onto a sugar cube to make it palatable to children.

Medical Moment

Dangerous Practice: Pox Parties

"Pox parties" is a term used to describe the purposeful exposure of children to a child known to be infected with the chickenpox virus. Many parents believe in this practice, despite the fact that there is now an effective vaccine for chickenpox. Parents bring their children to parties hosted in the home of an infected child and encourage the children to play together in the hope that their children will become infected with the virus.

In a new twist on pox parties, in 2011, it was discovered that parents were contacting each other via Facebook in order to mail items that had been contaminated by a child with chickenpox to mothers of children who had not yet had the disease. Infected items included lollipops that infected children had put in their mouth and items of clothing worn by children with chickenpox. Needless to say, this is a dangerous practice.

A Facebook site had the following post on its profile page in March 2014: "Shingles is caused by a lack of exposure to children with chickenpox. Find a pox party near you." The statement is blatantly false. Shingles is caused by *reactivation* of the chickenpox virus, which you are protected against if you receive the vaccination.

It is a well-accepted principle that memory B and T cells are only created from clones activated by a specific antigen. This provides a much quicker and more effective response on the second exposure and all exposures afterward. But researchers are now investigating a phenomenon that has been suspected for some time and confirmed in rigorous studies. It seems that exposure to a particular antigen can result in memory cells that will recognize antigens that are *chemically related* to it, even if those antigens have not been seen by the host. This might explain the well-known phenomenon, seen most clearly in developing countries, that vaccines against one disease can provide some protection against others. In Africa, for example, vaccinating against measles also cuts deaths from pneumonia, sepsis, and diarrhea by a third.

This realization upsets the long-held view that memory only exists because of specific exposures, but it makes sense if we consider how activation of specific immunity occurs, via recognition of epitopes, small pieces of macromolecules on the surfaces of microbes. If other microbes share those chemical signatures (epitopes), memory cells will react against them as well. This is a promising development, because it could result in using nonpathogenic microbes in vaccines to protect against more dangerous ones.

13.5 LEARNING OUTCOMES—Assess Your Progress

16. Diagram the steps in B-cell activation, including all types of cells produced.
17. Make a detailed drawing of an antibody molecule.
18. Explain the various end results of antibody binding to an antigen.
19. List the five types of antibodies and important facts about each.
20. Draw and label a graph—with time on the horizontal axis—that shows the development of the primary and secondary immune responses.

13.6 Specific Immunity and Vaccination

Specific immunity in humans and other mammals is categorized using two different sets of criteria, which, when combined, result in four specific descriptors of the immune state. Immunity can either be natural or artificial. Also, it can be either active or passive.

Natural immunity encompasses any immunity that is acquired during the normal biological experiences of an individual rather than through medical intervention.

Artificial immunity is protection from infection obtained through medical procedures. This type of immunity is induced by immunization with vaccines and immune serum.

Active immunity occurs when an individual receives an immune stimulus (antigen) that activates the B and T cells, causing the body to produce immune substances such as antibodies. Active immunity is marked by several characteristics: (1) It creates a memory that renders the person ready for quick action upon reexposure to that same antigen; (2) it requires several days to develop; and (3) it lasts for a relatively long time, sometimes for life. Active immunity can be stimulated by natural or artificial means.

Passive immunity occurs when an individual receives immune substances (usually antibodies) that were produced actively in the body of another human or animal donor. The recipient is protected for a short time, even though he or she has not had prior exposure to the antigen. It is characterized by (1) lack of memory for the original antigen; (2) lack of production of new antibodies against that disease; (3) immediate onset of protection; and (4) short-term effectiveness, because antibodies have a limited period of function and, ultimately, the recipient's body disposes of them. Passive immunity can also be natural or artificial in origin.

Table 13.10 illustrates the various possible combinations of acquired immunities.

Table 13.10 The Four Types of Acquired Immunity

Natural Immunity is acquired through the normal life experiences of a human and is not induced through medical means.

Active
After recovering from infectious disease, a person will generally be actively resistant to reinfection for a period that varies according to the disease. In the case of childhood viral infections such as measles, mumps, and rubella, this natural active stimulus provides nearly lifelong immunity. Other diseases result in a less extended immunity of a few months to years (such as pneumococcal pneumonia and shigellosis), and reinfection is possible. Even a subclinical infection can stimulate natural active immunity. This probably accounts for the fact that some people are immune to an infectious agent without ever having been noticeably infected with or vaccinated for it.

Passive
Natural, passively acquired immunity occurs only as a result of the prenatal and postnatal mother-child relationship. During fetal life, IgG antibodies circulating in the maternal bloodstream are small enough to pass or be actively transported across the placenta. This natural mechanism provides an infant with a mixture of many maternal antibodies that can protect it for the first few critical months outside the womb, while its own immune system is gradually developing active immunity. Depending on the microbe, passive protection lasts anywhere from a few months to a year.

Another source of natural passive immunity comes to the baby by way of the mother's milk. Although the human infant acquires 99% of natural passive immunity *in utero* and only about 1% through nursing, the milk-borne antibodies provide a special type of intestinal protection that is not available from transplacental antibodies.

Artificial Immunity is that produced purposefully through medical procedures.

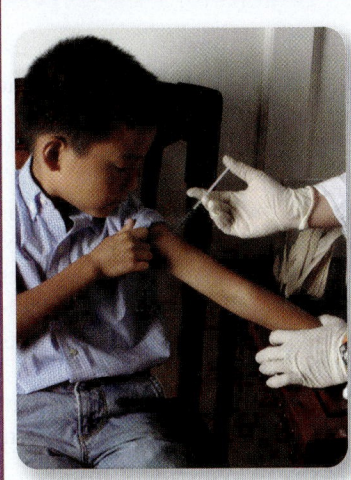

Active
Vaccination exposes a person to a specially prepared microbial (antigenic) stimulus, which then triggers the immune system to produce antibodies and lymphocytes to protect the person upon future exposure to that microbe. As with natural active immunity, the degree and length of protection vary.

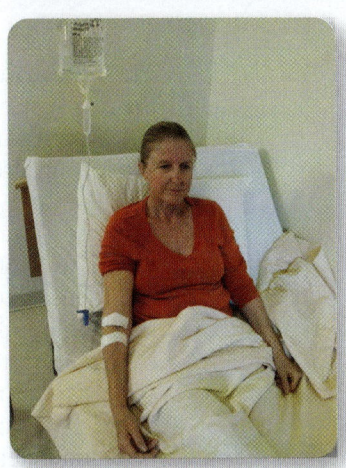

Passive
Passive immunotherapy involves a preparation that contains specific antibodies against a particular infectious agent. Pooled human serum from donor blood (gamma globulin) and immune serum globulins containing high quantities of antibodies are frequently used.

Immunization: A Lively History

The basic notion of immunization has existed for thousands of years. It probably stemmed from the observation that persons who had recovered from certain communicable diseases rarely got a second case. Undoubtedly, the earliest crude attempts involved bringing a susceptible person into contact with a diseased person or animal. The first recorded attempt at immunization occurred in sixth-century China. It consisted of drying and grinding up smallpox scabs and blowing them with a straw into the nostrils of vulnerable family members. By the 10th century, this practice had changed to the deliberate inoculation of dried pus from the smallpox pustules of one patient into the arm of a healthy person, a technique later called **variolation** (variola

Figure 13.9 **Ridicule of the new idea of vaccination.** This cartoon appeared during Edward Jenner's advocacy for vaccination against smallpox.

is the smallpox virus). This method was used in parts of the Far East for centuries before it was brought to England in 1721.

Although the principles of the technique had some merit, unfortunately many recipients and their contacts died of smallpox. This outcome vividly demonstrates a cardinal rule for a workable vaccine: It must contain an antigen that will provide protection but not cause the disease.

Eventually, this human experimentation paved the way for the first really effective vaccine, developed by the English physician Edward Jenner in 1796. Jenner's work gave rise to the words **vaccine** and *vaccination* (from L., *vacca*, "cow"), which now apply to any immunity obtained by inoculation with selected antigens. Jenner was inspired by the case of a dairymaid who had been infected by a pustular infection called cowpox. This related virus afflicts cattle but causes a milder condition in humans. She explained that she and other milkmaids had remained free of smallpox. Other residents of the region expressed a similar confidence in the cross-protection of cowpox. To test the effectiveness of this new vaccine, Jenner prepared material from human cowpox lesions and inoculated a young boy. When challenged 2 months later with an injection of crusts from a smallpox patient, the boy proved immune.

Jenner's discovery—that a less pathogenic agent could confer protection against a more pathogenic one—is especially remarkable in view of the fact that microscopy was still in its infancy and the nature of viruses was unknown. At first, the use of the vaccine was regarded with some fear and skepticism (**figure 13.9**). When Jenner's method proved successful and word of its significance spread, it was eventually adopted in many other countries. In 1979, the World Health Organization declared that smallpox had been eradicated. As a result, smallpox vaccination had been discontinued until recently, due to the threat of bioterrorism.

Passive Immunization

The only natural forms of passive immunization occur as (1) a fetus develops and encounters selected antibodies that are able to cross the placental barrier, and (2) a newborn nurses and receives IgA in breast milk that is secreted at birth and for a short time afterward.

The first attempts at artificial passive immunization involved the transfusion of horse serum containing antitoxins to prevent tetanus and to treat patients exposed to diphtheria. Since then, antisera from animals have been replaced with products of human origin that function with various degrees of specificity. Immune serum globulin (ISG), sometimes called *gamma globulin*, contains immunoglobulin extracted from the pooled blood of human donors. The method of processing ISG concentrates the antibodies to increase potency and eliminates potential pathogens (such as hepatitis B and HIV). It is a treatment of choice for replacing antibodies in immunodeficient patients.

A preparation called specific immune globulin (SIG) is derived from a more defined group of donors. Companies that prepare SIG obtain serum from patients who are convalescing and in a hyperimmune state after such infections as pertussis, tetanus, chickenpox, and hepatitis B. These globulins are preferable to ISG because they contain higher titers of specific antibodies obtained from a smaller pool of patients. Although donated immunities only last a relatively short time, they act immediately and can protect patients for whom no other useful medication or vaccine exists.

Artificial Active Immunity: Vaccination

The basic principle behind vaccination is to stimulate a primary response and a memory response that primes the immune system for future exposure to a virulent pathogen. If this pathogen enters the body, the immune response will be immediate, powerful, and sustained.

Vaccines have profoundly reduced the prevalence and impact of many infectious diseases that were once common and often deadly. Initially, the emphasis was on immunizing babies and children against formerly common childhood diseases, like measles, mumps, and rubella. Recent years have seen an additional push to immunize adolescents and adults against conditions such as human papilloma virus (HPV), *Streptococcus pneumoniae,* and shingles. Vaccines are also being developed for threats to human health that do not involve microbes at all.

In this section, we survey the principles of vaccine preparation and important considerations surrounding vaccine indication and safety. (Vaccines are also given specific consideration in later chapters on infectious diseases and organ systems.)

Principles of Vaccine Preparation

In natural immunity, an infectious agent stimulates a relatively long-term protective response. In artificial active immunity, the objective is to obtain this same response with a modified version of the microbe or its components. Qualities of an effective vaccine are as follows:

- It should protect against exposure to natural, wild forms of the pathogen.
- It should have a low level of adverse side effects or toxicity and not cause harm.
- It should stimulate both antibody (B-cell) response and cell-mediated (T-cell) response.
- It should have long-term, lasting effects (produce memory).
- It should not require numerous doses or boosters.
- It should be inexpensive, have a relatively long shelf life, and be easy to administer.

Vaccine preparations can be broadly categorized as either whole-organism or part-of-organism preparations. These categories also have subcategories:

1. Whole cells or viruses

    a. live, attenuated cells or viruses
    b. killed cells or inactivated viruses

2. Part-of-organism preparations: antigenic molecules derived from bacterial cells or viruses (subunits)

    a. subunits derived from cultures of cells or viruses
    b. subunits chemically synthesized to mimic natural molecules found on pathogens
    c. subunits manufactured via genetic engineering
    d. subunits conjugated with proteins (often from other microbes) to make them more immunogenic. These are called **conjugated vaccines.**

These categories are also shown in **table 13.11.**

Note: As of 2008, there were no vaccines available in the United States that consisted of killed whole bacteria. The last two available were those for cholera and plague.

Development of New Vaccines

Despite considerable successes, dozens of bacterial, viral, protozoal, and fungal diseases still remain without a functional vaccine. At the present time, no reliable vaccines are available for malaria, HIV/AIDS, various diarrheal diseases, respiratory diseases, and worm infections that affect over 200 million people per year worldwide. Worse than that, most existing vaccines are out of reach for much of the world's population.

Currently, much attention is being focused on newer strategies for vaccine preparation that employ antigen synthesis, recombinant DNA, and gene cloning technology. **DNA vaccines** are one promising new approach to immunization. The technique in these formulations is very similar to gene therapy as described in

Scientists are working on foods that can have vaccines genetically engineered into them to make vaccines more widely available in developing countries.

Table 13.11 Types of Vaccines

Whole-Cell Vaccines

Whole cells or viruses are very effective immunogens, since they are so large and complex. Depending on the vaccine, these are either killed or attenuated.

Killed vaccines (viruses are termed "inactivated" instead of "killed") are prepared by cultivating the desired strain or strains of a bacterium or virus and treating them with chemicals, radiation, heat, or some other agent that does not destroy antigenicity. The hepatitis A vaccine and three forms of the influenza vaccine contain inactivated viruses. Because the microbe does not multiply, killed vaccines often require a larger dose and more boosters to be effective.

Live attenuated vaccines contain live microbes whose virulence has been *attenuated*, or lessened/eliminated. This is usually achieved by modifying the growth conditions or manipulating microbial genes in a way that eliminates virulence factors. Vaccines for measles, mumps, polio (Sabin), and rubella contain live, nonvirulent viruses.

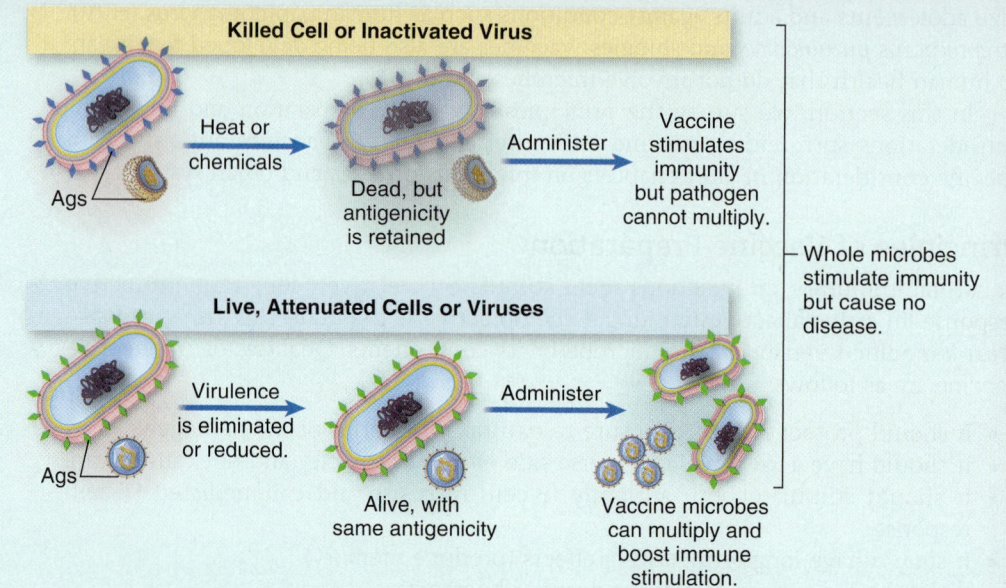

The advantages of live preparations are as follows:

1. Viable microorganisms can multiply and produce infection (but not disease) like the natural organism.
2. They confer long-lasting protection.
3. They usually require fewer doses and boosters than other types of vaccines.
4. They are particularly effective at inducing cell-mediated immunity.

Disadvantages of using live microbes in vaccines are that they require special storage facilities and can conceivably mutate back to become virulent again.

Subunit Vaccines (Parts of Organisms)

If the exact epitopes that stimulate immunity are known, it is possible to produce a vaccine based on a selected component of a microorganism. These vaccines for bacteria are called **subunit vaccines.** The antigens used in these vaccines may be taken from cultures of the microbes, produced by genetic engineering or synthesized chemically.

Examples of component antigens currently in use are the capsules of the pneumococcus and meningococcus, the protein surface antigen of anthrax, and the surface proteins of hepatitis B virus. A special type of vaccine is the **toxoid,** which consists of a purified bacterial exotoxin that has been chemically denatured. By eliciting the production of antitoxins that can neutralize the natural toxin, toxoid vaccines provide protection against diseases such as diphtheria, tetanus, and pertussis.

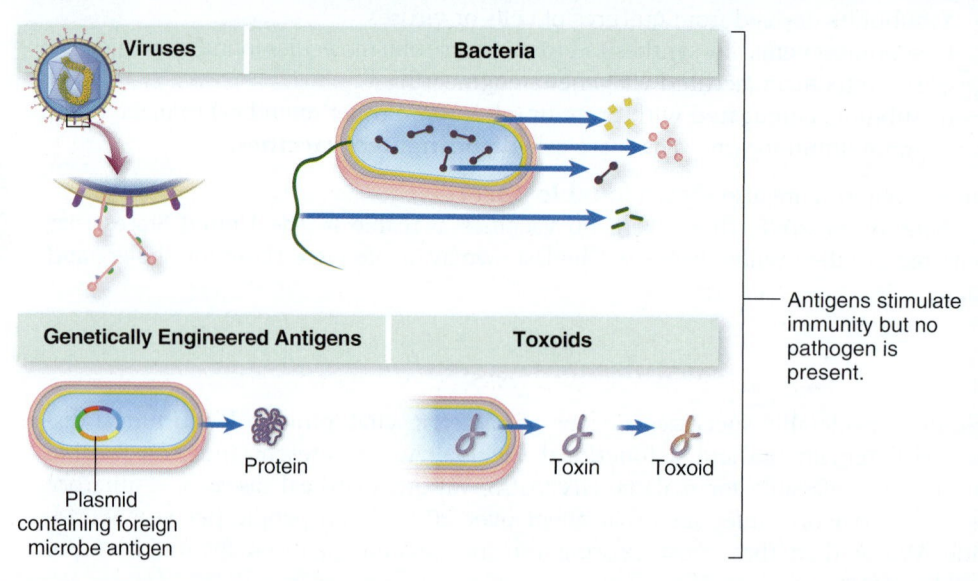

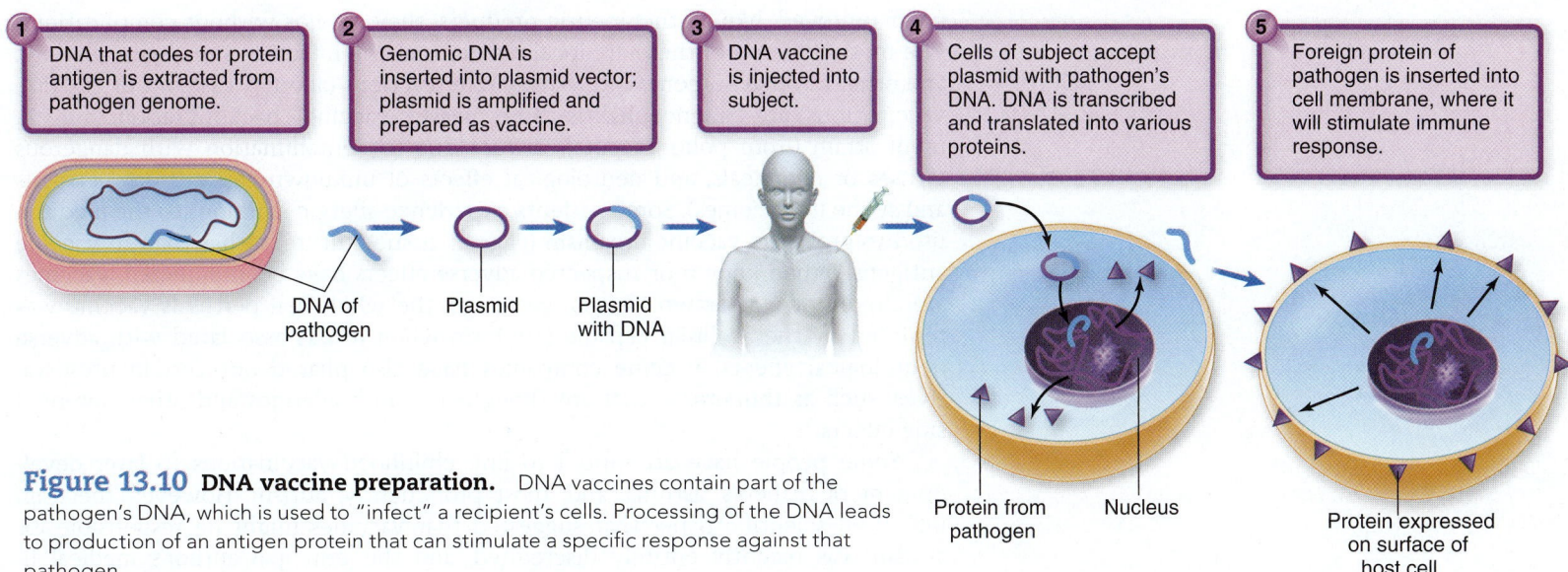

Figure 13.10 **DNA vaccine preparation.** DNA vaccines contain part of the pathogen's DNA, which is used to "infect" a recipient's cells. Processing of the DNA leads to production of an antigen protein that can stimulate a specific response against that pathogen.

chapter 8, except in this case, microbial (not human) DNA is inserted into a plasmid vector and inoculated into a recipient **(figure 13.10)**. The expectation is that the human cells will take up some of the plasmids and express the microbial DNA in the form of proteins. Because these proteins are foreign, they will be recognized during immune surveillance and cause B and T cells to be sensitized and form memory cells.

Vaccination strategies are now under intense investigation for the prevention and treatment of noninfectious diseases. One of the best examples is vaccination for Alzheimer's disease. Researchers are examining whether administering a peptide that is found in the brain plaques of Alzheimer's patients as a vaccine can lead to prevention of the disease. The principle is the same as with vaccines against microbial infection. If T and B memory cells that recognize these disease-inducing proteins can be created in a host via vaccination, once the proteins begin to form in the patient the memory response eliminates them before they cause damage. You can expect to see vaccines developed for a wide variety of conditions–infectious and noninfectious–in the future.

Route of Administration and Side Effects of Vaccines

Most vaccines are injected by subcutaneous, intramuscular, or intradermal routes. One type of the influenza vaccine comes in the form of a nasal spray. Oral (or nasal) vaccines are available for only a few diseases, but they have some distinct advantages. An oral or nasal dose of a vaccine can stimulate protection (IgA) on the mucous membrane of the portal of entry. Oral and nasal vaccines are also easier to give than injections, are more readily accepted, and are well tolerated.

Some vaccines require the addition of a special binding substance, or **adjuvant** (ad'-joo-vunt). An adjuvant is any compound that enhances immunogenicity and prolongs antigen retention at the injection site. The adjuvant precipitates the antigen and holds it in the tissues so that it will be released gradually. Its gradual release presumably facilitates contact with antigen-presenting cells and lymphocytes. The most common adjuvant is alum (aluminum hydroxide salts).

Vaccines must go through many years of trials in experimental animals and human volunteers before they are licensed for general use. Even after they have been approved, like all therapeutic products, they are not without complications. The most common of these are local reactions at the injection site, fever, allergies, and other adverse outcomes. Relatively rare reactions (about 1 case out of 220,000 vaccinations) are panencephalitis (from measles vaccine), back-mutation to a virulent strain (from polio vaccine), disease due to contamination with dangerous viruses or chemicals, and neurological effects of unknown cause (from pertussis and swine flu vaccines). Some patients experience allergic reactions to the medium used to grow the vaccine organism (eggs or tissue culture) rather than to vaccine antigens. When known or suspected adverse effects have been detected, vaccines are altered or withdrawn. Several years ago, the whole-cell pertussis vaccine was replaced by the acellular capsule (aP) form when it was associated with adverse neurological effects. Vaccine companies have also phased out certain preservatives, such as thimerosal, that are thought to cause allergies and other potential side effects.

Some people have attempted to link childhood vaccinations to later development of diabetes, asthma, and, most prominently, autism. However, the original 1998 scientific paper that suggested that vaccines might be responsible for autism was recently entirely discredited, and the principal author's medical license was revoked after authorities found the research and its claims fraudulent. Scores of experts have studied these negative associations and found that they are unsupportable.

In 2011, the Institute of Medicine, an independent nonprofit agency of the widely respected National Academies of Science, published the results of its comprehensive examination of childhood vaccines and stated unequivocally that the MMR vaccine does not cause autism. In contrast, the price of not being vaccinated has become painfully clear. Outbreaks of measles, mumps, diphtheria, polio, typhoid fever, and whooping cough have popped up all over this country in college dormitories, in antivaccination religious communities, and in airplanes.

These outbreaks are often attributed to a decrease in the level of herd immunity, a phenomenon in which a certain percentage of the population is vaccinated, which means that the microbe is unable to maintain its circulation through the population. Think about it—this means that getting vaccinated serves the common good as well as your individual good. At a time when most of the world's population is clamoring for vaccines, some in the developed world are refusing vaccination, essentially relying on others' willingness to be vaccinated to keep them and their children safe.

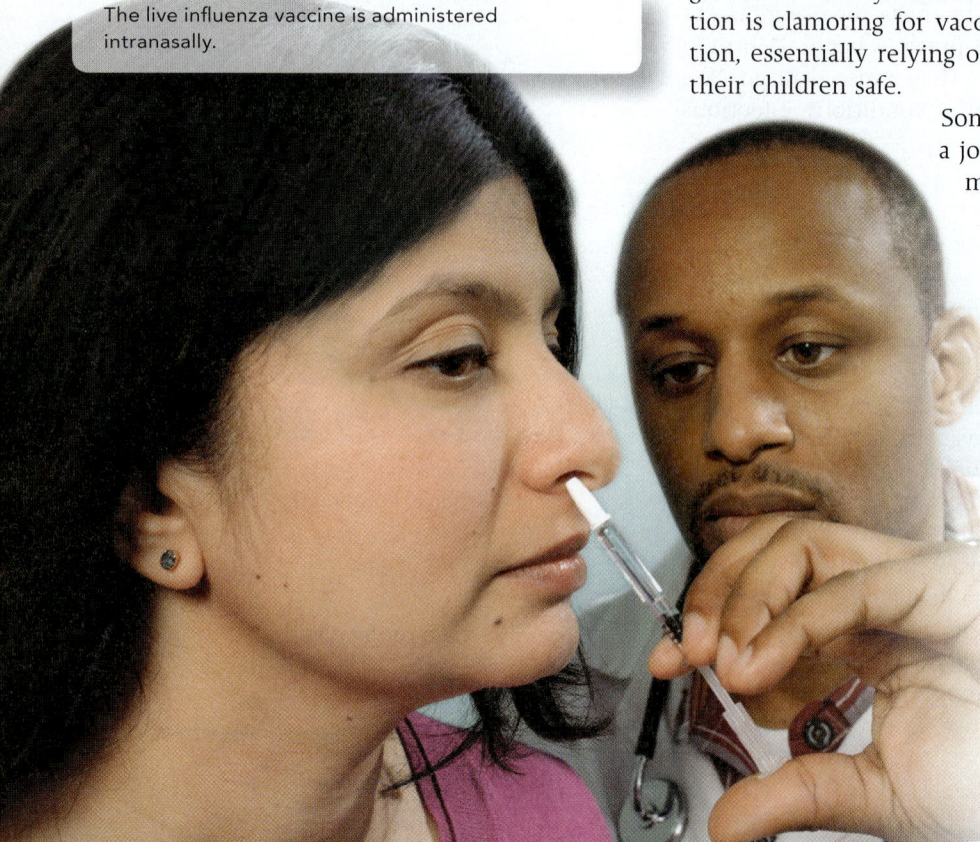

The live influenza vaccine is administered intranasally.

Some have speculated that vaccination has done too good of a job—at least in terms of being so effective for so long that many young parents have no memory of the prevaccination era and don't appreciate the much greater risk of not vaccinating compared to vaccinating. In the decade before measles vaccination began, 3 to 4 million cases occurred each year in the United States. Typically, 300 to 400 children died annually and 1,000 more were chronically disabled due to measles encephalitis. Put simply, childhood vaccines save the lives of 2.5 million children a year (worldwide), according to UNICEF.

Professionals involved in giving vaccinations must understand their inherent risks but also realize that the risks from the infectious disease almost always outweigh the chance of an adverse vaccine reaction. The greatest caution must be exercised in giving live vaccines to immunocompromised or pregnant patients, the latter because of possible risk to the fetus.

Vaccination: For Whom and When?

Until recently, vaccination was recommended for all typical childhood diseases for which a vaccine is available and for adults only in certain special circumstances (health workers, travelers, military personnel). It has become apparent to public health officials that vaccination of adults is often needed in order to boost an older immunization, protect against "adult" infections (e.g., pneumonia in elderly people), or provide special protection in people with certain medical conditions.

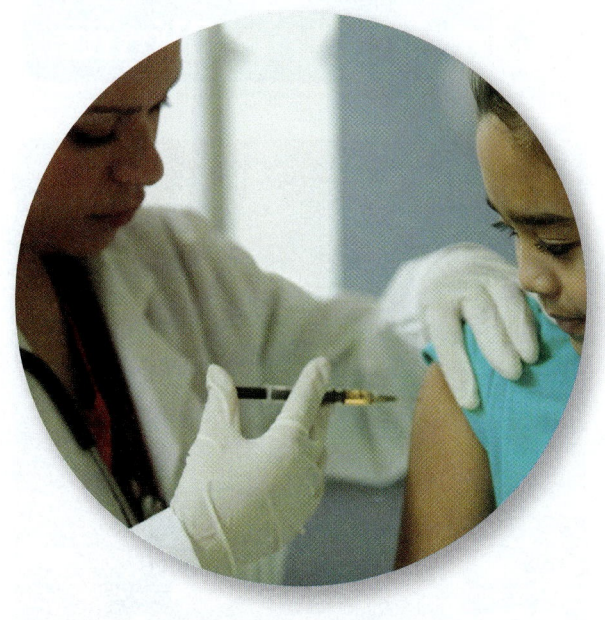

13.6 LEARNING OUTCOMES—Assess Your Progress

21. List and define the four different descriptors of specific immune states.

22. Discuss the qualities of an effective vaccine.

23. Name the two major categories of vaccines and then the subcategories under each.

24. Explain the principle of herd immunity and the risks that unfold when it is not maintained.

CASE FILE WRAP-UP

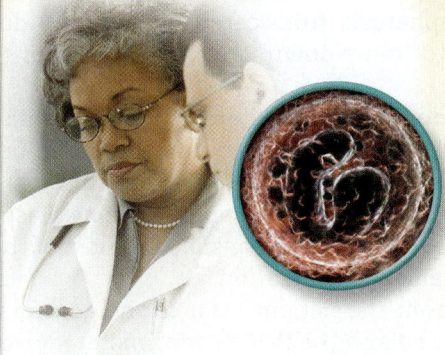

Lymphoma is the most common type of cancer affecting the blood. In lymphoma, lymphocytes multiply and grow in an uncontrolled fashion. These abnormal lymphocytes can travel to the spleen, the lymph nodes, the blood, the bone marrow, and other organs and form a tumor or mass. Both types of lymphocytes (B cells and T cells) can develop into lymphomas. Diffuse large B-cell lymphoma is the most common form of non-Hodgkin's lymphoma. Symptoms may include lymph node enlargement, night sweats, weight loss, and fever.

Because the lymphocytes are abnormal in lymphoma, they are unable to do their job effectively. Individuals with lymphoma are at increased risk of infection. Individuals who are immunosuppressed (like Mr. Campbell from the Case File at the beginning of the chapter) are at increased risk of developing lymphoma; in fact, immunosuppression is a known risk factor for B-cell lymphoma.

IVIG Therapy

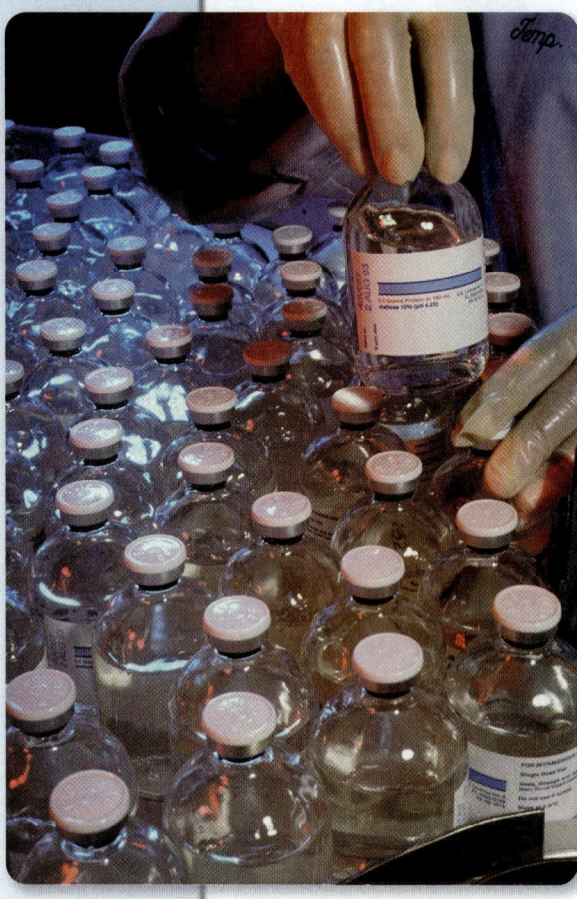

Intravenous immune globulin (IVIG) can be used to treat a variety of conditions affecting immunity. IVIG is comprised primarily of IgG, but other classes of immunoglobulins are also included in smaller amounts (e.g., IgA, IgM, IgD, and IgE). IVIG is used to treat the following autoimmune and immunodeficiency conditions:

- graft versus host disease
- Kawasaki syndrome
- chronic variable immune deficiency
- primary immune deficiency
- multiple sclerosis
- multifocal motor neuropathy
- dermatomyositis
- Guillain-Barré syndrome
- demyelinating inflammatory polyneuropathies
- idiopathic thrombocytopenia purpura (ITP)
- infections in premature infants

IVIG is also used for many off-label conditions for which this therapy has not yet been approved.

IVIG therapy can be thought of as a form of passive immunity for individuals whose immune system may be very immature (i.e., infants who are premature) or individuals who lack the ability to form antibodies. IVIG confers passive immunity through antibodies that are present in pools of donor plasma that has been harvested through the process of plasmapheresis from carefully screened donors. Immunoglobulins may also be used to "tamp down" the immune response in some forms of autoimmune disease.

Because IVIG is derived from human donor plasma, there is always a small risk of transmitting blood-borne pathogens. This risk is very low, however, because donors are strictly screened and the IVIG product itself is put through a variety of processes designed to inactivate any viruses present, including washing of the product to remove most of the IgA (which is responsible for most of the adverse reactions encountered), filtration, and pasteurization.

Informed consent must be obtained from the patient prior to the administration of IVIG. Intravenous access is obtained and the IVIG is administered over 2 to 4 hours. Special filtration tubing may be used. During the transfusion, the patient is monitored carefully for any adverse reactions. Most patients tolerate IVIG well without experiencing any adverse reactions. Rarely, IVIG administration will result in a reaction similar to a blood transfusion reaction (fever, chills, shortness of breath, hives), renal failure, or increased blood viscosity causing thrombotic complications (i.e., deep vein thrombosis, pulmonary embolism, stroke). For this reason, it is advisable for patients to receive their treatments in the hospital or treatment center where adequate staff are present to monitor for complications.

Chapter Summary

13.1 Specific Immunity: The Third and Final Line of Defense

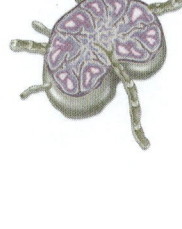

- The surfaces of all cell membranes contain protein markers or receptors. These function in identification, communication, and cell development. Some are also self identity markers.
- Human identity markers are genetically determined by major histocompatibility complexes (MHCs).
- Lymphocytes respond to a specific portion of an antigen called the epitope. A given microorganism has many such epitopes.
- Acquired specific immunity consists of four stages:
  - **Stage I.** Lymphocytes originate in hematopoietic tissue but go on to diverge into two distinct types: B cells and T cells.
  - **Stage II.** Antigen-presenting cells detect invading pathogens and present these antigens to lymphocytes, which recognize the antigen and initiate the specific immune response.
  - **Stage III.** Lymphocytes proliferate, producing clones of progeny that include groups of responder cells, regulator cells, and memory cells.
  - **Stage IV.** Activated T lymphocytes (one of three subtypes) regulate and participate directly in the specific immune responses. Activated B lymphocytes become plasma cells that produce and secrete large quantities of antibodies. Regulatory B cells are also produced.

13.2 Stage I: The Development of Lymphocyte Diversity

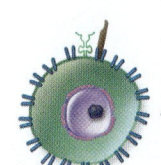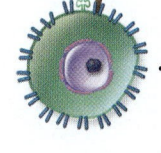

- Immature lymphocytes released from hematopoietic tissue migrate to one of two sites for further development. B cells mature in the stromal cells of the bone marrow. T cells mature in the thymus.
- During development, both B and T cells develop millions of genetically different clones. Together these clones possess enough genetic variability to respond to many millions of different antigens.
- Binding of antigen to a particular clone is called clonal selection. That clone is exclusively amplified in a process called clonal expansion.

13.3 Stage II: Presentation of Antigens

- Antigens or immunogens are proteins or other complex molecules of high molecular weight that trigger the immune response in the host.
- Antigens are molecular markers recognized as foreign to the individual host.

- Antigen-presenting cells (APCs), such as dendritic cells and macrophages, engulf and process foreign antigen and bind the epitope to MHC class II molecules on their cell surface for presentation to CD4 T lymphocytes.

13.4 Stages III and IV: T-Cell Response

- The three main classes of T cells are T helper cells, T regulatory cells, and T cytotoxic cells. Each subset of T cell produces a distinct set of cytokines that stimulate lymphocytes or destroy foreign cells.
- Gamma-delta T cells have T-cell receptors but can respond in a nonspecific manner.
- Natural killer T (NKT) cells contain natural killer (NK) markers and T-cell markers. Natural killer cells are not themselves specific, but they participate in specific immune responses.

13.5 Stages III and IV: B-Cell Response

- B cells produce five classes of antibody: IgM, IgG, IgA, IgD, and IgE.
- Antibodies bind physically to the specific portion of the antigen (the epitope) that stimulates their production, thereby immobilizing the antigen and enabling it to be destroyed by other components of the immune system.
- The memory response means that the second exposure to antigen calls forth a much faster and more vigorous response than the first.

13.6 Specific Immunity and Vaccination

- Active immunity means that your body produces antibodies to a disease agent. If you contract the disease, you can develop natural active immunity. If you are vaccinated, your body will produce artificial active immunity.
- In passive immunity, you receive antibodies from another person. Natural passive immunity comes from the mother. Artificial passive immunity is administered medically.
- Artificial passive immunity usually involves administration of antiserum. Antibodies collected from donors (human or otherwise) are injected into people who need protection immediately.
- Artificial active agents are vaccines that provoke a protective immune response in the recipient but do not cause the actual disease.

Multiple-Choice Questions   Bloom's Levels 1 and 2: Remember and Understand

Select the correct answer from the answers provided.

1. The primary B-cell receptor is
   - a. IgD.
   - b. IgA.
   - c. IgE.
   - d. IgG.

2. In humans, B cells mature in the ___ and T cells mature in the ___ .
   - a. GALT; liver
   - b. bursa; thymus
   - c. bone marrow; thymus
   - d. lymph nodes; spleen

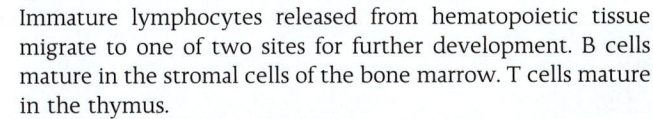
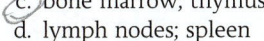

3. The cross-linkage of antigens by antibodies is known as
   a. opsonization.
   c. agglutination.
   b. a cross-reaction.
   d. complement fixation.

4. T-____ cells assist in the functions of certain B cells and other T cells.
   a. sensitized
   c. helper
   b. cytotoxic
   d. natural killer

5. T_C cells are important in controlling
   a. virus infections.
   c. autoimmunity.
   b. allergy.
   d. all of these.

6. Which of the following can serve as antigen-presenting cells (APCs)?
   a. T cells
   d. dendritic cells
   b. B cells
   e. b, c, and d
   c. macrophages

7. A vaccine that contains parts of viruses is called
   a. acellular.
   c. a subunit.
   b. recombinant.
   d. attenuated.

8. Conjugated vaccines combine antigens and
   a. antibodies.
   c. epitopes.
   b. adjuvants.
   d. foreign proteins.

Critical Thinking   Bloom's Levels 3, 4, and 5: Apply, Analyze, and Evaluate

Critical thinking is the ability to reason and solve problems using facts and concepts. These questions can be approached from a number of angles and, in most cases, they do not have a single correct answer.

1. Describe the major histocompatibility complex, and explain how it participates in immune reactions.

2. Explain the clonal selection theory of antibody specificity and diversity.

3. Describe three ways that B cells and T cells are similar and at least five major ways in which they are different.

4. Describe the actions of an antigen-presenting cell.

5. a. Describe the structure of immunoglobulin.
   b. What are the functions of the Fab and Fc portions?
   c. Describe four or five ways that antibodies function in immunity.

Visual Connections   Bloom's Level 5: Evaluate

This question connects previous images to a new concept.

1. **From table 13.9.** In this figure describing primary and secondary responses to antigen, indicate where a vaccination might be most effective, and also indicate where natural infection would play a role.

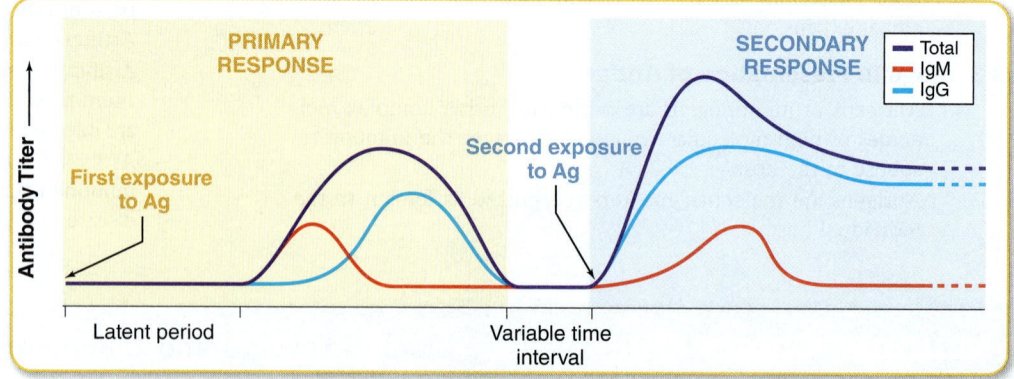

www.mcgrawhillconnect.com
Enhance your study of this chapter with study tools and practice tests. Also ask your instructor about the resources available through Connect, including the media-rich eBook, interactive learning tools, and animations.

Too Much of a Good Thing

I was working in a neurology unit when a young woman named Stacy was admitted. Stacy was 24 years old. She was married and had a 2-year-old at home. She was a full-time student, studying social work at the local university.

Stacy had been referred to one of our neurologists by her family doctor with some unusual symptoms. Over the past several months, Stacy had been experiencing increasing fatigue that she described as akin to "walking under water." She had to walk considerable distances on campus and she was finding it very difficult to climb stairs. She complained of blurred vision and her eyelids sagged. She had lost weight and stated that eating was "too much effort." She ate mainly soft foods that were easy to chew because eating meat and other foods that required a lot of chewing took too much effort. Her facial expression was fixed—even smiling took too much energy. Stacy and her family were very frightened by her symptoms.

She was examined by the neurologist, who immediately suspected myasthenia gravis. He admitted her to the unit and ordered several tests to confirm the diagnosis, including a blood test for acetylcholine receptor antibodies, an electromyography, and nerve conduction studies. A CT of the chest was also ordered. Fifteen percent of patients with myasthenia gravis have a thymoma, or a tumor on the thymus.

The CT showed the presence of a thymoma, and the other tests were all indicative of myasthenia gravis. Stacy was scheduled for a thymectomy. She was also started on prednisone and neostigmine.

Following surgery, Stacy gradually improved, although she stated that fatigue continued to be a problem; however, she noticed an overall improvement in muscle strength. She was able to continue her education on a part-time basis with scheduled rest periods.

- What is the underlying cause of myasthenia gravis?

- How do immunosuppressant drugs improve symptoms in myasthenia gravis?

Case File Wrap-Up appears on page 404.

Disorders in Immunity

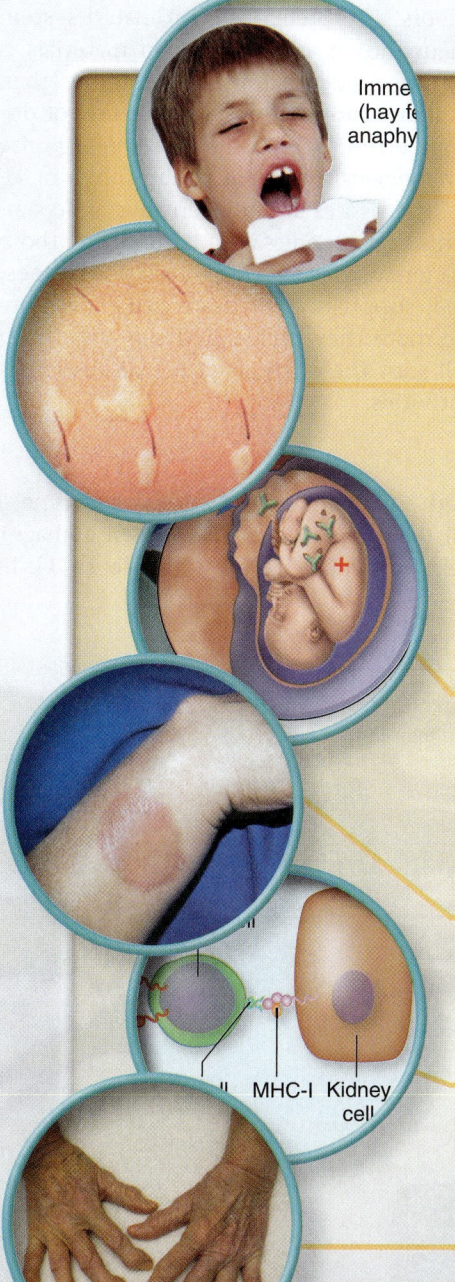

Imme
(hay fe
anaphy

IN THIS CHAPTER...

14.1 The Immune Response: A Two-Sided Coin

1. Define *immunopathology*, and describe the two major categories of immune dysfunction.
2. Identify the four major categories of hypersensitivities, or overreactions to antigens.

14.2 Type I Allergic Reactions: Atopy and Anaphylaxis

3. Summarize genetic and environmental factors that influence allergy development.
4. Identify three conditions caused by IgE-mediated allergic reactions.
5. Identify the two clinical forms of anaphylaxis, explaining why one is more often fatal than the other.
6. Explain the mode of action of two strategies for treating and preventing type I allergic reactions.

14.3 Type II Hypersensitivities: Reactions That Lyse Foreign Cells

7. List the three immune components causing cell lysis in type II hypersensitivity reactions.
8. Explain the role of Rh factor in hemolytic disease development and how it is prevented in newborns.

14.4 Type III Hypersensitivities: Immune Complex Reactions

9. Identify commonalities and differences between type II and type III hypersensitivities.

14.5 Type IV Hypersensitivities: Cell-Mediated (Delayed) Reactions

10. Describe one example of a type IV delayed hypersensitivity reaction.
11. List four classes of grafts, and explain how host versus graft and graft versus host diseases develop.

14.6 An Inappropriate Response to Self: Autoimmunity

12. List at least three autoimmune diseases and the common immunologic features in them.

14.7 Immunodeficiency Diseases: Hyposensitivity of the Immune System

13. Distinguish between primary and secondary immunodeficiencies, explaining how each develops.

MHC-I Kidney cell

14.1 The Immune Response: A Two-Sided Coin

Humans possess a powerful and intricate system of defense, which by its very nature also carries the potential to cause injury and disease. In most instances, a defect in immune function is expressed in commonplace but miserable symptoms such as those of hay fever and dermatitis. But abnormal or undesirable immune functions are also actively involved in debilitating or life-threatening diseases such as asthma, anaphylaxis, diabetes, rheumatoid arthritis, and graft rejection.

With few exceptions, our previous discussions of the immune response have centered around its numerous beneficial effects. The precisely coordinated system that seeks out, recognizes, and destroys an unending array of foreign materials is clearly protective, but it also has another side—a side that promotes rather than prevents disease. In this chapter, we survey **immunopathology,** the study of disease states associated with overreactivity or underreactivity of the immune response **(figure 14.1).** Overreactivity (also called hypersensitivity) takes the forms of allergy and autoimmunity. In these conditions, the tissues are innocent bystanders attacked by immunologic functions that can't distinguish one's own tissues from those expressing foreign material. In immunodeficiency or **hyposensitivity diseases,** immune function is incompletely developed, suppressed, or destroyed. The more scientists learn about immune disorders, the more they understand the underlying mechanisms of the immune system. We will start the chapter with a discussion of hypersensitivity, and then discuss hyposensitivities.

Hypersensitivity: Four Types

Hypersensitivity reactions can be classified into four major categories: type I ("common" allergy and anaphylaxis), type II (IgG- and IgM-mediated cell damage), type III (immune complex), and type IV (delayed hypersensitivity) **(table 14.1).** In

Figure 14.1 Overview of disorders of the immune system. Just as the system of T cells and B cells provides necessary protection against infection and disease, the same system can cause serious and debilitating conditions by overreacting (hypersensitivity) or underreacting (hyposensitivity) to immune stimuli.

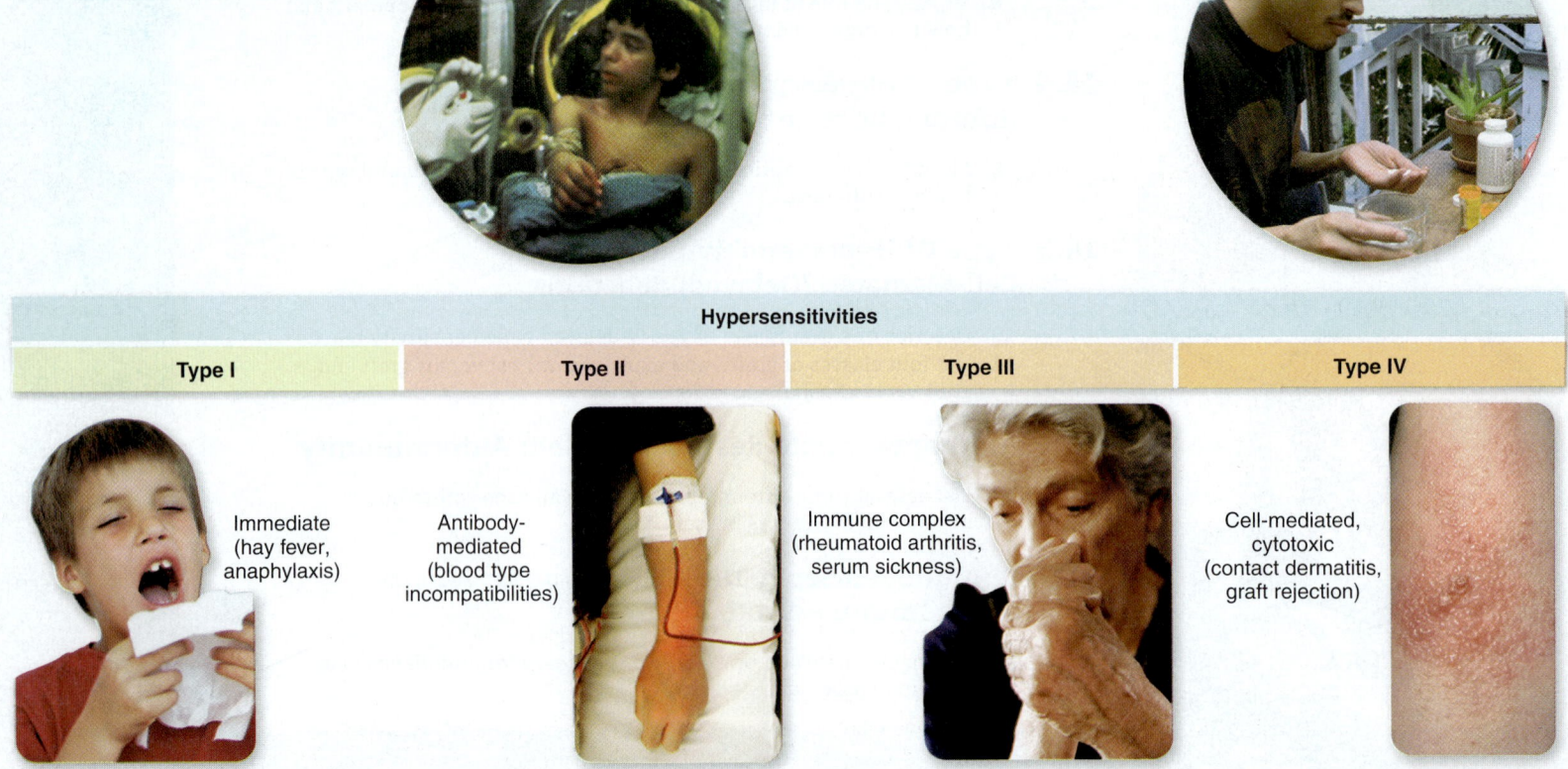

Hyposensitivities

Primary Immunodeficiency

Secondary Immunodeficiency

Hypersensitivities

Type I	Type II	Type III	Type IV
Immediate (hay fever, anaphylaxis)	Antibody-mediated (blood type incompatibilities)	Immune complex (rheumatoid arthritis, serum sickness)	Cell-mediated, cytotoxic (contact dermatitis, graft rejection)

Table 14.1 Hypersensitivity States

Type		Systems and Mechanisms Involved	Examples
I	Immediate hypersensitivity	IgE-mediated; involves mast cells, basophils, and allergic mediators	Anaphylaxis, allergies such as hay fever, asthma
II	Antibody-mediated	IgG, IgM antibodies act upon cells with complement and cause cell lysis; includes some autoimmune diseases	Blood group incompatibility; pernicious anemia; myasthenia gravis
III	Immune complex-mediated	Antibody-mediated inflammation; circulating IgG complexes deposited in basement membranes of target organs; includes some autoimmune diseases	Systemic lupus erythematosus; rheumatoid arthritis; serum sickness; rheumatic fever
IV	T-cell-mediated	Delayed hypersensitivity and cytotoxic reactions in tissues; includes some autoimmune diseases	Infection reactions; contact dermatitis; graft rejection

general, types I, II, and III involve a B-cell–immunoglobulin response, and type IV involves a T-cell response. The antigens that elicit these reactions can be exogenous, originating from outside the body (microbes, pollen grains, and foreign cells and proteins), or endogenous, arising from self tissue (autoimmunities).

One of the reasons allergies are easily mistaken for infections is that both involve tissue damage and thus trigger the inflammatory response, as described in chapter 12. Many symptoms and signs of inflammation (redness, heat, skin eruptions, edema, and granuloma) are prominent features of allergies.

14.1 LEARNING OUTCOMES—Assess Your Progress

1. Define *immunopathology*, and identify the two major categories of immune dysfunction.
2. Identify the four major categories of hypersensitivities, or overreactions to antigens.

14.2 Type I Allergic Reactions: Atopy and Anaphylaxis

The term **allergy** refers to an exaggerated immune response that is manifested by inflammation. Allergic individuals are acutely sensitive to repeated contact with antigens, called **allergens,** which do not noticeably affect nonallergic individuals. All type I reactions share a similar physiological mechanism, are immediate in onset, and are associated with exposure to specific antigens. However, there are two levels of severity: **Atopy** is any chronic local allergy such as hay fever or asthma; **anaphylaxis** (an″-uh-fih-lax′-us) is a systemic, sometimes fatal reaction that involves airway obstruction and circulatory collapse.

Although the general effects of hypersensitivity are detrimental, we must be aware that it involves the very same types of immune reactions as those at work in protective immunities. Based upon this fact, *all* humans have the potential to develop allergies under particular circumstances.

Who Is Affected?

In the United States, nearly half of the population is affected by airborne allergens, such as dust, pollen, and mold. Treatment of asthma, hay fever, and eczema associated with these allergens results in a price tag of about $21 million annually, making it the sixth most costly condition in the United States. Monetary loss due to employee debilitation and absenteeism is immeasurable, as is the loss of school and play

Asthma is a form of type I hypersensitivity.

time for affected children. The majority of type I allergies are relatively mild, but certain forms such as asthma and anaphylaxis may require hospitalization and can cause death especially in the youngest patients. In some individuals, atopic allergies last for a lifetime; others "outgrow" them; still others suddenly develop them later in life.

A predisposition to allergies seems to "run in families," that is, have a strong familial association. Be aware that what is hereditary is a *generalized susceptibility,* not the allergy to a specific substance. For example, a parent who is allergic to ragweed pollen can have a child who is allergic to cat hair. The prospect of a child's developing atopic allergy is at least 25% if one parent exhibits symptoms, and increases to nearly 50% if grandparents or siblings are also afflicted. The actual basis for atopy seems to be a genetic program that favors allergic antibody (IgE) production, increased reactivity of mast cells, and increased susceptibility of target tissue to allergic mediators.

The "hygiene hypothesis" provides one possible explanation for an environmental component to allergy development. This hypothesis suggests that the industrialized world has created a very hygienic environment, exemplified by antimicrobial products of all kinds and very well insulated homes, and that this has been bad for our immune systems. It seems that our immune systems need to be "trained" by interaction with microbes as we develop. In fact, children who grow up on farms have been found to have lower incidences of several types of allergies. Also, researchers have found that the combination of being delivered by cesarean section and a maternal history of allergy elevates the risk that a child will be allergic to foods by a factor of eight. Scientists suggest that delivery by cesarean section keeps the baby from being exposed to vaginal and stool bacteria. Additional work has shown that babies need to be exposed to commensal bacteria in order for the IgA system to develop normally.

Another possible factor affecting allergy development appears to be related to breast feeding. Newborn babies that are breast fed exclusively for the first 4 months of life have a lower risk of asthma and eczema, especially if they have a family history of allergy. This is thought to come from the presence of cytokines and growth factors in human milk that act on the baby's gut mucosa to induce tolerance, rather than reactivity, to allergens. New information from the Human Microbiome Project reveals that nearly 600 species of bacteria can be transferred to infants through breast milk. Combined with data from other studies showing that a disruption of microbial populations in the gut may influence the development of asthma, it is clear that these organisms play an important role in the development of tolerance to foreign antigens.

Might allergies reflect some beneficial evolutionary adaptation? Most allergy sufferers would answer with a resounding "No!" Why would humans and other mammals evolve an allergic response that causes suffering, tissue damage, and even death? One possible explanation may be that the components involved in an allergic response exist to defend against helminthic worms and other multicellular human parasites. It is only relatively recently in our evolutionary history that developed countries have seen dramatically fewer infections with these parasites. One hypothesis is that the part of the immune system that fights helminthic worms is left idle in a population that has recently been "scrubbed" of these parasites, and goes awry.

Finally, we have to remember that, as one author notes, "the current epidemic of allergy in industrialized countries is a small price to pay for the remarkable reduction of infant mortality provided by the elimination of pathogens through improved hygiene. Having too few microbes in our immediate environment seems to be problematic, but having many pathogens is far, far worse."[1]

The Nature of Allergens and Their Portals of Entry

As with all antigens, allergens have certain immunogenic characteristics. Proteins are more allergenic than carbohydrates, fats, or nucleic acids. Some allergens are haptens, nonproteinaceous substances with a molecular weight of less than 1,000 that can form complexes with carrier molecules in the body (shown in

---

[1] 2007, January-February. *American Scientist:* 28–35.

Figure 14.2 Common allergens, classified by portal of entry. (a) Common inhalants, or airborne environmental allergens, include pollen and insect parts. (b) Common ingestants, allergens that enter by mouth. (c) Common injectants, allergens that enter via the parenteral route. (d) Common contactants, allergens that enter through the skin.

figure 13.6). Organic and inorganic chemicals found in industrial and household products, cosmetics, food, and drugs are commonly of this type.

Allergens typically enter through epithelial portals in the respiratory tract, gastrointestinal tract, and skin **(figure 14.2).** The mucosal surfaces of the gut and respiratory system present a thin, moist surface that is normally quite penetrable. The dry, tough keratin coating of skin is less permeable, but access still occurs through tiny breaks, glands, and hair follicles.

Airborne environmental allergens such as pollen, house dust, dander (shed skin scales), or fungal spores are termed *inhalants* **(figure 14.2a).** Each geographic region harbors a particular combination of airborne substances that varies with the season and humidity. Pollen is given off seasonally by trees and other flowering plants, while mold spores are released throughout the year. Airborne animal hair and dander, feathers, and the saliva of dogs and cats are common sources of allergens. The component of house dust that appears to account for most dust allergies is not soil or other debris but the decomposed bodies and feces of tiny mites that commonly live in this dust.

Allergens that enter by mouth, called *ingestants*, often cause food allergies **(figure 14.2b).** *Injectant* allergies are triggered by drugs, vaccines, or hymenopteran (bee) venom **(figure 14.2c).** *Contactants* are allergens that enter through the skin **(figure 14.2d).** Many contact allergies are of the type IV (delayed) variety, discussed later in this chapter. It is also possible to be exposed to certain allergens, penicillin among them, during sexual intercourse due to the presence of allergens in the semen.

Mechanisms of Type I Allergy: Sensitization and Provocation

In general, type I allergies develop in stages. **Figure 14.3** tells the whole story, beginning with the initial encounter with allergen that sets up the conditions for the allergy to manifest on subsequent encounters.

The Role of Mast Cells and Basophils

Mast cells and basophils play an important role in allergy due to the following:

1. Their ubiquitous location in tissues. Mast cells are located in the connective tissue of virtually all organs, but particularly high concentrations exist in the lungs, skin, gastrointestinal tract, and genitourinary tract. Basophils circulate in the blood but migrate readily into tissues.
2. Their capacity to bind IgE during sensitization (**figure 14.3a**) and *degranulate*. Each cell carries 30,000 to 100,000 cell receptors, which trigger the release of inflammatory cytokines from cytoplasmic granules (secretory vesicles) when bound by allergen-associated IgE.

The symptoms of allergy are not caused by the direct action of allergen on tissues, but rather by the physiological effects of mast-cell-derived allergic mediators on target organs.

Cytokines, Target Organs, and Allergic Symptoms

Numerous substances involved in mediating allergy have been identified. The principal chemical mediators produced by mast cells and basophils are histamine, serotonin, leukotriene, platelet-activating factor, prostaglandins, and bradykinin (**figure 14.3c**). These inflammatory cytokines, acting alone or in combination, account for the tremendous scope of allergic symptoms.

Histamine, the most profuse and fastest-acting allergic mediator, is a potent stimulator of secretory glands and smooth muscle. Histamine's actions on smooth

Figure 14.3 A schematic view of cellular reactions during the type I allergic response. (a) Sensitization (initial contact with sensitizing dose), 1–6. (b) Provocation (later contacts with provocative dose), 7–9. (c) The spectrum of reactions to inflammatory cytokines released by mast cells and the common symptoms they elicit in target tissues and organs.

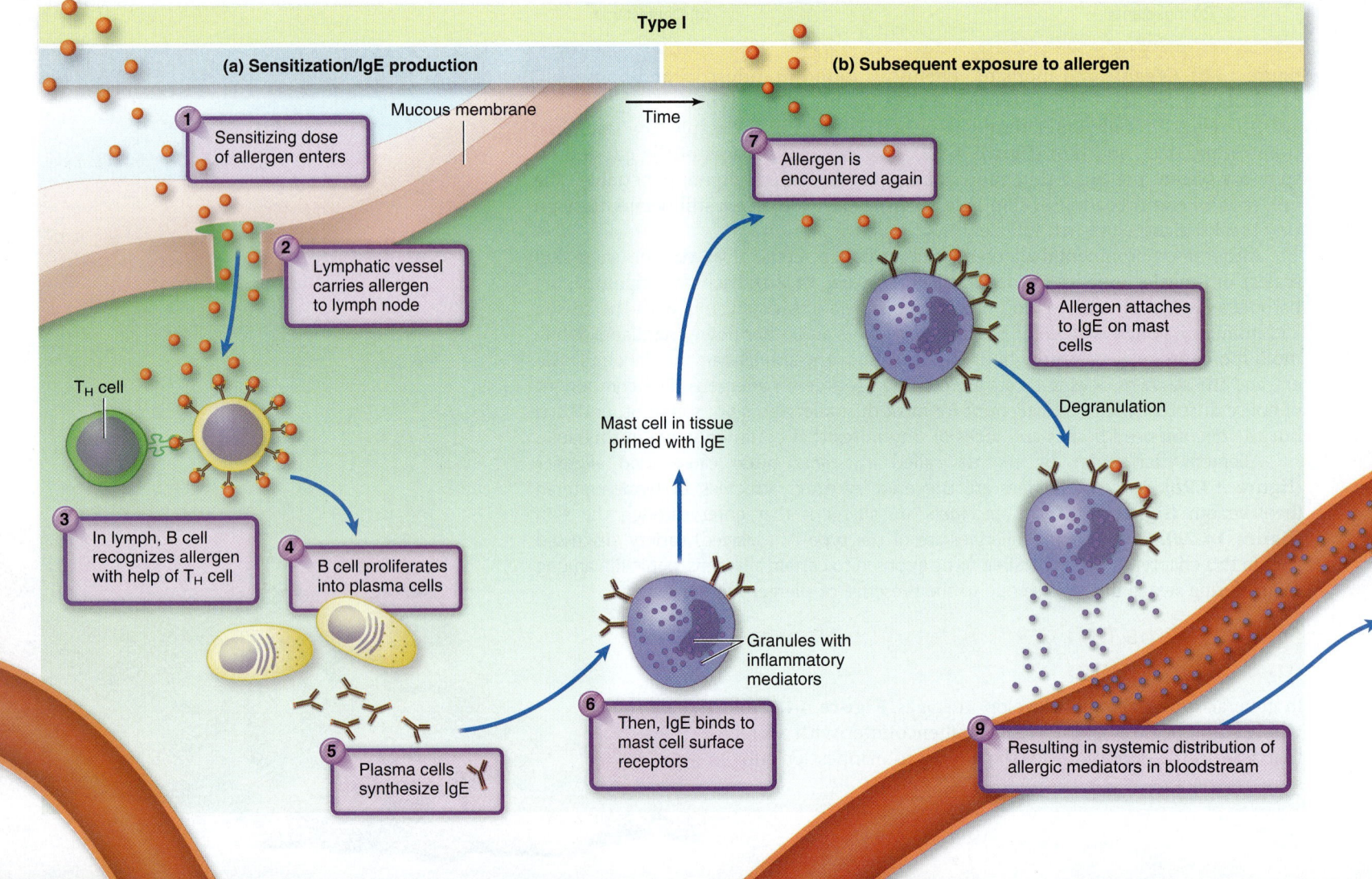

Type I

(a) Sensitization/IgE production

(b) Subsequent exposure to allergen

Mucous membrane

Time

1. Sensitizing dose of allergen enters
2. Lymphatic vessel carries allergen to lymph node

T_H cell

3. In lymph, B cell recognizes allergen with help of T_H cell
4. B cell proliferates into plasma cells
5. Plasma cells synthesize IgE
6. Then, IgE binds to mast cell surface receptors

Mast cell in tissue primed with IgE

Granules with inflammatory mediators

7. Allergen is encountered again
8. Allergen attaches to IgE on mast cells

Degranulation

9. Resulting in systemic distribution of allergic mediators in bloodstream

muscle vary with location. It *constricts* the smooth muscle layers of the small bronchi and intestine, thereby causing labored breathing and increased intestinal motility. In contrast, histamine *relaxes* vascular smooth muscle and dilates arterioles and venules, resulting in *wheal-and-flare* reactions in the skin and pruritus (itching). Histamine can also stimulate eosinophils to release inflammatory cytokines, escalating the symptoms.

In allergic reactions, **bradykinin** causes prolonged smooth muscle contraction of the bronchioles, dilatation of peripheral arterioles, increased capillary permeability, and increased mucus secretion. Although the exact role of **serotonin** in human allergy is uncertain, its effects appear to complement those of histamine and bradykinin.

Leukotriene (loo″-koh-try′-een) is known as the "slow-reacting substance of anaphylaxis" for its property of inducing gradual contraction of smooth muscle. This type of leukotriene is responsible for the prolonged bronchospasm, vascular permeability, and mucus secretion of the asthmatic individual. Other leukotrienes stimulate the activities of polymorphonuclear leukocytes, or granulocytes, which play a role in various immune functions (see chapter 12).

Prostaglandins are a group of powerful inflammatory agents. Normally, these substances regulate smooth muscle contraction (e.g., they stimulate uterine contractions during delivery). In allergic reactions, they are responsible for vasodilation, increased vascular permeability, increased sensitivity to pain, and bronchoconstriction. Nonsteroidal anti-inflammatory drugs (NSAIDs), such as aspirin and ibuprofen, work by preventing the actions of prostaglandins.

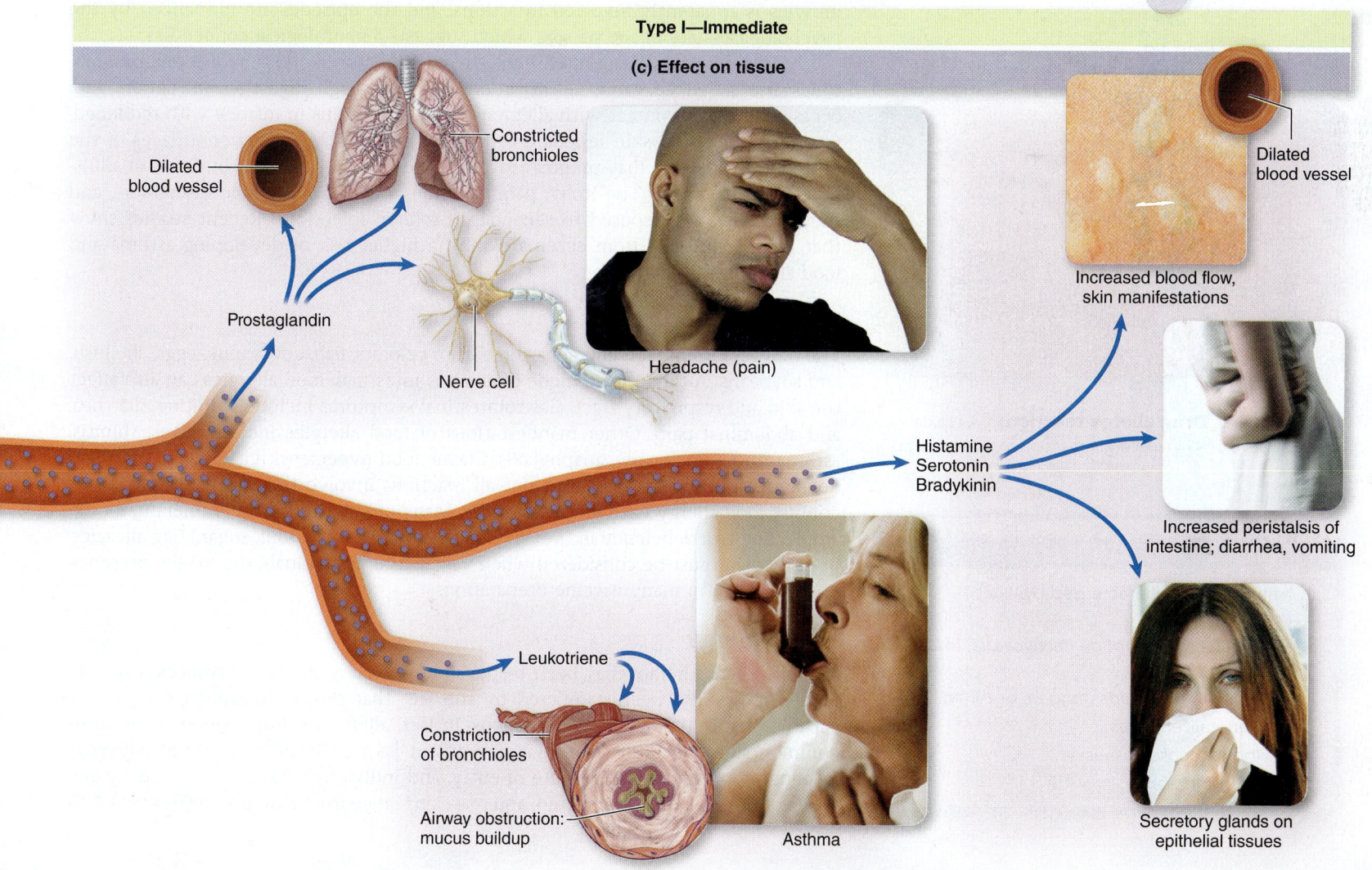

Type I—Immediate

(c) Effect on tissue

Dilated blood vessel

Constricted bronchioles

Prostaglandin

Nerve cell

Headache (pain)

Increased blood flow, skin manifestations

Dilated blood vessel

Histamine
Serotonin
Bradykinin

Increased peristalsis of intestine; diarrhea, vomiting

Leukotriene

Constriction of bronchioles

Airway obstruction: mucus buildup

Asthma

Secretory glands on epithelial tissues

Type I—Immediate

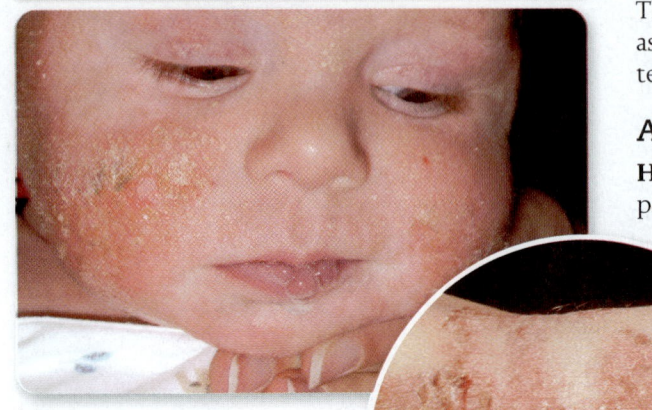

(a)

Figure 14.4 Atopic dermatitis, or eczema.
(a) Vesicular, weepy, encrusted lesions are typical in afflicted infants. **(b)** In adulthood, lesions are more likely to be dry, scaly, and thickened.

(b)

Type I—Immediate

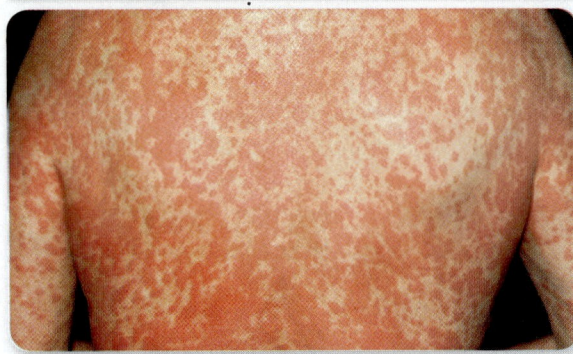

Figure 14.5 Drug allergy reaction.
A typical rash that develops in an allergic reaction to an antibiotic.

NCLEX® PREP

1. Anaphylaxis is characterized by what symptoms?
   a. *circulatory disruption (tachycardia, low blood pressure)*
   b. *swelling of the lips, tongue, or throat*
   c. *generalized hives*
   d. *loss of consciousness*
   e. *all of the above*

IgE- and Mast-Cell-Mediated Allergic Conditions

The mechanisms just described underlie the development of hay fever, allergic asthma, food allergy, drug allergy, and eczema; this section covers the main characteristics of these conditions.

Atopic Diseases

Hay fever is a generic term for allergic rhinitis, a seasonal reaction to inhaled plant pollen or molds, or a chronic, year-round reaction to a wide spectrum of airborne allergens or inhalants. The targets are typically respiratory membranes, and the symptoms include nasal congestion; sneezing; coughing; profuse mucus secretion; itchy, red, and teary eyes; and mild bronchoconstriction.

Asthma is a respiratory disease characterized by episodes of impaired breathing due to severe bronchoconstriction. The airways of asthmatic people are exquisitely responsive to minute amounts of inhalants, ingestants, or other stimuli, such as infectious agents. The symptoms of asthma range from occasional, annoying bouts of difficult breathing to fatal suffocation. Labored breathing, shortness of breath, wheezing, cough, and ventilatory **rales** are present to one degree or another. The respiratory tract of an asthmatic person is chronically inflamed and severely overreactive to allergic mediators, especially leukotrienes and serotonin from pulmonary mast cells. Upon activation of the allergic response, natural killer (NK) T cells are recruited and activated, adding to the cytokine storm brewing in the lungs. An imbalance in the nervous control of the respiratory smooth muscles is apparently involved in asthma, and the episodes are influenced by the psychological state of the person, which suggests a neurological connection.

Atopic dermatitis is an intensely itchy inflammatory condition of the skin, sometimes also called **eczema.** Sensitization occurs through ingestion, inhalation, and, occasionally, skin contact with allergens. It usually begins in infancy with reddened, weeping, encrusted skin lesions on the face, scalp, neck, and inner surfaces of the limbs and trunk that may progress to a dry, scaly, thickened skin condition in adulthood **(figure 14.4).** The itchy, painful lesions cause considerable discomfort, and they are often predisposed to secondary bacterial infections. Recent studies show that infants suffering from eczema exhibit a higher risk of developing asthma and food allergies as they age.

Food Allergy

The most common food allergens come from peanuts, fish, cow's milk, eggs, shellfish, and soybeans. Although the mode of entry is intestinal, food allergies can also affect the skin and respiratory tract. Gastrointestinal symptoms include vomiting, diarrhea, and abdominal pain. Other manifestations of food allergies include hives, rhinitis, asthma, and occasionally, anaphylaxis. Classic food hypersensitivity involves IgE and degranulation of mast cells, but not all reactions involve this mechanism. (Do not confuse food allergy with food *intolerance.* Many people are lactose intolerant, for example, due to a deficiency in the enzyme that degrades the milk sugar.) Egg allergies, in particular, must be considered when vaccinating individuals, due to the presence of egg protein in many vaccine preparations.

Drug Allergy

Modern chemotherapy has been responsible for many medical advances. Unfortunately, it has also been hampered by the fact that drugs are foreign compounds capable of stimulating allergic reactions. In fact, allergy to drugs is one of the most common side effects of treatment (present in 5% to 10% of hospitalized patients). Depending on the allergen, route of entry, and individual sensitivities, virtually any tissue of the body can be affected, and reactions range from a mild rash **(figure 14.5)** to fatal anaphylaxis. Compounds implicated most often are antibiotics (penicillin), synthetic antimicrobials (sulfa drugs), aspirin, opiates, and contrast dye used in X rays. The actual allergen is not the intact drug itself but a hapten given off when the liver processes the drug.

Anaphylaxis: An Overpowering IgE-Mediated Allergic Reaction

The term *anaphylaxis,* or anaphylactic shock, refers to a swift reaction to allergens. Two clinical types of anaphylaxis are seen in humans. *Cutaneous anaphylaxis* is the wheal-and-flare inflammatory reaction to the local injection of allergen. *Systemic anaphylaxis,* on the other hand, is characterized by sudden respiratory and circulatory disruption that can be fatal within minutes. The allergen type and route of entry causing anaphylaxis vary, though bee stings and injection of antibiotics or serum are implicated most often. Bee venom is a complex material containing several allergens and enzymes that can create a sensitivity that can last for decades after exposure.

The underlying physiological events in systemic anaphylaxis parallel those of atopy, but the concentration of chemical mediators and the strength of the response are greatly amplified. The immune system of a sensitized person exposed to a provocative dose of allergen responds with a sudden, massive release of chemicals into the tissues and blood, which act rapidly on the target organs. Anaphylactic persons have been known to die within 15 minutes from complete airway blockage.

Diagnosis of Allergy

Because allergy mimics infection and other conditions, it is important to determine if a person is actually allergic and to identify the specific allergen or allergens. Allergy diagnosis involves several levels of tests, including nonspecific, specific, *in vitro,* and *in vivo* methods.

The most widely used blood test is a radioallergosorbent test (RAST), which measures levels of IgE to specific allergens. A new test that can distinguish whether a patient has experienced an allergic attack measures elevated blood levels of tryptase, an enzyme released by mast cells that increases during an allergic response. Several types of specific *in vitro* tests can determine the allergic potential of a patient's blood sample. A differential blood cell count can reveal high levels of basophils and eosinophils, indicating allergy; the leukocyte histamine-release test measures the amount of histamine released from the patient's basophils when exposed to a specific allergen.

Skin Testing

A tried and true *in vivo* method to detect precise atopic or anaphylactic sensitivities is skin testing. With this technique, a patient's skin is injected, scratched, or pricked with a small amount of a pure allergen extract. There are hundreds of these allergen extracts containing common airborne allergens and more unusual allergens (mule dander, theater dust, bird feathers). Unfortunately, skin tests for food allergies using food extracts are unreliable in most cases. In patients with numerous allergies, the allergist maps the skin on the inner aspect of the forearms or back and injects the allergens intradermally according to this predetermined pattern **(figure 14.6*a*).** Approximately 20 minutes after antigenic challenge, each site is appraised for a wheal response indicative of histamine release. The diameter of the wheal is measured and rated on a scale of 0 (no reaction) to 4 (greater than 15 mm) **(figure 14.6*b*).**

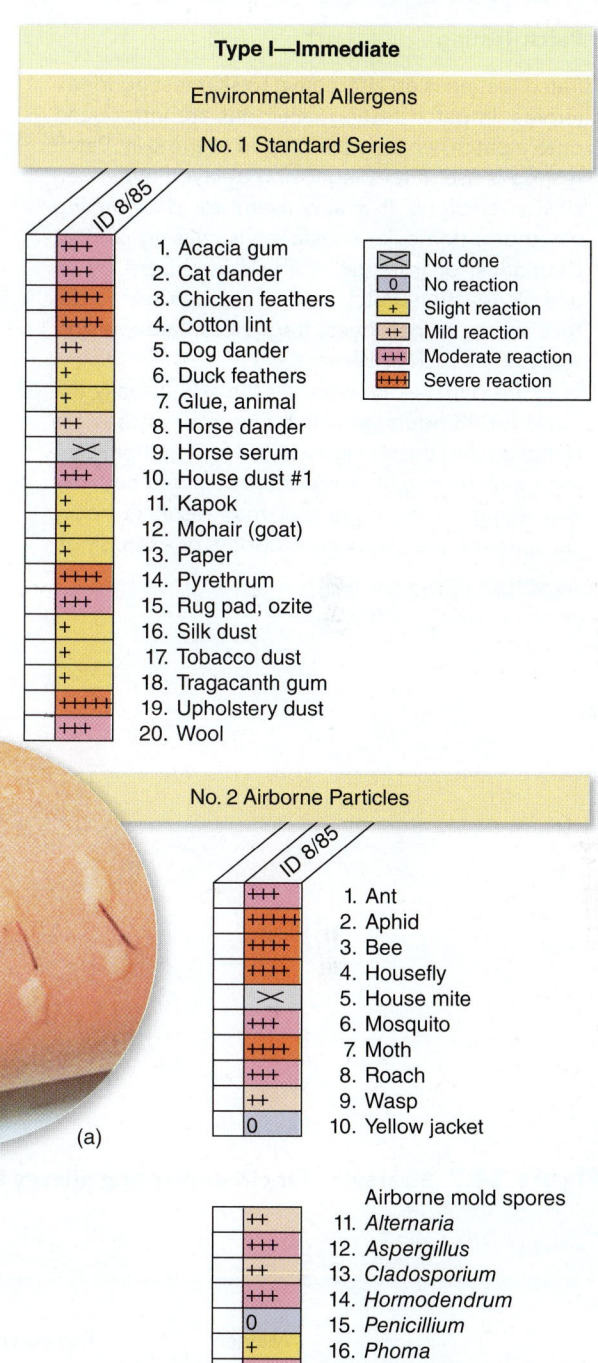

Type I—Immediate

Environmental Allergens

No. 1 Standard Series

ID 8/85

+++	1. Acacia gum
+++	2. Cat dander
++++	3. Chicken feathers
++++	4. Cotton lint
++	5. Dog dander
+	6. Duck feathers
+	7. Glue, animal
++	8. Horse dander
✕	9. Horse serum
+++	10. House dust #1
+	11. Kapok
+	12. Mohair (goat)
+	13. Paper
++++	14. Pyrethrum
+++	15. Rug pad, ozite
+	16. Silk dust
+	17. Tobacco dust
+	18. Tragacanth gum
+++++	19. Upholstery dust
+++	20. Wool

✕	Not done
0	No reaction
+	Slight reaction
++	Mild reaction
+++	Moderate reaction
++++	Severe reaction

No. 2 Airborne Particles

ID 8/85

+++	1. Ant
+++++	2. Aphid
++++	3. Bee
++++	4. Housefly
✕	5. House mite
+++	6. Mosquito
++++	7. Moth
+++	8. Roach
++	9. Wasp
0	10. Yellow jacket

Airborne mold spores

++	11. *Alternaria*
+++	12. *Aspergillus*
++	13. *Cladosporium*
+++	14. *Hormodendrum*
0	15. *Penicillium*
+	16. *Phoma*
+++	17. *Rhizopus*
	18. ....

(a)

(b)

Figure 14.6 A method for conducting an allergy skin test. The forearm (or back) is mapped and then injected with a selection of allergen extracts. The allergist must be very aware of the potential of anaphylactic attacks triggered by these injections. **(a)** Close-up of skin wheals showing a number of positive reactions (dark lines are measurer's marks). **(b)** An actual skin test record for some common environmental allergens [not related to **(a)**].

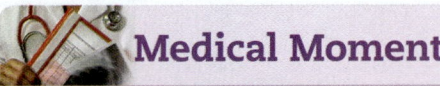

Medical Moment

Patch Testing

Patch testing is an alternative to skin testing for allergies. In patch testing, allergens are introduced onto a patch, which is then taped to the skin. Patch testing is useful for diagnosing delayed hypersensitivity reactions. It is also useful for determining whether a particular substance is causing contact dermatitis, or irritation of the skin caused by an allergic reaction. Patch testing can be used to detect allergy to hair dyes, fragrances, latex, resins, and metals such as silver.

The patches are worn on the skin (usually the back) for 48 hours, at which time the patches are removed and the skin is assessed for irritation. During patch testing, it is important to avoid bathing, showering, or activities that may result in heavy perspiration (i.e., working outdoors or exercising).

Treatment and Prevention of Allergy

In general, the methods of treating and preventing type I allergy involve the following:

1. avoiding the allergen, although this may be very difficult in many instances;
2. taking drugs that block the action of lymphocytes, mast cells, or chemical mediators; and
3. using injections to short-circuit the allergic reaction.

Taking Drugs to Block Allergy

The aim of antiallergy medication is to block the progress of the allergic response somewhere along the route between IgE production and the appearance of symptoms **(figure 14.7).** Oral anti-inflammatory drugs such as corticosteroids inhibit the activity of lymphocytes and thereby reduce the production of IgE, but they also have dangerous side effects and should not be taken for prolonged periods. Some drugs block the degranulation of mast cells and reduce the levels of inflammatory cytokines. Asthma and rhinitis sufferers can find relief with a drug that blocks synthesis of leukotriene and a monoclonal antibody that inactivates IgE (omalizumab [Xolair]).

Widely used medications for preventing symptoms of atopic allergy are **antihistamines,** the active ingredients in most over-the-counter allergy-control drugs. Antihistamines interfere with histamine activity by binding to histamine receptors on target organs. Other drugs that relieve inflammatory symptoms are aspirin and acetaminophen, which reduce pain by interfering with prostaglandin, and theophylline, a bronchodilator that reverses spasms in the respiratory smooth muscles. Persons who suffer from anaphylactic attacks are urged to carry at all times injectable or aerosolized epinephrine (adrenaline) and an identification tag indicating their sensitivity. Epinephrine reverses constriction of the airways and slows the release of allergic mediators. Although epinephrine works quickly and well, it has a very short half-life. It is very common to require more than one dose in anaphylactic reactions. Injectable epinephrine buys the individual time to get to a hospital for continuing treatment.

Allergy "Vaccines"

Approximately 70% of allergic patients benefit from controlled injections of specific allergens as determined by skin tests. This technique, called **desensitization,** or **hyposensitization,** is a therapeutic way to prevent reactions between allergen, IgE, and mast cells. The allergen preparations contain pure, preserved suspensions of plant antigens, venoms, dust mites, dander, and molds (but so far, hyposensitization for foods has not proved very effective). The immunologic basis of this treatment is open to differences in interpretation. One hypothesis suggests that injected allergens stimulate

Figure 14.7 **Strategies for circumventing allergy attacks.**

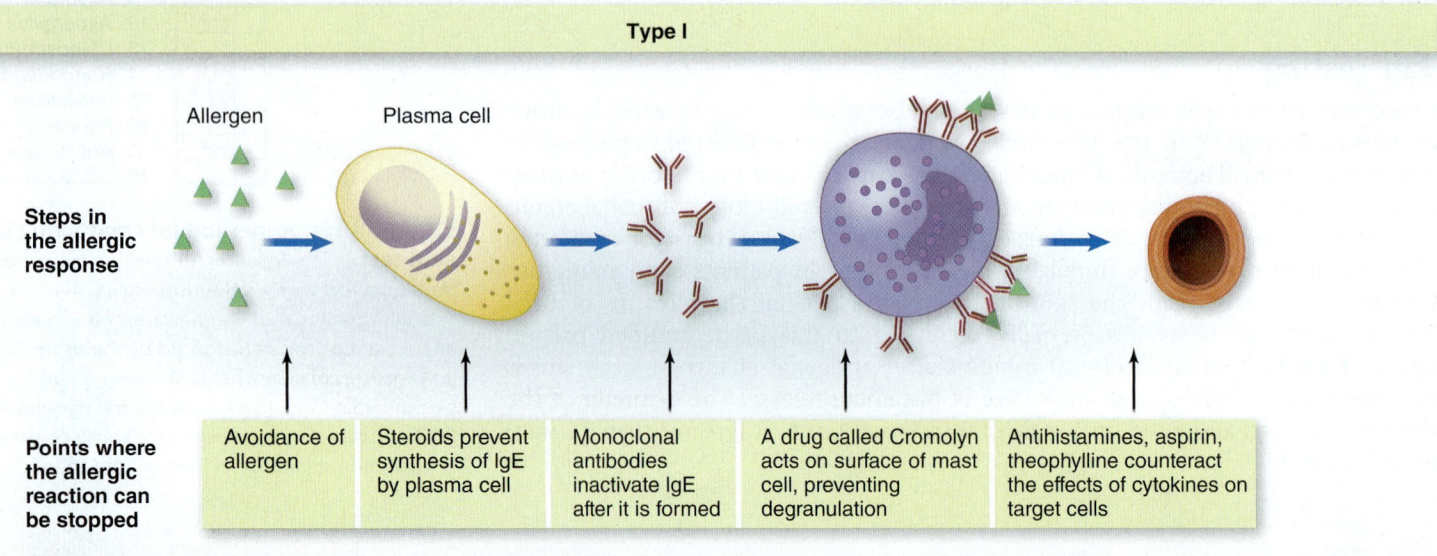

the formation of allergen-specific IgG–**blocking antibodies**–that can remove allergen from the system before it can bind to IgE (figure 14.7). It is also possible that allergen delivered in this fashion combines with the IgE itself and takes it from circulation.

A newer experimental therapy is an allergy shot that does not contain the allergen itself but instead contains a "decoy," an innocuous molecule that merely resembles a bacterium. It engages the components of the immune system that are active in allergy, causing them to stop reacting inappropriately to specific allergens. Just recently, a vaccine was developed to aid those suffering from cat allergies; ongoing research is leading to the development of other vaccines that work to alleviate allergies and at the same time protect against viral infection.

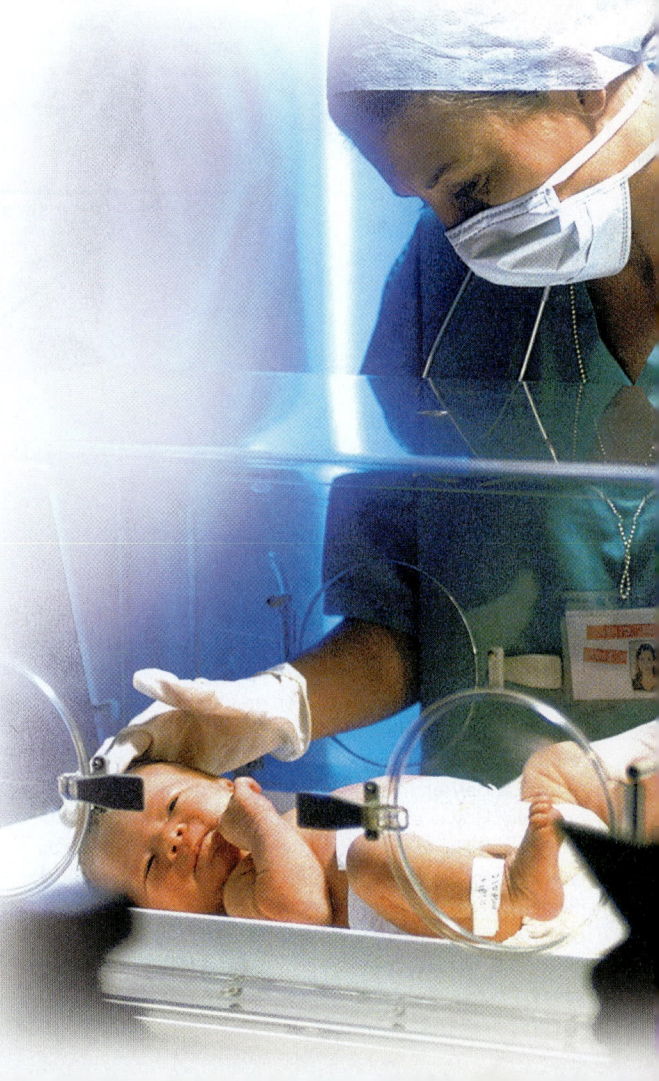

NCLEX® PREP

2. Antihistamines act to
   a. *make patients drowsy so that they are not bothered by symptoms.*
   b. *bind to histamine receptors on target organs.*
   c. *slow the release of allergic mediators.*
   d. *interfere with prostaglandins.*

> ### 14.2 LEARNING OUTCOMES—Assess Your Progress
>
> 3. Summarize genetic and environmental factors that influence allergy development.
> 4. Identify three conditions caused by IgE-mediated allergic reactions.
> 5. Identify the two clinical forms of anaphylaxis, explaining why one is more often fatal than the other.
> 6. Explain the mode of action of two strategies for treating and preventing type I allergic reactions.

14.3 Type II Hypersensitivities: Reactions That Lyse Foreign Cells

The diseases termed type II hypersensitivities are a complex group of syndromes that involve complement-assisted destruction (lysis) of foreign cells by antibodies (IgG and IgM) directed against those cells' surface antigens. This category includes transfusion reactions and some types of autoimmunities (discussed in a later section). The cells targeted for destruction are often red blood cells, but other cells can be involved.

Chapters 12 and 13 described the functions of unique surface markers on cell membranes. Ordinarily, these molecules play essential roles in transport, recognition, and development, but they become medically important when the tissues of one person are placed into the body of another person. Blood transfusions and organ donations introduce **alloantigens** (molecules that differ in the same species) on donor cells that are recognized by the lymphocytes of the recipient. These reactions are not really immune dysfunctions as allergy and autoimmunity are. The immune system is in fact working normally, but it is not equipped to distinguish between the desirable foreign cells of a transplanted tissue and the undesirable ones of a microbe.

The Rh Factor and Its Clinical Importance

We are all aware of the need to match blood types when transfusing blood into a person. Those blood types are matched on the basis of the A, B, and O antigens on the surfaces of red blood cells, as well as another antigen, called the **Rh factor** (or D antigen). Because the Rh factor can cause disease in natural circumstances (a normal pregnancy), we will use it to illustrate type II hypersensitivities. This factor was first discovered in experiments exploring the genetic relationships among animals. Rabbits inoculated with the red blood cells (RBCs) of rhesus monkeys produced an antibody that also reacted with human RBCs. Further tests showed that this monkey antigen (termed *Rh* for "rhesus") was present in about 85% of humans and absent in the other 15%. The details of Rh inheritance are complicated, but in simplest terms, a person's Rh type results from a combination of two possible alleles–a dominant one that codes for the factor and a recessive one that does not. The "+" or "−" that appears after a blood type (i.e., O+) reflects the Rh status of the person. Unlike with the ABO antigens, the only ways one can develop antibodies against this factor are through placental sensitization or transfusion. Although the Rh factor should be matched for a transfusion to avoid this situation, it is acceptable to transfuse Rh− blood if the Rh type is not known.

Hemolytic Disease of the Newborn and Rh Incompatibility

The potential for placental sensitization occurs when a mother is Rh− and her unborn child is Rh+. It is possible for fetal RBCs to leak into the mother's circulation during childbirth, when the detachment of the placenta creates avenues for fetal blood to enter the maternal circulation. The mother's immune system detects the foreign Rh factors on the fetal RBCs and is sensitized to them by producing antibodies and memory B cells. The first Rh+ child is usually not affected because the process begins so late in pregnancy that the child is born before maternal sensitization is completed. However, the mother's immune system has been strongly primed for a second contact with this factor in a subsequent pregnancy **(figure 14.8a)**.

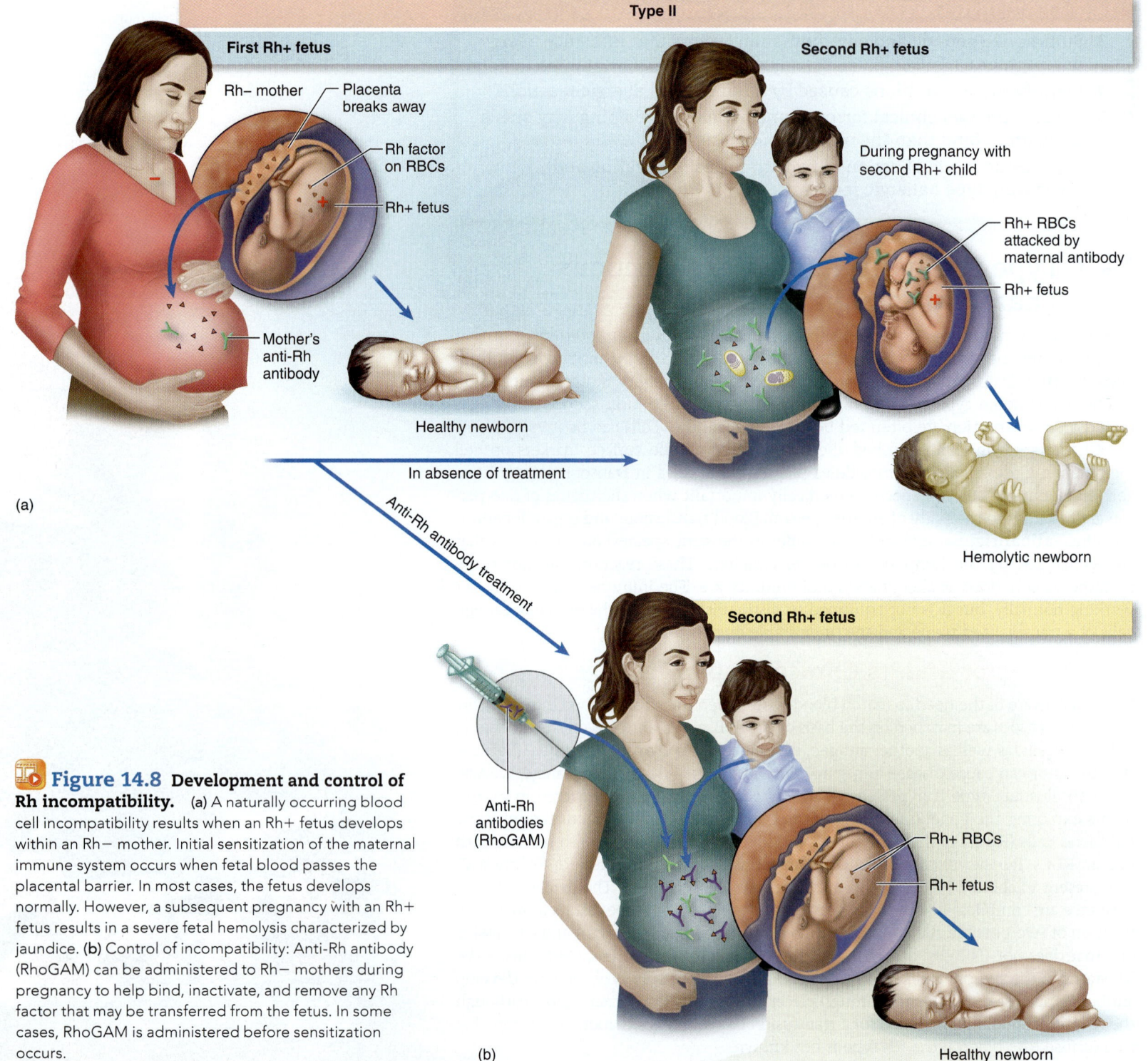

Figure 14.8 Development and control of Rh incompatibility. (a) A naturally occurring blood cell incompatibility results when an Rh+ fetus develops within an Rh− mother. Initial sensitization of the maternal immune system occurs when fetal blood passes the placental barrier. In most cases, the fetus develops normally. However, a subsequent pregnancy with an Rh+ fetus results in a severe fetal hemolysis characterized by jaundice. (b) Control of incompatibility: Anti-Rh antibody (RhoGAM) can be administered to Rh− mothers during pregnancy to help bind, inactivate, and remove any Rh factor that may be transferred from the fetus. In some cases, RhoGAM is administered before sensitization occurs.

In the next pregnancy with an Rh+ fetus, fetal blood cells escape into the maternal circulation late in pregnancy and elicit a memory response. Maternal anti-Rh antibodies then cross the placenta into the fetal circulation, where they affix to fetal RBCs and cause complement-mediated lysis. The outcome is a potentially fatal **hemolytic disease of the newborn (HDN),** characterized by severe anemia and jaundice. It is also called *erythroblastosis fetalis* (eh-rith″-roh-blas-toh′-sis fee-tal′-is), reflecting the release of immature nucleated RBCs (erythroblasts) into the blood to compensate for destroyed RBCs.

Maternal-fetal incompatibilities are also possible in the ABO blood group, but adverse reactions occur less frequently than with Rh sensitization because the antibodies to these blood group antigens are IgM rather than IgG and are unable to cross the placenta in large numbers. In fact, the maternal-fetal relationship is a fascinating instance of foreign tissue not being rejected, despite the extensive potential for contact.

Preventing Hemolytic Disease of the Newborn

Once sensitization of the mother to Rh factor has occurred, all other Rh+ fetuses will be at risk for hemolytic disease of the newborn. Prevention requires a careful family history of an Rh− pregnant woman. If the father is also Rh−, the child will be Rh− and free of risk; but if the father is Rh+, there is a possibility that the fetus may be Rh+. In this case, the mother must be passively immunized with antiserum containing antibodies against the Rh factor. This antiserum is called RhoGAM. It is the immunoglobulin fraction of human anti-Rh serum, prepared from pooled human sera. This antiserum, injected at 28 to 32 weeks and again immediately after delivery, reacts with any fetal RBCs that have escaped into the maternal circulation, thereby preventing the sensitization of the mother's immune system to Rh factor **(figure 14.8*b*).** Anti-Rh antibody must be given with each pregnancy that involves an Rh+ fetus, but it is ineffective if the mother has been previously sensitized.

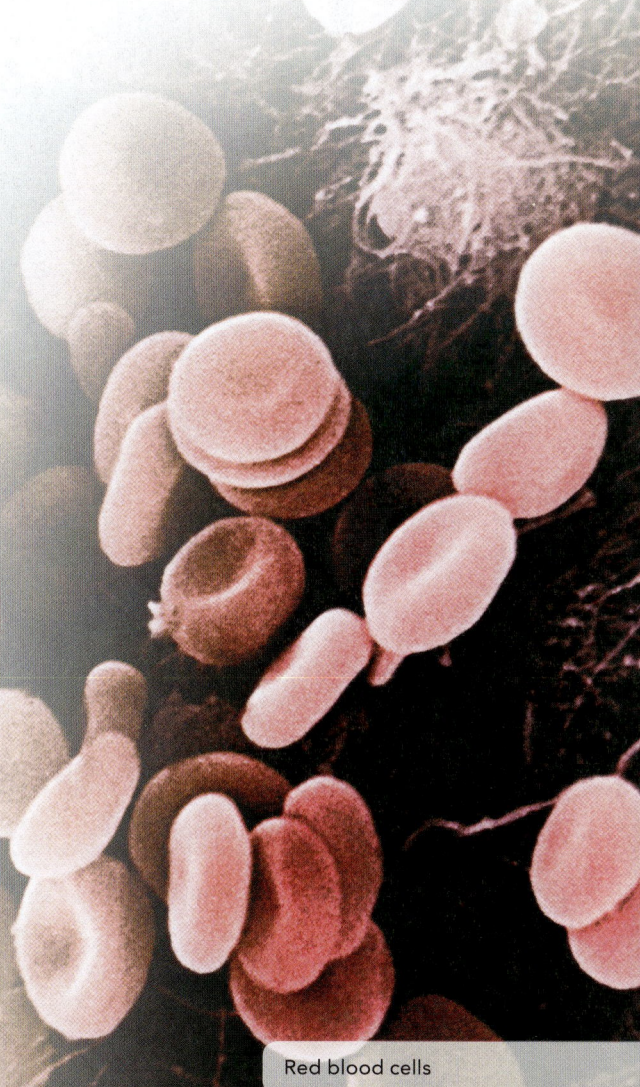

Red blood cells

14.3 LEARNING OUTCOMES—Assess Your Progress

7. List the three immune components causing cell lysis in type II hypersensitivity reactions.
8. Explain the role of Rh factor in hemolytic disease development and how it is prevented in newborns.

14.4 Type III Hypersensitivities: Immune Complex Reactions

Type III hypersensitivity involves the reaction of soluble antigen with antibody and the deposition of the resulting complexes in various tissues of the body. It is similar to type II, because it involves the production of IgG and IgM antibodies after repeated exposure to antigens and the activation of complement. Type III differs from type II because its antigens are not attached to the surface of a cell. The interaction of these antigens with antibodies produces free-floating complexes that can be deposited in the tissues, causing an **immune complex reaction,** or disease.

Mechanisms of Immune Complex Disease

After initial exposure to a profuse amount of antigen, the immune system produces large quantities of antibodies that circulate in the fluid compartments. When this

Type III

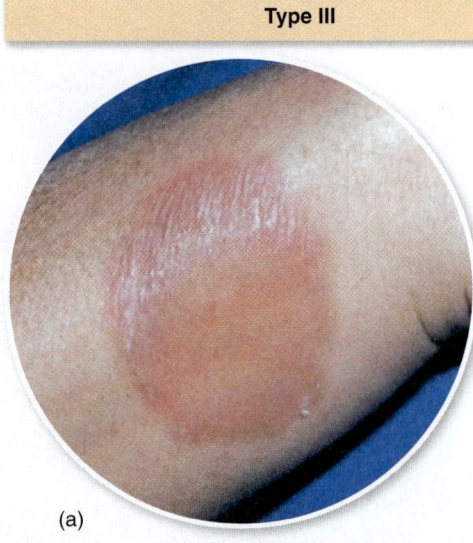

(a)

(b)

Figure 14.9 **Typical presentation of the Arthus reaction (a) and serum sickness (b), two immune complex diseases.**

antigen enters the system a second time, it reacts with the antibodies to form antigen-antibody complexes. These complexes recruit inflammatory components such as complement and neutrophils, which would ordinarily eliminate Ag-Ab complexes as part of the normal immune response. In an immune complex disease, however, these complexes are so abundant that they deposit in the **basement membranes** of epithelial tissues and become inaccessible. In response to these events, neutrophils release lysosomal granules that digest tissues and cause a destructive inflammatory condition.

Types of Immune Complex Disease

During the early tests of immunotherapy, hypersensitivity reactions to serum and vaccines were common. In addition to anaphylaxis, two syndromes were identified, the **Arthus reaction** and **serum sickness,** associated with certain types of passive immunization (especially with animal serum).

Serum sickness and the Arthus reaction are like anaphylaxis in that all of them require sensitization and preformed antibodies. Characteristics that set serum sickness and the Arthus reaction apart from anaphylaxis are as follows:

1. they depend on IgG, IgM, or IgA (precipitating antibodies) rather than IgE;
2. they require large doses of antigen (not a minuscule dose as in anaphylaxis); and
3. their symptoms are delayed (a few hours to days).

The Arthus reaction is a *localized* dermal injury due to inflamed blood vessels in the vicinity of any injected antigen. Serum sickness, however, is a *systemic* injury initiated by antigen-antibody complexes that circulate in the blood and settle into membranes at various sites.

The Arthus Reaction

The Arthus reaction is usually an acute response to a second injection of drugs or vaccines (boosters) at the same site as the first injection. In a few hours, the area becomes red, hot to the touch, swollen, and very painful **(figure 14.9a).** These symptoms are mainly due to the destruction of tissues in and around the blood vessels and the release of histamine from mast cells and basophils. Although the reaction is usually self-limiting and rapidly cleared, intravascular blood clotting can occasionally cause necrosis and loss of tissue.

Serum Sickness

Serum sickness was named for a condition that appeared in soldiers after repeated injections of horse serum to treat tetanus, though it is also caused by injections of animal hormones and drugs. The immune complexes enter the circulation; are carried throughout the body; and are eventually deposited in blood vessels of the kidney, heart, skin, and joints. The condition can become chronic, causing symptoms such as enlarged lymph nodes, rashes, painful joints, swelling, fever, and renal dysfunction **(figure 14.9b).**

14.4 LEARNING OUTCOMES—Assess Your Progress

9. Identify commonalities and differences between type II and type III hypersensitivities.

14.5 Type IV Hypersensitivities: Cell-Mediated (Delayed) Reactions

The adverse immune responses we have covered so far are explained primarily by B-cell involvement and antibodies. Type IV hypersensitivity is different; it involves primarily the T-cell branch of the immune system. In general, type IV diseases result when T cells respond to antigens displayed on self tissues or transplanted foreign cells. Type IV immune dysfunction has traditionally been known as delayed hypersensitivity because the symptoms arise one to several days following the second contact with an antigen.

Infectious Allergy

A classic example of delayed-type hypersensitivity occurs when a person sensitized by tuberculosis infection is injected with an extract (tuberculin) of the bacterium *Mycobacterium tuberculosis*. The so-called tuberculin reaction is an acute skin inflammation at the injection site appearing within 24 to 48 hours. So useful and diagnostic is this technique for detecting present or prior tuberculosis that it is the chosen screening strategy (see chapter 19). Other infections that use similar skin testing are leprosy, syphilis, histoplasmosis, toxoplasmosis, and candidiasis. This form of hypersensitivity arises from time-consuming cellular events involving a specific class of T cells ($T_H 1$) and their release of cytokines that recruit various inflammatory cells such as macrophages, neutrophils, and eosinophils. The buildup of fluid and cells at the site gives rise to a red bump **(figure 14.10*a*)**.

Contact Dermatitis

The most common delayed allergic reaction, contact dermatitis, is caused by exposure to resins in poison ivy or poison oak, to simple haptens in household and personal articles (jewelry, cosmetics, elasticized undergarments), and to certain drugs. Like immediate atopic dermatitis, the reaction to these allergens requires a sensitizing dose followed by a provocative dose. The allergen first penetrates the outer skin layers, is processed by Langerhans cells (skin dendritic cells), and presented to T cells. When subsequent exposures attract lymphocytes and macrophages to this area, these cells release enzymes and inflammatory cytokines that severely damage the epidermis in the immediate vicinity. This response accounts for the intensely itchy papules and blisters that are the early symptoms **(figure 14.10*b*)**. As healing progresses, the epidermis is replaced by a thick, keratinized layer.

T Cells and Their Role in Organ Transplantation

Transplantation or grafting of organs and tissues is a common medical procedure. Although it is life-saving, this technique is plagued by the natural tendency of lymphocytes to seek out foreign antigens and mount a campaign to destroy them. The bulk of the damage that occurs in graft rejections can be attributed to cytotoxic T-cell action.

The Genetic and Biochemical Basis for Graft Rejection

In chapter 13, we learned that the genes and markers in major histocompatibility (MHC or HLA) classes I and II are extremely important in recognizing self and in regulating the immune response. Although the cells of each person can exhibit variability in the pattern of these cell surface molecules, the pattern is identical in different cells of the same person. Similarity is seen among related siblings and parents, but the more distant the relationship, the less likely that the MHC genes and markers will be alike. When donor tissue (a graft) displays surface molecules of

Type IV

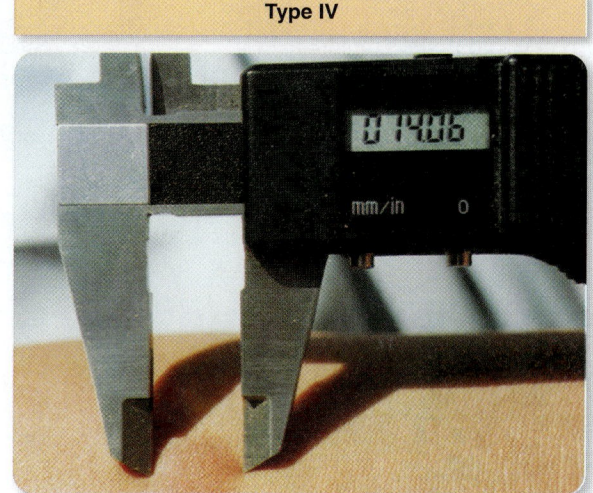

(a)

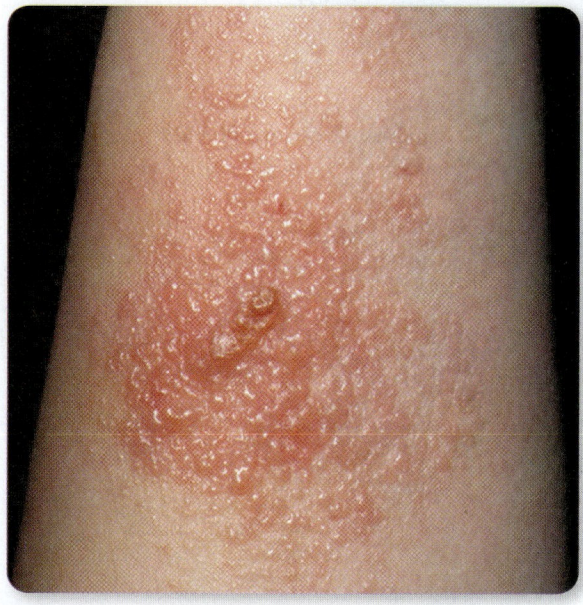

(b)

Figure 14.10 Type IV delayed reactions.
(a) Positive tuberculin test. Intradermal injection of tuberculin extract in a person sensitized to tuberculosis yields a slightly raised red bump greater than 10 mm in diameter. (b) Contact dermatitis from poison oak, showing various stages of involvement: blister, scales, and thickened patches.

Identical twins are the only people with exactly matching tissue antigens.

a different MHC class, the T cells of the recipient (called the host) will recognize its foreignness and react against it.

T-Cell-Mediated Recognition of Foreign MHC Receptors

Host Rejection of Graft   When the cytotoxic T cells of a host recognize foreign class I MHC markers on the surface of grafted cells, they release interleukin-2 as part of a general immune mobilization **(figure 14.11).** Antigen-specific helper and cytotoxic T cells bind to the grafted tissue and secrete lymphokines that begin the rejection process within 2 weeks of transplantation. Antibodies formed against the transplanted tissue contribute to the damage, resulting in the destruction of the vascular supply and death of the graft.

Graft Rejection of Host   Graft incompatibility is a two-way phenomenon. Some grafted tissues (especially bone marrow) contain an indigenous population called passenger lymphocytes (figure 14.11). This makes it quite possible for the graft to reject the host, causing **graft versus host disease (GVHD).** Because any host tissue bearing MHC markers foreign to the graft can be attacked, the effects of GVHD are widely systemic and toxic. A papular, peeling skin rash is the most common symptom though other organs are also affected. GVHD typically occurs within 100 to 300 days of the graft; overall, such reactions are declining due to better screening and a greater selection of tissues.

Classes of Grafts   Grafts are generally classified according to the genetic relationship between the donor and the recipient **(figure 14.12).** Tissue transplanted from one site on an individual's body to another site on his or her body is known as an **autograft.** Typical examples are skin replacement in burn repair and the use of a vein to fashion a coronary artery bypass. In an **isograft,** tissue from an identical twin is used. Because isografts do not contain foreign antigens, they are not rejected. **Allografts,** the most common type of grafts, are exchanges between

Figure 14.11 **Development of incompatible tissue graft reactions.**

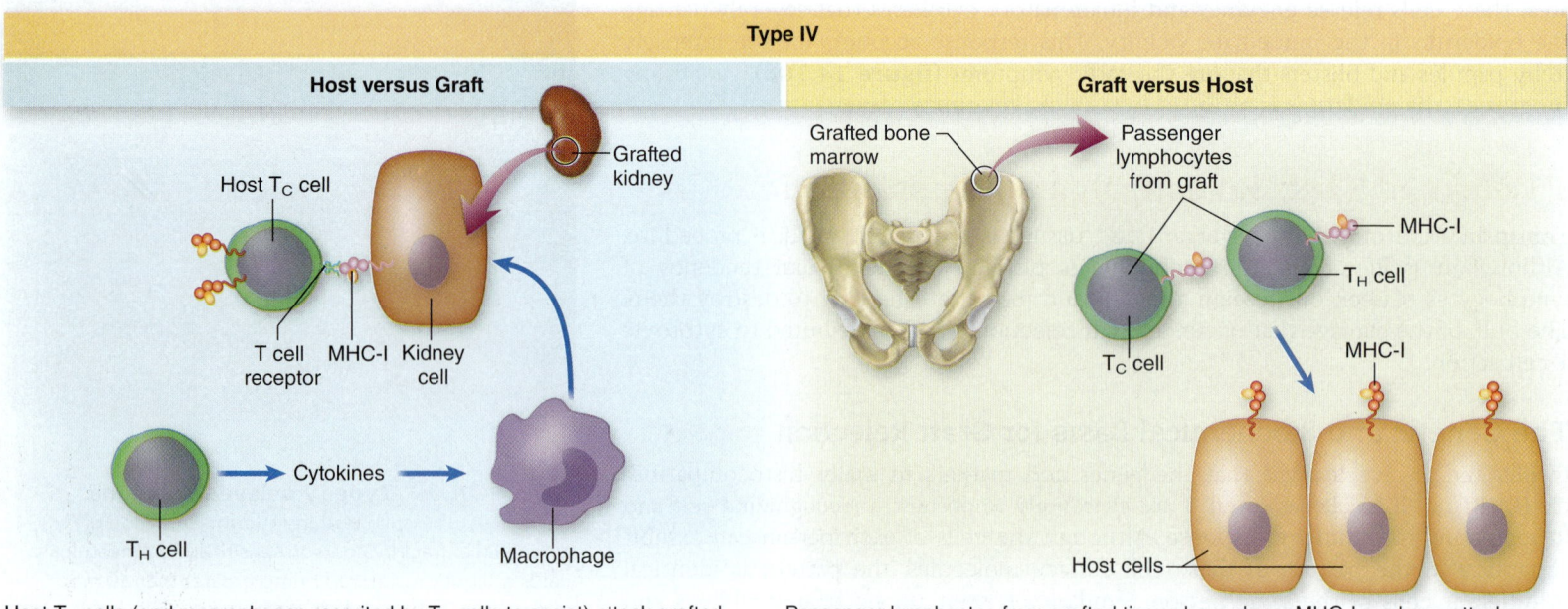

Host T$_C$ cells (and macrophages recruited by T$_H$ cells to assist) attack grafted cells with foreign MHC-I markers.

Passenger lymphoctes from grafted tissue have donor MHC-I markers; attack recipient cells with different MHC-I specificity.

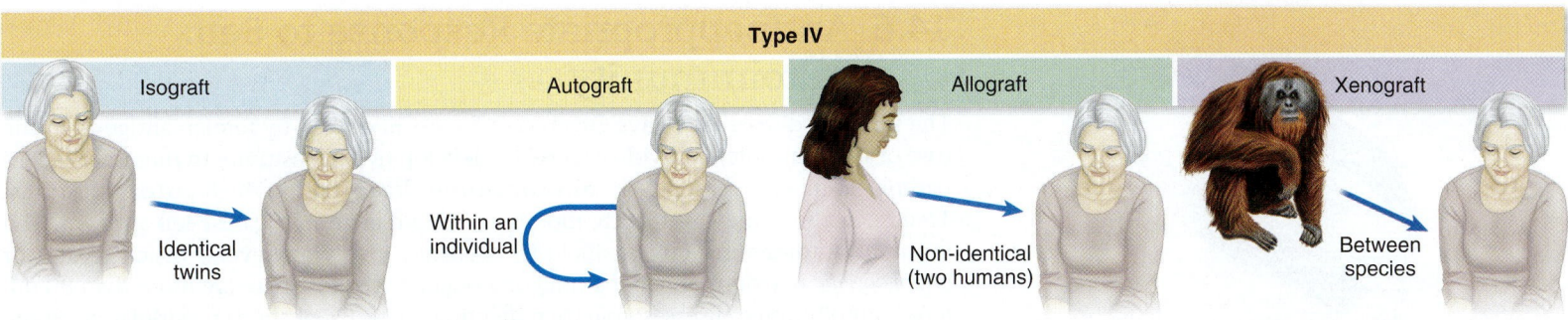

Figure 14.12 Four classes of tissue grafts.

genetically different individuals belonging to the same species (two humans). A close genetic correlation is sought for most allograft transplants (see next section). A **xenograft** is a tissue exchange between individuals of different species.

Types of Transplants

Over 28,000 people receive transplants each year in the United States, which reflects the beneficial nature of this medical procedure today. Transplantation involving every major organ, including parts of the brain, has been performed but most often involves skin, liver, heart, kidney, coronary artery, cornea, and bone marrow. The sources of organs and tissues are live donors (kidney, skin, bone marrow, liver), cadavers (heart, kidney, cornea), and fetal tissues.

In the past decade, advancements in transplantation science have expanded the possibilities for treatment and survival. Fetal tissues have been used in the treatment of diabetes and Parkinson disease, while parents have successfully donated portions of their organs to help save their children suffering from the effects of cystic fibrosis or liver disease. Recent advances in stem cell technology have made it possible to isolate stem cells more efficiently from blood donors, and the use of umbilical cord blood cells has furthered progress in this area of science. Though many hurdles still exist, scientists are using genetic engineering technology to develop an ample supply of immunologically compatible, safe tissues for xenotransplantation.

Bone marrow transplantation is used in patients with immune deficiencies, aplastic anemia, leukemia and other cancers, and radiation damage. Before closely matched donor marrow can be infused, the patient is pretreated with chemotherapy and whole-body irradiation to destroy the person's own blood stem cells, preventing rejection of the new marrow cells. The donor marrow cells are then dripped intravenously into the circulatory system, and the new cells settle automatically into the appropriate bone marrow regions. However, because donor lymphoid cells can still cause GVHD, antirejection drugs may be necessary. Interestingly, after transplantation, a recipient's blood type may change to the blood type of the donor!

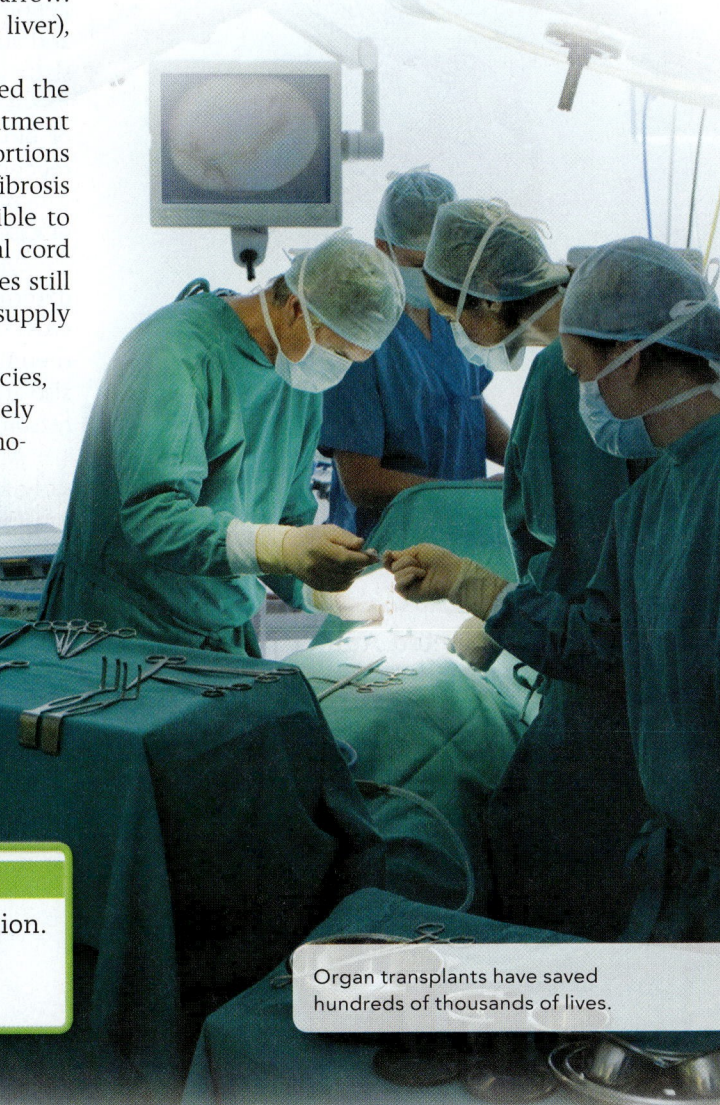

Organ transplants have saved hundreds of thousands of lives.

14.5 LEARNING OUTCOMES—Assess Your Progress

10. Describe one example of a type IV delayed hypersensitivity reaction.

11. List four classes of grafts, and explain how host versus graft and graft versus host diseases develop.

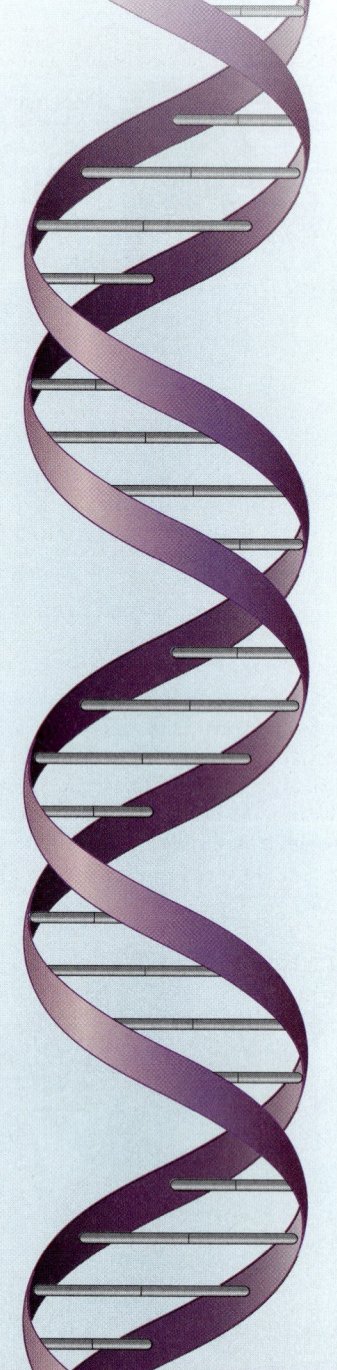

14.6 An Inappropriate Response to Self: Autoimmunity

The immune diseases we have covered so far are all caused by foreign antigens. In the case of autoimmunity, individuals actually develop hypersensitivity to themselves. This pathologic process accounts for **autoimmune diseases,** in which **autoantibodies,** T cells, and, in some cases, both, mount an abnormal attack against self antigens. The scope of autoimmune disease includes *systemic*, involving several major organs, or *organ-specific* reactions, involving only one organ or tissue. There are more than 80 different autoimmune diseases, together affecting 5% to 8% of the U.S. population. Some major diseases, their targets, and basic pathology are presented in **table 14.2.**

Genetic Correlation in Autoimmune Disease

In most cases, the precipitating cause of autoimmune disease remains obscure, but we do know that susceptibility can be influenced strongly by genetics and gender. Cases cluster in families, and even unaffected members tend to develop the autoantibodies for that disease. Studies show particular genes in the class I and II major histocompatibility complex coincide with certain autoimmune diseases. For example, autoimmune joint diseases such as rheumatoid arthritis and ankylosing spondylitis are more common in persons with the B-27 HLA type. With the expansion of genomic technology and the screening of whole genomes, many novel genes have recently been found to play a role in the pathway to autoimmunity. Sequencing of genomes may represent a new avenue for clinical diagnosis or treatment of disease, and studies have suggested that seemingly unrelated disorders, such as autism, may share a common genetic basis with autoimmune disease.

A moderate, regulated amount of autoimmunity is probably required to dispose of old cells and cellular debris. Disease apparently arises when this regulatory or recognition apparatus goes awry. Sometimes the processes go awry due to genetic irregularities or inherent errors in the host's physiological processes. In a large subset of cases, however, microbes are behind the malfunctioning.

Molecular mimicry is a process in which microbial antigens bearing molecular determinants similar to human cells induce the formation of antibodies that can cross-react with normal tissues. This is one possible explanation for the pathology of rheumatic fever. Similarly, T cells primed to react with streptococcal surface proteins also appear to react with keratin cells in the skin, causing them to proliferate. For this reason, psoriasis patients often report flare-ups after a strep throat infection.

Autoimmune disorders such as type 1 diabetes and multiple sclerosis are possibly triggered by *viral infection.* Viruses can noticeably alter cell receptors, thereby causing immune cells to attack the tissues bearing viral receptors.

Table 14.2 Selected Autoimmune Diseases

Disease	Target	Type of Hypersensitivity	Characteristics
Systemic lupus erythematosus (SLE)	Systemic	III	Inflammation of many organs; antibodies against red and white blood cells, platelets, clotting factors, nucleus DNA
Rheumatoid arthritis and ankylosing spondylitis	Systemic	II, III, and IV	Vasculitis; frequent target is joint lining; antibodies against other antibodies (rheumatoid factor), T-cell cytokine damage
Graves' disease	Thyroid	III	Antibodies against thyroid-stimulating hormone receptors
Myasthenia gravis	Muscle	III	Antibodies against the acetylcholine receptors on the nerve-muscle junction alter function
Type 1 diabetes	Pancreas	IV	T cells attack insulin-producing cells
Multiple sclerosis	Myelin	II and IV	T cells and antibodies sensitized to myelin sheath destroy neurons

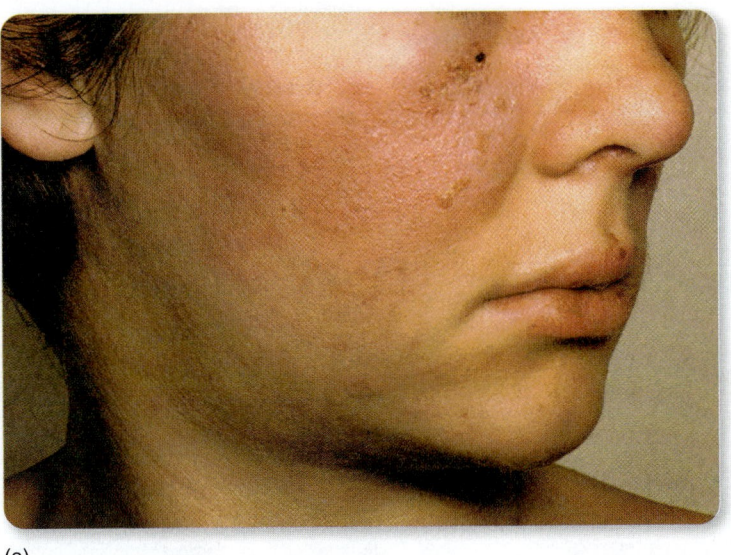

(a)

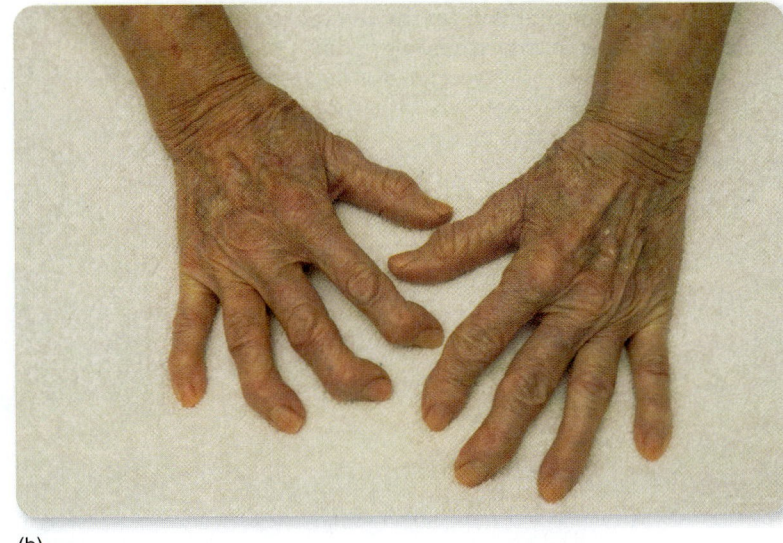

(b)

Examples of Autoimmune Disease

Systemic Autoimmunities

One of the most severe chronic autoimmune diseases is systemic lupus erythematosus (SLE, or lupus). This name originated from the characteristic butterfly-shaped rash that drapes across the nose and cheeks **(figure 14.13a),** as ancient physicians thought the rash resembled a wolf bite on the face (*lupus* is Latin for "wolf"). Although the manifestations of the disease vary considerably, all SLE patients produce autoantibodies against a variety of targets, including organs and tissues or intracellular materials, such as the nucleoprotein of the nucleus and mitochondria. It is not known how such a generalized loss of self-tolerance arises, though viral infection and loss of normal immune response suppression are suspected.

Rheumatoid arthritis, another systemic autoimmune disease, incurs progressive, debilitating damage to the joints and at times to the lungs, eyes, skin, and nervous system. In the joint form of the disease, autoantibodies form immune complexes that bind to the synovial membrane of the joints, which activates phagocytes to release cytokines. Chronic inflammation develops and leads to scar tissue and joint destruction. These cytokines (i.e., tumor necrosis factor [TNF]) can then trigger additional type IV delayed hypersensitivity responses. Epstein-Barr virus has been implicated as a precipitating cause, and the presence of an IgM antibody, called rheumatoid factor (RF), can be used in diagnosis of this disease. Treatment has recently involved the targeting of TNF or TNF-mediated pathways, but new drugs, targeting other immune system components, are now appearing.

Autoimmunities of the Endocrine Glands

The underlying cause of **Graves' disease** is the attachment of autoantibodies to receptors on the thyroxin-secreting follicle cells of the thyroid gland. The abnormal stimulation of these cells causes the overproduction of this hormone and the symptoms of hyperthyroidism, which affect nearly every body system.

Type 1 diabetes is another condition that may be a result of autoimmunity. Insulin, secreted by the beta cells in the pancreas, regulates the utilization of glucose by cells. Molecular mimicry has been implicated in the sensitization of cytotoxic T cells in type 1 diabetes, which leads to the lysis of beta cells. The reduced amount of insulin underlies the symptoms of this disease. A recent study showed permanent reversal of type 1 diabetes in patients that were re-infused with their own stem cells after complete immune suppression. We can expect to see more examples of such treatments as our understanding of autoimmunity increases.

Figure 14.13 Common autoimmune diseases. (a) Systemic lupus erythematosus. One symptom is a prominent rash across the bridge of the nose and on the cheeks. These papules and blotches can also occur on the chest and limbs. (b) Rheumatoid arthritis commonly targets the synovial membrane of joints. Over time, chronic inflammation causes thickening of this membrane, erosion of the articular cartilage, and fusion of the joint. These effects severely limit motion and can eventually swell and distort the joints.

 NCLEX® PREP

3. Which statement is true with regard to autoimmune diseases and hypersensitivity reactions?
   a. *Multiple sclerosis occurs more commonly in individuals who have the B-27 HLA type gene.*
   b. *Graves' disease is associated with type II hypersensitivity with abnormal T-cell function.*
   c. *Diabetes mellitus type 1 is associated with T-cell dysfunction.*
   d. *Systemic lupus erythematosus is associated with type IV hypersensitivity reaction.*

Figure 14.14 Mechanism for involvement of autoantibodies in myasthenia gravis.
Antibodies developed against receptors on the postsynaptic membrane block them so that acetylcholine cannot bind and muscle contraction is inhibited.

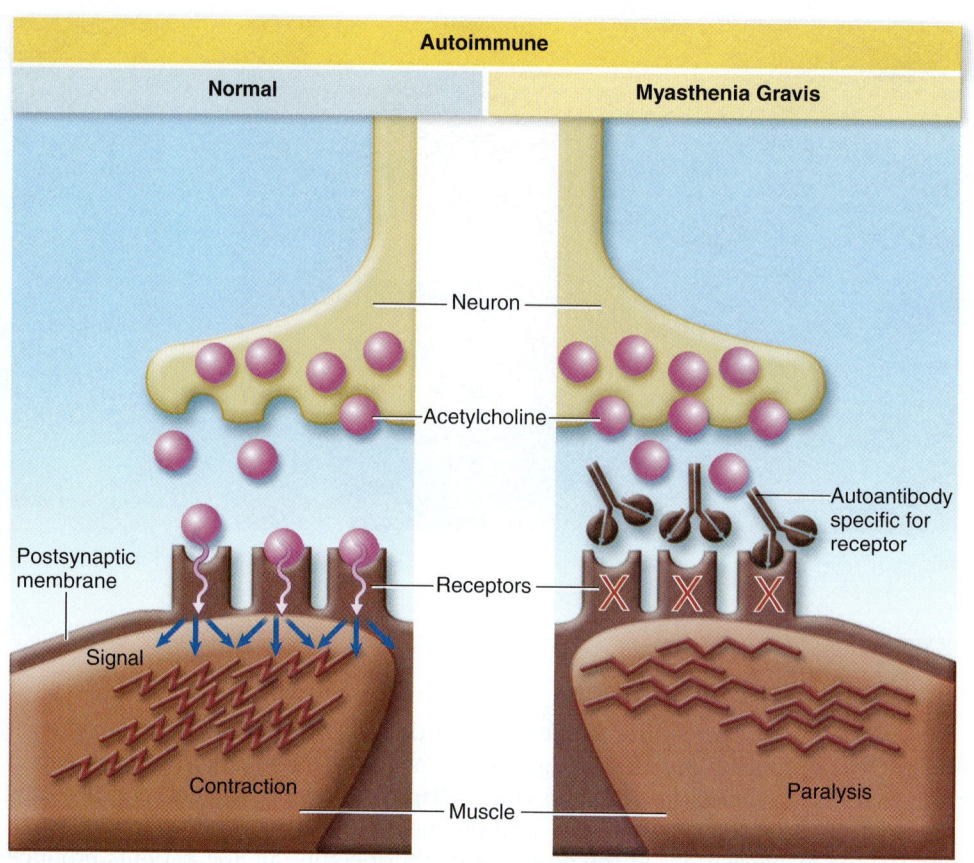

Neuromuscular Autoimmunities

Myasthenia gravis is a syndrome caused by autoantibodies binding to the receptors for acetylcholine, a chemical required to transmit a nerve impulse across the synaptic junction to a muscle **(figure 14.14).** The first effects are usually felt in the muscles of the eyes and throat, but the syndrome eventually progresses to complete loss of skeletal muscle function and death. Current treatment usually includes immunosuppressive drugs and therapy to remove the autoantibodies from the circulation.

Multiple sclerosis (MS) is a paralyzing neuromuscular disease associated with lesions in the insulating myelin sheath that surrounds neurons in the white matter of the central nervous system. T-cell and autoantibody-induced damage severely compromise the capacity of neurons to send impulses, resulting in muscular weakness and tremors, difficulties in speech and vision, and paralysis. Data suggest a possible association between infection with human herpesvirus 6 and the onset of disease. Immunosuppressants like cortisone and interferon beta alleviate symptoms, and the disease can be treated with monoclonal antibody therapy toward certain T-cell antigens.

14.6 LEARNING OUTCOMES—Assess Your Progress

12. List at least three autoimmune diseases and the common immunologic features in them.

Table 14.3 General Categories of Immunodeficiency Diseases with Selected Examples

Primary Immune Deficiencies (Genetic)	Secondary Immune Deficiencies (Acquired)

B-cell defects (low levels of B cells and antibodies)
    Agammaglobulinemia (X-linked, non-sex-linked)
    Hypogammaglobulinemia
    Selective immunoglobulin deficiencies
 	
T-cell defects (lack of all classes of T cells)
    Thymic aplasia (DiGeorge syndrome)
 	
Combined B-cell and T-cell defects (usually caused by lack or abnormality of lymphoid stem cell)
    Severe combined immunodeficiency (SCID) disease
    Adenosine deaminase (ADA) deficiency
 	
Complement defects
    Lacking one of C components
    Hereditary angioedema associated with rheumatoid diseases
	From natural causes
	    Infections (AIDS) or cancers
	    Nutrition deficiencies
	    Stress
	    Pregnancy
	    Aging
	
	From immunosuppressive agents
	    Irradiation
	    Severe burns
	    Steroids (cortisones)
	    Immunosuppressive drugs
	    Removal of spleen

14.7 Immunodeficiency Diseases: Hyposensitivity of the Immune System

Occasionally, errors occur in the development of the immune system, and a person is born with or develops weakened immune responses, called **immunodeficiencies.** The most obvious consequences of immunodeficiencies are recurrent, overwhelming infections, often with opportunistic microbes. Immunodeficiencies fall into two general categories: *primary diseases*, present at birth (congenital) and usually stemming from genetic errors, and *secondary diseases*, acquired after birth and caused by natural or artificial agents **(table 14.3).**

Primary Immunodeficiency Diseases

Primary deficiencies affect both specific immunities and less-specific ones such as phagocytosis. Consult **figure 14.15** to survey the places in the normal sequential development of lymphocytes, where defects can occur, and the possible consequences. In many cases, the deficiency is due to an inherited abnormality, though the exact nature of the abnormality is not known for a number of diseases. In some deficiencies, the lymphocyte in question is completely absent or is present at very low levels; in others, lymphocytes are present but do not function normally. Because the development of B cells and T cells diverges at some point, an individual can lack one or both cell lines. It must be emphasized, however, that some deficiencies affect other cell functions as well.

Clinical Deficiencies in B-Cell Development or Expression

Genetic deficiencies in B cells usually result in abnormal immunoglobulin expression. In some instances, only certain immunoglobulin classes are absent; in others, the levels of all types of immunoglobulin (Ig) are reduced.

   The term **agammaglobulinemia** literally means the absence of gamma globulin, the fraction of serum that contains immunoglobulins. Because it is very rare for Ig to be completely absent, some physicians prefer the term **hypogammaglobulinemia.** The symptoms of recurrent, serious bacterial infections usually appear about 6 months after birth. The bacteria most often implicated are pyogenic cocci, *Pseudomonas*, and *Haemophilus influenzae*; and the most common infection sites are the lungs, sinuses, meninges, and blood. Many Ig-deficient patients can have recurrent infections with viruses and protozoa, as well. The current treatment for this condition is passive immunotherapy with immune serum globulin and continuous antibiotic therapy.

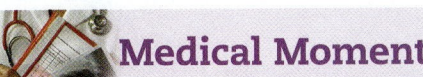

Medical Moment

Hand Washing

The importance of hand washing cannot be over-emphasized, particularly when caring for patients who have compromised immune systems. Such a simple step can be lifesaving for patients who are unable to fight off infection. Recognizing this fact, hospitals have made a conscious and concentrated effort to increase the number of hand washing stations available to increase compliance among health care workers.

   One of the most important skills that a health care worker will ever learn is proper hand washing technique. Alcohol-based hand sanitizers have increased hand washing compliance. Hand sanitizer should be applied liberally to the palm of one hand. The hands should then be rubbed together, dispersing the sanitizer over the surfaces of both hands, including the fingers and wrists. When hands are visibly soiled, clean running water and soap should be used.

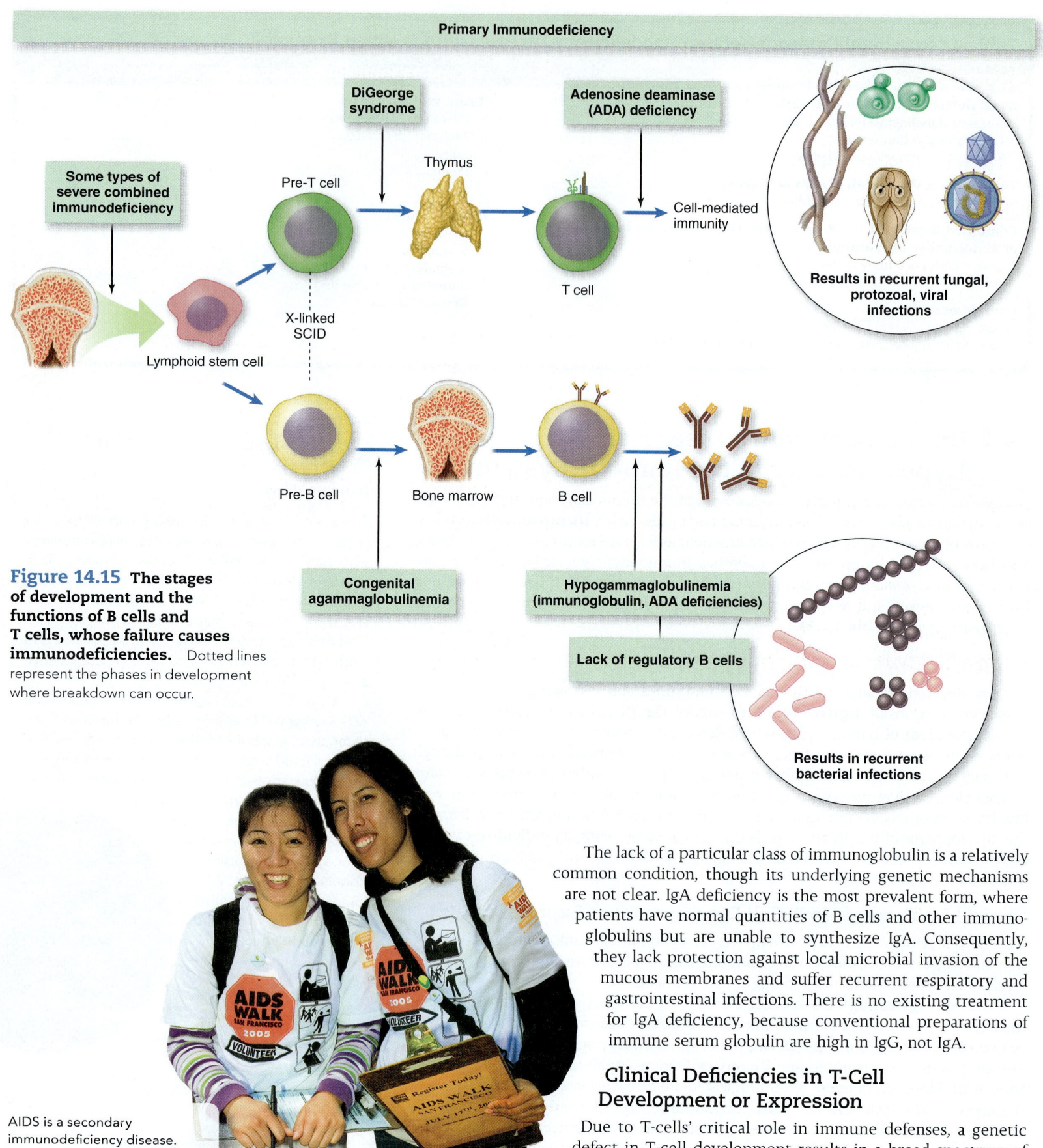

Primary Immunodeficiency

DiGeorge syndrome

Adenosine deaminase (ADA) deficiency

Some types of severe combined immunodeficiency

Pre-T cell

Thymus

Cell-mediated immunity

T cell

Lymphoid stem cell

X-linked SCID

Results in recurrent fungal, protozoal, viral infections

Pre-B cell

Bone marrow

B cell

Congenital agammaglobulinemia

Hypogammaglobulinemia (immunoglobulin, ADA deficiencies)

Lack of regulatory B cells

Results in recurrent bacterial infections

Figure 14.15 The stages of development and the functions of B cells and T cells, whose failure causes immunodeficiencies. Dotted lines represent the phases in development where breakdown can occur.

AIDS is a secondary immunodeficiency disease.

The lack of a particular class of immunoglobulin is a relatively common condition, though its underlying genetic mechanisms are not clear. IgA deficiency is the most prevalent form, where patients have normal quantities of B cells and other immunoglobulins but are unable to synthesize IgA. Consequently, they lack protection against local microbial invasion of the mucous membranes and suffer recurrent respiratory and gastrointestinal infections. There is no existing treatment for IgA deficiency, because conventional preparations of immune serum globulin are high in IgG, not IgA.

Clinical Deficiencies in T-Cell Development or Expression

Due to T-cells' critical role in immune defenses, a genetic defect in T-cell development results in a broad spectrum of disease, including severe opportunistic infections and cancer.

In fact, a dysfunctional T-cell line is usually more devastating than a defective B-cell line because T helper cells are required to assist in most specific immune reactions.

Abnormal Development of the Thymus The most severe of the T-cell deficiencies involves the congenital absence or immaturity of the thymus. Thymic aplasia, or **DiGeorge syndrome,** makes children highly susceptible to persistent infections by fungi, protozoa, and viruses. Vaccinations using live, attenuated microbes pose a danger, and common childhood infections such as chickenpox can be overwhelming and fatal in these children. Patients typically have reduced antibody levels as well.

Severe Combined Immunodeficiencies: Dysfunction in B and T Cells

Severe combined immunodeficiencies (SCIDs) are the most serious and potentially lethal forms of immunodeficiency disease because they involve dysfunction in both lymphocyte systems. Some SCIDs are due to the complete absence of the lymphocyte stem cell in the marrow; others are attributable to the dysfunction of B cells and T cells later in development. Infants with SCID usually manifest the T-cell deficiencies within days after birth by developing candidiasis, sepsis, pneumonia, or systemic viral infections.

In the two most common forms, Swiss-type agammaglobulinemia and thymic alymphoplasia, genetic defects in the development of the lymphoid cell line result in extremely low numbers of all lymphocyte types and poorly developed humoral and cellular immunity. A rare form of SCID, called **adenosine deaminase (ADA) deficiency,** is caused by an autosomal recessive defect in the metabolism of adenosine. In this case, lymphocytes develop but a metabolic product builds up abnormally and selectively destroys them. A small number of SCID cases are due to a developmental defect in receptors on B and T cells.

Because of their profound lack of specific adaptive immunities, SCID children require the most rigorous kinds of aseptic techniques to protect them from opportunistic infections. Aside from life in a sterile plastic bubble, exemplified by David Vetter **(figure 14.16a),** the only serious option for their longtime survival is total replacement or correction of dysfunctional lymphoid cells. Some infants can benefit from fetal liver or stem cell grafts, though transplantation is complicated by graft versus host disease. A more lasting treatment for both X-linked and ADA types of SCID is gene therapy–insertion of normal genes to replace the defective genes (see chapter 8).

Secondary Immunodeficiency Diseases

Secondary acquired deficiencies in B cells and T cells are caused by one of four general agents:

1. infection,
2. noninfectious metabolic disease,
3. chemotherapy, or
4. radiation.

The most recognized infection-induced immunodeficiency is **AIDS** (see chapter 18). This syndrome is caused when several types of immune cells, including T helper cells, monocytes, macrophages, and antigen-presenting cells, are infected by the human immunodeficiency virus (HIV). It is generally thought that the depletion of T helper cells and functional impairment of immune responses ultimately account for the cancers and opportunistic infections associated with this disease **(figure 14.16b).** Other infections that can deplete immunities are measles, leprosy, and malaria.

Primary Immunodeficiency

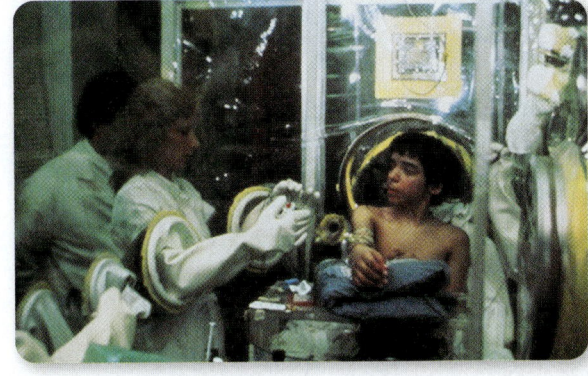

(a)

Secondary Immunodeficiency

(b)

Figure 14.16 The two types of immunodeficiency. (a) David Vetter, who lived from 1971 to his death in 1984 in a sterile bubble. (b) An AIDS patient.

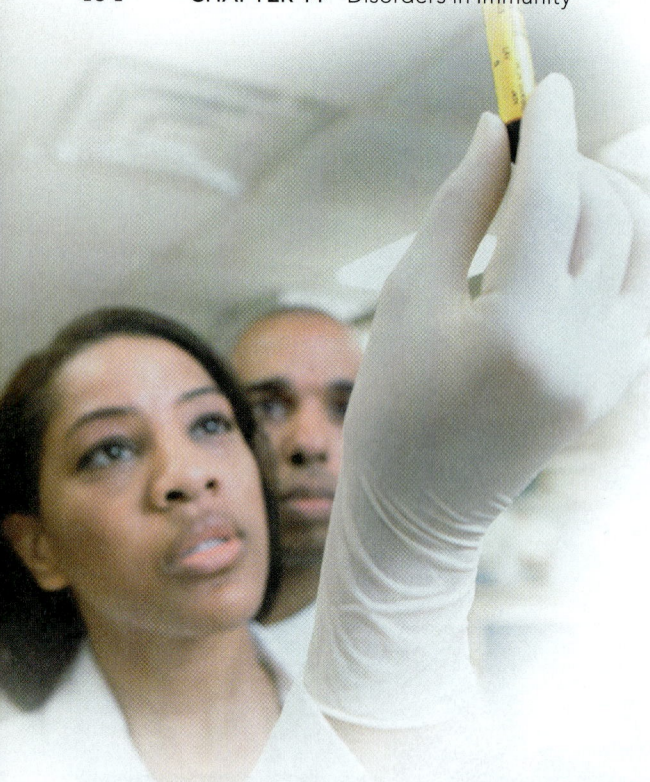

Cancers that target the bone marrow or lymphoid organs can be responsible for extreme malfunction of both humoral and cellular immunity. In leukemia, cancer cells outnumber normal cells, displacing them from bone marrow and blood. Plasma cell tumors produce large amounts of nonfunctional antibodies, while thymus tumors cause severe T-cell deficiencies.

An ironic outcome of lifesaving medical procedures is the possible suppression of a patient's immune system. Drugs that prevent graft rejection or decrease the symptoms of rheumatoid arthritis can likewise suppress beneficial immune responses, while radiation and anticancer drugs are damaging to the bone marrow and other body cells.

14.7 LEARNING OUTCOMES—Assess Your Progress

13. Distinguish between primary and secondary immunodeficiencies, explaining how each develops.

CASE FILE WRAP-UP

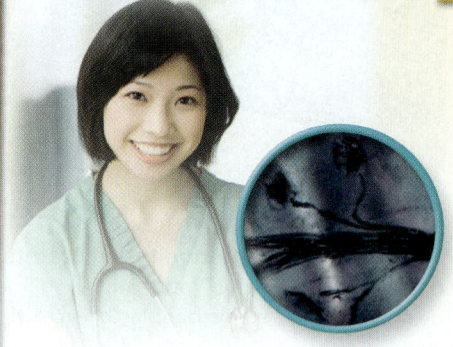

Myasthenia gravis commonly affects young women under the age of 40 and older men, but anyone can be affected. It is not directly inherited.

In myasthenia gravis, antibodies block or destroy acetylcholine receptors at the neuromuscular junction, preventing muscle contraction from occurring. Normally, nerve impulses travel down the nerve to the nerve ending, where the neurotransmitter acetylcholine binds to its receptors, which are then activated to generate a muscle contraction.

Immunosuppressants are used in myasthenia gravis to suppress the production of antibodies, thus improving muscle strength. Anticholinesterase agents, such as neostigmine, help to improve the transmission of nerve signals at the neuromuscular junction. Thymectomy (removal of the thymus) can help to reduce symptoms and may even cure some patients.

Two Types of Arthritis

Distinguishing between osteoarthritis and rheumatoid arthritis can be difficult, even for physicians. The following chart outlines some of the differences between the two conditions:

Distinguishing Factor	Osteoarthritis	Rheumatoid Arthritis
Age of onset	More common in older people–"wear and tear" arthritis	Typically between 30 and 60 years
Male/female predominance	More common in men before age 55, more common in women after age 55	Women are two to three times more likely to be affected, but men and children can get the disease
Time of day of worst symptoms	Pain generally worsens with activity; worse at end of the day	Prolonged morning stiffness (lasting longer than 30 minutes)
Cause	Age, genetic predisposition, obesity, injury, overuse of the joints	Inflammatory autoimmune disease, may be a triggering event or genetic cause
Joints affected	Most common in hips, knees, lower back (weight-bearing joints); can also affect smaller joints	Small bones of the hand and wrist, feet (may affect other joints as well); symmetrical joint involvement
Tests used in diagnosis	X ray, history, physical exam	X ray, physical exam, diagnostic tests (positive rheumatoid factor, anemia, elevated C-reactive protein, elevated erythrocyte sedimentation rate)
Frequency	Affects 21 million Americans	Affects over 2 million Americans
Other body systems affected	None–affects only the joints	May affect other organ systems
Onset	Gradual	May be sudden
Part of joint affected	Cartilage	Synovium (lining of the joint)

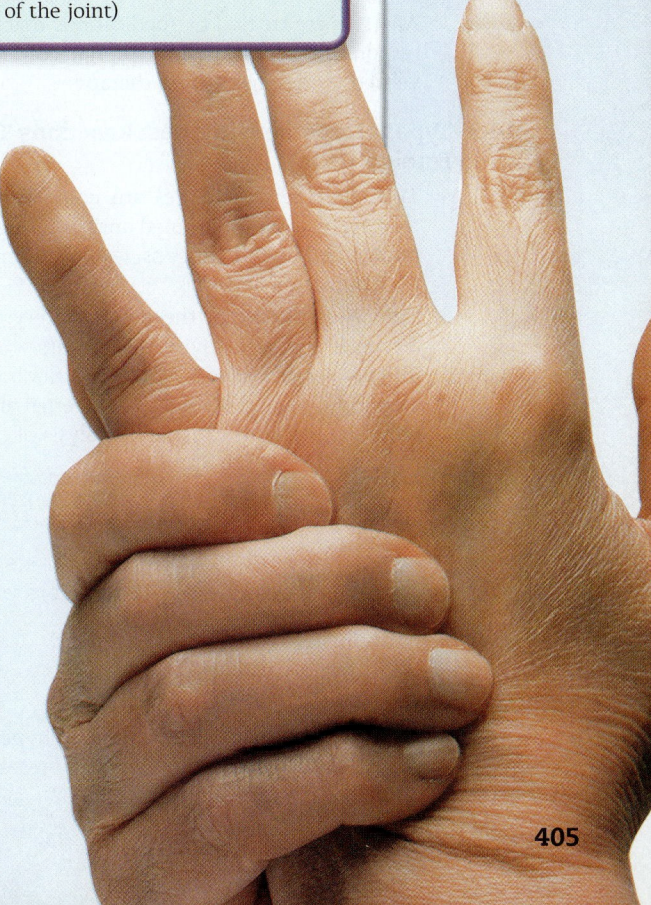

The term *arthritis* literally means "joint inflammation." Besides osteoarthritis and rheumatoid arthritis, the two most common types, there are many other types of arthritis. *Juvenile rheumatoid arthritis* affects children and can be debilitating. *Psoriatic arthritis* is a type of arthritis that affects people with psoriasis, an autoimmune skin condition. Psoriatic arthritis may precede the skin rash of psoriasis. *Gout* is another type of arthritis characterized by sudden and very painful attacks of redness and swelling affecting the joints, typically the big toe. Gout is caused by uric acid crystals that are deposited in the joints, causing pain and inflammation. *Ankylosing spondylitis* is a chronic inflammatory disease that results in eventual fusing of the spine. *Infectious arthritis* is inflammation of a joint caused by viruses, bacteria, or fungi. *Reiter's syndrome* (reactive arthritis) is a form of arthritis that affects not only the joints but the eyes and urinary tract as well. *Sjogren's syndrome* is characterized by dysfunction of the glands that produce moisture in the eyes and mouth. This syndrome may affect other parts of the body as well.

Fibromyalgia is a condition causing severe fatigue and joint/muscle pain. It can be confused with rheumatoid arthritis, but the joints of people with fibromyalgia do not become reddened or swollen. People with fibromyalgia have tender points within their muscles that cause pain. The cause of fibromyalgia is not known, but infection and trauma have both been implicated in causing the condition. Lab tests are usually normal.

Chapter Summary

14.1 The Immune Response: A Two-Sided Coin

- Immunopathology is the study of diseases associated with over-activity and underactivity of the immune response, including allergies, autoimmunity, transplantation and transfusion reactions, immunodeficiency diseases, and cancer.
- There are four categories of hypersensitivity reactions: type I (allergy and anaphylaxis), type II (complement, IgG- and IgM-mediated tissue destruction), type III (immune complex reactions), and type IV (delayed hypersensitivity reactions).

14.2 Type I Allergic Reactions: Atopy and Anaphylaxis

- Antigens that trigger hypersensitivity reactions are allergens, which are either exogenous (originate outside the host) or endogenous (involve the host's own tissue).
- Atopy is a chronic, local allergy, whereas anaphylaxis is a systemic, potentially fatal allergic response. Both result from excessive IgE production in response to exogenous antigens.
- Type I allergens include inhalants, ingestants, injectants, and contactants; potential portals of entry are the skin, respiratory tract, gastrointestinal tract, and genitourinary tract.
- Type I hypersensitivities are set up by a sensitizing dose of allergen and expressed when a second provocative dose triggers the allergic response.
- The primary participants in type I hypersensitivities are IgE, basophils, mast cells, and mediators of the inflammatory response.
- Allergies are diagnosed using *in vitro* and *in vivo* tests that assay for specific immune cells, IgE production, or local reactions to allergens.
- Allergies are treated by medications that interrupt the allergic response at certain points. Allergic reactions can often be prevented by desensitization therapy.

14.3 Type II Hypersensitivities: Reactions That Lyse Foreign Cells

- Type II hypersensitivities are complement-assisted reactions that occur when preformed antibodies (IgG or IgM) react with foreign cell-bound antigens, leading to membrane attack complex formation and lysis.
- Hemolytic disease of the newborn (erythroblastosis fetalis) is a type II hypersensitivity that occurs when Rh− mothers are initially sensitized to RBCs from a firstborn Rh+ baby. When carrying a second Rh+ fetus, maternal anti-Rh antibodies can cross the placenta, causing hemolysis of the fetal Rh+ RBCs.

14.4 Type III Hypersensitivities: Immune Complex Reactions

- Type III hypersensitivity reactions occur when large quantities of antigen react with host antibody to form insoluble immune complexes that settle in tissue cell membranes, causing chronic destructive inflammation.
- Like type II reactions, type III hypersensitivities involve the production of IgG and IgM and the activation of complement; they differ in that the antigen recognized in these reactions is soluble.
- The mediators of type III hypersensitivity reactions include soluble IgA, IgG, or IgM, and agents of the inflammatory response.
- Localized (Arthus) and systemic (serum sickness) reactions are two forms of type III hypersensitivities.

14.5 Type IV Hypersensitivities: Cell-Mediated (Delayed) Reactions

- Type IV or delayed hypersensitivity reactions, like the tuberculin reaction and transplant reactions (host rejection and GVHD), occur when cytotoxic T cells attack either self tissue or transplanted foreign cells.
- The four classes of transplants or grafts are determined by the degree of MHC similarity between graft and host. From most to least similar, these are autografts, isografts, allografts, and xenografts.

14.6 An Inappropriate Response to Self: Autoimmunity

- Autoimmune reactions occur when autoantibodies or host T cells mount an abnormal attack against self antigens, due to virus-induced alterations of immunity or failure to remove self-reactive clones.
- Examples of autoimmune diseases include systemic lupus erythematosus, rheumatoid arthritis, diabetes mellitus, myasthenia gravis, and multiple sclerosis.

14.7 Immunodeficiency Diseases: Hyposensitivity of the Immune System

- Immunodeficiency diseases occur when the immune response is reduced or absent.
- Primary immune diseases are genetically induced deficiencies of B cells, T cells, the thymus gland, or combinations of these. SCIDs are the most severe due to the loss of both humoral and cell-mediated immunity.
- Secondary immune diseases are caused by infection (i.e., AIDS), organic disease, chemotherapy, or radiation.

Multiple-Choice Questions Bloom's Levels 1 and 2: Remember and Understand

Select the correct answer from the answers provided.

1. Which statement is true of autoimmunity?
   a. It involves misshapen antibodies.
   b. It refers to "automatic immunity."
   c. It often manifests as types II, III, and IV hypersensitivities.
   d. It has an acute course and then usually resolves itself.

2. The T-cell branch of the immune system is primarily responsible for which hypersensitivities?
   a. type I   b. type II   c. type III   d. type IV

3. The contact with allergen that primes mast cells is the

    a. sensitizing dose.   c. provocative dose.

    b. degranulation dose.   d. desensitizing dose.

4. Select the correct pairing of the inflammatory cytokine and its function:

    a. prostaglandin/wheal-and-flare reaction   c. leukotriene/headache (pain)

    b. histamine/constricted bronchioles   d. estrogen/asthma

5. The hygiene hypothesis suggests that

    a. there are still too many microbes in our environment.

    b. we may need more contact with microbes as our immune systems mature.

    c. we may need more contact with antimicrobials as our immune systems mature.

    d. there are not enough microbes on farms.

6. Which of the following plays a role in the Arthus reaction?

    a. IgE antibodies   c. cell-associated antigens

    b. complement   d. cytotoxic T cells

7. Type II hypersensitivities are due to

    a. IgE reacting with mast cells.

    b. activation of cytotoxic T cells.

    c. IgG-allergen complexes that clog epithelial tissues.

    d. complement-induced lysis of cells in the presence of antibodies.

8. Which of the following statements is most correct regarding SCIDs?

    a. They are the least severe form of primary immunodeficiency disease.

    b. They result from viral infections.

    c. They result from the action of complement.

    d. They involve dysfunction of both lymphocyte (B- and T-cell) systems.

Critical Thinking   Bloom's Levels 3, 4, and 5: Apply, Analyze, and Evaluate

Critical thinking is the ability to reason and solve problems using facts and concepts. These questions can be approached from a number of angles and, in most cases, they do not have a single correct answer.

1. Trace the course of a pollen grain through sensitization and provocation in type I allergies, discussing the role of mast cells, basophils, IgE, and allergic mediators.

2. Why might it be advisable for an Rh− woman who has had an abortion, miscarriage, or an ectopic pregnancy to be immunized against the Rh factor?

3. Explain how tissue xenotransplantation can be successful in light of the immune system's robust ability to recognize foreign antigens.

4. A 31-year-old male develops a severe full body rash after receiving penicillin. He is certain that he has never received this antibiotic in the past. Summarize the immunologic reaction that is occurring in this patient, and discuss one hypothesis that would explain why the symptoms occurred when he never was prescribed the drug before.

5. a. Explain why babies with agammaglobulinemia do not develop opportunistic infections until about 6 months after birth.

    b. Explain why people with B-cell deficiencies can benefit from artificial passive immunotherapy, and discuss whether these individuals can be successfully vaccinated against various microbes.

Visual Connections   Bloom's Level 5: Evaluate

This question connects previous images to a new concept.

1. As you learned in **chapter 13,** B and T cells originate in tissues outside of the lymphatic system. With this in mind, provide at least one example of how an abnormality in these areas (i.e., in the bone marrow or thymus) can lead to immune deficiency.

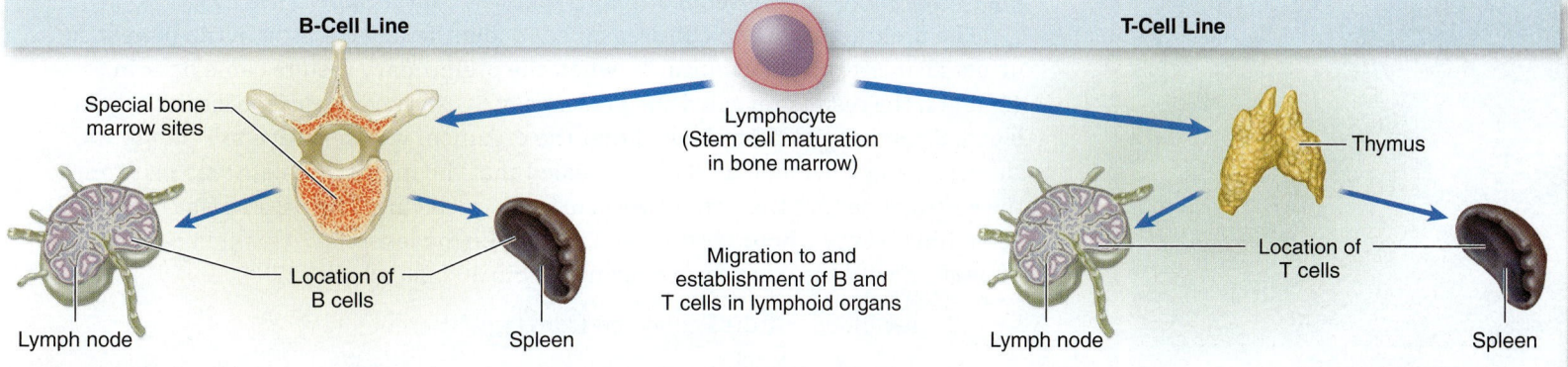

B-Cell Line — Special bone marrow sites — Location of B cells — Lymph node — Spleen — Lymphocyte (Stem cell maturation in bone marrow) — Migration to and establishment of B and T cells in lymphoid organs — T-Cell Line — Thymus — Location of T cells — Lymph node — Spleen

www.mcgrawhillconnect.com

Enhance your study of this chapter with study tools and practice tests. Also ask your instructor about the resources available through Connect, including the media-rich eBook, interactive learning tools, and animations.

CASE FILE

Tracing the Cause

When I was a lab tech student, my preceptor and I were asked to go to the emergency room to draw blood on a patient. The lab requisition filled out by the nurse caring for the patient stated the patient had "fever for 1 week with a decreased level of consciousness."

On reporting to the ER, we were directed to the patient, who was lying on a gurney in an exam room. The patient was a 78-year-old diabetic gentleman who appeared pale and thin. We introduced ourselves and told the patient that we were there to draw blood. The patient barely acknowledged our presence and seemed to have difficulty staying awake. The ER nurses had already collected a urine specimen from the patient's Foley catheter for urinalysis and culture. They handed us the specimens they had collected, which had been labeled correctly with the patient's name and the collection date and time.

We proceeded to draw blood according to the physician's order. We obtained blood for a complete blood count, electrolytes, blood glucose, liver function, and cardiac markers. We also obtained blood from two different sites for blood cultures. Once we had collected all of the samples, we notified the nursing staff that we were finished. We labeled all the collection vials with the necessary information and returned to the lab.

After running all of the tests that could be performed in-house, we sent copies of all of the reports to the ER. The patient's hemoglobin and hematocrit were slightly low. The white blood cell count was very high. The patient's potassium was high, and the sodium level was low. Liver function was normal, as were cardiac markers. The urinalysis showed ketones and protein in the urine, evidence of dehydration but not of infection. The patient's serum glucose was high but not dangerously so.

On the basis of these lab studies, the patient was rehydrated with intravenous fluids and his potassium and sodium levels were soon corrected. He was also given insulin to lower his blood glucose. The patient was started on broad-spectrum IV antibiotics pending the results of the urine and blood cultures. He continued to run a high fever and was only semiconscious following admission to the ICU.

The preliminary urine cultures failed to identify any bacteria in the patient's urine. That left the blood cultures; when the preliminary results came back in 48 hours, the results revealed the presence of *Staphylococcus epidermidis* in the patient's blood. Sensitivity studies showed that the organism was sensitive to Penicillin G, rifampin, and vancomycin. It also revealed that the microorganism was resistant to the ceftriaxone that the patient was currently receiving. The patient was switched to Penicillin G every 6 hours intravenously, and he made a relatively rapid recovery soon thereafter, with his fever disappearing within 36 hours.

- Why are blood cultures collected from two different sites?

- Why was it important for the laboratory staff to notify the nursing staff after the samples had been collected?

Case File Wrap-Up appears on page 432.

Diagnosing Infections

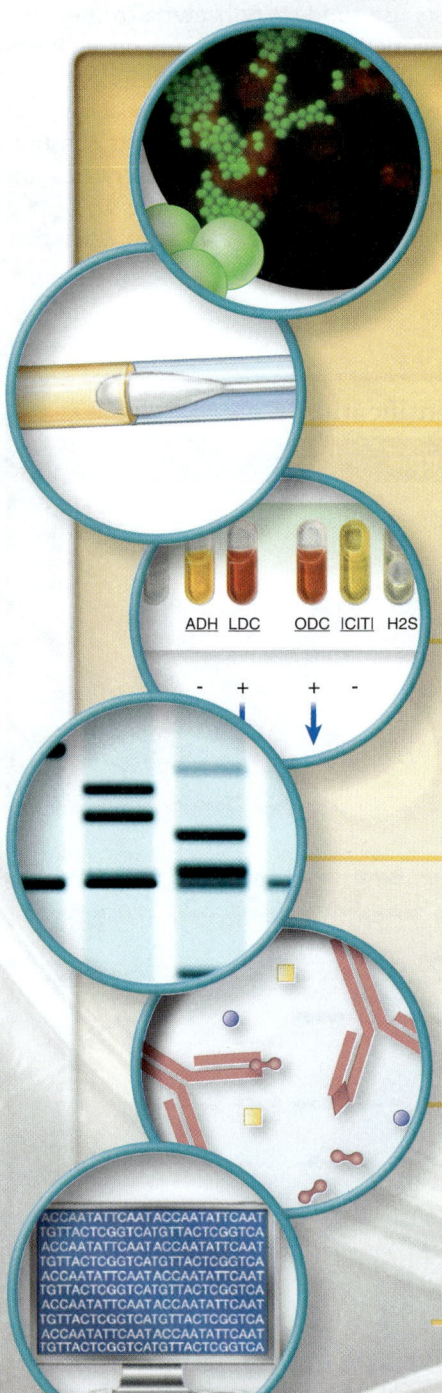

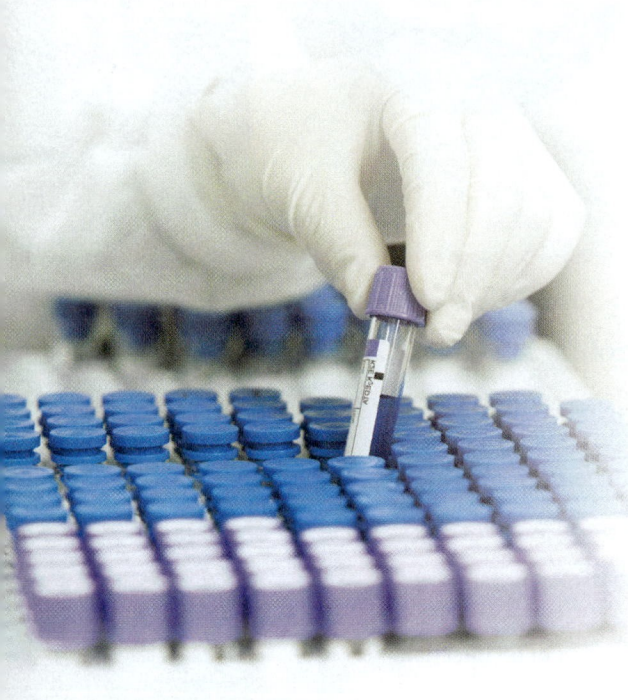

15.1 Preparation for the Survey of Microbial Diseases

Chapters 16 through 21 cover the most clinically significant bacterial, fungal, parasitic, and viral diseases. These chapters survey the most prevalent infectious conditions and the organisms that cause them. This chapter gets us started with an introduction to the how-to of diagnosing the infections.

For many students (and professionals), the most pressing topic in microbiology is *how to identify unknown bacteria* in patient specimens or in samples from nature. As seen in **table 15.1,** methods microbiologists use to identify bacteria to the level of genus and species fall into three main categories: **phenotypic,** which includes a consideration of morphology (microscopic and macroscopic) as well as bacterial physiology or biochemistry; **immunologic,** which entails serological analysis; and **genotypic** (or genetic) techniques. Data from a cross section of such tests can produce a unique profile for any bacterium. Increasingly, genetic means of identification are being used as a sole resource for identifying bacteria. We are on the verge of a revolution in infectious disease diagnosis. Within 5 years, most experts predict that

Table 15.1 Methods of Microbial Identification		
Category	**Description**	**Example**
Phenotypic	Observation of microbe's microscopic and macroscopic morphology, physiology, and biochemical properties	
Genotypic	Analysis of microbe's DNA or RNA	
Immunologic	Analysis of microbe using antibodies, or of patients' antibodies using prepackaged antigens	Agglutinated mat Nonagglutinated pellet Enlarged side view of wells

technology will allow large-scale and affordable adoption of genetic and other methods that will replace many of the phenotypic and immunologic methods. We will summarize these new technologies at the end of this chapter.

There are still many organisms, however, that must be identified in the "old-fashioned" way—via biochemical, serological, and morphological means. In fact, serology is so reliable for some diseases that it may never be replaced.

It should be noted that some other techniques can assist in the diagnosis. For example, computerized tomography (CT) scans are often used to diagnose peritonsillar abscesses, after which the causative agent can be identified using the techniques described in this chapter. Magnetic resonance imaging (MRI) and positron emission tomography (PET) scans are also used to find areas of deep tissue infection.

Phenotypic Methods

Microscopic Morphology

Light microscopy aids in the observation of microbial traits such as cell shape, size and arrangement, Gram stain reaction, acid-fast reaction, and special structures, including endospores, granules, and capsules. Electron microscopy studies can pinpoint additional structural features, such as the cell wall, flagella, pili, and fimbriae.

Macroscopic Morphology

Traits that can be assessed with the naked eye are also useful in identification. These include the appearance of colonies, including texture, size, shape, pigment, speed of growth, and patterns of growth in broth and gelatin media.

Physiological/Biochemical Characteristics

These have been the traditional mainstay of bacterial identification. Enzyme production and other biochemical properties of bacteria are fairly reliable and stable indicators of the identity of each species. Dozens of diagnostic tests exist for determining the presence of specific enzymes and to assess nutritional and metabolic activities. Examples include tests for fermentation of sugars; capacity to digest or metabolize complex polymers such as proteins and polysaccharides; production of gas; presence of enzymes such as catalase, oxidase, and decarboxylases; and sensitivity to antibiotics. Special rapid identification test systems have streamlined data collection and analysis in these types of tests.

Chemical Analysis

The analysis of specific structural substances, such as peptides in the cell wall and membrane lipid profiles, can also aid in microbial identification.

Genotypic Methods

Examining the genetic material itself (DNA and/or RNA) has revolutionized the identification and classification of bacteria. There are many advantages of genotypic methods over phenotypic methods, when they are available. The primary advantage is that culturing of the microorganisms is not always necessary. In recent decades, scientists have come to realize that there are many more microorganisms that we can't grow in the lab compared with those that we can. Numerous **viable nonculturable (VNC)** microbes are currently being identified in this manner, through studies such as the Human Microbiome Project. Another advantage is that genotypic methods are increasingly automated, producing rapid results that are often more precise than phenotypic methods.

Immunologic Methods

As you learned in chapter 13, one immune response to antigens is the production of antibodies, which are designed to bind tightly to specific antigens. The nature of the antibody response is exploited for diagnostic purposes when a patient sample is tested for the presence of specific antibodies to a suspected pathogen (antigen). This

is often easier than testing for the microbe itself. Laboratory kits based on this technique are available for immediate identification of a number of pathogens.

15.2 On the Track of the Infectious Agent: Specimen Collection

Regardless of the method of diagnosis, specimen collection is the common point that guides the health care decisions of every member of a clinical team. Indeed, the success of identification and treatment depends on how specimens are collected, handled, stored, and cultured. Specimens can be taken by a clinical laboratory scientist or medical technologist, nurse, physician, or even by the patient. However, it is imperative that general aseptic procedures be used, including sterile sample containers and other tools to prevent contamination from the environment or the patient. **Figure 15.1** delineates the most common sampling sites and procedures.

In sites that normally contain resident microbiota, care should be taken to sample only the infected site and not the surrounding areas. Saliva is an especially undesirable contaminant because it contains millions of bacteria per milliliter, most of which are normal biota. Sputum, the mucus secretion that coats the lower respiratory surfaces, especially the lungs, is discharged by coughing or taken by a thin tube called a catheter to avoid contamination with saliva. In addition, throat and nasopharyngeal swabs should not touch the tongue, cheeks, or saliva. On occasion saliva samples are needed for dental diagnosis and are obtained by having the patient expectorate into a container.

Urine is taken aseptically from the bladder with a catheter designed for that site. Another method, called a "clean catch," is taken by washing the external urethra and collecting the urine midstream. The latter method inevitably incorporates a few normal biota into the sample, but these can usually be differentiated from pathogens in an actual infection. Sometimes diagnostic techniques require first-voided "dirty catch" urine. The mucous lining of the urethra, vagina, or cervix can be sampled with a swab or applicator stick.

Depending on the nature of a skin lesion, skin can be swabbed or scraped with a scalpel to expose deeper layers. Wounds are cleansed prior to swabbing for culture to avoid collecting the many normal microbiota of the skin. Sterile materials such as blood, cerebrospinal fluid, and tissue fluids must be taken by sterile needle aspiration. Antisepsis of the puncture site is extremely important in these cases. Additional sources of specimens are the eye, ear canal, synovial fluid, nasal cavity (all by swab), and diseased tissue that has been surgically removed (biopsied).

After proper collection, the specimen is promptly transported to a lab and stored appropriately (usually refrigerated) if it must be held for a time. Nonsterile samples in particular, such as urine, feces, and sputum, are especially prone to deterioration at room temperature. Special swab and transport systems are designed to collect the specimen and maintain it in stable condition for several hours. These devices provide nonnutritive maintenance media (so microbes survive but do not grow), a buffering system, and an anaerobic environment to prevent destruction of oxygen-sensitive bacteria.

Knowing how to properly collect, transport, and store specimens is an important aspect of many health professionals' jobs. In addition, labeling and identifying specimens, as well as providing an accurate patient history, are crucial to obtaining timely and accurate results.

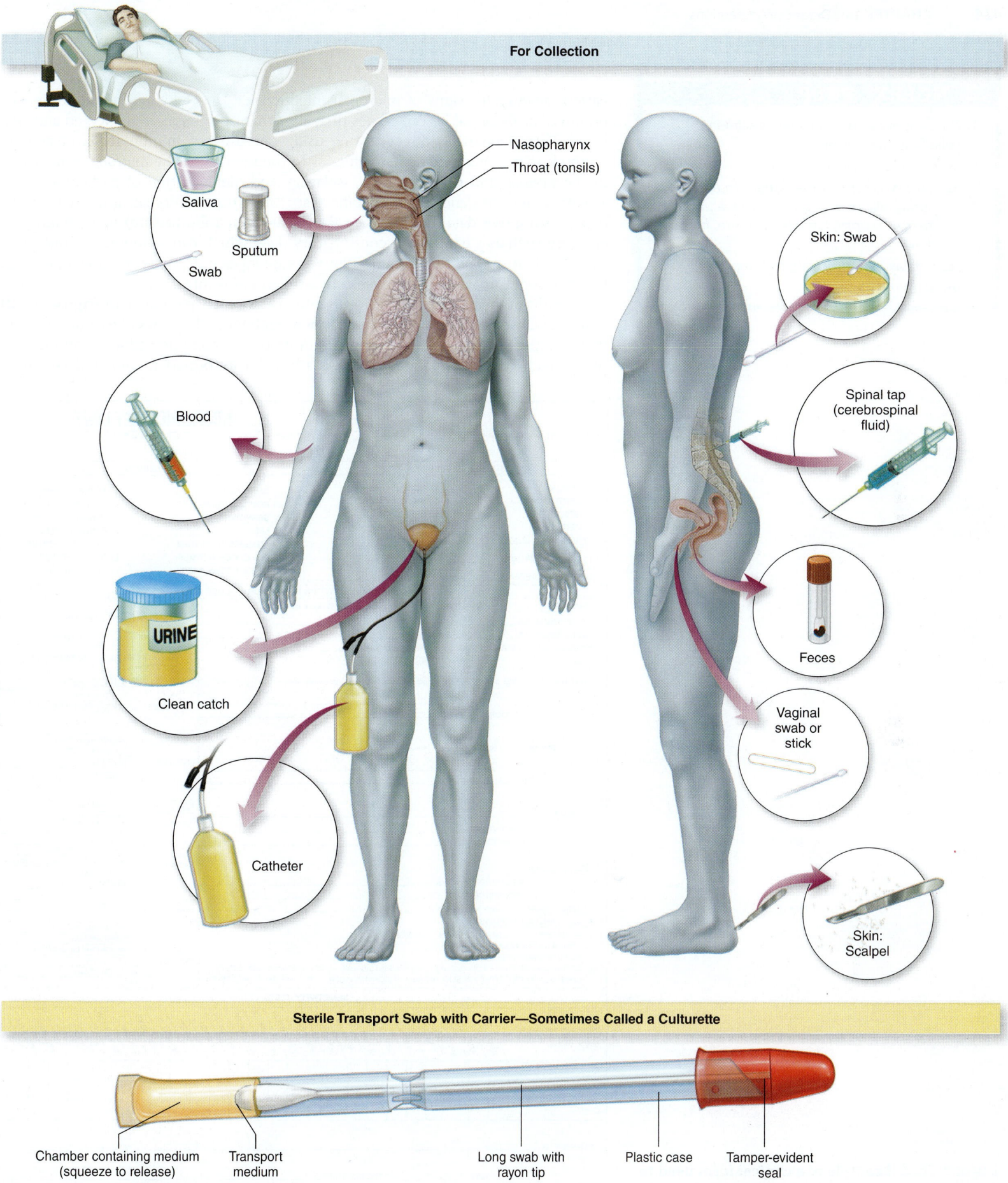

Nasopharynx

Throat (tonsils)

Saliva

Sputum

Swab

Skin: Swab

Blood

Spinal tap (cerebrospinal fluid)

URINE

Feces

Clean catch

Vaginal swab or stick

Catheter

Skin: Scalpel

Sterile Transport Swab with Carrier—Sometimes Called a Culturette

Chamber containing medium (squeeze to release)

Transport medium

Long swab with rayon tip

Plastic case

Tamper-evident seal

Figure 15.1 **Sampling sites and methods of collection for clinical laboratories.**

Overview of Laboratory Techniques

Patient analysis for signs of microbial infection (i.e., fever, wound exudate, mucus production, abnormal lesion) comes first; after that, specimens are collected and analyzed. This involves (1) direct testing using microscopic, immunologic, or genetic methods that provide immediate clues as to the identity of the microbe or microbes in the sample; and (2) cultivation, isolation, and identification of pathogens using a wide variety of general and specific tests. Most test results fall into two categories: presumptive data, which place the isolated microbe (isolate) in a preliminary category such as a genus, and confirmatory data, which can pinpoint the microbe's specific identity. The time required for testing ranges from a few minutes in a streptococcal sore throat to several weeks in a tuberculosis infection.

Results of specimen analysis are entered in a summary patient chart **(figure 15.2)** that can be used in assessment and treatment regimens. This looks like a boring form— but take the time to read it! You may notice that it compiles information on tests with which you are already familiar. As a health care provider, your understanding

MICROBIOLOGY UNIT

DATE, TIME & PERSON COLLECTING	SPECIMEN NUMBER	ANTIBIOTIC THERAPY	TENTATIVE DIAGNOSIS

SOURCE OF SPECIMEN		TEST REQUEST	
☐ THROAT	☐ BLOOD	☐ GRAM STAIN	☐ ACID FAST SMEAR
☐ SPUTUM	☐ URINE - CLEAN CATCH	☐ ROUTINE CULTURE	☐ ACID FAST CULTURE
☐ STOOL	☐ URINE - CATH	☐ SENSITIVITY	☐ FUNGUS WET MOUNT
☐ CERVIX	☐ BRONCHIAL WASHING	☐ MIC	☐ FUNGUS CULTURE
☐ AEROSOL INDUCED SPUTUM		☐ ANAEROBIC CULTURE	☐ PARASITE STUDIES
☐ WOUND - SPECIFY SITE _____		☐ R/O GROUP A STREP	☐ OCCULT BLOOD
☐ OTHER - SPECIFY _____		☐ WRIGHT STAIN (WBC)	☐ PCR ANALYSIS

DO NOT WRITE BELOW THIS LINE –# FOR LAB USE ONLY

GRAM STAIN   (4+ NUMEROUS; 3+ MANY; 2+ MODERATE; 1+FEW; 0 NONE SEEN)

COCCI: GRAM POS._____ GRAM NEG._____ W B C._____
BACILLI: GRAM POS._____ GRAM NEG._____ EPITHELIAL CELLS_____
INTRACELLULAR & EXTRACELLULAR GRAM-NEGATIVE DIPLOCOCCI_____
YEAST_____   ☐ No organisms seen.

CULTURE RESULTS

| 1+ FEW | 3+ MANY |
| 2+ MODERATE | 4+ NUMEROUS |

FUNGUS: WET MOUNT ☐ No mycotic elements or budding structures seen.
            ☐ _____
CULTURE   ☐ _____

AFB: SMEAR   ☐ No acid fast bacilli seen.
         ☐ _____
CULTURE ☐ _____

PARASITE   DIRECT:_____
STUDIES: CONCENTRATE:_____
             PERMANENT:_____

OCCULT BLOOD:
            APPEARANCE OF STOOL:_____
            OCCULT BLOOD:_____

COLONY COUNT: Urine organisms/ml.   ☐ _____ ☐ > 100,000

MISCELLANEOUS RESULTS:
☐ NO GROWTH IN: ☐ 2 DAYS ☐ 3 DAYS ☐ 5 DAYS ☐ 7 DAYS
☐ NORMAL FLORA ISOLATED
☐ NO ENTEROPATHOGENS ISOLATED
☐ SPUTUM UNACCEPTABLE FOR CULTURE — REPRESENTS SALIVA — NEW SPECIMEN REQUESTED
☐ URINE > 2 COLONY TYPES PRESENT REPRESENT CONTAMINATION — NEW SPECIMEN REQUESTED

ANAEROBES	☐ BACTEROIDES
	☐ CLOSTRIDIUM
	☐ PEPTOSTREPTOCOCCUS
	☐
ENTERICS	☐ ESCHERICHIA COLI
	☐ ENTEROBACTER
	☐ KLEBSIELLA
	☐ PROTEUS
	☐
STAPH-YLOCOCCUS	☐ AUREUS
	☐ EPIDERMIDIS
	☐ SAPROPHYTICUS
STREP-TOCOCCUS	☐ GROUP A
	☐ GROUP B
	☐ GROUP D ENTEROCOCCI
	☐ GROUP D NON ENTEROCOCCI
	☐ PNEUMONIAE
	☐ VIRIDANS
	☐
YEAST	☐ CANDIDA
	☐
OTHER ISOLATES	☐ PSEUDOMONAS
	☐ HAEMOPHILUS
	☐ GARDNERELLA VAGINALIS
	☐ NEISSERIA
	☐ CL. DIFFICILE
	☐

SENSITIVITY TESTS

NOTE: Bacteria with intermediate susceptibility may not respond satisfactorily to therapy.

	AMIKACIN	AMPICILLIN	BETA LACTAMASE PRODUCTION	CARBENICILLIN	CEFAZOLIN	CEFOTAXIME	CEFOXITIN	CEFUROXIME	CHLORAMPHENICOL	CLINDAMYCIN	ERYTHROMYCIN	GENTAMICIN	METHICILLIN	METRONIDAZOLE	NITROFURANTOIN	PENICILLIN	IMMUNOLOGY	TETRACYCLINE	TOBRAMYCIN	TRIMETHOPRIM SULFAMETHOXAZOLE	VANCOMYCIN	CIPROFLOX
A																						
B																						
C																						

☐ COMMENTS: _____

706-30A

DATE _____ TECHNOLOGIST _____

Figure 15.2 Example of a clinical form used to report data on a patient's specimens.

of and ability to communicate lab results are imperative to the successful treatment of patients.

Some diseases are diagnosed without analyzing actual microbes within specimens. Tests on patient serum provide indirect evidence for specific pathogens through analysis of the antibody response. Skin testing, for example, is important in identifying those in the general population who have had past exposures to infectious agents causing tuberculosis. Additionally, some pathogens are identified almost solely on patient signs and symptoms. AIDS, for example, is diagnosed by serological tests and a complex of signs and symptoms without ever isolating the virus. Other diseases, such as athlete's foot, are diagnosed purely on the typical presenting symptoms and may require no lab tests at all.

NCLEX® PREP

2. A clinical form used to report data on a patient's specimens may include
   a. *antibiotic history.*
   b. *patient symptoms.*
   c. *marital status.*
   d. *date and time of specimen collection.*
   e. *a, b, and d.*
   f. *b, c, and d.*

> **15.2 LEARNING OUTCOMES—Assess Your Progress**
>
> 3. Identify factors that may affect the identification of an infectious agent from a patient sample.
> 4. Compare the types of tests performed on microbial isolates versus those performed on patients themselves.

15.3 Phenotypic Methods

Immediate Direct Examination of Specimen

Direct microscopic observation of a fresh or stained specimen is one of the most rapid methods of determining presumptive and sometimes confirmatory microbial characteristics. The Gram stain and the acid-fast stain (see figure 2.18) are most often used for bacterial identification. As useful as these stains are, they can identify only a few organisms on their own. For that reason, we have a variety of other techniques.

Cultivation of Specimen

Isolation Media and Morphological Testing

Such a wide variety of media exist for microbial isolation that a certain amount of preselection must occur, based on the nature of the specimen. In cases in which the suspected pathogen is present in small numbers or is easily overgrown, the specimen can be initially enriched with specialized media. Nonsterile specimens containing a diversity of bacterial species, such as urine and feces, are cultured on selective media to encourage the growth of only the suspected pathogen. For example, approximately 80% of urinary tract infections are known to be caused by *Escherichia coli,* so selective media that will be sure to allow the growth of this common pathogen are chosen. Specimens are often inoculated into differential media to identify definitive characteristics, such as reactions in blood (blood agar) and fermentation patterns (mannitol salt and MacConkey agar). Working with a mixed or contaminated culture causes misleading and ambiguous results. So that subsequent steps in identification will be as accurate as possible, pure cultures of the microbe must be obtained from culturing on isolation media. Clinical microbiologists can then observe the suspected pathogen's microscopic morphology and staining reactions, cultural appearance, motility, and oxygen requirements, in addition to biochemical analysis and antibiotic sensitivity tests.

Biochemical Testing

The physiological reactions of bacteria to nutrients and other substrates provide excellent indirect evidence of the types of enzyme systems present in a particular species. Many of these tests are based on enzyme-mediated metabolic reactions that

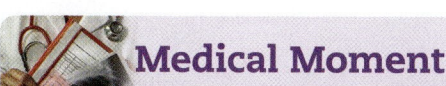

Medical Moment

Qualitative Versus Quantitative Diagnosis

Laboratory tests may be *qualitative* or *quantitative*. Qualitative tests are used to determine whether a substance is present in a sample, while quantitative tests measure the amounts of the substance that are present.

A pregnancy test is a perfect example of a test that may be qualitative or quantitative. Most home pregnancy tests are qualitative in that they test for the presence or absence of the hormone known as human chorionic gonadotropin (hCG), which will appear in the urine of pregnant women several days after conception.

A quantitative pregnancy test is determined from a blood sample. The hCG levels should double every 2 days in the first few weeks of pregnancy. When hCG levels do not rise as expected, there may be a problem with the pregnancy (such as miscarriage); extremely high levels may be associated with twins. Therefore, measuring the exact amount of hCG present in the bloodstream provides the ordering physician with much more information than a qualitative pregnancy test can, which only indicates whether or not a woman is pregnant.

are visualized by a color change. These types of reactions are particularly meaningful in bacteria, which are haploid organisms that generally express their genes for utilizing a given nutrient.

The microbe is cultured in a medium with a special substrate and then tested for a particular end product. Microbial expression of the enzyme is made visible by a colored dye; no coloration means it lacks the enzyme for utilizing the substrate in that particular way. Although routinely performed on cultured isolates, direct biochemical testing of patient samples now can be performed that produce results within hours instead of days.

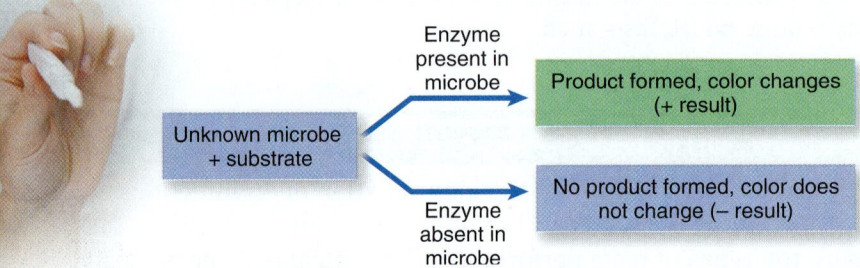

Antibodies are used to "diagnose" noninfectious conditions as well, such as pregnancy.

Among the prominent biochemical tests are carbohydrate fermentation (production of acid and/or gas); hydrolysis of gelatin, starch, and other polymers; enzyme actions such as catalase, oxidase, and coagulase; and various by-products of metabolism. Many are presently performed with rapid, miniaturized systems that can simultaneously determine up to 23 characteristics in small individual cups or spaces **(figure 15.3).** An important plus, given the complexity of biochemical profiles, is that such systems are readily adapted to computerized analysis.

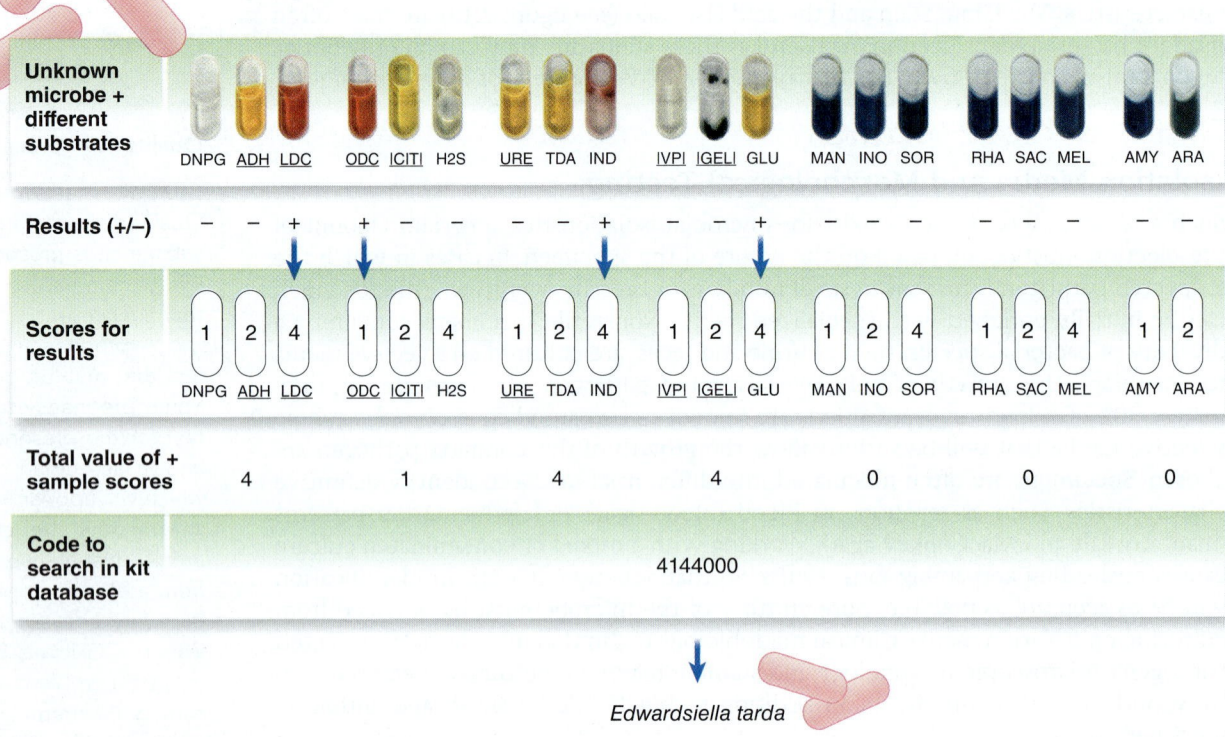

Figure 15.3 Rapid tests. A biochemical system for microbial identification. Samples of a single bacterial culture (an unknown gram-negative microbe) are placed in the different cups, which contain growth media and chemicals designed to test for a particular enzyme. The organism produces four positive results, as indicated. Scoring of the results is completed by adding the designated values of each positive result per set of cups indicated. This string of seven numbers creates a code that can be referenced in a manual to determine the identity of the unknown microbe.

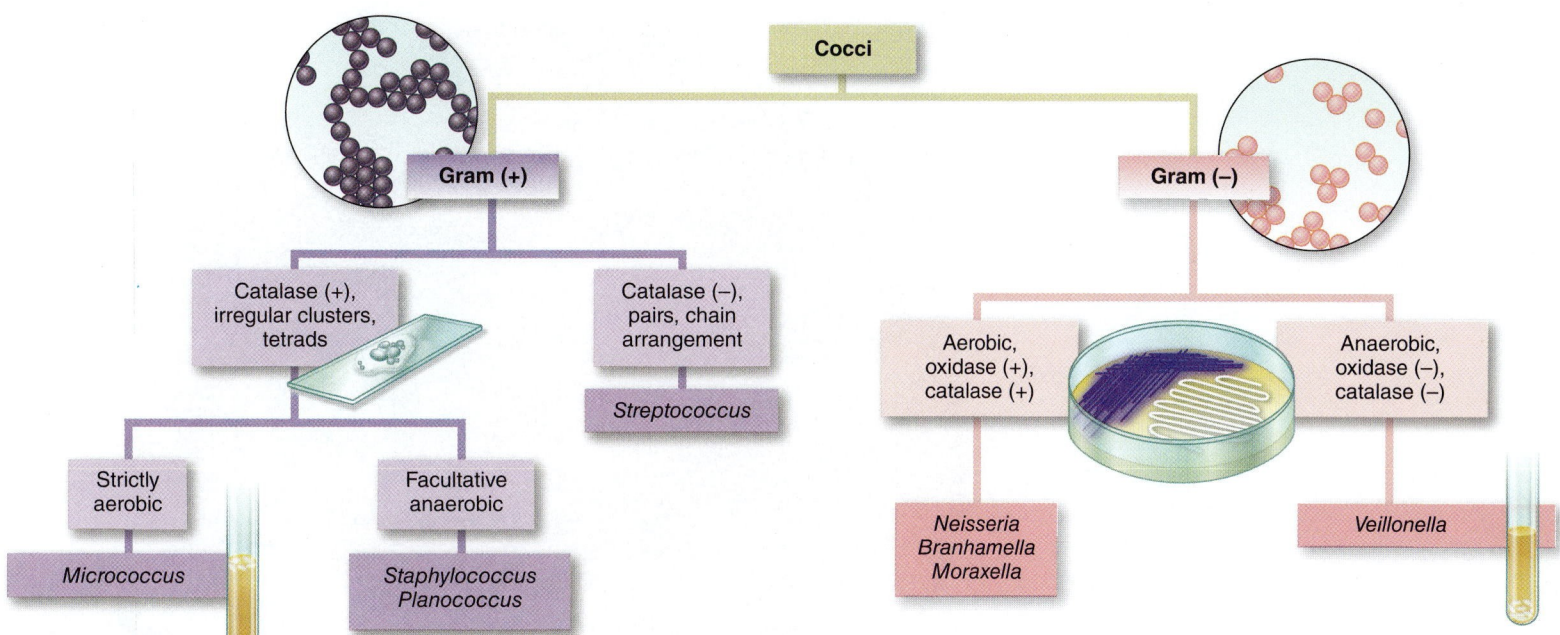

Figure 15.4 **Flowchart to separate primary genera of various cocci.**
Identification scheme for cocci commonly involved in human diseases.

Common schemes exist for identifying bacteria. These are based on easily recognizable characteristics such as motility, oxygen requirements, Gram stain reactions, morphology, spore formation, and various biochemical reactions. **Dichotomous keys** are flowcharts **(figure 15.4)** used to trace a route of identification by offering pairs of opposing characteristics (positive versus negative, for example) with two choices at each level from which to select. Eventually, an endpoint is reached, and the name of a genus or species that fits that particular combination of characteristics appears. Although dichotomous keys are useful in the student and research laboratory, *diagnostic tables,* which provide information on clinically important species in a condensed format, are preferred over dichotomous keys in most clinical laboratories.

Miscellaneous Tests

Phage typing relies on bacteriophages, viruses that attack bacteria in a very species-specific and strain-specific way. Such selection is useful in identifying some bacteria, primarily *Salmonella,* and is often used for tracing bacterial strains in epidemics. The technique of phage typing involves inoculating a lawn of bacterial cells onto agar, mapping it off into blocks, and applying a different phage to each sectioned area of growth **(figure 15.5).** Cleared areas corresponding to lysed cells indicate sensitivity to that phage, and a bacterial identification may be determined from this pattern.

Susceptible animals are needed to cultivate bacteria such as *Mycobacterium leprae* and significant quantities of *Treponema pallidum,* whereas avian embryos and cell cultures are used to grow host cell–dependent rickettsias, chlamydias, and viruses.

Antimicrobial sensitivity tests are not only important in determining the drugs to be used in treatment (see figure 10.3*b*), but the patterns of sensitivity can also be used in presumptive identification of some species, such as *Streptococcus, Pseudomonas,* and *Clostridium.*

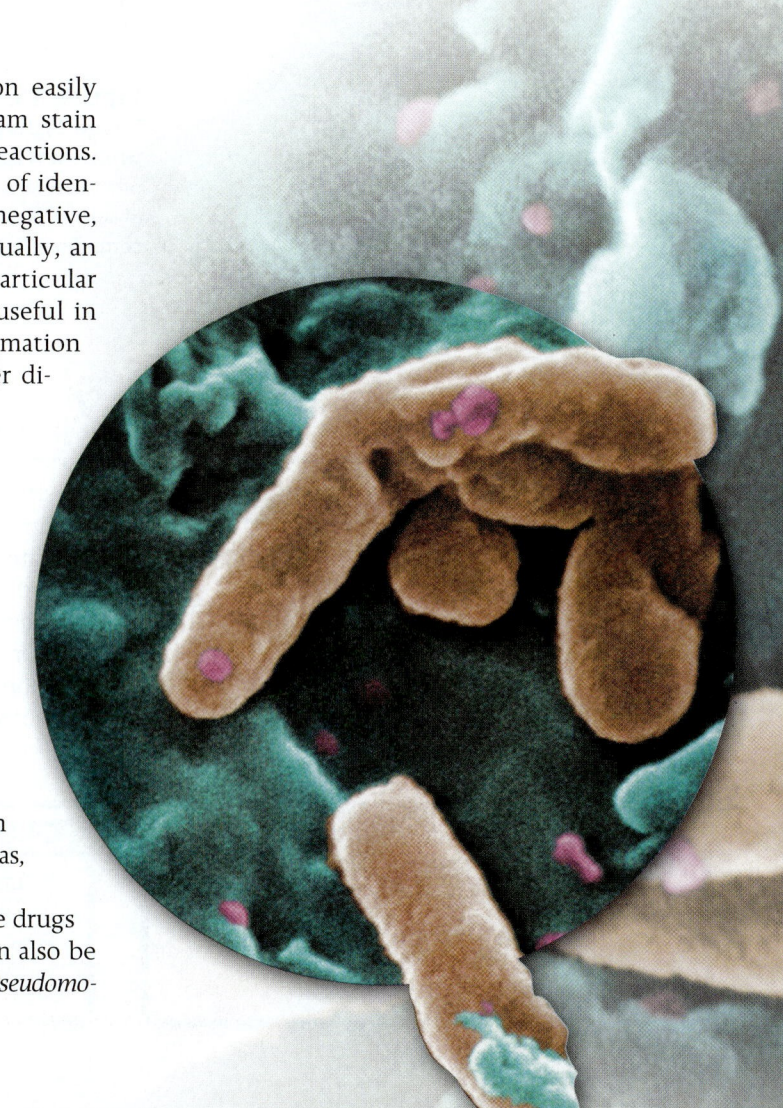

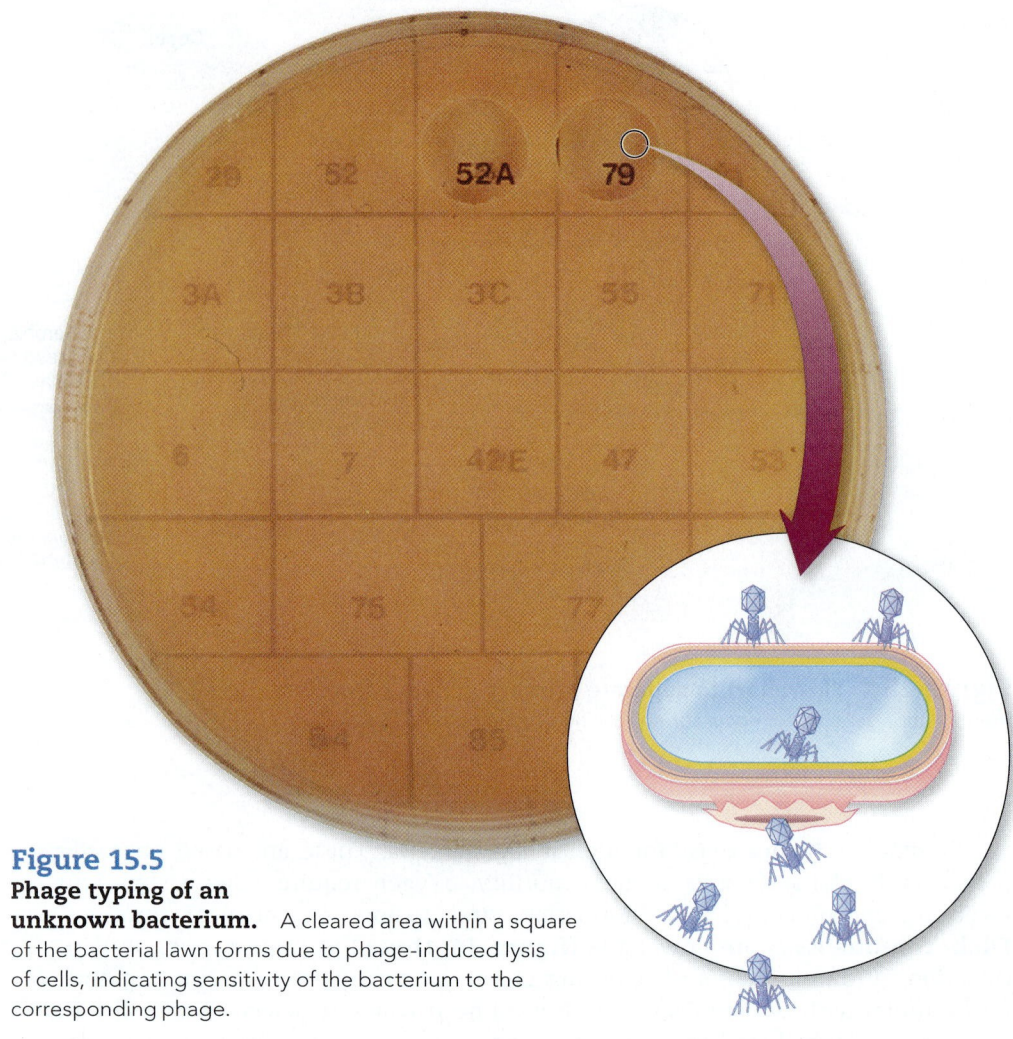

Figure 15.5
Phage typing of an
unknown bacterium. A cleared area within a square
of the bacterial lawn forms due to phage-induced lysis
of cells, indicating sensitivity of the bacterium to the
corresponding phage.

Determining Clinical Significance of Cultures

It is important to rapidly determine if an isolate from a specimen is clinically important or if it is merely a contaminant or normal biota. Although answering these questions may prove difficult, one can first focus on the number of microbes in a specimen. For example, a few colonies of *Escherichia coli* in a urine sample can simply indicate normal biota, whereas several hundred can mean active infection. In contrast, the presence of a single colony of a true pathogen, such as *Mycobacterium tuberculosis* in a sputum culture or an opportunist in sterile sites such as cerebrospinal fluid or blood, is highly suggestive of its role in disease. Furthermore, the repeated isolation of a relatively pure culture of any microorganism can mean it is an agent of disease.

NCLEX® PREP

3. When determining the clinical significance of cultures,
   a. the number of microbes is significant.
   b. the presence of a single colony of a true pathogen may indicate the presence of the disease if the culture comes from a site known to be sterile (i.e., cerebrospinal fluid).
   c. the repeated isolation of a relatively pure culture of any microorganism can mean it is an agent of disease, although this is not always the case.
   d. a range of tests may be needed to identify a pathogen.
   e. all of the above are true.

15.3 LEARNING OUTCOMES—Assess Your Progress

5. List at least three different tests that fall in the direct identification category.
6. Explain the main principle behind biochemical testing, and identify an example of such tests.

15.4 Genotypic Methods

The sequence of nitrogenous bases within DNA or RNA is unique to every microorganism. Because of this and the many recent advances in genomic technology, nucleic-acid-based tests have become a mainstay of microbial identification.

Polymerase Chain Reaction (PCR): Amplifying the Information

Many nucleic acid tests employ the use of the **polymerase chain reaction (PCR).** PCR results in the production of numerous identical copies of DNA or RNA molecules within hours (see table 8.9). This method can amplify even minute quantities of nucleic acids present in a sample, which greatly improves the sensitivity of these tests. PCR amplification can be performed on genetic material from a wide variety of bacteria, viruses, protozoa, and fungi. Metagenomic analysis of the human body and the environment also depends on the ability of PCR to amplify the amount of microbial information available for nucleic acid testing from these sites. In some cases, where the microbial populations are relatively unknown, a form of PCR called **random amplified polymorphic DNA (RAPD)** may be used because it employs primers of random sequence in an attempt to pick a microbial needle out of a haystack.

Hybridization: Probing for Identity

Hybridization is a technique that makes it possible to identify a microbe by analyzing segments of its genetic material. This requires small fragments of single-stranded DNA (or RNA) called **probes** that are known to be complementary to the specific sequences of nucleic acid isolated from a particular microbe. Base-pairing of the known probe to the nucleic acid can be observed providing evidence of the microbe's identity. Although hybridization techniques are quite specific, control probes must be used in order to rule out cross-reactivity. These tests have become more convenient and portable over the years. Probes are typically fluorescently labeled or attached to an enzyme that triggers a colorimetric change when hybridization occurs **(figure 15.6)**. This is based on the property of dyes such as fluorescein and rhodamine, which emit visible light in response to ultraviolet. This property of fluorescence has found numerous applications in diagnostic testing.

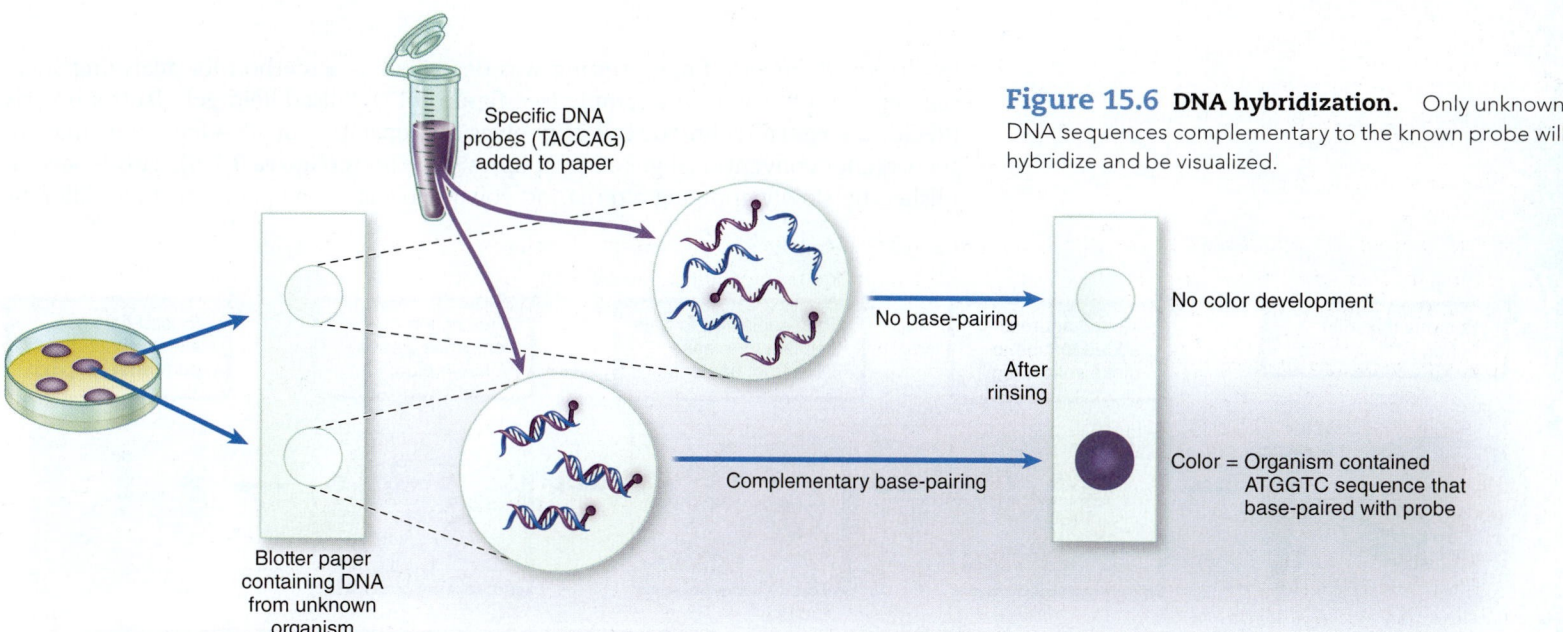

Specific DNA probes (TACCAG) added to paper

Blotter paper containing DNA from unknown organism

No base-pairing

After rinsing

Complementary base-pairing

No color development

Color = Organism contained ATGGTC sequence that base-paired with probe

Figure 15.6 DNA hybridization. Only unknown DNA sequences complementary to the known probe will hybridize and be visualized.

Figure 15.7 Peptide nucleic acid (PNA) FISH testing for S. aureus.
Testing can identify blood-borne pathogens more quickly than other methods.

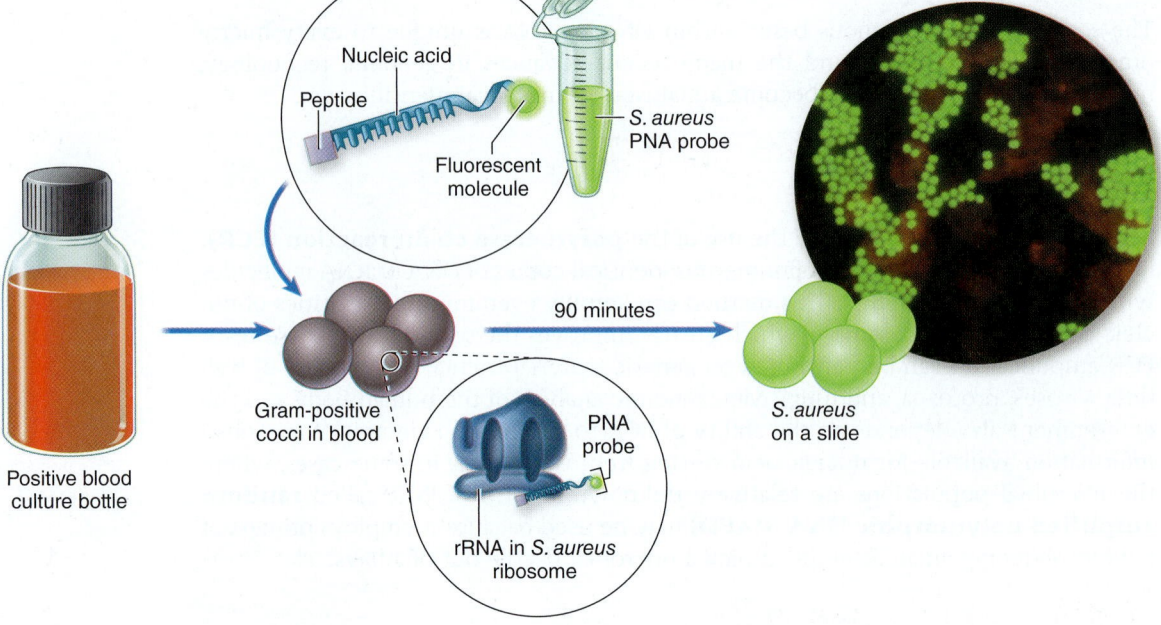

Several variations on the principle of hybridization are used in infectious disease diagnosis today. In the oldest method, unknown test DNA is extracted from cells in specimens or cultures and is bound to special blotter paper. After several different probes have been added to the blotter, it is observed for visible signs that the probes have become fixed (hybridized) to the test DNA. Such rapid hybridization test cards are still used routinely in the diagnosis of infection conditions such as vaginitis.

Fluorescent *in situ* hybridization (FISH) techniques involve the application of fluorescently labeled probes to intact cells within a patient specimen or an environmental sample **(figure 15.7)**. Microscopic analysis is used to locate "glowing" cells and conclude the identity of a specific microbe. FISH is often used to confirm a diagnosis or to identify the microbial components within a biofilm.

Pulsed-Field Gel Electrophoresis: Microbial Fingerprints

In chapter 8, genetic fingerprinting was described as a method for analyzing short segments of DNA within a sample (see figure 8.17). Pulsed-field gel electrophoresis (PFGE) is a similar technique but it involves the separation of DNA fragments that are too large for conventional gel electrophoresis methods **(figure 15.8)**. This is accomplished by slowly applying alternating voltage levels to the gel from three different

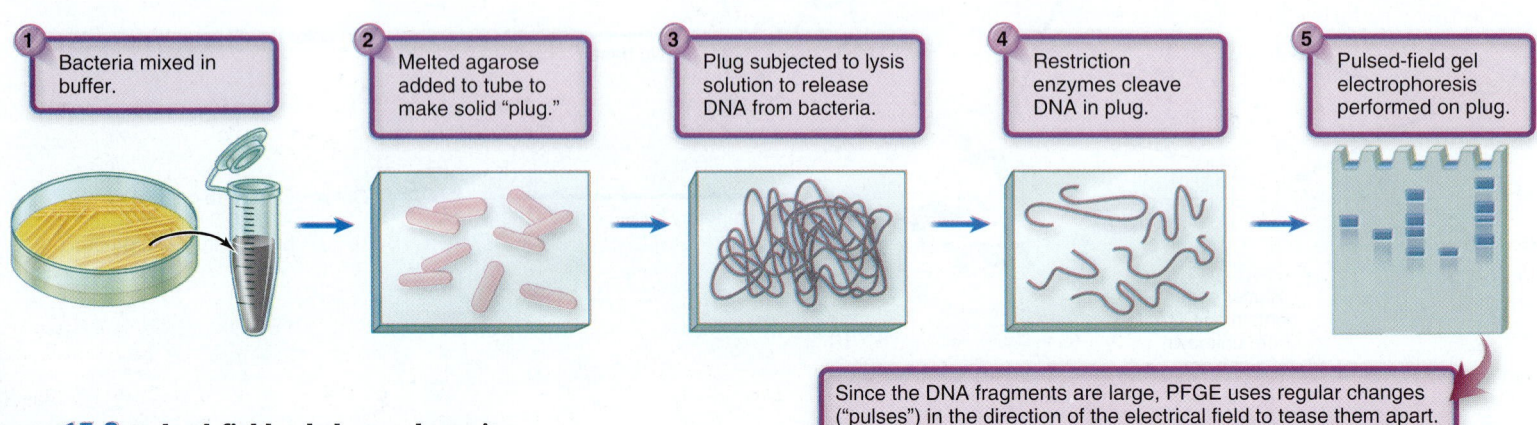

1. Bacteria mixed in buffer.
2. Melted agarose added to tube to make solid "plug."
3. Plug subjected to lysis solution to release DNA from bacteria.
4. Restriction enzymes cleave DNA in plug.
5. Pulsed-field gel electrophoresis performed on plug.

Since the DNA fragments are large, PFGE uses regular changes ("pulses") in the direction of the electrical field to tease them apart.

Figure 15.8 Pulsed-field gel electrophoresis.

directions, allowing even similarly sized DNA fragments to fully separate. This technique first came to prominence in 1993 when it was used by the CDC to determine that *E. coli* O157:H7 was the source of a food-borne outbreak in the United States. Since then, it has become an important method in epidemiological studies due to its accuracy in assessing microbial subtype and identification from patient samples. PulseNet is a program established by the CDC to assist in the investigation of possible infectious disease outbreaks, especially those caused by food-borne pathogens. Scientists from public health facilities across the country are able to rapidly communicate and compare PFGE data from patient specimens and other samples, allowing identification of outbreaks to occur within hours versus days or even weeks.

Ribotyping: rRNA Analysis

You may remember from chapter 1 that one of the most viable indicators of evolutionary relatedness and affiliation is comparison of the sequence of nitrogen bases in 16S ribosomal RNA (rRNA). The 16S ribosomal RNA is a component of the 30S subunit of bacterial and archaeal ribosomes. Other rRNA sequences vary between species, whereas 16S rRNA sequences are highly conserved across species and evolutionary time. This makes such RNA analysis, or ribotyping, perfectly suited for bacterial identification and subsequent diagnosis of infection. Ribosomal RNA is isolated, sequenced, and analyzed from cultured cells obtained from a patient site or environmental sample to obtain this information.

The FISH method previously discussed is also being used for rRNA analysis. It can rapidly identify 16S rRNA sequences without first culturing the organism. The turnaround time for identifying microorganisms present in blood cultures has been reduced from 24 hours to 90 minutes using this technique (see figure 15.7).

What's Next?

In the preceding sections, we have discussed widely used genomic diagnostic methods. An array of newer genomic methods have been developed, many of which will become standard tools for infectious disease diagnosis in the short-term future. These are discussed in section 15.6.

> ### 15.4 LEARNING OUTCOMES—Assess Your Progress
>
> **7.** Explain why PCR is useful for infectious disease diagnosis.
> **8.** List the major steps in a hybridization method of microbial identification.
> **9.** Explain how rRNA analysis has impacted the process of infectious disease diagnosis.

15.5 Immunologic Methods

The antibodies formed during an immune reaction are important in combating infection, but they hold additional practical value. Characteristics of antibodies (such as their quantity or specificity) can reveal the history of a patient's contact with microorganisms or other antigens. This is the underlying basis of serological testing. **Serology** is the branch of immunology that traditionally deals with *in vitro* diagnostic testing of serum. Serological testing is based on the principle that antibodies have extreme specificity for antigens, so when a particular antigen is exposed to its specific antibody, it will fit like a hand in a glove. The ability to visualize this interaction macroscopically or microscopically provides a powerful tool for detecting, identifying, and quantifying antibodies–or conversely, antigens. One can detect or identify an unknown antibody using a known antigen, or one can use an antibody of known specificity to help detect or identify an unknown antigen **(figure 15.9).** Modern serological methods have evolved beyond the ability to test sera, as urine, cerebrospinal fluid, whole tissues, and saliva can now be analyzed for the presence of specific antibodies. These and

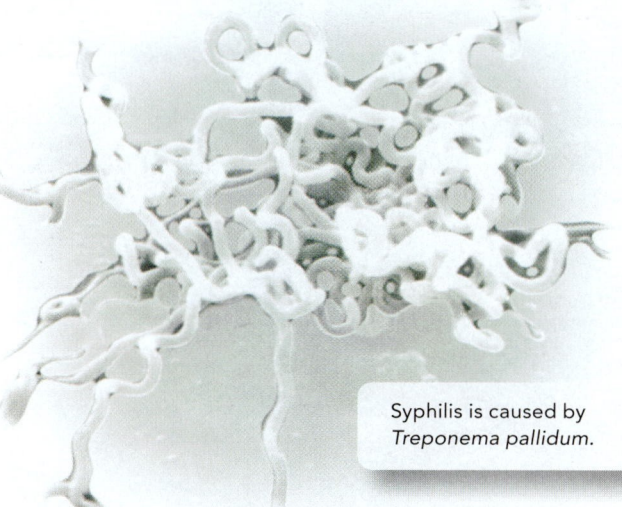

Syphilis is caused by *Treponema pallidum*.

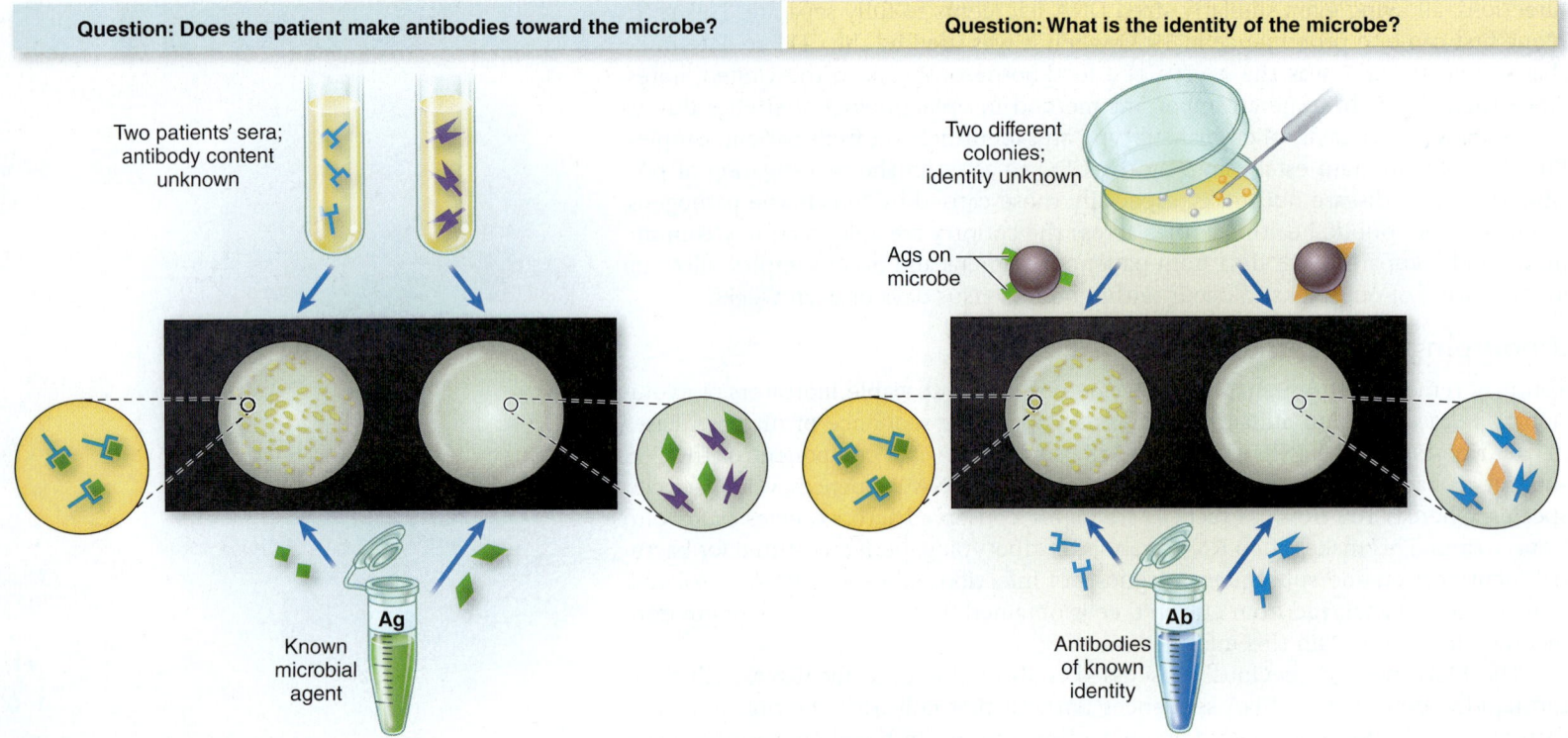

Question: Does the patient make antibodies toward the microbe?

Two patients' sera; antibody content unknown

Known microbial agent

Ag

Question: What is the identity of the microbe?

Two different colonies; identity unknown

Ags on microbe

Antibodies of known identity

Ab

Figure 15.9 **Basic principles of serological testing using antibodies and antigens.**

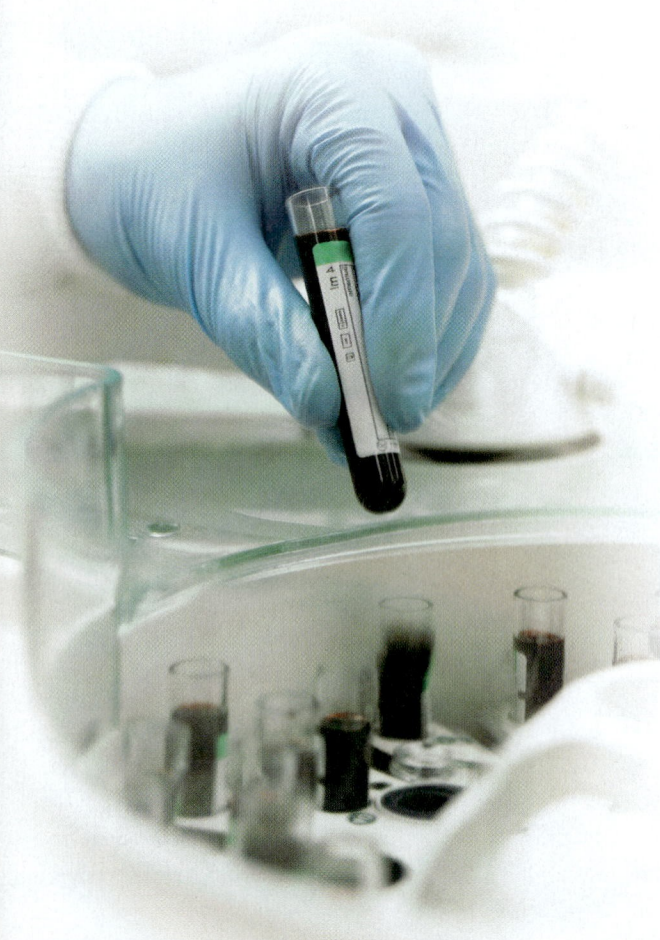

other immune tests help to determine the immunologic status of patients, confirm a suspected diagnosis, or screen individuals for disease.

General Features of Immune Testing

The applications of immunologic tests are diverse, and they demonstrate some of the brilliant and imaginative ways that antibodies and antigens can be used as tools. The most effective serological tests have a high degree of specificity and sensitivity. *Specificity* is the property of a test to focus on only a certain antibody or antigen and not to react with unrelated or distantly related ones. Better said, specificity is the degree to which a test does not (falsely) detect people who do not have a condition. *Sensitivity* refers to the detection of even minute quantities of antibodies or antigens in a specimen, and reflects the degree to which a test will detect every positive person.

Visualizing Antigen-Antibody Interactions

The molecular basis of immunologic testing is the binding of an antibody (Ab) to a specific site (epitope) on an antigen (Ag), and only electron microscopy provides adequate imaging of such complexes. To be useful in a clinical setting, serological tests were developed that produce a reaction visible to the naked eye or with magnification provided by light microscopy. When analyzing large cell surface antigens, antibody binding to these antigens creates large clumps or aggregates that are visible macroscopically or microscopically. Since the formation of smaller antigen-antibody complexes may not be readily observed, tests requiring special indicators are utilized that employ dyes or fluorescent reagents to visualize the endpoint of reactions.

Clinical Applications of Immune Testing

Agglutination and Precipitation Reactions

The essential differences between agglutination and precipitation as seen in **table 15.2** are in size, solubility, and location of the antigen. In agglutination, the antigens are whole cells or organisms such as red blood cells, bacteria, or viruses

Table 15.2 Summary of Immunologic Methods

Test	Description		Example
Agglutination	Involves antibody-mediated clumping of whole cells		Weil-Felix test, Ab titering
Precipitation	Produces antibody-antigen complexes in a cell-free system		RPR test, serotyping
Western blot	Separation of proteins followed by antibody-mediated detection		HIV verification test *Trichinella* diagnosis
Complement fixation	Testing involves lysin-mediated hemolysis of red blood cells		Histoplasmosis test
Fluorescence antibody	Monoclonal antibodies labeled with fluorescent dyes (FAbs)		
Direct		Unknown specimen is exposed to a known FAb solution	Meningitis, plague tests
Indirect		Fc region of patient's antibody binds with known FAb	FTA-ABS test for syphilis
Immunoassay	Sensitive, rapid tests for trace levels of antibody or antigen		
RIA		Measure of bound radioactively labeled antibodies or antigens	RAST or RIST for allergies
ELISA		Colorimetric test to detect unknown antigen or antibody	*Helicobacter*, HIV screening
In vivo methods	Antigen introduced into patient to elicit a reaction		Tuberculin reaction

Microscopic view

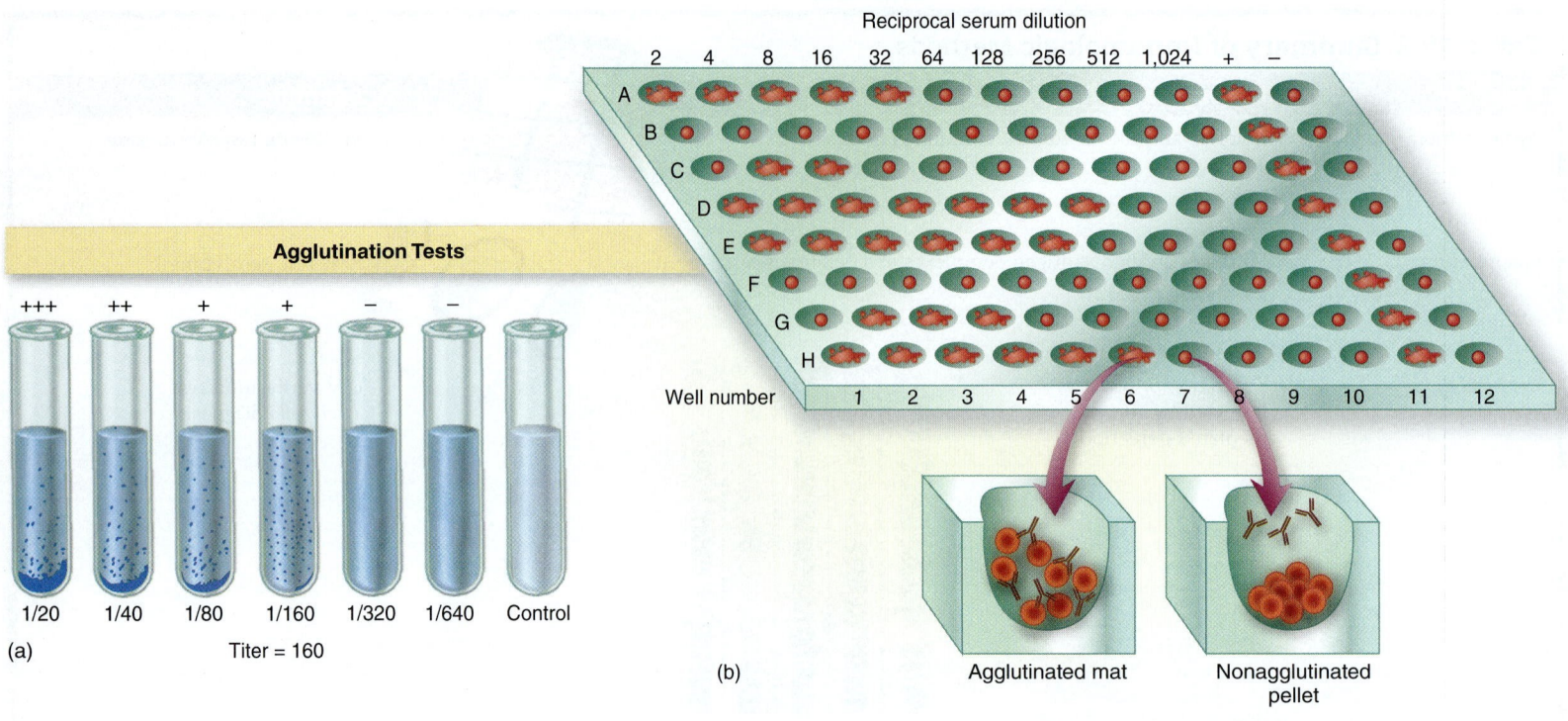

Agglutination Tests

+++ ++ + + − −

1/20 1/40 1/80 1/160 1/320 1/640 Control

(a)   Titer = 160

Reciprocal serum dilution

(b)

Agglutinated mat   Nonagglutinated pellet

Enlarged side view of wells

Figure 15.10 Agglutination tests. (a) Tube agglutination test for determining antibody titer. The same number of cells (antigen) is added to each tube, and the patient's serum is diluted in series. The titer in this example is 160 because there is no agglutination in the next tube in the dilution series (1/320). (b) A microtiter plate illustrating hemagglutination. The antibody is placed in wells 1–10. Positive controls (row 11) and negative controls (row 12) are included. Red blood cells are added to each well. If sufficient antibody is present to agglutinate the cells, they sink as a diffuse mat to the bottom of the well. If insufficient antibody is present, they form a tight pellet at the bottom.

displaying cell surface antigens **(figure 15.10)**; in precipitation, the antigen examined is a soluble molecule. In both reactions, when antigen and antibody concentrations are optimal, one antigen is interlinked by several antibodies to form insoluble aggregates that settle out in solution. Agglutination is easily seen because it forms visible clumps of cells, such as in the *Weil-Felix reaction* used in diagnosing rickettsial infections. Agglutination is also used to determine blood compatibility. Precipitation reactions, however, are more difficult to visualize because precipitates are easily disrupted in liquid media. New methods have been developed that produce results that resist disruption and are visible without magnification. The *rapid plasma reagin (RPR)* test, used to identify antibodies to syphilis, is one example. Although agglutination and precipitation tests for identifying infections have been largely replaced by fluorescent or genetic methods, they remain extremely useful in the developing world where newer technologies are not available.

Antibody Titers

An antigen-antibody reaction in liquid is read as a **titer,** or the concentration of antibodies in a sample (table 15.2). Titer is determined by serially diluting patient serum into test tubes or wells of a microtiter plate, all containing equal amounts of bacterial cells (antigen). Titer is defined as the highest dilution of serum that still produces agglutination. In general, the more a serum sample can be diluted and still react with antigen, the greater the concentration of antibodies and thus its titer. Antibody titers are often used to diagnose autoimmune disorders such as rheumatoid arthritis and lupus, and also to determine past exposure to certain diseases such as rubella.

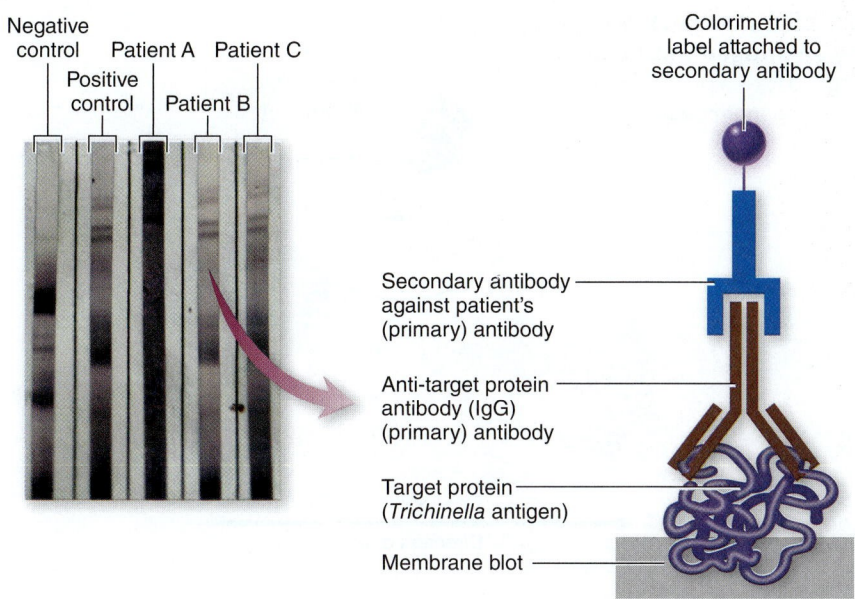

Figure 15.11 The Western blot procedure.
Major *Trichinella* surface proteins were separated via electrophoresis and transferred to a filter. The filter was incubated with the patient's (primary) and secondary antibody solutions and developed with a colorimetric label; *Trichinella* antigen-antibody complexes are visualized as bands. A positive control for all *Trichinella* antigens and a negative control serve as comparisons for patient sera A, B, and C. Patients B and C show positive results. The result for patient A is negative.

Serotyping

Serotyping is an antigen-antibody technique for identifying, classifying, and sub-grouping certain bacteria into categories called serotypes (table 15.2). This method employs antisera against cell antigens such as the capsule, flagellum, and cell wall. Serotyping is widely used in identifying *Salmonella* species and strains, and is the basis for differentiating the numerous pneumococcal and streptococcal serotypes. The Quellung test, which involves a precipitation reaction against capsular polysaccharide antigens, is one example.

The Western Blot Procedure

The **Western blot test** involves the electrophoretic separation of proteins, followed by antibody-mediated detection of these proteins **(figure 15.11).** First, a sample of proteins obtained from cells after lysing them is separated via electrical charge within a gel. The proteins embedded in the gel are then transferred and immobilized to a special filter. The filter is next incubated with antibody solutions, some of which have been labeled with radioactive, fluorescent, or luminescent molecules. After incubation, sites of specific antigen-antibody binding will appear as a pattern of bands that can be compared with known positive and negative controls. It is a highly specific and sensitive way to identify or verify the presence of microbial-specific antigens or antibodies in a patient sample. Western blotting is the second (verification) test for preliminary antibody-positive HIV screening tests.

Immunofluorescence Testing

The fundamental tool in immunofluorescence testing is a fluorescent antibody–a monoclonal antibody labeled by a fluorescent dye. Fluorescent antibodies (FAbs) can be used for diagnosis in two ways. In *direct testing*, an unknown test specimen or antigen is fixed to a slide and exposed to a FAb solution of known composition. If antibody-antigen complexes form, they will remain bound to the sample and will be visualized by fluorescence microscopy, thus indicating a positive result **(figure 15.12).** These tests are valuable for identifying and locating microbial antigens on cell surfaces or in tissues and in identifying the causative agents of syphilis, gonorrhea, and meningitis, among others.

In contrast, FAbs used in *indirect testing* recognize the Fc region of antibodies in patient sera (remember that antibodies are antigenic themselves!). Known antigen

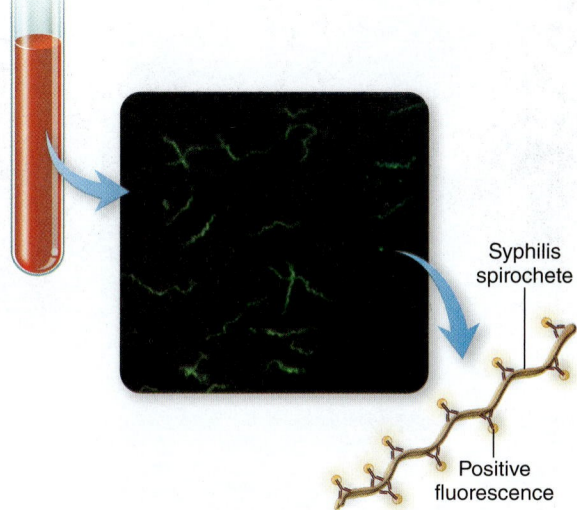

Figure 15.12 Direct fluorescence antigen test.
Photomicrograph of a direct fluorescence test for *Treponema pallidum*, the syphilis spirochete.

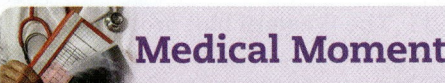

Medical Moment

Understanding Lab Results

If you are planning to work in health care, it will be important for you to be able to interpret lab results and apply your knowledge to your patient's clinical situation. For example, your patient was admitted with a fever and a wound on the lower leg, and he is being treated for a bacterial skin infection. Cultures were obtained from the infected area on admission and the patient was started empirically on broad-spectrum antibiotics. Over the next several hours, the infected area on the patient's lower leg increases in size and redness despite antibiotic therapy, and the patient's temperature remains elevated. You receive the patient's culture report during the evening shift and realize that the antibiotic is not appropriate for the microorganism that has been identified. You notify the physician immediately, and the physician orders a different antibiotic based on the sensitivity report. The patient begins to respond to the new therapy within 24 hours—his temperature decreases and the redness begins to recede. Your decision to notify the physician when you did may have prevented the patient from getting sicker and may have shortened his hospital stay.

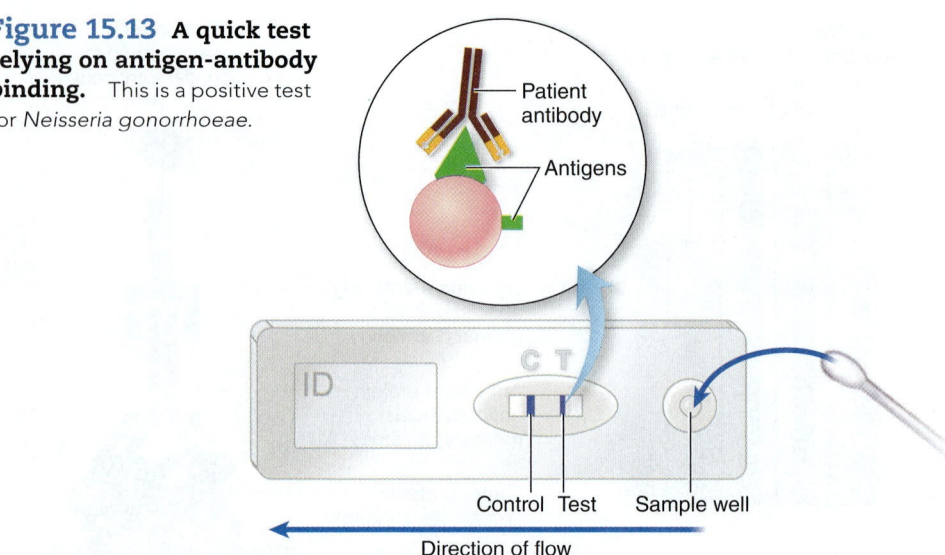

Figure 15.13 A quick test relying on antigen-antibody binding. This is a positive test for *Neisseria gonorrhoeae*.

(i.e., bacterial cells) is added to the test serum, and binding of the fluorescent antibody is visualized through fluorescence microscopy. Fluorescing aggregates or cells indicate that the FAbs have complexed with the microbe-specific antibodies in the test serum. This technique is frequently used to diagnose syphilis and various viral infections.

There is yet an easier application of antigen-antibody binding. Many easy-to-use kits have been created that utilize known antibodies, which are applied to patient specimens or bacterial isolates. In these reactions, the results can be seen with the naked eye because the manufacturer has added dye to the antibody. An example of this is shown in **figure 15.13.**

Immunoassays

The nature of the antibody response is being exploited today in athletics, criminology, government, and business to test for trace amounts of substances such as hormones, metabolites, and drugs. But traditional techniques in serology are not able to detect only a few molecules of these chemicals. **Immunoassays** are alternative methods that employ monoclonal antibodies and permit rapid, accurate measurement of trace antigen or antibody levels due to their enhanced sensitivity.

Radioimmunoassay (RIA)   Antibodies or antigens labeled with a **radioactive isotope** can be used to pinpoint minute quantities of a corresponding antigen or antibody (table 15.2). Radioimmunoassay can be used to detect hormone levels in samples, and to diagnose allergies in patients, chiefly by the radioimmunosorbent test (RIST) or radioallergosorbent test (RAST).

Enzyme-Linked Immunosorbent Assay (ELISA)   The **ELISA** test, also known as enzyme immunoassay (EIA), uses an enzyme-linked (most often the enzymes are horseradish peroxidase and alkaline phosphatase) indicator antibody to visualize antigen-antibody reactions. This technique also relies on a solid support such as a plastic microtiter plate that can *adsorb* (attract on its surface) the reactants **(figure 15.14).**

The *indirect ELISA* test detects microbe-specific antibodies in patient sera. A known antigen is adsorbed to the surface of a well and mixed with unknown antibody. If an antibody-antigen complex forms, an added indicator antibody will bind, and subsequent development will produce a color change indicating a positive result. This is the common test used for antibody screening for HIV, various rickettsial species, hepatitis A and C, and *Helicobacter*. Because false positives can occur, a verification test (Western blot) may be necessary.

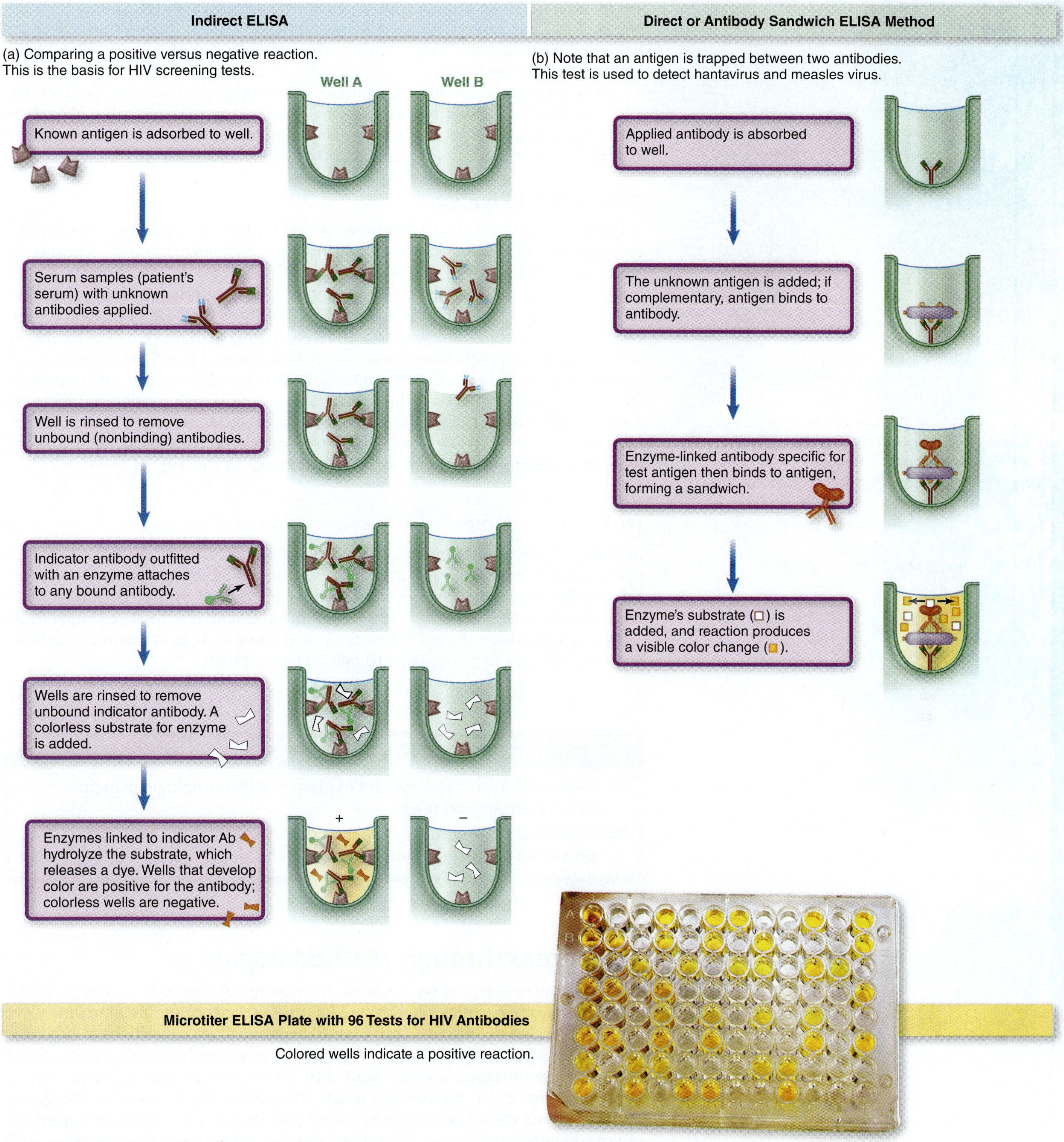

Indirect ELISA

(a) Comparing a positive versus negative reaction.
This is the basis for HIV screening tests.

Well A   Well B

Known antigen is adsorbed to well.

Serum samples (patient's serum) with unknown antibodies applied.

Well is rinsed to remove unbound (nonbinding) antibodies.

Indicator antibody outfitted with an enzyme attaches to any bound antibody.

Wells are rinsed to remove unbound indicator antibody. A colorless substrate for enzyme is added.

Enzymes linked to indicator Ab hydrolyze the substrate, which releases a dye. Wells that develop color are positive for the antibody; colorless wells are negative.

+   −

Direct or Antibody Sandwich ELISA Method

(b) Note that an antigen is trapped between two antibodies.
This test is used to detect hantavirus and measles virus.

Applied antibody is absorbed to well.

The unknown antigen is added; if complementary, antigen binds to antibody.

Enzyme-linked antibody specific for test antigen then binds to antigen, forming a sandwich.

Enzyme's substrate (□) is added, and reaction produces a visible color change (■).

Microtiter ELISA Plate with 96 Tests for HIV Antibodies

Colored wells indicate a positive reaction.

Figure 15.14   Methods of ELISA testing.   (a) Indirect ELISAs detect patient antibody. Here, both positive and negative tests are illustrated. (b) Direct ELISAs detect patient antigen (microbes). Here, only the positive reaction is illustrated.

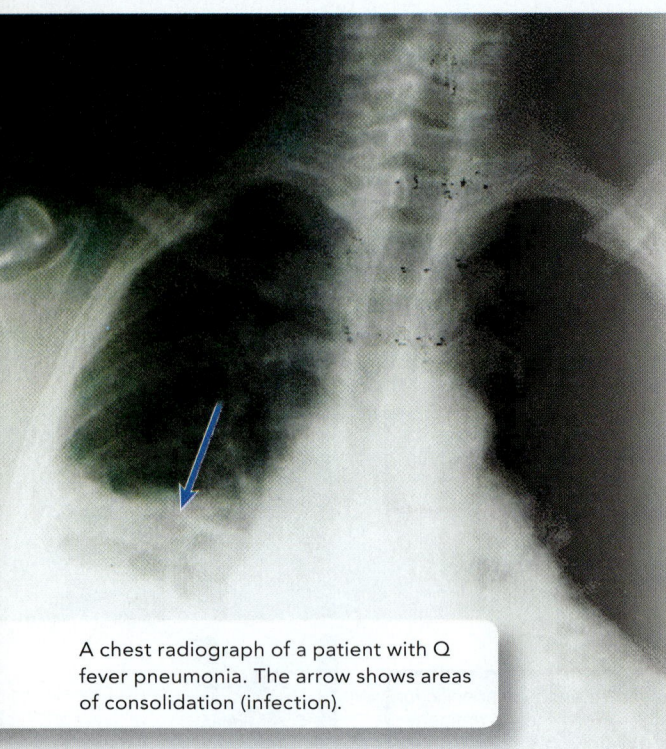

A chest radiograph of a patient with Q fever pneumonia. The arrow shows areas of consolidation (infection).

In *direct ELISA* (or sandwich) tests, a known antibody is adsorbed to the bottom of a well and incubated with an unknown antigen. If an antibody-antigen complex forms, it will attract the indicator antibody, and color will develop in these wells, indicating a positive result.

More frequently, alkaline phosphatase–based tests are used that produce visible light that is detected or quantified by machines and photographic films. New cutting-edge detection systems utilize computer chips that sense minute changes in electrical current that occur when antibody-antigen complexes are formed.

Complement Fixation

A **lysin** is an antibody that causes the lysis or rupture of target cells, and requires complement (see chapter 12 for a discussion of complement) to complete this destruction. *Hemo*lysins can interact with complement system components on red blood cells, causing the cells to hemolyze (lyse and release their hemoglobin). Since this lysin-mediated hemolysis is readily observable, it is the basis of a group of tests called complement fixation, or CF (table 15.2). Complement fixation tests are invaluable in diagnosing fungal diseases such as coccidioidomycosis and histoplasmosis, and in detecting antibodies to *Coxiella burnetii*, the cause of Q fever. The antistreptolysin O (ASO) titer test employs a technique related to complement fixation and measures antibody levels against streptolysin toxin, an important hemolysin of group A streptococci. This is an important verification test for scarlet fever, rheumatic fever, and other related streptococcal syndromes (see chapters 16 and 19).

In Vivo Testing

In practice, *in vivo* tests employ principles similar to serological tests, except in this case an antigen is introduced into a patient to elicit some sort of visible reaction. The **tuberculin reaction,** where a small amount of purified protein derivative (PPD) from *Mycobacterium tuberculosis* is injected into the skin, is a classic example (table 15.2). The appearance of a red, raised, thickened lesion in 48 to 72 hours can indicate previous exposure to tuberculosis (shown in figure 14.10).

> ### 15.5 LEARNING OUTCOMES—Assess Your Progress
> **10.** Define the term *serology*, and explain the immunologic principle behind serological tests.
> **11.** Explain the difference between a direct and an indirect ELISA, providing a clinical application for each.

15.6 Breakthrough Methodologies

Diagnostic microbiology is on the brink of a new era. A vast array of new technologies in the area of genetics, physics, and information science (in the form of massive databases) has led the medical profession to begin to adopt radically new diagnostic techniques. While the gold standard of diagnosis has always been growing an isolated culture of the offending microbe (and then identifying it), that technique has its downfalls. What if the microbe you isolate and grow is not the cause of the disease but just a bystander? Also, culturing takes time—18 hours minimum, and many organisms require much longer incubation times. In addition, more and more we realize that many infections are polymicrobial. Our single-minded efforts to isolate one disease-causing organism can result in serious misdiagnoses.

Take the very common example of infection of the bloodstream (septicemia). This is a condition that can kill very quickly. The critical time frame for appropriate management is estimated to be less than 6 hours. However, traditionally, the course of events is as follows:

- Blood sample is drawn and inoculated into blood culture tube.
- Broad-spectrum antimicrobials are begun.
- After 18 to 24 hours, identification of bacteria in blood culture is attempted, and antimicrobial sensitivities are determined.
- More appropriate antimicrobial therapy is instituted.

Studies show that during the 18- to 24-hour incubation period, if the patient is improving based on the broad-spectrum antimicrobial he or she is receiving, many physicians do not rewrite the antibiotic order for a more appropriate drug. In the case in which the patient does not improve, the delay in changing the antimicrobial is frequently fatal.

For these reasons, there is a call for more sophisticated diagnostic techniques that can occur immediately, and often at the point of care at the patient's bedside or in the doctor's office. Although many of the techniques described earlier in this chapter are relatively inexpensive compared to the newer methods, the consensus seems to be that improved patient outcomes with the new tests will soon drive hospitals and clinics to adopt the new technologies. After all, infectious disease specialists point out that even though diagnostic tests influence approximately 70% of health care treatment decisions, currently only 2% of U.S. health care costs are expended on them.

Thus follows an overview of the most likely new diagnostic techniques to be seen in the next 5 years in U.S. health care. **Figure 15.15** illustrates the techniques.

Microarrays

Microarrays designed for infectious disease diagnosis are "chips" (absorbent plates) that contain gene sequences from potentially thousands of different possible infectious agents, selected based on the syndrome being investigated (such as respiratory infection, or meningitis symptoms). The arrays are selected based on a very large differential diagnosis; in other words, what possible microbes could cause disease in this syndrome? Arrays can be made to contain bacterial, viral, and fungal genes in a single test. In this scenario, patient samples (sputum, cerebrospinal fluid) or the nucleic acids isolated from them are incubated with labeled gene sequences on the microarray. Matching sequences hybridize to the chip, and the label (in most cases it is fluorescence) is detected by a computer program, which provides the identity of the isolate or isolates **(figure 15.15a)**.

Nucleic Acid Sequencing: The Whole Story

The development of high-throughput nucleic acid sequencing has revolutionized the analysis of the human genome, as described in chapter 8. The cost of whole-genome sequencing, in terms of time and money, is becoming so low that this technique may become commonplace in clinical and epidemiological laboratories around the world. It is particularly useful for rapid analysis of outbreaks and drug-resistant organisms. It has also led to the creation of so-called "next generation sequencing" technologies. A technique like random amplified polymorphic DNA analysis uses random primers to allow for the identification of novel nucleic acid sequences. More importantly, a single genome can be scanned and analyzed multiple times in a process called "deep sequencing," which minimizes errors. Some scientists suspect that these types of sequencing will become so cheap and so routine that we will soon just "sequence everything" from a patient sample to find the one or more microbes causing symptoms **(figure 15.15b)**.

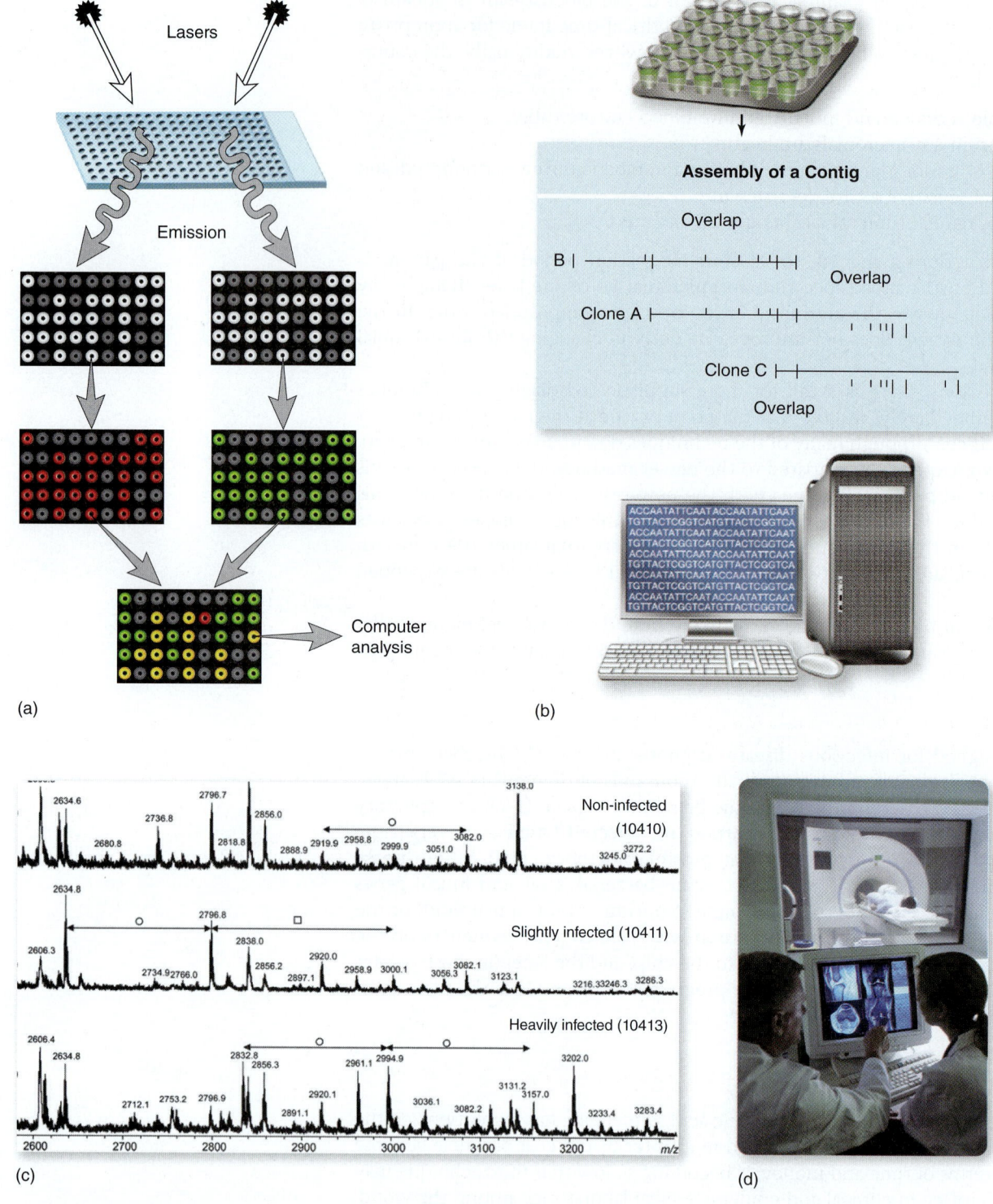

Figure 15.15 Overview of emerging diagnostic technologies. (a) Microarrays. (b) Nucleic acid sequencing. A contig is a set of overlapping DNA segments. (c) Mass spectometry. (d) A patient undergoing MRI (magnetic resonance imaging) testing.

Mass Spectometry

Mass spectrometry has been utilized for years to determine the structure and composition of various chemical compounds and biological molecules. Analysis of samples by mass spectrometry is poised to become the new cutting-edge technology for providing rapid and highly accurate microbial identification within just minutes. This technique, which is also often called MALDI-TOF, can be used to analyze a protein fingerprint from pure culture isolates or directly from patient specimens. It works by adding the patient sample to a metal plate and then striking it with a laser. This causes the sample to become ionized. The ions from the sample are guided into a machine that separates them and identifies them according to their mass-to-charge ratio. This technology has been applied to the identification of bacteria, viruses, and fungi, so far, and may become commonplace in many clinical and research laboratories due to its ability to produce rapid, precise, and cost-effective results compared to conventional phenotypic, genotypic, and immunologic methods.

Both the microarray and mass spectrometry technologies can also be used to simultaneously detect antibiotic susceptibilities–obviously, an important advantage **(figure 15.15c)**.

Imaging

An old way of diagnosing infections, which found use in only occasional infections, involves various imaging techniques. Infections associated with hip implants, for example, may be difficult to access through blood samples. The bacteria may be growing in biofilms on the implanted materials, or they may be growing in an abscess deep in the hip joint. Magnetic resonance imaging, computerized tomography (CT) scan, and positron emission tomography (PET) scans have been increasingly employed to find areas of localized infection in deep tissue, which can later be biopsied to aspirate samples for culture. In the event no infection is found on the image, the patient has been spared an invasive procedure. This seems very new and "high tech," but imaging in the form of X rays has been used for centuries in the diagnosis of tuberculosis **(figure 15.15d)**.

15.6 LEARNING OUTCOMES—Assess Your Progress

12. Explain why isolating a pathogen through standard culture methods may become an outdated diagnosis strategy.

13. Name at least three "breakthrough technologies" and the principles behind them.

CASE FILE WRAP-UP

Research has found that only 5% to 15% of blood cultures are found to be positive. Increasing the number of blood cultures obtained increases the likelihood of isolating the offending organism and can be lifesaving. When two sets of blood cultures are obtained and a pathogen is identified from both cultures, it is highly unlikely that the organism cultured is a contaminant. Both aerobic and anaerobic blood cultures are obtained to provide the best chance of isolating the organism.

Blood cultures are indicated before beginning parenteral antibiotic therapy in patients with fever and leukocytosis (elevated white blood cell count). It is important to obtain blood cultures before IV antibiotics are started because the antibiotics may inhibit the growth of the microorganism before it can be isolated.

The patient in the opening Case File had a fever with a decreased level of consciousness, both signs of sepsis. Blood cultures are used to identify microorganisms that have invaded the bloodstream, causing life-threatening septicemia. It was important for the lab workers to notify the nursing staff as soon as the cultures were collected so that the nursing staff could administer the initial dose of IV antibiotics.

Sampling Cerebrospinal Fluid via Lumbar Puncture

Lumbar puncture, sometimes referred to as a spinal tap, is performed to collect a sample of cerebrospinal fluid (CSF), the fluid that surrounds the brain and spinal cord. Conditions affecting the brain and/or spinal cord caused by infection with viruses, fungi, and bacteria can be detected by lumbar puncture. Performing a lumbar puncture is an extremely invasive procedure; it is therefore especially important to handle specimens carefully because repeating the test is not a desirable outcome for anyone, particularly the patient!

Although physicians perform the procedure, other health care professionals may be required to assist. The patient is positioned on his or her side with the neck flexed and knees drawn up to the chest. Positioning the patient in this way helps to increase the space between the spinal vertebrae, giving the physician extra room to work. Alternately, the patient may sit upright with his or her spine rounded and head and neck flexed forward. The area to be sampled is prepped using aseptic technique. The appropriate vertebrae are located (usually L3/L4 or L4/L5, as the spinal cord stops near L2). A local anesthetic is injected under the skin and along the intended path of the spinal needle. Once the area has been numbed, the spinal needle is inserted between the vertebrae until the needle is through the dura mater, which is indicated by a sensation of the tissue "giving." The needle is then advanced a little farther, through the arachnoid mater and into the subarachnoid space. Once the spinal needle is in the correct place, the stylet within the needle is removed and fluid can be collected.

Lumbar puncture is often performed when meningitis is suspected because there is no other reliable test for the condition. Lumbar puncture is also used in some cases of fever of unknown origin, especially in infants and small children. Lab staff must be immediately available to process the specimens collected. Lab staff may be tasked with being present during the procedure to take the samples immediately to the lab once they are collected, or this may be the responsibility of the nursing staff.

Patients are generally required to lie still for several hours following the procedure in order to reduce the likelihood of a "spinal headache," which occurs in up to 40% of patients. Tenderness in the lower back may also occur. Bleeding and brain stem herniation are rare side effects of lumbar puncture. Vital signs, neurological status, and the puncture site are monitored closely following the procedure.

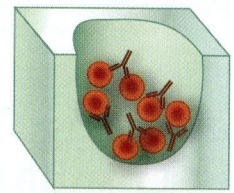

Chapter Summary

15.1 Preparation for the Survey of Microbial Diseases

- Microbiologists use three categories of techniques to diagnose infections: *phenotypic, genotypic,* and *immunologic.*

15.2 On the Track of the Infectious Agent: Specimen Collection

- The first step in clinical diagnosis (after patient observation) is obtaining a specimen. If this step is not performed correctly, specimen analysis will not be accurate no matter how "sensitive" the test.

15.3 Phenotypic Methods

- The main phenotypic methods include the direct examination of specimens, observing the growth of specimen cultures on special media, and biochemical testing of pure cultures.

15.4 Genotypic Methods

- The use of genotypic methods in microbial identification has grown exponentially.
  - Polymerase chain reaction is used on its own for diagnosis, and also as a precursor to other genetic methods.
  - Hybridization techniques exploit the base-pairing characteristics of nucleic acids.
  - Pulsed-field gel electrophoresis and ribotyping have specific applications in diagnosis.

15.5 Immunologic Methods

- Serological testing can be performed on a variety of body fluids or tissues and is based on the principle that antibodies have extreme specificity for antigens.
- Testing for microbial-specific antigens or antibodies is typically performed *in vitro*, and antigen-antibody interactions are made macroscopically or microscopically visible.
- Agglutination reactions occur between antibody and antigens bound to cells, resulting in visible clumping. It is the basis of determining titer, or antibody concentration, in patient sera.

- Precipitation reactions also occur between antibody and antigen and produce insoluble, visible precipitates, but they are typically made visible by adding radioactive or enzyme markers.
- Serotyping employs antisera against cellular antigens such as the capsule, flagellum, and cell wall to identify bacterial species and strains.
- In the Western blot procedure, proteins that have been separated by electrical current are identified by labeled antibodies.
- Direct fluorescence antibody tests indicate the presence of microbial antigens and are useful in identifying infectious agents; indirect fluorescence tests indicate the presence of microbe-specific antibodies and are used to diagnose infection.
- Immunoassays can detect very small quantities of antigen, antibody, or other substances and use dyes or radioactive isotopes to visualize antigen-antibody complexes.
- The ELISA test is widely used to detect antigens (direct method) or antibodies (indirect method) in patient samples.
- Complement fixation involves the complement-dependent action of lysins to detect antimicrobial antibodies and is used in diagnosing fungal and bacterial diseases.
- *In vivo* serological testing, such as the tuberculin reaction, involves injection of antigen to elicit a visible immune response in the host.

15.6 Breakthrough Methodologies

- The next 5 years are likely to bring many new technologies to the widespread diagnosis of infectious diseases.
- Whole-genome sequencing relies on DNA sequencing of microbes.
- Mass spectrometry detects microbes via their protein fingerprints.

Multiple-Choice Questions   Bloom's Levels 1 and 2: Remember and Understand

Select the correct answer from the answers provided.

1. PCR is used to do which of the following?
   a. Amplify the number of copies of genes.
   b. Identify the nucleic acid sequence of a gene.
   c. Determine the identity of a particular gene.
   d. Determine the function of a particular gene.

2. Mass spectrometry identifies microbes via
   a. fluorescence of antibodies.
   b. precipitation reactions.
   c. protein fingerprints.
   d. fluorescence of antigens.

3. Because they are sufficiently unique in their appearances, viruses can sometimes be identified at a family or genus level by which phenotypic method?
   a. cell culture
   b. PCR
   c. electron microscopy
   d. ELISA

4. Which of the following methods can identify different strains of a microbe?
   a. microscopic examination
   b. radioimmunoassay
   c. serotyping
   d. agglutination test

5. In agglutination reactions, the antigen is a ____; in precipitation reactions, it is a ____ .
   a. soluble molecule; whole cell
   b. whole cell; soluble molecule
   c. bacterium; virus
   d. protein; carbohydrate

6. Which type of methods is based on a microbe's utilization of nutrients?
   a. genotypic
   b. immunologic
   c. biochemical

7. Which of the following is an *in vivo* immunologic method?
   a. determination of antibody titer
   b. tuberculin reaction
   c. determination of blood type
   d. indirect ELISA

8. Which of the following is the correct pairing of an immunologic test and the subject to be identified?
   a. indirect ELISA test/an unknown microbial antigen
   b. direct fluorescence antibody test/an unknown antibody
   c. Western blot/an unknown microbial antigen or antibody
   d. agglutination test/an unknown soluble microbial toxin

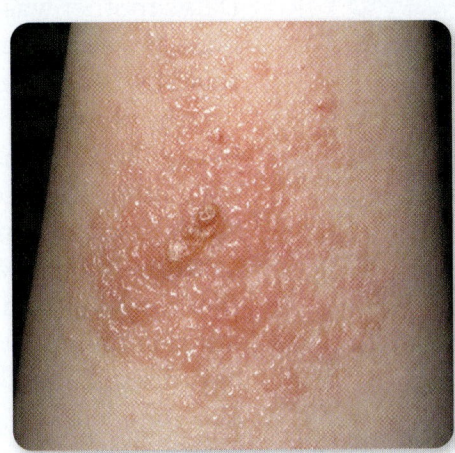

Critical Thinking   Bloom's Levels 3, 4, and 5: Apply, Analyze, and Evaluate

Critical thinking **is the ability to reason and solve problems using facts and concepts. These questions can be approached from a number of angles and, in most cases, they do not have a single correct answer.**

1. Explain why specimens should be taken aseptically and processed quickly to ensure accurate results, even when nonsterile sites are being sampled and selective media are to be used.

2. A patient arrives at a health center and the attending nurse suspects an infection. Describe the sequence of *general* steps he and the health center lab technicians might use to identify the exact identity of the infectious agent.

3. Differentiate between the serological tests used to identify isolated cultures of pathogens and those used to diagnose disease from the patient's serum.

4. a. Seropositivity means having a blood serum that tests positive. Explain why it may or may not develop at the same rate in all patients exposed to the same microbe.

   b. Would a high rate of false-positives decrease the sensitivity or specificity of a serological test? (False-positivity was discussed in this chapter, and also in chapter 11.)

5. Discuss how new information on the human microbiome may lead to the identification of specific microbes associated with diseases currently thought to be noninfectious.

Visual Connections   Bloom's Level 5: Evaluate

This question connects previous images to a new concept.

1. **From chapter 14, figure 14.10b.** Imagine that this patient is being seen by his or her physician for this unknown rash. What rapid phenotypic test could suggest that this condition is caused by a bacterium? Could a rapid immunoassay or fluorescent procedure be used to identify a specific viral cause? Explain your answer.

CASE FILE

A Rash of Symptoms

I was a newly graduated nurse working in the emergency room of a large urban children's hospital when Dale was brought in by his mother. Dale was a 6-year-old who, until a few days prior, had been very healthy. Dale's mother told the triage nurse that Dale was very sick and had a rash "all over." As was policy, Dale was given a mask to wear because he was coughing and was sent back out to the waiting room after having a brief history taken and a set of vital signs.

Dale had been triaged as urgent due to his cough, high fever, and elevated heart rate, so it wasn't long before I called Dale to an exam room to perform a more in-depth history and head-to-toe assessment. According to Dale's mother, Dale had initially complained of a sore throat. Within a day, he also complained of a headache and developed a high fever, which his mother treated appropriately with acetaminophen. He then developed a cough, which his mother assumed was part of a "bad cold." She was not overly concerned until Dale developed a rash, which started on his face and head and rapidly spread to his chest and back, and then to his arms and legs.

I had Dale's mother undress him and put him in an examination gown, and it was only then that I realized the extent of Dale's rash. It was winter and Dale had been covered in clothing from head to toe, having only his jacket removed when seen by the triage nurse. Dressed in a gown, I saw that Dale's mother was not exaggerating—Dale's rash covered almost all of his body. It was a maculopapular rash and did not look like any childhood rash that I had ever seen, before or since.

The physician on duty came in to see Dale and immediately asked Dale to open his mouth. She asked the mother about Dale's immunizations, and his mother admitted that she did not believe in immunizations and Dale had never been immunized. The physician then turned immediately to me and asked me to put Dale in an isolation room. Upon my return, she spoke to Dale's mother, and I will never forget what she said: "Well, it was a very bad decision to not have Dale immunized, because now he has the measles. Not only does your son have the measles, but by bringing him here you exposed everyone in the waiting room to what can be a life-threatening illness." She then ordered a chest X ray and blood work to confirm the diagnosis and said that she would immediately notify the state health department.

- Why was a chest X ray performed?

- Why was the state health department contacted?

Case File Wrap-Up appears on page 460.

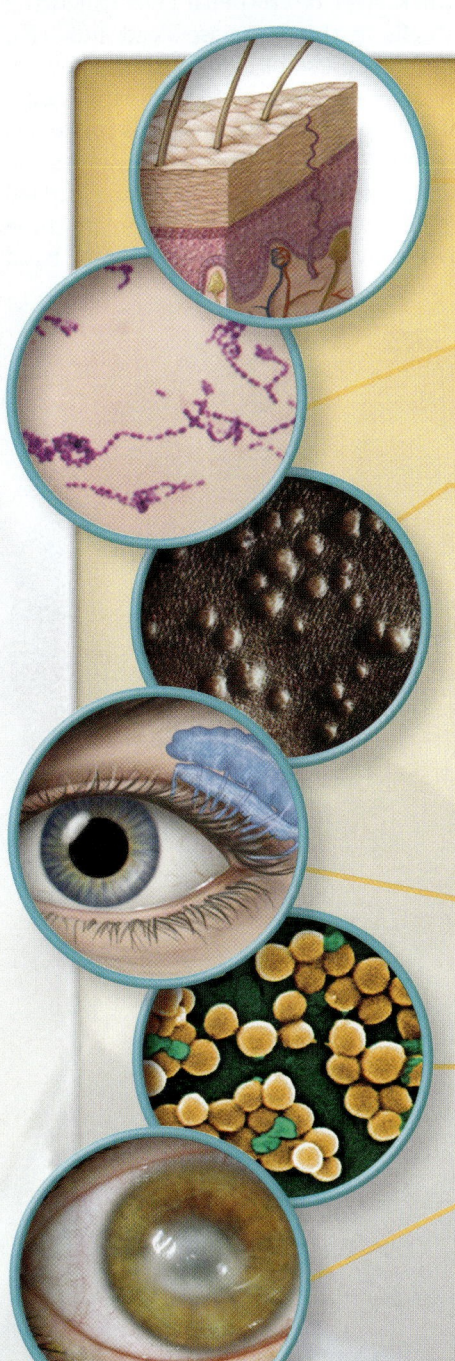

CHAPTER

16

Infectious Diseases Affecting the Skin and Eyes

IN THIS CHAPTER...

16.1 The Skin and Its Defenses

1. Describe the important anatomical features of the skin.
2. List the natural defenses present in the skin.

16.2 Normal Biota of the Skin

3. List characteristics of the skin's normal microbiota.

16.3 Skin Diseases Caused by Microorganisms

4. Explain the important features of the "Highlight Disease," MRSA skin and soft-tissue infection.
5. List the possible causative agents, modes of transmission, virulence factors, diagnostic techniques, and prevention/treatment for the "Highlight Disease," maculopapular rash diseases.
6. Discuss important features of the other infectious skin diseases. These are impetigo, cellulitis, staphylococcal scalded skin syndrome, vesicular/pustular rash diseases, large pustular skin lesions, and cutaneous and superficial mycoses.
7. Discuss the relative dangers of rubella and rubeola viruses in different populations.

16.4 The Surface of the Eye and Its Defenses

8. Describe the important anatomical features of the eye.
9. List the natural defenses present in the eye.

16.5 Normal Biota of the Eye

10. List the types of normal biota presently known to occupy the eye.

16.6 Eye Diseases Caused by Microorganisms

11. List the possible causative agents, modes of transmission, virulence factors, diagnostic techniques, and prevention/treatment for the "Highlight Disease," conjunctivitis.
12. Discuss important features of keratitis caused by either HSV or by *Acanthamoeba*.

Sweat is cooling and it has several antimicrobial properties.

16.1 The Skin and Its Defenses

The organs under consideration in this chapter—the skin and eyes—form the boundary between the human and the environment. The skin, together with the hair, nails, and sweat and oil glands, forms the **integument.** The skin has a total surface area of 1.5 to 2 square meters. Its thickness varies from 1.5 millimeters at places such as the eyelids to 4 millimeters on the soles of the feet. Several distinct layers can be found in this thickness, and we summarize them here. Follow **figure 16.1** as you read.

The outermost portion of the skin is the epidermis, which is further subdivided into four or five distinct layers. On top is a thick layer of epithelial cells called the stratum corneum, about 25 cells thick. The cells in this layer are dead and have migrated from the deeper layers during the normal course of cell division. They are packed with a protein called **keratin,** which the cells produce in very large quantities. Cells emerge from the deepest levels of the epidermis. Because this process is continuous, the entire epidermis is replaced every 25 to 45 days. Keratin gives the cells their ability to withstand damage, abrasion, and water penetration; the surface of the skin is termed *keratinized* for this reason. Below the stratum corneum are three or four more layers of epithelial cells. The lowest layer, the stratum basale, or basal

Figure 16.1 **A cross section of skin.**

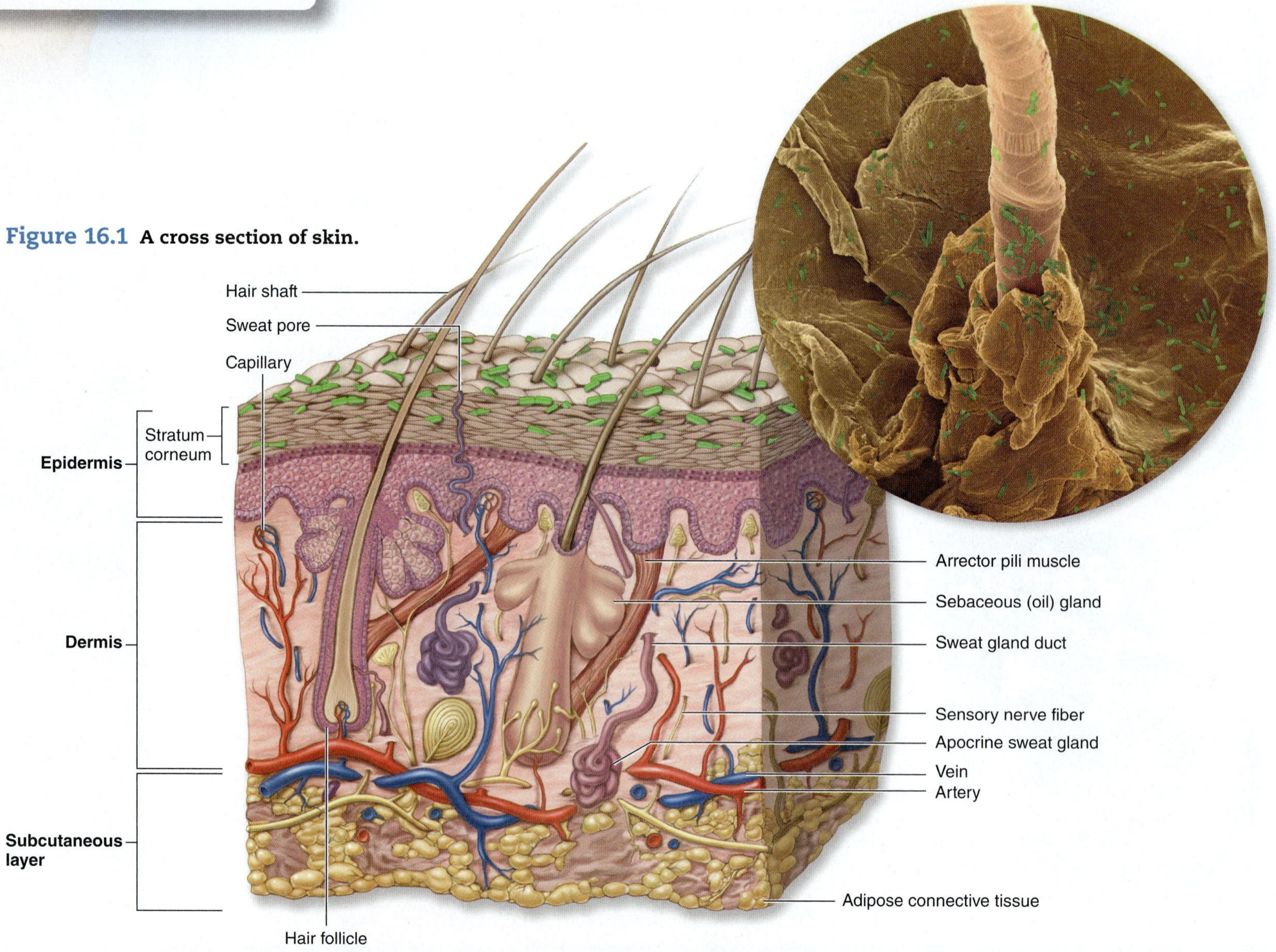

Hair shaft
Sweat pore
Capillary

Epidermis — Stratum corneum

Dermis

Subcutaneous layer

Hair follicle

Arrector pili muscle
Sebaceous (oil) gland
Sweat gland duct
Sensory nerve fiber
Apocrine sweat gland
Vein
Artery
Adipose connective tissue

layer, is attached to the underlying dermis and is the source for all of the cells that make up the epidermis.

The dermis, underneath the epidermis, is composed of connective tissue instead of epithelium. This means that it is a rich matrix of fibroblast cells and fibers such as collagen, and it contains macrophages and mast cells. The dermis also harbors a dense network of nerves, blood vessels, and lymphatic vessels. Damage to the epidermis generally does not result in bleeding, whereas damage deep enough to penetrate the dermis results in broken blood vessels. Blister formation, the result of friction trauma or burns, causes a separation between the dermis and epidermis.

The "roots" of hairs, called follicles, are in the dermis. **Sebaceous** (oil) **glands** and scent glands are associated with the hair follicle. Separate sweat glands are also found in this tissue. All of these glands have openings on the surface of the skin, so they pass through the epidermis as well.

Millions of cells from the stratum corneum slough off every day, and attached microorganisms slough off with them. The skin is also brimming with antimicrobial substances. Perhaps the most effective skin defense against infection is the one most recently discovered. In the past 20 years, small molecules called **antimicrobial peptides** have been identified in epithelial cells. These are positively charged chemicals that act by disrupting the negatively charged membranes of bacteria. There are many different types of these peptides, and they seem to be chiefly responsible for keeping the microbial count on skin relatively low.

The sebaceous glands' secretion, called **sebum,** has a low pH, which makes the skin inhospitable to many microorganisms. Sebum is oily due to its high concentration of lipids. The lipids can serve as nutrients for normal microbiota, but breakdown of the fatty acids contained in lipids leads to toxic by-products that inhibit the growth of microorganisms not adapted to the skin environment. This mechanism helps control the growth of potentially pathogenic bacteria. Sweat is also inhibitory to microorganisms, because of both its low pH and its high salt concentration. **Lysozyme** is an enzyme found in sweat (and tears and saliva) that specifically breaks down peptidoglycan, which you learned in chapter 3 is a unique component of bacterial cell walls.

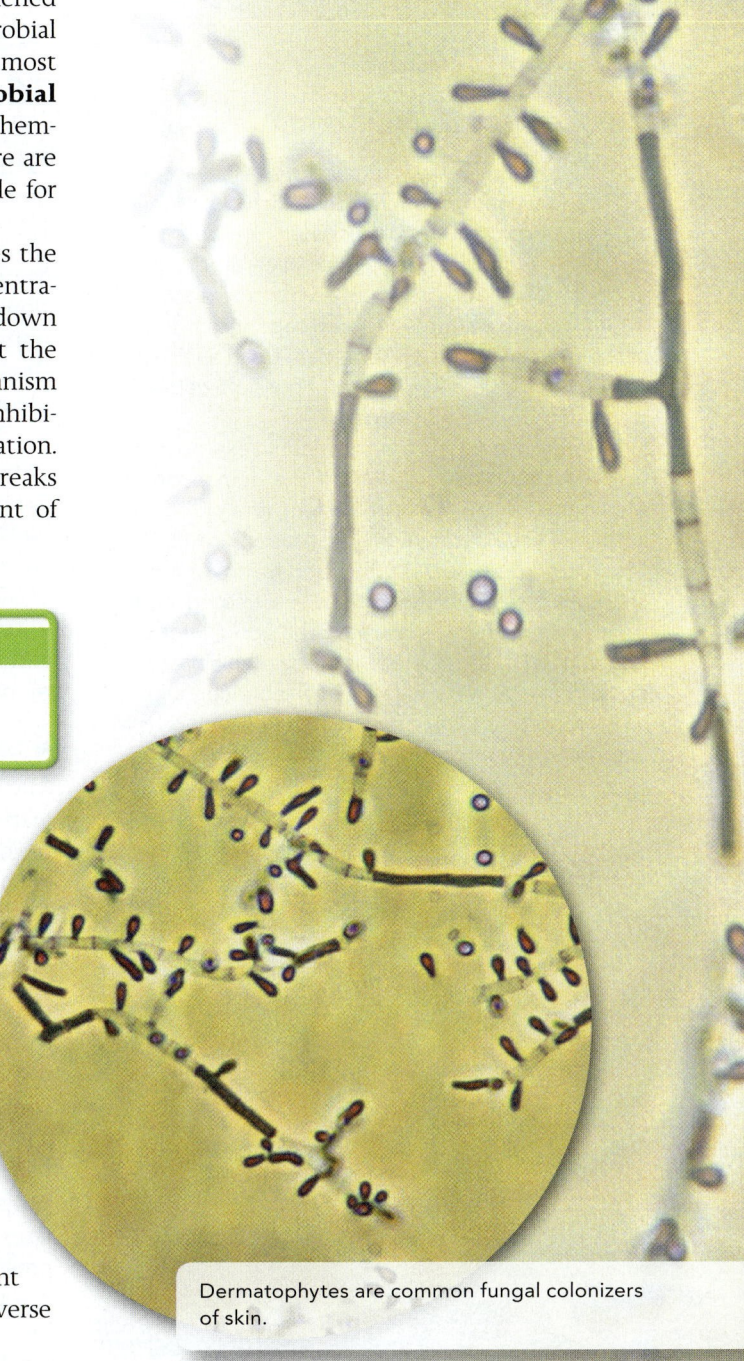

Dermatophytes are common fungal colonizers of skin.

16.1 LEARNING OUTCOMES—Assess Your Progress

1. Describe the important anatomical features of the skin.
2. List the natural defenses present in the skin.

16.2 Normal Biota of the Skin

Microbes that do live on the skin surface as normal biota must be capable of living in the dry, salty conditions they find there. Microbes are relatively sparsely distributed over dry, flat areas of the body, such as on the back, but they can grow into dense populations in moist areas and skin folds, such as the underarm and groin areas. The normal microbiota also live in the protected environment of the hair follicles and glandular ducts.

Relying on cultivation-independent techniques, the Human Microbiome Project (HMP) has begun to deliver fascinating information about our skin's biota. We are learning that hundreds of species of microbes, including some well-known pathogens, inhabit our epidermis, dermis, and subcutaneous skin layers. It is also common for different species to favor different areas of our bodies, and for different people to have different species. Last, in spite of this variation, it seems common for an individual's microbiota to remain relatively constant over time. In brief, the HMP is showing us that the skin microbiota is far more diverse than we imagined.

A Note About the Chapter Organization

In a clinical setting, patients present themselves to health care practitioners with a set of symptoms, and the health care team makes an "anatomical" diagnosis—such as a *generalized vesicular rash*. The anatomical diagnosis allows practitioners to narrow down the list of possible causes to microorganisms that are known to be capable of creating such a condition. Then the proper tests can be performed to arrive at an etiologic diagnosis (i.e., determining the exact microbial cause). So the order of events is as follows:

1. anatomical diagnosis,
2. differential diagnosis, and
3. etiologic diagnosis.

In this book, we organize diseases according to the anatomical diagnosis (which appears as a gold-colored heading). Then the agents in the differential diagnosis are each addressed, each of them appearing as turquoise headings. When we finish addressing each agent that could cause the condition, we sum them up in a Disease Table, whether there is only one possible cause or whether there are nine or ten.

Some conditions are truly iconic examples of how infectious agents work in the human body. Meningitis in the nervous system is a good example. These "paradigm" diseases are called "Highlight Diseases" at the beginning of these chapters. The other conditions are treated more briefly, since there is a wealth of good information available about these diseases; however, we do summarize every disease in a table containing all its important features, so that your textbook is a complete reference resource.

Defenses and Normal Biota of the Skin

	Defenses	Normal Biota
Skin	Keratinized surface, sloughing, low pH, high salt, lysozyme, antimicrobial peptides	*Streptococcus, Staphylococcus, Corynebacterium, Propionibacterium, Pseudomonas, Lactobacillus;* yeasts such as *Candida*

16.2 LEARNING OUTCOMES—Assess Your Progress

3. List characteristics of the skin's normal microbiota.

16.3 Skin Diseases Caused by Microorganisms

Highlight Disease

MRSA Skin and Soft-Tissue Infection

Methicillin-resistant *Staphylococcus aureus* (MRSA) is a common cause of skin lesions in non-hospitalized people. (The hospitalized population is more likely to acquire systemic–bloodstream–infections from MRSA, addressed in chapter 18.) Even though the name mentions "methicillin," these strains are usually resistant to multiple antibiotics.

Staphylococcus aureus is a gram-positive coccus that grows in clusters, like a bunch of grapes. It is nonmotile. Much of its destructiveness is due to its array of superantigens (see chapter 12). It can be highly virulent, but it also appears as "normal" biota on the skin of one-third of the population. Strains that are methicillin-resistant are also found on healthy people.

This species is considered the sturdiest of all non-endospore-forming pathogens, with well-developed capacities to withstand high salt (7.5% to 10%), extremes in pH, and high temperatures (up to 60°C for 60 minutes). *S. aureus* also remains viable after months of air drying and resists the effect of many disinfectants and antibiotics.

▶ Signs and Symptoms

MRSA infections of the skin tend to be raised, red, tender, localized lesions, often featuring pus and feeling hot to the touch **(figure 16.2).** They occur easily in breaks in the skin caused by injury, shaving, or even just abrading. They may localize around a hair follicle. Fever is a common feature.

▶ Transmission and Epidemiology

MRSA is a common contaminant of all kinds of surfaces you touch daily, especially if the surfaces are not routinely sanitized. Gym equipment, airplane tray tables, electronic devices, razors, and so on, are all sources of indirect contact infection. Persons with active MRSA skin infections should keep them covered in order to avoid direct contact transmission to others.

▶ Pathogenesis and Virulence Factors

All pathogenic *S. aureus* strains typically produce coagulase, an enzyme that coagulates plasma. Because 97% of all human isolates of *S. aureus* produce this enzyme, its presence is considered the most diagnostic species characteristic.

Other enzymes expressed by *S. aureus* include hyaluronidase, which digests the intercellular "glue" (hyaluronic acid) that binds connective tissue in host tissues; staphylokinase, which digests blood clots; a nuclease that digests DNA (DNase); and lipases that help the bacteria colonize oily skin surfaces.

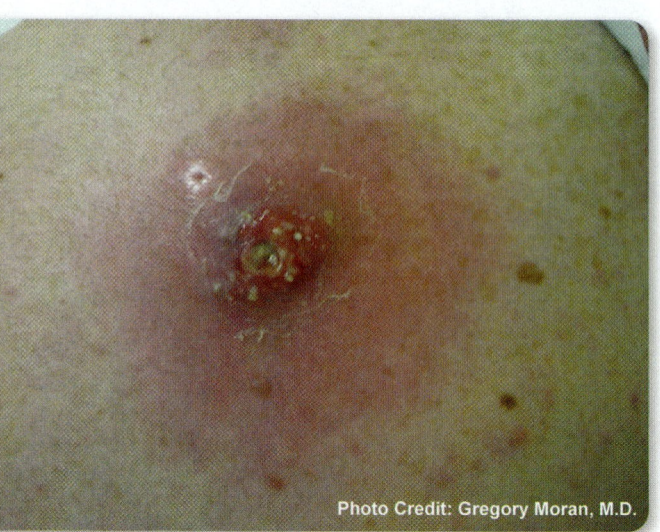
Photo Credit: Gregory Moran, M.D.

Figure 16.2 **A typical MRSA skin lesion.**

▶ Culture and/or Diagnosis

Polymerase chain reaction (PCR) is routinely used to diagnose MRSA. Alternatively, primary isolation of *S. aureus* is achieved by inoculation on blood agar **(figure 16.3)**. For heavily contaminated specimens, selective media such as mannitol salt agar are used. The production of catalase, an enzyme that breaks down hydrogen peroxide accumulated during oxidative metabolism, can be used to differentiate the staphylococci, which produce it, from the streptococci, which do not.

One key technique for separating *S. aureus* from other species of *Staphylococcus* is the coagulase test **(figure 16.4)**. By definition, any isolate that coagulates plasma is *S. aureus*; all others are coagulase-negative.

▶ Prevention and Treatment

Prevention is only possible with good hygiene. Treatment of these infections often starts with incision of the lesion and drainage of the pus. Antimicrobial treatment should include more than one antibiotic. Current recommendations in the United States are for the use of clindamycin + TMP/SMZ or doxycycline. These recommendations, of course, will change based on antibiotic-resistance patterns.

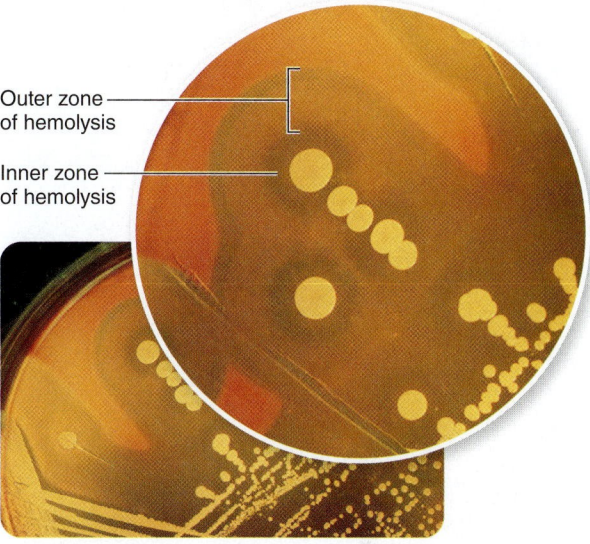

Figure 16.3 *Staphylococcus aureus.*   Blood agar plate growing *S. aureus*. Some strains show two zones of hemolysis, caused by two different hemolysins. The inner zone is clear, whereas the outer zone is fuzzy and appears only if the plate has been refrigerated after growth.

Outer zone of hemolysis

Inner zone of hemolysis

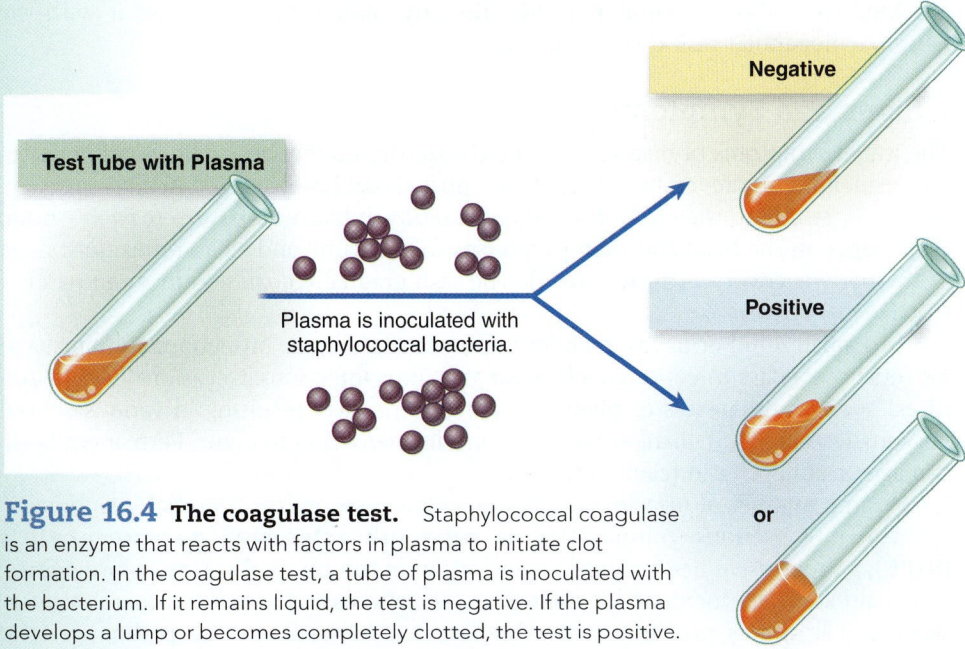

Negative

Test Tube with Plasma

Plasma is inoculated with staphylococcal bacteria.

Positive

Figure 16.4   **The coagulase test.**   Staphylococcal coagulase is an enzyme that reacts with factors in plasma to initiate clot formation. In the coagulase test, a tube of plasma is inoculated with the bacterium. If it remains liquid, the test is negative. If the plasma develops a lump or becomes completely clotted, the test is positive.

or

Disease Table 16.1 MRSA Skin and Soft-Tissue Infection

Causative Organism(s)	Methicillin-resistant *Staphylococcus aureus*
Most Common Modes of Transmission	Direct contact, indirect contact
Virulence Factors	Coagulase, other enzymes, superantigens
Culture/Diagnosis	PCR, culture and Gram stain, coagulase and catalase tests, multitest systems
Prevention	Hygiene practices
Treatment	Clindamycin + TMP/SMZ; in **Serious Threat** category in CDC Antibiotic Resistance Report
Epidemiology	Community-associated MRSA infections most common in children and young to middle-aged adults Incidence increasing in communities (decreasing in hospitals)

Maculopapular Rash Diseases

There are a variety of microbes that can cause the type of skin eruptions classified as *maculopapular,* a term denoting flat to slightly raised colored bumps.

Measles

Every year hundreds of thousands of children in the developing world die from this disease (about 385 a day), even though an extremely effective vaccine has been available since 1963. Health campaigns all over the world seek to make measles vaccine available to all and have been very effective. Since 2002, worldwide deaths from measles have dropped 74%. Ironically, it seems that more work and education need to be done in *developed* countries now. Many parents are opting to not have their children vaccinated, due to unfounded fears about the link between the vaccine and autism. We would do well to remember that before the vaccine was introduced, measles killed 6 million people worldwide each year.

Measles is also known as **rubeola.** Be very careful not to confuse it with the next maculopapular rash disease, rubella.

▶ Signs and Symptoms

The initial symptoms of measles are sore throat, dry cough, headache, conjunctivitis, lymphadenitis, and fever. In a short time, unusual oral lesions called *Koplik's spots* appear as a prelude to the characteristic red maculopapular **exanthem** (eg-zan'-thum) that erupts on the head and then progresses to the trunk and extremities until most of the body is covered **(figure 16.5).** The rash gradually coalesces into red patches that fade to brown.

In a small number of cases, children develop laryngitis, bronchopneumonia, and bacterial secondary infections such as ear and sinus infections. Occasionally (1 in 100 cases), measles progresses to pneumonia or encephalitis, resulting in various central nervous system (CNS) changes ranging from disorientation to coma. Permanent brain damage or epilepsy can result.

A large number of measles patients experience secondary bacterial infections. The most serious complication is **subacute sclerosing panencephalitis (SSPE),** a progressive neurological degeneration of the cerebral cortex, white matter, and brain stem. Its incidence is approximately one case in a million measles infections, and it afflicts primarily male children and adolescents. The pathogenesis of SSPE appears to involve a defective virus, one that has lost its ability to form a capsid and be released from an infected cell. Instead, it spreads unchecked through the brain by cell fusion, gradually destroying neurons and accessory cells and breaking down myelin. The disease causes profound intellectual and neurological impairment. The course of the disease invariably leads to coma and death in a matter of months or years.

▶ Pathogenesis and Virulence Factors

The virus implants in the respiratory mucosa and infects the tracheal and bronchial cells. From there, it travels to the lymphatic system, where it multiplies and then enters the bloodstream, a condition known as *viremia.* The bloodstream carries the virus to the skin and to various organs.

The measles virus induces the cell membranes of adjacent host cells to fuse into large **syncytia** (sin-sish'-uh), giant cells with many nuclei. These cells no longer perform their proper function. The virus seems proficient at disabling many aspects of the host immune response, especially cell-mediated immunity and delayed-type hypersensitivity.

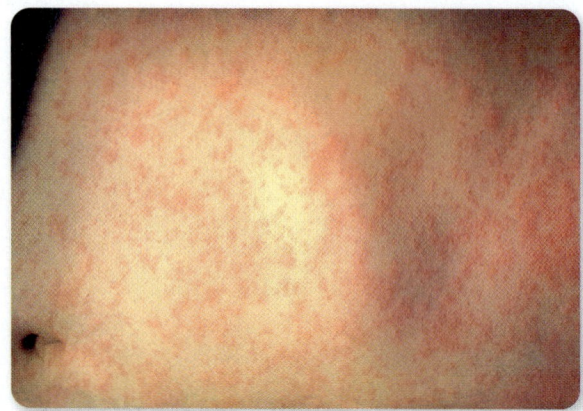

Figure 16.5 **The rash of measles.**

▶ Transmission and Epidemiology

Measles is one of the most contagious infectious diseases, transmitted principally by respiratory droplets. There is no reservoir other than humans, and a person is infectious during the periods of incubation, prodrome phase, and the skin rash but usually not during convalescence.

▶ Culture and Diagnosis

The disease can be diagnosed on clinical presentation alone; but if further identification is required, an ELISA test is available that tests for patient IgM to measles antigen, indicating a current infection.

▶ Prevention

The MMR vaccine (for measles, mumps, and rubella) contains live attenuated measles virus, which confers protection for up to 20 years. Measles immunization is recommended for all healthy children at the age of 12 to 15 months, with a booster before the child enters school.

▶ Treatment

Treatment relies on reducing fever, suppressing cough, and replacing lost fluid. Complications require additional remedies to relieve neurological and respiratory symptoms and to sustain nutrient, electrolyte, and fluid levels. Vitamin A supplements are recommended by some physicians; they have been found effective in reducing the symptoms and decreasing the rate of complications.

Rubella

This disease is also known as German measles. Rubella is derived from the Latin for "little red," and that is a good way to remember it because it causes a relatively minor rash disease with few complications. Sometimes it is called the 3-day measles. The only exception to this mild course of events is when a fetus is exposed to the virus while in its mother's womb (*in utero*). Serious damage can occur, and for that reason women of childbearing years must be sure to be vaccinated well before they plan to conceive.

▶ Signs and Symptoms

The two clinical forms of rubella are referred to as postnatal infection, which develops in children or adults, and **congenital** (prenatal) infection of the fetus, expressed in the newborn as various types of birth defects.

Postnatal Rubella   The rash of pink macules and papules first appears on the face and progresses down the trunk and toward the extremities, advancing and resolving in about 3 days. The rash is milder looking than the measles rash **(see Disease Table 16.2).** Adult rubella is often characterized by joint inflammation and pain rather than a rash.

Congenital Rubella   Rubella is a strongly **teratogenic** virus. Transmission of the rubella virus to a fetus *in utero* can result in a serious complication called **congenital rubella (figure 16.6).** The mother is able to transmit the virus even if she is asymptomatic. Infection in the first trimester is most likely to induce miscarriage or multiple permanent defects in the newborn. The most common of these is deafness and may be the only defect seen in some babies. Other babies may experience cardiac abnormalities, ocular lesions, deafness, and mental and physical retardation in varying combinations. Less drastic sequelae that usually resolve in time are anemia, hepatitis, pneumonia, carditis, and bone infection.

▶ Causative Agent

The rubella virus is a *Rubivirus,* in the family *Togaviridae.* The virus has the ability to stop mitosis, which is an important process in a rapidly developing embryo

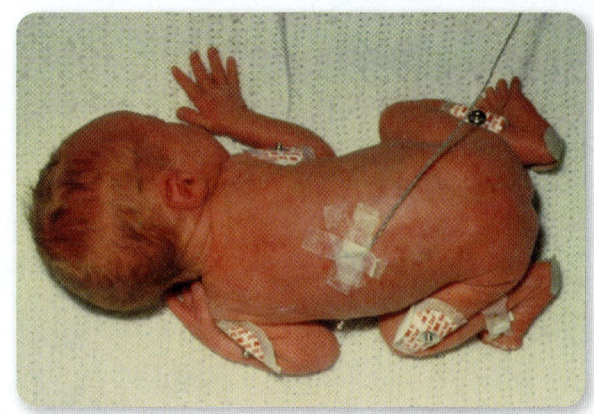

Figure 16.6 An infant born with congenital rubella can display a papular pink or purple rash.

A Note About Statistics in the Disease Tables

Each condition we study is summarized in a disease table. The last row of each table contains information about the epidemiology of the disease. The type of epidemiological information that is most relevant to a particular disease can vary. For example, it is vitally important to know the numbers of new cases (**incidence**) of some diseases. This is the case for measles in the United States (see Disease Table 16.2), as we track the resurgence of a disease once controlled by vaccination. For other conditions, such as fifth disease in the same table, it is more useful to know how many people have been affected by the time they reach a certain age. **Prevalence**, or the current number of people affected by the condition, is another common measure. And for some diseases the most informative statistic is how deadly they are (**mortality rate**). These tables contain the most useful information about each condition.

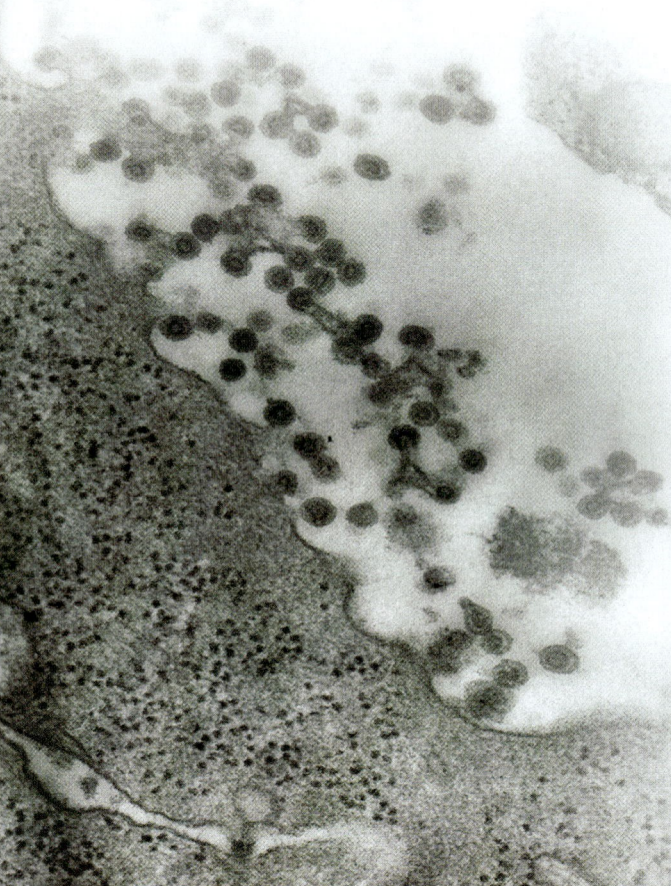

and fetus. It also induces apoptosis (programmed cell death) of normal tissue cells. This inappropriate cell death can do irreversible harm to organs it affects. And last, the virus damages vascular endothelium, leading to poor development of many organs.

▶ Transmission and Epidemiology

Rubella is a disease with worldwide distribution. Infection is initiated through contact with respiratory secretions and occasionally urine. The virus is shed during the prodromal phase and up to a week after the rash appears. Congenitally infected infants are contagious for a much longer period of time. Because the virus is only moderately communicable, close living conditions are required for its spread. This disease is well controlled in the United States, with fewer than 10 cases reported in each of the last several years. Most cases are reported among adolescents and young adults in military training camps, colleges, and summer camps.

▶ Culture and Diagnosis

Diagnosing rubella relies on the same twin techniques discussed earlier for measles. Because it mimics other diseases, rubella should not be diagnosed on clinical grounds alone. IgM antibody to rubella virus can be detected early using an ELISA technique or a latex-agglutination card. Women in developed countries routinely undergo antibody testing at the beginning of pregnancy to determine their immune status.

▶ Prevention and Treatment

The attenuated rubella virus vaccine is usually given to children in the combined form (MMR vaccination) at 12 to 15 months and a booster at 4 or 6 years of age. The vaccine for rubella can be administered on its own, without the measles and mumps components. Postnatal rubella is generally benign and requires only symptomatic treatment. No specific treatment is available for the congenital manifestations.

Fifth Disease

This disease, more precisely called *erythema infectiosum,* is so named because about 100 years ago it was the fifth of the diseases recognized by doctors to cause rashes in children. The first four were scarlet fever, measles, rubella, and another rash that was thought to be distinct but was probably not. Fifth disease is a very mild disease that often results in a characteristic "slapped-cheek" appearance because of a confluent reddish rash that begins on the face. Within 2 days, the rash spreads on the body but is most prominent on the arms, legs, and trunk. The rash of fifth disease may reoccur for several weeks and may be brought on by any activity that increases body heat (i.e., exercise, fever, sunlight, warm baths, and even high emotion).

The causative agent is parvovirus B19. You may have heard of "parvo" as a disease of dogs, but strains of this virus group infect humans. Fifth disease is usually diagnosed by the clinical presentation, but sometimes it is helpful to rule out rubella by testing for IgM against rubella.

This infection is very contagious. There is no vaccine and no treatment for this usually mild disease.

Roseola

This disease is common in young children and babies. It is sometimes known as "sixth disease." It can result in a maculopapular rash, but a high percentage (up to 70%) of cases proceed without the rash stage. Children sick with this disease exhibit a high fever (up to 41°C, or 105°F) that comes on quickly and lasts for up to 3 days. Seizures may occur during this period, but other than that patients remain alert and do not

Disease Table 16.2 Maculopapular Rash Diseases

Disease	Measles (Rubeola)	Rubella	Fifth Disease	Roseola	Other Conditions to Consider
Causative Organism(s)	Measles virus	Rubella virus	Parvovirus B19	Human herpesvirus 6	Other conditions resulting in a rash that may look similar to these maculopapular conditions should be considered:
Most Common Modes of Transmission	Droplet contact	Droplet contact	Droplet contact, direct contact	Unknown	
Virulence Factors	Syncytium formation, ability to suppress CMI	In fetuses: inhibition of mitosis, induction of apoptosis, and damage to vascular endothelium	–	Ability to remain latent	Scarlet fever (covered in chapter 19) Secondary syphilis (chapter 20)
Culture/Diagnosis	ELISA for IgM, acute/convalescent IgG	Acute IgM, acute/convalescent IgG	Usually diagnosed clinically	Usually diagnosed clinically	Rocky Mountain spotted fever (chapter 18)
Prevention	Live attenuated vaccine (MMR or MMRV)	Live attenuated vaccine (MMR or MMRV)	–	–	
Treatment	No antivirals; vitamin A, antibiotics for secondary bacterial infections	–	–	–	
Distinguishing Features of the Rashes	Starts on head, spreads to whole body, lasts over a week	Milder red rash, lasts approximately 3 days	"Slapped-face" rash first, spreads to limbs and trunk, tends to be confluent rather than distinct bumps	High fever precedes rash stage; rash not always present	
Epidemiological Features	Incidence increasing in North America; in developing countries incidence is 30 million cases/yr and 1 million deaths	3 cases reported in United States in 2009; worldwide: 100,000 infants/yr born with congenital rubella syndrome	60% of population seropositive by age 20	> 90% seropositive; 90% of disease cases occur before age of 2	
Appearance of Lesions					

act terribly ill. On the fourth day, the fever disappears, and it is at this point that a rash can appear, first on the chest and trunk and less prominently on the face and limbs. By the time the rash appears, the disease is almost over.

Roseola is caused by a human herpesvirus called HHV-6. Like all herpesviruses, it can remain latent in its host indefinitely after the disease has cleared. Very occasionally, the virus reactivates in childhood or adulthood, leading to mononucleosis-like or hepatitis-like symptoms. It is thought that 100% of the U.S. population is infected with this virus by adulthood. Some people experienced the disease roseola when they became infected, and some of them did not. The suggestion has been made that this virus causes other disease conditions later in life, such as multiple sclerosis. No vaccine and no treatment exist for roseola.

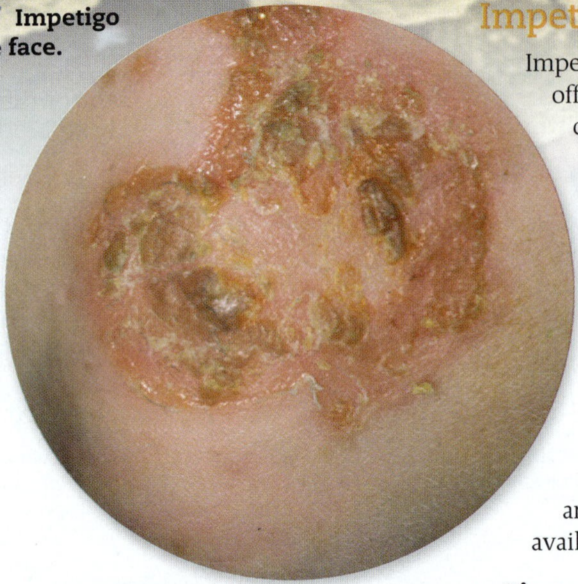

Figure 16.7 Impetigo
lesion on the face.

Impetigo

Impetigo is a superficial bacterial infection that causes the skin to flake or peel off **(figure 16.7).** It is not a serious disease but is highly contagious, and children are the primary victims. Impetigo can be caused by either *Staphylococcus aureus* or *Streptococcus pyogenes*, and some cases are probably caused by a mixture of the two. It has been suggested that *S. pyogenes* begins all cases of the disease, and in some cases *S. aureus* later takes over and becomes the predominant bacterium cultured from lesions. Because *S. aureus* produces a bacteriocin (toxin) that can destroy *S. pyogenes*, it is possible that *S. pyogenes* is often missed in culture-based diagnosis.

Impetigo, whether it is caused by *S. pyogenes*, *S. aureus*, or both, is highly contagious and transmitted through direct contact but also via fomites and mechanical vector transmission. The peak incidence is in the summer and fall.

The only current prevention for impetigo is good hygiene. Vaccines are in development for both of the etiologic agents, but none are currently available.

▶ Signs and Symptoms

The "lesion" of impetigo looks variously like peeling skin, crusty and flaky scabs, or honey-colored crusts. Lesions are most often found around the mouth, face, and extremities, though they can occur anywhere on the skin. It is very superficial and it itches. The symptomatology does not indicate whether the infection is caused by *Staphylococcus* or *Streptococcus*.

Impetigo Caused by *Staphylococcus aureus*

The most important virulence factors relevant to *S. aureus* impetigo are exotoxins called exfoliative toxins A and B, which are coded for by a phage that infects some *S. aureus* strains. At least one of the toxins attacks a protein that is very important for epithelial cell-to-cell binding in the outermost layer of the skin. Breaking up this protein leads to the characteristic blistering seen in the condition. The breakdown of skin architecture also facilitates the spread of the bacterium.

Impetigo Caused by *Streptococcus pyogenes*

Streptococcus pyogenes is thoroughly described in chapter 19. The important features are briefly summarized here, and the features pertinent to impetigo are listed in **Disease Table 16.3.**

S. pyogenes is a gram-positive coccus and is beta-hemolytic on blood agar. In addition to impetigo, it causes streptococcal pharyngitis (strep throat), scarlet fever, pneumonia, puerperal fever, necrotizing fasciitis, serious bloodstream infections, and poststreptococcal conditions such as rheumatic fever.

If the precise etiologic agent must be identified, there are well-established methods for identifying group A streptococci. Refer to chapter 19.

▶ Pathogenesis and Virulence Factors

Like *S. aureus*, this bacterium possesses a huge arsenal of enzymes and toxins. Some of these are listed in Disease Table 16.3.

Rarely, impetigo caused by *S. pyogenes* can be followed by acute poststreptococcal glomerulonephritis (see chapter 19). The strains that cause impetigo never cause rheumatic fever, however.

S. pyogenes is more often the cause of impetigo in newborns, and *S. aureus* is more often the cause of impetigo in older children, but both can cause infection in either age group.

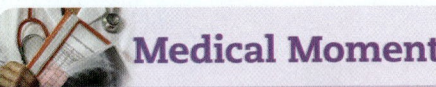

NCLEX® PREP

3. A client has been treated for impetigo but has not been compliant with medical therapy. What potential complication could ensue?
   a. development of renal pathology
   b. necrotizing fasciitis
   c. increased likelihood of developing rheumatic fever
   d. development of dental disease

Medical Moment

Hand, Foot, and Mouth Disease

Hand, foot, and mouth disease is a viral illness caused by an enterovirus, most often Coxsackie virus A16. It occurs most often in the summer and fall and is more common in children under the age of 10. When adults get the disease, their symptoms are generally milder. It is spread via the fecal-oral route, although contact with the lesions and droplet infection may also spread the virus efficiently. The disease often occurs in outbreaks.

In the prodrome, affected individuals may have fever, lose their appetite, and feel generally unwell. Soon after, lesions appear in the mouth, starting as macular lesions and progressing to erosions. These erosions may make eating and drinking very uncomfortable. Approximately 75% of patients also develop "itchy" lesions on the soles, palms, and between the toes and fingers. Lesions may also appear on the trunk, genitals, and/or buttocks and generally last 3 to 6 days.

Disease Table 16.3 Impetigo

Causative Organism(s)	*Staphylococcus aureus*	*Streptococcus pyogenes*
Most Common Modes of Transmission	Direct contact, indirect contact	Direct contact, indirect contact
Virulence Factors	Exfoliative toxin A, coagulase, other enzymes	Streptokinase, plasminogen-binding ability, hyaluronidase, M protein
Culture/Diagnosis	Routinely based on clinical signs, when necessary, culture and Gram stain, coagulase and catalase tests, multitest systems, PCR	Routinely based on clinical signs, when necessary, culture and Gram stain, coagulase and catalase tests, multitest systems, PCR
Prevention	Hygiene practices	Hygiene practices
Treatment	Topical mupirocin or retapamulin, oral dicloxacillin, cephalexin, or TMP-SMZ; (MRSA is in **Serious Threat** category in CDC Antibiotic Resistance Report)	Topical mupirocin or retapamulin
Distinguishing Features	Seen more often in older children, adults	Seen more often in newborns
Epidemiological Features	Prevalence approximately 1% of children in North America	

Disease Table 16.4 Cellulitis

Causative Organism(s)	*Staphylococcus aureus*	*Streptococcus pyogenes*	Other bacteria or fungi
Most Common Modes of Transmission	Parenteral implantation	Parenteral implantation	Parenteral implantation
Virulence Factors	Exfoliative toxin A, coagulase, other enzymes	Streptokinase, plasminogen-binding ability, hyaluronidase, M protein	–
Culture/Diagnosis	Based on clinical signs	Based on clinical signs	Based on clinical signs
Prevention	–	–	–
Treatment	Oral or IV antibiotic (dicloxacillin if sensitive, vancomycin if MRSA); surgery sometimes necessary; (MRSA is in **Serious Threat** category in CDC Antibiotic Resistance Report)	Oral or IV antibiotic (penicillin); surgery sometimes necessary	Aggressive treatment with oral or IV antibiotic; surgery sometimes necessary
Distinguishing Features	–	–	More common in immunocompromised
Epidemiological Features	Incidence highest among males 45–64		

Cellulitis

Cellulitis is a condition caused by a fast-spreading infection in the dermis and in the subcutaneous tissues below. It causes pain, tenderness, swelling, and warmth. Fever and swelling of the lymph nodes draining the area may also occur. Frequently, red lines leading away from the area are visible (a phenomenon called *lymphangitis*); this symptom is the result of microbes and inflammatory products being carried by the lymphatic system. Bacteremia could develop with this disease, but uncomplicated cellulitis has a good prognosis.

Cellulitis generally follows introduction of bacteria or fungi into the dermis, either through trauma or by subtle means (with no obvious break in the skin). The most common causes of the condition in healthy people are *Staphylococcus aureus* and *Streptococcus pyogenes*, although almost any bacterium and some fungi can cause this condition in an immunocompromised patient. In infants, group B streptococci are a frequent cause (see chapter 21).

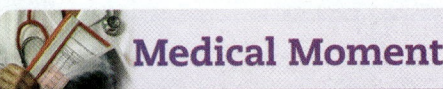

Medical Moment

Scrum Pox: Herpes Gladiatorum

Herpes gladiatorum is caused by herpes simplex virus type 1 (HSV-1). It often affects athletes such as wrestlers who come into close contact with each other (it is called "scrum pox" among rugby players or "mat herpes" among wrestlers). Outbreaks of the infection may occur in gyms or fitness centers.

The lesions occur in clusters of blisters that may or may not be painful. The lesions generally heal within a week to 10 days. Once a person has been infected with HSV-1, the person is infected for life and can pass the infection on to others. Periodic outbreaks can occur during times of stress, trauma, or sun exposure. During the first episode, the infected person may experience fever, sore throat, headache, and swollen glands. Some individuals can be infected with the virus but will fail to develop any skin lesions. These individuals carry the virus and can still pass the virus to others through skin-to-skin contact.

Mild cellulitis responds well to oral antibiotics chosen to be effective against both *S. aureus* and *S. pyogenes*. More involved infections and infections in immunocompromised people require intravenous antibiotics. If there are extensive areas of tissue damage, surgical debridement (duh-breed'-munt) is warranted **(Disease Table 16.4).**

Staphylococcal Scalded Skin Syndrome (SSSS)

This syndrome is another **dermolytic** condition caused by *Staphylococcus aureus*. It affects mostly newborns and babies, although children and adults can experience the infection. Transmission may occur when caregivers carry the bacterium from one baby to another. Adults in the nursery can also directly transfer *S. aureus* because approximately 30% of adults are asymptomatic carriers. Carriers can harbor the bacteria in the nasopharynx, axilla, perineum, and even the vagina. (Fortunately, only about 5% of *S. aureus* strains are lysogenized by the type of phage that codes for the toxins responsible for this disease.)

Like impetigo, this is an exotoxin-mediated disease. The phage-encoded exfoliative toxins A and B are responsible for the damage. Unlike impetigo, the toxins enter the bloodstream from some focus of infection (the throat, the eye, or sometimes an impetigo infection) and then travel to the skin throughout the body. These toxins cause **bullous lesions,** which often appear first around the umbilical cord (in neonates) or in the diaper or axilla area. A split occurs in the epidermal tissue layers just above the stratum basale (see figure 16.1). Widespread **desquamation** of the skin follows, leading to the burned appearance referred to in the name **(figure 16.8).**

Once a tentative diagnosis of SSSS is made, immediate antibiotic therapy should be instituted.

Vesicular or Pustular Rash Diseases

There are two diseases that present as generalized "rashes" over the body in which the individual lesions contain fluid. The lesions are often called *pox*, and the two diseases are chickenpox and smallpox **(Disease Table 16.6).** Chickenpox is very common and mostly benign, but even a single case of smallpox constitutes a public health emergency. Both are viral diseases.

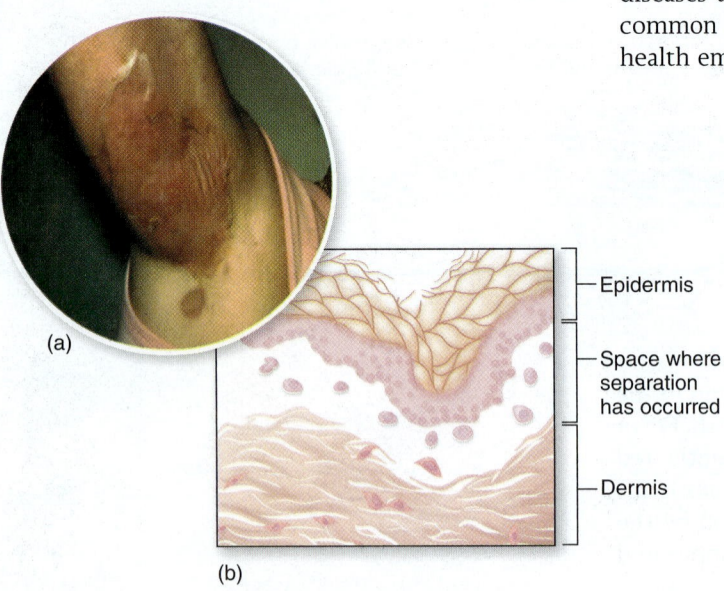

(a)

(b)

— Epidermis

— Space where separation has occurred

— Dermis

Figure 16.8 Staphylococcal scalded skin syndrome (SSSS) in the underarm area. **(a)** Exfoliative toxin produced in local infections causes blistering and peeling away of the outer layer of skin. **(b)** Realistic drawing of a segment of skin affected with SSSS. The point of epidermal shedding, or desquamation, is in the epidermis. The lesions will heal well because the level of separation is so superficial.

Disease Table 16.5 Scalded Skin Syndrome	
Causative Organism(s)	*Staphylococcus aureus*
Most Common Modes of Transmission	Direct contact, droplet contact
Virulence Factors	Exfoliative toxins A and B
Culture/Diagnosis	Histological sections; culture performed but false negatives common because toxins alone are sufficient for disease
Prevention	Eliminate carriers in contact with neonates
Treatment	Immediate systemic antibiotics (nafcillin if sensitive; vancomycin if MRSA); (MRSA is in **Serious Threat** category in CDC Antibiotic Resistance Report)
Distinguishing Features	Split in skin occurs *within* epidermis
Epidemiological Features	Mortality 1%–5% in children, 50%–60% in adults

Chickenpox

After an incubation period of 10 to 20 days, the first symptoms to appear are fever and an abundant rash that begins on the scalp, face, and trunk and radiates in sparse crops to the extremities. Skin lesions progress quickly from macules and papules to itchy vesicles filled with a clear fluid. In several days, they encrust and drop off, usually healing completely but sometimes leaving a tiny pit or scar. Lesions number from a few to hundreds and are more abundant in adolescents and adults than in young children. **Figure 16.9** contains images of the chickenpox lesions in a child and in an adult. The lesion distribution is *centripetal*, meaning that there are more in the center of the body and fewer on the extremities, in contrast to the distribution seen with smallpox. The illness usually lasts 4 to 7 days; new lesions stop appearing after about 5 days. Patients are considered contagious until all of the lesions have crusted over.

Approximately 0.1% of chickenpox cases are followed by encephalopathy, or inflammation of the brain caused by the virus. It can be fatal, but in most cases recovery is complete.

▶ Shingles

After recuperation from chickenpox, the virus enters into the sensory nerve endings of cutaneous spinal nerve branches, especially those that innervate the skin of the chest and head **(figure 16.10a)**. From there, it becomes latent in the ganglia and may reemerge as **shingles** (also known as **herpes zoster)** with its characteristic asymmetrical distribution on the skin of the trunk or head **(figure 16.10b)**.

Shingles develops abruptly after reactivation by such stimuli as psychological stress, X-ray treatments, immunosuppressive and other drug therapy, surgery, or

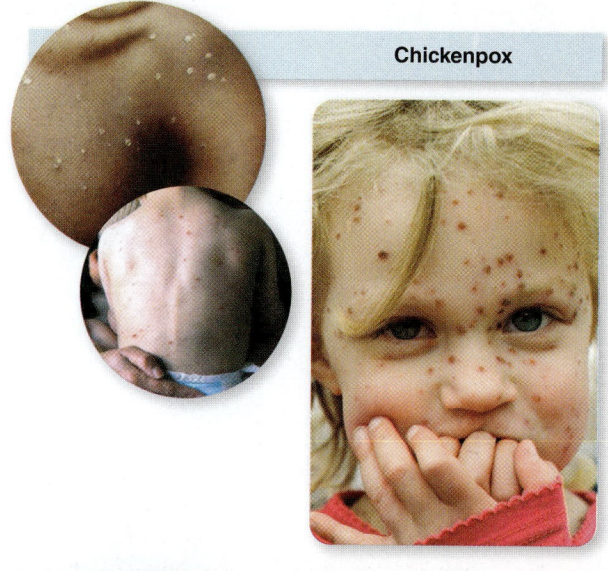

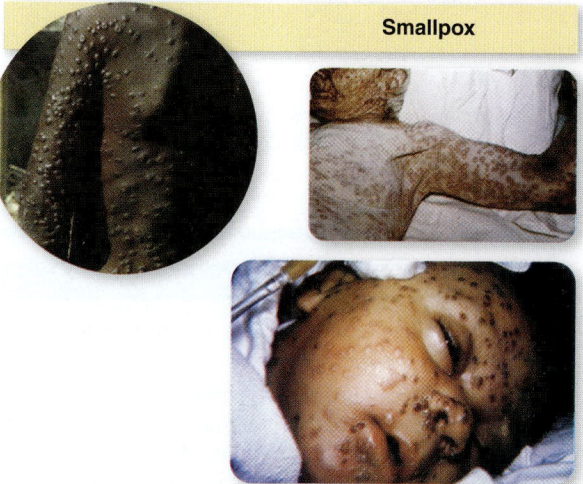

Figure 16.9 Images of chickenpox and smallpox.

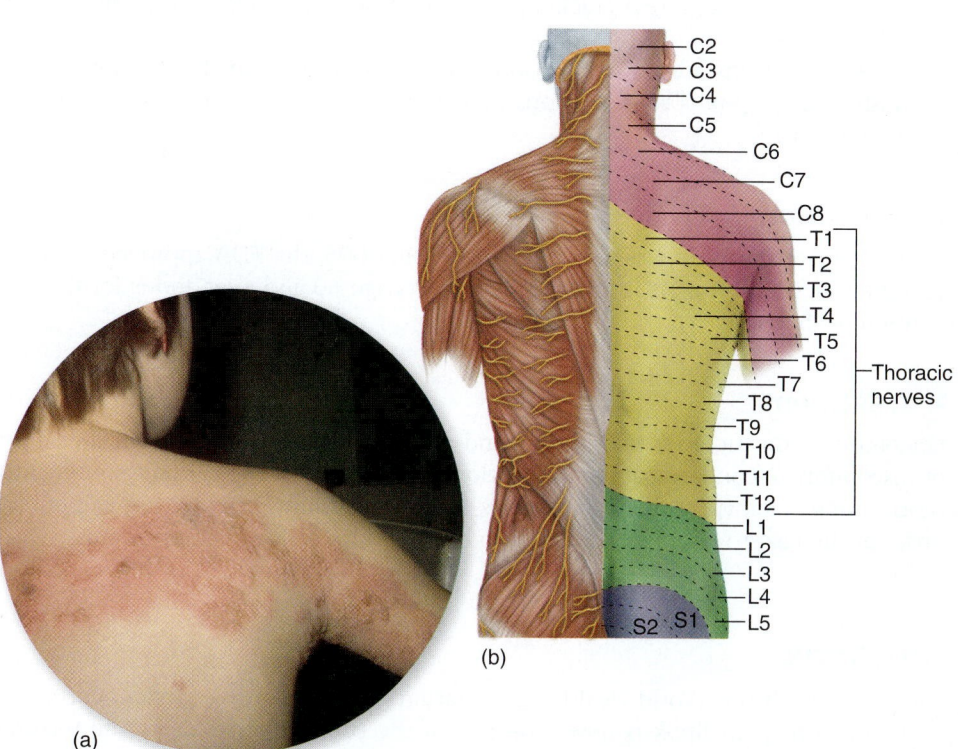

Figure 16.10 **Varicella-zoster virus reemergence as shingles.** (a) Clinical appearance of shingles lesions. (b) Dermatomes are areas of the skin served by a single ganglion. Shingles generally affects a single dermatome for this reason.

a developing malignancy. The virus is believed to migrate down the ganglion to the skin, where multiplication resumes and produces crops of tender, persistent vesicles. Inflammation of the ganglia and the pathways of nerves can cause pain and tenderness, known as postherpetic neuralgia, that can last for several months. Involvement of cranial nerves can lead to eye inflammation and ocular and facial paralysis.

▶ Causative Agent

Human herpesvirus 3 (HHV-3, also called **varicella** [var″-ih′sel′-ah]) causes chicken-pox, as well as the condition called herpes zoster or shingles. The virus is sometimes referred to as the varicella-zoster virus (VZV). Like other herpesviruses, it is an enveloped DNA virus.

▶ Pathogenesis and Virulence Factors

HHV-3 enters the respiratory tract, attaches to respiratory mucosa, and then invades and enters the bloodstream. The viremia disseminates the virus to the skin, where the virus causes adjacent cells to fuse and eventually lyse, resulting in the characteristic lesions. The virus enters the sensory nerves at this site, traveling to the dorsal root ganglia.

The ability of HHV-3 to remain latent in ganglia is an important virulence factor, because resting in this site protects it from attack by the immune system and provides a reservoir of virus for the reactivation condition of shingles.

NCLEX® PREP

4. The following statements are true regarding shingles, *except*
   a. *the condition is caused by a reactivation of the varicella virus that may remain latent in the ganglia of nerve fibers for years.*
   b. *you can "come down" with shingles by being exposed to the fluid of shingles lesions.*
   c. *psychological stress can reactivate the virus, which causes shingles.*
   d. *shingles may cause eye inflammation and facial paralysis.*

▶ Transmission and Epidemiology

Humans are the only natural hosts for HHV-3. The virus is harbored in the respiratory tract but is communicable from both respiratory droplets and the fluid of active skin lesions. People can acquire a chickenpox infection by being exposed to the fluid of shingles lesions.

Infected persons are most infectious a day or two prior to the development of the rash. Chickenpox is so contagious that if you are exposed to it you almost certainly will get it.

▶ Prevention

Live attenuated vaccine was licensed in 1995. In 2006, the FDA approved a unique vaccine called Zostavax. It is intended for adults age 60 and over and is for the prevention of shingles.

▶ Treatment

Uncomplicated varicella is self-limiting and requires no therapy aside from alleviation of discomfort. Secondary bacterial infection is treated with topical or systemic antibiotics. Oral cidofovir or related antivirals should be administered within 24 hours of onset of the rash to people considered to be at risk for serious complications.

Smallpox

Largely through the World Health Organization's comprehensive global efforts, naturally occurring smallpox is now a disease of the past. However, after the terrorist attacks on the United States on September 11, 2001, and the anthrax bioterrorism shortly thereafter, the U.S. government began taking the threat of smallpox bioterrorism very seriously. Vaccination, which had been discontinued, was once again offered to certain U.S. populations.

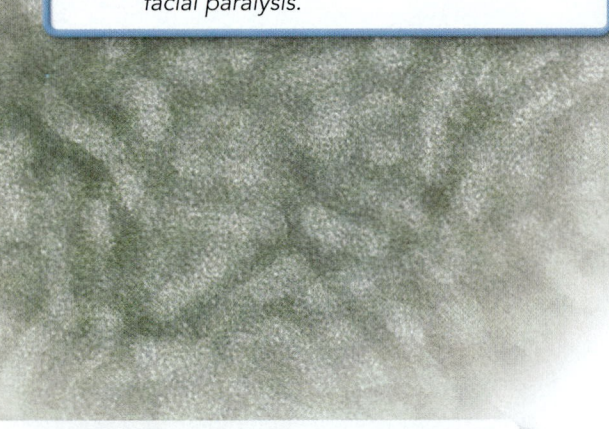

Vaccinia virus, the cause of cowpox, is used as a vaccine against smallpox, caused by the variola virus.

▶ Signs and Symptoms

Infection begins with fever and malaise, and later a rash begins in the pharynx, spreads to the face, and progresses to the extremities. Initially, the rash is *macular*, evolving in turn to *papular, vesicular,* and *pustular* before eventually crusting over, leaving nonpigmented sites pitted with scar tissue. There are two principal forms of smallpox: variola minor and variola major. Variola major is a highly virulent form that causes toxemia, shock, and intravascular coagulation. People who have survived any form of smallpox nearly always develop lifelong immunity.

It is vitally important for health care workers to be able to recognize the early signs of smallpox. The diagnosis of even a single suspected case must be treated as a health and law enforcement emergency. The symptoms of variola major progress as follows: After the prodrome period of high fever and malaise, a rash emerges, first in the mouth. Severe abdominal and back pain sometimes accompany this phase of the disease. A rash appears on the skin and spreads throughout the body within 24 hours. The rash typically occurs more on the extremities (a centrifugal distribution; figure 16.9).

By the third or fourth day of the rash, the bumps become larger and fill with a thick opaque fluid. A major distinguishing feature of this disease is that the pustules are indented in the middle. Also, patients report that the lesions feel as if they contain a BB pellet. Within a few days, these pustules begin to scab over. After 2 weeks, most of the lesions will have crusted over; the patient remains contagious until the last scabs fall off because the crusts contain the virus. During the entire rash phase, the patient is very ill. The lesions occur at the dermal level, which is the reason that scars remain after the lesions are healed.

A patient with variola minor has a rash that is less dense and is generally less ill than someone with variola major.

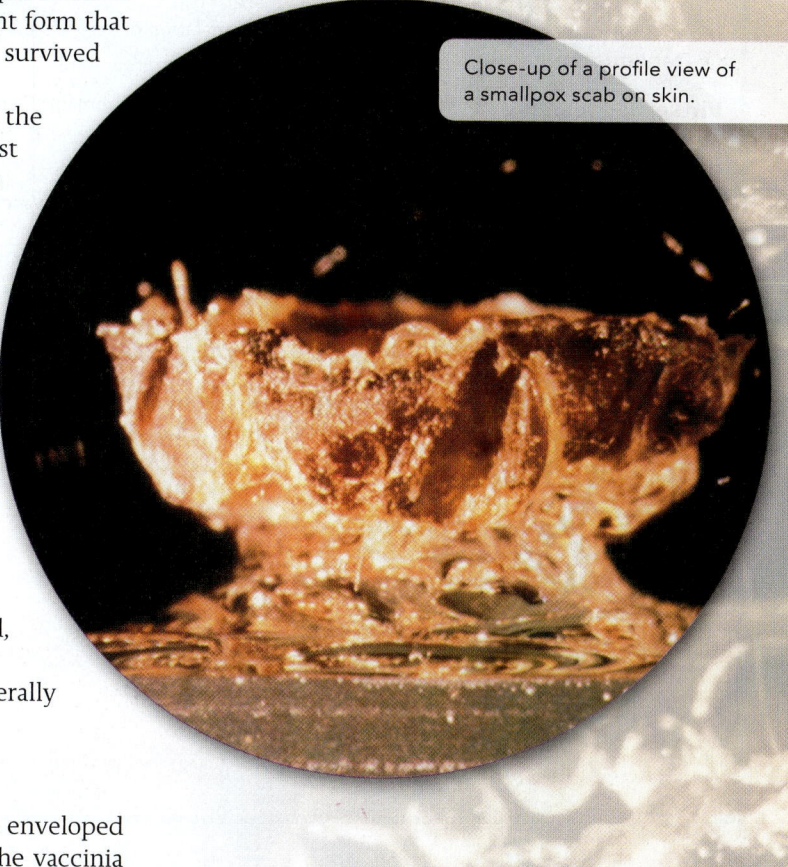

Close-up of a profile view of a smallpox scab on skin.

▶ Causative Agent

The causative agent of smallpox, the variola virus, is an orthopoxvirus, an enveloped DNA virus. Other members of this group are the monkeypox virus and the vaccinia virus from which smallpox vaccine is made. Variola is a hardy virus, surviving outside the host longer than most viruses.

▶ Transmission and Epidemiology

Smallpox is spread primarily through droplets, although fomites such as contaminated bedding and clothing can also spread it.

In the early 1970s, smallpox was endemic in 31 countries. Every year, 10 to 15 million people contracted the disease, and approximately 2 million people died from it. By 1977, after 11 years of intensive effort by the world health community, the last natural case occurred in Somalia.

▶ Prevention

In chapter 13, you read about Edward Jenner and his development of vaccinia virus to inoculate against smallpox. To this day, the vaccination for smallpox is based on the vaccinia virus. Immunizations were stopped in the United States in 1972. Since the terrorist events of 2001, most military branches are requiring that their personnel take the vaccination before deploying to certain parts of the world. In 2007, a new vaccine was approved by the Food and Drug Administration; it is called ACAM 2000.

Vaccination is also useful for postexposure prophylaxis, meaning that it can prevent or lessen the effects of the disease after you have already been infected with it.

▶ Treatment

There is no treatment for smallpox **(Disease Table 16.6)**.

Disease Table 16.6 Vesicular/Pustular Rash Diseases

Disease	Chickenpox	Smallpox
Causative Organism(s)	Human herpesvirus 3 (varicella-zoster virus)	Variola virus
Most Common Modes of Transmission	Droplet contact, inhalation of aerosolized lesion fluid	Droplet contact, indirect contact
Virulence Factors	Ability to fuse cells, ability to remain latent in ganglia	Ability to dampen, avoid immune response
Culture/Diagnosis	Based largely on clinical appearance	Based largely on clinical appearance
Prevention	Live attenuated vaccine; there is also vaccine to prevent reactivation of latent virus (shingles)	Live virus vaccine (vaccinia virus)
Treatment	None in uncomplicated cases; acyclovir for high risk	Cidofovir, immune globulin
Distinguishing Features	No fever prodrome; lesions are superficial; in centripetal distribution (more in center of body)	Fever precedes rash, lesions are deep and in centrifugal distribution (more on extremities)
Epidemiological Features	Chickenpox: vaccine decreased hospital visits by 88%, ambulatory visits by 59%; shingles: 1 million cases annually	Last natural case worldwide was in 1977
Appearance of Lesions		

Large Pustular Skin Lesions

Leishmaniasis

Two infections that result in large lesions (greater than a few millimeters across) deserve mention in this chapter on skin infections. The first is leishmaniasis, a zoonosis transmitted among various mammalian hosts by female sand flies. This infection can express itself in several different forms, depending on which species of the protozoan *Leishmania* is involved. Cutaneous leishmaniasis is a localized infection of the capillaries of the skin caused by *L. tropica,* found in Mediterranean, African, and Indian regions. A form of mucocutaneous leishmaniasis called espundia is caused by *L. brasiliensis,* endemic to parts of Central and South America. It affects both the skin and mucous membranes. Another form of this infection is systemic leishmaniasis.

Leishmania is transmitted to the mammalian host by the sand fly when it ingests the host's blood. The disease is endemic to equatorial regions that provide favorable conditions for the sand fly. At particular risk are travelers or immigrants who have never had contact with the protozoan and lack specific immunity.

There is no vaccine; avoiding the sand fly is the only prevention.

Cutaneous Anthrax

This form of anthrax is the most common and least dangerous version of infection with *Bacillus anthracis.* (The spectrum of anthrax disease is discussed fully in chapter 18.) It is caused by endospores entering the skin through small cuts or

Disease Table 16.7 Large Pustular Skin Lesions

Disease	Leishmaniasis	Cutaneous Anthrax
Causative Organism(s)	*Leishmania* spp.	*Bacillus anthracis*
Most Common Modes of Transmission	Biological vector	Direct contact with endospores
Virulence Factors	Multiplication within macrophages	Endospore formation; capsule, lethal factor, edema factor
Culture/Diagnosis	Culture of protozoa, microscopic visualization	Culture on blood agar; serology, PCR performed by CDC
Prevention	Avoiding sand fly	Avoid contact; vaccine available but not widely used
Treatment	Sodium stibogluconate, pentamidine	Ciprofloxacin, doxycycline, levofloxacin
Distinguishing Features	Mucocutaneous and systemic forms	Can be fatal
Epidemiological Features	Untreated visceral leishmaniasis mortality rate is 100%; 10% for cutaneous leishmaniasis	Untreated cutaneous anthrax mortality rate: 20%; treated mortality rate less than 1%
Appearance of Lesions		

abrasions. Germination and growth of the pathogen in the skin are marked by the production of a papule that becomes increasingly necrotic and later ruptures to form a painless, black **eschar** (ess′-kar) **(see Disease Table 16.7).** In the fall of 2001, 11 cases of cutaneous anthrax occurred in the United States as a result of bioterrorism (along with 11 cases of inhalational anthrax). Mail workers and others contracted the infection when endospores were sent through the mail. The infection can be naturally transmitted by contact with hides of infected animals (especially goats).

Left untreated, even the cutaneous form of anthrax is fatal approximately 20% of the time. A vaccine exists but is recommended only for high-risk persons and the military (Disease Table 16.7).

Cutaneous and Superficial Mycoses
Ringworm

A group of fungi that is collectively termed **dermatophytes** causes a variety of body surface conditions. These mycoses are strictly confined to the nonliving epidermal tissues (stratum corneum) and their derivatives (hair and nails). All these conditions have different names that begin with the word **tinea** (tin′-ee-ah), which derives from the erroneous belief that they were caused by worms. That misconception is also the reason these diseases are often called *ringworm*–ringworm of the scalp (tinea capitis), beard (tinea barbae), body (tinea corporis), groin (tinea cruris), foot (tinea pedis), and hand (tinea manuum). (Don't confuse these "tinea" terms with genus and species names. It is simply an old practice for naming the conditions.) Most of these conditions are caused by one of three different dermatophytes, which are discussed here.

The signs and symptoms of ringworm conditions are summarized in **table 16.1.**

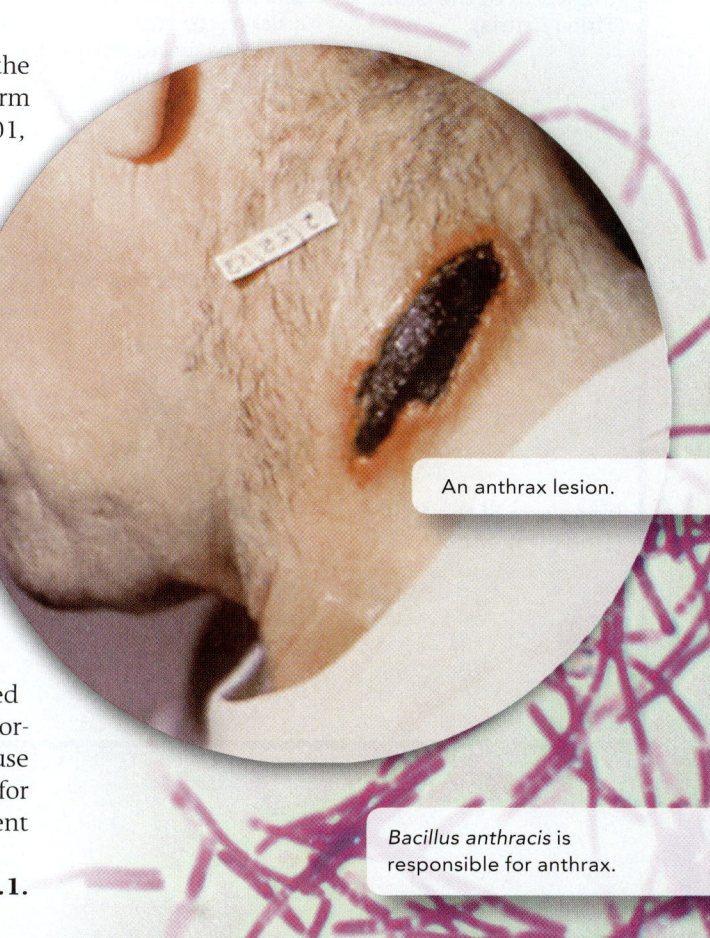

An anthrax lesion.

Bacillus anthracis is responsible for anthrax.

Table 16.1 Signs and Symptoms of Cutaneous Mycoses

Ringworm of the Scalp (Tinea Capitis)	This mycosis results from the fungal invasion of the scalp and the hair of the head, eyebrows, and eyelashes.
Ringworm of the Beard (Tinea Barbae)	This tinea, also called *barber's itch*, affects the chin and beard of adult males. Although once a common aftereffect of unhygienic barbering, it is now contracted mainly from animals.
Ringworm of the Body (Tinea Corporis)	This extremely prevalent infection of humans can appear nearly anywhere on the body's glabrous (smooth and bare) skin.
Ringworm of the Groin (Tinea Cruris)	Sometimes known as *jock itch*, crural ringworm occurs mainly in males on the groin, perianal skin, scrotum, and, occasionally, the penis. The fungus thrives under conditions of moisture and humidity created by sweating.
Ringworm of the Foot (Tinea Pedis)	Tinea pedis has more colorful names as well, including athlete's foot and jungle rot. Infections begin with blisters between the toes that burst, crust over, and can spread to the rest of the foot and nails.
Ringworm of the Nail (Tinea Unguium)	Fingernails and toenails, being masses of keratin, are often sites for persistent fungus colonization. The first symptoms are usually superficial white patches in the nail bed. A more invasive form causes thickening, distortion, and darkening of the nail.

► Causative Agents

There are about 39 species in the genera *Trichophyton, Microsporum,* and *Epidermophyton* that can cause the preceding conditions **(figure 16.11).** The causative agent of a given type of ringworm varies from one geographic location to another and is not restricted to a particular genus and species.

Diagnosis of tinea of the scalp caused by some species is aided by use of a long-wave ultraviolet lamp that causes infected hairs to fluoresce. Samples of hair, skin scrapings, and nail debris treated with heated potassium hydroxide (KOH) show a thin, branching fungal mycelium if infection is present.

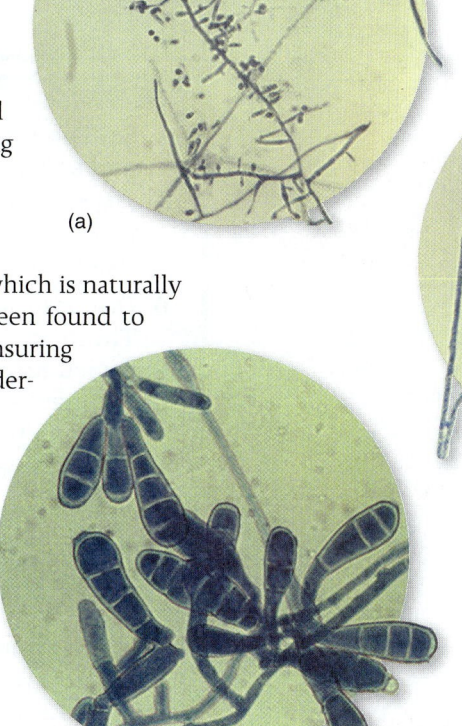

► Pathogenesis and Virulence Factors

The dermatophytes have the ability to invade and digest keratin, which is naturally abundant in the cells of the stratum corneum. They have also been found to suppress the ability of the immune system to respond to them, ensuring their long-term persistence. The fungi do not invade deeper epidermal layers.

► Transmission and Epidemiology

Transmission of the fungi that cause these diseases is via direct and indirect contact with other humans or with infected animals. Some of these fungi can be acquired from the soil.

Therapy is usually a topical antifungal agent. Ointments containing tolnaftate, miconazole, itraconazole, terbinafine, or thiabendazole are applied regularly for several weeks. Some drugs work by speeding up loss of the outer skin layer.

Figure 16.11 Examples of dermatophyte spores. (a) Regular, numerous microconidia of *Trichophyton*. (b) Macroconidia of *Microsporum canis,* a cause of ringworm in cats, dogs, and humans. (c) Smooth-surfaced macroconidia in clusters characteristic of *Epidermophyton*.

Superficial Mycoses

Agents of **superficial mycoses** involve the outer epidermal surface and are ordinarily innocuous infections with cosmetic rather than inflammatory effects. Tinea versicolor is caused by the yeast genus *Malassezia*, a genus that has at least 10 species living on human skin. The yeast feeds on the high oil content of the skin glands. Even though these yeasts are very common (carried by nearly 100% of

Disease Table 16.8 Cutaneous and Superficial Mycoses

Disease	Cutaneous Infections	Superficial Infections (Tinea Versicolor)
Causative Organism(s)	*Trichophyton, Microsporum, Epidermophyton*	*Malassezia* species
Most Common Modes of Transmission	Direct and indirect contact, vehicle (soil)	Endogenous "normal biota"
Virulence Factors	Ability to degrade keratin, invoke hypersensitivity	–
Culture/Diagnosis	Microscopic examination, KOH staining, culture	Usually clinical, KOH can be used
Prevention	Avoid contact	None
Treatment	Topical tolnaftate, itraconazole, terbinafine, miconazole, thiabendazole, oral terbinafine	Topical antifungals
Epidemiological Features	Among schoolchildren, 0%–19% prevalence, in humid climates up to 30%	Highest incidence among adolescents

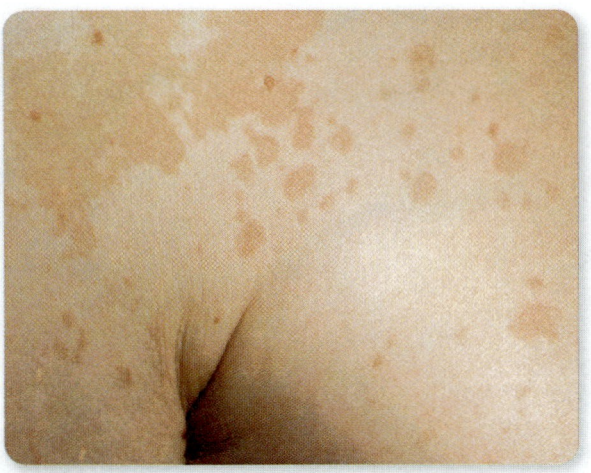

Figure 16.12 Tinea versicolor. Mottled, discolored skin pigmentation is characteristic of superficial skin infection by *Malassezia furfur*.

humans tested), in some people its growth elicits mild, chronic scaling and interferes with production of pigment by melanocytes. The trunk, face, and limbs may take on a mottled appearance **(figure 16.12)**. Other superficial skin conditions in which *Malassezia* is implicated are folliculitis, psoriasis, and seborrheic dermatitis (dandruff) **(Disease Table 16.8)**.

> ### 16.3 Learning Outcomes—Assess Your Progress
>
> 4. Explain the important features of the "Highlight Disease," MRSA skin and soft-tissue infection.
> 5. List the possible causative agents, modes of transmission, virulence factors, diagnostic techniques, and prevention/treatment for the "Highlight Disease," maculopapular rash diseases.
> 6. Discuss important features of the other infectious skin diseases. These are impetigo, cellulitis, staphylococcal scalded skin syndrome, vesicular/pustular rash diseases, large pustular skin lesions, and cutaneous and superficial mycoses.
> 7. Discuss the relative dangers of rubella and rubeola viruses in different populations.

16.4 The Surface of the Eye and Its Defenses

The eye is a complex organ with many different tissue types, but for the purposes of this chapter we consider only its exposed surfaces, the *conjunctiva* and the *cornea* **(figure 16.13)**. The **conjunctiva** is a very thin membranelike tissue that covers the eye (except for the cornea) and lines the eyelids. It secretes an oil- and mucus-containing fluid that lubricates and protects the eye surface. The **cornea** is the dome-shaped central portion of the eye lying over the iris (the colored part of the eye). It has five to six layers of epithelial cells that can regenerate quickly if they are superficially damaged. It has been called "the windshield of the eye."

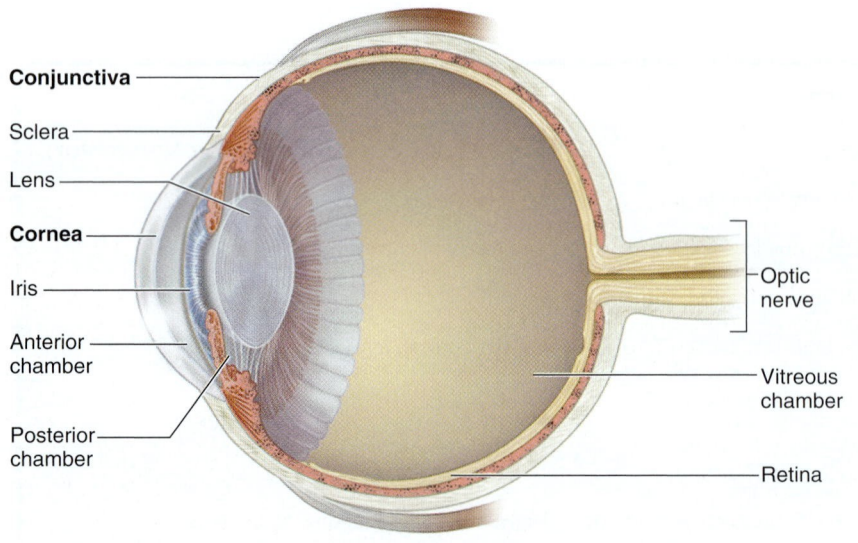

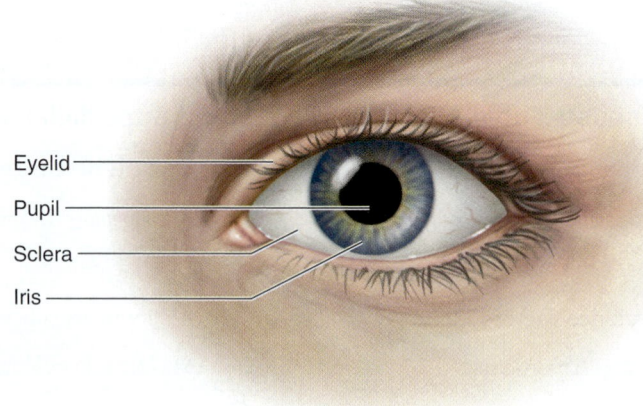

Figure 16.13 The anatomy of the eye.

The eye's best defense is the film of tears, which consists of an aqueous fluid, oil, and mucus. The tears are formed in the lacrimal gland at the outer and upper corner of each eye **(figure 16.14),** and they drain into the lacrimal duct at the inner corner. Tears contain sugars, lysozyme, and lactoferrin. These last two substances have antimicrobial properties. The mucous layer contains proteins and sugars and plays a protective role. And, of course, the flow of the tear film prevents the attachment of microorganisms to the eye surface.

Because the eye's primary function is vision, anything that hinders vision would be counterproductive. For that reason, inflammation does not occur in the eye as readily as it does elsewhere in the body. Flooding the eye with fluid containing a large number of light-diffracting objects, such as lymphocytes and phagocytes, in response to every irritant would mean almost constantly blurred vision. So even though the eyes are relatively vulnerable to infection (not being covered by keratinized epithelium), the evolution of the vertebrate eye has been toward reduced innate immunity and a corresponding reduction in inflammatory response. This characteristic is sometimes known as **immune privilege.**

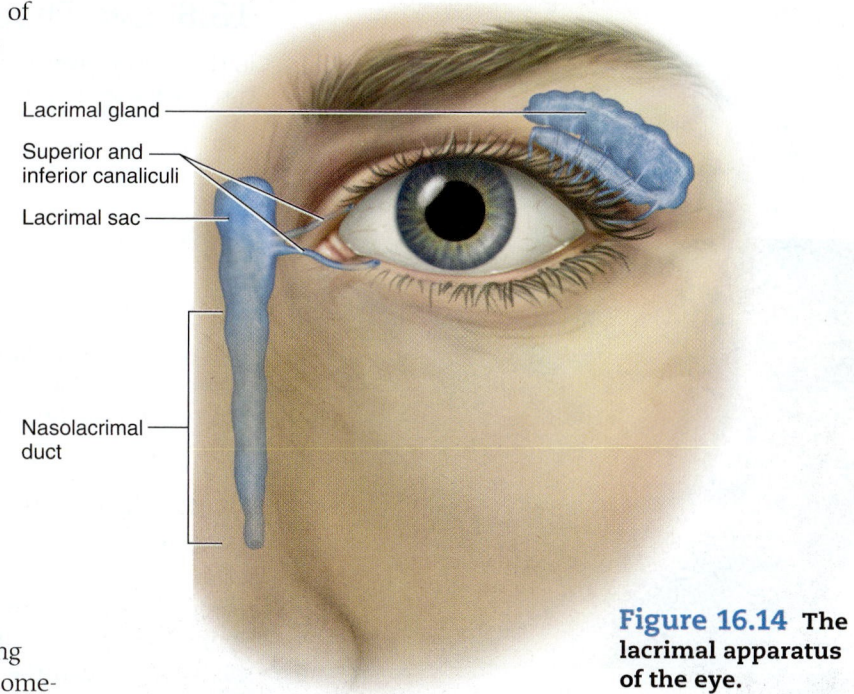

Lacrimal gland

Superior and inferior canaliculi

Lacrimal sac

Nasolacrimal duct

Figure 16.14 **The lacrimal apparatus of the eye.**

16.5 Normal Biota of the Eye

Like the surface of dry skin, the normal biota of the eye is relatively sparse. For instance, up to 20% of samples show no culturable bacteria, and the few bacteria that are found resemble the normal biota of the skin–namely, staphylococci, streptococci, *Corynebacterium*, and some yeasts. *Neisseria* species can also live on the surface of the eye. However, cultivation-independent techniques, such as those used by the Human Microbiome Project and the Ocular Microbiome Project, are revealing that the cornea alone is home to dozens of types of bacteria and viruses. Further, these bacteria and viruses appear to differ quite a bit from those on other parts of the body.

Defenses and Normal Biota of the Eyes		
	Defenses	**Normal Biota**
Eyes	Mucus in conjunctiva and in tears, lysozyme and lactoferrin in tears	Sparsely populated with *Staphylococcus aureus*, *Staphylococcus epidermidis*, and *Corynebacterium* species

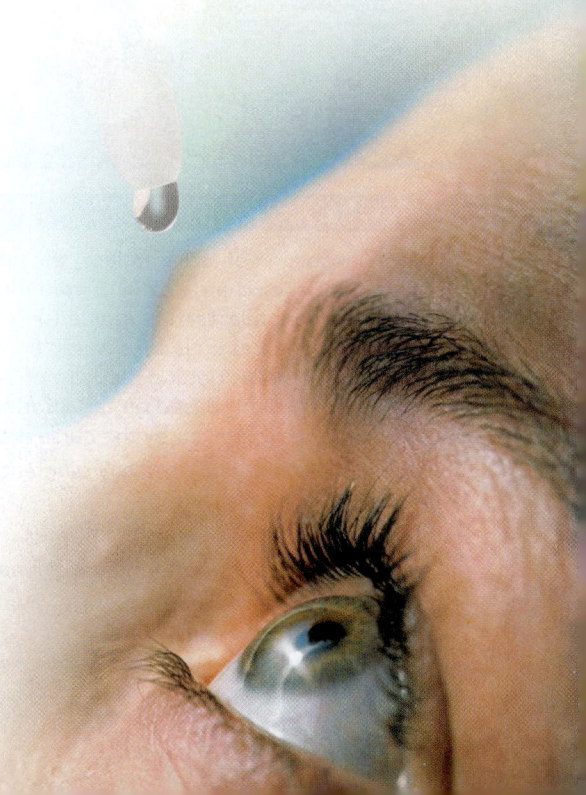

16.6 Eye Diseases Caused by Microorganisms

In this section, we cover the infectious agents that cause diseases of the surface structures of the eye—namely, the cornea and conjunctiva.

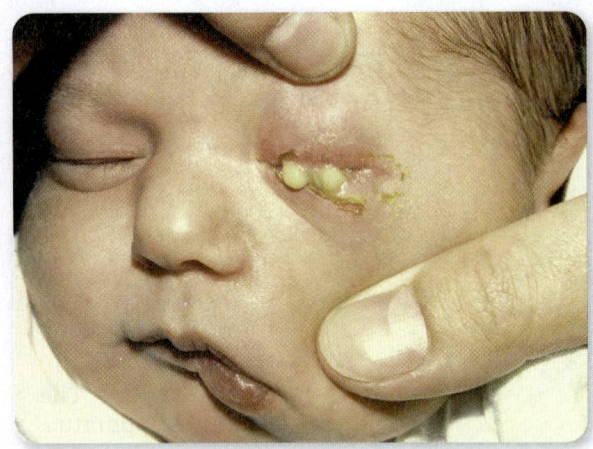

Figure 16.15 Neonatal conjunctivitis.

Highlight Disease

Conjunctivitis

Infection of the conjunctiva is relatively common. It can be caused by specific microorganisms that have a predilection for eye tissues, by contaminants that proliferate due to the presence of a contact lens or an eye injury, or by accidental inoculation of the eye by a traumatic event.

▶ Signs and Symptoms

Just as there are many different causes of conjunctivitis, there are many different clinical presentations. Most bacterial infections produce a milky discharge, whereas viral infections tend to produce a clear watery exudate. It is typical for a patient to wake up in the morning with an eye "glued" shut by secretions that have accumulated and solidified through the night. Some conjunctivitis cases are caused by an allergic response, and these often produce copious amounts of clear fluid as well. The informal name for common conjunctivitis is pinkeye.

▶ Causative Agents and Their Transmission

Cases of neonatal eye infection with *Neisseria gonorrhoeae* or *Chlamydia trachomatis* are usually transmitted vertically from a genital tract infection in the mother (discussed in chapter 21). Either one of these eye infections can lead to serious eye damage if not treated promptly **(figure 16.15)**. Note that herpes simplex can also cause neonatal conjunctivitis, but it is often accompanied by generalized herpes infection (covered in chapter 21).

Bacterial conjunctivitis in other age groups is most commonly caused by *Staphylococcus epidermidis*, *Streptococcus pyogenes*, or *Streptococcus pneumoniae*, although *Haemophilus influenzae* and *Moraxella* species are also frequent causes. *N. gonorrhoeae* and *C. trachomatis* can also cause conjunctivitis in adults. These infections may result from autoinoculation from a genital infection or from sexual activity, although *N. gonorrhoeae* can be part of the normal biota in the respiratory tract. A wide variety of bacteria, fungi, and protozoa can contaminate contact lenses and lens cases and then be transferred to the eye, resulting in disease that may be very serious. Viral conjunctivitis is commonly caused by adenoviruses, although other viruses may be responsible. Both bacterial and viral conjunctivitis are transmissible by direct and even indirect contact and are usually highly contagious.

▶ Prevention and Treatment

Newborn children in the United States are administered antimicrobials in their eyes after delivery to prevent neonatal conjunctivitis from either *N. gonorrhoeae* or *C. trachomatis*. Treatment of those infections, if they are suspected, is started before lab results are available and usually is accomplished with erythromycin, both topical and oral. If *N. gonorrhoeae* is confirmed, oral therapy is usually switched to ceftriaxone. If antibacterial therapy is prescribed for other conjunctivitis cases, it should cover all possible bacterial pathogens. Ciprofloxacin eyedrops are a common choice. Erythromycin or gentamicin are also often used. Because conjunctivitis is usually diagnosed based on clinical signs, a physician may prescribe prophylactic antibiotics even if a viral cause is suspected. If symptoms don't begin improving within 48 hours, more extensive diagnosis may be performed. **Disease Table 16.9** lists the most common causes of conjunctivitis; keep in mind that other microorganisms can also cause conjunctival infections.

NCLEX® PREP

5. Upon inspection of a client's face, the nurse notes a milky discharge coming from the right eye. The client relates that upon arising this morning, he felt as if his eyes were "tightly shut." The nurse is getting ready to physically examine the client. Based on this information, the priority action taken by the nurse is to

a. *ask the client if he used any different products on his face.*

b. *flush out the eye with sterile water.*

c. *obtain gloves to perform an examination.*

d. *note whether the client has allergies to food or medication.*

Disease Table 16.9 Conjunctivitis

Disease	Neonatal Conjunctivitis	Bacterial Conjunctivitis	Viral Conjunctivitis
Causative Organism(s)	*Chlamydia trachomatis* or *Neisseria gonorrhoeae*	*Streptococcus pneumoniae, Staphylococcus epidermidis, Staphylococcus aureus, Haemophilus influenzae, Moraxella,* and also *Neisseria gonorrhoeae, Chlamydia trachomatis*	Adenoviruses and others
Most Common Modes of Transmission	Vertical	Direct, indirect contact	Direct, indirect contact
Virulence Factors	–	–	–
Culture/Diagnosis	Gram stain and culture	Clinical diagnosis	Clinical diagnosis
Prevention	Screen mothers, apply antibiotic or silver nitrate to newborn eyes	Hygiene	Hygiene
Treatment	Topical and oral antibiotics; (antibiotic-resistant *N. gonorrhoeae* is in **Urgent Threat** category in CDC Antibiotic Resistance Report)	Gatifloxacin or levofloxacin ophthalmic solution	None, although antibiotics often given because type of infection not distinguished
Distinguishing Features	In babies <28 days old	Mucopurulent discharge	Serous (clear) discharge
Epidemiological Features	Less than 0.5% in developed world; higher incidence in developing world	More common in children	More common in adults

Keratitis

Keratitis is a more serious eye infection than conjunctivitis. Invasion of deeper eye tissues occurs and can lead to complete corneal destruction. Any microorganism can cause this condition, especially after trauma to the eye, but this section focuses on one of the more common causes: herpes simplex virus. It can cause keratitis in the absence of predisposing trauma.

The usual cause of herpetic keratitis is a "misdirected" reactivation of (oral) herpes simplex virus type 1 (HSV-1). The virus, upon reactivation, travels into the ophthalmic rather than the mandibular branch of the trigeminal nerve. Infections with HSV-2 can also occur as a result of a sexual encounter with the virus or transfer of the virus from the genital to eye area. Blindness due to herpes is the leading infectious cause of blindness in the United States. Bacterial and fungal causes of keratitis are more common in developing countries.

The viral condition is treated with trifluridine or acyclovir or both.

Disease Table 16.10 Keratitis

	Herpes simplex virus	Miscellaneous microorganisms
Causative Organism(s)	Herpes simplex virus	Miscellaneous microorganisms
Most Common Modes of Transmission	Reactivation of latent virus, although primary infections can occur in the eye	Often traumatic introduction (parenteral)
Virulence Factors	Latency	Various
Culture/Diagnosis	Usually clinical diagnosis; viral culture or PCR if needed	Various
Prevention	–	–
Treatment	Topical trifluridine +/– oral acyclovir	Specific antimicrobials
Epidemiological Features	One-third worldwide population infected; in United States, annual incidence of 500,000	

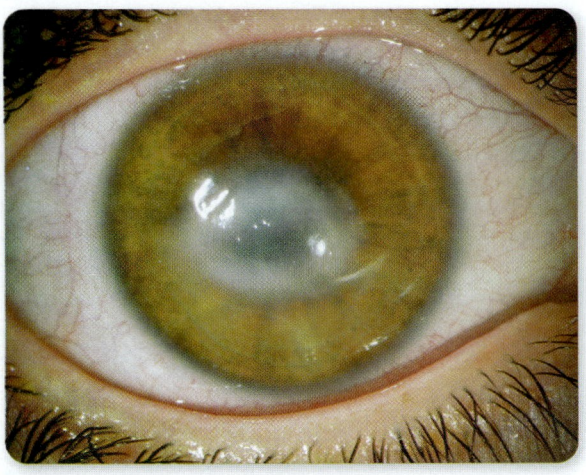

Figure 16.16 *Acanthamoeba* **infection of the eye.**

In the last few years, another form of keratitis has been increasing in incidence. An amoeba called *Acanthamoeba* has been causing serious keratitis cases, especially in people who wear contact lenses. This free-living amoeba is everywhere—it lives in tap water, freshwater lakes, and the like. The infections are usually associated with less-than-rigorous contact lens hygiene, or previous trauma to the eye **(figure 16.16) (Disease Table 16.10).**

16.6 LEARNING OUTCOMES—Assess Your Progress

11. List the possible causative agents, modes of transmission, virulence factors, diagnostic techniques, and prevention/treatment for the "Highlight Disease," conjunctivitis.

12. Discuss important features of keratitis caused by either HSV or by *Acanthamoeba*.

CASE FILE WRAP-UP

Dale, the patient in the chapter opening case file, was admitted to the hospital after his X ray confirmed a mild case of pneumonia. He was given IV antibiotics to treat his pneumonia and fluids to counter dehydration, as well as antipyretics to treat his fever and make him more comfortable. He was discharged home after 4 days and recovered completely.

The physician contacted the state health department, as must be done legally. State health department investigators contacted all of Dale's contacts to ensure they had not fallen ill. A few contacts were found who had never been vaccinated because their parents shared Dale's mother's erroneous view of vaccines. These children were quarantined at home to prevent the potential spread of the disease. Thankfully, all of the patients in the hospital waiting room were found to be immune to the disease and no one else became ill. Dale was assumed to have contracted measles while on a family vacation abroad.

▶ Summing Up

Taxonomic Organization Microorganisms Causing Diseases of the Skin and Eyes

Microorganism	Pronunciation	Location of Disease Table
Gram-positive bacteria		
Staphylococcus aureus	staf″-uh-lo-kok′-us are′-ee-us	MRSA skin and soft-tissue infection, p. 441
		Impetigo, p. 447
		Cellulitis, p. 447
		Scalded skin syndrome, p. 448
Streptococcus pyogenes	strep″-tuh-kok′-us pie′-ah″-gen-eez	Impetigo, p. 447
		Cellulitis, p. 447
		Scarlet fever, p. 455
Bacillus anthracis	buh-sill′-us an′-thray″-sus	Cutaneous anthrax, p. 453
Gram-negative bacteria		
Neisseria gonorrhoeae	nye-seer″-ee-uh′ gon′-uh-ree″-uh	Neonatal conjunctivitis, p. 459
Chlamydia trachomatis	kluh-mi″-dee-uh′ truh-koh′-muh-tis	Neonatal conjunctivitis, p. 459
DNA viruses		
Human herpesvirus 3	hew′-mun hur″-peez-vie′-russ	Chickenpox, shingles, p. 452
Variola virus	vayr′-ee-oh″-luh vie′-russ	Smallpox, p. 452
Parvovirus B19	par″-voh-vie′-russ	Fifth disease, p. 445
Human herpesvirus 6 and 7	hew′-mun hur″-peez-vie′-russ	Roseola, p. 445
Herpes simplex virus	hur″-peez sim′-plex vie′-russ	Keratitis, p. 459
RNA viruses		
Measles virus	mee′-zulls vie′-russ	Measles, p. 445
Rubella virus	roo′-bell″-uh vie′-russ	Rubella, p. 445
Fungi		
Trichophyton	try″-ko-fie′-tahn	Ringworm, p. 454
Microsporum	my″-krow′-spoor′-um	Ringworm, p. 454
Epidermophyton	ep′-uh-dur″-moh-fie′-tahn	Ringworm, p. 454
Malassezia spp.	mal′-uh-see″-zee-uh	Superficial mycoses, p. 455
Protozoa		
Leishmania spp.	leesh-mayn″-ee-uh	Leishmaniasis, p. 453
Acanthamoeba	ay-kanth″-uh-mee′-buh	Keratitis, p. 459

Erythema Multiforme

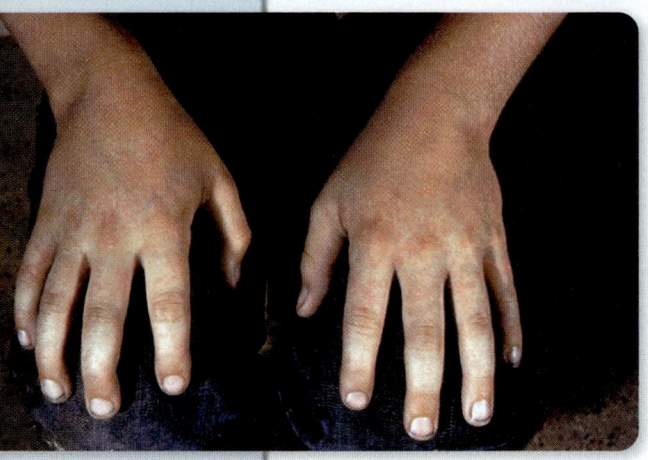

Erythema multiforme (EM) is a hypersensitivity reaction that can be triggered by an infection or by a reaction to a drug. EM becomes a skin rash that can also involve the mucous membranes. Although the rash can be widespread on the body, it often resolves without complications. Anyone can get EM, but young adults are most frequently affected, with males affected slightly more often.

Herpes simplex virus is the most common trigger of EM, most often genital herpes or herpes labialis, which causes common cold sores on the lip. The herpes infection usually occurs before EM develops. Another common trigger is *Mycoplasma pneumoniae,* which causes mild pneumonia. HIV, the varicella zoster virus, adenoviruses, and cytomegalovirus are occasional triggers of the disease.

Chills, fever, painful joints, and weakness may be experienced in the prodrome, but most individuals will have no warning of EM. Within a day, several to hundreds of skin lesions begin to develop, usually on the backs of the hands and the top of the feet to start, then advancing up the arms and legs to the trunk. The neck and face are often affected as well. Skin lesions may coalesce on the knees and elbows.

At first, the lesions are red/pink, flat, and round. They then become raised and spread until they are up to several centimeters in diameter. Thereafter, the center of the lesion darkens in color and may crust over or blister. Within 3 days, the lesions are usually fully evolved. Lesions may be found at various stages of development, hence the term *multiforme.* The lips of affected individuals may become swollen.

Mucous membrane involvement may be extensive in erythema multiforme major, in which one or more mucous membranes are affected, most often the oral mucosa (lips, tongue, inside of the cheeks, gums, or palate). The eyes, anus and genital areas, gastrointestinal tract, trachea, and bronchioles are less commonly affected. In erythema multiforme minor, mucous membranes are not involved or are only minimally affected (i.e., redness or minor ulceration of the lips).

Lesions of the mucosa involve redness, swelling, and blister formation. When the blisters rupture, they leave behind large shallow and painful ulcers that are covered with a whitish-colored membrane. Swallowing and speaking may be difficult due to pain. Episodes of EM may recur regularly several times a year.

Treatment of EM is supportive, as the condition often clears up on its own, although it often takes several weeks to do so. Patients who have difficulty eating or drinking due to pain may require treatment in a hospital. Recurrent EM may be treated with continuous acyclovir therapy (an antiviral agent) for 6 months.

Infectious Diseases Affecting
The Skin and Eyes

Keratitis

Herpes simplex virus
Acanthamoeba

Large Pustular Skin Lesions

Leishmania species
Bacillus anthracis

Staphylococcal Scalded Skin Syndrome

Staphylococcus aureus

Maculopapular Rash Diseases

Measles virus
Rubella virus
Parvovirus B19
Human herpesvirus 6 or 7

Impetigo

Staphylococcus aureus
Streptococcus pyogenes

Conjunctivitis

Neisseria gonorrhoeae
Chlamydia trachomatis
Various bacteria
Various viruses

Vesicular or Pustular Rash Disease

Human herpesvirus 3 (varicella)
Variola virus

Cellulitis

Staphylococcus aureus
Streptococcus pyogenes

Cutaneous and Superficial Mycoses

Trichophyton
Microsporum
Epidermophyton
Malassezia

MRSA Skin and Soft-Tissue Infection

Staphylococcus aureus

	Helminths
	Bacteria
	Viruses
	Protozoa
	Fungi

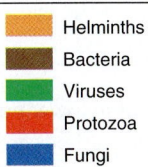

System Summary Figure 16.17

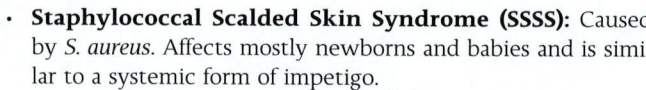

Chapter Summary

16.1 The Skin and Its Defenses

- The epidermal cells contain the protein keratin, which "waterproofs" the skin and protects it from microbial invasion.
- Other defenses include antimicrobial peptides, low pH sebum, high salt and lysozyme in sweat, and antimicrobial peptides.

16.2 Normal Biota of the Skin

- The Human Microbiome Project and other programs like it are revealing that the normal skin microbiota are far more diverse than we had previously understood from cultivation methods.
- The normal biota includes hundreds of species of microorganisms, differs from region to region of the body, differs between people, and yet can be constant over time on the same person.

16.3 Skin Diseases Caused by Microorganisms

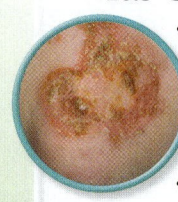

- **MRSA Skin and Soft-Tissue Infection**
  - Serious skin lesions caused by methicillin-resistant *S. aureus* are on the rise in the community environment.
  - MRSA isolates are commonly found on healthy people and on community surfaces.
- **Maculopapular Rash Diseases**
  - **Measles:** Measles or *rubeola* results in characteristic red maculopapular exanthem that erupts on the head and then progresses to the trunk and extremities until most of the body is covered. The MMR vaccine (measles, mumps, and rubella) contains attenuated measles virus.
  - **Rubella:** Also known as German measles, can appear in two forms: postnatal and congenital (prenatal) infection of the fetus. The MMR vaccination contains protection from rubella.
  - **Fifth Disease:** Also called *erythema infectiosum*, fifth disease is a very mild but highly contagious disease that often results in characteristic "slapped-cheek" appearance. Causative agent is parvovirus B19.
  - **Roseola:** Can result in a maculopapular rash; is caused by a human herpesvirus called HHV-6 and sometimes by HHV-7.
- **Impetigo:** A highly contagious superficial infection that can cause skin to peel or flake off. Causative organisms can be either *Staphylococcus aureus* or *Streptococcus pyogenes* or both.
- **Cellulitis:** Results from a fast-spreading infection of the dermis and subcutaneous tissue below. Most commonly caused by *S. aureus* or *S. pyogenes*.

- **Staphylococcal Scalded Skin Syndrome (SSSS):** Caused by *S. aureus*. Affects mostly newborns and babies and is similar to a systemic form of impetigo.
- **Vesicular or Pustular Rash Diseases**
  - **Chickenpox:** Skin lesions progress quickly from macules and papules to itchy vesicles filled with clear fluid.
    - **Shingles:** Chickenpox virus becomes latent in the ganglia and may reemerge to cause shingles. Human herpesvirus 3 causes chickenpox, as well as herpes zoster or shingles.
  - **Smallpox:** Naturally occurring smallpox has been eradicated from the world. Causative agent is the variola virus, an enveloped DNA virus.
- **Large Pustular Skin Lesions**
  - **Leishmaniasis:** A zoonosis transmitted by the female sand fly when it ingests the host's blood.
  - **Cutaneous Anthrax:** Most common and least dangerous version of infections with *Bacillus anthracis*.
- **Cutaneous and Superficial Mycoses**
  - **Ringworm:** A group of fungi collectively termed dermatophytes cause mycoses in the nonliving epidermal tissues, hair, and nails. Often called "ringworm." Species in the genera *Trichophyton*, *Microsporum*, and *Epidermophyton* are the cause.
  - **Superficial Mycoses:** Agents of superficial mycoses, such as *Malassezia* species, involve only the outer epidermis.

16.4 The Surface of the Eye and Its Defenses

- The flushing action of the tears, which contain lysozyme and lactoferrin, is the major protective feature of the eye.

16.5 Normal Biota of the Eye

- The eye has similar microbes as the skin but in lower numbers.
- As with the skin, cultivation-independent methods are showing that the eye is home to many more species than we previous thought.

16.6 Eye Diseases Caused by Microorganisms

- **Conjunctivitis:** Infection of the conjunctiva (commonly called pinkeye) can be caused by either bacteria or viruses. Both bacterial and viral conjunctivitis are highly contagious.
- **Keratitis:** A more serious eye infection than conjunctivitis. Herpes simplex viruses (HSV-1 and HSV-2) and *Acanthamoeba* cause two different forms of the disease.

Multiple-Choice Questions   Bloom's Levels 1 and 2: Remember and Understand

Select the correct answer from the answers provided.

1. Cultivation-independent methods of identifying microbes have recently found approximately how many different species of microbes on the skin?

   a. 0
   b. 1-9
   c. 10-90
   d. 100-900

2. *Staphylococcus aureus* is part of the differential diagnosis of which diseases?

   a. impetigo
   b. maculopapular rash
   c. MRSA infection
   d. all of the above
   e. two of the above

3. Which of the following is probably the most important defense factor for skin?

   a. phagocytes
   b. sebum
   c. dryness
   d. antimicrobial peptides

4. Name the organism(s) most commonly associated with cellulitis.

   a. *Staphylococcus aureus*
   b. *Propionibacterium acnes*
   c. *Streptococcus pyogenes*
   d. both a and b
   e. both a and c

5. Herpesviruses can cause all of the following diseases, *except*

   a. chickenpox.
   b. shingles.
   c. keratitis.
   d. smallpox.
   e. roseola.

6. Which disease is incorrectly matched with the causative agent?

   a. viral conjunctivitis/adenovirus
   b. shingles/varicella virus
   c. smallpox/variola virus
   d. measles/*Staphylococcus aureus*

7. Dermatophytes are fungi that infect the epidermal tissue by invading and attacking

   a. collagen.
   b. keratin.
   c. fibroblasts.
   d. sebaceous glands.

8. Poor contact lens hygiene is likely to get you a case of

   a. herpetic keratitis.
   b. *Wolbachia* infection.
   c. *Acanthamoeba* keratitis.
   d. ophthalmic gonorrhea.

Critical Thinking   Bloom's Levels 3, 4, and 5: Apply, Analyze, and Evaluate

Critical thinking is the ability to reason and solve problems using facts and concepts. These questions can be approached from a number of angles and, in most cases, they do not have a single correct answer.

1. With which pair would you see the biggest differences in the microbiota: your cheek *versus* your palm, your cheek *versus* your friend's cheek, your cheek today *versus* your cheek 3 months from now? With which pair would you be likely to see the least difference? Explain and justify your answers.

2. Who is more likely to experience shingles–your 10-year-old niece, or your 60-year-old aunt? Explain and justify your answer using your understanding of how we contract shingles.

3. Why would antibiotics in the penicillin family be ineffective in treating fungal infections?

4. Smallpox has been widely reported as a possible bioterror weapon. Given what you know about the etiology of the disease and the current state of the world's immunity to smallpox, discuss how effective (or ineffective) a smallpox weapon might be. What kind of defense could be mounted against such an attack?

5. Despite the availability of the measles vaccine, outbreaks of measles still occur. Discuss some of the reasons for these occurrences.

Visual Connections   Bloom's Level 5: Evaluate

This question connects previous images to a new concept.

1. **From chapter 11, figure 11.3.** How does this figure help explain impetigo caused by *Staphylococcus aureus* or *Streptococcus pyogenes*?

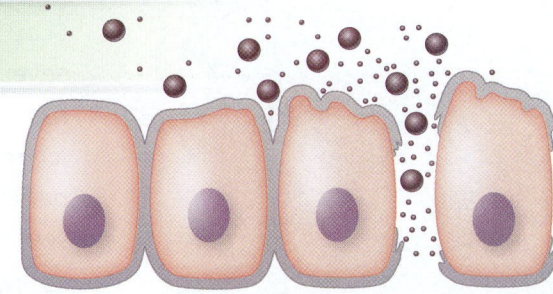

www.mcgrawhillconnect.com

Enhance your study of this chapter with study tools and practice tests. Also ask your instructor about the resources available through Connect, including the media-rich eBook, interactive learning tools, and animations.

Something New

I was working in a rural New York state hospital in the summer of 1999. Toward the end of that summer, we had seen our first cases of West Nile Virus (WNV). The patients who came in to see us in the hospital with WNV mostly complained of headache and fatigue. Some patients had a rash. For the most part, the disease was mild and self-limiting, and the affected patients required reassurance more than anything. The disease was new to us all, so the staff was learning about the disease and its transmission as quickly as we could so that we could care for patients who had contracted the disease and educate others on how to protect themselves from getting it.

Scott was 58 years of age and was taking corticosteroids to treat rheumatoid arthritis. He had no other health issues. He came in to the emergency room complaining of flulike symptoms. He had a fever and sore throat and complained of a severe headache with photophobia. He also complained of overwhelming fatigue and weakness. Scott's weakness was extreme. He had weakness in his lower extremities and found walking more than a couple of feet exhausting.

Because Scott's symptoms were so severe, he was admitted to the hospital for observation and supportive care. Blood was drawn and sent to the state laboratory for testing to be sure that WNV was the cause of his symptoms. A lumbar puncture was performed to rule out other causes of Scott's symptoms, such as bacterial or viral meningitis, but the results of the lumbar puncture were normal. His white blood cell count was low, which was indicative of a viral infection. It was determined that Scott was suffering from West Nile encephalitis.

Over the course of a week, Scott's weakness progressed to the point where his legs were essentially paralyzed. Physiotherapy was started and Scott gradually improved, although he remained weak and unable to work for several months after falling ill. He suffered no long-term residual effects, but his experience served as a reminder that, although most cases of WNV are asymptomatic or only mildly symptomatic, some patients, particularly older patients and patients who are immunocompromised, may experience a more severe and protracted version of the illness.

- How is West Nile virus spread?
- What factor may have put Scott at risk of contracting West Nile encephalitis?

Case File Wrap-Up appears on page 492.

Infectious Diseases Affecting the Nervous System

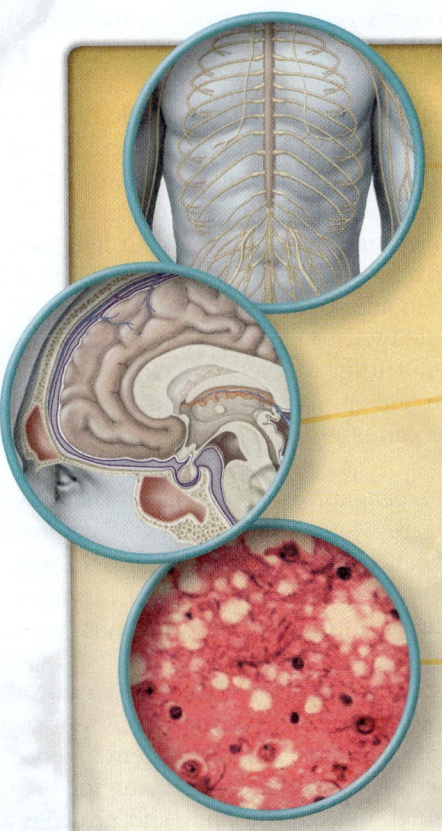

IN THIS CHAPTER...

17.1 The Nervous System and Its Defenses

1. Describe the important anatomical features of the nervous system.
2. List the natural defenses present in the nervous system.

17.2 Normal Biota of the Nervous System

3. Discuss the current state of knowledge of the normal microbiota of the nervous system.

17.3 Nervous System Diseases Caused by Microorganisms

4. List the possible causative agents, modes of transmission, virulence factors, diagnostic techniques, and prevention/treatment for the "Highlight Diseases," meningitis and poliomyelitis.
5. Identify the most common and also the most deadly of the multiple possible causes of meningitis.
6. Explain the difference between the oral polio vaccine and the inactivated polio vaccine, and under which circumstances each is appropriate.
7. Discuss important features of the diseases most directly involving the brain. These are meningoencephalitis, encephalitis, and subacute encephalitis.
8. Identify which encephalitis-causing viruses you should be aware of in your geographic area.
9. Discuss important features of the other diseases in the nervous system. These are rabies, poliomyelitis, tetanus, and botulism.

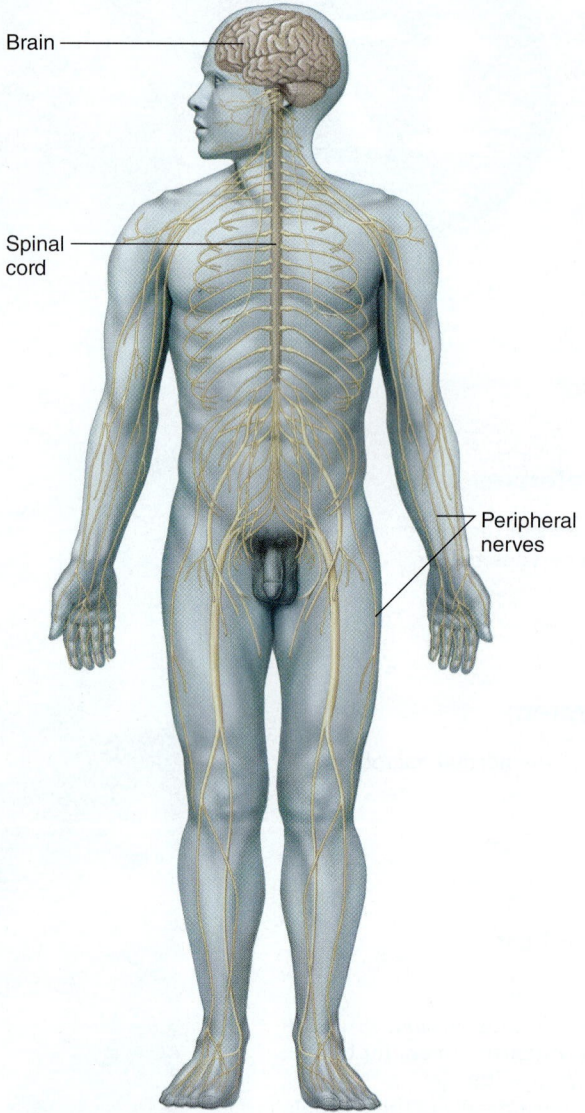

Brain

Spinal cord

Peripheral nerves

Figure 17.1 The two component parts of the nervous system. The central nervous system and the peripheral nerves.

17.1 The Nervous System and Its Defenses

The nervous system has two component parts: the central nervous system (CNS), consisting of the brain and spinal cord, and the peripheral nervous system (PNS), which contains the nerves that emanate from the brain and spinal cord to sense organs and to the periphery of the body **(figure 17.1).** The nervous system performs three important functions–sensory, integrative, and motor. The sensory function is fulfilled by sensory receptors at the ends of peripheral nerves. They generate nerve impulses that are transmitted to the central nervous system. There, the impulses are translated, or integrated, into sensation or thought, which in turn drives the motor function. The motor function necessarily involves structures outside of the nervous system, such as muscles and glands.

The brain and the spinal cord are dense structures made up of cells called **neurons.** They are both surrounded by bone. The brain is situated inside the skull, and the spinal cord lies within the spinal column **(figure 17.2),** which is composed of a stack of interconnected bones called vertebrae. The soft tissue of the brain and spinal cord is encased within a tough casing of three membranes called the **meninges.** The layers of these three membranes, from outermost to innermost position, are the dura mater, the arachnoid mater, and the pia mater. Between the arachnoid mater and pia mater is the subarachnoid space (i.e., the space under the arachnoid mater). The subarachnoid space is filled with a clear serumlike fluid called cerebrospinal fluid (CSF). The CSF provides nutrition to the CNS, while also providing a liquid cushion for the sensitive brain and spinal cord. The meninges are a common site of infection, and microorganisms can often be found in the CSF when meningeal infection **(meningitis)** occurs.

The PNS consists of nerves and ganglia. A ganglion is a swelling in the nerve where the cell bodies of the neurons aggregate. Nerves are bundles of neuronal axons that receive and transmit nerve signals. The axons and dendrites of adjacent neurons communicate with each other over a very small space, called a synapse. Chemicals called neurotransmitters are released from one cell and act on the next cell in the synapse.

The defenses of the nervous system are mainly structural. The bony casings of the brain and spinal cord protect them from traumatic injury. The cushion of surrounding CSF also serves a protective function. The entire nervous system is served by the vascular system, but the interface between the blood vessels serving the brain and the brain itself is different from that of other areas of the body and provides a third structural protection. The cells that make up the walls of the blood vessels allow very few molecules to pass through. In other parts of the body, there is freer passage of ions, sugars, and other metabolites through the walls of blood vessels. The restricted permeability of blood vessels in the brain is called the **blood-brain barrier,** and it prohibits most microorganisms from passing into the central nervous system. The drawback of this phenomenon is that drugs and antibiotics are difficult to introduce into the CNS also.

The CNS is considered an "immunologically privileged" site. These sites are able to mount only a partial, or at least a different, immune response when exposed to immunologic challenge. The functions of the CNS are so vital for the life of an organism that even temporary damage that could potentially result from "normal" immune responses would be very detrimental. The uterus and parts of the eye are other immunologically privileged sites. Specialized cells in the central nervous system perform defensive functions. Microglia are a type of cell having phagocytic capabilities, and brain macrophages also exist in the CNS, although the activity of both of these types of cells is thought to be less than that of phagocytic cells elsewhere in the body.

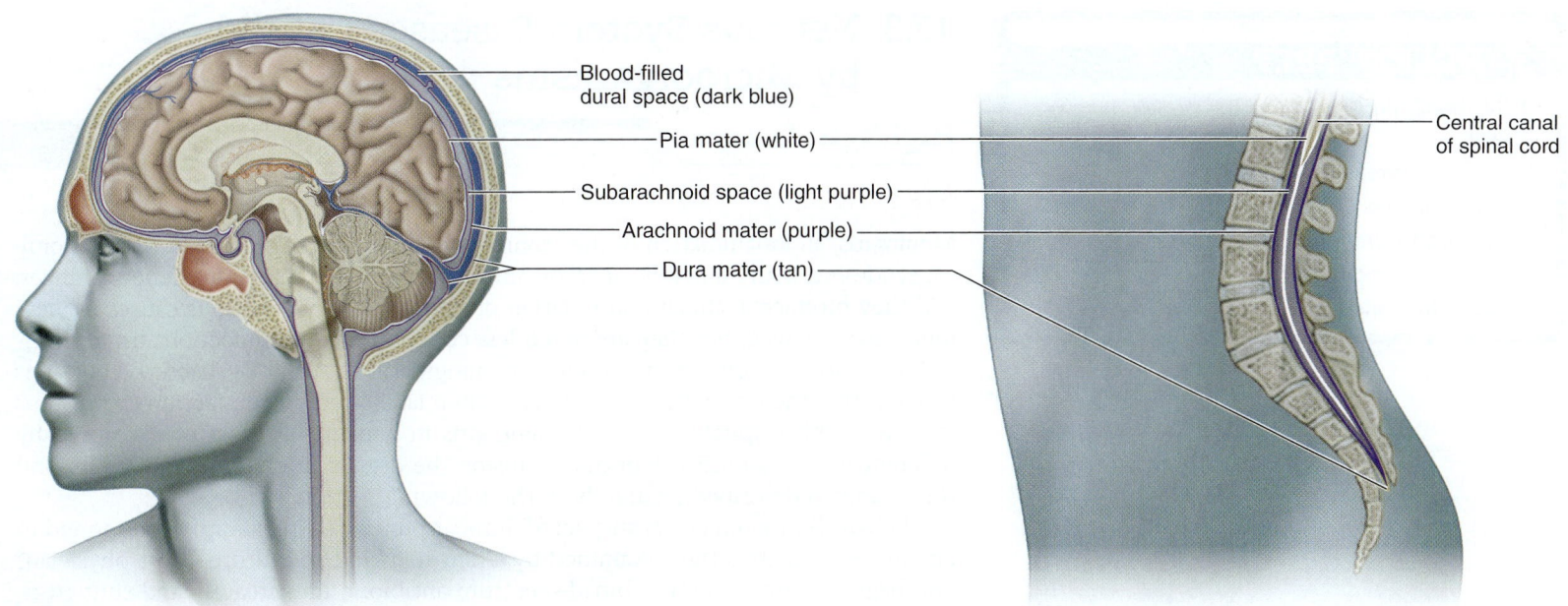

Blood-filled
dural space (dark blue)

Pia mater (white)

Subarachnoid space (light purple)

Arachnoid mater (purple)

Dura mater (tan)

Central canal
of spinal cord

Figure 17.2 **Detailed anatomy of the brain and spinal cord.**

17.1 LEARNING OUTCOMES—Assess Your Progress

1. Describe the important anatomical features of the nervous system.
2. List the natural defenses present in the nervous system.

17.2 Normal Biota of the Nervous System

It is still believed that there is no normal biota in either the CNS or PNS, and that finding microorganisms of any type in these tissues represents a deviation from the healthy state. Viruses such as herpes simplex live in a dormant state in the nervous system between episodes of acute disease, but they are not considered normal biota. The Human Microbiome Project is not sampling this system at the present time.

Nervous System Defenses and Normal Biota

	Defenses	Normal Biota
Nervous System	Bony structures, blood-brain barrier, microglial cells, and macrophages	None

---

17.2 LEARNING OUTCOMES—Assess Your Progress

3. Discuss the current state of knowledge of the normal microbiota of the nervous system.

NCLEX® PREP

2. What is the significance of the blood-brain barrier with respect to medication administration?

a. *It allows for the absorption of all medication therapies administered to clients in the clinical setting.*

b. *Medication therapies must be able to penetrate the blood-brain barrier to be effective.*

c. *The blood-brain barrier is a nonpermeable membrane and as such prevents absorption of all medications.*

d. *Increased permeability is a characteristic property of the blood-brain barrier.*

17.3 Nervous System Diseases Caused by Microorganisms

Highlight Disease

Meningitis

Meningitis, an inflammation of the meninges, is an excellent example of an anatomical syndrome. Many different microorganisms can cause an infection of the meninges, and they produce a similar constellation of symptoms. Noninfectious causes of meningitis exist as well, but they are much less common than the infections listed here.

The more serious forms of acute meningitis are caused by bacteria, but it is thought that their entrance to the CNS is often facilitated by coinfection or previous infection with respiratory viruses. Meningitis in neonates is most often caused by different microorganisms than those causing the disease in children and adults, and therefore it is described separately in the following section.

Whenever meningitis is suspected, lumbar puncture (spinal tap) is performed to obtain CSF, which is then examined by Gram stain and/or culture. Most physicians will begin treatment with a broad-spectrum antibiotic immediately and shift treatment if necessary after a diagnosis has been confirmed.

▶ Signs and Symptoms

No matter the cause, meningitis results in these typical symptoms: severe headache, painful or stiff neck, fever, and nausea and vomiting. Early symptoms may be mistaken for flu symptoms. Photophobia (sensitivity to light) may also be noted. Skin rashes may be present in specific types of meningitis. There is usually an increased number of white blood cells in the CSF. Specific microorganisms may cause additional, and sometimes characteristic, symptoms, which are described in the individual sections that follow.

Like many other infectious diseases, meningitis can manifest as acute or chronic disease. Some microorganisms are more likely to cause acute meningitis, and others are more likely to cause chronic disease.

In a normal healthy person, it is very difficult for microorganisms to gain access to the nervous system. Those that are successful usually have specific virulence factors.

Neisseria meningitidis

Neisseria meningitidis appears as gram-negative diplococci (round cells occurring in joined pairs) and is commonly known as the meningococcus. It is often associated with epidemic forms of meningitis. This organism causes the most serious form of acute meningitis and accounts for 15% to 20% of all meningitis cases. Most cases occur in young children, because vaccination of otherwise healthy children against this disease is not recommended until age 11. Although 12 different strains with different capsular antigens exist, serotypes A, B, and C are responsible for most cases of infection.

▶ Pathogenesis and Virulence Factors

The bacterium enters the body via the upper respiratory tract, moves into the blood, rapidly penetrates the meninges, and produces symptoms of meningitis. The most serious complications of meningococcal infection are due to meningococcemia **(figure 17.3),** which can accompany meningitis but can also occur on its own. The pathogen releases endotoxin into the generalized circulation, which is a potent stimulus for certain white blood cells. Damage to the blood vessels caused by cytokines released by the white blood cells leads to vascular collapse, hemorrhage, and crops of lesions called **petechiae** (pee-tee′-kee-eye) on the trunk and appendages. A *petechia* (singular) is a small, 1 to 2 mm red or purple spot that may occur anywhere on the body **(figure 17.4).**

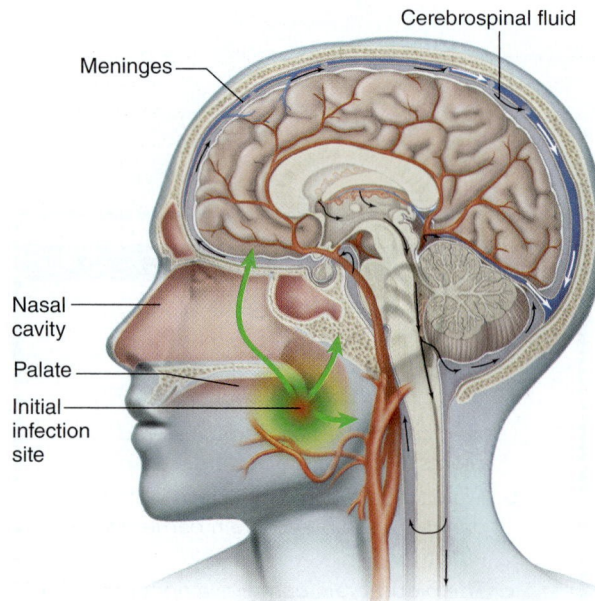

Cerebrospinal fluid

Meninges

Nasal cavity

Palate

Initial infection site

Figure 17.3 Dissemination of the meningococcus from a nasopharyngeal infection. Bacteria spread to the roof of the nasal cavity, which borders a highly vascular area at the base of the brain. From this location, they can enter the blood, causing meningococcemia, and escape into the cerebrospinal fluid, leading to infection of the meninges.

In a small number of cases, meningococcemia becomes an overwhelming disease with a high mortality rate.

The disease has a sudden onset, marked by fever higher than 40°C or 104°F, sore throat, chills, delirium, severe widespread areas of bleeding under the skin, shock, and coma. Generalized intravascular clotting, cardiac failure, damage to the adrenal glands, and death can occur within a few hours. The bacterium has an IgA protease and a capsule, both of which counter the body's defenses.

▶ Transmission and Epidemiology

Because meningococci do not survive long in the environment, these bacteria are usually acquired through close contact with secretions or droplets.

Meningococcal meningitis has a sporadic or epidemic incidence in late winter or early spring. The continuing reservoir of infection is humans who harbor the pathogen in the nasopharynx. The carriage state, which can last from a few days to several months, exists in 3% to 30% of the adult population and can exceed 50% in institutional settings. The scene is set for transmission when carriers live in close quarters with nonimmune individuals, as might be expected in families, day care facilities, college dormitories, and military barracks. The highest risk groups are young children (6 to 36 months old) and older children and young adults (10 to 20 years old).

▶ Culture and Diagnosis

Suspicion of bacterial meningitis constitutes a medical emergency, and differential diagnosis must be done with great haste and accuracy. It is most important to confirm (or rule out) meningococcal meningitis, because it can be rapidly fatal. Treatment is usually begun with this bacterium in mind until it can be ruled out. Cerebrospinal fluid, blood, or nasopharyngeal samples are stained and observed directly for the typical gram-negative diplococci. Cultivation may be necessary to differentiate the bacterium from other species. Specific rapid tests are also available for detecting the capsular polysaccharide or the cells directly from specimens without culturing.

It is usually necessary to differentiate this species from normal *Neisseria* that also live in the human body and can be present in infectious fluids. Immediately after collection, specimens are streaked on Modified Thayer-Martin medium (MTM) or chocolate agar and incubated in a high CO_2 atmosphere. Presumptive identification of the genus is obtained by a Gram stain and oxidase testing on isolated colonies **(figure 17.5).**

▶ Prevention and Treatment

The infection rate in most populations is about 1%, so well-developed natural immunity to the meningococcus appears to be the rule. A sort of natural immunization occurs during the early years of life as one is exposed to the meningococcus and its close relatives. Because even treated meningococcemial disease has a mortality rate of up to 15%, it is vital that chemotherapy begin as soon as possible with one or more drugs; it is generally given in high doses intravenously. Patients may also require treatment for shock and intravascular clotting.

When family members, medical personnel, or children in day care or school have come in close contact with infected people, preventive therapy with rifampin or tetracycline may be warranted. In the United States, immunization begins at the age of 11 followed by a booster dose. Vaccines are also available for younger children and for adults over the age of 55 who are at high risk for infection. The vaccine used in the United States contains two of the three most common serotypes of *N. meningitidis* found in the United States. In 2014, an outbreak at Princeton University was of the third type, serotype B; a vaccine containing that serotype, which is licensed in Europe but not in the United States, was used to protect the student population. Authorities took the unusual step of using the unlicensed vaccine due to the perceived danger of the outbreak.

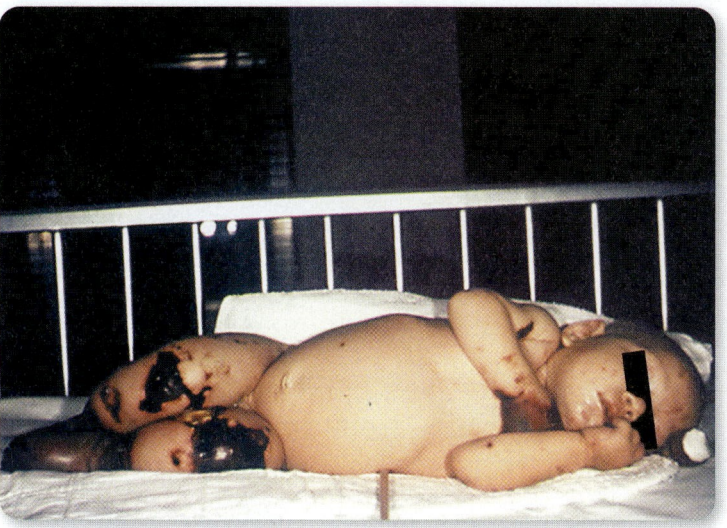

Figure 17.4 Vascular damage associated with meningococcal meningitis.

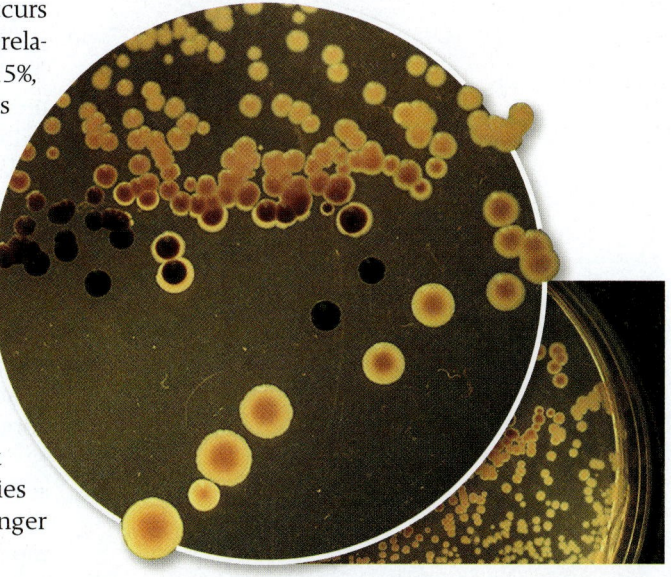

Figure 17.5 The oxidase test. A drop of oxidase reagent is placed on a suspected *Neisseria* or *Branhamella* colony. If the colony reacts with the chemical to produce a purple to black color, it is oxidase-positive; those that remain white to tan are oxidase-negative. Because several species of gram-negative rods are also oxidase-positive, this test is presumptive for these two genera only if a Gram stain has verified the presence of gram-negative cocci.

Disease Table 17.1 Meningitis

Causative Organism(s)	Neisseria meningitidis	Streptococcus pneumoniae	Haemophilus influenzae	
Most Common Modes of Transmission	Droplet contact	Droplet contact	Droplet contact	
Virulence Factors	Capsule, endotoxin, IgA protease	Capsule, induction of apoptosis, hemolysin and hydrogen peroxide production	Capsule	
Culture/Diagnosis	Gram stain/culture of CSF, blood, rapid antigenic tests, oxidase test	Gram stain/culture of CSF	Culture on chocolate agar	
Prevention	Conjugated vaccine; ciprofloxacin, rifampin, or ceftriaxone used to protect contacts	Two vaccines: PCV13 (children and adults), and PPSV23 (adults)	Hib vaccine, ciprofloxacin, rifampin, or ceftriaxone	
Treatment	Ceftriaxone, aztreonam, chloramphenicol	Vancomycin + ceftriaxone; in "Serious Threat" category in CDC Antibiotic Resistance Report	Ceftriaxone	
Distinctive Features	Petechiae, meningococcemia rapid decline	Serious, acute, most common meningitis in adults	Serious, acute, less common since vaccine became available	
Epidemiological Features	United States: 0.9–1.5 cases per 100,000 annually; meningitis belt: 1,000 cases per 100,000 annually	U.S. incidence before vaccine for children: 7.7 hospitalizations per 100,000. After vaccine for children: 2.6 per 100,000	Before vaccine, 300,000–400,000 deaths worldwide per year	

Streptococcus *pneumoniae*

You will see in chapter 19 that *Streptococcus pneumoniae*, also referred to as the **pneumococcus,** causes the majority of bacterial pneumonias. Meningitis is also caused by this bacterium; indeed, it is the most frequent cause of community-acquired meningitis and is also very severe. It does not cause the petechiae associated with meningococcal meningitis, and that difference is useful diagnostically. As many as 25% of pneumococcal meningitis patients will also have pneumococcal pneumonia. Pneumococcal meningitis is most likely to occur in patients with underlying susceptibility, such as alcoholic patients and patients with sickle-cell disease or those with absent or defective spleen function.

This bacterium is covered thoroughly in chapter 19, because it is a common cause of ear infections and pneumonia. It obviously has the potential to be highly pathogenic, while appearing as normal biota in many people. It can penetrate the respiratory mucosa; gain access to the bloodstream; and then, under certain conditions, enter the meninges.

Like the meningococcus, this bacterium has a polysaccharide capsule that protects it against phagocytosis. It also produces an alpha-hemolysin and hydrogen peroxide, both of which have been shown to induce damage in the CNS. It also appears capable of inducing brain cell apoptosis.

The bacterium is a small gram-positive flattened coccus that appears in end-to-end pairs. It has a distinctive appearance in a Gram stain of cerebrospinal fluid. Staining or culturing the nasopharynx is not useful because it is often normal biota there. Treatment requires a drug to which the bacterium is not resistant; penicillin is therefore not a good choice. Cefotaxime is often used, but drug susceptibilities must always be tested. It is recommended that a steroid be administered 20 minutes prior to antibiotic administration. This will dampen the inflammatory response to cell wall components that are released by antibiotic treatment of the gram-positive bacterium.

As mentioned in chapter 19, two vaccines are available for *S. pneumoniae:* a seven-valent conjugated vaccine (Prevnar), which is now recommended as part of the childhood immunization schedule, and a 23-valent polysaccharide vaccine (Pneumovax 23), which is available for adults.

Listeria monocytogenes	Cryptococcus neoformans	Coccidioides immitis	Viruses
Vehicle (food)	Vehicle (air, dust)	Vehicle (air, dust, soil)	Droplet contact
Intracellular growth	Capsule, melanin production	Granuloma (spherule) formation	Lytic infection of host cells
Cold enrichment, rapid methods	Negative staining, biochemical tests, DNA probes, cryptococcal antigen test	Identification of spherules, cultivation on Sabouraud's agar	Initially, absence of bacteria/fungi/protozoa, followed by viral culture or antigen tests
Cooking food, avoiding unpasteurized dairy products	–	Avoiding airborne endospores	–
Ampicillin, trimethoprim-sulfamethoxazole	Amphotericin B and fluconazole	Amphotericin B or oral or IV itraconazole	Usually none (unless specific virus identified and specific antiviral exists)
Asymptomatic in healthy adults; meningitis in neonates, elderly, and immunocompromised	Acute or chronic, most common in AIDS patients	Almost exclusively in endemic regions	Generally milder than bacterial or fungal
Mortality can be as much as 33%	Incidence before AIDS: >1 case per million per year; 66 cases per year in pre-HAART era; worldwide: 1 million new cases per year	Incidence in endemic areas: 200–300 annually	In United States, 4 of 5 meningitis cases caused by viruses: 26,000–42,000 hospitalizations/year

Haemophilus influenzae

The meningitis caused by this bacterium is severe. Before the vaccine was introduced in 1988, it was a very common cause of severe meningitis and death. In the course of the last 13 years, meningitis caused by this bacterium is much less common in the United States, a situation that can always change if a lower percentage of people get the vaccine and herd immunity is compromised.

Listeria monocytogenes

Listeria monocytogenes is a gram-positive bacterium that ranges in morphology from coccobacilli to long filaments in palisades formation **(figure 17.6).** Cells do not produce capsules or spores and have from one to four flagella. *Listeria* is not fastidious and is resistant to cold, heat, salt, pH extremes, and bile. It grows inside host cells and can move directly from an infected host cell to an adjacent healthy cell.

Listeriosis in healthy adults is often a mild or subclinical infection with nonspecific symptoms of fever, diarrhea, and sore throat. However, listeriosis in elderly or immunocompromised patients, fetuses, and neonates (described later) usually affects the brain and meninges and results in septicemia. (*Septicemia* is a term that means the multiplication of bacteria in the bloodstream.) The death rate is around 20%. Pregnant women are especially susceptible to infection, which can be transmitted to the infant prenatally when the microbe crosses the placenta or postnatally through the birth canal. Intrauterine infections usually result in premature abortion and fetal death.

Apparently, the primary reservoir is soil and water, and animals, plants, and food are secondary sources of infection. Most cases of listeriosis are associated with ingesting contaminated dairy products, poultry, and meat. Recent epidemics have spurred an in-depth investigation into the prevalence of *L. monocytogenes* in these sources. A 2003 U.S. government report concluded that consumers are exposed to low to moderate levels of *L. monocytogenes* on a regular basis. The pathogen has been isolated in 10% to 15% of ground beef and in 25% to 30% of chicken and turkey carcasses and is also present in 5% to 10% of luncheon meats, hot dogs, and cheeses. In 2011, contaminated cantaloupe from Colorado caused the third deadliest food-borne disease outbreak known in U.S.

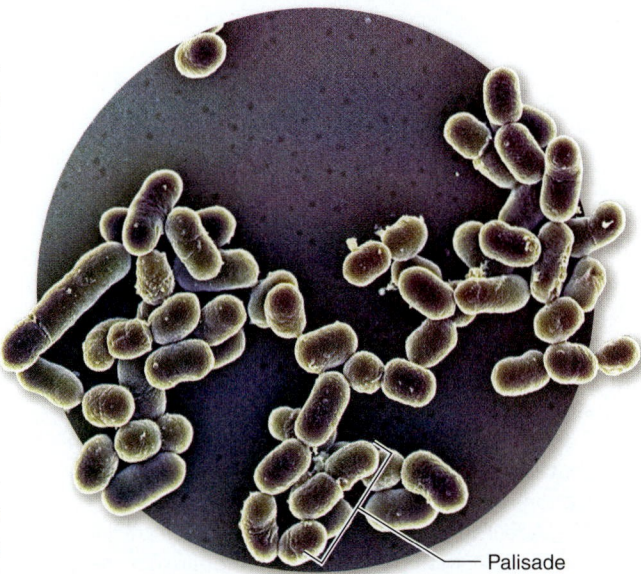

Palisade

Figure 17.6 **Listeria monocytogenes.** The bacterium is generally rod shaped. In Gram stains, individual cells tend to stack up in structures called palisades. That arrangement, pointed out here, is more obvious on a Gram stain where many more bacteria are seen.

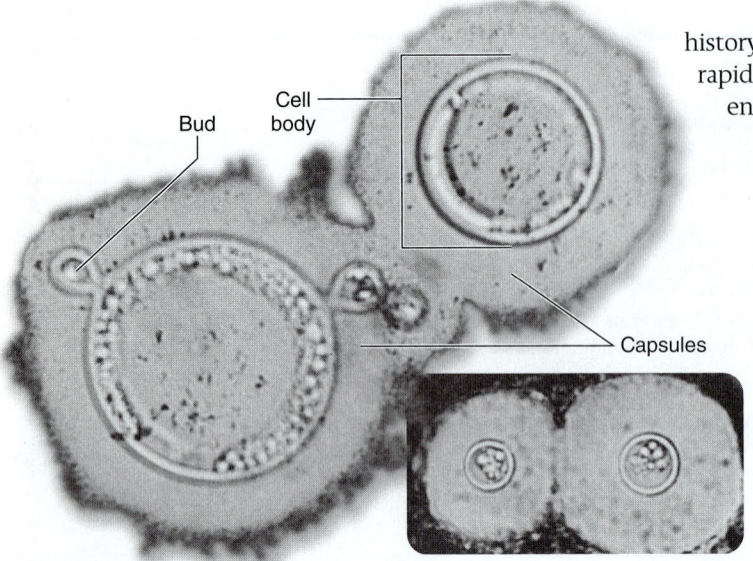

Bud

Cell body

Capsules

Figure 17.7 *Cryptococcus neoformans* **from infected spinal fluid stained negatively with India ink.**   Halos around the large spherical yeast cells are thick capsules. Also note the buds forming on one cell. Encapsulation is a useful diagnostic sign for cryptococcosis, although the capsule is fragile and may not show up in some preparations (150×).

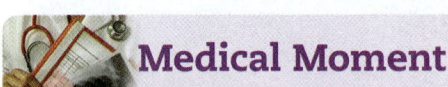

Medical Moment

Fungal Meningitis

In the summer and fall of 2012, almost 14,000 people in 20 different states were injected with tainted steroids from the New England Compounding Center. Although the facility has long since been shut down, patients who received the tainted injections are still suffering.

According to the CDC, which is still tracking the disease, there have been 233 confirmed cases of meningitis and 151 cases of meningitis combined with paraspinal or spinal infection. Thus far, 64 patients have died.

The fungus involved in these infections is *Exerohilum rostratum,* which is a common mold found in plants and in the soil. Like all molds, it thrives in humid and warm climates. It is not a common cause of infection in humans. Fungal meningitis is not spread from person to person. These infections can be very slow to develop, which has made tracking this unfortunate incident even more difficult for epidemiologists.

history, killing more than 40 people from listeriosis. It was also among the most rapid successful responses to food-borne illness. It spurred greater attention to and enforcement of food safety policies and regulations.

Diagnosing listeriosis is hampered by the difficulty in isolating it. The chances of isolation, however, can be improved by using a procedure called *cold enrichment,* in which the specimen is held at 4°C and periodically plated onto media, but this procedure can take 4 weeks. Rapid diagnostic kits using ELISA, immunofluorescence, and gene probe technology are now available for direct testing of dairy products and cultures. Antibiotic therapy should be started as soon as listeriosis is suspected. Ampicillin and trimethoprim-sulfamethoxazole are the first choices, followed by erythromycin. Prevention can be improved by adequate pasteurization temperatures and by proper washing, refrigeration, and cooking of foods that are suspected of being contaminated with animal manure or sewage. Pregnant women are cautioned by the U.S. Food and Drug Administration not to eat soft, unpasteurized cheeses.

Cryptococcus neoformans

The fungus *Cryptococcus neoformans* causes a more chronic form of meningitis with a more gradual onset of symptoms, although in AIDS patients the onset may be fast and the course of the disease more acute. It is sometimes classified as a meningoencephalitis (inflammation of both brain and meninges). Headache is the most common symptom, but nausea and neck stiffness are very common. This fungus is a widespread resident of human habitats. It has a spherical to ovoid shape, with small, constricted buds and a large capsule that is important in its pathogenesis **(figure 17.7).**

The primary ecological niche of *C. neoformans* is the bird population. It is prevalent in urban areas where pigeons congregate, and it proliferates in the high-nitrogen environment of droppings that accumulate on pigeon roosts. Masses of dried yeast cells are readily scattered into the air and dust. Its role as an opportunist is supported by evidence that healthy humans have strong resistance to it and that frank (obvious) infection occurs primarily in debilitated patients.

By far the highest rates of cryptococcal meningitis occur among patients with AIDS. This meningitis is frequently fatal. Other conditions that predispose individuals to infection are steroid treatment, diabetes, and cancer. It is not considered communicable among humans.

▶ Prevention and Treatment

Systemic cryptococcosis requires immediate treatment with amphotericin B and fluconazole over a period of weeks or months. There is no prevention.

Coccidioides species

This fungus causes a condition that is often called "Valley Fever" in the U.S. Southwest. The morphology of *Coccidioides* is very distinctive. At 25°C, it forms a moist white to brown colony with abundant, branching, septate hyphae. These hyphae fragment into thick-walled, blocklike **arthroconidia** (arthrospores) at maturity **(figure 17.8a).** On special media incubated at 37°C to 40°C, an arthrospore germinates into the parasitic phase, a small, spherical cell called a spherule **(figure 17.8b)** that can be found in infected tissues as well.

There are two species of *Coccidioides: C. immitis* is responsible for disease in California's San Joaquin valley, and *C. posadasil* is more widely distributed in the U.S. Southwest, Mexico, and South America.

▶ Pathogenesis and Virulence Factors

This is a true systemic fungal infection of high virulence, as opposed to an opportunistic infection. It usually begins with pulmonary infection but can disseminate quickly throughout the body. Coccidioidomycosis of the meninges is the most serious manifestation. All persons inhaling the arthrospores probably develop some degree of infection, but certain groups have a genetic susceptibility that gives rise to more serious disease. After the arthrospores are inhaled, they develop into spherules in the lungs. These spherules release scores of endospores into the lungs. (Unfortunately, *endospores* is the term used for this phase of the infection even though we have learned that endospores are *bacterial* structures.) At this point, the patient either experiences mild respiratory symptoms, which resolve themselves, or the endospores cause disseminated disease. Disseminated disease can include meningitis, osteomyelitis, and skin granulomas.

The highest incidence of coccidioidomycosis, estimated at 100,000 cases per year, occurs in the southwestern United States, although it also occurs in Mexico and parts of Central and South America. Especially concentrated reservoirs exist in the San Joaquin Valley of California and in southern Arizona. Outbreaks are usually associated with farming activity, archeological digs, construction, and mining. Climate change, which is drying out much of the southwestern and western United States, is predicted to increase the incidence and range of this disease in the western United States.

Viruses

It is estimated that four of five meningitis cases are caused by one of a wide variety of viruses. Because no bacteria or fungi are found in the CSF in viral meningitis, the condition is often called *aseptic meningitis*. Aseptic meningitis may occasionally have noninfectious causes.

The majority of cases of viral meningitis occur in children, and 90% are caused by enteroviruses. A common cause of viral meningitis is initial infection with HSV-2, concurrent with a genital infection. But many other viruses also gain access to the central nervous system on occasion.

Viral meningitis is generally milder than bacterial or fungal meningitis, and it is usually resolved within 2 weeks. The mortality rate is less than 1%. Diagnosis begins with the failure to find bacteria, fungi, or protozoa in CSF and can be confirmed, depending on the virus, by viral culture or specific antigen tests. In most cases, no treatment is indicated.

Meningitis-causing agents have their own research branch at the CDC.

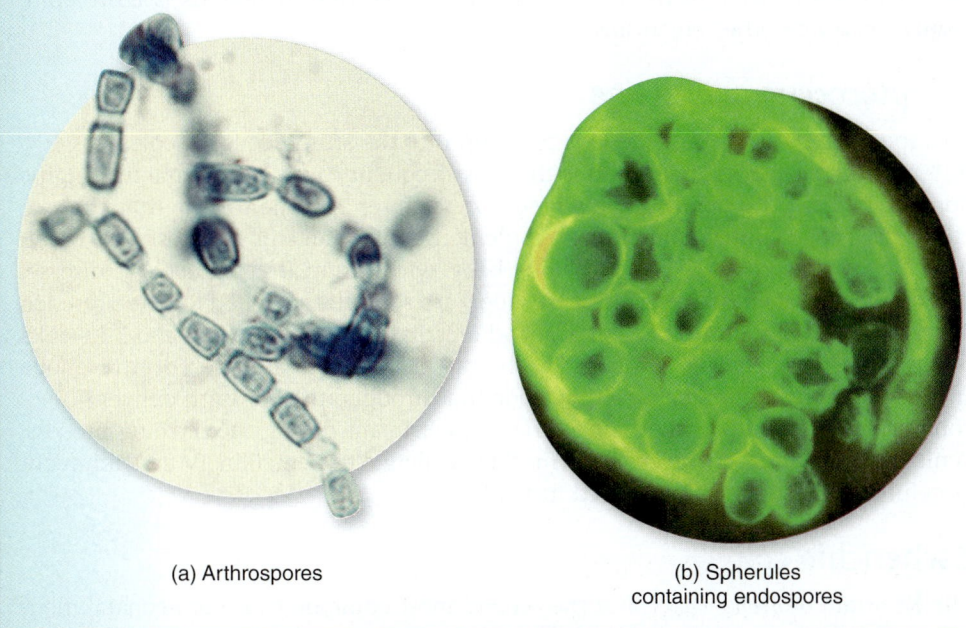

(a) Arthrospores

(b) Spherules containing endospores

Figure 17.8 Two phases of *Coccidioides* infection.   (a) Arthrospores are present in the environment and are inhaled. (b) In the lungs, the brain, or other tissues, arthrospores develop into spherules that are filled with endospores. Endospores are released and induce damage.

Disease Table 17.2 Neonatal and Infant Meningitis

Causative Organism(s)	*Streptococcus agalactiae*	*Escherichia coli*, strain K1	*Listeria monocytogenes*	*Cronobacter sakazakii*
Most Common Modes of Transmission	Vertical (during birth)	Vertical (during birth)	Vertical	Vehicle (baby formula)
Virulence Factors	Capsule	–	Intracellular growth	Ability to survive dry conditions
Culture/Diagnosis	Culture mother's genital tract on blood agar; CSF culture of neonate	CSF Gram stain/culture	Cold enrichment, rapid methods	Chromogenic differential agar, or rapid detection kits
Prevention	Culture and treatment of mother	–	Cooking food, avoiding unpasteurized dairy products	Safe preparation and use of, or avoidance of, powdered formula
Treatment	Penicillin G plus aminoglycosides	Ceftazidime or cefepime +/− gentamicin	Ampicillin, trimethoprim-sulfamethoxazole	Begin with broad-spectrum drugs until susceptibilities determined
Distinctive Features	Most common; positive culture of mother confirms diagnosis	Suspected if infant is premature		–
Epidemiological Features	Before intrapartum antibiotics in 1996: 1.8 cases per 1,000 live births After intrapartum antibiotics: 0.32 cases per 1,000 live births	Estimated at 0.2-5 per 1,000 live births; 20% of pregnant women colonized	Mortality can be as high as 33%	12 cases in United States in 2011

Neonatal Meningitis

Meningitis in newborns is almost always a result of infection transmitted by the mother, either *in utero* or (more frequently) during passage through the birth canal. As more premature babies survive, the rates of neonatal meningitis increase, because the condition is favored in patients with immature immune systems. In the United States, the two most common causes are *Streptococcus agalactiae* and *Escherichia coli*. *Listeria monocytogenes* is also found frequently in neonates. It has already been covered here but is included in **Disease Table 17.2** as a reminder that it can cause neonatal cases as well. In the developing world, neonatal meningitis is more commonly caused by other organisms.

Streptococcus agalactiae

This species of *Streptococcus* belongs to group B of the streptococci. It colonizes 10% to 30% of female genital tracts and is the most frequent cause of neonatal meningitis (for details about this condition in women, see chapter 21). The treatment for neonatal disease is Penicillin G, sometimes supplemented with an aminoglycoside. Women who are considered high risk (previous baby with group B streptococcal disease, early rupture of membranes, premature labor) are typically screened for the presence of these bacteria by means of a cervical and rectal swab at between 35 and 37 weeks of gestation. Women who are found to harbor the bacteria are offered intravenous antibiotics at the beginning of active labor and throughout labor until delivery is accomplished to avoid passing the bacteria to their infant during the birthing process. Penicillin is the drug of choice; if the mother is allergic to penicillin, IV erythromycin or cefuroxime may be administered instead.

Escherichia coli

The K1 strain of *Escherichia coli* is the second most common cause of neonatal meningitis. Most babies who suffer from this infection are premature, and their prognosis is poor. Twenty to thirty percent of them die, even with aggressive antibiotic treatment, and those who survive often have brain damage.

The bacterium is usually transmitted from the mother's birth canal. It causes no disease in the mother but can infect the vulnerable tissues of a neonate. It seems to have a predilection for the tissues of the central nervous system. Ceftazidime or cefepime +/− gentamicin is usually administered intravenously.

Cronobacter sakazakii

Cronobacter, formerly known as *Enterobacter sakazakii,* is found mainly in the environment but has been implicated in outbreaks of neonatal and infant meningitis transmitted through contaminated powdered infant formula. Although cases of *Cronobacter* meningitis are rare, mortality rates can reach 40%. The FDA and the CDC advise hospitals to use ready-to-feed or concentrated liquid formulas. They also recommend that home caregivers wash their hands and use clean feeding equipment when preparing formula, use fresh formula for each feeding, and discard any leftover formula.

Poliomyelitis

Poliomyelitis (poh″-lee-oh′my″-eh′ly′tis) (polio) is an acute enteroviral infection of the spinal cord that can cause neuromuscular paralysis. Because it often affects small children, in the past it was called *infantile paralysis.* No civilization or culture has escaped the devastation of polio. The efforts of a WHO campaign have significantly reduced the global incidence of polio. It was the campaign's goal to eradicate all of the remaining wild polioviruses by 2000, and then by 2005. It didn't happen. Eventually, billionaire Bill Gates got involved and contributed $700 million to help eradicate the disease. In 2014, India was declared polio-free. However, also during that year, the World Health Organization declared a public health emergency with respect to polio because it had spread to eight countries in Africa and the Middle East, threatening to undo progress toward its eradication.

▶ Signs and Symptoms

Most infections are contained as short-term, mild viremia. Some persons develop mild nonspecific symptoms of fever, headache, nausea, sore throat, and myalgia. If the viremia persists, viruses can be carried to the central nervous system through its blood supply. The virus then spreads along specific pathways in the spinal cord and brain. The virus is **neurotropic,** that is, it attacks the nervous system. It infiltrates the motor neurons of the anterior horn of the spinal cord, although it can also attack spinal ganglia, cranial nerves, and motor nuclei. Nonparalytic disease involves the invasion but not the destruction of nervous tissue. It gives rise to muscle pain and spasm, meningeal inflammation, and vague hypersensitivity.

In paralytic disease, invasion of motor neurons causes various degrees of flaccid paralysis over a period of a few hours to several days. Depending on the level of damage to motor neurons, paralysis of the muscles of the legs, abdomen, back, intercostals, diaphragm, pectoral girdle, and bladder can result. In rare cases of **bulbar poliomyelitis,** the brain stem, medulla, or even cranial nerves are affected. This situation leads to loss of control of cardiorespiratory regulatory centers, requiring mechanical respirators. In time, the unused muscles begin to atrophy, growth is slowed, and severe deformities of the trunk and limbs develop. Common sites of deformities are the spine, shoulder, hips, knees, and feet. Because motor function but not sensation is compromised, the crippled limbs are often very painful.

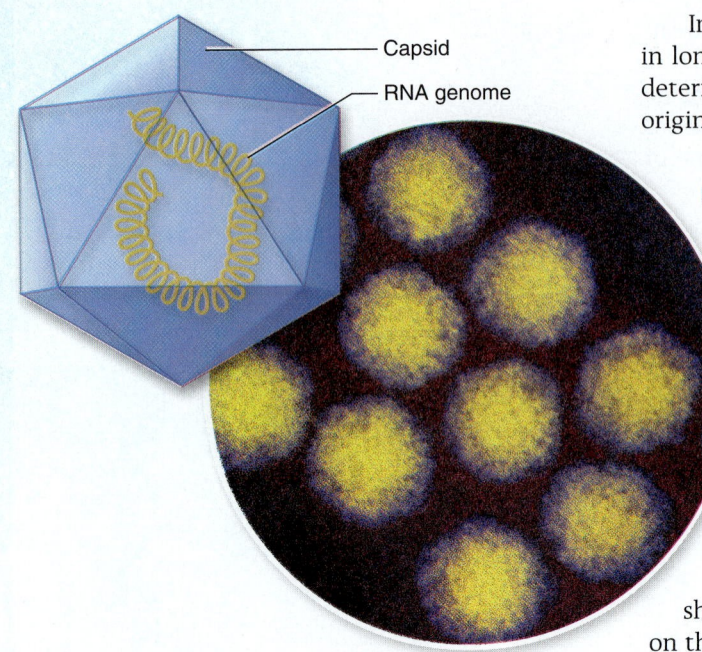

Capsid

RNA genome

Figure 17.9 **Typical structure of a picornavirus.** (a) A poliovirus, a type of picornavirus that is one of the simplest and smallest viruses (30 nm). It consists of an icosahedral capsid shell around a molecule of RNA. (b) A mass of stacked poliovirus particles in an infected host cell (300,000×).

In recent times, a condition called post-polio syndrome (PPS) has been diagnosed in long-term survivors of childhood infection. PPS manifests as a progressive muscle deterioration that develops in about 25% to 50% of patients several decades after their original polio attack.

▶ Causative Agent

The poliovirus is in the family *Picornaviridae*, genus *Enterovirus*–named for its small (*pico*) size and its RNA genome **(figure 17.9).** It is nonenveloped and nonsegmented. The naked capsid of the virus confers chemical stability and resistance to acid, bile, and detergents. By this means, the virus survives the gastric environment and other harsh conditions, which contributes to its ease of transmission.

▶ Pathogenesis and Virulence Factors

After being ingested, polioviruses adsorb to receptors of mucosal cells in the oropharynx and intestine. Here, they multiply in the mucosal epithelia and lymphoid tissue. Multiplication results in large numbers of viruses being shed into the throat and feces, and some of them leak into the blood. Depending on the number of viruses in the blood and their duration of stay there, an individual may exhibit no symptoms, mild nonspecific symptoms such as fever or short-term muscle pain, or devastating paralysis.

▶ Transmission and Epidemiology

Sporadic cases of polio can break out at any time of the year, but its incidence is more pronounced during the summer and fall. Humans are the only known reservoir; the virus is passed within the population through food, water, hands, objects contaminated with feces, and mechanical vectors. Although the 20th century saw a very large rise in paralytic polio cases, it was also the century during which effective vaccines were developed. The infection was eliminated from the Western Hemisphere in the late 20th century. Sadly, it is proving extremely difficulty to eradicate from the developing world.

▶ Prevention and Treatment

Treatment of polio rests largely on alleviating pain and suffering. During the acute phase, muscle spasm, headache, and associated discomfort can be alleviated by pain-relieving drugs. Respiratory failure may require artificial ventilation maintenance. Prompt physical therapy to diminish crippling deformities and to retrain muscles is recommended after the acute febrile phase subsides.

The mainstay of polio prevention is vaccination as early in life as possible, usually in four doses starting at about 2 months of age. Adult candidates for immunization are travelers and members of the armed forces. The two forms of vaccine currently in use are inactivated poliovirus vaccine (IPV), developed by Jonas Salk in 1954, and oral poliovirus vaccine (OPV), developed by Albert Sabin in the 1960s.

For many years, the oral vaccine was used in the United States because it is easily administered by mouth, but it is not free of medical complications. It contains an attenuated virus that can multiply in vaccinated people and be spread to others. In very rare instances, the attenuated virus reverts to a neurovirulent strain that causes disease rather than protects against it. For this reason, IPV, using killed virus, is the only vaccine used in the United States.

Disease Table 17.3 **Poliomyelitis**

Causative Organism(s)	Poliovirus
Most Common Modes of Transmission	Fecal-oral, vehicle
Virulence Factors	Attachment mechanisms
Culture/Diagnosis	Viral culture, serology
Prevention	Live attenuated (OPV) (developing world) or inactivated vaccine (IPV) (developed world)
Treatment	None, palliative, supportive
Epidemiological Features	Eighty percent of world's population lives in polio-free areas; still endemic in Pakistan, Afghanistan, and Nigeria as of 2014

Meningoencephalitis

Two microorganisms cause a distinct disease called *meningoencephalitis* (disease in both the meninges and brain), and they are both amoebas. *Naegleria fowleri* and *Acanthamoeba* are accidental parasites that invade the body only under unusual circumstances.

Naegleria fowleri

Most cases of *Naegleria* infection reported worldwide occur in people who have been swimming in warm, natural bodies of freshwater. Infection can begin when amoebas are forced into human nasal passages as a result of swimming, diving, or other aquatic activities. Once the amoeba is inoculated into the favorable habitat of the nasal mucosa, it burrows in, multiplies, and uses the olfactory nerve to migrate into the brain and surrounding structures. The result is primary amoebic meningo-encephalitis (PAM), a rapid, massive destruction of brain and spinal tissue that causes hemorrhage and coma and invariably ends in death within a week or so **(figure 17.10).** Note that this organism is very common—children often carry the amoeba as harmless biota, especially during the summer months, and the series of events leading to disease is exceedingly rare.

Unfortunately, *Naegleria* meningoencephalitis advances so rapidly that treatment usually proves futile. Studies have indicated that early therapy with amphotericin B, sulfadiazine, or tetracycline in some combination can be of some benefit. Because of the wide distribution of the amoeba and its hardiness, no general method of control exists. Public swimming pools and baths must be adequately chlorinated and checked periodically for the amoeba.

Acanthamoeba

This protozoan differs from *Naegleria* in its portal of entry; it invades broken skin, the conjunctiva, and occasionally the lungs and urogenital epithelia. Although it causes a meningoencephalitis somewhat similar to that of *Naegleria*, the course of infection is lengthier but nearly as deadly, with only a 2% to 3% survival rate. The disease is called granulomatous amoebic meningoencephalitis (GAM). At special risk for infection are people with traumatic eye injuries, contact lens wearers, and AIDS patients exposed to contaminated water. We discussed ocular infections in chapter 16. Cutaneous and CNS infections with this organism are occasional complications in AIDS.

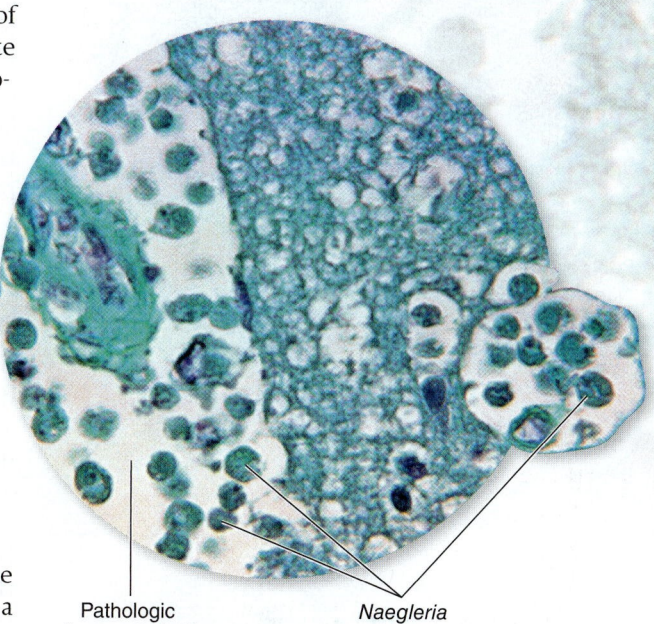

Pathologic changes in brain

Naegleria

Figure 17.10 ***Naegleria fowleri* in the brain.** The trophozoite form invades brain tissue, destroying it.

Disease Table 17.4 Meningoencephalitis		
Disease	**Primary Amoebic Meningoencephalitis**	**Granulomatous Amoebic Meningoencephalitis**
Causative Organism(s)	*Naegleria fowleri*	*Acanthamoeba*
Most Common Modes of Transmission	Vehicle (exposure while swimming in water)	Direct contact
Virulence Factors	Invasiveness	Invasiveness
Culture/Diagnosis	Examination of CSF; brain imaging, biopsy	Examination of CSF; brain imaging, biopsy
Prevention	Limit warm freshwater or untreated tap water entering nasal passages	–
Treatment	Amphotericin B; mostly ineffective	Surgical excision of granulomas; ketoconazole may help
Epidemiological Features	United States: 37 infections in 10-year period	Predominantly occurs in immunocompromised patients

Pools are chlorinated to prevent *Naeglaria* and other outbreaks.

Acute Encephalitis

Encephalitis (inflammation of the brain) can present as acute or **subacute.** It is always a serious condition, as the tissues of the brain are extremely sensitive to damage by inflammatory processes. Acute encephalitis is almost always caused by viral infection. One category of viral encephalitis is caused by viruses borne by insects (called *arboviruses,* which is short for *arthropod-borne viruses*), including West Nile virus. Alternatively, other viruses, such as members of the herpes family, are causative agents. Bacteria such as those covered under meningitis can also cause encephalitis, but the symptoms are usually more pronounced in the meninges than in the brain.

The signs and symptoms of encephalitis vary, but they may include behavior changes or confusion because of inflammation. Decreased consciousness and seizures frequently occur. Symptoms of meningitis are often also present. Few of these agents have specific treatments, but because swift initiation of acyclovir therapy can save the life of a patient suffering from herpesvirus encephalitis, most physicians will begin empiric therapy with acyclovir in all seriously ill neonates and most other patients showing evidence of encephalitis. Treatment will, in any case, do no harm in patients who are infected with other agents.

Arthropod-Borne Viruses (Arboviruses)

Most arthropods that serve as infectious disease vectors feed on the blood of hosts. Infections show a peak incidence when the arthropod is actively feeding and reproducing, usually from late spring through early fall. Warm-blooded vertebrates also maintain the virus during the cold and dry seasons. Humans can serve as dead-end, accidental hosts, as in equine encephalitis, or they can be a maintenance reservoir, as in yellow fever (discussed in chapter 18).

Arboviral diseases have a great impact on humans **(figure 17.11).** Although exact statistics are unavailable, it is believed that millions of people acquire infections each year and thousands of them die. One common outcome of arboviral infection is an acute fever, often accompanied by rash. Viruses that primarily cause these symptoms are covered in chapter 18.

The arboviruses discussed in this chapter can cause encephalitis, and we consider them as a group because the symptoms and management are similar. The

Figure 17.11
Worldwide distribution of major arboviral encephalitides.

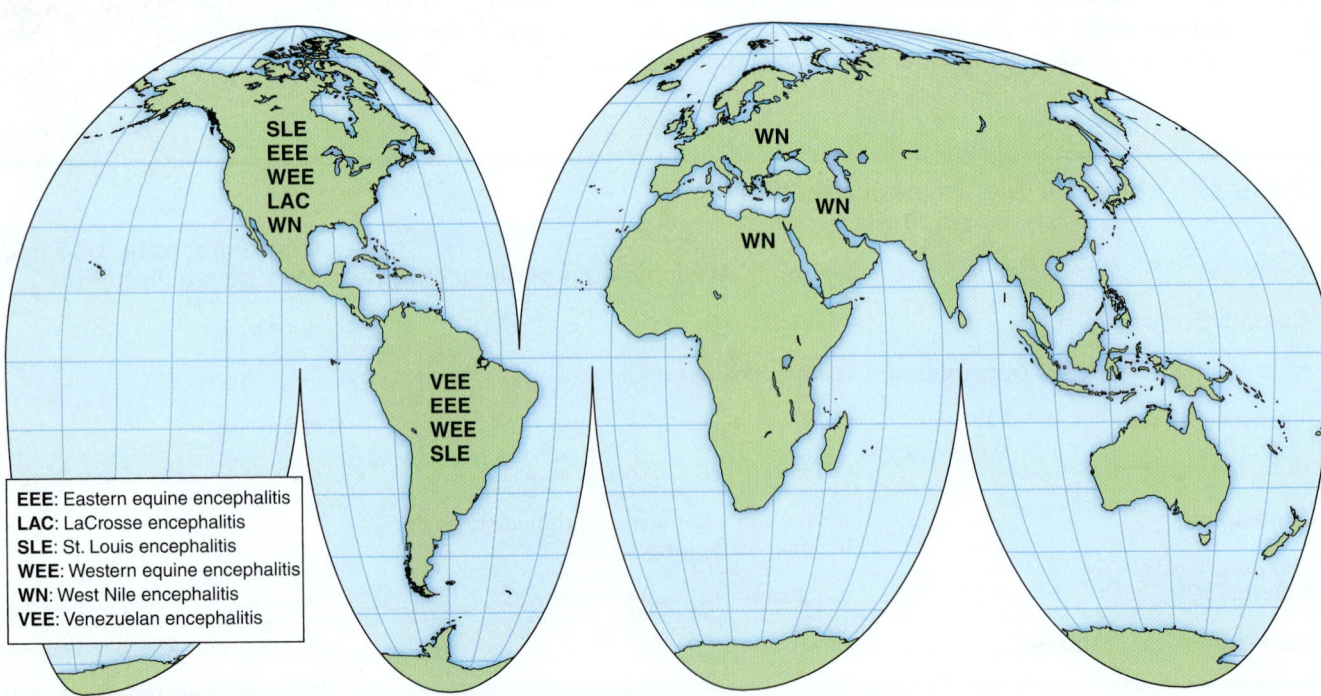

EEE: Eastern equine encephalitis
LAC: LaCrosse encephalitis
SLE: St. Louis encephalitis
WEE: Western equine encephalitis
WN: West Nile encephalitis
VEE: Venezuelan encephalitis

transmission and epidemiology of individual viruses are different, however, and are discussed for each virus in **table 17.1.** All of the infections here are transmitted by mosquitoes.

▶ Pathogenesis and Virulence Factors

Arboviral encephalitis begins with an arthropod bite, the release of the virus into tissues, and its replication in nearby lymphatic tissues. All the arboviruses we describe here are transmitted by mosquitoes. Prolonged viremia establishes the virus in the brain, where inflammation can cause swelling and damage to the brain, nerves, and meninges. Symptoms are extremely variable and can include coma, convulsions, paralysis, tremor, loss of coordination, memory deficits, changes in speech and personality, and heart disorders. In some cases, survivors experience some degree of permanent brain damage. Young children and the elderly are most sensitive to injury by arboviral encephalitis.

▶ Culture and Diagnosis

Except during epidemics, detecting arboviral infections can be difficult. The patient's history of travel to endemic areas or contact with vectors, along with serum analysis, is highly supportive of a diagnosis. Rapid serological and nucleic acid amplification tests are available for some of the viruses.

▶ Treatment

No satisfactory treatment exists for any of the arboviral encephalitides (plural of *encephalitis*). As mentioned earlier, empiric acyclovir treatment may be begun in case the infection is actually caused by either herpes simplex virus or varicella zoster. Treatment of the other infections relies entirely on support measures to control fever, convulsions, dehydration, shock, and edema.

Most of the control safeguards for arbovirus disease are aimed at the arthropod vectors. Mosquito abatement by eliminating breeding sites and by broadcast-spreading insecticides has been highly effective in restricted urban settings. Birds play a role as reservoirs of the virus, but direct transmission between birds and humans does not occur.

Public health officials regularly take samples of standing water in rain gutters, swimming pools, and elsewhere, looking for larval mosquitoes like those seen here.

Table 17.1 Arboviral Encephalitis

	Geographic Distribution	Notes
Western Equine Encephalitis (WEE)	This disease occurs sporadically in the western United States and Canada, appearing first in horses and later in humans.	The disease is extremely dangerous to infants and small children, with a case fatality rate of 3% to 7%.
Eastern Equine Encephalitis (EEE)	Endemic to an area along the eastern coasts of the United States and Canada. The usual pattern is sporadic, but occasional epidemics can occur in humans and horses.	The case fatality rate can be very high (70%).
California Encephalitis	The California strain occurs occasionally in the western United States and has little impact on humans. The LaCrosse strain is widely distributed in the eastern United States and Canada and is a prevalent cause of viral encephalitis in North America.	Two different virus strains; children living in rural areas are the primary target group, and most of them exhibit mild, transient symptoms. Fatalities are rare.
St. Louis Encephalitis (SLE)	Cases appear throughout North and South America, but epidemics in the United States occur most often in the Midwest and South.	May be the most common of all American viral encephalitides. Inapparent infection is very common, and the total number of cases is probably thousands of times greater than the 50 to 100 reported each year. The seasons of peak activity are spring and summer, depending on the region and species of mosquito.
West Nile Encephalitis	Commonly found in Africa, the Middle East, and parts of Asia, but after mid-1999 is now common in the Americas. The virus is known to infect a host of mammals (including humans), as well as birds and mosquitoes.	A close relative of the SLE virus. It emerged in the United States in 1999, and by 2008 the CDC was reporting that 1% of people in the United States—or approximately 3 million people—had evidence of past or present infection.

Disease Table 17.5 Acute Encephalitis

Causative Organism(s)	Arboviruses (viruses causing WEE, EEE, California encephalitis, SLE, West Nile encephalitis)	Herpes simplex 1 or 2	JC virus
Most Common Modes of Transmission	Vector (arthropod bites)	Vertical or reactivation of latent infection	? Ubiquitous
Virulence Factors	Attachment, fusion, invasion capabilities	–	–
Culture/Diagnosis	History, rapid serological tests, nucleic acid amplification tests	Clinical presentation, PCR, Ab tests, growth of virus in cell culture	PCR of cerebrospinal fluid
Prevention	Insect control; vaccines for WEE and EEE available	Maternal screening for HSV	None
Treatment	None	Acyclovir	Zidovudine or other antivirals
Distinctive Features	History of exposure to insect important	In infants, disseminated disease present; rare between 30 and 50 years	In severely immunocompromised, especially AIDS
Epidemiological Features	For West Nile virus: In Egypt, up to half infected during childhood and do not experience neurological disease In United States approx. 5,000 cases (half of them neuroinvasive) since appearance in 1999	HSV-1 most with common cause of encephalitis 2 cases per million per year	Affects 5% of adults with untreated AIDS

Herpes Simplex Virus (HSV)

Herpes simplex type 1 and 2 viruses can cause encephalitis in newborns born to HSV-positive mothers. In this case, the virus is disseminated and the prognosis is poor. Older children and young adults (ages 5 to 30), as well as older adults (over 50 years old), are also susceptible to herpes simplex encephalitis caused most commonly by HSV-1. In these cases, the HSV encephalitis usually represents a reactivation of dormant HSV from the trigeminal ganglion.

It should be noted the varicella-zoster virus (see chapter 16) can also reactivate from the dormant state, and it is responsible for rare cases of encephalitis.

JC Virus

The **JC virus (JCV)** gets its name from the initials of the patient in whom it was first diagnosed as the cause of illness. Serological studies indicate that asymptomatic infection with this polyoma virus is commonplace. In patients with immune dysfunction, especially in those with AIDS, it can cause a condition called **progressive multifocal leukoencephalopathy** (loo″-koh-en-sef″uh-lop′-uh-thee) **(PML).** This uncommon but generally fatal infection is a result of JC virus attack of accessory brain cells. The infection demyelinizes certain parts of the cerebrum. This virus should be considered when encephalitis symptoms are observed in AIDS patients. Recently a few deaths from this condition have been prevented with high doses of zidovudine.

Subacute Encephalitis

When encephalitis symptoms take longer to show up, and when the symptoms are less striking, the condition is termed *subacute encephalitis*. The most common cause of subacute encephalitis is the protozoan *Toxoplasma*. Another form of subacute encephalitis can be caused by persistent measles virus as many as 7 to 15 years after the initial infection. Finally, a class of infectious agents known as prions can cause a condition called spongiform encephalopathy.

Toxoplasma gondii

Infection in the fetus and in immunodeficient people, especially those with AIDS, is severe and often fatal. Although infection in otherwise healthy people is generally unnoticed, recent data suggest that it may have subtle but profound effects on their brain and the responses it controls. People with a history of *Toxoplasma* infection are often more likely to display thrill-seeking behaviors and seem to have slower reaction times.

▶ Signs and Symptoms

Most cases of toxoplasmosis are asymptomatic or marked by mild symptoms such as sore throat, lymph node enlargement, and low-grade fever. In patients whose immunity is suppressed by infection, cancer, or drugs, the outlook may be grim. The infection causes a more chronic or subacute form of encephalitis than do most viruses, often producing extensive brain lesions and fatal disruptions of the heart and lungs. A pregnant woman with toxoplasmosis has a 33% chance of transmitting the infection to her fetus. Congenital infection occurring in the first or second trimester is associated with stillbirth and severe abnormalities such as liver and spleen enlargement, liver failure, hydrocephalus, convulsions, and damage to the retina that can result in blindness.

▶ Pathogenesis and Virulence Factors

Toxoplasma is an obligate intracellular parasite, making its ability to invade host cells an important factor for virulence.

▶ Transmission and Epidemiology

T. gondii is a very successful parasite with so little host specificity that it can attack at least 200 species of birds and mammals. However, the parasite undergoes a sexual phase in the intestine of cats and is then released in feces, where it becomes an infective *oocyst* that survives in moist soil for several months. These forms eventually enter an asexual cyst state in tissues, called a *pseudocyst*. Most of the time, the parasite does not cycle in cats alone and is spread by oocysts to intermediate hosts, including rodents and birds. The cycle returns to cats when they eat these infected prey animals. Cattle and sheep can also be infected.

In 2007, scientists at Stanford University found that the protozoan crowds into a part of the rat brain that usually directs the rat to avoid the smell of cat urine (a natural defense against a domestic rat's major predator). When *Toxoplasma* infects rat brains, the rats lose their fear of cats. Infected rats are then easily eaten by cats, ensuring the continuing *Toxoplasma* life cycle. All other neurological functions in the rat are left intact.

Humans appear to be constantly exposed to the pathogen. The rate of prior infections, as detected through serological tests, can be as high as 90% in some populations. Many cases are caused by ingesting pseudocysts in undercooked contaminated meat, and other sources include contact with other mammals or even dirt and dust contaminated with oocysts. Fetuses may become infected when tachyzoites cross the placenta.

In view of the fact that the oocysts are so widespread and resistant, hygiene is of paramount importance in controlling toxoplasmosis. Adequate cooking or freezing below −20°C destroys both oocysts and tissue cysts. Oocysts can also be avoided by washing the hands after handling cats or soil possibly contaminated with cat feces, especially sandboxes and litter boxes. Pregnant women should be especially

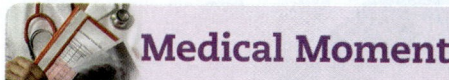

Medical Moment

Neglected Parasitic Infections

Toxoplasmosis is one of the five diseases on the CDC's Neglected Parasitic Infections list. This list highlights the five conditions that are under-diagnosed, underreported, and undertreated in the United States. The other four diseases are Chagas disease (chapter 18), toxocariasis (chapter 20), trichomoniasis (chapter 21), and cysticercosis (chapter 20). Cysticercosis also can cause symptoms in the nervous system. It is a tapeworm disease that comes from eating pork or from ingesting the eggs from an infected person via the fecal-oral route. When the route of transmission is the fecal-oral route, the larvae of the tapeworm become encysted in tissues throughout the body, often in the brain, leading to a condition called neurocysticercosis. Up to 300,000 people in the United States experience this condition.

It's up to health care providers to change the description of these five diseases so that they are no longer "underdiagnosed and undertreated." The only way to do this is to raise awareness about them, both among the public and among health care providers.

NCLEX® PREP

5. For which disease processes are immunizations available as a method of prevention? Select all that apply.
   a. *Cryptococcus neoformans*
   b. *Listeria monocytogenes*
   c. *Haemophilus influenzae*
   d. *Streptococcus pneumoniae*
   e. *Neisseria meningitidis*

attentive to these rules and should never clean the cat's litter box. *Toxoplasma* infection of the fetus can result in miscarriage, premature birth, or babies born with brain and/or vision problems.

Prions

As you read in chapter 5, prions are *proteinaceous infectious particles* containing, apparently, no genetic material. They are known to cause diseases called **transmissible spongiform encephalopathies (TSEs),** neurodegenerative diseases with long incubation periods but rapid progressions once they begin. The human TSEs are **Creutzfeldt-Jakob disease (CJD),** Gerstmann-Strussler-Scheinker disease, and fatal familial insomnia. TSEs are also found in animals and include a disease called scrapie in sheep and goats, transmissible mink encephalopathy, and bovine spongiform encephalopathy (BSE). This last disease is commonly known as mad cow disease and was in the headlines in the 1990s due to its apparent link to a variant form of Creutzfeldt-Jakob human disease in Great Britain. We consider CJD in this section.

Prions were found in the meat of cattle fed with animal scraps. The highest number of cases occurred in the 1990s in the United Kingdom but isolated cases have been found in the United States as recently as 2014.

▶ Signs and Symptoms of CJD

Symptoms of CJD include altered behavior, dementia, memory loss, impaired senses, delirium, and premature senility. Uncontrollable muscle contractions continue until death, which usually occurs within 1 year of diagnosis.

▶ Causative Agent of CJD

The transmissible agent in CJD is a prion. In some forms of the disease, it involved the transformation of a normal host protein (called PrP), a protein that is supposed to function to help the brain develop normally, and that has recently been found to protect against Alzheimer's disease. Once this happens, the abnormal PrP itself becomes catalytic and able to spontaneously convert other normal human PrP proteins into the abnormal form. This becomes a self-propagating chain reaction that creates a massive accumulation of altered PrP, leading to plaques, spongiform damage (i.e., holes in the brain) **(figure 17.12),** and severe loss of brain function.

Using the term *transmissible agent* may be a bit misleading, however, as some cases of CJD arise through genetic mutation of the *PrP* gene, which can be a heritable trait. So it seems that although one can acquire a defective PrP protein via transmission, one can also have an altered *PrP* gene passed on through heredity.

Prions are incredibly hardy "pathogens." They are highly resistant to chemicals, radiation, and heat. They can withstand prolonged autoclaving.

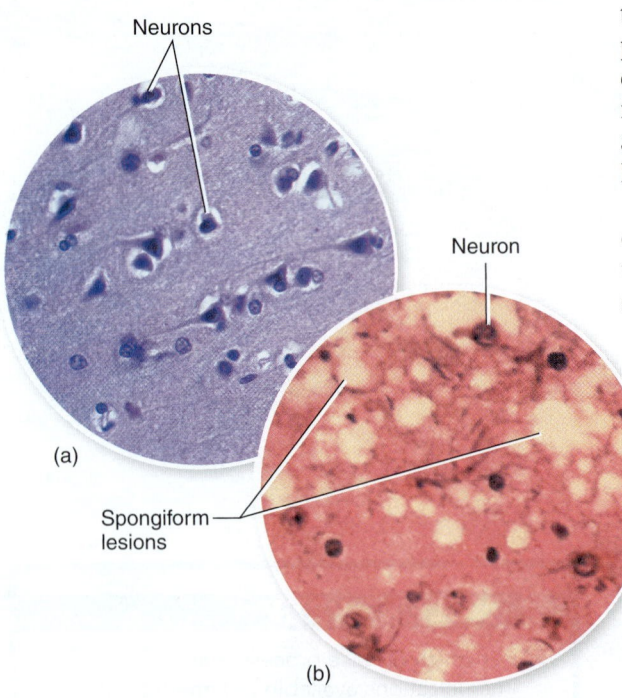

▶ Transmission and Epidemiology

In the late 1990s, it became apparent that humans were contracting a variant form of CJD (vCJD) after ingesting meat from cattle that had been afflicted by bovine spongiform encephalopathy. Presumably, meat products had been contaminated with fluid or tissues infected with the prion. Cases of this disease have centered around Great Britain, where many cows were found to have BSE. The median age at death of patients with vCJD is 28 years. In contrast, the median age at death of patients with other forms of CJD is 68 years.

Health care professionals should be aware of the possibility of CJD in patients, especially when surgical procedures are performed, as cases have been reported of transmission of CJD via contaminated surgical instruments. Due to the heat and chemical resistance of prions, normal disinfection and sterilization procedures are usually not sufficient to eliminate them from instruments and surfaces. The latest CDC guidelines for handling of CJD patients in a health care environment should be consulted. There is no known treatment for CJD, and mortality appears to be 100%; but there is active research into treatments for prion diseases.

Figure 17.12 The microscopic effects of spongiform encephalopathy. (a) Normal cerebral cortex section, showing neurons and glial cells. (b) Sectioned cortex in CJD patient shows numerous round holes, producing a "spongy" appearance. This destroys brain architecture and causes massive loss of neurons and glial cells.

Disease Table 17.6 Subacute Encephalitis

Causative Organism(s)	*Toxoplasma gondii*	Subacute sclerosing panencephalitis	Prions
Most Common Modes of Transmission	Vehicle (meat) or fecal-oral	Persistence of measles virus	CJD = direct/parenteral contact with infected tissue, or inherited vCJD = vehicle (meat, parenteral)
Virulence Factors	Intracellular growth	Cell fusion, evasion of immune system	Avoidance of host immune response
Culture/Diagnosis	Serological detection of IgM, culture, histology	EEGs, MRI, serology (Ab versus measles virus)	Biopsy, image of brain
Prevention	Personal hygiene, food hygiene	None	Avoiding tissue
Treatment	Pyrimethamine and/or leucovorin and/or sulfadiazine	None	None
Distinctive Features	Subacute, slower development of disease	History of measles	Long incubation period; fast progression once it begins
Epidemiological Features	15%–29% of U.S. population is seropositive; internationally, seroprevalence is to 90%; disease occurs in 3%–15% of AIDS patients	United States: fewer than 10 cases/year; incidence has declined 90% in countries who vaccinate against measles	CJD: 1 case per year per million worldwide; seen in older adults vCJD: 98% cases originated in United Kingdom

Rabies

Rabies is a slow, progressive zoonotic disease characterized by a fatal encephalitis. It is so distinctive in its pathogenesis and its symptoms that we discuss it separately from the other encephalitides. It is distributed nearly worldwide, except for perhaps two dozen countries that have remained rabies-free by practicing rigorous animal control.

▶ Signs and Symptoms

The average incubation period of rabies is 1 to 2 months or more, depending on the wound site, its severity, and the inoculation dose. The incubation period is shorter in facial, scalp, or neck wounds because of closer proximity to the brain. The prodromal phase begins with fever, nausea, vomiting, headache, fatigue, and other nonspecific symptoms.

Until recently, humans were never known to survive rabies. But a handful of patients have recovered in recent years after receiving intensive, long-term treatment.

▶ Pathogenesis and Virulence Factors

Infection with rabies virus typically begins when an infected animal's saliva enters a puncture site. The virus occasionally is inhaled or inoculated through the membranes of the eye. The rabies virus remains up to a week at the trauma site, where it multiplies. The virus then gradually enters nerve endings and advances toward the ganglia, spinal cord, and brain. Viral multiplication throughout the brain is eventually followed by migration to such diverse sites as the eye, heart, skin, and oral cavity. The infection cycle is completed when the virus replicates in the salivary gland and is shed into the saliva. Clinical rabies proceeds through several distinct stages that almost inevitably end in death, unless post-exposure vaccination is performed before symptoms begin.

Scientists have discovered that virulence is associated with an envelope glycoprotein that seems to give the virus its ability to spread in the CNS and to invade certain types of neural cells.

Bats are one of the vectors for rabies.

Disease Table 17.7	Rabies
Causative Organism(s)	Rabies virus
Most Common Modes of Transmission	Parenteral (bite trauma), droplet contact
Virulence Factors	Envelope glycoprotein
Culture/Diagnosis	RT-PCR of saliva; Ab detection of serum or CSF; skin biopsy
Prevention	Inactivated vaccine
Treatment	Postexposure passive and active immunization; induced coma and ventilator support
Epidemiological Features	United States: 1–5 cases per year Worldwide: 35,000–55,000 cases annually

► Transmission and Epidemiology

The primary reservoirs of the virus are wild mammals such as canines, skunks, raccoons, badgers, cats, and bats that can spread the infection to domestic dogs and cats. Both wild and domestic mammals can spread the disease to humans through bites, scratches, and inhalation of droplets. The annual worldwide total for human rabies is estimated to be about 35,000 to 50,000 cases, but only a tiny number of these cases occur in the United States, the majority of these transmitted to humans from bats. Most U.S. cases of rabies occur in wild animals (about 6,000 to 7,000 cases per year), while dog rabies has declined **(figure 17.13).**

The epidemiology of animal rabies in the United States varies. The most common wild animal reservoir hosts have changed from foxes to skunks to raccoons. Regional differences in the dominant reservoir also occur. Rats, skunks, and bobcats are the most common carriers of rabies in California, raccoons are the predominant carriers in the East, and coyotes dominate in Texas.

Diagnosis requires multiple tests. Reverse transcription PCR is used with saliva samples but must be accompanied by detection of antibodies to the virus in serum or spinal fluid. Skin biopsies are also used.

► Prevention and Treatment

A bite from a wild or stray animal demands assessment of the animal, meticulous care of the wound, and a specific treatment regimen. A wild mammal, especially a skunk, raccoon, fox, or coyote that bites without provocation, is presumed to be rabid, and therapy is immediately begun.

Rabies is one of the few infectious diseases for which a combination of passive and active postexposure immunization is indicated (and successful). Initially the wound is infused with human rabies immune globulin (HRIG) to impede the

Figure 17.13 Distribution of rabies in the United States. Rabies is found in 10 distinct geographic areas. In each area, a particular animal is the reservoir as illustrated by four different colors. The prevalence in bats is shown by the black dots.

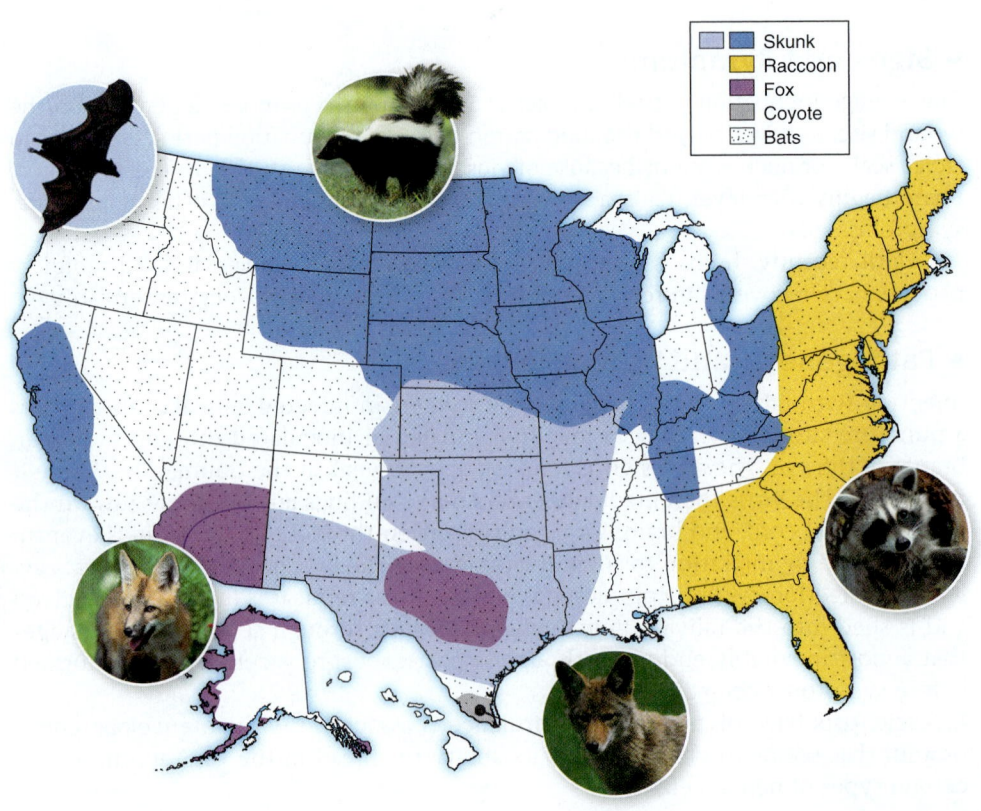

	Skunk
	Raccoon
	Fox
	Coyote
	Bats

spread of the virus, and globulin is also injected intramuscularly to provide immediate systemic protection. A full course of vaccination is started simultaneously. The current vaccine of choice is the **human diploid cell vaccine (HDCV).** The routine postexposure vaccination entails intramuscular or intradermal injection on days 0, 3, 7, and 14, sometimes with two additional boosters. High-risk groups such as veterinarians, animal handlers, laboratory personnel, and travelers should receive three doses to protect against possible exposure. A DNA vaccine for rabies is in development.

Tetanus

Tetanus is a neuromuscular disease whose alternate name, lockjaw, refers to an early effect of the disease on the jaw muscle. The etiologic agent, *Clostridium tetani*, is a common resident of cultivated soil and the gastrointestinal tracts of animals. It is a gram-positive, spore-forming rod. The endospores it produces often swell the vegetative cell **(figure 17.14).** Spores are produced only under anaerobic conditions.

▶ Signs and Symptoms

C. tetani releases a powerful neurotoxin, **tetanospasmin,** that binds to target sites on peripheral motor neurons, spinal cord and brain, and in the sympathetic nervous system. The toxin acts by blocking the inhibition of muscle contraction. Without inhibition of contraction, the muscles contract uncontrollably, resulting in spastic paralysis. The first symptoms are clenching of the jaw, followed in succession by extreme arching of the back, flexion of the arms, and extension of the legs **(figure 17.15).** Lockjaw confers the bizarre appearance of *risus sardonicus* (sardonic grin), which looks eerily as though the person is smiling **(figure 17.16).** Death most often occurs due to paralysis of the respiratory muscles and respiratory arrest.

▶ Pathogenesis and Virulence Factors

The mere presence of endospores in a wound is not sufficient to initiate infection because the bacterium is unable to invade damaged tissues readily. It is also a strict anaerobe, and the endospores cannot become established unless tissues at the site of the wound are necrotic and poorly supplied with blood, conditions that favor germination.

As the vegetative cells grow, the tetanospasmin toxin is released into the infection site. The toxin spreads to nearby motor nerve endings in the injured tissue, binds to them, and travels via axons to the ventral horns of the spinal cord (figure 17.16). The toxin blocks the release of neurotransmitter, and only a small amount is required to initiate the symptoms.

▶ Transmission and Epidemiology

Endospores usually enter the body through accidental puncture wounds, burns, umbilical stumps, frostbite, and crushed body parts. The incidence of tetanus is low in North America. Most cases occur among geriatric patients and intravenous drug abusers. Historically, however, the worldwide incidence of maternal and neonatal tetanus has been very high, causing 60,000 deaths each year. In response, the World Health Organization (WHO) has made dramatic global progress in reducing mortality through the promotion of more hygienic delivery practices and vaccination.

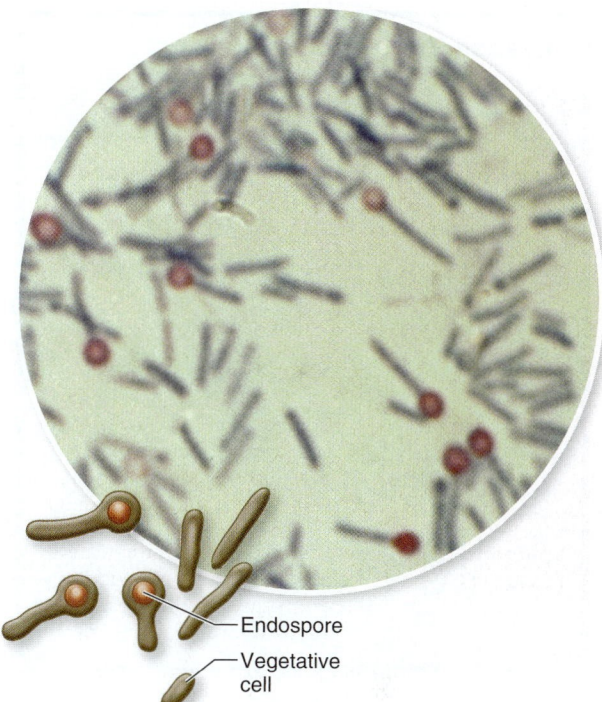

Endospore
Vegetative cell

Figure 17.14 *Clostridium tetani.*   Its typical tennis racket morphology is created by terminal endospores that swell the end of the cell (170×).

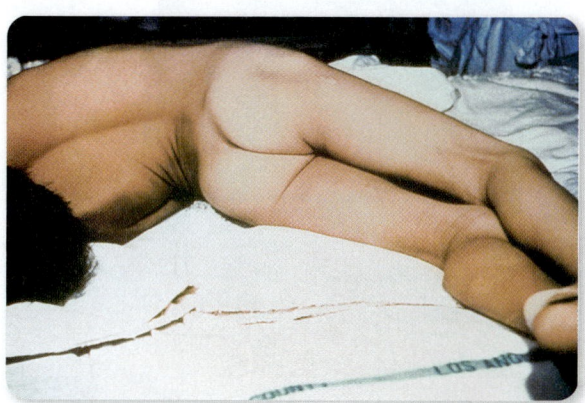

Figure 17.15 **Late-stage tetanus.**

Disease Table 17.8 Tetanus

Causative Organism(s)	*Clostridium tetani*
Most Common Modes of Transmission	Parenteral, direct contact
Virulence Factors	Tetanospasm exotoxin
Culture/Diagnosis	Symptomatic
Prevention	Tetanus toxoid immunization
Treatment	Combination of passive antitoxin and tetanus toxoid active immunization, Penicillin G, muscle relaxants
Epidemiological Features	United States: Approximately 30 cases/year; worldwide: estimated 1 million cases annually, 50% in newborns

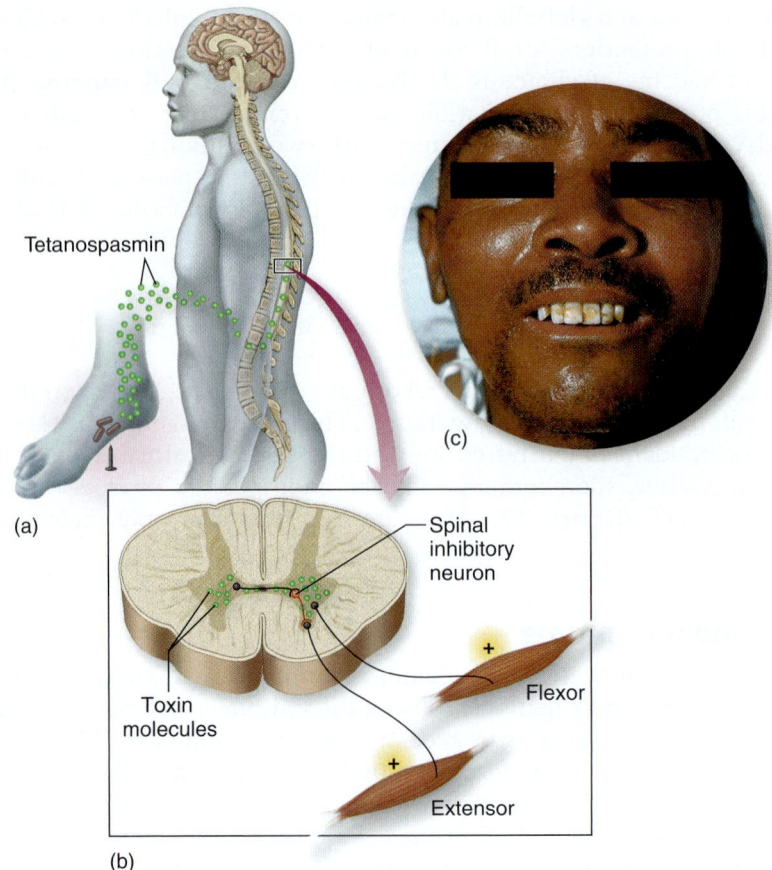

Figure 17.16 The events in tetanus.   (a) After traumatic injury, bacteria infecting the local tissues secrete tetanospasmin, which is absorbed by the peripheral axons and is carried to the target neurons in the spinal column. (b) In the spinal cord, the toxin attaches to the junctions of regulatory neurons that inhibit inappropriate contraction. Released from inhibition, the muscles, even opposing members of a muscle group, receive constant stimuli and contract uncontrollably. (c) Muscles contract spasmodically, without regard to regulatory mechanisms or conscious control. Note the clenched jaw, called *risus sardonicus*.

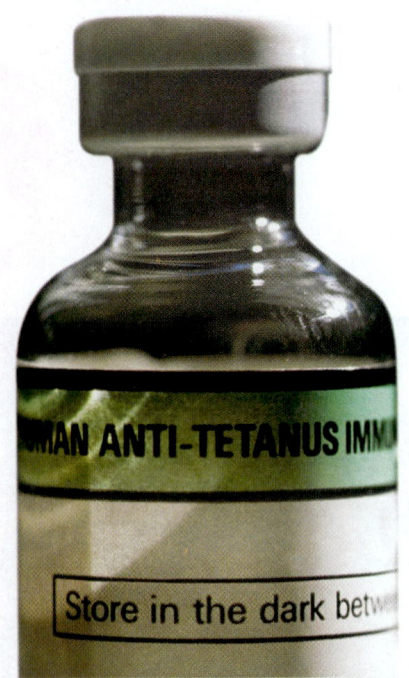

Passive tetanus immunoglobulin is given immediately to halt the progress of the toxin molecules.

▶ Prevention and Treatment

A patient with a clinical appearance suggestive of tetanus should immediately receive antitoxin therapy with human tetanus immune globulin (TIG) and Penicillin G.

The recommended vaccination series for 1- to 3-month-old babies consists of three injections of DTaP (diphtheria, tetanus, and acellular pertussis) given 2 months apart, followed by booster doses about 1 and 4 years later. Children thus immunized probably have protection for 10 years. At that point, and every 10 years thereafter, they should receive a dose of TD, tetanus-diphtheria vaccine. Additional protection against neonatal tetanus may be achieved by vaccinating pregnant women, whose antibodies will be passed to the fetus. Toxoid should also be given to injured persons who have never been immunized, have not completed the series, or whose last booster was received more than 10 years previously. The vaccine can be given simultaneously with passive TIG immunization to achieve immediate as well as long-term protection.

Botulism

Botulism is an **intoxication** (i.e., caused by an exotoxin) associated with eating poorly preserved foods, although it can also occur as a true infection. Until recent times, it was relatively common and frequently fatal, but modern techniques of food preservation and medical treatment have reduced both its incidence and its fatality rate.

▶ Signs and Symptoms

There are three major forms of botulism, distinguished by their means of transmission and the population they affect. **Table 17.2** summarizes these. The symptoms are largely the same in all three forms, however. From the circulatory system, an exotoxin called the **botulinum toxin** travels to its principal site of action, the neuromuscular junctions of skeletal muscles **(figure 17.17).** The effect of botulinum is to prevent the release of the neurotransmitter, acetylcholine, that initiates the signal for muscle contraction. The usual time before onset of symptoms is 12 to 72 hours, depending on the size of the dose. Neuromuscular symptoms first affect the muscles of the head and include double vision, difficulty in swallowing, and dizziness, but there is no sensory or mental lapse. Later symptoms are descending muscular paralysis and respiratory compromise. In the past, death resulted from respiratory arrest, but mechanical respirators have reduced the fatality rate to about 10%.

Table 17.2 Three Types of Botulism

		Transmission and Epidemiology	Culture and Diagnosis
Food-Borne Botulism	**Pure intoxication**	Many botulism outbreaks occur in home-processed foods, including canned vegetables, smoked meats, and cheese spreads. Several factors in food processing can lead to botulism. Endospores are present on the vegetables or meat at the time of gathering and are difficult to remove completely. When contaminated food is put in jars and steamed in a pressure cooker that does not reach reliable pressure and temperature, some endospores survive (botulinum endospores are highly heat resistant). At the same time, the pressure is sufficient to evacuate the air and create anaerobic conditions. Storage of the jars at room temperature favors endospore germination and vegetative growth, and one of the products of the cell's metabolism is botulinum, the most potent microbial toxin known. Bacterial growth may not be evident in the appearance of the jar or can or in the food's taste or texture, and only minute amounts of toxin may be present. Botulism is never transmitted from person to person.	Some laboratories attempt to identify the toxin in the offending food. Alternatively, if multiple patients present with the same symptoms after ingesting the same food, a presumptive diagnosis can be made. The cultivation of *C. botulinum* in feces is considered confirmation of the diagnosis since the carrier rate is very low.
Infant Botulism	**Infection followed by intoxication**	This is currently the most common type of botulism in the United States, with approximately 75 cases reported annually. The exact food source is not always known, although raw honey has been implicated in some cases, and the endospores are common in dust and soil. Apparently, the immature state of the neonatal intestine and microbial biota allows the endospores to gain a foothold, germinate, and give off neurotoxin. As in adults, babies exhibit flaccid paralysis, usually manifested as a weak sucking response, generalized loss of tone (the "floppy-baby syndrome"), and respiratory complications. Although adults can also ingest botulinum endospores in contaminated vegetables and other foods, the adult intestinal tract normally inhibits this sort of infection.	Finding the toxin or the organism in the feces confirms the diagnosis.
Wound Botulism	**Infection followed by intoxication**	Perhaps three or four cases of wound botulism occur each year in the United States. In this form of the disease, endospores enter a wound or puncture, much as in tetanus, but the symptoms are similar to those of food-borne botulism. Increased cases of this form of botulism are being reported in intravenous drug users as a result of needle puncture.	The toxin should be demonstrated in the serum, or the organism should be grown from the wound.

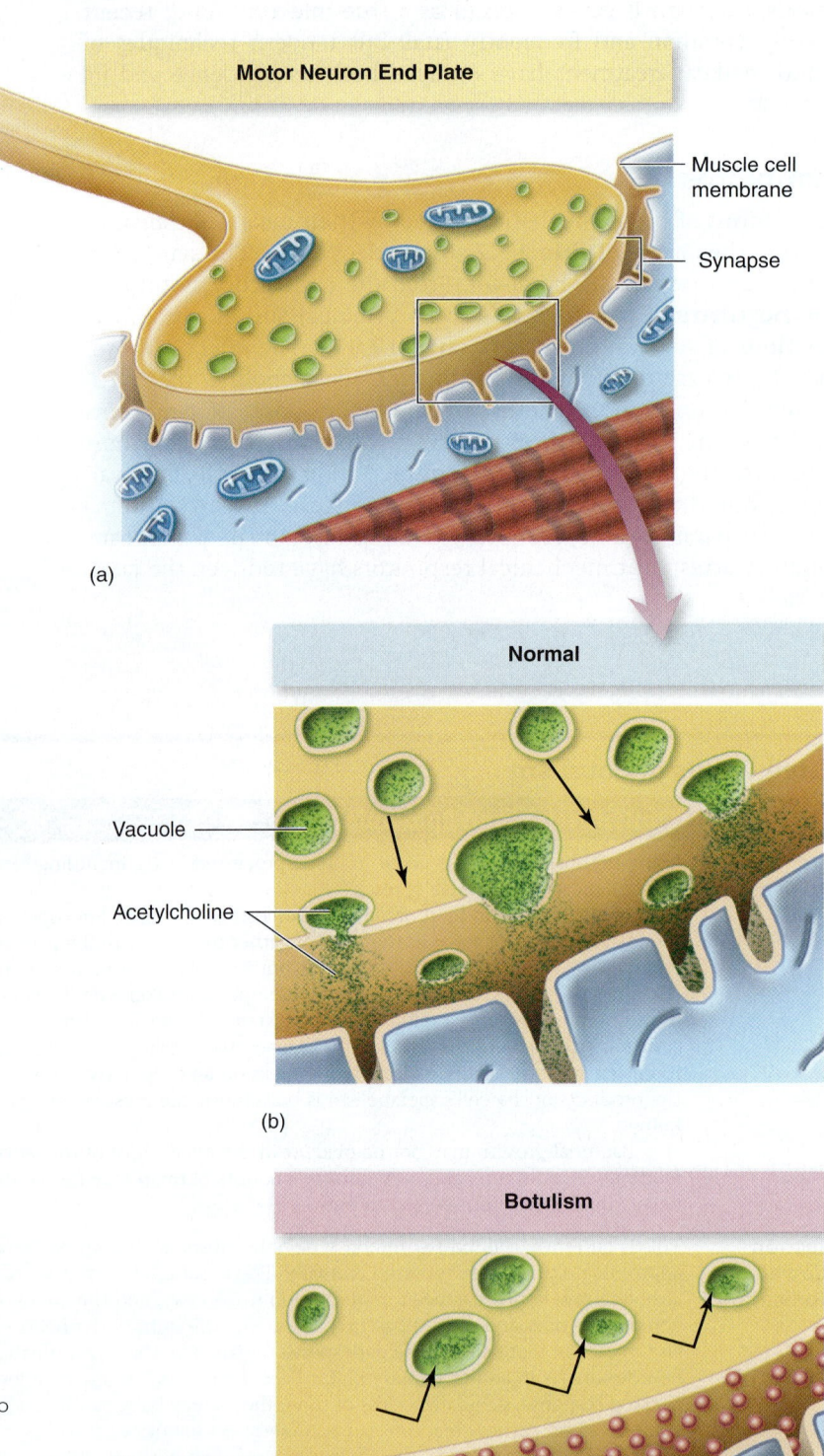

Figure 17.17 The physiological effects of botulism toxin (botulinum). (a) The relationship between the motor neuron and the muscle at the neuromuscular junction. (b) In the normal state, acetylcholine released at the synapse crosses to the muscle and creates an impulse that stimulates muscle contraction. (c) In botulism, the toxin enters the motor end plate and attaches to the presynaptic membrane, where it blocks release of the chemical. This prevents impulse transmission and keeps the muscle from contracting. This causes flaccid paralysis.

▶ Causative Agent

Clostridium botulinum, like *Clostridium tetani,* is an endospore-forming anaerobe that does its damage through the release of an exotoxin. *C. botulinum* commonly inhabits soil and water and occasionally the intestinal tract of animals. It is distributed worldwide but occurs most often in the Northern Hemisphere. The species has seven distinctly different types (designated A, B, C, D, E, F, and G) that vary in distribution among animals, regions of the world, and types of exotoxin. Human disease is usually associated with types A, B, E, and F, and animal disease with types A, B, C, D, and E.

Both *C. tetani* and *C. botulinum* produce neurotoxins; but tetanospasmin, the toxin made by *C. tetani,* results in spastic paralysis (uncontrolled muscle contraction). In contrast, botulinum, the *C. botulinum* neurotoxin, results in flaccid paralysis, a loss of ability to contract the muscles. The botulinum exotoxin is so potent, it is used (as Botox) to treat cross-eyes and uncontrollable blinking, and also for cosmetic purposes to reduce facial wrinkles caused by muscle contraction.

▶ Culture and Diagnosis

Diagnostic standards are slightly different for the three different presentations of botulism. Because minute amounts of the toxin are highly dangerous, laboratory testing should be performed only by experienced personnel. A suspected case of botulism should trigger a phone call to the state health department or the CDC before proceeding with diagnosis or treatment.

▶ Prevention and Treatment

The CDC maintains a supply of types A, B, and E trivalent horse antitoxin, which, when administered soon after diagnosis, can prevent the worst outcomes of the disease. Patients are also managed with respiratory and cardiac support systems. In all cases, hospitalization is required and recovery takes weeks. There is an overall 5% mortality rate.

Disease Table 17.9   Botulism

Causative Organism(s)	*Clostridium botulinum*
Most Common Modes of Transmission	Vehicle (food-borne toxin, airborne organism); direct contact (wound); parenteral (injection)
Virulence Factors	Botulinum exotoxin
Culture/ Diagnosis	Culture of organism; demonstration of toxin
Prevention	Food hygiene; toxoid immunization available for laboratory professionals
Treatment	Antitoxin, Penicillin G for wound botulism, supportive care
Epidemiological Features	United States: 75% of botulism is infant botulism; approximately 100–150 cases annually

---

17.3 LEARNING OUTCOMES—Assess Your Progress

4. List the possible causative agents, modes of transmission, virulence factors, diagnostic techniques, and prevention/treatment for the "Highlight Diseases," meningitis and poliomyelitis.

5. Identify the most common and also the most deadly of the multiple possible causes of meningitis.

6. Explain the difference between the oral polio vaccine and the inactivated polio vaccine, and under which circumstances each is appropriate.

7. Discuss important features of the diseases most directly involving the brain. These are meningoencephalitis, encephalitis, and subacute encephalitis.

8. Identify which encephalitis-causing viruses you should be aware of in your geographic area.

9. Discuss important features of the other diseases in the nervous system. These are rabies, poliomyelitis, tetanus, and botulism.

CASE FILE WRAP-UP

Scott, the victim of West Nile encephalitis in the opening Case File, contracted the disease through the bite of a mosquito. The disease can also be spread through blood transfusion and organ transplantation, but this is exceedingly rare. Infected mothers can pass on West Nile virus to their infants through the placenta prior to birth and via breast milk after birth.

In Scott's case, his age put him at increased risk of severe disease. Persons older than age 50 tend to be more severely affected by West Nile virus. In addition, Scott's use of steroids to control his rheumatoid arthritis might have suppressed his immune system sufficiently to have been a factor in his developing the severe form of the infection.

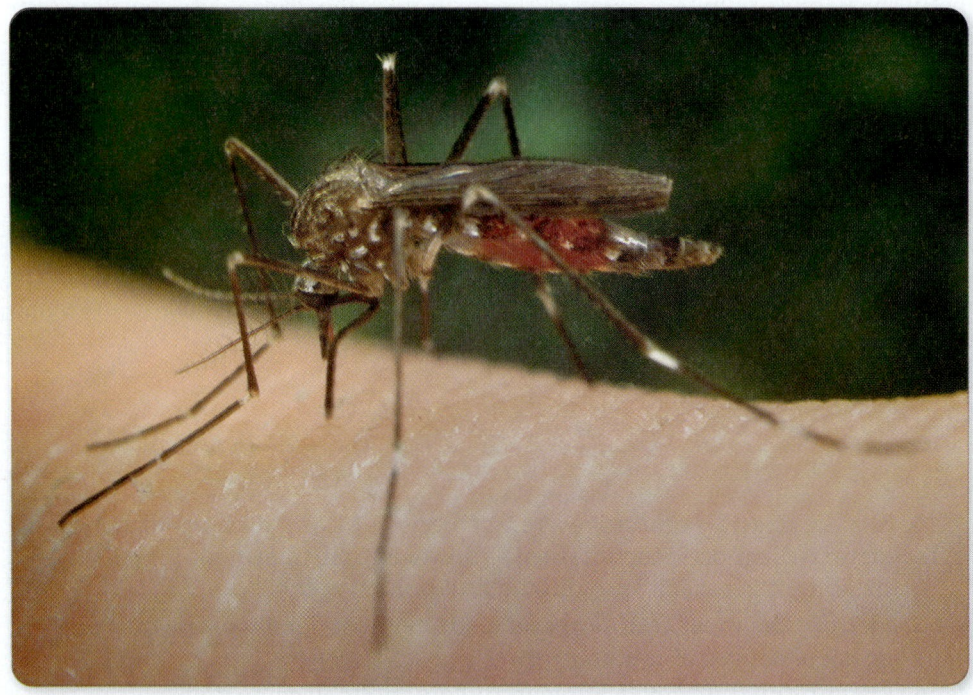

▶ Summing Up

Taxonomic Organization Microorganisms Causing Disease in the Nervous System

Microorganism	Pronunciation	Location of Disease Table
Gram-positive endospore-forming bacteria		
Clostridium botulinum	klos-trid″-ee-um bot′-yew-lin″-um	Botulism, p. 491
Clostridium tetani	klos-trid″-ee-um tet′-a-nie	Tetanus, p. 488
Gram-positive bacteria		
Streptococcus agalactiae	strep″-tuh-kok′-us ay-ga-lact′-tee-ay	Neonatal meningitis, p. 476
Streptococcus pneumoniae	strep″-tuh-kok′-us nu-mo′-nee-ay	Meningitis, p. 472
Listeria monocytogenes	lis-teer′-ee-uh mon′-oh-sy-toj″-eh-nees	Meningitis, p. 472
		Neonatal meningitis, p. 476
Gram-negative bacteria		
Cronobacter sakazakii	krow″-no-bak′-tur sock″-uh-zock′-ee	Neonatal meningitis, p. 476
Escherichia coli	esh′-shur-eesh″-ee-uh col′-eye	Neonatal meningitis, p. 476
Haemophilus influenzae	huh-mah′-fuh-luss in′-floo-en″-zay	Meningitis, p. 472
Neisseria meningitidis	nye-seer″-ee-uh′ men′-in-jit″-ih-dus	Meningitis, p. 472
DNA viruses		
Herpes simplex virus type 1 and 2	hur′-peez sim′-plex vie′-russ	Encephalitis, p. 482
JC virus	jay′-cee″ vie′-russ	Progressive multifocal leukoencephalopathy, p. 482
RNA viruses		
Arboviruses	ar′-bow-vie′-russ-suz	Encephalitis, p. 482
Western equine encephalitis virus, Eastern equine encephalitis virus, California encephalitis virus (California and LaCrosse strains), St. Louis encephalitis virus, West Nile virus	wes′-turn ee′-cwine en-sef′-ah-ly′-tiss vie′-russ, ee′-stern ee′-cwine en-sef′-ah ly″-tuss vie′-russ, cal′-i-for′-nee-uh en-sef′-ah-ly″-tiss vie′-russ, saynt lew′-iss en-cef′-ah-ly″-tiss vie′- russ, west ny′-il vie′-russ	
Poliovirus	poh′-lee-oh vie′-russ	Poliomyelitis, p. 478
Rabies virus	ray′-bees vie′-russ	Rabies, p. 486
Fungi		
Cryptococcus neoformans	crip-tuh-kok′-us nee′-oh-for″-mans	Meningitis, p. 472
Coccidioides	cox-sid″-ee-oid′-ees	Meningitis, p. 472
Prions		
Creutzfeldt-Jakob prion	croytz′-felt yaw′-cob pree′-on	Creutzfeldt-Jakob disease, p. 485
Protozoa		
Acanthamoeba	ay-kanth″-uh-mee′-buh	Meningoencephalitis, p. 479
Naegleria fowleri	nay-glar′-ee-uh fow-lahr′-ee	Meningoencephalitis, p. 479
Toxoplasma gondii	tox′-oh-plas″-mah gon′-dee-eye	Subacute encephalitis, p. 485

Surviving *Naegleria fowleri*

When 12-year-old Kayli Hardig went swimming in a water park in Arkansas in July, she and her family never foresaw the battle for Kayli's life that was to come. Kayli fell ill just days after swimming in the popular water park. Her symptoms included fever, headache, nausea and vomiting, and confusion, which eventually progressed to a coma. She was diagnosed with *Naegleria fowleri,* an amoebic infection that causes a deadly form of meningitis. Death can occur within 5 days of the onset of symptoms.

Naegleria fowleri is an almost-always fatal disease that kills 99% of the people who are diagnosed with it. Kayli is believed to be only the third person to have survived the infection in the United States. The infection is rare but not unheard of. There have even been at least two deaths in Minnesota in recent years. This is unusual since the organism is most common in southern states, and is thought to be a consequence of warming temperatures in North America. Kayli's battle is not over yet—she is still rehabilitating, learning to speak clearly and walk again.

The average age of victims infected with *N. fowleri* in the United States is 12. Children are more likely than adults to play in water. Activities that result in water forcefully entering the nose, such as diving and swimming under water, allow the amoeba access to the brain, where it quickly multiplies and destroys brain tissue in the frontal lobe. The infection results in massive swelling of the brain and eventual death. There is no cure, although antifungal medications such as amphotericin B are often given.

Although public panic is understandable following reports of an amoeba that is often referred to as a "brain-eating amoeba," note that the amoeba cannot survive in chlorinated water. Chlorinated swimming pools are perfectly safe to swim in; rather, warm, stagnant bodies of water are the ones that pose a risk, such as ponds, shallow lakes, and even slow-moving rivers. Swimming is also not to blame; rather, activities that result in water being forced up the nose are the high-risk activities. Given the numbers of children who swim in such areas during the hot summer months, few people become infected. This has led some scientists to speculate that many of us have antibodies to the disease.

In August 2013, Zachary Reyna of Florida was kneeboarding in a ditch with friends when he contracted the deadly amoeba. He fell ill soon afterward and was in a coma when his family made the difficult decision to remove him from life support on August 26. His family hoped that Zachary would survive as Kayli did, making him the fourth known person to survive the disease, but it was sadly not to be.

Pia mater

Abundance of inflammatory neutrophils

Infectious Diseases Affecting
The Nervous System

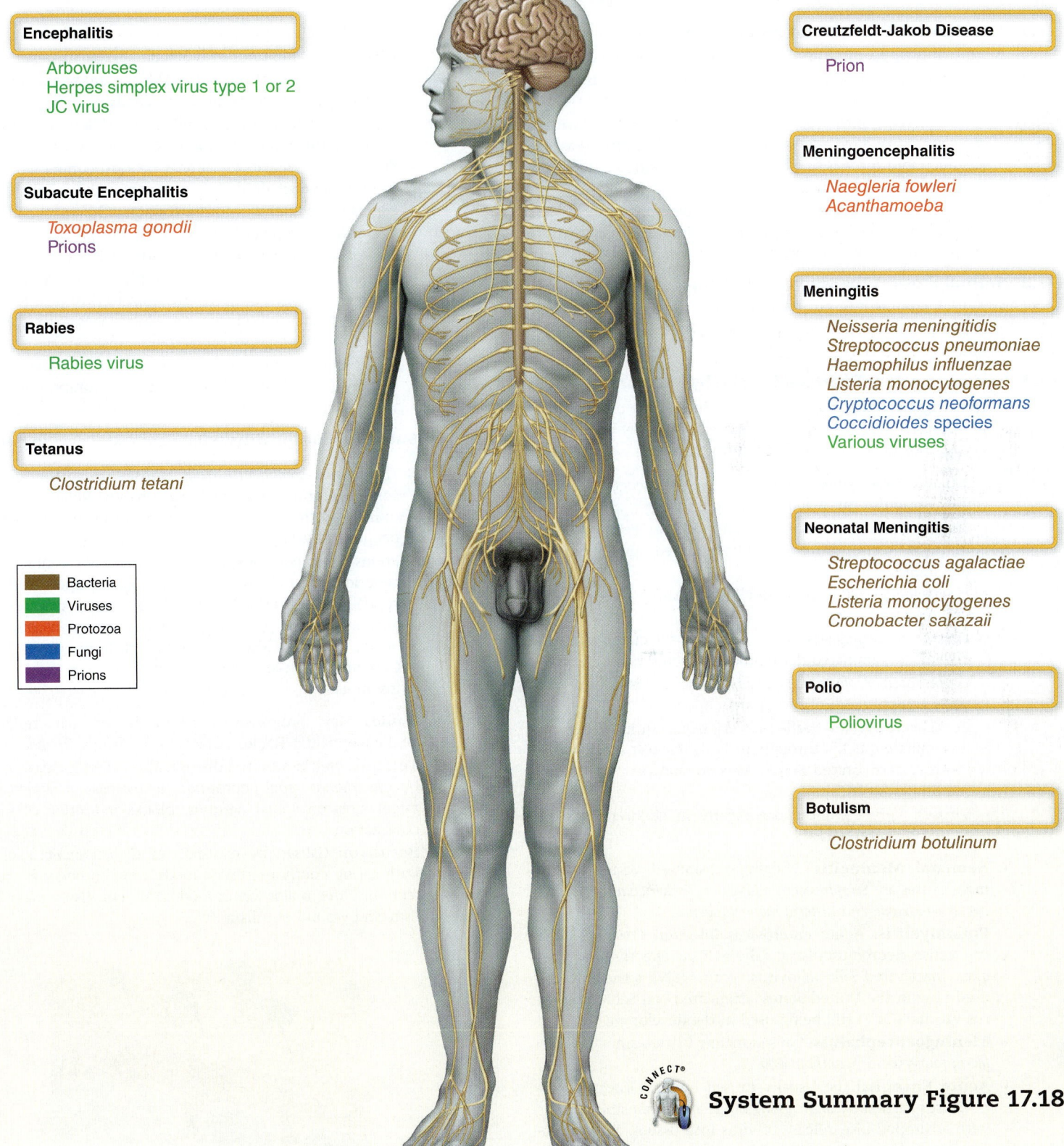

Encephalitis

Arboviruses
Herpes simplex virus type 1 or 2
JC virus

Subacute Encephalitis

Toxoplasma gondii
Prions

Rabies

Rabies virus

Tetanus

Clostridium tetani

▉	Bacteria
▉	Viruses
▉	Protozoa
▉	Fungi
▉	Prions

Creutzfeldt-Jakob Disease

Prion

Meningoencephalitis

Naegleria fowleri
Acanthamoeba

Meningitis

Neisseria meningitidis
Streptococcus pneumoniae
Haemophilus influenzae
Listeria monocytogenes
Cryptococcus neoformans
Coccidioides species
Various viruses

Neonatal Meningitis

Streptococcus agalactiae
Escherichia coli
Listeria monocytogenes
Cronobacter sakazaii

Polio

Poliovirus

Botulism

Clostridium botulinum

System Summary Figure 17.18

Chapter Summary

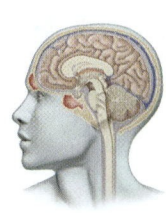

17.1 The Nervous System and Its Defenses

- The nervous system has two parts: the central nervous system (the brain and spinal cord), and the peripheral nervous system (nerves and ganglia).
- The soft tissue of the brain and spinal cord is encased within the tough casing of three membranes called the *meninges*. The subarachnoid space is filled with a clear serumlike fluid called cerebrospinal fluid (CSF).
- The nervous system is protected by the *blood-brain barrier*, which limits the passage of substances from the bloodstream to the brain and spinal cord.

17.2 Normal Biota of the Nervous System

- At the present time, we believe there is no normal biota in either the central nervous system (CNS) or the peripheral nervous system (PNS).

17.3 Nervous System Diseases Caused by Microorganisms

- **Meningitis:** Inflammation of the meninges. The more serious forms are caused by bacteria, often facilitated by coinfection or previous infection with respiratory viruses.

  - *Neisseria meningitidis:* Gram-negative diplococcus; causes most serious form of acute meningitis.
  - *Streptococcus pneumoniae:* Gram-positive coccus; most frequent cause of community-acquired pneumococcal meningitis.
  - *Haemophilus influenzae:* Declined sharply because of vaccination.
  - *Listeria monocytogenes:* Most cases are associated with ingesting contaminated dairy products, poultry, and meat.
  - *Cryptococcus neoformans:* Fungus; causes chronic form with more gradual onset of symptoms.
  - *Coccidioides* species: "Valley Fever"; begins in lungs but can disseminate quickly throughout body; highest incidence in southwestern United States, Mexico, and parts of Central and South America.
  - Viruses: Very common, particularly in children; 90% are caused by enteroviruses.

- **Neonatal Meningitis:** Usually transmitted vertically. Primary causes are *Streptococcus agalactiae, Escherichia coli, Cronobacter sakazakii,* and *Listeria monocytogenes.*
- **Poliomyelitis:** Acute enterovirus infection of spinal cord; can cause neuromuscular paralysis. Two effective vaccines exist: Inactivated Salk poliovirus vaccine (IPV) is the only one used now in the United States; attenuated oral Sabin poliovirus vaccine (OPV) still being used in the developing world.
- **Meningoencephalitis:** Caused mainly by two amoebas, *Naegleria fowleri* and *Acanthamoeba.*
- **Acute Encephalitis:** Usually caused by viral infection. Arboviruses carried by arthropods often are responsible. Begins with arthropod bite, release of virus into tissues, and replication in nearby lymphatic tissues.

- Western equine encephalitis (WEE): Occurs sporadically in western United States and Canada.
- Eastern equine encephalitis (EEE): Endemic to eastern coasts of the United States and Canada.
- California encephalitis: Caused by two different viral strains, the California strain and the LaCrosse strain.
- St. Louis encephalitis (SLE): May be most common of American viral encephalitides. Appears throughout North, South America; epidemics occur most often in Midwest and South.
- West Nile encephalitis: West Nile virus is close relative of SLE virus. Emerged in United States in 1999.
- Herpes simplex virus: Herpes simplex virus types 1 and 2 cause encephalitis in newborns born to HSV-positive mothers, older children and young adults (ages 5 to 30), older adults.
- JC virus: Can cause progressive multifocal leukoencephalopathy (PML), particularly in immunocompromised individuals. Fatal infection.

- **Subacute Encephalitis:** Symptoms take longer to manifest.

  - *Toxoplasma gondii:* Protozoan, causes toxoplasmosis, most common form of subacute encephalitis. Relatively asymptomatic in the healthy; can be severe in immunodeficient people and fetuses.
  - Prions: Proteinaceous infectious particles containing no genetic material. Cause transmissible spongiform encephalopathies (TSEs), neurodegenerative diseases with long incubation periods but rapid progressions once they begin. Human TSEs are Creutzfeldt-Jakob disease (CJD), Gerstmann-Strussler-Scheinker disease, and fatal familial insomnia.

- **Rabies:** Slow, progressive zoonotic disease characterized by fatal encephalitis. Rabies virus is in the family *Rhabdoviridae.*
- **Tetanus:** Neuromuscular disease, also called lockjaw; caused by *Clostridium tetani* neurotoxin, *tetanospasmin,* which binds target sites on spinal neurons, blocks inhibition of muscle contraction.
- **Botulism:** Caused by exotoxin of *C. botulinum;* associated with eating poorly preserved foods; can also occur as true infection. Three major forms: food-borne botulism, infant botulism, and wound botulism.

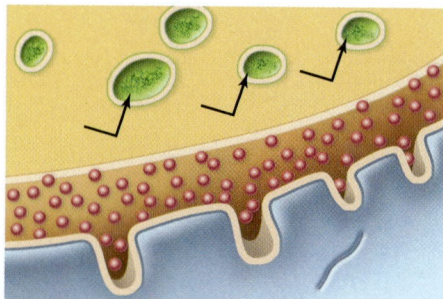

Multiple-Choice Questions   Bloom's Levels 1 and 2: Remember and Understand

Select the correct answer from the answers provided.

1. Cerebrospinal fluid is found in the
   - a. meninges.
   - b. dura mater.
   - c. subarachnoid space.
   - d. ganglia.
   - e. peripheral nerves.

2. The first choice antibiotic for bacterial meningitis is the broad-spectrum
   - a. cephalosporin.
   - b. penicillin.
   - c. ampicillin.
   - d. vancomycin.

3. Subacute encephalitis can be caused by
   - a. *Toxoplasma gondii.*
   - b. *Streptococcus agalactiae.*
   - c. *Naegleria fowleri.*
   - d. *Neisseria meningitidis.*
   - e. *Haemophilus influenzae.*

4. Which of the following neurological diseases is not caused by a prion?
   - a. Creutzfeldt-Jakob disease
   - b. scrapie
   - c. mad cow disease
   - d. St. Louis encephalitis

5. *Cryptococcus neoformans* is primarily transmitted by
   - a. direct contact.
   - b. bird droppings.
   - c. fomites.
   - d. sexual activity.

6. CJD is caused by a/an
   - a. arbovirus.
   - b. prion.
   - c. protozoan.
   - d. bacterium.

7. What food should you avoid feeding a child under 1 year old because of potential botulism?
   - a. honey
   - b. milk
   - c. apple juice
   - d. applesauce

8. Which of the following is *not* caused by an arbovirus?
   - a. St. Louis encephalitis
   - b. Eastern equine encephalitis
   - c. West Nile encephalitis
   - d. mad cow disease

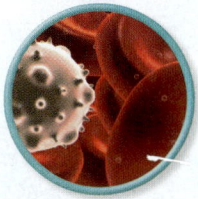

Critical Thinking   Bloom's Levels 3, 4, and 5: Apply, Analyze, and Evaluate

Critical thinking is the ability to reason and solve problems using facts and concepts. These questions can be approached from a number of angles and, in most cases, they do not have a single correct answer.

1. Why is encephalitis often difficult to diagnosis?

2. Propose a hypothesis to explain the demonstrated effect of *Toxoplasma gondii* on rats and its possible effects on humans. Include a consideration of the evolutionary benefits these effects might confer on the protozoan.

3. In the section on meningococcal meningitis, the following sentence appears: "If no samples were obtained prior to antibiotic treatment, a PCR test is the best bet for identifying the pathogen." Why?

4. Why is there no normal biota associated with the nervous system?

5. Even though the oral polio vaccine is not used in the developed world, it is still widely used in the developing world, in part because it confers what might be called "accidental" herd immunity. Can you speculate on what this is?

Visual Connections   Bloom's Level 5: Evaluate

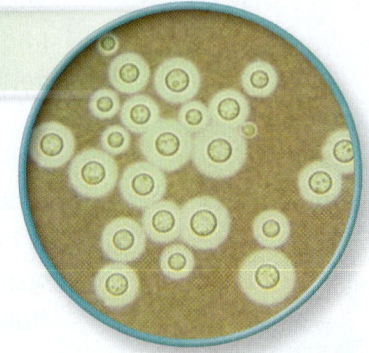

This question connects previous images to a new concept.

1. **From chapter 2, figure 2.19a.** Without looking back to the figure in chapter 2, speculate on which meningitis-causing organism you are seeing here. How could your presumptive diagnosis be confirmed?

www.mcgrawhillconnect.com

Enhance your study of this chapter with study tools and practice tests. Also ask your instructor about the resources available through Connect, including the media-rich eBook, interactive learning tools, and animations.

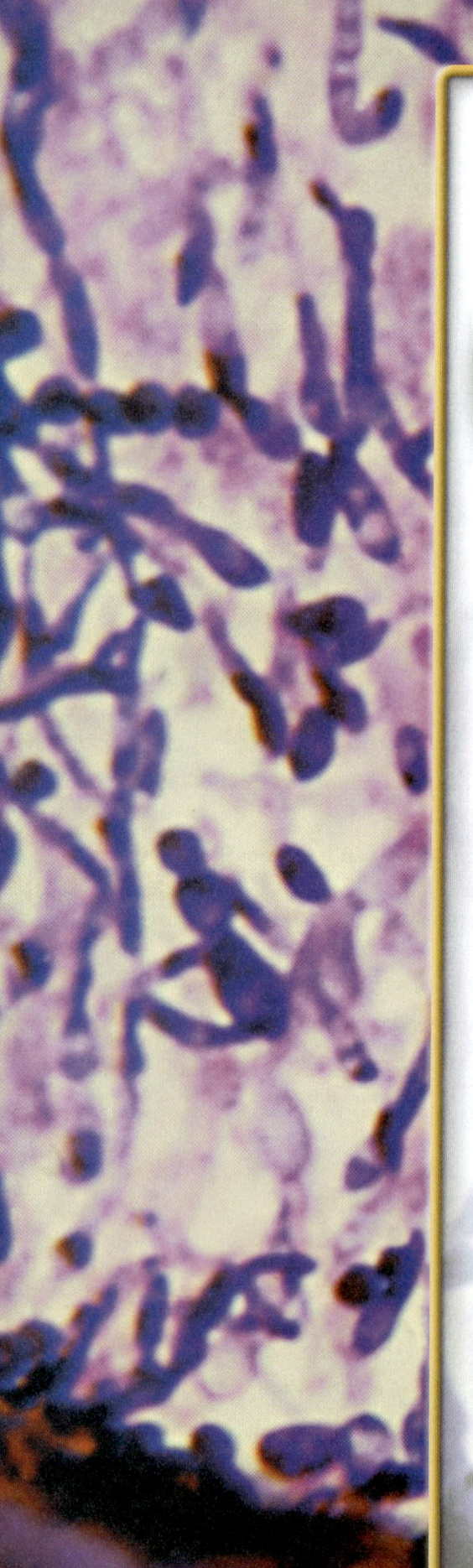

CASE FILE

Heartache

Donna, a 22-year-old university student, came to the emergency room on a Friday night at the beginning of the school break for Christmas holidays. She had just completed taking her final exams. When she got home to her parents' house, her mother took one look at her and insisted she go to the emergency room.

Donna had been complaining to her mother about feeling unwell for several weeks. When Donna had been home for Thanksgiving, she had been her usual self. She had had two wisdom teeth removed at that time but had healed quickly and denied any lingering oral symptoms.

Shortly after Thanksgiving, Donna had begun to experience an intermittent fever and weakness. She stated that her joints were often painful. In the past several days, she stated that she had also had palpitations, feeling as though her heart were racing and beating erratically. She complained of feeling exhausted. Donna had chalked up her symptoms to the stress of her workload, too much coffee, and final exams. Donna's mother stated that Donna looked as though she had lost weight since her last visit home a month earlier.

On exam, Donna had a temperature of 100.8°F (38.2°C). Her heart rate was abnormally fast at 112 beats per minute. Her blood pressure was low at 95/62 mmHg, and her oxygen saturation was 93%, which was a little lower than I expected for someone of Donna's age and condition.

I listened to Donna's lungs, which were clear. When I listened to Donna's heart, I thought I detected a murmur. When I questioned Donna and her mother, Donna's mother stated that Donna had had a heart murmur since she was an infant, but that she had been told it was nothing to be concerned about. Donna was very pale. When I checked her nail beds for signs of inadequate oxygenation, I noted that on several of Donna's fingernails there were dark lines running vertically. Both of Donna's palms had small red spots present.

The doctor on call came in to examine Donna. After examining her and noting the heart murmur, the lines on her nails, and the spots on her palms, he ordered an ECG (echocardiogram) and an array of blood tests, including a set of blood cultures. He spent time asking Donna and her mother about Donna's wisdom tooth extraction and her heart murmur. He told Donna and her mother that, based on her symptoms, he believed that Donna had subacute endocarditis.

- Which layer of the heart is affected in endocarditis?

- How might Donna have contracted endocarditis?

Case File Wrap-Up appears on page 526.

CHAPTER 18

Infectious Diseases Affecting the Cardiovascular and Lymphatic Systems

IN THIS CHAPTER...

18.1 The Cardiovascular and Lymphatic Systems and Their Defenses

1. Describe the important anatomical features of the cardiovascular and lymphatic systems.
2. List the natural defenses present in the cardiovascular and lymphatic systems.

18.2 Normal Biota of the Cardiovascular and Lymphatic Systems

3. Explain the "what" and the "why" of the normal biota of the cardiovascular and lymphatic systems.

18.3 Cardiovascular and Lymphatic System Diseases Caused by Microorganisms

4. List the possible causative agents, modes of transmission, virulence factors, diagnostic techniques, and prevention/treatment for the "Highlight Diseases" malaria and HIV.
5. Discuss the epidemiology of malaria.
6. Describe the epidemiology of HIV infection in the developing world.
7. Discuss the important features of infectious cardiovascular diseases that have more than one possible cause. These are the two forms of endocarditis, septicemia, hemorrhagic fever diseases, and nonhemorrhagic fever diseases.
8. Identify factors that distinguish hemorrhagic and nonhemorrhagic fever diseases.
9. Outline the series of events that may lead to septicemia and how it should be prevented and treated.
10. Discuss the important features of infectious cardiovascular diseases that have only one possible cause. These are plague, tularemia, Lyme disease, infectious mononucleosis, Chagas disease, and anthrax.
11. Describe what makes anthrax a good agent for bioterrorism, and list the important presenting signs to look for in patients.

18.1 The Cardiovascular and Lymphatic Systems and Their Defenses

The Cardiovascular System

The *cardiovascular system* is the pipeline of the body. It is composed of the blood vessels, which carry blood to and from all regions of the body, and the heart, which pumps the blood. This system moves the blood in a closed circuit, and it is therefore known as the *circulatory system*. The cardiovascular system provides tissues with oxygen and nutrients and carries away carbon dioxide and waste products, delivering them to the appropriate organs for removal. A closely related but largely separate system, the *lymphatic system*, is a major source of immune cells and fluids, and it serves as a one-way passage, returning fluid from the tissues to the cardiovascular system. **Figure 18.1** shows you how the two systems work together. You first saw this figure in chapter 12.

Figure 18.1 The anatomy of the cardiovascular and lymphatic systems.

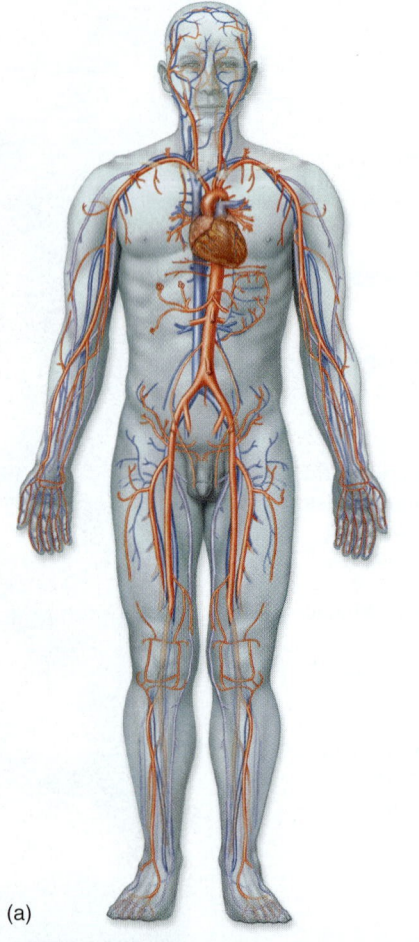

(a)

The Circulatory System: Surveillance

Body compartments are screened by circulating WBCs.

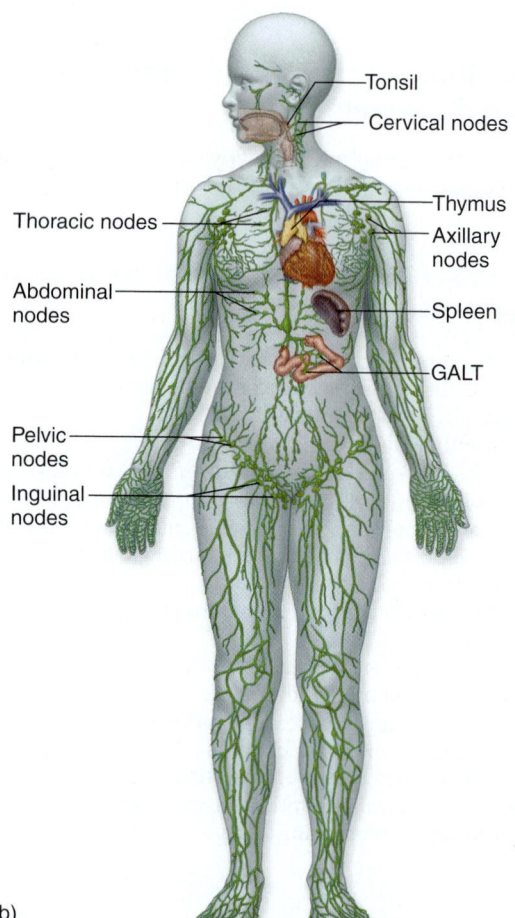

(b)

- Tonsil
- Cervical nodes
- Thymus
- Axillary nodes
- Spleen
- GALT

Thoracic nodes
Abdominal nodes
Pelvic nodes
Inguinal nodes

The Lymphatic System

The lymphatic system consists of a branching network of vessels that extend into most body areas. Note the higher density of lymphatic vessels in the "dead-end" areas of the hands, feet, and breast, which are frequent contact points for infections. Other lymphatic organs include the lymph nodes, spleen, gut-associated lymphoid tissue (GALT), the thymus, and the tonsils.

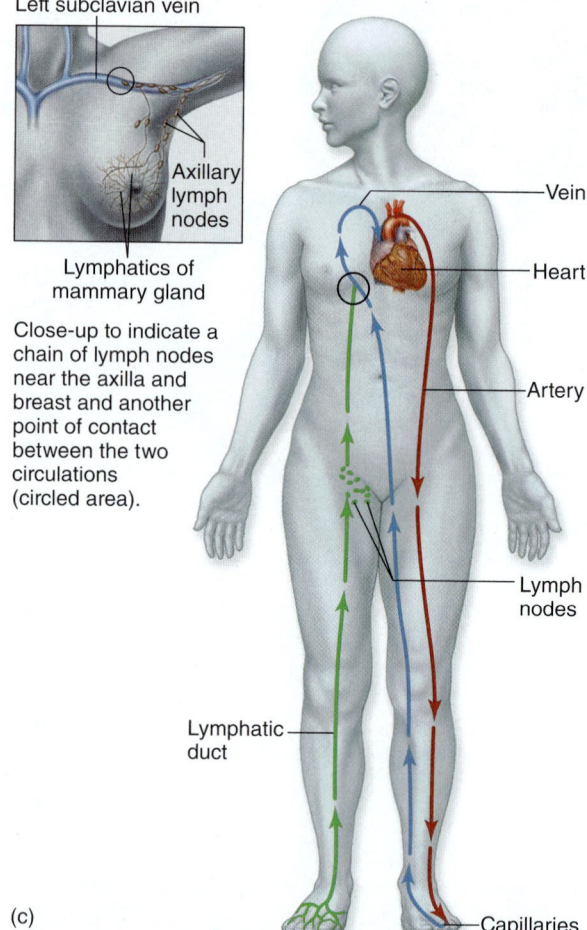

(c)

Left subclavian vein

Axillary lymph nodes

Lymphatics of mammary gland

Close-up to indicate a chain of lymph nodes near the axilla and breast and another point of contact between the two circulations (circled area).

- Vein
- Heart
- Artery
- Lymph nodes
- Capillaries

Lymphatic duct

The Lymphatic and Circulatory Systems

Comparison of the generalized circulation of the lymphatic system and the blood. Although the lymphatic vessels parallel the regular circulation, they transport in only one direction unlike the cyclic pattern of blood. Direct connection between the two circulations occurs at points near the heart where large lymph ducts empty their fluid into veins (circled area).

The heart is a fist-size muscular organ that pumps blood through the body. It is divided into two halves, each of which is divided into an upper and lower chamber **(figure 18.2).** The upper chambers are called atria (singular, *atrium*), and the lower are ventricles. The entire organ is encased in a fibrous covering, the pericardium, which is an occasional site of infection. The actual wall of the heart has three layers: from outer to inner, they are the epicardium, the myocardium, and the endocardium. The endocardium also covers the valves of the heart, and it is a relatively common target of microbial infection.

The blood vessels consist of *arteries, veins,* and *capillaries.* Arteries carry oxygenated blood away from the heart under relatively high pressure. They branch into smaller vessels called arterioles. Veins actually begin as smaller venules in the periphery of the body and coalesce into veins. The smallest blood vessels, the capillaries, connect arterioles to venules. Both arteries and veins have walls made of three layers of tissue. The innermost layer is composed of a smooth epithelium called endothelium. Its smooth surface encourages the smooth flow of cells and platelets through the system. The next layer is composed of connective tissue and muscle fibers. The outside layer is a thin layer of connective tissue. Capillaries, the smallest vessels, have walls made of only one layer of endothelium.

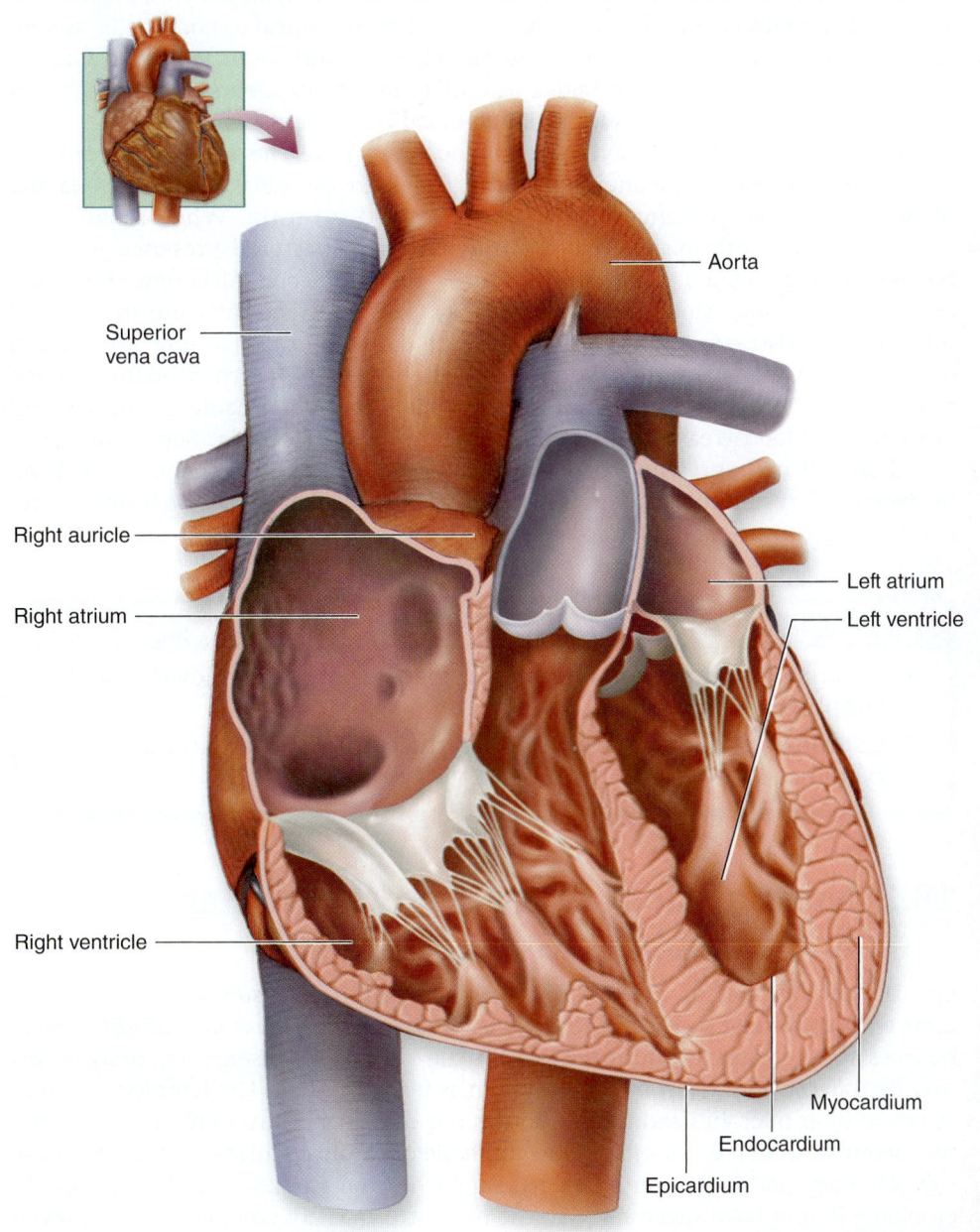

Conceptual image of blood cells in an artery.

Figure 18.2 **The heart.**

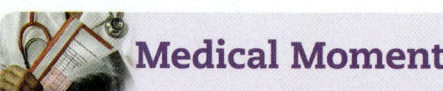

Medical Moment

Lymphangitis

Lymphangitis is an infection of the lymph vessels, which can occur as a result of bacterial infection. Staphylococcal and streptococcal skin infections are a common cause of lymphangitis. Lymphangitis can be a serious condition that may progress rapidly to sepsis if improperly treated.

Generally, an individual with a skin infection will notice a red streak spreading toward the armpit or groin area. Fever, pain, and flulike symptoms, as well as enlarged lymph nodes (lymphadenopathy) are all common symptoms of lymphangitis. Often the lymph nodes closest to the infected area are the ones affected.

Prompt recognition and treatment are essential. A swab of the affected area or blood cultures may yield the causative organism. Antibiotics should be started immediately. Anti-inflammatory drugs may help to reduce inflammation. In addition, warm compresses may help to decrease swelling and ease discomfort.

Lymphangitis may be confused with thrombophlebitis. However, lymphangitis can often be recognized by the presence of a skin infection and other signs of infection such as fever and chills. Most people recover from lymphangitis without any complications, but swelling of the affected area may continue for several weeks.

The Lymphatic System

The lymphatic system consists mainly of the lymph vessels, which roughly parallel the blood vessels; lymph nodes, which cluster at body sites such as the groin, neck, armpit, and intestines; and the spleen. It serves to collect fluid that has left the blood vessels and entered tissues, filter it of impurities and infectious agents, and return it to the blood.

Defenses of the Cardiovascular and Lymphatic Systems

The cardiovascular system is highly protected from microbial infection. Microbes that successfully invade the system, however, gain access to every part of the body, and every system may potentially be affected. For this reason, bloodstream infections are called **systemic** infections.

Multiple defenses against infection reside in the bloodstream. The blood is full of leukocytes, with approximately 5,000 to 10,000 white blood cells per milliliter of blood. The various types of white blood cells include the lymphocytes, responsible for specific immunity, and the phagocytes, which are so critical to nonspecific as well as specific immune responses. Very few microbes can survive in the blood with so many defensive elements. That said, a handful of infectious agents have nonetheless evolved exquisite mechanisms for avoiding blood-borne defenses.

Medical conditions involving the blood often have the suffix *-emia*. For instance, viruses that cause meningitis can travel to the nervous system via the bloodstream. Their presence in the blood is called **viremia.** When fungi are in the blood, the condition is termed **fungemia,** and bacterial presence is called **bacteremia,** a general term denoting only their *presence*. Although the blood contains no normal biota (see next section), bacteria frequently are introduced into the bloodstream during the course of daily living. Brushing your teeth or tearing a hangnail can introduce bacteria from the mouth or skin into the bloodstream; this situation is usually temporary. But when bacteria flourish and grow in the bloodstream, the condition is termed **septicemia.** Septicemia (also called sepsis) can very quickly lead to cascading immune responses, resulting in decreased systemic blood pressure, which can lead to **septic shock,** a life-threatening condition.

18.1 LEARNING OUTCOMES—Assess Your Progress

1. Describe the important anatomical features of the cardiovascular and lymphatic systems.
2. List the natural defenses present in the cardiovascular and lymphatic systems.

18.2 Normal Biota of the Cardiovascular and Lymphatic Systems

Like the nervous system, the cardiovascular and lymphatic systems are "closed" systems with no normal access to the external environment. Therefore, current science believes they possess no normal biota. In the absence of disease, microorganisms may be transiently present in either system as just described. The lymphatic system serves to filter microbes and their products out of tissues. Thus, in the healthy state, no microorganisms *colonize* either the lymphatic or cardiovascular systems. Of course, this is biology, and it is never quite that simple. Recent studies from the Human Microbiome Project have suggested that the bloodstream is not completely sterile, even during periods of apparent health. It is tempting to speculate that these low-level microbial "infections" may contribute to diseases for which no etiology has previously been found or for conditions currently thought to be noninfectious.

Cardiovascular and Lymphatic Systems Defenses and Normal Biota

	Defenses	Normal Biota
Cardiovascular System	Blood-borne components of nonspecific and specific immunity–including phagocytosis, specific immunity	None
Lymphatic System	Numerous immune defenses reside here.	None

18.2 LEARNING OUTCOMES—Assess Your Progress

3. Explain the "what" and the "why" of the normal biota of the cardiovascular and lymphatic systems.

18.3 Cardiovascular and Lymphatic System Diseases Caused by Microorganisms

Categorizing cardiovascular and lymphatic infections according to clinical presentation is somewhat difficult because most of these conditions are systemic, with effects on multiple organ systems. We start with two extremely important conditions, malaria and HIV.

Highlight Disease

Malaria

From prehistoric time until the present, malaria has been one of the greatest afflictions, in the same rank as bubonic plague, influenza, and tuberculosis. Even now, as the dominant protozoal disease, it threatens 40% of the world's population every year. The origin of the name is from the Italian words *mal,* "bad," and *aria,* "air."

▶ Signs and Symptoms

After a 10- to 16-day incubation period, the first symptoms are malaise, fatigue, vague aches, and nausea with or without diarrhea, followed by bouts of chills, fever, and sweating. These symptoms occur at 48- or 72-hour intervals, as a result of the synchronous rupturing of red blood cells. The interval, length, and regularity of symptoms reflect the type of malaria. Patients with falciparum malaria, the most virulent type, often display persistent fever, cough, and weakness for weeks without relief. Complications of malaria are hemolytic anemia from lysed blood cells and organ enlargement and rupture due to cellular debris that accumulates in the spleen, liver, and kidneys. One of the most serious complications of falciparum malaria is termed *cerebral malaria.* In this condition, small blood vessels in the brain become obstructed due to the increased ability of red blood cells (RBCs) to adhere to vessel walls (a condition called *cytoadherence* induced by the infecting protozoan). The resulting decrease in oxygen in brain tissue can result in coma and death. In general, malaria has the highest death rate in the acute phase, especially in children. Certain kinds of malaria (those caused by *Plasmodium vivax* and *P. ovale*) are subject to relapses because some infected liver cells harbor dormant protozoans for up to 5 years.

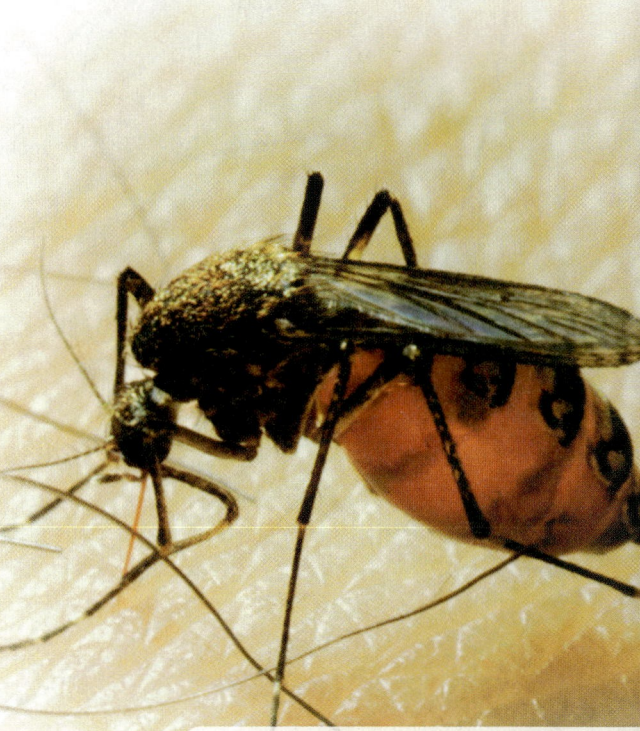

Mosquitos transmit the malaria protozoan.

▶ Causative Agent

Plasmodium species are protozoa in the sporozoan group. The genus *Plasmodium* contains over 200 species, but only five are known to commonly infect humans: *P. malariae, P. vivax, P. knowlesi, P. ovale,* and *P. falciparum.* The five species show variations in the pattern and severity of disease. For instance, *P. falciparum* is responsible for the vast majority of deaths.

Table 18.1 Life Cycle of the Malarial Parasite

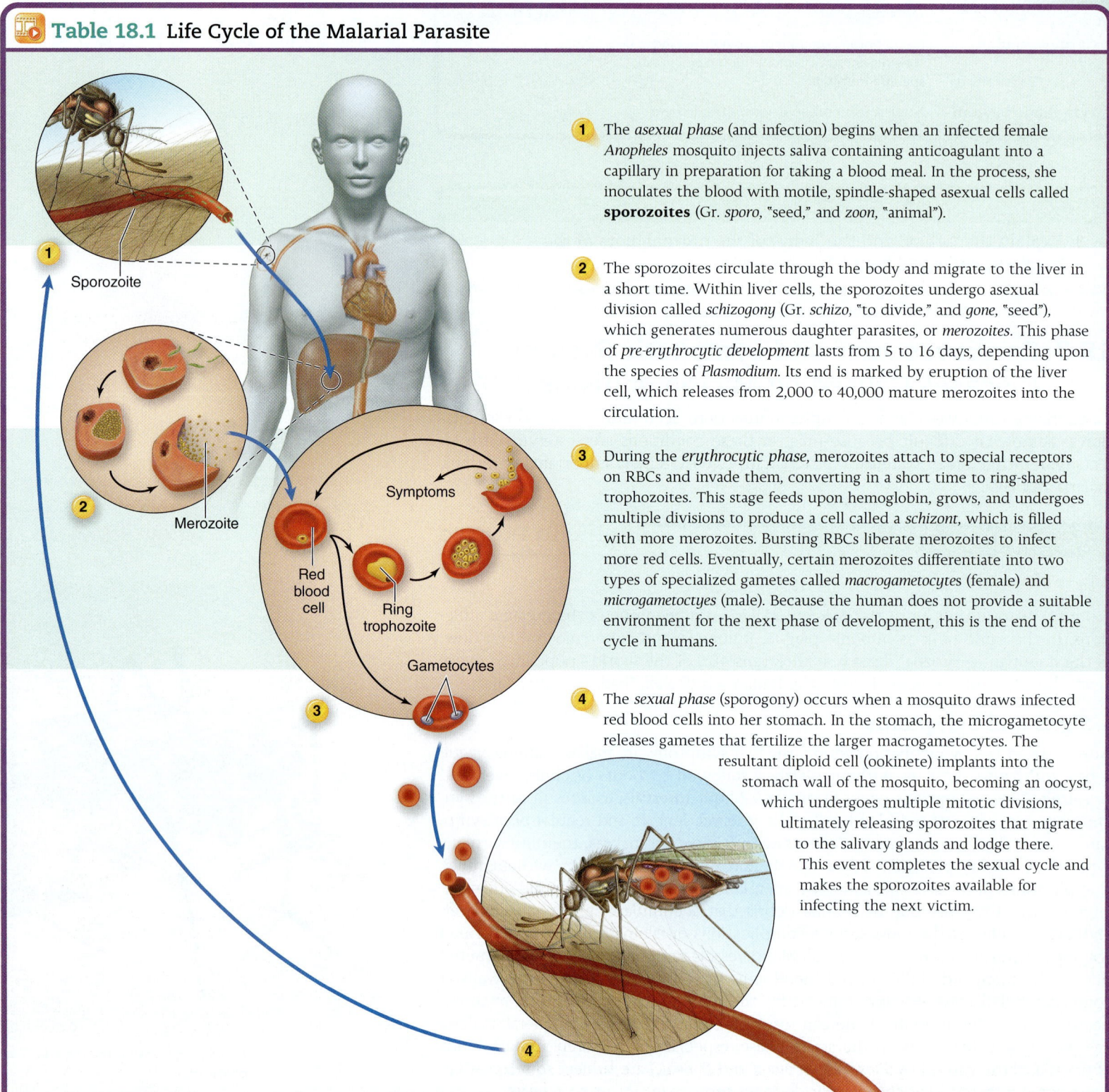

1 The *asexual phase* (and infection) begins when an infected female *Anopheles* mosquito injects saliva containing anticoagulant into a capillary in preparation for taking a blood meal. In the process, she inoculates the blood with motile, spindle-shaped asexual cells called **sporozoites** (Gr. *sporo,* "seed," and *zoon,* "animal").

2 The sporozoites circulate through the body and migrate to the liver in a short time. Within liver cells, the sporozoites undergo asexual division called *schizogony* (Gr. *schizo,* "to divide," and *gone,* "seed"), which generates numerous daughter parasites, or *merozoites.* This phase of *pre-erythrocytic development* lasts from 5 to 16 days, depending upon the species of *Plasmodium.* Its end is marked by eruption of the liver cell, which releases from 2,000 to 40,000 mature merozoites into the circulation.

3 During the *erythrocytic phase,* merozoites attach to special receptors on RBCs and invade them, converting in a short time to ring-shaped trophozoites. This stage feeds upon hemoglobin, grows, and undergoes multiple divisions to produce a cell called a *schizont,* which is filled with more merozoites. Bursting RBCs liberate merozoites to infect more red cells. Eventually, certain merozoites differentiate into two types of specialized gametes called *macrogametocytes* (female) and *microgametoctyes* (male). Because the human does not provide a suitable environment for the next phase of development, this is the end of the cycle in humans.

4 The *sexual phase* (sporogony) occurs when a mosquito draws infected red blood cells into her stomach. In the stomach, the microgametocyte releases gametes that fertilize the larger macrogametocytes. The resultant diploid cell (ookinete) implants into the stomach wall of the mosquito, becoming an oocyst, which undergoes multiple mitotic divisions, ultimately releasing sporozoites that migrate to the salivary glands and lodge there. This event completes the sexual cycle and makes the sporozoites available for infecting the next victim.

Development of the malarial parasite is divided into two distinct phases: the asexual phase, carried out in the human, and the sexual phase, carried out in the mosquito. **Table 18.1** lists the steps of the malarial life cycle.

Figure 18.3 illustrates the ring trophozoite stage in a malarial infection.

▶ Pathogenesis and Virulence Factors

The invasion of the merozoites into RBCs leads to the release of fever-inducing chemicals into the bloodstream. Chills and fevers often occur in a cyclic pattern. *Plasmodium* also metabolizes glucose at a very high rate, leading to hypoglycemia in the human host. The damage to RBCs results in anemia. The accumulation of malarial products in the liver and the immune stimulation in the spleen can lead to enlargement of these organs. Individual protozoa within a host can express distinctly different surface antigens, making it difficult for the vertebrate immune system to battle.

▶ Transmission and Epidemiology

All forms of malaria are spread primarily by the female *Anopheles* mosquito. Although malaria was once distributed throughout most of the world, the control of mosquitoes in temperate areas has successfully restricted it mostly to a belt extending around the equator **(figure 18.4)**. Despite this achievement, approximately 300 million new cases are still reported each year, about 90% of them in Africa. The most frequent victims are children and young adults, of whom at least 1 million die annually. A particular form of the malarial protozoan causes damage to the placenta in pregnant women, leading to excess mortality among fetuses and newborns. The total case rate in the United States is about 1,000 to 2,000 new cases a year. While most of these are found in people who acquired it in a known endemic area, locally transmitted infections are on the rise.

▶ Culture and Diagnosis

Malaria can be diagnosed definitively by the discovery of a typical stage of *Plasmodium* in stained blood smears (see figure 18.3). Newer serological procedures have made diagnosis more accurate while requiring less skill to perform. Other indications are knowledge of the patient's residence or travel in endemic areas and symptoms such as recurring chills, fever, and sweating.

Red blood cell   Ring trophozoites

Figure 18.3 The ring trophozoite stage in a *Plasmodium falciparum* infection. A smear of peripheral blood shows ring forms in red blood cells. Some RBCs have multiple trophozoites.

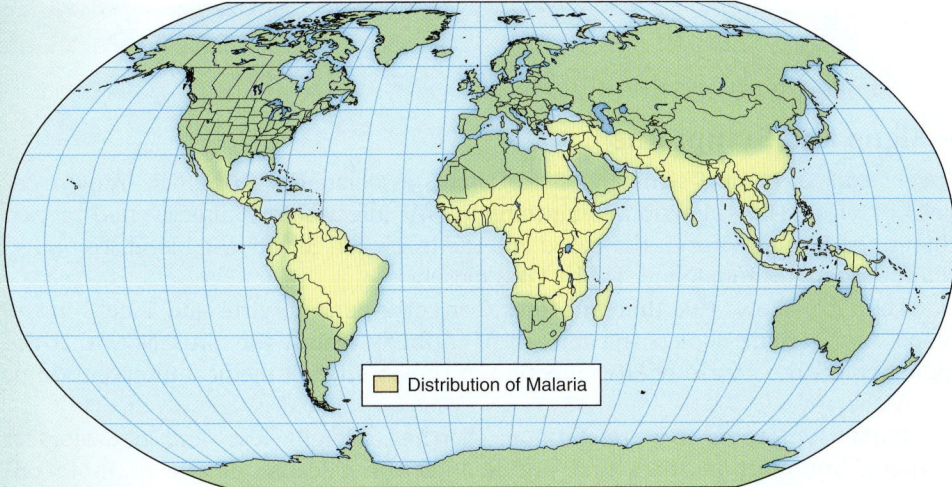

Distribution of Malaria

Figure 18.4 The malaria belt. Yellow zones outline the major regions that harbor malaria. The malaria belt corresponds to a band around the equator.

Disease Table 18.1 Malaria	
Causative Organism(s)	*Plasmodium falciparum, P. vivax, P. ovale, P. malariae, P. knowlesi*
Most Common Modes of Transmission	Biological vector (mosquito), vertical
Virulence Factors	Multiple life stages; multiple antigenic types; ability to scavenge glucose, GPI, cytoadherence
Culture/Diagnosis	Blood smear; serological methods
Prevention	Mosquito control; use of bed nets; no vaccine yet available; prophylactic antiprotozoal agents
Treatment	Chloroquine, mefloquine, artemisinin, pyrimethamine plus sulfadoxine (Fansidar), quinine, or proguanil; consult WHO
Epidemiological Features	United States: cases are generally in travelers or immigrants; internationally, 300 million cases in "malaria belt"; 1 million deaths per year; more deadly in children

Figure 18.5
An African family sits under a treated mosquito net from the UNICEF mosquito nets program.

▶ Prevention

Malaria prevention is attempted through long-term mosquito abatement and human chemoprophylaxis. Abatement includes elimination of standing water that could serve as a breeding site and spraying of insecticides to reduce populations of adult mosquitoes, especially in and near human dwellings. Scientists have also tried introducing sterile male mosquitoes into endemic areas in an attempt to decrease mosquito populations. Humans can reduce their risk of infection considerably by using netting, screens, and repellants; by remaining indoors at night; and by taking weekly doses of prophylactic drugs. (Western travelers to endemic areas are usually prescribed antimalarials for the duration of their trips.) People with a recent history of malaria are not allowed to give blood. The WHO and other international organizations focus on efforts to distribute bed nets and to teach people how to dip the nets into an insecticide **(figure 18.5).** The use of bed nets has been estimated to reduce childhood mortality from malaria by 20%. Here is an area where we can report some success: Bed-net use has tripled in 16 of 20 sub-Saharan African countries since 2000.

The best protection would come from a malaria vaccine, and scientists have struggled for decades to develop one. A successful malaria vaccine must be capable of striking a diverse and rapidly changing target. Scientists estimate that the parasite has 5,300 different antigens. Another potentially powerful strategy is the use of interfering RNAs in the mosquitoes to render them resistant to *Plasmodium* infection.

▶ Treatment

Quinine has long been a mainstay of malaria treatment. Chloroquine, the least toxic type, is used in nonresistant forms of the disease. In areas of the world where resistant strains of *P. falciparum* and *P. vivax* predominate, a course of mefloquine or pyrimethamine plus sulfadoxine (Fansidar) may be indicated, but more commonly artemisinin, another plant compound, has been most effective. Predictably, artemisinin resistance has now been found in Cambodia. The World Health Organization now recommends only administering artemisinin in combination with other antimalarials, in order to prevent resistance development.

Highlight Disease

HIV Infection and AIDS

▶ Signs and Symptoms

A spectrum of clinical signs and symptoms is associated with human immunodeficiency virus (HIV) infection. Symptoms in HIV infection are directly tied to two things: the level of virus in the blood and the level of T cells in the blood. To understand the progression, follow **table 18.2** closely.

The table shows two different lines that correspond to virus and T cells in the blood. Another line depicts the amount of antibody against the virus. Note that the table depicts the course of HIV infection in the absence of medical intervention or chemotherapy.

Initial symptoms may be fatigue, diarrhea, weight loss, and neurological changes, but most patients first notice infection because of one or more opportunistic infections or neoplasms (cancers). These conditions are known as AIDS-defining illnesses (ADIs) and are detailed in **table 18.3.** Other disease-related symptoms appear to accompany severe immune deregulation, hormone

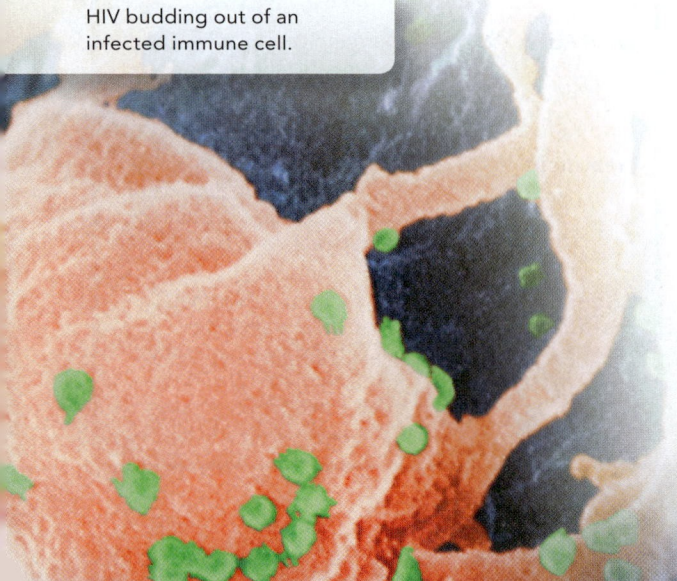

HIV budding out of an infected immune cell.

Table 18.2 Dynamics of Virus Antigen, Antibody, and T Cells in Circulation

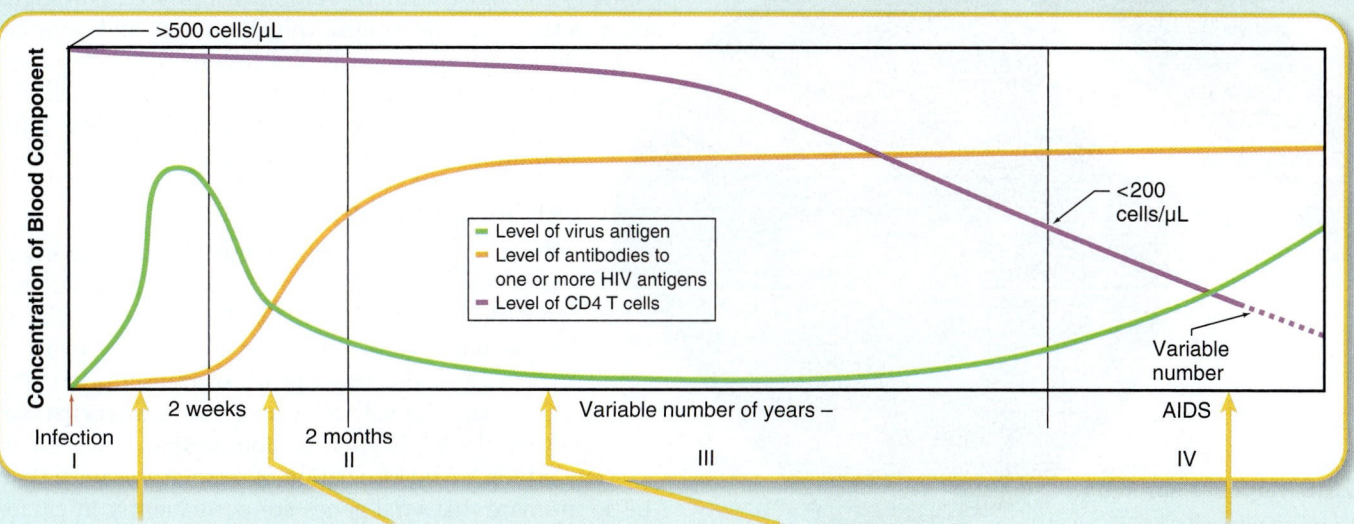

Initial infection is often attended by vague, mononucleosis-like symptoms that soon disappear. This phase corresponds to the initial high levels of virus (the green line above). Antibodies are not yet abundant.

In the second phase, virus numbers in blood drop dramatically and antibody begins to appear. CD4 T cells begin to decrease in number.

A long period of mostly asymptomatic infection ensues. During this time, which can last from 2 to 15 years, lymphadenopathy may be the prominent symptom. During the mid- to late-asymptomatic period, the number of T cells in the blood is steadily decreasing. Once the T-cell level reaches a (low) threshold, the symptoms of AIDS ensue.

Once T cells drop below 200 cells/μL, AIDS results. Note that even though antibody levels remain high, virus levels in the blood begin to rise.

Table 18.3 AIDS-Defining Illnesses

Skin and/or Mucous Membranes (Includes Eyes)	Nervous System	Cardiovascular and Lymphatic System or Multiple Organ Systems	Respiratory Tract	Gastrointestinal Tract	Genitourinary and/or Reproductive Tract
Cytomegalovirus retinitis (with loss of vision)	Cryptococcosis, extrapulmonary	Coccidioidomycosis, disseminated or extrapulmonary	Candidiasis of trachea, bronchi, or lungs	Candidiasis of esophagus, GI tract	Invasive cervical carcinoma (HPV)
Herpes simplex chronic ulcers (>1 month duration)	HIV encephalopathy	Cytomegalovirus (other than liver, spleen, nodes)	Herpes simplex bronchitis or pneumonitis	Herpes simplex chronic ulcers (>1 month duration) or esophagitis	Herpes simplex chronic ulcers (>1 month duration)
Kaposi's sarcoma	Lymphoma; primarily in brain	Histoplasmosis	*Mycobacterium avium* complex	Isosporiasis, intestinal	
	Progressive multifocal leukoencephalopathy	Burkitt's lymphoma	Tuberculosis (*Mycobacterium tuberculosis*)	Cryptosporidiosis, chronic intestinal (>1 month duration)	
	Toxoplasmosis of the brain	Immunoblastic lymphoma	*Pneumocystis jiroveci* pneumonia		
		Mycobacterium kansasii, disseminated or extrapulmonary	Pneumonia, recurrent		
		Mycobacterium tuberculosis, disseminated or extrapulmonary			
		Salmonella septicemia, recurrent			
		Wasting syndrome			

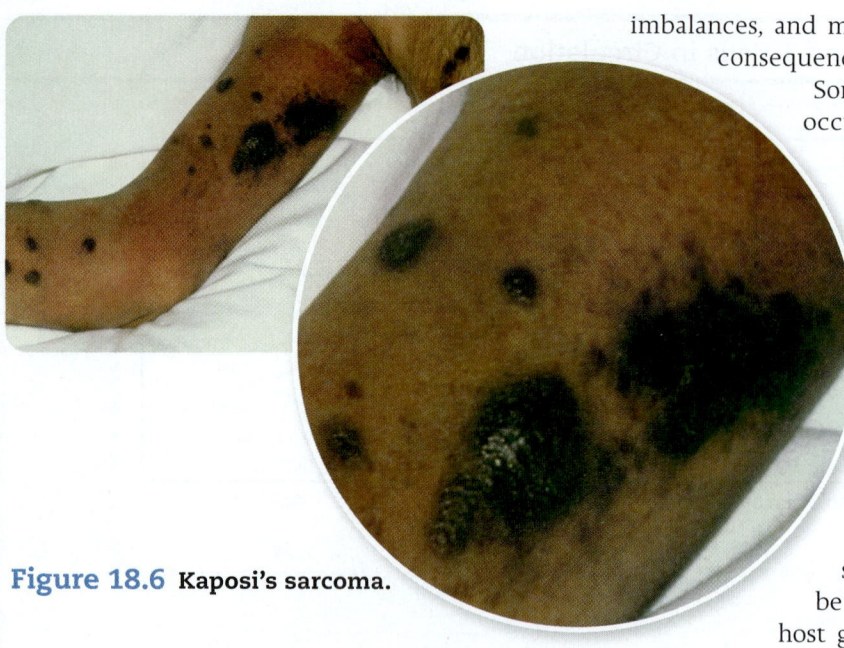

Figure 18.6 Kaposi's sarcoma.

imbalances, and metabolic disturbances. Pronounced wasting of body mass is a consequence of weight loss, diarrhea, and poor nutrient absorption.

Some of the most virulent complications are neurological. Lesions occur in the brain, meninges, spinal column, and peripheral nerves. Patients with nervous system involvement show some degree of withdrawal, persistent memory loss, spasticity, sensory loss, and progressive AIDS dementia. **Figure 18.6** depicts a common ADI, Kaposi's sarcoma.

▶ Causative Agent

HIV is a retrovirus, in the genus *Lentivirus*. Many retroviruses have the potential to cause cancer and produce dire, often fatal diseases and are capable of altering the host's DNA in profound ways. They are named "retroviruses" because they reverse the usual order of transcription. They contain an unusual enzyme called **reverse transcriptase (RT)** that catalyzes the replication of double-stranded DNA from single-stranded RNA. The association of retroviruses with their hosts can be so intimate that viral genes are permanently integrated into the host genome. Not only can this retroviral DNA be incorporated into the host genome as a provirus that can be passed on to progeny cells, but some retroviruses also transform cells (make them malignant) and regulate certain host genes.

HIV and other retroviruses display structural features typical of enveloped RNA viruses **(figure 18.7a)**. The outermost component is a lipid envelope with transmembrane glycoprotein spikes that mediate viral adsorption to the host cell. HIV can only infect host cells that present the required receptors, which is a combination receptor consisting of the CD4 marker plus a coreceptor called CCR-5. The virus uses these receptors to gain entrance to several types of leukocytes and tissue cells **(figure 18.7b)**.

▶ Pathogenesis and Virulence Factors

HIV enters a mucous membrane or the skin and travels to dendritic cells beneath the epithelium. In the dendritic cell, the virus grows and is shed from the cell without

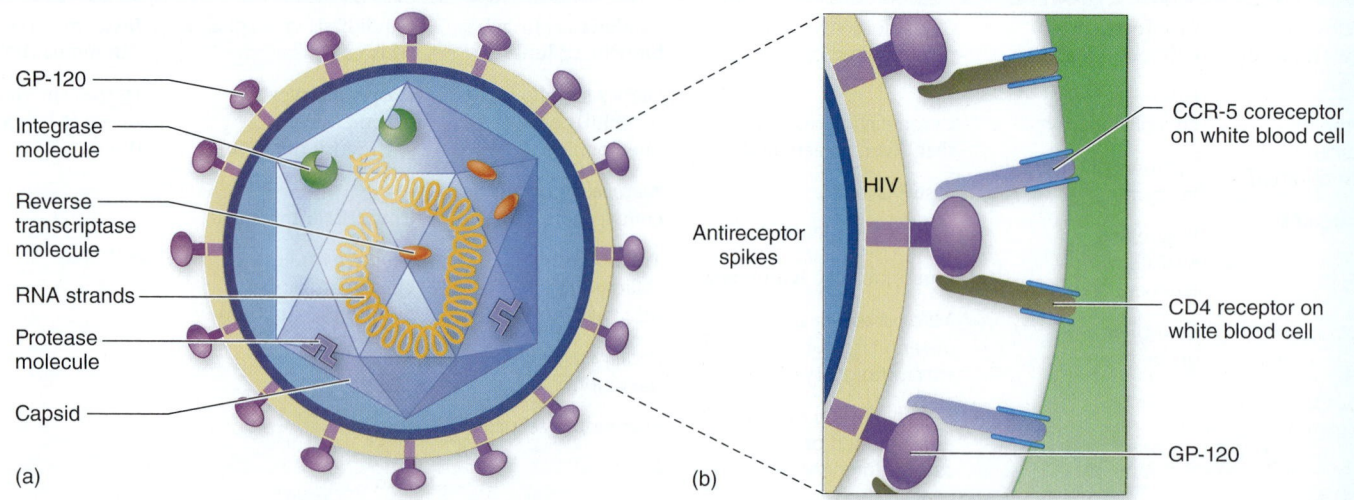

(a)

GP-120

Integrase molecule

Reverse transcriptase molecule

RNA strands

Protease molecule

Capsid

(b)

Antireceptor spikes

HIV

CCR-5 coreceptor on white blood cell

CD4 receptor on white blood cell

GP-120

Figure 18.7 The general structure of HIV. (a) The virus consists of glycoprotein (GP) spikes in the envelope, two identical RNA strands, and several molecules of reverse transcriptase, protease, and integrase encased in a protein capsid. (b) The snug attachment of HIV glycoprotein molecules to their specific receptors on a human cell membrane. These receptors are CD4 and a coreceptor called CCR-5 (fusin) that permit docking with the host cell and fusion with the cell membrane.

Table 18.4 The Multiplication Cycle of HIV

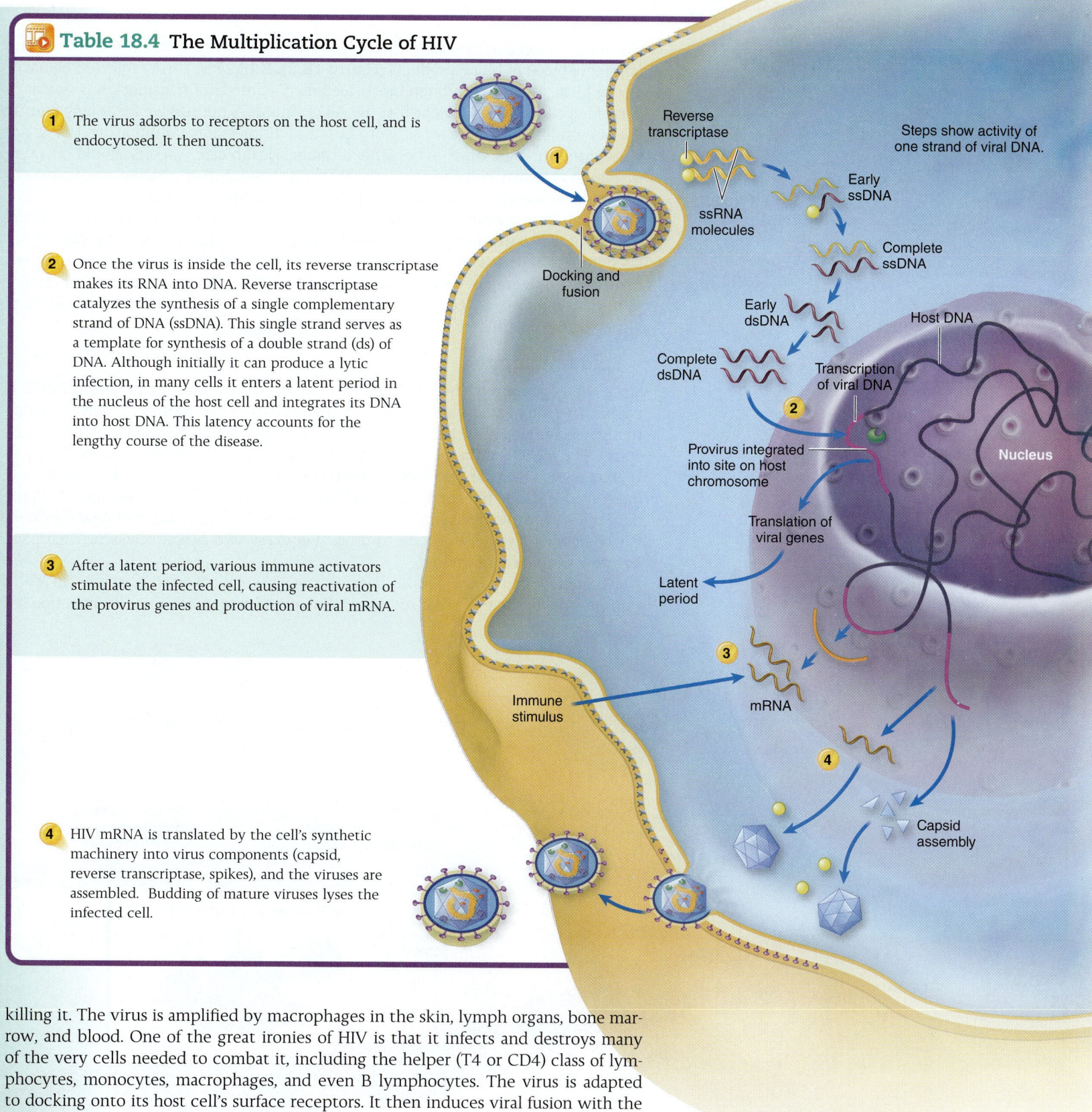

1 The virus adsorbs to receptors on the host cell, and is endocytosed. It then uncoats.

2 Once the virus is inside the cell, its reverse transcriptase makes its RNA into DNA. Reverse transcriptase catalyzes the synthesis of a single complementary strand of DNA (ssDNA). This single strand serves as a template for synthesis of a double strand (ds) of DNA. Although initially it can produce a lytic infection, in many cells it enters a latent period in the nucleus of the host cell and integrates its DNA into host DNA. This latency accounts for the lengthy course of the disease.

3 After a latent period, various immune activators stimulate the infected cell, causing reactivation of the provirus genes and production of viral mRNA.

4 HIV mRNA is translated by the cell's synthetic machinery into virus components (capsid, reverse transcriptase, spikes), and the viruses are assembled. Budding of mature viruses lyses the infected cell.

Reverse transcriptase

Steps show activity of one strand of viral DNA.

ssRNA molecules

Early ssDNA

Complete ssDNA

Early dsDNA

Host DNA

Complete dsDNA

Transcription of viral DNA

Docking and fusion

Provirus integrated into site on host chromosome

Nucleus

Translation of viral genes

Latent period

Immune stimulus

mRNA

Capsid assembly

killing it. The virus is amplified by macrophages in the skin, lymph organs, bone marrow, and blood. One of the great ironies of HIV is that it infects and destroys many of the very cells needed to combat it, including the helper (T4 or CD4) class of lymphocytes, monocytes, macrophages, and even B lymphocytes. The virus is adapted to docking onto its host cell's surface receptors. It then induces viral fusion with the cell membrane and creates syncytia.

Table 18.4 illustrates the life cycle of the virus.

Public health officials attribute the rapid spread of HIV in Africa in part to the sexual practices of long-haul truckers along major highways on the continent.

► Transmission

HIV transmission occurs mainly through two forms of contact: sexual intercourse and transfer of blood or blood products **(figure 18.8)**. Babies can also be infected before or during birth, as well as through breast feeding. The mode of transmission is similar to that of hepatitis B virus, except that the AIDS virus does not survive for as long outside the host and it is far more sensitive to heat and disinfectants. Additionally, HIV is not transmitted through saliva, as hepatitis B can be. Health care workers should be aware that fluids they may come in contact with during childbirth or invasive procedures can also transmit the virus. These are amniotic fluid, synovial fluid, and spinal fluid.

Semen and vaginal secretions also harbor free virus and infected white blood cells; thus, they are significant factors in sexual transmission. The virus can be isolated from urine, tears, sweat, and saliva in the laboratory—but in such small numbers that these fluids are not considered sources of infection. Because breast milk contains significant numbers of leukocytes, neonates who have escaped infection prior to and during birth can still become infected through nursing.

► Epidemiology

Since the beginning of the AIDS epidemic in the early 1980s, more than 30 million people have died worldwide. The best global estimate of the number of individuals currently infected with HIV is 32.2-38.8 million (2012), with over 1 million in the United States. The WHO estimates that 2.1 million new infections occurred in 2013. There is good news: Over the past three years there has been a 13 percent decrease in new infections. And, due to efforts of many global AIDS initiatives, many more people in the developing world are receiving lifesaving treatments. But the number of new infections is still growing faster than access to drugs: For every two people receiving treatment, five new people are diagnosed.

In most parts of the world, heterosexual intercourse is the primary mode of transmission. In the industrialized world, the overall rate of heterosexual infection has

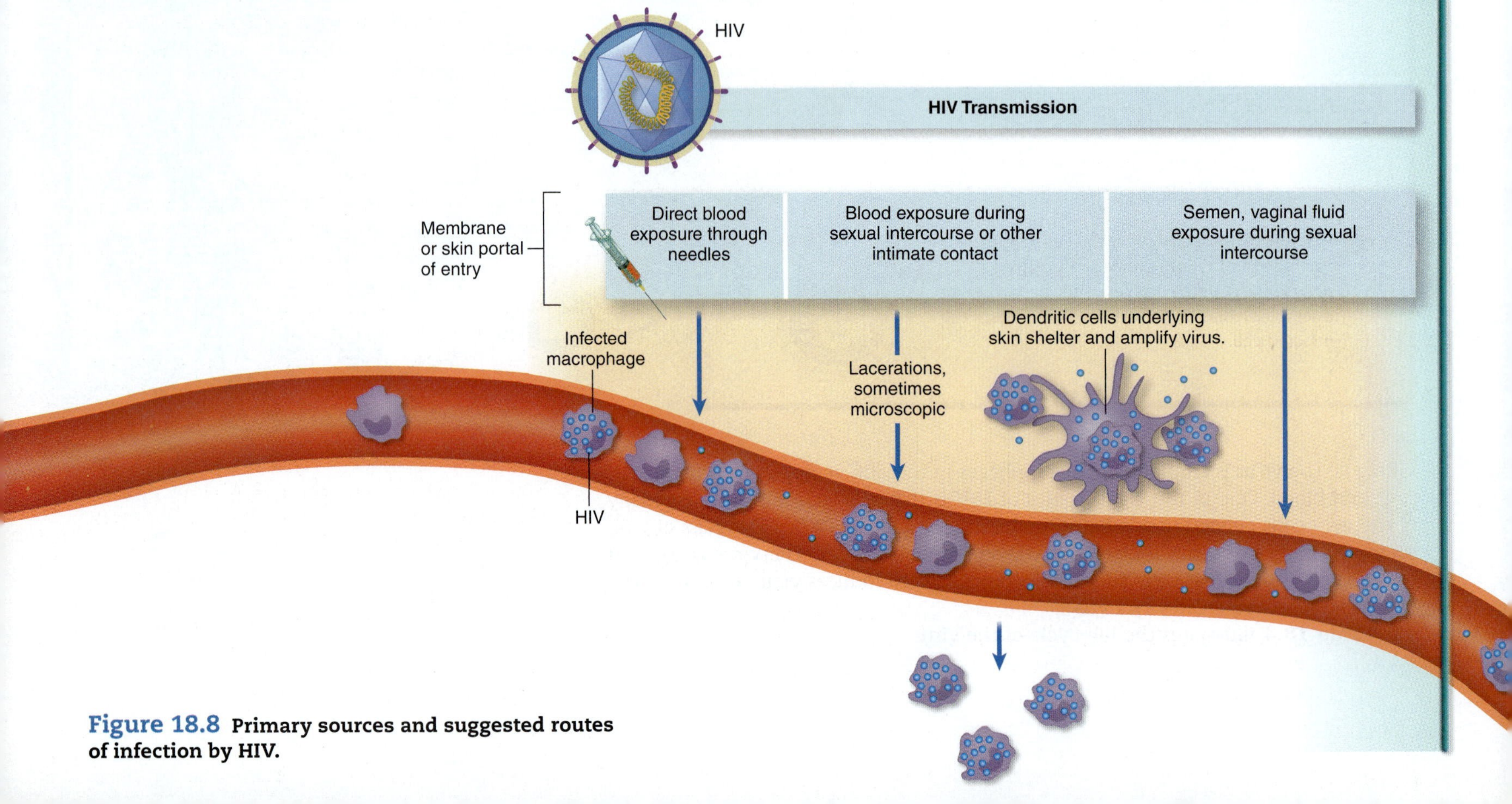

HIV

HIV Transmission

Membrane or skin portal of entry

| Direct blood exposure through needles | Blood exposure during sexual intercourse or other intimate contact | Semen, vaginal fluid exposure during sexual intercourse |

Dendritic cells underlying skin shelter and amplify virus.

Infected macrophage

Lacerations, sometimes microscopic

HIV

Figure 18.8 **Primary sources and suggested routes of infection by HIV.**

increased dramatically in the past several years, especially in adolescent and young adult women. In the United States, African-American women have a very high rate of infection. While they represent only 14% of the U.S. female population, they constitute 66% of HIV infections in females. The CDC comments that "while African-American women do not engage in more risky behaviors than other women, a complex range of social and environmental factors place them at greater risk for HIV."[1]

We should note that not everyone who becomes infected or is antibody-positive develops AIDS. About 1% of people who are antibody-positive remain free of disease, indicating that functioning immunity to the virus can develop. Any person who remains healthy despite HIV infection is termed a *nonprogressor*. These people are the object of intense scientific study. Some have been found to lack the cytokine receptors that HIV requires. Others are infected by a weakened virus mutant.

Treatment of HIV-infected mothers with an anti-HIV drug has dramatically decreased the rate of maternal-to-infant transmission of HIV during pregnancy. Current treatment regimens result in a transmission rate of approximately 11%, with some studies of multidrug regimens claiming rates as low as 5%. Evidence suggests that giving mothers protease inhibitors can reduce the transmission rate to around 1%. (Untreated mothers pass the virus to their babies at the rate of 33%.)

▶ Culture and Diagnosis

A person is diagnosed as having HIV infection, as opposed to having AIDS, if he or she has tested positive for the human immunodeficiency virus. In 2012, the U.S. Preventive Services Task Force recommended that all people between the ages of 15 and 64 be tested for HIV. People outside of that age group who are at high risk, as well as pregnant women, should also be tested.

Most viral testing is based on detection of antibodies specific to the virus in serum or other fluids. This allows for the inexpensive screening of large numbers of samples **(Figure 18.9).** The initial screening tests include the older ELISA and newer latex agglutination and rapid antibody tests.

False-negative results can occur when testing is performed before the onset of detectable antibody production (see table 18.2). To rule out this possibility, persons who test negative but feel they may have been exposed should be tested a second time 3 to 6 months later. Alternatively, testing for HIV antigens (rather than for HIV antibodies) can close the window of time between infection and detectable levels of antibodies during which infection could be missed by antibody tests.

Positive tests always require follow-up with a more specific test—usually a Western blot analysis. This test detects several different anti-HIV antibodies and can usually rule out false-positive results.

In the United States, people are diagnosed with AIDS, as opposed to HIV, if they meet the following criteria: (1) They are positive for the virus, *and* (2) they fulfill one of these additional criteria:

- They have a CD4 (helper T cell) count of fewer than 200 cells per microliter of blood.
- Their CD4 cells account for fewer than 14% of all lymphocytes.
- They experience one or more of a CDC-provided list of AIDS-defining illnesses (ADIs).

▶ Prevention

Avoidance of sexual contact with infected persons is a cornerstone of HIV prevention. A sexually active person should consider every partner to be infected unless proven otherwise. Barrier protection (condoms) should be used when having sex with anyone whose HIV status is not known with certainty to be negative. Although avoiding intravenous drugs is an important preventive measure, many drug addicts do not, or cannot, choose this option. In such cases, risk can be decreased by not sharing syringes or needles. Brand new research has shown that treating uninfected or newly infected people with antiretrovirals can prevent the progression to AIDS. This treatment is termed *pre-exposure prophylaxis* (*PrEP*).

1. Available from http://hivtest.cdc.gov/takecharge/.

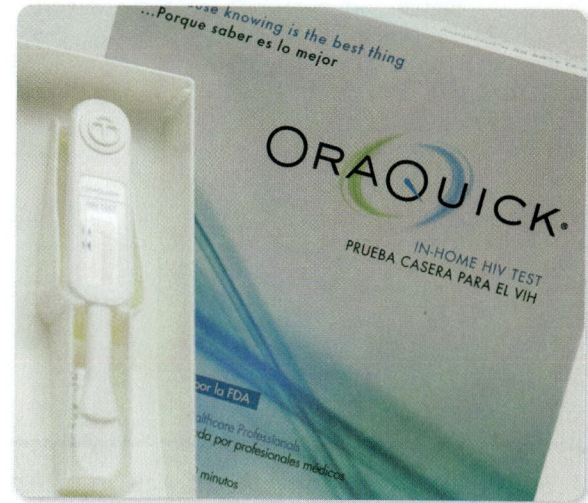

Figure 18.9 **At-home HIV tests may help increase awareness and promote early detection.**

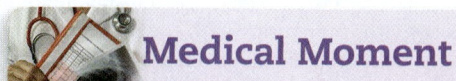

Medical Moment

Kaposi's Sarcoma

Prior to the 1980s, Kaposi's sarcoma was a cancer rarely seen in the United States. Kaposi's sarcoma primarily affected middle-aged men of Jewish or Mediterranean descent. The disease was also endemic to parts of Africa. When AIDS exploded on the scene in the 1980s, large numbers of Kaposi's sarcoma lesions began to be identified among the first AIDS patients in the United States among the male homosexual population.

Kaposi's sarcoma is not a true sarcoma but, rather, arises from lymphatic tissue in which channels form and blood cells aggregate, giving Kaposi's sarcoma lesions their deep color. Lesions may be dark purple, dark red, black, or brown and are typically raised (papular). Kaposi's sarcoma may also affect other body systems, such as the gastrointestinal and respiratory tracts.

Early AIDS researchers believed that Kaposi's sarcoma could be the cause of AIDS. Of course, this theory was negated when the HIV virus was identified. Kaposi's sarcoma is now known to be caused by a herpesvirus (the eighth human herpesvirus, or HHV-8).

Kaposi's sarcoma continues to be an opportunistic infection in AIDS patients and is one of the ADIs (AIDS-defining illnesses). It is sometimes the first presenting symptom of AIDS.

► Treatment

There is no cure for HIV. However, new drugs and new insights into their use have greatly improved the quality of life for some people living with HIV and AIDS. Unfortunately, much of the world is too poor to afford or properly use the latest medications. Two great challenges are (1) to make more widely available the drugs we do have and (2) to develop drugs and delivery mechanisms that are more affordable and practical.

Clear-cut guidelines exist for treating people who test HIV-positive. These guide-lines are updated regularly. The most recent update involves a new combination treatment called Triumeq, which contains one integrase inhibitor and two reverst transcriptase inhibitors. The newer recommendations call for treatment to begin soon after HIV diagnosis. In addition to antiviral chemotherapy, HIV-positive persons should

Table 18.5 Mechanisms of Action of Anti-HIV Drugs

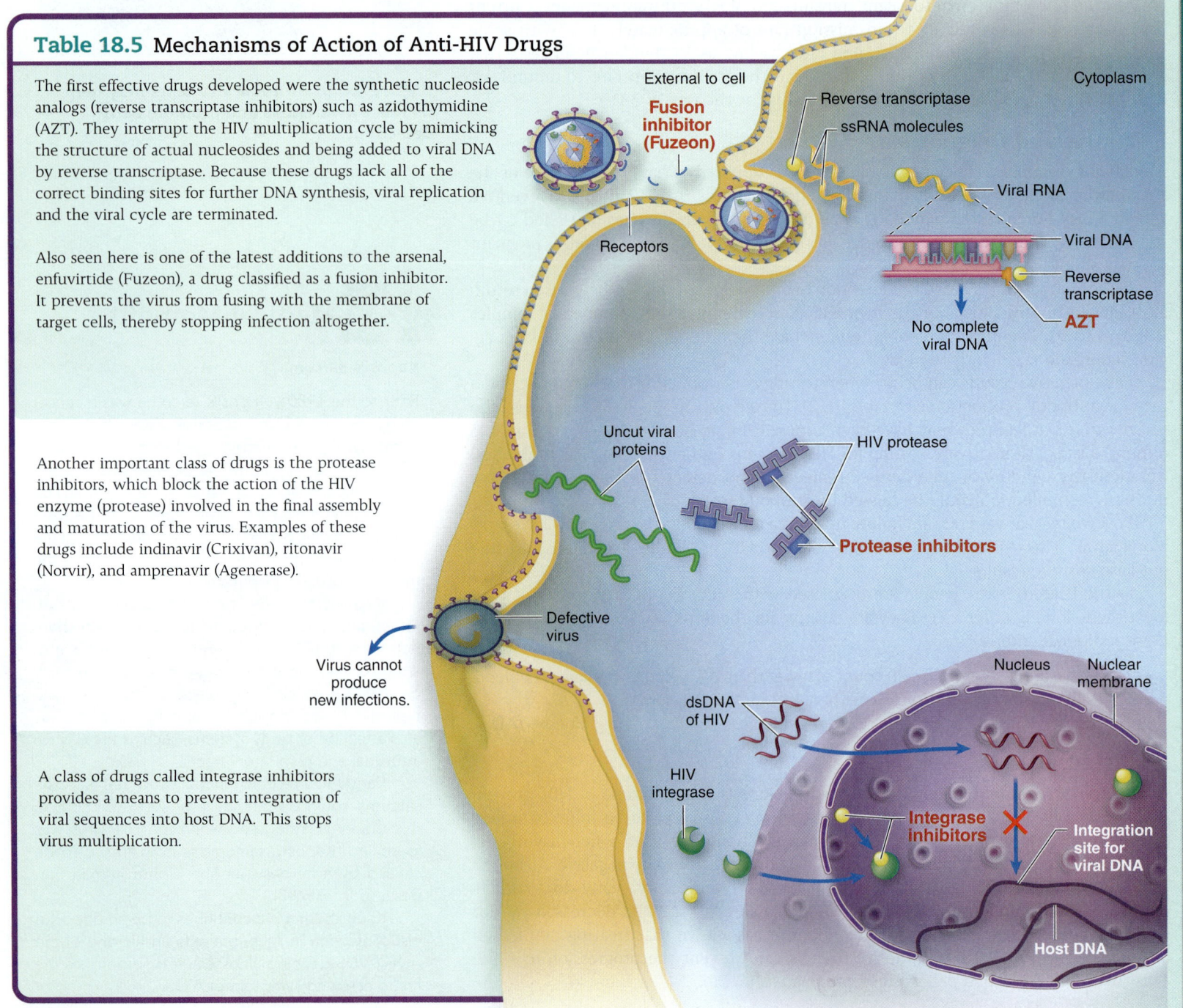

The first effective drugs developed were the synthetic nucleoside analogs (reverse transcriptase inhibitors) such as azidothymidine (AZT). They interrupt the HIV multiplication cycle by mimicking the structure of actual nucleosides and being added to viral DNA by reverse transcriptase. Because these drugs lack all of the correct binding sites for further DNA synthesis, viral replication and the viral cycle are terminated.

Also seen here is one of the latest additions to the arsenal, enfuvirtide (Fuzeon), a drug classified as a fusion inhibitor. It prevents the virus from fusing with the membrane of target cells, thereby stopping infection altogether.

Another important class of drugs is the protease inhibitors, which block the action of the HIV enzyme (protease) involved in the final assembly and maturation of the virus. Examples of these drugs include indinavir (Crixivan), ritonavir (Norvir), and amprenavir (Agenerase).

A class of drugs called integrase inhibitors provides a means to prevent integration of viral sequences into host DNA. This stops virus multiplication.

receive a wide array of drugs to prevent or treat a variety of opportunistic infections and other ADIs such as wasting disease. In **table 18.5,** the variety of drugs available to treat HIV is depicted.

Endocarditis

Endocarditis is an inflammation of the endocardium, or inner lining of the heart. Most of the time, endocarditis refers to an infection of the valves of the heart, often the mitral or aortic valve **(figure 18.10).** Two variations of infectious endocarditis have been described: acute and subacute. Each has distinct groups of possible causative agents, primarily bacterial. On rare occasions, infections may be caused by fungi and perhaps viruses. Physical trauma can also cause such inflammation.

The surgical innovation of prosthetic valves presents a new hazard for development of endocarditis. Patients with prosthetic valves can acquire acute endocarditis if bacteria are introduced during the surgical procedure, typically with high rates of morbidity and mortality. Alternatively, the prosthetic valves can serve as infection sites for the subacute form of endocarditis long after the surgical procedure.

Because the symptoms and the diagnostic procedures are similar for both forms of endocarditis, they are discussed first; then the specific aspects of acute and subacute endocarditis are addressed.

▶ Signs and Symptoms

The signs and symptoms are similar for both types of endocarditis, except that in the subacute condition they develop more slowly and are less pronounced than with the acute disease. Symptoms include fever, fatigue, joint pain, edema (swelling of feet, legs, and abdomen), weakness, anemia, abnormal heartbeat, and sometimes symptoms similar to myocardial infarction (heart attack, including shortness of breath or chills). Abdominal or side pain is sometimes reported. The patient may look very ill and may have petechiae (small red-to-purple discolorations) over the upper half of the body and under the fingernails (splinter hemorrhages). Red, painless skin spots on the palms and soles (Janeway lesions) and small painful nodes on the pads of fingers and toes (Osler's nodes) may also be apparent on examination. In subacute cases, an enlarged spleen may have developed over time; cases of extremely long duration (years) can lead to clubbed fingers and toes due to lack of oxygen in the blood.

Acute Endocarditis

Acute endocarditis is most often the result of an overwhelming bloodstream challenge with bacteria. Some of these bacteria seem to have the ability to colonize normal heart valves. Accumulations of bacteria on the valves (vegetations) hamper their function and can lead directly to cardiac malfunction and death. Alternatively, pieces of the bacterial vegetation can break off and create emboli (blockages) in vital organs. The bacterial colonies can also provide a constant source of blood-borne bacteria, with the accompanying systemic inflammatory response and shock.

▶ Causative Agents

The acute form of endocarditis is most often caused by *Staphylococcus aureus*. Other agents that cause it are *Streptococcus pyogenes, Streptococcus pneumoniae*, and *Neisseria gonorrhoeae*, as well as a host of other bacteria. Each of these bacteria is described elsewhere in this book.

▶ Transmission and Epidemiology

The most common route of transmission for acute endocarditis is parenteral—that is, via direct entry into the body. Intravenous or subcutaneous drug users have been a growing risk group for the condition. Traumatic injuries and surgical procedures can also introduce the large number of bacteria required for the acute form of endocarditis.

Disease Table 18.2 HIV Infection and AIDS	
Causative Organism(s)	Human immunodeficiency virus 1 or 2
Most Common Modes of Transmission	Direct contact (sexual), parenteral (blood-borne), vertical (perinatal and via breast milk)
Virulence Factors	Attachment, syncytia formation, reverse transcriptase, high mutation rate
Culture/ Diagnosis	Initial screening for antibody followed by Western blot confirmation of antibody
Prevention	Avoidance of contact with infected sex partner, contaminated blood, breast milk; pre-exposure prophylaxis (PrEP) for high-risk individuals
Treatment	Multiple simultaneous antiretroviral drugs
Epidemiological Features	Global infections: 32–39 million, 2.1 new infections in 2013. U.S. infections: approximately 1 million. One quarter of new U.S. infections in women (66% of these in African-American women).

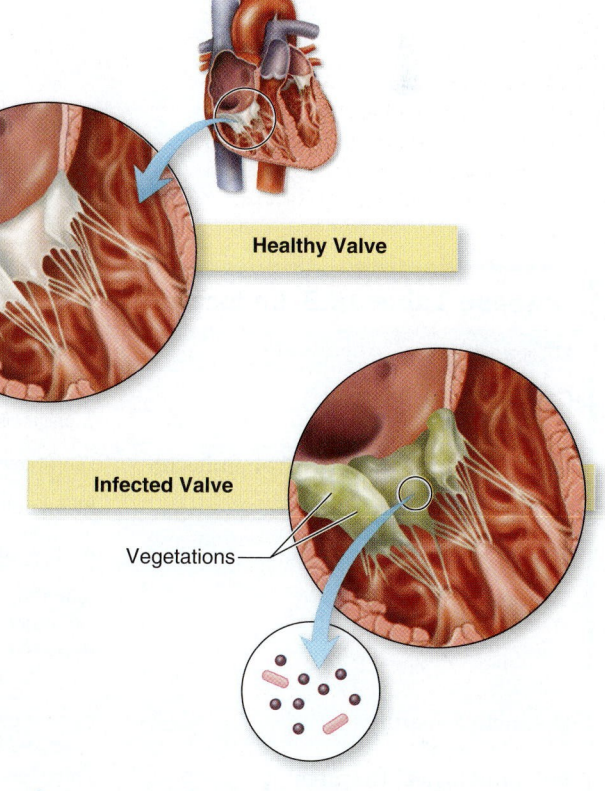

Healthy Valve

Infected Valve

Vegetations

Figure 18.10 Endocarditis. Infected valves don't work properly.

Subacute Endocarditis

Subacute forms of this condition are almost always preceded by some form of damage to the heart valves or by congenital malformation. Irregularities in the valves encourage the attachment of bacteria, which then form biofilms and impede normal function, as well as provide an ongoing source of bacteria to the bloodstream. People who have suffered rheumatic fever and the accompanying damage to heart valves are particularly susceptible to this condition (see chapter 19 for a complete discussion of rheumatic fever).

▶ Causative Agents

Most commonly, subacute endocarditis is caused by bacteria of low pathogenicity, often originating in the oral cavity. Alpha-hemolytic streptococci, such as *Streptococcus sanguis*, *S. oralis*, and *S. mutans*, are most often responsible, although normal biota from the skin and other bacteria can also colonize abnormal valves and lead to this condition.

▶ Transmission and Epidemiology

Minor disruptions in the skin or mucous membranes, such as those induced by overly vigorous toothbrushing, dental procedures, or relatively minor cuts and lacerations, can introduce bacteria into the bloodstream and lead to valve colonization. The bacteria are not, therefore, transmitted from other people or from the environment.

▶ Prevention

The practice of prophylactic antibiotic therapy in advance of surgical and dental procedures on patients with underlying valve irregularities has decreased the incidence of this infection. **Disease Table 18.3.**

Septicemia

Septicemia occurs when organisms are actively multiplying in the blood. Many different bacteria (and a few fungi) can cause this condition. Patients suffering from these infections are sometimes described as "septic."

▶ Signs and Symptoms

Fever is a prominent feature of septicemia. The patient appears very ill and may have an altered mental state, shaking chills, and gastrointestinal symptoms. Often an increased breathing rate is exhibited, accompanied by respiratory alkalosis (increased tissue pH due to breathing disorder). Low blood pressure is a hallmark of this condition and is caused by the inflammatory response to infectious agents in the

Disease Table 18.3 Endocarditis

Disease	Acute Endocarditis	Subacute Endocarditis
Causative Organism(s)	*Staphylococcus aureus, Streptococcus pyogenes, S. pneumoniae, Neisseria gonorrhoeae,* others	Alpha-hemolytic streptococci, others
Most Common Modes of Transmission	Parenteral	Endogenous transfer of normal biota to bloodstream
Culture/Diagnosis	Blood culture	Blood culture
Prevention	Aseptic surgery, injections	Prophylactic antibiotics before invasive procedures
Treatment	Nafcillin or oxacillin +/− gentamicin or tobramycin OR vancomycin + gentamicin; surgery may be necessary. Be aware that several of these bacteria are on CDC "Threat" list for antibiotic resistance.	Surgery may be necessary
Distinctive Features	Acute onset, high fatality rate	Slower onset
Epidemiological Features	Three times more common in males than females	–

bloodstream, which leads to a loss of fluid from the vasculature. This condition is the most dangerous feature of the disease, often culminating in death.

▶ Causative Agents

The vast majority of septicemias are caused by bacteria, and they are approximately evenly divided between gram-positives and gram-negatives. MRSA is a very common cause. Perhaps 10% are caused by fungal infections. Polymicrobial bloodstream infections increasingly are being identified in which more than one microorganism is causing the infection.

▶ Pathogenesis and Virulence Factors

Gram-negative bacteria multiplying in the blood release large amounts of endotoxin into the bloodstream, stimulating a massive inflammatory response mediated by a host of cytokines. This response invariably leads to a drastic drop in blood pressure, a condition called **endotoxic shock.** Gram-positive bacteria can instigate a similar cascade of events when fragments of their cell walls are released into the blood.

▶ Transmission and Epidemiology

In many cases, septicemias can be traced to parenteral introduction of the microorganisms via intravenous lines or surgical procedures. Other infections may arise from serious urinary tract infections or from renal, prostatic, pancreatic, or gallbladder abscesses. Patients with underlying spleen malfunction may be predisposed to multiplication of microbes in the bloodstream. Meningitis, osteomyelitis (bone infections), and pneumonia can all lead to sepsis. Hospitalization for sepsis has more than doubled in recent years. More alarming is the fact there is a 20% to 50% mortality rate, reflecting the high risks of this disease even when treatment is available. At least 200,000 cases occur each year in the United States, resulting in more than 100,000 deaths.

▶ Culture and Diagnosis

Because the infection is in the bloodstream, a blood culture is the obvious route to diagnosis. A full regimen of media should be inoculated to ensure isolation of the causative microorganism. Antibiotic susceptibilities should be assessed. Empiric therapy should be started immediately before culture and susceptibility results are available.

▶ Prevention and Treatment

Empiric therapy, which is begun immediately after blood cultures are taken, often begins with a broad-spectrum antibiotic. Once the organism is identified and its antibiotic susceptibility is known, treatment can be adjusted accordingly, as data show that rapid diagnosis and treatment are of paramount importance. **Disease Table 18.4.**

Plague

Although pandemics of plague have occurred since antiquity, the first one that was reliably chronicled killed an estimated 100 million people in the sixth century AD.

▶ Signs and Symptoms

Three possible manifestations of infection occur with the bacterium causing plague. **Pneumonic plague** is a respiratory disease, described in chapter 19. In **bubonic plague,** the bacterium, which is injected by the bite of a flea, enters the lymph and is filtered by a local lymph node. Infection causes inflammation and necrosis of the node, resulting in a swollen lesion called a **bubo,** usually in the groin or axilla **(figure 18.11).** The incubation period lasts 2 to 8 days, ending abruptly with the onset of fever, chills, headache, nausea, weakness, and tenderness of the bubo. Mortality rates, even with treatment, can reach up to 15%.

These cases often progress to massive bacterial growth in the blood termed **septicemic plague.** The presence of the bacteria in the blood results in disseminated intravascular coagulation, subcutaneous hemorrhage, and **purpura** that may degenerate into necrosis and gangrene. Mortality rates, once the disease has progressed to this

Disease Table 18.4	Septicemia
Causative Organism(s)	Bacteria or fungi, often MRSA
Most Common Modes of Transmission	Parenteral, endogenous transfer
Virulence Factors	Cell wall or membrane components
Culture/Diagnosis	Blood culture
Prevention	–
Treatment	Broad-spectrum antibiotic until identification and susceptibilities tested
Epidemiological Features	In United States: 200,000 cases and 100,000 deaths per year

Prairie dogs in the southwest United States sometimes harbor *Yersinia pestis.*

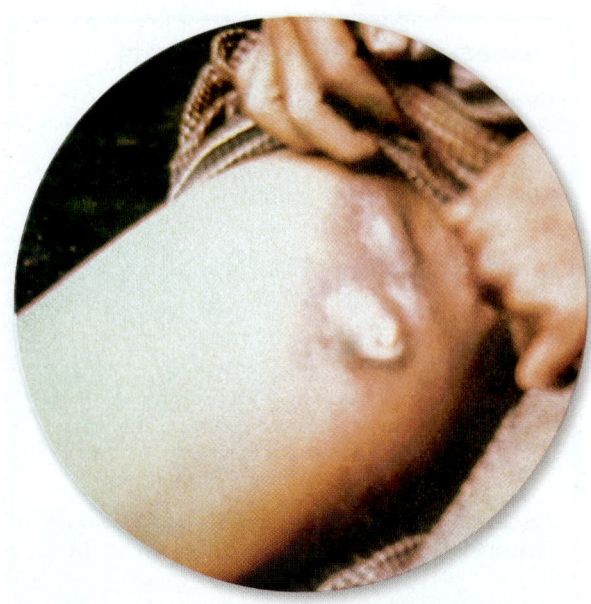

Figure 18.11 **A classic inguinal bubo of bubonic plague.** This hard nodule is very painful and can rupture onto the surface.

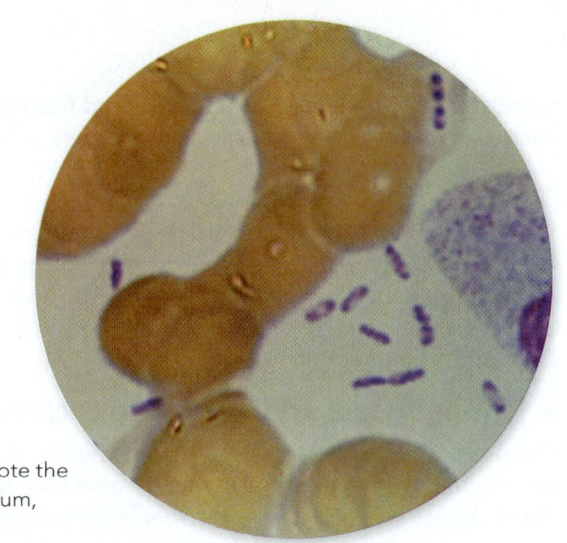

Figure 18.12 *Yersinia pestis.* Note the more darkly stained poles of the bacterium, lending it a "safety pin" appearance.

point, are 30% to 50% with treatment and 100% without treatment. Because of the visible darkening of the skin, the plague has often been called the "Black Death."

▶ Causative Agent

The cause of this dreadful disease is a tiny gram-negative rod, *Yersinia pestis*, a member of the family *Enterobacteriaceae*. *Y. pestis* displays unusual bipolar staining that makes it look like a safety pin **(figure 18.12).**

▶ Pathogenesis and Virulence Factors

The number of bacteria required to initiate a plague infection is small—perhaps only 3 to 50 cells.

▶ Transmission and Epidemiology

The plague bacterium resides in over 200 species of mammalian hosts. Some of these, such as mice and voles, serve as long-term endemic reservoirs, which are not affected by the disease. Other species, including rats and rabbits, are amplifying reservoirs, which get sick and tend to be closely connected to outbreaks of plague in humans.

The principal agents in the transmission of the plague bacterium are fleas. After a flea ingests a blood meal from an infected animal, the bacteria multiply in its gut. The bacterium promotes its spread by causing coagulation and blockage of the flea's esophagus. Being unable to feed properly, the ravenous flea jumps from animal to animal in a futile attempt to get nourishment. Regurgitated infectious material then is inoculated into the bite wound.

The distribution of plague is extensive. Although the incidence of disease has been reduced in the developed world, it has actually been increasing in Africa and other parts of the world. In the United States, sporadic cases (usually less than 10 per year) occur as a result of contact with wild and domestic animals. This disease is considered endemic in U.S. western and southwestern states. Persons most at risk for developing plague are veterinarians and people living and working near woodlands and forests. Dogs and cats can be infected with the plague, often from contact with infected wild animals such as prairie dogs. **Disease Table 18.5.**

Disease Table 18.5 Plague	
Causative Organism(s)	*Yersinia pestis*
Most Common Modes of Transmission	Vector, biological; also droplet contact (pneumonic) and direct contact with body fluids
Virulence Factors	Capsule, plasminogen activator
Culture/Diagnosis	Rapid genomic methods
Prevention	Flea and/or animal control; vaccine available for high-risk individuals
Treatment	Streptomycin or gentamicin
Epidemiological Features	United States: endemic in all western and southwestern states; internationally, 95% of human cases occur in Africa, including Madagascar

Tularemia

▶ Signs and Symptoms

After an incubation period ranging from a few days to 3 weeks, acute symptoms of headache, backache, fever, chills, coughing, and weakness appear. Further clinical manifestations are tied to the portal of entry. They include ulcerative skin lesions, swollen lymph glands, conjunctival inflammation, sore throat, intestinal disruption, and pulmonary involvement. Unfortunately, clinicians can misinterpret these signs and symptoms, delaying effective treatment. The death rate in the most serious forms of disease is 30%, but proper treatment with gentamicin or streptomycin reduces mortality to almost zero.

▶ Causative Agent

The causative agent of tularemia is a facultative intracellular gram-negative bacterium called *Francisella tularensis*. It is a zoonotic disease of assorted mammals endemic to the Northern Hemisphere. Because it has been associated with outbreaks of disease in wild rabbits, it is sometimes called rabbit fever. It is currently listed as a Category A bioterrorism agent, along with anthrax, plague, and others.

▶ Transmission and Epidemiology

Although rabbits and rodents (muskrats and ground squirrels) are the chief reservoirs, other wild animals (skunks, beavers, foxes, opossums) and some domestic animals are implicated as well. The chief route of transmission in the past had been through the activity of skinning rabbits, but with the decline of rabbit hunting, transmission via tick bites is more common. Ticks are the most frequent arthropod vector, followed by biting flies, mites, and mosquitoes. This has led to increased disease incidence during the summer.

With an estimated infective dose of between 10 and 50 organisms, *F. tularensis* is often considered one of the most infectious of all bacteria. Cases of tularemia have appeared in people who have accidentally run over rabbits while lawn mowing, presumably from inhaling aerosolized bacteria. In 2009, two different people in Alaska acquired tularemia after wresting infected rabbits from their dogs' mouths.

▶ Prevention

Because the intracellular persistence of *F. tularensis* can lead to relapses, antimicrobial therapy must not be discontinued prematurely. Protection is available in the form of a live attenuated vaccine. Laboratory workers and other occupationally exposed personnel must wear gloves, masks, and eyewear. **Disease Table 18.6.**

Lyme Disease

▶ Signs and Symptoms

Lyme disease is slow-acting, but it often evolves into a slowly progressive syndrome that mimics neuromuscular and rheumatoid conditions. An early symptom in 70% of cases is a rash at the site of a tick bite. The lesion, called *erythema migrans*, looks something like a bull's eye, with a raised erythematous (reddish) ring that gradually spreads outward and a pale central region **(figure 18.13)**. Other early symptoms are fever, headache, stiff neck, and dizziness. If not treated or if treated too late, the disease can advance to the second stage, during which cardiac and neurological symptoms, such as facial palsy, can develop. After several weeks or months, the third stage may occur. This involves a crippling arthritis; some people acquire chronic neurological complications that are severely disabling.

▶ Causative Agent

Borrelia burgdorferi are considered the cause of Lyme disease. They are unusual spirochetes in that they are large, ranging from 0.2 to 0.5 μm in width and from 10 to 20 μm in length, and contain 3 to 10 irregularly spaced and loose coils **(figure 18.14)**. A number of researchers are suggesting that other *Borrelia* species–and even other types of microbes–may be responsible for this condition. To be sure we still are in the dark about much concerning this disease.

▶ Pathogenesis and Virulence Factors

The bacterium is a master of immune evasion. It changes its surface antigens while it is in the tick and again after it has been transmitted to a mammalian host. It provokes a strong humoral and cellular immune response, but this response is mainly ineffective, perhaps because of the bacterium's ability to switch its antigens. Indeed, it is possible that the immune response contributes to the pathology of the infection.

Disease Table 18.6	Tularemia
Causative Organism(s)	*Francisella tularensis*
Most Common Modes of Transmission	Vector, biological; also direct contact with body fluids from infected animal; airborne
Virulence Factors	Intracellular growth
Culture/Diagnosis	Culture dangerous to lab workers and not reliable; serology most often used
Prevention	Live attenuated vaccine for high-risk individuals
Treatment	Gentamicin or streptomycin
Epidemiological Features	United States: several hundred cases per year; internationally, 500,000 cases per year

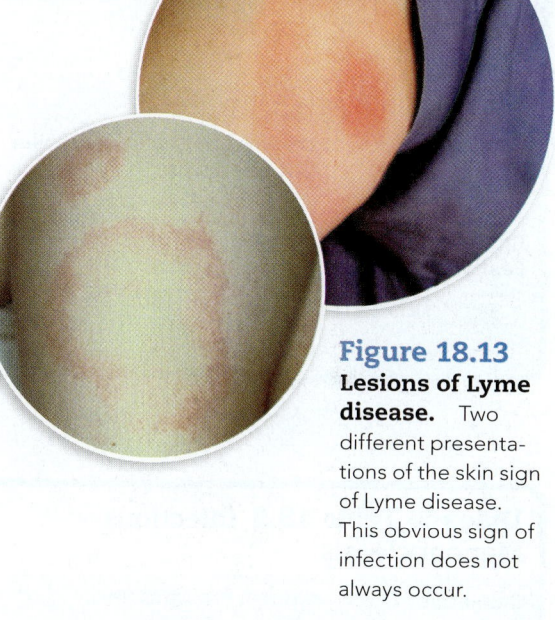

Figure 18.13 Lesions of Lyme disease. Two different presentations of the skin sign of Lyme disease. This obvious sign of infection does not always occur.

Hunting and cleaning rabbits is a common cause of tularemia.

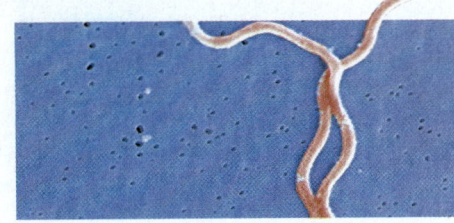

Figure 18.14 *Borrelia* has 3 to 10 loose, irregular coils.

Disease Table 18.7 Lyme Disease

Causative Organism(s)	*Borrelia burgdorferi*
Most Common Modes of Transmission	Vector, biological
Virulence Factors	Antigenic shifting, adhesins
Culture/ Diagnosis	ELISA for Ab, PCR
Prevention	Tick avoidance
Treatment	Doxycycline and/or amoxicillin (3–4 weeks), also cephalosporins and penicillin
Epidemiological Features	Endemic in North America, Europe, and Asia; approximately 300,000 new cases/year in the United States; only 10% diagnosed/reported

Disease Table 18.8 Infectious Mononucleosis

Causative Organism(s)	Epstein-Barr virus (EBV)
Most Common Modes of Transmission	Direct, indirect contact; parenteral
Virulence Factors	Latency, ability to incorporate into host DNA
Culture/ Diagnosis	Differential blood count, monospot test for heterophile antibody, specific ELISA
Prevention	–
Treatment	Supportive
Distinctive Features	Most common in teens
Epidemiological Features	United States: 500 cases per 100,000 per year

▶ Transmission and Epidemiology

B. burgdorferi is transmitted primarily by hard ticks of the genus *Ixodes*. In the northeastern part of the United States, *Ixodes scapularis* (the black-legged deer tick, **figure 18.15**) passes through a complex 2-year cycle that involves two principal hosts. In California, the transmission cycle involves *Ixodes pacificus*, another black-legged tick, and the dusky-footed woodrat as reservoir.

The greatest concentrations of Lyme disease are found in areas having large populations of both the intermediate and definitive hosts **(figure 18.16)**.

▶ Culture and Diagnosis

Diagnosis of Lyme disease can be difficult because of the range of symptoms it presents. Most suggestive are the ring-shaped lesions, isolation of spirochetes from the patient, and serological testing with an ELISA method that tracks a rising antibody titer. Tests for spirochete DNA in specimens is especially helpful for late-stage diagnosis.

▶ Prevention and Treatment

A vaccine for Lyme disease was available for a brief period of time, but it was withdrawn from the market in early 2002 due to poor sales. Other vaccines are in development. Anyone involved in outdoor activities should wear protective clothing, boots, leggings, and insect repellent containing DEET. Individuals exposed to heavy infestation should routinely inspect their bodies for ticks and remove ticks gently without crushing, preferably with forceps or fingers protected with gloves, because it is possible to become infected by tick feces or body fluids.

Early, prolonged (3 to 4 weeks) treatment with doxycycline and amoxicillin is effective, and other antibiotics such as ceftriaxone and penicillin are used in late Lyme disease therapy. Unfortunately, 10% to 20% of treated patients develop chronic Lyme disease. **Disease Table 18.7.**

Infectious Mononucleosis

This lymphatic system disease, which is often simply called "mono" or the "kissing disease," can be caused by a number of bacteria or viruses, but the vast majority of cases are caused by the **Epstein-Barr virus (EBV)**, a member of the herpesvirus family.

▶ Signs and Symptoms

The symptoms of mononucleosis are sore throat, high fever, and cervical lymphadenopathy, which develop after a long incubation period (30 to 50 days). Many patients also have a gray-white exudate in the throat, a skin rash, and enlarged spleen and liver. A notable sign of mononucleosis is sudden leukocytosis, consisting initially of infected B cells and later T cells. Fatigue is a hallmark of the disease. Patients remain fatigued for a period of weeks. During that time, they are advised not to engage in strenuous activity due to the possibility of injuring their enlarged spleen (or liver).

Eventually, the strong, cell-mediated immune response is decisive in controlling the infection and preventing complications. But after recovery, people usually remain chronically infected with EBV.

▶ Transmission and Epidemiology

More than 90% of the world's population is infected with EBV. In general, the virus causes no noticeable symptoms, but the time of life when the virus is first encountered seems to matter. In the case of EBV, infection during the teen years seems to result in disease, whereas infection before or after this period is usually asymptomatic.

Direct oral contact and contamination with saliva are the principal modes of transmission, although transfer through blood transfusions, sexual contact, and organ transplants is possible.

▶ Prevention and Treatment

The usual treatments for infectious mononucleosis are directed at symptomatic relief of fever and sore throat. Hospitalization is rarely needed. Occasionally, rupture of the spleen necessitates immediate surgery to remove it.

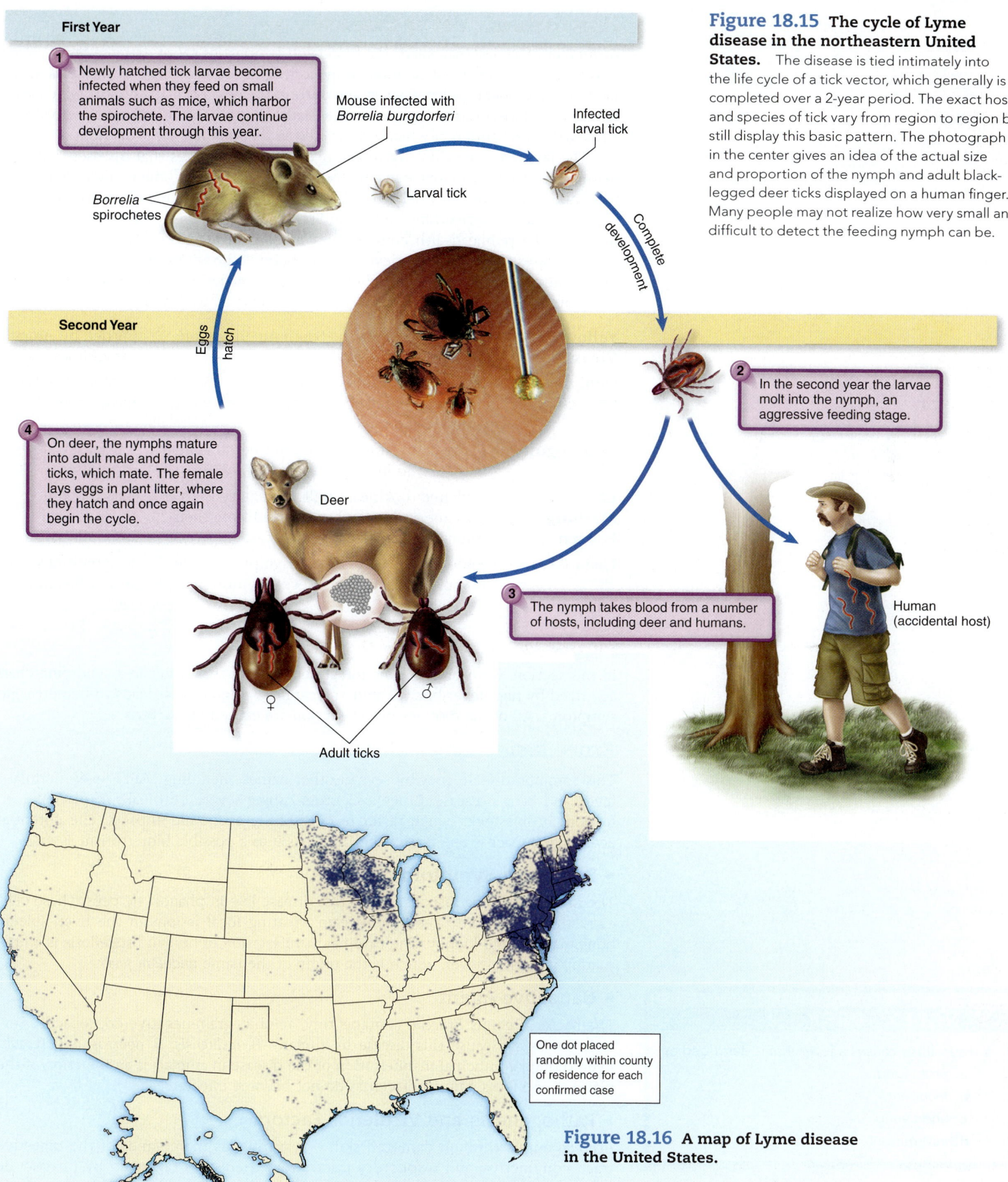

First Year

1 Newly hatched tick larvae become infected when they feed on small animals such as mice, which harbor the spirochete. The larvae continue development through this year.

Mouse infected with *Borrelia burgdorferi*

Infected larval tick

Borrelia spirochetes

Larval tick

Complete development

Second Year

Eggs hatch

4 On deer, the nymphs mature into adult male and female ticks, which mate. The female lays eggs in plant litter, where they hatch and once again begin the cycle.

Deer

2 In the second year the larvae molt into the nymph, an aggressive feeding stage.

3 The nymph takes blood from a number of hosts, including deer and humans.

Human (accidental host)

♀   ♂

Adult ticks

Figure 18.15 The cycle of Lyme disease in the northeastern United States. The disease is tied intimately into the life cycle of a tick vector, which generally is completed over a 2-year period. The exact hosts and species of tick vary from region to region but still display this basic pattern. The photograph in the center gives an idea of the actual size and proportion of the nymph and adult black-legged deer ticks displayed on a human finger. Many people may not realize how very small and difficult to detect the feeding nymph can be.

One dot placed randomly within county of residence for each confirmed case

Figure 18.16 A map of Lyme disease in the United States.

Hemorrhagic Fever Diseases

A number of agents that infect the blood and lymphatics cause extreme fevers, some of which are accompanied by internal hemorrhaging. The presence of the virus in the bloodstream causes capillary fragility and disrupts the blood-clotting system, which leads to various degrees of pathology, including death. All of these viruses are RNA enveloped viruses, the distribution of which is restricted to their natural host's distribution.

The fast-spreading Ebola outbreak of 2014 demonstrates that diseases that may get minimal amounts of space in a textbook can suddenly become worldwide threats. As of the writing of this book, the Ebola epidemic is outpacing efforts to control it in Africa, and has arrived in the United States. This has revealed a number of weaknesses in the public health community's preparedness for swiftly-spreading contagious diseases. Chikungunya, never before found to be transmitted in the Americas (only found in people returning from endemic countries) is, as of 2014, being locally transmitted by mosquitoes in the United States and Central America.

Yellow Fever Virus:	Endemic in Africa and South America; more frequent in rainy climates.	Carried by *Aedes* mosquitoes.
Dengue Fever:	Endemic in southeast Asia and India; epidemics have occurred in South America and Central America, and the Caribbean.	Carried by *Aedes* mosquitoes.
Chikungunya:	Endemic in Africa; arrived in Central America in 2013 and in U.S. and Europe in 2014.	Carried by *Aedes* mosquitoes.
Ebola and Marburg Fevers:	Endemic to Africa; capillary fragility is extreme and patients can bleed from their orifices and mucous membranes.	Bats thought to be natural reservoir of Ebola.
Lassa Fever:	Endemic to West Africa. Asymptomatic in 80% of cases. In others, severe symptoms develop.	Reservoir of virus is multimammate rat.

Nonhemorrhagic Fever Diseases

In this section, we examine some infectious diseases that result in a syndrome characterized by high fever but without the capillary fragility that leads to hemorrhagic symptoms. All of the diseases in this section are caused by bacteria.

Brucellosis

This common disease goes by several other names, including Malta fever, undulant fever, and Bang's disease. Brucellosis often causes severe outbreaks of placental infections in livestock, which result in devastating economic impacts. The potential economic impact is one reason the CDC lists it as a possible bioterrorism agent.

▶ Signs and Symptoms

The *Brucella* bacteria responsible for this disease live in phagocytic cells. These cells carry the bacteria into the bloodstream, creating focal lesions in the liver, spleen, bone marrow, and kidney. The cardinal manifestation of human brucellosis is a fluctuating pattern of fever, which is the origin of the name *undulant fever*.

▶ Causative Agent

The bacterial genus *Brucella* contains tiny, aerobic, gram-negative coccobacilli. Several species can cause this disease in humans: *B. melitensis, B. abortus,* and *B. suis.* Even though a principal manifestation of the disease in animals is an infection of the placenta and fetus, human placentas do not become infected.

▶ Pathogenesis and Virulence Factors

Brucella enters through damaged skin or via mucous membranes of the digestive tract, conjunctiva, and respiratory tract. From there it is taken up by phagocytic

NCLEX® PREP

4. Brucellosis causes a fever that is described as
   a. *prolonged.*
   b. *recurrent.*
   c. *undulating.*
   d. *intermittent.*

Disease Table 18.9 Hemorrhagic Fevers

Disease	Yellow Fever	Dengue Fever	Chikungunya	Ebola and/or Marburg	Lassa Fever
Causative Organism(s)	Yellow fever virus	Dengue fever virus	Chikungunya virus	Ebola virus, Marburg virus	Lassa fever virus
Most Common Modes of Transmission	Biological vector	Biological vector	Biological vector	Direct contact, body fluids	Droplet contact (aerosolized rodent excretions), direct contact with infected fluids
Virulence Factors	Disruption of clotting factors	Disruption of clotting factors	Disruption of clotting factors	Disruption of clotting factors	Disruption of clotting factors
Culture/Diagnosis	ELISA, PCR	Rise in IgM titers	PCR	PCR, viral culture (conducted at CDC)	ELISA
Prevention	Live attenuated vaccine available	–	–	–	Avoiding rats, safe food storage
Treatment	Supportive	Supportive	Supportive	Supportive	Ribavirin
Distinctive Features	Accompanied by jaundice	"Breakbone fever"–so named due to severe pain in some forms	Arthritic symptoms	Massive hemorrhage; rash sometimes present	Chest pain, deafness as long-term sequelae
Epidemiological Features	United States: only sporadic cases in travelers; internationally, 200,000 cases annually, 30,000 deaths; 90% of cases in Africa	United States: only sporadic cases in travelers; internationally, 50-100 million people infected every year and 22,000 deaths occur, mostly among children	United States: no reported cases; internationally, periodic epidemics	United States: infections only associated with facilities handling imported monkeys; internationally, sporadic outbreaks in Africa; major Ebola outbreak in 2014.	United States: no reported cases; internationally, occasional outbreaks in West Africa

cells. Because it is able to avoid destruction in the phagocytes, the bacterium is transported easily through the bloodstream and to various organs, such as the liver, kidney, breast tissue, or joints. Scientists suspect that the up-and-down nature of the fever is related to unusual properties of the bacterial lipopolysaccharide.

Q Fever

The name of this disease arose from the frustration created by not being able to identify its cause. The Q stands for "query." Its cause, a bacterium called *Coxiella burnetii*, was finally identified in the mid-1900s. The clinical manifestations of acute Q fever are abrupt onset of fever, chills, head and muscle ache, and, occasionally, a rash. The disease is sometimes complicated by pneumonitis (30% of cases), hepatitis, and endocarditis. About a quarter of the cases are chronic rather than acute and result in vascular damage and endocarditis-like symptoms.

C. burnetii is a very small pleomorphic (variously shaped) gram-negative bacterium, and for a time it was considered a rickettsia, in the same genus as the bacterium that causes Rocky Mountain spotted fever (below). *C. burnetii* is apparently harbored by a wide assortment of vertebrates and arthropods, especially ticks, which play an essential role in transmission between wild and domestic animals. Ticks do not transmit the disease to humans, however. Humans acquire infection largely by means of environmental contamination and airborne spread. Birth products, such as placentas, of infected domestic animals contain large numbers of bacteria. Other sources of infectious material include urine, feces, milk, and airborne particles from infected animals. The primary portals of entry are the lungs, skin, conjunctiva, and gastrointestinal tract.

People at highest risk are farm workers, meat cutters, veterinarians, laboratory technicians, and consumers of raw milk products.

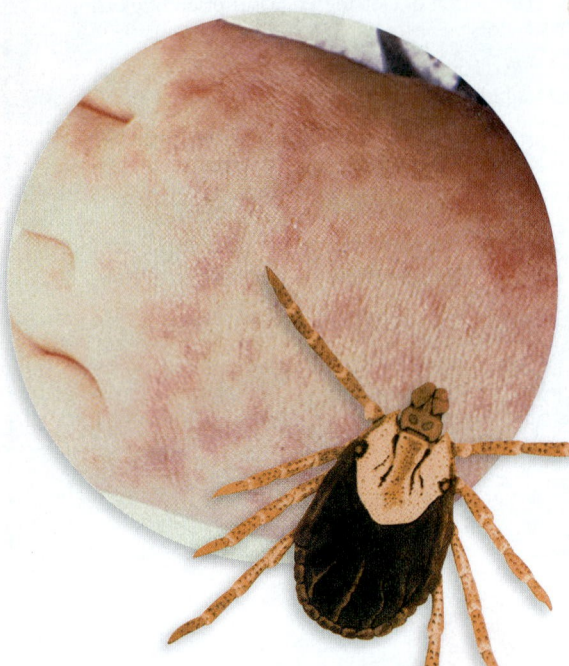

Figure 18.17 The rash in RMSF.
This case occurred in a child several days after the onset of fever.

Figure 18.18 A representative Triatomine bug, the carrier of *T. cruzi.*

Cat-Scratch Disease

This disease is one of a group of diseases caused by different species of the small gram-negative rod *Bartonella*. *Bartonella* species are considered to be emerging pathogens. They are fastidious but not obligate intracellular parasites, so they will grow on blood agar. In addition to cat-scratch disease, a new species of *Bartonella* that causes high fever and life-threatening anemia was identified in 2007.

Bartonella henselae is the agent of cat-scratch disease (CSD), an infection connected with being clawed or bitten by a cat. The pathogen is present in over 40% of cats, especially kittens. There are approximately 25,000 cases per year in the United States, 80% of them in children 2 to 14 years old. The symptoms start after 1 to 2 weeks, with a cluster of small papules at the site of inoculation. Most infections remain localized and resolve in a few weeks, but drugs such as azithromycin, erythromycin, and rifampin can be effective therapies. The disease can be prevented by thorough antiseptic cleansing of a cat bite or scratch.

Rocky Mountain Spotted Fever (RMSF)

This disease is named for the region in which it was first detected in the United States–the Rocky Mountains of Montana and Idaho. In spite of its name, the disease occurs infrequently in the western United States. The majority of cases are concentrated in the Southeast and Eastern Seaboard regions. It also occurs in Canada and Central and South America. Infections occur most frequently in the spring and summer, when the tick vector is most active. The yearly rate of RMSF is 20 to 40 cases per 10,000 population, with fluctuations coinciding with weather and tick infestations.

RMSF is caused by a bacterium called *Rickettsia rickettsii* transmitted by hard ticks such as the wood tick, the American dog tick, and the Lone Star tick. The dog tick is probably most responsible for transmission to humans because it is the major vector in the southeastern United States.

After 2 to 4 days of incubation, the first symptoms are sustained fever, chills, headache, and muscular pain. A distinctive spotted rash usually comes on within 2 to 4 days after the prodrome **(figure 18.17),** which usually appears first on the wrists, forearms, and ankles, before spreading. Early lesions are slightly mottled like measles, but later ones can change shape to look like other types of rashes. In the most severe untreated cases, the enlarged lesions merge and can become necrotic, predisposing to gangrene of the toes or fingertips.

Although the spots are the most obvious symptom of the disease, the most serious manifestations are cardiovascular disruption, including hypotension, thrombosis, and hemorrhage. Conditions of restlessness, delirium, convulsions, tremor, and coma are signs of the often overwhelming effects on the central nervous system. Fatalities occur in an average of 20% of untreated cases and 5% to 10% of treated cases.

Chagas Disease

Chagas disease is sometimes called "the American trypanosomiasis." The causative agent is the flagellated protozoan *Trypanosoma cruzi*. A different trypanosome, *T. brucei,* causes sleeping sickness on the African continent. Chagas disease has been called "the new AIDS of the Americas," because it has a long incubation time and is very difficult to cure.

▶ Signs and Symptoms

Once the trypanosomes are transmitted by a group of insects called the Triatomines **(figure 18.18),** they multiply in muscle and blood cells. From time to time, the blood cells rupture and large numbers of trypanosomes are released into the bloodstream. The disease manifestations are divided into acute and chronic phases. Soon after infection, the acute phase begins; symptoms are relatively nondescript and range from mild to severe fever, nausea, and fatigue. A swelling called a "chagoma"

Disease Table 18.10 Nonhemorrhagic Fever Diseases

Disease	Brucellosis	Q Fever	Cat-Scratch Disease	Rocky Mountain Spotted Fever
Causative Organism(s)	*Brucella* species	*Coxiella burnetii*	*Bartonella henselae*	*Rickettsia rickettsii*
Most Common Modes of Transmission	Direct contact, airborne, parenteral (needlesticks)	Airborne, direct contact, food-borne	Parenteral (cat scratch or bite)	Biological vector (tick)
Virulence Factors	Intracellular growth; avoidance of destruction by phagocytes	Endospore-like structure	Endotoxin	Induces apoptosis in cells lining blood vessels
Culture/Diagnosis	Gram stain of biopsy material	Serological tests for antibody	Biopsy of lymph nodes plus Gram staining; ELISA (performed by CDC)	Fluorescent antibody, PCR
Prevention	Animal control, pasteurization of milk	Vaccine for high-risk population	Clean wound sites	Avoid ticks
Treatment	Doxycycline plus gentamicin or streptomycin	Doxycycline	Azithromycin	Doxycycline
Distinctive Features	Undulating fever, muscle aches	Airborne route of transmission, variable disease presentation	History of cat bite or scratch; fever not always present	Most common in eastern and southeastern United States
Epidemiological Features	United States: fewer than 100 cases per year; internationally, 500,000 cases per year	United States: number of cases increased from 21 in 1999 to 169 in 2006	United States: estimated incidence is 9.3 cases per 100,000; internationally, seroprevalence from 0.6% to 37% depending on cat population	Only in Americas

at the site of the bug bite may be present. If the bug bite is close to the eyes, a distinct condition called Romana's sign, swelling of the eyelids, may appear. The acute phase lasts for weeks or months after which the condition becomes chronic, which is virtually asymptomatic for a period of years or indefinitely. Eventually, the trypanosomes are found in numerous sites around the body; in later years, this may lead to inflammation and disruption of function in organs such as the heart, the brain, and the intestinal tract.

▶ Transmission and Epidemiology

Estimates put the prevalence of this disease at 8 million people, 300,000 of whom live in the United States. Most U.S. cases were acquired in an endemic area by travelers or other people who have since immigrated to this country.

As already noted, the disease is transmitted through the bite of a Triatomine bug. The trypanosome can also be transmitted vertically because it crosses the placenta, and via blood transfusion with infected blood. Recently, the United States began screening all donated blood for this disease.

▶ Prevention

No vaccine exists for Chagas disease. In endemic areas, pesticides and improved building materials in houses are used to minimize the presence of the bug.

▶ Treatment

Treatment is most successful if begun during the acute phase. However, it is often not accomplished because the acute phase of the disease is not necessarily suggestive of Chagas. Drugs for treatment are only available through the CDC. During the chronic phase of the disease, symptomatic treatment of cardiac and other problems may be indicated.

Disease Table 18.11	Chagas Disease
Causative Organism	*Trypanosoma cruzi*
Most Common Modes of Transmission	Biological vector, vertical
Virulence Factors	Antioxidant enzymes, co-opting host antigens; induces autoimmunity
Culture/Diagnosis	Blood smear in acute phase; serological methods in later stages
Prevention	Insect control
Treatment	Consult CDC
Epidemiological Features	Endemic in Central and South America; 300,000 cases present in United States

Anthrax

Anthrax causes disease in the lungs and in the skin, and is addressed elsewhere in this book. We discuss anthrax in this chapter because it multiplies in large numbers in the blood and because septicemic anthrax is a possible outcome of all forms of anthrax.

For centuries, anthrax has been known as a zoonotic disease of herbivorous livestock (sheep, cattle, and goats). It has an important place in the history of medical microbiology because Robert Koch used anthrax as a model for developing his postulates in 1877–and, later, Louis Pasteur used the disease to prove the usefulness of vaccination.

▶ Signs and Symptoms

As just noted, anthrax infection can exhibit its primary symptoms in various locations of the body: on the skin (cutaneous anthrax), in the lungs (pulmonary anthrax), in the gastrointestinal tract (acquired through ingestion of contaminated foods), and in the central nervous system (anthrax meningitis). The cutaneous and pulmonary forms of the disease are the most common. In all of these forms, the anthrax bacterium gains access to the bloodstream, and death, if it occurs, is usually a result of an overwhelming septicemia. Pulmonary anthrax–and the accompanying pulmonary edema and hemorrhagic lung symptoms–can sometimes be the primary cause of death, although it is difficult to separate the effects of septicemia from the effects of pulmonary infection.

In addition to symptoms specific to the site of infection, septicemic anthrax results in headache, fever, and malaise. Bleeding in the intestine and from mucous membranes and orifices may occur in late stages of septicemia.

▶ Causative Agent

Bacillus anthracis is a gram-positive endospore-forming rod that is among the largest of all bacterial pathogens. It is composed of block-shaped, angular rods 3 to 5 μm long and 1 to 1.2 μm wide. Central endospores develop under all growth conditions except in the living body of the host **(figure 18.19)**. Because the primary habitat of many *Bacillus* species, including *B. anthracis,* is the soil, endospores are continuously dispersed by means of dust into water and onto the bodies of plants and animals.

▶ Pathogenesis and Virulence Factors

The main virulence factors of *B. anthracis* are its polypeptide capsule and what is referred to as a "tripartite" toxin–a protein complex composed of three separate exotoxins. The end result of exotoxin action is massive inflammation and initiation of shock.

Additional virulence factors for *B. anthracis* include hemolysins and other enzymes that damage host membranes.

▶ Transmission and Epidemiology

The anthrax bacillus undergoes its cycle of vegetative growth and sporulation in the soil. Animals become infected while grazing on grass contaminated with endospores.

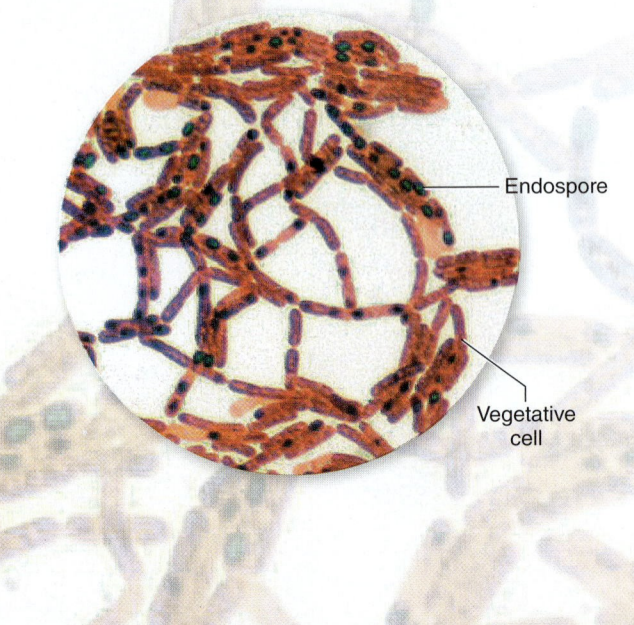

Figure 18.19 *Bacillus anthracis.* Note the centrally placed endospores and streptobacillus arrangement (600×).

Endospore

Vegetative cell

When the pathogen is returned to the soil in animal excrement or carcasses, it can sporulate and become a long-term reservoir of infection for the animal population. The majority of natural anthrax cases are reported in livestock from Africa, Asia, and the Middle East. Most recent (natural) cases in the United States have occurred in textile workers handling imported animal hair or hide or products made from them. Because of effective control procedures, the number of cases in the United States is extremely low (fewer than 10 per year).

The anthrax attacks of 2001 aimed at two senators and several media outlets focused a great deal of attention on the threat of bioterrorism. During that attack, 22 people acquired anthrax and 5 people died.

▶ Culture and Diagnosis

Diagnosis requires a high index of suspicion. This means that anthrax must be present as a possibility in the clinician's mind or it is likely not to be diagnosed, because it is such a rare disease in the developed world and because, in all of its manifestations, it can mimic other infections that are not so rare. First-level (presumptive) diagnosis begins with culturing the bacterium on blood agar and performing a Gram stain. Ultimately, samples should be handled by the Centers for Disease Control and Prevention, which will perform confirmatory tests, usually involving direct fluorescent antibody testing and phage typing tests.

▶ Prevention and Treatment

Humans should be vaccinated with the purified toxoid if they have occupational contact with livestock or products such as hides and bone or if they are members of the military. Effective vaccination requires six inoculations given over 1.5 years, with yearly boosters. The cumbersome nature of the vaccination has spurred research and development of more manageable vaccines. Persons who are suspected of being exposed to the bacterium are given prophylactic antibiotics, which seem to be effective at preventing disease even after exposure.

The recommended treatment for anthrax is usually penicillin, doxycycline, or ciprofloxacin. However, there is still debate about the best way to treat anthrax exposure, because antibiotic usage can sometimes worsen symptoms by releasing large amounts of toxin into the bloodstream. Treatment of human cases are conducted in consultation with the CDC.

Disease Table 18.12 Anthrax

Causative Organism(s)	*Bacillus anthracis*
Most Common Modes of Transmission	Vehicle (air, soil), indirect contact (animal hides), vehicle (food)
Virulence Factors	Triple exotoxin
Culture/Diagnosis	Culture, direct fluorescent antibody tests
Prevention	Vaccine for high-risk population
Treatment	In consultation with the CDC
Epidemiological Features	Internationally, 2,000–20,000 cases annually, most cutaneous

18.3 LEARNING OUTCOMES—Assess Your Progress

4. List the possible causative agents, modes of transmission, virulence factors, diagnostic techniques, and prevention/treatment for the "Highlight Diseases" malaria and HIV.

5. Discuss the epidemiology of malaria.

6. Describe the epidemiology of HIV infection in the developing world.

7. Discuss the important features of infectious cardiovascular diseases that have more than one possible cause. These are the two forms of endocarditis, septicemia, hemorrhagic fever diseases, and nonhemorrhagic fever diseases.

8. Identify factors that distinguish hemorrhagic and nonhemorrhagic fever diseases.

9. Outline the series of events that may lead to septicemia and how it should be prevented and treated.

10. Discuss the important features of infectious cardiovascular diseases that have only one possible cause. These are plague, tularemia, Lyme disease, infectious mononucleosis, Chagas disease, and anthrax.

11. Describe what makes anthrax a good agent for bioterrorism, and list the important presenting signs to look for in patients.

CASE FILE WRAP-UP

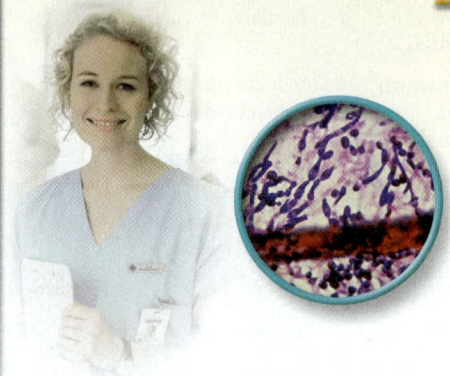

Endocarditis is an inflammation affecting the innermost layer of the heart. The mitral and aortic valves are most often affected. Endocarditis may be acute or subacute. While symptoms are mostly the same, subacute endocarditis has a slower onset.

In Donna's case, she had been diagnosed with a heart murmur as a child. Heart murmurs are abnormal or extra heart sounds resulting from the heart valves closing as blood is propelled through the heart. Unfortunately, Donna's heart murmur had never been adequately assessed and followed and was actually the result of a congenital abnormality of the mitral valve. Such abnormalities predispose the heart to infections, and Donna's dental work had allowed bacteria to enter Donna's bloodstream and travel to and infect Donna's mitral valve. Donna's cultures came back positive for streptococci, and she was diagnosed with subacute endocarditis. She required intravenous antibiotics to treat the infection and eventually underwent surgery to replace her mitral valve.

▶ Summing Up

Taxonomic Organization Summing up Microorganisms Causing Disease in the Cardiovascular and Lymphatic Systems

Microorganism	Pronunciation	Location of Disease Table
Gram-positive endospore-forming bacteria		
Bacillus anthracis	buh-sill′-us an-thray′-sus	Anthrax, p. 525
Gram-positive bacteria		
Staphylococcus aureus	staf″-uh-lo-kok′-us are′-ee-us	Acute endocarditis, p. 514; septicemia, p. 515
Streptococcus pyogenes	strep″-tuh-kok′-us pie′-ah″-gen-eez	Acute endocarditis, p. 514
Streptococcus pneumoniae	strep″-tuh-kok′-us nu-mo′-nee-ay	Acute endocarditis, p. 514
Gram-negative bacteria		
Yersinia pestis	yur-sin′-ee-uh pes′-tiss	Plague, p. 516
Francisella tularensis	fran-si′-sell″-uh tew′-luh-ren″-sis	Tularemia, p. 517
Borrelia burgdorferi	bor-rill′-ee-ah berg-dorf′-fur-eye	Lyme disease, p. 518
Brucella abortus, B. suis	bru-sell′-uh uh-bort′-us	Brucellosis p. 523
Coxiella burnetii	cox-ee-ell′-uh bur-net′-tee-eye	Q fever, p. 523
Bartonella henselae	bar-ton-nell′-uh hen′-sell-ay	Cat-scratch disease, p. 523
Neisseria gonorrhoeae	nye-seer″-ee-uh′ gon′-uh-ree″-uh	Acute endocarditis, p. 514
Rickettsia rickettsii	ri-ket′-see-uh ri-ket′-see-eye	Rocky Mountain spotted fever, p. 523
DNA viruses		
Epstein-Barr virus	ep′-steen bar″ vie′-russ	Infectious mononucleosis, p. 518
RNA viruses		
Yellow fever virus	yel′-loh fee′-ver vie′-russ	Yellow fever, p. 521
Dengue fever virus	den′-gay fee′-ver vie′-russ	Dengue fever, p. 521
Ebola and Marburg viruses	ee-bowl′-uh and mar′-berg vie′-russ-suz′	Ebola and Marburg hemorrhagic fevers, p. 521
Lassa fever virus	lass′-sah fee′-ver vie′-russ	Lassa fever, p. 521
Chikungunya virus	chick-un-goon′ yah vie′-russ	Hemorrhagic fevers, p. 521
Retroviruses		
Human immunodeficiency virus 1 and 2	hew′-mun im′-muh-noh-dee-fish″-shun-see vie′-russ	HIV infection and AIDS, p. 513
Protozoa		
Plasmodium falciparum, P. vivax, P. knowlesi, P. ovale, P. malariae	plas-moh′-dee-um fal-sip′-uh-rum, plas-moh′-dee-um vy′-vax, plas-moh′-dee-um nohles′-eye, plas-moh′-dee-um oh-val′-ee, plas-moh′-dee-um ma-lair′-ee-ay	Malaria, p. 505
Trypanosoma cruzi	tri-pan″-uh-sohm′-uh krewz′-ee-eye	Chagas disease, p. 524

Ebola

As you read this, you know more about the Ebola epidemic of 2014/2015 than the authors of this textbook do. We are writing this "Inside the Clinic" in October of 2014, and like everyone else around us, we are wondering what will happen in the next weeks and months.

First, we'll present the important features of the virus in more detail than we did in the section on hemorrhagic fevers, including relevant information from the epidemic.

Ebola Fever		Specific information from 2014/2015 epidemic
Causative Organism	Ebola virus, family *Filoviridae*	
Most Common Modes of Transmission	Direct contact, body fluids	
Virulence Factors	Ability to disrupt clotting, glycoprotein spikes for binding, a viral protein critical for viral replication	
Culture/Diagnosis	There are PCR and ELISA tests, as well as other antibody-mediated assays. Cell culture is also available.	A fever of > 101.5° F, after a known exposure, led to further testing
Prevention	Avoiding contact with patients and their fluids; two vaccines are in human trials	In U.S., quarantine precautions were instituted for 21 days for people exposed to patients or their fluids. There was some confusion about voluntary vs. mandatory quarantine in early days.
Treatment	Only supportive	In 2014 a few patients were treated with Z-MAPP, a "drug" containing monoclonal antibodies to the virus. Then the (still experimental) drug ran out.
Distinctive Features	High fatality rate: 30-90%	In Africa the infection displayed a 55% fatality rate
Epidemiology	Occurs in sporadic outbreaks in Africa, often in rural areas where outbreaks can be contained due to limited population.	2014 epidemic occurred mainly in three African countries and was harder to control because it hit urban areas of dense population.

Data in this table was current as of October 2014.

Epidemiologists classify infectious agents according to two important factors: their likelihood of infecting contacts (infectivity or infectiousness), and their virulence (deadliness). The two factors are weighed when trying to predict an outbreak's continuation and/or results. The figure on this page shows that Ebola (see arrow) has a low infectiousness. This may sound surprising, since we saw health care workers become infected even after wearing personal protective equipment. You see that four of three of the five diseases at the far right of the infectiousness scale (the highly contagious ones) are droplet or airborne diseases (mumps, whooping cough, and measles) and the fourth, malaria, is efficiently delivered to victims by the mosquito. Rotavirus (the fifth infection) is spread through diarrhea. The virus, even on a visibly "clean" object, can survive on surfaces for a long time.

The graph shows that Ebola is deadly, but less infectious than many, many microbes. The people at *most* risk are those who are caring for patients in the late stages of disease, where there might be copious amounts of body fluids (blood, vomit, diarrhea). The virus can also remain viable for several days in a corpse. In Africa, this can put family and community members and others at risk. In a first-world country like the United States, healthcare workers are at highest risk. In the first transmissions in the U.S., a man traveling from Liberia carrying

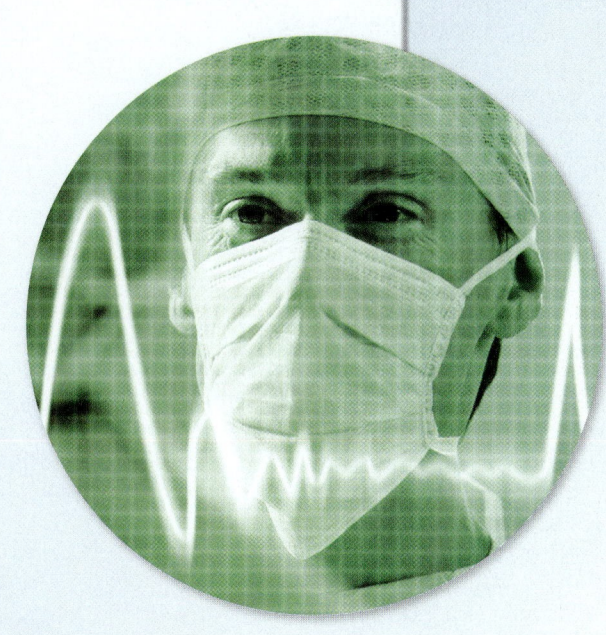

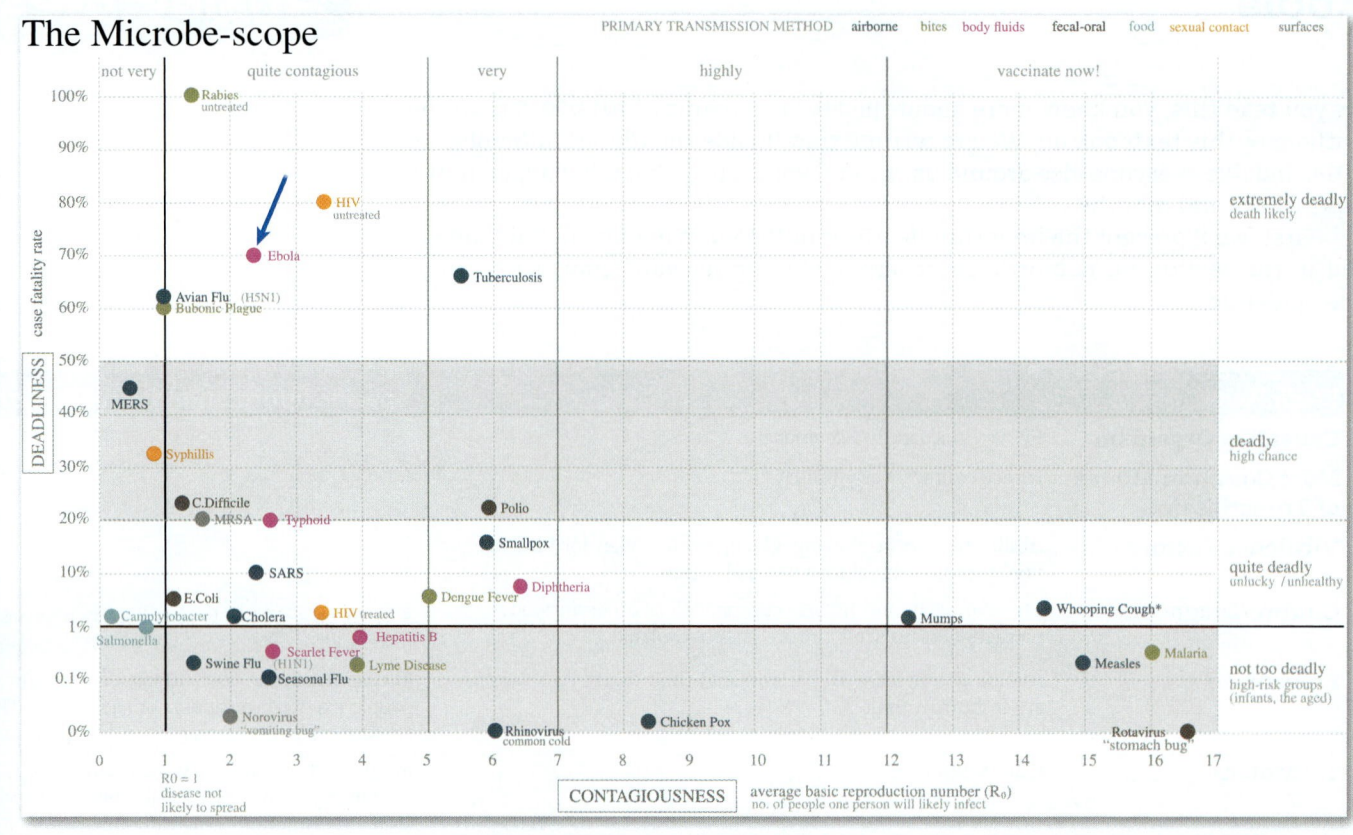

The Microbe-scope

PRIMARY TRANSMISSION METHOD   airborne   bites   body fluids   fecal-oral   food   sexual contact   surfaces

Source: CDC

the infection turned up at a hospital in Dallas, and was hospitalized. Two of his nurses became infected.

Because there is a now a high degree of awareness of the disease, it is quite possible that people with the disease will be quickly hospitalized, and that hospital personnel will quickly learn how to improve their protective practices, and that there will be no more spread in hospitals. Community transmission would also be avoided.

Of course, writing this, we don't know whether that was how it came to pass. But you do.

528

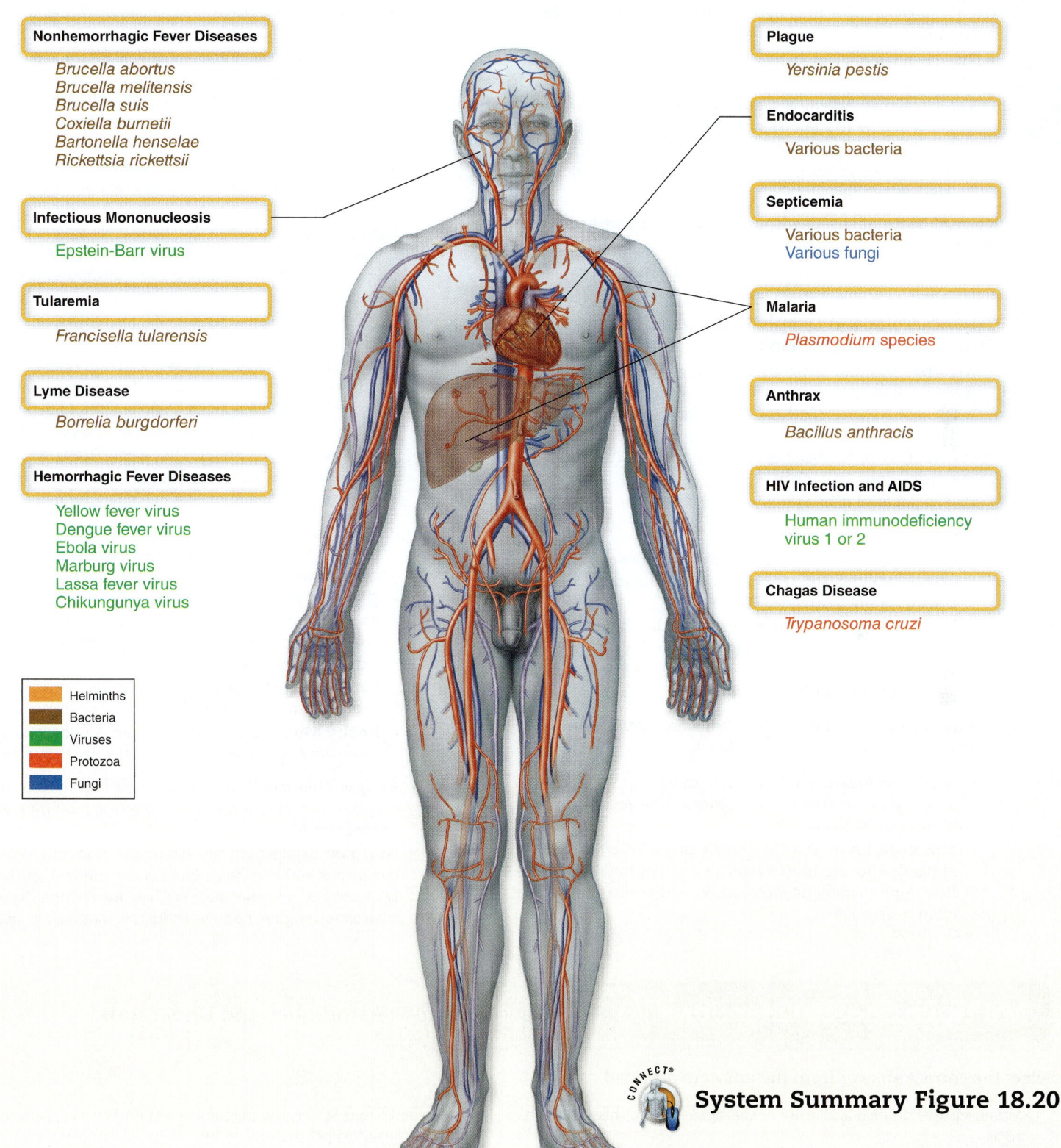

Nonhemorrhagic Fever Diseases

Brucella abortus
Brucella melitensis
Brucella suis
Coxiella burnetii
Bartonella henselae
Rickettsia rickettsii

Infectious Mononucleosis

Epstein-Barr virus

Tularemia

Francisella tularensis

Lyme Disease

Borrelia burgdorferi

Hemorrhagic Fever Diseases

Yellow fever virus
Dengue fever virus
Ebola virus
Marburg virus
Lassa fever virus
Chikungunya virus

Plague

Yersinia pestis

Endocarditis

Various bacteria

Septicemia

Various bacteria
Various fungi

Malaria

Plasmodium species

Anthrax

Bacillus anthracis

HIV Infection and AIDS

Human immunodeficiency
virus 1 or 2

Chagas Disease

Trypanosoma cruzi

■	Helminths
■	Bacteria
■	Viruses
■	Protozoa
■	Fungi

CONNECT®

System Summary Figure 18.20

Chapter Summary

18.1 The Cardiovascular and Lymphatic Systems and Their Defenses

- The cardiovascular system is composed of the blood vessels and the heart.
- The lymphatic system is a one-way passage, returning fluid from the tissues to the cardiovascular system.
- The systems are highly protected from microbial infection, as they are not an open body system and they contain many components of the host's immune system.

18.2 Normal Biota of the Cardiovascular and Lymphatic Systems

- At the present time, we believe that the cardiovascular and lymphatic systems contain no normal biota.

18.3 Cardiovascular and Lymphatic System Diseases Caused by Microorganisms

- **Malaria:** Symptoms are malaise, fatigue, vague aches, and nausea, followed by bouts of chills, fever, and sweating. Causative organisms are *Plasmodium* species: *P. malariae, P. vivax, P. falciparum, P. knowlesi,* and *P. ovale.* Carried by *Anopheles* mosquito.
- **HIV Infection and AIDS:** Symptoms directly tied to the level of virus in the blood versus the level of T cells in the blood.
  - HIV is a retrovirus. Contains *reverse transcriptase,* which catalyzes the replication of double-stranded DNA from single-stranded RNA.
  - Destruction of T4 lymphocytes paves the way for invasion by opportunistic agents and malignant cells.
  - Transmission occurs mainly through sexual intercourse and transfer of blood or blood products.
- **Endocarditis:** Inflammation of the endocardium, usually due to infection of the valves of the heart.
  - **Acute Endocarditis:** Most often caused by *Staphylococcus aureus,* group A streptococci, *Streptococcus pneumoniae,* and *Neisseria gonorrhoeae.*
  - **Subacute Endocarditis:** Usually preceded by some form of damage to the heart valves or by congenital malformation. Alpha-hemolytic streptococci and normal biota most often responsible.

- **Septicemia:** Organisms actively multiplying in the blood. Most caused by bacteria, often MRSA, a lesser extent by fungi.
- **Plague:** *Pneumonic plague* is a respiratory disease; *bubonic plague* causes inflammation and necrosis of the lymph nodes; *septicemic plague* is the result of multiplication of bacteria in the blood. *Yersinia pestis* is the causative organism. Fleas are principal agents in transmission of the bacterium.
- **Tularemia:** Causative agent is *Francisella tularensis,* a facultative intracellular gram-negative bacterium. Disease is often called rabbit fever.
- **Lyme Disease:** Caused by *Borrelia burgdorferi.* Syndrome mimics neuromuscular and rheumatoid conditions. *B. burgdorferi* is a spirochete transmitted primarily by *Ixodes* ticks.
- **Infectious Mononucleosis:** Vast majority of cases are caused by the herpesvirus *Epstein-Barr virus (EBV).*
- **Hemorrhagic Fever Diseases:** Extreme fevers often accompanied by internal hemorrhaging. Hemorrhagic fever diseases described here (Yellow fever, Dengue fever, Ebola and Marburg diseases, and Lassa fever) are caused by RNA enveloped viruses.
- **Nonhemorrhagic Fever Diseases:** Characterized by high fever without the capillary fragility that leads to hemorrhagic symptoms.
  - **Brucellosis:** Also called Malta fever, undulant fever, Bang's disease. Multiple species cause this disease in humans—among them *B. melitensis, B. abortus,* and *B. suis.*
  - **Q Fever:** Caused by *Coxiella burnetii,* a small pleomorphic gram-negative bacterium and intracellular parasite. *C. burnetii* is harbored by a wide assortment of vertebrates and arthropods, especially ticks.
  - **Cat-Scratch Disease:** Infection by *Bartonella henselae* connected with being clawed or bitten by a cat.
  - **Rocky Mountain Spotted Fever:** Another tick-borne disease; causes a distinctive rash. Caused by *Rickettsia rickettsii.*
- **Chagas Disease:** *Trypanosoma cruzi* transmitted by insects; endemic in South and Central America; now frequent in the United States.
- **Anthrax:** Exhibits primary symptoms in various locations: skin (cutaneous anthrax), lungs (pulmonary anthrax), gastrointestinal tract, central nervous system (anthrax meningitis). Caused by *Bacillus anthracis,* gram-positive, endospore-forming rod found in soil.

Multiple-Choice Questions   Bloom's Levels 1 and 2: Remember and Understand

Select the correct answer from the answers provided.

1. When viruses flourish and grow in the bloodstream, this is referred to as
   - a. viremia.
   - b. endocemia.
   - c. septicemia.
   - d. serocemia.

2. In the United States, the plague bacterium *Yersinia pestis* is transmitted to people mainly by
   - a. mosquitoes.
   - b. fleas.
   - c. rats.
   - d. prairie dogs.

3. Lyme disease is transmitted to people by

    a. ticks.
    d. woodrats.
    b. deer.
    e. fleas.
    c. mice.

4. Cat-scratch disease is effectively treated with

    a. rifampin.
    c. amoxicillin.
    b. penicillin.
    d. acyclovir.

5. Normal biota found in the oral cavity are most likely to cause

    a. acute endocarditis.
    c. malaria
    b. subacute endocarditis.
    d. tularemia.

6. A distinctive bull's-eye rash results from a tick bite transmitting

    a. Lyme disease.
    c. Q fever.
    b. tularemia.
    d. Rocky Mountain spotted fever.

7. Wool-sorter's disease is caused by

    a. *Brucella abortus.*
    c. *Coxiella burnetii.*
    b. *Bacillus anthracis.*
    d. rabies virus.

8. Which of the following is *not* a hemorrhagic fever?

    a. Lassa fever
    c. Ebola fever
    b. Marburg fever
    d. tularemia

Critical Thinking   Bloom's Levels 3, 4, and 5: Apply, Analyze, and Evaluate

Critical thinking is the ability to reason and solve problems using facts and concepts. These questions can be approached from a number of angles and, in most cases, they do not have a single correct answer.

1. In the Middle Ages, during a massive plague pandemic, one of the control measures officials instituted was the quarantine of infected people. Why was this not successful?

2. Why do you think that malarial infection is more often fatal in children than in adults in areas where it is endemic?

3. Explain the differences between the epidemiology of AIDS in the United States and in the developing world.

4. Use the terms *prevalence* and *incidence* (see chapter 11) to explain how better treatment options have led to a higher prevalence of AIDS in the world.

5. What characteristics make tularemia a potential bioweapon?

Visual Connections   Bloom's Level 5: Evaluate

This question connects previous images to a new concept.

1. a. **From chapter 12, figure 12.10a.** Imagine that the WBCs shown in this illustration are unable to control the microorganisms in the blood. Could the change that has occurred in the vessel wall help the organism spread to other locations? If so, how?

   b. If the organisms are able to survive phagocytosis, how could that impact the progress of this disease? Explain your answer.

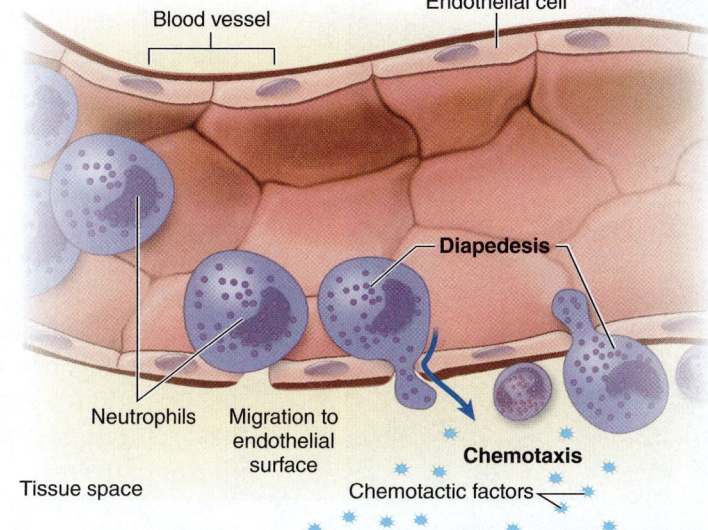

www.mcgrawhillconnect.com

Enhance your study of this chapter with study tools and practice tests. Also ask your instructor about the resources available through Connect, including the media-rich eBook, interactive learning tools, and animations.

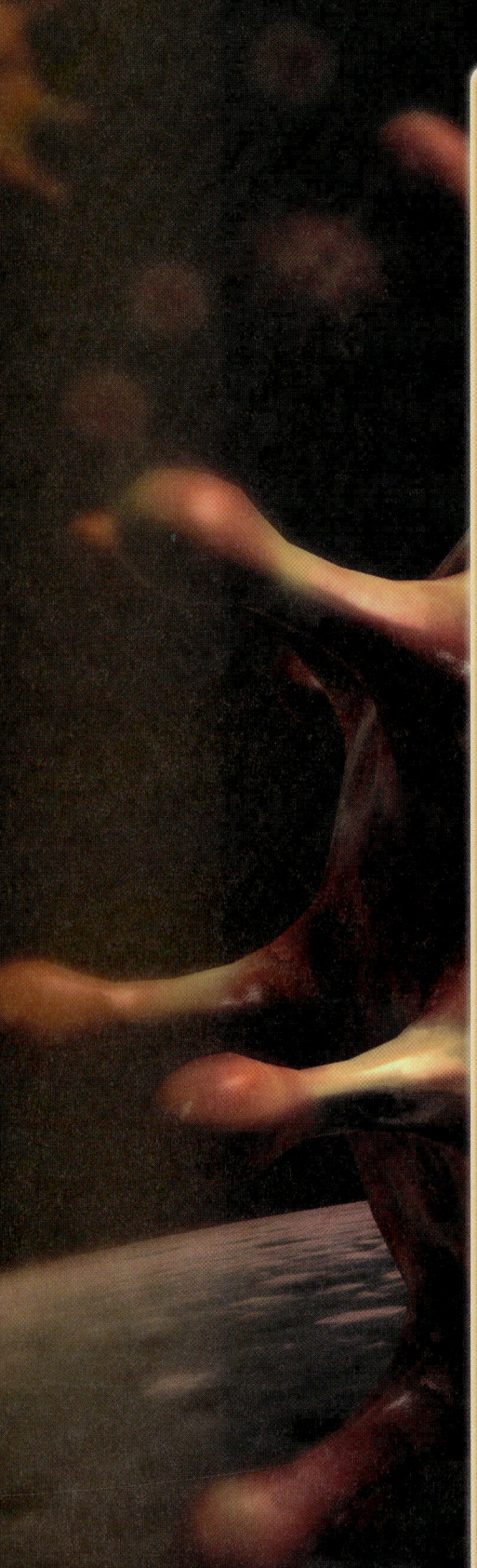

CASE FILE

Very Sick, Very Fast

I was a respiratory therapist student completing a rotation in pediatrics when I became involved in caring for Michael, a 6-month-old who was admitted to the pediatric unit with a respiratory infection.

Michael was a healthy infant who was born at term (39 weeks' gestation). He was current with all of his immunizations. Michael was the youngest of three children, with two older siblings who were school age. Michael attended day care on a part-time basis when his mother worked.

Michael and both his siblings had become ill during recent days with cold symptoms. At first, Michael had a runny nose and a slight cough. He was fussy and irritable and his mother stated he had been sleeping more than usual. Michael was being breast-fed. However, when he became ill, he fussed at the breast and fed less often, even though his mother offered him the breast frequently to encourage fluid intake and provide comfort. Having three children, Michael's mother was well versed in caring for a sick child and was not prone to panic over a common cold.

However, on the third day of his illness, Michael took an alarming turn for the worse. Although he acted as though he was hungry, he would stop feeding almost as soon as he started. Michael had copious nasal discharge, and his mother rightly tried to clear his nostrils using a bulb syringe, but Michael still fed poorly. He had only had one slightly wet diaper over the course of the night and had not urinated at all in the morning. He was running a high fever of 39.6°C (103.3°F). His eyes appeared sunken in his face and there were no tears when he cried, both signs of dehydration. He was coughing frequently. Most concerning, Michael's respiratory rate was 48 breaths/minute. His nail beds were cyanosed (blue) and he was very pale in color. His oxygen saturation was 89% on room air.

Michael was seen in the emergency room and was immediately admitted to the pediatric unit by the physician on call, who was very concerned. My preceptor and I were called to assess Michael on the unit. When we arrived, Michael had a chest X ray and blood work done. He was lying in a crib with a nasal cannula supplying humidified oxygen taped to his face. His color was still very pale but his nail beds were pink. His oxygen saturation had increased to 95% with oxygen flow at 5 L/minute. I listened to Michael's chest and heard wheezes throughout Michael's lungs. After I assessed Michael, my preceptor asked me what I thought might be going on with Michael. Hedging my bets, I replied that all I could tell for certain was that Michael had a respiratory tract illness and a virus was the likely culprit. My preceptor promptly replied, "You're right that Michael has a viral illness of the respiratory tract, but which one?" When I confessed that I did not know, my preceptor informed me that Michael's symptoms were consistent with respiratory syncytial virus (RSV). Michael's tests would prove him right.

- How is RSV spread?
- How might Michael have contracted the respiratory syncytial virus?

Case File Wrap-Up appears on page 555.

Infectious Diseases Affecting the Respiratory Systems

19

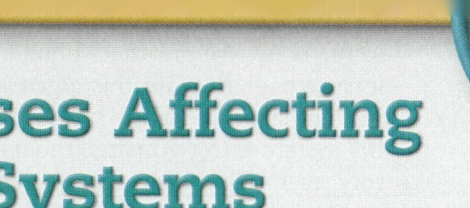

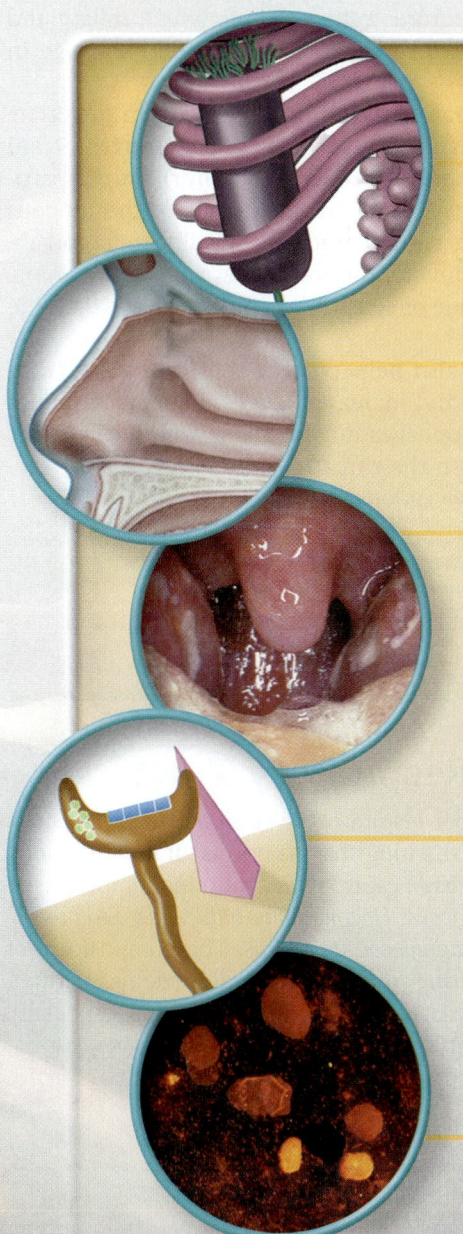

IN THIS CHAPTER...

19.1 The Respiratory Tract and Its Defenses

1. Draw or describe the anatomical features of the respiratory tract.
2. List the natural defenses present in the respiratory tract.

19.2 Normal Biota of the Respiratory Tract

3. List the types of normal biota presently known to occupy the respiratory tract.

19.3 Upper Respiratory Tract Diseases Caused by Microorganisms

4. List the possible causative agents, modes of transmission, virulence factors, diagnostic techniques, and prevention/treatment for the "Highlight Disease" pharyngitis.
5. Discuss important features of the other infectious diseases of the upper respiratory tract. These are rhinitis, sinusitis, acute otitis media, and diphtheria.
6. Identify two bacteria that can cause dangerous pharyngitis cases.

19.4 Diseases Caused by Microorganisms Affecting Both the Upper and Lower Respiratory Tract

7. List the possible causative agents, modes of transmission, virulence factors, diagnostic techniques, and prevention/treatment for the "Highlight Disease" influenza.
8. Compare and contrast *antigenic drift* and *antigenic shift* in influenza viruses.
9. Discuss important features of the other infectious diseases of the upper and lower respiratory tracts. These are pertussis and RSV disease.

19.5 Lower Respiratory Tract Diseases Caused by Microorganisms

10. List the possible causative agents, modes of transmission, virulence factors, diagnostic techniques, and prevention/treatment for the "Highlight Disease" tuberculosis.
11. Discuss the problems associated with MDR-TB and XDR-TB.
12. Discuss important features of the other lower respiratory tract diseases, community-acquired and healthcare-associated pneumonia.

19.1 The Respiratory Tract and Its Defenses

The respiratory tract is the most common place for infectious agents to gain access to the body. We breathe 24 hours a day, and anything in the air we breathe passes at least temporarily into this organ system.

The structure of the system is illustrated in **figure 19.1a.** Most clinicians divide the system into two parts, the *upper* and *lower respiratory tracts*. The upper respiratory tract includes the mouth, the nose, nasal cavity and sinuses above it, the throat or pharynx, and the epiglottis and larynx. The lower respiratory tract begins with the trachea, which feeds into the bronchi and bronchioles in the lungs. Attached to the bronchioles are small balloonlike structures called alveoli, which inflate and deflate with inhalation and exhalation. These are the sites of oxygen exchange in the lungs.

Several anatomical features of the respiratory system protect it from infection. As described in chapter 12, nasal hair serves to trap particles. Cilia **(figure 19.1b)** on the epithelium of the trachea and bronchi (the ciliary escalator) propel particles upward and out of the respiratory tract. Mucus on the surface of the mucous membranes lining the respiratory tract is a natural trap for invading microorganisms. Once the microorganisms are trapped, involuntary responses such as coughing, sneezing, and swallowing can move them out of sensitive areas. These are first-line defenses.

The second and third lines of defense also help protect the respiratory tract. Complement action, antimicrobial peptides, and increased levels of chemocytokines all help battle pathogens in the lungs. Macrophages inhabit the alveoli of the lungs and the clusters of lymphoid tissue (tonsils) in the throat. Secretory IgA against specific pathogens can be found in the mucus secretions as well.

> ### 19.1 LEARNING OUTCOMES—Assess Your Progress
> 1. Draw or describe the anatomical features of the respiratory tract.
> 2. List the natural defenses present in the respiratory tract.

19.2 Normal Biota of the Respiratory Tract

The latest research shows that a healthy upper respiratory system harbors thousands of commensal microorganisms and that even the lungs have a normal, if limited, biota. Part of this normal biota can cause serious disease, especially in immunocompromised individuals; these include *Streptococcus pyogenes*, *Haemophilus influenzae*, *Streptococcus pneumoniae*, *Neisseria meningitidis*, and *Staphylococcus aureus*. The composition of the lung microbiome differs in patients suffering from lung disorders such as chronic obstructive pulmonary disease (COPD), asthma, and cystic fibrosis. Yeasts, especially *Candida albicans*, also colonize the mucosal surfaces of the mouth in the upper respiratory tract. Smokers and nonsmokers also appear to have different microbiota.

Medical Moment

Epiglottitis

The epiglottis is a small piece of cartilage that partially covers the larynx. Its job is to help prevent the inhalation of food and fluids into your lungs. Epiglottitis occurs when an infection (often *Haemophilus influenzae* type b) causes the epiglottis to swell, which may result in the inability to draw air into the lungs.

Symptoms of epiglottitis may include fever, an extremely sore throat, a muffled or hoarse voice, stridor (a high-pitched sound that occurs during inhalation), and difficulty breathing and swallowing. Children may drool due to their inability to swallow oral secretions and may appear very anxious.

Epiglottitis is a medical emergency and may cause complete blockage of the airway. If you suspect that you or someone you know has epiglottitis, it is important to seek medical care immediately. Do not attempt to look at the affected person's throat, as this may result in laryngospasm. The affected person should be kept quiet in an upright position to facilitate breathing.

Respiratory Tract Defenses and Normal Biota

	Defenses	Normal Biota
Upper Respiratory Tract	Nasal hair, ciliary escalator, mucus, involuntary responses such as coughing and sneezing, secretory IgA	*Prevotella, Sphingomonas, Pseudomonas, Acinetobacter, Fusobacterium, Megasphaera, Veillonella, Staphylococcus,* and *Streptococcus*
Lower Respiratory Tract	Mucus, alveolar macrophages, secretory IgA	Still unclear; low levels of colonization by a few species probable

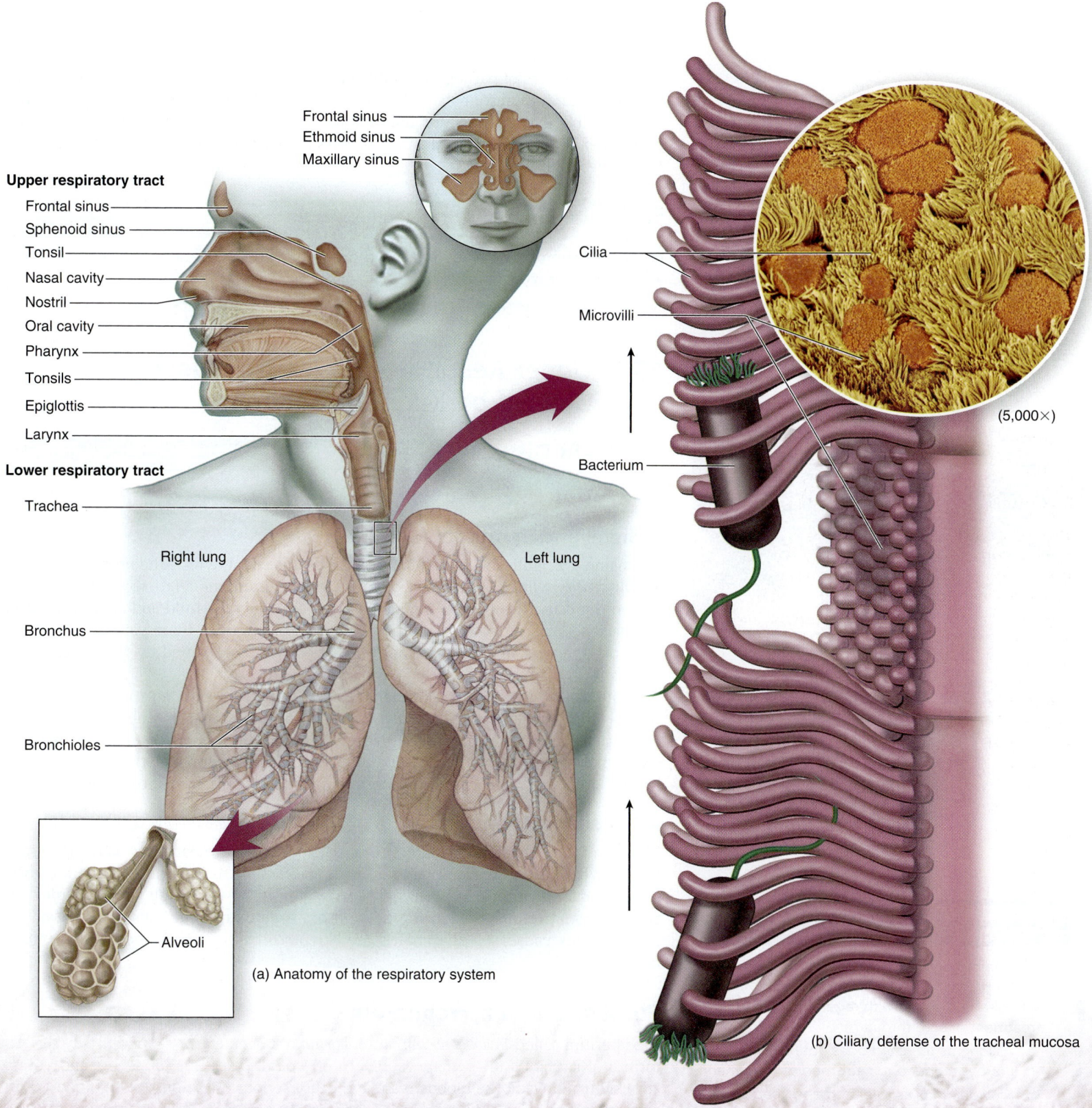

Upper respiratory tract

Frontal sinus
Sphenoid sinus
Tonsil
Nasal cavity
Nostril
Oral cavity
Pharynx
Tonsils
Epiglottis
Larynx

Lower respiratory tract

Trachea

Right lung

Left lung

Bronchus

Bronchioles

Alveoli

(a) Anatomy of the respiratory system

Frontal sinus
Ethmoid sinus
Maxillary sinus

Cilia

Microvilli

Bacterium

(5,000×)

(b) Ciliary defense of the tracheal mucosa

Figure 19.1 The respiratory tract.   (a) Important structures in the upper and lower respiratory tract. The four pairs of sinuses are pictured in the inset. (b) Ciliary defense of the respiratory tract.

In the respiratory system, as in some other organ systems, the normal biota performs the important function of microbial antagonism (see chapter 11). This reduces the chances of pathogens establishing themselves in the same area by competing with them for resources and space. To illustrate this point, *Lactobacillus sakei*, a known member of the sinus microbiome, can suppress the pathogenic potential of another normal biota organism *Corynebacterium tuberculostearicum*, reducing the incidence of sinus infection.

19.2 LEARNING OUTCOMES—Assess Your Progress

3. List the types of normal biota presently known to occupy the respiratory tract.

19.3 Upper Respiratory Tract Diseases Caused by Microorganisms

Highlight Disease

Pharyngitis

▶ Signs and Symptoms

The name says it all–this is an inflammation of the throat, which the host experiences as pain and swelling. The severity of pain can range from moderate to severe, depending on the causative agent. Viral sore throats are generally mild and sometimes lead to hoarseness. Sore throats caused by bacteria are generally more painful than those caused by viruses, and they are more likely to be accompanied by fever, headache, and nausea.

Clinical signs of a sore throat are reddened mucosa, swollen tonsils, and sometimes white packets of inflammatory products visible on the walls of the throat, especially in streptococcal disease **(figure 19.2).** The mucous membranes may be swollen, affecting speech and swallowing. Often pharyngitis results in foul-smelling breath. The incubation period for most sore throats is generally 2 to 5 days.

▶ Causative Agents

The same viruses causing the common cold are those most often causing a sore throat, which can also accompany other diseases, such as infectious mononucleosis (described in chapter 18). Pharyngitis may simply be the result of mechanical irritation from prolonged shouting or from drainage of an infected sinus cavity. The most serious causes of pharyngitis are *Streptococcus pyogenes* and *Fusobacterium necrophorum*.

Fusobacterium necrophorum

A bacterium called *Fusobacterium necrophorum* has been causing a potentially serious form of pharyngitis in as many as 10% to 30% of all cases in adolescents and young adults. It can invade the bloodstream and cause serious infections of the bloodstream and other organs, a condition called Lemierre's syndrome. *F. necrophorum* may be seen more often recently because physicians are using antibiotics more cautiously, only after screening to rule out viral infection. Unfortunately, they tend to screen only for *Streptococcus* ("strep") and in doing so may miss infection by *F. necrophorum*. This bacterium is sensitive to penicillin and related drugs, which are the first-line drugs for *S. pyogenes* as well. However, the second-line drugs for *Streptococcus*, such as azithromycin, have no effect on *F. necrophorum*. If *Fusobacterium* is identified, the

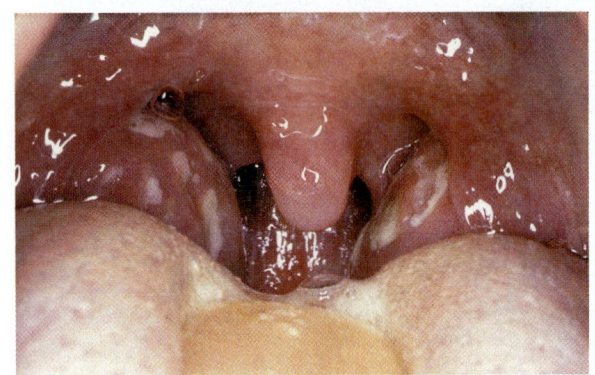

Figure 19.2 **The appearance of the throat in pharyngitis and tonsillitis.** The pharynx and tonsils become bright red and suppurative. Whitish pus nodules may also appear on the tonsils.

NCLEX® PREP

1. What information should be included in health promotion classes for parents of school-age children with regard to the treatment of sore throats?

  a. *As long as the child does not have a fever, there is no need to seek medical treatment.*

  b. *Provide fluids as tolerated and keep the child well hydrated.*

  c. *Parents should take their child to a health care provider so that a rapid throat culture can be obtained.*

  d. *If the child can swallow with minimal pain, all that is needed is increased fluid.*

first-choice drug is clindamycin. If left untreated, pharyngitis caused by this pathogen can lead to meningitis and has a significant mortality rate.

There are currently no rapid diagnostic tests for *F. necrophorum*. However, if a patient with pharyngitis is negative for a rapid strep test and displays neck swelling and rapidly worsening conditions, it is recommended to treat with penicillin and metronidazole or with clindamycin by itself to completely resolve a potential *F. necrophorum* infection.

Streptococcus pyogenes

S. pyogenes is a gram-positive coccus that grows in chains. It does not form endospores, is nonmotile, and forms capsules and slime layers. *S. pyogenes* is a facultative anaerobe that ferments a variety of sugars. It does not produce catalase, but it does have a peroxidase system for inactivating hydrogen peroxide, which allows its survival in the presence of oxygen.

▶ Pathogenesis

Untreated streptococcal throat infections can occasionally result in serious complications, either right away or days to weeks after the throat symptoms subside. The post-streptococcal conditions can be caused by the presence of an extra toxin (as in scarlet fever), or by the deposition of antigen-antibody complexes in the body (as in glomerulonephritis). Another cause of post-streptococal complications is immune system attack of self tissues triggered by streptococcal "superantigens" that cause immune activation to similar or even unrelated (human) proteins. This is responsible for rheumatic fever and for some types of obsessive-compulsive disorder (brain tissue is targeted) and psoriasis (skin is targeted). More details about scarlet fever and rheumatic fever are presented here.

Scarlet Fever   Scarlet fever is the result of infection with an *S. pyogenes* strain that is itself infected with a bacteriophage. This lysogenic virus gives the streptococcus the ability to produce erythrogenic toxin, described in the section on virulence. Scarlet fever is characterized by a sandpaper-like rash, most often on the neck, chest, elbows, and inner surfaces of the thighs. High fever accompanies the rash. It most often affects school-age children, and was a source of great suffering in the United States in the early part of the 20th century. In epidemic form, the disease can have a fatality rate of up to 95%. Most cases seen today are mild.

Rheumatic Fever   Rheumatic fever is thought to be due to an immunologic cross-reaction between the streptococcal M protein and heart muscle. It tends to occur approximately 3 weeks after pharyngitis has subsided. It can result in permanent damage to heart valves. Other symptoms include arthritis in multiple joints and the appearance of nodules over bony surfaces just under the skin.

▶ Virulence Factors

The virulence of *S. pyogenes* comes both from its surface antigens' mimicry of host proteins and from its superantigens.

Streptococci display numerous surface antigens **(figure 19.3).** Specialized polysaccharides on the surface of the cell wall help to protect the bacterium from being dissolved by the lysozyme of the host. Lipoteichoic acid (LTA) contributes to the adherence of *S. pyogenes* to epithelial cells in the pharynx. A spiky surface projection called *M protein* contributes to virulence by resisting phagocytosis and possibly by contributing to adherence. A capsule made of hyaluronic acid (HA) is formed by most *S. pyogenes* strains. It probably contributes to the bacterium's adhesiveness.

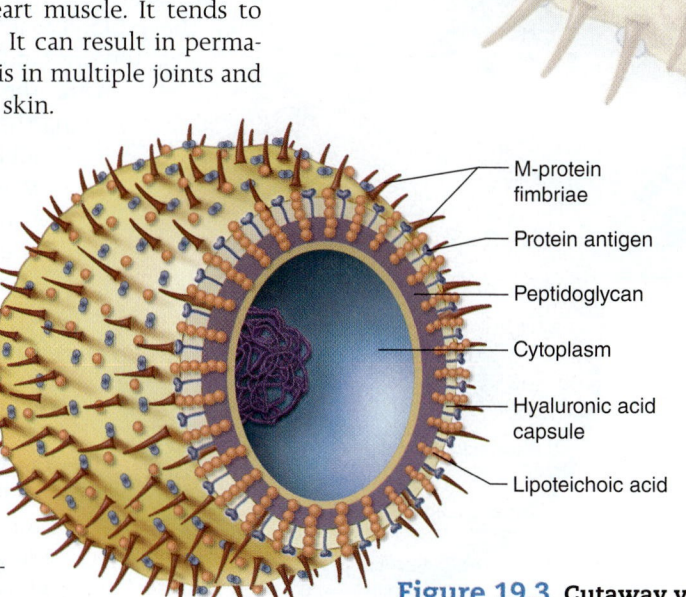

- M-protein fimbriae
- Protein antigen
- Peptidoglycan
- Cytoplasm
- Hyaluronic acid capsule
- Lipoteichoic acid

Figure 19.3 Cutaway view of group A *Streptococcus.*

Extracellular Toxins   Group A streptococci owe some of their virulence to the effects of hemolysins called **streptolysins.** The two types are streptolysin O (SLO) and streptolysin S (SLS). Both types cause beta-hemolysis of sheep blood agar (see "Culture and Diagnosis"). Both hemolysins rapidly injure many cells and tissues, including leukocytes and liver and heart muscle (in other forms of streptococcal disease).

A key toxin in the development of scarlet fever is **erythrogenic** (eh-rith′-roh-jen′-ik) **toxin.** This toxin is responsible for the bright red rash typical of this disease, and it also induces fever by acting upon the temperature regulatory center in the brain. Only lysogenic strains of *S. pyogenes* that contain genes from a temperate bacteriophage can synthesize this toxin. (For a review of the concept of lysogeny, see chapter 5.)

▶ Transmission and Epidemiology

Physicians estimate that 30% of sore throats may be caused by *S. pyogenes*, adding up to several million cases each year. Most transmission of *S. pyogenes* is via respiratory droplets or direct contact with mucus secretions. This bacterium is carried as "normal" biota by 15% of the population, but transmission from this reservoir is less likely than from a person who is experiencing active disease from the infection because of the higher number of bacteria present in the disease condition. It is less common but possible to transmit this infection via fomites. Humans are the only significant reservoir of *S. pyogenes*.

More than 80 serotypes of *S. pyogenes* exist, and thus people can experience multiple infections throughout their lives because immunity is serotype-specific. Even so, only a minority of encounters with the bacterium result in disease. An immunocompromised host is more likely to suffer from strep pharyngitis as well as serious sequelae of the throat infection.

▶ Culture and Diagnosis

The failure to recognize group A streptococcal infections can have devastating effects. Rapid cultivation and diagnostic techniques to ensure proper treatment and prevention measures are essential. Several different rapid diagnostic test kits are used in clinics and doctors' offices to detect group A streptococci from pharyngeal swab samples. These tests are based on antibodies that react with the outer carbohydrates of group A streptococci **(figure 19.4*a*).** Because the rapid tests have a significant possibility of returning a false-negative result, guidelines call for confirming the negative finding with a culture, which can be read the following day.

A culture is generally taken at the same time as the rapid swab and is plated on sheep blood agar. *S. pyogenes* displays a beta-hemolytic pattern due to its streptolysins (and hemolysins) **(figure 19.4*b*).** If the pharyngitis is caused by a virus, the blood agar dish will show a variety of colony types, representing the normal bacterial biota. Active infection with *S. pyogenes* will yield a plate with a majority of beta-hemolytic colonies. Group A streptococci are by far the most common beta-hemolytic isolates in human diseases, but lately an increased number of infections by group B streptococci (also beta-hemolytic), as well as the existence of beta-hemolytic enterococci, have made it important to use differentiation tests. A positive bacitracin disc test (figure 19.4*b*) provides additional evidence for group A.

▶ Prevention

No vaccine exists for group A streptococci, although many researchers are working on the problem. A vaccine against this bacterium would also be a vaccine against rheumatic fever, and thus it is in great demand. In the meantime, infection can be prevented by good hand washing, especially after coughing and sneezing and before preparing foods or eating.

Figure 19.4 Streptococcal tests.   (a) A rapid, immunologic test for diagnosis of group A infections. (b) Bacitracin disc test. With very few exceptions, only *Streptococcus pyogenes* is sensitive to a minute concentration (0.02 µg) of bacitracin. Any zone of inhibition around the B disc is interpreted as a presumptive indication of this species. (Note: Group A streptococci are negative for sulfamethoxazole-trimethoprim [SXT] sensitivity and the CAMP test.)

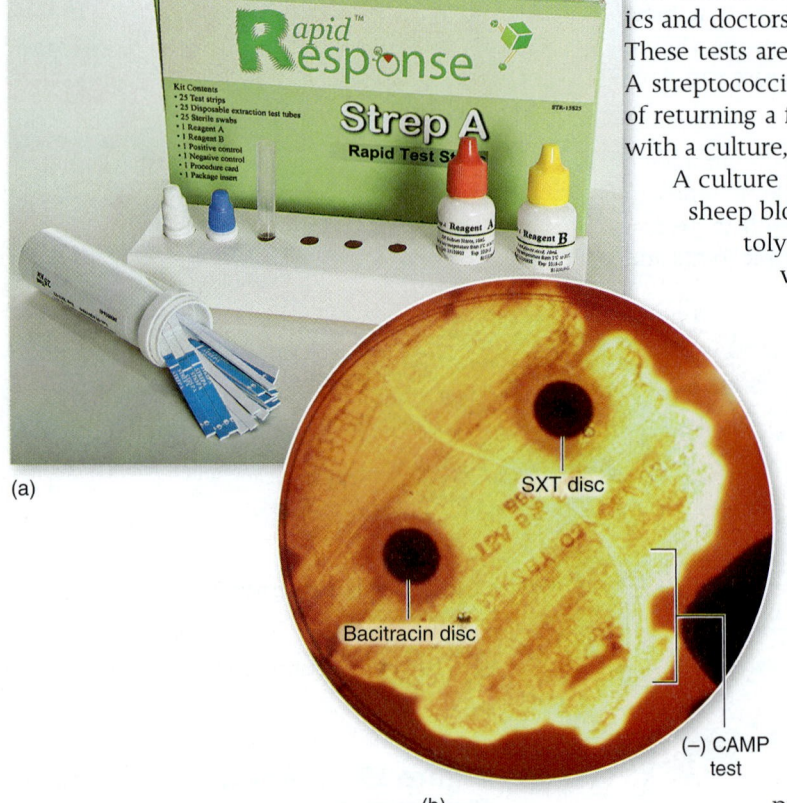

(a)

SXT disc

Bacitracin disc

(−) CAMP test

(b)

Disease Table 19.1 Pharyngitis

Causative Organism(s)	*Streptococcus pyogenes*	*Fusobacterium necrophorum*	Viruses
Most Common Modes of Transmission	Droplet or direct contact	Opportunistic	All forms of contact
Virulence Factors	LTA, M protein, hyaluronic acid capsule, SLS and SLO, superantigens, induction of autoimmunity	Invasiveness, endotoxin, leukotoxin	–
Culture/Diagnosis	Beta-hemolytic on blood agar, sensitive to bacitracin, rapid antigen tests	Growth on anaerobic agar	Goal is to rule out *S. pyogenes*, further diagnosis usually not performed
Prevention	Hygiene practices	Hygiene practices	Hygiene practices
Treatment	Penicillin, cephalexin in penicillin-allergic	Penicillin + metronidazole or clindamycin	Symptom relief only
Distinctive Features	Generally more severe than viral pharyngitis	Common in adolescents and young adults, neck swelling common; infections spread to cardiovascular system or deeper tissues	Hoarseness frequently accompanies viral pharyngitis
Epidemiological Features	United States: 20%–30% of all cases of pharyngitis	United States: 10%–30% of cases	Ubiquitous; responsible for 40%–60% of all pharyngitis

▶ Treatment

The antibiotic of choice for *S. pyogenes* is penicillin. In patients with penicillin allergies, a first-generation cephalosporin, such as cephalexin, is prescribed. Although most sore throats caused by *S. pyogenes* can resolve on their own, they should be treated with antibiotics because serious sequelae are a distinct possibility.

The Common Cold

Everyone is familiar with the symptoms of a common cold: sneezing, scratchy throat, and runny nose, which usually begin 2 or 3 days after infection. An uncomplicated cold generally is not accompanied by fever, although children can experience low fevers (less than 102°F). The incubation period is usually 2 to 5 days. People with asthma and other underlying respiratory conditions, such as asthma or chronic obstructive pulmonary disease (COPD), often suffer more severe symptoms triggered by the common cold.

The common cold is caused by one of over 200 different kinds of viruses. The most common type of virus leading to rhinitis is the group called rhinoviruses, of which there are 99 serotypes. Coronaviruses and adenoviruses are also major causes. Also, the respiratory syncytial virus (RSV) causes colds in most people, but in some, especially infants and children, they can lead to more serious respiratory tract symptoms (discussed later in the chapter).

Sinusitis

Commonly called a *sinus infection*, this inflammatory condition of any of the four pairs of sinuses in the skull (see figure 19.1c) can actually be caused by allergy (most common), infections, or simply by structural problems such as narrow passageways or a deviated nasal septum. The infectious agents that may be responsible for the condition commonly include a variety of viruses or bacteria and, less commonly, fungi.

Acute Otitis Media (Ear Infection)

Viral infections of the upper respiratory tract lead to inflammation of the eustachian tubes and the buildup of fluid in the middle ear, which can lead to bacterial multiplication in those fluids. Although the middle ear normally has no biota, bacteria can migrate along the eustachian tube from the upper respiratory

Disease Table 19.2 The Common Cold

Causative Organism(s)	Approximately 200 viruses
Most Common Modes of Transmission	Indirect contact, droplet contact
Virulence Factors	Attachment proteins; most symptoms induced by host response
Culture/Diagnosis	Not necessary
Prevention	Hygiene practices
Treatment	For symptoms only
Epidemiological Features	Highest incidence among preschool and elementary schoolchildren with average of three to eight colds per year; adults and adolescents: two to four colds per year

Disease Table 19.3 Sinusitis

Causative Organism(s)	Viruses	Various bacteria, often mixed infection	Various fungi
Most Common Modes of Transmission	Direct contact, indirect contact	Endogenous (opportunism)	Introduction by trauma or opportunistic overgrowth
Virulence Factors	–	–	–
Culture/Diagnosis	Culture not usually performed; diagnosis based on clinical presentation.	Culture not usually performed; diagnosis based on clinical presentation, occasionally X rays or other imaging technique used	Same
Prevention	Hygiene	–	–
Treatment	None	Broad-spectrum antibiotics or none	Physical removal of fungus; in severe cases, antifungals used
Distinctive Features	Viral and bacterial much more common than fungal	Viral and bacterial much more common than fungal	Suspect in immunocompromised patients
Epidemiological Features	–	United States: affects 1 of 7 adults; between 12 and 30 million diagnoses per year	Fungal sinusitis varies with geography; in United States, more common in SE and SW; internationally: more common in India, North Africa, Middle East

Figure 19.5 An infected middle ear.

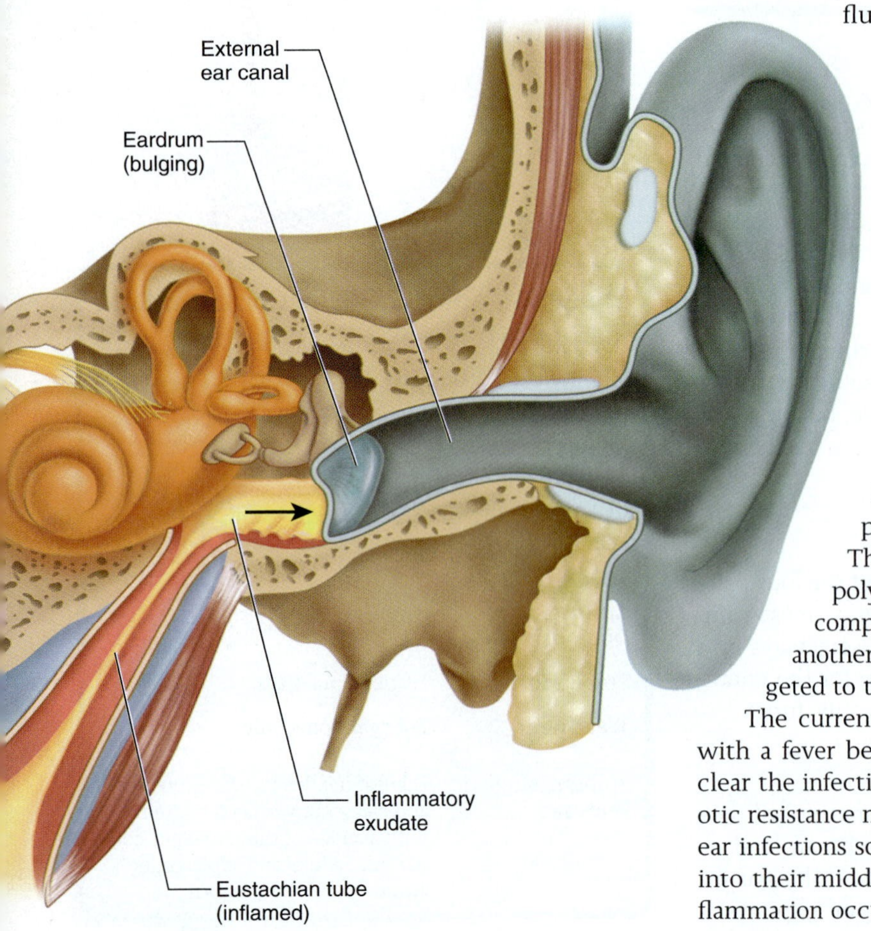

External ear canal

Eardrum (bulging)

Inflammatory exudate

Eustachian tube (inflamed)

tract **(figure 19.5).** When bacteria encounter mucus and fluid buildup in the middle ear, they multiply rapidly. Their presence increases the inflammatory response, leading to pus production and continued fluid secretion. This fluid is referred to as *effusion*.

Another condition, known as chronic otitis media, occurs when fluid remains in the middle ear for indefinite periods of time. Until recently, physicians considered it to be the result of a noninfectious immune reaction because they could not culture bacteria from the site and because antibiotics were not effective. New data suggest that this form of otitis media is caused by a mixed biofilm of bacteria that is attached to the mucosa of the middle ear. Biofilm bacteria generally are less susceptible to antibiotics (as discussed in chapter 3), and their presence in biofilm form would explain the inability to culture them from ear fluids. Scientists now believe that the majority of acute and chronic otitis media cases are mixed infections with viruses and bacteria acting together.

The single most common bacterium seen in acute otitis media is *Streptococcus pneumoniae*. A vaccine against *S. pneumoniae* has been a part of the recommended childhood vaccination schedule since 2000. The vaccine (Prevnar) is a conjugated vaccine (see chapter 13). It contains polysaccharide capsular material from 13 different strains of the bacterium complexed with a chemical that makes it more antigenic. It is distinct from another vaccine for the same bacterium (Pneumovax), which is primarily targeted to the older population to prevent pneumococcal pneumonia.

The current treatment recommendation for uncomplicated acute otitis media with a fever below 104°F is "watchful waiting" for 72 hours to allow the body to clear the infection, avoiding the use of antibiotics. When antibiotics are used, antibiotic resistance must be considered. Children who experience frequent recurrences of ear infections sometimes have small tubes placed through the tympanic membranes into their middle ears to provide a means of keeping fluid out of the site when inflammation occurs.

Disease Table 19.4 Otitis Media

Causative Organism(s)	Streptococcus pneumoniae	Haemophilus influenzae	Other bacteria/viruses
Most Common Modes of Transmission	Endogenous (may follow upper respiratory tract infection by S. pneumoniae or other microorganisms)	Endogenous (follows upper respiratory tract infection)	Endogenous
Virulence Factors	Capsule, hemolysin	Capsule, fimbriae	–
Culture/Diagnosis	Usually relies on clinical symptoms and failure to resolve within 72 hours	Same	Same
Prevention	Pneumococcal conjugate vaccine	Hib vaccine	None
Treatment	Wait for resolution; if needed, amoxicillin (high rates of resistance) or amoxicillin + clavulanate or cefuroxime; in **Serious Threat** category in CDC Antibiotic Resistance Report	Same as for S. pneumoniae	Wait for resolution; if needed, a broad-spectrum antibiotic (azithromycin) may be used in absence of etiologic diagnosis
Distinctive Features	–	–	Suspect if fully vaccinated against other two
Epidemiological Features	United States: 70% of children experience at least one case before age 2; in developing world: chronic otitis media results in significant hearing loss in 100s of millions and death in approx. 30,000 per year (in absence of treatment)		

Diphtheria

For hundreds of years, diphtheria was a significant cause of morbidity and mortality, but in the last 50 years, both the number of cases and the fatality rate have steadily declined throughout the world. In the United States in recent years, only one or two cases have been reported each year.

The disease is caused by *Corynebacterium diphtheriae*, a non-endospore-forming, gram-positive club-shaped bacterium. The most striking symptom of this disease is a characteristic membrane, usually referred to as a pseudomembrane, that forms on the tonsils or pharynx **(figure 19.6).**

The major virulence factor is an exotoxin encoded by a bacteriophage of *C. diphtheriae*. Strains of the bacterium that are not lysogenized by this phage do not cause serious disease. The release of diphtheria toxin in the blood leads to complications in distant organs, especially myocarditis and neuritis. Neuritis affects motor nerves and may result in temporary paralysis of limbs, the soft palate, and even the diaphragm, a condition that can predispose a patient to other lower respiratory tract infections. **(Disease Table 19.5)**

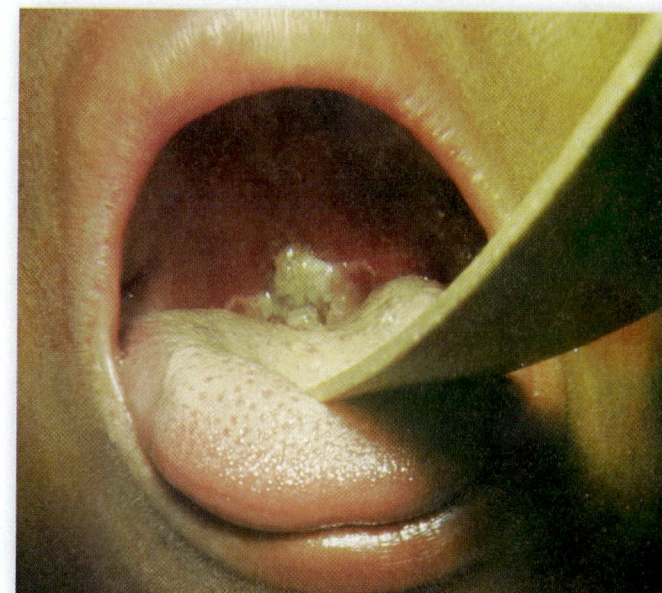

Figure 19.6 Diagnosing diphtheria. The clinical appearance in diphtheria infection includes gross inflammation of the pharynx and tonsils marked by grayish patches (a pseudomembrane) and swelling over the entire area.

Disease Table 19.5 Diphtheria

Causative Organism(s)	Corynebacterium diphtheriae
Most Common Modes of Transmission	Droplet contact, direct contact or indirect contact with contaminated fomites
Virulence Factors	Exotoxin: diphtheria toxin
Culture/Diagnosis	Tellurite medium–gray/black colonies, club-shaped morphology on Gram stain; treatment begun before definitive identification
Prevention	Diphtheria toxoid vaccine (part of DTaP, Tdap, and Td)
Treatment	Antitoxin plus penicillin or erythromycin
Epidemiological Features	United States: no cases since 2003; internationally: +/- 5,000 cases per year, even though there is 83% vaccine coverage

NCLEX® PREP

3. A mother brings her child into the clinic setting and tells the nurse that the "baby has an ear infection" and wants the physician to prescribe antibiotics. After the client sees the health care provider, no prescription is given. The nurse is then providing discharge instructions to the client. The mother is visibly upset and tells the nurse, "I don't know why I even came to the doctor if he is not going to treat my baby." What is the nurse's best response?

a. Ask the client why she feels this way and offer an empathetic ear.

b. Explain the concept of "watchful waiting" and developing antibiotic resistance.

c. Tell the client to wait, relate this concern to the physician, and have the health care provider go back and talk to the client.

d. Tell the client that her baby's symptoms will go away in a few days and therefore there is no need to start antibiotic therapy.

19.3 LEARNING OUTCOMES—Assess Your Progress

4. List the possible causative agents, modes of transmission, virulence factors, diagnostic techniques, and prevention/treatment for the "Highlight Disease" pharyngitis.

5. Discuss important features of the other infectious diseases of the upper respiratory tract. These are rhinitis, sinusitis, acute otitis media, and diphtheria.

6. Identify two bacteria that can cause dangerous pharyngitis cases.

19.4 Diseases Caused by Microorganisms Affecting Both the Upper and Lower Respiratory Tract

A number of infectious agents affect both the upper and lower respiratory tract regions. We address the more well-known diseases in this section; specifically, they are influenza, whooping cough, and respiratory syncytial virus (RSV).

Highlight Disease

Influenza

The "flu" is a very important disease to study for several reasons. First of all, everyone is familiar with the cyclical increase of influenza infections occurring during the winter months in the United States. Second, many conditions are erroneously termed the "flu," while in fact only diseases caused by influenza viruses are actually the flu. Third, the way that influenza viruses evolve provides an excellent illustration of the way other viruses can, and do, change to cause more serious diseases than they did previously.

Influenzas that occur every year are called "seasonal" flus. Often these are the only flus that circulate each year. Occasionally another flu strain appears, one that is new and may cause worldwide pandemics. In some years, such as in 2009, both of these flus were issues. They may have different symptoms, affect different age groups, and have separate vaccine protocols.

▶ Signs and Symptoms

Influenza begins in the upper respiratory tract but in serious cases may also affect the lower respiratory tract. There is a 1- to 4-day incubation period, after which symptoms begin very quickly. These include headache, chills, dry cough, body aches, fever, stuffy nose, and sore throat. Even the sum of all these symptoms can't describe how a person actually feels: lousy. The flu is known to "knock you off your feet." Extreme fatigue can last for a few days or even a few weeks. An infection with influenza can leave patients vulnerable to secondary infections, often bacterial. Influenza infection alone occasionally leads to a pneumonia that can cause rapid death, even in young healthy adults.

The latest pandemic virus, H1N1, or the swine flu of 2009, had similar symptoms but with a couple of differences. Not all patients had a fever (very unusual for influenza), and many patients had gastrointestinal distress.

▶ Causative Agent

All influenza is caused by one of three influenza viruses: A, B, or C. They belong to the family *Orthomyxoviridae*. They are spherical particles with an average diameter of 80 to 120 nanometers. Each virus has a lipoprotein envelope that is studded with glycoprotein spikes acquired during viral maturation **(figure 19.7)**. Also note that the envelope contains proteins that form a channel for ions into the virus. The two glycoproteins that make up the spikes of the envelope and contribute to virulence are called

hemagglutinin (H) and neuraminidase (N). The name hemagglutinin is derived from this glycoprotein's agglutinating action on red blood cells, which is the basis for viral assays used to identify the viruses. Hemagglutinin contributes to infectivity by binding to host cell receptors of the respiratory mucosa, a process that facilitates viral penetration. Neuraminidase breaks down the protective mucous coating of the respiratory tract, assists in viral budding and release, keeps viruses from sticking together, and participates in host cell fusion.

The ssRNA genome of the influenza virus is known for its extreme variability. It is subject to constant genetic changes that alter the structure of its envelope glycoproteins. Research has shown that genetic changes are very frequent in the area of the glycoproteins recognized by the host immune response but very rare in the areas of the glycoproteins used for attachment to the host cell **(figure 19.8)**. In this way, the virus can continue to attach to host cells while managing to decrease the effectiveness of the host response to its presence. This constant mutation of the glycoproteins is called **antigenic drift**— the antigens gradually change their amino acid composition, resulting in decreased ability of host memory cells to recognize them.

An even more serious phenomenon is known as **antigenic shift.** The genome of the virus consists of just 10 genes, encoded on 8 separate RNA strands. Antigenic shift is the swapping out of one of those genes or strands with a gene or strand from a different influenza virus. Some explanation is in order. For example, we know that certain influenza viruses infect both humans and swine. Other influenza viruses infect birds and swine. All of these viruses have 10 genes coding for the same important influenza proteins (including H and N)—but the actual sequence of the genes is different in the different types of viruses. Second, when the two viruses just described infect a single swine host, with both virus types infecting the same host cell, the viral packaging step can accidentally produce a human influenza virus that contains seven human influenza virus RNA strands plus a single duck influenza virus RNA strand **(figure 19.9).** When that virus infects a human, no immunologic recognition of the protein that came

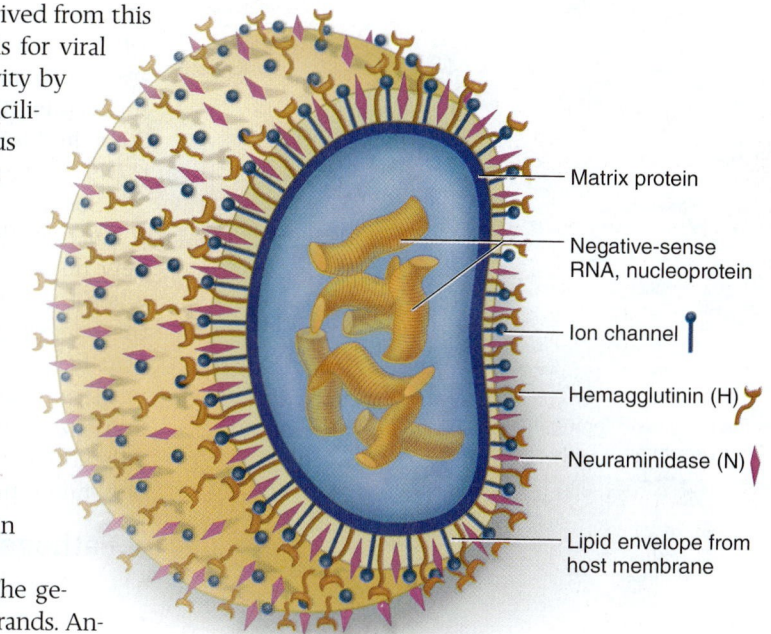

Figure 19.7 **Schematic drawing of influenza virus.**

- Matrix protein
- Negative-sense RNA, nucleoprotein
- Ion channel
- Hemagglutinin (H)
- Neuraminidase (N)
- Lipid envelope from host membrane

Figure 19.8 **Schematic drawing of hemagglutinin (H) of influenza virus.** Blue boxes depict site used to attach virus to host cells; green circles depict sites for anti-influenza antibody binding.

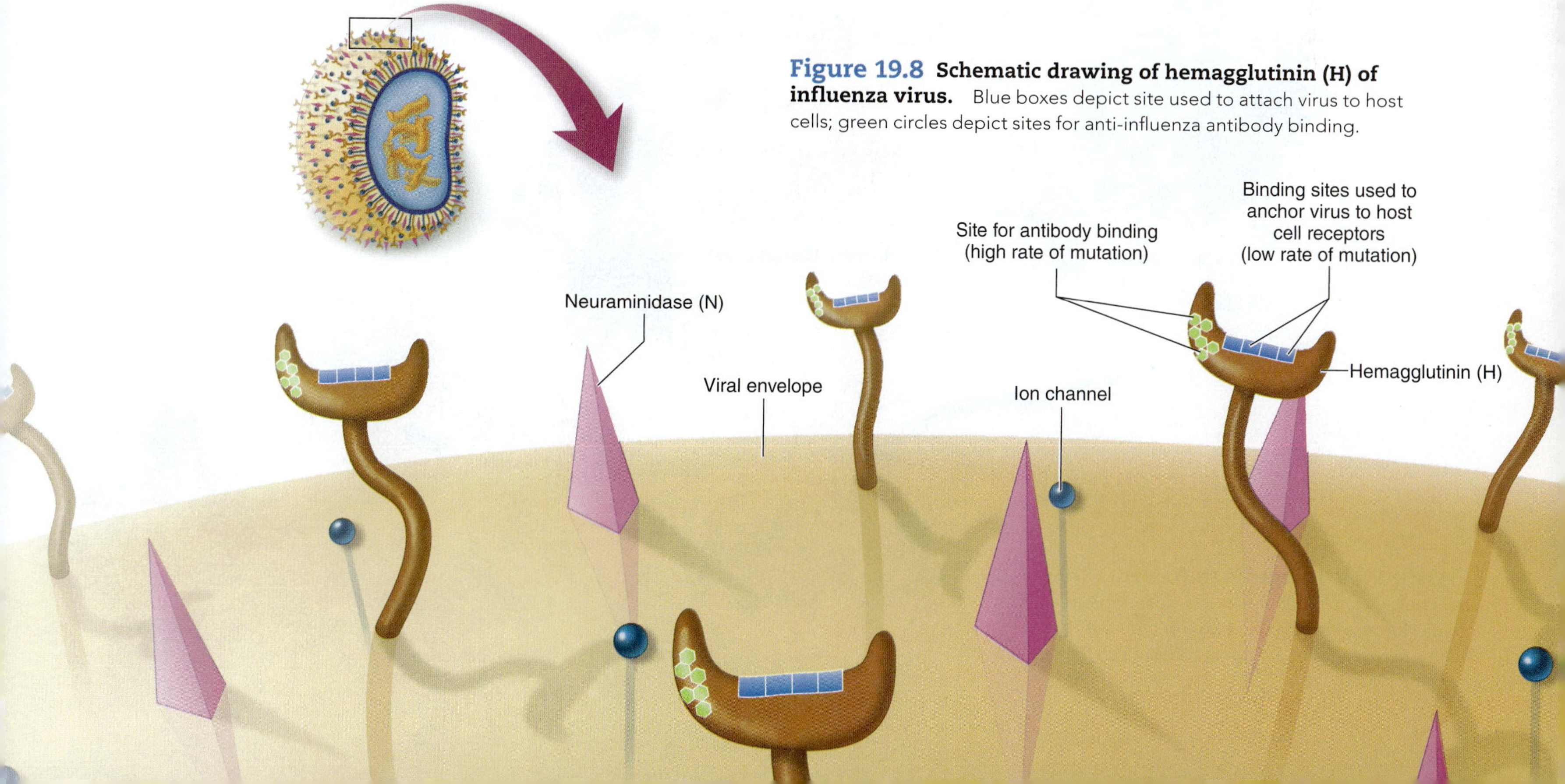

Neuraminidase (N)

Viral envelope

Ion channel

Site for antibody binding (high rate of mutation)

Binding sites used to anchor virus to host cell receptors (low rate of mutation)

Hemagglutinin (H)

Influenza viruses (orange) budding from a host cell (83,300×)

from the duck virus occurs. Experts have traced the flu pandemics of 1918, 1957, 1968, 1977, and 2009 to strains of a virus that came from pigs (swine flu). Influenza A viruses are named according to the different types of H and N spikes they display on their surfaces. In 2009, a swine flu called H1N1 caused a limited pandemic. It reappeared in 2014. In 2013, the main circulating virus was H3N2. This strain—along with H5N1 and H7N9—is called bird flu, because it originates in poultry.

Some people express frustration that public health officials seem to raise the alarm about new flu threats that don't pan out. Now that you have learned how easily a new flu strain can become lethal to a new species (i.e., humans), and knowing that when that happened in 1918 the flu killed 20-50 million people, you can appreciate the science—and the necessary precaution—behind these warnings. The warnings are usually accompanied by extra measures of preparation, such as additional vaccine, antivirals, and diagnostic kits being made available. "Better safe than sorry" is the thinking. Influenza B viruses are not divided into subtypes because they are thought to undergo only antigenic drift and not antigenic shift. Influenza C viruses are thought to cause only minor respiratory disease and are probably not involved in epidemics.

▶ Pathogenesis and Virulence Factors

The influenza virus binds primarily to ciliated cells of the respiratory mucosa. Infection causes the rapid shedding of these cells along with a load of viruses. Stripping the respiratory epithelium to the basal layer eliminates protective ciliary clearance. Sometimes the immune system responds too aggressively in a phenomenon called a "cytokine storm," and this leads to more severe irritation and inflammation in the lungs. The illness is further aggravated by fever, headache, and the other symptoms just described. The viruses tend to remain in the respiratory tract rather than spread to the bloodstream. As the normal ciliated epithelium is restored in a week or two, the symptoms subside.

As just noted, the glycoproteins and their structure are important virulence determinants. First of all, they mediate the adhesion of the virus to host cells. Second,

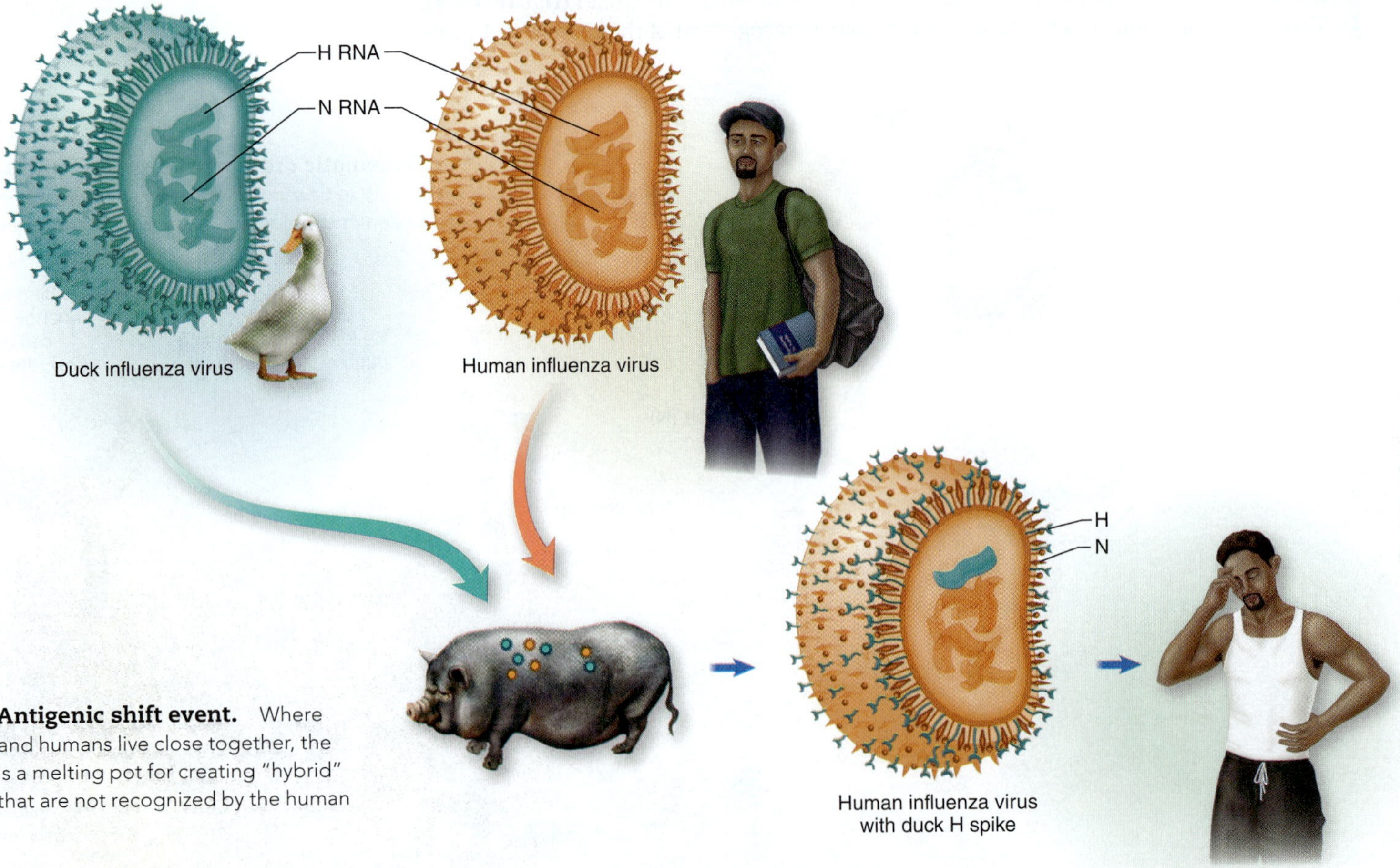

H RNA

N RNA

Duck influenza virus

Human influenza virus

H
N

Human influenza virus with duck H spike

Figure 19.9 Antigenic shift event. Where ducks and swine and humans live close together, the swine can serve as a melting pot for creating "hybrid" influenza viruses that are not recognized by the human immune system.

they change gradually and sometimes suddenly, evading immune recognition. One feature of the 2009 H1N1 virus is that it bound to cells lower in the respiratory tract and at a much higher rate, leading to massive damage—and often death—in the worst-affected patients.

▶ Transmission and Epidemiology

Inhalation of virus-laden aerosols and droplets constitutes the major route of influenza infection, although fomites can play a secondary role. Transmission is greatly facilitated by crowding and poor ventilation in classrooms, barracks, nursing homes, dormitories, and military installations in the late fall and winter. The drier air of winter facilitates the spread of the virus, as the moist particles expelled by sneezes and coughs become dry very quickly, helping the virus remain airborne for longer periods of time. In addition, the dry cold air makes respiratory tract mucous membranes more brittle, with micro-scopic cracks that facilitate invasion by viruses. Influenza is highly contagious and affects people of all ages. Annually, there are approximately 36,000 U.S. deaths from seasonal influenza and its complications, mainly among the very young and the very old.

The 2009 H1N1 virus took a particularly heavy toll on young people. Previously healthy children and teenagers formed a small but important risk group, with quite a few becoming ill within hours and dying within days.

▶ Culture and Diagnosis

There is a wide variety of culture-based and non-culture-based methods to diagnose the infection. Rapid influenza tests (such as PCR, ELISA-type assays, or immuno-fluorescence) provide results within 24 hours; viral culture provides results in 3 to 10 days. Cultures are not typically performed at the point of care; they must be sent to diagnostic laboratories, and they require up to 10 days for results. Despite these disadvantages, cultures can be useful to identify which subtype of influenza is causing infections, which is important for public health authorities to know. In 2009, officials did not often test for H1N1 but tested for influenza A or B virus, assuming if it was A then it was H1N1, since the circulating seasonal virus was influenza B. When speci-mens were tested, 100% of the influenza A isolates were in fact the H1N1.

▶ Prevention

Preventing influenza infections and epidemics is one of the top priorities for public health officials. There are currently four main types of influenza vaccines: inacti-vated vaccine designed for intramuscular injection; inactivated vaccine designed for intradermal injection; live attenuated virus vaccine, which is administered intrana-sally; and recombinant vaccine, approved in 2013, which is made through genetic engineering and not by growing the virus in chicken eggs as with the previous vac-cines. It is designed for intramuscular injection and approved for people 18 to 49 years old. The CDC recommends that everyone over the age of 6 months receive one of these vaccinations. Because of the changing nature of the antigens on the viral surface, annual vaccination is considered the best way to avoid infection.

One of the most promising new vaccine prospects is a vaccine that would protect against *all* flu viruses and not need to be given every year. This vaccine, in testing stages, would target the ion-channel proteins that are present on the envelope of influenza vi-ruses. Apparently these proteins are the same on all flu viruses, and they do not mutate readily. This discovery has the possibility of revolutionizing influenza prevention.

▶ Treatment

Influenza is one of the first viral diseases for which effective antiviral drugs became available. The drugs must be taken early in the infection, preferably by the second day. This requirement is an inherent difficulty because most people do not realize until later that they may have the flu. Amantadine and rimantadine can be used to treat and prevent some influenza type A infections, but they do not work against influenza type B viruses.

Disease Table 19.6 Influenza	
Causative Organism(s)	Influenza A, B, and C viruses
Most Common Modes of Transmission	Droplet contact, direct contact, or indirect contact
Virulence Factors	Glycoprotein spikes, overall ability to change genetically, ability to slow down immune system
Culture/Diagnosis	Viral culture (3–10 days) or rapid antigen-based or PCR tests
Prevention	Annual vaccination with one of four types of vaccines
Treatment	Oseltamivir or zanamivir
Epidemiological Features	For seasonal flu, deaths vary from year to year. United States: range from 17,000–52,000; internationally: range from 250,000–500,000

The 2009 H1N1 influenza virus caused an extreme—sometimes counterproductive—immune system response in previously healthy vigorous people.

Disease Table 19.7 Pertussis (Whooping Cough)

Causative Organism(s)	*Bordetella pertussis*
Most Common Modes of Transmission	Droplet contact
Virulence Factors	Fimbrial hemagglutinin (adhesion), pertussis toxin and tracheal cytotoxin, endotoxin
Culture/ Diagnosis	PCR or grown on B-G, charcoal, or potato-glycerol agar; diagnosis can be made on symptoms
Prevention	Acellular vaccine (DTaP), azithromycin, sulfamethoxazole for contacts
Treatment	Azithromycin
Epidemiological Features	United States: great increase in cases—more than 40,000 in 2012, fewer in 2013, but rose again in 2014 (data still incomplete); internationally: 140,000 cases in 2012

Disease Table 19.8 RSV Disease

Causative Organism(s)	Respiratory syncytial virus (RSV)
Most Common Modes of Transmission	Droplet and indirect contact
Virulence Factors	Syncytia formation
Culture/ Diagnosis	Direct antigen testing
Prevention	Passive antibody (humanized monoclonal) in high-risk children
Treatment	Ribavirin in severe cases
Epidemiological Features	United States: general population, less than 1% mortality rates, 3% to 5% mortality in premature infants or those with congenital heart defects; internationally: 7 times higher fatality rate in children in developing countries

Zanamivir (Relenza) is an inhaled drug that works against influenza A and B. Oseltamivir (Tamiflu) is available in capsules or as a powdered mix to be made into a drink. It can also be used for prevention of influenza A and B. Over the period of 2007–2009, different influenza viruses began to show resistance to one or more of these drugs, which called into question the practice of using the drugs preventively in epidemics. As we know with all antimicrobials, the more we use them, the more quickly we lose them (the more quickly they lose their effectiveness).

Whooping Cough (Pertussis)

This disease has two distinct symptom phases called the catarrhal and paroxysmal stages, which are followed by a long recovery (or convalescent) phase, during which a patient is particularly susceptible to other respiratory infections. After an incubation period of from 3 to 21 days, the **catarrhal** stage begins when bacteria present in the respiratory tract cause what appear to be cold symptoms, most notably a runny nose. This stage lasts 1 to 2 weeks. The disease worsens in the second **(paroxysmal)** stage, which is characterized by severe and uncontrollable coughing (a *paroxysm* can be thought of as a convulsive attack). The common name for the disease comes from the whooping sound a patient makes as he or she tries to grab a breath between uncontrollable bouts of coughing.

As in any disease, the convalescent phase is the time when numbers of bacteria are decreasing and no longer cause ongoing symptoms.

Over the the past 30 years, pertussis has been on the rise in the United States, and this appears to be due to three factors. First, we are learning that the vaccine does not provide lifelong immunity, and so adults are contracting and spreading it. Second, the virus appears to be evolving, rendering the older vaccine less effective. Last, some parents have the mistaken notion that the vaccine is more dangerous than pertussis itself and so choose to leave their children unvaccinated and unprotected. **Disease Table 19.7.**

Respiratory Syncytial Virus Infection

As its name indicates, respiratory syncytial virus (RSV) infects the respiratory tract and produces giant multinucleated cells (syncytia). Outbreaks of droplet-spread RSV disease occur regularly throughout the world, with peak incidence in the winter and early spring. Children 6 months of age or younger, as well as premature babies, are especially susceptible to serious disease caused by this virus. RSV is the most prevalent cause of respiratory infection in the newborn age group, and nearly all children have experienced it by age 2. An estimated 100,000 children are hospitalized with RSV infection each year in the United States. Infection in older children and adults usually manifests as a cold.

The virus is highly contagious and is transmitted through droplet contact but also through fomite contamination. Diagnosis of RSV infection is more critical in babies than in older children or adults. The afflicted child is conspicuously ill, with signs typical of pneumonia and bronchitis.

There is no RSV vaccine available yet, but an effective passive antibody preparation is used as prevention in high-risk children and babies born prematurely.

19.4 LEARNING OUTCOMES—Assess Your Progress

7. List the possible causative agents, modes of transmission, virulence factors, diagnostic techniques, and prevention/treatment for the "Highlight Disease" influenza.

8. Compare and contrast *antigenic drift* and *antigenic shift* in influenza viruses.

9. Discuss important features of the other infectious diseases of the upper and lower respiratory tracts. These are pertussis and RSV disease.

19.5 Lower Respiratory Tract Diseases Caused by Microorganisms

In this section, we consider microbial diseases that affect the lower respiratory tract primarily–namely, the bronchi, bronchioles, and lungs, with minimal involvement of the upper respiratory tract. Our discussion focuses on tuberculosis and pneumonia.

Highlight Disease

Tuberculosis

Mummies from the Stone Age, ancient Egypt, and Peru provide evidence that tuberculosis (TB) is an ancient human disease. In fact, historically it has been such a prevalent cause of death that it was called "Captain of the Men of Death" and "White Plague." After the discovery of streptomycin in 1943, the rates of tuberculosis in the developed world declined rapidly. But since the mid-1980s, it has reemerged as a serious threat. Worldwide, 2 billion people are currently infected. Two billion—that is over one-fourth of the world's population! The cause of tuberculosis is primarily the bacterial species *Mycobacterium tuberculosis*, informally called the tubercle bacillus. In this discussion, we will first address the general aspects of the infection and then turn to its most troubling form today, multidrug-resistant tuberculosis (MDR-TB).

▶ Signs and Symptoms

A clear-cut distinction can be made between infection with the TB bacterium and the disease it causes. In general, humans are rather easily infected with the bacterium but are resistant to the disease. Estimates project that only about 5% to 10% of infected people actually develop a clinical case of tuberculosis. Untreated tuberculosis progresses slowly, and people with the disease may have a normal life span, with periods of health alternating with episodes of morbidity. The majority (85%) of TB cases are contained in the lungs, even though disseminated TB bacteria can give rise to tuberculosis in any organ of the body. Clinical tuberculosis is divided into primary tuberculosis, secondary (reactivation or reinfection) tuberculosis, and disseminated or extrapulmonary tuberculosis.

Primary Tuberculosis   The minimum infectious dose for lung infection is low, around 10 bacterial cells. Alveolar macrophages phagocytose these cells, but they are not killed and continue to multiply inside the macrophages. This period of hidden infection is asymptomatic or is accompanied by mild fever. Some bacteria escape from the lungs into the blood and lymphatics. After 3 to 4 weeks, the immune system mounts a complex, cell-mediated assault against the bacteria. The large influx of mononuclear cells into the lungs plays a part in the formation of specific infection sites called **tubercles.** Tubercles are granulomas that consist of a central core containing TB bacteria in enlarged macrophages and an outer wall made of fibroblasts, lymphocytes, and macrophages **(figure 19.10)**. Although this response further checks spread of infection and helps prevent the disease, it also carries a potential for damage. Frequently, as neutrophils come on the scene and release their enzymes, the centers of tubercles break down into necrotic **caseous** (kay'-see-us) **lesions** that gradually heal by *calcification*–normal lung tissue is replaced by calcium deposits. The response of T cells to *M. tuberculosis* proteins also causes a cell-mediated immune response evident in the skin test called the **tuberculin reaction,** a valuable diagnostic and epidemiological tool.

TB infection outside of the lungs is more common in immunosuppressed patients and young children. Organs most commonly involved in **extrapulmonary TB** are the regional lymph nodes, kidneys, long bones, genital tract, brain, and meninges. Because of the debilitation of the patient and the high load of TB bacteria, these complications are usually grave.

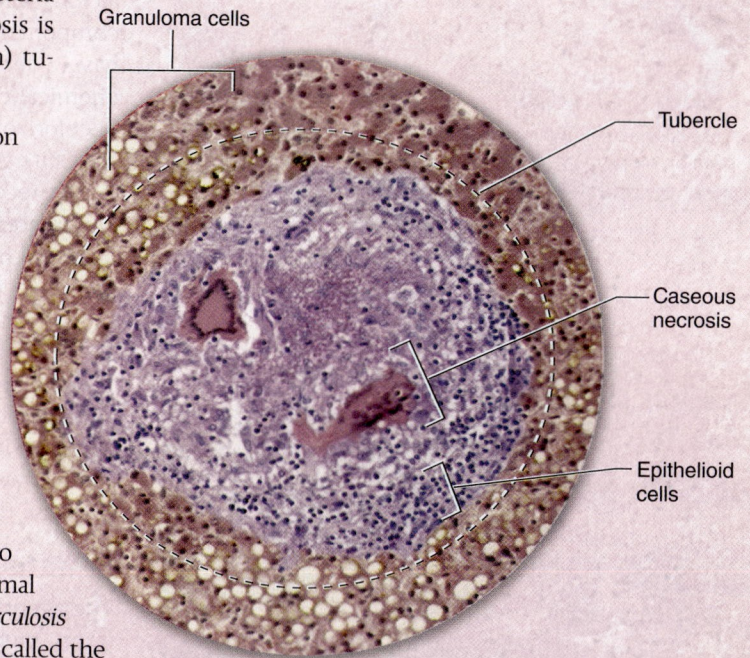

Granuloma cells

Tubercle

Caseous necrosis

Epithelioid cells

Figure 19.10 Tubercle formation. Photomicrograph of a tubercle. The massive granuloma infiltrate has obliterated the alveoli and set up a dense collar of fibroblasts, lymphocytes (granuloma cells), and epithelioid cells. The core of this tubercle is a caseous (cheesy) material containing the bacilli.

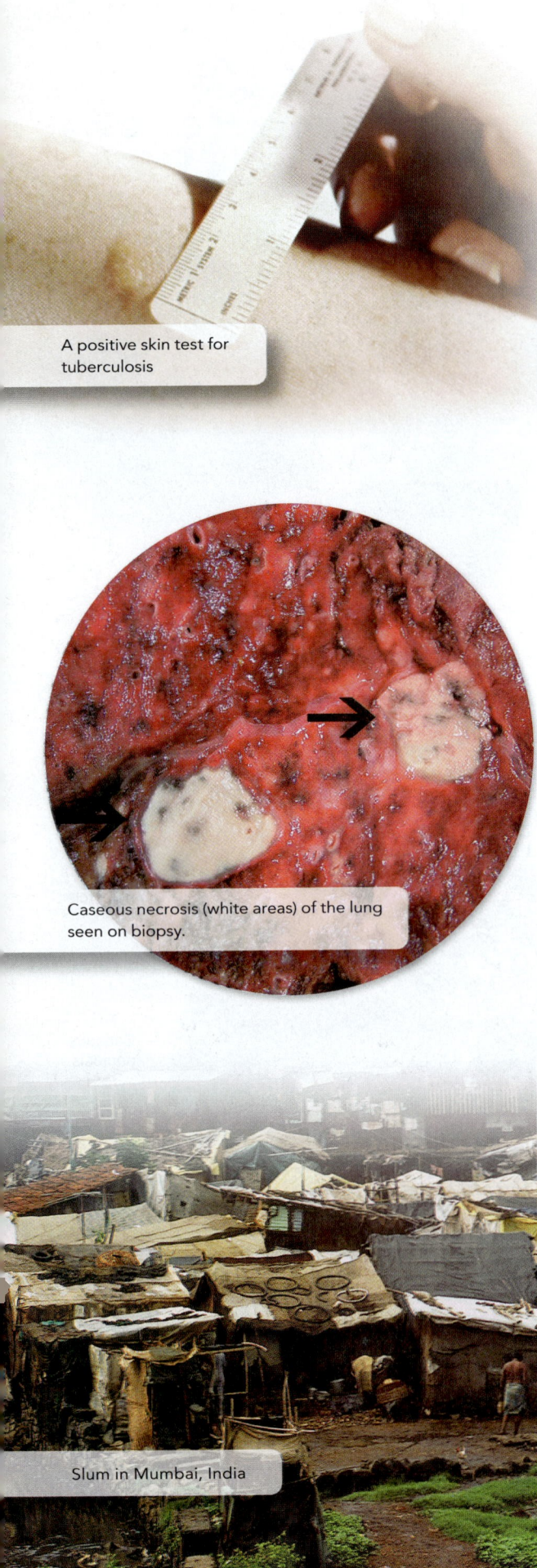

A positive skin test for tuberculosis

Caseous necrosis (white areas) of the lung seen on biopsy.

Slum in Mumbai, India

Tubercular meningitis is the result of an active brain lesion seeding bacteria into the meninges. Over a period of several weeks, the infection of the cranial compartments can create mental deterioration, permanent retardation, blindness, and deafness. Untreated tubercular meningitis is invariably fatal, and even treated cases can have a 30% to 50% mortality rate.

Secondary (Reactivation) Tuberculosis Although the majority of adequately treated TB patients recover more or less completely from the primary episode of infection, live bacteria can remain dormant and become reactivated weeks, months, or years later, especially in people with weakened immunity. In chronic tuberculosis, tubercles filled with masses of bacteria expand, cause cavities in the lungs, and drain into the bronchial tubes and upper respiratory tract. The patient gradually experiences more severe symptoms, including violent coughing, greenish or bloody sputum, low-grade fever, anorexia, weight loss, extreme fatigue, night sweats, and chest pain. It is the gradual wasting of the body that accounts for an older name for tuberculosis–*consumption*. Untreated secondary disease has nearly a 60% mortality rate.

▶ Causative Agents

M. tuberculosis, the cause of tuberculosis in most patients, is an acid-fast rod, long and thin. It is a strict aerobe, and technically speaking, there is still debate about whether it is a gram-positive or a gram-negative organism. It is rarely called gram anything, however, because its acid-fast nature is much more relevant in a clinical setting. It grows very slowly. With a generation time of 15 to 20 hours, a period of up to 6 weeks is required for colonies to appear in culture. (Note: The prefix *Myco-* might make you think of fungi, but this is a bacterium. The prefix in the name came from the mistaken impression that colonies growing on agar resembled fungal colonies. Be sure to differentiate this bacterium from *Mycoplasma*–they are unrelated.)

Robert Koch identified that *M. tuberculosis* often forms serpentine cords while growing, and he called the unknown substance causing this style of growth *cord factor*. Cord factor appears to be associated with virulent strains, and it is a lipid component of the mycobacterial cell wall. All mycobacterial species have walls that have a very high content of complex lipids, including mycolic acid and waxes. This chemical characteristic makes them relatively impermeable to stains and difficult to decolorize (acid-fast) once they are stained. The lipid wall of the bacterium also influences its virulence and makes it resistant to drying and disinfectants.

In recent decades, tuberculosis-like conditions caused by *Mycobacterium avium* and related mycobacterial species (sometimes referred to as the *M. avium* complex, or MAC) have been found in AIDS patients and other immunocompromised people. In this section, we consider only *M. tuberculosis*.

▶ Pathogenesis and Virulence Factors

The course of the infection–and all of its possible variations–was described under "Signs and Symptoms." Important characteristics of the bacterium that contribute to its virulence are its waxy surface (contributing both to its survival in the environment and its survival within macrophages) and its ability to stimulate a strong cell-mediated immune response that contributes to the pathology of the disease.

▶ Transmission and Epidemiology

The agent of tuberculosis is transmitted almost exclusively by fine droplets of respiratory mucus suspended in the air. The TB bacterium is highly resistant and can survive for 8 months in fine aerosol particles. Although larger particles become trapped in mucus and are expelled, tinier ones can be inhaled into the bronchioles and alveoli. This effect is especially pronounced among people sharing small closed rooms with limited access to sunlight and fresh air.

The epidemiological patterns of *M. tuberculosis* infection vary with the living conditions in a community or an area of the world. Factors that significantly affect people's susceptibility to tuberculosis are inadequate nutrition, debilitation of the immune system, poor access to medical care, lung damage, and their own genetics. Put simply, TB is an infection of poverty. People in developing countries are often infected as infants and harbor the microbe for many years until the disease is reactivated in young adulthood. In 2012, approximately 1.3 million people died from TB, or over 3,500 people every day.

Case rates have begun to drop in the United States, from a high in 2004, but some populations are at higher risk of infection or of developing life-threatening forms of the disease. For instance, about 60% of cases in the United States are among foreign-born persons. People who work or live in certain communities such as nursing homes, hospitals, or jails are also among those at greater risk. This is important to know as a health care provider so you can be alert for TB in certain populations.

▶ Culture and Diagnosis

You are probably familiar with several methods of detecting tuberculosis in humans. Clinical diagnosis of tuberculosis relies on four techniques: (1) tuberculin testing, (2) chest X rays, (3) detection of immune responsiveness in the blood, and (4) cultural isolation and antimicrobial susceptibility testing. Acid-fast staining is used as a supplement to these techniques.

Tuberculin Sensitivity and Testing   Because infection with the TB bacillus can lead to delayed hypersensitivity to tuberculoproteins, testing for hypersensitivity has been an important way to screen populations for tuberculosis infection and disease. Although there are newer methods available, the most widely used test is still the tuberculin skin test, called the **Mantoux test.** It involves local injection of purified protein derivative (PPD), a standardized solution taken from culture fluids of *M. tuberculosis*. The injection is done intradermally into the forearm to produce an immediate small bleb. After 48 and 72 hours, the site is observed for a red wheal called an **induration,** which is measured and interpreted as positive or negative according to size **(figure 19.11).**

Tuberculin testing is currently limited to selected groups known to have higher risk for tuberculosis infection. It is no longer used as a routine screening method among populations of children or adults who are not within the target groups. The reasoning behind this change is to allow more focused screening and to reduce expensive and unnecessary follow-up tests and treatments.

X Rays   Chest X rays can help verify TB when other tests have given indeterminate results, and they are generally used after a positive test for further verification. X-ray films reveal abnormal radiopaque patches, the appearance and location of which can be very indicative. Primary tubercular infection presents the appearance of fine areas of infiltration **(figure 19.12)** and enlarged lymph nodes in the lower and central areas of the lungs. Secondary tuberculosis films show more extensive infiltration in the upper lungs and bronchi and marked tubercles. Scars from older infections often show up on X rays and can furnish a basis for comparison when trying to identify newly active disease.

Blood Testing   Because the main immune response to *Mycobacterium* comes from T cells, the blood test for tuberculosis looks for T-cell activities in the form of cytokine release. The assay is called interferon gamma release assay (IGRA). It uses whole blood and can be completed within 24 hours.

Culture   Diagnosis that differentiates between *M. tuberculosis* and other mycobacteria must be accomplished as rapidly as possible so that appropriate treatment and isolation precautions can be instituted. Treatment and isolation should be instituted presumptively until culture results are available. Antibiotic sensitivities can be determined after the culture has grown. Even though newer cultivation schemes exist that shorten the incubation period from 6 weeks to several days, this delay is unacceptable for beginning treatment or isolation precautions. But a culture still must be performed because growing colonies are required to determine antibiotic sensitivities.

Acid-Fast Staining   The diagnosis of tuberculosis in people with positive skin tests or X rays can be backed up by acid-fast staining of sputum or other specimens. Several variations on the acid-fast stain are currently in use. The Ziehl-Neelsen stain produces bright red acid-fast bacilli (AFB) against a blue background **(figure 19.13).**

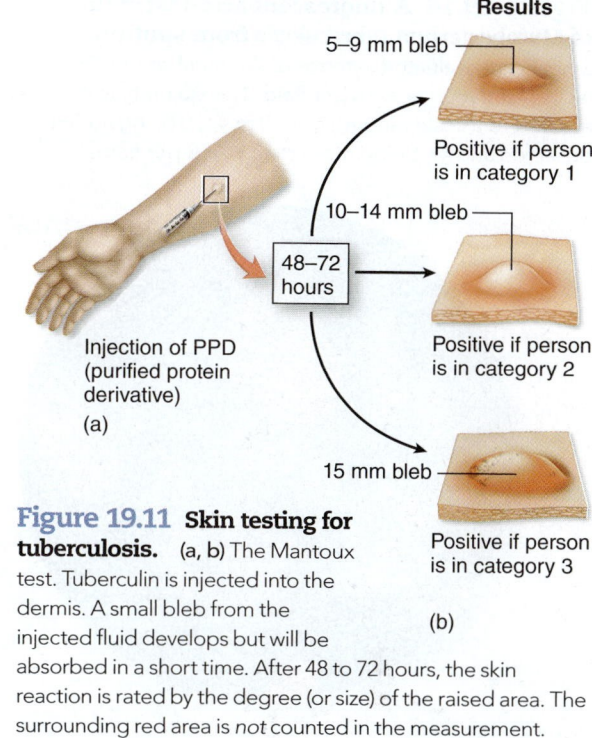

Figure 19.11   Skin testing for tuberculosis.   (a, b) The Mantoux test. Tuberculin is injected into the dermis. A small bleb from the injected fluid develops but will be absorbed in a short time. After 48 to 72 hours, the skin reaction is rated by the degree (or size) of the raised area. The surrounding red area is *not* counted in the measurement.

Results

5–9 mm bleb — Positive if person is in category 1

Injection of PPD (purified protein derivative)

(a)

48–72 hours

10–14 mm bleb — Positive if person is in category 2

15 mm bleb — Positive if person is in category 3

(b)

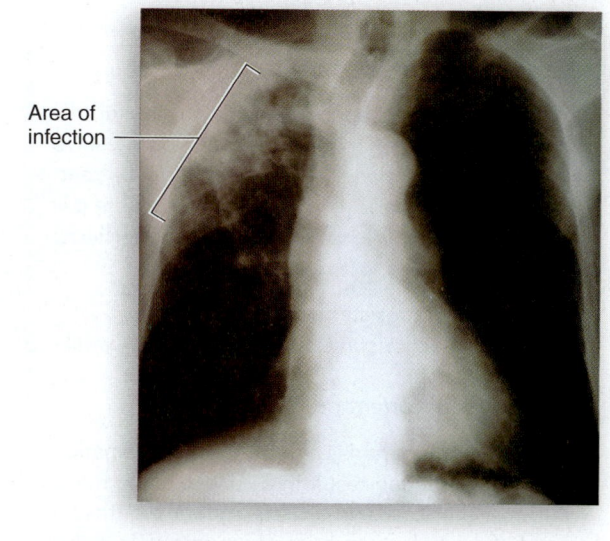

Figure 19.12   Primary tuberculosis.

Area of infection

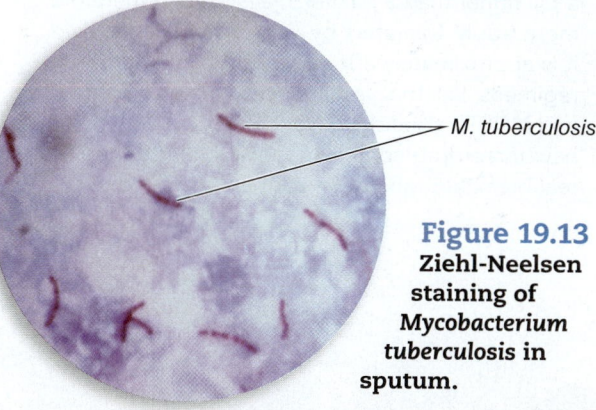

M. tuberculosis

Figure 19.13 Ziehl-Neelsen staining of Mycobacterium tuberculosis in sputum.

Figure 19.14 A fluorescent acid-fast stain of *Mycobacterium tuberculosis* from sputum. Smears are evaluated in terms of the number of AFB (acid-fast bacteria) seen per field. This quantity is then applied to a scale ranging from 0 to 4+, 0 being no AFB observed and 4+ being more than 9 AFB per field.

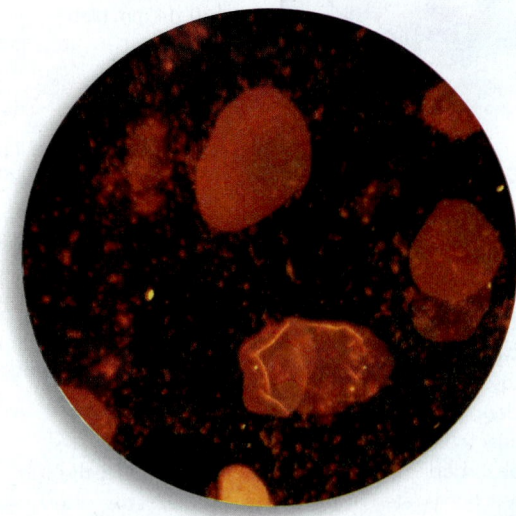

Fluorescence staining shows luminescent yellow-green bacteria against a dark background **(figure 19.14)**.

PCR for Simultaneous Identification and Susceptibility Testing

Recently, a PCR method has become available. It is known as the *Xpert® MTB/RIF assay* and simultaneously detects *M. tuberculosis* and determines its rifampin sensitivity within 100 minutes. The WHO started encouraging its use in 2010; as of 2014, many low- and middle-income countries had purchased the kits.

▶ Prevention

Preventing TB in the United States is accomplished by limiting exposure to infectious airborne particles. Extensive precautions, such as isolation in negative-pressure rooms, are used in health care settings when a person with active TB is identified. Vaccine is generally not used in the United States, although an attenuated vaccine called BCG is used in many countries. BCG stands for Bacille Calmette-Guerin, named for two French scientists who created the vaccine in the early 1900s. It is a live strain of a bovine tuberculosis bacterium that has been made avirulent by long passage through artificial media. Remember that persons vaccinated with BCG may respond positively to a tuberculin skin test.

In the past, prevention in the context of tuberculosis referred to preventing a person with latent TB from experiencing reactivation. This strategy is more accurately referred to as treatment of latent infection and is considered in the next section, "Treatment."

▶ Treatment

Treatment of latent TB infection is effective in preventing full-blown disease in persons who have positive skin tests and who are at risk for reactivated TB. Treatment of latent TB is with three drugs: isoniazid, rifampin, and rifapentine. The rifampin and rifapentine are taken for 4 and 3 months, respectively, and the isoniazid is continued for 9 months.

Treatment of active TB infection occurs in two phases. In the first, four drugs—rifampin, isoniazid, ethambutol, and pyrazinamide—are used for 2 months. The second phase uses only rifampin and isoniazid and lasts either 4 or 7 months, decided on a case-by-case basis.

One of the biggest problems with TB therapy is noncompliance on the part of the patient. It is very difficult, even under the best of circumstances, to keep to a regimen of multiple antibiotics daily for months—and most TB patients are living under conditions that are far from the best of circumstances. Failure to adhere to the antibiotic regimen leads to antibiotic resistance in the slow-growing microorganism; in fact, many *M. tuberculosis* isolates are now found to be **MDR-TB,** or multidrug-resistant TB.

Multidrug-Resistant Tuberculosis (MDR-TB)

Multidrug-Resistant Tuberculosis is defined as being resistant to at least isoniazid and rifampin. It requires treatment of 18 to 24 months with four to six drugs. It is particularly common in people who have been previously treated for tuberculosis. In this population, 20% of cases are MDR-TB overall. In some parts of the world, the rate of MDR-TB among previously treated TB patients is 50% to 60%. Among those being treated for TB for the first time, the MDR-TB rate is closer to 4%.

People with MDR-TB are generally sicker and have higher mortality rates than those infected with non-MDR-TB. This is true even when they are being treated, as the multiple-drug combination has severe side effects. In 2012, a drug called bedaquiline was approved for use for MDR-TB as the fourth drug in a combination therapy. Scientists are hopeful that a new drug combination called PaMZ will drastically improve treatment if it makes it to the market in 2014 or 2015 (see **Medical Moment**).

Extensively Drug-Resistant Tuberculosis (XDR-TB)

MDR-TB strains with resistance to two additional drugs are called **XDR-TB**. These strains have been reported in 84 countries. Worldwide, 9% of the MDR-TB cases also qualify as XDR-TB. Patients with XDR-TB have few treatment options and

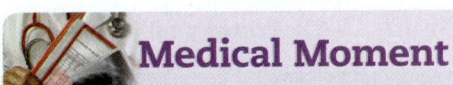

Medical Moment

Breakthrough TB Treatment

In 2014 researchers announced the successful trial of a new combination of drugs that could solve many of the problems associated with TB treatment, especially in developing countries where it is taking its most terrible toll. These problems include:

- the necessity of long treatment times
- multiply drug resistant strains
- painful injections required for already weakened patients
- high cost of current drugs

The new regimen is a three-drug combination called PaMZ. It can result in a cure, even in many drug-resistant cases, in as little as four months instead of the two years currently required. It is also a pill rather than a painful injection, and therefore more easily tolerated by very sick patients. And it is approximately 20x cheaper than current drug regimens. Doctors and agencies trying to fight TB and MDRTB around the world are hopeful that this new three-drug combination will provide the first real breakthrough in TB treatment in 60 years.

Disease Table 19.9 Tuberculosis

Causative Organism(s)	*Mycobacterium tuberculosis*	MDR-TB and XDR-TB
Most Common Modes of Transmission	Vehicle (airborne)	
Virulence Factors	Lipids in wall, ability to stimulate strong cell-mediated immunity (CMI)	
Culture/Diagnosis	Rapid methods plus culture; initial tests are skin testing, positive blood culture, chest X ray	
Prevention	Avoiding airborne *M. tuberculosis*; BCG vaccine in other countries	
Treatment	Isoniazid, rifampin, and pyrazinamide + ethambutol or streptomycin for varying lengths of time (always lengthy)	Multiple-drug regimen, which may include bedaquiline; in **Serious Threat** category in CDC Antibiotic Resistance Report
Distinctive Features	Responsible for nearly all non-MDR-TB except for some HIV-positive patients and severely immunosuppressed patients who have *Mycobacterium avium* complex (MAC)	Much higher fatality rate over shorter duration
Epidemiological Features	United States: approx. 10,000 cases/year, 16% of cases Whites, 84% ethnic minorities; internationally: 1.3 million deaths in 2012	United States: fewer than 100/year; worldwide: 94,000 with MDR-TB in 2012 (Note: probably greatly underdiagnosed)

their mortality rate is estimated to be about 70% within months of diagnosis. The epidemiology is difficult to document, but India and China have the highest burden of XDR-TB. A few cases are seen in the United States every year.

Pneumonia

Pneumonia is a classic example of a disease characterized by an *anatomical diagnosis*. It is defined as an inflammatory condition of the lung in which fluid fills the alveoli. The set of symptoms that we call pneumonia can be caused by a wide variety of different microorganisms. In a sense, the microorganisms need only to have appropriate characteristics to allow them to circumvent the host's defenses and to penetrate and survive in the lower respiratory tract. In particular, the microorganisms must avoid being phagocytosed by alveolar macrophages, or at least avoid being killed once inside the macrophage. Bacteria, fungi, and a wide variety of viruses can cause pneumonias, and there is a lot of variation in the virulence of different pathogenic agents. Pneumonia can be deadly, and across the globe, more children under the age of 5 die from pneumonia than any other infectious disease; over 4,000 children die each day. In the United States, residents experience 2 to 3 million cases of pneumonia and more than 45,000 deaths due to this condition every year. It is much more common in the winter.

Even though all pneumonias have similar kinds of symptoms, physicians often distinguish between two forms of pneumonia, characterized by different modes of transmission and pathogenic agents. *Community-acquired pneumonia (CAP)* is experienced by persons in the general population. *Healthcare-associated pneumonia (HCAP)* develops in individuals receiving treatment at health care facilities, including hospitals.

▶ Causative Agents of Community-Acquired Pneumonia

Streptococcus pneumoniae accounts for up to 40% of community-acquired bacterial pneumonia cases. It causes more lethal pneumonia cases than any other microorganism. *Legionella* is a less common but also serious cause of the disease. *Haemophilus influenzae* had been a major cause of community-acquired pneumonia, but the introduction of the Hib vaccine in 1988 has reduced its incidence. A number of bacteria cause a milder form of pneumonia that is often referred to as "walking pneumonia." Two of these are *Mycoplasma pneumoniae* and *Chlamydophila pneumoniae* (formerly

A Note About Directly Observed Therapy

Although it is highly labor intensive, directly observed therapy (DOT), in which ingestion of medications is observed by a responsible person, seems to be the most effective means of curbing infections and preventing further development of antibiotic resistance. The WHO estimates that 8 million deaths have been prevented by DOT over the last 15 years. Patients are referred for DOT if a physician suspects they will have trouble adhering to the very rigorous and lengthy antibiotic schedule. At that point, a public health worker is assigned to visit them at their home and/or workplace to watch them take their medicines. One innovative program to alleviate the labor-intensiveness of such an approach has been developed at the Massachusetts Institute of Technology. Patients receive a container of filter paper that dispenses a filter paper at timed intervals. They dip the paper in their urine and if the antibiotic is present in their urine, the filter paper reveals a code that the patients text to a central database. If they miss fewer than five pills a month, they receive free minutes for their cell phones.

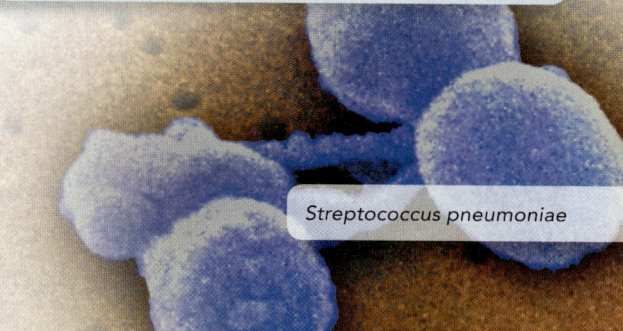

Streptococcus pneumoniae

Figure 19.15 Signature tents in Yosemite National Park. These tent types harbored hantavirus in the 2012 Yosemite outbreak.

Figure 19.16 Sign in wooded area in Kentucky. The sign is covered in bird droppings. Up to 90% of the population in the Ohio Valley show evidence of past infection with *Histoplasma*.

known as *Chlamydia pneumoniae*). *Histoplasma capsulatum* is a fungus that infects many people but causes a pneumonia-like disease in relatively few. One virus causes a type of pneumonia that can be very serious: hantavirus, which emerged in 1993 in the United States. Pneumonia may be a secondary effect of influenza disease.

Streptococcus pneumoniae   This bacterium, which is often simply called the pneumococcus, is a small gram-positive flattened coccus that often appears in pairs, lined up end to end. It is alpha-hemolytic on blood agar. Factors that favor the ability of the pneumococcus to cause disease are old age, the season (rate of infection is highest in the winter), underlying viral respiratory disease, diabetes, and chronic abuse of alcohol or narcotics. Healthy people commonly inhale this and other microorganisms into the respiratory tract without serious consequences because of the host defenses present there.

Because the pneumococcus is such a frequent cause of pneumonia in older adults, this population is encouraged to seek immunization with the older pneumococcal polysaccharide vaccine PPSV23, which stimulates immunity to the capsular polysaccharides of 23 different strains of the bacterium.

Legionella pneumophila   *Legionella* is a weakly gram-negative bacterium that has a range of shapes, from coccus to filaments. Several species or subtypes have been characterized, but *L. pneumophila* ("lung-loving") is the one most frequently isolated from infections.

Legionella's ability to survive and persist in natural habitats has been something of a mystery, yet it appears to be widely distributed in aqueous habitats as diverse as tap water, cooling towers, spas, ponds, and other freshwaters. It is resistant to chlorine. The bacterium can live in close association with free-living amoebas **(figure 19.15).** It is released during aerosol formation and can be carried for long distances. Cases have been traced to supermarket vegetable sprayers, hotel fountains, air-conditioning vents, and even the fallout from the Mount St. Helens volcano eruption in 1980.

Atypical Pneumonias   Pneumonias caused by *Mycoplasma* (as well as those caused by *Chlamydophila* and some other microorganisms) are often called atypical pneumonia– atypical in the sense that the symptoms do not resemble those of pneumococcal or other severe pneumonias. *Mycoplasma* pneumonia is transmitted by aerosol droplets among people confined in close living quarters, especially families, students, and the military. Lack of acute illness in most patients has given rise to the name "walking pneumonia."

Hantavirus   In 1993, hantavirus suddenly burst into the American consciousness. A cluster of unusual cases of severe lung edema among healthy young adults arose in the Four Corners area of New Mexico. Most of the patients died within a few days. They were later found to have been infected with hantavirus, an agent that had previously only been known to cause severe kidney disease and hemorrhagic fevers in other parts of the world. The new condition was named hantavirus pulmonary syndrome (HPS). Since 1993, the disease has occurred sporadically, but it has a mortality rate of at least 33%.

Very soon after the initial cases in 1993, it became clear that the virus was associated with the presence of mice in close proximity to the victims. Investigators eventually determined that the virus, an enveloped virus of the *Bunyaviridae* family, is transmitted via airborne dust contaminated with the urine, feces, or saliva of infected rodents. Deer mice and other rodents can carry the virus with few apparent symptoms. Small outbreaks of the disease are usually correlated with increases in the local rodent population. In 2012, hantavirus caused a localized outbreak among visitors to Yosemite National Park. Visitors staying in "signature tent cabins" were exposed to mouse droppings containing the virus. Three people died, and at least 10 were sickened. The signature tent cabins were of a new double-walled design that was meant to be safer, but the space between the walls turned out to be a perfect nesting space for the mice.

Histoplasma capsulatum   This organism is endemically distributed on all continents except Australia. Its highest rates of incidence occur in the eastern and central regions of the United States, especially in the Ohio Valley. This fungus appears to grow

Disease Table 19.10 Pneumonia

Causative Organism(s)	*Streptococcus pneumoniae*	*Legionella* species	*Mycoplasma pneumoniae*	Hantavirus	*Histoplasma capsulatum*	*Pneumocystis jiroveci*	Respiratory viruses
Most Common Modes of Transmission	Droplet contact or endogenous transfer	Vehicle (water droplets)	Droplet contact	Vehicle–airborne virus emitted from rodents	Vehicle– inhalation of contaminated soil	Droplet contact	Droplet contact or endogenous transfer
Virulence Factors	Capsule	–	Adhesins	Ability to induce inflammatory response	Survival in phagocytes	–	–
Culture/ Diagnosis	Gram stain often diagnostic, alpha-hemolytic on blood agar	Requires selective charcoal yeast extract agar; serology unreliable	Rule out other etiologic agents	Serology (IgM), PCR identification of antigen in tissue	Rapid antigen tests, microscopy	PCR	Failure to find bacteria or fungi
Prevention	Pneumococcal polysaccharide vaccine (PPSV23)	–	No vaccine, no permanent immunity	Avoid mouse habitats and droppings	Avoid contaminated soil/bat, bird droppings	Antifungals given to AIDS patients to prevent this	Hygiene
Treatment	Cefotaxime, ceftriaxone, with or without vancomycin; in **Serious Threat** category in CDC Antibiotic Resistance Report	Fluoroquinolone, azithromycin, clarithromycin	Doxycycline	Supportive	Itraconazole	Trimethoprim-sulfamethoxazole	None
Distinctive Features	Patient usually severely ill	Mild pneumonias in healthy people; can be severe in elderly or immuno-compromised	Usually mild; "walking pneumonia"	Rapid onset; high mortality rate	Many infections asymptomatic	Vast majority occur in AIDS patients	Usually mild
Epidemiological Features	40% of CAP cases; in 2009 H1N1 epidemic, 29% of fatalities were co-infected with this bacterium	United States: 8,000-10,000 cases per year; internationally: 2 million cases per year	20%-40% of CAP cases	United States: 10-40 cases per year; fatality rate 25%-75%	In United States, 250,000 infected per year; only 5%-10% have symptoms	80% of untreated AIDS patients are infected	30% of CAP cases

most abundantly in moist soils high in nitrogen content, especially those supplemented by bird and bat droppings **(figure 19.16).**

A useful tool for determining the distribution of *H. capsulatum* is to inject a fungal extract into the skin and monitor for allergic reactions (much like the TB skin test). Application of this test has verified the extremely widespread distribution of the fungus. In high-prevalence areas such as southern Ohio, Illinois, Missouri, Kentucky, Tennessee, Michigan, Georgia, and Arkansas, 80% to 90% of the population show signs of prior infection.

Pneumocystis (carinii) jiroveci

Although the fungus *Pneumocystis jiroveci* (formerly called *P. carinii*) was discovered in 1909, it remained relatively obscure until it was suddenly propelled into clinical prominence as the agent

A Note About SARS and MERS

In 2003, a virus in the *Coronavirus* family, previously known to cause only coldlike symptoms, burst onto the world stage as it started to cause pneumonias and death in Hong Kong. The SARS epidemic ended nearly as quickly as it started, and since 2004 new cases of SARS have not been detected anywhere on the planet. In 2013, another new coronavirus emerged from Saudi Arabia. It was named MERS (Middle East Respiratory Syndrome). It has a 30% fatality rate and so far does not seem to spread easily from human to human.

Disease Table 19.11 Healthcare-Associated Pneumonia

Causative Organism(s)	Gram-negative and gram-positive bacteria from upper respiratory tract or stomach; environmental contamination of ventilator
Most Common Modes of Transmission	Endogenous (aspiration)
Virulence Factors	–
Culture/Diagnosis	Culture of lung fluids
Prevention	Elevating patient's head, preoperative education, care of respiratory equipment
Treatment	Broad-spectrum antibiotics
Epidemiological Features	United States: 300,000 cases per year; occurs in 0.5%–1.0% of admitted patients; mortality rate in United States and internationally is 20%-50%

of *Pneumo cystis* pneumonia. PCP is one of the most frequent opportunistic infections in AIDS patients, most of whom will develop one or more episodes during their lifetimes. There is some debate about how *Pneumocystis* is acquired, but inhalation of spores is probably common, and healthy people may even harbor it as "normal biota" in their lungs.

Traditional antifungal drugs are ineffective against *Pneumocystis* pneumonia because the chemical makeup of the organism's cell wall differs from that of most fungi.

Respiratory Viruses Viruses are very common causes of community-acquired pneumonia. They are viruses that are either residents in the upper respiratory tract or acquired through our daily activities. Viral pneumonias are generally mild. **Disease Table 19.10.**

Healthcare-Associated Pneumonia

Up to 1% of hospitalized or institutionalized people experience the complication of pneumonia. It is the second most common healthcare-associated infection, preceded by urinary tract infections. The mortality rate is quite high, between 30% and 50%. The most frequent causes are *Pseudomonas aeruginosa* and *Acinetobacter baumannii* although infections with *Streptococcus pneumoniae* and *Klebsiella pneumoniae* are common as well. Pneumonia due to *S. aureus* infection arises frequently in HAP and is frequently caused by MRSA strains of the bacterium. Further complicating matters, many healthcare-associated pneumonias appear to be polymicrobial in origin–meaning that there are multiple microorganisms multiplying in the alveolar spaces.

In healthcare-associated infections, bacteria gain access to the lower respiratory tract through abnormal breathing and aspiration of the normal upper respiratory tract biota (and occasionally the stomach) into the lungs. Stroke victims have high rates of healthcare-associated pneumonia. Mechanical ventilation is another route of entry for microbes. Once there, the organisms take advantage of the usually lowered immune response in a hospitalized patient and cause pneumonia symptoms.

Culture of sputum or of tracheal swabs is not very useful in diagnosing healthcare-associated pneumonia, because the condition is usually caused by normal biota. Cultures of fluids obtained through endotracheal tubes or from bronchoalveolar lavage provide better information but are fairly intrusive. It is also important to remember that if the patient has already received antibiotics, culture results will be affected.

Because most healthcare-associated pneumonias are caused by microorganisms aspirated from the upper respiratory tract, measures that discourage the transfer of microbes into the lungs are very useful for preventing the condition. Elevating patients' heads to a 45-degree angle helps reduce aspiration of secretions. Good preoperative education of patients about the importance of deep breathing and frequent coughing can reduce postoperative infection rates. Proper care of mechanical ventilation and respiratory therapy equipment is essential as well.

Studies have shown that delaying antibiotic treatment of suspected healthcare-associated pneumonia leads to a greater likelihood of death. Even in this era of conservative antibiotic use, empiric therapy should be started as soon as healthcare-associated pneumonia is suspected, using multiple antibiotics that cover both gram-negative and gram-positive organisms.

19.5 LEARNING OUTCOMES—Assess Your Progress

10. List the possible causative agents, modes of transmission, virulence factors, diagnostic techniques, and prevention/treatment for the "Highlight Disease" tuberculosis.

11. Discuss the problems associated with MDR-TB and XDR-TB.

12. Discuss important features of the other lower respiratory tract diseases, community-acquired and healthcare-associated pneumonia.

CASE FILE WRAP-UP

Respiratory syncytial virus (RSV) is prevalent during the winter months and is a common cause of lower respiratory tract infection during infancy and childhood. Older children and adults generally experience mild disease, but infants ages 6 months and younger may develop bronchiolitis (inflammation of the smaller airways) or pneumonia as a consequence of the infection. Premature infants are particularly at risk and may develop life-threatening disease. There is no vaccine for RSV, but premature infants and/or infants born with cardiac or lung disease may be given a series of injections of the monoclonal antibody palivizumab (Synagis), which has proven to be moderately effective at preventing the disease.

RSV spreads easily by droplet contact. It can also survive for hours on inanimate objects and surfaces such as tables. Michael, the infant in the case file at the beginning of this chapter, may have contracted RSV at his day care, or he may have been exposed to the virus from his school-aged siblings. By 2 to 3 years of age, most children in the United States will have encountered RSV. Unfortunately, there is no vaccine for RSV, but researchers are getting closer to developing one.

▶ Summing Up

Taxonomic Organization Microorganisms Causing Disease in the Respiratory Tract

Microorganism	Pronunciation	Location of Disease Table
Gram-positive bacteria		
Streptococcus pneumoniae	strep′-tuh-kok″-us nu- moh′ nee-ay	Otitis media, p. 541
		Pneumonia, p. 553
S. pyogenes	strep′-tuh-kok″-us pie-ah′-gen-eez	Pharyngitis, p. 539
Corynebacterium diphtheriae	cor-eye′-nee-back-teer″-e-em dip-theer′-e-ay	Diphtheria, p. 541
Gram-negative bacteria		
Fusobacterium necrophorum	fuze′-oh″-back-teer″ee-em neck″-row-for′-em	Pharyngitis, p. 539
Bordetella pertussis	bor′-duh-tell′-uh per-tuss′-is	Whooping cough, p. 546
Mycobacterium tuberculosis*	my″-co-back-teer′-ee-em tuh-ber′-cue-loh-sis	Tuberculosis, p. 551
Legionella spp.	lee″-juhn-el′-uh	Pneumonia, p. 553
Pseudomonas aeruginosa	sue′-doe-moe″-ness air-roo′-gi-no″-sa	Healthcare-associated pneumonia, p. 554
Acinetobacter baumannii	a′-ci-nay′-toe-bak″-ter bow-mah″-nee-i	Healthcare-associated pneunomia, p. 554
Other bacteria		
Mycoplasma pneumoniae	my″-co-plazz′-muh nu-moh′-nee-ay	Pneumonia, p. 553
RNA viruses		
Respiratory syncytial virus	ress″-pur-uh-tor′-ee sin-sish′-ull vie′-russ	RSV disease, p. 546
Influenza virus A, B, and C	in″-floo-en′-zuh vie′-russ	Influenza, p. 545
Hantavirus	haun″-tuh vie′-russ	Pneumonia, p. 553
Fungi		
Pneumocystis jiroveci	new-moh-siss′-tiss yee″-row-vet′-zee	Pneumonia, p. 553
Histoplasma capsulatum	hiss″-toe-plazz′-muh cap″-sue-lah′-tum	Pneumonia, p. 553

*There is some debate about the gram status of the genus *Mycobacterium;* it is generally not considered gram-positive or gram-negative.

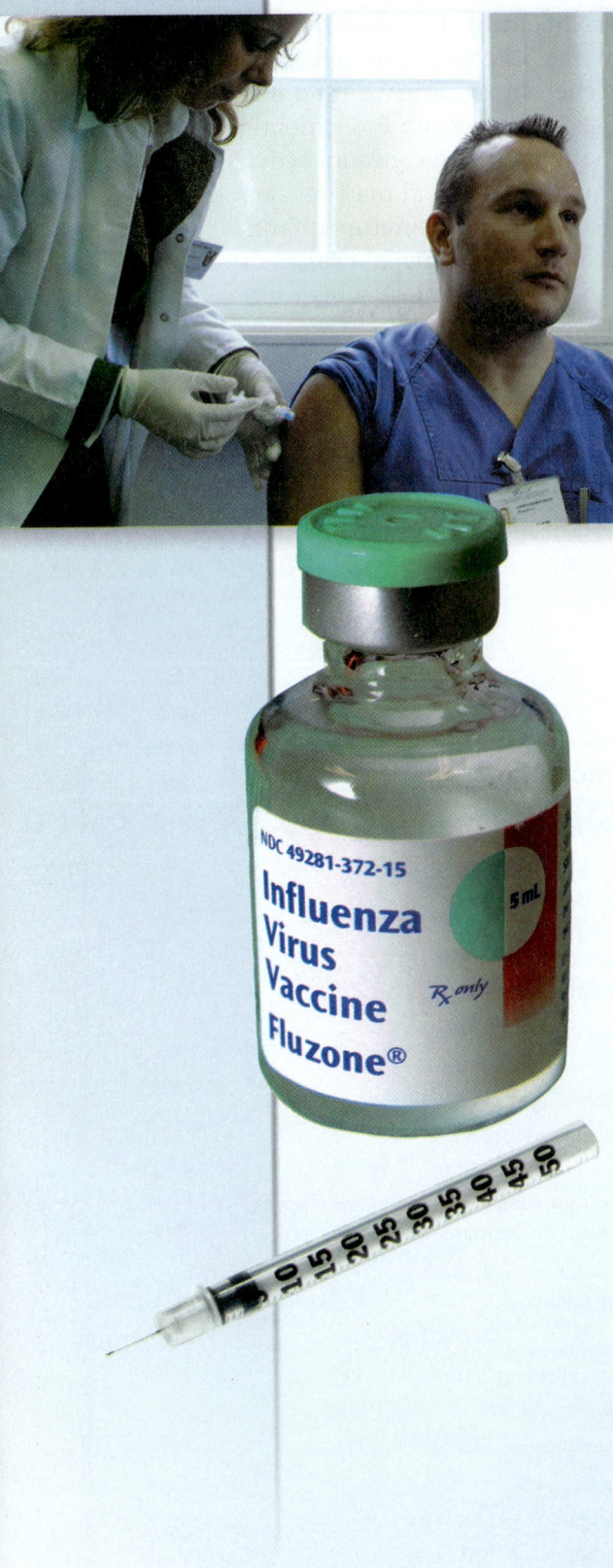

Mandatory Flu Shots for Health Care Workers: The Debate

Health care workers have notoriously poor track records when it comes to yearly immunization against seasonal influenza. According to the CDC, fewer than half (approximately 46%) of all health care workers were immunized against the seasonal flu in 2007–2008. Why were so many health care workers reluctant to get vaccinated against influenza? Many reasons are cited for poor compliance rates. Some health care workers have experienced adverse reactions to an influenza vaccination in the past, making them more reluctant to get vaccinated again. Some health care workers do not believe in the vaccine's ability to prevent influenza—others simply do not believe that they need to be vaccinated if they are healthy with no risk factors for serious disease should they contract the flu.

In 2009, during the height of panic regarding the novel H1N1 pandemic, which swept across the world within a few short months, the controversy surrounding the vaccine's rush to market and the exceptional media hype surrounding the pandemic forced some health care organizations to take a new approach to their yearly campaign to vaccinate their employees. The New York Health Department took the unprecedented step of mandating influenza vaccinations for health care workers, which caused a backlash of controversy as other states and organizations followed their lead. Health care workers balked at the concept of forced vaccination, with several organizations (for example, the Washington State Nurses Association) suing health care organizations over making forced vaccination a condition of employment.

There are two sides to every story, and mandatory vaccination is no exception. Those who support mandatory vaccination believe that health care workers have an obligation to protect their patients from a potentially deadly disease. They cite the fact that other vaccinations are mandatory: Health care workers must often provide proof of immunization against measles, diphtheria, rubella, and other diseases that were once major health concerns before beginning employment. In short, proponents of mandatory vaccination believe that an individual's rights are trumped by the rights of those they are charged with caring for—namely, their patients. Opponents of mandatory vaccination raise the point that individuals have the right to self-determination, and that forcing employees to submit to forced vaccinations violates human rights. Both sides raise valid points: The controversy that raged in 2009 (and is still raging today) will likely never be solved to everyone's satisfaction.

Several health care organizations found a compromise in this ongoing debate. In some hospitals, employees who have received the seasonal influenza wear color-coded badges that show that they have been vaccinated, while employees who do not receive the vaccine must wear masks at all times when engaged in patient care. Employees may be exempted from receiving the vaccine for valid medical reasons, such as a documented allergy to eggs (the vaccine is composed of inactivated viruses that are grown in the embryos of eggs); however, these employees are still required to wear a mask while working. Some employees are permitted to refuse the vaccine if they have a moral objection but are required to wear a mask at work at all times to protect others.

The 2009 influenza epidemic may have changed some minds. According to 2013 CDC data, the immunization rate among health care workers jumped from 46% in 2007–2008 to 63.5% in 2010–2011 and continued to climb to 72% in the 2012–2013 season. It appears that immunization rates are slowly climbing year by year among health care workers. Many hope that this trend will be mirrored by similar trends in other areas of public life.

Infectious Diseases Affecting
The Respiratory System

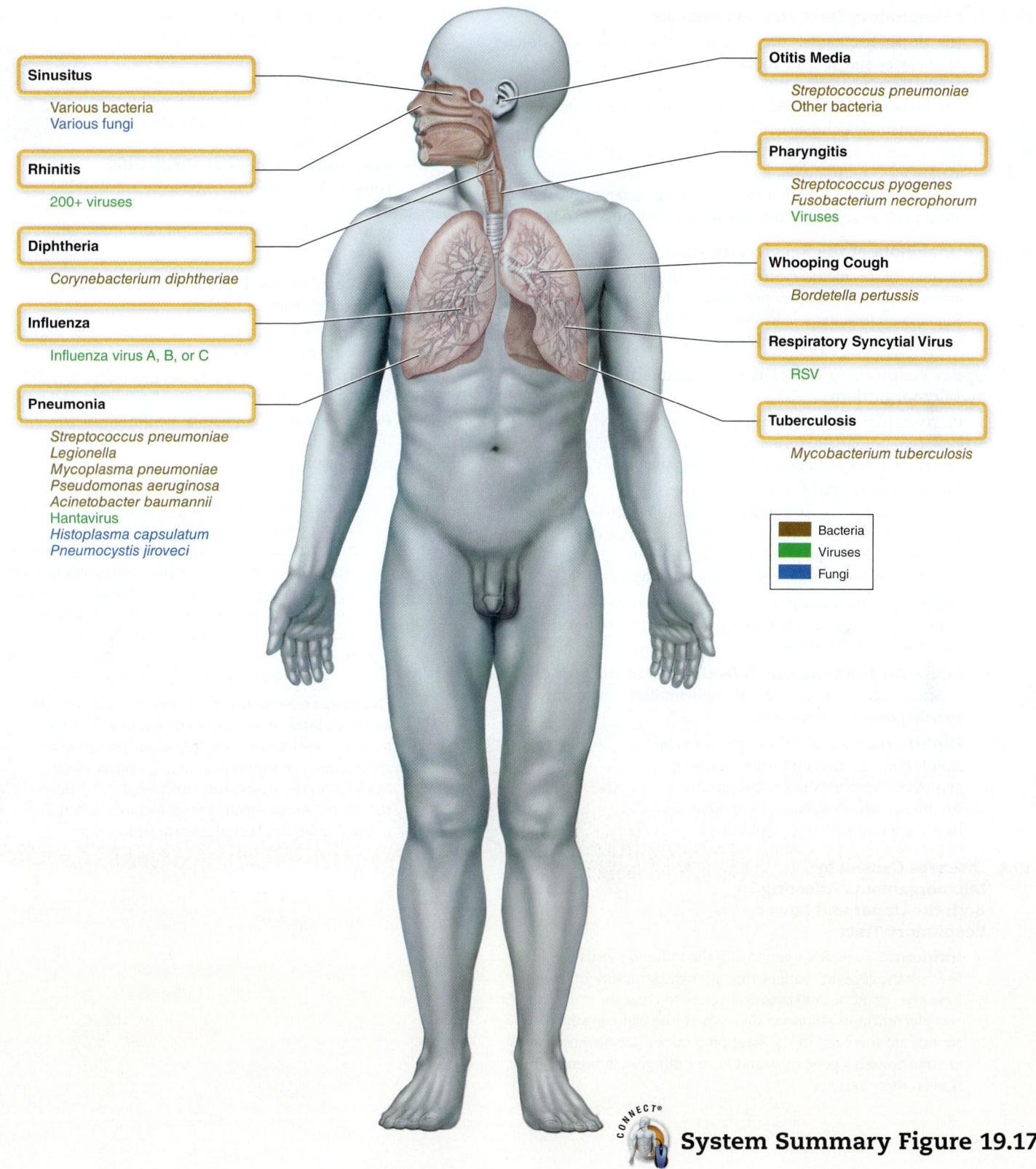

System Summary Figure 19.17

Chapter Summary

19.1 The Respiratory Tract and Its Defenses

- The upper respiratory tract includes the mouth, the nose, nasal cavity and sinuses above it, the throat or pharynx, and epiglottis and larynx.
- The lower respiratory tract consists of the trachea, the bronchi, bronchioles, and alveoli in the lungs.
- The ciliary escalator, mucus on the surface of the mucous membranes, and involuntary responses such as coughing, sneezing, and swallowing are structural defenses. Alveolar macrophages, and secretory IgA are also helpful.

19.2 Normal Biota of the Respiratory Tract

- Normal biota include *Streptococcus pyogenes*, *Haemophilus influenzae*, *Streptococcus pneumoniae*, *Neisseria meningitidis*, *Staphylococcus aureus*, *Lactobacillis sakei*, *Corynebacterium tuberculostearicum*, and *Candida albicans*.

19.3 Upper Respiratory Tract Diseases Caused by Microorganisms

- **Pharyngitis:** Viruses are common cause of pharyngitis. However, two potentially serious causes of pharyngitis are *Streptococcus pyogenes* and *Fusobacterium necrophorum*.

- **The Common Cold:** Caused by one of over 200 different kinds of viruses, most commonly the rhinoviruses, followed by the coronaviruses. Respiratory syncytial virus (RSV) causes colds in many people, but in some, especially children, they can lead to more serious respiratory tract symptoms.

- **Sinusitis:** Inflammatory condition of the sinuses in the skull, most commonly caused by allergy or infections by a variety of viruses or bacteria and, less commonly, fungi.

- **Acute Otitis Media (Ear Infection):** Most common cause is *Streptococcus pneumoniae*, though multiple organisms are usually present in infections.

- **Diphtheria:** Caused by *Corynebacterium diphtheriae*, a non-endospore-forming, gram-positive club-shaped bacterium. An important exotoxin is encoded by a bacteriophage of *C. diphtheriae*.

19.4 Diseases Caused by Microorganisms Affecting Both the Upper and Lower Respiratory Tract

- **Influenza:** The ssRNA genome of the influenza virus is subject to constant genetic changes that alter the structure of its envelope glycoprotein. *Antigenic drift* refers to constant mutation of this glycoprotein. *Antigenic shift*, where the eight separate RNA strands are involved in the swapping out of one of those genes or strands with a gene or strand from a different influenza virus, is even more serious.

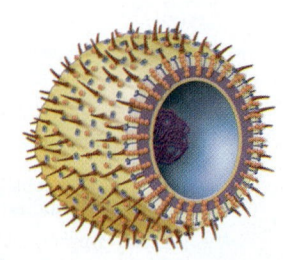

- **Whooping Cough:** Causative agent, *Bordetella pertussis*, releases multiple exotoxins *pertussis toxin* and *tracheal cytotoxin* that damage ciliated respiratory epithelial cells and cripple other components of the host defenses.

- **Respiratory Syncytial Virus Infection:** RSV infects the respiratory tract and produces giant multinucleated cells (syncytia). RSV is the most prevalent cause of respiratory infection in the newborn age group.

19.5 Lower Respiratory Tract Diseases Caused by Microorganisms

- **Tuberculosis:** Cause is primarily the bacterial species *Mycobacterium tuberculosis*. Multidrug-resistant tuberculosis (MDR-TB) and extensively drug-resistant tuberculosis (XDR-TB) are growing problems and have higher fatality rates.

- **Pneumonia:** Inflammatory condition of the lung in which fluid fills the alveoli. Caused by a wide variety of different microorganisms.

  - *Streptococcus pneumoniae:* The main agent for community-acquired bacterial pneumonia cases. *Legionella* is a less common but serious cause of the disease. Other bacteria: *Mycoplasma pneumoniae* and *Chlamydophila pneumoniae*. *Histoplasma capsulatum* is a fungus that can cause a pneumonia-like disease. A hantavirus causes a pneumonia-like condition named hantavirus pulmonary syndrome (HPS). Physicians may treat pneumonia empirically, meaning they do not determine the etiologic agent.

- **Healthcare-Associated Pneumonia:** *Pseudomonas aeruginosa*, *Acinetobacter baumannii*, *Streptococcus pneumoniae*, and *Klebsiella pneumoniae* are common. Pneumonia due to *S. aureus*, often MRSA strains, arises frequently in HAP. Furthermore, many healthcare-associated pneumonias appear to be polymicrobial in origin.

Multiple-Choice Questions   Bloom's Levels 1 and 2: Remember and Understand

Select the correct answer from the answers provided.

1. The common cold is caused by
   a. rhinoviruses.
   b. coronaviruses.
   c. *Staphylococcus aureus*.
   d. a and b.
   e. a, b, and c.

2. Which is not a characteristic of *Streptococcus pyogenes*?
   a. group A streptococcus
   b. alpha-hemolytic
   c. sensitive to bacitracin
   d. gram-positive

3. Which of the following techniques is/are used to diagnose tuberculosis?
   a. tuberculin testing
   b. chest X rays
   c. cultural isolation and antimicrobial testing
   d. all of the above

4. The DTaP vaccine provides protection against the following diseases, *except*
   a. diphtheria.
   b. pertussis.
   c. pneumonia.
   d. tetanus.

5. Which of the following infections often has/have a polymicrobial cause?
   a. otitis media
   b. healthcare-associated pneumonia
   c. sinusitis
   d. all of the above

6. The vast majority of pneumonias caused by this organism occur in AIDS patients.
   a. hantavirus
   b. *Histoplasma capsulatum*
   c. *Pneumocystis jiroveci*
   d. *Mycoplasma pneumoniae*

7. The beta-hemolysis of blood agar observed with *Streptococcus pyogenes* is due to the presence of
   a. streptolysin.
   b. M protein.
   c. hyaluronic acid.
   d. catalase.

8. Approximately what percentage of the world population are infected with *Mycobacterium tuberculosis*?
   a. 5%
   b. 10%
   c. 30%
   d. 50%

Critical Thinking   Bloom's Levels 3, 4, and 5: Apply, Analyze, and Evaluate

Critical thinking is the ability to reason and solve problems using facts and concepts. These questions can be approached from a number of angles and, in most cases, they do not have a single correct answer.

1. What characteristics of the bacterium *S. pneumoniae* do you think make it such a frequent pathogen?

2. Explain why it is unlikely that we will ever see a vaccine for sinusitis.

3. What are some of the likely explanations if you are not responding to antibiotic treatment for sinusitis?

4. Why is noncompliance during TB therapy such a big concern?

5. Why do we need to take the flu vaccine every year? Why does it not confer long-term immunity to the flu like other vaccines?

Visual Connections   Bloom's Level 5: Evaluate

This question connects previous images to a new concept.

1. **From chapter 2, figure 2.18b.** Although there are many different organisms present in the respiratory tract, an acid-fast stain of sputum like the one shown here along with patient symptoms can establish a presumptive diagnosis of tuberculosis. Explain why.

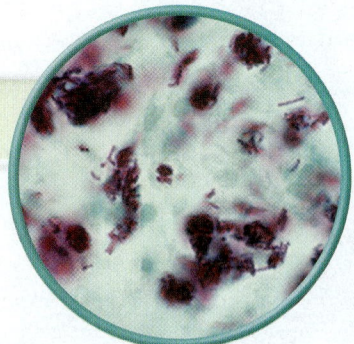

Acid-fast stain
Red cells are acid-fast.
Blue cells are non-acid-fast.

www.mcgrawhillconnect.com
Enhance your study of this chapter with study tools and practice tests. Also ask your instructor about the resources available through Connect, including the media-rich eBook, interactive learning tools, and animations.

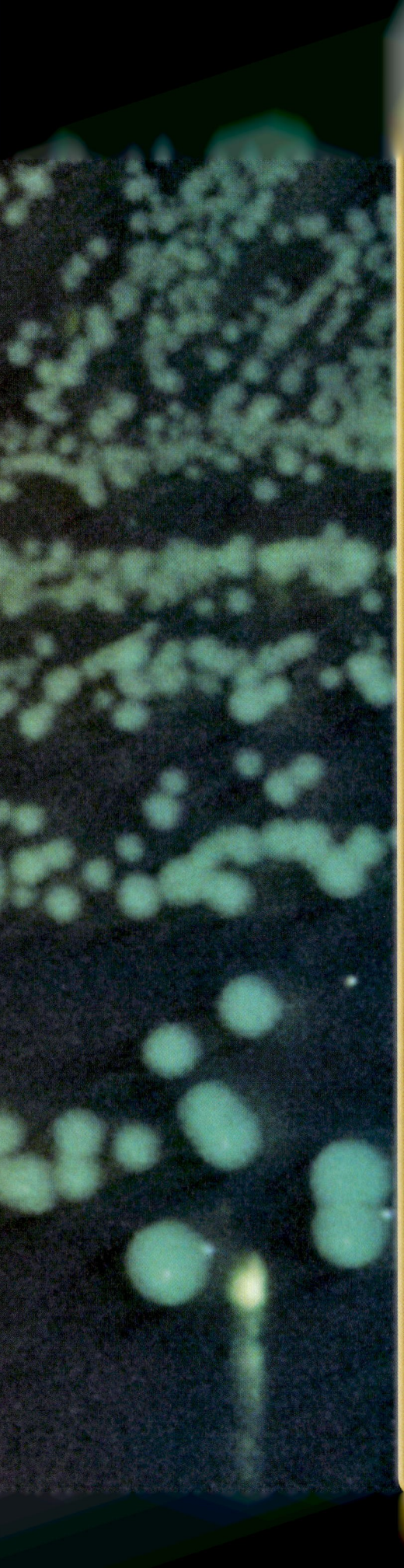

"Blood and Guts"

I was working as an LPN in a rural facility when Peter came into the emergency room one evening. My supervising RN and I realized almost immediately that Peter was very ill, and we worked together quickly to obtain Peter's history and vital signs. Peter stated that he had been ill since the morning, when he woke up earlier than normal with diarrhea. He stated that he had been well the night before when he went to bed and had awoken in the morning with severe abdominal cramping. The diarrhea started almost immediately after the cramps began. Peter estimated that he had had approximately 20 bouts of watery, foul-smelling diarrhea throughout the day. He also stated that there was fresh blood present in his stool.

I took Peter's vital signs. He had a fever of 38.4°C (101°F). He was tachycardic with a heart rate of 118 beats per minute. His blood pressure was low at 100/50 mmHg. He was pale and diaphoretic (sweaty) and complained of chills and a headache. He also complained of intermittent severe cramping. His abdomen was tender in all four quadrants on palpation and his bowel sounds were hyperactive on auscultation. His mucous membranes were dry and he complained of thirst. His skin turgor was poor.

My supervising RN called the physician to report Peter's symptoms and vital signs. The doctor stated that he would be coming in right away and to notify the lab. He asked us to start an intravenous on Peter.

We inserted a large-bore IV and started an infusion of normal saline at 250 mL/hour to rehydrate Peter. The physician and lab technician arrived almost simultaneously, and blood was drawn by the lab tech while the physician examined Peter and asked about his symptoms. Peter was asked to provide a stool sample for culture and an ova and parasites test (O&P test), which he was able to provide in a short time. The physician asked if Peter's wife and daughter were ill and if they had eaten the same food as Peter had in the last few days.

Peter's blood work came back. His white blood cell count was elevated and his potassium level was low. Peter was admitted to the hospital for rehydration and was started on broad-spectrum intravenous antibiotics for full coverage of potential pathogens while he awaited the stool culture results. The doctor felt that Peter likely had bacterial food poisoning.

Peter's stool cultures came back, revealing that Peter had *Shigella,* which was sensitive to sulfamethoxazole/trimethoprim and ciprofloxacin. Peter had an allergy to ciprofloxacin, so he was started on sulfamethoxazole/trimethoprim in oral form twice a day and IV fluids were continued. After 5 days in the hospital, Peter had recovered enough to go home. He continued the antibiotic therapy for another week and eventually fully recovered. The source of the infection was never determined.

- How is *Shigella* transmitted?
- How can *Shigella* be prevented?

Case File Wrap-Up appears on page 594.

Infectious Diseases Affecting the Gastrointestinal Tract

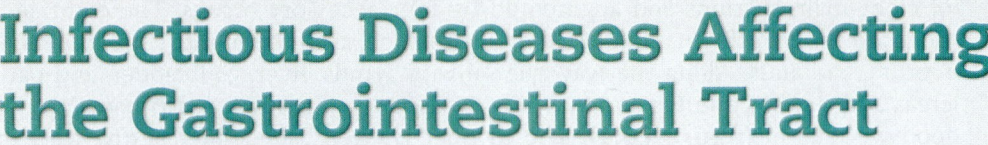

IN THIS CHAPTER...

20.1 The Gastrointestinal Tract and Its Defenses

1. Draw or describe the anatomical features of the gastrointestinal tract.
2. List the natural defenses present in the gastrointestinal tract.

20.2 Normal Biota of the Gastrointestinal Tract

3. List the types of normal biota presently known to occupy the gastrointestinal tract.
4. Describe how our view has changed of normal biota present in the stomach.

20.3 Gastrointestinal Tract Diseases Caused by Microorganisms (Nonhelminthic)

5. List the possible causative agents, modes of transmission, virulence factors, diagnostic techniques, and prevention/treatment for the "Highlight Disease" acute diarrhea.
6. Discuss important features of the conditions food poisoning and chronic diarrhea.
7. Discuss important features of the two categories of oral conditions: dental caries and periodontal diseases.
8. Identify the most important features of mumps, gastritis, and gastric ulcers.
9. Differentiate among the main types of hepatitis, and discuss the causative agents, mode of transmission, diagnostic techniques, prevention, and treatment of each.

20.4 Gastrointestinal Tract Diseases Caused by Helminths

10. Describe some distinguishing characteristics and commonalities seen in helminthic infections.
11. List four helminths that cause primarily intestinal symptoms, and identify which life cycle they follow and one unique fact about each one.
12. Explain reasons for and the consequences of the following statement, "Cysticercosis is underdiagnosed in the United States."

20.1 The Gastrointestinal Tract and Its Defenses

The gastrointestinal (GI) tract can be thought of as a long tube, extending from mouth to anus. It is a very sophisticated delivery system for nutrients, composed of eight main sections and augmented by four accessory organs. The eight sections are the mouth, pharynx, esophagus, stomach, small intestine, large intestine, rectum, and anus. Along the way, the salivary glands, liver, gallbladder, and pancreas add digestive fluids and enzymes to assist in digesting and processing the food we take in **(figure 20.1).** The GI tract is often called the *digestive tract* or the *enteric tract*.

The GI tract has a very heavy load of microorganisms, and it encounters millions of new ones every day. Because of this, defenses against infection are extremely important. All intestinal surfaces are coated with a layer of mucus, which confers mechanical protection. Secretory IgA can also be found on most intestinal surfaces. The muscular walls of the GI tract keep food (and microorganisms) moving through the system through the action of peristalsis. Various fluids in the GI tract have antimicrobial properties. Saliva contains the antimicrobial proteins lysozyme and lactoferrin. The stomach fluid is antimicrobial by virtue of its extremely high acidity. Bile is also antimicrobial.

The entire system is outfitted with cells of the immune system, collectively called gut-associated lymphoid tissue (GALT). The tonsils and adenoids in the oral cavity and pharynx, small areas of lymphoid tissue in the esophagus, Peyer's patches in the small intestine, and the appendix are all packets of lymphoid tissue consisting of T and B cells as well as cells of nonspecific immunity. One of their jobs is to produce IgA, but they perform a variety of other immune functions.

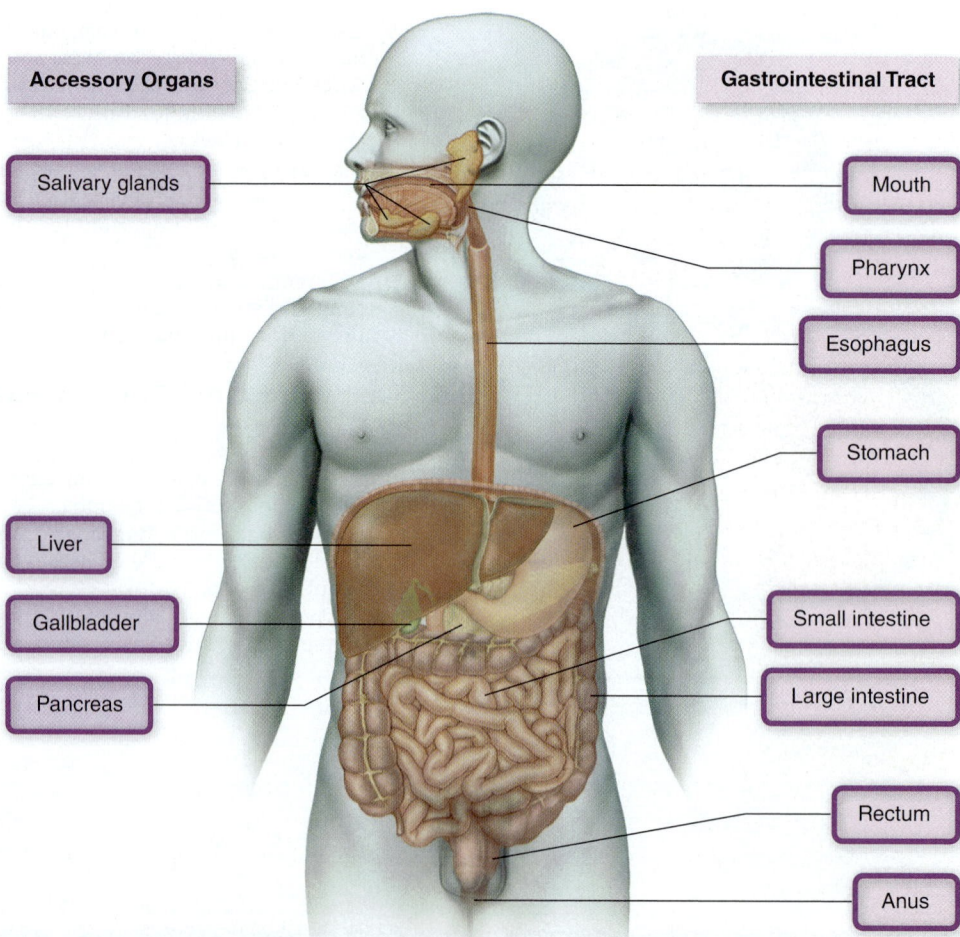

Figure 20.1 Major organs of the digestive system.

A huge population of commensal organisms lives in this system, especially in the large intestine. They avoid immune destruction through various mechanisms, including cloaking themselves with host sugars they find on the intestinal walls.

> ### 20.1 LEARNING OUTCOMES—Assess Your Progress
>
> 1. Draw or describe the anatomical features of the gastrointestinal tract.
> 2. List the natural defenses present in the gastrointestinal tract.

 NCLEX® PREP

1. Gut-associated lymphoid tissue (GALT) includes
   a. *Peyer's patches in the small intestine.*
   b. *the appendix.*
   c. *tonsils.*
   d. *adenoids.*
   e. *all of the above.*
   f. *none of the above.*

20.2 Normal Biota of the Gastrointestinal Tract

As just mentioned, the GI tract is home to a large variety of normal biota. The oral cavity alone is populated by more than 600 known species of microorganisms, including *Actinomyces, Lactobacillis, Neisseria, Prevatella, Streptococcus, Treponema,* and *Veillonella* species. Fungi such as *Candida albicans* are also numerous. A few protozoa (*Trichomonas tenax, Entamoeba gingivalis*) also call the mouth "home." Bacteria live on the teeth as well as the soft structures in the mouth. Numerous species of normal biota bacteria live on the teeth in a synergistic community called dental plaque, which is a type of biofilm (see chapter 3). Bacteria are held in the biofilm by specific recognition molecules. Alpha-hemolytic streptococci are generally the first colonizers of the tooth surface after it has been cleaned. The streptococci attach specifically to proteins in the **pellicle,** a mucinous glycoprotein covering on the tooth. Then other species attach specifically to proteins or sugars on the surface of the streptococci, and so on.

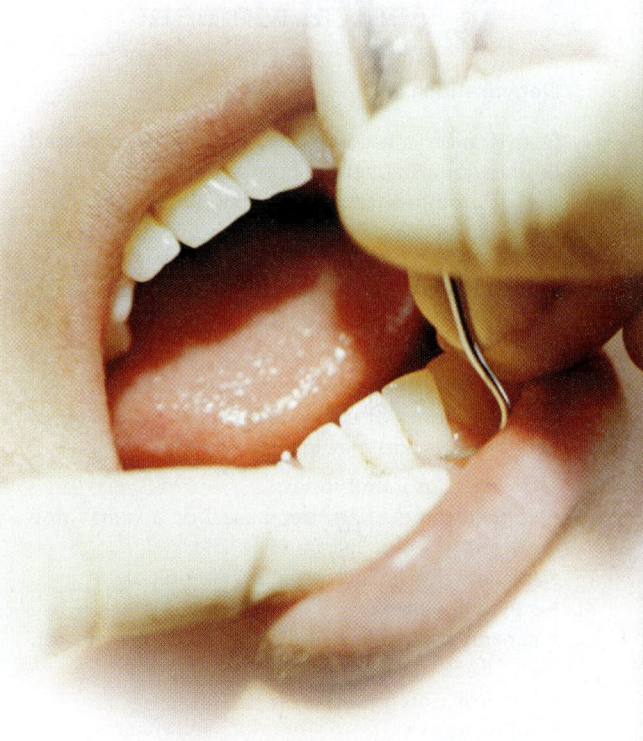

The pharynx contains a variety of microorganisms, which were described in chapter 19. Although the stomach was previously thought to be sterile due to its very low pH, researchers in 2008 found the molecular signatures of 128 different species of microorganisms in the stomach. Though many of these are likely to be "just passing through," a few, including *Bacillus, Clostridium, Staphylococcus,* and *Streptococcus,* are permanent residents. The large intestine has always been known to be a haven for billions of microorganisms (10^{11} per gram of contents), including the bacteria *Bacteroides, Bifidobacterium, Clostridium, Enterobacter, Escherichia, Fusobacterium, Lactobacillus, Peptostreptococcus, Staphylococcus,* and *Streptococcus;* the fungus *Candida;* and several protozoa as well. Researchers have also found archaea species there.

The normal biota in the gut provide a protective function, but they perform other jobs as well. Some of them help with digestion. Some provide nutrients that we can't produce ourselves. *E. coli,* for instance, synthesizes vitamin K. Its mere presence in the large intestine seems to be important for the proper formation of epithelial cell structure. And the normal biota in the gut plays an important role in "teaching" our immune system to react to microbial antigens. Scientists believe that the mix of microbiota in the healthy gut can influence a host's chances for obesity or autoimmune diseases.

The accessory organs (salivary glands, gallbladder, liver, and pancreas) are free of microorganisms, just as all internal organs are.

> ### 20.2 LEARNING OUTCOMES—Assess Your Progress
>
> 3. List the types of normal biota presently known to occupy the gastrointestinal tract.
> 4. Describe how our view has changed of normal biota present in the stomach.

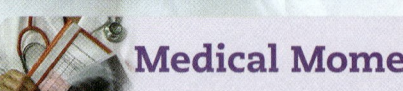

Chickens and their eggs can be colonized by *Salmonella*.

Medical Moment

Dehydration

Dehydration is a common symptom experienced by individuals suffering from conditions affecting the gastrointestinal tract. Recognizing and treating dehydration is a common task of health care workers.

Dehydration can be thought of as an imbalance of fluids resulting from excessive fluid loss (as in vomiting and diarrhea) or inadequate fluid intake (due to nausea or loss of appetite).

Symptoms of dehydration may be subtle or very obvious and may include the following:

- dry mucous membranes;
- concentrated urine, decreased or absent urine output;
- lack of tears (in infants and children);
- sunken eyes;
- sunken fontanelles (the "soft spot" on the top and back of the infant's skull);
- lethargy or fatigue;
- weakness;
- tachycardia (rapid heart rate) and tachypnea (rapid breathing);
- hypotension (low blood pressure);
- poor skin turgor (skin that does not spring back into position when gently pinched);
- altered blood glucose;
- capillary refill (assessment of nail beds);
- thirst;
- headache; and
- seizures.

Health care workers classify dehydration as mild, moderate, or severe—based on the estimated amount of weight lost. This is often expressed in percentages. Capillary refill, skin turgor, and breathing are considered to be the most accurate signs in estimating degree of dehydration.

20.3 Gastrointestinal Tract Diseases Caused by Microorganisms (Nonhelminthic)

In this section, we first address microbes that cause diarrhea of various types. Then we discuss oral diseases (dental caries and periodontitis), stomach conditions (gastritis and gastric ulcers), and the spectrum of hepatitis infections.

Highlight Disease

Acute Diarrhea

Diarrhea—usually defined as three or more loose stools in a 24-hour period—needs little explanation. In recent years, on average, citizens of the United States experienced 1.2 to 1.9 cases of diarrhea per person per year, and among children that number is twice as high. In tropical countries, children may experience more than 10 episodes of diarrhea a year. In fact, more than 3 million children a year, mostly in developing countries, die from a diarrheal disease. In developing countries, the high mortality rate is not the only issue. Children who survive dozens of bouts with diarrhea during their developmental years are likely to have permanent physical and cognitive effects. Diarrheal illnesses are often accompanied by fever, abdominal pain and/or cramping, nausea, vomiting, and dehydration.

In the United States, up to a third of all acute diarrhea is transmitted by contaminated food. In recent years, consumers have become much more aware of the possibility of *E. coli*-contaminated hamburgers or *Salmonella*-contaminated ice cream. New food safety measures are being implemented all the time, but it is still necessary for the consumer to be aware and to practice good food handling.

Although most diarrhea episodes are self-limiting and therefore do not require treatment, others (such as *E. coli* O157:H7) can have devastating effects. In most diarrheal illnesses, antimicrobial treatment is contraindicated (inadvisable), but some, such as shigellosis, call for quick treatment with antibiotics. For public health reasons, it is important to know which agents are causing diarrhea in the community, but in most cases identification of the agent is not performed.

Some species of bacteria are subdivided into serotypes or subtypes. Many gram-negative enteric bacteria are named and designated according to the following antigens: H, the flagellar antigen; K, the capsular antigen; and O, the cell wall antigen. Not all enteric bacteria carry the H and K antigens, but all have O, the polysaccharide portion of the lipopolysaccharide implicated in endotoxic shock (see chapter 18). Most species of gram-negative enterics exhibit a variety of subspecies, variants, or serotypes caused by slight variations in the chemical structure of the HKO antigens. Some bacteria in this chapter (e.g., *E. coli* O157:H7) are named according to their surface antigens; however, we use Latin variant names for *Salmonella*.

In this section, we describe acute diarrhea having infectious agents as the cause. In the sections following this one, we discuss acute diarrhea and vomiting caused by toxins, commonly known as food poisoning, and chronic diarrhea and its causes.

Salmonella

A decade ago, one of every three chickens destined for human consumption was contaminated with *Salmonella*, but the rate is now about 10%. Other poultry, such as ducks and turkeys, is also affected. Eggs are infected as well because the bacteria may actually enter the egg while the shell is being formed in the chicken. In 2007, peanut butter was found to be the source of a *Salmonella* outbreak in the United States. **Figure 20.2** depicts the top five food-borne microbes that land people in the hospital. You see that *Salmonella* is the number one cause. *Salmonella* is a very large genus of bacteria but with a complicated nomenclature. The disease condition in which we are interested is caused by *Salmonella enterica* subspecies *enterica*. This subspecies is further subdivided into serogroups that have formal names such as "*Salmonella enterica* subspecies *enterica* serovar Typhi"; in clinical situations, this latter organism is usually just called "*Salmonella typhi*."

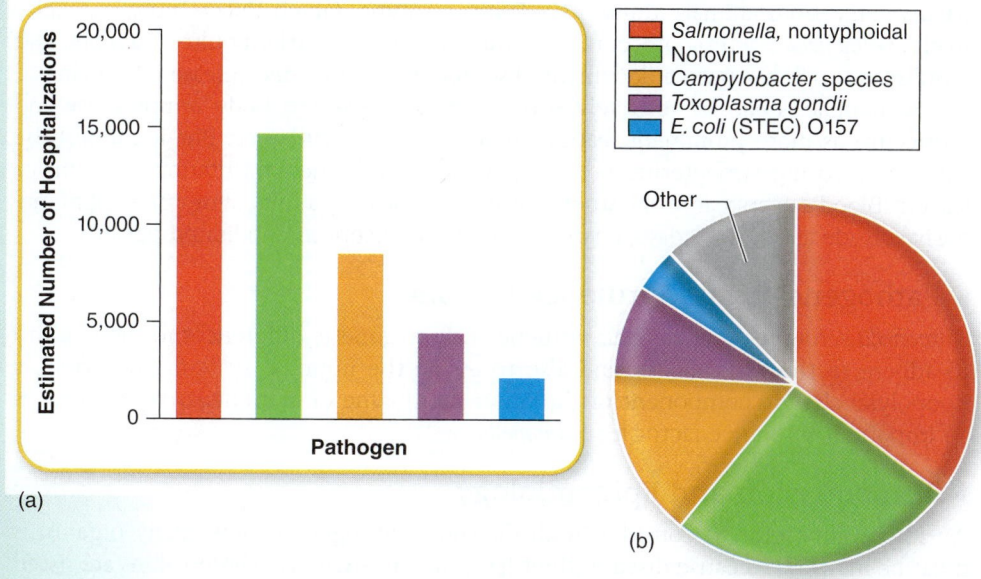

Legend:
- *Salmonella*, nontyphoidal
- Norovirus
- *Campylobacter* species
- *Toxoplasma gondii*
- *E. coli* (STEC) O157

Figure 20.2   Top five food-borne causes of hospitalization in the United States.
(a) *Salmonella*, norovirus, and *Campylobacter* are in our "Acute Diarrhea" section. *Toxoplasma* is discussed in chapter 17. (b) The same data presented as a pie chart, illustrating that "other" causes represent a significant proportion of the total. Data are from 2011 (CDC).

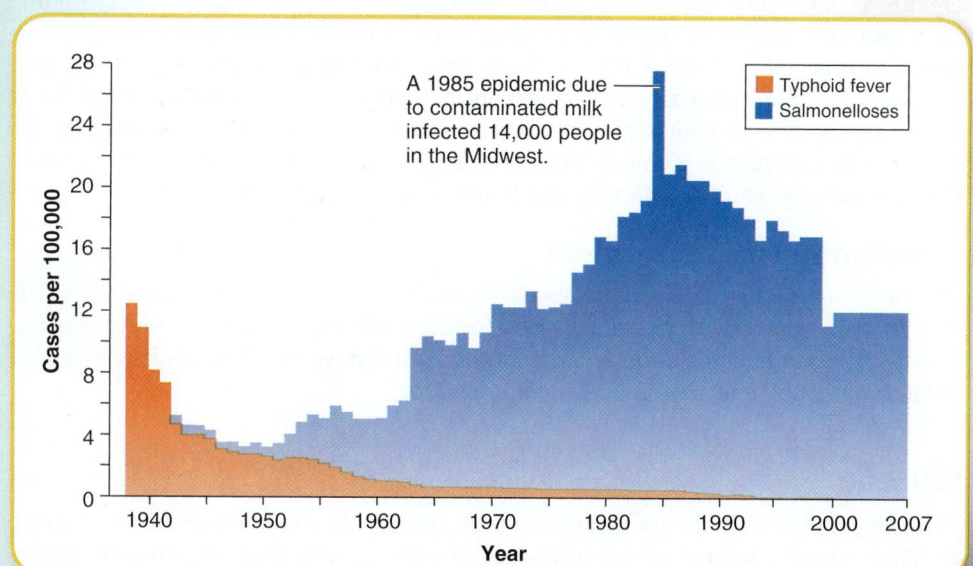

Figure 20.3   Data on the prevalence of typhoid fever and other salmonelloses from 1940 to 2007.   Nontyphoidal salmonelloses did occur before 1940, but the statistics are not available.

Pet reptiles have occasionally been the source of *Salmonella* infections in humans.

Salmonellae are motile; they ferment glucose with acid and sometimes gas; and most of them produce hydrogen sulfide (H_2S) but not urease. They grow readily on most laboratory media and can survive outside the host in inhospitable environments such as freshwater and freezing temperatures. These pathogens are resistant to chemicals such as bile and dyes, which are the basis for isolation on selective media.

► Signs and Symptoms

The genus *Salmonella* causes a variety of illnesses in the GI tract and beyond. Until fairly recently, its most severe manifestation was typhoid fever. Since the mid-1900s, a milder disease usually called salmonellosis has been much more common **(figure 20.3).** Sometimes the condition is also called enteric fever or gastroenteritis. Whereas typhoid fever is caused by the Typhi serotype, gastroenteritises are generally caused by the serotypes known as Typhimurium, Enteritidis, Heidelberg, Newport, and Javiana. Most of these strains come from animals, unlike the Typhi serotype, which infects humans exclusively. *Salmonella* bacteria are normal intestinal biota in cattle, poultry, rodents, and reptiles and each has been a documented source of infection and disease in humans.

Salmonellosis can be relatively severe, with an elevated body temperature and septicemia as more prominent features than GI tract disturbance. But it can also be fairly mild, with gastroenteritis—vomiting, diarrhea, and mucosal irritation—as its major feature. Blood can appear in the stool. In otherwise healthy adults, symptoms spontaneously subside after 2 to 5 days; death is infrequent except in debilitated persons.

▶ Pathogenesis and Virulence Factors

Salmonella serotypes vary in their virulence due to genetic differences in their ability to adhere to the gut mucosa, and also to evade the immune system. Endotoxin, a lipopolysaccharide component of the outer membrane of gram-negative bacteria, is an important virulence factor for *Salmonella*.

▶ Transmission and Epidemiology

An important factor to consider in all diarrheal pathogens is how many organisms must be ingested to cause disease (their ID_{50}). It varies widely. These values are listed in a special row in **Disease Table 20.1.** *Salmonella* has a high ID_{50}, meaning a lot of organisms have to be ingested in order for disease to result. Animal products such as meat and milk can be readily contaminated with *Salmonella* during slaughter, collection, and processing. A 2001 U.S. outbreak was traced to green grapes.

Most cases are traceable to a common food source such as milk or eggs. Some cases may be due to poor sanitation. In one outbreak, about 60 people became infected after visiting the Komodo dragon exhibit at the Denver zoo. They picked up the infection by handling the rails and fence of the dragon's cage.

▶ Prevention and Treatment

The only prevention for salmonellosis is avoiding contact with the bacterium. Uncomplicated cases of salmonellosis are treated with fluid and electrolyte replacement; if the patient has underlying immunocompromise or if the disease is severe, trimethoprim-sulfamethoxazole is recommended.

Shigella

The *Shigella* bacteria are gram-negative rods, nonmotile and non-endospore-forming. They do not produce urease or hydrogen sulfide, traits that help in their identification. They are primarily human parasites, though they can infect apes. All produce a similar disease that can vary in intensity. These bacteria resemble some types of pathogenic *E. coli* very closely.

▶ Signs and Symptoms

The symptoms of shigellosis include frequent, watery stools, as well as fever, and often intense abdominal pain. Nausea and vomiting are common. Stools often contain obvious blood, and even more often are found to have occult (not visible to the naked eye) blood. Diarrhea containing blood is also called **dysentery.** Mucus from the GI tract will also be present in the stools.

▶ Pathogenesis and Virulence Factors

Shigellosis is different from many GI tract infections in that *Shigella* invades the villus cells of the large intestine rather than the small intestine. In addition, it is not as invasive as *Salmonella* and does not perforate the intestine or invade the blood. It enters the intestinal mucosa by means of special cells in Peyer's patches. Once in the mucosa, *Shigella* instigates an inflammatory response that causes extensive tissue destruction. The release of endotoxin causes fever.

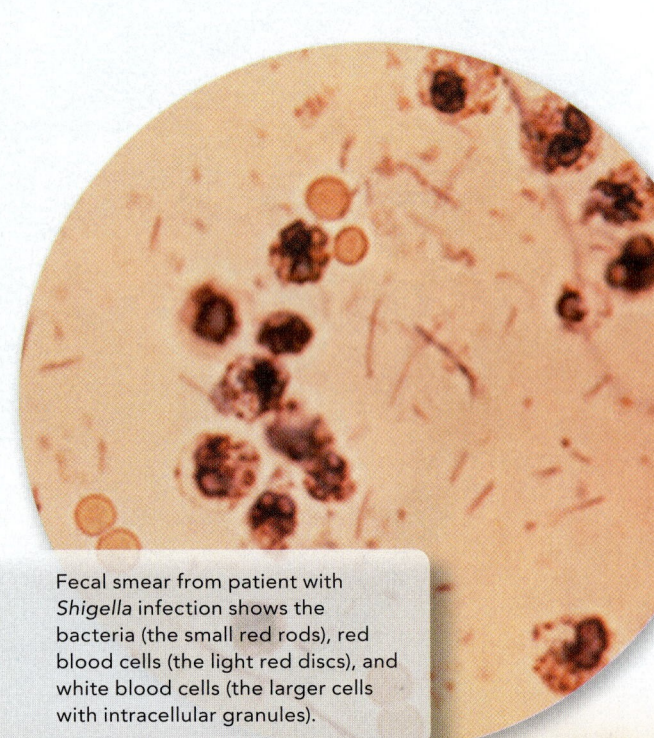

Fecal smear from patient with *Shigella* infection shows the bacteria (the small red rods), red blood cells (the light red discs), and white blood cells (the larger cells with intracellular granules).

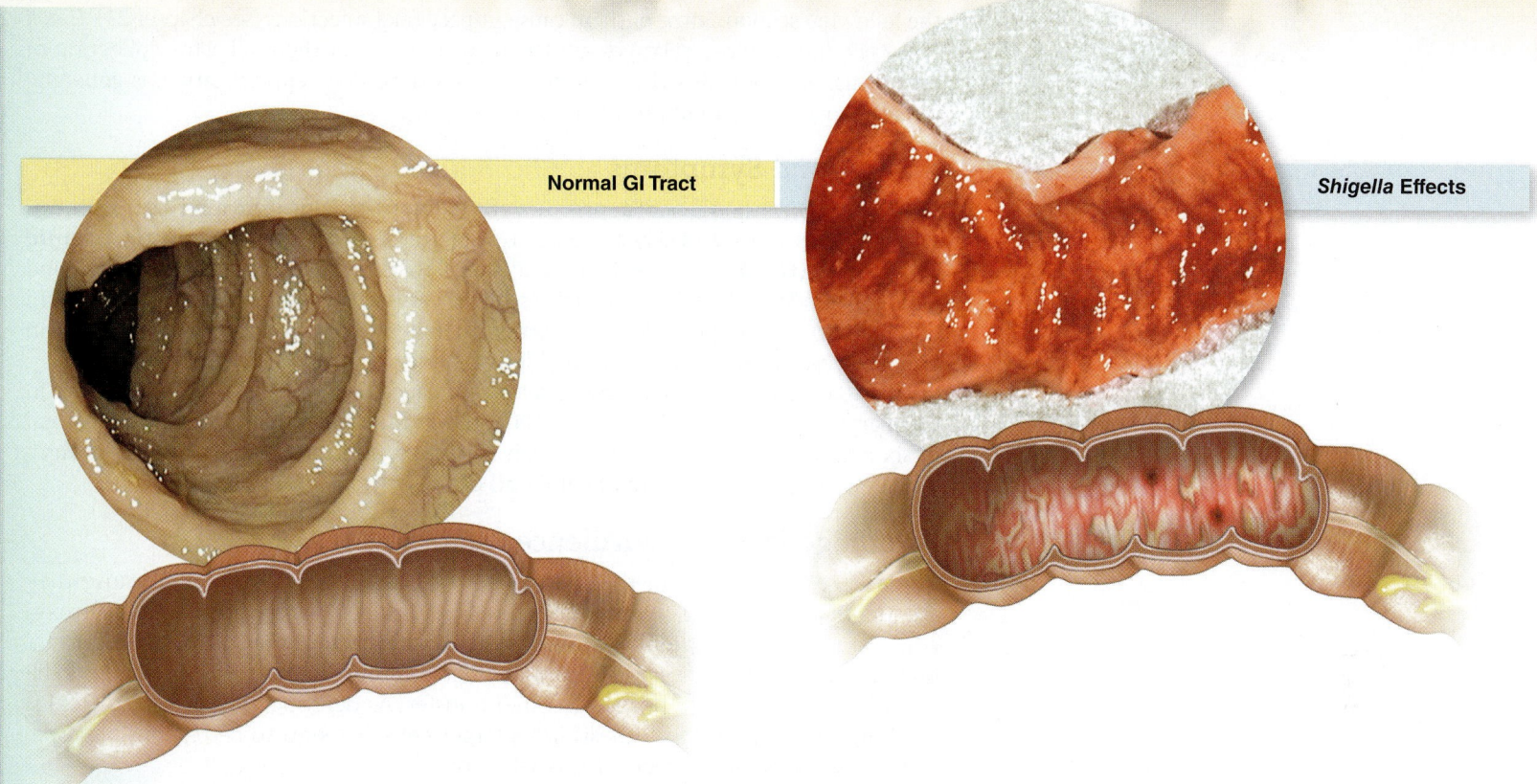

Normal GI Tract

Shigella **Effects**

Figure 20.4 **The appearance of the large intestinal mucosa in *Shigella* dysentery.** Note the patches of blood and mucus, the erosion of the lining, and the absence of perforation.

Enterotoxin, an exotoxin that affects the enteric (or GI) tract, damages the mucosa and villi. Local areas of erosion give rise to bleeding and heavy secretion of mucus **(figure 20.4)**. *Shigella dysenteriae* (and perhaps some of the other species) produces a heat-labile exotoxin called **shiga toxin,** which seems to be responsible for the more serious damage to the intestine as well as any systemic effects, including injury to nerve cells.

▶ Transmission and Epidemiology

In addition to the usual oral route, shigellosis is also acquired through direct person-to-person contact, largely because of the small infectious dose required (from 10 to 200 bacteria). The disease is mostly associated with lax sanitation, malnutrition, and crowding; and it is spread epidemically in day care centers, prisons, mental institutions, nursing homes, and military camps. *Shigella* can establish a chronic carrier condition in some people that lasts several months.

▶ Prevention and Treatment

The only prevention of this and most other diarrheal diseases is good hygiene and avoiding contact with infected persons. Most physicians recommend prompt treatment of shigellosis with ciprofloxacin.

Shiga-Toxin-Producing *E. coli* (STEC)

Dozens of different strains of *E. coli* exist, most of which cause no disease at all. A handful of them cause various degrees of intestinal symptoms, as described in this and the following section. Some of them cause urinary tract infections (see chapter 21). *E. coli* O157:H7 and its close relatives are the most virulent of them all. This collection of organisms, of which this *E. coli* strain is the most famous representative, is generally referred to as shiga-toxin-producing *E. coli* (STEC).

▶ Signs and Symptoms

E. coli O157:H7 is the agent of a spectrum of conditions, ranging from mild gastroenteritis with fever to bloody diarrhea. About 10% of patients develop **hemolytic uremic syndrome (HUS),** a severe hemolytic anemia that can cause kidney damage and failure. Neurological symptoms such as blindness, seizure, and stroke (and long-term debilitation) are also possible. These serious manifestations are most likely to occur in children younger than age 5 and in elderly people.

In 2011, a new HUS-causing *E. coli* strain caused a large and deadly outbreak in Germany. It was named *E. coli* O104:H4 and was identified as an STEC strain. A total of six additional STEC strains have been identified, and the U.S. Department of Agriculture started testing ground beef for all of these strains in 2012.

▶ Pathogenesis and Virulence Factors

These *E. coli* owe much of their virulence to shiga toxins (so named because they are identical to the shiga exotoxin secreted by virulent *Shigella* species). Shiga toxin genes are present on prophage genes donated by bacteriophage in *E. coli* but are on the chromosome of *Shigella dysenteriae*, suggesting that the *E. coli* acquired the virulence factor through phage-mediated transfer. As described for *Shigella*, the shiga toxin interrupts protein synthesis in its target cells. It seems to be responsible especially for the systemic effects of this infection.

Another important virulence determinant for STEC is the ability to efface (rub out or destroy) enterocytes, which are gut epithelial cells.

The net effect is a lesion in the gut (effacement), usually in the large intestine. The microvilli are lost from the gut epithelium, and the lesions produce bloody diarrhea.

▶ Transmission and Epidemiology

The most common mode of transmission for STEC is the ingestion of contaminated and undercooked beef, although other foods and beverages can be contaminated as well.

Any farm product may also become contaminated by cattle feces. Products that are eaten raw, such as lettuce, vegetables, and apples used in unpasteurized cider, are particularly problematic. The disease can also be spread via the fecal-oral route of transmission, especially among young children in group situations. Even touching surfaces contaminated with cattle feces can cause disease, since ingesting as few as 10 organisms has been found to be sufficient to initiate this disease.

▶ Culture and Diagnosis

Infection with this type of *E. coli* should be confirmed with stool culture or with ELISA, PCR, or pulsed-field gel electrophoresis.

▶ Prevention and Treatment

The best prevention for this disease is never to eat raw or even rare hamburger and to wash raw vegetables well. The shiga toxin is heat-labile and the *E. coli* is killed by heat as well.

No vaccine exists for *E. coli* O157:H7 or other STEC strains. However, some countries vaccinate cattle against *E. coli* O157:H7 as a means to protect human populations.

Antibiotics may be contraindicated for this infection, as they may increase the pathology by releasing more toxin, leading to HUS. Supportive therapy, including plasma transfusions to dilute toxin in the blood, is a good option.

Other E. coli

At least five other categories of *E. coli* can cause diarrheal diseases. Scientists call these **enterotoxigenic** *E. coli* (ETEC), **enteroinvasive** *E. coli* (EIEC), **enteropathogenic** *E. coli* (EPEC), diffusely adherent *E. coli* (DAEC), and **enteroaggregative** *E. coli* (EAEC). In clinical practice, most physicians are interested in differentiating shiga-toxin-producing *E. coli* (STEC) from all the others. In Disease Table 20.1, the non-shiga-toxin-producing *E. coli* are grouped together in one column.

Campylobacter

Although you may never have heard of *Campylobacter*, it is considered to be the most common bacterial cause of diarrhea in the United States. It probably causes more diarrhea than *Salmonella* and *Shigella* combined, with 2.4 million cases of diarrhea credited to it per year. Even though it appears on the "food-borne" list for hospitalizations in figure 20.2, just as often it is not serious enough to result in hospitalization or even a doctor's visit.

The symptoms of campylobacteriosis are frequent watery stools, fever, vomiting, headaches, and abdominal pain. The symptoms may last longer than most acute diarrheal episodes, sometimes extending beyond 2 weeks. They may subside and then recur over a period of weeks.

Campylobacter jejuni is the most common cause, although there are other pathogenic *Campylobacter* species. Campylobacters are slender, curved, or spiral gram-negative bacteria propelled by polar flagella at one or both poles, often appearing in S-shaped or gull-winged pairs. These bacteria tend to be microaerophilic inhabitants of the intestinal tract, genitourinary tract, and oral cavity of humans and animals. Transmission of this pathogen takes place via the ingestion of contaminated beverages and food, especially water, milk, meat, and chicken.

Once ingested, *C. jejuni* cells reach the mucosa at the last segment of the small intestine (ileum) near its junction with the colon; they adhere, burrow through the mucus, and multiply. Symptoms commence after an incubation period of 1 to 7 days. The mechanisms of pathology appear to involve a heat-labile enterotoxin that stimulates a secretory diarrhea like that of cholera. In a small number of cases, infection with this bacterium can lead to a serious neuromuscular paralysis called Guillain-Barré syndrome.

Guillain-Barré syndrome (GBS) (pronounced gee"-luhn-buh-ray') is the leading cause of acute paralysis in the United States since the eradication of polio there. The good news is that many patients recover completely from this paralysis. The condition is still mysterious in many ways, but it seems to be an autoimmune reaction that can be brought on by infection with viruses and bacteria, by vaccination in rare cases, and even by surgery. The single most common precipitating event for the onset of GBS is *Campylobacter* infection. Twenty to forty percent of GBS cases are preceded by infection with *Campylobacter*. The reasons for this are not clear. (Note that even though 20% to 40% of GBS cases are preceded by *Campylobacter* infection, only about 1 in 1,000 cases of *Campylobacter* infection results in GBS.)

Resolution of infection occurs in most instances with simple, nonspecific rehydration and electrolyte balance therapy. In more severely affected patients, it may be necessary to administer azithromycin. Antibiotic resistance is increasing in these bacteria, in large part due to the use of fluoroquinolones in the poultry industry. Because vaccines are yet to be developed, prevention depends on rigid sanitary control of water and milk supplies and care in food preparation.

Yersinia Species

Yersinia is a genus of gram-negative bacteria that includes the infamous plague bacterium, *Yersinia pestis* (discussed in chapter 18). There are two species that cause GI tract disease: *Y. enterocolitica* and *Y. pseudotuberculosis*. The infections are most notable for the high degree of abdominal pain they cause. This symptom is accompanied by fever. Often the symptoms are mistaken for appendicitis.

Hand washing is a very effective way to reduce transmission of fecal bacteria.

Clostridium difficile

Clostridium difficile is a gram-positive endospore-forming rod found as normal biota in the intestine. It was once considered relatively harmless but now is known to cause a condition called pseudomembranous colitis, also called antibiotic-associated colitis. In most cases, this infection seems to be precipitated by therapy with broad-spectrum antibiotics such as ampicillin, clindamycin, or cephalosporins. It is a major cause of diarrhea in hospitals, although community-acquired infections have been on the rise in the last few years. Also, new studies suggest that the use of gastric acid inhibitors for the treatment of heartburn can predispose people to this infection. Although *C. difficile* is relatively noninvasive, it is able to superinfect the large intestine when drugs have disrupted the normal biota. It produces two enterotoxins, toxins A and B, that cause areas of necrosis in the wall of the intestine. The predominant symptom is diarrhea commencing late in antibiotic therapy or even after therapy has stopped. More severe cases exhibit abdominal cramps, fever, and leukocytosis. The colon is inflamed and gradually sloughs off loose, membranelike patches called pseudomembranes consisting of fibrin and cells **(figure 20.5).** If the condition is not stopped, perforation of the cecum and death can result.

Nearly 15,000 people in the United States die each year from *C. difficile* infection, and both disease incidence and drug resistance are on the rise. Mild, uncomplicated cases respond to the withdrawal of antibiotics and replacement therapy for lost fluids and electrolytes. More severe infections are treated with metronidazole, vancomycin, or fidaxomicin (Dificid) for several weeks until the intestinal biota returns to normal. Researchers and patients have also experimented with a variety of methods, such as fecal, transplants, to provide the GI tract with a new healthy microbiome.

Vibrio cholerae

Cholera has been a devastating disease for centuries. It is not an exaggeration to say that the disease has shaped a good deal of human history in Asia and Latin America, where it has been endemic. These days we have come to expect outbreaks of cholera to occur after natural disasters, war, or large refugee movements, especially in underdeveloped parts of the world.

These bacteria are rods with a single polar flagellum. They belong to the family *Vibrionaceae*. A freshly isolated specimen of *Vibrio cholerae* contains quick, darting cells that slightly resemble a comma. *Vibrio* shares many characteristics with members of the *Enterobacteriaceae* family. Vibrios are fermentative and grow on ordinary or selective media containing bile at 37°C. They possess unique O and H antigens and membrane receptor antigens that provide some basis for classifying members of the family. There are two major types, called classic and *El Tor*.

▶ Signs and Symptoms

After an incubation period of a few hours to a few days, symptoms begin abruptly with vomiting, followed by copious watery feces called secretory diarrhea. The intestinal contents are lost very quickly, leaving only secreted fluids. This voided fluid contains flecks of mucus—hence, the description "rice-water stool." Fluid losses of nearly 1 liter per hour have been reported in severe cases, and an untreated patient can lose up to 50% of body weight during the course of this disease. The diarrhea causes loss of blood volume, acidosis from bicarbonate loss, and potassium depletion, which manifest in muscle cramps, severe thirst, flaccid skin, sunken eyes, and, in young children, coma and convulsions. Secondary circulatory consequences can include hypotension, tachycardia, cyanosis, and collapse from shock within 18 to 24 hours. If cholera is left untreated, death can occur in less than 48 hours, and the mortality rate is between 55% and 70%.

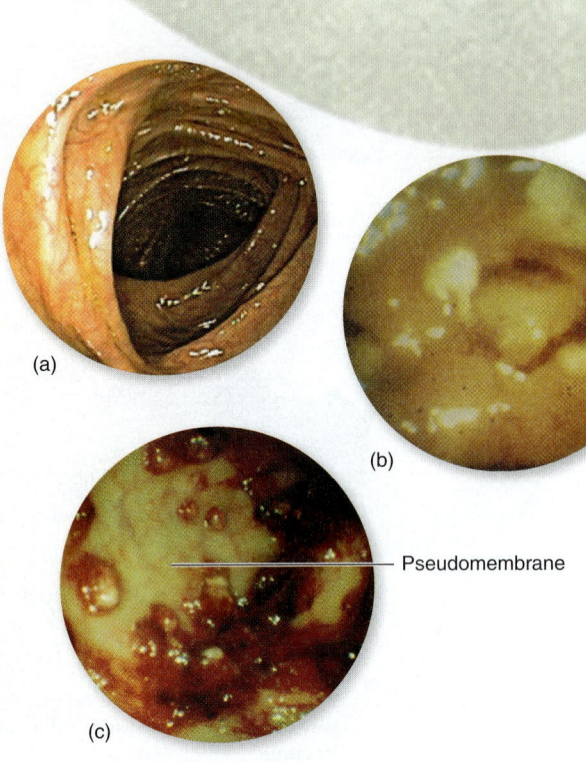

(a)

(b)

Pseudomembrane

(c)

Figure 20.5 **Antibiotic-associated colitis.**
(a) Normal colon. (b) A mild form of colitis with diffuse, inflammatory patches. (c) Heavy yellow plaques, or pseudomembranes, typical of more severe cases. Photographs were made by a sigmoidoscope, an instrument capable of photographing the interior of the colon.

▶ Pathogenesis and Virulence Factors

V. cholerae has a relatively high infectious dose (10^8 cells). At the junction of the duodenum and jejunum, the vibrios penetrate the mucous barrier using their flagella, adhere to the microvilli of the epithelial cells, and multiply there. The bacteria never enter the host cells or invade the mucosa. The virulence of *V. cholerae* is due to an enterotoxin called cholera toxin (CT), which disrupts the normal physiology of intestinal cells. Under the influence of this system, the cells shed large amounts of electrolytes into the intestine, an event accompanied by profuse water loss.

▶ Transmission and Epidemiology

The pattern of cholera transmission and the onset of epidemics are greatly influenced by the season of the year and the climate. Cold, acidic, dry environments inhibit the migration and survival of *Vibrio*, whereas warm, monsoon, alkaline, and saline conditions favor them. The bacteria survive in water sources for long periods of time. Recent outbreaks in several parts of the world have been traced to giant cargo ships that pick up ballast water in one port and empty it in another elsewhere in the world. Cholera ranks among the top seven causes of morbidity and mortality, affecting several million people in endemic regions of Asia and Africa. See some interesting history about this disease in chapter 22.

In nonendemic areas such as the United States, the microbe is spread by water and food contaminated by asymptomatic carriers, but it is relatively uncommon.

▶ Prevention and Treatment

Effective prevention is contingent on proper sewage treatment and water purification. Vaccines are available for travelers and people living in endemic regions. One vaccine contains killed *V. cholerae* but protects for only 6 months or less. An oral vaccine containing live, attenuated bacteria was developed to be a more effective alternative, but evidence suggests it also confers only short-term immunity. It is not routinely used in the United States.

The key to cholera therapy is prompt replacement of water and electrolytes, because their loss accounts for the severe morbidity and mortality. This therapy can be accomplished by various rehydration techniques that replace the lost fluid and electrolytes. One of these, oral rehydration therapy (ORT), is incredibly simple and astonishingly effective. Until the 1970s, the treatment, if one could access it, was rehydration through an IV drip. This treatment usually required traveling to the nearest clinic, often miles or days away. Most affected children received no treatment at all, and 3 million of them died every year. Then scientists tested a simple sugar-salt solution that patients could drink.

The relatively simple solution, developed by the World Health Organization (WHO), consists of a mixture of the electrolytes sodium chloride, sodium bicarbonate, potassium chloride, and glucose or sucrose dissolved in water. When administered early in amounts ranging from 100 to 400 milliliters per hour, the solution can restore patients in 4 hours, often bringing them literally back from the brink of death. Infants and small children, who once would have died, now survive so often that the mortality rate for treated cases of cholera is near zero. This therapy has several advantages, especially for countries with few resources. It does not require medical facilities, high-technology equipment, or complex medication protocols. It also eliminates the need for clean needles, which is a pressing issue in many parts of the world.

Oral antibiotics, such as doxycycline, are given orally along with rehydration. These can shorten periods of both diarrhea and bacterial excretion.

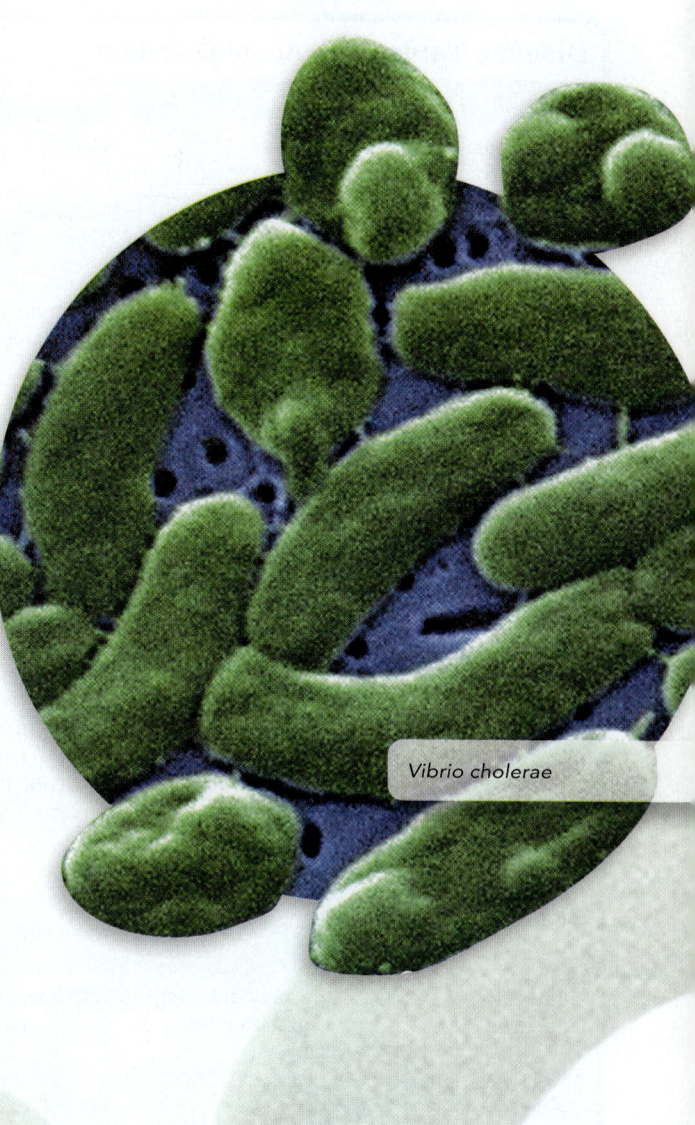

Vibrio cholerae

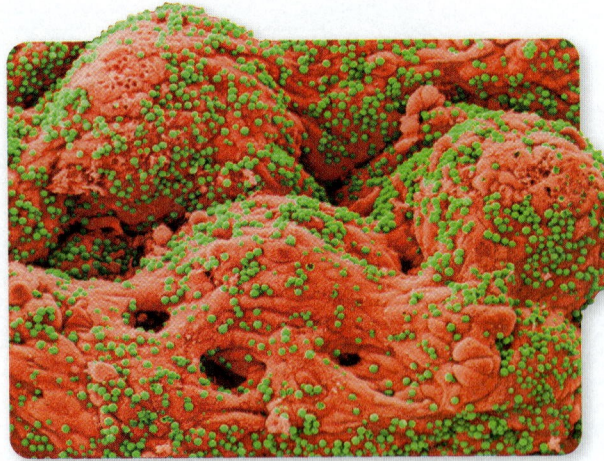

Figure 20.6 Scanning electron micrograph of *Cryptosporidium* (green) attached to the intestinal epithelium.

Cryptosporidium

Cryptosporidium is an intestinal protozoan of the apicomplexan type (see chapter 4) that infects a variety of mammals, birds, and reptiles. For many years, cryptosporidiosis was considered an intestinal ailment exclusive to calves, pigs, chickens, and other poultry, but it is clearly a zoonosis as well. The organism's life cycle includes a hardy intestinal oocyst as well as a tissue phase. Humans accidentally ingest the oocysts with water or food that has been contaminated by feces from infected animals. The oocyst "excysts" once it reaches the intestines and releases sporozoites that attach to the epithelium of the small intestine **(figure 20.6)**. The organism penetrates the intestinal cells and lives intracellularly in them. It undergoes asexual and sexual reproduction in these cells, produces more oocysts, which are released into the gut lumen, excreted from the host, and after a short time become infective again. The oocysts are highly infectious and extremely resistant to treatment with chlorine and other disinfectants.

Disease Table 20.1 Acute Diarrhea

	Bacterial Causes				
Causative Organism(s)	*Salmonella*	*Shigella*	Shiga-toxin-producing *E. coli* (STEC)	Other *E. coli* (non-shiga-toxin producing)	*Campylobacter*
Most Common Modes of Transmission	Vehicle (food, beverage), fecal-oral	Fecal-oral, direct contact	Vehicle (food, beverage), fecal-oral	Vehicle, fecal-oral	Vehicle (food, water), fecal-oral
Virulence Factors	Adhesins, endotoxin	Endotoxin, enterotoxin, shiga toxins in some strains	Shiga toxins; proteins for attachment, secretion, effacement	Various: proteins for attachment, secretion, effacement; heat-labile and/or heat-stable exotoxins; invasiveness	Adhesins, exotoxin, induction of autoimmunity
Culture/ Diagnosis	Stool culture, not usually necessary	Stool culture; antigen testing for shiga toxin	Stool culture, antigen testing for shiga toxin	Stool culture not usually necessary in absence of blood, fever	Stool culture not usually necessary; dark-field microscopy
Prevention	Food hygiene and personal hygiene	Food hygiene and personal hygiene	Avoid live *E. coli* (cook meat and clean vegetables)	Food and personal hygiene	Food and personal hygiene
Treatment	Rehydration; no antibiotic for uncomplicated disease	Ciprofloxacin in severe cases, rehydration; in **Serious Threat** category in CDC Antibiotic Resistance Report	Antibiotics contraindicated, supportive measures	Rehydration, antimotility agent	Rehydration, azithromycin in severe cases (antibiotic resistance rising); in **Serious Threat** category in CDC Antibiotic Resistance Report
Fever Present?	Usually	Often	Often	Sometimes	Usually
Blood in Stool?	Sometimes	Often	Usually	Sometimes	No
Distinctive Features	Often associated with chickens, reptiles	Very low ID_{50}	Hemolytic uremic syndrome	EIEC, ETEC, EPEC	Guillain-Barré syndrome
Epidemiological Features	United States: 20% of all cases require hospitalization; death rate of 0.6%	United States: estimated 450,000 cases per year; internationally: 165 million cases per year	Internationally: causes HUS in 10% of patients; 25% of HUS patients suffer neurological complications; 50% have chronic renal sequelae	–	United States: 2.4 million cases per year; internationally: 400 million cases per year

The prominent symptoms mimic other types of gastroenteritis, with headache, sweating, vomiting, severe abdominal cramps, and diarrhea. AIDS patients may experience chronic persistent cryptosporidial diarrhea that can be used as a criterion to help diagnose AIDS. The agent can be detected in fecal samples or in biopsies (**figure 20.7**) using ELISA or acid-fast staining. Stool cultures should be performed to rule out other (bacterial) causes of infection.

Half of the outbreaks of diarrhea associated with swimming pools are caused by *Cryptosporidium*. Because chlorination is not entirely successful in eradicating the cysts, most treatment plants use filtration to remove them, but even this method can fail.

Treatment is not usually required for otherwise healthy patients. Antidiarrheal agents (antimotility drugs) may be used. Although no curative antimicrobial agent exists for *Cryptosporidium*, physicians will often try nitazoxanide, which can be effective against protozoa in immunocompetent patients.

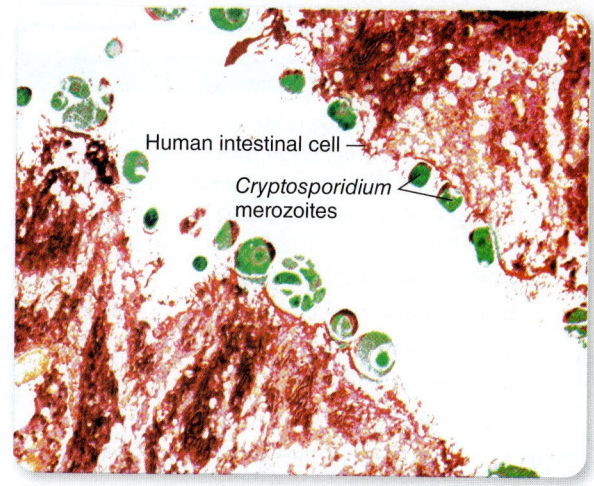

Human intestinal cell

Cryptosporidium merozoites

Figure 20.7 A micrograph of a *Cryptosporidium* merozoite that has penetrated the intestinal mucosa.

			Nonbacterial Causes		
Yersinia	*Clostridium difficile*	*Vibrio cholerae*	*Cryptosporidium*	Rotavirus	Norovirus
Vehicle (food, water), fecal-oral, indirect contact	Endogenous (normal biota)	Vehicle (water and some foods), fecal-oral	Vehicle (water, food), fecal-oral	Fecal-oral, vehicle, fomite	Fecal-oral, vehicle
Intracellular growth	Enterotoxins A and B	Cholera toxin (CT)	Intracellular growth	–	–
Cold-enrichment stool culture	Stool culture, PCR, ELISA demonstration of toxins in stool	Clinical diagnosis, microscopic techniques, serological detection of antitoxin	Fluorescence microscopy	Usually not performed	Rapid antigen test
Food and personal hygiene	–	Water hygiene	Water treatment, proper food handling	Oral live virus vaccine	Hygiene
None in most cases; doxycycline, gentamicin, or TMP-SMZ for bacteremia	Withdrawal of antibiotic; in severe cases metronidazole or fidaxomicin (Dificid)	Rehydration; in severe cases, doxycycline	None; nitazoxanide used sometimes	Rehydration	Rehydration
Usually	Sometimes	No	Often	Often	Sometimes
Occasionally	Not usually; mucus prominent	No	Not usually	No	No
Severe abdominal pain	Antibiotic-associated diarrhea	Rice-water stools	Resistant to chlorine disinfection	Severe in babies	–
Uncommon; more likely in children than adults and in winter than in other seasons	United States: 3 million cases per year	Global estimate: 100,000–130,000 deaths annually	United States: estimated 748,000 cases per year; 30% seropositive	United States: 2–3 million cases per year; internationally: 125 million cases of infantile diarrhea annually	United States: most common cause of diarrhea in <18-year-olds

NCLEX® PREP

2. A nurse is preparing to assess a client for whom a report has been given noting 6% dehydration. Which signs/symptoms would the nurse anticipate as being present? *Select all that apply.*

  a. *increased respiratory rate*

  b. *hypotension*

  c. *headache*

  d. *extreme thirst*

  e. *sustained tachycardia*

Rotavirus

Rotavirus is a member of the *Reovirus* group, which consists of an unusual double-stranded RNA genome with both an inner and an outer capsid. Globally, rotavirus is the primary viral cause of morbidity and mortality resulting from diarrhea, accounting for nearly 50% of all cases. It is estimated that there are 1 million cases of rotavirus infection in the United States every year, leading to 70,000 hospitalizations. Peak occurrences of this infection are seasonal–in the U.S. Southwest, the peak is often in the late fall, and in the Northeast, the peak comes in the spring. The virus gets its name from its physical appearance, which is said to resemble a spoked wheel.

The virus is transmitted by the fecal-oral route, including through contaminated food, water, and fomites. For this reason, disease is most prevalent in areas of the world with poor sanitation. In the United States, rotavirus infection is relatively common, but its course is generally mild.

The effects of infection vary with the age, nutritional state, general health, and living conditions of the patient. Babies from 6 to 24 months of age lacking maternal antibodies have the greatest risk for fatal disease. These children present symptoms of watery diarrhea, fever, vomiting, dehydration, and shock. The intestinal mucosa can be damaged in a way that chronically compromises nutrition, and long-term or repeated infections can retard growth. Newborns seem to be protected by maternal antibodies. Adults can also acquire this infection, but it is generally mild and self-limiting.

Children are treated with oral replacement fluid and electrolytes. An oral live virus vaccine has been available since 2006, and hospital admissions have declined by nearly 90% since then.

Norovirus

A bewildering array of viruses can cause gastroenteritis, including adenoviruses, astroviruses, and noroviruses. As you see in figure 20.2, noroviruses are the second most common cause of hospitalizations from food-borne diseases in the United States.

Transmission is fecal-oral or via contamination of food and water. Viruses generally cause a profuse, watery diarrhea of 3 to 5 days' duration. Vomiting may accompany the disease, especially in the early phases. Mild fever is often seen.

Treatment of these infections always focuses on rehydration.

Food Poisoning

If a patient presents with severe nausea and frequent vomiting accompanied by diarrhea, and reports that companions with whom he or she shared a recent meal (within the last 1 to 6 hours) are suffering the same fate, food poisoning should be suspected. **Food poisoning** refers to symptoms in the gut that are caused by a preformed toxin of some sort. In many cases, the toxin comes from *Staphylococcus aureus*. In others, the source of the toxin is *Bacillus cereus* or *Clostridium perfringens*. The toxin occasionally comes from nonmicrobial sources such as fish, shellfish, or mushrooms. In any case, if the symptoms are violent and the incubation period is very short, this condition, which is an *intoxication* (the effects of a toxin) rather than an *infection*, should be considered.

Staphylococcus aureus Exotoxin

This illness is associated with eating foods such as custards, sauces, cream pastries, processed meats, chicken salad, or ham that have been contaminated by handling and then left unrefrigerated for a few hours. Because of the high salt tolerance of *S. aureus*, even foods containing salt as a preservative are not exempt. The toxins produced by the multiplying bacteria do not noticeably alter the food's taste or smell. The exotoxin (which is an enterotoxin) is heat-stable; inactivation requires

100°C for at least 30 minutes. Thus, heating the food after toxin production may not prevent disease. The ingested toxin acts upon the gastrointestinal epithelium and stimulates nerves, with acute symptoms of cramping, nausea, vomiting, and diarrhea. Recovery is also rapid, usually within 24 hours. Most often, the *S. aureus* comes from a food handler's skin or nose.

This condition is almost always self-limiting, and antibiotics are definitely not warranted.

Bacillus cereus Exotoxin

Bacillus cereus is a sporulating gram-positive bacterium that is naturally present in soil. As a result, it is a common resident on vegetables and other products in close contact with soil. It produces two exotoxins, one of which causes a diarrheal-type disease, the other of which causes an **emetic** (ee-met'-ik) or vomiting disease. The type of disease that takes place is influenced by the type of food that is contaminated by the bacterium. The emetic form is most frequently linked to fried rice, especially when it has been cooked and kept warm for long periods of time. These conditions are apparently ideal for the expression of the low-molecular-weight, heat-stable exotoxin having an emetic effect. The diarrheal form of the disease is usually associated with cooked meats or vegetables that are held at a warm temperature for long periods of time. These conditions apparently favor the production of the high-molecular-weight, heat-labile exotoxin. The symptom in these cases is a watery, profuse diarrhea that lasts only for about 24 hours.

In both cases, the only prevention is the proper handling of food.

Clostridium perfringens Exotoxin

Another sporulating gram-positive bacterium that causes intestinal symptoms is *Clostridium perfringens*. Endospores from *C. perfringens* can contaminate many kinds of foods. Those most frequently implicated in disease are animal flesh (meat, fish) and vegetables such as beans that have not been cooked thoroughly enough to destroy endospores. When these foods are cooled, endospores germinate, and the germinated cells multiply, especially if the food is left unrefrigerated. If the food is eaten without adequate reheating, live *C. perfringens* cells enter the small intestine and release exotoxin. The toxin, acting upon epithelial cells, initiates acute abdominal pain, diarrhea, and nausea in 8 to 16 hours. Recovery is rapid, and deaths are extremely rare.

Disease Table 20.2 Food Poisoning

Causative Organism(s)	*Staphylococcus aureus* exotoxin	*Bacillus cereus*	*Clostridium perfringens*
Most Common Modes of Transmission	Vehicle (food)	Vehicle (food)	Vehicle (food)
Virulence Factors	Heat-stable exotoxin	Heat-stable toxin, heat-labile toxin	Heat-labile toxin
Culture/Diagnosis	Usually based on epidemiological evidence	Microscopic analysis of food or stool	Detection of toxin in stool
Prevention	Proper food handling	Proper food handling	Proper food handling
Treatment	Supportive	Supportive	Supportive
Fever Present	Not usually	Not usually	Not usually
Blood in Stool	No	No	No
Distinctive Features	Suspect in foods with high salt or sugar content	Two forms: emetic and diarrheal	Acute abdominal pain
Epidemiological Features	United States: estimated 240,000 cases per year	United States: estimated 63,000 cases per year	United States: estimated 966,000 cases per year

C. perfringens also causes an enterocolitis infection similar to that caused by *C. difficile*. This infectious type of diarrhea is acquired from contaminated food, or it may be transmissible by inanimate objects.

Chronic Diarrhea

Chronic diarrhea is defined as lasting longer than 14 days. It can have infectious causes or can reflect noninfectious conditions. Most of us are familiar with diseases that present a constellation of bowel syndromes, such as irritable bowel syndrome and ulcerative colitis, neither of which is directly caused by a microorganism as far as we know. Increasing evidence suggests that a chronically disrupted intestinal biota (from long-term use of antibiotics, for example) can predispose people to these conditions.

People suffering from AIDS almost universally suffer from chronic diarrhea. Most of the patients who are not taking antiretroviral drugs have diarrhea caused by a variety of opportunistic microorganisms, including *Cryptosporidium*, *Mycobacterium avium*, and so forth. A patient's HIV status should be considered if he or she presents with chronic diarrhea.

Here we examine a few of the microbes that can be responsible for chronic diarrhea in otherwise healthy people.

Enteroaggregative E. coli (EAEC)

In the section on acute diarrhea, you read about the various categories of *E. coli* that can cause disease in the gut. One type, the enteroaggregative *E. coli* (EAEC), is particularly associated with chronic disease, especially in children. This bacterium was first recognized in 1987 and can be difficult to diagnose in a clinical lab. It is distinguished by its ability to adhere to human cells in aggregates rather than as single cells **(figure 20.8)**. Its presence appears to stimulate secretion of large amounts of mucus in the gut, which may be part of its role in causing chronic diarrhea.

Cyclospora

Cyclospora cayetanensis is an emerging protozoal pathogen. Since the first occurrence in 1979, hundreds of outbreaks have been reported in the United States and Canada. Its mode of transmission is fecal-oral, and most cases have been associated with consumption of fresh produce and water presumably contaminated with feces. This disease occurs worldwide, and although primarily of human origin, it is not spread directly from person to person. Outbreaks have been traced to imported raspberries, salad made with fresh greens, and drinking water. A major outbreak of this organism occurred on a cruise ship in April of 2009, wherein 135 of 1,318 passengers, and 25 crew members, became ill with *Cyclospora*.

Giardia

Giardia lamblia (also known as *Giardia intestinalis*) is a pathogenic flagellated protozoan first observed by Antonie van Leeuwenhoek in his own feces. For 200 years, it was considered a harmless or weak intestinal pathogen; and only since the 1950s has its importance as a cause of diarrhea been recognized. In fact, it is the most common flagellate isolated in clinical specimens. Observed straight on, the trophozoite has a unique symmetrical heart shape with organelles positioned in such a way that it resembles a face **(figure 20.9)**. Four pairs of flagella emerge from the ventral surface, which is concave and acts like a suction cup for attachment to a substrate. *Giardia* cysts are small and compact and contain four nuclei.

Typical symptoms include diarrhea of long duration, abdominal pain, and flatulence. Stools have a greasy, foul-smelling quality to them. Fever is usually not present.

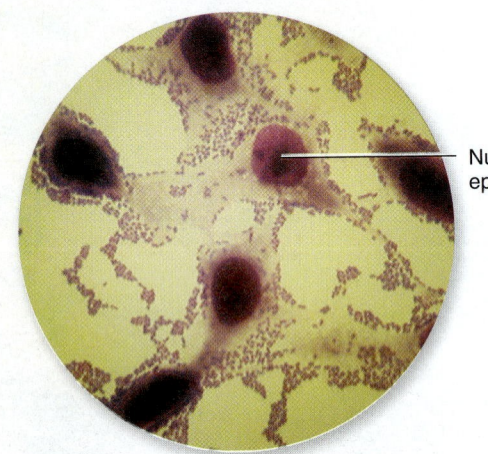

Nucleus of epithelial cell

Figure 20.8 **Enteroaggregative E. coli adhering to epithelial cells.**

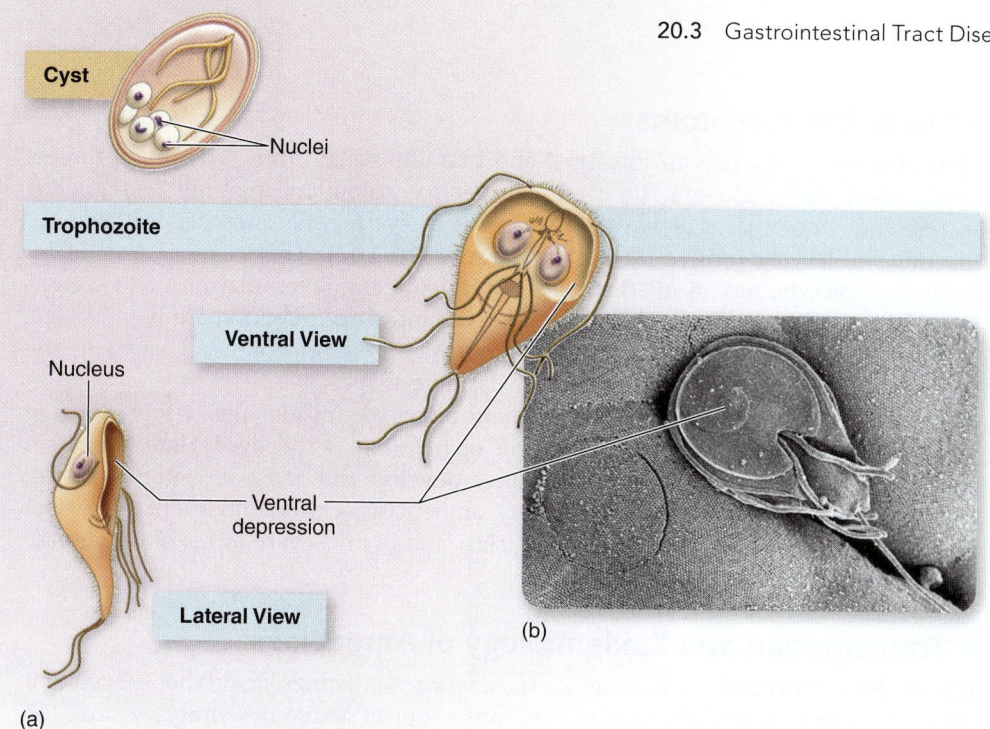

(a)

(b)

Figure 20.9 *Giardia lamblia* **trophozoite.**
(a) Schematic drawing. (b) Scanning electron micrograph of intestinal surface, revealing (on the left) the lesion left behind by adhesive disc of a *Giardia* that has detached. The trophozoite on the right is lying on its "back" and is revealing its adhesive disc.

▶ Transmission and Epidemiology of Giardiasis

Giardiasis has a complex epidemiological pattern. The protozoan has been isolated from the intestines of beavers, cattle, coyotes, cats, and human carriers, but the precise reservoir is unclear at this time. Although both trophozoites and cysts escape in the stool, the cysts play a greater role in transmission. Unlike other pathogenic flagellates, *Giardia* cysts can survive for 2 months in the environment. Cysts are usually ingested with water and food or swallowed after close contact with infected people or contaminated objects. Infection can occur with a dose of only 10 to 100 cysts.

▶ Prevention and Treatment

There is a vaccine against *Giardia* that can be given to animals, including dogs. No human vaccine is available. Avoiding drinking from freshwater sources is the major preventive measure that can be taken.

Treatment is with tinidazole or nitazoxanide.

Entamoeba

Amoebas are widely distributed in aqueous habitats and are frequent parasites of animals, but only a small number of them have the necessary virulence to invade tissues and cause serious pathology. One of the most significant pathogenic amoebas is *Entamoeba histolytica* (en"-tah-mee'-bah his"-toh-lit'-ih-kuh). The relatively simple life cycle of this parasite alternates between a large trophozoite that is motile by means of pseudopods and a smaller, compact, nonmotile cyst **(figure 20.10a–c).** The trophozoite lacks most of the organelles of other eukaryotes, and it has a large single nucleus that contains a prominent nucleolus called a *karyosome*. The mature cyst is encased in a thin yet tough wall and contains four nuclei as well as distinctive cigar-shaped bodies called *chromatoidal bodies,* which are actually dense clusters of ribosomes.

▶ Signs and Symptoms

Clinical amoebiasis exists in intestinal and extraintestinal forms. The initial targets of intestinal amoebiasis are the cecum, appendix, colon, and rectum. The amoeba secretes enzymes that dissolve tissues, and it actively penetrates deeper layers of the mucosa, leaving erosive ulcerations **(figure 20.10d).** This phase is marked by dysentery (bloody, mucus-filled stools), abdominal pain, fever, diarrhea, and weight loss. The most life-threatening manifestations of intestinal infection are hemorrhage, perforation, appendicitis, and tumorlike growths called amoebomas. Lesions in the mucosa of the colon have a characteristic flasklike shape.

Extraintestinal infection occurs when amoebas invade the viscera of the peritoneal cavity. The most common site of invasion is the liver. Here, abscesses containing necrotic tissue and trophozoites develop and cause amoebic hepatitis. Another rarer complication is pulmonary amoebiasis. Other infrequent targets of infection are the spleen, adrenals, kidney, skin, and brain. Severe forms of the disease result in about a 10% fatality rate.

▶ Transmission and Epidemiology of Amoebiasis

Entamoeba is harbored by chronic carriers whose intestines favor the encystment stage of the life cycle. Cyst formation cannot occur in active dysentery because the feces are so rapidly flushed from the body; but after recuperation, cysts are continuously shed in feces.

Infection is usually acquired by ingesting food or drink contaminated with cysts released by an asymptomatic carrier. The amoeba is thought to be carried in the intestines of one-tenth of the world's population, and it kills up to 100,000 people a year. Occurrence is highest in tropical regions (Africa, Asia, and Latin America), where "night soil" (human excrement) or untreated sewage is used to fertilize crops, and sanitation of water and food can be substandard. Although the prevalence of the disease is lower in the United States, as many as 10 million people could harbor the agent.

▶ Prevention and Treatment

Prevention of the disease relies on purification of water because no vaccine currently exists. Because regular chlorination of water supplies does not kill cysts, more rigorous methods such as boiling or iodine are required.

Effective treatment usually involves the use of metronidazole (Flagyl) or chloroquine. Dehydroemetine is

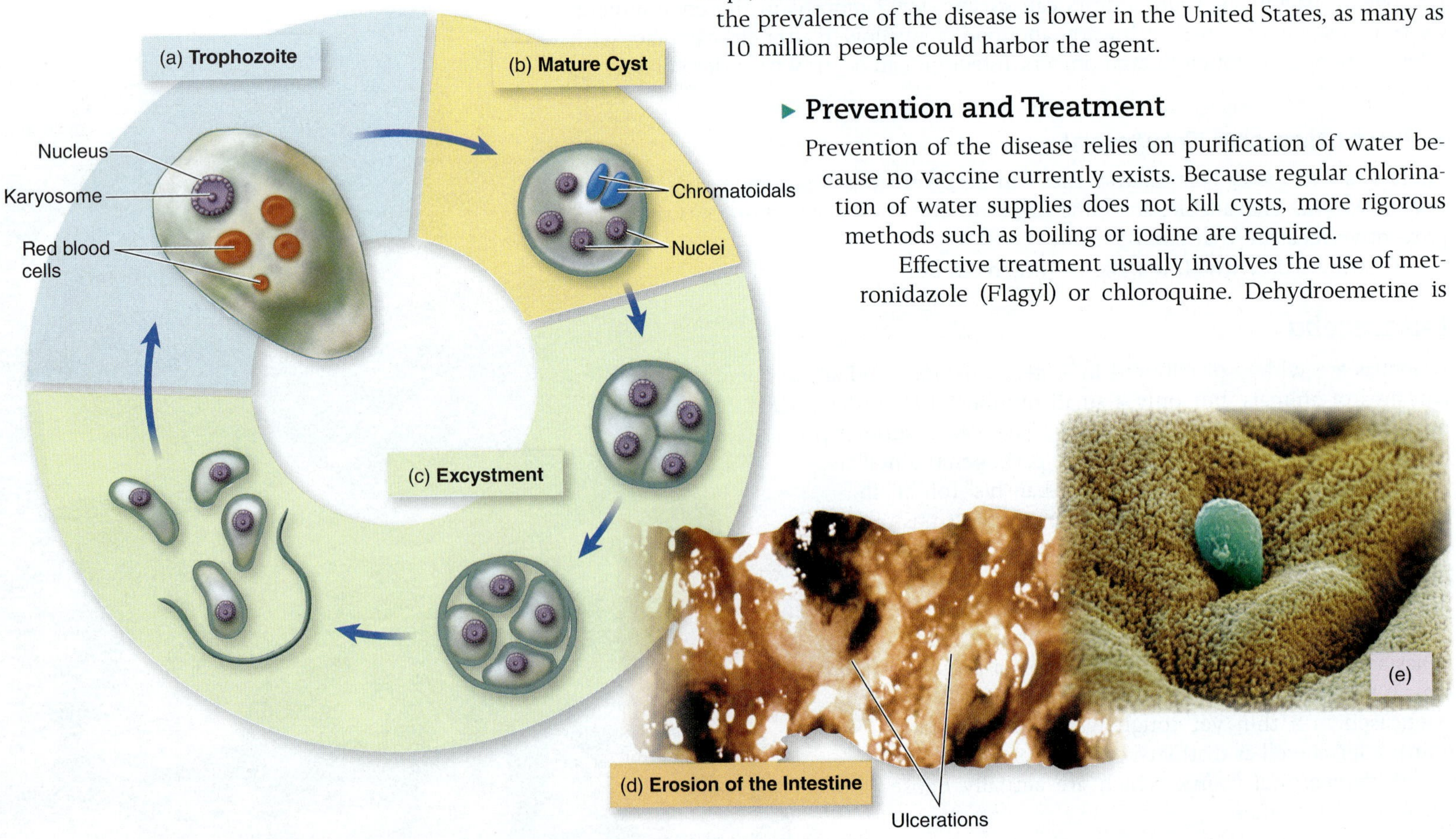

Figure 20.10 *Entamoeba histolytica.*
(a) A trophozoite containing a single nucleus, a karyosome, and red blood cells. (b) A mature cyst with four nuclei and two blocky chromatoidals. (c) Stages in excystment. Divisions in the cyst create four separate cells, or metacysts, that differentiate into trophozoites and are released. (d) Intestinal amoebiasis and dysentery of the cecum. Red patches are sites of amoebic damage to the intestinal mucosa. (e) Trophozoite of *Entamoeba histolytica.* Note the fringe of very fine pseudopods it uses to invade and feed on tissue.

(a) **Trophozoite**

Nucleus
Karyosome
Red blood cells

(b) **Mature Cyst**

Chromatoidals
Nuclei

(c) **Excystment**

(d) **Erosion of the Intestine**

Ulcerations

(e)

Disease Table 20.3 Chronic Diarrhea

Causative Organism(s)	Enteroaggregative E. coli (EAEC)	Cyclospora cayetanensis	Giardia lamblia	Entamoeba histolytica
Most Common Modes of Transmission	Vehicle (food, water), fecal-oral	Fecal-oral, vehicle	Vehicle, fecal-oral, direct and indirect contact	Vehicle, fecal-oral
Virulence Factors	?	Invasiveness	Attachment to intestines alters mucosa	Lytic enzymes, induction of apoptosis, invasiveness
Culture/Diagnosis	Difficult to distinguish from other E. coli	Stool examination, PCR	Stool examination, ELISA	Stool examination, ELISA, serologya
Prevention	?	Washing, cooking food, personal hygiene	Water hygiene, personal hygiene	Water hygiene, personal hygiene
Treatment	None, or ciprofloxacin	TMP-SMZ	Tinidazole, nitazoxanide	Metronidazole or chloroquine
Fever Present	No	Usually	Not usually	Yes
Blood in Stool	Sometimes, mucus also	No	No, mucus present (greasy and malodorous)	Yes
Distinctive Features	Chronic in the malnourished	–	Frequently occurs in backpackers, campers	–
Epidemiological Features	Developing countries: 87% of chronic diarrhea in children >2 years old	United States: estimated 16,000 cases per year; internationally: endemic in 27 countries, mostly tropical	United States: estimated 1.2 million cases per year; internationally: prevalence rates from 2% to 5% in industrialized world; 40-50 million cases per year	Internationally: 40,000–100,000 deaths annually

used to control symptoms, but it will not cure the disease. Other drugs are given to relieve diarrhea and cramps, while lost fluid and electrolytes are replaced by oral or intravenous therapy. Infection with *E. histolytica* provokes antibody formation against several antigens, but permanent immunity is unlikely and reinfection can occur.

Tooth and Gum Infections

It is difficult to pinpoint exactly when the "normal biota biofilm" described for the oral environment becomes a "pathogenic biofilm." If left undisturbed, the biofilm structure eventually contains anaerobic bacteria that can damage the soft tissues and bones (referred to as the periodontium) surrounding the teeth. Also, the introduction of carbohydrates to the oral cavity can result in breakdown of hard tooth structure (the dentition) due to the production of acid by certain oral streptococci in the biofilm. These two separate circumstances are discussed here.

Dental Caries (Tooth Decay)

Dental caries is the most common infectious disease of human beings. The process involves the dissolution of solid tooth surface due to the metabolic action of bacteria. **(Figure 20.11** depicts the structure of a tooth.) The symptoms are often not noticeable but range from minor disruption in the outer (enamel) surface of the tooth to complete destruction of the enamel and then destruction of deeper

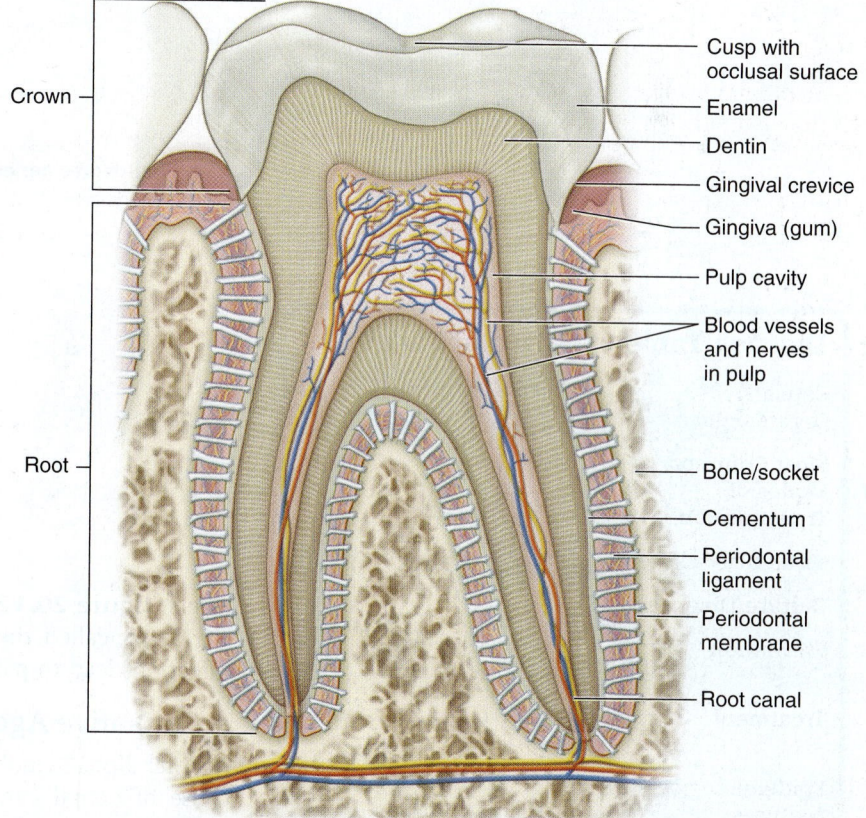

Crown

Root

Cusp with occlusal surface
Enamel
Dentin
Gingival crevice
Gingiva (gum)
Pulp cavity
Blood vessels and nerves in pulp
Bone/socket
Cementum
Periodontal ligament
Periodontal membrane
Root canal

Figure 20.11 The anatomy of a tooth.

Figure 20.12 Stages in plaque development and cariogenesis. (a) A microscopic view of pellicle and plaque formation, acidification, and destruction of tooth enamel. (b) Progress and degrees of cariogenesis.

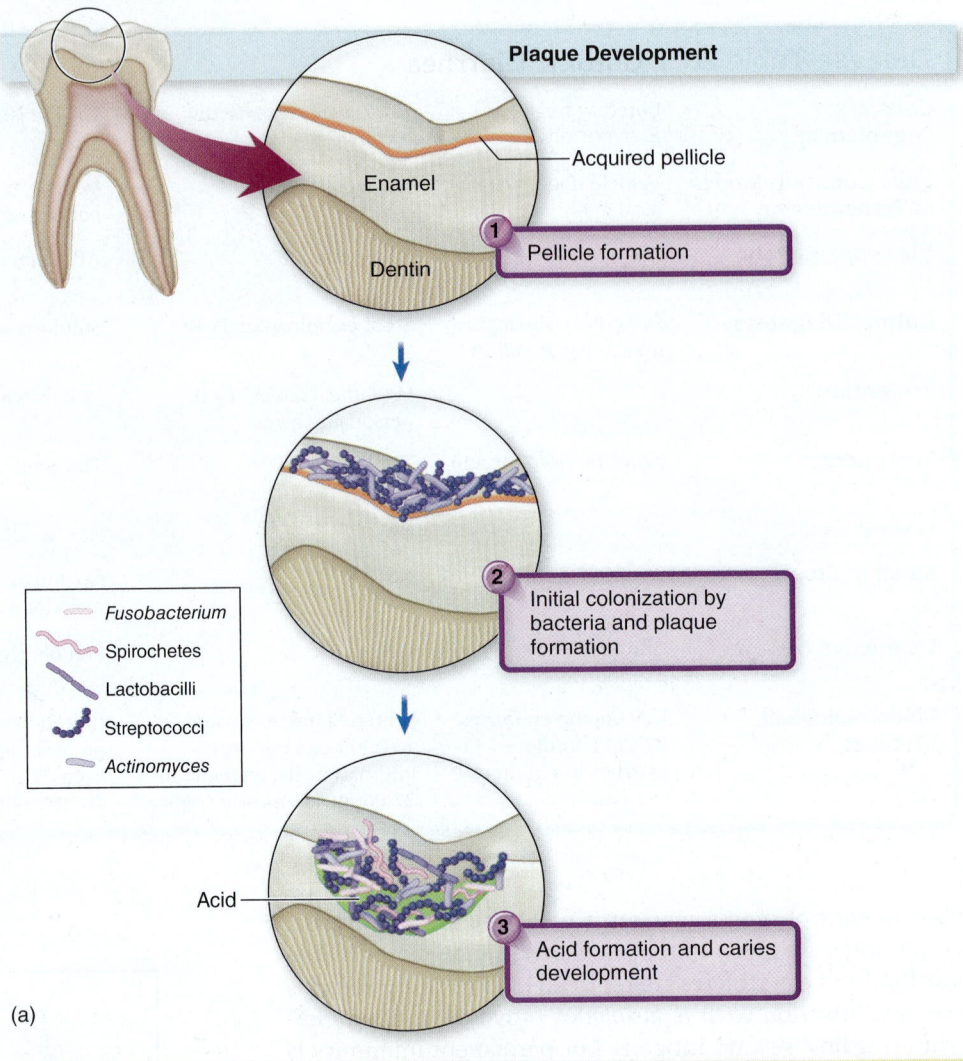

Plaque Development

Acquired pellicle

Enamel

Dentin

1 Pellicle formation

Fusobacterium
Spirochetes
Lactobacilli
Streptococci
Actinomyces

2 Initial colonization by bacteria and plaque formation

Acid

3 Acid formation and caries development

(a)

Cariogenesis

First-degree caries	Second-degree caries	Third-degree caries

Enamel affected

Dentin penetrated

Exposure of pulp

(b)

Disease Table 20.4 Dental Caries

Causative Organism(s)	*Streptococcus mutans, Streptococcus sobrinus,* others
Most Common Modes of Transmission	Direct contact
Virulence Factors	Adhesion, acid production
Culture/Diagnosis	–
Prevention	Oral hygiene, fluoride supplementation
Treatment	Removal of diseased tooth material
Epidemiological Features	Globally, 60%–90% prevalence in school-age children

layers **(figure 20.12)**. Deeper lesions can result in infection to the soft tissue inside the tooth, called the pulp, which contains blood vessels and nerves. These deeper infections lead to pain, referred to as a "toothache."

▶ Causative Agent

An oral alpha-hemolytic streptococcus, *Streptococcus mutans*, seems to be the main cause of dental caries, although a mixed species consortium, consisting of other *Streptococcus* species and some lactobacilli, is probably the best route to caries. A specific condition called early childhood caries may also be caused by a newly identified species, *Scardovia wiggsiae*. Note that in the absence of dietary carbohydrates, bacteria do not cause decay.

▶ Pathogenesis and Virulence Factors

In the presence of sucrose and, to a lesser extent, other carbohydrates, *S. mutans* and other streptococci produce sticky polymers of glucose called fructans and glucans. These adhesives help bind them to the smooth enamel surfaces and contribute to the sticky bulk of the plaque biofilm **(figure 20.13)**. If mature plaque is not removed from sites that readily trap food, it can result in a carious lesion. This is due to the action of the streptococci and other bacteria that produce acid as they ferment the carbohydrates. If the acid is immediately flushed from the plaque and diluted in the mouth, it has little effect. However, in the denser regions of plaque, the acid can accumulate in direct contact with the enamel surface and lower the pH to below 5, which is acidic enough to begin to dissolve (decalcify) the calcium phosphate of the enamel in that spot. This initial lesion can remain localized in the enamel and can be repaired with various inert materials (fillings). Once the deterioration has reached the level of the dentin, tooth destruction speeds up and the tooth can be rapidly destroyed.

▶ Transmission and Epidemiology

The bacteria that cause dental caries are transmitted to babies and children by their close contacts, especially the mother or closest caregiver. There is evidence for transfer of oral bacteria between children in day care centers, as well.

▶ Culture and Diagnosis

Dental professionals diagnose caries based on the tooth condition. Culture of the lesion is not routinely performed.

▶ Prevention and Treatment

The best way to prevent dental caries is through dietary restriction of sucrose and other refined carbohydrates. Regular brushing and flossing to remove plaque are also important. Most municipal communities in the United States add trace amounts of fluoride to their drinking water, because fluoride, when incorporated into the tooth structure, can increase tooth (as well as bone) hardness. Fluoride can also encourage the remineralization of teeth that have begun the demineralization process. These and other proposed actions of fluoride could make teeth less susceptible to decay. Fluoride is also added to toothpastes and mouth rinses and can be applied in gel form. Many European countries do not fluoridate their water due to concerns over additives in drinking water.

Treatment of a carious lesion involves removal of the affected part of the tooth (or the whole tooth in the case of advanced caries), followed by restoration of the tooth structure with an artificial material.

Periodontal Diseases

Periodontal disease is so common that 97% to 100% of the population has some manifestation of it by age 45. Most kinds are due to bacterial colonization and varying degrees of inflammation that occur in response to gingival damage.

Periodontitis

The initial stage of periodontal disease is **gingivitis,** the signs of which are swelling, loss of normal contour, patches of redness, and increased bleeding of the gums (gingiva). Spaces or pockets of varying depth also develop between the tooth and the gingiva. If this condition persists, a more serious disease called periodontitis results. The deeper involvement increases the size of the pockets and can cause bone resorption severe enough to loosen the tooth in its socket. If the condition is allowed to progress, the tooth can be lost **(figure 20.14).**

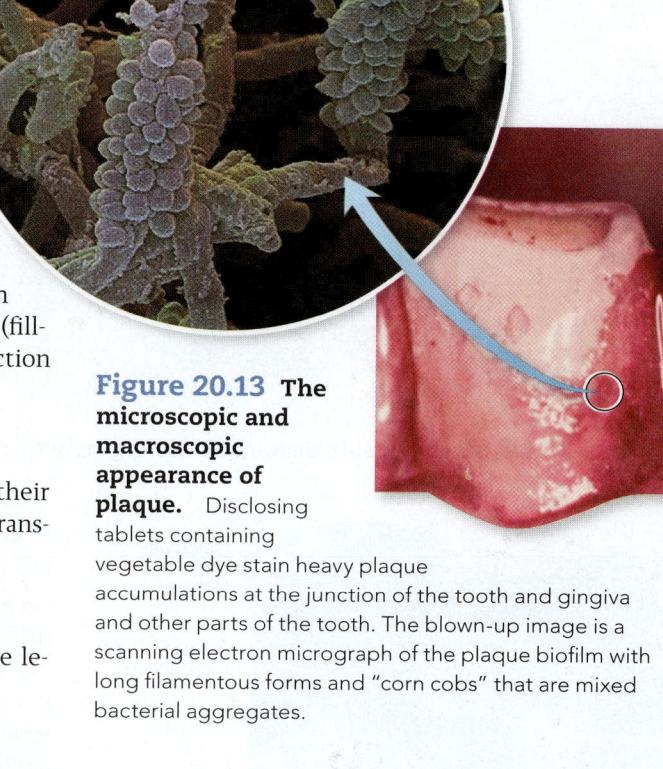

Figure 20.13   The microscopic and macroscopic appearance of plaque.   Disclosing tablets containing vegetable dye stain heavy plaque accumulations at the junction of the tooth and gingiva and other parts of the tooth. The blown-up image is a scanning electron micrograph of the plaque biofilm with long filamentous forms and "corn cobs" that are mixed bacterial aggregates.

Disease Table 20.5   Periodontitis

Causative Organism(s)	Polymicrobial community including some or all of: *Tannerella forsythia, Aggregatibacter actinomycetemcomitans, Porphyromonas gingivalis*, others
Most Common Modes of Transmission	–
Virulence Factors	Induction of inflammation, enzymatic destruction of tissues
Culture/ Diagnosis	–
Prevention	Oral hygiene
Treatment	Removal of plaque and calculus, gum reconstruction, possibly anti-inflammatory treatments
Epidemiological Features	United States: smokers = 11%, nonsmokers = 2%; internationally: 10%–15% of adults

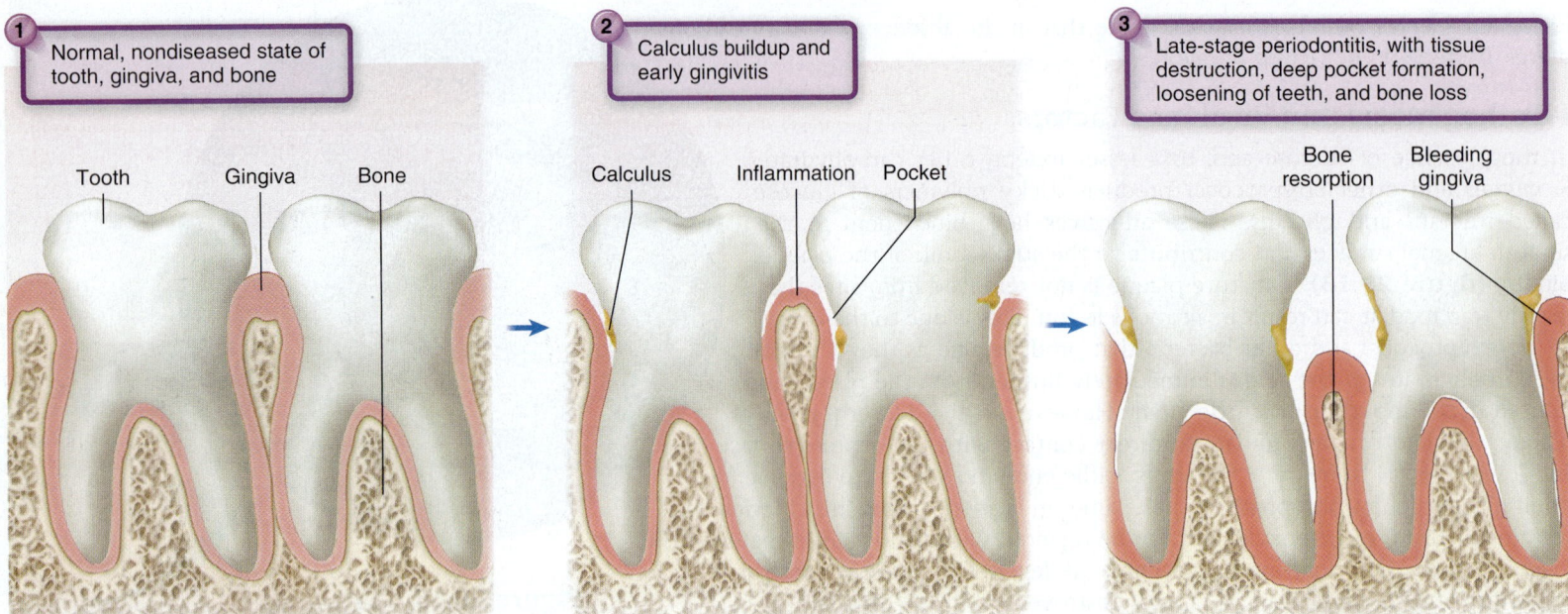

Figure 20.14 **Stages in soft-tissue infection, gingivitis, and periodontitis.**

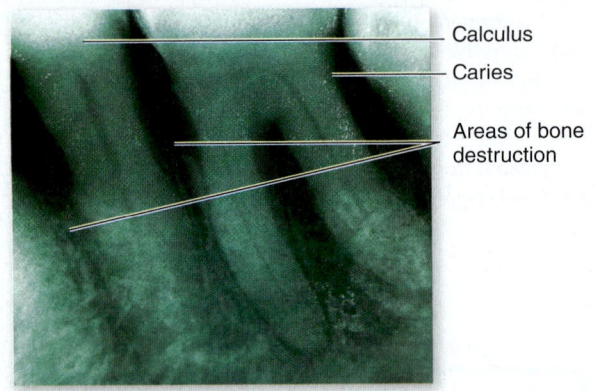

Figure 20.15 **The nature of calculus.**
Radiograph of the lower premolar and molar, showing calculus on the top and a caries lesion on the right. Bony defects caused by periodontitis affect both teeth.

▶ Causative Agents

Dental scientists stop short of stating that particular bacteria cause periodontal disease, because not all of the criteria for establishing causation have been satisfied. In fact, data from the Human Microbiome Project reveal that the composition of the microbial *community*, rather than single organisms, relates directly to risk of dental caries or periodontitis. When the polymicrobial biofilms consist of the right combination of bacteria, such as the anaerobes *Tannerella forsythia* (formerly *Bacteroides forsythus*), *Aggregatibacter actinomycetemcomitans, Porphyromonas gingivalis,* and perhaps *Fusobacterium* and spirochete species, the periodontal destruction process begins. The most common predisposing condition occurs when the plaque becomes mineralized (calcified) with calcium and phosphate crystals. This process produces a hard, porous substance called **calculus** above and below the gingival margin (edge) that can induce varying degrees of periodontal damage **(figure 20.15)**. The presence of calculus leads to a series of inflammatory events that probably allow the bacteria to cause disease.

Recent research has also shown that some of these microbes can escape into the bloodstream and increase the risk of coronary diseases, autoimmune diseases, and even meningitis.

Most periodontal disease is treated by removal of calculus and plaque and maintenance of good oral hygiene. Often, surgery to reduce the depth of periodontal pockets is required. Antibiotic therapy, either systemic or applied in periodontal packings, may also be utilized.

Mumps

The word *mumps* is Old English for "lump" or "bump." The symptoms of this viral disease are so distinctive that Hippocrates clearly characterized it in the fifth century BC as a self-limited, mildly epidemic illness associated with painful swelling at the angle of the jaw **(figure 20.16)**.

▶ Signs and Symptoms

After an average incubation period of 2 to 3 weeks, symptoms of fever, nasal discharge, muscle pain, and

Disease Table 20.6 Mumps	
Causative Organism(s)	Mumps virus (genus *Paramyxovirus*)
Most Common Modes of Transmission	Droplet contact
Virulence Factors	Spike-induced syncytium formation
Culture/Diagnosis	Clinical, fluorescent Ag tests, ELISA for Ab
Prevention	MMR live attenuated vaccine
Treatment	Supportive
Epidemiological Features	United States: fluctuates between a few hundred cases a year and a few thousand; internationally: are epidemic peaks every 2-5 years

malaise develop. These may be followed by inflammation of the salivary glands (especially the parotids), producing the classic gopherlike swelling of the cheeks on one or both sides. Swelling of the gland is called parotitis, and it can cause considerable discomfort. Viral multiplication in salivary glands is followed by invasion of other organs, especially the testes, ovaries, thyroid gland, pancreas, meninges, heart, and kidney. Despite the invasion of multiple organs, the prognosis of most infections is complete, uncomplicated recovery with permanent immunity.

Complications in Mumps   In 20% to 30% of young adult males, mumps infection localizes in the epididymis and testis, usually on one side only. The resultant syndrome of orchitis and epididymitis may be rather painful, but no permanent damage usually occurs.

▶ Transmission and Epidemiology of Mumps Virus

Humans are the exclusive natural hosts for the mumps virus. It is communicated primarily through salivary and respiratory secretions. Most cases occur in children under the age of 15, and as many as 40% are subclinical. Because lasting immunity follows any form of mumps infection, no long-term carrier reservoir exists in the population. The incidence of mumps in the United States spiked in 2006 and 2010 (with 2,600 and 1,500 cases reported).

▶ Prevention and Treatment

The general pathology of mumps is mild enough that symptomatic treatment to relieve fever, dehydration, and pain is usually adequate. Vaccine recommendations call for a dose of MMR at 12 to 15 months and a second dose at 4 to 6 years. Health care workers and college students who haven't already had both doses are advised to do so.

Gastritis and Gastric Ulcers

The curved cells of *Helicobacter* were first detected by J. Robin Warren in 1979 in stomach biopsies from ulcer patients. He and a partner, Barry J. Marshall, isolated the microbe in culture and even served as guinea pigs by swallowing a large inoculum to prove that it would cause gastric ulcers. Warren and Marshall won the Nobel Prize in Medicine in 2005 for their discovery.

▶ Signs and Symptoms

Gastritis is experienced as sharp or burning pain emanating from the abdomen. Gastric or peptic ulcers are actual lesions in the mucosa of the stomach (gastric ulcers) or in the uppermost portion of the small intestine (duodenal ulcers). Severe ulcers can be accompanied by bloody stools, vomiting, or both. The symptoms are often worse at night, after eating, or under conditions of psychological stress.

The second most common cancer in the world is stomach cancer (although it has been declining in the United States), and ample evidence suggests that long-term infection with *Helicobacter pylori* is a major contributing factor.

▶ Causative Agent

Helicobacter pylori is a curved gram-negative rod, closely related to *Campylobacter*, which we studied earlier in this chapter.

▶ Pathogenesis and Virulence Factors

Once the bacterium passes into the gastrointestinal tract, it bores through the outermost mucous layer that lines the stomach epithelial tissue. Then it attaches to specific binding sites on the cells and entrenches itself.

Before the bacterium was discovered, spicy foods, high-sugar diets (which increase acid levels in the stomach), and psychological stress were considered to be the cause of gastritis and ulcers. Now it appears that these factors merely aggravate the underlying infection.

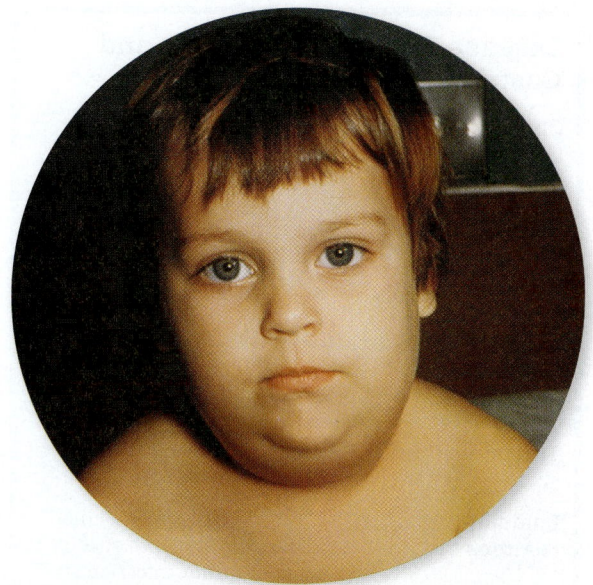

Figure 20.16 The external appearance of swollen parotid glands in mumps (parotitis).

 NCLEX® PREP

3. A client has been diagnosed with Guillain-Barré syndrome (GBS). Which observation found in the client's past medical history is relevant?
   a. *congestive heart failure*
   b. Campylobacter *infection*
   c. *dehydration*
   d. *diabetes mellitus*

Disease Table 20.7 Gastritis and Gastric Ulcers

Causative Organism(s)	*Helicobacter pylori*
Most Common Modes of Transmission	?
Virulence Factors	Adhesins, urease
Culture/Diagnosis	Endoscopy, urea breath test, stool antigen test
Prevention	None
Treatment	Amoxicillin followed by clarithromycin + tinidazole
Epidemiological Features	United States: infection (not disease) rates at 35% of adults; internationally: infection rates at 50%

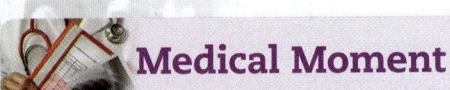

Medical Moment

Assessing Jaundice

As a health care professional, you will likely encounter a patient with jaundice at some point in your career. Jaundice is caused by elevated levels of bilirubin in the blood. How is jaundice recognized in a patient? Jaundice progresses caudally from face to trunk and extremities, so you should begin by looking at the patient's face. In some patients, jaundice will be easily recognized, as the patient's sclerae (the white portions of the eye) will be yellow—sometimes startlingly so! The skin may also have a yellowish discoloration. In dark-skinned individuals, especially older adults, the sclerae may not be reliable indicators as the sclerae may have a yellowish tint even in patients without jaundice. You can assess the palms and soles of the feet in patients with dark skin. Blanching of the skin (applying pressure to the skin with your fingertip) also helps to reveal the underlying skin color. The mucous membranes of affected individuals will also reveal a yellow discoloration.

▶ Transmission and Epidemiology

The mode of transmission of this bacterium remains a mystery. Studies have revealed that the pathogen is present in a large proportion of the human population. It occurs in the stomachs of 25% of healthy middle-age adults and in more than 60% of adults over 60 years of age. *H. pylori* is probably transmitted from person to person by the oral-oral or fecal-oral route. It seems to be acquired early in life and carried asymptomatically until its activities begin to damage the digestive mucosa. Because other animals are also susceptible to *H. pylori* and even develop chronic gastritis, it has been proposed that the disease is a zoonosis transmitted from an animal reservoir. The bacterium has also been found in water sources, suggesting that perhaps proper sanitation may reduce transmission.

▶ Prevention and Treatment

The only preventive approaches available currently are those that diminish some of the aggravating factors just mentioned. Many over-the-counter remedies offer symptom relief; most of them act to neutralize stomach acid. The best treatment is a course of antibiotics augmented by acid suppressors. The antibiotics most prescribed are clarithromycin or metronidazole.

Hepatitis

When certain viruses infect the liver, they cause **hepatitis,** an inflammatory disease marked by necrosis of hepatocytes and a mononuclear response that swells and disrupts the liver architecture. This pathologic change interferes with the liver's excretion of bile pigments such as bilirubin into the intestine. When bilirubin, a greenish-yellow pigment, accumulates in the blood and tissues, it causes **jaundice,** a yellow tinge in the skin and eyes. The condition can be caused by a variety of different viruses, including cytomegalovirus and Epstein-Barr virus. The others are all called "hepatitis viruses" but only because they all can cause this inflammatory condition in the liver. They are quite different from one another. While there are some recently discovered hepatitis viruses, they are not yet well characterized so we will cover the five that are well understood, named hepatitis A–E.

Note that noninfectious conditions can also cause inflammation and disease in the liver, including some autoimmune conditions, drugs, and alcohol overuse.

Hepatitis A and E Viruses

Hepatitis A virus (HAV) and hepatitis E virus (HEV) are both single-stranded nonenveloped RNA viruses. These two viruses are considered together since they are both transmitted through the fecal-oral route, and both cause relatively minor, self-limited hepatitis. The important exception to this is when HEV infects pregnant women, in whom it causes a 15% to 25% fatality rate.

▶ Signs and Symptoms

Most infections by these viruses are either subclinical or accompanied by vague, flulike symptoms. In more overt cases, the presenting symptoms may include jaundice and swollen liver. The viruses are not oncogenic (cancer causing), and in most everyone besides pregnant women, complete uncomplicated recovery results.

▶ Transmission and Epidemiology

In general, the disease is associated with deficient personal hygiene and lack of public health measures. In countries with inadequate sewage control, most outbreaks are associated with fecally contaminated water and food. Most infections result from close institutional contact, unhygienic food handling, eating shellfish, sexual transmission, or travel to other countries.

Hepatitis A occasionally can be spread by blood or blood products, but this is the exception rather than the rule. In developing countries, children are the most common victims, because exposure to the virus tends to occur early in life, whereas in North America and Europe, more cases appear in adults. Because the virus is not carried chronically, the principal reservoirs are asymptomatic, short-term carriers (often children) or people with clinical disease.

▶ Prevention and Treatment

Prevention of hepatitis A is based primarily on immunization. An inactivated viral vaccine (Havrix) has been in use since the mid-1990s. Short-term protection can be conferred by passive immune globulin. This treatment is useful for people who have come in contact with HAV-infected individuals, or who have eaten at a restaurant that was the source of a recent outbreak. It has also recently been discovered that administering Havrix after exposure can prevent symptoms. A combined hepatitis A/hepatitis B vaccine, called Twinrix, is recommended for people who may be at risk for both diseases, such as people with chronic liver dysfunction, intravenous drug users, and anyone engaging in anal-oral intercourse. Travelers to areas with high rates of both diseases should obtain vaccine coverage as well. Hepatitis E has no vaccine.

No specific medicine is available for hepatitis A or E once the symptoms begin. Drinking lots of fluids and avoiding liver irritants such as aspirin or alcohol will speed recovery.

Hepatitis B Virus (and Hepatitis D)

Hepatitis B virus (HBV) is an enveloped DNA virus in the family *Hepadnaviridae*. Intact viruses are often called Dane particles. The genome is partly double-stranded and partly single-stranded. **Hepatitis D** is an enveloped RNA virus. It is actually a *subvirus satellite* of HBV; that is, it can propagate only in the presence of HBV.

▶ Signs and Symptoms

In addition to the direct damage to liver cells just outlined, the spectrum of hepatitis disease may include fever, chills, malaise, anorexia, abdominal discomfort, diarrhea, and nausea. Rashes may appear and arthritis may occur. Hepatitis B infection can be very serious, even life-threatening. A small number of patients develop glomerulonephritis and arterial inflammation. Complete liver regeneration and restored function occur in most patients; however, a small number of patients develop chronic liver disease in the form of necrosis or cirrhosis (permanent liver scarring and loss of tissue). In some cases, chronic HBV infection can lead to liver cancer.

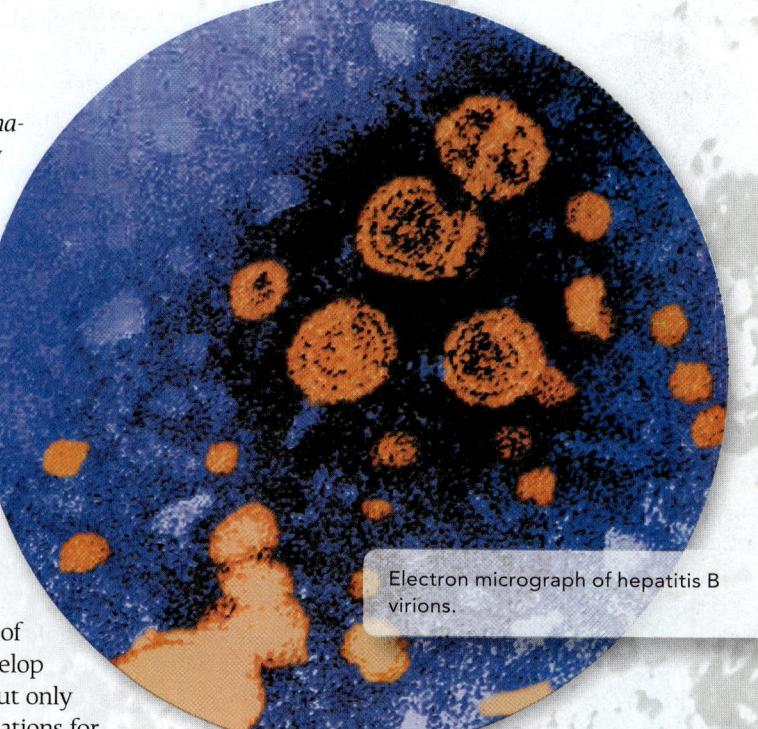

Electron micrograph of hepatitis B virions.

Patients who become infected as children have significantly higher risks of long-term infection and disease. In fact, 90% of neonates infected at birth develop chronic infection, as do 30% of children infected between the ages of 1 and 5, but only 6% of persons infected after the age of 5. This finding is one of the major justifications for the routine vaccination of children. Also, infection becomes chronic more often in men than in women. The mortality rate is 15% to 25% for people with chronic infection.

When patients are infected with both HBV and hepatitis D virus (HDV), the disease can become more severe and is more likely to progress to permanent liver damage.

▶ Pathogenesis and Virulence Factors

The hepatitis B virus enters the body through a break in the skin or mucous membrane or by injection into the bloodstream. Eventually, it reaches the liver cells (hepatocytes) where it multiplies and releases viruses into the blood during an incubation period of 4 to 24 weeks (7 weeks average). Surprisingly, the majority of those infected exhibit few overt symptoms and eventually develop an immunity to HBV, but some people experience the symptoms described earlier.

▶ Transmission and Epidemiology

An important factor in the transmission pattern of hepatitis B virus is that it multiplies exclusively in the liver, which continuously seeds the blood with viruses. Electron microscopic studies have revealed up to 10^7 virions per milliliter of infected blood. Even a

---

NCLEX® PREP

4. Which conditions may be associated with jaundice? Select all that apply.
   a. *cholecystitis*
   b. *pruritus*
   c. *clay-colored stools*
   d. *decreased bilirubin levels*
   e. *concentrated dark urine*

minute amount of blood (a *millionth* of a milliliter) can transmit infection. The abundance of circulating virions is so high and the minimal dose so low that such simple practices as sharing a toothbrush or a razor can transmit the infection. HBV can be transmitted via semen and vaginal secretions. Growing concerns about virus spread through donated organs and tissue are prompting increased testing prior to surgery. Vertical transmission is possible, and it predisposes the child to development of the carrier state and increased risk of liver cancer. It is sometimes known as *serum hepatitis.*

This virus is one of the major infectious concerns for health care workers. Needlesticks can easily transmit the virus, and therefore most workers are required to have the full series of HBV vaccinations. Unlike the more notorious HIV, HBV remains infective for days in dried blood, for months when stored in serum at room temperature, and for decades if frozen. Although it is not inactivated after 4 hours of exposure to 60°C, boiling for the same period can destroy it. Disinfectants containing chlorine, iodine, and glutaraldehyde show potent anti-hepatitis B activity.

▶ Culture and Diagnosis

Serological tests can detect either virus antigen or antibodies. Radioimmunoassay and ELISA testing permit detection of the important surface antigen (S antigen) of HBV very early in infection. Antibody tests are most valuable in patients who are negative for the antigen.

▶ Prevention and Treatment

The primary prevention for HBV infection is vaccination. The most widely used vaccines are recombinant, containing the pure surface antigen cloned in yeast cells. Vaccination is a must for medical and dental workers and students, patients receiving multiple transfusions, immunodeficient persons, and cancer patients. The vaccine is also now strongly recommended for all newborns as part of a routine immunization schedule. As just mentioned, a combined vaccine for HAV/HBV may be appropriate for certain people.

Passive immunization with hepatitis B immune globulin (HBIG) gives significant immediate protection to people who have been exposed to the virus through needle puncture, broken blood containers, or skin and mucosal contact with blood. Another group for whom passive immunization is highly recommended is neonates born to infected mothers.

Mild cases of hepatitis B are managed by symptomatic treatment and supportive care. Chronic infection can be controlled with recombinant human interferon, tenofovir, or entevir. Each of these can help to slow virus multiplication and prevent liver damage in many but not all patients. None of the drugs is considered curative.

Hepatitis C Virus

Hepatitis C virus (HCV) is sometimes referred to as the "silent epidemic" because more than 4 million Americans are infected with the virus, but it takes many years to cause noticeable symptoms. In the United States, its incidence fell between 1992 and 2003, but no further decreases have been seen since then. Liver failure from hepatitis C is one of the most common reasons for liver transplants in this country. Hepatitis C is an RNA virus in the *Flaviviridae* family. It used to be known as "non-A non-B" virus. It is usually diagnosed with a blood test for antibodies to the virus.

▶ Signs and Symptoms

People have widely varying experiences with this infection. It shares many characteristics of hepatitis B disease, but it is much more likely to become chronic. Of those infected, 75% to 85% will remain infected indefinitely. (In contrast, only about 6% of persons who acquire hepatitis B after the age of 5 will be chronically infected.) With HCV infection, it is possible to have severe symptoms without permanent liver damage, but it is more common to have chronic liver disease even if there are no overt symptoms. Cancer may also result from chronic hepatitis C virus (HCV) infection. Worldwide, HBV infection

Disease Table 20.8 Hepatitis

Causative Organism(s)	Hepatitis A or E virus	Hepatitis B virus	Hepatitis C virus
Most Common Modes of Transmission	Fecal-oral, vehicle	Parenteral (blood contact), direct contact (especially sexual), vertical	Parenteral (blood contact), vertical
Virulence Factors	-	Latency	Core protein suppresses immune function?
Culture/Diagnosis	IgM serology	Serology (ELISA, radioimmunoassay)	Serology
Prevention	Hepatitis A vaccine or combined; HAV/HBV vaccine	HBV recombinant vaccine	-
Treatment	HAV: hepatitis A vaccine or immune globulin; HEV: immune globulin	Interferon, tenofovir, or entecavir	(Pegylated) interferon, with or without ribavirin
Incubation Period	2-7 weeks	1-6 months	2-8 weeks
Epidemiological Features	Hepatitis A, United States: 20,000 cases annually and 40% of adults show evidence of prior infection; internationally: 1.4 million cases per year; hepatitis E, internationally: 20 million infections per year; 60% in East and Southeast Asia	United States: prevalence rate 1.5 per 100,000; 800,000 to 1.4 million have chronic infection; internationally: 240 million	United States: estimated 17,000 new cases per year; 2.7 million with chronic HCV; internationally: 150 million chronically infected

is the most common cause of liver cancer, but in the United States it is more likely to be caused by HCV. In 2012, the CDC recommended that all baby boomers (those born between 1945 and 1965) be tested for HCV.

▶ Transmission and Epidemiology

This virus is acquired in similar ways to HBV. It is more commonly transmitted through blood contact (both "sanctioned," such as in blood transfusions, and "unsanctioned," such as needle sharing by injecting drug users) than through transfer of other body fluids. Vertical transmission is also possible.

Before a test was available to test blood products for this virus, it seems to have been frequently transmitted through blood transfusions. Hemophiliacs who were treated with clotting factor prior to 1985 were infected with HCV at a high rate. Once blood began to be tested for HIV (in 1985) and screened for so-called "non-A non-B" hepatitis, the risk of contracting HCV from blood was greatly reduced.

▶ Prevention and Treatment

There is currently no vaccine for hepatitis C. The current treatment regimen is ribavirin plus a form of interferon called pegylated interferon. The treatments are not curative, but they may prevent or lessen damage to the liver. In 2011, two new protease inhibitor drugs were approved for treating hepatitis C.

20.3 LEARNING OUTCOMES—Assess Your Progress

5. List the possible causative agents, modes of transmission, virulence factors, diagnostic techniques, and prevention/treatment for the "Highlight Disease" acute diarrhea.

6. Discuss important features of the conditions food poisoning and chronic diarrhea.

7. Discuss important features of the two categories of oral conditions: dental caries and periodontal diseases.

8. Identify the most important features of mumps, gastritis, and gastric ulcers.

9. Differentiate among the main types of hepatitis, and discuss the causative agents, mode of transmission, diagnostic techniques, prevention, and treatment of each.

20.4 Gastrointestinal Tract Diseases Caused by Helminths

Helminths that parasitize humans are amazingly diverse, ranging from barely visible roundworms (0.3 mm) to huge tapeworms (25 m long). In the introduction to these organisms in chapter 4, we grouped them into three categories: nematodes (round-worms), trematodes (flukes), and cestodes (tapeworms), and we discussed basic characteristics of each group. You may wish to review those sections before continuing. In this section, we examine the intestinal diseases caused by helminths. Although they can cause symptoms that may be mistaken for some of the diseases discussed elsewhere in this chapter, helminthic diseases are usually accompanied by an additional set of symptoms that arise from the host response to helminths. Worm infection usually provokes an increase in granular leukocytes called eosinophils, which have a specialized capacity to destroy multicellular parasites. This increase, termed **eosinophilia,** is a hallmark of helminthic infection and is detectable in blood counts. If the following symptoms occur coupled with eosinophilia, helminthic infection should be suspected.

Helminthic infections may be acquired through the fecal-oral route or through penetration of the skin, but most of these organisms spend part of their lives in the intestinal tract. **Figure 20.17** depicts the four different types of life cycles of the helminths. While the worms are in the intestines, they can produce a gamut of intestinal symptoms. Some of them also produce symptoms outside of the intestines.

Clinical Considerations

There is a very wide variety of helminthic diseases that afflict humans (and animals, for that matter). We don't have space in this book to describe them all, but we will

Trained sushi chefs are skilled at detecting helminth infection in the raw fish they prepare.

Figure 20.17 Four basic helminth life and transmission cycles.

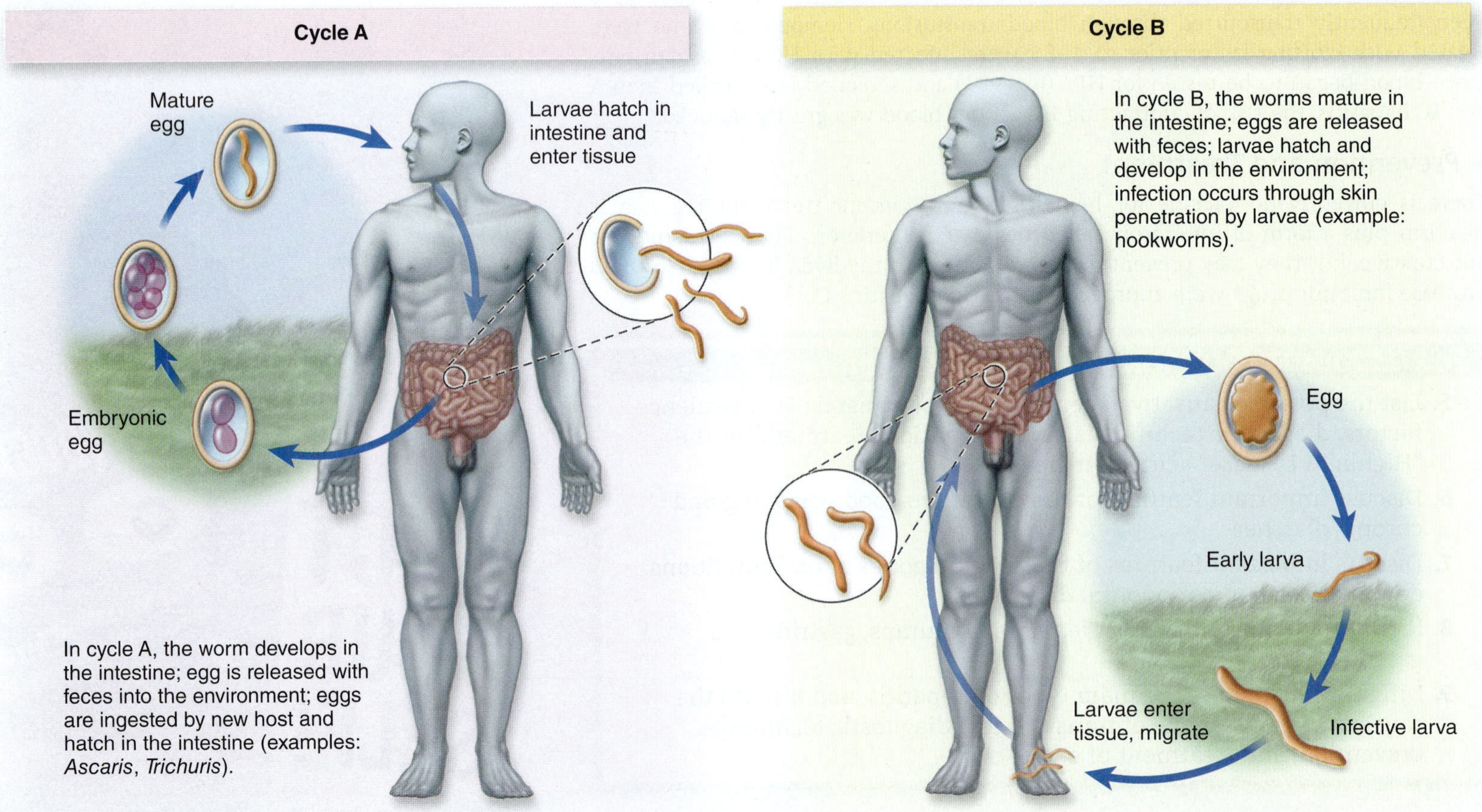

Cycle A

Mature egg

Larvae hatch in intestine and enter tissue

Embryonic egg

In cycle A, the worm develops in the intestine; egg is released with feces into the environment; eggs are ingested by new host and hatch in the intestine (examples: *Ascaris, Trichuris*).

Cycle B

In cycle B, the worms mature in the intestine; eggs are released with feces; larvae hatch and develop in the environment; infection occurs through skin penetration by larvae (example: hookworms).

Egg

Early larva

Larvae enter tissue, migrate

Infective larva

present two classic types of infections in some detail and provide you disease tables for several more.

We talk about diagnosis, pathogenesis and prevention, and treatment of the helminths as a group in the next subsections. We'll then highlight one of the most common forms of helminthic disease, and finish with summaries of the others.

▶ Pathogenesis and Virulence Factors in General

Helminths have numerous adaptations that allow them to survive in their hosts. They have specialized mouthparts for attaching to tissues and for feeding, enzymes with which they liquefy and penetrate tissues, and a cuticle or other covering to protect them from host defenses. In addition, their organ systems are usually reduced to the essentials: getting food and processing it, moving, and reproducing. The damage they cause in the host is very often the result of the host's response to the presence of the invader.

Many helminths have more than one host during their lifetimes. If this is the case, the host in which the adult worm is found is called the **definitive host** (usually a vertebrate). Sometimes the actual definitive host is not the host usually used by the parasite but an accidental bystander. Humans often become the accidental definitive hosts for helminths whose normal definitive host is a cow, pig, or fish. Larval stages of helminths are found in intermediate hosts. Humans can serve as intermediate hosts, too. Helminths may require no intermediate host at all or may need one or more intermediate hosts for their entire life cycle.

▶ Diagnosis in General

Diagnosis of almost all helminthic infections follows a similar series of steps. A differential blood count showing increased eosinophils and serological tests indicating

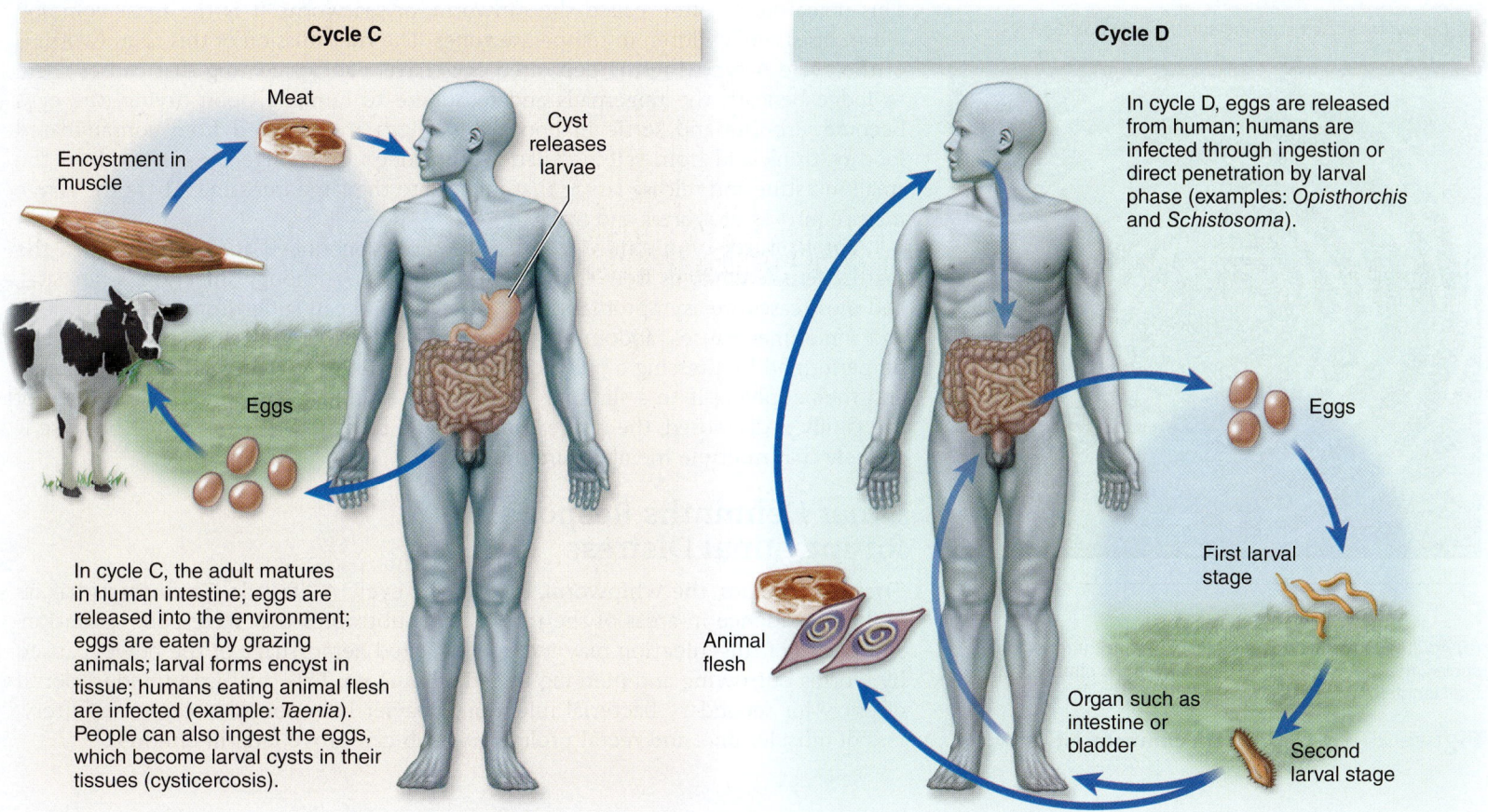

Cycle C

Meat

Encystment in muscle

Cyst releases larvae

Eggs

In cycle C, the adult matures in human intestine; eggs are released into the environment; eggs are eaten by grazing animals; larval forms encyst in tissue; humans eating animal flesh are infected (example: *Taenia*). People can also ingest the eggs, which become larval cysts in their tissues (cysticercosis).

Cycle D

In cycle D, eggs are released from human; humans are infected through ingestion or direct penetration by larval phase (examples: *Opisthorchis* and *Schistosoma*).

Eggs

First larval stage

Animal flesh

Organ such as intestine or bladder

Second larval stage

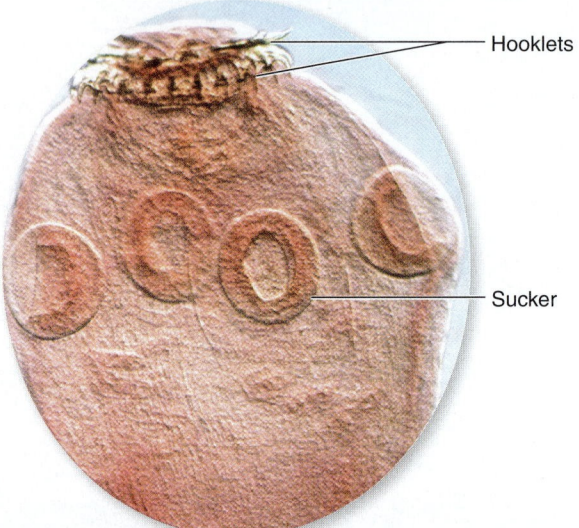

Hooklets

Sucker

(a) Tapeworm scolex showing sucker and hooklets.

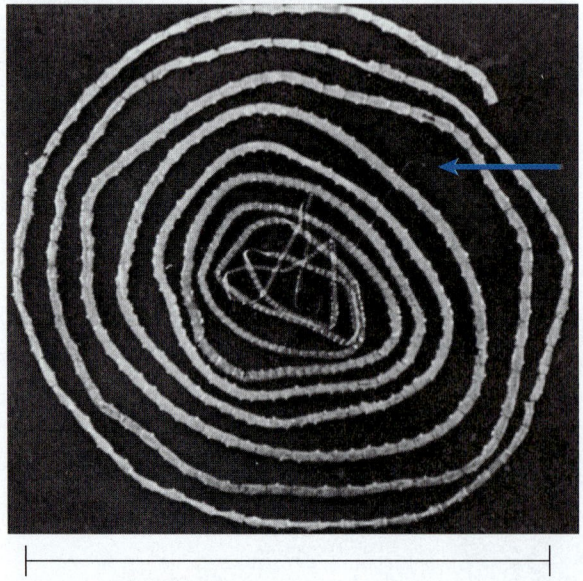

24 in

(b) Adult tapeworm. The arrow points to the (barely visible) scolex; the remainder of the tape, called the strobila, has a total length of 5 meters.

Figure 20.18 Tapeworm characteristics.

sensitivity to helminthic antigens all provide indirect evidence of worm infection. A history of travel to the tropics or immigration from those regions is also helpful, even if it occurred years ago, because some flukes and nematodes persist for decades. The most definitive evidence, however, is the discovery of eggs, larvae, or adult worms in stools or other tissues. The worms are sufficiently distinct in morphology that positive identification can be based on any stage, including eggs. That said, not all of these diseases result in eggs or larval stages that can easily be found in stool.

▶ Prevention and Treatment in General

There are no vaccines for the helminthic infections described here. In populations in which the infections are common, prophylactic treatment twice a year with antihelminthic drugs has been shown to keep people healthy.

Although several useful antihelminthic medications exist, the cellular physiology of the eukaryotic parasites resembles that of humans, and drugs toxic to them can also be toxic to us. Some antihelminthic drugs suppress a metabolic process that is more important to the worm than to the human. Others inhibit the worm's movement and prevent it from maintaining its position in a certain organ. Note that some helminths have developed resistance to the drugs used to treat them. In some cases, surgery may be necessary to remove worms or larvae.

==Highlight Disease==

Helminthic Infections: Intestinal Distress as the Primary Symptom

Both tapeworms and roundworms can infect the intestinal tract in such a way as to cause primary symptoms there. We will highlight one nematode and one tapeworm, and include the rest of the probable agents in our summary table.

Enterobius vermicularis

This nematode is often called the pinworm, or seatworm. It is the most common worm disease of children in temperate zones. The transmission of this roundworm is of the cycle A type. Freshly deposited eggs have a sticky coating that causes them to lodge beneath the fingernails and to adhere to fomites. Upon drying, the eggs become airborne and settle in house dust. Eggs are ingested from contaminated food or drink and from self-inoculation from one's own fingers. Eggs hatch in the small intestine and release larvae that migrate to the large intestine. There the larvae mature into adult worms and mate.

The hallmark symptom of this condition is pronounced anal itching when the mature female emerges from the anus and lays eggs. Although infection is not fatal, and most cases are asymptomatic, the afflicted child can suffer from disrupted sleep and sometimes nausea, abdominal discomfort, and diarrhea. A simple rapid test can be performed by pressing a piece of transparent adhesive tape against the anal skin and then applying it to a slide for microscopic examination. When one member of the family is diagnosed, the entire family should be tested and/or treated because it is likely that multiple members are infected.

Other Helminths Responsible for Intestinal Distress

Trichuris trichiura, the whipworm, follows the cycle A lifestyle. Trichuriasis has its highest incidence in areas of the tropics and subtropics that have poor sanitation. Symptoms of this infection may include localized hemorrhage of the bowel caused by worms burrowing and piercing intestinal mucosa. This can also provide a portal of entry for secondary bacterial infection. Heavier infections can cause dysentery, loss of muscle tone, and rectal prolapse, which can prove fatal in children.

The tapeworm *Diphyllobothrium latum* has an intermediate host in fish. It follows cycle C just as *Taenia solium* does. It is common in the Great Lakes, Alaska, and Canada. Mammals, including humans, act as definitive hosts. It develops in the intestine and can cause long-term symptoms. It can be transmitted in raw food such as sushi and sashimi made from salmon.

Hymenolepis species are small tapeworms and are the most common human tapeworm infections in the world. They follow cycle C. There are two species: *Hymenolepis nana*, known as the dwarf tapeworm because it is only 15 to 40 mm in length, and *H. diminuta*, the rat tapeworm, which is usually 20 to 60 cm in length as an adult.

Disease Table 20.9 Intestinal Distress

Causative Organism(s)	*Enterobius vermicularis* (pinworm)	*Trichuris trichiura* (whipworm)	*Diphyllobothrium latum* (fish tapeworm)	*Hymenolepis nana* and *H. diminuta*
Most Common Modes of Transmission	Cycle A: vehicle (food, water), fomites, self-inoculation	Cycle A: vehicle (soil), fecal-oral	Cycle C: vehicle (seafood)	Cycle C: vehicle (ingesting insects), fecal-oral
Virulence Factors	–	Burrowing and invasiveness	Vitamin B_{12} usage	–
Culture/Diagnosis	Adhesive tape + microscopy	Blood count, serology, egg or worm detection	Blood count, serology, egg or worm detection	Blood count, serology, egg or worm detection
Prevention	Hygiene	Hygiene, sanitation	Cook meat	Hygienic environment
Treatment	Mebendazole, piperazine	Mebendazole	Praziquantel	Praziquantel
Distinctive Features	Common in United States	Humans sole host	Large tapeworm; anemia	Most common tapeworm infection
Epidemiological Features	United States: prevalence in children, 0.2%-20%; higher in the South	United States: prevalence approx. 0.1%; internationally: prevalence as high as 80% in Southeast Asia, Africa, the Caribbean, and Central and South America	Estimated 20 million infections worldwide	United States: prevalence approximately 0.4%; internationally: the single most prevalent tapeworm infection

Helminthic Infections: Intestinal Distress Accompanied by Migratory Symptoms

A diverse group of helminths enter the body as larvae or eggs, mature to the worm stage in the intestine, and then migrate into the circulatory and lymphatic systems, after which they travel to the heart and lungs, migrate up the respiratory tree to the throat, and are swallowed. This journey returns the mature worms to the intestinal tract where they then take up residence. All of these conditions, in addition to causing symptoms in the digestive tract, may induce inflammatory reactions along their migratory routes, resulting in eosinophilia and, during their lung stage, pneumonia. Examples of this type of infection appear in **Disease Table 20.10**.

Cysticercosis

Taenia solium is a tapeworm. Adult worms are usually around 5 meters long and have a scolex with hooklets and suckers to attach to the intestine (**figure 20.19**). This helminth follows cycle C in figure 20.17, in which humans are infected by eating animal flesh that contains the worm eggs, or even the worms themselves.

Disease Table 20.10 Intestinal Distress Plus Migratory Symptoms

Causative Organism(s)	*Toxocara* species	*Ascaris lumbricoides* (intestinal roundworm)	*Necator americanus* and *Ancylostoma duodenale* (hookworms)
Most Common Modes of Transmission	Cycle A: dog or cat feces	Cycle A: vehicle (soil/fecal-oral), fomites, self-inoculation	Cycle B: vehicle (soil), fomite
Virulence Factors	–	Induction of hypersensitivity, adult worm migration, abdominal obstruction	Induction of hypersensitivity, adult worm migration, abdominal obstruction
Culture/Diagnosis	Blood count, serology, egg or worm detection	Blood count, serology, egg or worm detection	Blood count, serology, egg or worm detection
Prevention	Hygiene	Hygiene	Sanitation
Treatment	Albendazole	Albendazole	Albendazole
Distinctive Features	Can cause migration symptoms or blindness	Most cases mild, unnoticed	Penetrates skin, serious intestinal symptoms
Epidemiological Features	Nearly 100% of newborn puppies in United States infected; 14% of people in United States have been infected	Internationally: up to 25% prevalence, 80,000–100,000 deaths per year	United States: widespread in Southeast until early 1900s; internationally: 800 million infected

Disease Table 20.11 Cysticercosis

Causative Organism(s)	*Taenia solium* (pork tapeworm)
Most Common Modes of Transmission	Cycle C: vehicle (pork), fecal-oral
Virulence Factors	–
Culture/Diagnosis	Blood count, serology, egg or worm detection
Prevention	Cook meat, avoid pig feces
Treatment	Praziquantel
Distinctive Features	Ingesting larvae embedded in pork leads to intestinal tapeworms: ingesting eggs (fecal-oral route) causes cysticercosis, larval cysts embedded in tissue of new host
Epidemiological Features	United States: considered a neglected parasitic infection, common cause of seizures; internationally: very common in Latin America and Asia

Figure 20.19 Cysticerci in the brain caused by *Taenia solium*.

Disease caused by *T. solium* (the pig tapeworm) is distributed worldwide but is mainly concentrated in areas where humans live in close proximity with pigs or eat undercooked pork. In pigs, the eggs hatch in the small intestine and the released larvae migrate throughout the organs. Ultimately, they encyst in the muscles, becoming *cysticerci,* young tapeworms that are the infective stage for humans. When humans ingest a live cysticercus in pork, the coat is digested and the organism is flushed into the intestine, where it firmly attaches by the scolex and develops into an adult tapeworm. Infection with *T. solium* can take another form when humans ingest the tapeworm eggs rather than infected meat. Then the human, instead of the tapeworm in the gut, becomes the host of the encysted larvae, or cysticerci, leading to a condition called **cysticercosis,** one of the five neglected parasitic infections (NPIs) in the United States (see Inside the Clinic at the end of this chapter). It is estimated that tens of thousands of Latinos living in the United States are affected by cysticercosis, but it is not often recognized because American physician may not know to look for it. A particularly nasty form of this condition is neurocysticerosis, in which the larvae encyst in the brain. It is estimated to be responsible for 10% of seizures requiring emergency room visits in some U.S. cities.

Schistosomiasis: Liver Disease

When liver swelling or malfunction is accompanied by eosinophilia, **schistosomiasis** should be suspected. Schistosomiasis has afflicted humans for thousands of years. The disease described here is caused by the blood flukes *Schistosoma mansoni* or *S. japonicum*, species that are morphologically and geographically distinct but share similar life cycles, transmission methods, and general disease manifestations. It is one of the few infectious agents that can invade intact skin.

Another species called *Schistosoma haematobium* causes disease in the bladder.

▶ Signs and Symptoms

The most severe consequences associated with chronic infection are hepatomegaly, liver disease, and splenomegaly. Occasionally, eggs from the worms are carried into the central nervous system and heart, and create a severe granulomatous response. Adult flukes can live for many years and, by eluding the immune defenses, cause a chronic affliction.

▶ Causative Agent

Schistosomes are trematodes, or flukes (see chapter 4), but they are more cylindrical than flat **(figure 20.20).** They are often called blood flukes. Humans are the definitive hosts for the blood fluke, and snails are the intermediate host.

▶ Pathogenesis and Virulence Factors

This parasite is clever indeed. Once inside the host, it coats its outer surface with proteins from the host's bloodstream, basically "cloaking" itself from the host defense system. This coat reduces its surface antigenicity and allows it to remain in the host indefinitely.

▶ Transmission and Epidemiology

The life cycle of the schistosome is of the "D" type, and is very complex (figure 20.20). The cycle begins when infected humans release eggs into irrigated fields or ponds, either by deliberate fertilization with excreta or by defecating or urinating directly into the water.

The disease is endemic to 74 countries located in Africa, South America, the Middle East, and the Far East. Schistosomiasis (including the urinary tract form) is the second most prominent parasitic disease after malaria, probably affecting 200 million people at any one time worldwide.

Disease Table 20.12 Schistosomiasis Liver Disease	
Causative Organism(s)	*Schistosoma mansoni, S. japonicum*
Most Common Modes of Transmission	Cycle D: vehicle (contaminated water)
Virulence Factors	Antigenic "cloaking"
Culture/ Diagnosis	Identification of eggs in feces, scarring of intestines detected by endoscopy
Prevention	Avoiding contaminated vehicles
Treatment	Praziquantel
Distinctive Features	Penetrates skin, lodges in blood vessels of intestine, damages liver
Epidemiological Features	Internationally: 230 million new infections per year by these and the urinary schistosome

20.4 LEARNING OUTCOMES—Assess Your Progress

10. Describe some distinguishing characteristics and commonalities seen in helminthic infections.
11. List four helminths that cause primarily intestinal symptoms, and identify which life cycle they follow and one unique fact about each one.
12. Explain reasons for and the consequences of the following statement, "Cysticercosis is underdiagnosed in the United States."

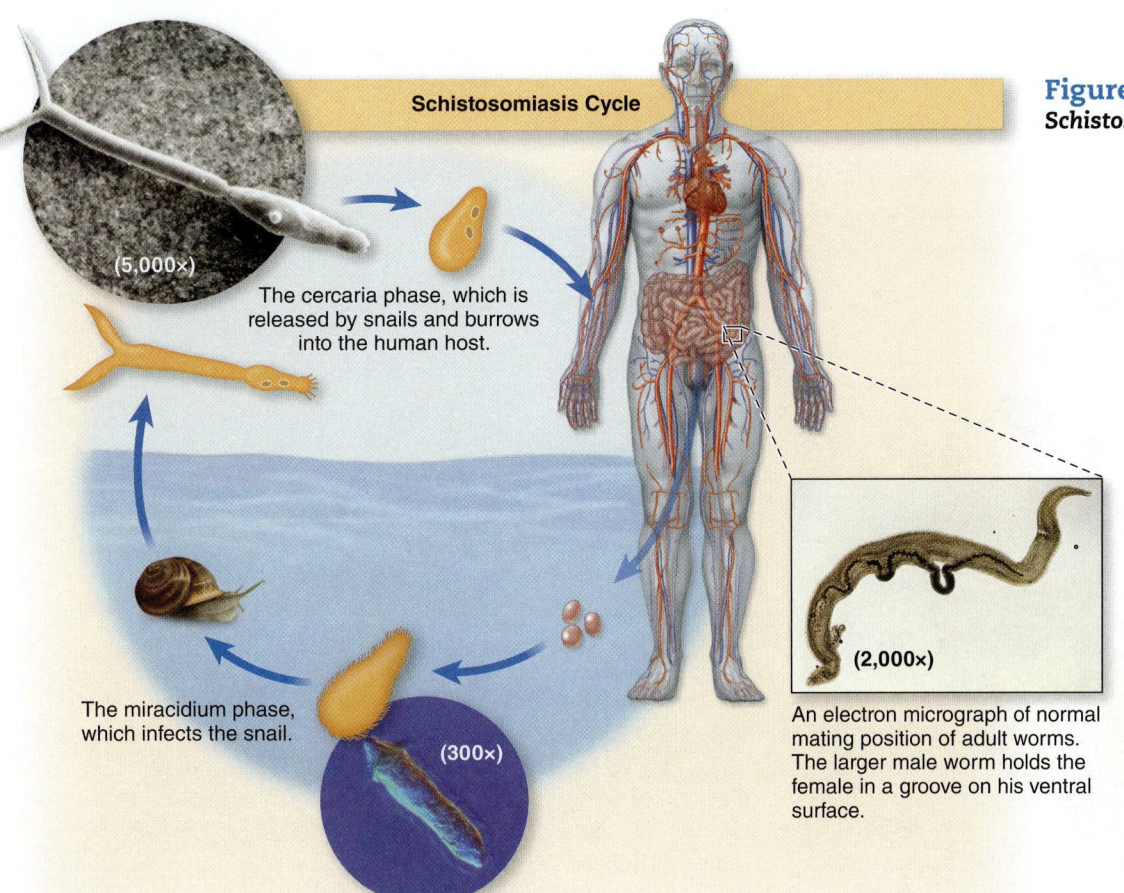

Schistosomiasis Cycle

(5,000×)

The cercaria phase, which is released by snails and burrows into the human host.

The miracidium phase, which infects the snail.

(300×)

An electron micrograph of normal mating position of adult worms. The larger male worm holds the female in a groove on his ventral surface.

(2,000×)

Figure 20.20 **Stages in the life cycle of** *Schistosoma.*

CASE FILE WRAP-UP

Shigellosis is the disease caused by a group of bacteria called *Shigella*. Symptoms of *Shigella* include fever, abdominal cramps, and diarrhea that may be bloody. Some people with shigellosis are asymptomatic. Illness requiring hospitalization is uncommon—most people recover without treatment within a week. However, one form of *Shigella* (*Shigella dysenteriae* type 1) causes deadly epidemics in developing countries.

Shigellosis is spread through direct contact with someone who is carrying the bacteria. Poor hygiene is often a factor. Shigella may be found in contaminated food and water. Proper food handling can help prevent shigellosis, as can frequent hand washing, especially following a trip to the bathroom. Peter, the patient diagnosed with *Shigella* in the case file at the beginning of this chapter, may have come in contact with someone who was infected and did not wash his or her hands after a trip to the bathroom; or he may have been exposed to contaminated food, although no one else in his family fell ill.

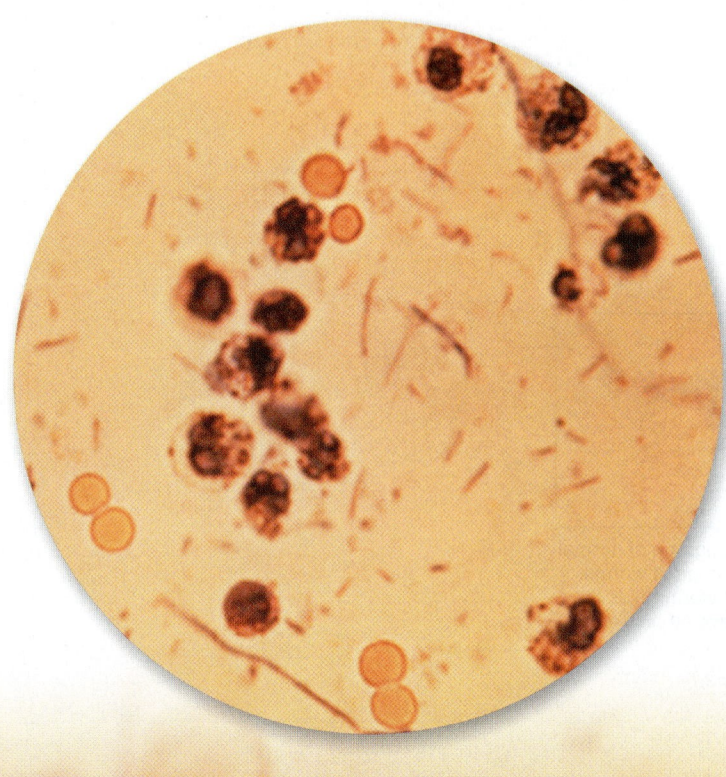

▶ Summing Up

Taxonomic Organization Microorganisms Causing Disease in the GI Tract

Microorganism	Pronunciation	Location of Disease Table
Gram-positive endospore-forming bacteria		
Clostridium difficile	klos-trid"-ee-um dif"-i-sil	Acute diarrhea, p. 572
Clostridium perfringens	klos-trid"-ee-um purr-frinj'-unz	Food poisoning, p. 575
Bacillus cereus	buh-sill'-us seer'-ee-uhs	Food poisoning, p. 575
Gram-positive bacteria		
Streptococcus mutans	strep"-tuh-kok'-us mew'-tans	Dental caries, p. 580
Streptococcus sobrinus	strep"-tuh-kok'-us so-brin'-us	Dental caries, p. 580
Staphylococcus aureus	staf"-uh-lo-kok'-us are'-ee-us	Food poisoning, p. 575
Scardovia wiggsiae	skar-doh"-vee-uh' wig'-zee-ay	Dental caries, p. 580
Gram-negative bacteria		
Campylobacter jejuni	cam"-plo-bac'-ter juh-june'-ee	Acute diarrhea, p. 572
Helicobacter pylori	heel"-i-coe-back'-tur pie-lor'-ee	Gastritis/gastric ulcers, p. 584
Shiga-toxin-producing *Escherichia coli*	esh'-shur-eesh" ee -uh-col'-eye	Acute diarrhea plus hemolytic syndrome, p. 572
Other *E. coli*		Acute diarrhea, p. 572
		Chronic diarrhea, p. 579
Salmonella	sal'-muh-nel"-luh	Acute diarrhea, p. 572
Shigella	shi-gel'-luh	Acute diarrhea, p. 572
Vibrio cholerae	vib'-ree-oh col'-er-ee	Acute diarrhea, p. 572
Yersinia enterocolitica and *Y. pseudotuberculosis*	yur-sin'-ee-ah en-ter-oh-coe-lit'-i-cuh, yur-sin'-ee-ah soo'-doh-tuh-bur'-cue-loh" sis	Acute diarrhea, p. 572
Tannerella forsythia, Aggregatibacter actinomycetemcomitans, Porphyromonas gingivalis, Treponema vincentii, Prevotella intermedia, Fusobacterium	tan-er-rel'-ah for-sye"-thee-ah, ag-gruh-ga"-ti-bac'-tur ack-tin'-oh-my-see"-tem-cah'-mi-tans, por-fuhr'-oh-moan"-as jin-ji-vall'-is, trep'-oh-nee-ma vin-cen'-tee-eye, prev'-oh-tell-ah in-ter-meed'-ee-ah, few'-zo-bac-teer'-ee'-um	Periodontal disease, p. 581
DNA viruses		
Hepatitis B virus	hep-uh-tie'-tis B vie'-russ	Hepatitis, p. 587
RNA viruses		
Hepatitis A virus	hep-uh-tie'-tis A vie'-russ	Hepatitis, p. 587
Hepatitis C virus	hep-uh-tie'-tis C vie'-russ	Hepatitis, p. 587
Hepatitis E virus	hep-uh-tie'-tis E vie'-russ	Hepatitis, p. 587
Mumps virus	mumps vie'-russ	Mumps, p. 582
Rotavirus	ro'-ta-vie'-russ	Acute diarrhea, p. 572
Norovirus	no"-row-vie'-russ	Acute diarrhea, p. 572
Protozoa		
Entamoeba histolytica	en"-tah-mee'-ba his"-toh-lit'-ih-kuh	Chronic diarrhea, p. 579
Cryptosporidium	crip'-toe-spor-id"-ee-um	Acute diarrhea, p. 572
Cyclospora	Sie"-clo-spor'-ah	Chronic diarrhea, p. 579
Giardia lamblia	jee-ar'-dee-ah lam'-blee-ah	Chronic diarrhea, p. 579
Helminths—nematodes		
Ascaris lumbricoides	a-scare'-is lum'-bri-coi"-dees	Intestinal distress plus migratory symptoms, p. 592
Enterobius vermicularis	en'-ter-oh"-bee-us ver-mick"-u-lar'-is	Intestinal distress, p. 591
Trichuris trichiura	tri-cur'-is trick-ee-ur'-ah	Intestinal distress, p. 591
Necator americanus and *Ancylostoma duodenale*	neh-cay'-ter a-mer'-i-can"-us, an"-sy-lo-sto"-mah dew-ah'-den-al"-ee	Intestinal distress plus migratory symptoms, p. 592
Toxocara species	tox'-uh-kair"-uh	Intestinal distress plus migratory symptoms, p. 592
Helminths—cestodes		
Hymenolepis	hie'-men-oh"-lep-is	Intestinal distress, p. 591
Taenia solium	te'-ne-ah so'-lee-um	Cysticercosis, p. 592
Diphyllobothrium latum	dif'-oh-lo-both"-ree-um lah'-tum	Intestinal distress, p. 591
Helminths—trematodes		
Schistosoma mansoni and *S. japonicum*	shis'-toh-so-my"-a-sis man-sohn"-ee, ja-pawn'-i-cum	Helminthic liver disease, p. 593

Right Here at Home: Neglected Parasitic Infections

This chapter has presented us with a variety of "unsavory" infections, including ones caused by worms. Maybe it came as a surprise to you that up to one-fourth of the world's population is infected with intestinal roundworms, for example. Or maybe it is easier to stomach because we are confident that protozoal and helminthic infections are relatively rare in the United States.

That's a mistake. The CDC has begun a campaign against five neglected parasitic infections (NPIs) in the United States. The five are

- Chagas disease—the trypanosome disease caused by *Trypanosoma cruzi* (chapter 18)
- Neurocysticercosis—caused by the tapeworm *Taenia solium* (this chapter)
- Toxocariasis—caused by worms that travel through tissues and can cause blindness (this chapter)
- Toxoplasmosis—the protozoan with which 60 million people in the United States are infected (chapter 17)
- Trichomoniasis—a protozoal infection of the genital tract that leaves those infected more vulnerable to other sexually transmitted infections, including HIV; also leads to premature births by infected mothers (chapter 21)

These diseases can affect anyone, but in the United States they are much more likely to be found in residents of poor neighborhoods and in immigrants from countries in which the diseases are prevalent. These are also the people least likely to seek or have access to medical care. One of the consequences of that circumstance is that it is difficult to estimate infection rates, but new attempts have begun.

These are a few pertinent statistics from the CDC:

- Neurocysticercosis is the single most common infectious cause of seizures in some areas of the United States.
- Toxocariasis is caused by dog and cat roundworms; up to 14% of the U.S. population has been exposed. About 70 people a year are blinded by this infection.
- Up to 300,000 people in the United States are currently infected with the protozoan that causes Chagas disease.
- In the United States, 3.7 million people are currently infected with *Trichomonas*.

It is time to stop thinking of these infections as "other people's problems."

Infectious Diseases Affecting
The Gastrointestinal Tract

Cysticercosis

Taenia solium

Dental Caries

Streptococcus mutans
Streptococcus sobrinus
Scardovia wiggsiae
Other bacteria

Periodontitis and Necrotizing Ulcerative Diseases

Tannerella forsythia
Aggregatibacter actinomycetemcomitans
Porphyromonas gingivalis
Treponema vincentii
Prevotella intermedia
Fusobacterium

Schistosomiasis

Schistosoma mansoni
Schistosoma japonicum

Hepatitis

Hepatitis A or E
Hepatitis B or C

Helminthic Infections with Intestinal and Migratory Symptoms

Ascaris lumbricoides
Necator americanus
Ancylostoma duodenale
Toxocara species

Helminths
Bacteria
Viruses
Protozoa

Mumps

Mumps virus

Gastritis and Gastric Ulcer

Helicobacter pylori

Acute Diarrhea

Salmonella
Shigella
Shiga-toxin-producing E. coli
Other E. coli
Campylobacter
Yersinia enterocolitica
Yersinia pseudotuberculosis
Clostridium difficile
Vibrio cholerae
Cryptosporidium
Rotavirus
Norovirus

Chronic Diarrhea

EAEC
Cyclospora cayetanensis
Giardia lamblia
Entamoeba histolytica

Food Poisoning

Staphylococcus aureus
Bacillus cereus
Clostridium perfringens

Tract Infections Causing Intestinal Distress

Trichuris trichiura
Enterobius vermicularis
Diphyllobothrium latum

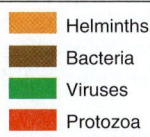

System Summary Figure 20.21

Chapter Summary

20.1 The Gastrointestinal Tract and Its Defenses

- The GI tract has a very heavy load of microorganisms, and it encounters millions of new ones every day. There are significant mechanical, chemical, and antimicrobial defenses to combat microbial invasion.

20.2 Normal Biota of the Gastrointestinal Tract

- Bacteria abound in all of the eight main sections of the gastrointestinal tract. Even the highly acidic stomach is colonized.

20.3 Gastrointestinal Tract Diseases Caused by Microorganisms (Nonhelminthic)

- **Acute Diarrhea:** In the United States, a third of all acute diarrhea is transmitted by contaminated food.
  - *Salmonella:* The *Salmonella* species that causes gastrointestinal disease is divided into many serotypes, based on major surface antigens.
  - *Shigella:* Frequent, watery, bloody stools, fever, and often intense abdominal pain. The bacterium *Shigella dysenteriae* produces a heat-labile exotoxin called shiga toxin.
  - Shiga-Toxin-Producing *E. coli* (STEC): This group contains *E. coli* O157:H7 and other virulent strains that cause hemolytic uremic syndrome and other debilitating effects.
  - Other *E. coli:* Enterotoxigenic *E. coli* (traveler's diarrhea), enteroinvasive *E. coli*, enteropathogenic *E. coli*, diffusely adherent *E. coli*, and enteroaggregative *E. coli* also cause diarrhea.
  - *Campylobacter:* Frequent, watery stools, fever, vomiting, headaches, and severe abdominal pain. Infrequently, infection can lead to serious neuromuscular paralysis called *Guillain-Barré syndrome.*
  - *Yersinia* Species: Both *Y. enterocolitica* and *Y. pseudotuberculosis* are agents of GI disease via food and beverage contamination.
  - *Clostridium difficile*: Pseudomembranous colitis (antibiotic-associated colitis), precipitated by therapy with broad-spectrum antibiotics.
  - *Vibrio cholerae:* Symptoms of secretory diarrhea and severe fluid loss can lead to death in less than 48 hours.
  - *Cryptosporidium:* Intestinal waterborne protozoan that infects mammals, birds, and reptiles.
  - Rotavirus: Primary viral cause of morbidity and mortality resulting from diarrhea, accounting for 50% of all cases.
  - Norovirus: Second most common cause of U.S. hospitalizations from food-borne diseases.
- **Food Poisoning:** Refers to symptoms in the gut that are caused by a preformed toxin. Caused most often by exotoxins from *Staphylococcus aureus*, *Bacillus cereus*, and *Clostridium perfringens*.
- **Chronic Diarrhea:** Caused most often by enteroaggregative *E. coli* (EAEC), or the protozoa *Cyclospora cayetanensis* or *Entamoeba histolytica.*
- **Tooth and Gum Infections**
  - Dental Caries: Alpha-hemolytic *Streptococcus mutans* is main cause.

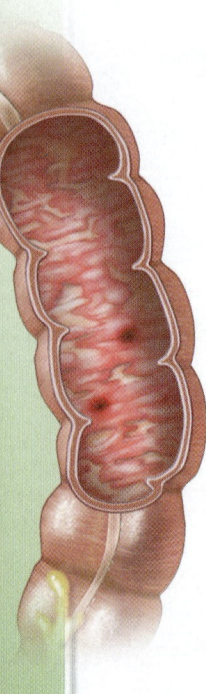

- Periodontal Diseases: The anaerobic bacteria *Tannerella forsythia, Aggregatibacter actinomycetemcomitans, Porphyromonas, Fusobacterium,* and spirochete species are causative agents.
- **Mumps:** Caused by an enveloped, single-stranded RNA virus (mumps virus) from the genus *Paramyxovirus.*
- **Gastritis and Gastric Ulcers:** *Helicobacter pylori*, a curved gram-negative rod, is the causative agent.
- **Hepatitis:** Inflammatory disease marked by necrosis of hepatocytes and a mononuclear response that swells and disrupts the liver, causing jaundice. Can be caused by a variety of different viruses.
  - Hepatitis A virus (HAV): A nonenveloped, single-stranded RNA enterovirus of low virulence. Spread through fecal-oral route. Inactivated vaccine available.
  - Hepatitis B virus (HBV): Enveloped DNA virus in the family *Hepadnaviridae.* Can be very serious, even life-threatening; some patients develop chronic liver disease in the form of necrosis, cirrhosis, or liver cancer. Transmitted by blood and other bodily fluids. Virus is major infectious concern for health care workers.
  - Hepatitis C Virus: RNA virus in *Flaviviridae* family. Shares characteristics of hepatitis B disease but is much more likely to become chronic. More commonly transmitted through blood contact than through other body fluids.

20.4 Gastrointestinal Tract Diseases Caused by Helminths

- **Helminthic Infections: Intestinal Distress as the Primary Symptom**
  - *Enterobius vermicularis:* "Pinworm"; most common worm disease of children in temperate zones. Not fatal, and most cases are asymptomatic.
- **Helminthic Infections: Intestinal Distress Accompanied by Migratory Symptoms:** The intestinal roundworm is acquired by the fecal-oral route. *Ascaris lumbricoides* can obstruct the GI tract or leave the GI tract and migrate throughout the body. Two worms known as hookworms, *Necator americanus* and *Anclyostoma duodenale*, penetrate intact skin and travel through the bloodstream to the lungs and then to the GI tract.
- **Cysticercosis:** Larval tapeworm cysts embedded in brain and other tissues, often causing seizures.
- **Schistosomiasis: Liver Disease:** *Schistosomiasis* in intestines is caused by blood flukes *Schistosoma mansoni* and *S. japonicum*. Symptoms include fever, chills, diarrhea, and liver and spleen disease.

Multiple-Choice Questions   Bloom's Levels 1 and 2: Remember and Understand

Select the correct answer from the answers provided.

1. Which of the following is not typically considered a defense against infection within the GI tract?
   a. keratinized layer of cells
   b. mucous coat
   c. secretory IgA
   d. continuous passage of cells

2. Gastric ulcers are caused by
   a. *Treponema vincentii.*
   b. *Prevotella intermedia.*
   c. *Helicobacter pylori.*
   d. all of the above.

3. The normal biota of the GI tract is most diverse (has the most different types of microbes) in the
   a. pharynx.
   b. stomach.
   c. small intestine.
   d. large intestine.

4. Which of these microorganisms is associated with Guillain-Barré syndrome?
   a. *E. coli*
   b. *Salmonella*
   c. *Campylobacter*
   d. *Shigella*

Critical Thinking   Bloom's Levels 3, 4, and 5: Apply, Analyze, and Evaluate

Critical thinking is the ability to reason and solve problems using facts and concepts. These questions can be approached from a number of angles and, in most cases, they do not have a single correct answer.

1. The normal biota of the GI tract seems to include a lot of disease-causing organisms. How can this be? That is, if they are normal, why are they also consider pathogens? If they are pathogens, why are they considered normal?

2. Why is it thought that the shiga toxin found in *E. coli* originated in *Shigella,* and not vice versa?

3. Why is heating food contaminated with *Staphylococcus aureus* no guarantee that the associated food poisoning will be prevented?

4. What are some of the ways we can prevent or slow down the spread of helminthic diseases?

5. Which members of the population are most at risk for hepatitis C? Why?

Visual Connections   Bloom's Level 5: Evaluate

This question connects previous images to a new concept.

1. **From chapter 11, figure 11.3*b*.** Imagine for a minute that the organism in this illustration is a shiga-toxin-producing *E. coli*. What would be one reason not to treat a patient having this infection with powerful antibiotics?

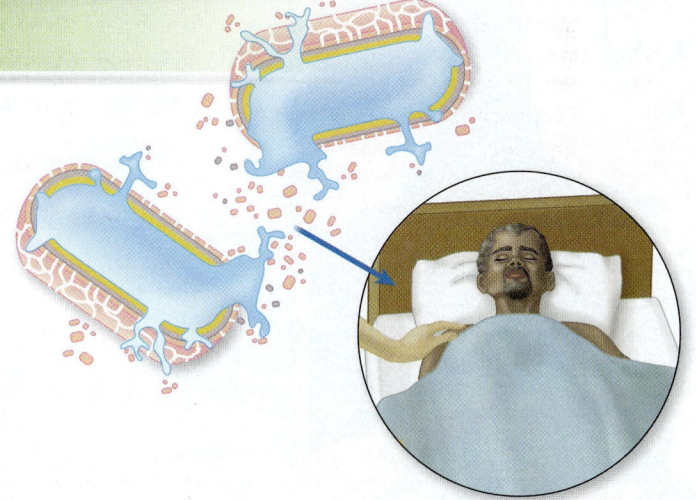

Outer membrane component causes fever, malaise, aches, and shock.

www.mcgrawhillconnect.com
Enhance your study of this chapter with study tools and practice tests. Also ask your instructor about the resources available through Connect, including the media-rich eBook, interactive learning tools, and animations.

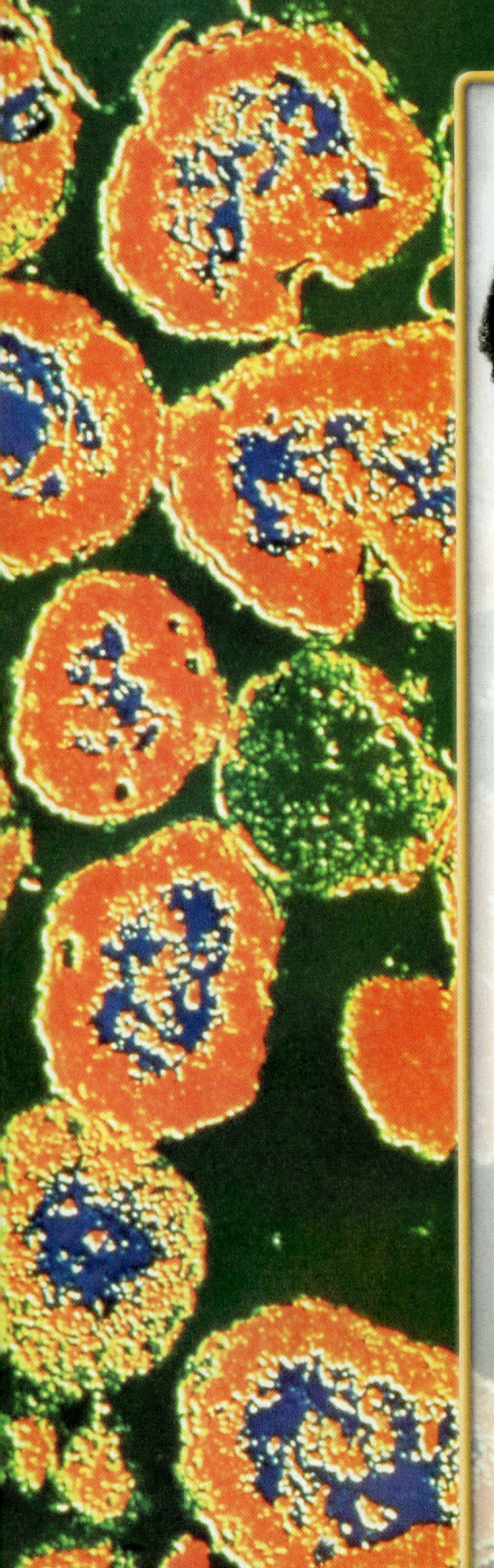

It's All in the Walk

One of the most disturbing and gut-wrenching cases I was ever involved in concerned a 10-year-old girl who was brought to the emergency room by her mother, who was afraid her daughter had appendicitis. What her daughter turned out to have was far more troubling.

The patient had developed a high fever (39.4°C [103°F]) and was complaining of severe lower abdominal pain. Upon questioning, the patient also admitted to painful urination. As I led her into an examining room, I noticed that she walked taking very small steps. She did not lift her feet off the floor and seemed reluctant or unable to straighten up fully. Her odd gait raised a red flag in my mind.

After performing a quick abdominal examination and obtaining vital signs, I found the physician on call and told her about the patient's symptoms. I also described how the patient moved with an odd shuffling gait. Although I could not recall what I had learned about the symptom, I knew that it was a diagnostic feature. The physician asked me how old the patient was and asked if I had had the opportunity to ask the patient whether she was sexually active. I had not done so due to the patient's age. To my surprise, the physician asked me to prepare for an internal (vaginal) exam in another room while she examined the patient and spoke to her mother.

- What condition did the physician suspect?

- Is the condition reportable? If so, to whom would you report it?

Case File Wrap-Up appears on page 625.

Infectious Diseases Affecting the Genitourinary System

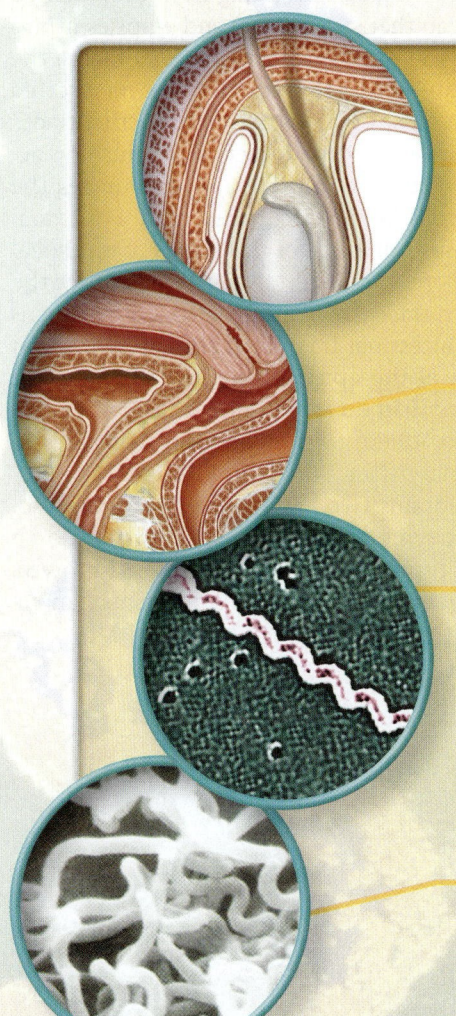

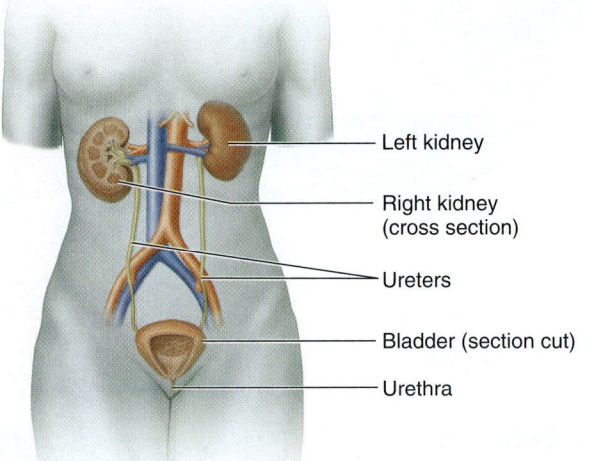

Figure 21.1 The urinary system.

21.1 The Genitourinary Tract and Its Defenses

As suggested by the name, the structures considered in this chapter are really two distinct organ systems. The *urinary tract* has the job of removing substances from the blood, regulating certain body processes, and forming urine and transporting it out of the body. The *genital system* has reproduction as its major function. It is also called the *reproductive system.*

The urinary tract includes the kidneys, ureters, bladder, and the urethra **(figure 21.1).** The kidneys remove metabolic wastes from the blood, acting as a sophisticated filtration system. Ureters are tubular organs extending from each kidney to the bladder. The bladder is a collapsible organ that stores urine and empties it into the urethra, which is the conduit of urine to the exterior of the body. In males, the urethra is also the terminal organ of the reproductive tract, but in females the urethra is separate from the vagina, which is the outermost organ of the reproductive tract.

The most obvious defensive mechanism in the urinary tract is the flushing action of the urine flowing out of the system. The flow of urine also encourages the **desquamation** (shedding) of the epithelial cells lining the urinary tract. For example, each time a person urinates, he or she loses hundreds of thousands of epithelial cells! Any microorganisms attached to them are also shed, of course. Probably the most common microbial threat to the urinary tract is the group of microorganisms that comprise the normal biota in the gastrointestinal tract, because the two organ systems are in close proximity. But the cells of the epithelial lining of the urinary tract have different chemicals on their surfaces than do those lining the GI tract. For that reason, most bacteria that are adapted to adhere to the chemical structures in the GI tract cannot gain a foothold in the urinary tract.

Urine, in addition to being acidic, also contains two antibacterial proteins, lysozyme and lactoferrin. You may recall that lysozyme is an enzyme that breaks down peptidoglycan. Lactoferrin is an iron-binding protein that inhibits bacterial growth. Finally, secretory IgA specific for previously encountered microorganisms can be found in the urine.

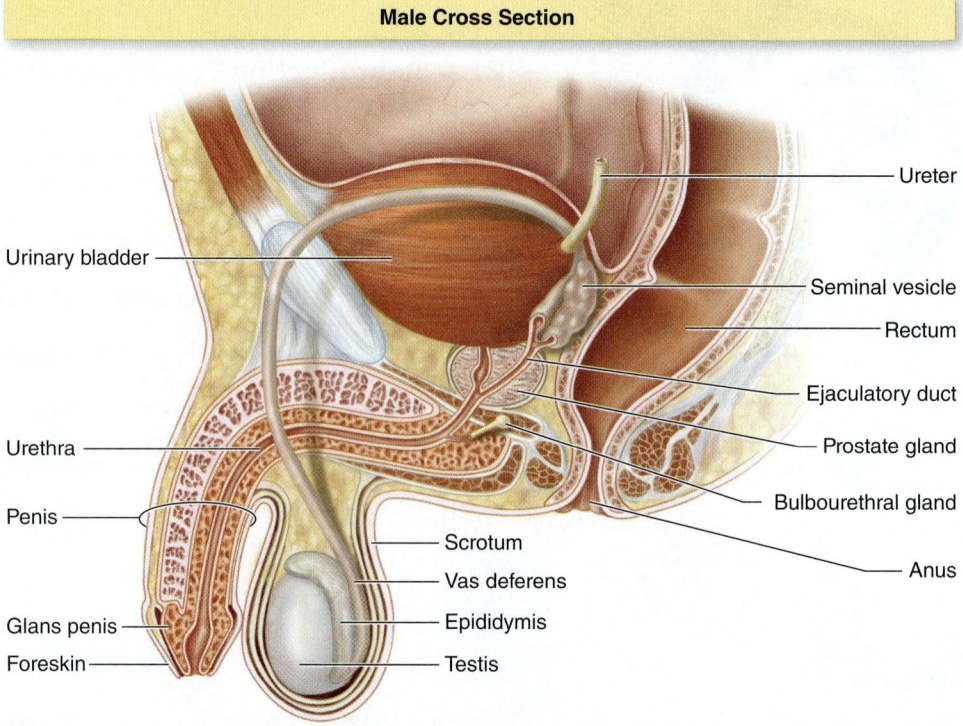

Figure 21.2 The male reproductive system.

The male reproductive system produces, maintains, and transports sperm cells and is the source of male sex hormones. It consists of the *testes*, which produce sperm cells and hormones, and the *epididymides*, which are coiled tubes leading out of the testes. Each epididymis terminates in a *vas deferens*, which combines with the seminal vesicle and terminates in the ejaculatory duct **(figure 21.2).** The contents of the ejaculatory duct empty into the urethra during ejaculation. The *prostate gland* is a walnut-shaped structure at the base of the urethra. It also contributes to the released fluid (semen). The external organs are the scrotum, containing the testes, and the *penis*, a cylindrical organ that houses the urethra. As for its innate defenses, the male reproductive system also benefits from the flushing action of the urine, which helps move microorganisms out of the system.

The female reproductive system consists of the *uterus*, the *fallopian tubes* (also called uterine tubes), *ovaries*, and *vagina* **(figure 21.3).** One very important tissue of the female reproductive tract is the *cervix*, which is the lower one-third of the uterus and the part that connects to the vagina. The cervix serves as the opening to the uterus. The cervix is a common site of infection in the female reproductive tract.

The natural defenses of the female reproductive tract vary over the lifetime of the woman. The vagina is lined with mucous membranes and, thus, has the protective covering of secreted mucus. During childhood and after menopause, this mucus is the major nonspecific defense of this system. Secretory IgA antibodies specific for any previously encountered infections would be present on these surfaces. During a woman's reproductive years, a major portion of the defense is provided by changes in the pH of the vagina brought about by the release of estrogen. This hormone stimulates the vaginal mucosa to secrete glycogen, which certain bacteria can ferment into acid, lowering the pH of the vagina to about 4.5. Before puberty, a girl produces little estrogen and little glycogen and has a vaginal pH of about 7. The change in pH beginning in adolescence results in a vastly different normal biota in the vagina, described later. The biota of women in their childbearing years is thought to prevent the establishment and invasion of microbes that might have the potential to harm a developing fetus.

Female Cross Section

Fallopian (uterine) tube

Ovary

Vertebral column

Uterus

Urinary bladder

Cervix of uterus

Symphysis pubis

Vagina

Mons pubis

Urethra

Rectum

Clitoris

Anus

Urethral orifice

Vaginal orifice

Labia minora

Labia majora

Figure 21.3 **The female reproductive system.**

The genitourinary tract includes the reproductive organs.

Normal biota in the female genital tract is different in women of childbearing age than in younger or older women.

21.2 Normal Biota of the Genitourinary Tract

As with all other organ systems, recent research is showing that the urinary system harbors a more diverse microbiota than we used to think. The lower urethra has a well-established microbiota, while the upper urinary tract appears to have fewer types and lower abundance. The exact microbial composition varies between men and women—and also among individuals. Because the urethra in women is so short (about 3.5 cm long) and is in such close proximity to the anus, it can act as a pipeline for bacteria from the GI tract to the bladder, resulting in urinary tract infections. The outer surface of the penis is colonized by *Pseudomonas* and *Staphylococcus* species—aerobic bacteria. In an uncircumcised penis, the area under the foreskin is colonized by anaerobic gram-negatives.

Normal Biota of the Male Genital Tract

Because the terminal "tube" of the male genital tract is the urethra, the normal biota of the male genital tract (i.e., in the urethra) is comprised of the same residents just described. However, after sexual activity begins, microbes associated with sexually transmitted infections (STIs) can sometimes become long-term residents.

Normal Biota of the Female Genital Tract

In the female genital tract, only the vagina harbors a large population of normal biota. As just mentioned, before puberty and after menopause, the pH of the vagina is close to neutral and the vagina harbors a biota that is similar to that found in the urethra. After the onset of puberty, estrogen production leads to glycogen release in the vagina, resulting in an acidic pH. *Lactobacillus* species thrive in the acidic environment and contribute to it, converting sugars to acid. Their predominance in the vagina, combined with the acidic environment, discourages the growth of many microorganisms. The estrogen-glycogen effect continues, with minor disruptions, throughout the childbearing years until menopause, when the biota gradually returns to a mixed population similar to that of prepuberty. Note that the fungus *Candida albicans* is also present at low levels in the healthy female reproductive tract.

Genitourinary Tract Defenses and Normal Biota

	Defenses	Normal Biota
Urinary Tract (Both Genders)	Flushing action of urine; specific attachment sites not recognized by most nonnormal biota; shedding of urinary tract epithelial cells, secretory IgA, lysozyme, and lactoferrin in urine	Nonhemolytic *Streptococcus, Staphylococcus, Corynebacterium, Lactobacillus, Prevotella, Veillonella, Gardnerella*
Female Genital Tract (Childhood and Postmenopause)	Mucus secretions, secretory IgA	Same as for urinary tract
Female Genital Tract (Childbearing Years)	Acidic pH, mucus secretions, secretory IgA	Variable, but often *Lactobacillus* predominates; also *Prevotella, Sneathia, Streptococcus,* and *Candida albicans*
Male Genital Tract	Same as for urinary tract	Urethra: same as for urinary tract Outer surface of penis: *Pseudomonas* and *Staphylococcus* Sulcus of uncircumcised penis: anaerobic gram-negatives

When the normal vaginal microbiota is altered and is unable to keep *Candida albicans* in check, an overgrowth of the fungus may occur, resulting in a symptomatic yeast infection.

21.3 Urinary Tract Diseases Caused by Microorganisms

We consider two types of diseases in this section. **Urinary tract infections (UTIs)** result from invasion of the urinary system by bacteria or other microorganisms. **Leptospirosis,** by contrast, is a spirochete-caused disease transmitted by contact of broken skin or mucous membranes with contaminated animal urine.

Highlight Disease

Urinary Tract Infections (UTIs)

Even though the flushing action of urine helps to keep infections to a minimum in the urinary tract, urine itself is a good growth medium for many microorganisms. When urine flow is reduced, or bacteria are accidentally introduced into the bladder, an infection of that organ (known as *cystitis*) can occur. Occasionally, the infection can also affect the kidneys, in which case it is called *pyelonephritis*. If an infection is limited to the urethra, it is called *urethritis*.

▶ Signs and Symptoms

Cystitis is a disease of sudden onset. Symptoms include pain, frequent urges to urinate even when the bladder is empty, and burning pain accompanying urination (called *dysuria*). The urine can be cloudy due to the presence of bacteria and white blood cells. It may have an orange tinge from the presence of red blood cells (*hematuria*). Low-grade fever and nausea are frequently present. If back pain is present and fever is high, it is an indication that the kidneys may also be involved (pyelonephritis). Pyelonephritis is a serious infection that can result in permanent damage to the kidneys if improperly or inadequately treated. If only the bladder is involved, the condition is sometimes called acute uncomplicated UTI.

▶ Causative Agents

As we saw in the discussion of pneumonia, it is important to distinguish between UTIs that are acquired in health care facilities and those acquired outside of the health care setting. When they occur in health care facilities, they are almost always a result of catheterization and are therefore called *catheter-associated UTIs (CA-UTIs)*. (Be careful! The abbreviation "CA" in other infections often refers to "community-acquired"–just the opposite of what is meant here! We will spell out "community" in referring to non-healthcare-associated UTIs.)

In 95% of UTIs, the cause is bacteria that are normal biota in the gastrointestinal tract. *Escherichia coli* is by far the most common of these, accounting for approximately 80% of community-acquired urinary tract infections. *Staphylococcus saprophyticus* and *Enterococcus* are also common culprits. These last two are only referenced in **Disease Table 21.1** following the discussion of *E. coli*.

The *E. coli* species that cause UTIs are ones that exist as normal biota in the gastrointestinal tract. They are not the ones that cause diarrhea and other digestive tract diseases.

Disease Table 21.1 Urinary Tract Infections

Causative Organism(s)	*Escherichia coli*	*Staphylococcus saprophyticus*	*Enterococcus*
Most Common Modes of Transmission	Opportunism: transfer from GI tract (community-acquired) or environment or GI tract (via catheter)		
Virulence Factors	Adhesins, motility	-	-
Culture/Diagnosis	Usually culture-based; antimicrobial susceptibilities always checked		
Prevention	Hygiene practices; in case of CA-UTIs, limit catheter usage		
Treatment	Based on susceptibility testing	Based on susceptibility testing	Based on susceptibility testing; vancomycin-resistant *Enterococcus* is in **Serious Threat** category in CDC Antibiotic Resistance Report
Epidemiological Features	Causes 90% of community UTIs and 50%-70% of CA-UTIs	Causes small percentage of community UTIs and even lower percentage of CA-UTIs	Frequent cause of CA-UTIs

▶ Transmission and Epidemiology

Community-acquired UTIs are nearly always "transmitted" *not* from one person to another but from one organ system to another, namely from the GI tract to the urinary system. They are much more common in women than in men because of the nearness of the female urethral opening to the anus (see figure 21.3). Many women experience what have been referred to as "recurrent urinary tract infections," although it is now known that some *E. coli* can invade the deeper tissue of the urinary tract and therefore avoid being destroyed by antibiotics. They can emerge later to cause symptoms again. It is not clear how many "recurrent" infections are actually infections that reactivate in this way.

Catheter-associated UTIs are also most commonly caused by *E. coli*, *S. saprophyticus*, and *Enterococcus*. *Klebsiella* species are another common cause. The National Healthcare Safety Network is now recommending minimizing the use of urinary catheters as much as possible to limit the incidence of these infections.

▶ Treatment

Sulfa drugs such as trimethoprimsulfa-methoxazole are most often used for UTIs of various etiologies. If there is a lot of resistance to this treatment in the local area, other drugs must be used. Often another nonantibiotic drug called phenazopyridine (Pyridium) is administered simultaneously. This drug relieves the very uncomfortable symptoms of burning and urgency. However, some physicians are reluctant to administer this medication for fear that it may mask worsening symptoms; when Pyridium is used, it should be used only for a maximum of 2 days. Pyridium is an azo dye and causes the urine to turn a dark orange to red color. It may also color contact lenses. A large percentage of *E. coli* strains is resistant to penicillin derivatives, so these should be avoided. Also, a new strain of *E. coli* (ST131) has arisen, which is highly virulent and, more troubling, resistant to multiple antibiotics. Medical professionals are ringing alarm bells about this strain, saying that if it acquires resistance to one more class of antibiotics it will become virtually untreatable.

Leptospirosis

This infection is a zoonosis associated with wild animals and domesticated animals. It can affect the kidneys, liver, brain, and eyes. It is considered in this section because it can have its major effects on the kidneys and because its presence in animal urinary tracts causes it to be shed into the environment through animal urine.

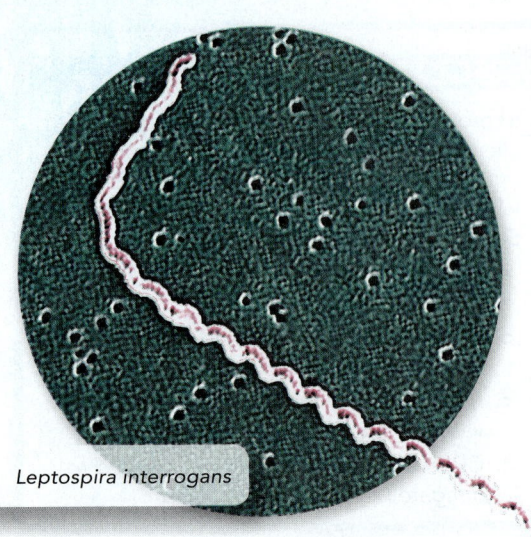

Leptospira interrogans

▶ Signs and Symptoms

Leptospirosis has two phases. During the early, or leptospiremic, phase, the pathogen appears in the blood and cerebrospinal fluid. Symptoms are sudden high fever, chills, headache, muscle aches, conjunctivitis, and vomiting. During the second phase (called the immune phase), the blood infection is cleared by natural defenses. This period is marked by milder fever; headache due to leptospiral meningitis; and *Weil's syndrome*, a cluster of symptoms characterized by kidney invasion, hepatic disease, jaundice, anemia, and neurological disturbances. Long-term disability and even death can result from damage to the kidneys and liver, but they occur primarily with the most virulent strains and in elderly persons.

▶ Causative Agent

Leptospires are typical spirochete bacteria marked by tight, regular, individual coils with a bend or hook at one or both ends. *Leptospira interrogans* (lep″-toh-spy′-rah in-terr′-oh-ganz) is the species that causes leptospirosis in humans and animals. There are nearly 200 different serotypes of this species distributed among various animal groups, which accounts for extreme variations in the disease manifestations in humans.

▶ Transmission and Epidemiology

Infection occurs almost entirely through contact of skin abrasions or mucous membranes with animal urine or some environmental source containing urine. In 1998, dozens of athletes competing in the swimming phase of a triathlon in Illinois contracted leptospirosis from the water.

▶ Treatment

Early treatment with doxycycline, penicillin G, or ceftriazone rapidly reduces symptoms and shortens the course of disease, but delayed therapy is less effective. Other spirochete diseases, such as syphilis (described later), exhibit this same pattern of being susceptible to antibiotics early in the infection but less so later on.

> **21.3 LEARNING OUTCOMES—Assess Your Progress**
>
> **4.** List the possible causative agents, modes of transmission, virulence factors, diagnostic techniques, and prevention/treatment for the Highlight Disease, urinary tract infection.
>
> **5.** Discuss important features of leptospirosis.

21.4 Reproductive Tract Diseases Caused by Microorganisms

Not all reproductive tract diseases are sexually transmitted, though many are. Vaginitis/vaginosis may or may not be; prostatitis probably is not.

It was very difficult to choose a "highlight" disease for this section, though we were sure we wanted it to be a sexually transmitted disease (now often called *sexually transmitted infections*, due to the fact that so many of them are silent, while still being serious). Sexually transmitted infections (STIs) now are more common in the United States than in any other industrialized country. The discharge diseases are responsible for unprecedented numbers of infertility cases. Herpes and human papillomavirus (HPV) infections are incurable and therefore simply increase in their prevalence over time. Americans have become more familiar with the dangers of HPV infection since the introduction of the HPV vaccine, so, in a split decision, we have decided to highlight the dangers of curable–but usually undetected–discharge diseases.

Disease Table 21.2 Leptospirosis	
Causative Organism(s)	*Leptospira interrogans*
Most Common Modes of Transmission	Vehicle: contaminated soil or water
Virulence Factors	Adhesins, invasion proteins
Culture/Diagnosis	Slide agglutination test of patient's blood for antibodies
Prevention	Avoiding contaminated vehicles
Treatment	Doxycycline, Penicillin G, or ceftriaxone
Epidemiological Features	United States: 100–200 cases per year, half in Hawaii; internationally: 80% of people in tropical areas are seropositive

A Note About STI Statistics

It is difficult to compare the incidence of different STIs to one another, for several reasons. The first is that many, many infections are "silent," and therefore infected people don't access the health care system and don't get counted. Of course, we know that many silent infections are actually causing damage that won't be noticed for years—and when it is, the original causative organism is almost never sought out. The second reason is that only some STIs are officially reportable to the CDC, and states' regulations vary. *Chlamydia* infection and gonorrhea are nationally reportable, for example, but herpes and HPV infections are not. In each section, we present estimates of the prevalence and/or incidence of the diseases wherever possible.

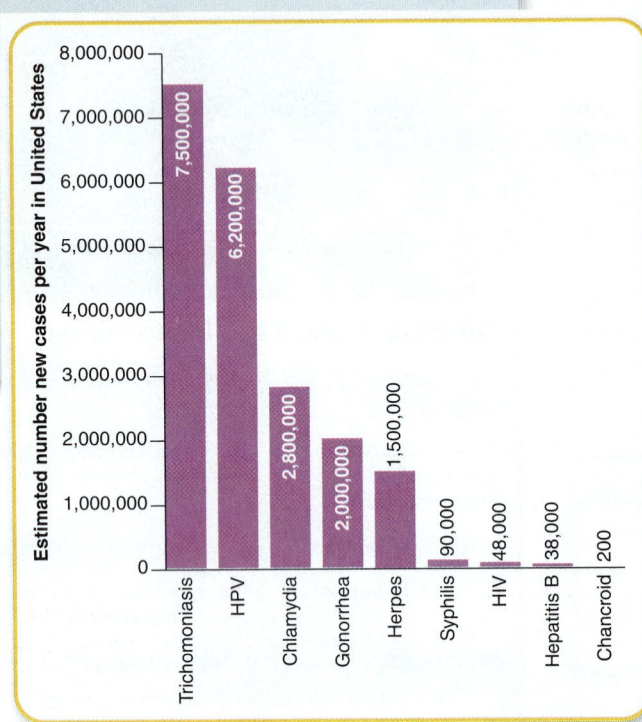

Discharge Diseases with Major Manifestation in the Genitourinary Tract

Discharge diseases are those in which the infectious agent causes an increase in fluid discharge in the male and female reproductive tracts. Examples are trichomoniasis, gonorrhea, and *Chlamydia* infection. The causative agents are transferred to new hosts when the fluids in which they live contact the mucosal surfaces of the receiving partner.

Gonorrhea

Gonorrhea has been known as a sexually transmitted disease since ancient times. It was named by the Greek physician Claudius Galen, who thought that it was caused by an excess flow of semen.

▶ Signs and Symptoms

In the male, infection of the urethra elicits urethritis, painful urination, and a yellowish discharge, although a relatively large number of cases are asymptomatic. In most cases, infection is limited to the distal urogenital tract, but it can occasionally spread from the urethra to the prostate gland and epididymis (refer to figure 21.2). Scar tissue formed in the spermatic ducts during healing of an invasive infection can render a man infertile.

In the female, it is likely that both the urinary and genital tracts will be infected during sexual intercourse. A mucopurulent (containing mucus and pus) or bloody vaginal discharge occurs in about half of the cases, along with painful urination if the urethra is affected. Major complications occur when the infection ascends from the vagina and cervix to higher reproductive structures such as the uterus and fallopian tubes **(figure 21.4).** One disease resulting from this progression is **salpingitis**

Figure 21.4 Invasive gonorrhea in women. *(Left)* Normal state. *(Right)* In ascending gonorrhea, the gonococcus is carried from the cervical opening up through the uterus and into the fallopian tubes. Pelvic inflammatory disease (PID) is a serious complication that can lead to scarring in the fallopian tubes, ectopic pregnancies, and mixed anaerobic infections.

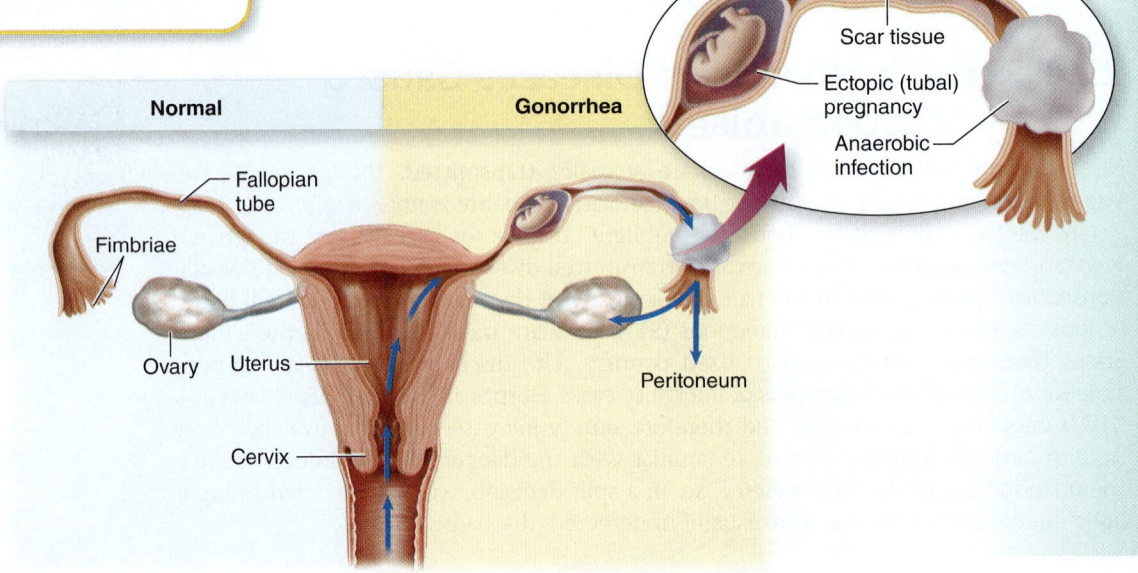

(sal"-pin-jy'-tis). This inflammation of the fallopian tubes may be isolated, or it may also include inflammation of other parts of the upper reproductive tract, termed *pelvic inflammatory disease (PID)*. It is not unusual for the microbe that initiates PID to become involved in mixed infections with anaerobic bacteria. The buildup of scar tissue from PID can block the fallopian tubes, causing sterility or ectopic pregnancies.

Serious consequences of gonorrhea can occur outside of the reproductive tract. In a small number of cases, the gonococcus enters the bloodstream and is disseminated to the joints and skin. Involvement of the wrist and ankle can lead to chronic arthritis and a painful, sporadic, papular rash on the limbs. Rare complications of gonococcal bacteremia are meningitis and endocarditis.

Children born to gonococcus carriers are also in danger of being infected as they pass through the birth canal. Because of the potential harm to the fetus, physicians usually screen pregnant mothers for its presence. Gonococcal eye infections are very serious and often result in keratitis, ophthalmia neonatorum, and even blindness **(figure 21.5)**. A universal precaution to prevent such complications is the use of antibiotic eyedrops or ointments (usually erythromycin) for newborn babies. The pathogen may also infect the pharynx and respiratory tract of neonates. Finding gonorrhea in children other than neonates is strong evidence of sexual abuse by infected adults, and it calls for child welfare consultation along with thorough bacteriologic analysis.

▶ Causative Agent

N. gonorrhoeae is a pyogenic gram-negative diplococcus. It appears as pairs of kidney bean–shaped bacteria, with their flat sides touching **(figure 21.6)**.

▶ Pathogenesis and Virulence Factors

Successful attachment is key to the organism's ability to cause disease. Gonococci use specific chemical groups on the tips of fimbriae to anchor themselves to mucosal epithelial cells. Once the bacterium attaches, it invades the cells and multiplies within the basement membrane.

The fimbriae may also play a role in slowing down effective immunity. The fimbrial proteins are controlled by genes that can be turned on or off, depending on the bacterium's situation. This phenotypic change is called phase variation. In addition, the genes can rearrange themselves to put together fimbriae of different configurations. This antigenic variation confuses the body's immune system. Antibodies that previously recognized fimbrial proteins may not recognize them once they are rearranged.

The gonococcus also possesses an enzyme called IgA protease, which can cleave IgA molecules stationed for protection on mucosal surfaces. In addition, pieces of its outer membrane are shed during growth. These "blebs," containing endotoxin, probably play a role in pathogenesis because they can stimulate portions of the nonspecific defense response, resulting in localized damage.

▶ Transmission and Epidemiology

Except for neonatal infections, the gonococcus is spread through some form of sexual contact.

Gonorrhea is a strictly human infection that occurs worldwide and ranks among the most common sexually transmitted diseases. About 350,000 cases are reported in the United States each year, while the number of asymptomatic infections is in the millions. The good news is that these numbers have been declining since 2006.

It is important to consider the reservoir of asymptomatic males and females when discussing the transmission of the infection. Because approximately 10% of infected males and 50% of infected females experience no symptoms, it is often spread unknowingly.

▶ Culture and Diagnosis

In males, it is easy to diagnose this disease; a Gram stain of urethral discharge is diagnostic. The normal biota of the male urethra is so sparse that it is easy to see the

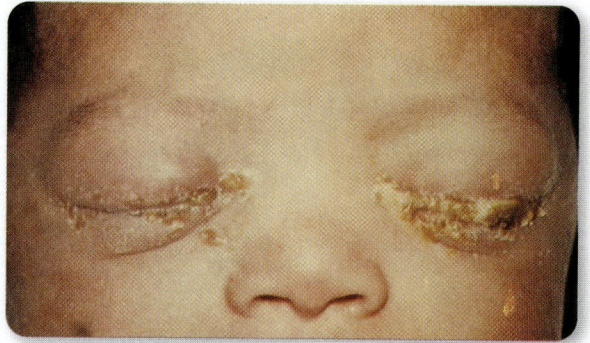

Figure 21.5 Gonococcal ophthalmia neonatorum in a week-old infant. The infection is marked by intense inflammation and edema; if allowed to progress, it causes damage that can lead to blindness. Fortunately, this infection is completely preventable and treatable.

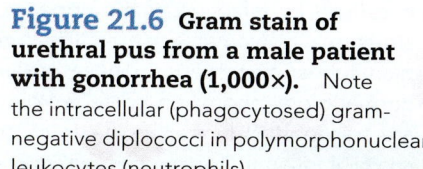

Figure 21.6 Gram stain of urethral pus from a male patient with gonorrhea (1,000×). Note the intracellular (phagocytosed) gramnegative diplococci in polymorphonuclear leukocytes (neutrophils).

Neutrophils   Gonococci

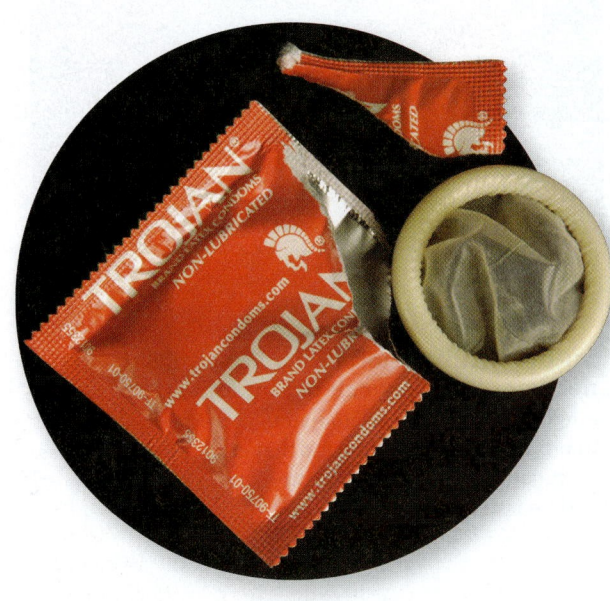

diplococcus inside of neutrophils (figure 21.6). In females, other methods, such as ELISA or PCR tests, are called for. Alternatively, the bacterium can be cultured on Thayer-Martin agar, a rich chocolate agar base with added antibiotics that inhibit competing bacteria.

N. gonorrhoeae grows best in an atmosphere containing increased CO_2. Because *Neisseria* is so fragile, it is best to inoculate it onto media directly from the patient rather than using a transport tube. Gonococci produce catalase, enzymes for fermenting various carbohydrates, and the enzyme cytochrome oxidase, which can be used for identification as well. Gonorrhea is a reportable disease.

▶ Prevention

No vaccine is yet available for gonorrhea. Using condoms is an effective way to avoid transmission of this and other discharge diseases.

▶ Treatment

The CDC runs a program called the Gonococcal Isolate Surveillance Project (GISP) to monitor the growing prevalence of antibiotic-resistant strains of *N. gonorrhoeae*. Every month in 28 local STD clinics around the country, *N. gonorrhoeae* isolates from the first 25 males diagnosed with the infection are sent to regional testing labs, their antibiotic sensitivities are determined, and the data are provided to the GISP program at the CDC. Although the most highly resistant strains have not yet been observed in the United States, the CDC has recently changed its overall recommendation for the treatment of gonorrhea worldwide. It now advises the use of ceftriaxone + azithromycin or doxycycline. This bacterium is at the highest threat level on the CDC's list of antibiotic-resistant organisms.

Chlamydia

Genital *Chlamydia* infection is the most common reportable infectious disease in the United States. Annually, more than 1 million cases are reported, but the actual infection rate may be five to seven times that number. The overall prevalence among sexually active young women ages 14 to 19 is 6.8% according to the CDC. It is at least two to three times as common as gonorrhea. The vast majority of cases are asymptomatic. When we consider the serious consequences that may follow *Chlamydia* infection, those facts are very disturbing.

▶ Signs and Symptoms

In males who experience symptoms of *Chlamydia* infection, the bacterium causes an inflammation of the urethra. The symptoms mimic gonorrhea—namely, discharge and painful urination. Untreated infections may lead to epididymitis. Females who experience symptoms have cervicitis, a discharge, and often salpingitis. Pelvic inflammatory disease is a frequent sequela of female chlamydial infection. A woman is even more likely to experience PID as a result of a *Chlamydia* infection than as a result of gonorrhea. Up to 75% of *Chlamydia* infections are asymptomatic, which puts women at risk for developing PID because they don't seek treatment for initial infections. The PID itself may be acute and painful, or it may be relatively asymptomatic, allowing damage to the upper reproductive tract to continue unchecked.

Certain strains of *C. trachomatis* can invade the lymphatic tissues, resulting in another condition called lymphogranuloma venereum. This condition is accompanied by headache, fever, and muscle aches. The lymph nodes near the lesion begin to fill with granuloma cells and become enlarged and tender. These "nodes" can cause long-term lymphatic obstruction that leads to chronic, deforming edema of the genitalia or anus. The disease is endemic to South America, Africa, and Asia, but occasionally occurs in other parts of the world. Its incidence in the United States is about 500 cases per year.

A Note About Pelvic Inflammatory Disease (PID) and Infertility

The National Center for Health Statistics estimates that more than 6 million women in the United States have impaired fertility. There are many different reasons for infertility, but the leading cause is pelvic inflammatory disease, or PID. PID is caused by infection of the upper reproductive structures in women—namely, the uterus, fallopian tubes, and ovaries. These organs have no normal biota, and when bacteria from the vagina are transported higher in the tract, they start a chain of inflammatory events that may or may not be noticeable to the patient. The inflammation can be acute, resulting in pain, abnormal vaginal discharge, fever, and nausea, or it can be chronic, with less noticeable symptoms. In acute cases, women usually seek care; in some ways, these can be considered the lucky ones. If the inflammation is curbed at an early stage by using antibiotics to kill the bacteria, chances are better that the long-term sequelae of PID can be avoided. *Chlamydia* infection is the leading cause of PID, followed closely by *N. gonorrhoeae* infection. But other bacteria, perhaps also including normal biota of the reproductive tract, can also cause PID if they are traumatically introduced into the uterus.

Babies born to mothers with *Chlamydia* infections can develop eye infections and also pneumonia if they become infected during passage through the birth canal. Infant conjunctivitis caused by contact with maternal *Chlamydia* infection is the most prevalent form of conjunctivitis in the United States (100,000 cases per year). Antibiotic drops or ointment applied to newborns' eyes are chosen to eliminate both *Chlamydia* and *N. gonorrhoeae*.

▶ Causative Agent

C. trachomatis is a very small gram-negative bacterium. It lives inside host cells as an obligate intracellular parasite. All *Chlamydia* species alternate between two distinct stages, illustrated in **table 21.1.**

▶ Pathogenesis and Virulence Factors

Chlamydia's ability to grow intracellularly contributes to its virulence because it escapes certain aspects of the host's immune response. Also, the bacterium has a unique cell wall that apparently prevents the phagosome from fusing with the lysosome inside phagocytes. The presence of the bacteria inside cells causes the release of cytokines that provoke intense inflammation. This defensive response leads to most of the actual tissue damage in *Chlamydia* infection. Of course, the last step of inflammation is repair, which often results in scarring. This can have disastrous effects on a narrow-diameter structure like the fallopian tube.

▶ Transmission and Epidemiology

The reservoir of pathogenic strains of *C. trachomatis* is the human body. The microbe shows an astoundingly broad distribution within the population, and incidence is rising. Adolescent women are more likely than older women to harbor the bacterium because it prefers to infect cells that are particularly prevalent on the adolescent cervix. It is transmitted through sexual contact and also vertically. Fifty percent of (untreated) babies born to infected mothers will acquire chlamydial conjunctivitis or pneumonia.

▶ Culture and Diagnosis

Infection with this microorganism is usually detected initially using a rapid technique such as PCR or ELISA. Direct fluorescent antibody detection is also used. Serology is not always reliable. In addition, antibody to *Chlamydia* is very common in adults and often indicates past, not present, infection. A urine test is available, which has definite advantages for widespread screening, but it is slightly less accurate for females than males.

▶ Prevention

Avoiding contact with infected tissues and secretions through abstinence or barrier protection (condoms) is the only means of prevention.

▶ Treatment

The CDC recommends annual screening of young women for this often asymptomatic infection. It is also recommended that older women with some risk factor (new sexual partner, for instance) also be screened. Treatment is usually with doxycycline or azithromycin. Coinfection with gonorrhea should be assumed and treated similarly. Many patients become reinfected soon after treatment; therefore, the recommendation is that patients be rechecked for *Chlamydia* infection 3 to 4 months after treatment. Treatment of all sexual partners of the patient is also recommended to prevent reinfection. Repeated infections with *Chlamydia* increase the likelihood of PID and other serious sequelae.

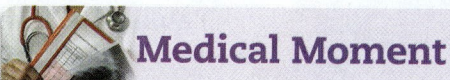

Medical Moment

Female Condoms

Condoms are a reliable way to prevent the spread of STIs. Although not 100% effective, they greatly reduce the risk of contracting a sexually transmitted infection. The female condom was introduced in the 1990s to provide an alternative for women.

Why are more women not embracing the female condom? Female condoms are more expensive and may be less readily available. They are also more awkward to insert and may reduce sensation for some women (a common complaint of men regarding the male condom).

The female condom has a higher failure rate than male condoms. An estimated 21 out of 100 women will become pregnant in the first year of typical use—this may be due to failure to use the condom correctly. Other causes of failure may include breakage of the condom, the condom slipping out of place during intercourse, or the penis slipping between the condom and the vagina.

Table 21.1 The Life Cycle of *Chlamydia*

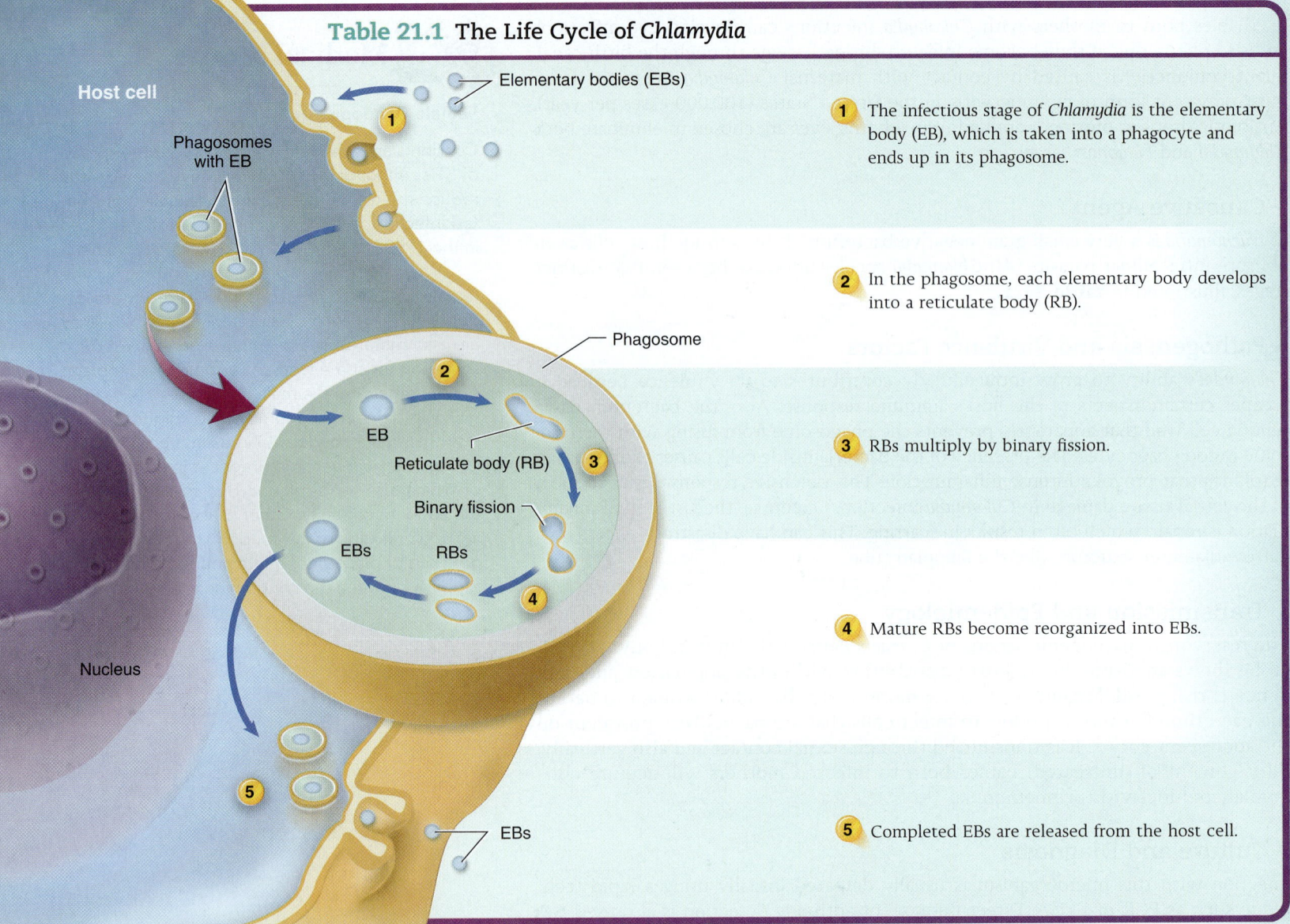

1. The infectious stage of *Chlamydia* is the elementary body (EB), which is taken into a phagocyte and ends up in its phagosome.

2. In the phagosome, each elementary body develops into a reticulate body (RB).

3. RBs multiply by binary fission.

4. Mature RBs become reorganized into EBs.

5. Completed EBs are released from the host cell.

Disease Table 21.3 Genital Discharge Diseases (in Addition to Vaginitis/Vaginosis)

	Gonorrhea	Chlamydia
Causative Organism(s)	*Neisseria gonorrhoeae*	*Chlamydia trachomatis*
Most Common Modes of Transmission	Direct contact (STI), also vertical	Direct contact (STI), vertical
Virulence Factors	Fimbrial adhesins, antigenic variation, IgA protease, membrane blebs/endotoxin	Intracellular growth resulting in avoiding immune system and cytokine release, unusual cell wall preventing phagolysosome fusion
Culture/Diagnosis	Gram stain in males, rapid tests (PCR, ELISA) for females, culture on Thayer-Martin agar	PCR or ELISA, can be followed by cell culture
Prevention	Avoid contact; condom use	Avoid contact; condom use
Treatment	Coinfection with gonorrhea and *C. trachomatis* should be assumed; treat with doxycycline or azithromycin. Be on alert for multidrug-resistant *N. gonorrhoeae*, which is in **Urgent Threat** category in CDC Antibiotic Resistance Report.	
Distinctive Features	Rare complications including arthritis, meningitis, endocarditis	More commonly asymptomatic than gonorrhea
Effects on Fetus	Eye infections, blindness	Eye infections, pneumonia
Epidemiological Features	United States: rates decreased 16% between 2006 and 2010; internationally: 26 million cases	United States: 2.8 million new infections per year; internationally: eye infection (trachoma) has 90% prevalence rate in developing world

Vaginitis and Vaginosis

▶ Signs and Symptoms

Vaginitis, an inflammation of the vagina, is a condition characterized by some degree of vaginal itching, depending on the etiologic agent. Symptoms may also include burning and sometimes a discharge, which may take different forms as well. Vaginosis is similar but does not include significant inflammation.

▶ Causative Agents

While a variety of bacteria and even protozoa can cause vaginitis, the most well-known agent is the fungus *Candida albicans*. The vaginal condition caused by this fungus is known as a *yeast infection*.

Candida albicans

C. albicans is a dimorphic fungus that is normal biota in from 50% to 100% of humans, living in low numbers on many mucosal surfaces such as the mouth, gastrointestinal tract, vagina, and so on. The vaginal condition it causes is also called vulvovaginal candidiasis. The yeast is easily detectable on a wet prep or a Gram stain of material obtained during a pelvic exam **(figure 21.7)**. The presence of pseudohyphae in the smear is a clear indication that the yeast is growing rapidly and causing a yeast infection.

In otherwise healthy people, the fungus is not invasive and limits itself to this surface infection. Please note, however, that *Candida* infections of the bloodstream do occur and they have high mortality rates. They do not normally stem from vaginal infections with the fungus, however, but are seen most frequently in hospitalized patients. AIDS patients are also at risk of developing systemic *Candida* infections.

▶ Transmission and Epidemiology

Vaginal infections with this organism are nearly always opportunistic. Disruptions of the normal bacterial biota or even minor damage to the mucosal epithelium in the vagina can lead to overgrowth by this fungus. Disruptions may be mechanical, such as trauma to the vagina, or they may be chemical, as when broad-spectrum antibiotics taken for some other purpose temporarily diminish the vaginal bacterial population. Diabetics and pregnant women are also predisposed to vaginal yeast overgrowths. Some women are prone to this condition during menstruation.

It is possible to transmit this yeast through sexual contact, especially if a woman is experiencing an overgrowth of it. The recipient's immune system may well subdue the yeast so that it acts as normal biota in them. But the yeast may be passed back to the original partner during further sexual contact after treatment. Women with HIV infection experience frequently recurring yeast infections. Also, a small percentage of women with no underlying immune disease experience chronic or recurrent vaginal infection with *Candida* for reasons that are not clear.

▶ Prevention and Treatment

No vaccine is available for *C. albicans*. Topical and oral azole drugs are used to treat vaginal candidiasis, and many of them are now available over the counter. Many women experience this condition multiple times in their lives, but if infections recur frequently or fail to resolve, it is important to see a physician.

Gardnerella Species

The bacterium *Gardnerella* is associated with a particularly common condition in women in their childbearing years. This condition is usually called *vaginosis* rather than *vaginitis* because it doesn't appear to induce inflammation in the vagina. It is also known as BV, or bacterial vaginosis. Despite the absence of an inflammatory response, a vaginal discharge is associated with the condition, which often has a very

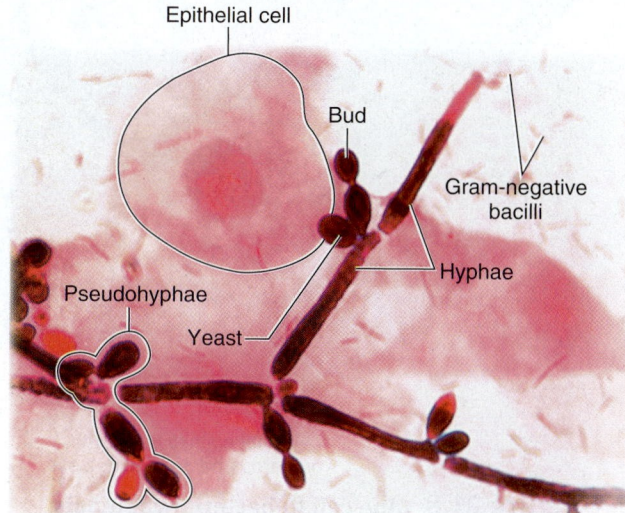

Figure 21.7 Gram stain of *Candida albicans* in a vaginal smear.

Disease Table 21.4 Vaginitis/Vaginosis

Causative Organism(s)	*Candida albicans*	Mixed infection, usually including *Gardnerella*	*Trichomonas vaginalis*
Most Common Modes of Transmission	Opportunism	Opportunism or STI?	Direct contact (STI)
Virulence Factors	–	–	–
Culture/Diagnosis	Wet prep or Gram stain	Visual exam of vagina, or clue cells seen in Pap smear or other smear	Protozoa seen on Pap smear or Gram stain
Prevention	–	–	Barrier use during intercourse
Treatment	Topical or oral azole drugs, some over-the-counter drugs	Metronidazole or clindamycin	Metronidazole, tinidazole
Distinctive Features	White curdlike discharge	Discharge may have fishy smell	Discharge may be greenish
Epidemiological Features	United States: 20% of all vaginitis; 75% of women report at least one case during lifetime	United States: estimated 7.4 million cases per year; internationally: prevalence rates vary from 20% to 50%	United States: 3–8 million people infected; is one of CDC's "neglected parasitic infections"

fishy odor. Itching is also common, but it is also true that many women have this condition with no noticeable symptoms.

Vaginosis is most likely a result of a shift from a predominance of "good" bacteria (lactobacilli) in the vagina to a predominance of "bad" bacteria, and one of those is *Gardnerella vaginalis*. This genus of bacteria is a facultative anaerobe and gram-positive, although in a Gram stain it usually appears gram-negative. (Some texts refer to it as *gram-variable* for this reason.) Probably a mixed infection leads to the condition, however. Anaerobic streptococci and other bacteria, particularly a genus known as *Mobiluncus*, that are normally found in low numbers in a healthy vagina can also often be found in high numbers in this condition. The often-mentioned fishy odor comes from the metabolic by-products of anaerobic metabolism by these bacteria.

▶ Pathogenesis and Virulence Factors

The mechanism of damage in this disease is not well understood, but some of the outcomes are. Besides the symptoms just mentioned, vaginosis can lead to complications such as pelvic inflammatory disease, infertility, and, more rarely, ectopic pregnancies. Babies born to some mothers with vaginosis have low birth weights.

▶ Transmission and Epidemiology

This mixed infection is not necessarily considered to be sexually transmitted, although women who have never had sex rarely develop the condition. It is very common in sexually active women. It may be that the condition is associated with sex but not transmitted by it. This situation could occur if the act of penetration or the presence of semen (or saliva) causes changes in the vaginal epithelium or in the vaginal biota. We do not know exactly what causes the increased numbers of *Gardnerella* and other normally rare biota. The low pH typical of the vagina is usually higher in vaginosis. It is not clear whether this causes or is caused by the change in bacterial biota.

▶ Culture and Diagnosis

The condition can be diagnosed by a variety of methods. Sometimes a simple stain of vaginal secretions is used to examine sloughed vaginal epithelial cells. In vaginosis, some cells will appear to be nearly covered with adherent bacteria. In normal times, vaginal epithelial cells are sparsely covered with bacteria. These cells are called clue cells and are a helpful diagnostic indicator **(figure 21.8).** They can also be found on Pap smears, but sometimes genomic analyses are needed.

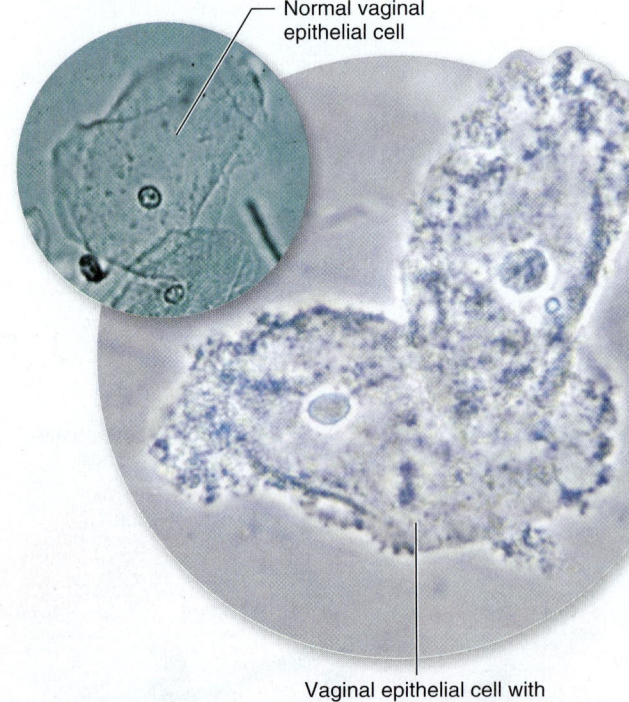

Normal vaginal epithelial cell

Vaginal epithelial cell with numerous bacteria (clue cell)

Figure 21.8 Clue cell in bacterial vaginosis. These epithelial cells came from a pelvic exam. The cells in the large circle have an abundance of bacteria attached to them.

▶ Prevention and Treatment

Women who find the condition uncomfortable or who are planning on becoming pregnant should be treated. Women who use intrauterine devices (IUDs) for contraception should also be treated because IUDs can provide a passageway for the bacteria to gain access to the upper reproductive tract. The usual treatment is oral or topical metronidazole or clindamycin.

Trichomonas vaginalis

Trichomonads are small, pear-shaped protozoa with four anterior flagella and an undulating membrane. *Trichomonas vaginalis* seems to cause asymptomatic infections. In approximately 50% of its victims, Trichomonads are considered asymptomatic infectious agents rather than normal biota because of evidence that some people experience long-term negative effects. Even though *Trichomonas* is a protozoan, it has no cyst form, and it does not survive long outside of the host.

Many cases are asymptomatic, and men seldom have symptoms. Women often have vaginitis symptoms, which can include a white to green frothy discharge. Chronic infection can make a person more susceptible to other infections, including HIV. Also, women who become infected during pregnancy are predisposed to premature labor and low-birth-weight infants. Chronic infection may also lead to infertility.

Because *Trichomonas* is common biota in so many people, it is easily transmitted through sexual contact. It has been called the most common nonviral sexually transmitted infection. It does not appear to undergo opportunistic shifts within its host (i.e., to become symptomatic under certain conditions), but rather, the protozoan causes symptoms when transmitted to a noncarrier.

Prostatitis

Prostatitis is an inflammation of the prostate gland (see figure 21.2). It can be acute or chronic. Acute prostatitis is virtually always caused by bacterial infection. The bacteria are usually normal biota from the intestinal tract or may have caused a previous urinary tract infection. Chronic prostatitis is also often caused by bacteria. Researchers have found that chronic prostatitis, often unresponsive to antibiotic treatment, can be caused by mixed biofilms of bacteria in the prostate. Some forms of chronic prostatitis have no known microbial cause, though many infectious disease specialists feel that one or more bacteria are involved, but they are simply not culturable with current techniques.

Symptoms may include pain in the groin and lower back, frequent urge to urinate, difficulty in urinating, blood in the urine, and painful ejaculation. Treatment is with broad-spectrum antibiotics.

NCLEX® PREP

2. A wet prep indicates the presence of pseudophyphae. Based on this finding, the nurse suspects that the client has
   a. *gonorrhea.*
   b. *Chlamydia.*
   c. *Candida albicans.*
   d. *pyelonephritis.*

Disease Table 21.5 Prostatitis

Causative Organism(s)	GI tract biota
Most Common Modes of Transmission	Endogenous transfer from GI tract; otherwise unknown
Virulence Factors	Various
Culture/Diagnosis	Digital rectal exam to examine prostate; culture of urine or semen
Prevention	None
Treatment	Antibiotics, muscle relaxers, alpha blockers
Distinctive Features	Pain in genital area and/or back, difficulty urinating
Epidemiological Features	United States: 50% of men experience during lifetime

Genital Ulcer Diseases

Three common infectious conditions can result in lesions (ulcers) on the genitals: syphilis, chancroid, and genital herpes. In this section, we consider each of these. One very important fact to remember about the ulcer diseases is that having one of them increases the chances of infection with HIV because of the open lesions.

Syphilis

Untreated syphilis is marked by distinct clinical stages designated as *primary, secondary,* and *tertiary syphilis.* The disease also has latent periods of varying duration during which it is quiescent. The spirochete appears in the lesions and blood during the primary and secondary stages and, thus, is transmissible at these times. During the early latency period between secondary and tertiary syphilis, it is also transmissible. Syphilis is largely nontransmissible during the "late latent" and tertiary stages.

▶ Primary Syphilis

The earliest indication of syphilis infection is the appearance of a hard **chancre** (shang'-ker) at the site of entry of the pathogen. Because these ulcers tend to be painless, they may escape notice, especially when they are on internal surfaces. The chancre heals spontaneously without scarring in 3 to 6 weeks, but the healing is deceptive because the spirochete has escaped into the circulation and is entering a period of tremendous activity.

▶ Secondary Syphilis

About 3 weeks to 6 months after the chancre heals, the secondary stage appears. By then, many systems of the body have been invaded, and the signs and symptoms are more profuse and intense. Initial symptoms are fever, headache, and sore throat, followed by lymphadenopathy and a peculiar red or brown rash that breaks out on all skin surfaces, including the palms of the hands and the soles of the feet. A person's hair often falls out. Like the chancre, the lesions contain viable spirochetes and disappear spontaneously in a few weeks. The major complications of this stage, occurring in the bones, hair follicles, joints, liver, eyes, and brain, can linger for months and years.

▶ Latency and Tertiary Syphilis

After resolution of secondary syphilis, about 30% of infections enter a highly varied latent period that can last for 20 years or longer. During latency, although antibodies to the bacterium are readily detected, the bacterium itself is not. The final stage of the disease, tertiary syphilis, is relatively rare today because of widespread use of antibiotics. But it is so damaging that it is important to recognize. By the time a patient reaches this phase, numerous pathologic complications occur in susceptible tissues and organs. Cardiovascular syphilis results from damage to the small arteries in the aortic wall. As the fibers in the wall weaken, the aorta is subject to distension and fatal rupture. The same pathologic process can damage the aortic valves, resulting in heart failure.

In one form of tertiary syphilis, painful swollen syphilitic tumors called **gummas** (goo-mahz') develop in tissues such as the liver, skin, bone, and cartilage **(figure 21.9).** Gummas are usually benign and only occasionally lead to death, but they can impair function. Neurosyphilis can involve any part of the nervous system, but it shows particular affinity for the blood vessels in the brain, cranial nerves, and dorsal roots of the spinal cord. The diverse results include

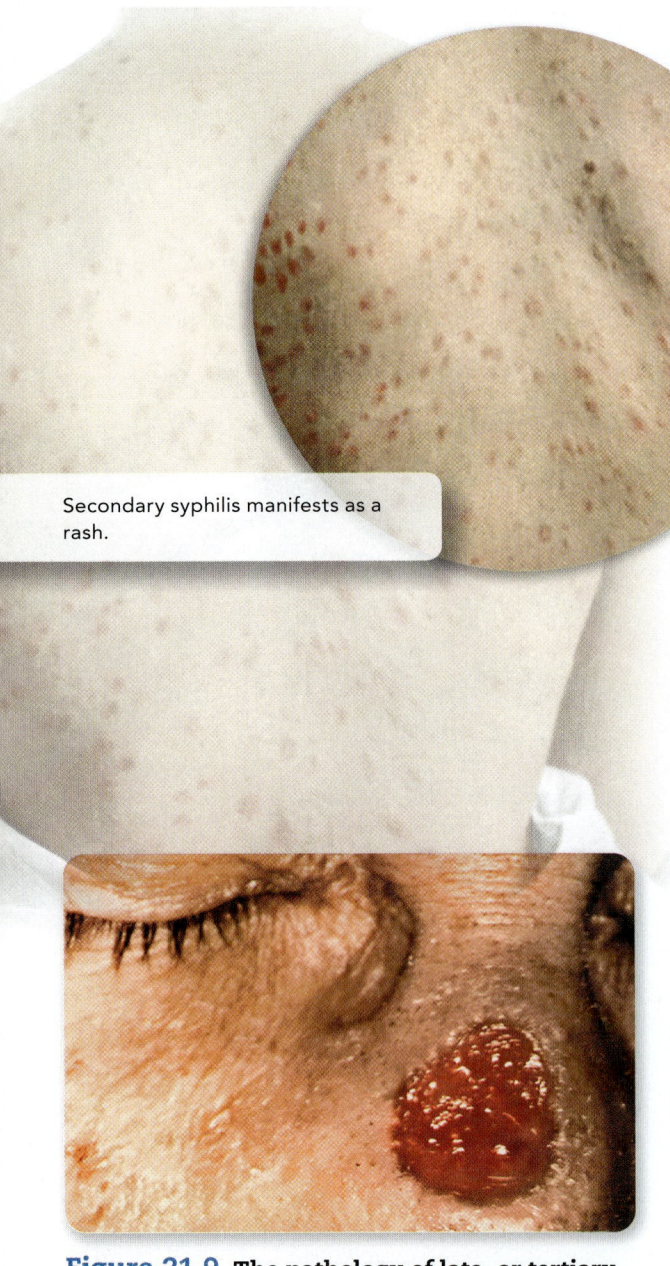

Secondary syphilis manifests as a rash.

Figure 21.9 **The pathology of late, or tertiary, syphilis.** An ulcerating syphilis tumor, or gumma, appears on the nose of this patient. Other gummas can be internal.

severe headaches, convulsions, atrophy of the optic nerve, blindness, dementia, and a sign called the Argyll-Robertson pupil–a condition caused by adhesions along the inner edge of the iris that fix the pupil's position into a small irregular circle.

▶ Congenital Syphilis

The syphilis bacterium can pass from a pregnant woman's circulation into the placenta and can be carried throughout the fetal tissues. An infection leading to congenital syphilis can occur in any of the three trimesters, but it is most common in the second and third. The pathogen inhibits fetal growth and disrupts critical periods of development with varied consequences, ranging from mild to the extremes of spontaneous miscarriage or stillbirth. Early congenital syphilis encompasses the period from birth to 2 years of age and is usually first detected 3 to 8 weeks after birth. Infants often demonstrate such signs as profuse nasal discharge **(figure 21.10a),** skin eruptions, bone deformation, and nervous system abnormalities. The late form gives rise to an unusual assortment of problems in the bones, eyes, inner ear, and joints, and causes the formation of Hutchinson's teeth **(figure 21.10b).**

▶ Causative Agent

Treponema pallidum, a spirochete, is a thin, regularly coiled cell with a gram-negative cell wall. It is a strict parasite with complex growth requirements that necessitate cultivating it in living host cells.

▶ Pathogenesis and Virulence Factors

Brought into direct contact with mucous membranes or abraded skin, *T. pallidum* binds avidly by its hooked tip to the epithelium **(figure 21.11).** At the binding site, the spirochete multiplies and penetrates the capillaries nearby. Within a short time, it moves into the circulation, and the body is literally transformed into a large receptacle for incubating the pathogen. Virtually any tissue is a potential target.

T. pallidum produces no toxins and does not appear to kill cells directly. Studies have shown that, although phagocytes seem to act against it and several types of antitreponemal antibodies are formed, immune responses are unable to contain it.

▶ Transmission and Epidemiology

Humans are evidently the sole natural hosts and source of *T. pallidum.* The bacterium is extremely fastidious and sensitive and cannot survive for long outside the host, being rapidly destroyed by heat, drying, disinfectants, soap, high oxygen tension, and pH changes. It survives a few minutes to hours when protected by body secretions and about 36 hours in stored blood. The risk of infection from an infected sexual partner is 12% to 30% per encounter.

For centuries, syphilis was a common and devastating disease in the United States, so much so that major medical centers had "Departments of Syphilology." Its effect on social life was enormous. This effect diminished quickly when antibiotics were discovered. But since 2003, the rates have been increasing again in the United States. And syphilis continues to be a serious problem worldwide, especially in Africa and Asia. As mentioned previously, persons with syphilis often suffer concurrent infections with other STIs. Coinfection with the AIDS virus can be an especially deadly combination with a rapidly fatal course.

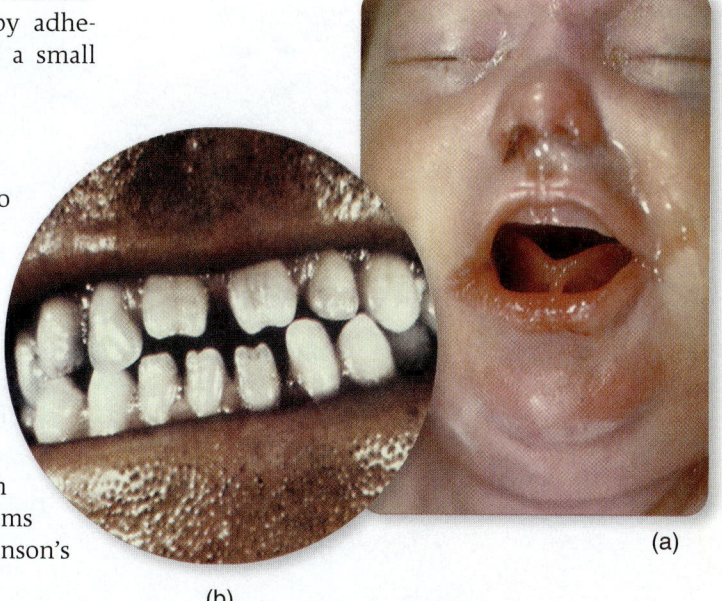

(a)

(b)

Figure 21.10 **Congenital syphilis.** (a) An early sign is snuffles, a profuse nasal discharge that obstructs breathing. (b) A common characteristic of late congenital syphilis is notched, barrel-shaped incisors (Hutchinson's teeth).

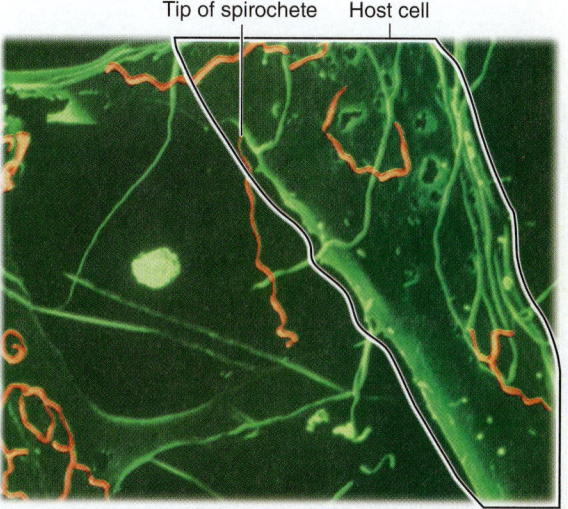

Tip of spirochete   Host cell

Figure 21.11 **Electron micrograph of the syphilis spirochete attached to cells.**

Spirochete

Tissue cells

Figure 21.12 *Treponema pallidum* from a syphilitic chancre, viewed with dark-field illumination. Its tight spirals are highlighted next to human cells and tissue debris.

▶ Culture and Diagnosis

Syphilis can be detected in patients most rapidly by using dark-field microscopy of a suspected lesion **(figure 21.12)**. A single negative test is not enough to exclude syphilis because the patient may have removed the organisms by washing, so follow-up tests are recommended.

Very commonly, blood tests are used for this diagnosis. The best test is that which specifically reacts with treponemal antigens. Additional specific tests are available when considered necessary. One of these is the indirect immunofluorescent method called the FTA-ABS (fluorescent treponemal antibody absorption) test. The test serum is first allowed to react with treponemal cells and then reacted with antihuman globulin antibody labeled with fluorescent dyes. If antibodies to the treponeme are present, the fluorescently labeled antibody will bind to the human antibody bound to the treponemal cells. The result is highly visible with a fluorescence microscope.

▶ Prevention

The core of an effective prevention program depends on detection and treatment of the sexual contacts of syphilitic patients. Public health departments and physicians are charged with the task of questioning patients and tracing their contacts. All individuals identified as being at risk, even if they show no signs of infection, are given immediate prophylactic penicillin in a single long-acting dose.

The barrier effect of a condom provides excellent protection during the primary phase. Protective immunity apparently does arise in humans, allowing the prospect of an effective immunization program in the future, although no vaccine exists currently.

▶ Treatment

Throughout most of history, the treatment for syphilis was a dose of mercury or even a "mercurial rub" applied to external lesions. In 1906, Paul Ehrlich discovered that a derivative of arsenic called salvarsan could be very effective. The fact that toxic compounds like mercury and arsenic were used to treat syphilis gives some indication of how dreaded the disease was and to what lengths people would go to rid themselves of it.

Once penicillin became available, it replaced all other treatments, and Penicillin G retains its status as a wonder drug in the treatment of all stages and forms of syphilis.

Chancroid

This ulcerative disease usually begins as a soft papule, or bump, at the point of contact. It develops into a "soft chancre" (in contrast to the hard syphilis chancre), which is very painful in men, but may be unnoticed in women **(Disease Table 21.6)**. Inguinal lymph nodes can become very swollen and tender.

Chancroid is caused by a pleomorphic gram-negative rod called *Haemophilus ducreyi*. Recent research indicates that a hemolysin (exotoxin) is important in the pathogenesis of chancroid disease. It is very common in the tropics and subtropics and is becoming more common in the United States. Chancroid is transmitted exclusively through direct contact, especially sexually. This disease is associated with sex workers and poor hygiene; uncircumcised men seem to be more commonly infected than those who have been circumcised. People may carry this bacterium asymptomatically.

Genital Herpes

Genital herpes is much more common than most people think. It is caused by herpes simplex viruses (HSVs). Two types of HSV have been identified, HSV-1 and HSV-2.

▶ Signs and Symptoms

Genital herpes infection has multiple presentations. After initial infection, a person may notice no symptoms. Alternatively, herpes could cause the appearance of single or multiple vesicles on the genitalia, perineum, thigh, and buttocks. The vesicles are small and are filled with a clear fluid (see Disease Table 21.6). They can be intensely painful to the touch. The appearance of lesions the first time you get them can be accompanied by malaise, anorexia, fever, and bilateral swelling and tenderness in the groin. Occasionally central nervous system symptoms such as meningitis or encephalitis can develop. Thus, we see that initial infection can either be completely asymptomatic or be serious enough to require hospitalization.

After recovery from initial infection, a person may have recurrent episodes of lesions. They are generally less severe than the original symptoms, although the whole gamut of possible severity is seen here as well. Some people never have recurrent lesions. Others have nearly constant outbreaks with little recovery time between them. On average, the number of recurrences is four or five a year. Their frequency tends to decrease over the course of years.

In most cases, patients remain asymptomatic or experience recurrent "surface" infections indefinitely. Very rarely, complications can occur. Every year, one or two persons per million with chronic herpes infections develop encephalitis. The virus disseminates along nerve pathways to the brain (although it can also infect the spinal cord). The effects on the central nervous system begin with headache and stiff neck and can progress to mental disturbances and coma. The fatality rate in untreated encephalitis cases is 70%, although treatment with acyclovir is effective. Patients with underlying immunodeficiency are more prone to severe, disseminated herpes infection than are immunocompetent patients.

▶ Herpes of the Newborn

Although HSV infections in healthy adults are annoying and unpleasant, only rarely are they life-threatening. However, in the neonate and the fetus **(figure 21.13),** HSV infections are very destructive and can be fatal. Most cases occur when infants are contaminated by the mother's reproductive tract immediately before or during birth, but they have also been traced to hand transmission from the mother's lesions to the baby. In infants whose disease is confined to the mouth, skin, or eyes, the mortality rate is 30%, but disease affecting the central nervous system has a 50% to 80% mortality rate.

Pregnant women with a history of recurrent infections must be monitored for any signs of viral shedding, especially in the last 4 weeks of pregnancy. If no evidence of recurrence is seen, vaginal birth is indicated, but any evidence of an outbreak at the time of delivery necessitates a cesarean section.

▶ Causative Agent

Both HSV-1 and HSV-2 can cause genital herpes if the virus contacts the genital epithelium, although HSV-1 is thought of as a virus that infects the oral mucosa, resulting in "cold sores" or "fever blisters" **(figure 21.14),** and HSV-2 is thought of as the genital virus. In reality, either virus can infect either region, depending on the type of contact.

▶ Pathogenesis and Virulence Factors

Herpesviruses have a strong tendency to become latent. The molecular basis of latency is not entirely clear. During latency, some type of signal causes most of the HSV genome not to be transcribed. This allows the virus to be maintained within cells of the nervous system between episodes. Recent research has found that microRNAs are in part responsible for the latency of HSV-1. It is further suggested that in some peripheral cells, viral replication takes place at a constant, slow rate, resulting in constant low-level shedding of the virus without lesion production.

Reactivation of the virus can be triggered by a variety of stimuli, including stress, UV radiation (sunlight), injury, menstruation, or another microbial infection. At that point, the virus begins manufacturing large numbers of entire virions, which cause new lesions on the surface of the body served by the neuron, usually in the same site as previous lesions.

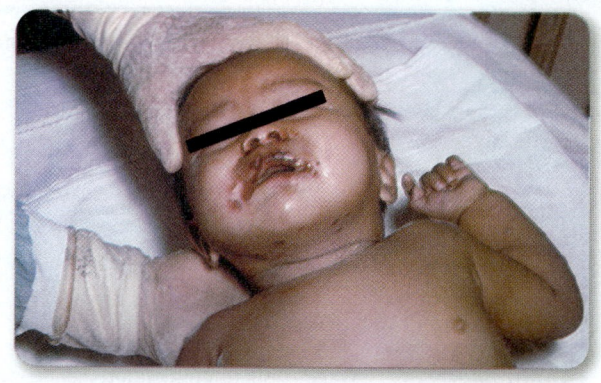

Figure 21.13 Neonatal herpes simplex.   This premature infant was born with the classic "cigarette burn" pattern of HSV infection. Babies can be born with the lesions or develop them 1 to 2 weeks after birth.

Disease Table 21.6 Genital Ulcer Diseases

	Syphilis	Chancroid	Herpes
Causative Organism(s)	*Treponema pallidum*	*Haemophilus ducreyi*	Herpes simplex 1 and 2
Most Common Modes of Transmission	Direct contact and vertical	Direct contact (vertical transmission not documented)	Direct contact, vertical
Virulence Factors	Lipoproteins	Hemolysin (exotoxin)	Latency
Culture/ Diagnosis	Direct tests (immunofluorescence, dark-field microscopy), blood tests for treponemal and nontreponemal antibodies, PCR	Culture from lesion	Clinical presentation, PCR, antibody tests, growth of virus in cell culture
Prevention	Antibiotic treatment of all possible contacts, avoiding contact	Avoiding contact	Avoiding contact, antivirals can reduce recurrences
Treatment	Penicillin G	Ceftriaxone or azithromycin	Acyclovir and derivatives
Distinctive Features	Three stages of disease plus latent period, possibly fatal	No systemic effects	Ranges from asymptomatic to frequent recurrences
Effects on Fetus	Congenital syphilis	None	Blindness, disseminated herpes infection
Appearance of Lesions			Vesicles
Epidemiological Features	United States: estimated 90,000 new cases per year; internationally: estimated 12 million new infections per year	United States: no more than 200 per year; internationally: estimated 7 million cases annually	United States: 20% prevalence in adults; internationally: estimated 536 million infected in 15-49 age group

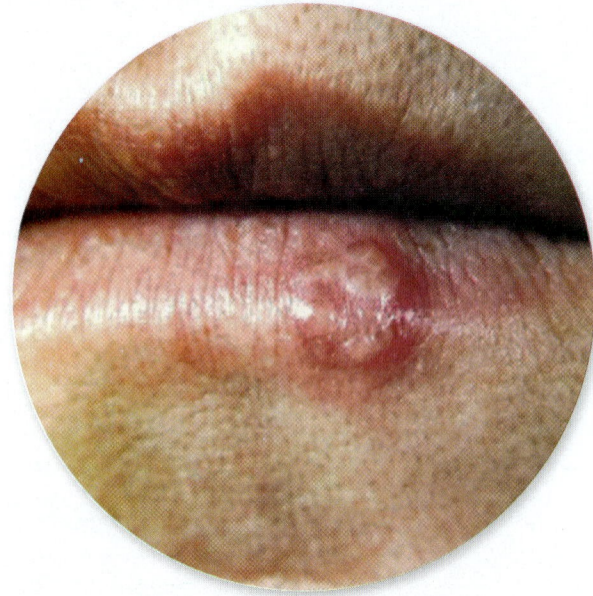

Figure 21.14 Oral herpes infection. Tender itchy papules erupt around the mouth and progress to vesicles that burst, drain, and scab over. These sores and fluid are highly infectious and should not be touched.

▶ Transmission and Epidemiology

Herpes simplex infection occurs globally in all seasons and among all age groups. Because these viruses are relatively sensitive to the environment, transmission is primarily through direct exposure to secretions containing the virus. People with active lesions are the most significant source of infection, but studies indicate that genital herpes can be transmitted even when no lesions are present (due to the constant shedding just discussed).

Earlier in this chapter, you read that *Chlamydia* infection is the most common *reported* infectious disease in the United States. Elsewhere you might hear that gonorrhea is one of the most common reportable *STIs* in the United States. Both statements are true. It is also true that genital herpes is much more common than either of these diseases. Herpes, however, is not a nationally *reportable* disease.

It is estimated that about 20% of American adults have genital herpes. Fifty to ninety percent of people who are infected don't even know it, either because they have rare symptoms that they fail to recognize or because they have no symptoms at all.

▶ Culture and Diagnosis

These two viruses are sometimes diagnosed based on the characteristic lesions alone. PCR tests are available to test for these viruses directly from lesions. Alternatively, antibody to either of the viruses can be detected from blood samples. Detecting antibody to either HSV-1 or HSV-2 in blood does not necessarily indicate whether the infection is oral or genital, or whether the infection is new or preexisting.

Herpes-infected mucosal cells display notable characteristics in a Pap smear **(figure 21.15).** Laboratory culture and specific tests are essential for diagnosing severe or complicated herpes infections.

▶ Prevention

No vaccine is currently licensed for HSV, but more than one is being tested in clinical trials, meaning that vaccines may become available very soon. In the meantime, avoiding contact with infected body surfaces is the only way to avoid HSV. Condoms provide good protection when they actually cover the site where the lesion is, but lesions can occur outside of the area covered by a condom.

Mothers with cold sores should be careful in handling their newborns; they should never kiss their infants on the mouth.

▶ Treatment

Several agents are available for treatment. These agents often result in reduced viral shedding and a decrease in the frequency of lesion occurrence. They are not curative. Acyclovir and its derivatives (famciclovir or valacyclovir) are very effective. Topical formulations can be applied directly to lesions, and pills are available as well. Sometimes medicines are prescribed on an ongoing basis to decrease the frequency of recurrences, and sometimes they are prescribed to be taken at the beginning of a recurrence to shorten it.

Wart Diseases

In this section, we describe two viral STIs that cause wartlike growths. The more serious disease is caused by the *human papillomavirus (HPV)*; the other condition, called *molluscum contagiosum*, apparently has no serious effects.

Human Papillomavirus Infection

These viruses are the causative agents of genital warts. But an individual can be infected with these viruses without having any warts, while still risking serious consequences.

▶ Signs and Symptoms

Symptoms, if present, may manifest as warts—outgrowths of tissue on the genitals **(Disease Table 21.7).** In females, these growths can occur on the vulva and in and around the vagina. In males, the warts can occur in or on the penis and the scrotum. In both sexes, the warts can appear in or on the anus and even on the skin around the groin, such as the area between the thigh and the pelvis. The warts themselves range from tiny, flat, inconspicuous bumps to extensively branching, cauliflower-like masses called **condyloma acuminata.** The warts are unsightly and can be obstructive, but they don't generally lead to more serious symptoms.

Other types of HPV can lead to more subtle symptoms. Certain types of the virus infect cells on the female cervix. This infection may be "silent," or it may lead to abnormal cell changes and malignancies of the cervix. Approximately 4,000 women die each year in the United States from cervical cancer, and the vast majority of these are caused by HPV.

Males can also get cancer from infection with these viruses. The sites most often affected are the penis and the anus. These cases are much less common than cervical cancer. Mouth and throat cancers in both genders have also been more recently associated with HPV infection and are thought to be a consequence of oral sex. (see Inside the Clinic at the end of this chapter)

▶ Causative Agent

The human papillomaviruses are a group of nonenveloped DNA viruses belonging to the *Papovaviridae* family. There are more than 100 different types of HPV. Some types

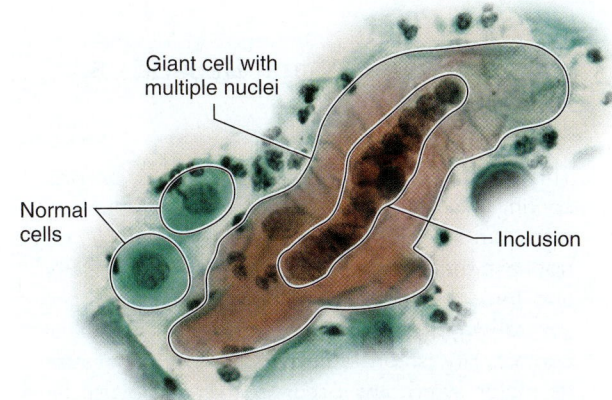

Giant cell with multiple nuclei

Normal cells

Inclusion

Figure 21.15 The appearance of herpesvirus infection in a Pap smear. A Pap smear of a cervical scraping shows enlarged, multinucleate, giant cells and intranuclear inclusions typical of herpes simplex type 2. This appearance is not specific for HSV, but most other herpesviruses do not infect the reproductive mucosa.

Medical Moment

Crabs

Pediculosis pubis, commonly known as crabs, is caused by infestation of the pubic hair by *Phthirus pubis*, a tiny (1 mm or less) insect that can multiply rapidly. The condition can be spread from person to person through sexual contact and can therefore be considered a sexually transmitted condition. The eyebrows, eyelashes, chest hair, scalp hair, and facial hair may also be affected, generally through oral sexual activities.

Although these tiny creatures do not pose much of a risk to human health, infestation with crabs may lead to a higher risk for other STIs. It is estimated that approximately 3 million people are infected yearly in the United States.

Symptoms include intense itching of the affected areas. Some people do not experience itching but while they are bathing may observe the insects, as the lice are visible to the naked eye.

Treatment is applied topically in the form of creams, shampoos, or lotions. Manual removal of adult insects and their eggs is recommended using a fine-toothed metal comb. Clothing and bedding should be laundered in hot water to prevent reinfestation. Abstinence is the only way to reliably prevent the condition.

A Note About HIV and Hepatitis B and C

This chapter is about diseases *whose major (presenting) symptoms occur in the genitourinary tract.* But some sexually transmitted diseases do not have their major symptoms in this system. HIV and hepatitis B and C can all be transmitted in several ways, one of them being through sexual contact. HIV is considered in chapter 18 because its major symptoms occur in the cardiovascular and lymphatic systems. Because the major disease manifestations of hepatitis B and C occur in the gastrointestinal tract, these diseases are discussed in chapter 20. Anyone diagnosed with any sexually transmitted disease should also be tested for HIV.

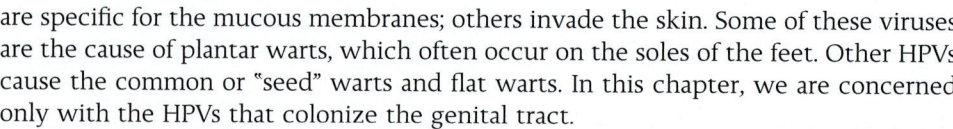

are specific for the mucous membranes; others invade the skin. Some of these viruses are the cause of plantar warts, which often occur on the soles of the feet. Other HPVs cause the common or "seed" warts and flat warts. In this chapter, we are concerned only with the HPVs that colonize the genital tract.

Among the HPVs that infect the genital tract, some are more likely to cause the appearance of warts. Others that have a preference for growing on the cervix can lead to cancerous changes. Five types in particular, HPV-16, -18, -31, -33, and -35, are closely associated with development of cervical cancer. Other types put you at higher risk for vulvar or penile cancer.

▶ Pathogenesis and Virulence Factors

Scientists are working hard to understand how viruses cause the growths we know as warts and also how some of them can cause cancer. The major virulence factor for cancer-causing HPVs is their **oncogenes,** which code for proteins that interfere with normal host cell function, resulting in uncontrolled growth.

▶ Transmission and Epidemiology

Young women have the highest rate of HPV infections; 25% to 46% of women under the age of 25 are infected with genital HPV. It is estimated that 14% of female college students become infected with this incurable condition each year. Overall, about 15% of people between ages 15 and 49 are HPV-positive. It is difficult to know whether genital herpes or HPV is more common, but it is probably safe to assume that any unprotected sex carries a good chance of encountering either HSV or HPV.

The mode of transmission is direct contact. Autoinoculation is also possible— meaning that the virus can be spread to other parts of the body by touching warts. Indirect transmission occurs but is more common for nongenital warts caused by HPV.

▶ Culture and Diagnosis

PCR-based screening tests can be used to test samples from a pelvic exam for the presence of dangerous HPV types. These tests are now recommended for women over the age of 30. The Pap smear is still the single best screening procedure available for cervical cancer; it will detect nearly all precancerous cells if conducted on the recommended schedule.

▶ Prevention

When discussing HPV prevention, we must consider two possibilities. One of these is infection with the viruses, which is prevented the same way other sexually transmitted infections are prevented—by avoiding direct, unprotected contact, but also by one of two vaccines that are now available (Gardasil and Cervarix). The vaccines prevent infection by a small number of types of HPV and are recommended in both girls and boys as young as age 9. Despite the fears of some parents, being vaccinated against the virus does not encourage young people to become sexually active but instead causes them to realize the dangers of sex, according to a study conducted in 2009 among 553 teenage girls in Britain.

The second issue is the prevention of cervical cancer. Even though women now have access to the vaccines, cancer can still result from HPV types not included in the vaccines. The good news is that cervical cancer is slow in developing, so that even if a woman is infected with a malignant HPV type, regular screening of the cervix can detect abnormal changes early. The standardized screen for cervical cell changes is the Pap smear. Precancerous changes show up very early, and the development process can be stopped by removal of the affected tissue. Women should have their first Pap smear by age 21 or within 3 years of their first sexual activity, whichever comes first. New Pap smear technologies have been developed; and depending on which one your physician uses, it is now possible that you need to be screened only once every 2 or 3 years. But you should base your screening practices on the sound advice of a physician.

► Treatment

Infection with any HPV is incurable. Genital warts can be removed through a variety of methods, some of which can be used at home, but the virus causing them will most likely remain with you. It is possible for the viral infection to resolve itself, but this is very unpredictable.

Molluscum Contagiosum

An unclassified virus in the family *Poxviridae* can cause a condition called molluscum contagiosum. This disease can take the form of skin lesions, and it can also be transmitted sexually. The wartlike growths that result from this infection can be found on the mucous membranes or the skin of the genital area (see Disease Table 21.7). Few problems are associated with these growths beyond the warts themselves. In severely immunocompromised people, the disease can be more extensive.

The virus causing these growths can also be transmitted through fomites such as clothing or towels and through autoinoculation. For a more detailed description of this condition, see chapter 16.

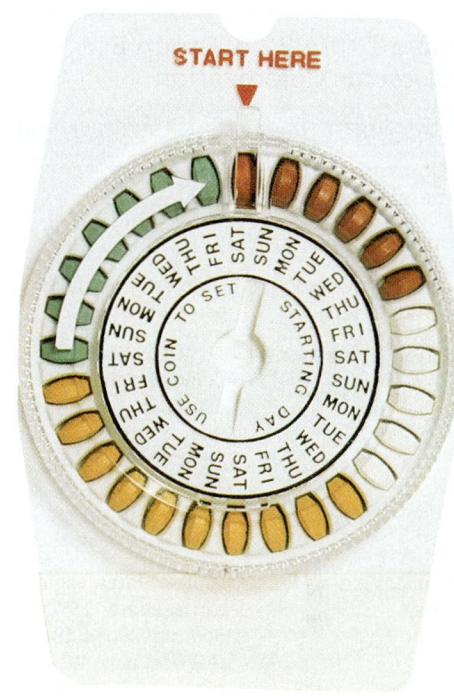

Disease Table 21.7 Wart Diseases

	HPV	Molluscum Contagiosum
Causative Organism(s)	Human papillomaviruses	Poxvirus, sometimes called the molluscum contagiosum virus (MCV)
Most Common Modes of Transmission	Direct contact (STI), also autoinoculation, indirect contact	Direct contact (STI), also indirect and autoinoculation
Virulence Factors	Oncogenes (in the case of malignant types of HPV)	–
Culture/Diagnosis	PCR tests for certain HPV types, clinical diagnosis	Clinical diagnosis, also histology, PCR
Prevention	Vaccines available; avoid direct contact; prevent cancer by screening cervix	Avoid direct contact
Treatment	Warts or precancerous tissue can be removed; virus not treatable	Warts can be removed; virus not treatable
Distinguishing Features	Infection may or may not result in warts; infection may result in malignancy	Wartlike growths are only known consequence of infection
Effects on Fetus	May cause laryngeal warts	–
Appearance of Lesions		
Epidemiological Features	United States: estimated 6 million new infections per year; 12,000 new cases of HPV-associated cervical cancer	United States: affects 2%-10% of children annually

Group B *Streptococcus* "Colonization"—Neonatal Disease

Ten to forty percent of women in the United States are colonized, asymptomatically, by a beta-hemolytic *Streptococcus* in Lancefield group B. Nonpregnant women experience no ill effects from this colonization. But colonization of pregnant women with this organism is associated with preterm delivery. Additionally, about half of their infants become colonized by the bacterium during passage through the birth canal or by ascension of the bacteria through ruptured membranes; thus, this colonization is considered a reproductive tract disease.

A small percentage of infected infants experience life-threatening bloodstream infections, meningitis, or pneumonia. If they recover from these acute conditions, they may have permanent disabilities, such as developmental disabilities, hearing loss, or impaired vision. In some cases, the mothers also experience disease, such as amniotic infection or subsequent stillbirths. Although group B *Streptococcus* infections have declined sharply in the United States, they remain a major threat to infant morbidity and mortality worldwide.

In 2002, the CDC recommended that all pregnant women be screened for group B *Streptococcus* colonization at 35 to 37 weeks of pregnancy. Recommendations for earlier testing are sometimes warranted because colonization has been associated with preterm birth. Women positive for the bacterium should be treated with penicillin or ampicillin, unless the bacterium is found to be resistant to these and unless allergy to penicillin is present, in which case erythromycin may be used.

Disease Table 21.8 Group B *Streptococcus* Colonization

Causative Organism(s)	Group B *Streptococcus*
Most Common Modes of Transmission	Vertical
Virulence Factors	–
Culture/Diagnosis	Culture of mother's genital tract
Prevention/Treatment	Treat mother with penicillin/ampicillin; watch for clindamycin-resistant strains as they are in **Concerning Threats** category on CDC Antibiotic Resistance Report
Epidemiological Features	United States: vaginal carriage rates 15%-45%; neonatal sepsis due to this occurs in 1.8-3.2 per 1,000 live births; internationally: vaginal carriage rates 12%-27%

21.4 LEARNING OUTCOMES—Assess Your Progress

6. List the possible causative agents, modes of transmission, virulence factors, and prevention/treatment for gonorrhea and *Chlamydia* infection.
7. Distinguish between vaginitis and vaginosis.
8. Discuss prostatitis.
9. Name three diseases that result in genital ulcers, and discuss their important features.
10. Differentiate between the two diseases causing warts in the reproductive tract.
11. Provide some detail about the first "cancer vaccine" and how it works.
12. Identify the most important risk group for group B *Streptococcus* infection and explain why that group is important.

CASE FILE WRAP-UP

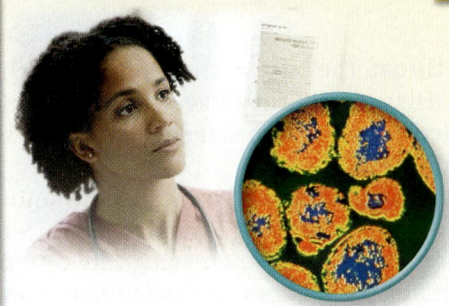

Transmission electron micrograph of Neisseria gonorrhoeae.

The young patient featured in the opening case study was given a pelvic exam after consent was obtained from the patient's mother. The patient was found to have a purulent and foul-smelling vaginal discharge, in addition to fever and lower abdominal pain. Cultures of the vaginal discharge and the cervix were obtained. Blood work and an ultrasound were also ordered. The blood work revealed an elevated white blood cell count and erythrocyte sedimentation rate (ESR), while the ultrasound revealed an abscess near the left ovary. The patient was admitted with a diagnosis of pelvic inflammatory disease (PID) for IV antibiotics and possible surgical drainage of the abscess. Cultures eventually yielded the specific causative agent, *Neisseria gonorrhoeae*. The "PID shuffle" is a term used to describe the typical gait of a patient with PID, in which the feet are advanced in a shuffling manner to avoid jarring the pelvic organs, which results in severe pain.

The patient was only 10 years of age, which should raise suspicion of child abuse. Not only should the case be reported to the appropriate state health authority, but it should also be reported to social services or the police in order to determine whether the patient was a victim of abuse.

Oral Cancer and Sex

Cancers of the tonsils, the oral part of the throat, the base of the tongue, and the soft palate are called *oropharyngeal cancers*. Historically they have been relatively rare, but in the past several years their incidence has increased dramatically in Europe and North America.

In the past, these cancers were most closely associated with cigarette smoking. So, predictably, their occurrence had been falling as smoking rates fell. However, scientists at Johns Hopkins started noticing an increasing rate of oropharyngeal cancers in nonsmokers. They tested the tumors and found that many of them contained human papillomavirus type 16. As they studied the issue more, they found that being exposed to HPV-16 meant that you were *32 times* more likely to suffer oropharyngeal cancer than people who were not exposed. That is an astronomical risk percentage. For example, people who are heavy smokers have a 3 times greater risk of oropharyngeal cancer than those who do not smoke.

The surge in oral cancers is most striking among men. In the period from 1998 to 2008, a 36% increase in oropharyngeal cancer occurred (this at the same time that men's smoking declined by 13%).

How is it transmitted? Through direct contact with the virus. Where is the virus most common? On genitals. Oral sex is a major culprit for transmitting HPVs of all types to the oral cavity.

The bad news? The CDC says that the vast majority of sexually active adults will become infected with one or more types of HPV during their lifetimes. The good news? Most of these HPVs are not pathogenic. A few types cause warts (both genital and oral), and only a few appear to be cancer-causing. Oral cancer seems to be associated most strongly with HPV-16.

These are a few of the reasons the HPV vaccines are recommended for girls *and* boys—women *and* men.

▶ Summing Up

Taxonomic Organization Microorganisms Causing Disease in the Genitourinary Tract

Microorganism	Pronunciation	Location of Disease Table
Gram-positive bacteria		
Staphylococcus saprophyticus	staf″-uh-lo-kok′-us sap′-pro-fit″-uh-cus	Urinary tract infection, p. 606
Gardnerella (note: stains gram-negative)	gard′-ner-el″-uh	Vaginitis or vaginosis, p. 614
Group B *Streptococcus*	groop bee′ strep′-tuh-kok″-us	Group B *Streptococcus* colonization, p. 624
Gram-negative bacteria		
Escherichia coli	esh′shur-eesh″-ee-uh col′-eye	Urinary tract infection, p. 606
Leptospira interrogans (spirochete)	lep′-toh-spy′-ruh in-ter′-ruh-ganz	Leptospirosis, p. 607
Enterococcus	ent′-ter-oh-kok″-us	Urinary tract infection, p. 606
Neisseria gonorrhoeae	nye-seer″-ee-uh′ gon′-uh-ree″-uh	Genital discharge disease, p. 612
Chlamydia trachomatis	kluh-mi″-dee-uh′ truh-koh′-muh-tis	Genital discharge disease, p. 612
Treponema pallidum (spirochete)	trep′-oh-nee″-ma pal′-uh-dum	Genital ulcer disease, p. 620
Haemophilus ducreyi	huh-mah′-fuh-luss doo-cray′-ee-eye	Genital ulcer disease, p. 620
DNA viruses		
Herpes simplex virus 1 and 2	hur′-peez sim′-plex vie′-russ	Genital ulcer disease, p. 620
Human papillomavirus	hew′-mun pap′-uh-loh″-muh-vie′-russ	Wart diseases, p. 623
Poxvirus	pox′-vie′-russ	Wart diseases, p. 623
Fungi		
Candida albicans	can′-duh-duh al″-buh-cans′	Vaginitis or vaginosis, p. 614
Protozoa		
Trichomonas vaginalis	trick″-uh-mon′-us vaj′-ih-nal″-us	Vaginitis or vaginosis, p. 614

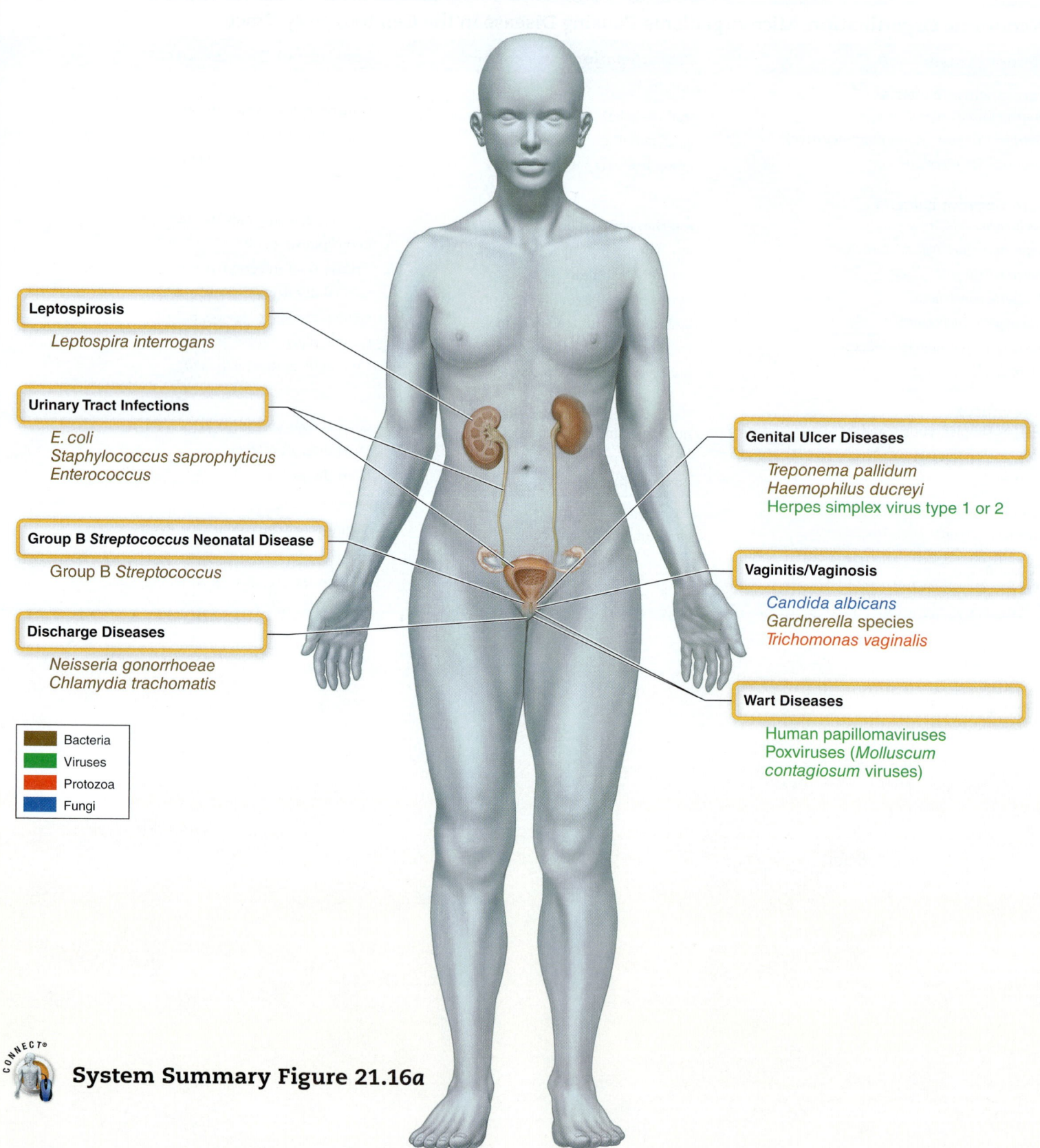

Infectious Diseases Affecting
The Genitourinary System

Leptospirosis

Leptospira interrogans

Urinary Tract Infections

E. coli
Staphylococcus saprophyticus
Enterococcus

Group B *Streptococcus* Neonatal Disease

Group B *Streptococcus*

Discharge Diseases

Neisseria gonorrhoeae
Chlamydia trachomatis

Genital Ulcer Diseases

Treponema pallidum
Haemophilus ducreyi
Herpes simplex virus type 1 or 2

Vaginitis/Vaginosis

Candida albicans
Gardnerella species
Trichomonas vaginalis

Wart Diseases

Human papillomaviruses
Poxviruses (*Molluscum contagiosum* viruses)

■	Bacteria
■	Viruses
■	Protozoa
■	Fungi

System Summary Figure 21.16*a*

Infectious Diseases Affecting
The Genitourinary System

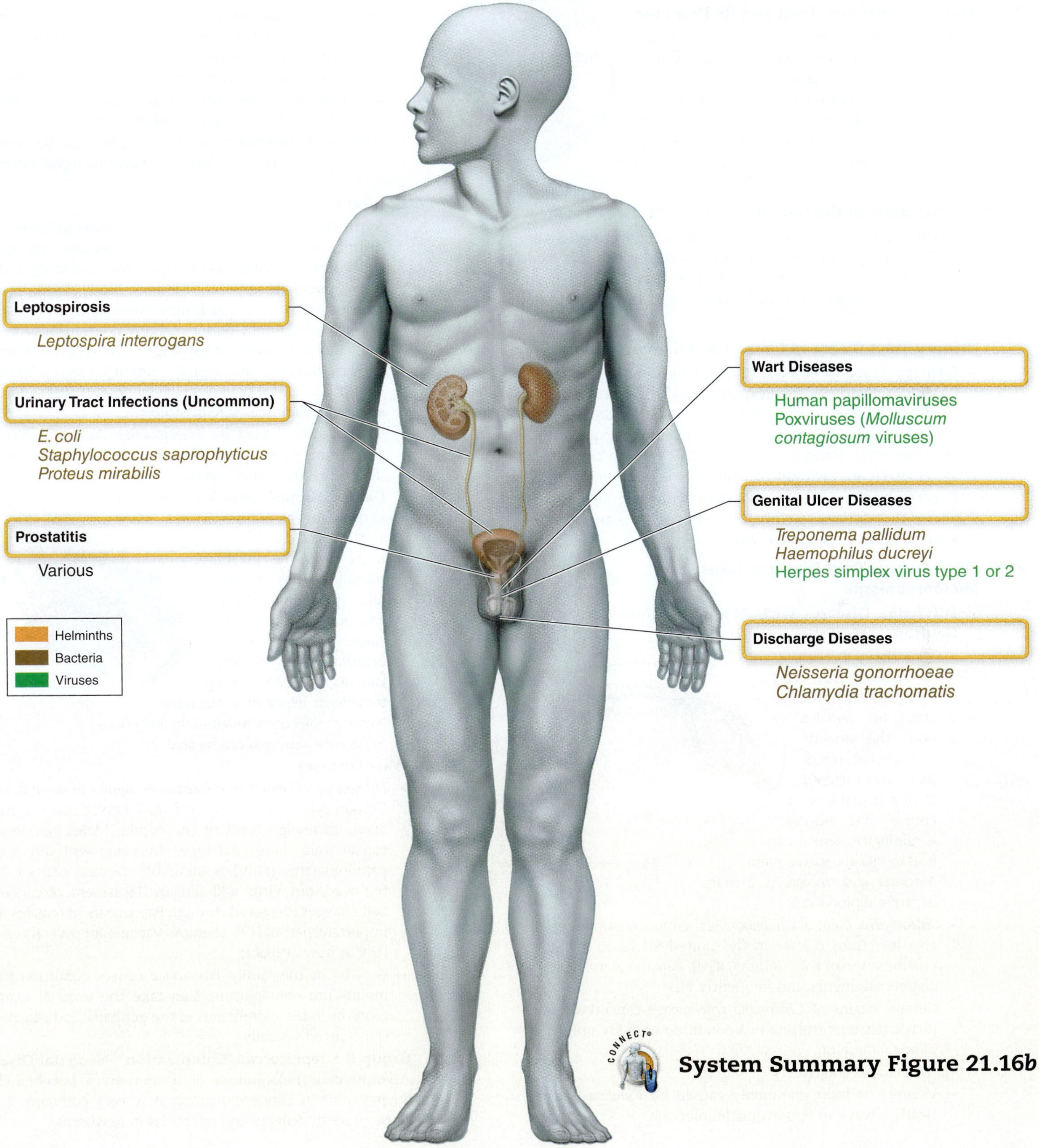

Leptospirosis

Leptospira interrogans

Urinary Tract Infections (Uncommon)

E. coli
Staphylococcus saprophyticus
Proteus mirabilis

Prostatitis

Various

Helminths
Bacteria
Viruses

Wart Diseases

Human papillomaviruses
Poxviruses (*Molluscum contagiosum* viruses)

Genital Ulcer Diseases

Treponema pallidum
Haemophilus ducreyi
Herpes simplex virus type 1 or 2

Discharge Diseases

Neisseria gonorrhoeae
Chlamydia trachomatis

System Summary Figure 21.16b

Chapter Summary

21.1 The Genitourinary Tract and Its Defenses

- The urinary system allows excretion of fluid and wastes from the body. It has mechanical, chemical defense mechanisms.
- The reproductive tract is composed of structures and substances that allow for sexual intercourse and the creation of a new fetus; protected by normal mucosal defenses and specialized features (such as low pH).

21.2 Normal Biota of the Genitourinary Tract

- Normal biota in the male reproductive and urinary systems are in the distal part of the urethra and resemble skin biota. Same is generally true for the female urinary system. Normal biota in the female reproductive tract changes over the course of her lifetime.

21.3 Urinary Tract Diseases Caused by Microorganisms

- **Urinary Tract Infections (UTIs):** Can occur in the bladder (*cystitis*), the kidneys (*pyelonephritis*), and the urethra (*urethritis*). Most common causes are *Escherichia coli*, *Staphylococcus saprophyticus*, and *Proteus mirabilis*. Community-acquired UTIs are most often transmitted from the GI tract to the urinary system. UTIs are the most common nosocomial infection.
- **Leptospirosis:** Zoonosis associated with wild animals that affects the kidneys, liver, brain, and eyes. Causative agent is spirochete *Leptospira interrogans*.

21.4 Reproductive Tract Diseases Caused by Microorganisms

- **Discharge Diseases with Major Manifestation in the Genitourinary Tract**
  - Gonorrhea can elicit urethritis in males, but many cases are asymptomatic. In females, both the urinary and genital tracts may be infected during sexual intercourse. One sequela is salpingitis, which can lead to PID. Causative agent, *Neisseria gonorrhoeae*, is a gram-negative diplococcus.
  - *Chlamydia:* Genital *Chlamydia* is the most common reportable infectious disease in the United States. In males: an inflammation of the urethra (NGU). Females: cervicitis, discharge, salpingitis, and frequently PID.
  - Certain strains of *Chlamydia trachomatis* can invade lymphatic tissues, resulting in a condition called lymphogranuloma venereum.
- **Vaginitis and Vaginosis**
  - Vaginitis is most commonly caused by *Candida albicans*. Nearly always an opportunistic infection.
  - *Gardnerella* is associated with vaginosis; has a discharge but no inflammation. Could lead to complications such as pelvic inflammatory disease (PID).
  - *Trichomonas vaginalis* causes mostly asymptomatic infections in females and males. *Trichomonas*, a flagellated protozoan, is easily transmitted through sexual contact.
- **Prostatitis:** Inflammation of the prostate; can be acute or chronic. Not all cases established to have microbial cause, but most do.
- **Genital Ulcer Diseases**
  - **Syphilis:** Caused by spirochete *Treponema pallidum*. Three distinct clinical stages: *primary, secondary,* and *tertiary syphilis,* with a latent period between secondary and tertiary. Spirochete appears in lesions, blood during primary and secondary stages; is transmissible at these times, also during early latency period. Largely nontransmissible during "late latent" and tertiary stages. The syphilis bacterium can lead to congenital syphilis, inhibiting fetal growth and disrupting critical periods of development. This can lead to spontaneous miscarriage or stillbirth.
  - **Chancroid:** Caused by *Haemophilus ducreyi*, a pleomorphic gram-negative rod. Transmitted exclusively through direct–mainly sexual–contact.
  - **Genital Herpes:** Caused by two types of herpes simplex viruses (HSVs): HSV-1 and HSV-2. May be no symptoms, or may be fluid-filled, painful vesicles on genitalia, perineum, thigh, and buttocks. In severe cases, meningitis or encephalitis can develop. Patients remain asymptomatic or experience recurrent "surface" infections indefinitely. Infections in neonate and fetus can be fatal.
- **Wart Diseases**
  - *Human papillomaviruses:* Causative agents of genital warts. Certain types infect cells on female cervix that eventually result in malignancies of the cervix. Males can also get cancer from these viral types. Infection with any human papillomavirus (HPV) is incurable. Genital warts can be removed, but virus will remain. Treatment of cancerous cell changes–detected through Pap smears in females–is an important part of HPV therapy. Vaccine for several types of HPV is now available.
  - A virus in the family *Poxviridae* causes condition called molluscum contagiosum. Can take the form of wartlike growths in the membranes of the genitalia, and can also be transmitted sexually.
- **Group B *Streptococcus* "Colonization"–Neonatal Disease:** Asymptomatic colonization of women by a beta-hemolytic *Streptococcus* in Lancefield group B is very common. It can cause preterm delivery and infections in newborns.

Multiple-Choice Questions   Bloom's Levels 1 and 2: Remember and Understand

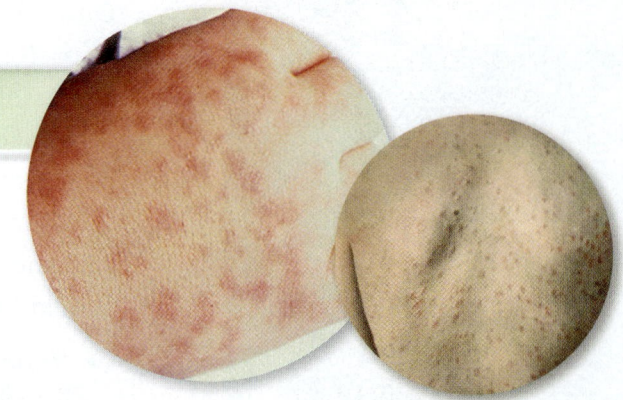

Select the correct answer from the answers provided.

1. The genitourinary system can be thought of as how many different, largely independent organ systems?

   a. 1   c. 3
   b. 2   d. 4

2. Which part of the urinary tract has the most diverse normal microbiota?

   a. lower urethra   c. bladder
   b. upper urethra   d. kidneys

3. Syphilis is caused by

   a *Treponema pallidum.*   c. *Trichomonas vaginalis.*
   b. *Neisseria gonorrhoeae.* d. *Haemophilus ducreyi.*

4. Bacterial vaginosis is commonly associated with the following organism:

   a. *Candida albicans*   d. all of the above
   b. *Gardnerella*   e. none of the above
   c. *Trichomonas*

5. This dimorphic fungus is a common cause of vaginitis.

   a. *Candida albicans*   c. *Trichomonas*
   b. *Gardnerella*   d. all of the above

6. Genital herpes transmission can be reduced or prevented by all of the following *except*

   a. a condom.
   b. abstinence.
   c. the contraceptive pill.
   d. a female condom.

7. This protozoan can be treated with the drug metronidazole (Flagyl).

   a. *Neisseria gonorrhoeae*
   b. *Chlamydia trachomatis*
   c. *Treponema pallidum*
   d. *Trichomonas vaginalis*

8. Which group has the highest rate of HPV infection?

   a. female college students
   b. male college students
   c. college professors of either gender
   d. baby boomers

Critical Thinking   Bloom's Levels 3, 4, and 5: Apply, Analyze, and Evaluate

Critical thinking is the ability to reason and solve problems using facts and concepts. These questions can be approached from a number of angles and, in most cases, they do not have a single correct answer.

1. Infection with *Gardnerella* does not induce inflammation in the vagina. Can you speculate on characteristics of the infection or the organism that could result in no inflammation?.

2. What advantages does the life cycle of *Chlamydia* confer on the pathogen?

3. What are some of the stimuli that can trigger reactivation of a latent herpesvirus infection? Speculate on why.

4. Why do you suppose a urine screening test for *Chlamydia* is more accurate for males than for females?

5. It has been stated that the actual number of people in the United States who have genital herpes may be a lot higher than official statistics depict. What are some possible reasons for this discrepancy?

Visual Connections   Bloom's Level 5: Evaluate

This question connects previous images to a new concept.

1. **From chapters 18 and 21, figure 18.17 and the photo on page 616 in this chapter.** Compare these two rashes. What kind of information would help you determine the diagnosis in both cases?

 **www.mcgrawhillconnect.com**

Enhance your study of this chapter with study tools and practice tests. Also ask your instructor about the resources available through Connect, including the media-rich eBook, interactive learning tools, and animations.

Leona's Beloved Cheese

While I was in nursing school, I was working as a nurse's aide in a long-term care facility. One of my favorite patients was Leona. She was 88 years old and generally healthy, though frail. She loved to tell stories of her youth in Mexico in the 1930s and 1940s before she immigrated to the United States. She spent the first 3 weeks of August looking forward to a visit from her extended family during their family reunion, which was being held locally.

The family visit came and went, and Leona talked about this cousin or that nephew for days afterward. One day when I checked in on Leona at the beginning of my shift, she was quite ill. As soon as I walked into her room, she leaned over the side of the bed to vomit, and I immediately detected the telltale odor of diarrhea. Then she began to cry. She was a mess. I called for a doctor and cleaned her up. I tried to comfort her, but she was cramping and experiencing severe pain. When the doctor arrived, she ordered rehydration and stool cultures and tried to make Leona comfortable. Then she questioned Leona about possible exposures. Because Leona never leaves the facility, we were worried that she had acquired an infection there. We needed to know how it was transmitted. Different measures would need to be taken if it was from the environment, from a caregiver, or from food, for example. Any of those possibilities suggested that more residents could be affected.

Eventually, the doctor established that her family had brought some of her favorite delicacies to her from their large spread of food at the reunion. One of her favorite foods was *queso blanco,* a Mexican cheese made from unpasteurized milk. The family had brought her a large plate of it, and she had eaten it 3 days previously. Upon questioning, the family revealed that two other people had reported mild GI tract symptoms as well but that dozens of people at the reunion had eaten the cheese.

Leona's stool cultures came back positive for *Campylobacter.* The doctor said she was fairly certain that the queso blanco was the culprit and that we should have no further cases of diarrhea at the facility.

- If the source of *Campylobacter* was, indeed, the cheese, why did only three people (of dozens) become ill?

- Why is most milk in the United States pasteurized?

Case File Wrap-Up appears on page 652.

One Health

The Interconnected Health of the Environment, Humans, and Other Animals

By Ronald M. Atlas, University of Louisville

22

IN THIS CHAPTER...

22.1 One Health

1. Draw your own representation or logo depicting how animal health, human health, and environmental conditions interact.

22.2 Animals and Infectious Disease: Zoonoses

2. What features of rabies make it a typical zoonosis?
3. Describe the path (through organisms and environments) that the West Nile virus took from Africa to New York City.
4. Discuss the effects of deforestation and reforestation on the eventual emergence of Lyme disease in the northeastern United States.
5. Explain how HIV transferred from animals to humans.

22.3 The Environment and Infectious Disease

6. Differentiate between coliforms and *E. coli*.
7. Identify three different modes of transmission for cholera.
8. Name at least two diseases whose incidence may be affected by increased rainfall.
9. Discuss the chain of environmental and biological events leading to the outbreak of hantavirus pulmonary syndrome in the southwestern United States in 1993.

22.4 Microbes to the Rescue

10. Define *biological oxygen demand*.
11. Outline the three phases in wastewater treatment.
12. Provide an example of a xenobiotic that has been made biodegradable.

22.1 One Health

We live in a complex and ever-changing world in which human health, animal health, and environmental health are tightly connected. So much today is global in nature–global travel, global economy, global movement of livestock, global climate change. It is no wonder, then, that human health, animal health, and environmental health are inextricably interrelated–and that we are beginning to appreciate that the health of all life on earth is connected.

The concept that the health of humans, animals, and the environment should be viewed in a holistic (all-inclusive) way is called *one health*. This is not a new idea for microbiologists who have always needed to consider all the causes that lead to the spread of infectious diseases and the ability to control them. Pioneering microbiologists like Louis Pasteur and Robert Koch recognized the interconnection between animal and human health. They carried out research on diseases such as anthrax and rabies, which affected both humans and other animals. Medical and veterinary educators of that era also freely crossed the boundaries of human and animal health. In the words of Rudolf Virchow, a leading medical educator of the late 1800s, "Between animal and human medicine there are no dividing lines–nor should there be."

The contemporary revival of the *one health* concept has expanded beyond the human–animal health interface to encompass the health and sustainability of all the world's ecosystems. Microorganisms circulate among human hosts, animal hosts, and environmental reservoirs. Disruption of the environment can lead to transmission of pathogens to animals and humans; evolution of new microbial traits can occur in response to changes in the environment. In addition, reservoirs of pathogens and virulence traits can persist in the environment, poised to enter humans and other animals at an opportune time.

Global Mixing Bowl

One way to think about *one health* is to picture three overlapping spheres representing humans, animals, and the environment **(figure 22.1).** A change in any one of these spheres impacts the others and with that the microbes that each contains. The mixing of microbes in different animal hosts and under different environmental conditions can foster the evolution of new and potentially deadly pathogens. Human activities in particular can promote the emergence of infectious diseases, for example, through ecological disturbances and movement of animals. This occurs frequently for some microbes. Look at the example of influenza viruses, in which the mixing of different strains of influenza viruses in birds, swine, and humans results in the evolution of new recombinant strains with the potential to spread globally almost every year (see chapter 19 for the details of how this works).

Essentially, we have a global mixing bowl in which microbes have greater opportunities to establish new niches, cross species lines, be transported globally, become resistant to antimicrobial drugs and vaccines, and very quickly create new exposures and challenges in the populations of people and animals and in the environment. The result has been the dawn of a new era of emerging and reemerging diseases **(table 22.1).**

The patterns of newly emergent infectious diseases, such as SARS (in 2003) and West Nile fever, reflect contemporary demographic and environmental changes–more people living closer together and in greater contact with wild and domesticated animals, and habitat destruction. Another issue is the multipurpose use of resources–such as using the same water for waste disposal, irrigation, and human consumption. These changes have been driving repeated pathogen spillover from wildlife and the spread of newly evolved pathogens in dense human populations. Over the past four decades, the rate of infectious disease emergence has increased in both humans and animals. We are likely to see new infectious diseases emerge that will be the plagues of our modern world.

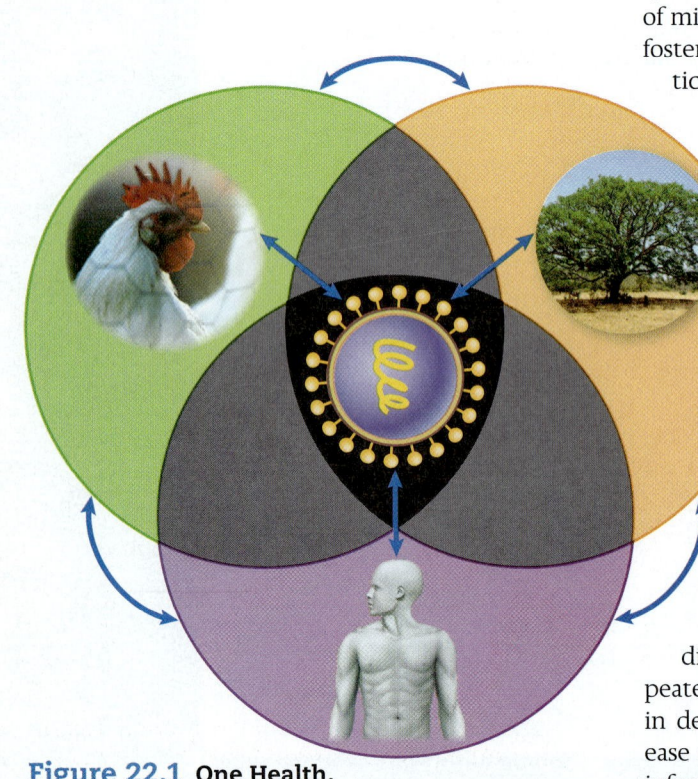

Figure 22.1 One Health.

Table 22.1 Emerging and Reemerging Diseases

Group 1: Pathogens *Newly* Recognized in Past Two Decades	Group II: Reemerging Pathogens
Acanthamebiasis	*Clostridium difficile*
Australian bat lyssavirus	Enterovirus 71
Babesia, atypical	Mumps virus
Bartonella henselae	*Staphylococcus aureus*
Ehrlichiosis	*Streptococcus*, group A
Encephalitozoon cuniculi	
Encephalitozoon hellem	
Enterocytozoon bieneusi	
Hendra or equine morbilli virus	
Human herpesvirus 6	
Human herpesvirus 8	
Lyme disease	
Parvovirus B19	

22.1 LEARNING OUTCOMES—Assess Your Progress

1. Draw your own representation or logo depicting how animal health, human health, and environmental conditions interact.

22.2 Animals and Infectious Disease: Zoonoses

The term **zoonosis** was initially coined to describe the transfer of disease-causing microorganisms from vertebrate animals to humans. Worldwide, diseases that humans acquire from animals result in millions of cases of illness and perhaps a million deaths each year **(figure 22.2)**. Today, however, we also recognize that microbes can transfer back *and* forth between animals and humans. If you are trekking with gorillas in Rwanda, you may want to worry more about the danger of transmitting disease-causing microbes from yourself to the gorillas than about your own safety.

In addition to diseases that are routinely transmitted from an animal to a human (such as rabies), we have to consider another type of zoonosis: the infectious agent that is transferred from an animal to a human and then becomes adapted to human-to-human transfer (such as HIV). Over the last three decades, approximately 75% of newly emerging human diseases have been zoonotic and many have come from or through wildlife. Given that there are 50,000 known vertebrate species and assuming that each has 20 endemic viruses (which is likely to be an underestimate; bats alone harbor 20,000), there probably are more than 1 million vertebrate viruses. Only 2,000 or so viruses have been described, so 99.8% of vertebrate viruses remain to be discovered. These calculations begin to give you an appreciation of the enormous potential for future zoonotic diseases to emerge.

Once introduced into human populations, some zoonotic agents can then spread from human to human—in some cases causing global pandemics. For example, the HIV/AIDS epidemic, which has resulted in more than 40 million human cases worldwide, had its origin as a chimpanzee retrovirus that jumped species and then adapted itself to human-to-human transmission. Such epidemics are extremely difficult to control and can have high rates of morbidity and mortality.

In 2015, the ever-changing influenza virus has started raising alarm again. The fear is that bird flu, which has started to infect humans through bird-to-human transmission in Asia, is on the verge of becoming transmissible from one person to the next (human-to-human transmission). If it does become easily transmissible, the stage could be set for a major pandemic that has the potential to kill millions—like the 1917–1918 influenza pandemic, which is estimated to have killed 50 million people worldwide.

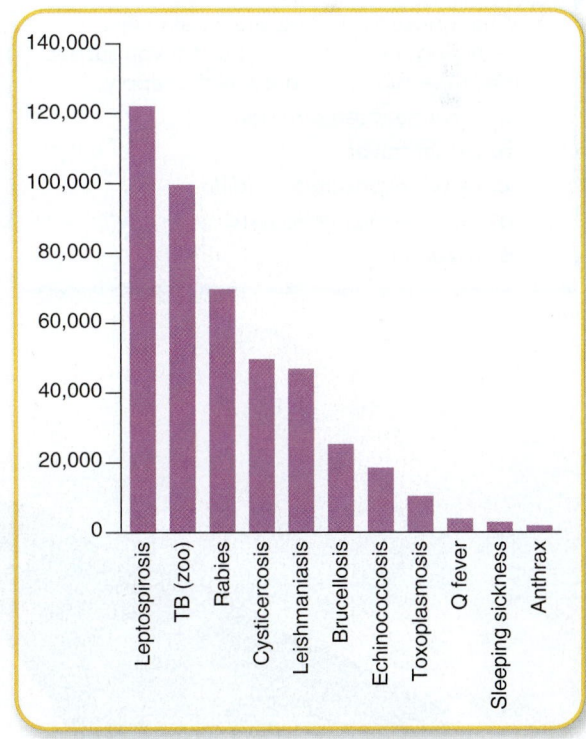

Figure 22.2 Annual number of human deaths worldwide caused by zoonoses. Data from the International Livestock Research Institute

In the next section, we will consider some examples of zoonotic diseases and examine the interactions between animals, humans, and the environment that can lead to their spread and potential control.

Rabies: The Classic Zoonotic Disease

Rabies typically is transmitted to humans through exposure to the virus-containing saliva of an infected animal—mostly through scratches or bites from rabid dogs or by inhaling the virus from aerosols in caves with high bat populations. Rabies is nearly 100% fatal in humans, leaving little opportunity to develop a human-to-human mode of transmission. In recent years, a few people in the United States have survived rabies as a result of experimental, intensive interventions. The cases were well publicized, but the interventions would not be available to most rabies victims in the world.

Over the last 100 years, the epidemiology of rabies in the United States has changed dramatically. More than 90% of all animal cases reported annually to the CDC now occur in wildlife; before 1960, the majority were in domestic animals. The principal rabies hosts in the United States today are wild carnivores and bats (see figure 17.13). Raccoons are the most frequently reported rabid wildlife species in the United States (36% of all animal cases during 2010), followed by skunks (24%), bats (23%), foxes (7%), and other wild animals, including mice, rats, and rabbits (2%).

The number of rabies-related human deaths in the United States has declined from more than 100 annually at the turn of the 20th century to one or two per year currently. The great reduction in human cases is largely the result of vaccinating dogs against rabies. Millions of dogs are vaccinated each year in the United States at a cost of over $300 million per year.

Worldwide, rabies still causes over 50,000 fatal human cases per year. Rabies can be thought of as a neglected disease of poverty, affecting people in underprivileged communities—especially children under age 15 (accounting for 30% to 50% of all exposures). According to the World Health Organization (WHO), more than 2.5 billion people are at risk in over 100 countries. Rabid dogs still cause over 99% of human deaths from rabies worldwide. India, with its estimated 35 million stray dogs, has the highest prevalence of human rabies (approximately 20,000 fatalities per year). Postexposure use of human rabies vaccine can reduce the number of human cases of rabies, but the vaccine is expensive and not adequately used in developing countries.

West Nile Virus—The Emergence of a Contemporary Zoonosis

Mosquitoes are pivotal in the emergence of many zoonotic diseases. They were responsible for the emergence of West Nile virus in the United States in 1999. Almost simultaneously, several elderly people in New York City became deathly ill with signs of encephalitis, and crows began dying in large numbers. A cormorant, several flamingos, and a bald eagle at the Bronx Zoo also died. This raised serious concerns for Dr. Tracey McNamara who was the head veterinary pathologist at the zoo. Initially, there was no thought of a connection because of the traditional separation of human public health and animal disease diagnostic laboratories. McNamara helped connect the dots and establish the connection between the animal and human deaths.

Analyses of human blood specimens by the Centers for Disease Control and Prevention (CDC) initially suggested that the cause of the human cases was St. Louis encephalitis (SLE), a disease that is transmitted from infected birds to humans by mosquitoes. However, analysis of samples from the dead zoo birds by the U.S. National Veterinary Services Lab revealed a virus too small to be SLE virus. It soon became clear that the human and bird deaths were being caused by the same virus and that this

NCLEX® PREP

1. What question(s) is/are vital to ask when gathering the history of a patient you suspect may have rabies? Choose all that apply.
   a. *recent exposure to dogs*
   b. *recent travel*
   c. *recent exposure to wildlife*
   d. *recent exposure to bats*
   e. *occupation*

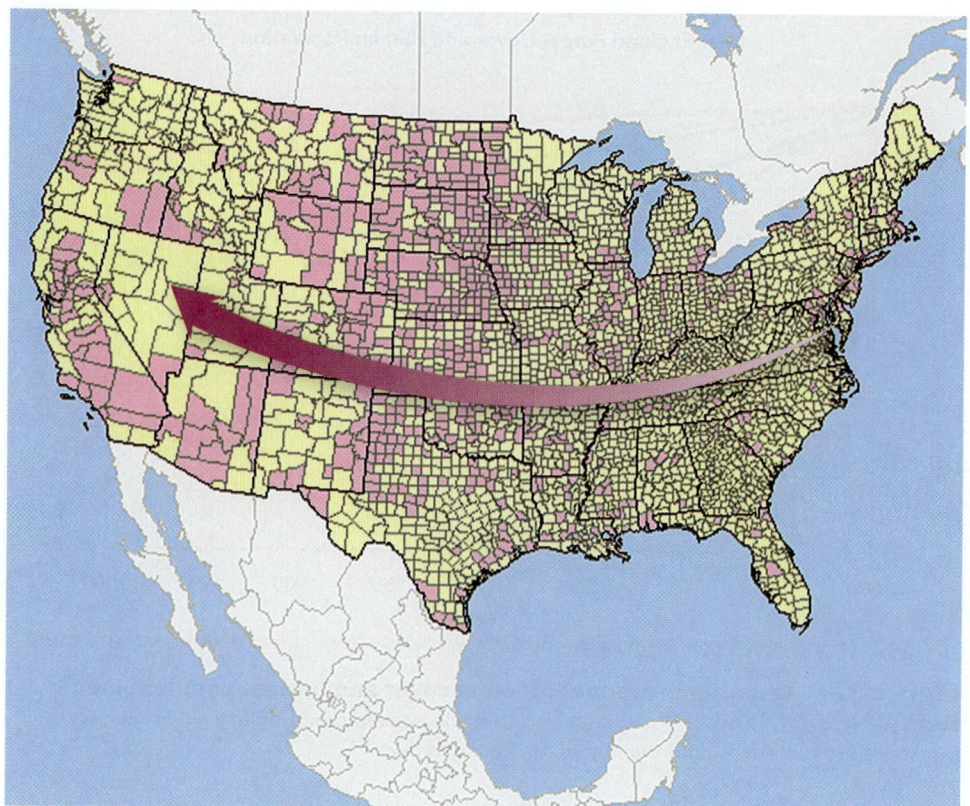

Figure 22.3 **Map of counties in which human cases of West Nile fever occurred in 2013.**

was a newly emerging disease. Nearly 3 months after the initial outbreak, government scientists announced that the disease was caused by West Nile virus, a disease that had never before been found in the Western Hemisphere.

Studies on the genes of West Nile virus suggest that it first evolved in Africa and that as birds migrated from Africa to other continents, they spread the virus to new bird species and eventually to mosquito vectors, which transferred the virus to other birds and humans. Since its introduction into the eastern United States, the disease has spread westward. Currently, West Nile virus occurs in all states except Hawaii and Alaska **(figure 22.3).** Had there been a *one health* approach to surveillance and diagnosis in 1999, the etiology of the disease might have been revealed much sooner and its spread might have been contained.

Lyme Disease—Deer, Ticks, and Environmental Change

During the European colonization of North America, woodland in New England was cleared for farming **(figure 22.4).** At the same time, deer were hunted almost to extinction. However, during the 20th century, environmental conditions changed in the northeastern United States. People abandoned farming, and migration to cities increased. Farmland was converted back to woodland. Deer, protected from hunters and predators, proliferated. White-footed mice also flourished in the forests. With the increase in deer and mouse populations, the deer tick, which is the reservoir host for *Borrelia burgdorferi,* also thrived.

In 1976, a number of children living near Lyme, Connecticut, began developing arthritis-like symptoms. Because rheumatoid arthritis is not an infectious disease, the cluster of children with these symptoms was unusual, suggesting an underlying microbial cause. Soon this disease would be recognized as a bacterial infection caused by

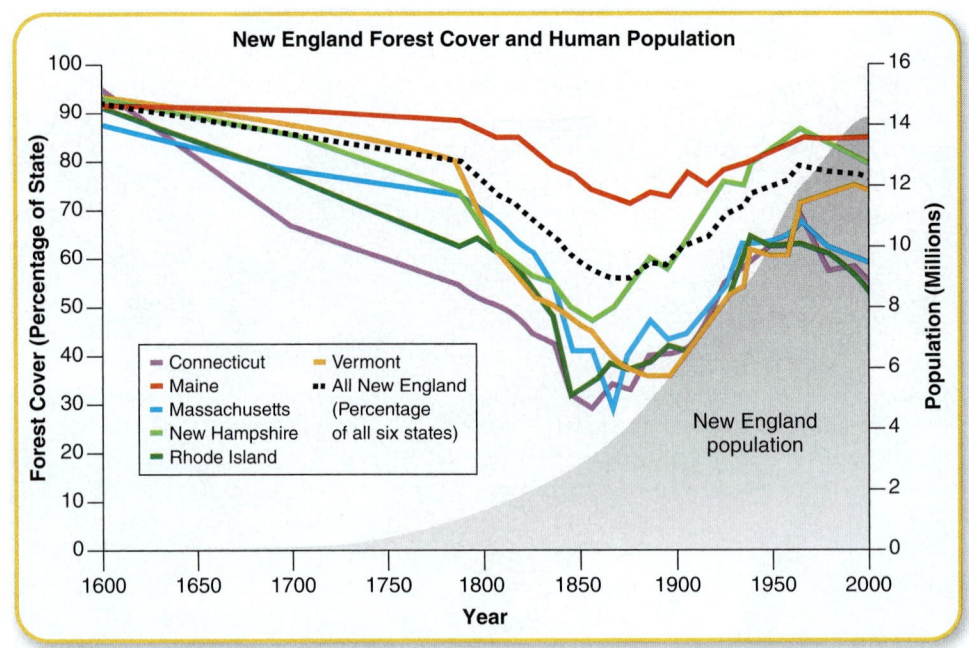

Figure 22.4 Changes in amount of forest cover and human population over time in New England. Note that forest cover decreased greatly in the 1700s and 1800s but started to recover around the turn of the 20th century.

the spirochete *Borrelia burgdorfori* and given the name *Lyme disease*. It is now the most commonly reported arthropod-borne illness in the United States and Europe and is also found in Asia **(figure 22.5)**. The life cycle of the bacterium, depicted in figure 18.15, involves ticks, small mammals, and large mammals. The reasons for the emergence of Lyme disease reflect changing demographics and human behavior, including efforts to restore disrupted environments.

As suburban areas of the northeastern United States became heavily populated, and growing numbers of housing developments backed up to woodland areas, conditions became ideal for *B. burgdorferi*-infected ticks to come into frequent contact with humans.

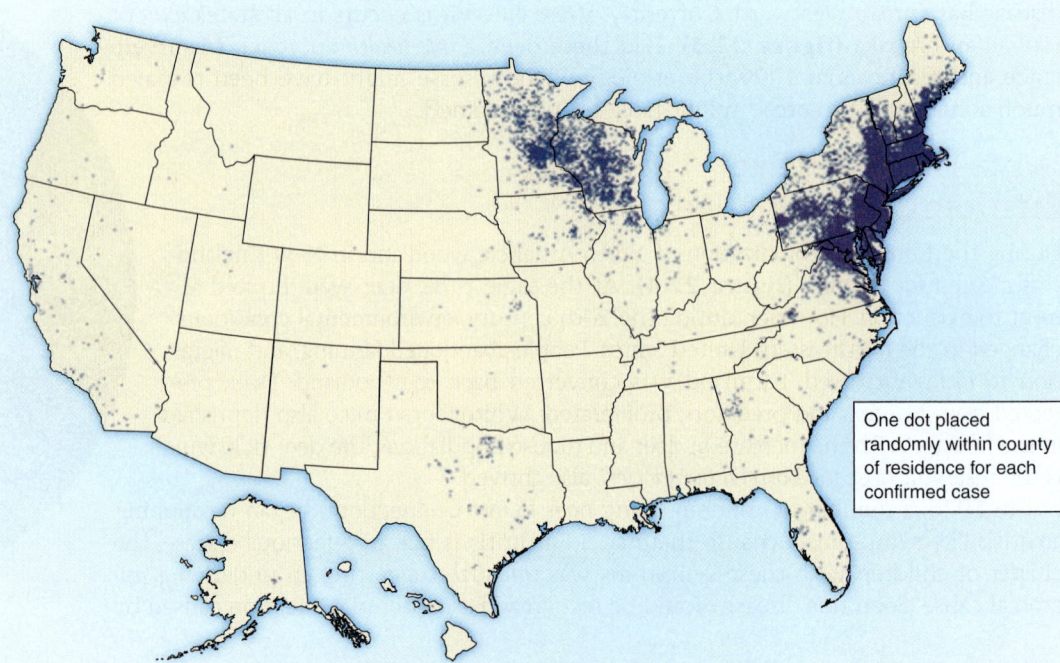

One dot placed randomly within county of residence for each confirmed case

Figure 22.5 A map of Lyme disease in the United States.

During the past 40 years, the infection has continued to spread in the northeastern United States. Climate change (covered later in this chapter) is also influencing the epidemiology of Lyme disease. Scientists mapped the places where temperatures have increased in North America between 1971 and 2010, and found a corresponding increase in the tick population in those areas. Warmer conditions mean that larval ticks are more likely to survive to maturity. This is another reason that Lyme disease–and a host of other tick-borne diseases–are on the rise in temperate climates.

HIV/AIDS—Our Contemporary Plague

Since the beginning in the 1980s, almost 70 million people have been infected with the human immunodeficiency virus (HIV) and about 35 million people have died of acquired immunodeficiency syndrome (AIDS). Globally, about 35 million people are living with HIV infections today.

The development of the contemporary HIV/AIDS pandemic has a complex history that began over a century ago **(figure 22.6).** Molecular evidence shows that this modern pandemic actually began as a zoonotic disease in the late 1800s in central Africa. The HIV-1 group M, which is responsible for most cases of human AIDS, probably originated before 1900 from a simian immunodeficiency virus (SIV) found in chimpanzees in Cameroon. A second lineage of HIV-2 spilled over into humans later from SIV-infected sooty mangabeys (Old World monkeys).

The SIVs most likely were transferred into the blood of hunters harvesting bush meat. Hunters who killed and butchered chimps and monkeys were regularly exposed to animal blood with high concentrations of SIVs. Cuts, bites, and scratches are quite common among bush meat hunters, and these provide an easy portal of entry for the virus.

Colonization and urbanization in Africa undoubtedly contributed to the spread of HIV. Medical interventions of the time, particularly by French colonial physicians, used intravenous blood transfusions, some of which were carried out with reused syringes. This fostered the transfer of HIV. Eventually, enough people were infected to initiate wider spread.

Infected individuals probably moved down the Sangha and Congo rivers toward Kinshasa in the Congo, spreading HIV along the way. During colonial times, efforts

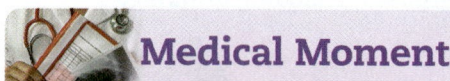

Medical Moment

The Evolution of Virulence: HIV

Did you know that the virulence of microbes can change according to how we humans behave? This is a principle that evolutionary biologists have studied for a long time. It seems that *how a microbe is transmitted* can determine how virulent it is. Here's how it sometimes (but not always) happens: If a microbe is transmitted through intimate human contact, as in the case of sexually transmitted infections (STIs), then it will be more effectively and frequently transmitted if infection does not result in a quick death or debilitation of its host. That means that the variants of the virus that continue (i.e., are passed on) are the ones that cause milder symptoms. Over the course of time (decades), then, STIs become less acutely virulent through this process of natural selection.

On the other hand, what seems to have happened in the case of HIV in the early 1980s is just the opposite. For different reasons on the two continents, the rate of sexual partner exchange was high in certain populations in both Africa and the United States. In such a scenario, random mutations that made the virus highly virulent were maintained in the virus population because even the strains making people very sick, very quickly were being passed on due to the frequency of sexual contact. This phenomenon seems to have fueled the explosion in the 1980s of a high-virulence pandemic caused by a virus that had been infecting people more slowly and with longer survival times for decades before that.

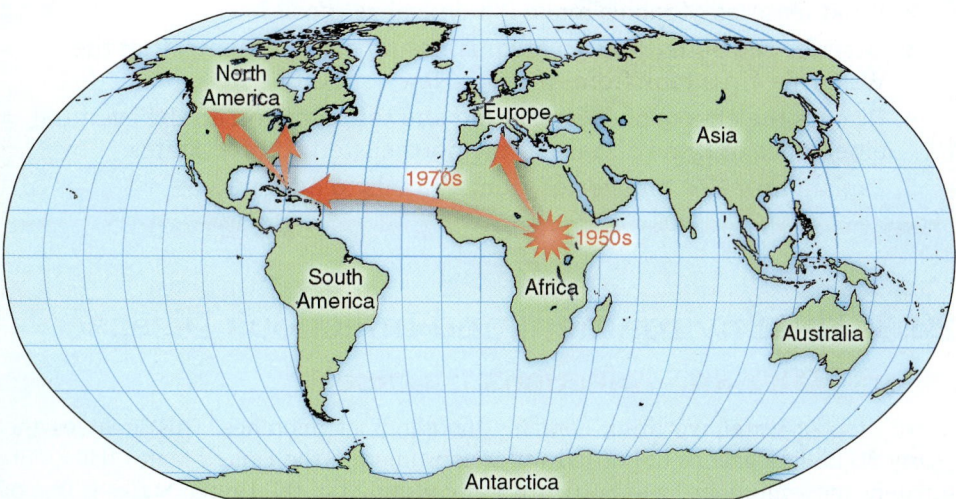

Figure 22.6 Path of HIV-1 after being introduced into humans. The virus probably jumped to humans in the late 1800s/early 1900s. In the 1950s and 1960s, it had infected many in Central Africa. It probably was transferred to the Caribbean in the 1970s. It either traveled from there or directly from Central Africa to large urban centers in the United States in the 1980s.

The NAMES project created and curates a massive quilt, in which each panel commemorates a person lost to AIDS.

Figure 22.7 Much of the world's population uses the same source of water for many different purposes, including drinking.

to exploit rubber and ivory in Central Africa created routes to transport these resources that became pathways for infectious disease propagation. Syphilis reached epidemic proportions along porter routes and riverside trading posts in Cameroon and throughout the Congo Basin. The human immunodeficiency virus presumably moved along the same routes. Once it arrived in Kinshasa, the evolving HIV appears to have entered a wide urban sexual network and spread quickly. Conditions were right for the amplification of HIV (see Medical Moment). The stage was set for the AIDS epidemic.

Based upon analyses of stored blood samples, it appears that about 2,000 people in Africa might have been infected with HIV by the 1960s. Urban prostitution increased in Kinshasa. Belgians and other Europeans fled the Congo and were replaced by thousands of Haitians. By 1970, there was a surge in opportunistic infections such as cryptococcal meningitis and Kaposi's sarcoma. Although it was not recognized at the time, this was the first sign of the HIV/AIDS epidemic.

It is not known exactly how the virus was transferred to the United States, but in 1980, the detection of an abnormal cluster of Kaposi's sarcoma among gay men in San Francisco triggered public health investigations leading to the identification of AIDS and its cause, HIV. It had become a global pandemic. Once a disease of chimpanzees, it had spilled over into humans and become human-to-human transmissible—a zoonotic disease had evolved into a major human killer.

> ### 22.2 LEARNING OUTCOMES—Assess Your Progress
>
> **2.** What features of rabies make it a typical zoonosis?
> **3.** Describe the path (through organisms and environments) that the West Nile virus took from Africa to New York City.
> **4.** Discuss the effects of deforestation and reforestation on the eventual emergence of Lyme disease in the northeastern United States.
> **5.** Explain how HIV transferred from animals to humans.

22.3 The Environment and Infectious Disease

Clean Water and Infectious Disease

Every drop of water we drink contains thousands of microbes. This includes the nearly 30 million gallons of bottled water consumed in the United States daily. Fortunately, most municipal and bottled water consumed in the United States is free of human pathogens—but this is not the case everywhere.

While the United States and many other nations take great care to ensure the safety of potable (drinkable) water, this is not the case worldwide—particularly in rural communities of developing countries. In fact, most of the world's population lack access to safe drinking water. According to the World Health Organization (WHO), more than 1 billion people lack access to clean water, and 2.6 billion people lack access to basic sanitation **(figure 22.7).**

Poor water sanitation and a lack of safe drinking water take a greater human toll than war and terrorism. Waterborne diarrheal disease alone accounts for 4.1% of the daily global disease burden, and more than 3.4 million deaths each year are attributable to unsafe drinking water–primarily among children in developing countries. According to a U.N. report, 4,000 children die each day as a result of infectious diseases acquired by drinking contaminated water. The poorest and most vulnerable members of society bear the greatest burden of waterborne and diarrheal disease.

Drinking Water Quality—Microbiological Safety of Potable Water

One of the reasons potable water is generally safe in the United States and many developed nations is that great care is taken to disinfect the water and test it for bacterial content before it is consumed. Water is generally disinfected by filtration or by treatment with chlorine to ensure that it is free of potential pathogens. When there is a breach in the water delivery system and the quality of the water cannot be ensured, a boil-water advisory is issued; that is, the public is told to boil the water to kill most potential pathogens before drinking the water.

Routine bacterial monitoring of potable waters involves testing for **indicator bacteria,** that is, microorganisms normally found in mammalian gastrointestinal tracts whose presence would indicate likely contamination with fecal matter **(figure 22.8).** Testing of the actual pathogens of concern (protozoa like *Giardia* and *Cryptosporidium*; bacteria like *Salmonella, Shigella, Vibrio,* and *Campylobacter;* and

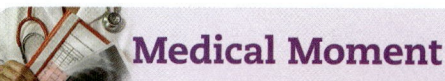

Medical Moment

Plastic Bottles for Clean Water

Every week around the world, 30,000 people die from lack of clean water. Ninety percent of these are children under 5 years old. Clean water—taken for granted in the developed world—is a resource more precious than gold on the rest of the planet. Even though we take it for granted, the processes and infrastructure used to deliver it to us are complex and expensive. How can we export those to other settings? Maybe we don't have to. Solar water disinfection is a method of safely disinfecting drinking water by simply placing contaminated water in a transparent plastic bottle and leaving it in the sun for 6 hours. Ultraviolet light kills bacteria and parasites and inactivates viruses, making the water safe. This technique has been used all over the world in impoverished nations where citizens have no access to clean drinking water, and it has proven to be an effective way of preventing diarrheal disease.

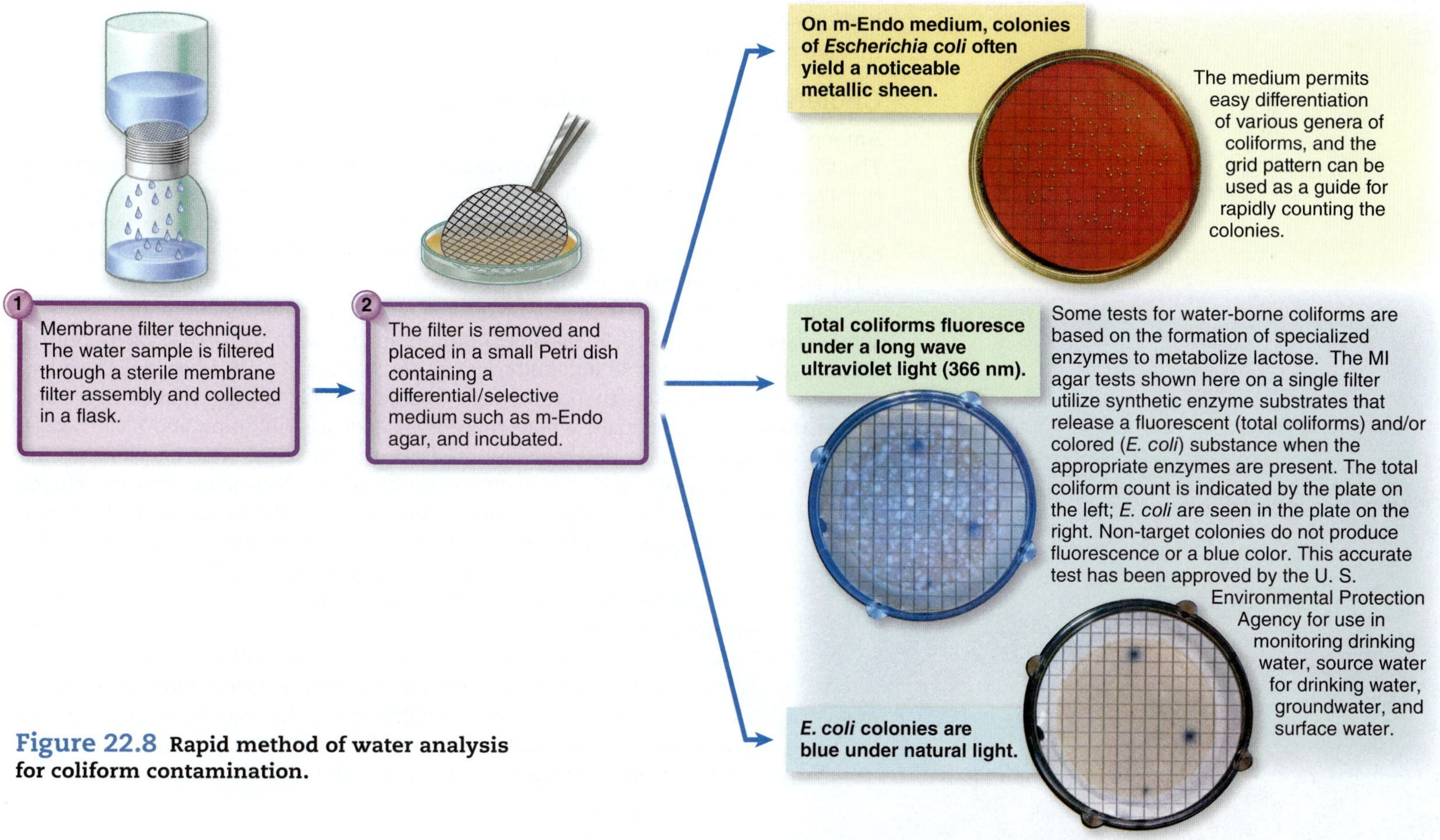

On m-Endo medium, colonies of *Escherichia coli* often yield a noticeable metallic sheen.

The medium permits easy differentiation of various genera of coliforms, and the grid pattern can be used as a guide for rapidly counting the colonies.

① Membrane filter technique. The water sample is filtered through a sterile membrane filter assembly and collected in a flask.

② The filter is removed and placed in a small Petri dish containing a differential/selective medium such as m-Endo agar, and incubated.

Total coliforms fluoresce under a long wave ultraviolet light (366 nm).

Some tests for water-borne coliforms are based on the formation of specialized enzymes to metabolize lactose. The MI agar tests shown here on a single filter utilize synthetic enzyme substrates that release a fluorescent (total coliforms) and/or colored (*E. coli*) substance when the appropriate enzymes are present. The total coliform count is indicated by the plate on the left; *E. coli* are seen in the plate on the right. Non-target colonies do not produce fluorescence or a blue color. This accurate test has been approved by the U. S. Environmental Protection Agency for use in monitoring drinking water, source water for drinking water, groundwater, and surface water.

***E. coli* colonies are blue under natural light.**

Figure 22.8 Rapid method of water analysis for coliform contamination.

viruses like hepatitis A and Norwalk) would be difficult and may not provide the necessary margin of safety given the large volumes of water that are distributed. The intestinal bacteria most useful in the routine monitoring of microbial pollution are gram-negative rods called **coliforms** (*E. coli* and similar bacteria), which are gram-negative bacteria that ferment lactose and produce gas. These coliform bacteria, as well as fecal streptococci, which also are useful indicator bacteria, survive in natural waters but do not multiply there. Finding them in high numbers thus indicates recent fecal contamination and the possible presence of intestinal pathogens. When significant numbers of coliform bacteria or fecal streptococci are detected, the water is not considered safe to drink.

Although total coliform counts are useful indicators for ensuring water safety, they can be misleading. In many circumstances, it is important to differentiate between coliforms that can be naturally found in soils and uncontaminated waters (*Enterobacter, Klebsiella, Citrobacter*) and **fecal coliforms** (*E. coli*) that live mainly in human and animal intestines. Although most *E. coli* strains are not pathogenic, they almost always come from a mammal's intestinal tract, so their presence in a sample is a clear indicator of fecal contamination.

Sometimes the public and the media confuse *coliform* with *E. coli*. This can cause unwarranted concerns. For example, in 1995, there was a minor panic when media outlets reported that iced tea from restaurants in Cincinnati, Ohio, contained significant numbers of "fecal coliforms." One headline read, "Iced Tea Worse Than River Water." Despite such alarmist reports, no one became sick from drinking the iced tea. When scientists did more detailed testing, they found that the predominant species were *Klebsiella* and *Enterobacter,* both of which are commonly found on plants, including tea leaves. Clearly, the total coliform count was misleading in this situation.

No doubt the coliform test that has been used for decades will soon be replaced with more direct testing for fecal contamination, either through specific testing for *E. coli* or through new technologies being developed such as **biosensors,** machines that detect structures on specific microbes and transmit electrical or physical signals that are easily read.

Cholera and Water Contamination

Among the waterborne diseases, cholera has a particularly rich and deadly history. The bacterium *Vibrio cholerae* had existed in the Ganges river delta in India for centuries. During the 1800s, it began spreading across the world from its original reservoir. Six subsequent pandemics killed millions of people across all of the continents. The current (seventh) pandemic, which started in South Asia in 1961, reached Africa in 1971 and the Americas in 1991.

Cholera transmission is closely linked to inadequate environmental management. *Vibrio cholerae* is released in massive amounts in the watery diarrhea of its human victims and remains viable indefinitely in water sources. At-risk areas include urban slums, where basic infrastructure is not available, as well as camps for refugees and other displaced individuals, where there is inadequate sanitation and clean water. For example, cholera infected hundreds of thousands of individuals who fled Rwanda in 1994 during that country's civil war. Rwandans fled to refugee camps across the border in Goma, Zaire, to avoid the massacre in their own country **(figure 22.9).** The water in the camps became contaminated due to inadequate sanitation. Twenty thousand individuals were dying each day at the height of the epidemic, making it the worst cholera epidemic in many centuries.

Water companies today still collect their water from rivers that are also used for sewage disposal. For example, in Louisville, Kentucky, drinking water comes from the Ohio River that has been used upriver by Cincinnati, Ohio, for its sewage disposal. Because Cincinnati treats its sewage and Louisville disinfects and regularly tests its water for pathogens, the Ohio River is not a source for the spread of waterborne disease. In fact, Louisville water was recently named the best municipal drinking water in the United States.

Our earliest recognition that water can serve as a conduit for the spread of cholera and other infectious diseases comes from the work of John Snow in England in the mid-19th century. Snow postulated that an 1854 outbreak of cholera came from the water supply. This was a serious and terrifying outbreak that killed hundreds and sent the most well-to-do residents of the Soho area in London fleeing to avoid being stricken with cholera. Of course, this left those who could not afford to leave the city to suffer the consequences of the disease.

Snow carefully mapped where the cases were occurring and soon determined that an unusually high number of deaths were taking place near a water pump on Broad Street. It appeared that buckets of waste being disposed from a nearby tenement where a child had cholera entered the water being pumped from this well. Shutting down the well quickly led to a dramatic decline in the number of deaths due to cholera. By establishing the pattern of disease, Snow was able to pinpoint the source and the way of stopping the epidemic. In doing so, he established the fundamentals of epidemiology, the study of disease spread and control in populations. (Note that after the immediate problem of the epidemic was solved, public officials still balked at accepting the general principle–that feces contaminating the water supplies were responsible for disease outbreaks. Apparently, it was just too unsavory to consider.)

Today, molecular methods are providing valuable tools for determining the sources of infectious disease outbreaks. Such tools were used to help identify the source of an outbreak of cholera in Haiti that occurred in the aftermath of the devastating earthquake that struck that country in 2010. These molecular analyses identified the strain of *Vibrio cholerae* responsible for the outbreak as an Asian strain that had not previously been seen in the Western Hemisphere.

As of 2014, the outbreak in Haiti had yet to be brought under control as the water infrastructure was still being rebuilt **(figure 22.10).** In mid-2014, close to 700,000 cases and more than 8,500 deaths had been reported since the onset of the cholera epidemic. Like John Snow's work a century and a half earlier, geographic clusters were used to map the epidemic's movement, pointing to the source of the

John Snow's original map indicating (with black boxes) where cases of cholera had occurred in London.

Figure 22.9 **Rwandan refugees in Zaire in 1994.**

Figure 22.10 **The aftermath of the 2010 earthquake in Haiti.**

outbreak as a U.N. relief camp housing soldiers from Nepal. The Centers for Disease Control and Prevention conducted molecular tests on strains from the U.N. base and from afflicted individuals, concluding that the most likely source of the outbreak was one of the U.N. soldiers and that the disease was spread by the poor sanitary conditions and floods that contaminated Haitian drinking water following the earthquake and subsequent hurricane.

Like the situation in Haiti, control of cholera is a major problem in several Asian countries, as well as in Africa. Since 2005, the reemergence of cholera has been noted in parallel with the ever-increasing size of vulnerable populations living in unsanitary conditions. Bangladesh experiences major cholera epidemics annually; outbreaks often occur following monsoons that wash human fecal matter into waters used for drinking water. In such endemic areas of Asia, the death rate is normally 5% to 15%. The number of cholera cases reported to the World Health Organization continues to rise in countries with inadequate sanitation and water treatment. The true burden of the disease is estimated to be 3 to 5 million cases and 100,000 to 120,000 deaths annually. Not surprisingly, governments frequently downplay the extent of a cholera epidemic, especially if fisheries or tourism are important to their economies. In 1991, during a massive cholera epidemic in Peru, the country's president famously appeared on television eating ceviche (a raw fish dish) and serving it to his cabinet ministers. Predictably, the hospitalization rate for cholera in Peru skyrocketed in the ensuing weeks.

Microbial Contamination of the Food Chain

Cholera is an example of an infectious disease of the "environment" (water) that can contaminate food that we consume (shellfish). There are many other ways that our environment and the animals with whom we share it impact the food we eat.

Hepatitis A virus is another pathogen that can be transmitted via water contaminated with human fecal matter. Hepatitis A virus is widespread throughout the world. Transmission rates are very high in areas where there is no sewage treatment. Hepatitis A viral infections have an especially high incidence in developing countries and rural areas. In rural areas of South Africa, the seroprevalence is 100%; that is, all of the people in those areas have been infected with hepatitis A, although many are asymptomatic and show no signs of disease even though they are carriers of the hepatitis A virus.

Hepatitis A infections also occur in developed countries. Approximately 25,000 cases of hepatitis A viral infections are reported in the United States every year—but because many cases are asymptomatic or not reported, the actual annual incidence of hepatitis A infections is estimated to be more than 260,000.

You have only to look at the news to find the latest food-borne outbreak of disease, and many of these are described in chapter 20. *E. coli* outbreaks from beef are caused when cattle are infected, and *E. coli* outbreaks associated with vegetables are generally triggered when crops are watered with unclean water containing fecal matter or, possibly, when infected animals or birds have access to the crops. Many helminth diseases are acquired through a contaminated environment (organisms burrowing through skin or being being swallowed) or from ingesting infected food. As has been stated already, the concept of *one health* is an old one indeed; but in many parts of the research and regulatory communities, the last few decades have seen increasing compartmentalization so that the connections were not as well appreciated as they have now become.

Global Climate Change and Biology

Over the past 40 years, the concentrations of carbon dioxide and methane in the atmosphere have increased. They are now at their highest levels since the first humans set foot on earth **(figure 22.11).** As a result of the increased concentrations of carbon dioxide and methane, the climate is changing. Because these compounds absorb long-wavelength energy coming from the earth's surface and reflect some of it back to earth, they cause a rise in temperatures—so things get warmer, much like in a greenhouse.

Climate change can be detected as changes in weather—more rainfall and generally higher average temperatures over the year with more variable weather patterns **(figure 22.12)**. Most scientists argue that human activities are the root cause and that if we do not change course there will be devastating impacts on human, animal, and environmental health.

We hear quite frequently about rising sea levels and the threat of devastating storms if we do not halt the atmospheric buildup of greenhouse gases. We should also understand that climate change may increase the global burden of disease. Drought and lack of water are likely to get worse in some places while in other areas increased rainfall likely will lead to increases in some pathogens and vectors of infectious disease. *Vibrio cholerae*, for example, thrives in warm waters. There are predictions that the global incidence of cholera will increase as a result of global warming. Mosquito populations likely also will increase in areas of increased rainfall and with them the global incidence of infectious diseases like yellow fever and malaria. Mathematical models show that a 2- to 3-degree rise in temperature could increase the number of people affected by malaria by several hundred million. Already, dengue fever is moving into Florida.

The United States saw an elegant example of how weather patterns can lead to a sudden outbreak of disease in 1993, when hantavirus suddenly began killing healthy young adults in the Four Corners area of the American Southwest. The virus had never before been known to cause the severe, overwhelming respiratory distress and death seen in that outbreak (previously, it was known as a kidney pathogen). After the outbreak was contained and research conducted into the causes of the outbreak, the weather phenomenon El Niño, a warming trend that originates over the Pacific Ocean, looked like the ultimate culprit. Here's how: In the year preceding this outbreak, El Niño caused wetter winter conditions in the Southwest. This led to a great increase in the pine nut population, a favorite delicacy of deer mice. By the spring and summer of 1993, there was a very large deer mouse population. Deer mice are the reservoir for hantavirus, and the people who became infected all had exposure to large amounts of deer mouse feces and/or urine. When these substances dried out and were aerosolized, they delivered lethal doses of the virus to the lungs of humans.

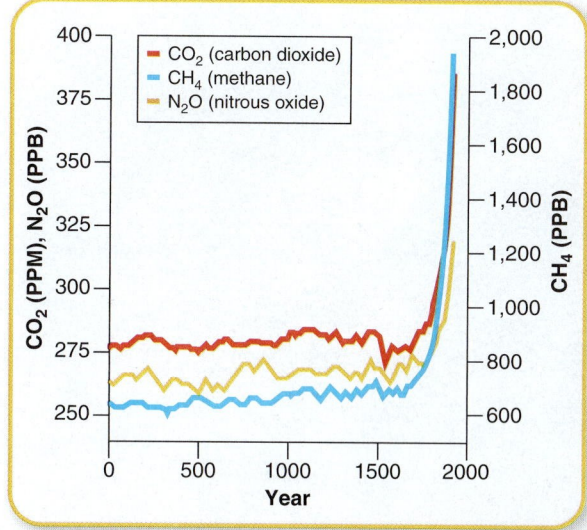

Figure 22.11 **The increase in greenhouse gas concentrations over the past 2,000 years.**

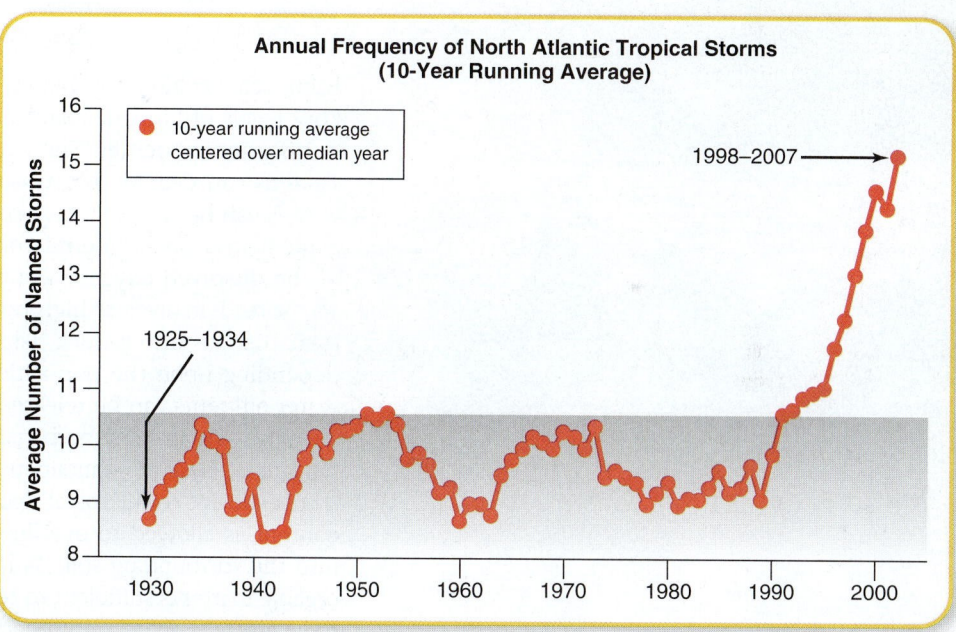

Figure 22.12 **Number of named tropical storms in the North Atlantic.**

Climate change will undeniably have far-reaching effects, where the change in temperature of even a few degrees can start a cascade of events that can easily lead to unusual occurrences of infectious disease. It is not just humans that are being affected by climate change; plants and animals are also being impacted. Bluetongue virus, which causes bluetongue disease in domestic and wild ruminants, has reached Britain from Africa; the likely cause is that warm winds resulting from climate change have carried midges, which are small flies that are the vectors of the disease, northward from Africa to Europe. We can expect future movements of vectors and the diseases they carry to result in increased disease outbreaks if climate change continues on its current course.

22.4 Microbes to the Rescue

While the presence of pathogens in the environment is clearly a threat to human and animal health, non-disease-causing microbes are of extreme importance in *saving us* from the wastes and pollutants that we produce. In New York City alone, wastewater treatment plants, with the help of microbes, treat over 1 billion gallons of sewage a day, removing an average of 65% of the organic matter. Urban solid waste in the United States amounts to roughly 150 million tons per year. Microbes also help us by cleaning up our pollutants, such as industrial chemicals and oil spills, both when we specifically "ask them to" and when we do not.

Liquid Waste Treatment—Sewage

Each year, trillions of gallons of human waste are released into the environment. Low levels of human wastes can be accommodated because natural waters have self-purification capacities. The naturally occurring nonpathogenic microbes in those ecosystems can degrade the wastes. However, in areas of high population densities, the wastes can be too great. When waters are overwhelmed by concentrated inputs of organic matter (including microbes) they exhibit a high demand for oxygen. Exhaustion of the dissolved oxygen occurs and the water becomes putrid and septic. Pathogens also spread. In areas of high population densities, therefore, it is generally necessary to treat the waste to reduce risk. Several different sewage treatment processes are used, depending upon the population density and the environment into which the wastewater effluents can be released.

Many rural and suburban areas with relatively low population densities rely on septic tanks. These are containers into which the sewage flows. The solid material settles and is subject to microbial decomposition. The liquid, with its greatly reduced organic content, is allowed to overflow and is distributed through a series of perforated pipes into the surrounding soil. As long as the houses are far enough apart, the reduction of organic matter is sufficient to reduce the concentrations of organic matter being released to levels that can be accommodated without causing environmental harm. Such systems, however, are inadequate to handle the wastes from densely populated communities.

For some municipalities, the answer is to pump the sewage away–sometimes off-shore into the ocean away from beaches and shellfish beds and sometimes to agricultural fields away from the city. Mexico City, for example, only treated 15% of the wastewater it collected in 2008 and still only has a maximum capacity for treating 25% of the sewage generated by the 25 million inhabitants of the city. The remaining untreated sewage is pumped to agricultural areas–largely alfalfa but also barley, corn, and wheat fields–that collectively cover an area the size of the state of Rhode Island.

Most other major cities–from New York City to Johannesburg, South Africa–have extensive sewage treatment plants. The border cities of Tijuana in Mexico and San Diego in the United States have developed an international agreement for jointly treating their wastewater. Sewage from Tijuana is piped across the border to a newly constructed treatment plant on the U.S. side of the border. This plant also treats the wastewater for San Diego.

The sewage that is piped to wastewater treatment plants is subjected to physical, biological (microbial), and sometimes chemical processes **(figure 22.13).** The

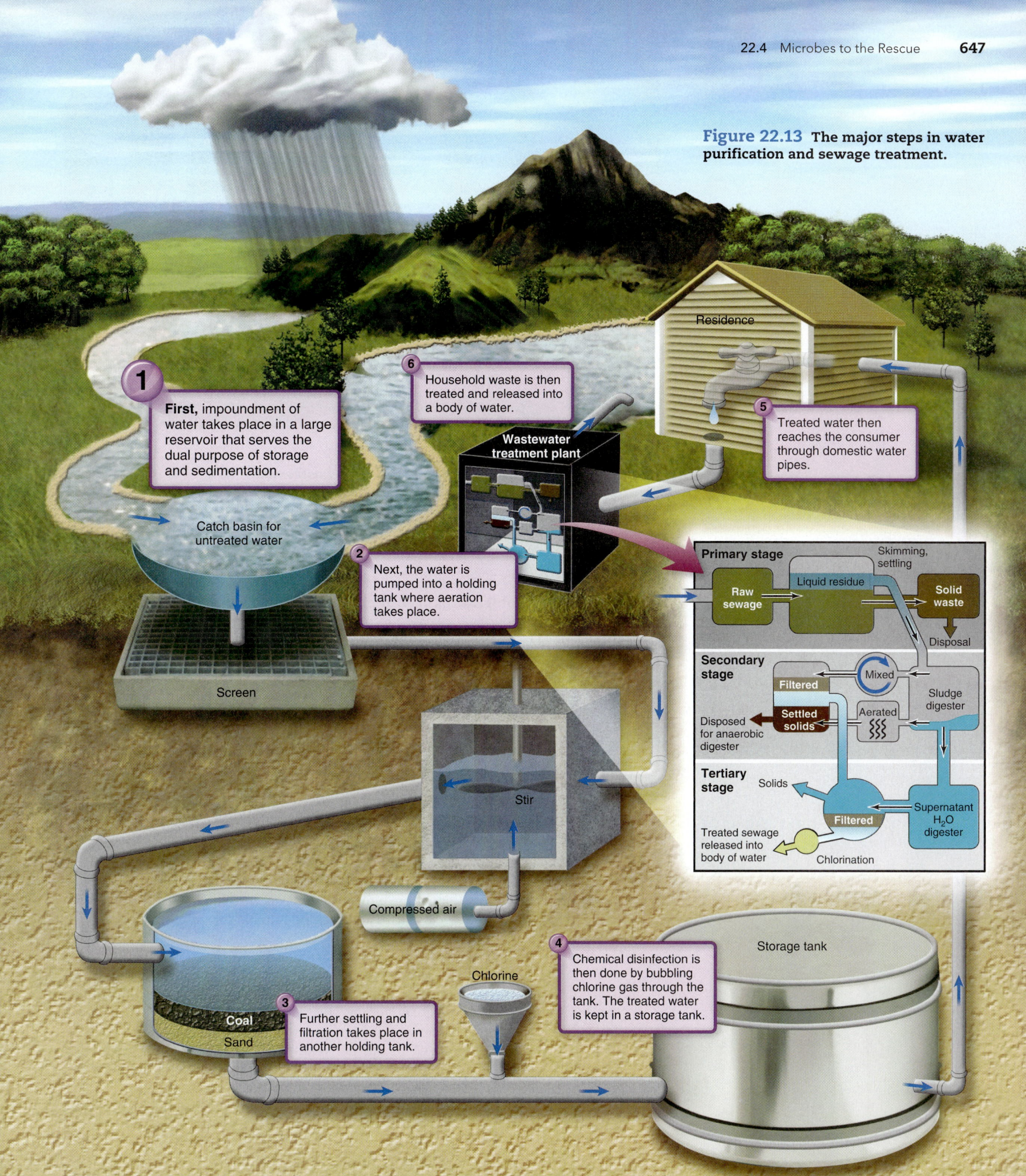

Figure 22.13 The major steps in water purification and sewage treatment.

1 **First,** impoundment of water takes place in a large reservoir that serves the dual purpose of storage and sedimentation.

Catch basin for untreated water

Screen

2 Next, the water is pumped into a holding tank where aeration takes place.

Stir

Compressed air

Coal

Sand

3 Further settling and filtration takes place in another holding tank.

Chlorine

4 Chemical disinfection is then done by bubbling chlorine gas through the tank. The treated water is kept in a storage tank.

Storage tank

6 Household waste is then treated and released into a body of water.

Wastewater treatment plant

Residence

5 Treated water then reaches the consumer through domestic water pipes.

Primary stage

Raw sewage → Liquid residue → Skimming, settling → Solid waste

Disposal

Secondary stage

Filtered   Mixed

Settled solids   Aerated   Sludge digester

Disposed for anaerobic digester

Tertiary stage

Solids

Filtered   Supernatant H_2O digester

Treated sewage released into body of water   Chlorination

(b)

(a)

Figure 22.14 Treatment of sewage and wastewater. (a) Digester tanks used in the primary phase of treatment; each tank can process several million gallons of raw sewage a day. (b) View inside the secondary reactor shows the large stirring paddle that mixes the sludge to aerate it to encourage microbial decomposition.

main aim of wastewater treatment is to reduce the organic matter content, that is, to lessen the **biological oxygen demand,** or requirement for oxygen, when the wastewater is eventually released into the environment. Normally, the numbers of pathogens also decline during the treatment processes so that the water released from the plant is safer.

The processes involved in wastewater treatment are divided into (1) the primary phase **(figure 22.14a),** which involves physical separation of solid materials, largely through settling; (2) a secondary phase in which the liquid portion of the waste is subjected to microbial decomposition that is largely aerobic **(figure 22.14b);** and (3) an optional tertiary phase, which may be chemical, physical, or biological, to remove additional inorganic substances like ammonia, nitrate, and phosphate and to eliminate pathogens.

The *trickling filter system* is a relatively simple and inexpensive secondary treatment system. Liquid waste is sprayed over a porous bed of rocks and allowed to flow downward for collection or directly into the groundwater. The rocks are covered by a biofilm of microorganisms. As the wastewater slowly percolates downward, the microorganisms in the biofilms aerobically degrade the organic compounds in the wastewater, reducing the biological oxygen demand. These trickling filters, though, have limited capacities and are not suitable for very high volumes of wastewater.

For treating high volumes of wastewater, the *activated sludge process* is used for secondary treatment of the sewage. In this process, the liquid waste is placed into large tanks. Air is forced into the tanks and the wastewater is rigorously mixed to ensure aerobic conditions. An active microbial community develops that degrades the organic compounds in the wastewater. A sludge forms that is allowed to settle. Most of the sludge is removed for further treatment, but a portion of the sludge containing high numbers of microbes (called *activated sludge* because the microbes are already adapted to degrading the organic compounds in the wastewater) is reintroduced into the aerated treatment tank along with the next batch of wastewater to be treated.

The activated sludge process generally reduces the biological oxygen demand by 85% to 90%. It also greatly reduces the numbers of pathogens, which are largely outcompeted by the nonpathogens and tend to settle in the sludge. The concentrations of *Salmonella, Shigella,* enteroviruses, and other pathogens are generally 90% to 99% lower in the effluent water that is discharged from the plant than in the water that comes into the treatment facility. The water from the secondary treatment process often is released into nearby water bodies.

Sometimes, prior to release, the water from the secondary treatment tanks is subjected to tertiary treatment. Tertiary treatment removes nutrients that could support algal blooms and is important if the wastewater is going to be released into a pristine lake. The tertiary phase may also involve disinfection of the wastewater—for example, by filtration or chlorine treatment—to eliminate pathogens. Thus, cities like Cincinnati and Louisville can safely discharge the effluents from their wastewater treatment plants into the Ohio River.

The sludge from primary settling tanks and activated sludge treatment tanks can be treated in **anaerobic digestors.** Here, anaerobic bacteria and archaea further decompose the solid wastes, producing a stable solid material and methane that can be collected and used as a fuel. The gas produced from the anaerobic digestor treatment of the sludge can be used to heat the treatment facility or can be purified and sold. Thus, waste treatment and energy generation can be coupled to improve environmental health.

Solid Waste Treatment—Composting and Landfills

Urban solid waste (what most of us know as "garbage") production in the United States amounts to roughly 150 million tons per year. Much of this material is glass, metal, and plastic that is not subject to microbial degradation. The rest is decomposable organic compounds, such as food scraps and paper.

Today, we attempt to remove the nonbiodegradable material from the solid waste and recycle it; this includes paper that is relatively difficult for microbes to decompose. In some cases, the organic wastes are dumped offshore or simply discarded on land, but excessive dumping can cause serious problems. Most municipalities in the United States, therefore, use a combination of sanitary landfills and composting to treat solid wastes.

Composting is an aerobic process in which air and inorganic nutrients support the growth of diverse aerobic microbial communities of bacteria and fungi that are able to decompose the organic wastes. The wastes are generally arranged in a heap or pile that can be managed to ensure adequate aeration and moisture retention **(figure 22.15).** Sometimes, homeowners compost leaves and kitchen and garden wastes to get rid of them. Initially, the decomposition process is carried out by mesophilic microorganisms; but as the decomposition proceeds, heat builds up and there is a shift to thermophiles. The end result of the decomposition of the organic wastes is a material that can be added to soils as a soil conditioner or fertilizer.

In 2013, the mayor of New York City, Michael Bloomberg, proposed that residents be required to separate food scraps from other solid wastes and that these organic wastes would then be subject to "composting." However, his use of the term *composting* did not accurately describe what he was proposing. He wanted the food scraps to undergo anaerobic decomposition to produce biogas (approximately 70% methane and 30% carbon dioxide) that could be purified and used as a fuel. This process would essentially be the same as the one used for anaerobic digestion of sewage sludge.

The alternative to composting or decomposition in an anaerobic digestor is to send the wastes to a landfill. Generally, organic and inorganic solid wastes are deposited together on land that has minimal real estate value. Because exposed waste

Figure 22.15 A backyard composting bin.

Figure 22.16 A typical landfill.

can cause esthetic and odor issues and also attract insects and rodents, each day's waste is covered with a layer of soil–creating a sanitary landfill **(figure 22.16).** Eventually, the waste and soil layers build to form a hill. As the buried waste is decomposed, largely by anaerobic microorganisms, the land mass subsides. After 30 to 50 years, the land often is stable and suitable for other uses like housing development. One of the campuses of your author Kelly Cowan's university is built on top of a landfill!

Biodegradation and Bioremediation—Oil and Chemical Spills

Microorganisms have an extraordinary capacity to degrade and to transform chemicals that enter our environment either through natural processes or human activities. Without these microbial activities, all life would cease on earth within a few weeks. Yet, human exploitation of fossil fuels that sometimes results in major oil spills and the release into the environment of xenobiotics–novel, chemically synthesized compounds that do not naturally occur–is proving challenging to microbial **biodegradation** and the sustainability of environmental quality.

Some xenobiotic compounds are totally resistant–recalcitrant–to microbial attack. When these compounds enter the environment, they accumulate and can cause environmental harm. In some cases, it has been possible to redesign these compounds so that they are biodegradable. Laundry detergents, for example, now have linear alkyl benzyl sulfonates, which are readily biodegradable, instead of the previously used branched (nonlinear) alkyl benzyl sulfonates. It was discovered that these were recalcitrant and accumulated in the environment. The story of laundry detergent design is significant because it was one of the first instances in which a xenobiotic chemical was redesigned to make it biodegradable. Today, we also have biodegradable plastics and various other substances that have been designed to be readily degraded by microorganisms.

In contrast to recalcitrant xenobiotic compounds for which microorganisms have not had sufficient time to evolve the capacity to degrade them, microorganisms have long been exposed to petroleum hydrocarbons that routinely seep into the environment. Over 10 million tons of oil pollutants enter the world's oceans each year from natural seepages, accidental spillages, and the disposal of oily wastes. Microorganisms that naturally occur in the oceans have the ability to degrade most of the compounds in petroleum, which is why the oceans are not covered with a layer of oil. However, as we periodically witness, the sudden release of large amounts of oil from a supertanker accident or well blowout can overwhelm the microbial biodegradative capacity.

In the case of the 1989 *Exxon Valdez* spill, far more oil washed up on the shorelines of Prince William Sound, Alaska, than the microbes could quickly biodegrade. In particular, there was a lack of sufficient inorganic nutrients to support the microbial growth needed to consume the hydrocarbons in the oil rapidly. To overcome this limitation, inorganic nitrogen- and phosphate-containing fertilizers were added to stimulate the growth of the naturally occurring oil-degrading microbes—a process called **bioremediation.** No microbes were added—they were already there and just needed the added fertilizer to allow them to grow faster.

Likewise, within 2 to 6 days, microbes in the Gulf of Mexico were estimated to biodegrade half of the dispersed oil released from the BP *Deepwater Horizon* well explosion in 2010 **(figure 22.17).** Even in the cold, deep waters of the Gulf, microbes consumed over 90% of the oil within a month of its release. Molecular analyses showed that diverse microbes, particularly *Oceanospirillum* and *Colwellia*, were responsible for the rapid biodegradation of this oil. Nature has an amazing self-cleaning capacity—one that is fully dependent upon microorganisms.

Figure 22.17 **The 2011 Gulf oil spill.**

Counteracting Climate Change?

Some scientists think that it may be possible to manage microbial activities in soils and oceans to combat global warming. Microbes are responsible for the production and consumption of the major greenhouse gases, including carbon dioxide and methane. Some microbes produce carbon dioxide, whereas others, like cyanobacteria and algae, consume it. Some microbes, like the archaea in the guts of herbivorous animals, produce large amounts of methane. In fact, according to the U.S. Environmental Protection Agency, cows release more methane into the atmosphere than automobiles. Thus, some microorganisms contribute to the production of greenhouse gases, whereas others tend to reduce atmospheric concentrations of carbon dioxide and methane and thereby lessen the impact of human activities.

Whether changes in microbial processes can lead to a net positive or negative feedback for greenhouse gas emissions is unclear. It is also unclear whether the balance of microbial production and consumption of carbon dioxide and methane can be controlled. What does seem certain is that if the concentrations of greenhouse gases continue to increase and climate changes continue, human, animal, and environmental health will also change—and the ultimate outcome may well depend upon the microbes.

22.4 LEARNING OUTCOMES—Assess Your Progress

10. Define *biological oxygen demand*.
11. Outline the three phases in wastewater treatment.
12. Provide an example of a xenobiotic that has been made biodegradable.

CASE FILE WRAP-UP

Raw milk can harbor microbes such as *Campylobacter, Listeria, Salmonella,* and *Escherichia coli* from fecal contamination or infection of the cow's udders, skin biota of cows, or contamination of the milking equipment. Leona was sickened by *Campylobacter,* presumably from the raw cheese she consumed. *Campylobacter* is likely to cause only slight symptoms (a few loose bowel movements, for example) in healthy people, but it can be much more virulent in people with underlying immunosuppression. Elderly people are particularly at risk. So, if this was a food-borne outbreak, most of the "victims" probably didn't even notice their disease.

By the 1950s, **pasteurization** was implemented as a standard procedure in the United States, which greatly reduced the number of infections and diseases transmitted by cow's milk and other dairy products. It is estimated that the risk of disease from raw milk and milk products is 150 times greater than the risk of disease from pasteurized milk. Additionally, disease outbreaks caused by raw milk result in more *severe* disease than those caused by pasteurized milk. Even though the culprit in this case caused relatively mild disease, it can be much worse. According to the CDC, many of the outbreaks associated with raw milk are caused by *E. coli* O157:H7, which leads to hemolytic uremic syndrome (HUS), kidney failure, and death. In January of 2012, the family of a 5-year-old child in Washington State filed a lawsuit against the Cozy Valley Dairy after the child developed HUS after drinking raw milk from the dairy.

Anthracimycin: Ocean Mud Yields New Antibiotic

In 2013, researchers in San Diego, California, published an article about an exciting discovery they had made in the least likely of all places—mud. Specifically, this mud was from the ocean off the coast of Santa Barbara, California. What they discovered was a bacterium that produces a substance they have named *anthracimycin*. As you may surmise from the name, the substance has promise for being a new antibiotic in our arsenal.

Culture extracts from the new species have been shown to be capable of destroying the bacterium responsible for anthrax, a deadly disease caused by *Bacillus anthracis*, a bacterium that is common in soil and persists for a long time there due to its endospore-forming capabilities. It made the news in a big way in 2001, when five people died and 22 people were sickened after handling mail containing *B. anthracis* endospores. Although an anthrax vaccine currently exists, it is not widely available to the public. A new antibiotic that is effective against anthrax will be a welcome addition, given the current world and political climate.

The marine microorganism that was discovered in the mud is a new species of *Streptomyces* that has been found to be effective against several gram-positive species, including *Streptococcus, Enterococcus,* and *Staphylococcus,* and even, it is thought, against methicillin-resistant *Staphyloccus aureus* (MRSA). If this turns out to be the case, many lives will potentially be saved. As the problem of antibiotic resistance is still growing, the discovery of a new antibiotic is both welcome and timely.

Chapter Summary

22.1 One Health

- The health of the environment, the health of animals, and the health of humans are intertwined. Scientists are calling this concept *one health*.

22.2 Animals and Infectious Disease: Zoonoses

- Human infectious diseases that originated in animals are called zoonoses (singular, *zoonosis*). Human diseases can also be transmitted to animals.
- Rabies spreads from animals to humans and is close to 100% fatal.
- West Nile virus was carried by birds from Africa to the United States, where it is easily transferred from birds to humans via mosquitoes.
- Lyme disease is transmitted by ticks among small and large mammals, including humans. Encroaching on the habitat of wildlife puts us in the chain of transmission.
- HIV originated as SIV, or simian immunodeficiency virus, in monkeys. It crossed species and then became human-to-human transmissible.

22.3 The Environment and Infectious Disease

- Drinking water that does not contain microbial pathogens is fundamental to life. Most of the world's population struggles to obtain clean water.
- The presence of coliforms is used as a sign that water has been exposed to fecal material, although it is a less-than-perfect test.

- Cholera is caused by a human pathogen that survives well in water. Outbreaks often occur in places disrupted by war or natural disasters.
- Food is another part of our environment that is easily contaminated with microbes. Even in industrialized countries, keeping pathogens out of the food chain is a constant struggle.
- Climate change results in changes in our weather patterns and the environment and plants and animals that are affected by it. This can lead to changes in the transmission of microbes to humans.

22.4 Microbes to the Rescue

- Humans have learned how to take advantage of microbial processes to help us maintain a clean environment and clean up pollution that we create.
- Sewage treatment is assisted in multiple ways by using microbes to break down toxic substances.
- Solid waste (garbage) can also be degraded by microbes via composting or anaerobic digestion.
- When humans create problems by dumping oils or chemicals into the environment, we often rely on naturally occuring microbes to degrade the oils or chemicals and clean up the spills. This is called bioremediation.
- Xenobiotics are human-made compounds that do not occur naturally on our planet. Often, they cannot be broken down in the environment. Changing the chemical structures of these chemicals can sometimes make them biodegradable.

Multiple-Choice Questions   Bloom's Levels 1 and 2: Remember and Understand

Select the correct answer from the answers provided.

1. During sewage treatment, microbial action on a large scale first takes place

   a. in the primary phase.
   b. in the secondary phase.
   c. after the secondary phase is completed.
   d. Microbial action is not a part of sewage treatment.

2. Contaminated water has

   a. lower oxygen content than pure water.
   b. higher oxygen content than pure water.
   c. the same oxygen content as pure water.
   d. no oxygen.

3. In which way can infectious agents pass between humans and other animals?

   a. from animals to humans
   b. from humans to animals
   c. Both of the above are possible.
   d. Neither *a* nor *b* is possible.

4. Which of these infections involves a life cycle that requires more than one host species *other than* humans?

   a. rabies
   b. cholera
   c. West Nile fever
   d. all of the above

5. SIV probably jumped into the human species for the first time
   a. during the 1980s.
   b. during the 1970s.
   c. during the Dark Ages.
   d. in the early part of the 20th century.
   e. in the early part of the 21st century.

6. Which of the following is a true statement about food-borne infections?
   a. Hepatits B is frequently food-borne.
   b. *E. coli* outbreaks are nearly always caused by tainted ground beef.
   c. All *E. coli* are pathogenic.
   d. Shellfish can be contaminated with *Vibrio cholera* from their water source.

7. Which of the following pairs of words are most clearly opposites?
   a. biodegradable, recalcitrant
   b. activated sludge, sewage
   c. carbon dioxide, methane
   d. SIV, HIV

8. Which of the following refers to a process in which microbial activity is *dis*couraged?
   a. composting
   b. trickling filter system
   c. bioremediation
   d. none of the above

Critical Thinking   Bloom's Levels 3, 4, and 5: Apply, Analzye, and Evaluate

Critical thinking is the ability to reason and solve problems using facts and concepts. These questions can be approached from a number of angles and, in most cases, they do not have a single correct answer.

1. Construct a diagram of how human activity can affect the environment in a way that ultimately harms humans. Include at least one microbial agent.

2. Provide details about how bird flu (H5N1) could transform from a strictly avian pathogen to a human pandemic strain.

3. How would you go about determining how many people have been infected with West Nile virus in a particular state?

4. Create an infographic or visual representation that describes the difference between *E. coli* and a coliform.

5. Defend or refute this statement with information from this chapter: The environment would be a safer place for human health if all microbes could be eliminated from it.

Visual Connections   Bloom's Level 5: Evaluate

This question connects previous images to a new concept.

1. **From chapter 18, figure 18.7a.** What part of the viral anatomy is most likely to have changed when SIV became HIV? Justify your answer.

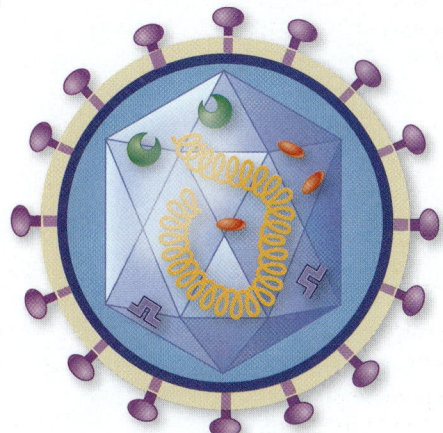

www.mcgrawhillconnect.com

Enhance your study of this chapter with study tools and practice tests. Also ask your instructor about the resources available through Connect, including the media-rich eBook, interactive learning tools, and animations.

Answers to NCLEX® Prep and Multiple-Choice Questions

Chapter 1

NCLEX® Prep
1. a, b, d, e
2. c

Multiple-Choice
1. d
2. d
3. a
4. c
5. c
6. d
7. d
8. a

Chapter 2

NCLEX® Prep
1. b
2. a, b, c, d, e
3. c

Multiple-Choice
1. c
2. b
3. c
4. c
5. d
6. a
7. b
8. d

Chapter 3

NCLEX® Prep
1. a
2. b
3. b

Multiple-Choice
1. d
2. c
3. b
4. b
5. b
6. d
7. c
8. d

Chapter 4

NCLEX® Prep
1. b
2. c
3. c

Multiple-Choice
1. b
2. c
3. b
4. c
5. d
6. c
7. c
8. c

Chapter 5

NCLEX® Prep
1. c, d, e
2. a
3. b, c, e
4. d
5. c

Multiple-Choice
1. c
2. a
3. d
4. a
5. a
6. b
7. d
8. a

Chapter 6

NCLEX® Prep
1. d
2. e
3. a
4. b, c, e

Multiple-Choice
1. a
2. b
3. b
4. a
5. a
6. b
7. c
8. b

Chapter 7

NCLEX® Prep
1. b, c
2. c
3. b

Multiple-Choice
1. b
2. d
3. e
4. b
5. b
6. d
7. c
8. c

Chapter 8

NCLEX® Prep
1. b
2. c
3. c
4. d
5. b

Multiple-Choice
1. d
2. b
3. b
4. c
5. a
6. b
7. a
8. c

Chapter 9

NCLEX® Prep
1. a
2. b
3. b

Multiple-Choice
1. c
2. a
3. c
4. b
5. d
6. b
7. d
8. d

Chapter 10

NCLEX® Prep
1. b
2. d
3. d
4. c
5. a

Multiple-Choice
1. b
2. e
3. b
4. d
5. d
6. d
7. c
8. b

Chapter 11

NCLEX® Prep
1. b, c, e, f
2. d
3. d
4. b, c, e
5. c
6. b

Multiple-Choice
1. d
2. b
3. e
4. d
5. b
6. c
7. a
8. c

Chapter 12

NCLEX® Prep
1. a
2. c, d, e
3. c
4. a
5. c

Multiple-Choice
1. b
2. e
3. b
4. a
5. d
6. b
7. d
8. c

Chapter 13

NCLEX® Prep
1. a
2. b
3. c

Multiple-Choice
1. b
2. e
3. b
4. d
5. d
6. b
7. c
8. b

Chapter 14

NCLEX® Prep
1. e
2. b
3. c
4. d

Multiple-Choice
1. c
2. d
3. a
4. b
5. b
6. b
7. d
8. d

Chapter 15

NCLEX® Prep
1. d
2. e
3. e

Multiple-Choice
1. a
2. c
3. c
4. c
5. a
6. e
7. c
8. d

Chapter 16

NCLEX® Prep
1. c
2. b
3. a
4. b
5. c

Chapter 17

NCLEX® Prep
1. a
2. b
3. d
4. d
5. c, d, e

Multiple-Choice
1. c
2. a
3. a
4. d
5. b
6. b
7. a
8. d

Chapter 18

NCLEX® Prep
1. b
2. c
3. d
4. c
5. a

Multiple-Choice
1. a
2. b
3. a
4. a
5. b
6. a
7. b
8. d

Chapter 19

NCLEX® Prep
1. c
2. d
3. b
4. a, b, d, e

Multiple-Choice
1. d
2. e
3. d
4. e
5. d
6. d
7. b
8. c

Chapter 20

NCLEX® Prep
1. e
2. a, b, c, d, e
3. b
4. a, b, c, e
5. f

Multiple-Choice
1. a
2. c
3. d
4. c

Chapter 21

NCLEX® Prep
1. a, b, c
2. c
3. d

Multiple-Choice
1. b
2. a
3. a
4. b
5. a
6. c
7. d
8. a

Chapter 22

NCLEX® Prep
1. a, b, c, d, e

Multiple-Choice
1. b
2. a
3. c
4. c
5. d
6. d
7. a
8. d

Multiple-Choice
1. d
2. b
3. d
4. c
5. d
6. c
7. a
8. c

Displaying Disease Statistics

Infectious disease specialists use a number of different methods to visually represent the numbers of disease cases or deaths.

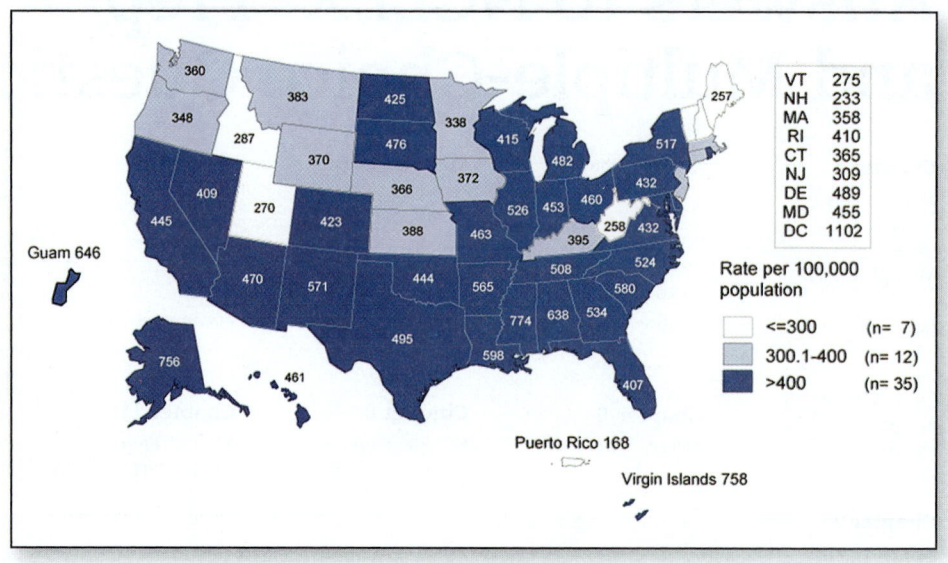

Chlamydia—Rates by State, United States and Outlying Areas, 2012

VT	275
NH	233
MA	358
RI	410
CT	365
NJ	309
DE	489
MD	455
DC	1102

Guam 646

Rate per 100,000 population

	<=300	(n= 7)
	300.1–400	(n= 12)
	>400	(n= 35)

Puerto Rico 168

Virgin Islands 758

NOTE: The total rate of *Chlamydia* for the United States and outlying areas (Guam, Puerto Rico, and Virgin Islands) was 453.5 per 100,000 population.

SOURCE: CDC.

Chlamydia—Rates by County, United States, 2012

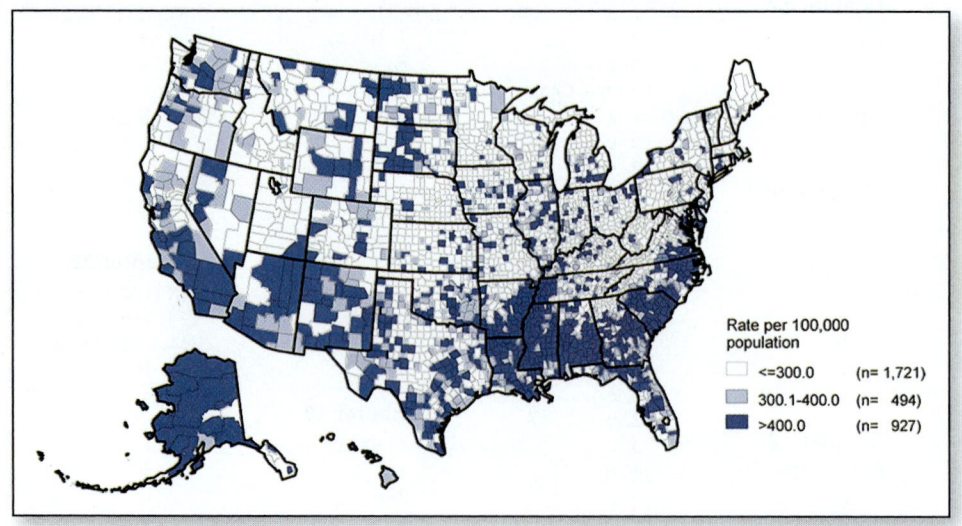

Rate per 100,000 population

	<=300.0	(n= 1,721)
	300.1–400.0	(n= 494)
	>400.0	(n= 927)

Some methods emphasize the geographic distribution of disease.

SOURCE: CDC.

Chlamydia—Rates by Sex, United States, 1992–2012

Other methods may represent trends by sex.

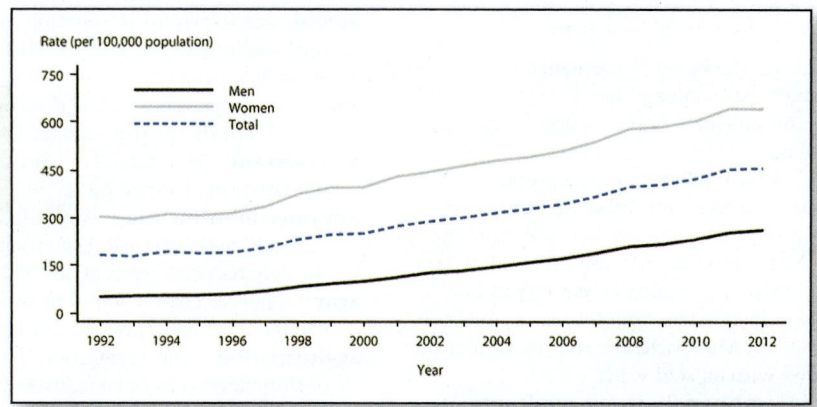

NOTE: As of January 2000, all 50 states and the District of Columbia have regulations that require the reporting of *Chlamydia* cases.

SOURCE: CDC.

Chlamydia—Rates by Age and Sex, United States, 2012

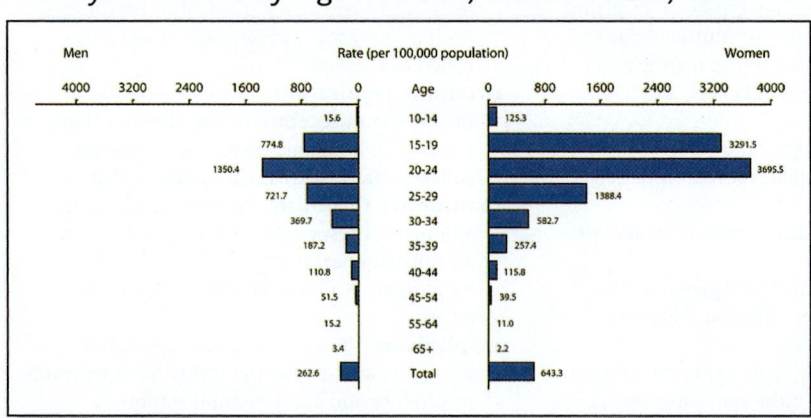

Or, the image may depict information by age and sex.

SOURCE: CDC.

Visit **www.cdc.gov/mmwr** for more in-depth discussions
and visual representations of disease statistics.

Glossary

A

abiogenesis The belief in spontaneous generation as a source of life.

abscess An inflamed, fibrous lesion enclosing a core of pus.

acid-fast A term referring to the property of mycobacteria to retain carbol fuchsin even in the presence of acid alcohol. The staining procedure is used to diagnose tuberculosis.

acid-fast stain A solution containing carbol fuchsin, which, when bound to lipids in the envelopes of *Mycobacterium* species, cannot be removed with an acid wash.

acquired immunodeficiency syndrome See AIDS.

actin Protein component of long filaments of protein arranged under the cytoplasmic membrane of bacteria; contributes to cell shape and division.

active immunity Immunity acquired through direct stimulation of the immune system by antigen.

active site The specific region on an apoenzyme that binds substrate. The site for reaction catalysis. Also called *catalytic site*.

active transport Nutrient transport method that requires carrier proteins in the membranes of the living cells and the expenditure of energy.

acute Characterized by rapid onset and short duration.

adenine (A) One of the nitrogen bases found in DNA and RNA with a purine form.

adenosine deaminase (ADA) deficiency An immunodeficiency disorder and one type of SCID that is caused by an inborn error in the metabolism of adenine. The accumulation of adenine destroys both B and T lymphocytes.

adenosine triphosphate (ATP) A nucleotide that is the primary source of energy to cells.

adhesion The process by which microbes gain a more stable foothold at the portal of entry; often involves a specific interaction between the molecules on the microbial surface and the receptors on the host cell.

adjuvant In immunology, a chemical vehicle that enhances antigenicity, presumably by prolonging antigen retention at the injection site.

adsorption A process of adhering one molecule onto the surface of another molecule.

aerobe A microorganism that lives and grows in the presence of free gaseous oxygen (O_2).

aerobic respiration Respiration in which the final electron acceptor in the electron transport chain is oxygen (O_2).

aerosols Suspensions of fine dust or moisture particles in the air that contain live pathogens.

aerotolerant The state of not utilizing oxygen but not being harmed by it.

agammaglobulinemia Also called *hypogammaglobulinemia*. The absence of or severely reduced levels of antibodies in serum.

agar A polysaccharide found in seaweed and commonly used to prepare solid culture media.

agglutination The aggregation by antibodies of suspended cells or similar-size particles (agglutinogens) into clumps that settle.

agranulocyte One form of leukocyte (white blood cell) having globular, nonlobed nuclei and lacking prominent cytoplasmic granules.

AIDS Acquired immunodeficiency syndrome. The complex of signs and symptoms characteristic of the late phase of human immunodeficiency virus (HIV) infection.

alcoholic fermentation An anaerobic degradation of pyruvic acid that results in alcohol production.

algae Photosynthetic, plantlike organisms that generally lack the complex structure of plants; they may be single-celled or multicellular and inhabit diverse habitats such as marine and freshwater environments, glaciers, and hot springs.

allele A gene that occupies the same location as other alternative (allelic) genes on paired chromosomes.

allergen A substance that provokes an allergic response.

allergy The altered, usually exaggerated, immune response to an allergen. Also called *hypersensitivity*.

alloantigen An antigen that is present in some but not all members of the same species.

allograft Relatively compatible tissue exchange between nonidentical members of the same species. Also called *homograft*.

allosteric Pertaining to the altered activity of an enzyme due to the binding of a molecule to a region other than the enzyme's active site.

alternative splicing The ability of eukaryotic organisms to create variant mRNAs from a single genetic sequence by cutting it in different places.

amino acids The building blocks of protein. Amino acids exist in 20 naturally occurring forms that impart different characteristics to the various proteins they compose.

aminoglycoside A complex group of drugs derived from soil actinomycetes that impairs ribosome function and has antibiotic potential. Example: streptomycin.

ammonification Phase of the nitrogen cycle in which ammonia is released from decomposing organic material.

amphibolism Pertaining to the metabolic pathways that serve multiple functions in the breakdown, synthesis, and conversion of metabolites.

amphipathic Relating to a compound that has contrasting characteristics, such as hydrophilichydrophobic or acid-base.

amphitrichous Having a single flagellum or a tuft of flagella at opposite poles of a microbial cell.

amplicon DNA strand that has been primed for replication during polymerase chain reaction.

anabolism The energy-consuming process of incorporating nutrients into protoplasm through biosynthesis.

anaerobe A microorganism that grows best, or exclusively, in the absence of oxygen.

anaerobic digesters Closed chambers used in a microbial process that converts organic sludge from waste treatment plants into useful fuels such as methane and hydrogen gases. Also called *bioreactors*.

anaerobic respiration Respiration in which the final electron acceptor in the electron transport chain is an inorganic molecule containing sulfate, nitrate, nitrite, carbonate, and so on.

anamnestic response In immunology, an augmented response or memory related to a prior stimulation of the immune system by antigen. It boosts the levels of immune substances.

anaphylaxis The unusual or exaggerated allergic reaction to antigen that leads to severe respiratory and cardiac complications.

anion A negatively charged ion.

annotating In the context of genome sequencing, it is the process of assigning biological function to genetic sequence.

antagonism Relationship in which microorganisms compete for survival in a common environment by taking actions that inhibit or destroy another organism.

antibiotic A chemical substance from one microorganism that can inhibit or kill another microbe even in minute amounts.

antibody A large protein molecule evoked in response to an antigen that interacts specifically with that antigen.

antibody-mediated immunity   Specific protection from disease provided by the products of B cells.

anticodon   The trinucleotide sequence of transfer RNA that is complementary to the trinucleotide sequence of messenger RNA (the codon).

antigen   Any cell, particle, or chemical that induces a specific immune response by B cells or T cells and can stimulate resistance to an infection or a toxin. See **immunogen.**

antigen binding site   Specific region at the ends of the antibody molecule that recognize specific antigens. These sites have numerous shapes to fit a wide variety of antigens.

antigenic drift   Minor antigenic changes in the influenza A virus due to mutations in the spikes' genes.

antigenic shift   Major changes in the influenza A virus due to recombination of viral strains from two different host species.

antigenicity   The property of a substance to stimulate a specific immune response, such as antibody formation.

antigen-presenting cells (APCs)   Cells of the immune system that digest foreign cells and particles and place pieces of them on their own surfaces in such a way that other cells of the immune system recognize them.

antihistamine   A drug that counters the action of histamine and is useful in allergy treatment.

antimicrobial   A special class of compounds capable of destroying or inhibiting microorganisms.

antimicrobial peptides   Short protein molecules found in epithelial cells; have the ability to kill bacteria.

antiparallel   A description of the two strands of DNA, which are parallel to each other, but the orientation of the deoxyribose and phosphate groups run in the opposite directions, with the 5' carbon at the top of the leading strand and the 3' carbon at the top of the lagging strand.

antisense DNA   A DNA oligonucleotide that binds to a specific piece of RNA, thereby inhibiting translation; used in gene therapy.

antisense RNA   An RNA oligonucleotide that binds to a specific piece of RNA, thereby inhibiting translation; used in gene therapy.

antisepsis   Chemical treatments to kill or inhibit the growth of all vegetative microorganisms on body surfaces.

antiseptic   A growth-inhibiting agent used on tissues to prevent infection.

antiserum   Antibody-rich serum derived from the blood of animals (deliberately immunized against infectious or toxic antigen) or from people who have recovered from specific infections.

antitoxin   Globulin fraction of serum that neutralizes a specific toxin. Also refers to the specific antitoxin antibody itself.

apicomplexans   A group of protozoa that lack locomotion in the mature state.

apoenzyme   The protein part of an enzyme, as opposed to the nonprotein or inorganic cofactors.

apoptosis   The genetically programmed death of cells that is both a natural process of development and the body's means of destroying abnormal or infected cells.

appendages   Accessory structures that sprout from the surface of bacteria. They can be divided into two major groups: those that provide motility and those that enable adhesion.

applied microbiology   The study of the practical uses of microorganisms.

aqueous   Referring to solutions in which water is used as the solvent.

archaea   Prokaryotic single-celled organisms of primitive origin that have unusual anatomy, physiology, and genetics and live in harsh habitats; when capitalized **(Archaea),** the term refers to one of the three domains of living organisms as proposed by Woese.

arthroconidia   Reproductive body of *Coccidioides immitis;* also *arthrospore.*

Arthus reaction   An immune complex phenomenon that develops after repeat injection. This localized inflammation results from aggregates of antigen and antibody that bind, complement, and attract neutrophils.

artificial immunity   Immunity that is induced as a medical intervention, either by exposing an individual to an antigen or administering immune substances to him or her.

ascospore   A spore formed within a saclike cell (ascus) of Ascomycota following nuclear fusion and meiosis.

ascus   Special fungal sac in which haploid spores are created.

asepsis   A condition free of viable pathogenic microorganisms.

aseptic technique   Method of handling microbial cultures, patient specimens, and other sources of microbes in a way that prevents infection of the handler and others who may be exposed.

assay medium   Microbiological medium used to test the effects of specific treatments to bacteria, such as antibiotic or disinfectant treatment.

assembly (viral)   The step in viral multiplication in which capsids and genetic material are packaged into virions.

asthma   A type of chronic allergy in which the airways become constricted and produce excess mucus in reaction to allergens, exercise, stress, or cold temperatures.

astromicrobiology   A branch of microbiology that studies the potential for and the possible role of microorganisms in space and on other planets.

asymptomatic   An infection that produces no noticeable symptoms even though the microbe is active in the host tissue.

asymptomatic carrier   A person with an inapparent infection who shows no symptoms of being infected yet is able to pass the disease agent on to others.

atmosphere   That part of the biosphere that includes the gaseous envelope up to 14 miles above the earth's surface. It contains gases such as carbon dioxide, nitrogen, and oxygen.

atom   The smallest particle of an element to retain all the properties of that element.

atomic weight   The average of the mass numbers of all the isotopic forms for a particular element.

atopy   Allergic reaction classified as type I, with a strong familial relationship; caused by allergens such as pollen, insect venom, food, and dander; involves IgE antibody; includes symptoms of hay fever, asthma, and skin rash.

ATP synthase   A unique enzyme located in the mitochondrial cristae and chloroplast grana that harnesses the flux of hydrogen ions to the synthesis of ATP.

attenuate   To reduce the virulence of a pathogenic bacterium or virus by passing it through a nonnative host or by long-term subculture.

AUG (start codon)   The codon that signals the point at which translation of a messenger RNA molecule is to begin.

autoantibody   An "anti-self" antibody having an affinity for tissue antigens of the subject in which it is formed.

autoclave   A sterilization chamber that allows the use of steam under pressure to sterilize materials. The most common temperature/pressure combination for an autoclave is 121°C and 15 psi.

autograft   Tissue or organ surgically transplanted to another site on the same subject.

autoimmune disease   The pathologic condition arising from the production of antibodies against autoantigens. Example: rheumatoid arthritis. Also called *autoimmunity.*

autotroph   A microorganism that requires only inorganic nutrients and whose sole source of carbon is carbon dioxide.

axenic   A sterile state such as a pure culture. An axenic animal is born and raised in a germ-free environment. See **gnotobiotic.**

axial filament   A type of flagellum (called an endoflagellum) that lies in the periplasmic space of spirochetes and is responsible for locomotion. Also called *periplasmic flagellum.*

azole   Five-membered heterocyclic compounds typical of histidine, which are used in antifungal therapy.

B

B lymphocyte (B cell)   A white blood cell that gives rise to plasma cells and antibodies.

bacillus   Bacterial cell shape that is cylindrical (longer than it is wide).

back-mutation   A mutation that counteracts an earlier mutation, resulting in the restoration of the original DNA sequence.

bacteremia   The presence of viable bacteria in circulating blood.

Bacteria   When capitalized, can refer to one of the three domains of living organisms proposed by Woese containing all nonarchaea prokaryotes.

bacteria (singular, **bacterium**)   Category of prokaryotes with peptidoglycan in their cell walls and circular chromosome(s). This group of small cells is widely distributed in the earth's habitats.

bacterial chromosome   A circular body in bacteria that contains the primary genetic material. Also called *nucleoid*.

bactericide   An agent that kills bacteria.

bacteriophage   A virus that specifically infects bacteria.

bacteristatic   Any process or agent that inhibits bacterial growth.

barophile   A microorganism that thrives under high (usually hydrostatic) pressure.

basement membrane   A thin layer (1–6 μm) of protein and polysaccharide found at the base of epithelial tissues.

beta oxidation   The degradation of long-chain fatty acids. Two-carbon fragments are formed as a result of enzymatic attack directed against the second or beta carbon of the hydrocarbon chain. Aided by coenzyme A, the fragments enter the Krebs cycle and are processed for ATP synthesis.

beta-lactamase   An enzyme secreted by certain bacteria that cleaves the beta-lactam ring of penicillin and cephalosporin and thus provides for resistance against the antibiotic. See **penicillinase.**

binary fission   The formation of two new cells of approximately equal size as the result of parent cell division.

binomial system   Scientific method of assigning names to organisms that employs two names to identify every organism–genus name plus species name.

biodegradation   The breaking down of materials through the action of microbes or insects.

biofilm   A complex association that arises from a mixture of microorganisms growing together on the surface of a habitat.

biogenesis   Belief that living things can only arise from others of the same kind.

biological oxygen demand   Requirement for oxygen by microbes in a niche; used to measure drinking water quality.

biological vector   An animal that not only transports an infectious agent but plays a role in the life cycle of the pathogen, serving as a site in which it can multiply or complete its life cycle. It is usually an alternate host to the pathogen.

bioremediation   Decomposition of harmful chemicals by microbes or consortia of microbes.

biosensor   A device used to detect microbes or trace amounts of compounds through PCR, genome techniques, or electrochemical signaling.

biota   Beneficial or harmless resident bacteria commonly found on and/or in the human body.

biotechnology   The intentional use by humans of living organisms or their products to accomplish a goal related to health or the environment.

blast cell   An immature precursor cell of B and T lymphocytes. Also called a *lymphoblast.*

blocking antibody   The IgG class of immunoglobulins that competes with IgE antibody for allergens, thus blocking the degranulation of basophils and mast cells.

blood cells   Cellular components of the blood consisting of red blood cells, primarily responsible for the transport of oxygen and carbon dioxide, and white blood cells, primarily responsible for host defense and immune reactions.

blood-brain barrier   Decreased permeability of the walls of blood vessels in the brain, restricting access to that compartment.

botulinum toxin   *Clostridium botulinum* toxin. Ingestion of this potent exotoxin leads to flaccid paralysis.

bradykinin   An active polypeptide that is a potent vasodilator released from IgE-coated mast cells during anaphylaxis.

broad-spectrum   Denotes drugs that have an effect on a wide variety of microorganisms.

Brownian movement   The passive, erratic, nondirectional motion exhibited by microscopic particles. The jostling comes from being randomly bumped by submicroscopic particles, usually water molecules, in which the visible particles are suspended.

brucellosis   A zoonosis transmitted to humans from infected animals or animal products; causes a fluctuating pattern of severe fever in humans as well as muscle pain, weakness, headache, weight loss, and profuse sweating. Also called *undulant fever.*

bubo   The swelling of one or more lymph nodes due to inflammation.

bubonic plague   The form of plague in which bacterial growth is primarily restricted to the lymph and is characterized by the appearance of a swollen lymph node referred to as a *bubo.*

budding   See **exocytosis.**

bulbar poliomyelitis   Complication of polio infection in which the brain stem, medulla, or cranial nerves are affected. Leads to loss of respiratory control and paralysis of the trunk and limbs.

bullous   Consisting of fluid-filled blisters.

C

calculus   Dental deposit formed when plaque becomes mineralized with calcium and phosphate crystals. Also called *tartar.*

cancer   Any malignant neoplasm that invades surrounding tissue and can metastasize to other locations. A carcinoma is derived from epithelial tissue, and a sarcoma arises from proliferating mesodermal cells of connective tissue.

capsid   The protein covering of a virus's nucleic acid core. Capsids exhibit symmetry due to the regular arrangement of subunits called *capsomers.* See **icosahedron.**

capsomer   A subunit of the virus capsid shaped as a triangle or disc.

capsular staining   Any staining method that highlights the outermost polysaccharide and/or protein structure on a bacterial, fungal, or protozoal cell.

capsule   In bacteria, the loose, gel-like covering or slime made chiefly of polysaccharides. This layer is protective and can be associated with virulence.

carbohydrate   A compound containing primarily carbon, hydrogen, and oxygen in a 1:2:1 ratio.

carbohydrate fermentation medium   A growth medium that contains sugars that are converted to acids through fermentation. Usually contains a pH indicator to detect acid protection.

carrier   A person who harbors infections and inconspicuously spreads them to others. Also, a chemical agent that can accept an atom, chemical radical, or subatomic particle from one compound and pass it on to another.

caseous lesion   Necrotic area of lung tubercle superficially resembling cheese. Typical of tuberculosis.

catabolism   The chemical breakdown of complex compounds into simpler units to be used in cell metabolism.

catalyst   A substance that alters the rate of a reaction without being consumed or permanently changed by it. In cells, enzymes are catalysts.

catalytic site   The niche in an enzyme where the substrate is converted to the product. Also called *active site.*

catarrhal   A term referring to the secretion of mucus or fluids; term for the first stage of pertussis.

cation   A positively charged ion.

cell   An individual membrane-bound living entity; the smallest unit capable of an independent existence.

cell wall   In bacteria, a rigid structure made of peptidoglycan that lies just outside the cytoplasmic membrane; eukaryotes also have a cell wall but may be composed of a variety of materials.

cell-mediated immunity   The type of immune responses brought about by T cells, such as cytotoxic and helper effects.

cellulitis   The spread of bacteria within necrotic tissue.

cellulose   A long, fibrous polymer composed of β-glucose; one of the most common substances on earth.

cephalosporins   A group of broad-spectrum antibiotics isolated from the fungus *Cephalosporium.*

cercaria   The free-swimming larva of the schistosome trematode that emerges from the snail host and can penetrate human skin, causing schistosomiasis.

cestode The common name for tapeworms that parasitize humans and domestic animals.

chancre The primary sore of syphilis that forms at the site of penetration by *Treponema pallidum*. It begins as a hard, dull red, painless papule that erodes from the center.

chemical bond A link formed between molecules when two or more atoms share, donate, or accept electrons.

chemical mediators Small molecules that are released during inflammation and specific immune reactions that allow communication between the cells of the immune system and facilitate surveillance, recognition, and attack.

chemiosmosis The generation of a concentration gradient of hydrogen ions (called the *proton motive force*) by the pumping of hydrogen ions to the outer side of the membrane during electron transport.

chemoautotroph An organism that relies upon inorganic chemicals for its energy and carbon dioxide for its carbon. Also called a *chemolithotroph*.

chemoheterotroph Microorganisms that derive their nutritional needs from organic compounds.

chemokine Chemical mediators (cytokines) that stimulate the movement and migration of white blood cells.

chemostat A growth chamber with an outflow that is equal to the continuous inflow of nutrient media. This steady-state growth device is used to study such events as cell division, mutation rates, and enzyme regulation.

chemotactic factors Chemical mediators that stimulate the movement of white blood cells. See **chemokine.**

chemotaxis The tendency of organisms to move in response to a chemical gradient (toward an attractant or to avoid adverse stimuli).

chemotherapy The use of chemical substances or drugs to treat or prevent disease.

chemotroph Organism that oxidizes compounds to feed on nutrients.

chitin A polysaccharide similar to cellulose in chemical structure. This polymer makes up the horny substance of the exoskeletons of arthropods and certain fungi.

chlorophyll A group of mostly green pigments that are used by photosynthetic eukaryotic organisms and cyanobacteria to trap light energy to use in making chemical bonds.

chloroplast An organelle containing chlorophyll that is found in photosynthetic eukaryotes.

cholesterol Best-known member of a group of lipids called *steroids*. Cholesterol is commonly found in cell membranes and animal hormones.

chromatin The genetic material of the nucleus. Chromatin is made up of nucleic acid and stains readily with certain dyes.

chromosome The tightly coiled bodies in cells that are the primary sites of genes.

chronic Any process or disease that persists over a long duration.

cilium (plural, *cilia*) Eukaryotic structure similar to a flagellum that propels a protozoan through the environment.

class In the levels of classification, the division of organisms that follows phylum.

classical pathway Pathway of complement activation initiated by a specific antigen-antibody interaction.

clonal selection theory A conceptual explanation for the development of lymphocyte specificity and variety during immune maturation.

clone A colony of cells (or group of organisms) derived from a single cell (or single organism) by asexual reproduction. All units share identical characteristics. Also used as a verb to refer to the process of producing a genetically identical population of cells or genes.

cloning host An organism such as a bacterium or a yeast that receives and replicates a foreign piece of DNA inserted during a genetic engineering experiment.

coagulase A plasma-clotting enzyme secreted by *Staphylococcus aureus*. It contributes to virulence and is involved in forming a fibrin wall that surrounds staphylococcal lesions.

coccobacillus An elongated coccus; a short, thick, oval-shaped bacterial rod.

coccus A spherical-shaped bacterial cell.

codon A specific sequence of three nucleotides in mRNA (or the sense strand of DNA) that constitutes the genetic code for a particular amino acid.

coenzyme A complex organic molecule, several of which are derived from vitamins (e.g., nicotinamide, riboflavin). A coenzyme operates in conjunction with an enzyme. Coenzymes serve as transient carriers of specific atoms or functional groups during metabolic reactions.

coevolution A biological process whereby a change in the genetic composition in one organism leads to a change in the genetics of another organism.

cofactor An enzyme accessory. It can be organic, such as coenzymes, or inorganic, such as Fe^{2+}, Mn^{2+}, or Zn^{2+} ions.

cold sterilization The use of nonheating methods such as radiation or filtration to sterilize materials.

coliform A collective term that includes normal enteric bacteria that are gram-negative and lactose-fermenting.

colony A macroscopic cluster of cells appearing on a solid medium, each arising from the multiplication of a single cell.

colostrum The clear yellow early product of breast milk that is very high in secretory antibodies. Provides passive intestinal protection.

commensalism An unequal relationship in which one species derives benefit without harming the other.

common-source epidemic An outbreak of disease in which all affected individuals were exposed to a single source of the pathogen, even if they were exposed at different times.

communicable infection Capable of being transmitted from one individual to another.

competent Referring to bacterial cells that are capable of absorbing free DNA in their environment either naturally or through induction by exposure to chemicals or electrical currents.

competitive inhibition Control process that relies on the ability of metabolic analogs to control microbial growth by successfully competing with a necessary enzyme to halt the growth of bacterial cells.

complement In immunology, serum protein components that act in a definite sequence when set in motion either by an antigen-antibody complex or by factors of the alternative (properdin) pathway.

complementary DNA (cDNA) DNA created by using reverse transcriptase to synthesize DNA from RNA templates.

compounds Molecules that are a combination of two or more different elements.

concentration The expression of the amount of a solute dissolved in a certain amount of solvent. It may be defined by weight, volume, or percentage.

condyloma acuminata Extensive, branched masses of genital warts caused by infection with human papillomavirus.

congenital Transmission of an infection from mother to fetus.

congenital rubella Transmission of the rubella virus to a fetus *in utero*. Injury to the fetus is generally much more serious than it is to the mother.

congenital syphilis A syphilis infection of the fetus or newborn acquired from maternal infection *in utero*.

conidia Asexual fungal spores shed as free units from the tips of fertile hyphae.

conidiospore A type of asexual spore in fungi; not enclosed in a sac.

conjugated vaccines Subunit vaccines combined with carrier proteins, often from other microbes, to make them more immunogenic.

conjugation In bacteria, the contact between donor and recipient cells associated with the transfer of genetic material such as plasmids. Can involve special (sex) pili. Also a form of sexual recombination in ciliated protozoa.

conjunctiva The thin fluid-secreting tissue that covers the eye and lines the eyelid.

constitutive enzyme An enzyme present in bacterial cells in constant amounts, regardless of the presence of substrate. Enzymes of the central catabolic pathways are typical examples.

contagious Communicable; transmissible by direct contact with infected people and their fresh secretions or excretions.

contaminant An impurity; any undesirable material or organism.

contaminated culture A medium that once held a pure (single or mixed) culture but now contains unwanted microorganisms.

convalescence Recovery; the period between the end of a disease and the complete restoration of health in a patient.

corepressor A molecule that combines with inactive repressor to form active repressor, which attaches to the operator gene site and inhibits the activity of structural genes subordinate to the operator.

cornea The transparent, dome-shaped tissue covering the iris, pupil, and anterior chamber of the eye composed of five to six layers of quickly regenerating epithelial cells.

covalent A type of chemical bond that involves the sharing of electrons between two atoms. Jeanne - can delete one of these

covalent bond A chemical bond formed by the sharing of electrons between two atoms.

Creutzfeldt-Jakob disease (CJD) A spongiform encephalopathy caused by infection with a prion. The disease is marked by dementia, impaired senses, and uncontrollable muscle contractions.

crista The infolded inner membrane of a mitochondrion that is the site of the respiratory chain and oxidative phosphorylation.

cryptosporidiosis A gastrointestinal disease caused by *Cryptosporidium parvum,* a protozoan.

culture The visible accumulation of microorganisms in or on a nutrient medium. Also, the propagation of microorganisms with various media.

cutaneous Second level of skin, including the stratum corneum and occasionally the upper dermis.

cyanosis Blue discoloration of the skin or mucous membranes indicative of decreased oxygen concentration in blood.

cyst The resistant, dormant but infectious form of protozoa. Can be important in spread of infectious agents such as *Entamoeba histolytica* and *Giardia lamblia.*

cysteine A sulfide-containing amino acid that usually produces covalent disulfide bonds in an amino acid sequence, contributing to the tertiary structure of the protein.

cysticercosis Condition caused by *Taenia solium* tapeworm. Larval cysts embed in brain, muscle, and other tissue, causing seizures and other symptoms.

cystine An amino acid, $HOOC-CH(NH_2)-CH_2-S-S-CH_2-CH(NH_2)COOH$. An oxidation product of two cysteine molecules in which the OSH (sulfhydryl) groups form a disulfide union. Also called *dicysteine.*

cytochrome A group of heme protein compounds whose chief role is in electron and/or hydrogen transport occurring in the last phase of aerobic respiration.

cytokine Regulatory chemical released by cells of the immune system that serves as signal between different cells.

cytopathic effect The degenerative changes in cells associated with virus infection. Examples: the formation of multinucleate giant cells (Negri bodies), the prominent cytoplasmic inclusions of nerve cells infected by rabies virus.

cytoplasm Dense fluid encased by the cytoplasmic membrane; the site of many of the cell's biochemical and synthetic activities.

cytoplasmic membrane Lipid bilayer that encloses the cytoplasm of bacterial cells.

cytosine (C) One of the nitrogen bases found in DNA and RNA, with a pyrimidine form.

cytotoxicity The ability to kill cells; in immunology, certain T cells are called *cytotoxic T cells* because they kill other cells.

D

daptomycin A lipopetide antibiotic that disrupts the cytoplasmic membrane.

deamination The removal of an amino group from an amino acid.

death phase End of the cell growth due to lack of nutrition, depletion of environment, and accumulation of wastes. Population of cells begins to die.

debridement Trimming away devitalized tissue and foreign matter from a wound.

decomposer A consumer that feeds on organic matter from the bodies of dead organisms. These microorganisms feed from all levels of the food pyramid and are responsible for recycling elements (also called *saprobes*).

decomposition The breakdown of dead matter and wastes into simple compounds that can be directed back into the natural cycle of living things.

decontamination The removal or neutralization of an infectious, poisonous, or injurious agent from a site.

deductive approach Method of investigation that uses deduction.

definitive host The organism in which a parasite develops into its adult or sexually mature stage. Also called the *final host.*

degerm To physically remove surface oils, debris, and soil from skin to reduce the microbial load.

degranulation The release of cytoplasmic granules, as when cytokines are secreted from mast cell granules.

dehydration synthesis During the formation of a carbohydrate bond, the step in which one carbon molecule gives up its OH group and the other loses the H from its OH group, thereby producing a water molecule. This process is common to all polymerization reactions.

denaturation The loss of normal characteristics resulting from some molecular alteration. Usually in reference to the action of heat or chemicals on proteins whose function depends upon an unaltered tertiary structure.

dendritic cell A large, antigen-processing cell characterized by long, branchlike extensions of the cell membrane.

denitrification The end of the nitrogen cycle when nitrogen compounds are returned to the reservoir in the air.

dental caries A mixed infection of the tooth surface that gradually destroys the enamel and may lead to destruction of the deeper tissue.

deoxyribonucleic acid (DNA) The nucleic acid often referred to as the "double helix." DNA carries the master plan for an organism's heredity.

deoxyribose A 5-carbon sugar that is an important component of DNA.

dermatophytes A group of fungi that cause infections of the skin and other integument components. They survive by metabolizing keratin.

dermolytic Capable of damaging the skin.

desensitization See **hyposensitization.**

desiccation To dry thoroughly. To preserve by drying.

desquamate To shed the cuticle in scales; to peel off the outer layer of a surface.

diabetes mellitus A disease involving compromise in insulin function. In one form, the pancreatic cells that produce insulin are destroyed by autoantibodies; in another, the pancreas does not produce sufficient insulin.

diapedesis The migration of intact blood cells between endothelial cells of a blood vessel such as a venule.

dichotomous keys Flowcharts that offer two choices or pathways at each level.

differential medium A single substrate that discriminates between groups of microorganisms on the basis of differences in their appearance due to different chemical reactions.

differential stain A technique that utilizes two dyes to distinguish between different microbial groups or cell parts by color reaction.

diffusion The dispersal of molecules, ions, or microscopic particles propelled down a concentration gradient by spontaneous random motion to achieve a uniform distribution.

DiGeorge syndrome A birth defect usually caused by a missing or incomplete thymus that results in abnormally low or absent T cells and other developmental abnormalities.

dimorphic In mycology, the tendency of some pathogens to alter their growth form from mold to yeast in response to rising temperature.

diplococci Spherical or oval-shaped bacteria, typically found in pairs.

direct or total cell count 1. Counting total numbers of individual cells being viewed with magnification. 2. Counting isolated colonies of organisms growing on a plate of media as a way to determine population size.

disaccharide A sugar containing two monosaccharides. Example: sucrose (fructose + glucose).

disease Any deviation from health, as when the effects of microbial infection damage or disrupt tissues and organs.

disinfection The destruction of pathogenic nonsporulating microbes or their toxins, usually on inanimate surfaces.

division In the levels of classification, an alternate term for *phylum.*

DNA See **deoxyribonucleic acid.**

DNA polymerase Enzyme responsible for the replication of DNA. Several versions of the enzyme exist, each completing a unique portion of the replication process.

DNA profiling A pattern of restriction enzyme fragments that is unique for an individual organism.

DNA sequencing Determining the exact order of nucleotides in a fragment of DNA. Most commonly done using the Sanger dideoxy sequencing method.

DNA vaccine A newer vaccine preparation based on inserting DNA from pathogens into host cells to encourage them to express the foreign protein and stimulate immunity.

domain In the levels of classification, the broadest general category to which an organism is assigned. Members of a domain share only one or a few general characteristics.

doubling time Time required for a complete fission cycle–from parent cell to two new daughter cells. Also called *generation time.*

droplet nuclei The dried residue of fine droplets produced by mucus and saliva sprayed while sneezing and coughing. Droplet nuclei are less than 5 μm in diameter (large enough to bear a single bacterium and small enough to remain airborne for a long time) and can be carried by air currents. Droplet nuclei are drawn deep into the air passages.

drug resistance An adaptive response in which microorganisms begin to tolerate an amount of drug that would ordinarily be inhibitory.

dysentery Diarrheal illness in which stools contain blood and/or mucus.

dyspnea Difficulty in breathing.

E

echinocandins Antifungal drugs that inhibit the manufacture of fungal cell walls.

ecosystem A collection of organisms together with its surrounding physical and chemical factors.

ectoplasm The outer, more viscous region of the cytoplasm of a phagocytic cell such as an amoeba. It contains microtubules, but not granules or organelles.

eczema An acute or chronic allergy of the skin associated with itching and burning sensations. Typically, red, edematous, vesicular lesions erupt, leaving the skin scaly and sometimes hyperpigmented.

edema The accumulation of excess fluid in cells, tissues, or serous cavities. Also called *swelling.*

electrolyte Any compound that ionizes in solution and conducts current in an electrical field.

electron A negatively charged subatomic particle that is distributed around the nucleus in an atom.

electrophoresis The separation of molecules by size and charge through exposure to an electrical current.

electrostatic Relating to the attraction of opposite charges and the repulsion of like charges. Electrical charge remains stationary as opposed to electrical flow or current.

element A substance comprising only one kind of atom that cannot be degraded into two or more substances without losing its chemical characteristics.

ELISA Abbreviation for **e**nzyme-**l**inked **i**mmuno**s**orbent **a**ssay, a very sensitive serological test used to detect antibodies in diseases such as AIDS.

emetic Inducing to vomit.

encephalitis An inflammation of the brain, usually caused by infection.

endemic disease A native disease that prevails continuously in a geographic region.

endergonic reaction A chemical reaction that occurs with the absorption and storage of surrounding energy. See **exergonic reaction.**

endocytosis The process whereby solid and liquid materials are taken into the cell through membrane invagination and engulfment into a vesicle.

endogenous Originating or produced within an organism or one of its parts.

endoplasmic reticulum (ER) An intracellular network of flattened sacs or tubules with or without ribosomes on their surfaces.

endospore A small, dormant, resistant derivative of a bacterial cell that germinates under favorable growth conditions into a vegetative cell. The bacterial genera *Bacillus* and *Clostridium* are typical endospore formers.

endotoxic shock A massive drop in blood pressure caused by the release of endotoxin from gram-negative bacteria multiplying in the bloodstream.

endotoxin A bacterial toxin that is not ordinarily released (as is exotoxin). Endotoxin is composed of a phospholipid-polysaccharide complex that is an integral part of gram-negative bacterial cell walls. Endotoxins can cause severe shock and fever.

enriched medium A nutrient medium supplemented with blood, serum, or some growth factor to promote the multiplication of fastidious microorganisms.

enteric Pertaining to the intestine.

enteroaggregative The term used to describe certain types of intestinal bacteria that tend to stick to each other in large clumps.

enteroinvasive Predisposed to invade the intestinal tissues.

enteropathogenic Pathogenic to the alimentary canal.

enterotoxigenic Having the capacity to produce toxins that act on the intestinal tract.

enterotoxin A bacterial toxin that specifically targets intestinal mucous membrane cells. Enterotoxigenic strains of *Escherichia coli* and *Staphylococcus aureus* are typical sources.

enumeration medium Microbiological medium that does not encourage growth and allows for the counting of microbes in food, water, or environmental samples.

enveloped virus A virus whose nucleocapsid is enclosed by a membrane derived in part from the host cell. It usually contains exposed glycoprotein spikes specific for the virus.

enzyme A protein biocatalyst that facilitates metabolic reactions.

enzyme induction One of the controls on enzyme synthesis. This occurs when enzymes appear only when suitable substrates are present.

enzyme repression The inhibition of enzyme synthesis by the end product of a catabolic pathway.

eosinophil A leukocyte whose cytoplasmic granules readily stain with red eosin dye.

eosinophilia An increase in eosinophil concentration in the bloodstream, often in response to helminthic infection.

epidemic A sudden and simultaneous outbreak or increase in the number of cases of disease in a community.

epidemiology The study of the factors affecting the prevalence and spread of disease within a community.

epitope The precise molecular group of an antigen that defines its specificity and triggers the immune response.

Epstein-Barr virus (EBV) Herpesvirus linked to infectious mononucleosis, Burkitt's lymphoma, and nasopharyngeal carcinoma.

erysipelas An acute, sharply defined inflammatory disease specifically caused by hemolytic *Streptococcus.* The eruption is limited to the skin but can be complicated by serious systemic symptoms.

erythrogenic toxin An exotoxin produced by lysogenized group A strains of β-hemolytic streptococci that is responsible for the severe fever and rash of scarlet fever in the nonimmune individual. Also called a *pyrogenic toxin.*

eschar A dark, sloughing scab that is the lesion of anthrax and certain rickettsioses.

essential nutrient Any ingredient such as a certain amino acid, fatty acid, vitamin, or mineral that cannot be formed by an organism and must be supplied in the diet. A growth factor.

ethylene oxide A potent, highly water-soluble gas invaluable for gaseous sterilization of heat-sensitive objects such as plastics, surgical and diagnostic appliances, and spices.

etiologic agent The microbial cause of disease; the pathogen.

eubacteria Term used for nonarchaea prokaryotes, means "true bacteria."

Eukarya One of the three domains (sometimes called superkingdoms) of living organisms, as proposed by Woese; contains all eukaryotes.

eukaryotic cell A cell that differs from a prokaryotic cell chiefly by having a nuclear membrane (a well-defined nucleus), membrane-bound subcellular organelles, and mitotic cell division.

evolution Scientific principle that states that living things change gradually through hundreds of millions of years, and these changes are expressed in structural and functional adaptations in each organism. Evolution presumes that those traits that favor survival are preserved and passed on to following generations, and those traits that do not favor survival are lost.

exanthem   An eruption or rash of the skin.

exergonic reaction   A chemical reaction associated with the release of energy to the surroundings. See **endergonic reaction.**

exfoliative toxin   A poisonous substance that causes superficial cells of an epithelium to detach and be shed. Example: staphylococcal exfoliatin. Also called an *epidermolytic toxin.*

exocytosis   The process that releases enveloped viruses from the membrane of the host's cytoplasm.

exoenzyme   An extracellular enzyme chiefly for hydrolysis of nutrient macromolecules that are otherwise impervious to the cell membrane. It functions in saprobic decomposition of organic debris and can be a factor in invasiveness of pathogens.

exogenous   Originating outside the body.

exon   A stretch of eukaryotic DNA coding for a corresponding portion of mRNA that is translated into peptides. Intervening stretches of DNA that are not expressed are called *introns.* During transcription, exons are separated from introns and are spliced together into a continuous mRNA transcript.

exotoxin   A toxin (usually protein) that is secreted and acts upon a specific cellular target. Examples: botulin, tetanospasmin, diphtheria toxin, and erythrogenic toxin.

exponential   Pertaining to the use of exponents, numbers that are typically written as a superscript to indicate how many times a factor is to be multiplied. Exponents are used in scientific notation to render large, cumbersome numbers into small workable quantities.

exponential growth phase   The period of maximum growth rate in a growth curve. Cell population increases logarithmically.

extrapulmonary tuberculosis   A condition in which tuberculosis bacilli have spread to organs other than the lungs.

extremophiles   Organisms capable of living in harsh environments, such as extreme heat or cold.

exudate   Fluid that escapes cells into the extracellular spaces during the inflammatory response.

F

facilitated diffusion   The passive movement of a substance across a plasma membrane from an area of higher concentration to an area of lower concentration utilizing specialized carrier proteins.

facultative   Pertaining to the capacity of microbes to adapt or adjust to variations; not obligate. Example: The presence of oxygen is not obligatory for a facultative anaerobe to grow. See **obligate.**

family   In the levels of classification, a mid-level division of organisms that groups more closely related organisms than previous levels. An order is divided into families.

fastidious   Requiring special nutritional or environmental conditions for growth; said of bacteria.

fecal coliforms   Any species of gram-negative lactose-positive bacteria (primarily *Escherichia coli*) that live primarily in the intestinal tract and not the environment. Finding evidence of these bacteria in a water or food sample is substantial evidence of fecal contamination and potential for infection. See **coliform.**

feedback inhibition   Temporary end to enzyme action caused by an end-product molecule binding to the regulatory site and preventing the enzyme's active site from binding to its substrate.

fermentation   The extraction of energy through anaerobic degradation of substrates into simpler, reduced metabolites. In large industrial processes, fermentation can mean any use of microbial metabolism to manufacture organic chemicals or other products.

fermentor   A large tank used in industrial microbiology to grow mass quantities of microbes that can synthesize desired products. These devices are equipped with means to stir, monitor, and harvest products such as drugs, enzymes, and proteins in very large quantities.

fertility (F) factor   Donor plasmid that allows synthesis of a pilus in bacterial conjugation. Presence of the factor is indicated by F$^+$, and lack of the factor is indicated by F$^-$.

filament   A helical structure composed of proteins that is part of bacterial flagella.

fimbria   A short, numerous-surface appendage on some bacteria that provides adhesion but not locomotion.

Firmicutes   Taxonomic category of bacteria that have gram-positive cell envelopes.

flagellar staining   A staining method that highlights the flagellum of a bacterium.

flagellum   A structure that is used to propel the organism through a fluid environment.

fluid mosaic model   A conceptualization of the molecular architecture of cellular membranes as a bilipid layer containing proteins. Membrane proteins are embedded to some degree in this bilayer, where they float freely about.

fluorescence   The property possessed by certain minerals and dyes to emit visible light when excited by ultraviolet radiation. A fluorescent dye combined with specific antibody provides a sensitive test for the presence of antigen.

fluoroquinolones   Synthetic antimicrobial drugs chemically related to quinine. They are broad-spectrum and easily adsorbed from the intestine.

focal infection   Occurs when an infectious agent breaks loose from a localized infection and is carried by the circulation to other tissues.

fomite   Virtually any inanimate object an infected individual has contact with that can serve as a vehicle for the spread of disease.

food poisoning   Symptoms in the intestines (which may include vomiting) induced by preformed exotoxin from bacteria.

frameshift mutation   An insertion or deletion mutation that changes the codon reading frame from the point of the mutation to the final codon. Almost always leads to a nonfunctional protein.

fructose   One of the carbohydrates commonly referred to as sugars. Fructose is commonly fruit sugars.

functional group   In chemistry, a particular molecular combination that reacts in predictable ways and confers particular properties on a compound. Examples: $-COOH$, $-OH$, $-CHO$.

fungemia   The condition of fungi multiplying in the bloodstream.

fungi   Macroscopic and microscopic heterotrophic eukaryotic organisms that can be uni- or multicellular.

fungus   Heterotrophic unicellular or multicellular eukaryotic organism that may take the form of a larger macroscopic organism, as in the case of mushrooms, or a smaller microscopic organism, as in the case of yeasts and molds.

fuzeon   Anti-HIV drug that inhibits viral attachment to host cells.

G

gamma globulin   The fraction of plasma proteins high in immunoglobulins (antibodies). Preparations from pooled human plasma containing normal antibodies make useful passive immunizing agents against pertussis, polio, measles, and several other diseases.

gas gangrene   Disease caused by a clostridial infection of soft tissue or wound. The name refers to the gas produced by the bacteria growing in the tissue. Unless treated early, it is fatal. Also called *myonecrosis.*

gastritis   Pain and/or nausea, usually experienced after eating; result of inflammation of the lining of the stomach.

gel electrophoresis   A laboratory technique for separating DNA fragments according to length by employing electricity to force the DNA through a gel-like matrix typically made of agarose. Smaller DNA fragments move more quickly through the gel, thereby moving farther than larger fragments during the same period of time.

gene   A site on a chromosome that provides information for a certain cell function. A specific segment of DNA that contains the necessary code to make a protein or RNA molecule.

gene probe   Short strand of single-stranded nucleic acid that hybridizes specifically with complementary stretches of nucleotides on test samples and thereby serves as a tagging and identification device.

gene therapy   The introduction of normal functional genes into people with genetic diseases such as sickle-cell anemia and cystic fibrosis. This is usually accomplished by a virus vector.

generation time　Time required for a complete fission cycle–from parent cell to two new daughter cells. Also called *doubling time*.

genetic engineering　A field involving deliberate alterations (recombinations) of the genomes of microbes, plants, and animals through special technological processes.

genetics　The science of heredity.

genital warts　A prevalent STI linked to some forms of cancer of the reproductive organs. Caused by infection with human papillomavirus.

genome　The complete set of chromosomes and genes in an organism.

genomic libraries　Collections of DNA fragments representing the entire genome of an organism inserted into plasmids and stored in vectors such as bacteria or yeasts.

genomics　The systematic study of an organism's genes and their functions.

genotype　The genetic makeup of an organism. The genotype is ultimately responsible for an organism's phenotype, or expressed characteristics.

genus　In the levels of classification, the second most specific level. A family is divided into several genera.

geomicrobiology　A branch of microbiology that studies the role of microorganisms in the earth's crust.

germ free　See **axenic**.

germ theory of disease　A theory first originating in the 1800s that proposed that microorganisms can be the cause of diseases. The concept is actually so well established in the present time that it is considered a fact.

germicide　An agent lethal to non-endospore-forming pathogens.

giardiasis　Infection by the *Giardia* flagellate. Most common mode of transmission is contaminated food and water. Symptoms include diarrhea, abdominal pain, and flatulence.

gingivitis　Inflammation of the gum tissue in contact with the roots of the teeth.

gluconeogenesis　The formation of glucose (or glycogen) from noncarbohydrate sources such as protein or fat. Also called *glyconeogenesis*.

glucose　One of the carbohydrates commonly referred to as sugars. Glucose is characterized by its 6-carbon structure.

glycerol　A 3-carbon alcohol, with three OH groups that serve as binding sites.

glycocalyx　A filamentous network of carbohydrate-rich molecules that coats cells.

glycogen　A glucose polymer stored by cells.

glycolysis　The energy-yielding breakdown (fermentation) of glucose to pyruvic or lactic acid. It is often called *anaerobic glycolysis* because no molecular oxygen is consumed in the degradation.

glycosidic bond　A bond that joins monosaccharides to form disaccharides and polymers.

gnotobiotic　Referring to experiments performed on germ-free animals.

Golgi apparatus　An organelle of eukaryotes that participates in packaging and secretion of molecules.

gonococcus　Common name for *Neisseria gonorrhoeae*, the agent of gonorrhea.

Gracilicutes　Taxonomic category of bacteria that have gram-negative envelopes.

graft　Live tissue taken from a donor and transplanted into a recipient to replace damaged or missing tissues such as skin, bone, blood vessels.

graft versus host disease (GVHD)　A condition associated with a bone marrow transplant in which T cells in the transplanted tissue mount an immune response against the recipient's (host) normal tissues.

Gram stain　A differential stain for bacteria useful in identification and taxonomy. Gram-positive organisms appear purple from crystal violet mordant retention, whereas gram-negative organisms appear red after loss of crystal violet and absorbance of the safranin counterstain.

gram-negative　A category of bacterial cells that describes bacteria with an outer membrane, a cytoplasmic membrane, and a thin cell wall.

gram-positive　A category of bacterial cells that describes bacteria with a thick cell wall and no outer membrane.

grana　Discrete stacks of chlorophyll-containing thylakoids within chloroplasts.

granulocyte　A mature leukocyte that contains noticeable granules in a Wright stain. Examples: neutrophils, eosinophils, and basophils.

granuloma　A solid mass or nodule of inflammatory tissue containing modified macrophages and lymphocytes. Usually a chronic pathologic process of diseases such as tuberculosis or syphilis.

granzymes　Enzymes secreted by cytotoxic T cells that damage proteins of target cells.

Graves' disease　A malfunction of the thyroid gland in which autoantibodies directed at thyroid cells stimulate an overproduction of thyroid hormone (hyperthyroidism).

greenhouse effect　The capacity to retain solar energy by a blanket of atmospheric gases that redirects heat waves back toward the earth.

group translocation　A form of active transport in which the substance being transported is altered during transfer across a plasma membrane.

growth curve　A graphical representation of the change in population size over time. This graph has four periods known as lag phase, exponential or log phase, stationary phase, and death phase.

growth factor　An organic compound such as a vitamin or amino acid that must be provided in the diet to facilitate growth. An essential nutrient.

guanine (G)　One of the nitrogen bases found in DNA and RNA in the purine form.

Guillain-Barré syndrome　A neurological complication of infection or vaccination.

gumma　A nodular, infectious granuloma characteristic of tertiary syphilis.

gut-associated lymphoid tissue (GALT)　A collection of lymphoid tissue in the gastrointestinal tract that includes the appendix, the lacteals, and Peyer's patches.

gyrase　The enzyme responsible for supercoiling DNA into tight bundles; a type of topoisomerase.

H

HAART　**H**ighly **a**ctive **a**nti**r**etroviral **t**herapy; three-antiviral treatment for HIV infection.

habitat　The environment to which an organism is adapted.

halogens　A group of related chemicals with antimicrobial applications. The halogens most often used in disinfectants and antiseptics are chlorine and iodine.

halophile　A microbe whose growth is either stimulated by salt or requires a high concentration of salt for growth.

Hansen's disease　A chronic, progressive disease of the skin and nerves caused by infection by a *Mycobacterium* species that is a slow-growing, strict parasite. *Hansen's disease* is the preferred name for leprosy.

hapten　An incomplete or partial antigen. Although it constitutes the determinative group and can bind antigen, hapten cannot stimulate a full immune response without being carried by a larger protein molecule.

Hashimoto's thyroiditis　An autoimmune disease of the thyroid gland that damages the thyroid follicle cells and results in decreased production of thyroid hormone (hypothyroidism).

hay fever　A form of atopic allergy marked by seasonal acute inflammation of the conjunctiva and mucous membranes of the respiratory passages. Symptoms are irritative itching and rhinitis.

healthcare-associated infection　Formerly referred to as *nosocomial infection*, any infection acquired as a direct result of a patient's presence in a hospital or health care setting.

helical　Having a spiral or coiled shape. Said of certain virus capsids and bacteria.

helminth　A term that designates all parasitic worms.

helper T cell　A class of thymus-stimulated lymphocytes that facilitate various immune activities such as assisting B cells and macrophages. Also called a *T helper cell*.

hematopoiesis　The process by which the various types of blood cells are formed, such as in the bone marrow.

hemoglobin　A protein in red blood cells that carries iron.

hemolysin　Any biological agent that is capable of destroying red blood cells and causing the release of hemoglobin. Many bacterial pathogens produce exotoxins that act as hemolysins.

hemolytic disease   Incompatible Rh factor between mother and fetus causes maternal antibodies to attack the fetus and trigger complement-mediated lysis in the fetus.

hemolytic uremic syndrome (HUS)   Severe hemolytic anemia leading to kidney damage or failure; can accompany *E. coli* O157:H7 intestinal infection.

hemolyze   When red blood cells burst and release hemoglobin pigment.

hepatitis   Inflammation and necrosis of the liver, often the result of viral infection.

hepatitis D   The delta agent; a defective RNA virus that cannot reproduce on its own unless a cell is also infected with the hepatitis B virus.

hepatocellular carcinoma   A liver cancer associated with infection with hepatitis B virus.

herd immunity   The status of collective acquired immunity in a population that reduces the likelihood that nonimmune individuals will contract and spread infection. One aim of vaccination is to induce herd immunity.

heredity   Genetic inheritance.

hermaphroditic   Containing the sex organs for both male and female in one individual.

herpes zoster   A recurrent infection caused by latent chickenpox virus. Its manifestation on the skin tends to correspond to dermatomes and to occur in patches that "girdle" the trunk. Also called *shingles*.

heterotroph   An organism that relies upon organic compounds for its carbon and energy needs.

hexose   A 6-carbon sugar such as glucose and fructose.

hierarchies   Levels of power. Arrangement in order of rank.

histamine   A cytokine released when mast cells and basophils release their granules. An important mediator of allergy, its effects include smooth muscle contraction, increased vascular permeability, and increased mucus secretion.

histiocyte   Another term for *macrophage*.

histone   Proteins associated with eukaryotic DNA. These simple proteins serve as winding spools to compact and condense the chromosomes.

HLA   An abbreviation for **h**uman **l**eukocyte **a**ntigens. This closely linked cluster of genes programs for cell surface glycoproteins that control immune interactions between cells and is involved in rejection of allografts. Also called the *major histocompatibility complex (MHC)*.

holoenzyme   An enzyme complete with its apoenzyme and cofactors.

horizontal gene transfer   Transmission of genetic material from one cell to another through nonreproductive mechanisms, that is, from one organism to another living in the same habitat.

host   Organism in which smaller organisms or viruses live, feed, and reproduce.

host range   The limitation imposed by the characteristics of the host cell on the type of virus that can successfully invade it.

human diploid cell vaccine (HDCV)   A vaccine made using cell culture that is currently the vaccine of choice for preventing infection by rabies virus.

human immunodeficiency virus (HIV)   A retrovirus that causes acquired immunodeficiency syndrome (AIDS).

human microbiome   The complete complement of microorganisms that live in or on humans.

Human Microbiome Project (HMP)   A project of the National Institutes of Health to identify microbial inhabitants of the human body and their role in health and disease; uses metagenomic techniques instead of culturing.

human papillomavirus (HPV)   A group of DNA viruses whose members are responsible for common, plantar, and genital warts.

humoral immunity   Protective molecules (mostly B lymphocytes) carried in the fluids of the body.

hybridization   A process that matches complementary strands of nucleic acid (DNA-DNA, RNA-DNA, RNA-RNA). Used for locating specific sites or types of nucleic acids.

hydration   The addition of water as in the coating of ions with water molecules as ions enter into aqueous solution.

hydrogen bond   A weak chemical bond formed by the attraction of forces between molecules or atoms–in this case, hydrogen and either oxygen or nitrogen. In this type of bond, electrons are not shared, lost, or gained.

hydrolysis   A process in which water is used to break bonds in molecules. Usually occurs in conjunction with an enzyme.

hydrophilic   The property of attracting water. Molecules that attract water to their surface are called *hydrophilic*.

hydrophobic   The property of repelling water. Molecules that repel water are called *hydrophobic*.

hypertonic   Having a greater osmotic pressure than a reference solution.

hyphae   The tubular threads that make up filamentous fungi (molds). This web of branched and intertwining fibers is called a *mycelium*.

hypogammaglobulinemia   An inborn disease in which the gamma globulin (antibody) fraction of serum is greatly reduced. The condition is associated with a high susceptibility to pyogenic infections.

hyposensitivity diseases   Diseases in which there is a diminished or lack of immune reaction to pathogens due to incomplete immune system development, immune suppression, or destruction of the immune system.

hyposensitization   A therapeutic exposure to known allergens designed to build tolerance and eventually prevent allergic reaction.

hypothesis   A tentative explanation of what has been observed or measured.

hypotonic   Having a lower osmotic pressure than a reference solution.

I

icosahedron   A regular geometric figure having 20 surfaces that meet to form 12 corners. Some virions have capsids that resemble icosahedral crystals.

immune complex reaction   Type III hypersensitivity of the immune system. It is characterized by the reaction of soluble antigen with antibody and the deposition of the resulting complexes in basement membranes of epithelial tissue.

immune privilege   The restriction or reduction of immune response in certain areas of the body that reduces the potential damage to tissues that a normal inflammatory response could cause.

immune tolerance   Tolerance to self; the inability of one's immune system to react to self proteins or antigens.

immunity   An acquired resistance to an infectious agent due to prior contact with that agent.

immunoassays   Extremely sensitive tests that permit rapid and accurate measurement of trace antigen or antibody.

immunocompetence   The ability of the body to recognize and react with multiple foreign substances.

immunodeficiency   Immune function is incompletely developed, suppressed, or destroyed.

immunodeficiency disease   A form of immunopathology in which white blood cells are unable to mount a complete, effective immune response, which results in recurrent infections. Examples would be AIDS and agammaglobulinemia.

immunogen   Any substance that induces a state of sensitivity or resistance after processing by the immune system of the body.

immunoglobulin (Ig)   The chemical class of proteins to which antibodies belong.

immunology   The study of the system of body defenses that protect against infection.

immunopathology   The study of disease states associated with overreactivity or underreactivity of the immune response.

immunotherapy   Preventing or treating infectious diseases by administering substances that produce artificial immunity. May be active or passive.

in utero   Literally means "in the uterus"; pertains to events or developments occurring before birth.

in vitro   Literally means "in glass," signifying a process or reaction occurring in an artificial environment, as in a test tube or culture medium.

in vivo   Literally means "in a living being," signifying a process or reaction occurring in a living thing.

incidence   In epidemiology, the number of new cases of a disease occurring during a period.

incineration   Destruction of microbes by subjecting them to extremes of dry heat.

Microbes are reduced to ashes and gas by this process.

inclusion A relatively inert body in the cytoplasm such as storage granules, glycogen, fat, or some other aggregated metabolic product.

inclusion body One of a variety of different storage compartments in bacterial cells.

incubate To isolate a sample culture in a temperature-controlled environment to encourage growth.

incubation period The period from the initial contact with an infectious agent to the appearance of the first symptoms.

index case The first case of a disease identified in an outbreak or epidemic.

indicator bacteria In water analysis, any easily cultured bacteria that may be found in the intestine and can be used as an index of fecal contamination. The category includes coliforms and enterococci. Discovery of these bacteria in a sample means that pathogens may also be present.

induced mutation Any alteration in DNA that occurs as a consequence of exposure to chemical or physical mutagens.

inducible enzyme An enzyme that increases in amount in direct proportion to the amount of substrate present.

inducible operon An operon that under normal circumstances is not transcribed. The presence of a specific inducer molecule can cause transcription of the operon to begin.

induction The process whereby a bacteriophage in the prophage state is activated and begins replication and enters the lytic cycle.

induration Area of hardened, reddened tissue associated with the tuberculin test.

infection The entry, establishment, and multiplication of pathogenic organisms within a host.

infectious disease The state of damage or toxicity in the body caused by an infectious agent.

inflammasome A large protein in phagocytic cells that contains pattern recognition receptors (PRRs) to help these cells initiate the inflammatory response.

inflammation A natural, nonspecific response to tissue injury that protects the host from further damage. It stimulates immune reactivity and blocks the spread of an infectious agent.

inoculation The implantation of microorganisms into or upon culture media.

inorganic chemicals Molecules that lack the basic framework of the elements of carbon and hydrogen.

integument The outer surfaces of the body: skin, hair, nails, sweat glands, and oil glands.

interferon (IFN) Natural human chemical that inhibits viral replication; used therapeutically to combat viral infections and cancer.

interferon gamma A protein produced by a virally infected cell that induces production of antiviral substances in neighboring cells. This defense prevents the production and maturation of viruses and thus terminates the viral infection.

interleukins A class of chemicals released from host cells that have potent effects on immunity.

intermediate filament Proteinaceous fibers in eukaryotic cells that help provide support to the cells and their organelles.

intoxication Poisoning that results from the introduction of a toxin into body tissues through ingestion or injection.

intron The segments on split genes of eukaryotes that do not code for polypeptide. They can have regulatory functions. See **exon.**

iodophor A combination of iodine and an organic carrier that is a moderate-level disinfectant and antiseptic.

ion An unattached, charged particle.

ionic bond A chemical bond in which electrons are transferred and not shared between atoms.

ionization The aqueous dissociation of an electrolyte into ions.

ionizing radiation Radiant energy consisting of short-wave electromagnetic rays (X rays) or high-speed electrons that cause dislodgment of electrons on target molecules and create ions.

irradiation The application of radiant energy for diagnosis, therapy, disinfection, or sterilization.

isograft Transplanted tissue from one monozygotic twin to the other; transplants between highly inbred animals that are genetically identical.

isolation The separation of microbial cells by serial dilution or mechanical dispersion on solid media to create discrete colonies.

isoniazid Older drug that targets the bacterial cell wall; used against *M. tuberculosis.*

isotonic Two solutions having the same osmotic pressure such that, when separated by a semipermeable membrane, there is no net movement of solvent in either direction.

isotope A version of an element that is virtually identical in all chemical properties to another version except that their atoms have slightly different atomic masses.

J

jaundice The yellowish pigmentation of skin, mucous membranes, sclera, deeper tissues, and excretions due to abnormal deposition of bile pigments. Jaundice is associated with liver infection, as with hepatitis B virus and leptospirosis.

JC virus (JCV) Causes a form of encephalitis (progressive multifocal leukoencephalopathy), especially in AIDS patients.

K

Kaposi's sarcoma A malignant or benign neoplasm that appears as multiple hemorrhagic sites on the skin, lymph nodes, and viscera, and apparently involves the metastasis of abnormal blood vessel cells. It is a clinical feature of AIDS.

keratin Protein produced by outermost skin cells provides protection from trauma and moisture loss.

killed or inactivated vaccine A whole cell or intact virus preparation in which the microbes are dead or preserved and cannot multiply but are still capable of conferring immunity.

killer T cells A T lymphocyte programmed to directly affix cells and kill them. See **cytotoxicity.**

kingdom In the levels of classification, the second division from more general to more specific. Each domain is divided into kingdoms.

Koch's postulates A procedure to establish the specific cause of disease. In all cases of infection: (1) The agent must be found; (2) inoculations of a pure culture must reproduce the same disease in animals; (3) the agent must again be present in the experimental animal; and (4) a pure culture must again be obtained.

Koplik's spots Tiny red blisters with central white specks on the mucosal lining of the cheeks. Symptomatic of measles.

Krebs cycle or **tricarboxylic acid (TCA) cycle** The second pathway of the three pathways that complete the process of primary catabolism. Also called the *citric acid cycle.*

L

L form A stage in the lives of some bacteria in which they have no peptidoglycan.

labile In chemistry, molecules or compounds that are chemically unstable in the presence of environmental changes.

lactoferrin A protein in mucosal secretions, tears, and milk that contains iron molecules and has antimicrobial activity.

lactose One of the carbohydrates commonly referred to as sugars. Lactose is commonly found in milk.

lactose (*lac*) operon Control system that manages the regulation of lactose metabolism. It is composed of three DNA segments, including a regulator, a control locus, and a structural locus.

lag phase The early phase of population growth during which no signs of growth occur.

lagging strand The newly forming 5' DNA strand that is discontinuously replicated in segments (Okazaki fragments).

lantibiotics Short peptides produced by bacteria that inhibit the growth of other bacteria.

latency The state of being inactive. Example: a latent virus or latent infection.

leading strand The newly forming 3' DNA strand that is replicated in a continuous fashion without segments.

Legionnaire's disease Infection by *Legionella* bacterium. Weakly gram-negative rods are able to survive in aquatic habitats. Some forms may be fatal.

lepromas Skin nodules seen on the face of persons suffering from lepromatous leprosy. The skin folds and thickenings are caused by the overgrowth of *Mycobacterium leprae.*

lepromatous leprosy Severe, disfiguring leprosy characterized by widespread dissemination of the leprosy bacillus in deeper lesions.

leprosy See **Hansen's disease.**

lesion A wound, injury, or some other pathologic change in tissues.

leukocidin A heat-labile substance formed by some pyogenic cocci that impairs and sometimes lyses leukocytes.

leukocytes White blood cells. The primary infection-fighting blood cells.

leukocytosis An abnormally large number of leukocytes in the blood, which can be indicative of acute infection.

leukopenia A lower-than-normal leukocyte count in the blood that can be indicative of blood infection or disease.

leukotriene An unsaturated fatty acid derivative of arachidonic acid. Leukotriene functions in chemotactic activity, smooth muscle contractility, mucus secretion, and capillary permeability.

ligase An enzyme required to seal the sticky ends of DNA pieces after splicing.

lipase A fat-splitting enzyme. Example: Triacylglycerol lipase separates the fatty acid chains from the glycerol backbone of triglycerides.

lipid A term used to describe a variety of substances that are not soluble in polar solvents, such as water, but will dissolve in nonpolar solvents such as benzene and chloroform. Lipids include triglycerides, phospholipids, steroids, and waxes.

lipopolysaccharide A molecular complex of lipid and carbohydrate found in the bacterial cell wall. The lipopolysaccharide (LPS) of gram-negative bacteria is an endotoxin with generalized pathologic effects such as fever.

lipoteichoic acid Anionic polymers containing glycerol that are anchored in the cytoplasmic membranes of gram-positive bacteria.

liquid [media] Growth-supporting substance in fluid form.

lithoautotroph Bacteria that rely on inorganic minerals to supply their nutritional needs. Sometimes referred to as *chemoautotrophs.*

lithotroph An autotrophic microbe that derives energy from reduced inorganic compounds such as N_2S.

live attenuated vaccines Vaccines composed of living organisms that have been weakened and cannot cause disease.

lobar pneumonia Infection involving whole segments (lobes) of the lungs, which may lead to consolidation and plugging of the alveoli and extreme difficulty in breathing.

localized infection Occurs when a microbe enters a specific tissue, infects it, and remains confined there.

locus (plural, *loci*) A site on a chromosome occupied by a gene.

logarithmic or log phase Maximum rate of cell division during which growth is geometric in its rate of increase. Also called *exponential growth phase.*

lophotrichous Describing bacteria having a tuft of flagella at one or both poles.

lumen The cavity within a tubular organ.

lymphadenitis Inflammation of one or more lymph nodes. Also called *lymphadenopathy.*

lymphatic system A system of vessels and organs that serve as sites for development of immune cells and immune reactions. It includes the spleen, thymus, lymph nodes, and gut-associated lymphoid tissue (GALT).

lymphocyte The second most common form of white blood cells.

lyophilization A method for preserving microorganisms (and other substances) by freezing and then drying them directly from the frozen state.

lyse To burst.

lysin A complement-fixing antibody that destroys specific targeted cells; examples: hemolysin and bacteriolysin.

lysis The physical rupture or deterioration of a cell.

lysogenic conversion A bacterium acquires a new genetic trait due to the presence of genetic material from an infecting phage.

lysogeny The indefinite persistence of bacteriophage DNA in a host without bringing about the production of virions.

lysosome A cytoplasmic organelle containing lysozyme and other hydrolytic enzymes.

lysozyme An enzyme found in sweat, tears, and saliva that breaks down bacterial peptidoglycan.

M

macromolecules Large, molecular compounds assembled from smaller subunits, most notably biochemicals.

macronutrient A chemical substance required in large quantities (phosphate, for example).

macrophage A white blood cell derived from a monocyte that leaves the circulation and enters tissues. These cells are important in nonspecific phagocytosis and in regulating, stimulating, and cleaning up after immune responses.

macroscopic Visible to the naked eye.

major histocompatibility complex (MHC) A set of genes in mammals that produces molecules on surfaces of cells that differentiate among different individuals in the species. See **HLA.**

maltose One of the carbohydrates referred to as sugars. A fermentable sugar formed from starch.

Mantoux test An intradermal screening test for tuberculin hypersensitivity. A red, firm patch of skin at the injection site greater than 10 mm in diameter after 48 hours is a positive result that indicates current or prior exposure to the TB bacillus.

mapping Determining the location of loci and other qualities of genomic DNA.

marine microbiology A branch of microbiology that studies the role of microorganisms in the oceans.

marker Any trait or factor of a cell, virus, or molecule that makes it distinct and recognizable; example: a genetic marker.

mass number (MN) Measurement that reflects the number of protons and neutrons in an atom of a particular element.

mast cell A nonmotile connective tissue cell implanted along capillaries, especially in the lungs, skin, gastrointestinal tract, and genitourinary tract. Like a basophil, its granules store mediators of allergy.

matrix The dense ground substance between the cristae of a mitochondrion that serves as a site for metabolic reactions.

matter All tangible materials that occupy space and have mass.

maximum temperature The highest temperature at which an organism will grow.

MDR-TB Multidrug-resistant tuberculosis.

mechanical vector An animal that transports an infectious agent but is not infected by it, such as a housefly whose feet become contaminated with feces.

medium (plural, *media*) A nutrient used to grow organisms outside of their natural habitats.

meiosis The type of cell division necessary for producing gametes in diploid organisms. Two nuclear divisions in rapid succession produce four gametocytes, each containing a haploid number of chromosomes.

membrane In a single cell, a thin double-layered sheet composed of lipids such as phospholipids and sterols and proteins.

memory (immunologic memory) The capacity of the immune system to recognize and act against an antigen upon second and subsequent encounters.

memory cell The long-lived progeny of a sensitized lymphocyte that remains in circulation and is genetically programmed to react rapidly with its antigen.

Mendosicutes Taxonomic category of bacteria that have unusual cell walls; archaea.

meninges The tough tri-layer membrane covering the brain and spinal cord. Consists of the dura mater, arachnoid mater, and pia mater.

meningitis An inflammation of the membranes (meninges) that surround and protect the brain. It is often caused by bacteria such as *Neisseria meningitidis* (the meningococcus) and *Haemophilus influenzae.*

merozoite The motile, infective stage of an apicomplexan parasite that comes from a liver or red blood cell undergoing multiple fission.

mesophile Microorganisms that grow at intermediate temperatures.

messenger RNA (mRNA) A single-stranded transcript that is a copy of the DNA template that corresponds to a gene.

metabolic analog Enzyme that mimics the natural substrate of an enzyme and vies for its active site.

metabolism A general term for the totality of chemical and physical processes occurring in a cell.

metabolites Small organic molecules that are intermediates in the stepwise biosynthesis or breakdown of macromolecules.

metabolomics The study of the complete complement of small chemicals present in a cell at any given time.

metachromatic granules A type of inclusion in storage compartments of some bacteria that stain a contrasting color when treated with colored dyes.

methanogens Methane producers.

MHC Major histocompatibility complex.

MIC Abbreviation for **m**inimum **i**nhibitory **c**oncentration. The lowest concentration of antibiotic needed to inhibit bacterial growth in a test system.

microaerophile An aerobic bacterium that requires oxygen at a concentration less than that in the atmosphere.

microbe See **microorganism.**

microbial antagonism Relationship in which microorganisms compete for survival in a common environment by taking actions that inhibit or destroy another organism.

microbial ecology The study of microbes in their natural habitats.

microbicides Chemicals that kill microorganisms.

microbiology A specialized area of biology that deals with living things ordinarily too small to be seen without magnification, including bacteria, archaea, fungi, protozoa, and viruses.

microbistatic The quality of inhibiting the growth of microbes.

microfilaments Cellular cytoskeletal element formed by thin protein strands that attach to cell membrane and form a network through the cytoplasm. Responsible for movement of cytoplasm.

micronutrient A chemical substance required in small quantities (trace metals, for example).

microorganism A living thing ordinarily too small to be seen without magnification; an organism of microscopic size.

microscopic Invisible to the naked eye.

microscopy Science that studies structure, magnification, lenses, and techniques related to use of a microscope.

microtubules Long hollow tubes in eukaryotic cells; maintain the shape of the cell and transport substances from one part of cell to another; involved in separating chromosomes in mitosis.

miliary tuberculosis Rapidly fatal tuberculosis due to dissemination of mycobacteria in the blood and formation of tiny granules in various organs and tissues. The term *miliary* means "resembling a millet seed."

mineralization The process by which decomposers (bacteria and fungi) convert organic debris into inorganic and elemental form. It is part of the recycling process.

minimum inhibitory concentration (MIC) The smallest concentration of drug needed to visibly control microbial growth.

minimum temperature The lowest temperature at which an organism will grow.

miracidium The ciliated first-stage larva of a trematode. This form is infective for a corresponding intermediate host snail.

missense mutation A mutation in which a change in the DNA sequence results in a different amino acid being incorporated into a protein, with varying results.

mitochondrion A double-membrane organelle of eukaryotes that is the main site for aerobic respiration.

mitosis Somatic cell division that preserves the somatic chromosome number.

mixed acid fermentation An anaerobic degradation of pyruvic acid that results in more than one organic acid being produced (e.g., acetic acid, lactic acid, succinic acid).

mixed culture A container growing two or more different, known species of microbes.

mixed infection Occurs when several different pathogens interact simultaneously to produce an infection. Also called a *synergistic infection.*

molecule A distinct chemical substance that results from the combination of two or more atoms.

monoclonal antibodies (MAbs) Antibodies that have a single specificity for a single antigen and are produced in the laboratory from a single clone of B cells.

monocyte A large mononuclear leukocyte normally found in the lymph nodes, spleen, bone marrow, and loose connective tissue. This type of cell makes up 3% to 7% of circulating leukocytes.

monomer A simple molecule that can be linked by chemical bonds to form larger molecules.

mononuclear phagocyte system (MPS) A collection of monocytes and macrophages scattered throughout the extracellular spaces that function to engulf and degrade foreign molecules.

monosaccharide A simple sugar such as glucose that is a basic building block for more complex carbohydrates.

monotrichous Describing a microorganism that bears a single flagellum.

morbidity A diseased condition.

morbidity rate The number of persons afflicted with an illness under question or with illness in general, expressed as a numerator, with the denominator being some unit of population (as in $x/100{,}000$).

mordant A chemical that fixes a dye in or on cells by forming an insoluble compound and thereby promoting retention of that dye. Example: Gram's iodine in the Gram stain.

morphology The study of organismic structure.

mortality rate The number of persons who have died as the result of a particular cause or due to all causes, expressed as a numerator, with the denominator being some unit of population (as in $x/100{,}000$).

most probable number (MPN) Test used to detect the concentration of contaminants in water and other fluids.

motility Self-propulsion.

mumps Viral disease characterized by inflammation of the parotid glands.

mutagen Any agent that induces genetic mutation; examples: certain chemical substances, ultraviolet light, radioactivity.

mutant strain A subspecies of microorganism that has undergone a mutation, causing expression of a trait that differs from other members of that species.

mutation A permanent inheritable alteration in the DNA sequence or content of a cell.

mutualism Organisms living in an obligatory but mutually beneficial relationship.

mycelium The filamentous mass that makes up a mold. Composed of hyphae.

mycoplasma A genus of bacteria that contain no peptidoglycan/cell wall, but the cytoplasmic membrane is stabilized by sterols.

mycosis Any disease caused by a fungus.

myonecrosis Death of muscle tissue.

N

NAD/NADH Abbreviation for the oxidized/reduced form of nicotinamide adenine dinucleotide, an electron carrier. Also known as the vitamin niacin.

nanobacteria (also *nanobes*) Bacteria that are up to 100 times smaller than average bacteria.

nanobes Cell-like particles found in sediments and other geologic deposits that some scientists speculate are the smallest bacteria. Short for *nanobacteria.*

narrow-spectrum Denotes drugs that are selective and limited in their effects. For example, they inhibit either gram-negative or gram-positive bacteria, but not both.

natural immunity Any immunity that arises naturally in an organism via previous experience with the antigen.

natural selection A process in which the environment places pressure on organisms to adapt and survive changing conditions. Only the survivors will be around to continue the life cycle and contribute their genes to future generations. This is considered a major factor in evolution of species.

necrosis A pathologic process in which cells and tissues die and disintegrate.

negative stain A staining technique that renders the background opaque or colored and leaves the object unstained so that it is outlined as a colorless area.

nematode A common name for helminths called roundworms.

neurons Cells that make up the tissues of the brain and spinal cord that receive and transmit signals to and from the peripheral nervous system and central nervous system.

neurotropic Having an affinity for the nervous system. Most likely to affect the spinal cord.

neutralization   The process of combining an acid and a base until they reach a balanced proportion, with a pH value close to 7.

neutron   An electrically neutral particle in the nuclei of all atoms except hydrogen.

neutrophil   A mature granulocyte present in peripheral circulation, exhibiting a multilobular nucleus and numerous cytoplasmic granules that retain a neutral stain. The neutrophil is an active phagocytic cell in bacterial infection.

nitrogen base   A ringed compound of which pyrimidines and purines are types.

nitrogenous base   A nitrogen-containing molecule found in DNA and RNA that provides the basis for the genetic code. Adenine, guanine, and cytosine are found in both DNA and RNA, while thymine is found exclusively in DNA and uracil is found exclusively in RNA.

nomenclature   A set system for scientifically naming organisms, enzymes, anatomical structures, and so on.

noncommunicable   An infectious disease that does not arrive through transmission of an infectious agent from host to host.

noncompetitive inhibition   Form of enzyme inhibition that involves binding of a regulatory molecule to a site other than the active site.

nonionizing radiation   Method of microbial control, best exemplified by ultraviolet light, that causes the formation of abnormal bonds within the DNA of microbes, increasing the rate of mutation. The primary limitation of nonionizing radiation is its inability to penetrate beyond the surface of an object.

nonpolar   A term used to describe an electrically neutral molecule formed by covalent bonds between atoms that have the same or similar electronegativity.

nonself   Molecules recognized by the immune system as containing foreign markers, indicating a need for immune response.

nonsense codon   A triplet of mRNA bases that does not specify an amino acid but signals the end of a polypeptide chain.

nonsense mutation   A mutation that changes an amino-acid-producing codon into a stop codon, leading to premature termination of a protein.

nosocomial infection   An infection not present upon admission to a hospital but incurred while being treated there.

nucleocapsid   In viruses, the close physical combination of the nucleic acid with its protective covering.

nucleoid   The basophilic nuclear region or nuclear body that contains the bacterial chromosome.

nucleolus   A granular mass containing RNA that is contained within the nucleus of a eukaryotic cell.

nucleosome   Structure in the packaging of DNA. Formed by the DNA strands wrapping around the histone protein to form nucleus bodies arranged like beads on a chain.

nucleotide   The basic structural unit of DNA and RNA; each nucleotide consists of a phosphate, a sugar (ribose in RNA, deoxyribose in DNA), and a nitrogenous base such as adenine, guanine, cytosine, thymine (DNA only), or uracil (RNA only).

numerical aperture   In microscopy, the amount of light passing from the object and into the object in order to maximize optical clarity and resolution.

nutrient   Any chemical substance that must be provided to a cell for normal metabolism and growth. Macronutrients are required in large amounts, and micronutrients in small amounts.

nutrition   The acquisition of chemical substances by a cell or organism for use as an energy source or as building blocks of cellular structures.

O

obligate   Without alternative; restricted to a particular characteristic. Example: An *obligate parasite* survives and grows only in a host; an *obligate aerobe* must have oxygen to grow; an *obligate anaerobe* is destroyed by oxygen.

Okazaki fragment   In replication of DNA, a segment formed on the lagging strand in which biosynthesis is conducted in a discontinuous manner dictated by the $5' \rightarrow 3'$ DNA polymerase orientation.

oligodynamic action   A chemical having antimicrobial activity in minuscule amounts. Example: Certain heavy metals are effective in a few parts per billion.

oligonucleotides   Short pieces of DNA or RNA that are easier to handle than long segments.

oncogene   A naturally occurring type of gene that when activated can transform a normal cell into a cancer cell.

oncovirus   Mammalian virus capable of causing malignant tumors.

oocyst   The encysted form of a fertilized macrogamete or zygote; typical in the life cycles of apicomplexan parasites.

operator   In an operon sequence, the DNA segment where transcription of structural genes is initiated.

operon   A genetic operational unit that regulates metabolism by controlling mRNA production. In sequence, the unit consists of a regulatory gene, inducer or repressor control sites, and structural genes.

opportunistic   In infection, ordinarily nonpathogenic or weakly pathogenic microbes that cause disease primarily in an immunologically compromised host.

opsonization   The process of stimulating phagocytosis by affixing molecules (opsonins such as antibodies and complement) to the surfaces of foreign cells or particles.

optimum temperature   The temperature at which a species shows the most rapid growth rate.

order   In the levels of classification, the division of organisms that follows class. Increasing similarity may be noticed among organisms assigned to the same order.

organelle   A small component of eukaryotic cells that is bounded by a membrane and specialized in function.

organic chemicals   Molecules that contain the basic framework of the elements carbon and hydrogen.

osmophile   A microorganism that thrives in a medium having high osmotic pressure.

osmosis   The diffusion of water across a selectively permeable membrane in the direction of lower water concentration.

osteomyelitis   A focal infection of the internal structures of long bones, leading to pain and inflammation. Often caused by *Staphylococcus aureus*.

outer membrane (OM)   An additional membrane possessed by gram-negative bacteria; a lipid bilayer containing specialized proteins and polysaccharides. It lies outside of the cell wall.

oxidation   In chemical reactions, the loss of electrons by one reactant.

oxidation-reduction   Redox reactions in which paired sets of molecules participate in electron transfers.

oxidative phosphorylation   The synthesis of ATP using energy given off during the electron transport phase of respiration.

oxidizing agent   An atom or a compound that can receive electrons from another in a chemical reaction.

oxygenic   Any reaction that gives off oxygen; usually in reference to the result of photosynthesis in eukaryotes and cyanobacteria.

P

palindrome   A word, verse, number, or sentence that reads the same forward or backward. Palindromes of nitrogen bases in DNA have genetic significance as transposable elements, as regulatory protein targets, and in DNA splicing.

palisades   The characteristic arrangement of *Corynebacterium* cells resembling a row of fence posts and created by snapping.

PAMPs   Pathogen-associated molecular patterns. Chemical signatures present on many different microorganisms, but not on host, which are recognized by host as foreign.

pandemic   A disease afflicting an increased proportion of the population over a wide geographic area (often worldwide).

parasite   An organism that lives on or within another organism (the host) from which it obtains nutrients and enjoys protection. The parasite produces some degree of harm in the host.

parasitism   A relationship between two organisms in which the host is harmed in some way while the colonizer benefits.

parenteral   Administering a substance into a body compartment other than through the gastrointestinal tract, such as via intravenous, subcutaneous, intramuscular, or intramedullary injection.

paroxysmal   Events characterized by sharp spasms or convulsions; sudden onset of a symptom such as fever and chills.

passive carrier   Persons who mechanically transfer a pathogen without ever being infected by it. For example, a health care worker who doesn't wash his/her hands adequately between patients.

passive immunity   Specific resistance that is acquired indirectly by donation of preformed immune substances (antibodies) produced in the body of another individual.

passive transport   Nutrient transport method that follows basic physical laws and does not require direct energy input from the cell.

pasteurization   Heat treatment of perishable fluids such as milk, fruit juices, or wine to destroy heat-sensitive vegetative cells, followed by rapid chilling to inhibit growth of survivors and germination of spores. It prevents infection and spoilage.

pathogen   Any agent (usually a virus, bacterium, fungus, protozoan, or helminth) that causes disease.

pathogen-associated molecular patterns (PAMPs)   Molecules on the surfaces of many types of microbes that are not present on host cells that mark the microbes as foreign.

pathogenicity   The capacity of microbes to cause disease.

pathogenicity islands   Areas of the genome containing multiple genes, which contribute to a new trait for the organism that increases its ability to cause disease.

pathognomic   Distinctive and particular to a single disease; suggestive of a diagnosis.

pathologic   Capable of inducing physical damage on the host.

pathology   The structural and physiological effects of disease on the body.

pattern recognition receptors (PRRs)   Molecules on the surface of host defense cells that recognize pathogen-associated molecular patterns on microbes.

pellicle   A membranous cover; a thin skin, film, or scum on a liquid surface; a thin film of salivary glycoproteins that forms over newly cleaned tooth enamel when exposed to saliva.

pelvic inflammatory disease (PID)   An infection of the uterus and fallopian tubes that has ascended from the lower reproductive tract. Caused by gonococci and chlamydias.

penetration (viral)   The step in viral multiplication in which virus enters the host cell.

penicillinase   An enzyme that hydrolyzes penicillin; found in penicillin-resistant strains of bacteria.

penicillins   A large group of naturally occurring and synthetic antibiotics produced by *Penicillium* mold and active against the cell wall of bacteria.

pentose   A monosaccharide with five carbon atoms per molecule; examples: arabinose, ribose, xylose.

peptide   Molecule composed of short chains of amino acids, such as a dipeptide (two amino acids), a tripeptide (three), and a tetrapeptide (four).

peptide bond   The covalent union between two amino acids that forms between the amine group of one and the carboxyl group of the other. The basic bond of proteins.

peptidoglycan (PG)   A network of polysaccharide chains cross-linked by short peptides that forms the rigid part of bacterial cell walls. Gram-negative bacteria have a smaller amount of this rigid structure than do gram-positive bacteria.

perforin   Proteins released by cytotoxic T cells that produce pores in target cells.

perinatal   In childbirth, occurring before, during, or after delivery.

period of invasion   The period during a clinical infection when the infectious agent multiplies at high levels, exhibits its greatest toxicity, and becomes well established in the target tissues.

periplasmic space   The region between the cell wall and cell membrane of the cell envelopes of gram-negative bacteria.

peritrichous   In bacterial morphology, having flagella distributed over the entire cell.

petechiae   Minute hemorrhagic spots in the skin that range from pinpoint- to pinhead-size.

Peyer's patches   Oblong lymphoid aggregates of the gut located chiefly in the wall of the terminal and small intestine. Along with the tonsils and appendix, Peyer's patches make up the gut-associated lymphoid tissue that responds to local invasion by infectious agents.

pH   The symbol for the negative logarithm of the H ion concentration; p (power) or $[H^+]_{10}$. A system for rating acidity and alkalinity.

phagocyte   A class of white blood cells capable of engulfing other cells and particles.

phagocytosis   A type of endocytosis in which the cell membrane actively engulfs large particles or cells into vesicles.

phagolysosome   A body formed in a phagocyte, consisting of a union between a vesicle containing the ingested particle (the phagosome) and a vacuole of hydrolytic enzymes (the lysosome).

phase variation   The process of bacteria turning on or off a group of genes that changes its phenotype in a heritable manner.

phenotype   The observable characteristics of an organism produced by the interaction between its genetic potential (genotype) and the environment.

phosphate   An acidic salt containing phosphorus and oxygen that is an essential inorganic component of DNA, RNA, and ATP.

phospholipid   A class of lipids that compose a major structural component of cell membranes.

phosphorylation   Process in which inorganic phosphate is added to a compound.

photoautotroph   An organism that utilizes light for its energy and carbon dioxide chiefly for its carbon needs.

photon   A subatomic particle released by electromagnetic sources such as radiant energy (sunlight). Photons are the ultimate source of energy for photosynthesis.

photophosphorylation   The process of electron transport during photosynthesis that results in the synthesis of ATP from ADP.

photosynthesis   A process occurring in plants, algae, and some bacteria that traps the sun's energy and converts it to ATP in the cell. This energy is used to fix CO_2 into organic compounds.

phototrophs   Microbes that use photosynthesis to feed.

phylum   In the levels of classification, the third level of classification from general to more specific. Each kingdom is divided into numerous phyla. Sometimes referred to as a *division.*

physiology   The study of the function of an organism.

phytoplankton   The collection of photosynthetic microorganisms (mainly algae and cyanobacteria) that float in the upper layers of aquatic habitats where sun penetrates. These microbes are the basis of aquatic food pyramids and, together with zooplankton, make up the plankton.

pili (singular, *pilus*)   Long, tubular structures made of pilin protein produced by gram-negative bacteria and used for conjugation.

pinocytosis   The engulfment, or endocytosis, of liquids by extensions of the cell membrane.

plague   Zoonotic disease caused by infection with *Yersinia pestis*. The pathogen is spread by flea vectors and harbored by various rodents.

plaque   In virus propagation methods, the clear zone of lysed cells in tissue culture or chick embryo membrane that corresponds to the area containing viruses. In dental application, the filamentous mass of microbes that adheres tenaciously to the tooth and predisposes to caries, calculus, or inflammation.

plasma   The carrier fluid element of blood.

plasma cell   A progeny of an activated B cell that actively produces and secretes antibodies.

plasmids   Extra chromosomal genetic units characterized by several features. A plasmid is a double-stranded DNA that is smaller than and replicates independently of the cell chromosome; it bears genes that are not essential for cell growth; it can bear genes that code for adaptive traits; and it is transmissible to other bacteria.

platelet-activating factor   A substance released from basophils that causes release of allergic mediators and the aggregation of platelets.

platelets   Formed elements in the blood that develop when megakaryocytes disintegrate.

Platelets are involved in hemostasis and blood clotting.

pleomorphism   Normal variability of cell shapes in a single species.

pluripotential   Stem cells having the developmental plasticity to give rise to more than one type; example: undifferentiated blood cells in the bone marrow.

pneumococcus   Common name for *Streptococcus pneumoniae*, the major cause of bacterial pneumonia.

pneumonia   An inflammation of the lung leading to accumulation of fluid and respiratory compromise.

pneumonic plague   The acute, frequently fatal form of pneumonia caused by *Yersinia pestis*.

point mutation   A change that involves the loss, substitution, or addition of one or a few nucleotides.

point-source epidemic   An outbreak of disease in which all affected individuals were exposed to a single source of the pathogen at a single point in time.

polar   Term to describe a molecule with an asymmetrical distribution of charges. Such a molecule has a negative pole and a positive pole.

poliomyelitis   An acute enteroviral infection of the spinal cord that can cause neuromuscular paralysis.

polyclonal   In reference to a collection of antibodies with mixed specificities that arose from more than one clone of B cells.

polyclonal antibodies   A mixture of antibodies that were stimulated by a complex antigen with more than one antigenic determinant.

polymer   A macromolecule made up of a chain of repeating units; examples: starch, protein, DNA.

polymerase   An enzyme that produces polymers through catalyzing bond formation between building blocks (polymerization).

polymerase chain reaction (PCR)   A technique that amplifies segments of DNA for testing. Using denaturation, primers, and heat-resistant DNA polymerase, the number can be increased several-million-fold.

polymicrobial   Involving multiple distinct microorganisms.

polymorphonuclear leukocytes (PMNLs)   White blood cells with variously shaped nuclei. Although this term commonly denotes all granulocytes, it is used especially for the neutrophils.

polymyxin   A mixture of antibiotic polypeptides from *Bacillus polymyxa* that are particularly effective against gram-negative bacteria.

polypeptide   A relatively large chain of amino acids linked by peptide bonds.

polyribosomal complex   An assembly line for mass production of proteins composed of a chain of ribosomes involved in mRNA transcription.

polysaccharide   A carbohydrate that can be hydrolyzed into a number of monosaccharides; examples: cellulose, starch, glycogen.

population   A group of organisms of the same species living simultaneously in the same habitat. A group of different populations living together constitutes the community level.

porin   Transmembrane proteins of the outer membrane of gram-negative cells that permit transport of small molecules into the periplasmic space but bar the penetration of larger molecules.

portal of entry   Route of entry for an infectious agent; typically a cutaneous or membranous route.

portal of exit   Route through which a pathogen departs from the host organism.

positive stain   A method for coloring microbial specimens that involves a chemical that sticks to the specimen to give it color.

posttranslational   Referring to modifications to the protein structure that occur after protein synthesis is complete, including removal of formyl methionine, further folding of the protein, addition of functional groups, or addition of the protein to a quaternary structure.

potable   Describing water that is relatively clear, odor-free, and safe to drink.

PPNG   Penicillinase-producing *Neisseria gonorrhoeae*.

prebiotics   Nutrients used to stimulate the growth of favorable biota in the intestine.

prevalence   The total number of cases of a disease in a certain area and time period.

primary infection   An initial infection in a previously healthy individual that is later complicated by an additional (secondary) infection.

primary response   The first response of the immune system when exposed to an antigen.

primary structure   Initial protein organization described by type, number, and order of amino acids in the chain. The primary structure varies extensively from protein to protein.

primers   Synthetic oligonucleotides of known sequence that serve as landmarks to indicate where DNA amplification will begin.

prion   A concocted word to denote "proteinaceous infectious agent"; a cytopathic protein associated with the slow-virus spongiform encephalopathies of humans and animals.

probes   Small fragments of single-stranded DNA (RNA) that are known to be complementary to the specific sequence of DNA being studied.

probiotics   Preparations of live microbes used as a preventive or therapeutic measure to displace or compete with potential pathogens.

prodromal stage   A short period of mild symptoms occurring at the end of the period of incubation. It indicates the onset of disease.

proglottid   The egg-generating segment of a tapeworm that contains both male and female organs.

progressive multifocal leukoencephalopathy (PML)   An uncommon, fatal complication of infection with JC virus (polyoma virus).

prokaryotic cells   Small cells, lacking special structures such as a nucleus and organelles. All **prokaryotes** are microorganisms.

promastigote   A morphological variation of the trypanosome parasite responsible for leishmaniasis.

promoter   Part of an operon sequence. The DNA segment that is recognized by RNA polymerase as the starting site for transcription.

promoter region   The site composed of a short signaling DNA sequence that RNA polymerase recognizes and binds to commence transcription.

propagated epidemic   An outbreak of disease in which the causative agent is passed from affected persons to new persons over the course of time.

prophage   A lysogenized bacteriophage; a phage that is latently incorporated into the host chromosome instead of undergoing viral replication and lysis.

prophylactic   Any device, method, or substance used to prevent disease.

prostaglandin   A hormonelike substance that regulates many body functions. Prostaglandin comes from a family of organic acids containing 5-carbon rings that are essential to the human diet.

protease   Enzymes that act on proteins, breaking them down into component parts.

protease inhibitors   Drugs that act to prevent the assembly of functioning viral particles.

protein   Predominant organic molecule in cells, formed by long chains of amino acids.

proteomics   The study of an organism's complement of proteins (its *proteome*) and functions mediated by the proteins.

proton   An elementary particle that carries a positive charge. It is identical to the nucleus of the hydrogen atom.

protozoa   A group of single-celled, eukaryotic organisms.

provirus   The genome of a virus when it is integrated into a host cell's DNA.

PRRs   Pattern recognition receptors. Molecules on the surface of host cells that recognize pathogen-associated molecular patterns (PAMPs) on microbial cells.

pseudohypha   A chain of easily separated, spherical to sausage-shaped yeast cells partitioned by constrictions rather than by septa.

pseudomembrane   A tenacious, noncellular mucous exudate containing cellular debris that tightly blankets the mucosal surface in infections such as diphtheria and pseudomembranous enterocolitis.

pseudopodium   A temporary extension of the protoplasm of an amoeboid cell. It serves both in amoeboid motion and for food gathering (phagocytosis).

pseudopods   Protozoal appendage responsible for motility. Also called *false feet*.

psychrophile   A microorganism that thrives at low temperature (0°C–20°C), with a temperature optimum of 0°C–15°C.

pulmonary Occurring in the lungs. Examples include pulmonary anthrax and pulmonary nocardiosis.

pure culture A container growing a single species of microbe whose identity is known.

purine A nitrogen base that is an important encoding component of DNA and RNA. The two most common purines are adenine and guanine.

purpura Purple-colored spots or blotches on the skin.

pus The viscous, opaque, usually yellowish matter formed by an inflammatory infection. It consists of serum exudate, tissue debris, leukocytes, and microorganisms.

pyogenic Pertains to pus formers, especially the pyogenic cocci: pneumococci, streptococci, staphylococci, and neisseriae.

pyrimidine Nitrogen bases that help form the genetic code on DNA and RNA. Uracil, thymine, and cytosine are the most important pyrimidines.

pyrimidine dimer The union of two adjacent pyrimidines on the same DNA strand, brought about by exposure to ultraviolet light. It is a form of mutation.

pyrogen A substance that causes a rise in body temperature. It can come from pyrogenic microorganisms or from polymorphonuclear leukocytes (endogenous pyrogens).

Q

quaternary structure Most complex protein structure characterized by the formation of large, multiunit proteins by more than one of the polypeptides. This structure is typical of antibodies and some enzymes that act in cell synthesis.

quats A word that pertains to a family of surfactants called *quaternary ammonium compounds*. These detergents are only weakly microbicidal and are used as sanitizers and preservatives.

quinine A substance derived from cinchona trees that was used as an antimalarial treatment; has been replaced by synthetic derivatives.

quinolone A class of synthetic antimicrobic drugs with broad-spectrum effects.

quorum sensing The ability of bacteria to regulate their gene expression in response to sensing bacterial density.

R

rabies The only rhabdovirus that infects humans. Zoonotic disease characterized by fatal meningoencephalitis.

radiation Electromagnetic waves or rays, such as those of light given off from an energy source.

radioactive isotopes Unstable isotopes whose nuclei emit particles of radiation. This emission is called *radioactivity* or *radioactive decay*. Three naturally occurring emissions are alpha, beta, and gamma radiation.

rales Sounds in the lung, ranging from clicking to rattling; indicate respiratory illness.

random amplified polymorphic DNA (RAPD) A type of PCR that utilizes primers with random sequences in order to identify microbial populations that are relatively unknown.

reactants Molecules entering or starting a chemical reaction.

real image An image formed at the focal plane of a convex lens. In the compound light microscope, it is the image created by the objective lens.

receptor Cell surface molecules involved in recognition, binding, and intracellular signaling.

recombinant An organism that contains genes that originated in another organism, whether through deliberate laboratory manipulation or natural processes.

recombinant DNA technology A technology, also known as *genetic engineering*, that deliberately modifies the genetic structure of an organism to create novel products, microbes, animals, plants, and viruses.

recombination A type of genetic transfer in which DNA from one organism is donated to another.

recycling A process that converts unusable organic matter from dead organisms back into their essential inorganic elements and returns them to their nonliving reservoirs to make them available again for living organisms. This is a common term that means the same as *mineralization* and *decomposition*.

redox Denoting an oxidation-reduction reaction.

reducing agent An atom or a compound that can donate electrons in a chemical reaction.

reducing medium A growth medium that absorbs oxygen and allows anaerobic bacteria to grow.

reduction In chemistry, the gain of electrons.

redundancy The property of the genetic code that allows an amino acid to be specified by several different codons.

refractive index The measurement of the degree of light that is bent, or refracted, as it passes between two substances such as air, water, or glass.

regulated enzymes Enzymes whose extent of transcription or translation is influenced by changes in the environment.

regulator DNA segment that codes for a protein capable of repressing an operon.

regulatory B cells (B$_{reg}$ cells) A type of activated B cell that controls the immune response.

regulatory site The location on an enzyme where a certain substance can bind and block the enzyme's activity.

release The final step in the multiplication cycle of viruses in which the assembled virus particle exits the host cell and moves on to infect another cell.

replication In DNA synthesis, the semiconservative mechanisms that ensure precise duplication of the parent DNA strands.

replication fork The Y-shaped point on a replicating DNA molecule where the DNA polymerase is synthesizing new strands of DNA.

reportable disease Those diseases that must be reported to health authorities by law.

repressible operon An operon that under normal circumstances is transcribed. The buildup of the operon's amino acid product causes transcription of the operon to stop.

repressor The protein product of a repressor gene that combines with the operator and arrests the transcription and translation of structural genes.

reservoir In disease communication, the natural host or habitat of a pathogen.

resistance (R) factor Plasmids, typically shared among bacteria by conjugation, that provide resistance to the effects of antibiotics.

resolving power The capacity of a microscope lens system to accurately distinguish between two separate entities that lie close to each other. Also called *resolution*.

respiratory chain A series of enzymes that transfer electrons from one to another, resulting in the formation of ATP. It is also known as the *electron transport chain*. The chain is located in the cytoplasmic membrane of bacteria and in the inner mitochondrial membrane of eukaryotes.

respiratory syncytial virus (RSV) An RNA virus that infects the respiratory tract. RSV is the most prevalent cause of respiratory infection in newborns.

restriction endonuclease An enzyme present naturally in cells that cleaves specific locations on DNA. It is an important means of inactivating viral genomes, and it is also used to splice genes in genetic engineering.

restriction fragment length polymorphisms (RFLPs) Variations in the lengths of DNA fragments produced when a specific restriction endonuclease acts on different DNA sequences.

restriction fragments Short pieces of DNA produced when DNA is exposed to restriction endonucleases.

retrovirus A group of RNA viruses (including HIV) that have the mechanisms for converting their genome into a double strand of DNA that can be inserted on a host's chromosome.

reverse transcriptase (RT) The enzyme possessed by retroviruses that carries out the reversion of RNA to DNA–a form of reverse transcription.

Reye's syndrome A sudden, usually fatal neurological condition that occurs in children after a viral infection. Autopsy shows cerebral edema and marked fatty change in the liver and renal tubules.

Rh factor An isoantigen that can trigger hemolytic disease in newborns due to incompatibility between maternal and infant blood factors.

ribonucleic acid (RNA) The nucleic acid responsible for carrying out the hereditary program transmitted by an organism's DNA.

ribose   A 5-carbon monosaccharide found in RNA.

ribosomal RNA (rRNA)   A single-stranded transcript that is a copy of part of the DNA template.

ribosome   A bilobed macromolecular complex of ribonucleoprotein that coordinates the codons of mRNA with tRNA anticodons and, in so doing, constitutes the peptide assembly site.

ribozyme   A part of an RNA-containing enzyme in eukaryotes that removes intervening sequences of RNA (introns) and splices together the true coding sequences (exons) to form a mature messenger RNA.

rickettsias   Medically important family of bacteria, commonly carried by ticks, lice, and fleas. Significant cause of important emerging diseases.

ringworm   A superficial mycosis caused by various dermatophytic fungi. This common name is actually a misnomer.

RNA editing   The alteration of RNA molecules before translation, found only in eukaryotes.

RNA polymerase   Enzyme process that translates the code of DNA to RNA.

rolling circle   An intermediate stage in viral replication of circular DNA into linear DNA.

rough endoplasmic reticulum (RER)   Microscopic series of tunnels that originates in the outer membrane of the nuclear envelope and is used in transport and storage. Large numbers of ribosomes, partly attached to the membrane, give the rough appearance.

rubeola (red measles)   Acute disease caused by infection with *Morbillivirus*.

S

S layer   Single layer of thousands of copies of a single type of protein linked together on the surface of a bacterial cell that is produced when the cell is in a hostile environment.

saccharide   Scientific term for sugar. Refers to a simple carbohydrate with a sweet taste.

salpingitis   Inflammation of the fallopian tubes.

sanitize   To clean inanimate objects using soap and degerming agents so that they are safe and free of high levels of microorganisms.

saprobe   A microbe that decomposes organic remains from dead organisms. Also known as a *saprophyte* or *saprotroph*.

sarcina   A cubical packet of 8, 16, or more cells; the cellular arrangement of the genus *Sarcina* in the family *Micrococcaceae*.

satellitism   A commensal interaction between two microbes in which one can grow in the vicinity of the other due to nutrients or protective factors released by that microbe.

saturation   The complete occupation of the active site of a carrier protein or enzyme by the substrate.

schistosomiasis   Infection by blood fluke, often as a result of contact with contaminated water in rivers and streams. Symptoms appear in liver, spleen, or urinary system depending on species of *Schistosoma*. Infection may be chronic.

schizogony   A process of multiple fission whereby first the nucleus divides several times and, subsequently, the cytoplasm is subdivided for each new nucleus during cell division.

scientific method   Principles and procedures for the systematic pursuit of knowledge, involving the recognition and formulation of a problem, the collection of data through observation and experimentation, and the formulation and testing of a hypothesis.

scolex   The anterior end of a tapeworm characterized by hooks and/or suckers for attachment to the host.

sebaceous glands   The sebum- (oily, fatty) secreting glands of the skin.

sebum   Low pH, oil-based secretion of the sebaceous glands.

secondary infection   An infection that compounds a preexisting one.

secondary response   The rapid rise in antibody titer following a repeat exposure to an antigen that has been recognized from a previous exposure. This response is brought about by memory cells produced as a result of the primary exposure.

secondary structure   Protein structure that occurs when the functional groups on the outer surface of the molecule interact by forming hydrogen bonds. These bonds cause the amino acid chain to either twist, forming a helix, or to pleat into an accordion pattern called a *β-pleated sheet*.

secretory antibody   The immunoglobulin (IgA) that is found in secretions of mucous membranes and serves as a local immediate protection against infection.

selective media   Nutrient media designed to favor the growth of certain microbes and to inhibit undesirable competitors.

selectively toxic   Property of an antimicrobial agent to be highly toxic against its target microbe while being far less toxic to other cells, particularly those of the host organism.

self   Natural markers of the body that are recognized by the immune system.

self-limited   Applies to an infection that runs its course without disease or residual effects.

semiconservative replication   In DNA replication, the synthesis of paired daughter strands, each retaining a parent strand template.

semisolid [media]   Nutrient media with a firmness midway between that of a broth (a liquid medium) and an ordinary solid medium; motility media.

semisynthetic   Drugs that, after being naturally produced by bacteria, fungi, or other living sources, are chemically modified in the laboratory. Compare to **synthetic.**

sensitizing dose   The initial effective exposure to an antigen or an allergen that stimulates an immune response. Often applies to allergies.

sepsis   The state of putrefaction; the presence of pathogenic organisms or their toxins in tissue or blood.

septic shock   Blood infection resulting in a pathological state of low blood pressure accompanied by a reduced amount of blood circulating to vital organs. Endotoxins of all gram-negative bacteria can cause shock, but most clinical cases are due to gram-negative enteric rods.

septicemia   Systemic infection associated with microorganisms multiplying in circulating blood.

septicemic plague   A form of infection with *Yersinia pestis* occurring mainly in the bloodstream and leading to high mortality rates.

septum (plural, *septa*)   A partition or cellular cross wall, as in certain fungal hyphae.

sequela   A morbid complication that follows a disease.

sequencing   Determining the actual order and types of bases in a segment of DNA.

serology   The branch of immunology that deals with *in vitro* diagnostic testing of serum.

seropositive   Showing the presence of specific antibody in a serological test. Indicates ongoing infection.

serotonin   A vasoconstrictor that inhibits gastric secretion and stimulates smooth muscle.

serotyping   The subdivision of a species or subspecies into an immunologic type, based upon antigenic characteristics.

serous   Referring to serum, the clear fluid that escapes cells during the inflammatory response.

serum   The clear fluid expressed from clotted blood that contains dissolved nutrients, antibodies, and hormones but not cells or clotting factors.

serum sickness   A type of immune complex disease in which immune complexes enter circulation, are carried throughout the body, and are deposited in the blood vessels of the kidney, heart, skin, and joints. The condition may become chronic.

severe acute respiratory syndrome (SARS)   A severe respiratory disease caused by infection with a newly described coronavirus.

severe combined immunodeficiency   A collection of syndromes occurring in newborns caused by a genetic defect that knocks out both B- and T-cell types of immunity. There are several versions of this disease, termed *SCID* for short.

sex pilus   A conjugative pilus.

sexually transmitted infections (STIs)   Infections resulting from pathogens that enter the body via sexual intercourse or intimate, direct contact.

shiga toxin   Heat-labile exotoxin released by some *Shigella* species and by *E. coli* O157:H7; responsible for worst symptoms of these infections.

shiga-toxin-producing *E. coli* (STEC)   A strain of *E. coli* that produces the shiga toxin.

shingles   Lesions produced by reactivated human herpesvirus 3 (chickenpox) infection; also known as herpes zoster.

siderophores Low-molecular-weight molecules produced by many microorganisms that can bind iron very tightly.

sign Any abnormality uncovered upon physical diagnosis that indicates the presence of disease. A *sign* is an objective assessment of disease, as opposed to a *symptom*, which is the subjective assessment perceived by the patient.

silent mutation A mutation that, because of the degeneracy of the genetic code, results in a nucleotide change in both the DNA and mRNA but not the resultant amino acid and, thus, not the protein.

simple stain Type of positive staining technique that uses a single dye to add color to cells so that they are easier to see. This technique tends to color all cells the same color.

slime layer A diffuse, unorganized layer of polysaccharides and/or proteins on the outside of some bacteria.

smooth endoplasmic reticulum (SER) A microscopic series of tunnels lacking ribosomes that functions in the nutrient processing function of a cell.

solute A substance that is uniformly dispersed in a dissolving medium or solvent.

solution A mixture of one or more substances (solutes) that cannot be separated by filtration or ordinary settling.

solvent A dissolving medium.

somatic (O or cell wall antigen) One of the three major antigens commonly used to differentiate gram-negative enteric bacteria.

source The person or item from which an infection is directly acquired. See **reservoir.**

Southern blot A technique that separates fragments of DNA using electrophoresis and identifies them by hybridization.

species In the levels of classification, the most specific level of organization.

specificity In immunity, the concept that some parts of the immune system only react with antigens that originally activated them.

spike A receptor on the surface of certain enveloped viruses that facilitates specific attachment to the host cell.

spirillum A type of bacterial cell with a rigid spiral shape and external flagella.

spirochete A coiled, spiral-shaped bacterium that has endoflagella and flexes as it moves.

spliceosome A molecule composed of RNA and protein that removes introns from eukaryotic mRNA before it is translated by forming a loop in the intron, cutting it from the mRNA, and joining exons together.

spontaneous generation Early belief that living things arose from vital forces present in nonliving, or decomposing, matter.

spontaneous mutation A mutation in DNA caused by random mistakes in replication and not known to be influenced by any mutagenic agent. These mutations give rise to an organism's natural, or background, rate of mutation.

sporadic Description of a disease that exhibits new cases at irregular intervals in unpredictable geographic locales.

sporangiospore A form of asexual spore in fungi; enclosed in a sac.

sporangium A fungal cell in which asexual spores are formed by multiple cell cleavage.

spore A differentiated, specialized cell form that can be used for dissemination, for survival in times of adverse conditions, and/or for reproduction. Spores are usually unicellular and may develop into gametes or vegetative organisms.

sporicide A chemical agent capable of destroying bacterial endospores.

sporozoite One of many minute elongated bodies generated by multiple division of the oocyst. It is the infectious form of the malarial parasite that is harbored in the salivary gland of the mosquito and inoculated into the victim during feeding.

sporulation The process of spore formation.

start codon The nucleotide triplet AUG that codes for the first amino acid in protein sequences.

stasis A state of rest or inactivity; applied to nongrowing microbial cultures. Also called *microbistasis.*

stationary growth phase Survival mode in which cells either stop growing or grow very slowly.

stem cells Pluripotent, undifferentiated cells.

sterile Completely free of all life forms, including spores and viruses.

sterilization Any process that completely removes or destroys all viable microorganisms, including viruses, from an object or habitat. Material so treated is sterile.

strain In microbiology, a set of descendants cloned from a common ancestor that retains the original characteristics. Any deviation from the original is a different strain.

streptolysin A hemolysin produced by streptococci.

strict or obligate anaerobe An organism that does not use oxygen gas in metabolism and cannot survive in oxygen's presence.

structural gene A gene that codes for the amino acid sequence (peptide structure) of a protein.

subacute Indicates an intermediate status between acute and chronic disease.

subacute sclerosing panencephalitis (SSPE) A complication of measles infection in which progressive neurological degeneration of the cerebral cortex invariably leads to coma and death.

subcellular vaccine A vaccine preparation that contains specific antigens, such as the capsule or toxin from a pathogen, and not the whole microbe.

subclinical A period of inapparent manifestations that occurs before symptoms and signs of disease appear.

subculture To make a second-generation culture from a well-established colony of organisms.

subcutaneous The deepest level of the skin structure.

substrate The specific molecule upon which an enzyme acts.

subunit vaccine A vaccine preparation that contains only antigenic fragments such as surface receptors from the microbe. Usually in reference to virus vaccines.

sucrose One of the carbohydrates commonly referred to as sugars. Common table or cane sugar.

sulfonamide Antimicrobial drugs that interfere with the essential metabolic process of bacteria and some fungi.

superantigens Bacterial toxins that are potent stimuli for T cells and can be a factor in diseases such as toxic shock.

superficial mycosis A fungal infection located in hair, nails, and the epidermis of the skin.

superinfection An infection occurring during antimicrobial therapy that is caused by an overgrowth of drug-resistant microorganisms.

superoxide A toxic derivative of oxygen; (O_2^-).

surfactant A surface-active agent that forms a water-soluble interface; examples: detergents, wetting agents, dispersing agents, and surface tension depressants.

sylvatic Denotes the natural presence of disease among wild animal populations; examples: sylvatic (sylvan) plague, rabies.

symbiosis An intimate association between individuals from two species; used as a synonym for *mutualism.*

symptom The subjective evidence of infection and disease as perceived by the patient.

syncytium (plural, *syncytia*) A multinucleated protoplasmic mass formed by consolidation of individual cells.

syndrome The collection of signs and symptoms that, taken together, paint a portrait of the disease.

synergism The coordinated or correlated action by two or more drugs or microbes that results in a heightened response or greater activity.

synthesis (viral) The step in viral multiplication in which viral genetic material and proteins are made through replication and transcription/translation.

synthetic Referring to a chemotherapeutic agent manufactured entirely through chemical processes in the laboratory that mimics the actions of antibiotics. Compare to **semisynthetic.**

synthetic biology The use of known genes to produce new applications.

syphilis A sexually transmitted bacterial disease caused by the spirochete *Treponema pallidum.*

systemic Occurring throughout the body; said of infections that invade many compartments and organs via the circulation.

T

T lymphocyte (T cell) A white blood cell that is processed in the thymus and is involved in cell-mediated immunity.

***Taq* polymerase** DNA polymerase from the thermophilic bacterium *Thermus aquaticus* that enables high-temperature replication of DNA required for the polymerase chain reaction.

tartar   See **calculus.**

taxa   Taxonomic categories.

taxonomy   The formal system for organizing, classifying, and naming living things.

teichoic acid   Anionic polymers containing glycerol that appear in the walls of gram-positive bacteria.

temperate phage   A bacteriophage that enters into a less virulent state by becoming incorporated into the host genome as a prophage instead of in the vegetative or lytic form that eventually destroys the cell.

template strand   The strand in a double-stranded DNA molecule that is used as a model to synthesize a complementary strand of DNA or RNA during replication or transcription.

Tenericutes   Taxonomic category of bacteria that lack cell walls.

teratogenic   Causing abnormal fetal development.

tertiary structure   Protein structure that results from additional bonds forming between functional groups in a secondary structure, creating a three-dimensional mass.

tetanospasmin   The neurotoxin of *Clostridium tetani,* the agent of tetanus. Its chief action is directed upon the inhibitory synapses of the anterior horn motor neurons.

tetracyclines   A group of broad-spectrum antibiotics with a complex four-ring structure.

tetrads   Groups of four.

theory   A collection of statements, propositions, or concepts that explains or accounts for a natural event.

theory of evolution   The evidence cited to explain how evolution occurs.

therapeutic index   The ratio of the toxic dose to the effective therapeutic dose that is used to assess the safety and reliability of the drug.

thermal death point   The lowest temperature that achieves sterilization in a given quantity of broth culture upon a 10-minute exposure. Examples: 55°C for *Escherichia coli,* 60°C for *Mycobacterium tuberculosis,* and 120°C for endospores.

thermal death time   The least time required to kill all cells of a culture at a specified temperature.

thermoduric   Resistant to the harmful effects of high temperature.

thermophile   A microorganism that thrives at a temperature of 50°C or higher.

thrush   *Candida albicans* infection of the oral cavity.

thymine (T)   One of the nitrogen bases found in DNA but not in RNA. Thymine is in a pyrimidine form.

thymus   Butterfly-shaped organ near the tip of the sternum that is the site of T-cell maturation.

tincture   A medicinal substance dissolved in an alcoholic solvent.

tinea   Ringworm; a fungal infection of the hair, skin, or nails.

tinea versicolor   A condition of the skin appearing as mottled and discolored skin pigmentation as a result of infection by the yeast *Malassezia furfur.*

titer   In immunochemistry, a measure of antibody level in a patient, determined by agglutination methods.

toll-like receptors (TLRs)   A category of pattern recognition receptors that bind to pathogen-associated molecular patterns on microbes.

tonsils   A ring of lymphoid tissue in the pharynx that acts as a repository for lymphocytes.

topoisomerases   Enzymes that can add or remove DNA twists and thus regulate the degree of supercoiling.

toxemia   Condition in which a toxin (microbial or otherwise) is spread throughout the bloodstream.

toxigenicity   The tendency for a pathogen to produce toxins. It is an important factor in bacterial virulence.

toxin   A specific chemical product of microbes, plants, and some animals that is poisonous to other organisms.

toxinosis   Disease whose adverse effects are primarily due to the production and release of toxins.

toxoid   A toxin that has been rendered nontoxic but is still capable of eliciting the formation of protective antitoxin antibodies; used in vaccines.

trace elements   Micronutrients (zinc, nickel, and manganese) that occur in small amounts and are involved in enzyme function and maintenance of protein structure.

transamination   The transfer of an amino group from an amino acid to a carbohydrate fragment.

transcript   A newly transcribed RNA molecule.

transcription   Messenger RNA (mRNA) synthesis; the process by which a strand of RNA is produced against a DNA template.

transduction   The transfer of genetic material from one bacterium to another by means of a bacteriophage vector.

transfection   The introduction of DNA into eukaryotic cells from the environment by exposing cells to chemicals or electrical currents; similar to **transformation.**

transferrin   A protein in the plasma fraction of blood that transports iron.

transfer RNA (tRNA)   A transcript of DNA that specializes in converting RNA language into protein language.

transformation   In microbial genetics, the transfer of genetic material contained in "naked" DNA fragments from a donor cell to a competent recipient cell.

transfusion   Infusion of whole blood, red blood cells, or platelets directly into a patient's circulation.

translation   Protein synthesis; the process of decoding the messenger RNA code into a polypeptide.

transmissible spongiform encephalopathies (TSEs)   Diseases caused by proteinaceous infectious particles (also known as *prions*).

transport medium   Microbiological medium that is used to transport specimens.

transposon   A DNA segment with an insertion sequence at each end, enabling it to migrate to another plasmid, to the bacterial chromosome, or to a bacteriophage.

trematode   A category of helminth; also known as *flatworm* or *fluke.*

trichomoniasis   Sexually transmitted disease caused by infection by the trichomonads, a group of protozoa. Symptoms include urinary pain and frequency and foul-smelling vaginal discharge in females, or recurring urethritis, with a thin milky discharge, in males.

triglyceride   A type of lipid composed of a glycerol molecule bound to three fatty acids.

triplet   See **codon.**

trophozoite   A vegetative protozoan (feeding form) as opposed to a resting (cyst) form.

true pathogen   A microbe capable of causing infection and disease in healthy persons with normal immune defenses.

trypomastigote   The infective morphological stage transmitted by the tsetse fly or the reduviid bug in African trypanosomiasis and Chagas disease.

tubercle   In tuberculosis, the granulomatous well-defined lung lesion that can serve as a focus for latent infection.

tuberculin   A glycerinated broth culture of *Mycobacterium tuberculosis* that is evaporated and filtered. Formerly used to treat tuberculosis, tuberculin is now used chiefly for diagnostic tests.

tuberculin reaction   A diagnostic test in which PPD, or purified protein derivative (of *M. tuberculosis*), is injected superficially under the skin and the area of reaction measured; also called the *Mantoux test.*

tuberculoid leprosy   A superficial form of leprosy characterized by asymmetrical, shallow skin lesions containing few bacterial cells.

tumor necrosis factor (TNF)   A cytokine involved in amplifying the immune response; is also implicated in autoimmune diseases and cancer.

turbid   Cloudy appearance of nutrient solution in a test tube due to growth of microbe population.

typhoid fever   Form of salmonellosis. It is highly contagious. Primary symptoms include fever, diarrhea, and abdominal pain. Typhoid fever can be fatal if untreated.

U

ubiquitous   Present everywhere at the same time.

ultraviolet (UV) radiation   Radiation with an effective wavelength from 240 nm to 260 nm. UV radiation induces mutations readily but has very poor penetrating power.

uncoating   The process of removal of the viral coat and release of the viral genome by its newly invaded host cell.

undulant fever   See **brucellosis.**

universal donor   In blood grouping and transfusion, a group O individual whose erythrocytes bear neither agglutinogen A nor B.

universal precautions (UPs)   Centers for Disease Control and Prevention guidelines for health care workers regarding the prevention of disease transmission when handling patients and body substances.

uracil (U)   One of the nitrogen bases in RNA but not in DNA. Uracil is in a pyrimidine form.

urinary tract infection (UTI)   Invasion and infection of the urethra and bladder by bacterial residents, most often *E. coli.*

V

vaccination   Exposing a person to the antigenic components of a microbe without its pathogenic effects for the purpose of inducing a future protective response.

vaccine   Originally used in reference to inoculation with the cowpox or vaccinia virus to protect against smallpox. In general, the term now pertains to injection of whole microbes (killed or attenuated), toxoids, or parts of microbes as a prevention or cure for disease.

vacuoles   In the cell, membrane-bounded sacs containing fluids or solid particles to be digested, excreted, or stored.

valence   The combining power of an atom based upon the number of electrons it can either take on or give up.

van der Waals forces   Weak attractive interactions between molecules of low polarity.

vancomycin   Antibiotic that targets the bacterial cell wall; used often in antibiotic-resistant infections.

variable region   The antigen binding fragment of an immunoglobulin molecule, consisting of a combination of heavy and light chains whose molecular conformation is specific for the antigen.

varicella   Informal name for virus responsible for chickenpox as well as shingles; also known as *human herpesvirus 3* (HHV-3).

variolation   A hazardous, outmoded process of deliberately introducing smallpox material scraped from a victim into the nonimmune subject in the hope of inducing resistance.

vasoactive   Referring to chemical mediators involved in the immune response that act on endothelial cells or the smooth muscle of blood vessels causing them to either restrict or relax.

vector   An animal that transmits infectious agents from one host to another, usually a biting or piercing arthropod like the tick, mosquito, or fly. Infectious agents can be conveyed mechanically by simple contact or biologically whereby the parasite develops in the vector. A genetic element such as a plasmid or a bacteriophage used to introduce genetic material into a cloning host during recombinant DNA experiments.

vegetative   In describing microbial developmental stages, a metabolically active feeding and dividing form, as opposed to a dormant, seemingly inert, nondividing form. Examples: a bacterial cell versus its spore; a protozoal trophozoite versus its cyst.

vehicle   An inanimate material (solid object, liquid, or air) that serves as a transmission agent for pathogens.

vesicle   A blister characterized by a thin-skinned, elevated, superficial pocket filled with serum.

viable nonculturable (VNC)   A description of a state in which bacteria are alive but are not metabolizing at an appreciable rate and will not grow when inoculated onto laboratory medium.

vibrio   A curved, rod-shaped bacterial cell.

viremia   The presence of viruses in the bloodstream.

virion   An elementary virus particle in its complete morphological and thus infectious form. A virion consists of the nucleic acid core surrounded by a capsid, which can be enclosed in an envelope.

viroid   An infectious agent that, unlike a virion, lacks a capsid and consists of a closed circular RNA molecule. Although known viroids are all plant pathogens, it is conceivable that animal versions exist.

virtual image   In optics, an image formed by diverging light rays; in the compound light microscope, the second, magnified visual impression formed by the ocular from the real image formed by the objective.

virucide   A chemical agent that inactivates viruses, especially on living tissue.

virulence   In infection, the relative capacity of a pathogen to invade and harm host cells.

virulence factors   A microbe's structures or capabilities that allow it to establish itself in a host and cause damage.

virus   Microscopic, acellular agent composed of nucleic acid surrounded by a protein coat.

virus particle   A more specific name for a virus when it is outside of its host cells.

vitamins   A component of coenzymes critical to nutrition and the metabolic function of coenzyme complexes.

W

wart   An epidermal tumor caused by papillomaviruses. Also called a *verruca.*

Western blot test   A procedure for separating and identifying antigen or antibody mixtures by two-dimensional electrophoresis in polyacrylamide gel, followed by immune labeling.

wheal   A welt; a marked, slightly red, usually itchy area of the skin that changes in size and shape as it extends to adjacent area. The reaction is triggered by cutaneous contact or intradermal injection of allergens in sensitive individuals.

white blood cells   In contrast to erythrocytes, or red blood cells, white blood cells are clear or colorless and include granulocytes (neutrophils, eosinophils, and basophils) and agranulocytes (lymphocytes and monocytes).

whitlow   A deep inflammation of the finger or toe, especially near the tip or around the nail. Whitlow is a painful herpes simplex virus infection that can last several weeks and is most common among health care personnel who come in contact with the virus in patients.

whole blood   A liquid connective tissue consisting of blood cells suspended in plasma.

Widal test   An agglutination test for diagnosing typhoid.

wild type   The natural, nonmutated form of a genetic trait.

wobble   A characteristic of amino acid codons in which the third base of a codon can be altered without changing the code for the amino acid.

X

XDR-TB   Extensively drug-resistant tuberculosis (worse than multidrug-resistant tuberculosis).

xenograft   The transfer of a tissue or an organ from an animal of one species to a recipient of another species.

Z

zoonosis   An infectious disease indigenous to animals that humans can acquire through direct or indirect contact with infected animals.

zooplankton   The collection of nonphotosynthetic microorganisms (protozoa, tiny animals) that float in the upper regions of aquatic habitat and together with phytoplankton comprise the plankton.

zygospore   A thick-walled sexual spore produced by the zygomycete fungi. It develops from the union of two hyphae, each bearing nuclei of opposite mating types.

Photo Credits

Front Matter

Page iv (Cowan): Courtesy Kelly Cowan; p. iv (Bunn): Courtesy Patrick Osachuk; p. iv (Atlas): Courtesy Michel Cohen Atlas; p. v: ©C Squared Studios/Getty Images; All background photo collages (clockwise from top left): CDC/Janice Haney Carr; CDC/Janice Haney Carr; CDC/Dr. Erskine Palmer & Byron Skinner; CDC/Dr. Stan Erlandsen; ©Science Photo Library/Getty Images RF; ©Science Photo Library/Getty Images RF; NIAID, NIH/Rocky Mountain Laboratories; p. viii (inset): ©Steve Gschmeissner/Science Source; p. ix (inset): CDC/Janice Haney Carr; p. x (inset): ©Steve Gschmeissner/Science Source; p. xi (inset): ©Jose Luis Pelaez Inc/Blend Images LLC RF; p. xii (inset): Centers for Disease Control; xiii (inset): ©Martin Oeggerli/Science Source; CO1: CDC/Dr. Erskine Palmer & Byron Skinner; CO1: CDC/Dr. Stan Erlandsen; CO1: ©Science Photo Library/Getty Images RF; CO2: ©Nathan Reading; CO3: CDC/Dr. Gilda Jones; CO4: CDC/Dr. Lucille K. George; CO6: CDC/Don Stalons; CO7: ©McGraw-Hill Education; CO8: ©Purestock/SuperStock; CO9: CDC/Janice Haney Carr; CO10: ©Glow Images RF; CO11: ©MedicalRF.com/Getty Images RF; CO12: ©The McGraw-Hill Companies, Inc./Al Telser, photographer; CO13: ©MedicalRF.com/Getty Images RF; CO14: ©McGraw-Hill Education/Al Telser; CO15: ©McGraw-Hill Education/Lisa Burgess; CO16: ©Ingram Publishing RF; CO17: ©Scott Camazine/Science Source; CO18: Centers for Disease Control; CO19: ©MedicalRF.com; CO20: Centers for Disease Control; CO21: ©Meredun Animal Health Ltd/Science Source; CO22: USDA/Photo by De Wood, digital colorization by Chris Pooley.

Chapter 1

Opener (clockwise from top left): CDC/Janice Haney Carr; CDC/Janice Haney Carr; CDC/Dr. Erskine Palmer & Byron Skinner; CDC/Dr. Stan Erlandsen; ©Science Photo Library/Getty Images RF; ©Science Photo Library/Getty Images RF; NIAID, NIH/Rocky Mountain Laboratories; p. 2 (auhor): ©Michael Williams; p. 3: ©Jerome Wexler/Science Source; p. 4 (bacteria): ©MedicalRF.com/Getty Images RF; p. 4 (microbe): NASA; 1.1: NASA GSFC image by Robert Simmon and Reto Stöckli; 1.2: ©Jerome Wexler/Science Source; p. 6: ©McGraw-Hill Education/Don Rubbelke; 1.3: U.S. Coast Guard/Chief Petty Officer John Kepsimelis, Atlantic Strike Team; 1.4 (Haloquadratum): ©Dr. Mike Dyall-Smith; 1.4 (Aspergillus): ©Michael Abbey/Science Source; 1.4 (E coli): CDC/Janice Haney Carr; 1.4 (Vorticella): ©Nuridsany et Perennou/Science Source; 1.4 (Herpes): CDC/Dr. Erskine Palmer; 1.4 (Taenia): CDC/Dr. Mae Melvin; p. 8: ©PhotoLink/Getty Images RF; 1.6 (both): ©Kathleen Talaro; 1.7: ©Bettmann/Corbis; p. 12: ©Steve Gschmeissner/SPL/Getty Images RF; p. 14: ©Stephen Durr RF; 1.8: ©H Lansdown/Alamy RF; 1.9: ©Stockbyte/PunchStock RF; p. 18: ©Guy Crittenden/Getty Images RF; p. 20: ©Comstock/Jupiter Images RF; p. 21, p. 24: ©Ingram Publishing RF; p. 25: CDC/Janice Haney Carr; p. 28 (web): ©Tony Sweet/Digital Vision/Getty Images RF; p. 28 (soil): ©Pixtal/age fotostock RF; p. 30 (author): ©Michael Williams; p. 30 (cell): ©Science Photo Library/Getty Images RF; p. 30 (background): ©Stephen Durr RF; p. 31: ©Roll Call/Getty Images; p. 33: NASA GSFC image by Robert Simmon and Reto StöckliP.

Chapter 2

Opener: This image was originally posted to Flickr by Nathan Reading at http://flickr.com/photos/54976525@N08/5835135777. It was reviewed on 1 April 2012 by the FlickreviewR robot and was confirmed to be licensed under the terms of the cc-by-2.0; p. 34 (doctor): ©Terry Vine/Blend Images, LLC RF; p. 35: ©Peter Skinner/Science Source; p. 37: Centers for Disease Control; 2.2 (all): ©Kathleen Talaro; p. 38: ©Dan Ippolito RF; 2.3 (all): ©Kathleen Talaro; p. 40: ©Image Source/Corbis RF; 2.4a-b, 2.5: ©Kathleen Talaro; 2.7: ©McGraw-Hill Education/Lisa Burgess; p. 44-45: ©Ryan McVay/Getty Images RF; 2.9 (all): ©Kathleen Talaro and Harold Benson; 2.11: ©Peter Skinner/Science Source; p. 48-49: Centers for Disease Control; 2.13 (both): Courtesy Nikon Instruments Inc.; p. 50 (bright field): M. I. Walker/Science Source; p. 50 (dark field): ©Kent Wood/Science Source; p. 50 (pase-contrast): ©Roland Birke/Photolibrary/Getty Images; p. 50 (diff. int.): ©Kent Wood/Science Source; p. 51 (fluorescence): ©Dr. Dan Kalman/Science Source; p. 51 (confocal): ©Dr. Gopal MurtScience Source; p. 51 (TEM): CDC/Dr. Erskine Palmer; p. 50-51, p. 51 (SEM): ©Science Photo Library/Getty Images RF; p. 52: ©Digital Vision/Getty Images RF; p. 53: CDC/DPDx–Melanie Moser; 2.17a: ©Kathleen Talaro; 2.17b: Centers for Disease Control; 2.18a: ©ASM/Science Source; 2.18b: ©Richard J. Green/Science Source; 2.18c: CDC/Courtesy Larry Stauffer, Oregon State Public Health Laboratory; 2.19a: CDC/Dr. Leanor Haley; 2.19b: ©David Fankhauser; p. 55 (dish): ©Flying Colours Ltd/Getty Images RF; p. 56 (doctor): ©Terry Vine/Blend Images, LLC RF; p. 56 (culture): This image was originally posted to Flickr by Nathan Reading at http://flickr.com/photos/54976525@N08/5835135777. It was reviewed on 1 April 2012 by the FlickreviewR robot and was confirmed to be licensed under the terms of the cc-by-2.0; p. 56 (bottles): ©Corbis RF; p. 57: ©Custom Medical Stock Photo/Newscom; p. 59: ©Kathleen Talaro and Harold Benson.

Chapter 3

Opener: CDC/Dr. Gilda Jones; p. 60 (doctor): ©JGI/Blend Images RF; p. 61 (second): ©Eye of Science/Science Source; p. 61 (third): ©Dennis Kunkel Microscopy, Inc./Phototake; p. 61 (fifth): ©Corbis RF; p. 61 (sixth): Courtesy Bergey's Manual Trust; p. 62-63: ©Science Photo Library/Getty Images RF; 3.2: ©Max Planck Institute/AFP/Newscom; 3.3: CDC/Billie Ruth Bird; p. 65 (coccus, rod): CDC/Janice Haney Carr; p. 65 (vibrio): ©Juergen Berger/Science Source; p. 65 (spirillum): USDA/Photo by De Wood. Digital colorization by Chris Pooley; p. 65 (spirochete): ©VEM/Science Source; p. 65 (branching): ©Eye of Science/Science Source; 3.5: ©De Agostini/Getty Images; 3.6b: Sarkar MK1, Paul K, Blair D., "Chemotaxis signaling protein CheY binds to the rotor protein FliN to control the direction of flagellar rotation in Escherichia coli," PNAS May 18, 2010 vol. 107 no. 20 9370-9375; 3.10a: ©Eye of Science/Science Source; 3.11: ©L. Caro/SPL/Science Source; 3.12: ©Russell Kightley/Science Source; 3.13a: ©Michael Abbey/Science Source; 3.13b: CDC/Courtesy Larry Stauffer, Oregon State Public Health Laboratory; 3.14b: ©Scimat/Science Source; 3.15 (gram-pos): ©Dr. Kari Lounatmaa/Science Source; 3.15 (gram-neg): ©Dennis Kunkel Microscopy, Inc./Phototake; 3.17: ©McGraw-Hill Education; p. 73: ©Steven p. Lynch RF; p. 76: ©Digital Vision/Getty Images RF; 3.19: ©Dennis Kunkel Microscopy, Inc./Phototake; 3.20, 3.21: ©Science Source; p. 78-79: CDC/Laura Rose & Janice Haney Carr; p. 80: ©Corbis RF; p. 81 (manual): Courtesy Bergey's Manual Trust; p. 81 (boy): ©Purestock/Getty Images RF; p. 82 (bacteria): CDC/Dr. Gilda Jones; p. 82 (doctor): ©JGI/Blend Images RF; p. 83: ©Steve Gschmeissner/Science Source; p. 84: ©Dennis Kunkel Microscopy, Inc./Phototake; p. 85: ©Eye of Science/Science Source; p. 85: ©ASM/Science Source.

Chapter 4

Opener: CDC/Dr. Lucille K. George; p. 86 (doctor): ©Jose Luis Pelaez Inc/Blend Images LLC RF; p. 87 (Aspergillus): ©BSIP/Universal Images Group/Getty Images; p. 87 (Trichomonas, liver fluke): ©Eye of Science/Science Source; p. 88: ©EM Research Services, Newcastle University RF; 4.2: ©Aaron J. Bell/Science Source; 4.3: ©Thomas Deerinck, NCMIR/Science Source; 4.4: ©D Spector/Photolibrary/Getty Images; 4.5: ©Don W. Fawcett/Science Source; 4.6: ©EM Research Services, Newcastle University RF; p. 94-95: ©Science Photo Library/Getty Images RF; 4.9: ©Kallista Images/Getty Images; p. 96: ©EM Research Services, Newcastle University RF; 4.10, 97: ©Dr. Torsten Wittmann/Science Source; p. 98: ©Science Photo Library/Getty Images RF; p. 99: ©BSIP/UIG/Getty Images; 4.11 (both): Courtesy Dr. Judy A. Murphy, Murphy Consuitancy Microscopy & Digital Imaging, Stockton, CA; 4.12: ©Science Photo Library/Getty Images RF; p. 102 (top): ©Kathleen Talaro; p. 102-103: ©William Marin, Jr./The Image Works; p. 104, 105: ©Stephen Durr RF; 4.14: CDC/Dr. Stan Erlandsen; p. 105: ©Dr. Tony Brain/Science Source; p. 105 (amoeboid): ©Stephen Durr RF; p. 105 (ciliated): ©J. R. Factor/Science Source; p. 105 (flagelated): ©Eye of Science/Science Source; p. 105 (apicomplexan): ©BSIP/Universal Images Group/Getty Images; 4.15 (liver fluke): ©Eye of Science/Science Source; 4.15 (tape worm): ©Geoff Brightling/Dorling Kindersley/Getty Images; p. 108 (flatworm): ©NHPA/M. I. Walker RF; 4.16b: Centers for Disease Control; p. 110 (cocci): CDC/Dr. Lucille K. George; p. 110 (doctor): ©Jose Luis Pelaez Inc/Blend Images LLC RF; p. 110 (farm): ©Robert Glusic/Exactostock/Superstock RF; p. 111: CDC/James Gathany.

Chapter 5

Opener (doctor): ©DreamPictures/Pam Ostrow/Blend Images LLC RF; p. 115 (first): ©Eye of Science/Science Source; p. 115 (fourth): Bakonyi T, Lussy H, Weissenböck H, Hornyák A, Nowotny N. "In vitro host-cell susceptibility to Usutu virus." Emerging Infectious Diseases, Vol. 11, No. 2, Feb. 2005. Available from http://wwwnc.cdc.gov/eid/article/11/2/04-1016.htm; p. 115 (fifth): CDC/Dr. Al Jenny; p. 115 (sixth): CDC/Cynthia Goldsmith; p. 116: ©Anna Yu/Getty Images RF; p. 117: CDC/Cynthia Goldsmith; 5.2a: CDC/J. Nakano; 5.2b: ©Phototake; 5.2c: ©A.B. Dowsette/SPL/Science Source; p. 119: CDC/Dr. Fred Murphy; p. 120 (tob. mosaic): ©Science Source; p. 120 (influenza): CDC/Dr. Fred Murphy; p. 120 (adenovirus), p. 121 (hep B): ©Dr. Linda M. Stannard, University of Cape Town/Science Source; p. 121 (herpes): ©Eye of Science/Science Source; p. 121 (capsid): ©Ami Images/Science Source; p. 122: ©Science Photo Library/Getty Images RF; p. 123: CDC/Dr. Fred Murphy; p. 124: CDC/Cynthia Goldsmith; p. 127: CDC/James Gathany; 5.6b: ©Chris Bjornberg/Science Source; 5.7a: ©Patricia Barber/Custom Medical Stock; 5.7b: Courtesy Massimo Battaglia, INeMM CNR, Rome Italy; 5.11: ©Lee D. Simon/Science Source; 5.12a (both): ©Corbis RF; 5.12a (both): Bakonyi T, Lussy H, Weissenböck H, Hornyák A, Nowotny N. "In vitro host-cell susceptibility to Usutu virus." Emerging Infectious Diseases, Vol. 11, No. 2, Feb. 2005. Available from http://wwwnc.cdc.gov/eid/article/11/2/04-1016.htm; p. 134 (left): ©M. Abbey/Science Source; p. 134 (right): U.S. Dept. of Agriculture–Animal and Plant Health Inspection Service, APHIS/DR. Al Jenny; p. 135: CDC/C. S. Goldsmith and A. Balish; p. 136 (virus): CDC/Cynthia Goldsmith; p. 136 (doctor): ©DreamPictures/Pam Ostrow/Blend Images LLC RF; p. 137 (rash): Centers for Disease Control; p. 137 (virus): ©Science Photo Library/Getty Images RF; p. 137 (child): ©Brand X Pictures/PunchStock RF.

Chapter 6

Opener: CDC/Don Stalons; p. 140 (doctor): ©Jose Luis Pelaez Inc/Blend Images LLC RF; p. 141: ©Terese M. Barta, Ph.D.; p. 142 (fish): ©McGraw-Hill Education/Barry Barker; p. 142 (coral): ©Michael Aw/Photodisc/Getty Images RF; p. 143: CDC/Peggy S. Hayes & Elizabeth H. White, M.S.; 6.1a: ©Steve Gschmeissner/Science Source; 6.1b: ©Martin Oeggerli/Science Source; p. 149: Hugh Threlfall/Alamy RF; 6.5a: ©Francois Gohier/Science Source; 6.5b: Courtesy Nozomu Takeuchi; p. 151: Courtesy Submarine Ring of Fire 2004 Exploration, NOAA Vents Program; p. 152 (tubes): ©Terese M. Barta, Ph.D.; p. 152 (anthrax): CDC/Laura Rose & Janice Haney Carr; p. 152 (H. pylori): ©Science Photo Library/Getty Images RF; p. 152 (E. coli, C. difficile, C. strep): CDC/Janice Haney Carr; p. 153: ©Photodisc/Alamy RF; p. 154: ©Getty Images RF; 6.7: Courtesy Ellen Swogger and Garth James, Center for Biofilm Engineering, Montana State University; p. 157, p. 158 (all): CDC/Janice Haney Carr; p. 158: ©Getty Images/Jonelle Weaver RF; p. 159: ©Corbis RF; p. 160: ©Kathleen Talaro; p. 161: Centers for Disease Control; p. 162 (doctor): ©Jose Luis Pelaez Inc/Blend Images LLC RF; p. 162 (micrograph): CDC/Don Stalons; p. 162 (chamber): ©ERproductions Ltd/Blend Images/Getty Images; p. 163: ©Digital Vision/SuperStock RF.

Chapter 7

Opener: ©McGraw-Hill Education; p. 166 (doctor): ©Purestock/SuperStock RF; p. 172: ©McGraw-Hill Education/Jill Braaten; p. 173: ©Frederick Bass/Getty Images RF; p. 175: ©Digital Vision/PunchStock RF; p. 177: ©Lars A. Niki RF; p. 179: ©imageshop–zefa visual media uk ltd/Alamy RF; p. 183 (grapes): ©Imagestate Media (John Foxx)/Imagestate RF; p. 183 (barrels): ©Ingram Publishing/SuperStock RF; p. 184: ©Chris Ryan/AGE Fotostock RF; p. 186: ©Joseph Sohm/Visions of America/Corbis RF; p. 187: USDA/Scott Bauer; p. 188 (doctor): ©Purestock/SuperStock RF; p. 188 (micrograph): ©McGraw-Hill Education; p. 188 (food): ©McGraw-Hill Education/John Thoeming; p. 188 (monitor): ©Realistic Reflections RF; p. 189: ©moodboard/SuperStock RF; p. 191 (gram-pos): ©Dr. Kari Lounatmaa/Science Source; p. 191 (gram-neg): ©Dennis Kunkel Microscopy, Inc./Phototake.

Chapter 8

Opener: ©Purestock/SuperStock RF; p. 192 (doctor): ©Juice Images/Glow Images RF; p. 193: ©Kathleen Talaro; p. 194: ©Adrian Neal/Getty Images RF; 8.2: ©Dr. Gopal Murti/Science Source; p. 197: ©Blend Images/Getty Images RF; p. 202 (background): ©EM Research Services, Newcastle University RF; p. 202, p. 203 (ribosome): ©Center for Molecular Biology of RNA, UC-Santa Cruz; 8.9: Courtesy Steven McKnight, PhD; p. 209: ©Ingram Publishing RF; p. 211 (cow): ©Imagestate Media (John Foxx)/Imagestate RF; p. 211 (snake): ©Ingram Publishing/Fotosearch RF; p. 212-213: CDC/Janice Haney Carr; p. 217: ©Science Source; p. 218: ©McGraw-Hill Education/Barry Barker; p. 220: ©ERproductions Ltd/Blend Images RF; p. 222: ©Chad Baker/Getty Images RF; 8.17b: ©Kathleen Talaro; p. 223 (doctor): ©Rob Melnychuk/Getty Images RF; p. 225: ©Science Photo Library/Getty Images RF; p. 226: CDC/Hsi Liu, Ph.D., MBA, James Gathany; p. 228 (doctor): ©Juice Images/Glow Images RF; p. 228 (cells): ©Purestock/SuperStock RF; p. 228 (x-ray): ©Corbis RF; p. 229: ©Evan Oto/Science Source; p. 231: ©Eye of Science/Science Source.

Chapter 9

Opener: CDC/Janice Haney Carr; p. 232: ©Pixtal/AGE Fotostock RF; p. 233 (second): ©McGraw-Hill Education/Charles D. Winters; p. 233 (third): ©Kathleen Talaro; p. 234: ©Corbis RF; p. 236: ©Doug Sherman/Geofile RF; p. 237: USDA/Ken Hammond; p. 240: ©Bruce Chambers, The Orange County Register/Newscom; p. 241 (man): ©Fuse/Getty Images RF; p. 241 (sardines): ©Burke/Triolo Productions/Getty Images RF; p. 242 (pot): ©McGraw-Hill Education/Charles D. Winters; p. 242 (pasteurization): ©James King-Holmes/Science Source; p. 242 (beer): ©John A. Rizzo/Getty Images RF; p. 242-243: ©Corbis RF; p. 243: ©iStock/360/Getty Images RF;

p. 244 (flame): ©UIG via Getty Images; p. 244 (oven): ©RayArt Graphics/Alamy RF; p. 244 (desert): ©PhotoAlto/PunchStock RF; p. 245: USDA–Agricultural Research Service; p. 246 (gamma ray): ©Adam Hart-Davis/Science Source; p. 246 (uv): ©Tom Pantages; 9.6, p. 246–247: ©Kallista Images/Getty Images; p. 248: ©Ingram Publishing RF; p. 249: ©Stockbyte/PunchStock RF; p. 250 (bleach): ©Richard Hutchings RF; p. 250 (peroxide): ©McGraw-Hill Education/Jill Braaten; p. 250, p. 252 (alcohol): ©Richard Hutchings RF; p. 252 (heavy metal): ©McGraw-Hill Education/Stephen Frisch; p. 245 (wound): ©McGraw-Hill Education; p. 254 (doctor): ©Pixtal/AGE Fotostock RF; p. 254 (bacteria): CDC/Janice Haney Carr; p. 255: ©Hans Neleman/Getty Images.

Chapter 10

Opener: ©Glow Images RF; p. 258 (doctor): ©Ryan McVay/Getty Images RF; p. 259: CDC/Dr. Richard Facklam; p. 260: Armed Forces Institute of Pathology; p. 260: CDC/Janice Haney Carr; p. 261: Library of Congress Prints and Photographs Division; p. 262-263: CDC/Janice Haney Carr; 10.1b: ©Kathleen Talaro; 10.2: CDC/Dr. Richard Facklam; 10.3b: Courtesy David Ellis; p. 264: ©Creatas/PunchStock RF; p. 265 (YeastOne): Courtesy David Ellis; p. 265 (pills): ©Thinkstock/Jupiterimages RF; p. 266-267: ©MedicalRF.com/Getty Images; p. 268: ©Jon Feingersh Photography/Getty Images RF; p. 268-269: ©Peter Dazeley/Photographer's Choice/Getty Images RF; p. 270: ©Image Source/Jupiterimages RF; p. 272: ©Forest & Kim Starr RF; p. 273: ©NHPA/M. I. Walker RF; p. 275: ©L. Caro/SPL/Science Source; p. 278: ©Dimitri Vervits/ImageState RF; 10.6: ©Kathleen Talaro; p. 280, p. 283: CDC; p. 284 (doctor): ©Ryan McVay/Getty Images RF; p. 284 (pill inset): ©Glow Images RF; p. 284 (background pills): ©Design Pics/Kristy-Anne Glubish RF; p. 284 (red pills): ©Markos Dolopikos/Alamy RF; p. 285 (doctor): ©Jose Luis Pelaez, Inc./Blend Images LLC/Getty Images RF; p. 285 (pills): ©Vstock LLC/Getty Images RF; p. 286: CDC/Janice Haney Carr; p. 287: ©Stockbyte/Getty Images RF.

Chapter 11

Opener: ©MedicalRF.com/Getty Images; p. 288 (doctor): ©ERproductions Ltd/Blend Images LLC RF; p. 289 (first): NIH Roadmap for Medical Research website; p. 289 (second): ©McGraw-Hill Education/Christopher Kerrigan; p. 290: NIH Roadmap for Medical Research website; p. 291: ©Brand X Pictures/Jupiterimages RF; p. 293: ©Ingram Publishing RF; p. 295-296: NIAID, NIH/Rocky Mountain Laboratories; p. 297: ©Brand X Pictures RF; 11.4: ©McGraw-Hill Education/Lisa Burgess; p. 301: ©Rubberball/Getty Images RF; p. 304 (writing): ©Thinkstock/Getty Images RF; p. 304 (man): ©Brand X Pictures/PunchStock RF; p. 304 (mosquito): CDC; p. 304 (breath): ©McGraw-Hill Education/Christopher Kerrigan; p. 306 (raccoons): ©imagebroker/Alamy RF; p. 306 (pig): ©Creatas/PunchStock RF; p. 306 (cat): ©Ingram Publishing/SuperStock RF; p. 306 (mouse): ©Punchstock/BananaStock RF; p. 307 (airplane): ©Tony Cordoza/Alamy RF; p. 307 (passengers): ©ColorBlind Images/Getty Images RF; p. 309: ©Dynamic Graphics/JupiterImages RF; p. 314: ©Donald Nausbaum/Getty Images RF; p. 315: NIAID, NIH/Arthur Friedlander; p. 318 (micrograph): CDC/Fred Murphy and Sylvia Whitfield; p. 318 (doctor): ©ERproductions Ltd/Blend Images LLC RF; p. 318 (kidney): ©MedicalRF.com/Getty Images; p. 319: ©ERproductions Ltd/Blend Images LLC RF; p. 321: ©Kathleen Talaro.

Chapter 12

Opener: ©The McGraw-Hill Companies, Inc./Al Telser, photographer; p. 322 (doctor): ©Pixtal/age fotostock RF; p. 323: ©Susumu Nishinaga/Science Source; p. 324: ©Ingram Publishing RF; p. 325: ©Andersen Ross/Blend Images RF; 12.3: ©Susumu Nishinaga/Science Source; p. 326-327: ©Science Photo Library/Getty Images RF; p. 329: ©MedicalRF.com/Getty Images; p. 330: ©Graham Bell/Corbis RF; p. 331: ©fStop/PunchStock RF; 12.8a: ©Eye of Science/Science Source; 12.9: ©SPL/Science Source;

p. 336: ©Brand X Pictures/PunchStock RF; 12.10b: Courtesy Steve Kunkel; p. 338: ©Brand X Pictures/Punchstock RF; p. 339: ©Science Photo Library/Getty Images RF; 12.12 (both): ©Sucharit Bhakdi; p. 344 (doctor): ©Pixtal/age fotostock RF; p. 344 (stomach): ©McGraw-Hill Education/Al Telser; p. 344 (antacids): ©McGraw-Hill Education/Mark Dierker; p. 345: ©Sam Edwards/age fotostock RF.

Chapter 13

Opener: ©MedicalRF.com/Getty Images RF; p. 348 (doctor): ©Jose Luis Pelaez Inc/Blend Images LLC RF; p. 349: ©Steve Gschmeissner/Science Source; p. 350-351: ©MedicalRF.com/Getty Images RF; p. 352: ©Comstock Images/PictureQuest RF; p. 353: ©MedicalRF.com/Getty Images RF; p. 360-361: ©Photographer's Choice/Getty Images RF; 13.7: ©Steve Gschmeissner/Science Source; 13.8b: ©Kenneth Eward/Science Source; p. 367: ©Blend Images/Getty Images RF; p. 368: ©Image Source/Getty Images RF; p. 369 (girl): ©Floresco Productions/Corbis RF; p. 369 (mother): ©Ingram Publishing/Superstock RF; p. 369 (boy): ©McGraw-Hill Education/Jill Braaten; p. 369 (IV fluids): ©McGraw-Hill Education/Mark Dieker; 13.9: Library of Congress Prints and Photographs Division; p. 371: ©Ingram Publishing/SuperStock RF; p. 374: CDC/James Gathany; p. 375 (shot): ©Blend Images/Jupiterimages RF; p. 375 (doctor): ©Jose Luis Pelaez Inc/Blend Images LLC RF; p. 375 (monocytes): ©MedicalRF.com/Getty Images RF; p. 376: ©Tracy Montana/PhotoLink/Getty Images RF.

Chapter 14

Opener: ©McGraw-Hill Education/Al Telser; p. 380 (doctor): ©Jose Luis Pelaez Inc/Blend Images LLC RF; p. 381 (first): ©Pixtal/age fotostock RF; p. 381 (second): ©STU/Custom Medical Stock; p. 381 (fourth): ©Phototake; p. 381 (sixth): ©McGraw-Hill Education/Aaron Roeth; p. 381 (seventh): Courtesy Baylor College of Medicine, Public Affairs; 14.1 (boy in hosp.): Courtesy Baylor College of Medicine, Public Affairs; 14.1 (taking pills): ©McGraw-Hill Education/Christopher Kerrigan; 14.1 (sneeze): ©Pixtal/age fotostock RF; 14.1 (transfusion): ©Roc Canals Photography/Getty Images; 14.1 (arthritis): ©Dynamic Graphics/JupiterImages RF; 14.1 (dermatitis): ©John Kaprielian/Science Source; p. 383: ©Science Photo Library/Getty Images RF; p. 384: ©Ivan Hunter/Getty Images RF; p. 385: ©Jupiterimages/Photos.com/Alamy RF; 14.3c (headache): ©Ingram Publishing RF; 14.3c (skin): ©STU/Custom Medical Stock; 14.3c (stomache ache): ©Brand X Pictures RF; 14.3c (blowing nose): ©Ingram Publishing RF; 14.3c (inhaler): ©Science Photo Library RF; 14.4 (infant): ©Dr. P. Marazzi/Science Source; 14.4 (arm): ©Biophoto Associates/Science Source; 14.5: ©Dr. P. Marazzi/Science Source; 14.6a: ©STU/Custom Medical Stock; p. 391: ©Flying Colours Ltd/Getty Images RF; p. 393: ©Image Source/Getty Images RF; 14.9a: ©Phototake; 14.9b: Courtesy Gary P. Wiliams, M.D.; 14.10a: ©Kathleen Talaro; 14.10b: ©John Kaprielian/Science Source; p. 396: ©Rubberball/Getty Images RF; p. 397: ©Getty Images/OJO Images RF; 14.13a: ©ISM/Phototake; 14.13b: ©McGraw-Hill Education/Aaron Roeth; 14.21: ©Realistic Reflections/Getty Images RF; p. 401: ©Scimat/Science Source; 14.23: ©McGraw-Hill Education/Christopher Kerrigan; 14.16a: Courtesy Baylor College of Medicine, Public Affairs; 14.16b: ©McGraw-Hill Education/Christopher Kerrigan; p. 404 (woman): ©ERproductions Ltd/Blend Images LLC RF; p. 404 (doctor): ©Jose Luis Pelaez Inc/Blend Images LLC RF; p. 404 (micrograph): ©McGraw-Hill Education/Al Telser; p. 405: ©Comstock/Alamy RF; p. 406: ©Pixtal/age fotostock RF.

Chapter 15

Opener: ©McGraw-Hill Education/Lisa Burgess; p. 408: ©McGraw-Hill Education/Trish Ofenlock; p. 409 (first): Image provided by AdvanDx, Inc.; p. 409 (fourth): ©Steven Puetzer/Getty Images RF; p. 410 (scientist): ©Adam Gault/Getty Images RF; p. 410 (hyphae): CDC/Arthur F. DiSalvo, M.D.; p. 410 (DNA): ©Steven

Puetzer/Getty Images RF; p. 411: ©Don Carstens/Artville RF; p. 412: ©Ingram Publishing RF; p. 416: ©Punchstock/BananaStock RF; p. 417: CDC/Janice Haney Carr; 15.5: Courtesy The University of Leeds; 15.7: Image provided by AdvanDx, Inc.; p. 421: CDC; p. 422: ©Adam Gault/Getty Images RF; p. 423 (aglutination): ©Image Source/Getty Images RF; p. 423 (precipitation): ©Mauro Fermariello/SPL/Science Source; p. 423 (western blot): ©Phanie/Science Source; p. 423 (comp. fixation): ©LeBeau/Custom Medical Stock Photo; p. 423 (direct fluor.): CDC; p. 423 (ELISA): ©Hank Morgan/Science Source/Science Source; p. 423 (in vivo): CDC/Gabrielle Benenson; 15.11: ©Phanie/Science Source; 15.12: CDC/Russell; 15.14: ©Hank Morgan/Science Source/Science Source; p. 428: Courtesy EM Scrimgeour et el. First report of Q fever in Oman. Emerging Infectious Diseases. Jan-Feb 2000. vol. 6, no 1; 15.15d: ©Pixtal/AGE Fotostock RF; p. 432 (doctor): ©McGraw-Hill Education/Trish Ofenlock; p. 432 (test): ©McGraw-Hill Education/Lisa Burgess; p. 432 (hospital): ©Adam Gault/Getty Images RF; p. 433: ©Don Carstens/Artville RF; p. 435: ©John Kaprielian/Science Source.

Chapter 16

Opener: ©Ingram Publishing RF; p. 436 (doctor): ©Pixtal/AGE Fotostock RF; p. 437 (second): Centers for Disease Control; p. 437 (third): CDC/Courtesy World Health Organization Diagnosis of Smallpox Slide Series; p. 437 (fifth): CDC/Janice Haney Carr/Jeff Hageman, M.H.S.; p. 437 (sixth): ©ISM/Phototake; p. 438: ©Corbis RF; 16.1: ©Dennis Kunkel Microscopy, Inc.; p. 439: CDC; 16.2: CDC/Gregory Moran, M.D.; 16.3: ©Kathleen Talaro; 16.5: CDC; 16.6: ©James Stevenson/Science Source; p. 444: CDC/Dr. Fred Murphy; Sylvia Whitfield; p. 445 (measles): CDC; p. 445 (rubella): Centers for Disease Control; p. 445 (slapped cheek): ©Dr. p. Marazzi/Science Source; p. 445 (roseola): ©Scott Camazine/Science Source; 16.7: ©Hercules Robinson/Alamy; p. 446-447: National Institute of Allergy and Infectious Diseases (NIAID)/NIH/USHHS; 16.8a: ©DermPics/Science Source; 16.9 (chickenpox face): ©David White/Alamy; 16.9 (chickenpox back): ©Picture Partners/Alamy; 16.9 (chickenpox torso): CDC; 16.9 (smallpox arm): CDC/Dr. Robinson; 16.9 (smallpox back): CDC/Dr. John Noble, Jr.; 16.9 (smallpox face): ©Everett Collection Historical/Alamy; 16.10a: ©Logical Images, Inc. All rights reserved; p. 450, p. 451: CDC; p. 452 (chickenpox): ©Gabriel Blaj/Alamy; p. 452 (smallpox): CDC/Dr. Charles Farmer, Jr.; p. 453 (leishmaniasis): Centers for Disease Control; p. 453 (cutaneous anthrax): ©Medical-on-Line/Alamy; p. 453 (anthrax on neck): CDC; p. 453 (anthrax micrograph): CDC/Dr. William A. Clark; p. 454 (scalp): CDC; p. 454 (face): Centers for Disease Control; p. 454 (body): ©Biophoto Associates/Science Source; p. 454 (groin): ©Dr. Harout Tanielian/Science Source; p. 454 (foot): ©Dr. p. Marazzi/Science Source; p. 454 (nails): CDC/Dr Edwin P Ewing, Jr.; 16.11 (all): CDC/Dr. Lucille K. George; 16.12: ©Biophoto Associates/Science Source; p. 457: ©Comstock Images/PictureQuest RF; 16.15: ©Medical-on-Line/Alamy; p. 458: CDC/Janice Haney Carr/Jeff Hageman, M.H.S.; 16.16: ©ISM/Phototake; p. 459 (doctor): ©Pixtal/AGE Fotostock RF; p. 459 (needles): ©Ingram Publishing RF; p. 459-460 (micrograph): CDC/Cynthia Goldsmith; p. 462: CDC; p. 464 (impetigo): ©Hercules Robinson/Alamy; p. 464 (chickenpox): CDC/John Noble, Jr., MD.

Chapter 17

Opener: ©Scott Camazine/Science Source; p. 466: ©Chris Ryan/age fotostock; p. 467: CDC/DR. Al Jenny; 17.4: CDC/Mr. Gust; 17.5: ©Kathleen Talaro; 17.6, p. 472-473: ©Dr. Gary Gaugler/Science Source; 17.7 (both): ©Gordon Love, M.D. VA, North CA Healthcare System, Martinez, CA; p. 475: CDC/Maryam I. Daneshvar, Ph.D.; 17.8a: CDC/Dr. Hardin; 17.8b: Science Source; 17.9, p. 476-477: ©NIBSC/Science Source; 17.10, p. 479: CDC; p. 480: ©Corbis RF; p. 481: CDC; p. 484: ©Tom Pepeira/Iconotec RF; 17.12a: ©M. Abbey/Science Source; 17.12b: CDC/DR. Al Jenny; p. 485, 17.13 (bat): ©PhotoLink/Getty Images RF; 17.13 (skunk): ©Stockbyte/Getty Images RF; 17.13 (raccoon): ©Corbis RF; 17.13 (tx fox): ©Ingram Publishing RF; 17.13 (az fox): ©Corbis RF; 17.14: ©Biophoto Associates/Science Source; 17.15: Centers for Disease Control; 17.16c: CDC/Dr. Thomas F. Sellers/Emory University; p. 488: ©Image Source/Getty Images RF; p. 492 (doctor): ©Chris Ryan/age fotostock; p. 492 (virus): ©Scott Camazine/Science Source; p. 492 (mosquito): CDC/James Gathany; p. 494 (micrograph): CDC/Dr. Govinda S. Visvesvara; p. 494 (pond): ©McGraw-Hill Education; p. 497 (cells): ©MedicalRF.com/Getty Images RF; p. 497 (capsule stain): CDC/Dr. Leanor Haley.

Chapter 18

Opener: Centers for Disease Control; p. 498 (doctor): ©Chris Ryan/Getty Images RF; p. 500-501: ©Purestock/Getty Images RF; p. 503: CDC; 18.3, p. 505: Courtesy Stephen B. Aley, PhD., University of Texas at El Paso; 18.5: ©Anthony Asael/Art in All of Us/Corbis; p. 506: CDC/Cynthia Goldsmith; 18.6: ©SPL/Science Source; p. 510: ©Comstock Images/Alamy RF; 18.9: ©Kristoffer Tripplaar/Alamy; p. 515: ©Corbis RF; 18.11: Centers for Disease Control; 18.12: ©Science Source; p. 517: ©Ingram Publishing/SuperStock RF; 18.13 (top): CDC/James Gathany; 18.13 (bottom): Centers for Disease Control; 18.14: CDC/Claudia Molins; 18.15: ©Scott Camazine/Science Source; p. 521: ©Comstock Images/Alamy RF; 18.17 (hand): Centers for Disease Control; 18.17 (tick): ©PhotoLink/Getty Images RF; 18.18: ©The Natural History Museum/The Image Works; 18.19, p. 524: CDC/Courtesy Larry Stauffer, Oregon State Public Health Laboratory; p. 526 (doctor): ©Chris Ryan/Getty Images RF; p. 526 (micrograph): Centers for Disease Control; p. 528: ©Corbis RF.

Chapter 19

Opener: ©MedicalRF.com; p. 532 (doctor): ©Purestock/SuperStock RF; p. 533 (third): ©Pulse Picture Library/CMP Images/PhotoTake; p. 533 (fifth): CDC/Ronald W. Smithwick; p. 534-535: ©Science Photo Library/Getty Images RF; 19.1b: ©Susumu Nishinaga/Science Source; 19.2: ©Pulse Picture Library/CMP Images/PhotoTake; p. 536: ©Photodisc Collection/Getty Images RF; 19.4a: ©Science Source; 19.4b: ©Van Bucher/SPL/Science Source; p. 539: ©ballyscanlon/Getty Images RF; 19.6: Centers for Disease Control; p. 542: ©Getty Images RF; p. 544: ©Science Photo Library/Getty Images RF; p. 545 (virus): CDC/C. S. Goldsmith and A. Balish; p. 545 (man): ©RubberBall/SuperStock RF; p. 547 (pills): ©Thinkstock/Jupiterimages RF; 547 (micrograph): ©Steve Gschmeissner/Science Source; p. 548 (skin test): CDC/Donald Kopanoff; p. 548 (lung): ©McGraw-Hill Education; p. 548 (slum): ©McGraw-Hill Education/Barry Barker; 19.12: ©SIU Bio Med/Custom Medical Stock Photo; 19.13, p. 550 (top): CDC/Dr. George p. Kubica; p. 550 (bottom): CDC/Ronald W. Smithwick; p. 551: CDC/Janice Haney Carr; 19.15: ©Jebb Harris/ZUMA Press/Newscom; 19.16: ©Tom Volk; p. 553: ©Travel Ink/Getty Images; p. 555 (doctor): ©Purestock/SuperStock RF; p. 555 (virus): ©MedicalRF.com; p. 556 (doctors): ©Getty Images; p. 556 (bottle): CDC/Jim Gathany; p. 556 (shot): ©Creatas/PunchStock RF; p. 558: ©Thinkstock/Jupiterimages RF; p. 559: ©Jack Bostrack/Visuals Unlimited.

Chapter 20

Opener: Centers for Disease Control; p. 560 (doctor): ©Purestock/SuperStock RF; p. 561 (second): ©Ingram Publishing RF; p. 561 (third): ©Moredun Animal Health Ltd./Science Source; p. 561 (fourth): ©Eric Grave/SPL/Getty Images RF; p. 562: ©Kate Mitchell/Corbis RF; p. 563: ©Ingram Publishing RF; p. 564: ©Kent Knudson/PhotoLink/Getty Images RF; p. 565: ©Ryan McVay/Getty Images RF; p. 566-567: Centers for Disease Control; 20.4 (left): ©Gastrolab/Science Source; 20.4 (right): Centers for Disease Control; p. 569: ©BananaStock/PunchStock RF; 20.5a: ©David Musher/Science Source; 20.5b: Courtesy Fred Pittman; 20.5c: ©Benjamin/Custom Medical Stock Photo; p. 570-571: CDC/Janice Haney Carr; 20.6: ©Moredun Animal Health Ltd./Science Source; 20.7: ©Custom Medical Stock Photo; p. 574-575: ©McGraw-Hill Education/Erica Simone Leeds; 20.8: ©Iruka Okeke; 20.9: CDC/Dr. Stan Erlandsen; p. 577: ©McGraw-Hill Education/Barry Barker; 20.10d: Centers for Disease Control; 20.10e: ©Eye of Science/Science Source; 20.13 (plaque): ©David Scharf/Science Source; 20.13 (tooth): ©BSIP SA/Alamy; 20.15: ©E.H. Gill/Custom Medical Stock Photo/Newscom; 20.16: Centers for Disease Control; p. 584-585: CDC/Dr. Erskine Palmer; p. 586: ©Ingram Publishing RF; p. 587: ©McGraw-Hill Education/Christopher Kerrigan; p. 588: ©Getty Images/Stockbyte RF; 20.18a: ©Eric Grave/SPL/Getty Images RF; 20.18b: Courtesy Dickson Despommier; p. 591: ©John A. Rizzo/Getty Images RF; 20.19 (brain): ©PR Bouree/age fotostock; 20.19 (scan): ©Science Photo Library/Getty Images RF; 20.20 (cercaria): ©Dr. Harvey Blankespoor; 20.20 (mating): CDC/Dr. Shirley Maddison; 20.20 (miracidium): ©Sinclair Stammers/Science Source; p. 594 (doctor): ©Purestock/SuperStock RF; p. 594 (agar & micrograph): Centers for Disease Control; p. 596: ©Ingram Publishing/SuperStock RF; p. 598: Centers for Disease Control.

Chapter 21

Opener: ©Meredun Animal Health Ltd/Science Source; p. 600 (doctor): ©Jose Luis Pelaez Inc./Getty Images RF; p. 601 (third & fourth): Centers for Disease Control; p. 603: ©Andersen Ross/Getty Images RF; p. 604: ©Ariel Skelley/Blend Images RF; p. 605: ©Photodisc/Getty Images RF; p. 606: Centers for Disease Control; p. 607: ©Corbis RF; 21.5: CDC/J. Pledger; 21.6, p. 609 (background): CDC/Dr. Norman Jacobs; p. 610: ©McGraw-Hill Education/Christopher Kerrigan; 21.7: Courtesy Danny L. Wiedbrauk; 21.8 (both): Centers for Disease Control; p. 615: ©Eye of Science/Science Source; p. 616: ©DoubleVision/Science Source; 21.9: CDC/J. Pledger; 21.10a: CDC/Dr. Norman Cole; 21.10b: CDC/Robert Sumpter; 21.11: ©Custom Medical Stock Photo; 21.12: CDC/Dr. Edwin P. Ewing, Jr.; 21.13: CDC/J.D. Millar; p. 620 (syphilis): ©ISM/Phototake; p. 620 (chancroid): ©Dr. M.A. Ansary/Science Source; p. 620 (herpes): ©Dr. P. Marazzi/Science Source; 21.14: Centers for Disease Control; 21.15: ©Dr. Walker/Science Source; p. 622 (woman): ©McGraw-Hill Education/Jill Braaten; p. 622 (both virus): ©Science Photo Library/Getty Images RF; p. 623 (pills): ©Comstock/Alamy RF; p. 623 (HPV): ©BSIP/age fotostock; p. 623 (molluscum): ©Dr. Jack Jerjian/Phototake; p. 624-625(virus): ©Meredun Animal Health Ltd/Science Source; p. 625 (doctor): ©Jose Luis Pelaez Inc./Getty Images RF; p. 626: ©Hybrid Images/Cultura/Getty Images RF; p. 630: ©Dr. Walker/Science Source; p. 631 (lip & hand): Centers for Disease Control; p. 631 (back): ©DoubleVision/Science Source.

Chapter 22

Opener: USDA-ARS/Photo by De Wood, digital colorization by Chris Pooley; p. 632 (doctor): ©Juice Images/Glow Images RF; 22.1: ©MedicalRF.com RF; p. 633 (third): ©Dr. Parvinder Sethi RF; p. 633 (fourth): US Coast Guard/Chief Petty Officer John Kepsimelis; p. 634 (terraces): ©BambooSIL/SuperStock RF; p. 634 (geese): U.S. Fish & Wildlife Service/Wyman Meinzer; 22.1 (rooster): ©Kent Knudson/PhotoLink/Getty Images RF; 22.1 (tree): ©OGphoto/Getty Images RF; p. 636 (raccoons): ©McGraw-Hill Education/Barry Barker; p. 636 (bird): ©Stockbyte RF; p. 637: ©William Leaman/Alamy RF; p. 639: Centers for Disease Control; p. 640: ©Corbis RF; 22.7: ©McGraw-Hill Education/Dr. Parvinder Sethi; 22.8 (top): ©Kathleen Talaro; 22.8 (middle & bottom): Reprinted from EPA Method 1604 (EPA-821-R-02-024) Courtesy Dr. Kristen Brenner from the Microbial Exposure Research Branch, Microbiological and Chemical Exposure Assessment Research Division, National Exposure Research Laboratory, Office of Research and Development, U.S Environmental Protection Agency; p. 642 (man): ©Corbis RF; p. 642 (river): ©Tom Uhlman/Alamy RF; p. 643: ©Wellcome Library, London; 22.9: ©Pascal Guyot/AFP/Getty Images; 22.10: USGS/Walter D. Mooney; p. 644 (salsa): ©Raul Taborda/Alamy RF; p. 644 (sprouts): ©Shawna Lemay/Moment Open/Getty Images RF; p. 645: ©Creatas/PunchStock RF; p. 646: ©Andrew Penner/Getty Images RF; 22.14 (both): Courtesy Sanitation Districts of Los Angeles County; p. 649: ©PhotoAlto RF; 22.15: ©Wave Royalty Free/Alamy RF; 22.16: ©Comstock Images/Jupiterimages RF; p. 650: ©Vanessa Vick/Science Source; 22.17: US Coast Guard/Chief Petty Officer John Kepsimelis; p. 652 (doctor): ©Juice Images/Glow Images RF; p. 652 (bacteria): USDA-ARS/Photo by De Wood, digital colorization by Chris Pooley; p. 653: © PIXTAL/PunchStock RF; p. 654: ©Corbis RF.

Line Art: Chapter 9 Table 9.5: From Perkins, *Principles and Methods of Sterilization in Health Science*, 2nd ed.

Index

Note: Page numbers followed by t and f refer to tables and figures, respectively. Page numbers in **boldface** refer to boxed material, and page numbers in *italics* refer to definitions or introductory discussions.

A

AAV (adeno-associated virus), 134
Abiogenesis, *9*
Abscesses, *301*
ACAM 2000 vaccine, 451
Acanthamoeba, 105t, 460, 479, 479t
Acetaminophen, 163, 390
Acetic acid, 253
Acid-fast stains, 54f, *55,* **74,** 549–550
Acidophiles, 153
Acids, as antimicrobial agents, 253
Acinetobacter baumannii, 554
Acriflavine dyes, 253
ACT (artemisinin combination therapy), 272
Actin, 77
Actin filaments, *96,* 96f
Activated sludge process, 647, 649
Active immunity, 368, 369t
Active site, of enzymes, 172, 173f
Active transport, **147,** 148, 148t
Active viruses, 116
Acute encephalitis, 480–482, 482t
Acute endocarditis, 513, 514t
Acute infections, defined, 300t
Acute otitis media, 539–540, 540f, 541t
Acyclovir, 137, 274t, 281t, 459, 462, 480, 621
ADA (adenosine deaminase) deficiency, 403
Adenine, 15t, 22, 196
Adeno-associated virus (AAV), 134
Adenoids, **331**
Adenosine deaminase (ADA) deficiency, 403
Adenosine diphosphate (ADP), 23, 177
Adenosine monophosphate (AMP), 177
Adenosine triphosphate (ATP), 23, 23f, 175, 177, 177f, 182
Adenoviruses, 120t, 458, 539
Adhesion, *296,* 334t
ADIs (AIDS-defining illnesses), 506, 507t, 508, **511**
Adjuvants, *373*
Administration, of vaccines, 373
ADP (adenosine diphosphate), 23, 177
Adsorption, in viral multiplication cycle, 124, 124f, 125t
Aerobes, 152t
Aerobic respiration, *179–182*

Aerosols, 308t
Aerotolerant anaerobes, 152t
African Americans
  HIV/AIDS among, 511
  vitamin D synthesis by, 189
Agammaglobulinemia, *401–402*
Agar, *16,* 39–43
Agar diffusion tests, 262–263
Agglutination reactions, 365t, 422, 423t, 424f
Aggregatibacter actinomycetemcomitans, 582
Agricultural industry, 99
AIDS. *See* HIV/AIDS
AIDS-defining illnesses (ADIs), 506, 507t, 508, **511**
Air
  environmental allergens in, 385
  as reservoir for infectious disease, 304t
  transmission of disease through, 308t
Airplanes, spread of disease through, 307
Albendazole, 272
Alcohol, as germicidal agent, 251t, 252
Alcohol-based hand cleansers, **248**
Alcoholic beverages, fermentation of, 184
Aldehydes, 250t
Alkalinophiles, 153
Alkalis, as antimicrobial agents, 253
Allergens, 359, 383, 384–385, 385f
Allergic reactions, 383–391
  anaphylaxis, 383, 389
  atopic diseases, 383, 384, 388–389
  case study, 258
  diagnosis of, 389, 389f, **390**
  to medications, 280, 282, 388–389, 388f
  prevalence of, 383–384
  sensitization and provocation of, 385–387f
  side effects vs., **283**
  symptom mediation, 386–387
  treatment and prevention of, 390–391, 390f
Allergic rhinitis, 383, 388
Allergies, defined, *383*
Allergy shots, 390–391
Alloantigens, 358, 391

Allografts, 396–397, 397f
Alpha helix, 20
Alpha toxin, 162
Alternative pathway, of complement cascade, 341, 342f
Alveoli, 534
Alzheimer's disease, 373
Amantadine, 274t, 281t, 545
Amikacin, 281t
Amino acids, 15t, 19, 19t, **187,** 204
Aminoglycosides, 265, 268t, 272, 281t
Ammonium hydroxide, 253
Amoebiasis, 105t, 577–579
Amoeboid protozoa, 105t
Amoxicillin, 268t, 271t, 344, 518
AMP (adenosine monophosphate), 177
Amphibolism, 185, 185t
Amphitrichous flagellar arrangements, 67, 67f
Amphotericin B, 265, 273t, 281t, 474, 479
Ampicillin, 268t, 271t, 281t, 474
Amprenavir, 512t
Amylase, 170
Anabolic enzymes, 209
Anabolism, *168,* 185–187
Anaerobes, 152t
Anaerobic digestors, 649
Anaerobic respiration, *179–183*
Anamnestic response, 367t
Anaphylaxis, 383, 389
Anatomical diagnosis, **440,** 551
Aniline dyes, 253
Animalia (kingdom), 26, 27f
Animals and animal viruses. *See also* Zoonotic infections; *specific animals*
  cultivation and identification techniques, 132–133
  multiplication cycles in, 124–128, 124f, 125t, 126f, 127t, 128f
  as reservoirs and sources of disease, 306
  treatment of, 136
Ankylosing spondylitis, 398, 398t, 405
Anopheles mosquito, 111, 505
Anoxygenic photosynthesis, 5
Antacids, 265
Antagonism, microbial, *154,* 291
Anthracimycin, 653

Anthrax
  as bioterrorism agent, 234, 450, 453, 525
  causative agent, 13, 524, 524f
  culture and diagnosis, 525
  cutaneous, 452–453, 453t, 524
  encapsulated bacterial cells in, 70
  endospore form of, 79
  overview, 525t
  pathogenesis and virulence factors, 524
  pulmonary, 524
  septicemic, 524
  signs and symptoms, 524
  S layers in, 69
  transmission and epidemiology, 524–525
  treatment and prevention of, 525
  as zoonotic infection, 306t, 524
Antibacterial drugs, 281t
Antibiograms, 262
Antibiosis, 154
Antibiotic-associated colitis, 282, 319, 570, 570f
Antibiotic resistance
  consumer's role in, 285
  dangers of, 277
  defined, *274*
  development of, 275
  fitness cost of, **220**
  hazard level classification, 277–278
  mechanisms of, 276, 276t
  natural selection and, 277, 277f
  plasmids and, 211
Antibiotics. *See also specific names and types of antibiotics*
  allergic reactions to, 258, 280, 282, **283**
  defined, 261t
  diarrhea as side effect of, **280**
  resistance to (*See* Antibiotic resistance)
  viral infections, ineffectiveness against, **122**
Antibodies
  antigen interactions with, 365
  defined, *21*
  diversity of, 356, 356f
  functions of, 365, 365t
  primary and secondary responses to antigens, 367–368, 367t